GROUP THEORY FOR PHYSICISTS

Second Edition

GROUP THEORY FOR PHYSICISTS

Second Edition

Zhong-Qi Ma
Institute of High Energy Physics, Beijing, China

World Scientific

NEW JERSEY • LONDON • SINGAPORE • BEIJING • SHANGHAI • HONG KONG • TAIPEI • CHENNAI • TOKYO

Published by

World Scientific Publishing Co. Pte. Ltd.

5 Toh Tuck Link, Singapore 596224

USA office: 27 Warren Street, Suite 401-402, Hackensack, NJ 07601

UK office: 57 Shelton Street, Covent Garden, London WC2H 9HE

Library of Congress Cataloging-in-Publication Data

Names: Ma, Zhongqi, 1940– author.

Title: Group theory for physicists / by Zhong-Qi Ma (Institute of High Energy Physics, Beijing, China).

Description: Second edition. | New Jersey : World Scientific, 2019. | Includes bibliographical references and index.

Identifiers: LCCN 2018041647| ISBN 9789813277380 (hardcover : alk. paper) | ISBN 9789813277960 (pbk : alk. paper)

Subjects: LCSH: Group theory. | Group theory--Textbooks. | Mathematical physics. | Mathematical physics--Textbooks.

Classification: LCC QC20.7.G76 M29 2019 | DDC 512/.2--dc23

LC record available at https://lccn.loc.gov/2018041647

British Library Cataloguing-in-Publication Data

A catalogue record for this book is available from the British Library.

For any available supplementary material, please visit
https://www.worldscientific.com/worldscibooks/10.1142/11187#t=suppl

Printed in Singapore

Preface

Ten years has passed since the textbook "Group theory for Physicists" published in 2007. During these ten years, when I was retired and I had more time to study, I concentrate my effort on some unsolved problems of the calculation method of group theory in physics, including some questions asked and discussed by my friends and students. I am pleased to share my new progress with readers in the second edition of this textbook.

1. The generalized Gel'fand's method.

One of the main tasks of group theory is to find the inequivalent irreducible representations of the groups used in physics. For the simple Lie algebras $\mathcal{L}$ with dimension more than one, its irreducible representation can be calculated by the method of the block weight diagram in principle. The difficulty is that the method of the block weight diagram does not give a certain rule for choosing the orthonormal basis states with multiple weights. However, the calculation quantity depends greatly on how to choose the suitable orthonormal basis states, especially when the multiplicities of weights are high. The Gel'fand's method [Gel'fand–Tsetlin (1950)] gives a certain rule to choose the orthonormal basis states of the Lie algebras A_ℓ and greatly simplifies the calculation.

The Gel'fand's method is a recursive way based on $A_\ell \supset A_{\ell-1} \oplus \mathcal{A}_\ell$. We develop the Gel'fand's method from A_ℓ to $\mathcal{L}$ (except for F_4, see Chap. 8) based on $\mathcal{L} \supset A_{\ell-1} \oplus \mathcal{A}_\ell$ and give a certain rule to choose the orthonormal basis states of the Lie algebras $\mathcal{L}$. We summarize the generalized Gel'fand's method in subsection 8.3.1 and explain this method through some examples in detail (see section 8.3). This method is also effective for calculating the Clebsch–Gordan coefficients in $\mathcal{L}$ (see section 8.5) and for calculating the wave functions of the basis states (see Chapters 9–12).

2. The irreducible bases of the symmetry group **I** of an icosahedron.

The irreducible bases of a finite group G give the main information of G. Since the order g of **I** is 60, the traditional calculation for the explicit forms of the irreducible bases of **I** is too long to write in a textbook. Based on the isomorphism between **I** and the alternating subgroup S_5' of the permutation group S_5, the parameters in an icosahedron and the irreducible bases of **I** are calculated in detail (see subsection 5.4.6).

3. The irreducible representations of the space groups $\mathcal{S}$.

Although the representations of the space groups can be looked over on the web: http://www.cryst.ehu.es/rep/repres.html, usually physicists prefer to understand the calculation method if it is not too difficult. In the first edition of this book we only explain the calculation method of the single-valued representations of $\mathcal{S}$ for the symmorphic space groups or for the non-symmorphic space group $\mathcal{S}$ where the star of wave vectors is not on the boundary of the Brillouin zone. Now, we also explain the calculation method for both the single-valued and the double-valued representations of $\mathcal{S}$ and for the non-symmorphic space group $\mathcal{S}$ where the star of wave vectors is on the boundary of the Brillouin zone (see subsection 6.2.4 and section 6.5).

4. Antisymmetric wave functions of fermions.

The spinor wave function of the system of n electrons is described by a tensor of rank n of SU(2). How to combine the spinor wave function with the orbital wave function to be totally antisymmetric (or symmetric) is a fundamental problem in physics because the wave function of a system of identical fermions should be totally antisymmetric. We introduce a simple method to calculate this problem in subsection 9.1.6.

I welcome any reader to discuss the problems on group theory through email: mazq@ihep.ac.cn. I express my heartfelt gratitude to Prof. Chen Ning Yang at Tsinghua University for his guidance in my research work and support in my study on group theory. I am indebted to Prof. Jing Zhang at Imperial College London of UK for his suggestion on English in this book. I am very grateful to my wife, Ms Xian Li, for her continuous support.

(Retired from Institute of High Energy Physics)
Taikang Community, Yue Garden
Guangdong Province, China *Zhong-Qi Ma*
March 2018

Preface to the First Edition

Group theory is a powerful tool for studying the symmetry of a physical system, especially the symmetry of a quantum system. Since the exact solution of the dynamic equation in the quantum theory is generally difficult to obtain, one has to find other methods to analyze the property of the system. Group theory provides an effective method by analyzing the symmetry of the system to obtain some precise information of the system verifiable with observations. Now, Group Theory is a required course for graduate students majored in physics and theoretical chemistry.

The course of Group Theory for the students majored in physics is very different from the same course for those majored in mathematics. A graduate student in physics needs to know the theoretical framework of group theory and more importantly to master the techniques in the application of group theory to various fields of physics, which is actually his main objective for taking the course. However, no course or textbook on group theory can be expected to include explicitly the solution to every problem of group theory in his research field of physics. A student of physics has to know the fundamental theory of group theory, otherwise he may not be able to apply the techniques creatively. On the other hand, the student of physics is not expected to completely grasp all the mathematics behind group theory due to the breadth of the knowledge required.

I first taught the group theory course in 1962. Since 1986, I have been teaching for 20 years the course of Group Theory to graduate students mainly majored in physics at the Graduate School of Chinese Academy of Sciences. In addition, most of my research work has been related to applications of group theory to physics. In 1996, the Chinese Academy of Sciences decided to publish a series of textbooks for graduate students. I was invited to write a textbook on Group Theory for the series. In the textbook, based on my experience in teaching and research work, I explained the fundamental concepts and techniques of group theory progressively and systematically using the language familiar to physicists, and also emphasized the ways with which group theory is applied to physics. The textbook (in Chinese) has been widely used for the Group Theory course in China since it was published by Science Press (Beijing) in 1998. The second edition of this textbook was published in 2006 after systematic revision. Up to the seventh printing in December 2006, 16800 copies of this textbook were printed altogether. This textbook is the foundation for the present textbook, although some new materials are included and some are moved to another exercise book [Ma and Gu (2004)].

By the request of the readers, an exercise book on group theory, by the same author, was published in 2002 by Science Press (Beijing) to form a complete set of textbooks on group theory. In order to make the exercise book self-contained, a brief review of the main concepts and techniques is given before the problems in each section. The reviews can be used as a concise textbook on group theory. Cooperated with my student, the English version of the exercise book was published in 2004 by World Scientific named "Problems and Solutions in Group Theory for Physicists" ([Ma and Gu (2004)]). A great deal of new materials drawn from teaching and research experiences is included. The reviews of each chapter has been extensively revised. Last four chapters are essentially new. Some useful results are listed in the exercise book for reference. Some materials are moved from the textbook to the exercise book such that the textbook can concentrate the main subjects on group theory and include some new developments in group theory. This exercise book serves as a supplement for the present textbook.

The present textbook consists of 10 chapters. Chap. 1 is a short review on linear algebra. The reader is required not only to be familiar with its basic concepts but also to master its applications, especially the similarity transformation method. In Chap. 2, the concepts of a group and its subsets are introduced from the physical problems and explained through examples of some finite groups. The importance of the group table of a finite group is emphasized. The group table of the symmetric groups of regular polyhedrons are introduced in a new way (see §2.5.2). The theory of representations of a group is studied in Chap. 3. The transformation operator P_R for the scalar functions bridges the gap between the representation theory and physical application. The subduced and induced representations of groups are used to construct the character tables of finite groups in Chap. 3, and to calculate the outer product of representations of the permutation groups in Chap. 6. The method of idempotents is systematically studied in §3.7. Based on this method the standard irreducible basis vectors in the group algebras of some finite groups in common use in physics are calculated. The method of Young operators studied in Chap. 6 is its development to the permutation groups.

The classification and representations of semisimple Lie algebras are introduced in Chap. 7 and partly in Chap. 4 by the language familiar to physicists. The methods of block weight diagrams and dominant weight diagrams are recommended for calculating the representation matrices of the generators and the Clebsch–Gordan series in a simple Lie algebra. The readers who are interested in the strict mathematical definitions and proofs in the theory of semisimple Lie algebras are recommended to read the more mathematically oriented books (e.g., [Bourbaki (1989)]).

The remaining part of the book is devoted to the important symmetric groups of physical systems. In Chap. 4 the symmetric group SO(3) of a spherically symmetric system in three dimensions is studied. The study on the symmetry of crystals is a typical example of the physical application of group theory. In Chap. 5, through a systematic analysis by the method of group theory, only based on the translation symmetry of crystals, the crystals are classified completely that there are 11 proper crystallographic point groups, 32 crystallographic point groups, 7 crystal systems, 14 Bravais lattices, 73 symmorphic space groups, and 230 space groups. The international symbols of the space groups are recommended in Chap. 5. The analysis method for the symmetry of a crystal from its international symbols are emphasized (see §5.4.6). The permutation groups S_n are the symmetric groups of the identical particles and are widely used in the decomposition of the tensor spaces. In Chap. 6 the permutation groups S_n are studied by the method of Young operators. The irreducible basis vectors in the group space of S_n are explicitly given based on the Young operators, and the calculating method for the similarity transformations from them to the orthonormal basis vectors is proved and demonstrated by examples. The matrix groups in common use in physics, such as the $SU(N)$ groups, the $SO(N)$ groups, the $USp(2\ell)$ groups, and the Lorentz group L_p are studied in some detail in the last three chapters.

The systematic examination of Young operators is an important characteristic of this book. We calculate the characters, the representation matrices, and the outer product of the irreducible representations of the permutation groups using the method of Young operators. For the matrix groups $SU(N)$, $SO(N)$, and $USp(2\ell)$, which are related to four classical Lie algebras, the basis states of the irreducible representations can be explicitly calculated using the method of Young operators. The dimensions of the irreducible representations of the permutation groups, the $SU(N)$ groups, the SO(N) groups, and the USp(2ℓ) groups are all calculated by the hook rule, a method based on the Young patterns.

An isolated quantum n-body system is invariant in the translation of space-time and the spatial rotation so that the energy, the momentum, and the total angular momentum of the system are conserved. The motion of the center-of-mass and the global rotation of the system should be separated from its internal motion, and its Schrödinger equation is reduced to the radial equation, depending only on the internal degrees of freedom. The method of the Jacobi coordinate vectors is summarized in §4.9.1 to separate the motion of the center-of-mass. The generalized harmonic polynomials are presented to separate the global rotation of an isolated quantum n-body system from the internal motion in §4.9 and generalized to arbitrary N-dimensional space in §9.4, where the Dirac equation in $(N+1)$-dimensional space-time is also studied.

In conclusion, I must express my cordial gratitude to my supervisor Prof. Ning Hu for his guidance, under which I entered the research field of the symmetric theory of particle physics. I am indebted to Prof. Yi-Shi Duan who made me an abecedarian to study and to teach the group theory. I am very grateful to my wife, Ms Xian Li, for her continuous support. This book was supported by the National Natural Science Foundation of China under Grants Nos. 10475082 and 10675050.

Institute of High Energy Physics
Beijing, China *Zhong-Qi Ma*
June 2007

List of Symbols

$\boldsymbol{a}$	vector,
$\hat{\boldsymbol{n}}$, $\hat{\boldsymbol{x}}$, $\boldsymbol{e}_\mu$	unit vector,
$\underline{a}$	column matrix,
Γ, Λ	diagonal matrix,
$D(R)^{-1}$, $D(R)^T$	inverse and transpose of a matrix $D(R)$,
$D(R)^*$, $D(R)^\dagger$	complex matrix and conjugate matrix,
$\mathcal{L}$, $\mathcal{L}_1$	linear space or subspace,
$\mathcal{L}_1 + \mathcal{L}_2$	sum of two subspaces,
$\mathcal{L}_1 \oplus \mathcal{L}_2$	direct sum of two subspaces,
AB	product of two digits, two matrices etc.,
$\times$	direct product of matrices, representations,
$\otimes$	direct product of groups, outer product,
E, e	identity of a group,
R, S	transformations, elements of a group,
$\mathcal{R}$	set, complex of elements,
$\mathcal{C}_\alpha$	class in a group,
$n(\alpha)$	number of elements in $\mathcal{C}_\alpha$,
C_α	class operator,
g	order of a group,
g_c	number of classes in a group,
$\rho \bmod n$	the integer $\rho + n$ is assumed to be the same as ρ,
$\mathrm{G} \approx \mathrm{G}'$	G is isomorphic onto G$'$,
$\mathrm{G} \sim \mathrm{G}'$	G is homomorphic onto G$'$,
C_N, D_N	cyclic group, dihedral group,
T, **O**, **I**	symmetric groups of polyhedrons,
$D(G) \simeq \overline{D}(G)$	equivalent representations,
e_a, e^j_μ	idempotent,
$\phi^j_{\mu\nu}$	the irreducible bases in the group algebra,
$\mathcal{L}^{(j)}_\mu$, $\mathcal{R}^{(j)}_\mu$, $\mathcal{I}^{(j)}$	left ideal, right ideal, two-side ideal,
T_a, $T^{(r)}_{ab}$, T_{ab}	generators in self representation,
I_j, I_A, $D(I_A)$	generators of a Lie group,
$Y^\ell_m(\hat{\boldsymbol{x}})$	spherical harmonic function,
$\mathcal{Y}^\ell_m(\boldsymbol{x})$	harmonic polynomial,
$Q^{\ell\lambda}_q(\boldsymbol{R}_1, \boldsymbol{R}_2)$	generalized harmonic polynomial,

$C^{jk}_{\mu\nu,J(r)M}$	Clebsch–Gordan coefficients,
$\langle j,\mu,k,\nu\vert J,r,M\rangle$	
$C^{\boldsymbol{M}^{(1)}\boldsymbol{M}^{(2)}}_{\boldsymbol{m}^{(1)},\boldsymbol{m}^{(2)};\boldsymbol{M},r,\boldsymbol{m}}$	
$\vert j,\mu\rangle\vert k,\nu\rangle$, $\vert\mu\rangle\vert\nu\rangle$,	basis vectors before similarity transformation,
$\vert\boldsymbol{M}_1,\boldsymbol{m}_1\rangle\vert\boldsymbol{M}_2,\boldsymbol{m}_2\rangle$	
$\vert\boldsymbol{m}_1\rangle\vert\boldsymbol{m}_2\rangle$,	
$\vert\vert\boldsymbol{M},\boldsymbol{m}\rangle_r$	basis vectors after similarity transformation,
$C_{jk}^{\ \ \ell}$, $C_{AB}^{\ \ \ D}$	structure constant of a Lie group,
$\boldsymbol{V}(x)_a$, $\boldsymbol{T}(x)_{a_1\ldots a_n}$	components of vector and tensor,
$\boldsymbol{\theta}_d$, $\boldsymbol{\theta}_{d_1\ldots d_n}$	basis vectors, basis tensors,
$\mathcal{T}$, $\mathcal{T}^*$, $\mathcal{T}^{[\lambda]}_\mu$	tensor space and subspace,
P_R, Q_R,	transformations operators for scalars, tensors
$O_R = P_R Q_R$	and spinors,
P and Q	horizontal and vertical permutations,
P_0 and Q_0	horizontal and vertical transpositions,
$\mathcal{Y}$, $\mathcal{Y}^{[\lambda]}_\mu$	Young operator,
$\mathcal{P}$, $\mathcal{Q}$	horizontal and vertical operators,
$[\lambda]$	Young pattern,
$\mathcal{L}$, $\mathcal{L}_R$	Lie algebra, real Lie algebra,
$\mathcal{H}$,	Cartan subalgebra of a Lie algebra,
g_{AB}, g_{jk}	Killing form,
H_j and $E_{\boldsymbol{\alpha}}$	Cartan–Weyl bases,
H_μ, E_μ and F_μ	Chevalley bases,
$\boldsymbol{r}_\mu$	simple roots,
$\boldsymbol{M}$, $\boldsymbol{m}$	the highest weight, weights,
$\boldsymbol{w}_\mu$	fundamental dominant weight,
$A_{\mu\nu}$	Cartan matrix,
$(\boldsymbol{r}_\nu)_\mu = A_{\mu\nu}$	component of a simple root,
$\vert\boldsymbol{M},\boldsymbol{m}\rangle$	basis state in the representation $D^{\boldsymbol{M}}$,
$L(\boldsymbol{M},\boldsymbol{m})$	level of a weight $\boldsymbol{m}$ in the representation $D^{\boldsymbol{M}}$,
$C_2(\boldsymbol{M})$	Casimir invariant of order two,
$[(\pm)\lambda]$	self-dual and anti-self-dual representations of SO(N),
$[s,\lambda]$	spinor representation of SO($2\ell+1$),
$[\pm s,\lambda]$	spinor representation of SO(2ℓ),
Δ	set of all roots in a simple Lie algebra,
Δ_+	set of all positive roots in a simple Lie algebra.

Contents

Chapter 1

REVIEW ON LINEAR ALGEBRAS

The main mathematical tool in group theory is linear algebras. In this chapter, we will review some fundamental concepts and calculation methods in linear algebras, which are often used in group theory.

1.1 Linear Space and Basis Vectors

Let $H(x)$ denote the Hamiltonian of a system. Suppose that the eigenvalue E of $H(x)$ is m-degenerate. Namely, there are m linear independent functions $\psi_\mu(x)$ with the same eigenvalue E:

$$H(x)\psi_\mu(x) = E\psi_\mu(x), \qquad \mu = 1,\ 2,\ \cdots,\ m, \tag{1.1}$$

where x briefly denotes the set of coordinates for all degrees of freedom. Any linear combination $\phi(x)$ of $\psi_\mu(x)$ is an eigenfunction of $H(x)$ with the same eigenvalue E:

$$\phi(x) = \sum_{\mu=1}^{m} \psi_\mu(x)a_\mu, \qquad H(x)\phi(x) = E\phi(x). \tag{1.2}$$

Two eigenfunctions satisfy the linear rule:

$$c\left(\sum_{\mu=1}^{m} \psi_\mu(x)a_\mu + \sum_{\mu=1}^{m} \psi_\mu(x)b_\mu\right) = \sum_{\mu=1}^{m} \psi_\mu(x)\left(c\, a_\mu + c\, b_\mu\right). \tag{1.3}$$

Conversely, any eigenfunction of $H(x)$ with the eigenvalue E can be expressed as a linear combination of $\psi_\mu(x)$ like (1.2). The set of $\phi(x)$ is called a linear space $\mathcal{L}$ of dimension m, generated by m **basis functions** $\psi_\mu(x)$. $\phi(x)$ is called a **vector** in $\mathcal{L}$ and a_μ is the μth-component of the vector $\phi(x)$ with respect to the basis functions $\psi_\mu(x)$.

In linear algebras, m objects $\boldsymbol{e}_\mu$ are said to be linearly dependent if there exist m coefficients c_μ which are not vanishing simultaneously such that

$$\sum_{\mu=1}^{m} \boldsymbol{e}_\mu c_\mu = 0. \tag{1.4}$$

Otherwise, $\boldsymbol{e}_\mu$ are **linearly independent** to each other. Assume that $\boldsymbol{e}_\mu$ satisfy the following linear formulas:

$$\begin{aligned} \boldsymbol{e}_\mu a_\mu + \boldsymbol{e}_\nu a_\nu &= \boldsymbol{e}_\nu a_\nu + \boldsymbol{e}_\mu a_\mu, \\ c\left(\sum_\mu \boldsymbol{e}_\mu a_\mu + \sum_\mu \boldsymbol{e}_\mu b_\mu\right) &= \sum_\mu \boldsymbol{e}_\mu \left(c a_\mu + c b_\mu\right), \end{aligned} \tag{1.5}$$

where c, a_μ, a_ν, and b_μ are arbitrary complex numbers. Let $\mathcal{L}$ denote the set of all possible complex combinations $\boldsymbol{a}$ of $\boldsymbol{e}_\mu$:

$$\boldsymbol{a} = \sum_{\mu=1}^{m} \boldsymbol{e}_\mu a_\mu. \tag{1.6}$$

$\mathcal{L}$ is called a m-dimensional **linear space** generated by m **basis vectors** $\boldsymbol{e}_\mu$. $\boldsymbol{a}$ is a **vector** in $\mathcal{L}$ and a_μ is its μth component. The basis vector is a special vector whose components all are zero except for one which is equal to one,

$$\left(\boldsymbol{e}_\mu\right)_\nu = \delta_{\mu\nu} = \begin{cases} 1 & \text{when } \mu = \nu, \\ 0 & \text{when } \mu \neq \nu, \end{cases} \tag{1.7}$$

where $\delta_{\mu\nu}$ is the Kronecker δ function. All possible real combinations $\boldsymbol{a}$ of $\boldsymbol{e}_\mu$ span a **real linear space**. A vector is called a **null vector** if its components all vanish. Two vectors $\boldsymbol{a}$ and $\boldsymbol{b}$ are said to be equal to each other if and only if their components are respectively equal, $a_\mu = b_\mu$. In linear algebras, the concepts of vectors and a linear space are independent of the physical content of the objects.

For a given linear space $\mathcal{L}$ and a given set of basis vectors, a vector $\boldsymbol{a}$ is completely described by the m components a_μ. Usually, the m ordered numbers are arranged as a column-matrix $\underline{a}$,

$$\underline{a} = \begin{pmatrix} a_1 \\ a_2 \\ \vdots \\ a_m \end{pmatrix}. \tag{1.8}$$

The column-matrix $\underline{a}$ is another form to denote a vector $\boldsymbol{a}$. For a given set of basis vectors $\boldsymbol{e}_\mu$, there is a one-to-one correspondence between $\underline{a}$ and $\boldsymbol{a}$.

n vectors $\boldsymbol{a}^{(1)}$, $\boldsymbol{a}^{(2)}$, $\cdots$, $\boldsymbol{a}^{(n)}$ are linearly independent if there does not exist a linear relation: $\sum_{i=1}^{n} \boldsymbol{a}^{(i)} c_i = 0$, where n coefficients c_i are not vanishing simultaneously. In an m-dimensional space $\mathcal{L}$ the number n of linearly independent vectors is not larger than m. In $\mathcal{L}$, n linearly independent vectors generate a subspace $\mathcal{L}_1$ of dimension n. A subspace is called a null space $\emptyset$ if it contains only the null vector. **The whole space $\mathcal{L}$ and the null space $\emptyset$ are two trivial subspaces.**

The sum of two subspaces $\mathcal{L}_1$ and $\mathcal{L}_2$ in $\mathcal{L}$ is a subspace, denoted by $\mathcal{L}_1 + \mathcal{L}_2$, which contains all linear combinations of the vectors belonging to $\mathcal{L}_1$ and $\mathcal{L}_2$. The intersection of two subspaces is a subspace, denoted by $\mathcal{L}_1 \cap \mathcal{L}_2$, which contains all vectors belonging to both subspaces.

$\mathcal{L}$ is said to be the direct sum of two subspaces, $\mathcal{L} = \mathcal{L}_1 \oplus \mathcal{L}_2$, if $\mathcal{L} = \mathcal{L}_1 + \mathcal{L}_2$, and one of the following three equivalent conditions is satisfied.

(1) The intersection of $\mathcal{L}_1$ and $\mathcal{L}_2$ is a null space.

(2) The dimension of $\mathcal{L}$ is equal to the sum of the dimensions of $\mathcal{L}_1$ and $\mathcal{L}_2$.

(3) Each vector in $\mathcal{L}$ can be expressed uniquely as a sum of two vectors, respectively belonging to two subspaces $\mathcal{L}_1$ and $\mathcal{L}_2$.

$\mathcal{L}_2$ is called the **complement** of $\mathcal{L}_1$ in $\mathcal{L}$ if $\mathcal{L} = \mathcal{L}_1 \oplus \mathcal{L}_2$. $\mathcal{L}_1$ is also the complement of $\mathcal{L}_2$. The complement of $\mathcal{L}_1$ in $\mathcal{L}$ is not unique. A space $\mathcal{L}$ can be decomposed as a direct sum of some subspaces more than two.

1.2 Matrices

Matrices are widely used in group theory. X is called an $m \times n$ matrix if there are mn numbers $X_{\mu\nu}$ arranged as a rectangular:

$$X = \begin{pmatrix} X_{11} & X_{12} & \cdots & X_{1n} \\ X_{21} & X_{22} & \cdots & X_{2n} \\ \vdots & \vdots & \vdots & \vdots \\ X_{m1} & X_{m2} & \cdots & X_{mn} \end{pmatrix}, \tag{1.9}$$

where $X_{\mu\nu}$ is the element of the matrix at the ν-column of the μ-row, and μ and ν are its row index and its column index, respectively. Y is called the transpose of X, $Y = X^T$, if $Y_{\mu\nu} = X_{\nu\mu}$. Y is called the complex conjugate of X, $Y = X^*$, if $Y_{\mu\nu} = X^*_{\mu\nu}$. Y is called the conjugate of X, $Y = X^\dagger$, if $Y_{\mu\nu} = X^*_{\nu\mu}$. X is called a null matrix, $X = 0$, if all elements of X vanish.

The product SR of two matrices S and R can be defined only if the column number k of the left-multiplying matrix S is equal to the row number of the right-multiplying matrix R:

$$(SR)_{\mu\nu} = \sum_{\rho=1}^{k} S_{\mu\rho} R_{\rho\nu}. \tag{1.10}$$

X is a column matrix if its column number n is one, and $X^\dagger$ is a row matrix. The product $X^\dagger X$ is a non-negative real number, called the square module of a column matrix X. X is called normalized if $X^\dagger X = 1$. Two column matrices X and Y are called orthogonal to each other if $Y^\dagger X = 0$. m column matrices are called orthonormal if $X_a^\dagger X_b = \delta_{ab}$. From the definitions (1.7), (1.8) and (1.10), the basis vectors $\boldsymbol{e}_\mu$ are orthonormal.

An $m \times m$ matrix X is called a square matrix with dimension m, or simply called a matrix. For a square matrix X, $X_{\mu\mu}$ is called its diagonal element. A matrix X is called diagonal if its non-diagonal elements all vanish: $X_{\mu\nu} = 0$ if $\mu \neq \nu$. A diagonal matrix is called constant if its diagonal elements are equal to each other, and is called the unit matrix $\mathbf{1}$ if its diagonal elements all are one. A constant matrix is a product of a constant c and a unit matrix: $c\mathbf{1}$.

Generally, the order of the product of two matrices with the same dimension cannot be exchanged:

$$[X,\ Y] \equiv XY - YX \neq 0, \qquad XY \neq YX. \tag{1.11}$$

$[X,\ Y]$ is called the commutator of X and Y. Y is called commutable with X if $[X,\ Y] = 0$.

The sum of all diagonal elements of a matrix X is called its trace, $\text{Tr}\ X = \sum_{\mu=1}^{m} X_{\mu\mu}$. The determinant of X is defined as

$$\begin{aligned} \det X &= \sum_{\mu_1\mu_2\cdots\mu_m} \epsilon_{\mu_1\mu_2\cdots\mu_m} X_{1\mu_1} X_{2\mu_2} \cdots X_{m\mu_m}, \\ \epsilon_{\nu_1\nu_2\cdots\nu_m} \det X &= \sum_{\mu_1\mu_2\cdots\mu_m} \epsilon_{\mu_1\mu_2\cdots\mu_m} X_{\nu_1\mu_1} X_{\nu_2\mu_2} \cdots X_{\nu_m\mu_m}, \end{aligned} \tag{1.12}$$

where $\epsilon_{\mu_1\mu_2\cdots\mu_m}$ is called the totally antisymmetric unit tensor of dimension m, which is antisymmetric with respect to the transposition of any pair of indices, and satisfies $\epsilon_{12\dots m} = 1$. It is easy to see

$$\sum_d \epsilon_{abd}\epsilon_{rsd} = \delta_{ar}\delta_{bs} - \delta_{as}\delta_{br}.$$

Generally,

$$\frac{1}{(m-n)!}\sum_{a_{n+1}\cdots a_m}\epsilon_{a_1\cdots a_n a_{n+1}\cdots a_m}\epsilon_{b_1\cdots b_n a_{n+1}\cdots a_m} = \sum_{p_1\cdots p_n}\epsilon_{p_1\cdots p_n}\delta_{a_1 b_{p_1}}\delta_{a_2 b_{p_2}}\cdots\delta_{a_n b_{p_n}}, \tag{1.13}$$

where $\epsilon_{p_1\cdots p_n}$ is an antisymmetric unit tensor of dimension n.

X is called a **non-singular matrix** if $\det X \neq 0$. From the definition (1.12), the sufficient and necessary condition for a matrix X to be non-singular is that the m column matrices of X are linearly independent, so its m row matrices. For a non-singular matrix X there exists its **inverse** X^{-1} satisfying

$$X^{-1}X = XX^{-1} = \mathbf{1}. \tag{1.14}$$

A matrix u is called **unitary** if its conjugate $u^\dagger$ is equal to its inverse:

$$u^\dagger u = uu^\dagger = \mathbf{1}, \qquad \sum_{\rho=1}^{m} u_{\rho\mu}^* u_{\rho\nu} = \sum_{\rho=1}^{m} u_{\mu\rho}u_{\nu\rho}^* = \delta_{\mu\nu}. \tag{1.15}$$

Namely, the column matrices $\left(\underline{u_{\cdot\nu}}\right)_\mu$ of u are orthonormal, so its row matrices $\left(\underline{u_{\mu\cdot}}\right)_\nu$. A real unitary matrix R is called **real orthogonal**: $R^* = R$ and $R^T = R^{-1}$. A matrix H is called **hermitian** if

$$H^\dagger = H, \qquad H_{\mu\nu} = H_{\nu\mu}^*. \tag{1.16}$$

The diagonal element of a hermitian matrix H must be real. A real hermitian matrix S is called **real symmetric**: $S_{\mu\nu} = S_{\mu\nu}^* = S_{\nu\mu}$.

1.3 Linear Transformations and Linear Operators

A transformation gives a rule to change a function $\psi(x)$ into another function $\phi(x)$. An operator $R(x)$ is the mathematical symbol to denote a transformation, $R(x)\psi(x) = \phi(x)$. An operator $R(x)$ is linear if it satisfies

$$R(x)\left\{c_1\phi_1(x) + c_2\phi_2(x)\right\} = c_1R(x)\phi_1(x) + c_2R(x)\phi_2(x), \tag{1.17}$$

where the coefficients c_1 and c_2 are constant. A linear operator describes a linear transformation. The operators used in this textbook are always linear unless specified otherwise. The multiplication $S(x)R(x)$ of two operators is an operator, defined as a successive application to the function first with $R(x)$ and then with $S(x)$. Namely, if $R(x)\psi(x) = \phi(x)$, then

$S(x)R(x)\psi(x) = S(x)\phi(x)$. Two operators are generally non-commutable: $R(x)S(x) \neq S(x)R(x)$.

If an operator $R(x)$ commutes with the Hamiltonian $H(x)$,

$$[H(x), R(x)] \equiv H(x)R(x) - R(x)H(x) = 0, \tag{1.18}$$

the application of $R(x)$ to the eigenfunction $\psi_\mu(x)$ of $H(x)$ is still an eigenfunction of $H(x)$ with the same eigenvalue E,

$$H(x)\{R(x)\psi_\mu(x)\} = R(x)\{H(x)\psi_\mu(x)\} = E\{R(x)\psi_\mu(x)\}. \tag{1.19}$$

Thus, $R(x)\psi_\mu(x)$ belongs to the linear space $\mathcal{L}$ generated by m eigenfunctions $\psi_\mu(x)$ of $H(x)$ with the eigenvalue E such that

$$R(x)\psi_\mu(x) = \sum_\nu \psi_\nu(x)D_{\nu\mu}(R). \tag{1.20}$$

$\mathcal{L}$ is called an **invariant space** to $R(x)$ if $R(x)\phi(x) \in \mathcal{L}$, $\forall \phi(x) \in \mathcal{L}$. The coefficients $D_{\nu\mu}(R)$ are arranged as a matrix $D(R)$ of dimension m, called the matrix of an operator $R(x)$ in its invariant space $\mathcal{L}$ with respect to the basis functions $\psi_\mu(x)$, or simply called the matrix of $R(x)$ in $\mathcal{L}$. $D(R)$ **depends on the operator** $R(x)$, **but not on** x.

In linear algebras, a linear operator R describes a transformation of vectors in a linear space $\mathcal{L}$ satisfying

$$R\{c_1\boldsymbol{a} + c_2\boldsymbol{b}\} = c_1R\,\boldsymbol{a} + c_2R\,\boldsymbol{b}. \tag{1.21}$$

$\mathcal{L}$ is called invariant to R if $R\boldsymbol{a} \in \mathcal{L}$, $\forall \boldsymbol{a} \in \mathcal{L}$. The matrix $D(R)$ of R in its invariant space $\mathcal{L}$ with respect to the basis vectors $\boldsymbol{e}_\mu$ is calculated as:

$$R\,\boldsymbol{e}_\mu = \sum_\nu \boldsymbol{e}_\nu D_{\nu\mu}(R). \tag{1.22}$$

The action of R on any vector $\boldsymbol{a} = \sum_\mu \boldsymbol{e}_\mu a_\mu \in \mathcal{L}$ can be calculated by $D(R)$:

$$R\boldsymbol{a} = \sum_\mu (R\boldsymbol{e}_\mu)\,a_\mu = \sum_\nu \boldsymbol{e}_\nu\left[\sum_\mu D_{\nu\mu}(R)a_\mu\right] = \boldsymbol{b} = \sum_\nu \boldsymbol{e}_\nu b_\nu,$$

$$(R\boldsymbol{a})_\nu = b_\nu = \sum_\mu D_{\nu\mu}(R)a_\mu, \qquad \underline{b} = D(R)\,\underline{a}. \tag{1.23}$$

It is worthy to emphasize the difference between Eqs. (1.22) and (1.23). Both equations describe the transformation of vectors in the action of an operator R. Equation (1.22) is a vector equation where a basis vector $\boldsymbol{e}_\mu$ transforms to a combination of basis vectors. The summed index ν is the order index of the basis vector and it is the **row index** of $D_{\nu\mu}(R)$. Equation (1.23) is a component equation where a vector $\boldsymbol{a}$ transforms to another vector $\boldsymbol{b}$. The summed index μ is the index of the vector component and it is the **column index** of $D_{\nu\mu}(R)$. Since a basis vector is also a vector, two equations are consistent:

$$(R\boldsymbol{e}_\mu)_\rho = \sum_\lambda D_{\rho\lambda}(R)\,(\boldsymbol{e}_\mu)_\lambda = D_{\rho\mu}(R) = \sum_\nu (\boldsymbol{e}_\nu)_\rho\, D_{\nu\mu}(R). \qquad (1.24)$$

1.4 Similarity Transformation

For a given set of basis vectors $\boldsymbol{e}_\mu$ in a linear space $\mathcal{L}$ of dimension m, there is **one-to-one correspondence (1.6) between a vector $\boldsymbol{a}$ and its column-matrix** $\underline{a}$, and there is **one-to-one correspondence (1.22) between an operator R and its matrix** $D(R)$. However, the choice of the basis vectors in $\mathcal{L}$ is not unique. Any set of m linearly independent vectors can be chosen to be the basis vectors. In this section we will discuss how the column-matrix of a vector and the matrix of an operator change when the basis vectors are changed.

Let $\boldsymbol{e}'_\nu$ denote m linearly independent vectors with the components $S_{\mu\nu}$ in the original basis vectors $\boldsymbol{e}_\mu$,

$$\boldsymbol{e}'_\nu = \sum_\mu \boldsymbol{e}_\mu S_{\mu\nu}, \qquad \underline{e'_\nu} = \underline{S_{\cdot\nu}}. \qquad (1.25)$$

Since $\boldsymbol{e}'_\nu$ are linearly independent to each other, S is a nonsingular matrix ($\det S \neq 0$) and has its inverse matrix S^{-1}.

$$\boldsymbol{e}_\mu = \sum_\nu \boldsymbol{e}'_\nu \left(S^{-1}\right)_{\nu\mu}. \qquad (1.26)$$

Choosing $\boldsymbol{e}'_\nu$ to be new basis vectors, the components a'_ν of the vector $\boldsymbol{a}$ and the matrix $\overline{D}(R)$ of the operator R can be calculated as follows:

$$\boldsymbol{a} = \sum_\mu \boldsymbol{e}_\mu a_\mu = \sum_\nu \boldsymbol{e}'_\nu \sum_\mu \left(S^{-1}\right)_{\nu\mu} a_\mu = \sum_\nu \boldsymbol{e}'_\nu a'_\nu,$$

$$a'_\nu = \sum_\mu \left(S^{-1}\right)_{\nu\mu} a_\mu, \qquad \underline{a}' = S^{-1}\underline{a}. \qquad (1.27)$$

$$R\boldsymbol{e}'_\nu = \sum_\rho (R\boldsymbol{e}_\rho)\, S_{\rho\nu} = \sum_\mu \boldsymbol{e}_\mu \sum_\rho D_{\mu\rho}(R) S_{\rho\nu},$$

$$Re'_\nu = \sum_\rho e'_\rho \overline{D}_{\rho\nu}(R) = \sum_\mu e_\mu \sum_\rho S_{\mu\rho}\overline{D}_{\rho\nu}(R),$$

$$\sum_\rho D_{\mu\rho}(R)S_{\rho\nu} = \sum_\rho S_{\mu\rho}\overline{D}_{\rho\nu}(R), \qquad \overline{D}(R) = S^{-1}D(R)S. \tag{1.28}$$

Eq. (1.28) can be re-written as

$$D(R)\ \underline{S_{\cdot\nu}} = \sum_\rho \underline{S_{\cdot\rho}}\ \overline{D}_{\rho\nu}(R), \qquad Re'_\nu = \sum_\rho e'_\rho \overline{D}_{\rho\nu}(R). \tag{1.29}$$

It is nothing but the definition of the matrix of R in the new basis vectors e'_ν. The relation (1.28) between $\overline{D}(R)$ and $D(R)$ is called a **similarity transformation** and two matrices $\overline{D}(R)$ and $D(R)$ are said to be **equivalent** to each other. S is called the matrix of the similarity transformation. Note that, the matrix of the similarity transformation for two equivalent matrices $D(R)$ and $\overline{D}(R)$ is **not unique**. If X commutes with $D(R)$ and Y commutes with $\overline{D}(R)$, XSY is also a matrix of the similarity transformation between $D(R)$ and $\overline{D}(R)$. The similarity transformation is called the **simple** one, if the new set of basis vectors is the same as the original one except for the order of basis vectors. **Two matrices are equivalent if they are equivalent to the same matrix**:

$$\begin{aligned} X^{-1}D(R)X &= M = Y^{-1}\overline{D}(R)Y, \\ \overline{D}(R) &= \left(XY^{-1}\right)^{-1} D(R)\left(XY^{-1}\right). \end{aligned} \tag{1.30}$$

If $\boldsymbol{b} = R\boldsymbol{a}$, one has $\underline{b} = D(R)\underline{a}$ in the original basis vectors e_μ. In the new set of basis vectors e'_ν one has

$$\underline{b}' = S^{-1}\underline{b} = S^{-1}D(R)\underline{a} = S^{-1}D(R)S\underline{a}' = \overline{D}(R)\underline{a}'. \tag{1.31}$$

Namely, **the different choices of the basis vectors do not change the action of an operator on a vector.**

If the new basis vectors $\underline{e'_\nu} = \underline{S_{\cdot\nu}}$ are orthonormal, too,

$$\sum_{\rho=1}^{m} \left(\underline{e'_\nu}\right)^*_\rho \left(\underline{e'_\mu}\right)_\rho = \sum_{\rho=1}^{m} S^*_{\rho\nu}S_{\rho\mu} = \delta_{\nu\mu}, \tag{1.32}$$

S is unitary due to Eq. (1.15). **Through a unitary similarity transformation S, a unitary matrix u becomes a unitary matrix**, and a **hermitian matrix H becomes a hermitian matrix**:

$$\left(S^{-1}uS\right)^\dagger = \left(S^{-1}uS\right)^{-1}, \qquad \left(S^{-1}HS\right)^\dagger = S^{-1}HS. \tag{1.33}$$

Let $\mathcal{L}$ denote an m-dimensional space, $\mathcal{L}_1$ be its n-dimensional subspace, invariant to the operator R, and $\mathcal{L}_2$ be the complement of $\mathcal{L}_1$ in $\mathcal{L}$. Choose a new set of basis vectors in $\mathcal{L}$ such that the first n basis vectors belong to $\mathcal{L}_1$ and the next $(m-n)$ ones belong to $\mathcal{L}_2$. Arrange the new basis vectors e'_ν to be the column matrices of S, $\underline{S_{\cdot\nu}} = \underline{e'_\nu}$. Through the similarity transformation S the matrix $D(R)$ of R is changed to $\overline{D}(R)$. Since $\mathcal{L}_1$ is invariant to R, one has

$$Re'_\mu = \sum_{\nu=1}^{n} e'_\nu \overline{D}_{\nu\mu}(R), \qquad 1 \leq \mu \leq n. \tag{1.34}$$

Namely, the down-left corner of $\overline{D}(R)$ is vanishing,

$$S^{-1}D(R)S = \overline{D}(R) = \begin{pmatrix} D^{(1)}(R) & M \\ 0 & D^{(2)}(R) \end{pmatrix}. \tag{1.35}$$

This matrix $\overline{D}(R)$ is called a **ladder matrix**. Furthermore, if $\mathcal{L}_2$ is also invariant to R, one has $M = 0$,

$$\overline{D}(R) = \begin{pmatrix} D^{(1)}(R) & 0 \\ 0 & D^{(2)}(R) \end{pmatrix} = D^{(1)}(R) \oplus D^{(2)}(R). \tag{1.36}$$

This matrix $\overline{D}(R)$ is a direct sum of two submatrices, and is called a **block matrix** in the type $[n, (m-n)]$. Generally, a matrix is also called a block one if it can be changed into the direct sum of submatrices by a simple similarity transformation. In other words, X is a block matrix if $X\Lambda = \Lambda X$ where Λ is diagonal.

1.5 Eigenvectors and Diagonalization of a Matrix

In quantum mechanics, the eigenequation of a physical operator $R(x)$ is

$$R(x)\psi(x) = \lambda\psi(x).$$

The eigenvalue λ is the possible observed value for the physical quantity. The eigenvalues describe the characteristic of the physical quantity and are independent of the choice of basis functions. In linear algebras, the eigenequation of an operator R is

$$R\, \boldsymbol{a} = \lambda\, \boldsymbol{a}. \tag{1.37}$$

If $\boldsymbol{a} \in \mathcal{L}$ and $\mathcal{L}$ is an m-dimensional space invariant to R, one has

$$D(R)\,\underline{a} = \lambda\,\underline{a}, \qquad \sum_{\nu} D_{\mu\nu}(R)\,a_\nu = \lambda\,a_\mu. \tag{1.38}$$

The eigenequation (1.38) is a set of coupled linear homogeneous equations for m variables a_ν. The condition for existence of a non-vanishing solution is that the coefficient determinant of the coupled equations vanishes:

$$\det\left[D(R) - \lambda \mathbf{1}\right] = 0. \tag{1.39}$$

Equation (1.39) is called the secular equation of $D(R)$, and its roots are the eigenvalues of $D(R)$. The secular equation is invariant in similarity transformation of $D(R)$, so that the eigenvalues are independent of the choice of the basis vectors. From Eq. (1.39), the sum and the product of the eigenvalues of $D(R)$ are its trace, Tr $D(R)$, and its determinant, $\det D(R)$, respectively (see Prob. 3).

Equation (1.37) shows that the eigenvector generates a one-dimensional subspace which is invariant to R. **The eigenvectors with different eigenvalues are linearly independent.** If there exist m linearly independent eigenvectors $\boldsymbol{a}^{(\nu)}$ of R in its invariant space $\mathcal{L}$ of dimension m, one may choose a new set of basis vectors $\boldsymbol{e}'_\nu = \boldsymbol{a}^{(\nu)}$ such that the matrix of R with respect to the new basis vectors $\boldsymbol{e}'_\nu$ is a diagonal matrix with the diagonal elements λ_ν:

$$\begin{aligned} &R\boldsymbol{a}^{(\nu)} = D(R)\boldsymbol{a}^{(\nu)} = \lambda_\nu \boldsymbol{a}^{(\nu)}, \qquad && \underline{S_{\cdot\nu}} = \underline{a^{(\nu)}}, \\ &S^{-1}D(R)S = \Lambda, && \Lambda_{\nu\mu} = \delta_{\nu\mu}\lambda_\nu. \end{aligned} \tag{1.40}$$

Therefore, **the key for diagonalizing an m-dimensional matrix is to find its m linearly independent eigenvectors.** Since the eigenequation (1.38) is linearly homogeneous with respect to the eigenvector $\boldsymbol{a}^{(\nu)}$, $\boldsymbol{a}^{(\nu)}$ can be multiplied by a constant c_ν. When the eigenvalue is degenerate, its eigenvectors can be made a nonsingular linear combination. This is the reason why the similarity transformation matrix S is not unique. The number of the arbitrary parameters in S is equal to $\sum_\nu n_\nu^2$, where n_ν is the multiplicity of the eigenvalue λ_ν. Those parameters play an important role in calculating a common similarity transformation for a few pairs of equivalent matrices, which is often used in group theory (see Prob. 8).

Substituting an eigenvalue into Eq. (1.38), one is able to solve at least one eigenvector. However, when an eigenvalue is a root of Eq. (1.39) with multiplicity n, it is not certain to find n linearly independent eigenvectors with the given eigenvalue by solving Eq. (1.38). The following matrix is the simplest example which has the eigenvalue 1 with multiplicity two, but only has one linearly independent eigenvector,

$$\begin{pmatrix} 1 & b \\ 0 & 1 \end{pmatrix} \begin{pmatrix} 1 \\ 0 \end{pmatrix} = \begin{pmatrix} 1 \\ 0 \end{pmatrix}, \qquad b \neq 0. \tag{1.41}$$

This matrix cannot be diagonalized by a similarity transformation. Generally, if all eigenvalues λ of a matrix R are the same, R can be written in the standard Jordan forms by a similarity transformation:

$$R_{ab} = \begin{cases} \lambda & \text{when } \ a = b, \\ 0 \ \text{ or } \ 1 & \text{when } \ a + 1 = b, \\ 0 & \text{the remaining cases.} \end{cases} \tag{1.42}$$

A hermitian matrix and a unitary matrix both can be diagonalized by a unitary similarity transformation (see Prob. 10). Namely, there exist m orthonormal eigenvectors for a hermitian matrix and for a unitary matrix with dimension m. Thus, the module of the eigenvalue of a unitary matrix is one, and the eigenvalue of a hermitian matrix is real. A real symmetric matrix can be diagonalized by a real orthogonal similarity transformation. A real orthogonal matrix can be diagonalized by a unitary similarity transformation, but generally not by a real orthogonal one, because its eigenvalue may not be real.

1.6 Inner Product of Vectors

There are three types of product of vectors where the product may be a scalar, a vector, or a tensor, respectively. **The product of two vectors is called the inner product if the product is a scalar.** In quantum mechanics the inner product of two wave functions is defined as

$$\langle \phi(x) | \psi(x) \rangle = \int (dx) \phi(x)^* \psi(x), \tag{1.43}$$

where the sign $\int$ denotes an integral for the continuous coordinate and a sum for the discrete coordinate. The form on the left-hand side of Eq. (1.43) is called the Dirac symbol. Generally, the inner product of two vectors $\langle \boldsymbol{a} | \boldsymbol{b} \rangle$ in linear algebras satisfies

$$\begin{aligned} &\langle c_1 \boldsymbol{a}^{(1)} + c_2 \boldsymbol{a}^{(2)} | \boldsymbol{b} \rangle = c_1^* \langle \boldsymbol{a}^{(1)} | \boldsymbol{b} \rangle + c_2^* \langle \boldsymbol{a}^{(2)} | \boldsymbol{b} \rangle, \\ &\langle \boldsymbol{a} | c_1 \boldsymbol{b}^{(1)} + c_2 \boldsymbol{b}^{(2)} \rangle = c_1 \langle \boldsymbol{a} | \boldsymbol{b}^{(1)} \rangle + c_2 \langle \boldsymbol{a} | \boldsymbol{b}^{(2)} \rangle, \\ &\langle \boldsymbol{b} | \boldsymbol{a} \rangle = \langle \boldsymbol{a} | \boldsymbol{b} \rangle^*, \qquad \langle \boldsymbol{a} | \boldsymbol{a} \rangle = |\boldsymbol{a}|^2 > 0, \ \text{ if } \boldsymbol{a} \neq 0. \end{aligned} \tag{1.44}$$

The inner product is linear for the second vector and antilinear for the first vector. The inner product becomes its complex conjugate if changing the order of two vectors. The self-inner product of a non-vanishing vector is a positive real number, called the square module of the vector. A vector is normal if its module is one. Two vectors $\boldsymbol{a}$ and $\boldsymbol{b}$ are orthogonal to each other if $\langle \boldsymbol{a}|\boldsymbol{b}\rangle = 0$. In quantum mechanics one gets used to that the basis vectors $\boldsymbol{e}_\mu$ are always chosen to be orthonormal. However, sometimes in group theory, one has to face the basis vectors not to be orthonormal.

Let Ω denote the inner product of two basis vectors:

$$\langle \boldsymbol{e}_\mu|\boldsymbol{e}_\nu\rangle = \Omega_{\mu\nu}, \qquad \Omega_{\nu\mu} = \Omega_{\mu\nu}^* = \left(\Omega^\dagger\right)_{\nu\mu}. \tag{1.45}$$

Ω is a hermitian matrix. The inner product of two arbitrary vectors can be calculated by Ω:

$$\boldsymbol{a} = \sum_\mu \boldsymbol{e}_\mu a_\mu, \qquad \boldsymbol{b} = \sum_\nu \boldsymbol{e}_\nu b_\nu, \qquad \langle \boldsymbol{a}|\boldsymbol{b}\rangle = \sum_{\mu\nu} a_\mu^* \Omega_{\mu\nu} b_\nu, \tag{1.46}$$

Let $\boldsymbol{a}$ denote an non-vanishing eigenvector of Ω with the eigenvalue λ,

$$\begin{gathered}\sum_\nu \Omega_{\mu\nu} a_\nu = \lambda a_\mu, \qquad \boldsymbol{a} = \sum_\nu \boldsymbol{e}_\nu a_\nu \neq 0, \\ \lambda \sum_\mu |a_\mu|^2 = \sum_{\mu\nu} a_\mu^* \Omega_{\mu\nu} a_\nu = \sum_{\mu\nu} a_\mu^* \langle \boldsymbol{e}_\mu|\boldsymbol{e}_\nu\rangle a_\nu = \langle \boldsymbol{a}|\boldsymbol{a}\rangle > 0.\end{gathered} \tag{1.47}$$

Hence, the eigenvalue λ of Ω is positive. **A hermitian matrix is called positive (negative) definite if its eigenvalues are all positive (negative).** Thus, Ω is positive definite.

The matrix of an operator R in an invariant linear space $\mathcal{L}$ with respect to the basis vectors $\boldsymbol{e}_\mu$ is defined in Eq. (1.22) which is independent of the definition of inner product. However, if the basis vectors $\boldsymbol{e}_\mu$ in $\mathcal{L}$ are not orthonormal, one has

$$\langle \boldsymbol{e}_\mu|R\boldsymbol{e}_\nu\rangle = \sum_\rho \langle \boldsymbol{e}_\mu|\boldsymbol{e}_\rho\rangle D_{\rho\nu} = \sum_\rho \Omega_{\mu\rho} D_{\rho\nu}. \tag{1.48}$$

The definition for the adjoint operator $R^\dagger$ in quantum mechanics is generalized from the definition of the conjugate matrix $D(R)^\dagger$:

$$\begin{aligned}&\left[\underline{a}^\dagger\ D(R)\ \underline{b}\right]^* = \left[D(R)\underline{b}\right]^\dagger\ \underline{a} = \underline{b}^\dagger\ D^\dagger(R)\ \underline{a}, \\ &\langle \boldsymbol{a}|R\boldsymbol{b}\rangle^* = \langle R\boldsymbol{b}|\boldsymbol{a}\rangle = \langle \boldsymbol{b}|R^\dagger \boldsymbol{a}\rangle.\end{aligned} \tag{1.49}$$

The adjoint relation between two operators is mutual. An operator is called hermitian if $H^\dagger = H$, and an operator is called unitary if $U^\dagger U = \mathbf{1}$. Note that the matrices of two adjoint operators are not conjugate to each other if the basis vectors are not orthonormal. Let $D(R)$ and X denote the matrices of two operators R and $R^\dagger$, respectively

$$Re_\mu = \sum_\rho e_\rho D_{\rho\mu}(R), \qquad R^\dagger e_\nu = \sum_\rho e_\rho X_{\rho\nu},$$

$$\begin{gathered}\sum_\rho D^*_{\rho\mu}(R)\langle e_\rho|e_\nu\rangle = \langle Re_\mu|e_\nu\rangle = \langle e_\mu|R^\dagger e_\nu\rangle = \sum_\rho \langle e_\mu|e_\rho\rangle X_{\rho\nu}, \\ D^\dagger(R)\Omega = \Omega X, \qquad X = \Omega^{-1}D^\dagger(R)\Omega.\end{gathered} \tag{1.50}$$

Only for the orthonormal basis vectors e_μ, $\Omega_{\mu\nu} = \delta_{\mu\nu}$, $\langle a|b\rangle = \underline{a}^\dagger \underline{b}$, $\langle e_\mu|Re_\nu\rangle = D_{\mu\nu}(R)$, and a hermitian (or an unitary) operator corresponds to a hermitian (or a unitary) matrix.

At last, there are a few different definitions for the inner product of vectors. Another inner product is defined to be linear for both factors,

$$\begin{aligned}&\langle c_1 a_1 + c_2 a_2|b\rangle = c_1\langle a_1|b\rangle + c_2\langle a_2|b\rangle, \\ &\langle a|c_1 b_1 + c_2 b_2\rangle = c_1\langle a|b_1\rangle + c_2\langle a|b_2\rangle, \\ &\langle e_\mu|e_\nu\rangle = \Omega_{\mu\nu} = \Omega_{\nu\mu}, \qquad \det\Omega \neq 0.\end{aligned} \tag{1.51}$$

In this definition, the self-inner product of a vector may not be real.

1.7 The Direct Product of Matrices

Discuss a quantum system composed by two subsystems. Let two functional spaces $\mathcal{L}_1$ and $\mathcal{L}_2$ for two subsystems, respectively, both be invariant to the operator R:

$$\begin{aligned}&\psi_\mu(x^{(1)}) \in \mathcal{L}_1, \qquad R\psi_\mu(x^{(1)}) = \sum_{\nu=1}^m \psi_\nu(x^{(1)})D^{(1)}_{\nu\mu}(R), \\ &\psi_\mu(x^{(2)}) \in \mathcal{L}_2, \qquad R\phi_\rho(x^{(2)}) = \sum_{\lambda=1}^n \phi_\lambda(x^{(2)})D^{(2)}_{\lambda\rho}(R).\end{aligned} \tag{1.52}$$

The total wave functions of the composed system are expressed as the products of two wave functions of the subsystems or their combination:

$$\psi_\mu(x^{(1)})\phi_\rho(x^{(2)}), \qquad 1 \leq \mu \leq m, \ \ 1 \leq \rho \leq n, \tag{1.53}$$

and generate the functional space $\mathcal{L}$ of the composed system. The (mn)-dimensional space $\mathcal{L}$ is called the direct product of two subspaces, $\mathcal{L} = \mathcal{L}_1 \otimes \mathcal{L}_2$ and is also invariant to R,

$$
\begin{aligned}
R\left[\psi_\mu(x^{(1)})\phi_\rho(x^{(2)})\right] &= \sum_{\nu=1}^{m}\sum_{\lambda=1}^{n} \psi_\nu(x^{(1)})D^{(1)}_{\nu\mu}(R)\phi_\lambda(x^{(2)})D^{(2)}_{\lambda\rho}(R), \\
&= \sum_{\nu=1}^{m}\sum_{\lambda=1}^{n} \psi_\nu(x^{(1)})\phi_\lambda(x^{(2)})\left[D^{(1)}(R)\times D^{(2)}(R)\right]_{\nu\lambda,\mu\rho}, \\
D^{(1)}_{\nu\mu}(R)D^{(2)}_{\lambda\rho}(R) &= \left[D^{(1)}(R)\times D^{(2)}(R)\right]_{\nu\lambda,\mu\rho}.
\end{aligned}
\tag{1.54}
$$

The (mn)-dimensional matrix $D^{(1)}(R)\times D^{(2)}(R)$ of R in $\mathcal{L}$ is called the direct product of two submatrices $D^{(1)}(R)$ and $D^{(2)}(R)$. The row and column of the direct product matrix are denoted by two indices (ν,λ) and (μ,ρ), respectively. The order of indices is usually arranged such that the second index ρ increases for a given μ, and then the first index μ increases. For example, the direct product of two 2-dimensional matrices X and Y is

$$
X\times Y = \begin{pmatrix} X_{11}Y & X_{12}Y \\ X_{21}Y & X_{22}Y \end{pmatrix} = \begin{pmatrix} X_{11}Y_{11} & X_{11}Y_{12} & X_{12}Y_{11} & X_{12}Y_{12} \\ X_{11}Y_{21} & X_{11}Y_{22} & X_{12}Y_{21} & X_{12}Y_{22} \\ X_{21}Y_{11} & X_{21}Y_{12} & X_{22}Y_{11} & X_{22}Y_{12} \\ X_{21}Y_{21} & X_{21}Y_{22} & X_{22}Y_{21} & X_{22}Y_{22} \end{pmatrix}.
$$

The product of two matrices $D^{(1)}(R)\times D^{(2)}(R)$ and $D^{(1)}(S)\times D^{(2)}(S)$ is

$$
\begin{aligned}
&\left[D^{(1)}(R)\times D^{(2)}(R)\right]\left[D^{(1)}(S)\times D^{(2)}(S)\right] \\
&\quad = \left[D^{(1)}(R)D^{(1)}(S)\right]\times\left[D^{(2)}(R)D^{(2)}(S)\right].
\end{aligned}
\tag{1.55}
$$

Thus,

$$
\begin{aligned}
\left[D^{(1)}(R)\times D^{(2)}(R)\right]^{-1} &= D^{(1)}(R)^{-1}\times D^{(2)}(R)^{-1}, \\
\left[D^{(1)}(R)\times D^{(2)}(R)\right]^{T} &= D^{(1)}(R)^{T}\times D^{(2)}(R)^{T}, \\
\left[D^{(1)}(R)\times D^{(2)}(R)\right]^{\dagger} &= D^{(1)}(R)^{\dagger}\times D^{(2)}(R)^{\dagger}.
\end{aligned}
\tag{1.56}
$$

The trace and the determinant of direct product $D^{(1)}(R)\times D^{(2)}(R)$ are

$$
\begin{aligned}
\text{Tr}\left[D^{(1)}(R)\times D^{(2)}(R)\right] &= \left[\text{Tr}\ D^{(1)}(R)\right]\left[\text{Tr}\ D^{(2)}(R)\right], \\
\det\left[D^{(1)}(R)\times D^{(2)}(R)\right] &= \left[\det\ D^{(1)}(R)\right]^{n}\left[\det\ D^{(2)}(R)\right]^{m},
\end{aligned}
\tag{1.57}
$$

where m and n are the dimensions of $D^{(1)}(R)$ and $D^{(2)}(R)$, respectively. If two matrices $D^{(1)}(R)$ and $D^{(2)}(R)$ depend on a continuous parameter α, one has

$$
\begin{aligned}
&\frac{d}{d\alpha}\left[D^{(1)}(R)\times D^{(2)}(R)\right] \\
&\quad = \left[\frac{dD^{(1)}(R)}{d\alpha}\right]\times D^{(2)}(R) + D^{(1)}(R)\times\left[\frac{dD^{(2)}(R)}{d\alpha}\right].
\end{aligned}
\tag{1.58}
$$

The direct product reduces to the product of a number and a matrix if one of two factor matrices is one-dimensional. Generally, $D^{(1)}(R) \times D^{(2)}(R)$ is not equal to $D^{(2)}(R) \times D^{(1)}(R)$, but their difference is only a simple similarity transformation.

1.8 Exercises

1. Prove: (1) $RR^\dagger = \mathbf{1}$ if $R^\dagger R = \mathbf{1}$;

(2) $RR^{-1} = \mathbf{1}$ if $R^{-1}R = \mathbf{1}$;

(3) $RR^T = \mathbf{1}$ if $R^T R = \mathbf{1}$.

2. Find the independent real parameters respectively in a unitary matrix, a real orthogonal matrix, and a hermitian matrix of two dimensions, respectively. Give the general expressions for those matrices.

3. Prove that the sum of the eigenvalues of a matrix is equal to the trace of the matrix, and the product of eigenvalues is equal to the determinant of the matrix.

4. Calculate the eigenvalues and eigenvectors of three Pauli matrices, respectively

$$\sigma_1 = \begin{pmatrix} 0 & 1 \\ 1 & 0 \end{pmatrix}, \qquad \sigma_2 = \begin{pmatrix} 0 & -\mathrm{i} \\ \mathrm{i} & 0 \end{pmatrix}, \qquad \sigma_3 = \begin{pmatrix} 1 & 0 \\ 0 & -1 \end{pmatrix}.$$

5. Find the similarity transformation to diagonalize the following matrices:

$$(1)\ R = \begin{pmatrix} 0 & 0 & 1 \\ 1 & 0 & 0 \\ 0 & 1 & 0 \end{pmatrix}, \qquad (2)\ S = \begin{pmatrix} 0 & 0 & 0 & 1 \\ 0 & 0 & 1 & 0 \\ 0 & 1 & 0 & 0 \\ 1 & 0 & 0 & 0 \end{pmatrix},$$

$$(3)\ \begin{pmatrix} 1 & -\sqrt{2} & 1 \\ \sqrt{2} & 0 & -\sqrt{2} \\ 1 & \sqrt{2} & 1 \end{pmatrix}, \qquad (4)\ \begin{pmatrix} \cos\theta & -\sin\theta \\ \sin\theta & \cos\theta \end{pmatrix}.$$

6. Find a similarity transformation matrix M which satisfies

$$M^{-1} \begin{pmatrix} 0 & -\cos\theta & \sin\theta\sin\varphi \\ \cos\theta & 0 & -\sin\theta\cos\varphi \\ -\sin\theta\sin\varphi & \sin\theta\cos\varphi & 0 \end{pmatrix} M = \begin{pmatrix} 0 & -1 & 0 \\ 1 & 0 & 0 \\ 0 & 0 & 0 \end{pmatrix}.$$

7. Find the common similarity transformation matrix M which satisfies the following three equations simultaneously:

$$M^{-1}\begin{pmatrix} 0 & -\mathrm{i} & 0 \\ \mathrm{i} & 0 & 0 \\ 0 & 0 & 0 \end{pmatrix} M = \begin{pmatrix} 1 & 0 & 0 \\ 0 & 0 & 0 \\ 0 & 0 & -1 \end{pmatrix},$$

$$M^{-1}\begin{pmatrix} 0 & 0 & 0 \\ 0 & 0 & -\mathrm{i} \\ 0 & \mathrm{i} & 0 \end{pmatrix} M = \frac{1}{\sqrt{2}}\begin{pmatrix} 0 & 1 & 0 \\ 1 & 0 & 1 \\ 0 & 1 & 0 \end{pmatrix},$$

$$M^{-1}\begin{pmatrix} 0 & 0 & \mathrm{i} \\ 0 & 0 & 0 \\ -\mathrm{i} & 0 & 0 \end{pmatrix} M = \frac{\mathrm{i}}{\sqrt{2}}\begin{pmatrix} 0 & -1 & 0 \\ 1 & 0 & -1 \\ 0 & 1 & 0 \end{pmatrix}.$$

8. Find the common similarity transformation matrix X satisfying

$$R = \begin{pmatrix} 1 & 0 \\ 0 & -1 \end{pmatrix}, \qquad X^{-1}(R \times R)X = \begin{pmatrix} 1 & 0 & 0 & 0 \\ 0 & -1 & 0 & 0 \\ 0 & 0 & 1 & 0 \\ 0 & 0 & 0 & -1 \end{pmatrix},$$

$$S = \frac{1}{2}\begin{pmatrix} -1 & -\sqrt{3} \\ \sqrt{3} & -1 \end{pmatrix}, \qquad X^{-1}(S \times S)X = \frac{1}{2}\begin{pmatrix} 2 & 0 & 0 & 0 \\ 0 & 2 & 0 & 0 \\ 0 & 0 & -1 & -\sqrt{3} \\ 0 & 0 & \sqrt{3} & -1 \end{pmatrix}.$$

9. Prove that any unitary matrix u can be diagonalized by a unitary similarity transformation, and any hermitian matrix H can be diagonalized by a unitary similarity transformation.

10. Show the general form of an $m \times m$ matrix which are unitary and hermitian simultaneously.

11. Prove that $D(R)$ and $D(R)^\dagger$ can be diagonalized by a common unitary similarity transformation if and only if $D(R)^\dagger$ commutes with $D(R)$.

12. If $\det R \neq 0$, prove that both $R^\dagger R$ and $RR^\dagger$ are positive definite hermitian matrices.

13. Prove that any matrix can be changed by a similarity transformation into a direct sum of the submatrices in the forms (the Jordan forms):

$$R_{ab} = \begin{cases} \lambda & \text{when } a = b, \\ 0 \text{ or } 1 & \text{when } a + 1 = b, \\ 0 & \text{the remaining cases.} \end{cases}$$

Chapter 2

GROUP AND ITS SUBSETS

Group theory is a powerful tool for studying the symmetry of a physical system. In this chapter, the mathematical definition of a group will be abstracted from the common property of the sets of the symmetry transformations of physical systems. Some simple examples of groups are explained to give the readers a concrete understanding on groups. The subsets in a group, the isomorphism and homomorphism of two groups, and the direct product of groups will be discussed.

2.1 Symmetry

The symmetry becomes more and more important in the modern sciences. What is the symmetry? One often says that a scalene triangle is asymmetric, an isosceles triangle and a regular triangle are more symmetric, and a circle is the most symmetric among them. How does one estimate the symmetry of a system?

The symmetry is tied up with transformations. A transformation which leaves the system invariant is called a **symmetry transformation** of the system. The symmetry of a system is described by the set of all its symmetry transformations. The identical transformation which leaves anything invariant is a trivial symmetry transformation of any system. A scalene triangle is not invariant under any transformation except for the identical transformation. An isosceles triangle is left invariant under the inversion with respect to the perpendicular bisectrix plane to its bottom edge. A regular triangle is left invariant under three inversions with respect to the perpendicular bisectrix planes to its edges and under the rotations through $\pm 2\pi/3$ about the axis perpendicular to the triangle at its center. A circle is left invariant under any perpendicular plane containing its diameter and under any rotation about the axis perpendicular to the circle at its center.

A quantum system is described by its Hamiltonian. A symmetry transformation of a quantum system leaves its Hamiltonian invariant. For example, the Hamiltonian of an isolated n-body system is

$$H = -\frac{\hbar^2}{2} \sum_{j=1}^{n} m_j^{-1} \nabla_j^2 + \sum_{i<j} U(|\boldsymbol{r}_i - \boldsymbol{r}_j|),$$

where $\boldsymbol{r}_j$ and m_j respectively are the position vector and the mass of the jth particle, ∇_j^2 is the Laplace operator with respect to $\boldsymbol{r}_j$, and U is the interaction potential depending on the distance between two particles. Obviously, the Hamiltonian is left invariant under the space translation, rotations, and inversion. For the system of identical particles, the Hamiltonian is left invariant under the permutations among particles. Those symmetry transformations characterize the symmetry of the quantum system.

In terms of the method of group theory, one is able to analyze symmetry of the system and to obtain some precise information of the system verifiable by observations, even without knowing the detail of the system. First of all, we would like to give the readers an intuitive estimation on the power of group theory through a very simple example: a quantum system with the symmetry of space inversion. Let P denote the transformation operator of the wave function in the space inversion, and let ψ denote an eigenfunction of the Hamiltonian of the system with the eigenvalue E,

$$P\psi(\boldsymbol{r}_1,\ \boldsymbol{r}_2,\ \ldots) = \psi(-\boldsymbol{r}_1,\ -\boldsymbol{r}_2,\ \ldots).$$

Since the Hamiltonian is left invariant under the inversion, $P\psi$ is also an eigenfunction of the Hamiltonian with the same eigenvalue, so any linear combination of ψ and $P\psi$. Take the following combinations:

$$\begin{array}{ll} \phi_S \sim \psi + P\psi, & \phi_A \sim \psi - P\psi, \\ P\phi_S = \phi_S, & P\phi_A = -\phi_A. \end{array} \tag{2.1}$$

In the space inversion, ϕ_S and ϕ_A are the wave functions with the even and odd parities, respectively. No matter how the detail of a system is, only if the system is left invariant under the space inversion, the wave function of any stationary state of the system (the eigenfunction of the Hamiltonian) can be chosen with a definite parity, namely, the parity is conserved and can be used to classify the wave functions of stationary states of the system. Furthermore, the electric dipole transition between two states with the same parity is suppressed. This is the selection rule for the electric dipole transition in quantum mechanics.

This simple example shows that **some precise information of a system can be obtained from its symmetry even though the Schrödinger equation is hard to be solved.**

2.2 Group and its Multiplication Table

2.2.1 *Definition of a Group*

A transformation is called the symmetry transformation of a system if it leaves the system invariant. The set of the symmetry transformations of a system characterizes the symmetry of the system. There are some common properties in the sets of symmetry transformations of different systems. In physics, the multiplication RS of two transformations R and S is usually defined as successive applications first by S and then by R. Thus, the multiplication of two symmetry transformations is still a symmetry one of the system. The multiplication law satisfies the associative law. The identical transformation E is a symmetry transformation of any system. For any symmetry transformation R of the system, the multiplication ER is equal to R. The inverse of a symmetry transformation is also a symmetry one of the system. Those properties are essential and common for the set of symmetry transformations of every system, and can be used to define a group.

Definition 2.1 A group G is a set of elements R satisfying the following four axioms with respect to the given multiplication law of elements.

(a) The set is **closed** to this multiplication,

$$RS \in G, \qquad \forall\ R \text{ and } S \in G. \tag{2.2}$$

(b) The multiplication law of elements satisfies the **associative law**,

$$R(ST) = (RS)T, \qquad \forall\ R,\ S \text{ and } T \in G. \tag{2.3}$$

(c) The set contains an **identity** $E \in$ G satisfying

$$ER = R, \qquad \forall\ R \in \mathrm{G}. \tag{2.4}$$

(d) For any element $R \in$ G, the set contains its **inverse** R^{-1} satisfying

$$R^{-1}R = E, \qquad \forall\ R \in \mathrm{G}. \tag{2.5}$$

In Definition 2.1, the elements can be any objects and the multiplication law of elements can be defined arbitrarily. The main thing for a group is that the set satisfies four axioms with respect to the given multiplication law. The set of symmetry transformations of any system with respect to the multiplication law of transformations satisfies four axioms, so that the set is a group, called the **symmetry group** of the system. In most cases in physics, the elements of a group are transformations, operators, or matrices. If without specification, the multiplication law is defined as successive applications when the elements are transformations or their operators, and as the matrix multiplication when the elements are matrices.

It can be shown from Definition 2.1 (see Prob. 1) that

$$\begin{aligned} &RE = R, \qquad RR^{-1} = E, \qquad (RS)^{-1} = S^{-1}R^{-1}, \\ &T = E, \qquad \text{if } TR = R \text{ or } RT = R, \\ &S = R^{-1}, \qquad \text{if } SR = E \text{ or } RS = E. \end{aligned} \tag{2.6}$$

Thus, the identity in a group is unique, the inverse of any element $R \in G$ is unique, and the inverse relation is mutual. In the proof one cannot use any calculation rule except for Definition 2.1. From now on, the conclusions (2.6) can be used because they have been proved in Prob. 1.

The multiplication of elements is generally non-commutable, $RS \neq SR$. A group is called abelian if the multiplication of every two elements in the group is commutable. A finite group contains finite number g of elements. g is called the order of a finite group. An infinite group contains infinite number of elements. If the elements in an infinite group can be described by a set of continuous parameters, the group is called a continuous group.

In Definition 2.1 the multiplication law of elements is very important for a group. From viewpoint of group theory, the property of objects, which are chosen to be elements in the group, is less important. Namely, one pays more attention to how the elements construct the group. If there is a one-to-one correspondence between elements of two groups, and the correspondence is invariant to the multiplication of elements, two groups are called **isomorphic**, and regard as the same in group theory.

Definition 2.2 A group G′ is called isomorphic onto another group G if there is a one-to-one correspondence between elements of two groups, and the multiplication of two elements in G′ is in the same way to map onto the multiplication of two corresponding elements in G. In mathematical symbols, G′ ≈ G if

$$\begin{aligned} &R' \longleftrightarrow R, \quad S' \longleftrightarrow S, \quad \text{and} \quad R'S' \longleftrightarrow RS, \\ &\forall\ R \in \mathrm{G}, \quad S \in \mathrm{G}, \quad R' \in \mathrm{G}', \quad \text{and} \quad S' \in \mathrm{G}', \end{aligned} \tag{2.7}$$

where "$\longleftrightarrow$" denotes one-to-one correspondence.

There are different one-to-one correspondences between elements in two groups. Two groups are isomorphic if there exists a one-to-one correspondence which is invariant to the multiplication of elements. One cannot make conclusion that two groups are NOT isomorphic only based on that there is a one-to-one correspondence between elements in two groups which is not invariant to the multiplication of elements.

A subset in a group is defined to be a complex of elements. Two complexes are equal to each other, $\mathcal{R} = \mathcal{S}$, if $\mathcal{R} \subset \mathcal{S}$ and $\mathcal{S} \subset \mathcal{R}$. Let $\mathcal{R} = \{R_1,\ R_2,\ \cdots,\ R_m\}$ and $\mathcal{S} = \{S_1,\ S_2,\ \cdots,\ S_n\}$ be two complexes of elements in a group G. The multiplication $\mathcal{RS}$ of two complexes $\mathcal{R}$ and $\mathcal{S}$ is a complex which is the set of all elements R_jS_k. Similarly, the multiplication $T\mathcal{R}$ (or $\mathcal{R}T$) of an element T and a complex $\mathcal{R}$ is a complex which consists of the elements TR_j (or R_jT). The set of complexes forms a group if the set satisfies the four axioms for a group with respect to the multiplication law of complexes.

Theorem 2.1 (Rearrangement theorem) For any element T in a group $\mathrm{G} = \{E,\ R,\ S,\ \cdots\}$, each of the following three sets is the same as the original group G:

$$\begin{aligned} T\mathrm{G} &= \{T,\ TR,\ TS,\ \cdots\} = G, \\ \mathrm{G}T &= \{T,\ RT,\ ST,\ \cdots\} = G, \\ \mathrm{G}^{-1} &= \{E,\ R^{-1},\ S^{-1},\ \cdots\} = G. \end{aligned} \tag{2.8}$$

Proof The proofs for three equalities are similar. We prove $T\mathrm{G} = \mathrm{G}$ as example. Due to Eq. (2.2) any element in the set TG is an element in the group G, so $T\mathrm{G} \subset \mathrm{G}$. Conversely, any element R in G can be expressed as $R = T(T^{-1}R)$. Since $T^{-1}R \in \mathrm{G}$, $R \in T\mathrm{G}$ and $\mathrm{G} \subset T\mathrm{G}$. □

For a finite group, the multiplication of each pair of elements can be listed in a table, called the multiplication table of a group, or briefly the **group table**. In a group table, the right-multiplication elements S are listed on the top, the left-multiplication elements R are listed on the left, and the multiplications RS are filled in the content of the table. Rearrangement theorem shows that **there are no repetitive elements in each row and in each column of the group table**. Two groups are isomorphic if their multiplication tables are the same. Conversely, two isomorphic groups may have different group tables if the enumerations are unsuitable. Theorem 2.1 restricts strongly the number of non-isomorphic groups with a given order g. For example, Theorem 2.1 completely determines the group tables with the order $g = 2$ or 3, as given in Tables 2.1 and 2.2. V_2, called the inversion group of order two, is the symmetry group of a system which is invariant in the space inversion σ. V_2 is isomorphic onto C_2, composed of two numbers 1 and -1 with respect to the multiplication law of numbers. The group C_3 is composed of three complex numbers $e = 1$, $\omega = \exp(-i2\pi/3)$, and $\omega' = \exp(i2\pi/3)$ with respect to the multiplication law of complex numbers.

Table 2.1 The group table of V_2

V_2	e	σ
e	e	σ
σ	σ	e

Table 2.2 The group table of C_3

C_3	e	ω	ω'
e	e	ω	ω'
ω	ω	ω'	e
ω'	ω'	e	ω

In V_2 one has $e = \sigma^2$. Each element in C_3 is equal to a power of ω: $\omega' = \omega^2$ and $e = \omega^3$. Generally, a finite group generated by the powers of one element R is called the cyclic group C_N

$$C_N = \{E,\ R,\ R^2,\ \cdots,\ R^{N-1}\}, \qquad R^N = E. \tag{2.9}$$

R, as well as R^{N-1}, is called the generator of C_N. C_N is an abelian group. The group table of C_N can be filled as follows. Arrange the elements both at the first column (left-multiplication elements) and at the top line (right-multiplication elements) of the group table in the order of powers: E, R, R^2, $\cdots$, R^{N-1}. The elements in each row of the content of the group table are obtained by moving the elements in the preceding row leftward by one box and moving the leftmost element to the rightmost position. For example, see Table 2.3 for C_4.

Table 2.3 The group table of C_4

C_4	E	R	R^2	R^3
E	E	R	R^2	R^3
R	R	R^2	R^3	E
R^2	R^2	R^3	E	R
R^3	R^3	E	R	R^2

Let $R(\hat{\boldsymbol{n}}, \omega)$ denote a rotation, by using the corkscrew rule, about an axis with the direction $\hat{\boldsymbol{n}}(\theta, \varphi)$ through the angle ω, where θ and φ are the polar angle and the azimuthal angle of the unit vector $\hat{\boldsymbol{n}}(\theta, \varphi)$, respectively. If a rotation $R(\hat{\boldsymbol{n}}, 2\pi/N)$ is the symmetry transformation of a physical system, $R(\hat{\boldsymbol{n}}, 2\pi/N)$, usually denoted by C_N, is called the N-fold proper rotation, or briefly called the N-**fold rotation**. The rotational axis of C_N along the direction $\hat{\boldsymbol{n}}$ is called the N-fold proper axis, or the N-**fold axis**. The 1-fold rotation is the identity E, and the 1-fold axis is trivial.

If $S_N = \sigma C_N = C_N \sigma$, where σ is the space inversion, is the symmetry transformation of a physical system, S_N is called the N-fold improper rotation and its rotational axis is called the the N-fold improper axis. The cyclic group generated by S_N is denoted by S_N or $\overline{\mathrm{C}}_N$, whose order is N or $2N$, depending on whether N is even or odd.

Now, we turn to a group G with order four. If the square of one element R in G does not equal to the identity E, due to the rearrangement theorem, R^3 does not equal to E, either. Thus, G is the cyclic group C_4. If the square of each element in G is equal to the identity, due to the rearrangement theorem, its group table has to be Table 2.4. A typical example of a group with the group table 2.4 is the **inversion group V_4 of order four**, which is composed of the identity e, the space inversion σ, the time inversion τ, and the space-time inversion ρ. Up to isomorphism there are only two different groups with order four, the group C_4 and V_4, depending on whether it contains an element whose square is not equal to the identity. Both C_4 and V_4 are abelian.

Table 2.4 The group table of V_4

V_4	e	σ	τ	ρ
e	e	σ	τ	ρ
σ	σ	e	ρ	τ
τ	τ	ρ	e	σ
ρ	ρ	τ	σ	e

2.2.2 *Subgroup*

A subset H $\in$ G is a subgroup of G if H forms a group under the same multiplication law of elements as used in G. Note that H **is not a subgroup of G if the multiplication law in H is different to that in** G. There are two trivial subgroups for any group G: The identity E and the whole group G. We will pay more attention to nontrivial subgroups. C_4 contains one nontrivial subgroup $\{E,\ R^2\}$. V_4 contains three nontrivial subgroup $\{e,\ \sigma\}$, $\{e,\ \tau\}$, and $\{e,\ \rho\}$. If $N = nm$, C_N contains two subgroups C_m and C_n, generated by C_N^n and by C_N^m, respectively.

Since the number of elements in a finite group G is limited, the powers of each element R in G must be repeated when the power is high enough, say $R^{a+n} = R^a$, namely, $R^n = E$. The lowest n for $R^n = E$ is called the **order of the element** R. The order of the identity E is one, and the order of other element $R \neq E$ must be larger than one. The subset of the powers of R in G is called the period of R, which forms a subgroup of G, called the cyclic subgroup. **Do not confuse the nomenclature "order" for a group and for an element**. Only for the cyclic group C_N generated by R, the order of R is equal to the order of the group C_N.

A subgroup must contain the whole period of its every element. A subset is not a subgroup if it does not contain the identity E. **For a finite group G, a subset H in G is a subgroup if H is closed** with respect to the multiplication law in G. In fact, if it is closed, H contains the period of its every element R, including the inverse $R^{-1} = R^{n-1}$ and the identity $E = R^n$, where n is the order of R. The associative law is satisfied because both H and G obey the same multiplication law of elements.

If the period of an element $R \in$ G is not equal to the whole group G, take an element $S \in$ G which does not belong to the period of R and construct a larger subset by multiplying R and S arbitrarily. For a finite group, one can enlarge the subset by adding more elements until the subset is equal to G. The elements are called the generators of a finite group G if each element in G can be expressed as a multiplication of the generators, and each generator is not equal to the multiplication of other generators. The choice of generators of a finite group is not unique. Choose the generators whose number is as small as possible. **The least number of generators is called the rank of a finite group**. The rank of the cyclic group C_N is one, and both R and R^{-1} can be chosen to be its generator.

2.2.3 *Symmetry Group of a Regular Polygon*

Discuss the symmetry transformations of a regular polygon with N sides. Put the regular polygon on a planar coordinates flame OXY where its center coincides with the origin O and one vertex A is located on the X-axis. The Z-axis is an N-fold proper axis of the polygon, where the N-fold rotation is denoted by C_N. When N is odd, any axis connecting a vertex and the midpoint of its opposite edge is a 2-fold axis. When N is even, any axis connecting two opposite vertices and any axis connecting the midpoints of two opposite edges both are the 2-fold axes. The 2-fold rotation about an axis with the angle $j\pi/N$ to the X-axis is denoted by $C_2^{(j)}$. The set of the 2-fold rotations $C_2^{(j)}$ and the powers $(C_N)^j$, $0 \leqslant j \leqslant N-1$, forms the symmetry group of the polygon with the order $g = 2N$, called by the Nth-dihedral group D_N.

Calculate the multiplication $C_N C_2^{(j)}$ in D_N. Let OB denote a 2-fold axis and let $C_2^{(j)}$ denote the 2-fold rotation about OB. The point B is left invariant under the rotation $C_2^{(j)}$ and moves to D under the rotation C_N. The angle between two axes OB and OD is $2\pi/N$ (see Fig. 2.1). On the other hand, D moves to the symmetric position P with respect to the axis OB under $C_2^{(j)}$, and then, moves to B under C_N. Since two points B and D transform to each other, $C_N C_2^{(j)}$ is the 2-fold rotation about the axis along the angular bisector of $\angle BOD$,

$$C_N C_2^{(j)} = C_2^{(j+1)}, \qquad j \bmod N. \tag{2.10}$$

where "j mod N" is a mathematical symbol, denoting $j+N$ is the same as j.

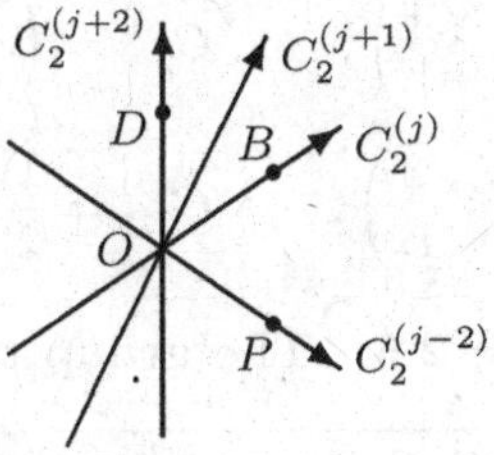

Fig. 2.1 The calculation for $C_N C_2^{(j)} = C_2^{(j+1)}$

Thus, one obtains the multiplication law in D_N as

$$\begin{aligned} &(C_N)^N = \left(C_2^{(j)}\right)^2 = E, && (C_N)^a (C_N)^b = (C_N)^{a+b}, \\ &(C_N)^a C_2^{(j)} = C_2^{(j+a)}, && (C_N)^a = C_2^{(j+a)} C_2^{(j)} = C_2^{(j)} C_2^{(j-a)}, \\ &C_2^{(j)} (C_N)^a = C_2^{(j-a)}, && j \text{ and } a \bmod N. \end{aligned} \tag{2.11}$$

This is another way than the group table to show the multiplication rule of elements in G. The generators of D_N are C_N and $C_2^{(0)}$.

The third way to show the multiplication law in D_N is the method of coordinate transformation. Let $D(R)$ denote the transformation matrix of vertices in the planar coordinate system under an element R of D_N :

$$\begin{pmatrix} x' \\ y' \end{pmatrix} = D(R)\begin{pmatrix} x \\ y \end{pmatrix} = \begin{pmatrix} a & b \\ c & d \end{pmatrix}\begin{pmatrix} x \\ y \end{pmatrix}. \tag{2.12}$$

For the vertex on the X-axis and its neighboring vertices, we have

$$\begin{pmatrix} \cos(2\pi/N) \\ \sin(2\pi/N) \end{pmatrix} = D(C_N)\begin{pmatrix} 1 \\ 0 \end{pmatrix}, \quad \begin{pmatrix} 1 \\ 0 \end{pmatrix} = D(C_N)\begin{pmatrix} \cos(2\pi/N) \\ -\sin(2\pi/N) \end{pmatrix},$$

$$\begin{pmatrix} 1 \\ 0 \end{pmatrix} = D(C_2^{(0)})\begin{pmatrix} 1 \\ 0 \end{pmatrix}, \quad \begin{pmatrix} \cos(2\pi/N) \\ -\sin(2\pi/N) \end{pmatrix} = D(C_2^{(0)})\begin{pmatrix} \cos(2\pi/N) \\ \sin(2\pi/N) \end{pmatrix}.$$

Thus,

$$D(C_N) = \begin{pmatrix} \cos(2\pi/N) & -\sin(2\pi/N) \\ \sin(2\pi/N) & \cos(2\pi/N) \end{pmatrix}, \quad D(C_2^{(0)}) = \begin{pmatrix} 1 & 0 \\ 0 & -1 \end{pmatrix}. \tag{2.13}$$

An important example of D_N is the symmetry group D_3 of a regular triangle which is the simplest non-abelian group. The elements in D_3 can be expressed by the following matrices:

$$\begin{aligned} E &= \begin{pmatrix} 1 & 0 \\ 0 & 1 \end{pmatrix}, & C_3 &= \frac{1}{2}\begin{pmatrix} -1 & -\sqrt{3} \\ \sqrt{3} & -1 \end{pmatrix}, \\ C_3^2 &= \frac{1}{2}\begin{pmatrix} -1 & \sqrt{3} \\ -\sqrt{3} & -1 \end{pmatrix}, & C_2^{(0)} &= \begin{pmatrix} 1 & 0 \\ 0 & -1 \end{pmatrix}, \\ C_2^{(1)} &= \frac{1}{2}\begin{pmatrix} -1 & \sqrt{3} \\ \sqrt{3} & 1 \end{pmatrix}, & C_2^{(2)} &= \frac{1}{2}\begin{pmatrix} -1 & -\sqrt{3} \\ -\sqrt{3} & 1 \end{pmatrix}. \end{aligned} \tag{2.14}$$

Table 2.5 The group table of D_3

	E	C_3	C_3^2	$C_2^{(0)}$	$C_2^{(1)}$	$C_2^{(2)}$
E	E	C_3	C_3^2	$C_2^{(0)}$	$C_2^{(1)}$	$C_2^{(2)}$
C_3	C_3	C_3^2	E	$C_2^{(1)}$	$C_2^{(2)}$	$C_2^{(0)}$
C_3^2	C_3^2	E	C_3	$C_2^{(2)}$	$C_2^{(0)}$	$C_2^{(1)}$
$C_2^{(0)}$	$C_2^{(0)}$	$C_2^{(2)}$	$C_2^{(1)}$	E	C_3^2	C_3
$C_2^{(1)}$	$C_2^{(1)}$	$C_2^{(0)}$	$C_2^{(2)}$	C_3	E	C_3^2
$C_2^{(2)}$	$C_2^{(2)}$	$C_2^{(1)}$	$C_2^{(0)}$	C_3^2	C_3	E

In other words, the matrix group constructed by six matrices given in Eq. (2.14) is isomorphic onto D_3. The group table of D_3, shown in Table 2.5, can be obtained by Eq. (2.11) or by multiplying those six matrices. It can be proved (see Prob. 9) that up to isomorphism there are only two different groups with order six: The group is isomorphic onto D_3 if there are three elements whose order is two, otherwise, it is isomorphic onto C_6.

The group D_3 contains four cyclic subgroups: $\{E,\ C_2^{(0)}\}$, $\{E,\ C_2^{(1)}\}$, $\{E,\ C_2^{(2)}\}$, and $\{E,\ C_3,\ C_3^2\}$. The group D_6 contains nine cyclic subgroups, three D_2 subgroups, and two D_3 subgroups:

$$\begin{aligned}
&D_2 = \{E,\ C_6^3,\ C_2^{(0)},\ C_2^{(3)}\}, \quad D_2' = \{E,\ C_6^3,\ C_2^{(1)},\ C_2^{(4)}\},\\
&D_2'' = \{E,\ C_6^3,\ C_2^{(2)},\ C_2^{(5)}\}, \quad D_3 = \{E,\ C_6^2,\ C_6^4,\ C_2^{(0)},\ C_2^{(2)},\ C_2^{(4)}\},\\
&D_3' = \{E,\ C_6^2,\ C_6^4,\ C_2^{(1)},\ C_2^{(3)},\ C_2^{(5)}\}.
\end{aligned}$$

Although there is no regular polygon with only two sides, one still can define the group $D_2 = \{E,\ C_2,\ C_2^{(0)},\ C_2^{(1)}\}$. Usually in the literature, the elements of D_2 is denoted as the 2-fold rotations about the coordinate axes: $C_2 = T_z$, $C_2^{(0)} = T_x$, and $C_2^{(1)} = T_y$. Obviously, $D_2 \approx V_4$.

2.2.4 *Symmetry of System of Identical Particles*

Study the symmetry of the system of n identical particles. A rearrangement of n objects is called a **permutation**. A permutation R which moves the jth object to the r_jth position is denoted by a $2 \times n$ matrix,

$$R = \begin{pmatrix} 1 & 2 & \ldots & n \\ r_1 & r_2 & \ldots & r_n \end{pmatrix}. \tag{2.15}$$

A digit j describes an object located in the jth position. Hereafter, a permutation of objects is said as a permutation of digits for simplicity. Usually, this simplification does not causes any confusion. In the expression (2.15), the corresponding relation between two digits in each column is essential, but R does not matter with the order of the columns, e.g.,

$$R = \begin{pmatrix} 1 & 2 & 3 & 4 & 5 \\ 3 & 4 & 5 & 2 & 1 \end{pmatrix} = \begin{pmatrix} 5 & 4 & 1 & 2 & 3 \\ 1 & 2 & 3 & 4 & 5 \end{pmatrix}. \tag{2.16}$$

The multiplication of two permutations does not satisfy the multiplication law of matrices, although a permutation is expressed by a matrix. The multiplication SR of two permutations is defined as successive applications with R and then with S. In the permutation SR, the jth object is transformed into the r_jth position by R, and then, into the s_{r_j}th position by S. Note that in the second permutation S, the jth object is transformed into the s_jth position, **no matter in what position it was located before the first permutation** R. In the practical calculation, one may change the order of columns in S or R such that the first row of S is the same as the second row of R, then, the matrix form of SR is obtained by combining the first row of R and the second row of S. For example, letting

$$S = \begin{pmatrix} 1 & 2 & 3 & 4 & 5 \\ 3 & 1 & 2 & 4 & 5 \end{pmatrix} = \begin{pmatrix} 3 & 4 & 5 & 2 & 1 \\ 2 & 4 & 5 & 1 & 3 \end{pmatrix}, \tag{2.17}$$

one has

$$SR = \begin{pmatrix} 3 & 4 & 5 & 2 & 1 \\ 2 & 4 & 5 & 1 & 3 \end{pmatrix} \begin{pmatrix} 1 & 2 & 3 & 4 & 5 \\ 3 & 4 & 5 & 2 & 1 \end{pmatrix} = \begin{pmatrix} 1 & 2 & 3 & 4 & 5 \\ 2 & 4 & 5 & 1 & 3 \end{pmatrix}, \tag{2.18}$$

$$SR = \begin{pmatrix} 1 & 2 & 3 & 4 & 5 \\ 3 & 1 & 2 & 4 & 5 \end{pmatrix} \begin{pmatrix} 5 & 4 & 1 & 2 & 3 \\ 1 & 2 & 3 & 4 & 5 \end{pmatrix} = \begin{pmatrix} 5 & 4 & 1 & 2 & 3 \\ 3 & 1 & 2 & 4 & 5 \end{pmatrix}. \tag{2.19}$$

The multiplication of two permutations satisfies the associative law, $(RS)T = R(ST)$, but it is generally not commutable. The identity E is described by a matrix whose two rows are exactly the same,

$$E = \begin{pmatrix} 1 & 2 & 3 & \ldots & n \\ 1 & 2 & 3 & \ldots & n \end{pmatrix}. \tag{2.20}$$

From the multiplication law one has $ER = R$. The matrix of R^{-1} is obtained from that of R by switching two rows,

$$R^{-1} = \begin{pmatrix} r_1 & r_2 & \ldots & r_n \\ 1 & 2 & \ldots & n \end{pmatrix}. \tag{2.21}$$

Due to Eq. (2.18) one has $R^{-1}R = E$.

There are $n!$ different permutations among n objects. In the multiplication law the set of $n!$ permutations satisfies the four axioms for a group, and forms the **permutation group of n objects**, denoted by S_n. The order of S_n is $g = n!$. Choosing arbitrarily m objects among n objects, one obtains a permutation group S_m by the permutations of m objects. S_m is a subgroup of S_n. Each row in the group table of a finite group shows a permutation of the group elements such that **a finite group G with order g is a subgroup of the permutation group** S_g. However, the order of S_g is too large to be used in practice.

If a permutation S leaves $(n - \ell)$ objects invariant and changes the remaining ℓ objects sequently, S is called a **cycle** with cycle length ℓ, characterized by a one-row matrix:

$$\begin{aligned} S &= \begin{pmatrix} a_1 & a_2 & \ldots & a_{\ell-1} & a_\ell & b_1 & \ldots & b_{n-\ell} \\ a_2 & a_3 & \ldots & a_\ell & a_1 & b_1 & \ldots & b_{n-\ell} \end{pmatrix} \\ &= (a_1 \ a_2 \ \ldots \ a_{\ell-1} \ a_\ell). \end{aligned} \tag{2.22}$$

In the one-row matrix for a cycle, the order of digits is essential, while the transformation of digits in sequence is permitted:

$$(q\ a\ b\ c\ \cdots\ p) = (a\ b\ c\ \cdots\ p\ q) = (b\ c\ \cdots\ p\ q\ a)\,. \tag{2.23}$$

For example,

$$\begin{aligned} &(1\ 2\ 3) = (2\ 3\ 1) = (3\ 1\ 2) = \begin{pmatrix} 1 & 2 & 3 \\ 2 & 3 & 1 \end{pmatrix} \\ &\neq (2\ 1\ 3) = (1\ 3\ 2) = (3\ 2\ 1) = \begin{pmatrix} 1 & 2 & 3 \\ 3 & 1 & 2 \end{pmatrix} = (1\ 2\ 3)^{-1}\,. \end{aligned} \tag{2.24}$$

A cycle with length 1 is the identity E. A cycle with length 2 is called a **transposition**,

$$(a\ b) = (b\ a)\,, \qquad (a\ b)\,(a\ b) = E. \tag{2.25}$$

The order of a cycle S with length ℓ is ℓ (see Prob. 6), namely, $S^{\ell} = E$.

Two cycles are called **independent** if they do not contain any common digit (object). The product of two independent cycles is commutable. **Any permutation can be decomposed into a product of independent cycles uniquely up to the product order.** In fact, checking an arbitrary digit a_1 in a given permutation R, one finds that under the action of R, a_1 changes to a_2, a_2 changes to a_3, and so on. In this chain there must exist an object, say a_ℓ, such that a_ℓ changes to a_1. Thus, R contains a cycle $(a_1,\ a_2,\ \ldots,\ a_\ell)$ with cycle length ℓ. In the remaining objects one checks another object, say b_1, and finds another chain of objects which gives the second cycle independent of the first one. This process can be done until all objects are contained in the cycles. Thus, the permutation is decomposed into a product of independent cycles. For example,

$$\begin{aligned} R &= \begin{pmatrix} 1 & 2 & 3 & 4 & 5 \\ 3 & 4 & 5 & 2 & 1 \end{pmatrix} = (1\ 3\ 5)\,(2\ 4) = (2\ 4)\,(1\ 3\ 5)\,, \\ S &= \begin{pmatrix} 1 & 2 & 3 & 4 & 5 \\ 3 & 1 & 2 & 4 & 5 \end{pmatrix} = (1\ 3\ 2)\,(4)\,(5) = (1\ 3\ 2)\,. \end{aligned} \tag{2.26}$$

The set of cycle lengths of the independent cycles in a permutation R is called its **cycle structure**. For example, the cycle structures of R and S in Eq. (2.26) are (3 2) and $(3,1,1) \equiv (3,1^2)$, respectively. Generally, the cycle structure of a permutation R is shown as

$$(\ell_1,\ \ell_2,\ \cdots,\ \ell_m), \qquad \sum_{j=1}^{m} \ell_j = n. \tag{2.27}$$

The arrangement order of ℓ_j in the cycle structure of a permutation is arbitrary and the repetitive ℓ_j can be expressed as an exponent. It is a common sense that the simplest form of a permutation is expressed as a product of independent cycles. A set of positive integers ℓ_j, whose sum is equal to an integer n, is called a partition of n. **The cycle structure of a permutation R in S_n is characterized by a partition of n**

In the calculation of the product of two permutations one has to deal with the product of two cycles with a few common objects. The product of two cycles with one common object can be calculated as

$$(a\ b\ c\ d)\,(d\ e\ f) = \begin{pmatrix} a\ b\ c\ e\ f\ d \\ b\ c\ d\ e\ f\ a \end{pmatrix} \begin{pmatrix} a\ b\ c\ d\ e\ f \\ a\ b\ c\ e\ f\ d \end{pmatrix}$$
$$= \begin{pmatrix} a\ b\ c\ d\ e\ f \\ b\ c\ d\ e\ f\ a \end{pmatrix} = (a\ b\ c\ d\ e\ f)\,.$$

Generally, **two cycles with one common object can be connected into one,**

$$(a\ \cdots\ b\ c)(c\ d\ \cdots\ f) = (a\ \cdots\ b\ c\ d\ \cdots\ f). \tag{2.28}$$

Conversely, **a cycle can be decomposed into a product of two cycles by cutting it at any digit and repeating the digit.** As generalization, the product of two cycles with more than one common objects can be calculated by "cutting" and "connecting", say

$$\begin{aligned}
&(a_1\ \ldots a_i\ c\ a_{i+1}\ \ldots\ a_j\ d)\,(d\ b_1\ \ldots b_r\ c\ b_{r+1}\ \ldots\ b_s) \\
&= (a_1\ \ldots a_i\ c)\,(c\ a_{i+1}\ \ldots\ a_j\ d)\,(d\ b_1\ \ldots b_r\ c).(c\ b_{r+1}\ \ldots\ b_s) \\
&= (a_1\ \ldots a_i\ c)\,(a_{i+1}\ \ldots\ a_j\ d\ c)\,(c\ d\ b_1\ \ldots b_r)\,(c\ b_{r+1}\ \ldots\ b_s) \\
&= (a_1\ \ldots a_i\ c)\,(a_{i+1}\ \ldots\ a_j\ d)\,(d\ c)\,(c\ d)\,(d\ b_1\ \ldots b_r)\,(c\ b_{r+1}\ \ldots\ b_s) \\
&= (a_1\ \ldots a_i\ c)\,(a_{i+1}\ \ldots\ a_j\ d)\,(d\ b_1\ \ldots b_r)\,(c\ b_{r+1}\ \ldots\ b_s) \\
&= (a_1\ \ldots a_i\ c\ b_{r+1}\ \ldots\ b_s)\,(a_{i+1}\ \ldots\ a_j\ d\ b_1\ \ldots b_r)\,.
\end{aligned}$$

Denoting three vertices of a regular triangle by digits 1, 2 and 3, one may express the symmetry transformation of the regular triangle as a permutation in S_3,

$$\begin{aligned}
&E = E, \qquad &&C_3 = (3\ 2\ 1), \qquad &&(C_3)^2 = (1\ 2\ 3), \\
&C_2^{(0)} = (2\ 3), \qquad &&C_2^{(1)} = (1\ 3), \qquad &&C_2^{(2)} = (1\ 2).
\end{aligned} \tag{2.29}$$

The group table of D_3 (see Table. 2.5) can be obtained again by multiplications of six permutations, for example, $C_3C_2^{(0)} = (1\,3\,2)(2\,3) = (1\,3) = C_2^{(1)}$. Namely, $S_3 \approx D_3$.

2.3 Subsets in a Group

2.3.1 *Cosets and Invariant Subgroup*

Let H of order h be a subgroup of the group G of order g:

$$\text{H} = \{S_1,\ S_2,\ S_3,\ \cdots,\ S_h\} \subset \text{G}, \qquad S_1 = E.$$

Taking an arbitrary element $R_j \in \text{G}$, which does not belong to H, one obtains two subsets by left- and right-multiplying R_j to H:

$$\begin{aligned} R_j\text{H} &= \{R_j,\ R_jS_2,\ R_jS_3,\ \cdots,\ R_jS_h\}, \\ \text{H}R_j &= \{R_j,\ S_2R_j,\ S_3R_j,\ \cdots,\ S_hR_j\}, \end{aligned} \qquad R_j \in \text{G}, \quad R_j \overline{\in} \text{H}. \tag{2.30}$$

$R_j\text{H}$ is called the left coset of H, and $\text{H}R_j$ the right coset.

The following conclusions for cosets are very easy to prove, and are left to the readers as exercise. A left coset does not contain any element belonging to the subgroup H. A left coset is not a subgroup. Two left cosets must be the same if there is at least one common element in them, namely, there is no common element between two different left cosets. Similar conclusions hold for the right cosets. The necessary and sufficient condition for two elements R and S belonging to the same left coset is $R^{-1}S \in \text{H}$, and that for them belonging to the same right coset is $RS^{-1} \in \text{H}$. For a finite group, the number of elements in any coset is equal to the order h of the subgroup H, and the group G is equal to the union of the subgroup H and its left cosets (or right cosets),

$$\text{G} = \bigcup_{j=1}^{d} R_j\text{H} = \bigcup_{j=1}^{d} \text{H}R_j, \qquad R_1 = E. \tag{2.31}$$

Thus, the ratio of the orders of the group G and its subgroup H is an integer: $g/h = d$ (Lagrange Theorem). d is called the index of the subgroup H in G. A group G does not contain any nontrivial subgroups if its order g is a prime number. It leads that **a group whose order n is a prime number is isomorphic onto the cyclic group C_n.**

Let H denote a subgroup of a group G. The left coset $R_j\text{H}$ of H in G is not necessary to be equal to its corresponding right coset $\text{H}R_j$. A subgroup is called invariant or normal if

$$R_j\text{H} = \text{H}R_j, \qquad R_j\text{H}R_j^{-1} = \text{H}, \qquad \forall R_j \in \text{G}. \tag{2.32}$$

Note that Eq. (2.32) means that $R_jS_\mu R_j^{-1} = S_\nu \in \text{H}$. The element $R_j \in \text{G}$ is not necessary to commute with any element S_μ in the invariant subgroup

H. Any subgroup in an abelian group is invariant. **The subgroup with index 2 is invariant**, because it has only one coset such that its left coset has to be equal to its right coset.

As the complexes of elements, the set of an invariant subgroup H of G and its cosets forms a group with respect to the multiplication law of the complexes. This group is called the quotient group G/H of an invariant subgroup H in G. In fact, from Eq. (2.32) one has

$$\begin{aligned} &R_j\mathrm{H}R_k\mathrm{H} = R_jR_k\mathrm{HH} = (R_jR_k)\,\mathrm{H}, \\ &\mathrm{H}R_j\mathrm{H} = R_j\mathrm{HH} = R_j\mathrm{H}, \\ &R_j^{-1}\mathrm{H}R_j\mathrm{H} = R_j^{-1}R_j\mathrm{HH} = \mathrm{H}. \end{aligned}$$

The identity in the quotient group is H and the order of G/H is the index $n = g/h$ of H in G. One cannot define a quotient group G/H if H is not an invariant subgroup of G.

H is called the maximal invariant subgroup of G if H is an non-trivial invariant subgroup of G and there is no non-trivial invariant subgroup in G with order larger than that of H. A subgroup with index 2 is certainly a maximal invariant subgroup. C_N is the maximal invariant subgroup of the group D_N.

Each cycle as well as each permutation in S_n can be decomposed into a product of transpositions by cutting the digits in the cycle:

$$(a\ b\ c\ \ldots\ p\ q) = (a\ b)\,(b\ c)\ldots(p\ q)\,. \tag{2.33}$$

This decomposition is not unique. However, it is fixed for a given permutation whether the number of the transpositions in its decomposition is even or odd. This conclusion can be proved by making use of the Vandermonde determinant, which is defined as

$$D(x_1, x_2, \ldots, x_n) = \begin{vmatrix} 1 & 1 & \ldots & 1 \\ x_1 & x_2 & \ldots & x_n \\ x_1^2 & x_2^2 & \ldots & x_n^2 \\ \ldots & \ldots & \ldots & \ldots \\ x_1^{n-1} & x_2^{n-1} & \ldots & x_n^{n-1} \end{vmatrix} = \prod_{i>j}(x_i - x_j).$$

The determinant changes its sign when the variables x_j are made a transposition. When the variables x_j are made a permutation R, it is completely determined by R whether the determinant changes its sign or not. On the other hand, R can be decomposed into a product of transpositions. The determinant changes its sign in R if the number of transpositions in the decomposition of R is odd. Conversely, the determinant does not change if it is even. Thus, the conclusion is proved.

A permutation is called even (or odd) if it is decomposed into a product of even (or odd) transpositions. Define the permutation parity $\delta(R)$ of a permutation R to denote whether R is even or odd

$$\delta(R) = \begin{cases} 1, & R \text{ is even,} \\ -1, & R \text{ is odd.} \end{cases} \tag{2.34}$$

The identity E is an even permutation. The permutation parity of a cycle with length ℓ is $(-1)^{\ell+1}$. Since $\delta(RS) = \delta(R)\delta(S)$, the product of two even permutations or two odd permutations is an even permutation, and the product of an even permutation and an odd permutation is an odd permutation. When $n > 1$, the subset of all even permutations in S_n forms a subgroup with index 2, called the alternating subgroup S'_n, which is the maximal invariant subgroup of S_n.

2.3.2 *Conjugate Elements and Class*

The element $S' = TST^{-1}$ is said to be conjugate to S where S, S', and T all belong to G. The conjugate relation of two elements is mutual. Two elements are conjugate if they both are conjugate to one element:

$$S' = TST^{-1}, \qquad S'' = RSR^{-1} = \left(RT^{-1}\right) S' \left(RT^{-1}\right)^{-1}.$$

The subset of all mutually conjugate elements in a group G is called a class $\mathcal{C}_\alpha$ in G,

$$\mathcal{C}_\alpha = \left\{S_1,\ S_2,\ \cdots,\ S_{n(\alpha)}\right\} = \left\{S_b | S_b = TS_aT^{-1}, T \in G\right\}, \tag{2.35}$$

where $n(\alpha)$ is the number of elements contained in the class $\mathcal{C}_\alpha$. **There is no common element in two different classes.** The identity E itself forms a class, denoted by $\mathcal{C}_1$. The other class $\mathcal{C}_\alpha$, $\alpha \neq 1$, does not contain the identity and is not a subgroup. Let g_c denote the number of classes in G,

$$\sum_{\alpha=1}^{g_c} n(\alpha) = g. \tag{2.36}$$

For an abelian group, each element itself forms a class, so that $g_c = g$.

For a given element $T \in$ G, $TS_aT^{-1} \neq TS_bT^{-1}$ if $S_a \neq S_b$. Thus,

$$T\mathcal{C}_\alpha T^{-1} = \mathcal{C}_\alpha, \qquad \forall\, T \in \text{G}. \tag{2.37}$$

Theorem 2.2 Let $n(\alpha)$ denote the number of elements in a class $\mathcal{C}_\alpha$ of a finite group G with the order g, and let $m(\alpha)$ denote the number of $T \in$ G satisfying $TS_aT^{-1} = S_b$, where S_a and S_b are two given elements in $\mathcal{C}_\alpha$, then $g = n(\alpha)m(\alpha)$, and $m(\alpha)$ is independent of S_a and S_b.

The key in the proof is to define a subgroup H of G whose elements commute with an element S_a in $\mathcal{C}_\alpha$. Any element T_μ in the left coset TH satisfies $T_\mu S_a T_\mu^{-1} = TS_aT^{-1} = S_b$. (see Prob. 16).

If S_a and S_b are two elements in a class $\mathcal{C}_\alpha$, $S_b = TS_aT^{-1}$, then S_a^{-1} and S_b^{-1} are also conjugate to each other, $S_b^{-1} = TS_a^{-1}T^{-1}$, so that the set of S_a^{-1} forms a class, called the **reciprocal class** $\mathcal{C}_\alpha^{(-1)}$ of $\mathcal{C}_\alpha$. $\mathcal{C}_\alpha^{(-1)}$ contains the same number of elements as $\mathcal{C}_\alpha$. The reciprocal relation of two classes is mutual. If $S_a \in \mathcal{C}_\alpha$ is conjugate to its inverse S_a^{-1}, then $\mathcal{C}_\alpha = \mathcal{C}_\alpha^{(-1)}$, called the **self-reciprocal class**.

There are some common properties among elements in one class. For example, **two conjugate elements in a finite group have the same order**. In fact, if $S^n = E$, then $(TST^{-1})^n = TS^nT^{-1} = E$. However, two elements with the same order are not necessary to be conjugate. Two elements are conjugate to each other if and only if **they can be expressed as two different products of other two elements**, TR and RT. In fact, $TR = T(RT)T^{-1}$, and two conjugate elements S and S' satisfy $S' = T(ST^{-1})$ and $S = (ST^{-1})T$. If we take the same arrangement order for both the left-multiplication elements and the right-multiplication elements in the group table of a finite group, two elements are conjugate to each other if and only if they appear at least once at the two symmetric positions with respect to the diagonal line of the group table.

G is called **a proper point group** of a system (e.g., a crystal) if the symmetry group G of the system consists of proper rotations about the axes with a common point, usually taken as the origin. Let an axis along the direction $\hat{\boldsymbol{n}}$ be an N-fold proper axis of the system, whose generator is denoted by $R = R(\hat{\boldsymbol{n}}, 2\pi/N)$, and let $S \in$ G change the vector $\hat{\boldsymbol{n}}$ to the vector $\hat{\boldsymbol{m}}$. Thus the conjugate element $R' = SRS^{-1}$ is an N-fold proper axis of the system about the axis along the direction $\hat{\boldsymbol{m}}$:

$$R' = SR(\hat{\boldsymbol{n}}, 2\pi/N)S^{-1} = R(\hat{\boldsymbol{m}}, 2\pi/N), \qquad \text{if } \ S\hat{\boldsymbol{n}} = \hat{\boldsymbol{m}}. \tag{2.38}$$

In fact, S^{-1} changes $\hat{\boldsymbol{m}}$ to $\hat{\boldsymbol{n}}$ first, then R makes a rotation about $\hat{\boldsymbol{n}}$ through $2\pi/N$, and at last S changes $\hat{\boldsymbol{n}}$ back to $\hat{\boldsymbol{m}}$. Thus, the axis along $\hat{\boldsymbol{m}}$ does not change in the rotation R', and it is also an N-fold proper axis of the system just like the axis along $\hat{\boldsymbol{n}}$.

Generally, if a symmetry transformation S of a system relates two directions and there is an N-fold proper axis along one direction, there must be a proper axis of the same fold along the other direction. Those two axes are called the **equivalent axes**. Two symmetry transformations about two equivalent axes through the same angle are conjugate to each other. An N-fold proper axis is called **nonpolar** if there is a symmetry transformation changes the axis to its opposite direction, such that two symmetry transformations about a nonpolar axis through two opposite angles $\pm\omega$ are conjugate to each other. In this meaning, a 2-fold proper axis is always nonpolar due to $C_2 = C_2^{-1}$. **No symmetry transformation of the system changes one proper axis to another with different folds.** Two symmetry transformations through the same non-zero angle are not conjugate to each other if they are proper rotations about two axes of different folds.

From Eq. (2.32) an invariant subgroup H⊂G contains all elements in G which are conjugate to any element in H. Namely, **H is composed of a few whole classes**. The standard steps to analyze a finite group G is as follows. First, find the order of every element in G. Then, judge whether the elements with the same order are conjugate or not, so that the classes in G are found. Third, combine a few classes as a subset and judge whether the subset forms a subgroup. Namely, judge whether the subset contains the identity, whether the number of elements in the subset is a factor of the order g of G, and whether the subset is closed to the multiplication of elements. If the subset forms a subgroup of G, it is an invariant subgroup, and then, its quotient group G/H can be found.

The symmetry group D_N of a regular polygon with N sides contains a nonpolar N-fold axis, called the principal axis, and N proper 2-fold axes, uniformly-distributing in the plane perpendicular to the principal axis. When $N = 2n + 1$ is odd, all 2-fold axes, each of which connects a vertex and the midpoint of its opposite edge, are equivalent to each other and there are $n + 2$ self-reciprocal classes in D_{2n+1}:

$$\begin{aligned} &\{E\}, \quad \{C_{2n+1}^m, C_{2n+1}^{-m}\}, \quad 1 \leqslant m \leqslant n \\ &\{C_2^{(0)}, C_2^{(1)}, \cdots, C_2^{(2n)}\}. \end{aligned} \tag{2.39}$$

When $N = 2n$ is even, the 2-fold axes are divided into two classes depending on whether the axis connect two opposite vertices or two midpoints of the opposite edges. The D_{2n} group contains $n + 3$ self-reciprocal classes,

$$\begin{aligned} &\{E\}, \quad \{C_{2n}^n\}, \quad \{C_{2n}^m, C_{2n}^{-m}\}, \quad 1 \leqslant m \leqslant n-1, \\ &\{C_2^{(0)}, C_2^{(2)}, \cdots, C_2^{(2n-2)}\}, \quad \{C_2^{(1)}, C_2^{(3)}, \cdots, C_2^{(2n-1)}\}. \end{aligned} \tag{2.40}$$

In addition to the cyclic subgroup C_{2n}, D_{2n} contains another two maximal invariant subgroups,

$$\begin{aligned} D_n &= \{E, C_{2n}^2, C_{2n}^4, \ldots, C_{2n}^{2n-2}, C_2^{(0)}, C_2^{(2)}, \ldots, C_2^{(2n-2)}\}, \\ D'_n &= \{E, C_{2n}^2, C_{2n}^4, \ldots, C_{2n}^{2n-2}, C_2^{(1)}, C_2^{(3)}, \ldots, C_2^{(2n-1)}\}. \end{aligned}$$

Discuss the classes in a permutation group S_n. The multiplication law of permutations can be understood as follows. In Eq. (2.18) when a permutation R is left-multiplied by a permutation S, the digits (objects) in the second row of R are made a S permutation. In Eq. (2.19) when S is right-multiplied by R, the digits in the first row of S are made an R^{-1} permutation. Therefore, **the conjugate permutation** SRS^{-1} of R can be calculated **by transforming all digits in both rows of R with the permutation** S. Especially, when R is a cycle, one has

$$\begin{aligned} S\,(a\ b\ c\ \ldots\ d)\,S^{-1} &= (s_a\ s_b\ s_c\ \ldots\ s_d)\,, \\ S\,(a\ b\ c\ \ldots\ d) &= (s_a\ s_b\ s_c\ \ldots s_d)\,S. \end{aligned} \tag{2.41}$$

When a permutation S is moved from the left-side of a permutation R to its right-side, the digits (objects) in R are made an S permutation. Conversely, when S is moved from the right-side of R to its left-side, the digits in R are made an S^{-1} permutation. This multiplication rule (2.41) of permutations are called the **interchanging rule**. Thus, two permutations are conjugate to each other if and only if their cycle structures are the same. The class in a permutation group S_n is characterized **by the cycle structure** of the permutations, namely, is characterized **by a partition of** n,

$$(\ell_1, \ell_2, \ldots, \ell_m)\,, \qquad \sum_{i=1}^{m} \ell_i = n. \tag{2.42}$$

The number $g_c(n)$ of classes in S_n is equal to the number of partitions of n. There is no analytic formula for $g_c(n)$ (see Chap. 1 of [Angrews (1976)]), but all partitions for a given positive integer n can be listed without dropping (see subsection 4.2.1). Some $g_c(n)$ are given as follows.

$$\begin{aligned} &g_c(1) = 1, \quad g_c(2) = 2, \quad g_c(3) = 3, \quad g_c(4) = 5, \\ &g_c(5) = 7, \quad g_c(6) = 11, \quad g_c(7) = 15, \quad g_c(8) = 22, \\ &g_c(9) = 30, \quad g_c(10) = 42, \quad g_c(20) = 627, \quad g_c(50) = 204226, \\ &g_c(100) = 190569292, \quad g_c(200) = 3972999029388. \end{aligned}$$

However, there is an analytic formula for the number $n(\alpha)$ of the elements in a class $\mathcal{C}_\alpha$ of S_n. In the following, without loss of generality, we calculate

$n(\alpha)$ in $\mathcal{C}_\alpha$ with the cycle structure

$$(k^{m_k}, (k-1)^{m_{k-1}}, \ldots, 2^{m_2}, 1^{m_1}), \qquad \sum_{\ell=1}^{k} \ell m_\ell = n, \tag{2.43}$$

where m_ℓ denotes the number of cycles with cycle length ℓ contained in the class $\mathcal{C}_\alpha$, and k denotes the largest cycle length among cycles in $\mathcal{C}_\alpha$. Define a parameter T_ℓ by

$$T_\ell = \sum_{a=1}^{\ell} a m_a = n - \sum_{b=\ell+1}^{k} b m_b, \qquad T_k = n, \qquad T_1 = m_1.$$

The number of elements in the class $(k^{m_k}, 1^{T_{k-1}})$ of S_n is $N[n,k,m_k]$:

$$N[n,k,m_k] = \{m_k!\}^{-1} \{(k-1)!\}^{m_k} \prod_{a=0}^{m_k-1} \binom{n-ak}{k} = \frac{n!}{k^{m_k} m_k! T_{k-1}!}.$$

The number of elements in the class $(k^{m_k}, (k-1)^{m_{k-1}}, 1^{T_{k-2}})$ is $N[n,k,m_k]N[T_{k-1}, k-1, m_{k-1}]$. Generally, The number of elements in the class $\mathcal{C}_\alpha$ is

$$n(\alpha) = \prod_{\ell=1}^{k} N[T_\ell, \ell, m_\ell] = n! \prod_{\ell=1}^{k} \{m_\ell! \ell^{m_\ell}\}^{-1}. \tag{2.44}$$

From the interchanging rule, one is able to calculate the rank of the permutation group S_n. Let $P_a = (a\ a+1)$ denote the transposition of two neighboring objects a and $(a+1)$. P_a satisfies

$$\begin{aligned} &P_a^2 = E, \qquad P_a P_{a+1} P_a = P_{a+1} P_a P_{a+1} = (a\ a+2), \\ &P_a P_b = P_b P_a \quad \text{when } |a-b| \geqslant 2. \end{aligned} \tag{2.45}$$

Each transposition can be expressed as a product of the transpositions of neighboring objects, so can a permutation,

$$\begin{aligned} (a\ \ d) = (d\ \ a) &= P_{a-1} P_{a-2} \cdots P_{d+1} P_d P_{d+1} \cdots P_{a-2} P_{a-1} \\ &= P_d P_{d+1} \cdots P_{a-2} P_{a-1} P_{a-2} \cdots P_{d+1} P_d, \qquad d < a. \end{aligned} \tag{2.46}$$

By making use of a cycle $W = (1\ 2\ \ldots\ n)$ of length n, one has

$$P_a = W P_{a-1} W^{-1} = W^{a-1} P_1 W^{-a+1}. \tag{2.47}$$

Thus, W and P_1 are the generators of a permutation group S_n. The rank of S_n is two.

2.4 Homomorphism

If there is a one-to-one correspondence between elements of two groups, and the correspondence is left invariant under the multiplication law of elements, two groups are **isomorphic**. If the correspondence between elements of two groups is not one-to-one, but many-to-one, the relation of the two groups becomes **homomorphic**.

Definition 2.3 A group G is said to be homomorphic onto another group G′, $G \sim G'$, if every element R' in G′ maps to at least one element R in G, every element R in G maps to one and only one element R' in G′, and the product of two elements in G′ is in the same way to map to the product of two corresponding elements in G and vice versa:

$$\begin{aligned} G \sim G' \ \text{if} \quad & R \longrightarrow R', \quad S \longrightarrow S', \quad \text{and} \quad RS \longrightarrow R'S', \\ & \forall \ R \in G, \quad S \in G, \quad R' \in G', \quad \text{and} \quad S' \in G', \end{aligned}$$

where the symbol "$R \longrightarrow R'$" denotes the many-to-one correspondence between R and R'.

There are different many-to-one correspondences between elements of two groups. A group G is homomorphic onto another group G′ if and only if there exists a many-to-one correspondence which is left invariant under the multiplication law of elements. One cannot make conclusion that two groups are NOT homomorphic only based on that there is a many-to-one correspondence between elements of two groups which is not invariant under the multiplication law of elements.

If two groups are isomorphic, each group describes the full property of another group. If a group G is homomorphic onto a group G′, the group G′ describes only part property of the group G. Theorem 2.3 will show what part property of G was described by G′ if $G \sim G'$.

Theorem 2.3 If $G \sim G'$, the subset H of elements in G which maps to the identity E' in G′ forms an invariant subgroup of G, and the subset of elements in G which maps to an element R' in G′ is a coset of H in G. The quotient group G/H of H in G is isomorphic onto the group G′, $G/H \approx G'$. H is called the **kernel** of homomorphism.

Proof The key of the proof is that the element R' in G′ to which the element R in G maps is **unique**. We will first prove that the subset H of elements in G which maps to the identity E' in G′ forms a subgroup of G. Then, prove that H is an invariant subgroup of G. At last, prove that the subset of elements in G which maps to an element R' in G$'$ forms a coset RH of H. Thus, it is obvious from Definition 2.2 that the quotient group G/H is isomorphic onto G$'$.

Let H $= \{S_1, S_2, \cdots, S_h\}$ be the subset of all elements in G which map to the identity E' in G$'$, $S_\mu \longrightarrow E'$. Because G $\sim$ G$'$, $S_\mu S_\nu \longrightarrow E'E' = E'$. Thus, H is closed to the multiplication law of elements. If the identity E in G maps to an element T' in G$'$, $ES_\mu \longrightarrow T'E' = T'$. At the same time one has $ES_\mu = S_\mu \longrightarrow E'$. From the uniqueness, $T' = E'$, so that E belongs to H. For an arbitrary element R in G and its inverse R^{-1}, letting $R \longrightarrow R'$ and $R^{-1} \longrightarrow P'$, one has $R^{-1}R = E$ which maps to both $P'R'$ and E'. From the uniqueness, $P' = R'^{-1}$. The inverse S_μ^{-1} of S_μ maps to $E'^{-1} = E'$ in G$'$, so that S_μ^{-1} belongs to H, too. Therefore, H satisfies four axioms and is a subgroup of G. Since $RS_\mu R^{-1}$ maps to $R'E'R'^{-1} = E'$, H is an invariant subgroup of G.

Each element in a coset RH of H maps to $R'E' = R'$ in G$'$. Conversely, if $R_\mu \longrightarrow R'$, then, $R^{-1}R_\mu \longrightarrow R'^{-1}R' = E'$. Namely, $R^{-1}R_\mu \in$ H, and both R and R_μ belong to the same coset RH. Therefore, there is a one-to-one correspondence between the complexes of elements in G and elements in G$'$: $H \longleftrightarrow E'$ and $RH \longleftrightarrow R'$. Their multiplication satisfies the same one-to-one correspondence. Thus, G/H $\approx$ G$'$. □

Theorem 2.3 shows that if G $\sim$ G$'$, G$'$ describes the property of the quotient group G/H, but the difference among elements in the kernel H of homomorphism is not described.

Theorem 2.4 Let G be a group and G$'$ be a set which is closed with respect to the given multiplication law of elements in G$'$. If there is a correspondence that every element R in G maps to one element R' in G$'$ uniquely and every element R' in G$'$ maps to at least one element R in G, and if this correspondence is left invariant under the multiplication law of elements, then G$'$ is a group and G is isomorphic or homomorphic onto G$'$, depending on whether the correspondence is one-to-one or many-to-one.

Proof The key of the proof is uniqueness of the element R' in G$'$ to which the element R in G maps. The proof is the same both for G $\sim$ G$'$ and for G $\approx$ G$'$. If $R \longrightarrow R'$, $S \longrightarrow S'$, and $T \longrightarrow T'$, then $RS \longrightarrow R'S'$, $ST \longrightarrow S'T'$, $(RS)T \longrightarrow (R'S')T'$, and $R(ST) \longrightarrow R'(S'T')$. From $(RS)T = R(ST)$, one has $(R'S')T' = R'(S'T')$ and the multiplication law of elements in G$'$ satisfies the associative law. From $E \longrightarrow E'$, $R \longrightarrow R'$, and $ER = R \longrightarrow E'R'$ one has $E'R' = R'$, and the set G$'$ contains the identity E'. From $R^{-1} \longrightarrow P'$, and $R^{-1}R = E \longrightarrow P'R'$, one has $P'R' = E'$, and the inverse of R' is P' which belongs to G$'$. □

2.5 Symmetry of a Regular Polyhedron

2.5.1 *Regular Polyhedrons*

A regular polyhedron is a geometric figure with high symmetry. Let the origin of the coordinate frame coincide with the center of the regular polyhedron. The symmetry transformations of the polyhedron leaves the origin invariant and the proper symmetry group of a regular polyhedron is a proper point group.

Consider a regular polyhedron with N side-faces, L edges, and V vertices. Each side-face in the polyhedron is a regular polygon with n edges. m edges as well as m side-faces meet at one vertex. Since each edge connects two vertices and is shared by two side-faces, one has

$$nN = 2L = mV. \tag{2.48}$$

The internal angle of a regular polygon with n edges is $(n-2)\pi/n$. For a solid figure, m has to be larger than 2 and $m(n-2)\pi/n < 2\pi$. Thus, $2 < m < 2n/(n-2)$. When $n = 3$, the regular polyhedrons are tetrahedron ($m = 3$), octahedron ($m = 4$), and icosahedron ($m = 5$), respectively. When $n = 4$ or 5, m has to be 3 and the regular polyhedrons are cube ($n = 4$) and dodecahedron ($n = 5$), respectively. There is no regular polyhedron with $n > 5$. The parameters of regular polyhedrons are listed in Table 2.6.

Table 2.6 The parameters of regular polyhedrons

N: No. of side-faces in polyhedron	4	6	8	12	20
V: No. of vertices in polyhedron	4	8	6	20	12
L: No. of edges in polyhedron	6	12	12	30	30
n: No. of edges in a side-face	3	4	3	5	3
m: No. of edges meeting at a vertex	3	3	4	3	5
G: proper point group of polyhedron	**T**	**O**	**O**	**I**	**I**
$g = 2L$: order of G	12	24	24	60	60

Connecting the centers of all side-faces of a regular polyhedron, one obtains another regular polyhedron. Two polyhedrons are called to be dual to each other. For two dual polyhedrons, N **interchanges with** V, n **interchanges with** m**, and they have the same** L **and the same symmetry group.** A dodecahedron is dual to an icosahedron and their proper symmetry group is denoted by **I**. A cube is dual to an octahedron and their proper symmetry group is denoted by **O**. The tetrahedron is a self-dual figure and its proper symmetry group is denoted by **T**.

Since the origin coincides with the center of the polyhedron, the position of the polyhedron is determined by the position of only one edge. In a symmetry transformation this edge may change to the position of another edge with two different direction so that the order g of the proper symmetry group is equal to $2L$. Thus, the orders of **T**, **O**, and **I** are 12, 24 and 60, respectively.

2.5.2 *Symmetry of a Cube*

A cube is shown in Fig. 2.2 where the origin of the rectangular coordinate frame coincides with the center of the cube, and the coordinate axes point from the origin to the centers of three neighboring squares of the cube. Four vertices above the XY plane are denoted by A_j sequently, $1 \leqslant j \leqslant 4$, and their opposite vertices are denoted by B_j, respectively. A_1 has positive components of X, Y, and Z. An octahedron is obtained by connecting the centers of six squares in the cube. A tetrahedron is obtained by connecting four un-neighboring vertices A_1, B_2, A_3, and B_4 in the cube.

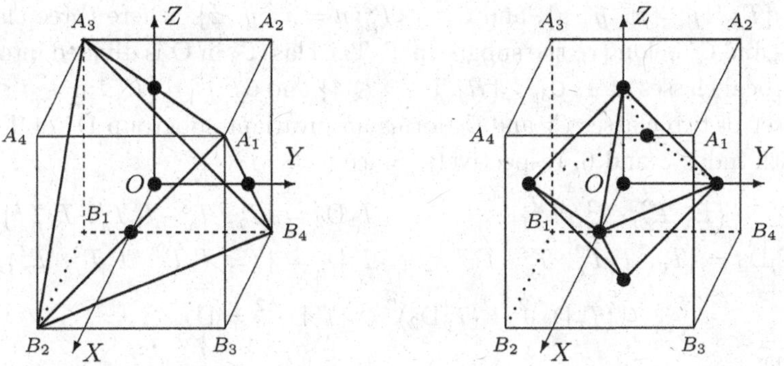

Fig. 2.2 A cube, a tetrahedron, and an octahedron.

From Fig. 2.2, three coordinate axes are the 4-fold axes of the cube, but the 2-fold axes of the tetrahedron. Let T_μ, $\mu = x,\ y,\ z$, denote the rotations about the coordinate axes through the angle $\pi/2$, respectively. Only T_μ^2, not T_μ, belongs to the group **T**. Four diagonals, respectively connecting two opposite vertices of the cube, are the 3-fold axes of both the cube and the tetrahedron. The direction of each 3-fold axis is defined from the center of one triangle to the opposite vertex in a tetrahedron:

$$\begin{aligned} R_1 : &\quad \text{from } B_1 \text{ to } A_1, \quad (e_x + e_y + e_z)/\sqrt{3}, \\ R_2 : &\quad \text{from } A_2 \text{ to } B_2, \quad (e_x - e_y - e_z)/\sqrt{3}, \\ R_3 : &\quad \text{from } B_3 \text{ to } A_3, \quad (-e_x - e_y + e_z)/\sqrt{3}, \\ R_4 : &\quad \text{from } A_4 \text{ to } B_4, \quad (-e_x + e_y - e_z)/\sqrt{3}. \end{aligned} \tag{2.49}$$

R_j and R_j^{-1}, $1 \leqslant j \leqslant 4$, belong to both group **O** and group **T**. Six lines, respectively connecting the midpoints of two opposite edges of the cube, are the 2-fold axes of the cube, but not of the tetrahedron. Let S_k, $1 \leqslant k \leqslant 6$, denote six rotations about those 2-fold axes through π angle:

$$\begin{aligned} S_1 : &\quad (e_x + e_y)/\sqrt{2}, \qquad S_2 : \quad (e_x - e_y)/\sqrt{2}, \\ S_3 : &\quad (e_y + e_z)/\sqrt{2}, \qquad S_4 : \quad (e_y - e_z)/\sqrt{2}, \\ S_5 : &\quad (e_x + e_z)/\sqrt{2}, \qquad S_6 : \quad (e_x - e_z)/\sqrt{2}. \end{aligned} \tag{2.50}$$

A cube has three 4-fold axes, four 3-fold axes, and six 2-fold axes. All axes are non-polar, and any two axes with the same fold are equivalent to each other. A tetrahedron has three 2-fold axes and four 3-fold axes, where 2-fold axes are non-polar and 3-fold axes are polar. The proper symmetry group **O** of a cube contains 24 elements and five self-reciprocal classes: $\mathcal{C}_1 = \{E\}$, $\mathcal{C}_2 = \{S_k; 1 \leqslant k \leqslant 6\}$, $\mathcal{C}_3 = \{R_j^{\pm 1}; 1 \leqslant j \leqslant 4\}$, $\mathcal{C}_4 = \{T_\mu^{\pm 1}; \mu = x,\ y,\ z\}$, and $\mathcal{C}_5 = \{T_\mu^2; \mu = x,\ y,\ z\}$, where three classes $\mathcal{C}_1$, $\mathcal{C}_3$ and $\mathcal{C}_5$ belong to the subgroup **T**. The class $\mathcal{C}_3$ in **O** is divided into two reciprocal classes of **T**: $\mathcal{C}_3 = \{R_j; 1 \leqslant j \leqslant 4\}$ and $\mathcal{C}_3^{(-1)} = \{R_j^{-1}; 1 \leqslant j \leqslant 4\}$. The set of two classes $\mathcal{C}_1$ and $\mathcal{C}_5$ forms an invariant subgroup D_2 of **T** and **O** with index 3 and 6, respectively. Since

$$\begin{aligned} \mathrm{D}_2 &= \{E,\ T_x^2,\ T_y^2,\ T_z^2\}, \qquad & T_x\mathrm{D}_2 &= \{T_x,\ T_x^3,\ T_xT_y^2,\ T_xT_z^2\}, \\ T_y\mathrm{D}_2 &= \{T_y,\ T_yT_x^2,\ T_y^3,\ T_yT_z^2\}, \qquad & T_z\mathrm{D}_2 &= \{T_z,\ T_zT_x^2,\ T_zT_y^2,\ T_z^3\}, \end{aligned}$$

$$(T_x\mathrm{D}_2)^2 = (T_y\mathrm{D}_2)^2 = (T_z\mathrm{D}_2)^2 = \mathrm{D}_2,$$

one has

$$\mathbf{T}/\mathrm{D}_2 \approx \mathrm{C}_3, \qquad \mathbf{O}/\mathrm{D}_2 \approx \mathrm{D}_3. \tag{2.51}$$

Let (2θ) denote the angle between two neighboring 3-fold axes, whose angular bisector is a 4-fold axis in the cube. The angle between a 3-fold axis and a neighboring 2-fold axis in the cube is $\pi/2 - \theta$. If the length of the edge of the cube is taken to be unit, the radius R of the circumcircle, the radius r_O of the inscribed circle of the cube, the radius r_T of the inscribed circle of the tetrahedron, and θ are listed as follows:

$$\begin{aligned} &R = \sqrt{3}/2, \qquad r_O = 1/2, \qquad r_T = \sqrt{1/12}, \\ &\cos\theta = \sqrt{1/3}, \quad \cos(2\theta) = -1/3, \quad \theta = 54.73^\circ. \end{aligned} \tag{2.52}$$

In comparison with **O**, the permutation group S_4 contains 24 elements and five self-reciprocal classes, too. $\overline{\mathcal{C}}_1$ contains only the identity E. $\overline{\mathcal{C}}_2$ with the cycle structure $(2,1,1)$ contains 6 elements with order 2. $\overline{\mathcal{C}}_3$ with the cycle structure $(3,1)$ contains 8 elements with order 3. $\overline{\mathcal{C}}_4$ with the cycle structure (4) contains 6 elements with order 4. $\overline{\mathcal{C}}_5$ with the cycle structure $(2,2)$ contains 3 elements with order 2. S_4 has an invariant subgroups S_4' with index two and an invariant subgroups D_2 with index six. S_4' consists of $\overline{\mathcal{C}}_1$, $\overline{\mathcal{C}}_3$, and $\overline{\mathcal{C}}_5$. D_2 consists of $\overline{\mathcal{C}}_1$ and $\overline{\mathcal{C}}_5$, and $S_4/D_2 \approx D_3$. Thus, S_4 has the same structure of classes as **O**. The structure of classes of a finite group includes how many classes the group contains, how many elements each class contains, to what the order of the element in each class is equal, the union of which classes forms an invariant subgroup, and onto what group the quotient group of the invariant subgroup is isomorphic. Therefore,

$$\mathbf{O} \approx S_4, \qquad \mathbf{T} \approx S_4'. \tag{2.53}$$

For two isomorphic groups, one has to find the one-to-one correspondence between elements of two groups and to show that the correspondence is invariant under multiplications of elements. One has to make the correspondence between each pair of classes in two groups **O** and S_4,

$$\mathcal{C}_j \longleftrightarrow \overline{\mathcal{C}}_j, \qquad 1 \leqslant j \leqslant 5. \tag{2.54}$$

There are three kinds of multiplications of elements. One is the powers of the element. The second is the conjugate relation (2.38) which corresponds to the interchanging rule (2.41) in S_n. The correspondence (2.54) is left invariant under these two kinds of multiplications.

Let us study the conjugate relation (2.38) in some detail. Since T_z is a 4-fold rotation, the remaining elements in **O**, except for the powers of T_z, can be divided into six subsets around T_z:

$$\begin{aligned} &\{R_1,\ R_2^2,\ R_3,\ R_4^2\}, \qquad &&\{R_1^2,\ R_2,\ R_3^2,\ R_4\}, \\ &\{S_5,\ S_3,\ S_6,\ S_4\}, &&\{S_1,\ S_2,\ S_1,\ S_2\}, \\ &\{T_x,\ T_y,\ T_x^{-1},\ T_y^{-1}\}, &&\{T_x^2,\ T_y^2,\ T_x^2\ T_y^2\}. \end{aligned} \tag{2.55}$$

The elements in each subsets satisfy sequently the conjugate relation (2.38), where S is taken to be T_z. For example, the elements in the first subset satisfy

$$\begin{aligned} T_zR_1T_z^{-1} &= R_2^2, \qquad T_zR_2^2T_z^{-1} = R_3, \\ T_zR_3T_z^{-1} &= R_4^2, \qquad T_zR_4^2T_z^{-1} = R_1. \end{aligned} \tag{2.56}$$

In addition to Eq. (2.56), there are more conjugate relations to relate the elements in different subsets (2.55): $T_x = R_1T_zR_1^{-1}$ and $S_1 = R_1S_5R_1^{-1}$.

The third kind of multiplications of elements is the multiplication of two elements belonging to different classes. From Fig. 2.2 one obtains

$$A_1 \xrightarrow{T_z} A_2 \xrightarrow{R_1} A_4, \qquad A_4 \xrightarrow{T_z} A_1 \xrightarrow{R_1} A_1,$$

Namely, the product R_1T_z is equal to S_5:

$$R_1T_z = S_5. \tag{2.57}$$

Equation (2.57) in $\mathbf{O}$ just likes Eq. (2.10) in D_N. Due to Eq. (2.54), T_z, R_1 and S_5 have to correspond to the permutations with the cycle structures (4), (3, 1) and (2, 1, 1), respectively. Without loss of generality, let $T_z \leftrightarrow (1\ 2\ 3\ 4)$. There are two choices of R_1 for correspondence: $(1\ 2\ 3)$ or $(2\ 1\ 3)$. Substituting them into Eq. (2.57), one has

$$\begin{aligned} R_1 &\leftrightarrow (1\ 2\ 3), \qquad S_5 \leftrightarrow (1\ 2\ 3)(1\ 2\ 3\ 4) = (2\ 1\ 3\ 4), \\ R_1 &\leftrightarrow (2\ 1\ 3), \qquad S_5 \leftrightarrow (2\ 1\ 3)(1\ 2\ 3\ 4) = (3\ 4). \end{aligned}$$

The first choice violates the correspondence (2.54) and the second one is suitable. One may choose correspondence such as $R_1 \leftrightarrow (3\ 2\ 4)$, $(4\ 3\ 1)$, or $(1\ 4\ 2)$, but they are different only formally due to Eq. (2.56). Based on $T_z \leftrightarrow (1\ 2\ 3\ 4)$, $R_1 \leftrightarrow (2\ 1\ 3)$ and $S_5 \leftrightarrow (3\ 4)$, the one-to-one correspondences between the elements of $\mathbf{O}$ and S_4 are calculated from $T_x = R_1T_zR_1^{-1}$, $S_1 = R_1S_5R_1^{-1}$ and the conjugate relations (2.56):

$$\begin{aligned} &T_z \leftrightarrow (1\ 2\ 3\ 4), \quad T_x \leftrightarrow (1\ 2\ 4\ 3), \quad T_y \leftrightarrow (2\ 3\ 1\ 4), \\ &R_1 \leftrightarrow (3\ 2\ 1), \quad R_2^2 \leftrightarrow (4\ 3\ 2), \quad R_3 \leftrightarrow (1\ 4\ 3), \\ &R_4^2 \leftrightarrow (2\ 1\ 4), \quad S_5 \leftrightarrow (3\ 4), \quad S_3 \leftrightarrow (4\ 1), \\ &S_6 \leftrightarrow (1\ 2), \quad S_4 \leftrightarrow (2\ 3), \quad S_1 \leftrightarrow (2\ 4), \\ &S_2 \leftrightarrow (3\ 1), \quad E \leftrightarrow E. \end{aligned} \tag{2.58}$$

and

$$(a\ b\ c\ d)^2 = (a\ c)(b\ d), \quad (a\ b\ c\ d)^3 = (d\ c\ b\ a), \quad (a\ b\ c)^2 = (c\ b\ a). \tag{2.59}$$

In terms of the multiplication law (2.28) of two permutations, any product of two elements in $\mathbf{O}$, as well as its group table, is easy to calculate through the correspondence (2.58). For example,

$$
\begin{aligned}
R_1T_x^2 &\leftrightarrow (3\ 2\ 1)(1\ 4)(2\ 3) = (1\ 4\ 3) \leftrightarrow R_3, \\
R_1T_y^2 &\leftrightarrow (3\ 2\ 1)(1\ 2)(3\ 4) = (2\ 3\ 4) \leftrightarrow R_2, \\
R_1T_z^2 &\leftrightarrow (3\ 2\ 1)(1\ 3)(2\ 4) = (1\ 2\ 4) \leftrightarrow R_4, \\
R_1^2T_x^2 &\leftrightarrow (1\ 2\ 3)(1\ 4)(2\ 3) = (1\ 4\ 2) \leftrightarrow R_4^2, \\
R_1^2T_y^2 &\leftrightarrow (1\ 2\ 3)(1\ 2)(3\ 4) = (1\ 3\ 4) \leftrightarrow R_3^2, \\
R_1^2T_z^2 &\leftrightarrow (1\ 2\ 3)(1\ 3)(2\ 4) = (3\ 2\ 4) \leftrightarrow R_2^2, \\
T_zT_x^2 &\leftrightarrow (1\ 2\ 3\ 4)(1\ 4)(2\ 3) = (4\ 2) \leftrightarrow S_1, \\
T_zT_y^2 &\leftrightarrow (1\ 2\ 3\ 4)(1\ 2)(3\ 4) = (1\ 3) \leftrightarrow S_2, \\
T_yT_x^2 &\leftrightarrow (1\ 4\ 2\ 3)(1\ 4)(2\ 3) = (1\ 2) \leftrightarrow S_6, \\
T_yT_z^2 &\leftrightarrow (1\ 4\ 2\ 3)(1\ 3)(2\ 4) = (3\ 4) \leftrightarrow S_5, \\
T_xT_y^2 &\leftrightarrow (1\ 2\ 4\ 3)(1\ 2)(3\ 4) = (1\ 4) \leftrightarrow S_3, \\
T_xT_z^2 &\leftrightarrow (1\ 2\ 4\ 3)(1\ 3)(2\ 4) = (3\ 2) \leftrightarrow S_4.
\end{aligned} \tag{2.60}
$$

Comparing $R_1T_z = S_5$ with $C_3C_2^{(0)} = C_2^{(1)}$ (see Eq. (2.10)), one obtains the one-to-one correspondence between elements of $\mathbf{O}/D_2$ and D_3:

$$
\mathbf{O}/D_2 \approx \mathrm{D}_3: \quad
\begin{aligned}
\mathrm{D}_2 &= \{E,\ T_x^2,\ T_y^2,\ T_z^2\} \leftrightarrow E, \\
R_1\mathrm{D}_2 &= \{R_1,\ R_3,\ R_2,\ R_4\} \leftrightarrow C_3, \\
R_1^2\mathrm{D}_2 &= \{R_1^2,\ R_4^2,\ R_3^2,\ R_2^2\} \leftrightarrow C_3^2, \\
T_z\mathrm{D}_2 &= \{T_z,\ S_1,\ S_2,\ T_z^3\} \leftrightarrow C_2^{(0)}, \\
T_y\mathrm{D}_2 &= \{T_y,\ S_6,\ T_y^3,\ S_5\} \leftrightarrow C_2^{(1)}, \\
T_x\mathrm{D}_2 &= \{T_x,\ T_x^3,\ S_3,\ S_4\} \leftrightarrow C_2^{(2)}.
\end{aligned} \tag{2.61}
$$

2.5.3 *Symmetry of an Icosahedron*

A regular icosahedron is shown in Fig. 2.3, where the origin O of the coordinate frame coincides with the center of the icosahedron and the Z-axis is in the direction from the origin O to one vertex denoted by A_0. The remaining five vertices above the XY plane are denoted by A_j sequently, $1 \leqslant j \leqslant 5$, where A_1 is on the XZ plane. The opposite vertex of A_j, including A_0, is denoted by B_j. The Y-axis is in the direction from O to the midpoint of the edge A_2B_5.

The regular icosahedron contains six 5-fold axes, ten 3-fold axis, and fifteen 2-fold axes. The 5-fold axes are along the directions from B_j to A_j with the generators T_j, $0 \leqslant j \leqslant 5$. Except for one 5-fold axis ($j = 0$) along the positive Z-axis, the polar angles of the remaining 5-fold axes all are θ_1, and their azimuthal angles are $2(j-1)\pi/5$, respectively. The 3-fold axes are along the lines connecting the centers of two opposite triangles with the generators R_j, $1 \leqslant j \leqslant 10$. The polar angles of the 3-fold axes are θ_2 when $1 \leqslant j \leqslant 5$, and θ_3 when $6 \leqslant j \leqslant 10$. Their azimuthal angles are $(2j-1)\pi/5$, respectively. The 2-fold axes are along the lines connecting the midpoints of two opposite edges with the generators S_j, $1 \leqslant j \leqslant 15$. The polar angles of the 2-fold axes are θ_4 when $1 \leqslant j \leqslant 5$, θ_5 when $6 \leqslant j \leqslant 10$, and $\pi/2$ when $11 \leqslant j \leqslant 15$. Their azimuthal angles are $2(j-1)\pi/5$ when $1 \leqslant j \leqslant 5$, $(2j-1)\pi/5$ when $6 \leqslant j \leqslant 10$, and $(4j-3)\pi/10$ when $11 \leqslant j \leqslant 15$, respectively. **All axes are non-polar and any two axes with the same fold are equivalent to each other.** θ_1 will be calculated by group theory in Eq. (5.96). Then, $2\theta_4 = \theta_1$ and $2\theta_5 = \theta_2 + \theta_3 = \pi - \theta_1$. If the length of the edge of the icosahedron is taken to be unit, the radius R of its circumcircle, the radius r of its inscribed circle, and θ_2 are calculable.

$$\begin{aligned}
&\tan\theta_1 = 2, \quad \tan\theta_4 = (\sqrt{5}-1)/2, \quad \tan\theta_5 = (\sqrt{5}+1)/2, \\
&R = \frac{1}{2\sin\theta_4} = \sqrt{\frac{5+\sqrt{5}}{8}} = 0.9511, \quad r = \sqrt{R^2 - \frac{1}{3}} = \frac{3+\sqrt{5}}{4\sqrt{3}} = 0.7558 \\
&\tan\theta_2 = (r\sqrt{3})^{-1} = 3-\sqrt{5}, \quad \tan\theta_3 = -\tan(\theta_1+\theta_2) = 3+\sqrt{5}, \\
&\theta_1 = 2\theta_4 \approx 63.43^\circ, \quad \theta_2 \approx 37.38^\circ, \quad \theta_3 \approx 79.19^\circ, \quad \theta_5 \approx 58.28^\circ.
\end{aligned} \tag{2.62}$$

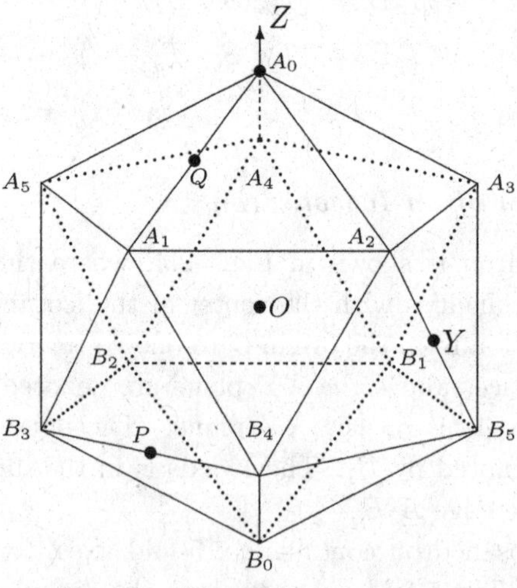

Fig. 2.3 The regular icosahedron.

The proper symmetry group **I** of an icosahedron contains 60 elements and five self-reciprocal classes: $\mathcal{C}_1 = \{E\}$, $\mathcal{C}_2 = \{S_j|1 \leqslant j \leqslant 15\}$, $\mathcal{C}_3 = \{R_j^{\pm 1}|1 \leqslant j \leqslant 10\}$, $\mathcal{C}_4 = \{T_j^{\pm 1}|0 \leqslant j \leqslant 5\}$, and $\mathcal{C}_5 = \{T_j^{\pm 2}|0 \leqslant j \leqslant 5\}$. The numbers of elements in those classes are 1, 15, 20, 12, and 12, respectively. **There is no nontrivial invariant subgroup in I.**

In comparison with **I**, the alternating subgroup S_5' of the permutation group S_5 contains 60 elements and five self-reciprocal classes, too. $\overline{\mathcal{C}}_1$ contains only the identity E. $\overline{\mathcal{C}}_2$ with the cycle structure $(2,2,1)$ contains 15 elements with order 2. $\overline{\mathcal{C}}_3$ with the cycle structure $(3,1,1)$ contains 20 elements with order 3. The class $\mathcal{C}_\alpha$ with the cycle structure (5) in S_5 contains 24 elements with order 5. In the following we will prove that, for the subgroup S_5', the class $\mathcal{C}_\alpha$ is divided into two classes $\overline{\mathcal{C}}_4 = \{T|T = SRS^{-1}, S \in \mathrm{S}_5'\}$ and $\overline{\mathcal{C}}_5 = \{T|T = SR^2S^{-1}, S \in \mathrm{S}_5'\}$, where $R = (1\ 2\ 3\ 4\ 5)$. Both $\overline{\mathcal{C}}_4$ and $\overline{\mathcal{C}}_5$ in S_5' are the self-reciprocal classes and contain 12 elements with order 5.

In fact, from Theorem 2.2, for given two elements T and $R = (1\,2\,3\,4\,5)$ in $\mathcal{C}_\alpha$, the number of S in S_5 satisfying $T = SRS^{-1}$ is $m(\alpha) = g/n(\alpha) = 120/24 = 5$. The five elements S can be expressed as PR^a with $0 \leqslant a \leqslant 4$ if $T = PRP^{-1}$. Since the permutation R is even, we have $\delta(PR^a) = \delta(P)$, and the class $\mathcal{C}_\alpha$ in S_5 is divided into two classes for the subgroup S_5', depending on whether P is even or odd. Because

$$\begin{aligned}
&(a\ b\ c\ d\ e)^{-1} = (a\ e\ d\ c\ b) = [(b\ e)(c\ d)](a\ b\ c\ d\ e)[(b\ e)(c\ d)], \\
&(a\ b\ c\ d\ e)^2 = (a\ c\ e\ b\ d) = (b\ c\ e\ d)(a\ b\ c\ d\ e)(b\ c\ e\ d)^{-1},
\end{aligned}$$

where $(b\ e)(c\ d)$ is even and $(b\ c\ e\ d)$ is odd. For S_5', $R \in \mathcal{C}_\alpha$ is conjugate with R^{-1}, but not conjugate with R^2. Thus, the class $\mathcal{C}_\alpha$ of S_5 is divided into two self-reciprocal classes $\overline{\mathcal{C}}_4$ and $\overline{\mathcal{C}}_5$ in the subgroup S_5'.

There is no nontrivial invariant subgroup in S_5' (see Prob. 19). S_5' has the same structure of classes as **I** such that

$$\mathbf{I} \approx \mathrm{S}_5'. \tag{2.63}$$

Since T_0 is a 5-fold rotation, the remaining elements in **I**, except for the powers of T_0, are divided into eleven subsets around T_0:

$$\begin{aligned}
&\left(T_j^a|1 \leqslant j \leqslant 5\right), \qquad 1 \leqslant a \leqslant 4, \\
&\left(R_{j+b}^a|1 \leqslant j \leqslant 5\right), \qquad a = 1,\ 2, \qquad b = 0,\ 5, \\
&\left(S_{j+b}|1 \leqslant j \leqslant 5\right), \qquad b = 0,\ 5,\ 10.
\end{aligned} \tag{2.64}$$

The elements in each subset satisfy the following conjugate relation where a and b are given in Eq. (2.64) and $1 \leqslant j \leqslant 4$,

$$\begin{aligned} &T_0T_j^aT_0^{-1} = T_{1+j}^a, && T_0T_5^aT_0^{-1} = T_1^a, \\ &T_0R_{b+j}^aT_0^{-1} = R_{b+1+j}^a, && T_0R_{b+5}^aT_0^{-1} = R_{b+1}^a, \\ &T_0S_{b+j}T_0^{-1} = S_{b+1+j}, && T_0S_{b+5}T_0^{-1} = S_{b+1}, \end{aligned} \tag{2.65}$$

From Fig. 2.3 one obtains

$$A_0 \xrightarrow{T_0} A_0 \xrightarrow{R_1} A_1, \qquad A_1 \xrightarrow{T_0} A_2 \xrightarrow{R_1} A_0,$$

Namely, the product R_1T_0 is equal to S_1:

$$R_1T_0 = S_1. \tag{2.66}$$

Equation (2.66) in **I** just likes Eq. (2.53) in **O**. Make the correspondence between each pair of classes in two groups **I** and S_5',

$$\mathcal{C}_j \longleftrightarrow \overline{\mathcal{C}}_j, \qquad 1 \leqslant j \leqslant 5. \tag{2.67}$$

Thus, T_0, R_1 and S_1 have to correspond to the permutations with the cycle structures (5), $(3,1,1)$ and $(2,2,1)$, respectively. Without loss of generality, let $T_0 \leftrightarrow (1\ 2\ 3\ 4\ 5)$. There are four choices for R_1 which are different in essence. From Eq. (2.66) one has

$$\begin{aligned} &R_1 \leftrightarrow (1\ 2\ 3), \quad && S_1 \leftrightarrow (1\ 2\ 3)(1\ 2\ 3\ 4\ 5) = (2\ 1\ 3\ 4\ 5), \\ &R_1 \leftrightarrow (2\ 1\ 3), && S_1 \leftrightarrow (2\ 1\ 3)(1\ 2\ 3\ 4\ 5) = (3\ 4\ 5), \\ &R_1 \leftrightarrow (1\ 2\ 4), && S_1 \leftrightarrow (1\ 2\ 4)(1\ 2\ 3\ 4\ 5) = (2\ 3\ 1\ 4\ 5), \\ &R_1 \leftrightarrow (2\ 1\ 4), && S_1 \leftrightarrow (2\ 1\ 4)(1\ 2\ 3\ 4\ 5) = (2\ 3)(4\ 5). \end{aligned}$$

The first three choices violate the correspondence (2.67), and the last one is suitable. In addition to Eq. (2.65), there are more conjugate relations for calculating the first element in each subsets in Eq. (2.64):

$$\begin{aligned} T_1 &= R_1T_0R_1^{-1} \leftrightarrow (2\ 1\ 4)(1\ 2\ 3\ 4\ 5)(4\ 1\ 2) = (4\ 1\ 3\ 2\ 5), \\ S_6 &= R_1S_1R_1^{-1} \leftrightarrow (2\ 1\ 4)[(2\ 3)(4\ 5)](4\ 2\ 1) = (1\ 3)(2\ 5), \\ R_6 &= S_6R_1S_6 \leftrightarrow [(1\ 3)(2\ 5)](2\ 1\ 4)[(1\ 3)(2\ 5)] = (5\ 3\ 4), \\ S_{11} &= R_6S_6R_6^{-1} \leftrightarrow (5\ 3\ 4)[(1\ 3)(2\ 5)](4\ 3\ 5) = (1\ 4)(2\ 3). \end{aligned} \tag{2.68}$$

Based on $T_0 \leftrightarrow (1\ 2\ 3\ 4\ 5)$, $R_1 \leftrightarrow (2\ 1\ 4)$, $S_1 \leftrightarrow (2\ 3)(4\ 5)$, the one-to-one correspondence between the elements of **I** and S_5' is calculated from the conjugate relations (2.64), (2.65) and (2.68):

$$\begin{array}{lll}
T_0 \leftrightarrow (1\,2\,3\,4\,5), & T_1 \leftrightarrow (1\,3\,2\,5\,4), & T_2 \leftrightarrow (2\,4\,3\,1\,5), \\
T_3 \leftrightarrow (3\,5\,4\,2\,1), & T_4 \leftrightarrow (4\,1\,5\,3\,2), & T_5 \leftrightarrow (5\,2\,1\,4\,3), \\
R_1 \leftrightarrow (2\,1\,4), & R_2 \leftrightarrow (3\,2\,5), & R_3 \leftrightarrow (4\,3\,1), \\
R_4 \leftrightarrow (5\,4\,2), & R_5 \leftrightarrow (1\,5\,3), & R_6 \leftrightarrow (5\,3\,4), \\
R_7 \leftrightarrow (1\,4\,5), & R_8 \leftrightarrow (2\,5\,1), & R_9 \leftrightarrow (3\,1\,2), \\
R_{10} \leftrightarrow (4\,2\,3), & S_1 \leftrightarrow (2\,3)(4\,5), & S_2 \leftrightarrow (3\,4)(5\,1), \\
S_3 \leftrightarrow (4\,5)(1\,2), & S_4 \leftrightarrow (5\,1)(2\,3), & S_5 \leftrightarrow (1\,2)(3\,4), \\
S_6 \leftrightarrow (1\,3)(2\,5), & S_7 \leftrightarrow (2\,4)(3\,1), & S_8 \leftrightarrow (3\,5)(4\,2), \\
S_9 \leftrightarrow (4\,1)(5\,3), & S_{10} \leftrightarrow (5\,2)(1\,4), & S_{11} \leftrightarrow (1\,4)(2\,3), \\
S_{12} \leftrightarrow (2\,5)(3\,4), & S_{13} \leftrightarrow (3\,1)(4\,5), & S_{14} \leftrightarrow (4\,2)(5\,1), \\
S_{15} \leftrightarrow (5\,3)(1\,2), & E \leftrightarrow E, &
\end{array} \tag{2.69}$$

and

$$\begin{array}{ll}
(a\,b\,c)^2 = (c\,b\,a), & (a\,b\,c\,d\,e)^2 = (a\,c\,e\,b\,d), \\
(a\,b\,c\,d\,e)^3 = (d\,b\,e\,c\,a), & (a\,b\,c\,d\,e)^4 = (e\,d\,c\,b\,a).
\end{array} \tag{2.70}$$

Any product of two elements in **I**, as well as its group table, is easy to calculate through the correspondence (2.69) and the multiplication law (2.28) of permutations.

2.6 Direct Product of Groups and Improper Point Groups

2.6.1 *Direct Product of Two Groups*

Definition 2.4 A group G is called the direct product $G = H_1 \otimes H_2$ of its two subgroups H_1 and H_2,

$$H_1 = \{R_1,\ R_2,\ \cdots,\ R_{h_1}\}, \qquad H_2 = \{S_1,\ S_2,\ \cdots,\ S_{h_2}\}, \tag{2.71}$$

if the following conditions are satisfied.

a) Except for the identity $E = R_1 = S_1$, there is no other common element in H_1 and H_2.

b) Any two elements respectively belonging to two subgroups H_1 and H_2 are commutable, namely $R_jS_\mu = S_\mu R_j$, $\forall$ $R_j \in H_1$ and $S_\mu \in H_2$.

c) The group G consists of all products R_jS_μ.

Obviously, if $G = H_1 \otimes H_2$, H_1 and H_2 both are the invariant subgroups of G. Since the identity E is the only common element in two subgroups H_1 and H_2, there is no repetitive element in the set $\{R_jS_\mu\}$. In fact, if $R_jS_\mu = R_kS_\nu$, then $R_k^{-1}R_j = S_\nu S_\mu^{-1}$ is the identity E, namely, $R_j = R_k$ and $S_\mu = S_\nu$. Thus, the order of G is $g = h_1h_2$.

The practical case of the direct product of two groups one often meets is that the groups H_1 and H_2 are two sets of operators respectively acting on two different subsystems so that the elements in two groups are commutable, $R_jS_\mu = S_\mu R_j$. Define two groups $\mathrm{H}_1S_1 \approx \mathrm{H}_1$ and $R_1\mathrm{H}_2 \approx H_2$, where R_1 and S_1 are the identities of H_1 and H_2, respectively. The element R_1S_1 is the only common element in two groups H_1S_1 and $R_1\mathrm{H}_2$. The set

$$\mathrm{G} = \{R_jS_\mu | R_j \in \mathrm{H}_1, S_\mu \in \mathrm{H}_2\} \tag{2.72}$$

satisfies four axioms under the multiplication laws of elements: $(R_jS_\mu)(R_kS_\nu) = (R_jR_k)(S_\mu S_\nu)$. This group G is the direct product of two groups H_1 and H_2: $\mathrm{G} = \mathrm{H}_1 \otimes \mathrm{H}_2$.

2.6.2 *Improper Point Groups*

Any improper rotation S' is the product of a proper rotation S and a space inversion σ. σ commutes with any rotation and its square is equal to the identity E,

$$S' = \sigma S = S\sigma, \qquad \sigma^2 = E. \tag{2.73}$$

The product of two proper rotations or two improper rotations is a proper rotation, and the product of one proper rotation and one improper rotation is an improper rotation. The inverse R^{-1} of a proper rotation R is a proper rotation. An improper point group G contains not only improper rotations, but also proper rotations. The subset H composed of all proper rotations in G is an invariant subgroup of G with index 2. H is called the proper subgroup of an improper point group G.

There are two types of improper point groups depending on whether it contains the space inversion σ. An improper point group G of I-type contains the space inversion σ, and it is the direct product of its proper subgroup H and the inversion group V_2 of order 2,

$$G = H \otimes \mathrm{V}_2, \qquad \mathrm{V}_2 = \{E, \sigma\}. \tag{2.74}$$

An improper point group G of P-type does not contain the space inversion σ. Let R_k and S_j denote the proper rotation and the improper rotation in G, respectively. $\sigma S_j = S_j'$ is a proper rotation and is not equal to any proper rotation R_k in G. In fact, if $\sigma S_j = R_k$, $\sigma = R_kS_j^{-1} \in G$. It is in conflict with the assumption. Let G′ denote a proper point group composed of those $S_j' = \sigma S_j$ and R_k. G′ is isomorphic onto G and also contains the invariant subgroup H with index 2. Conversely, from a proper point group

G′ containing an invariant subgroup H with index 2, one can construct an improper point group G of P-type by leaving H invariant and multiplying the elements in the coset of H in G′ with σ. G is isomorphic onto G′.

The Schönflies notations is widely used to denote improper point groups G. Choose the coordinate system such that the Z-axis is along a proper or improper axis with the highest fold in G, called the principal axis. For the groups **T** and **I**, the Z-axis is chosen to be along a 2-fold axis. The inversion with respect to the XY plane is nothing but the 2-fold improper rotation about the Z-axis, denoted by S_2. An improper point group G is denoted with a subscript "h" (horizontal) if and only if G contains S_2. There is only one exception $\mathrm{C}_s = \{E, S_2\}$ which is an improper point group of P-type. When an improper point group G does not contain S_2, G is denoted with a subscript "d" (dihedral) if G contains both the proper and improper 2-fold axes in the XY plane, G is denoted with a subscript "v" (vertical) if G contains only the improper 2-fold axes in the XY plane, and G is denoted by S_{4n} (improper $4n$-fold axis), C_i and $\mathrm{C}_{(2n+1)i}$ (I-type), if G contains no 2-fold axes in the XY plane.

According to the Schönflies notations, some improper point groups of I-type are listed as follows.

$$\begin{array}{ll} \mathrm{C}_i \approx \mathrm{C}_1 \otimes \mathrm{V}_2, & \mathrm{C}_{(2n)h} \approx \mathrm{C}_{2n} \otimes \mathrm{V}_2, \\ \mathrm{C}_{(2n+1)i} \approx \mathrm{C}_{(2n+1)} \otimes \mathrm{V}_2, & \mathrm{D}_{(2n)h} \approx \mathrm{D}_{2n} \otimes \mathrm{V}_2, \\ \mathrm{D}_{(2n+1)d} \approx \mathrm{D}_{(2n+1)} \otimes \mathrm{V}_2, & \mathbf{T}_h \approx \mathbf{T} \otimes \mathrm{V}_2, \\ \mathbf{O}_h \approx \mathbf{O} \otimes \mathrm{V}_2, & \mathbf{I}_h \approx \mathbf{I} \otimes \mathrm{V}_2, \end{array} \tag{2.75}$$

Some improper point groups of P-type are listed as follows.

$$\begin{array}{ll} \mathrm{S}_{4n} \approx \mathrm{C}_{4n}, \ \text{subgroup} \ \mathrm{C}_{2n}, & \mathrm{C}_{(2n+1)h} \approx \mathrm{C}_{4n+2}, \ \text{subgroup} \ \mathrm{C}_{2n+1}, \\ \mathrm{C}_s \approx \mathrm{C}_2, \ \text{subgroup} \ \mathrm{C}_1, & \mathrm{C}_{Nv} \approx \mathrm{D}_N, \ \text{subgroup} \ \mathrm{C}_N, \\ \mathrm{D}_{(2n)d} \approx \mathrm{D}_{4n}, \ \text{subgroup} \ \mathrm{D}_{2n}, & \mathrm{D}_{(2n+1)h} \approx \mathrm{D}_{4n+2}, \ \text{subgroup} \ \mathrm{D}_{2n+1}, \\ \mathbf{T}_d \approx \mathbf{O}, \ \text{subgroup} \ \mathbf{T}. & \end{array} \tag{2.76}$$

Since the space inversion σ is a symmetry transformation of both a cube and an icosahedron, but not a symmetry transformation of a tetrahedron, the symmetry groups of a cube and an icosahedron are $\mathbf{O}_h$ and $\mathbf{I}_h$, respectively. The axis connecting the midpoints of two opposite edges in a tetrahedron is its improper 4-fold axes, so that the symmetry groups of a tetrahedron is $\mathbf{T}_d$.

2.7 Exercises

1. Let E be the identity of a group G, R and S be any two elements in G with their inverses R^{-1} and S^{-1}, respectively. Try to show from Definition 2.1: (a) $RR^{-1} = E$; (b) $RE = R$; (c) $T = E$ if $TR = R$ or $RT = R$; (d) $T = R^{-1}$ if $TR = E$ or $RT = E$; (e) The inverse of (RS) is $S^{-1}R^{-1}$.
2. Show that the multiplicative group is isomorphic onto the additive group. The multiplicative group is composed of all positive real numbers where the multiplication law of elements is the product of numbers. The additive group is composed of all real numbers where the multiplication law of elements is the addition of numbers.
3. If H_1 and H_2 are two subgroups of a group G, prove that the common elements in H_1 and H_2 also form a subgroup of G.
4. Prove that a group must be abelian if the square of every element in the group is equal to the identity.
5. Simplify the following permutations as products of independent cycles:

$$(1): \ (1\ 2)(2\ 3)(1\ 2), \qquad (2): \ (1\ 2\ 3)(1\ 3\ 4)(3\ 2\ 1),$$

$$(3): \ (1\ 2\ 3\ 4)^{-1}, \qquad (4): \ (1\ 2\ 4\ 5)(4\ 3\ 2\ 6),$$

$$(5): \ (1\ 2\ 3)(4\ 2\ 6)(3\ 4\ 5\ 6).$$

6. Prove that the order of a cycle with cycle length ℓ is ℓ.
7. Calculate explicitly the number $n(\alpha)$ of elements contained in every class $\mathcal{C}_\alpha$ with the cycle structure $(\ell) = (\ell_1, \ell_2, \cdots, \ell_m)$ in the permutation groups S_4, S_5, and S_6, respectively.
8. Prove that a group whose order g is a prime number must be the cyclic group C_g.
9. Show that up to isomorphism, there are only two different groups of order 6: The cyclic group C_6 and the symmetry group D_3 of a regular triangle.
10. Show that up to isomorphism, there are only two different groups of order $2n$ where n is a prime number: The cyclic group C_{2n} and the symmetry group D_n of a regular polygon with n sides.
11. Study all groups with order 9 which are not isomorphic onto each other.

12. Show that the set of all possible products of the Pauli matrices σ_1 and σ_2 (see Prob. 4 in Chap. 1) forms a group G. Fill the multiplication table of G. Calculate the orders of this group G, the classes, the invariant subgroups and their quotient groups in G, respectively. Show that this group is isomorphic onto the symmetry group D_4 of a square.

13. Show that the set of all possible products of $i\sigma_1$ and $i\sigma_2$ forms a group Q_8. Fill the multiplication table of this group Q_8. Calculate the orders of this group, the classes, the invariant subgroups and their quotient groups in Q_8, respectively. Show that this group Q_8 is not isomorphic onto the group D_4.

14. Up to isomorphism, prove that there are only five different groups of order 8: The cyclic group C_8, $C_{4h} = C_4 \otimes V_2$, the symmetry group D_4 of a square, the quaternion group Q_8 (see Prob. 13), and $D_{2h} = D_2 \otimes V_2$.

15. Give a counter-example to show that the invariant subgroup of an invariant subgroup of a group G is not necessary to be an invariant subgroup of G. However, an invariant subgroup H_1 of G is also an invariant subgroup of a subgroup H of G if $H_1 \subset H$.

16. Let $\mathcal{C}_\alpha = \{S_1, S_2, \ldots, S_{n(\alpha)}\}$ be a class, containing $n(\alpha)$ elements, of a finite group G of order g. For any two given elements S_a and S_b in the class $\mathcal{C}_\alpha$, which may be the same or different, show that the number $m(\alpha)$ of the elements $T \in G$ satisfying $S_a = TS_bT^{-1}$ is $g/n(\alpha)$.

17. Let $\mathcal{C}_\alpha$ be a class of S_n with the cycle structure $(\ell) = (\ell_1, \ell_2, \cdots, \ell_m)$, where all ℓ_j are different. Any element R in $\mathcal{C}_\alpha$ is a product of m independent cycles R_j with cycle lengths ℓ_j: $R = \prod_{j=1}^m R_j$. Let C_{ℓ_j} denote the cyclic subgroup of S_n generated by R_j. Prove that $T \in S_n$ commutes with R, $TR = RT$, if and only if T belongs to the direct product of cyclic subgroups $\prod_{j=1}^m C_{\ell_j}$.

18. Let $\mathcal{C}_\alpha$ be a class of S_n with cycle structure $(\ell) = (\ell_1, \ell_2, \cdots, \ell_m)$, where all ℓ_j are different odd integers. Prove that, for the alternating subgroup S'_n of S_n, $\mathcal{C}_\alpha$ is decomposed into two classes, containing the same number of elements.

19. Prove that the alternating subgroup S'_5 of the permutation group S_5 does not contain any non-trivial invariant subgroup.

20. Based on $\mathbf{T} \approx S'_4$, calculate the multiplication table of the group $\mathbf{T}$ by the multiplication rule (2.28) of permutations.

21. Let $\mathcal{C}_\alpha$ and $\mathcal{C}_\beta$ be two classes in a group G. Let $\mathcal{C}_\alpha\mathcal{C}_\beta$ denote the set composed of all possible products $S_aT_b \in$ G, where $S_a \in \mathcal{C}_\alpha$ and $T_b \in \mathcal{C}_\beta$. Prove that the set $\mathcal{C}_\alpha\mathcal{C}_\beta \subset$ G contains every element conjugate to each given element S_aT_b in G.

22. There are 52 pieces of playing cards in a set of poker. The order of cards is changed in each shuffle, namely, the cards are made a permutation. If the shuffle is "strictly" done in the following rule: first separate the cards into two parts with equal number, then pick up one card from each part sequently. After the strict shuffle the first and the last cards do not change their positions, while the remaining cards are rearranged. Try to decompose this permutation into a product of independent cycles, to find the cycle structure of this permutation, and to explain after at least how many times of the strict shuffles the order of cards will change back to its original one.

23. The multiplication table of a finite group G is as follows.

	E	*A*	*B*	*C*	*D*	*F*	*I*	*J*	*K*	*L*	*M*	*N*
E	*E*	*A*	*B*	*C*	*D*	*F*	*I*	*J*	*K*	*L*	*M*	*N*
A	*A*	*E*	*F*	*I*	*J*	*B*	*C*	*D*	*M*	*N*	*K*	*L*
B	*B*	*F*	*A*	*K*	*L*	*E*	*M*	*N*	*I*	*J*	*C*	*D*
C	*C*	*I*	*L*	*A*	*K*	*N*	*E*	*M*	*J*	*F*	*D*	*B*
D	*D*	*J*	*K*	*L*	*A*	*M*	*N*	*E*	*F*	*I*	*B*	*C*
F	*F*	*B*	*E*	*M*	*N*	*A*	*K*	*L*	*C*	*D*	*I*	*J*
I	*I*	*C*	*N*	*E*	*M*	*L*	*A*	*K*	*D*	*B*	*J*	*F*
J	*J*	*D*	*M*	*N*	*E*	*K*	*L*	*A*	*B*	*C*	*F*	*I*
K	*K*	*M*	*J*	*F*	*I*	*D*	*B*	*C*	*N*	*E*	*L*	*A*
L	*L*	*N*	*I*	*J*	*F*	*C*	*D*	*B*	*E*	*M*	*A*	*K*
M	*M*	*K*	*D*	*B*	*C*	*J*	*F*	*I*	*L*	*A*	*N*	*E*
N	*N*	*L*	*C*	*D*	*B*	*I*	*J*	*F*	*A*	*K*	*E*	*M*

(a) Find the inverse of each element in G;

(b) Point out the elements commutable with every element in G;

(c) List the period and order of each element;

(d) Find the elements in each class of G;

(e) Find all invariant subgroups in G. Calculate the cosets of each invariant subgroup and show what group its quotient group is isomorphic onto;

(f) Make a judgment whether G is isomorphic onto the proper symmetry group **T** of a tetrahedral, or isomorphic onto the symmetry group D_6 of the regular polygon with six sides.

Chapter 3

THEORY OF REPRESENTATIONS

The theory of linear representations of groups is the foundation of group theory. In this chapter we will first introduce the definition of the representation of a group and the concepts of the equivalent representations and the irreducible representations. Then, through a few theorems we will study the methods for finding all inequivalent irreducible representations of a finite group, and demonstrate the fundamental steps for application of group theory to physics by an example. At last, we will calculate the irreducible bases in the group algebra of a finite group.

3.1 Linear Representations of a Group

3.1.1 *Definition of Linear Representation*

If a matrix group $D(\mathrm{G})$ is the isomorphic or homomorphic image of a given group G, the matrix group $D(\mathrm{G})$ describes the property of G at least partly and is called a linear representation of G, or briefly a representation.

Definition 3.1 A matrix group $D(\mathrm{G})$, composed of nonsingular $m \times m$ matrices $D(R)$, is called an m-dimensional representation of a group G, if $D(\mathrm{G})$ is the isomorphic or homomorphic image of G. The matrix $D(R)$, to which an element R in G maps, is called the representation matrix of R in the representation, and Tr $D(R) = \chi(R)$ is the character of R.

The representation matrix of the identity E is a unit matrix, $D(E) = \mathbf{1}$. The representation matrices of R and its inverse R^{-1} are mutually inverse matrices, $D(R^{-1}) = D(R)^{-1}$. Thus, $D(SRS^{-1}) = D(S)D(R)D(S)^{-1}$. **The characters of two elements belonging the same class $\mathcal{C}_\alpha$ are the same.** The character $\chi(R)$ of an element R is also said to be the character of a class $\mathcal{C}_\alpha$: $\chi(R) = \chi(\mathcal{C}_\alpha) = \chi_\alpha$ if $R \in \mathcal{C}_\alpha$.

Any group has an identical representation, or called the trivial one, where $D(R) = 1$ for every element R in G. The representation $D(\mathrm{G})$ is called faithful if G is isomorphic onto $D(\mathrm{G})$, and it is unfaithful if $\mathrm{G} \sim D(\mathrm{G})$. A matrix group is its own representation, called the self-representation. A representation is called unitary (or real orthogonal) if each representation matrix $D(R)$ is unitary (or real orthogonal). In this textbook we only study the representation with a finite dimension unless specified otherwise. An example of an infinite-dimensional unitary representation is studied in Prob. 25 of Chap. 4 of [Ma and Gu (2004)].

3.1.2 *Group Algebra and Regular Representation*

In a group G, the multiplication of two elements has been defined, but the sum of two elements is not defined. Now, we define the sum of elements in G. The elements in G are assumed to be linearly independent in the sum, and the sum of elements satisfies the following fundamental axioms:

$$\begin{aligned} c_1 R + c_2 R &= (c_1 + c_2) R, \\ c_1 R + c_2 S &= c_2 S + c_1 R, \\ c_3 (c_1 R + c_2 S) &= c_3 c_1 R + c_3 c_2 S. \end{aligned} \tag{3.1}$$

For a finite group G, the linear space $\mathcal{L}$ spanned by the elements R in G is called the **group space** of G, and the elements are called its **natural bases.** Any linear combination of the elements in G is a vector in $\mathcal{L}$,

$$X = \sum_{R \in \mathrm{G}} F(R) R, \qquad Y = \sum_{R \in \mathrm{G}} F_1(R) R. \tag{3.2}$$

A vector X is described completely by a **group function** $F(R)$, which is a map of the elements R in G onto a complex number $F(R)$. For a finite group G of order g, the dimension of $\mathcal{L}$ is g and only g values can be taken for a group function $F(R)$ so that there are only g linearly independent group functions for G. Define the inner product of vectors in $\mathcal{L}$ such that the natural bases are orthonormal:

$$\langle R|S\rangle = \delta_{RS}, \qquad \langle X|Y\rangle = \sum_{R \in \mathrm{G}} F(R)^* F_1(R). \tag{3.3}$$

Each matrix element $D_{\mu\nu}(R)$ of a representation $D(\mathrm{G})$ is a group function of G, and $D(\mathrm{G})$ is a matrix function. The character $\chi(R)$ is also a group function. In fact, $\chi(R)$ is a class function because $\chi(R) = \chi(TRT^{-1})$.

In a linear space $\mathcal{L}$, the sum of vectors and the product of a vector and a complex number, as well as the inner product of two vectors, have been defined. A linear space $\mathcal{L}$ is called an **algebra** if another product of vectors in $\mathcal{L}$ is defined to satisfy

$$XY \in \mathcal{L}, \quad Z(X+Y) = ZX + ZY, \quad \forall\, X,\, Y,\, Z \in \mathcal{L}. \tag{3.4}$$

In a group space $\mathcal{L}$, the product of two vectors can be defined such that the coefficients are multiplied like two numbers and the group elements are multiplied according to the multiplication rule in the group G:

$$\begin{aligned} XY &= \left\{\sum_{R\in G} F(R)R\right\}\left\{\sum_{S\in G} F_1(S)S\right\} = \sum_{R\in G}\sum_{S\in G} \{F(R)F_1(S)\}(RS) \\ &= \sum_{T\in G}\left\{\sum_{R\in G} F(R)F_1(R^{-1}T)\right\}T = \sum_{T\in G}\left\{\sum_{S\in G} F(TS^{-1})F_1(S)\right\}T. \end{aligned}$$

Thus, the group space $\mathcal{L}$ becomes a **group algebra**, still denoted by $\mathcal{L}$.

In the group algebra $\mathcal{L}$, a vector changes to another vector if an element S in G left-multiplies or right-multiplies on it. Namely, the element S in G, as well as a vector in $\mathcal{L}$, plays a role of a linear operator in $\mathcal{L}$.

First, we study the matrix form $D(S)$ of the left-multiplying operator S in the natural bases of $\mathcal{L}$:

$$SR = \sum_{P\in G} P\, D_{PR}(S). \tag{3.5}$$

In fact, there is only one term in the sum of Eq. (3.5), namely, there is only one non-vanishing element in each column of $D(S)$,

$$D_{PR}(S) = \begin{cases} 1, & P = SR, \\ 0, & P \neq SR. \end{cases} \tag{3.6}$$

The rows and columns of the matrix $D(S)$ are enumerated by the group elements. Due to Theorem 2.1, there is only one non-vanishing element in each row of $D(S)$, too. Equation (3.5) gives a one-to-one correspondence between the matrices $D(S)$ in $D(\mathrm{G})$ and the elements S in G, and the correspondence is left invariant under the multiplication rule of elements:

$$\begin{aligned} T(SR) &= \sum_{P\in G} TPD_{PR}(S) = \sum_{Q\in G} Q\left\{\sum_{P\in G} D_{QP}(T)D_{PR}(S)\right\} \\ &= (TS)R = \sum_{Q\in G} QD_{QR}(TS). \end{aligned}$$

$$D(S) \longleftrightarrow S, \qquad D(T)D(S) = D(TS) \longleftrightarrow TS.$$

From Theorem 2.4, the matrices $D(S)$ form a group $D(\mathrm{G})$, isomorphic onto the group G. Thus, $D(\mathrm{G})$ is a faithful representation of G, called the **regular representation**. Any finite group has the regular representation. The character $\chi(S)$ in the regular representation is

$$\chi(S) = \mathrm{Tr}\ D(S) = \begin{cases} g, & S = E, \\ 0, & S \neq E. \end{cases} \tag{3.7}$$

The regular representation matrix $D(S)$ is easy to calculate from the group table. The position (the Tth row) of the non-vanishing element in the Rth column of $D(S)$ is given by the product element ($SR = T$) in the R column of the Sth row of the group table. For example, the order of the rows (and columns) in the regular representation of D_3 is taken as: E, C_3, C_3^2, $C_2^{(0)}$, $C_2^{(1)}$, and $C_2^{(2)}$. From Table 2.5 one writes $D(C_3)$ and $D(C_2^{(0)})$ as follows,

$$D(C_3) = \begin{pmatrix} 0&0&1&0&0&0 \\ 1&0&0&0&0&0 \\ 0&1&0&0&0&0 \\ 0&0&0&0&0&1 \\ 0&0&0&1&0&0 \\ 0&0&0&0&1&0 \end{pmatrix}, \qquad D(C_2^{(0)}) = \begin{pmatrix} 0&0&0&1&0&0 \\ 0&0&0&0&0&1 \\ 0&0&0&0&1&0 \\ 1&0&0&0&0&0 \\ 0&0&1&0&0&0 \\ 0&1&0&0&0&0 \end{pmatrix}. \tag{3.8}$$

Second, we study the right-multiplying operator $\overline{S}$, where the bar is used to distinguish it from the left-multiplying operator S:

$$\overline{S}R \equiv RS. \tag{3.9}$$

The multiplication rule for the right-multiplying operator $\overline{S}$ is different from that for the left-multiplying operator S because the product of $\overline{T}$ and $\overline{S}$ is not equal to $\overline{(TS)}$, but $\overline{(ST)}$:

$$\overline{T}\,(\overline{S}R) = \overline{T}\,(RS) = (RS)\,T = R\,(ST) = \overline{(ST)}R. \tag{3.10}$$

Under this multiplication rule, the right-multiplying operators $\overline{S}$ form a group $\overline{\mathrm{G}}$, called the **intrinsic group** of G. $\overline{\mathrm{G}}$ is isomorphic onto G by a one-to-one correspondence $\overline{S} \longleftrightarrow S^{-1}$, and is usually called anti-isomorphic. The matrix forms of the right-multiplying operators $\overline{S}$ in the natural bases of $\mathcal{L}$ form a representation of the intrinsic group $\overline{G}$, and their transposes, denoted by $\overline{D}(S)$, form a representation of G:

$$\overline{S}R = RS = \sum_{P\in G} \overline{D}_{RP}(S)P, \tag{3.11}$$

because

$$\begin{aligned}\overline{T}(\overline{S}R) &= \sum_{P\in G} \overline{D}_{RP}(S)PT = \sum_{Q\in G} \left\{\sum_{P\in G} \overline{D}_{RP}(S)\overline{D}_{PQ}(T)\right\} Q \\ &= R(ST) = \sum_{Q\in G} \overline{D}_{RQ}(ST)\, Q, \\ \overline{D}(S) &\longleftrightarrow S, \qquad \overline{D}(S)\overline{D}(T) = \overline{D}(ST) \longleftrightarrow ST.\end{aligned}$$

From Eq. (3.11), one has

$$\overline{D}_{RP}(S) = \begin{cases} 1, & P = RS, \\ 0, & P \neq RS, \end{cases} \tag{3.12}$$

$$\overline{\chi}(S) = \text{Tr}\ \overline{D}(S) = \begin{cases} g, & S = E, \\ 0, & S \neq E. \end{cases} \tag{3.13}$$

The matrices $\overline{D}(S)$ form another faithful representation of G. **The character $\overline{\chi}(S)$ is the same as $\chi(S)$ given in** Eq. (3.7). The representation matrix $\overline{D}(S)$ is easy to calculate from the group table, too. The position (the Tth column) of the non-vanishing element in the Rth row of $\overline{D}(S)$ is given by the product element ($RS = T$) in the Rth row of the Sth column of the group table. For example, in the same order for the rows (and columns) as Eq. (3.8), one writes $\overline{D}(C_3)$ and $\overline{D}(C_2^{(0)})$ from Table 2.5 as follows,

$$\overline{D}(C_3) = \begin{pmatrix} 0&1&0&0&0&0 \\ 0&0&1&0&0&0 \\ 1&0&0&0&0&0 \\ 0&0&0&0&0&1 \\ 0&0&0&1&0&0 \\ 0&0&0&0&1&0 \end{pmatrix}, \qquad \overline{D}(C_2^{(0)}) = \begin{pmatrix} 0&0&0&1&0&0 \\ 0&0&0&0&1&0 \\ 0&0&0&0&0&1 \\ 1&0&0&0&0&0 \\ 0&1&0&0&0&0 \\ 0&0&1&0&0&0 \end{pmatrix}. \tag{3.14}$$

3.1.3 *Class Operator*

Let a finite group G of order g contain g_c classes, and let a class $\mathcal{C}_\alpha$ in G consist of $n(\alpha)$ elements:

$$\mathcal{C}_\alpha = \left\{S_1,\ S_2,\ \cdots,\ S_{n(\alpha)}\right\} = \left\{S_b | S_b = TS_aT^{-1}, T \in \text{G}\right\}. \tag{3.15}$$

The sum of elements S_a belonging to the class $\mathcal{C}_\alpha$ is a vector in the group algebra $\mathcal{L}$, and is also a linear operator in $\mathcal{L}$, called the class operator:

$$\mathsf{C}_\alpha = \sum_{S_a \in \mathcal{C}_\alpha} S_a. \tag{3.16}$$

From Eq. (2.37) and Theorem 2.2 one has

$$\begin{aligned} &T\mathsf{C}_\alpha T^{-1} = \mathsf{C}_\alpha, \qquad [T,\ \mathsf{C}_\alpha] = 0, \qquad && \forall\, T \in \mathrm{G}, \\ &\sum_{T\in G} TS_aT^{-1} = \frac{g}{n(\alpha)}\mathsf{C}_\alpha, && \forall\, S_a \in \mathcal{C}_\alpha. \end{aligned} \tag{3.17}$$

A class operator commutes with every element in G, as well as every class operator. Conversely, any X, which commutes with every element in G, must be a linear combination of the class operators. In fact, letting

$$X = \sum_{\alpha=1}^{g_c} \sum_{S_a \in \mathcal{C}_\alpha} F(S_a)S_a = TXT^{-1} = \frac{1}{g}\sum_{T\in G} TXT^{-1}, \tag{3.18}$$

one has from Eq. (3.17)

$$\begin{aligned} X &= \frac{1}{g}\sum_{\alpha=1}^{g_c} \sum_{S_a \in \mathcal{C}_\alpha} F(S_a) \sum_{T\in G} TS_aT^{-1} = \sum_{\alpha=1}^{g_c} \mathsf{C}_\alpha F_\alpha, \\ F_\alpha &= n(\alpha)^{-1} \sum_{S_a \in \mathcal{C}_\alpha} F(S_a). \end{aligned} \tag{3.19}$$

In comparison with Eq. (3.18), $F(S_a) = F_\alpha$ depends on the class $\mathcal{C}_\alpha$ but not on the element S_a. Since the product $\mathsf{C}_\alpha \mathsf{C}_\beta$ commutes with every element T in G, one has

$$\mathsf{C}_\alpha \mathsf{C}_\beta = \sum_{\gamma=1}^{g_c} f(\alpha,\beta,\gamma)\mathsf{C}_\gamma, \qquad f(\alpha,\beta,\gamma) = f(\beta,\alpha,\gamma), \tag{3.20}$$

where $f(\alpha,\beta,\gamma)$ is a non-negative integer and can be calculated directly from the group table (see Prob. 7).

The class operators are linearly independent and span a linear space $\mathcal{L}_c$ of dimension g_c, called the class space. Equation (3.20) shows that the class space $\mathcal{L}_c$ is closed for the multiplication of its vectors so that $\mathcal{L}_c$ is an algebra, called the class algebra.

3.2 Transformation Operators for Scalar Functions

Let x simply denote all the coordinates of degrees of freedom in a quantum system, and let $\psi(x)$ denote the scalar wave function. Let R denote a linear transformation of the system, which may be either a space-time transformation, such as a translation, a rotation, or an inversion etc., or an internal transformation, such as a rotation in the space of the isotopic spin, and so on. Under the transformation R, x is changed to $x' = Rx$ and the wave function $\psi(x)$ is changed to $\psi'(x')$. In order to show the dependence of ψ' on the transformation R explicitly, we introduce an operator P_R, $\psi'(x') \equiv P_R\psi(x')$.

Being a scalar wave function, the value of the transformed wave function $P_R\psi$ at the point Rx should be equal to the value of the original wave function ψ at the point x, namely

$$\begin{aligned} &x \xrightarrow{R} x' = Rx , \qquad x = R^{-1}x', \\ &\psi(x) \xrightarrow{R} \psi'(x') = P_R\psi(x') = P_R\psi(Rx) = \psi(x). \end{aligned} \tag{3.21}$$

Replacing the argument (Rx) in $P_R\psi(Rx)$ with x, one has

$$P_R\psi(x) = \psi(R^{-1}x). \tag{3.22}$$

$\psi(x)$ and $P_R\psi(x)$ are two different functions of x. Equation (3.22) shows the relation between the values of two functions at different points. At the same time, the relation gives the method to calculate the transformed function $P_R\psi(x)$ from the original function $\psi(x)$. Namely, **first replace the argument x in $\psi(x)$ with $R^{-1}x$, and then, regard $\psi(R^{-1}x)$ as a function of x**, which is nothing but the transformed function $P_R\psi(x)$.

Obviously, P_R is a linear operator

$$\begin{aligned} P_R\left\{a\psi(x) + b\phi(x)\right\} &= a\psi(R^{-1}x) + b\phi(R^{-1}x) \\ &= aP_R\psi(x) + bP_R\phi(x). \end{aligned} \tag{3.23}$$

Equation (3.22) shows a one-to-one correspondence between the operator P_R and the transformation R. This correspondence is invariant in the product of transformations,

$$\begin{aligned} &x' \xrightarrow{S} x'' = Sx' = (SR)x, \\ &\psi'(x') \xrightarrow{S} \psi''(x'') = P_S\psi'(x'') = P_SP_R\psi(x''), \\ &x \xrightarrow{SR} x'' = (SR)x, \\ &\psi(x) \xrightarrow{SR} P_{SR}\psi(x'') = P_{SR}\psi\left[(SR)x\right] = \psi(x). \end{aligned}$$

Because

$$P_S\psi'(x'') = P_S\psi'(Sx') = \psi'(x') = P_R\psi(x') = P_R\psi(Rx) = \psi(x),$$

$P_S P_R = P_{SR}$ corresponds to SR in the same rule. If the transformations R form a group G, then P_R form a group $P_{\rm G}$ isomorphic onto G. It is worthy to emphasize that

$$P_S\left[P_R\psi(x)\right] \neq P_S\psi(R^{-1}x) = \psi(S^{-1}R^{-1}x). \tag{3.24}$$

P_S has to act on the function $\psi' = P_R\psi$, not on the function ψ.

Ex. 1 Discuss the space translation of a system in one dimension:

$$x \xrightarrow{T(a)} x' = T(a)x = x + a, \qquad x = T(a)^{-1}x' = x' - a,$$

$$\begin{aligned} P_{T(a)}\psi(x) &= \psi[T(a)^{-1}x] = \psi(x-a) = \sum_{n=0}^{\infty} \frac{(-a)^n}{n!}\frac{d^n}{dx^n}\psi(x) \\ &= \exp\left\{-a\left(\frac{d}{dx}\right)\right\}\psi(x) = \exp\left\{-iap_x/\hbar\right\}\psi(x). \end{aligned} \tag{3.25}$$

Namely, the translation operator $P_{T(a)}$ for a scalar function is expressed as an exponential function of the momentum operator, $p_x = -i\hbar d/dx$.

Ex. 2 Discuss the rotation of a system in two dimensions. Let K be a laboratory-fixed coordinate frame, denoted by X and Y, and K' be a body-fixed one, denoted by X' and Y'. $R(\omega)$ is a rotation about the axis perpendicular to the coordinate plane through an angle ω, which changes K' from K to the present position, as shown in Fig. 3.1. The point $P(x,y)$ in the system changes to P' in the rotation $R(\omega)$. The coordinates of P' is (x', y') in the K frame, but is still (x, y) in the K' frame.

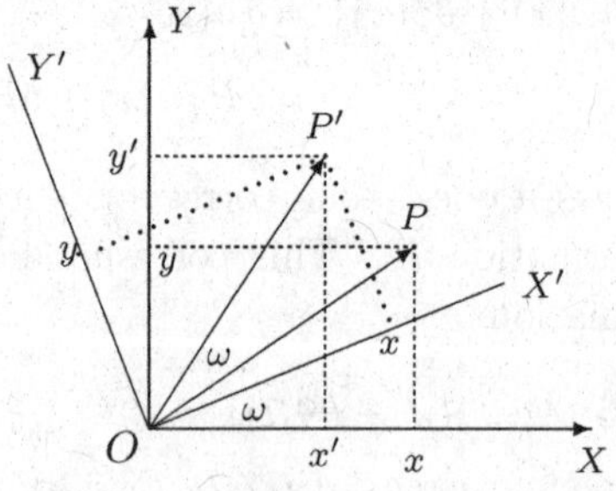

Fig. 3.1 A rotation about the Z axis.

From Fig. 3.1 one has

$$\begin{pmatrix} x' \\ y' \end{pmatrix} = R(\omega)\begin{pmatrix} x \\ y \end{pmatrix} = \begin{pmatrix} x\cos\omega - y\sin\omega \\ x\sin\omega + y\cos\omega \end{pmatrix}. \tag{3.26}$$

The inverse of $R(\omega)$ is $R(-\omega)$. There are three independent basis functions for the homogeneous polynomials of order two:

$$\psi_1(x,y) = x^2, \qquad \psi_2(x,y) = xy, \qquad \psi_3(x,y) = y^2,$$

which span a functional space $\mathcal{L}$ of dimension three, invariant under the rotation $R(\omega)$, as well as under the operator $P_{R(\omega)}$:

$$\begin{aligned}
P_{R(\omega)}\psi_1(x,y) &= \psi_1[R^{-1}(x,y)] = (x\cos\omega + y\sin\omega)^2 \\
&= \psi_1(x,y)\cos^2\omega + \psi_2(x,y)\sin(2\omega) + \psi_3(x,y)\sin^2\omega, \\
P_{R(\omega)}\psi_2(x,y) &= \psi_2[R^{-1}(x,y)] = (x\cos\omega + y\sin\omega)(-x\sin\omega + y\cos\omega) \\
&= -\psi_1(x,y)\sin\omega\cos\omega + \psi_2(x,y)\cos(2\omega) + \psi_3(x,y)\sin\omega\cos\omega, \\
P_{R(\omega)}\psi_3(x,y) &= \psi_3[R^{-1}(x,y)] = (-x\sin\omega + y\cos\omega)^2 \\
&= \psi_1(x,y)\sin^2\omega - \psi_2(x,y)\sin(2\omega) + \psi_3(x,y)\cos^2\omega.
\end{aligned}$$

The matrix form $D(\omega)$ of $P_{R(\omega)}$ in $\mathcal{L}$ is

$$P_{R(\omega)}\psi_\mu(x,y) = \psi_\mu[R^{-1}(x,y)] = \sum_\nu \psi_\nu(x,y) D_{\nu\mu}(\omega),$$

$$D(\omega) = \begin{pmatrix} \cos^2\omega & -\sin\omega\cos\omega & \sin^2\omega \\ \sin(2\omega) & \cos(2\omega) & -\sin(2\omega) \\ \sin^2\omega & \sin\omega\cos\omega & \cos^2\omega \end{pmatrix}. \tag{3.27}$$

$D(\omega)$ is not real orthogonal. However, it is easy to check that $D(\omega)$ will be real orthogonal if the bases $\psi_2(x,y)$ changes to $\sqrt{2}xy$.

Let a linear operator $L(x)$ change a state A to another state B,

$$\psi_B(x) = L(x)\psi_A(x),$$

where $\psi_A(x)$ and $\psi_B(x)$ are the wave functions of two states, respectively. After the transformation R, the wave functions are changed,

$$\begin{aligned}
\psi_A(x) &\xrightarrow{R} \psi'_A(x') = P_R\psi_A(x'), \\
\psi_B(x) &\xrightarrow{R} \psi'_B(x') = P_R\psi_B(x').
\end{aligned}$$

The operator $L(x)$ is changed to $L'(x')$ such that the operator still describes a transformation changing the state A to the state B:

$$L(x) \xrightarrow{R} L'(x'), \qquad \psi'_B(x') = L'(x')\psi'_A(x').$$

Replacing the argument x' with x, and noting that ψ_A is an arbitrary function, one has

$$L'(x)\left\{P_R\psi_A(x)\right\} = P_R\psi_B(x) = P_R\left\{L(x)\psi_A(x)\right\}.$$

$$L'(x) = P_R L(x) P_R^{-1}. \tag{3.28}$$

This is the transformation formula of a linear operator $L(x)$ under the transformation R. This formula can be understood as follows:

$$P_R \left[L(x)\psi(x)\right] = \left[P_R L(x) P_R^{-1}\right] \left[P_R \psi(x)\right] = L'(x)\psi'(x). \tag{3.29}$$

The transformation (3.28) for an operator seems to be different from Eq. (1.28), but they are consistent:

$$\begin{array}{lll} \underline{b} = D(R)\underline{a}, & \underline{a}' = S^{-1}\underline{a}, & D'(R) = S^{-1}D(R)S, \\ \psi_B(x) = L(x)\psi_A(x), & \psi'_A(x) = P_R\psi_A(x), & L'(x) = P_R L(x) P_R^{-1}. \end{array} \tag{3.30}$$

R is a symmetry transformation of a system if the Hamiltonian $H(x)$ of the system is invariant under R. According to Eq. (3.28), one has

$$H(x) \xrightarrow{R} P_R H(x) P_R^{-1} = H(x), \qquad [H(x),\ P_R] = 0. \tag{3.31}$$

Namely, **the symmetric transformation operator P_R commutes with the Hamiltonian** $H(x)$. If the energy level E is degeneracy with multiplicity m, there are m linearly independent eigenfunctions $\psi_\mu(x)$,

$$H(x)\psi_\mu(x) = E\psi_\mu(x), \qquad \mu = 1,\ 2,\ \cdots,\ m. \tag{3.32}$$

The basis functions $\psi_\mu(x)$ span an m-dimensional space $\mathcal{L}$ such that any function in $\mathcal{L}$ is the eigenfunction of $H(x)$ with the eigenvalue E and any eigenfunction of $H(x)$ with the eigenvalue E belongs to $\mathcal{L}$. From Eq. (3.31) one has

$$H(x)\left[P_R\psi_\mu(x)\right] = P_R H(x)\psi_\mu(x) = E\left[P_R\psi_\mu(x)\right]. \tag{3.33}$$

Hence, the space $\mathcal{L}$ is invariant with respect to P_R. Let $D(R)$ denote the matrix of P_R with respect to the basis functions $\psi_\mu(x)$,

$$P_R\psi_\mu(x) = \sum_{\nu=1}^{m} \psi_\nu(x) D_{\nu\mu}(R). \tag{3.34}$$

Equation (3.34) gives a one-to-one or a one-to-many correspondence between $D(R)$ and P_R, which is invariant under the product of the symmetric transformations. Thus, the matrix group $D(\mathrm{G})$ is the isomorphic or homomorphic image of the symmetry group G of the system,

$$\mathrm{G} \approx P_G \approx D(\mathrm{G}), \quad \text{or} \quad \mathrm{G} \approx P_G \sim D(\mathrm{G}). \tag{3.35}$$

$D(\mathrm{G})$ is an m-dimensional representation of G, and describes the transformation property of the static wave functions $\psi_\mu(x)$ under the symmetric transformations. In order to study the symmetric property of a system, one has to find all representations of the symmetry group G. It is one of the main tasks of group theory to find the representations of the symmetry groups used in physics. However, this task can be simplified.

It is worthy to notice the difference between two equations (3.22) and (3.34). The arguments of two functions in Eq. (3.22) are different. They are the scalar wave functions of the system after and before the transformation R. **Any scalar function should satisfy the transformation rule** (3.22). Equation (3.34) shows that a function $P_R\psi_\mu(x)$ in $\mathcal{L}$ is expanded in the basis functions $\psi_\nu(x)$, where the arguments of two functions are the same. **Equation (3.34) holds only when the functional space $\mathcal{L}$ spanned by $\psi_\nu(x)$ is invariant under the operator P_R.**

In quantum mechanics the inner product of two wave functions is defined as

$$\langle\psi(x)|\phi(x)\rangle = \int \psi(x)^*\phi(x)\mathrm{d}x.$$

For the space transformation R, such as the space rotation and the space translation, the Jacobi determinant from x to $R^{-1}x$ equals to one, so that

$$\langle P_R\psi(x)|P_R\phi(x)\rangle = \int \psi(R^{-1}x)^*\phi(R^{-1}x)\mathrm{d}x = \langle\psi(x)|\phi(x)\rangle. \tag{3.36}$$

Therefore, the transformation operator P_R is unitary and a hermitian operator becomes a hermitian one under P_R. If the basis functions $\psi_\mu(x)$ are orthonormal, the representation matrix $D(R)$ of P_R is unitary. There is an important exception which is the Lorentz transformation A. The transformation operator P_A is not unitary because the Jacobi determinant for the Lorentz transformation does not equal to one. Even though the basis functions $\psi_\mu(x)$ are orthonormal, the representation of the Lorentz group is not unitary generally.

3.3 Equivalent Representations

A representation $D(\mathrm{G})$ of a group G is a matrix group, whose representation matrix $D(R)$ acts on vectors in a linear space. This linear space is called the **representation space**. The group algebra of a finite group is the representation space of the regular representation. The eigenfunctions $\psi_\mu(x)$ of $H(x)$ with the energy E span a linear space $\mathcal{L}$, which is invariant under the symmetric transformation operators P_R. The matrices $D(R)$ of P_R form a representation $D(\mathrm{G})$ of the symmetry group G, and $\mathcal{L}$ is its representation space.

The choice of the basis vectors in the representation space is not unique. When the basis vectors make a linear combination,

$$\phi_\mu = \sum_\nu \psi_\nu X_{\nu\mu}, \qquad \det X \neq 0, \tag{3.37}$$

the matrix of P_R makes a similarity transformation

$$P_R\psi_\mu = \sum_\rho \psi_\rho D_{\rho\mu}(R), \qquad P_R\phi_\nu = \sum_\lambda \phi_\lambda \overline{D}_{\lambda\nu}(R),$$

$$\overline{D}(R) = X^{-1}D(R)X. \tag{3.38}$$

$\overline{D}(R)$ forms another representation of G. Both $D(R)$ and $\overline{D}(R)$ are the matrices of the same operator P_R in the same linear space $\mathcal{L}$, but with different basis vectors. These two representations are said to be equivalent.

Definition 3.2 Two representations $D(\mathrm{G})$ and $\overline{D}(\mathrm{G})$ of a group G is called equivalent to each other, $D(\mathrm{G}) \simeq \overline{D}(\mathrm{G})$, if two representation matrices $D(R)$ and $\overline{D}(R)$ for every element R in G satisfy the same similarity transformation (3.38).

The dimensions of two equivalent representations are same. **The characters of every element R in two equivalent representations $D(\mathrm{G})$ and $\overline{D}(\mathrm{G})$ are equal to each other**:

$$\chi(R) = \overline{\chi}(R), \qquad \forall R \in G. \tag{3.39}$$

Conversely, it will be shown (see Corollary 3.3.2) that two representations of a finite group G are equivalent if Eq. (3.39) holds.

The representation $\overline{D}(\mathrm{G})$ given in Eq. (3.12) have the same character as that in the regular representation $D(\mathrm{G})$ given in Eq. (3.6) for every element, respectively. They are equivalent and the similarity transformation matrix X is calculated to be

$$\overline{D}(R)X = XD(R), \qquad X_{PQ} = \begin{cases} 1, & PQ = T, \\ 0, & PQ \neq T, \end{cases} \tag{3.40}$$

where T is an arbitrarily given element in G (see Prob. 6).

The set of the complex conjugate matrices $D(R)^*$ of a representation $D(\mathrm{G})$ forms a representation $D(\mathrm{G})^*$ of G, called the conjugate representation of $D(\mathrm{G})$. A representation $D(\mathrm{G})$ is called **self-conjugate** if

$D(G)^* \simeq D(G)$, otherwise, non-self-conjugate. **The characters in a self-conjugate representation must be real.**

The representation $D(G)$ is called real if it is equivalent to a representation composed of real representation matrices. A real representation must be self-conjugate, but **a self-conjugate representation may not be real**, i.e., no similarity transformation exists to change this self-conjugate representation into that composed of real representation matrices. The criteria to distinguish those two kinds of self-conjugate representations will be given in Theorem 3.5.

From any representation $D(G)$ of G one is able to obtain infinite number of equivalent representations $\overline{D}(G)$ by different similarity transformations. **There is no essential difference between equivalent representations.** We only needs to study one in all equivalent representations.

Theorem 3.1 Any representation of a finite group is equivalent to a unitary representation, and two equivalent unitary representations can be related through a unitary similarity transformation.

Proof The first part of this Theorem means that for any representation $D(G)$ of a finite group G there is a matrix X satisfying

$$\overline{D}(R) = X^{-1}D(R)X, \qquad \overline{D}(R)^\dagger\overline{D}(R) = \mathbf{1}, \qquad \forall R \in G. \tag{3.41}$$

Namely,

$$D(R)^\dagger \left(XX^\dagger\right)^{-1} D(R) = \left(XX^\dagger\right)^{-1}. \tag{3.42}$$

Due to Theorem 2.1, the matrix $\left(XX^\dagger\right)^{-1}$ can be chosen to be H, where

$$\begin{aligned} H &\equiv \sum_{S\in G} D(S)^\dagger D(S) = \sum_{S\in G} D(SR)^\dagger D(SR) \\ &= D(R)^\dagger \left(\sum_{S\in G} D(S)^\dagger D(S)\right) D(R) = D(R)^\dagger H D(R). \end{aligned} \tag{3.43}$$

The problem is how to solve X from

$$\left(XX^\dagger\right)^{-1} = H. \tag{3.44}$$

In fact, H is a positive definite hermitian matrix. H can be diagonalized by a unitary similarity transformation U:

$$\Gamma = U^{-1}HU, \qquad U^\dagger U = \mathbf{1}, \qquad \Gamma_{\mu\mu} > 0.$$

Hence, $X = U\Gamma' U^{-1}$ with $\Gamma'_{\mu\nu} = \delta_{\mu\nu}(\Gamma_{\mu\mu})^{-1/2}$ satisfies Eq. (3.44).

Now, turn to the second part of this theorem. If two unitary representations $D(\mathrm{G})$ and $\overline{D}(\mathrm{G})$ are related by a non-unitary similarity transformation X, $X\overline{D}(R) = D(R)X$, one wants to find a matrix Y which commutes with $\overline{D}(R)$ and (XY) is unitary. Define a hermitian matrix $H_1 = X^\dagger X = \left(YY^\dagger\right)^{-1}$ which commutes with $\overline{D}(R)$:

$$\overline{D}(R)^{-1} H_1 \overline{D}(R) = \left[X\overline{D}(R)\right]^\dagger \left[X\overline{D}(R)\right] = \left[D(R)X\right]^\dagger \left[D(R)X\right] = H_1.$$

Y can be solved from H_1 just like X is solved from H. Diagonalizing H_1 by a unitary similarity transformation V, $\Gamma_1 = V^{-1}H_1V$, where the diagonal elements $(\Gamma_1)_{\mu\mu}$ are positive, one has $Y = V\Gamma_1' V^{-1}$ with $(\Gamma_1')_{\mu\nu} = \delta_{\mu\nu}(\Gamma_1)_{\mu\mu}^{-1/2}$. Thus, Y commutes with $\overline{D}(R)$. □

The key in the proof of Theorem 3.1 is Eq. (3.43) where Theorem 2.1 is used that the summation of a group function $F(R)$ on R is equal to its summation on RS. In other words, the average $\overline{F}(\mathrm{G})$ of a group function of a finite group G is invariant in multiplying with any group element S,

$$\overline{F}(\mathrm{G}) = \frac{1}{g}\sum_{S\in \mathrm{G}} F(S) = \frac{1}{g}\sum_{S\in \mathrm{G}} F(RS) = \frac{1}{g}\sum_{S\in \mathrm{G}} F(SR). \tag{3.45}$$

Hereafter, due to Theorem 3.1, we will only discuss the unitary representations and the unitary similarity transformations for a finite group G. For a finite group, $\chi(R^{-1}) = \chi(R)^*$. If the representation $D(\mathrm{G})$ is real, only real matrices appear in the proof of Theorem 3.1.

Corollary 3.1.1 Any real representation of a finite group is equivalent to a real orthogonal representation, and two equivalent real orthogonal representations can be related by a real orthogonal similarity transformation.

Corollary 3.1.2 In a finite group, the character of a self-reciprocal class is real, and two characters of a pair of reciprocal classes are complex conjugate to each other.

3.4 Inequivalent Irreducible Representations

3.4.1 *Irreducible Representations*

Definition 3.3 A representation $D(\mathrm{G})$ of a group G is said to be reducible if the representation matrix $D(R)$ of every element R of G can be changed to the same type of the echelon matrix through a common similarity transformation X,

$$X^{-1}D(R)X = \begin{pmatrix} D^{(1)}(R) & T(R) \\ 0 & D^{(2)}(R) \end{pmatrix}. \tag{3.46}$$

Otherwise, it is called an irreducible representation.

It is easy to show that both $D^{(1)}(R)$ and $D^{(2)}(R)$ form the representations of G. The character of R in the reducible representation $D(\mathrm{G})$ is the sum of the characters in these two representations

$$\chi(R) = \chi^{(1)}(R) + \chi^{(2)}(R). \tag{3.47}$$

From Definition 3.3, the representation space of $D^{(1)}(\mathrm{G})$ is a nontrivial invariant subspace in the representation space of a reducible representation $D(\mathrm{G})$. Conversely, if there is a nontrivial invariant subspace in the representation space of $D(\mathrm{G})$, we choose new basis vectors respectively belonging to this subspace and its complementary subspace. The representation matrices take the form (3.46) in the new basis vectors. Namely, **a representation of G is irreducible if and only if in the representation space there exists no nontrivial invariant subspace to G.**

If both complementary subspaces are invariant under $D(\mathrm{G})$, there exists a common similarity transformation X such that the representation matrix $D(R)$ of every element R of G takes the same type of the block matrix, which is the direct sum of two submatrices $D^{(1)}(R)$ and $D^{(2)}(R)$,

$$X^{-1}D(R)X = \begin{pmatrix} D^{(1)}(R) & 0 \\ 0 & D^{(2)}(R) \end{pmatrix} = D^{(1)}(R) \oplus D^{(2)}(R). \tag{3.48}$$

This representation $D(\mathrm{G})$ is called **completely reducible**. The representation $D^{(1)}(\mathrm{G}) \oplus D^{(2)}(\mathrm{G})$ is called the **reduced representation.**

A typical example for existence of the incompletely reducible representation is the following representation of the translation group $\mathcal{T}$ in one-dimensional space which is an infinite abelian group,

$$x \xrightarrow{T(a)} x' = x + a, \qquad T(a)T(b) = T(a+b),$$
$$D(a) = \begin{pmatrix} 1 & a \\ 0 & 1 \end{pmatrix}, \qquad D(a)D(b) = D(a+b).$$

Any reducible representation of a finite group G is completely reducible. In fact, the representation space $\mathcal{L}$ of a reducible unitary representation $D(\mathrm{G})$ of G must contain a nontrivial invariant subspace $\mathcal{L}_1$, say m_1-dimensional. Make a unitary similarity transformation X where the first m_1 column matrices belong to $\mathcal{L}_1$ and the remaining do not. Thus,

$X^{-1}D(R)X$ is in the form (3.46). Since $X^{-1}D(R)X$ is unitary, $D^{(1)}(R)$ is unitary, too, so that $T(R)$ has to be vanishing.

From irreducible representations of G, one is able to construct infinite number of reducible representations by direct summing them. One of the main tasks of group theory is to find all inequivalent and irreducible representations of the symmetry groups of physical systems.

3.4.2 *Schur Theorem*

Theorem 3.2 (The second Schur Theorem) For two inequivalent irreducible representations $D^{(1)}(\mathrm{G})$ and $D^{(2)}(\mathrm{G})$ of a group G, $X = 0$ if

$$D^{(1)}(R)X = XD^{(2)}(R), \qquad \forall\, R \in G. \tag{3.49}$$

Proof The key in the proof is that there is **no nontrivial invariant subspace in the representation space of an irreducible representation**. Let m_1 and m_2 denote the dimensions of $D^{(1)}(\mathrm{G})$ and $D^{(2)}(\mathrm{G})$, respectively. X is an $m_1 \times m_2$ matrix. We will prove for three cases that the space spanned by the column (or row) matrices of X is a null space.

(a) $m_1 > m_2$. Due to Eq. (3.49), the column matrices in X, $(\underline{X_{\cdot\mu}})_\lambda = X_{\lambda\mu}$, span an invariant subspace under $D^{(1)}(R)$ with the dimension less than m_1 so that the subspace is a null space:

$$D^{(1)}(R)\underline{X_{\cdot\mu}} = \sum_\rho \underline{X_{\cdot\rho}} D^{(2)}(R)_{\rho\mu}. \tag{3.50}$$

(b) $m_1 = m_2$. If $\det X \neq 0$, due to Eq. (3.49), $D^{(1)}(\mathrm{G})$ and $D^{(2)}(\mathrm{G})$ are equivalent which is in contradiction to the hypothesis. If $\det X = 0$, the dimension of the invariant subspace under $D^{(1)}(R)$, spanned by the column matrices of X, is less than m_1 so that the subspace is a null space.

(c) $m_1 < m_2$. Transposing Eq. (3.49), one has

$$D^{(2)}(R)^T X^T = X^T D^{(1)}(R)^T.$$

It means that the column matrices of X^T span an invariant subspace under $D^{(2)}(R)^T$ with the dimension less than m_2. By transposing, there is an invariant subspace in the representation space of $D^{(2)}(R)$ with the same dimension less than m_2 so that the subspace is a null space. □

Corollary 3.2.1 (The first Schur Theorem) For an irreducible representation $D(\mathrm{G})$ of a group G, $X = \lambda \mathbf{1}$ if

$$XD(R) = D(R)X, \qquad \forall R \in G. \tag{3.51}$$

Proof Letting λ be an eigenvalue of X, we define $Y = X - \lambda\mathbf{1}$, and then, $\det Y = 0$ and $YD(R) = D(R)Y$. Thus, through the similar proof to the case (b), $Y = 0$ and $X = \lambda\mathbf{1}$. □

The Schur theorems hold for any group. Since any reducible representation of a finite group is completely reducible, there exists a nonconstant matrix commutable with its every representation matrix.

Corollary 3.2.2 A representation of a finite group is reducible if and only if there exists a nonconstant matrix commutable with its every representation matrix.

3.4.3 *Orthogonal Relation*

For a finite group G of order g, any matrix element $D_{\mu\nu}(R)$ of a representation D(G) is a group function and describes a vector in the group space with respect to the natural basis vectors. The inner product of two group functions is defined through the inner product (3.3) of two vectors in the group space,

$$\sum_{R \in G} F_1(R)^* F_2(R). \tag{3.52}$$

Two group functions are orthogonal if their inner product is vanishing. The inner self-product of one group function is the square of the module of the function.

Theorem 3.3 The representation matrix elements $D^i_{\mu\rho}(R)$ and $D^j_{\nu\lambda}(R)$ in two inequivalent and irreducible unitary representations of a finite group G satisfy the orthogonal relation of the group functions:

$$\sum_{R \in G} D^i_{\mu\rho}(R)^* D^j_{\nu\lambda}(R) = \frac{g}{m_j}\delta_{ij}\delta_{\mu\nu}\delta_{\rho\lambda}, \tag{3.53}$$

where g is the order of G, m_j denotes the dimension of D^j(G), and $D^i(R) = D^j(R)$ when $i = j$.

Proof Define two $m_i \times m_j$ matrices $Y(\mu\nu)$ and $X(\mu\nu)$, where $Y(\mu\nu)$ contains only one non-vanishing matrix element, $Y(\mu\nu)_{\rho\lambda} = \delta_{\mu\rho}\delta_{\nu\lambda}$, and

$$X(\mu\nu) = \sum_R D^i(R^{-1})Y(\mu\nu)D^j(R).$$

Then, $X(\mu\nu)_{\rho\lambda}$ is nothing but the left-hand side of Eq. (3.53):

$$
\begin{aligned}
X(\mu\nu)_{\rho\lambda} &= \sum_{R\in G} \sum_{\tau\sigma} D^i_{\rho\tau}(R^{-1}) Y(\mu\nu)_{\tau\sigma} D^j_{\sigma\lambda}(R) \\
&= \sum_{R\in G} D^i_{\mu\rho}(R)^* D^j_{\nu\lambda}(R).
\end{aligned} \tag{3.54}
$$

Due to Theorem 2.1, $X(\mu\nu)$ satisfies

$$
X(\mu\nu)D^j(S) = \sum_{R\in G} D^i(S)D^i(RS)^{-1}Y(\mu\nu)D^j(RS) = D^i(S)X(\mu\nu).
$$

From the Schur theorem, $X(\mu\nu) = 0$ when $i \neq j$, and $X(\mu\nu)_{\rho\lambda} = C(\mu\nu)\delta_{\rho\lambda}$ when $i = j$, where $C(\mu\nu)$ is a constant depending on $Y(\mu\nu)$. Substituting $X(\mu\nu)_{\rho\lambda}$ into Eq. (3.54) with $i = j$ and $\rho = \lambda$ and summing over λ, one has

$$
m_j C(\mu\nu) = \sum_{R\in G} \sum_{\lambda} D^j_{\lambda\mu}(R^{-1}) D^j_{\nu\lambda}(R) = \sum_{R\in G} D^j_{\nu\mu}(RR^{-1}) = g\delta_{\mu\nu}.
$$

Thus, Eq. (3.53) follows. □

From Theorem 3.3, all matrix elements $D^i_{\mu\rho}(R)$ and $D^j_{\nu\lambda}(R)$ with $i \neq j$ are orthogonal to each other. Therefore, the orthogonality holds if two representations are not unitary because any representation of a finite group can be changed to a unitary one by a similarity transformation. From Eq. (3.53) with $i = j$, each unitary irreducible representation $D^j(\mathrm{G})$ provides m_j^2 group functions $D^j_{\nu\lambda}(R)$ of G, which are orthogonal to each other and are normalized to a common number g/m_j. Since there are only g linearly independent group functions in the group space of a finite group G of order g, **the square sum of the dimensions** m_j of all inequivalent irreducible representations of a finite group G of order g **is not larger than** g.

Taking $\mu = \rho$ and $\nu = \lambda$ in Eq. (3.53) and summing over μ and ν, one has

$$
\sum_{R\in G} \chi^i(R)^* \chi^j(R) = \sum_{\alpha=1}^{g_c} n(\alpha) \left(\chi^i_\alpha\right)^* \chi^j_\alpha = g\delta_{ij}, \tag{3.55}
$$

where $n(\alpha)$ is the number of elements contained in a class $\mathcal{C}_\alpha$ and $\chi^j_\alpha = \chi^j(\mathcal{C}_\alpha)$. $[n(\alpha)/g]^{1/2}\chi^j_\alpha$ are orthonormal in the class space:

$$
\sum_{\alpha=1}^{g_c} [n(\alpha)/g]^{1/2} \left(\chi^i_\alpha\right)^* [n(\alpha)/g]^{1/2} \chi^j_\alpha = \delta_{ij}. \tag{3.56}
$$

Corollary 3.3.1 Being group functions, the characters of the inequivalent irreducible representations of a finite group G of order g are orthogonal to each other and normalized to g.

The number of the inequivalent irreducible representations of a finite group G is not larger than the number g_c of classes contained in G.

A reducible representation $D(\mathrm{G})$ of a finite group G can be reduced to a direct sum of some irreducible representations through a similarity transformation X

$$X^{-1}D(R)X = \bigoplus_j a_j D^j(R), \qquad \chi(R) = \sum_j a_j \chi^j(R), \tag{3.57}$$

where $\chi(R)$ is the character of R in $D(\mathrm{G})$. a_j is called the **multiplicity** of an irreducible representation $D^j(\mathrm{G})$ in the representation $D(\mathrm{G})$. From Eq. (3.55), one has

$$a_j = \frac{1}{g} \sum_{R \in G} \chi^j(R)^* \chi(R). \tag{3.58}$$

If all inequivalent irreducible representations of a finite group G are known, the multiplicity a_j can be calculated from the character $\chi(R)$, so that the similarity transformation matrix X is able to calculate from Eq. (3.57). **It is the main method for reducing a representation of a finite group.** Two representations of a finite group G are equivalent if the multiplicities a_j of all inequivalent and irreducible representations $D^j(\mathrm{G})$ in them are equal to each other, respectively. From Eqs. (3.55) and (3.57), one has

$$\sum_{R \in G} |\chi(R)|^2 = g \sum_j a_j^2 \geqslant g. \tag{3.59}$$

The sum will be equal to g if the representation is irreducible.

Corollary 3.3.2 Two representations of a finite group are equivalent if and only if the characters of every element in them are equal to each other, respectively.

Corollary 3.3.3 A representation of a finite group of order g is irreducible if and only if the character $\chi(R)$ in the representation satisfies

$$\sum_{R \in G} |\chi(R)|^2 = g. \tag{3.60}$$

3.4.4 *Completeness of Representations*

From the character $\chi(R)$ of the regular representation $D(\mathrm{G})$ of a finite group G (see Eq. (3.6)), the multiplicity a_j of each irreducible representation

$D^j(\mathrm{G})$ in the regular representation $D(\mathrm{G})$ is equal to its dimension m_j:

$$a_j = \frac{1}{g} \sum_{R \in G} \chi^j(R)^* \chi(R) = \chi^j(E)^* = m_j.$$

Namely,

$$X^{-1} D(R) X = \bigoplus_j m_j D^j(R), \qquad \chi(R) = \sum_j m_j \chi^j(R). \tag{3.61}$$

Taking the character of the identity E, one has

$$g = \chi(E) = \sum_j m_j^2. \tag{3.62}$$

Theorem 3.4 The square sum of the dimensions of all inequivalent irreducible representations of a finite group G is equal to the order g of G.

Corollary 3.4.1 The representation matrix elements $D^j_{\nu\lambda}(R)$ of all inequivalent irreducible representations of a finite group G construct a complete set of orthogonal basis vectors in the group space, and any group function $F(R)$ can be expanded with respect to the basis vectors:

$$F(R) = \sum_{j\mu\nu} C^j_{\mu\nu} D^j_{\mu\nu}(R), \qquad C^j_{\mu\nu} = \frac{m_j}{g} \sum_{R \in G} D^j_{\mu\nu}(R)^* F(R). \tag{3.63}$$

The class function is a special group function, where the functional values for the elements in the same class are equal to each other. Expanding a class function with respect to $D^j_{\mu\nu}(R)$, one has

$$\begin{aligned} F(R) &= F(SRS^{-1}) = \frac{1}{g} \sum_{S \in G} F(SRS^{-1}) \\ &= \sum_{j\mu\nu} \frac{C^j_{\mu\nu}}{g} \sum_{S \in G} \sum_{\rho\lambda} D^j_{\mu\rho}(S) D^j_{\rho\lambda}(R) D^j_{\nu\lambda}(S)^* \\ &= \sum_j \left(\frac{1}{m_j} \sum_\mu C^j_{\mu\mu} \right) \chi^j(R). \end{aligned}$$

Thus, the characters $\chi^j(R)$ construct a complete set of basis vectors in the class space.

Corollary 3.4.2 The characters $\chi^j(R)$ of all inequivalent irreducible representations of a finite group G construct a complete set of basis vectors in the class space, and any class function $F(R)$ can be expanded with respect to the basis vectors:

$$F(R) = F(SRS^{-1}) = \sum_j C_j \chi^j(R),$$
$$C_j = \frac{1}{g} \sum_{R\in G} \chi^j(R)^* F(R). \tag{3.64}$$

Corollary 3.4.3 The number of the inequivalent and irreducible representations of a finite group G is equal to its class number g_c,

$$\sum_j 1 = g_c. \tag{3.65}$$

Define a $g \times g$ matrix U and a $g_c \times g_c$ matrix V:

$$U_{R,j\mu\nu} = \left(\frac{m_j}{g}\right)^{1/2} D^j_{\mu\nu}(R), \qquad V_{\alpha,j} = \left(\frac{n(\alpha)}{g}\right)^{1/2} \chi^j_\alpha. \tag{3.66}$$

The orthogonal relations (3.53) and (3.56) show that U and V are both unitary. Then, the complete relations of the representation matrix elements $D^j_{\mu\nu}(R)$ and the characters $\chi^j(\mathcal{C}_\alpha)$ are

$$\sum_{j\mu\nu} m_j D^j_{\mu\nu}(R) D^j_{\mu\nu}(S)^* = g\delta_{RS}, \tag{3.67}$$

$$\sum_j \chi^j_\alpha \left(\chi^j_\beta\right)^* = \frac{g}{n(\alpha)} \delta_{\alpha\beta}. \tag{3.68}$$

Let a finite group G contain g_c classes $\mathcal{C}_\alpha$, among which g_c' classes are self-reciprocal, denoted by $\mathcal{C}_\beta$, and the remaining in pair are reciprocal, denoted by $\mathcal{C}_\sigma$ and $\mathcal{C}_\sigma^{(-1)}$ with $\mathcal{C}_\sigma \neq \mathcal{C}_\sigma^{(-1)}$. Define a complete set of basis functions in the class space $\mathcal{L}_c$: g_c' basis functions are $F^{(1)}_\beta(\mathcal{C}_\alpha) = \delta_{\beta\alpha}$; $(g_c - g_c')/2$ basis functions are $F^{(2)}_\sigma(\mathcal{C}_\alpha)$ which are vanishing except for $F^{(2)}_\sigma(\mathcal{C}_\sigma) = F^{(2)}_\sigma(\mathcal{C}_\sigma^{(-1)}) = 1$; and $(g_c - g_c')/2$ basis functions are $F^{(3)}_\sigma(\mathcal{C}_\alpha)$ which are vanishing except for $F^{(3)}_\sigma(\mathcal{C}_\sigma) = -F^{(3)}_\sigma(\mathcal{C}_\sigma^{(-1)}) = 1$. Thus, $\mathcal{L}_c$ is decomposed into a direct sum of two subspaces, $\mathcal{L}_c = \mathcal{L}_{c1} \oplus \mathcal{L}_{c2}$, where the subspace $\mathcal{L}_{c1}$ with dimension $(g_c + g_c')/2$ is spanned by the basis functions $F^{(1)}_\beta(\mathcal{C}_\alpha)$ and $F^{(2)}_\sigma(\mathcal{C}_\alpha)$, and the subspace $\mathcal{L}_{c2}$ with dimension $(g_c - g_c')/2$ is spanned by the basis functions $F^{(3)}_\sigma(\mathcal{C}_\alpha)$. Note that the function $\Phi(\mathcal{C}_\alpha)$ in $\mathcal{L}_{c1}$ satisfies $\Phi(\mathcal{C}_\sigma) = \Phi(\mathcal{C}_\sigma^{(-1)})$, and the function $\Psi(\mathcal{C}_\alpha)$ in $\mathcal{L}_{c2}$ satisfies $\Psi(\mathcal{C}_\beta) = 0$ and $\Psi(\mathcal{C}_\sigma) = -\Psi(\mathcal{C}_\sigma^{(-1)})$.

From Corollary 3.4.2, the characters of g_c inequivalent irreducible representations $D^j(\text{G})$ of G span the class space $\mathcal{L}_c$, too. Assume that among $D^j(\text{G})$, m representations $D^k(\text{G})$ are self-conjugate, $D^k(\text{G})^* \simeq D^k(\text{G})$, and

$(g_c - m)/2$ pairs of representations are non-self-conjugate, $D^\ell(\mathrm{G})^* \simeq D^{\ell'}(\mathrm{G})$ in each pair. Due to Corollary 3.1.2, $\chi^k(\mathcal{C}_\alpha)$ of m self-conjugate representations and $\chi^\ell(\mathcal{C}_\alpha) + \chi^{\ell'}(\mathcal{C}_\alpha)$ of $(g_c - m)/2$ pairs of non-self-conjugate representations belong to $\mathcal{L}_{c1}$, and $\chi^\ell(\mathcal{C}_\alpha) - \chi^{\ell'}(\mathcal{C}_\alpha)$ of $(g_c - m)/2$ pairs of non-self-conjugate representations belong to $\mathcal{L}_{c2}$. Therefore, $m = g'_c$.

Corollary 3.4.4 The number m of the inequivalent irreducible self-conjugate representations of a finite group G is equal to the number g'_c of the self-reciprocal classes in G, $m = g_{c'}$.

3.4.5 *Character Tables of Finite Groups*

It is easier to calculate the characters $\chi(\mathcal{C}_\alpha)$ of the representations of a group G than to calculate the representation matrices $D(R)$. The characters of all inequivalent irreducible representations of a finite group G can be listed in a table, called the **character table of** G.

The characters of inequivalent irreducible representations of a finite group G satisfy **four necessary conditions**. The conditions (3.62) and (3.65) restrict the number of inequivalent and irreducible representations of G and their dimensions. The conditions (3.55) and (3.68) show orthogonality, normalization, and completeness of characters. However, **four conditions are not sufficient for characters**. A simplest example is that two groups C_4 and V_4 of order four have different character tables (see Tables 3.3 and 3.5), which both satisfy four necessary conditions.

A group has some faithful irreducible representations as well as some unfaithful ones, including the identical representation. An important step to calculate the character table of a finite group G is to find all invariant subgroups H of G. The **representations of the quotient groups** G/H are the unfaithful ones of G. Evidently, the character table of a finite group with lower order is easier to find than that with higher order.

Corollary 3.1.2 shows that the character of a self-reciprocal class in G is real, and two characters of a pair of reciprocal classes are complex conjugate to each other. Thus, only the characters of reciprocal classes in non-self-conjugate representations may **not be real**. The number of pairs of inequivalent irreducible non-self-conjugate representations of G is equal to the number of pairs of reciprocal classes in G (see Corollary 3.4.4).

The direct product of two representations of G is a representation of G. Especially, the direct product of an irreducible representation and a one-dimensional representation of G is an irreducible representation of G.

The irreducible representation $D^{jk}(\mathrm{G})$ of the direct product of two groups, $\mathrm{G} = \mathrm{H}_1 \otimes \mathrm{H}_2$, is equal the direct product of the irreducible representations $D^j(\mathrm{H}_1)$ and $D^k(\mathrm{H}_2)$ of two subgroups:

$$D^{jk}(RS) = D^j(R) \times D^k(S), \qquad \forall R \in \mathrm{H}_1, \quad S \in \mathrm{H}_2. \tag{3.69}$$

The above properties are helpful in calculating the character table of a finite group G. In principle, it is the task of mathematicians to find the character tables of G. However, understanding the basic techniques and the results of mathematicians in calculation is useful to a physicist.

A cyclic group $\mathrm{C}_N = \{E,\ R,\ R^2,\ \cdots,\ R^{N-1}\}$, where $R^N = E$, is an abelian group. The class number g_c of C_N is equal to its order N. So that there are N inequivalent one-dimensional representations $D^j(\mathrm{C}_N)$ of C_N. Solving $D^j(R)^N = D^j(E) = 1$, one obtains

$$D^j(R^m) = \exp\{-\mathrm{i}2\pi mj/N\}, \quad 0 \leqslant m \leqslant N-1, \quad 0 \leqslant j \leqslant N-1. \tag{3.70}$$

Tables 3.1 to 3.4 list the character tables of some cyclic groups where the names of representations come from the theory of crystals.

Table 3.1 The character table of C_2

C_2	E	R
A	1	1
B	1	-1

Table 3.2 The character table of C_3 ($\omega = \exp\{-\mathrm{i}2\pi/3\}$)

C_3	E	R	R^2
A	1	1	1
E	1	ω	ω^*
E'	1	ω^*	ω

Table 3.3 The character table of C_4

C_4	E	R	R^2	R^3
A	1	1	1	1
B	1	-1	1	-1
E	1	$-\mathrm{i}$	-1	i
E'	1	i	-1	$-\mathrm{i}$

The representation $D^B(\mathrm{C}_2)$ is usually called the antisymmetric one. The group C_4 contains an invariant subgroup $\mathrm{C}_2 = \{E, R^2\}$. The representation $D^B(\mathrm{C}_4)$ is the antisymmetric representation of the quotient group $\mathrm{C}_4/\mathrm{C}_2$, and $D^{E'}(\mathrm{C}_4) = D^B(\mathrm{C}_4) \times D^E(\mathrm{C}_4)$.

Table 3.4 The character table of C_5 ($\eta = \exp\{-i2\pi/5\}$)

C_5	E	R	R^2	R^3	R^4
A	1	1	1	1	1
E_1	1	η	η^2	η^3	η^4
E_1'	1	η^4	η^3	η^2	η
E_2	1	η^2	η^4	η	η^3
E_2'	1	η^3	η	η^4	η^2

In the inversion group V_4 of order 4, four elements are denoted by four diagonal matrices in a space of four dimensions:

$$\begin{aligned} &e = \text{diag}\{1,\ 1,\ 1,\ 1\}, \qquad &&\sigma = \text{diag}\{-1,\ -1,\ -1,\ 1\}, \\ &\tau = \text{diag}\{1,\ 1,\ 1,\ -1\}, \qquad &&\rho = \text{diag}\{-1,\ -1,\ -1,\ -1\}. \end{aligned} \tag{3.71}$$

Table 3.5 shows the character table of V_4 where D^{A_1} is the identical representation, D^{B_1} and D^{A_2} are the space and time components of elements in V_4, and $D^{B_2} = D^{A_2} \times D^{B_1}$ is the determinant of elements. The identity e and any other element construct a subgroup C_2 in V_4. Three representations D^{A_2}, D^{B_1}, and D^{B_2} are the antisymmetric representations of three quotient groups V_4/C_2, respectively. The group D_2 is isomorphic onto V_4 so that they have the same character table.

Table 3.5 The character table of V_4 and D_2

V_4	e	σ	τ	ρ
A_1	1	1	1	1
A_2	1	1	−1	−1
B_1	1	−1	1	−1
B_2	1	−1	−1	1

D_2	E	C_2	C_2'	C_2''
A_1	1	1	1	1
A_2	1	1	−1	−1
B_1	1	−1	1	−1
B_2	1	−1	−1	1

The simplest non-Abelian group D_3 was discussed in section 2.2. D_3 contains six elements and three classes. The identity E itself is a class $\mathcal{C}_1$, two 3-fold rotations C_3 and C_3^2 construct the class $\mathcal{C}_2$, and three 2-fold rotations $C_2^{(0)}$, $C_2^{(1)}$, $C_2^{(2)}$ construct the class $\mathcal{C}_3$. Two classes $\mathcal{C}_1$ and $\mathcal{C}_2$ forms an invariant subgroup C_3 of index 2, and the class $\mathcal{C}_3$ is its coset. Since $g = 6$, $g_c = 3$, and $1^2 + 1^2 + 2^2 = 6$, the D_3 group has two inequivalent representations D^A and D^B of one dimension, and one irreducible representation D^E of two dimensions. D^A and D^B are the representations of the quotient group D_3/C_3. There are a few methods to calculate the characters χ^E in the representation D^E. From Eq. (3.68) one has $\chi^E(C_2') = 0$. In fact, if $\chi^E(C_2') \neq 0$, the representation $D^B \times D^E$ would be a two-dimensional irreducible representation, inequivalent to D^E, which is in contradiction to Eq. (3.62). From Eq. (3.55) one obtains $\chi^E(C_3) = -1$. The two-dimensional representation D^E of D_3, as well as its characters, has been obtained by the method of coordinate transformation (see Eq. (2.12)). The character table of D_3 is listed in Table 3.6.

In the theory of crystals, a class $\mathcal{C}_\alpha$ in the character table of point groups is usually denoted by one element in $\mathcal{C}_\alpha$ and the coefficient in front of the element denotes the number $n(\alpha)$ of elements contained in $\mathcal{C}_\alpha$, if $n(\alpha)$ is larger than one. The N-fold proper rotation, $N > 2$, is denoted by C_N if its axis is along Z-axis, and otherwise by C'_N. The 2-fold proper rotation is denoted by C_2 if its axis is along Z-axis, by C'_2 if its axis is along X-axis, and otherwise by C''_2.

Table 3.6 The character table of D_3

D_3	E	$2C_3$	$3C'_2$
A	1	1	1
B	1	1	-1
E	2	-1	0

If the quotient group in a finite group G is isomorphic onto one of above groups, the listed character tables are helpful in calculating the unfaithful irreducible representations of G.

3.4.6 *Self-conjugate Representations*

The characters $\chi(R)$ of a self-conjugate representation is real, but a self-conjugate representation may not be real. Namely, a self-conjugate representation is not always equivalent to a representation composed of real representation matrices. The following theorem gives a criterion of a real representation.

Theorem 3.5 The similarity transformation matrix X, which relates the irreducible unitary self-conjugate representation of a finite group G and its conjugate one, must be symmetric or antisymmetric. X is symmetric if and only if the representation is real.

Proof Let $D(\mathrm{G})$ denote an irreducible unitary self-conjugate representation of a finite group G, and $X^{-1}D(R)X = D(R)^*$. Due to the Schur theorem, X is determined up to a constant factor. Without loss of generality, X can be assumed to be unitary,

$$D(R) = X^T D(R)^* X^* = \left(X^T X^{-1}\right) D(R) \left(X X^*\right).$$

From Corollary 3.2.1, $X^T X^{-1} = \tau \mathbf{1}$, $X^T = \tau X$, and $X = \tau X^T = \tau^2 X$.

Thus, $\tau = \pm 1$, namely, X must be symmetric or antisymmetric.

τ can be expressed by the characters such that τ is independent of a similarity transformation.

$$\begin{aligned}
\sum_{R\in G} \chi(R^2) &= \sum_{\mu\nu} \sum_{R\in G} \left\{D_{\mu\nu}(R)^*\right\}^* D_{\nu\mu}(R) \\
&= \sum_{\mu\nu} \sum_{R\in G} \left\{X^{-1}D(R)X\right\}^*_{\mu\nu} D_{\nu\mu}(R) \\
&= \sum_{\mu\nu\rho\lambda} \tau \sum_{R\in G} X_{\mu\rho} D_{\rho\lambda}(R)^* \left(X^{-1}\right)_{\nu\lambda} D_{\nu\mu}(R) \\
&= \frac{g\tau}{m} \sum_{\mu\nu} X_{\mu\nu} \left(X^{-1}\right)_{\nu\mu} = g\tau.
\end{aligned}$$

If the representation is not self-conjugate, the first line of the above equation is equal to zero owing to Theorem 3.3.

If $D(R)$ is real, X is a unit matrix and $\tau = 1$. Conversely, if X is a symmetric unitary matrix, $X^\dagger = X^* = X^{-1}$, we are going to show that X can be decomposed to Y^2, where Y is a symmetric unitary matrix. In fact, if $X\boldsymbol{a} = \lambda \boldsymbol{a}$, then $X^{-1}\boldsymbol{a}^* = \lambda^{-1}\boldsymbol{a}^*$. Both $\boldsymbol{a}$ and $\boldsymbol{a}^*$ are the eigenvectors of X with the same eigenvalue. Namely, the eigenvectors of X can be chosen to be real such that X can be diagonalized through a real orthogonal similarity transformation M: $M^{-1}XM = \Gamma' = \Gamma^2$ is diagonal and unitary. Thus, $Y = M\Gamma M^{-1}$ is unitary, $X = Y^2$, and $Y^\dagger = Y^{-1} = M\Gamma^* M^{-1} = Y^*$. Due to $Y^{-2}D(R)Y^2 = D(R)^*$, we have $Y^{-1}D(R)Y = YD(R)^*Y^{-1} = \left(Y^{-1}D(R)Y\right)^*$. $D(R)$ is a real representation. □

Corollary 3.5.1 The character in an irreducible representation of a finite group satisfies

$$\frac{1}{g}\sum_{R\in G} \chi(R^2) = \begin{cases} 1, & \text{for real representation,} \\ -1, & \text{for self-conjugate but not real,} \\ 0, & \text{for non-self-conjugate representation.} \end{cases} \tag{3.72}$$

3.5 Subduced and Induced Representations

3.5.1 *Frobenius Theorem*

Let $D^j(\mathrm{G})$ with dimension m_j and $\overline{D}^k(\mathrm{H})$ with dimension $\overline{m}_k$ denote the irreducible representations of G and its subgroup H, respectively. The set of representation matrices $D^j(T_t)$ with $T_t \in H$ forms a representation $D^j(\mathrm{H})$ of H, which is called the **subduced representation** of the irreducible representation $D^j(\mathrm{G})$ of G with respect to its subgroup H. Generally, the subduced representation $D^j(\mathrm{H})$ is a reducible one of H, and can be decomposed with respect to the irreducible representations $\overline{D}^k(\mathrm{H})$ of H:

$$\chi^j(T_t) = \sum_k a_{jk}\overline{\chi}^k(T_t), \qquad m_j = \sum_k a_{jk}\overline{m}_k. \tag{3.73}$$

where $\chi^j(T_t) = \chi^j(\overline{\mathcal{C}}_\beta)$ and $\overline{\chi}^k(T_t) = \overline{\chi}^k(\overline{\mathcal{C}}_\beta)$ are the characters of $T_t \in \overline{\mathcal{C}}_\beta \subset \mathrm{H}$ in the representations $D^j(\mathrm{H})$ and $\overline{D}^k(\mathrm{H})$, respectively. a_{jk} is the multiplicity of $\overline{D}^k(\mathrm{H})$ in $D^j(\mathrm{H})$.

Assume that H **is a finite group** of order h and the class $\overline{\mathcal{C}}_\beta$ in H contains $\overline{n}(\beta)$ elements. Thus,

$$a_{jk} = \frac{1}{h}\sum_{T_t\in\mathrm{H}} \overline{\chi}^k(T_t)^*\chi^j(T_t) = \frac{1}{h}\sum_\beta \overline{n}(\beta)\overline{\chi}^k(\overline{\mathcal{C}}_\beta)^*\chi^j(\overline{\mathcal{C}}_\beta), \tag{3.74}$$

Let ψ_μ denote the basis functions in the representation space of $\overline{D}^k(H)$,

$$P_{T_t}\psi_\mu = \sum_\nu \psi_\nu \overline{D}^k_{\nu\mu}(T_t), \qquad T_t \in \mathrm{H}. \tag{3.75}$$

Generally, the representation space of $\overline{D}^k(H)$ is not invariant under G. Now, let G **be also a finite group** of order $g = nh$, where n is the index of H in G. The left cosets of H in G are denoted by R_rH. Although R_r is not unique, we assume that R_r has been chosen, where $R_1 = E$. Namely, any element in G can be expressed as R_rT_t **uniquely**, where $T_t \in \mathrm{H}$. Define an extended space of dimension $n\overline{m}_k$ with the basis functions $\psi_{r\mu} = P_{R_r}\psi_\mu$, where $\psi_{1\mu} = \psi_\mu$. The extended space is invariant under G, and corresponds to a representation $\Delta^k(\mathrm{G})$ of G with dimension $n\overline{m}_k$. For any given element S in G and for each R_r, SR_r can be expressed as R_uT_t, where u and t are completely determined by S and r. Since

$$P_S\psi_{r\mu} = P_{SR_r}\psi_\mu = P_{R_u}P_{T_t}\psi_\mu = \sum_\nu \psi_{u\nu}\overline{D}^k_{\nu\mu}(T_t),$$

we obtain

$$\Delta^k_{u\nu,r\mu}(S) = \overline{D}^k_{\nu\mu}(T_t), \qquad \chi^k(S) = \sum_{r\mu} \Delta^k_{r\mu,r\mu}(S). \tag{3.76}$$

This representation $\Delta^k(\mathrm{G})$ is called the **induced representation** of the irreducible representation $\overline{D}^k(\mathrm{H})$ of the subgroup H with respect to G. Generally, the induced representation $\Delta^k(\mathrm{G})$ is a reducible one of G:

$$X^{-1}\Delta^k(S)X = \bigoplus_j b_{jk}D^j(S), \qquad (g/h)\,\overline{m}_k = \sum_j b_{jk}m_j,$$
$$b_{jk} = \frac{1}{g}\sum_{S\in G} \chi^j(S)^*\chi^k(S) = \frac{1}{g}\sum_\alpha n(\alpha)\chi^j(\mathcal{C}_\alpha)^*\chi^k(\mathcal{C}_\alpha), \tag{3.77}$$

where $\mathcal{C}_\alpha$ is the class in G.

Theorem 3.6 (Frobenius theorem) Let H be a subgroup of a finite group G. The multiplicity a_{jk} of the irreducible representation $\overline{D}^k(\mathrm{H})$ in the subduced representation $D^j(\mathrm{H})$ of the irreducible representation $D^j(\mathrm{G})$ of G with respect to H is equal to the multiplicity b_{jk} of the irreducible representation $D^j(\mathrm{G})$ in the induced representation $\Delta^k(\mathrm{G})$ of an irreducible representation $\overline{D}^k(\mathrm{H})$ of H with respect to G:

$$a_{jk} = b_{jk}. \tag{3.78}$$

Proof: In the class $\mathcal{C}_\alpha$ of G, some elements may belong to the subgroup H, and some may not. Those elements in $\mathcal{C}_\alpha$ belonging to H, if exist, constitute a few whole classes of H, denoted by $\overline{\mathcal{C}}_\beta$. From Eq. (3.76), the diagonal element of $\Delta^k(S)$ appears only when $u = r$, i.e., $SR_r = R_rT_t$. Thus, $\chi^k(S) = \chi^k(\mathcal{C}_\alpha) = 0$ if $S \in \mathcal{C}_\alpha$ and $\mathcal{C}_\alpha$ does not contain any element belonging to H. $\chi^k(\mathcal{C}_\alpha)$ is nonvanishing only when the class $\mathcal{C}_\alpha$ contains a few $\overline{\mathcal{C}}_\beta$ in H. For a given S, denoting by κ_β the number of different R_r satisfying $R_r^{-1}SR_r \in \overline{\mathcal{C}}_\beta$, one has $\chi^k(\mathcal{C}_\alpha) = \sum_\beta \kappa_\beta \overline{\chi}^k(\overline{\mathcal{C}}_\beta)$.

From Prob. 16 of Chap. 2, the number of elements R in G satisfying $R^{-1}SR = T_t \in \overline{\mathcal{C}}_\beta$ is $m(\alpha) = g/n(\alpha)$, and the number of elements T_y in the subgroup H satisfying $T_yT_tT_y^{-1} = T_t$ is $\overline{m}(\beta) = h/\overline{n}(\beta)$. Expressing R by R_rT_x, one has $R_r^{-1}SR_r = T_xT_tT_x^{-1} \in \overline{\mathcal{C}}_\beta$. If R_rT_x satisfies $S(R_rT_x) = (R_rT_x)T_t$, then $R_rT_xT_y$ satisfies $S(R_rT_xT_y) = (R_rT_xT_y)T_t$. However, the latter does not make any new contribution to the characters $\chi^k(S)$ because $R_r^{-1}SR_r = T_xT_yT_tT_y^{-1}T_x^{-1} = T_xT_tT_x^{-1}$. Therefore,

$$\kappa_\beta = \frac{m(\alpha)}{\overline{m}(\beta)} = \frac{g\,\overline{n}(\beta)}{h\,n(\alpha)}, \qquad \chi^k(\mathcal{C}_\alpha) = \frac{g}{h\,n(\alpha)}\sum_\beta \overline{n}(\beta)\,\overline{\chi}^k(\overline{\mathcal{C}}_\beta). \tag{3.79}$$

Substituting Eq. (3.79) into Eq. (3.77), where $\chi^j(\mathcal{C}_\alpha) = \chi^j(\overline{\mathcal{C}}_\beta)$ because the class $\mathcal{C}_\alpha$ in G contains a few classes $\overline{\mathcal{C}}_\beta$ of H, one obtains

$$b_{jk} = \frac{1}{g}\sum_\alpha n(\alpha)\chi^j(\mathcal{C}_\alpha)^*\chi^k(\mathcal{C}_\alpha) = \frac{1}{h}\sum_\beta \overline{n}(\beta)\chi^j(\overline{\mathcal{C}}_\beta)^*\overline{\chi}^k(\overline{\mathcal{C}}_\beta) = a_{jk}. \tag{3.80}$$

The sum over the classes $\mathcal{C}_\alpha$ in Eq. (3.80) is equivalent to the sum over the classes $\overline{\mathcal{C}}_\beta$ in H, because each class $\overline{\mathcal{C}}_\beta$ of H is contained in one and only one class $\mathcal{C}_\alpha$ of G. Thus, the multiplicity b_{jk} in Eq. (3.77) is equal to the multiplicity a_{jk} in Eq. (3.74). □

3.5.2 *Irreducible Representations of* D_N

The group D_N contains one N-fold axis, denoted as Z axis and called the principal axis, and N 2-fold axes located in the XY plane. The order of D_N is $g = 2N$. Two generators of D_N are the N-fold rotation C_N about the Z axis and one 2-fold rotation $C_2^{(0)}$ about the X axis.

The group D_N has an invariant subgroup C_N with index two. The cyclic group C_N has N inequivalent one-dimensional representations $D^j(\mathrm{C}_N)$ (see Eq. (3.70)). Denoting by $\psi^{(j)}$ the basis in the representation space of $D^j(\mathrm{C}_N)$, one has

$$P_{C_N}\psi^{(j)} = \mathrm{e}^{-\mathrm{i}2\pi j/N}\psi^{(j)}, \qquad 0 \leqslant j \leqslant N-1. \tag{3.81}$$

The evident form of the wave function $\psi^{(j)}$ depends on the physical problem that one deals with. Due to Eq. (3.81) we may assume $\psi^{(j)}(\varphi) = \mathrm{e}^{\mathrm{i}j\varphi}$ which is an planar function depending on the azimuth angle φ:

$$P_{C_N}\psi^{(j)}(\varphi) = \psi^{(j)}(C_N^{-1}\varphi) = \mathrm{e}^{\mathrm{i}j(\varphi-2\pi/N)} = \mathrm{e}^{-\mathrm{i}2\pi j/N}\psi^{(j)}(\varphi). \tag{3.82}$$

Extending the representation space by ϕ^j, we have

$$P_{C_2^{(0)}}\psi^j = \phi^j, \qquad P_{C_2^{(0)}}\phi^j = \psi^j, \qquad P_{C_N}\phi^j = \mathrm{e}^{\mathrm{i}2\pi j/N}\phi^j,$$

where from Eq. (2.11), $C_N C_2^{(0)} = C_2^{(1)} = C_2^{(0)} C_N^{-1}$ is used. Then, we obtain the induced two-diemensional representations $D^j(\mathrm{D}_N)$:

$$\begin{aligned} D^{E_j}(C_N) &= \begin{pmatrix} \mathrm{e}^{-\mathrm{i}2\pi j/N} & 0 \\ 0 & \mathrm{e}^{\mathrm{i}2\pi j/N} \end{pmatrix}, \qquad & D^{E_j}(C_2^{(0)}) &= \begin{pmatrix} 0 & 1 \\ 1 & 0 \end{pmatrix}, \\ \chi^{E_j}(C_N) &= 2\cos\left(\frac{2\pi j}{N}\right), & \chi^{E_j}(C_2^{(0)}) &= 0. \end{aligned} \tag{3.83}$$

Among those representations, some representations are equivalent:

$$\sigma_1^{-1} D^{E_{N-j}}(R)\sigma_1 = D^{E_j}(R), \qquad R \in \mathrm{D}_N, \quad \sigma_1 = \begin{pmatrix} 0 & 1 \\ 1 & 0 \end{pmatrix}. \tag{3.84}$$

$D^{E_0}(C_N)$ is a unit matrix such that $D^{E_0}(\mathrm{D}_N)$ is reducible.

When $N = 2n+1$, from Eq. (2.39) there are $g_c = n+2$ classes in D_{2n+1}. D_{2n+1} has n inequivalent irreducible representations of two dimensions as given in Eq. (3.83) where $1 \leqslant j \leqslant n$. The representation $D^{E_0}(\mathrm{D}_{2n+1})$ is reducible. Through a similarity transformation $X = \sqrt{\frac{1}{2}}\begin{pmatrix} 1 & -\mathrm{i} \\ 1 & \mathrm{i} \end{pmatrix}$ we obtained another useful form of those representations (see Eq. (2.13)):

$$\begin{aligned}
\overline{D}^{E_j}(C_{2n+1}^m) &= \begin{pmatrix} \cos\left[2\pi mj/(2n+1)\right] & -\sin\left[2\pi mj/(2n+1)\right] \\ \sin\left[2\pi mj/(2n+1)\right] & \cos\left[2\pi mj/(2n+1)\right] \end{pmatrix}, \\
\overline{D}^{E_j}(C_2^{(m)}) &= \begin{pmatrix} \cos\left[2\pi mj/(2n+1)\right] & \sin\left[2\pi mj/(2n+1)\right] \\ \sin\left[2\pi mj/(2n+1)\right] & -\cos\left[2\pi mj/(2n+1)\right] \end{pmatrix}.
\end{aligned} \tag{3.85}$$

where $0 \leqslant m \leqslant 2n$ and $1 \leqslant j \leqslant n$, and

$$\begin{aligned}
X^{-1}D^{E_0}(R)X &= \begin{pmatrix} D^A(R) & 0 \\ 0 & D^B(R) \end{pmatrix}, \qquad R \in \mathrm{D}_{2n+1}, \\
D^A(C_{2n+1}) &= D^A(C_2^{(0)}) = D^B(C_{2n+1}) = -D^B(C_2^{(0)}) = 1.
\end{aligned} \tag{3.86}$$

For the group D_3, Eq. (3.85) coincides with Eq. (2.14).

When $N = 2n$, from Eq. (2.40) there are $g_c = n + 3$ classes in D_{2n}. D_{2n} has $(n-1)$ inequivalent irreducible representations of two dimensions as given in Eq. (3.83) where $1 \leqslant j \leqslant n-1$. The representation $D^{E_0}(\mathrm{D}_{2n})$ and $D^{E_n}(\mathrm{D}_{2n})$ are reducible. Through a similarity transformation X we obtained another useful form of those representations:

$$\begin{aligned}
\overline{D}^{E_j}(C_{2n}^m) &= \begin{pmatrix} \cos\left[\pi mj/n\right] & -\sin\left[\pi mj/n\right] \\ \sin\left[\pi mj/n\right] & \cos\left[\pi mj/n\right] \end{pmatrix}, \\
\overline{D}^{E_j}(C_2^{(m)}) &= \begin{pmatrix} \cos\left[\pi mj/n\right] & \sin\left[\pi mj/n\right] \\ \sin\left[\pi mj/n\right] & -\cos\left[\pi mj/n\right] \end{pmatrix}.
\end{aligned} \tag{3.87}$$

where $0 \leqslant m \leqslant 2n-1$ and $1 \leqslant j \leqslant n-1$, and

$$\begin{aligned}
X^{-1}D^{E_0}(R)X &= \begin{pmatrix} D^{A_1}(R) & 0 \\ 0 & D^{A_2}(R) \end{pmatrix}, \qquad R \in \mathrm{D}_{2n}, \\
X^{-1}D^{E_n}(R)X &= \begin{pmatrix} D^{B_1}(R) & 0 \\ 0 & D^{B_2}(R) \end{pmatrix}, \\
D^{A_1}(C_{2n}) &= D^{A_2}(C_{2n}) = 1, \quad D^{B_1}(C_{2n}) = D^{B_2}(C_{2n}) = -1, \\
D^{A_1}(C_2^{(0)}) &= -D^{A_2}(C_2^{(0)}) = 1, \quad D^{B_1}(C_2^{(0)}) = -D^{B_2}(C_2^{(0)}) = 1.
\end{aligned} \tag{3.88}$$

Four inequivalent representations $D^{A_1}(\mathrm{D}_{2n})$, $D^{A_2}(\mathrm{D}_{2n})$, $D^{B_1}(\mathrm{D}_{2n})$, and $D^{B_2}(\mathrm{D}_{2n})$ of one dimension are those of the quotient group $\mathrm{D}_{2n}/\mathrm{C}_n \approx \mathrm{D}_2$ (see Table 3.5).

In Tables 3.7-16 we list the character tables of some D_N groups where $\eta_N = e^{-i2\pi/N}$, and $q_a^{(N)} = q_{N-a}^{(N)} = (\eta_N)^a + (\eta_N)^{-a} = 2\cos(2\pi a/N)$. The parameters satisfy some relations, for example

$$\sum_{a=0}^{N-1} (\eta_N)^a = 0, \qquad \sum_{a=1}^{n} \left(q_a^{(2n+1)}\right)^2 = 2n - 1,$$
$$q_{2a}^{(2n)} = q_a^{(n)}, \qquad q_a^{(2n)} = -q_{n-a}^{(2n)}.$$

Table 3.7 The character table of D_5

D_5	E	$2C_5$	$2C_5^2$	$5C_2'$
A	1	1	1	1
B	1	1	1	-1
E_1	2	$(\sqrt{5}-1)/2$	$-(\sqrt{5}+1)/2$	0
E_2	2	$-(\sqrt{5}+1)/2$	$(\sqrt{5}-1)/2$	0

Table 3.8 The character table of D_7

D_7	E	$2C_7$	$2C_7^2$	$2C_7^3$	$7C_2'$
A	1	1	1	1	1
B	1	1	1	1	-1
E_1	2	$q_1^{(7)}$	$q_2^{(7)}$	$q_3^{(7)}$	0
E_2	2	$q_2^{(7)}$	$q_3^{(7)}$	$q_1^{(7)}$	0
E_3	2	$q_3^{(7)}$	$q_1^{(7)}$	$q_2^{(7)}$	0

Table 3.9 The character table of D_9

D_9	E	$2C_7$	$2C_7^2$	$2C_7^3$	$2C_7^4$	$9C_2'$
A	1	1	1	1	1	1
B	1	1	1	1	1	-1
E_1	2	$q_1^{(9)}$	$q_2^{(9)}$	-1	$q_4^{(9)}$	0
E_2	2	$q_2^{(9)}$	$q_4^{(9)}$	-1	$q_1^{(9)}$	0
E_3	2	-1	-1	2	-1	0
E_4	2	$q_4^{(9)}$	$q_1^{(9)}$	-1	$q_2^{(9)}$	0

Table 3.10 The character table of D_4

D_4	E	$2C_4$	C_4^2	$4C_2'$	$4C_2''$
A_1	1	1	1	1	1
A_2	1	1	1	-1	-1
B_1	1	-1	1	1	-1
B_2	1	-1	1	-1	1
E	2	0	-2	0	0

Table 3.11 The character table of D_6

D_6	E	$2C_6$	$2C_6^2$	C_6^3	$4C_2'$	$4C_2''$
A_1	1	1	1	1	1	1
A_2	1	1	1	1	-1	-1
B_1	1	-1	1	-1	1	-1
B_2	1	-1	1	-1	-1	1
E_1	2	1	-1	-2	0	0
E_2	2	-1	-1	2	0	0

Table 3.12 The character table of D_8

D_8	E	$2C_8$	$2C_8^2$	$2C_8^3$	C_8^4	$4C_2'$	$4C_2''$
A_1	1	1	1	1	1	1	1
A_2	1	1	1	1	1	-1	-1
B_1	1	-1	1	-1	1	1	-1
B_2	1	-1	1	-1	1	-1	1
E_1	2	$\sqrt{2}$	0	$-\sqrt{2}$	-2	0	0
E_2	2	0	-2	0	2	0	0
E_3	2	$-\sqrt{2}$	0	$\sqrt{2}$	-2	0	0

Table 3.13 The character table of D_{10}

D_{10}	E	$2C_{10}$	$2C_{10}^2$	$2C_{10}^3$	$2C_{10}^4$	C_{10}^5	$5C_2'$	$5C_2''$
A_1	1	1	1	1	1	1	1	1
A_2	1	1	1	1	1	1	-1	-1
B_1	1	-1	1	-1	1	-1	1	-1
B_2	1	-1	1	-1	1	-1	-1	1
E_1	2	$-q_2^{(5)}$	$q_1^{(5)}$	$-q_1^{(5)}$	$q_2^{(5)}$	-2	0	0
E_2	2	$q_1^{(5)}$	$q_2^{(5)}$	$q_2^{(5)}$	$q_1^{(5)}$	2	0	0
E_3	2	$-q_1^{(5)}$	$q_2^{(5)}$	$-q_2^{(5)}$	$q_1^{(5)}$	-2	0	0
E_4	2	$q_2^{(5)}$	$q_1^{(5)}$	$q_1^{(5)}$	$q_2^{(5)}$	2	0	0

$$q_1^{(10)} = -q_2^{(5)} = (\sqrt{5}+1)/2, \quad q_2^{(10)} = q_1^{(5)} = (\sqrt{5}-1)/2.$$

Table 3.14 The character table of D_{12}

D_{12}	E	$2C_{12}$	$2C_{12}^2$	$2C_{12}^3$	$2C_{12}^4$	$2C_{12}^5$	C_{12}^6	$6C_2'$	$6C_2''$
A_1	1	1	1	1	1	1	1	1	1
A_2	1	1	1	1	1	1	1	-1	-1
B_1	1	-1	1	-1	1	-1	1	1	-1
B_2	1	-1	1	-1	1	-1	1	-1	1
E_1	2	$\sqrt{3}$	1	0	-1	$-\sqrt{3}$	-2	0	0
E_2	2	1	-1	-2	-1	1	2	0	0
E_3	2	0	-2	0	2	0	-2	0	0
E_4	2	-1	-1	2	-1	-1	2	0	0
E_5	2	$-\sqrt{3}$	1	0	-1	$\sqrt{3}$	-2	0	0

Table 3.15 The character table of D_{14}

D_{14}	E	$2C_{12}$	$2C_{12}^2$	$2C_{12}^3$	$2C_{12}^4$	$2C_{12}^5$	$2C_{12}^6$	C_{12}^7	$7C_2'$	$7C_2''$
A_1	1	1	1	1	1	1	1	1	1	1
A_2	1	1	1	1	1	1	1	1	−1	−1
B_1	1	−1	1	−1	1	−1	1	−1	1	−1
B_2	1	−1	1	−1	1	−1	1	−1	−1	1
E_1	2	$-q_3^{(7)}$	$q_1^{(7)}$	$-q_2^{(7)}$	$q_2^{(7)}$	$-q_1^{(7)}$	$q_3^{(7)}$	−2	0	0
E_2	2	$q_1^{(7)}$	$q_2^{(7)}$	$q_3^{(7)}$	$q_3^{(7)}$	$q_2^{(7)}$	$q_1^{(7)}$	2	0	0
E_3	2	$-q_2^{(7)}$	$q_3^{(7)}$	$-q_1^{(7)}$	$q_1^{(7)}$	$-q_3^{(7)}$	$q_2^{(7)}$	−2	0	0
E_4	2	$q_2^{(7)}$	$q_3^{(7)}$	$q_1^{(7)}$	$q_1^{(7)}$	$q_3^{(7)}$	$q_2^{(7)}$	2	0	0
E_5	2	$-q_1^{(7)}$	$q_2^{(7)}$	$-q_3^{(7)}$	$q_3^{(7)}$	$-q_2^{(7)}$	$q_1^{(7)}$	−2	0	0
E_6	2	$q_3^{(7)}$	$q_1^{(7)}$	$q_2^{(7)}$	$q_2^{(7)}$	$q_1^{(7)}$	$q_3^{(7)}$	2	0	0

$$q_1^{(14)} = -q_3^{(7)}, \quad q_2^{(14)} = q_1^{(7)}, \quad q_3^{(14)} = -q_2^{(7)}.$$

Table 3.16 The character table of D_{18}

D_{18}	E	$2C_{18}$	$2C_{18}^2$	$2C_{18}^3$	$2C_{18}^4$	$2C_{18}^5$	$2C_{18}^6$	$2C_{18}^7$	$2C_{18}^8$	C_{18}^9	$9C_2'$	$9C_2''$
A_1	1	1	1	1	1	1	1	1	1	1	1	1
A_2	1	1	1	1	1	1	1	1	1	1	−1	−1
B_1	1	−1	1	−1	1	−1	1	−1	1	−1	1	−1
B_2	1	−1	1	−1	1	−1	1	−1	1	−1	−1	1
E_1	2	$-q_4^{(9)}$	$q_1^{(9)}$	1	$q_2^{(9)}$	$-q_2^{(9)}$	−1	$-q_1^{(9)}$	$q_4^{(9)}$	−2	0	0
E_2	2	$q_1^{(9)}$	$q_2^{(9)}$	−1	$q_4^{(9)}$	$q_4^{(9)}$	−1	$q_2^{(9)}$	$q_1^{(9)}$	2	0	0
E_3	2	1	−1	−2	−1	1	2	1	−1	−2	0	0
E_4	2	$q_2^{(9)}$	$q_4^{(9)}$	−1	$q_1^{(9)}$	$q_1^{(9)}$	−1	$q_4^{(9)}$	$q_2^{(9)}$	2	0	0
E_5	2	$-q_2^{(9)}$	$q_4^{(9)}$	1	$q_1^{(9)}$	$-q_1^{(9)}$	−1	$-q_4^{(9)}$	$q_2^{(9)}$	−2	0	0
E_6	2	−1	−1	2	−1	−1	2	−1	−1	2	0	0
E_7	2	$-q_1^{(9)}$	$q_2^{(9)}$	1	$q_4^{(9)}$	$-q_4^{(9)}$	−1	$-q_2^{(9)}$	$q_1^{(9)}$	−2	0	0
E_8	2	$q_4^{(9)}$	$q_1^{(9)}$	−1	$q_2^{(9)}$	$q_2^{(9)}$	−1	$q_1^{(9)}$	$q_4^{(9)}$	2	0	0

$$q_1^{(18)} = -q_4^{(9)}, \quad q_2^{(18)} = q_1^{(9)}, \quad q_4^{(18)} = q_2^{(9)}, \quad q_3^{(18)} = -q_6^{(18)} = 1.$$

3.6 Applications in Physics

At the beginning of Chapter 2, before studying group theory, we raised a simple example to see how to obtain some precise information of the system through analyzing its symmetry. Now, we have studied the fundamental concepts on group theory and the theory of representations. It is time to discuss the typical applications of group theory to physics.

3.6.1 *Classification of Static Wave Functions*

The first step in the application of group theory to physics is to find the symmetric transformations of a given quantum system with the Hamiltonian $H(x)$. A symmetric transformation R leaves the

Hamiltonian invariant,

$$H(x) \xrightarrow{R} P_R H(x) P_R = H(x), \qquad [P_R,\ H(x)] = 0. \tag{3.89}$$

P_R is the transformation operator for scalar wave functions corresponding to the symmetric transformation R. The set of the symmetric transformations forms the symmetry group G of the system.

Second, **find the inequivalent irreducible representations and their characters of the symmetry group of the system.** This is a task of group theory. Usually, one chooses the convenient forms of the representation matrices in an irreducible representation $D^j(\mathrm{G})$ such that the representation matrices $D^j(A)$ of the generators A of G, as many as possible, are diagonal. There exist some generators B whose representation matrices $D^j(B)$ are not diagonal if G is not abelian. The representation matrices of any element in G can be calculated from those of the generators.

Third, if the energy level E is m degeneracy, there are m linearly independent eigenfunctions $\psi_\mu(x)$ of $H(x)$ with the eigenvalue E:

$$H(x)\psi_\mu(x) = E\psi_\mu(x), \qquad \mu = 1, 2, \ldots, m. \tag{3.90}$$

$\psi_\mu(x)$ span an m-dimensional functional space $\mathcal{L}$. Any function $\phi(x)$ in $\mathcal{L}$ is the eigenfunction of $H(x)$ with the eigenvalue E, and any eigenfunction of $H(x)$ with the eigenvalue E belongs to $\mathcal{L}$. Due to Eq. (3.89), $P_R\psi_\mu(x)$ is an eigenfunction of $H(x)$ with the same energy E. Namely, **$\mathcal{L}$ is invariant under the symmetric transformations P_R**. The matrix $D(R)$ of P_R in $\mathcal{L}$ with respect to the basis function $\psi_\mu(x)$ is calculated by

$$P_R\psi_\mu(x) = \psi_\mu(R^{-1}x) = \sum_{\nu=1}^{m} \psi_\nu(x) D_{\nu\mu}(R). \tag{3.91}$$

The set of $D(R)$ forms a representation of the symmetry group G of the system, called the representation corresponding to the energy E. $D(R)$ **describes the transformation rule of the eigenfunctions of $H(x)$ with E in the symmetric transformation R.** The character of R in $D(R)$ is $\chi(R) = \mathrm{Tr}D(R)$. Generally, the representation $D(\mathrm{G})$ is reducible and is not in the convenient form. Through a similarity transformation X, $D(\mathrm{G})$ can be decomposed into the direct sum of irreducible representations in the convenient forms,

$$X^{-1}D(R)X = \bigoplus_j a_j D^j(R), \qquad \chi(R) = \sum_j a_j \chi^j(R). \tag{3.92}$$

The multiplicity a_j of the irreducible representation $D^j(G)$ in the representation $D(G)$ can be calculated by the orthogonal relation (3.55):

$$a_j = \frac{1}{g}\sum_{R\in G} \chi^j(R)^*\chi(R) = \frac{1}{g}\sum_{\alpha} n(\alpha)\left(\chi^j_\alpha\right)^* \chi_\alpha. \qquad (3.93)$$

Then, X can be calculated from Eq. (3.92) in the following way. When R in Eq. (3.92) is taken to be the generators A where $D^j(A)$ are diagonal, **X is the similarity transformation matrix to diagonalize $D(A)$.** Namely, the column matrices of X are the eigenvectors of $D(A)$ with different eigenvalues, respectively. The solution X contains some parameters to be determined. Substituting X into Eq. (3.92), where R is taken to be the remaining generators B, one is able to determine part of parameters. In the calculation, X^{-1} in Eq. (3.92) should be moved to the right-hand side of the equation **to avoid the complicated calculation for the inverse matrix**. The calculated X, satisfying Eq. (3.92) for all generators, still contains some undetermined parameters, whose number is $\sum_j a_j^2$. **Those parameters should be chosen to make X as simple as possible.**

From Eq. (3.92), the row index of X is the same as the column index of $D(R)$, denoted by μ, and the column index of X is the same as the column index of the block matrix on the right-hand side of Eq. (3.92), which is enumerated by three indices j, ρ, and r. Two indices j and ρ are denoted the irreducible representation $D^j(R)$ and its row and the additional index r is needed to distinguish different D^j in the decomposition (3.92), when $a_j > 1$.

New basis functions $\Phi^j_{\rho r}(x)$ are the combinations of $\psi_\mu(x)$ by X:

$$\begin{aligned} \Phi^j_{\rho r}(x) &= \sum_\mu \psi_\mu(x) X_{\mu, j\rho r}, \\ P_R \Phi^j_{\rho r}(x) &= \sum_\lambda \Phi^j_{\lambda r}(x) D^j_{\lambda\rho}(R). \end{aligned} \qquad (3.94)$$

$\Phi^j_{\rho r}(x)$ are the basis functions chosen by group theory. The function $\Phi^j_{\rho r}(x)$ is called **a function belonging to the ρth row of the irreducible representation D^j** of G. This is the so-called classification of the static wave functions according to the irreducible representations of the symmetry group of the system. When $a_j > 1$, there are a_j sets of basis functions $\Psi^j_{\rho r}(x)$ which belong to the same representation D^j of G and are distinguished by the parameter r. Any linear combination of the functions $\Phi^j_{\rho r}(x)$ with the same j and ρ is the eigenfunction belonging to the ρth row of D^j

$$\Phi^j_{\rho s}(x) = \sum_r \Psi^j_{\rho r}(x) Y^j_{rs}, \tag{3.95}$$

where **the combination coefficients Y^j_{rs} are independent of ρ**. The combination matrix Y^j is related to the undetermined parameters in X.

The physical meaning of the function $\Psi^j_{\rho r}(x)$ depends on the given group G and the chosen representation $D^j(\mathrm{G})$. Since the representation matrices $D^j(A)$ of some generators A in G are diagonal, $\Psi^j_{\rho r}(x)$ is **the common eigenfunction of the operators P_A** as given in Eq. (3.94).

3.6.2 *Clebsch–Gordan Series and Coefficients*

If a quantum system consists of two subsystems (see section 1.6), the wave function of the system is expressed as the product of two wave functions of the subsystems or their combinations. Suppose that two functional spaces $\mathcal{L}_j$ and $\mathcal{L}_k$ of two subsystems are the representation spaces of irreducible representations $D^j(\mathrm{G})$ and $D^k(\mathrm{G})$, respectively, then

$$P_R \psi^j_\mu(x) = \sum_\rho \psi^j_\rho(x) D^j_{\rho\mu}(R), \qquad P_R \phi^k_\nu(y) = \sum_\lambda \phi^k_\lambda(y) D^k_{\lambda\nu}(R).$$

The functional space $\mathcal{L}$ of the composed system is the direct product of $\mathcal{L}_j$ and $\mathcal{L}_k$, and is spanned by $\Psi^{jk}_{\mu\nu}(x,y) = \psi^j_\mu(x)\phi^k_\nu(y)$ where $1 \leqslant \mu \leqslant m_j$ and $1 \leqslant \nu \leqslant m_k$. In the symmetric transformation, $\Psi^{jk}_{\mu\nu}(x,y)$ transforms according to the direct product representation

$$\begin{aligned} P_R \Psi^{jk}_{\mu\nu}(x,y) &= \sum_{\rho\lambda} \Psi^{jk}_{\rho\lambda}(x,y) \left[D^j(R) \times D^k(R)\right]_{\rho\lambda,\mu\nu}, \\ \left[D^j(R) \times D^k(R)\right]_{\rho\lambda,\mu\nu} &= D^j_{\rho\mu}(R) D^k_{\lambda\nu}(R). \end{aligned} \tag{3.96}$$

Generally, the direct product representation is deducible and can be decomposed through a similarity transformation C^{jk}:

$$\left(C^{jk}\right)^{-1} \left[D^j(R) \times D^k(R)\right] C^{jk} = \bigoplus_J a_J D^J(R). \tag{3.97}$$

The series on the right-hand side of Eq. (3.97) is called the Clebsch–Gordan series. Taking the trace of Eq. (3.97), one has

$$\chi^j(R)\chi^k(R) = \sum_J a_J \chi^J(R). \tag{3.98}$$

The multiplicity a_J as well as the matrix C^{jk} can be calculated from Eqs. (3.98) and (3.97). Similar to the discussion in the preceding subsection, the row and column indices of C^{jk} are denoted respectively by $\mu\nu$ and JMr, where the additional index r is needed when $a_J > 1$. New basis functions $\Phi^J_{Mr}(x,y)$ are combined from $\Psi^{jk}_{\mu\nu}(x,y)$ through C^{jk} such that they belong to the representation D^J of G,

$$\begin{aligned} \Phi^J_{Mr}(x,y) &= \sum_{\mu\nu} \Psi^{jk}_{\mu\nu}(x,y) C^{jk}_{\mu\nu,JMr}, \\ P_R \Phi^J_{Mr}(x,y) &= \sum_{M'} \Phi^J_{M'r}(x,y) D^J_{M'M}(R). \end{aligned} \tag{3.99}$$

The matrix elements $C^{jk}_{\mu\nu,JMr}$ are called the Clebsch–Gordan coefficients. The Clebsch–Gordan coefficients depend upon the group G and its chosen representations. In addition, as discussed in the preceding subsection, C^{jk} contains $\sum_J a_J^2$ undetermined parameters, which are chosen to make C^{jk} as simple as possible. Some examples for the calculations of the Clebsch–Gordan coefficients of point groups are given in Probs. 24–27 of Chap. 3 of [Ma and Gu (2004)].

Theorem 3.8 The Clebsch–Gordan series (3.97) for the direct product $D^j(\mathrm{G}) \times D^k(\mathrm{G})$ of two irreducible representations of a finite group G contains the identical representation if and only if $D^j(\mathrm{G})$ and $D^k(\mathrm{G})$ are conjugate to each other. The Clebsch–Gordan series contains one and only one identical representation if $D^j(\mathrm{G})$ and $D^k(\mathrm{G})$ are conjugate.

Proof Since the character of the identical representation is 1, we calculate a_J by Eq. (3.58):

$$a_J = \frac{1}{g} \sum_{R\in G} \chi^j(R)\chi^k(R).$$

Due to Eq. (3.55) $a_J = 1$ if $D^j(\mathrm{G})$ and $D^k(\mathrm{G})$ are conjugate to each other, and $a_J = 0$ if they are not. □

3.6.3 *Wigner–Eckart Theorem*

Theorem 3.7 (Wigner–Eckart Theorem) The functions belonging to two inequivalent irreducible representations of a unitary operator group P_G are orthogonal to each other. The functions belonging to different rows of a unitary irreducible representation of P_G are orthogonal to each other, and the self-inner products of those functions are independent of the row number.

Proof First of all, it is assumed that one can define an inner product of two functions such that the operators P_R are unitary:

$$\langle \phi(x)|\psi(x)\rangle = \langle P_R\phi(x)|P_R\psi(x)\rangle. \tag{3.100}$$

If P_G is the symmetry group of a physical system such as a finite group or the rotational group, the inner product usually used in physics makes P_R unitary. An important counter-example in physics is the proper Lorentz transformation.

Let $D^j(\mathrm{G})$ and $D^k(\mathrm{G})$ denote two inequivalent irreducible unitary representations of the group P_G. $\psi^j_\mu(x)$ belongs to the μth row of $D^j(\mathrm{G})$ and $\phi^k_\nu(x)$ belongs to the νth row of $D^k(\mathrm{G})$,

$$P_R\psi^j_\mu(x) = \sum_\rho \psi^j_\rho(x)D^j_{\rho\mu}(R), \qquad P_R\phi^k_\nu(x) = \sum_\lambda \phi^k_\lambda(x)D^k_{\lambda\nu}(R).$$

Letting $\langle \phi^k_\nu(x)|\psi^j_\mu(x)\rangle = X^{kj}_{\nu\mu}$, one has

$$\begin{aligned}\langle \phi^k_\nu(x)|P_R\psi^j_\mu(x)\rangle &= \sum_\rho \langle \phi^k_\nu(x)|\psi^j_\rho(x)\rangle D^j_{\rho\mu}(R) = \sum_\rho X^{kj}_{\nu\rho}D^j_{\rho\mu}(R)\\ = \langle P_R^{-1}\phi^k_\nu(x)|\psi^j_\mu(x)\rangle &= \sum_\lambda D^k_{\lambda\nu}(R^{-1})^*\langle \phi^k_\lambda(x)|\psi^j_\mu(x)\rangle = \sum_\lambda D^k_{\nu\lambda}(R)X^{kj}_{\lambda\mu}.\end{aligned} \tag{3.101}$$

From the Schur Theorem 3.3, $X^{kj}_{\nu\mu} = \delta_{kj}\delta_{\nu\mu}\ \langle \phi^k||\psi^j\rangle$, where $\langle \phi^k||\psi^j\rangle$ is called the **reduced matrix element**, which is a constant independent of the subscript μ. Thus, the Theorem is proved:

$$\langle \phi^k_\nu(x)|\psi^j_\mu(x)\rangle = \delta_{kj}\delta_{\nu\mu}\langle \phi^k||\psi^j\rangle. \tag{3.102}$$

The functions belonging to two inequivalent irreducible representations of P_G are orthogonal to each other, even for non-unitary representations. □

In quantum mechanics, most physical observables are calculated through elements of the matrix. When the static wave functions belong to given rows of a unitary irreducible representation, due to the Wigner–Eckart Theorem, **the problem of calculating $m_k m_j$ matrix elements $\langle \phi^k_\nu(x)|\psi^j_\mu(x)\rangle$ is reduced to the problem of calculating only one reduced matrix element**. Second, the static wave functions in the practical problems usually are hard to solve such that even one element of the matrix also cannot be calculated. However, **the ratios of the matrix elements can be obtained by eliminating the reduced matrix element as a parameter**. Third, some matrix elements vanish due to the symmetry as shown in Eq. (3.102). **These vanishing matrix elements are called**

the selection rule in physics. Thus, from analyzing the symmetry of the system, one can obtain some precise and observable information of the system.

If a set of operators $L^k_\rho(x)$ of mechanical quantities transforms as follows in the symmetric transformations P_R,

$$P_R L^k_\rho(x) P_R^{-1} = \sum_\lambda L^k_\lambda(x) D^k_{\lambda\rho}(R), \tag{3.103}$$

$L^k_\rho(x)$ are called the **irreducible tensor operators**, then,

$$P_R L^k_\rho(x)\psi^j_\mu(x) = \sum_{\lambda\tau} L^k_\lambda(x)\psi^j_\tau \left[D^k(R)\times D^j(R)\right]_{\lambda\tau,\rho\mu}. \tag{3.104}$$

$L^k_\rho(x)\psi^j_\mu(x)$ can be combined by the Clebsch–Gordan coefficients such that the combination function $F^J_{Mr}(x)$ belongs to the Mth row of D^J:

$$\begin{aligned} F^J_{Mr}(x) &= \sum_{\rho\mu} L^k_\rho(x)\psi^j_\mu(x) C^{kj}_{\rho\mu,JMr}, \\ P_R F^J_{Mr}(x) &= \sum_{M'} F^J_{M'r}(x) D^J_{M'M}(R), \\ L^k_\rho(x)\psi^j_\mu(x) &= \sum_{JMr} F^J_{Mr}(x)\left[\left(C^{kj}\right)^{-1}\right]_{JMr,\rho\mu}. \end{aligned} \tag{3.105}$$

There are $(m_{j'}m_k m_j)$ matrix elements $\langle\phi^{j'}_\nu(x)|L^k_\rho(x)|\psi^j_\mu(x)\rangle$ for the mechanical quantities $L^k_\rho(x)$ between two static wave functions $\phi^{j'}_\nu(x)$ and $\psi^j_\mu(x)$. The Wigner–Eckart Theorem greatly simplifies the calculation problems,

$$\begin{aligned} \langle\phi^{j'}_\nu(x)|L^k_\rho(x)|\psi^j_\mu(x)\rangle &= \sum_{JMr} \langle\phi^{j'}_\nu(x)|F^J_{Mr}(x)\rangle\left[\left(C^{kj}\right)^{-1}\right]_{JMr,\rho\mu} \\ &= \sum_r \langle\phi^{j'}||L^k||\psi^j\rangle_r\left[\left(C^{kj}\right)^{-1}\right]_{j'\nu r,\rho\mu}. \end{aligned} \tag{3.106}$$

Those matrix elements are expressed as the combination of the products of the Clebsch–Gordan coefficients and the reduced matrix elements $\langle\phi^{j'}||L^k||\psi^j\rangle_r$. The Clebsch–Gordan coefficients describe the property of those matrix elements related with the symmetry of the system. The reduced matrix elements, which are independent of the row indices ν, ρ, and μ, are related with the detail of the system. **The number of the reduced matrix elements is equal to the multiplicity of the irreducible representation $D^{j'}$ in the direct product representation $D^k \times D^j$.**

3.6.4 *Normal Degeneracy and Accidental Degeneracy*

Let $D(\mathrm{G})$ denote the representation of the symmetry group G, corresponding to the energy level E of the original Hamiltonian $H_0(x)$. The degeneracy for the energy E is called **normal** if $D(\mathrm{G})$ is irreducible and called **accidental** if $D(\mathrm{G})$ is reducible.

We begin with the energy level of the original system with normal degeneracy. Introduce a "symmetry perturbation" $\lambda H_1(x)$ which does not disturb the symmetry group of $H_0(x)$. Namely, $H_0(x)$ and $H_1(x)$ both commute with the symmetric transformation operators P_R:

$$[P_R,\ H_0(x)] = 0, \qquad [P_R,\ H_1(x)] = 0. \tag{3.107}$$

Let $\psi^j_\mu(x)$ denote the eigenfunctions of $H_0(x)$ with E belonging to the μth row of an irreducible representation $D^j(\mathrm{G})$:

$$P_R\psi^j_\mu(x) = \sum_\nu^m \psi^j_\nu(x)D^j_{\nu\mu}(R), \qquad 1 \leqslant \mu \leqslant m. \tag{3.108}$$

For the first approximation, the degeneracy does not split, and the energy shift ΔE^j is calculated by

$$\lambda\langle\psi^j_\nu(x)|H_1(x)|\psi^j_\mu(x)\rangle = \delta_{\nu\mu}\left(\Delta E^j\right). \tag{3.109}$$

It is a non-perturbation conclusion. The degeneracy does not split even in arbitrarily high approximation. In fact, let the perturbation is introduced continuously as the parameter λ increases from 0 to 1. The wave function $\psi^j_\mu(x)$, as well as any physical observable, changes continuously, too. Due to the Wigner-Eckart theorem, the functions belonging to inequivalent irreducible representations are orthogonal to each other. Since the symmetry does not change as λ increases, the wave function cannot change to that belonging to other irreducible representation of G suddenly. So that as λ increases continuously, the wave functions belong to the same irreducible representation of G, and the degeneracy is left invariant. **A symmetry perturbation cannot split an energy with a normal degeneracy.**

Now, we turn to an energy level E of the original Hamiltonian with an accidental degeneracy, where the representation corresponding to E is reducible, but does not contain an irreducible representation with the multiplicity $a_j > 1$. Under the symmetry perturbation, the energy splits only between eigenfunctions belonging to the inequivalent irreducible representations, and **the energies of eigenfunctions belonging to one irreducible representation only make the same shift as a whole**:

$$\lambda\langle\psi_{\nu}^{k}(x)|H_1(x)|\psi_{\mu}^{j}(x)\rangle = \delta_{kj}\delta_{\nu\mu}\left(\Delta E^{j}\right). \tag{3.110}$$

This conclusion is also non-perturbative. If the representation corresponding to E contains an irreducible representation $D^j(\mathrm{G})$ with the multiplicity $a_j > 1$, the sets of eigenfunctions belonging to $D^j(\mathrm{G})$ are combined

$$\lambda\langle\psi_{\nu r}^{k}(x)|H_1(x)|\psi_{\mu s}^{j}(x)\rangle = \delta_{kj}\delta_{\nu\mu}\left(\Delta E^{j}\right)_{rs}, \tag{3.111}$$

where $r,\ s = 1, 2, \ldots, a_j$. By making use of the symmetry of the system, the method of group theory greatly simplifies the calculation.

If the original Hamiltonian $H_0(x)$ and the perturbation Hamiltonian $H_1(x)$ have different symmetries, one may choose the common symmetric transformations of $H_0(x)$ and $H_1(x)$ to be the symmetry group of the system so that the perturbation Hamiltonian $H_1(x)$ is the "symmetry perturbation". It is common viewpoint that the energy level of the original system is normal degeneracy if G contains all symmetric transformations of the system. The accidental degeneracy is related to the existence of some undiscovered symmetric transformations of the system [Zou and Huang (1995)].

3.6.5 *Example of Physical Application*

We are going to raise a physical example to demonstrate the steps of application of group theory to physics. Discuss a quantum system with a square well potential in two dimensions. The Hamiltonian equation is ($\hbar = 2m = 1$)

$$\begin{aligned} H(x,y)\psi(x,y) &= -\frac{d^2\psi(x,y)}{dx^2} - \frac{d^2\psi(x,y)}{dy^2} + V(x,y)\psi(x,y) \\ &= E\psi(x,y), \end{aligned} \tag{3.112}$$
$$V(x,y) = \begin{cases} 0, & \text{when } |x| < \pi, \quad |y| < \pi, \\ \infty, & \text{the remaining cases.} \end{cases}$$

First, it is evident that the symmetry group of the system with the square well potential in two dimensions is the group D_4. The generators of D_4 are the 4-fold rotation C_4 about the Z-axis and the 2-fold rotation $C_2^{(0)}$ about the X-axis. As given in Eqs. (3.87) and (3.88), the inequivalent irreducible representations of D_4 are

$$\begin{aligned}
&D^{A_1}(C_4) = D^{A_2}(C_4) = 1, && D^{A_1}(C_2^{(0)}) = -D^{A_2}(C_2^{(0)}) = 1, \\
&D^{B_1}(C_4) = D^{B_2}(C_4) = -1, && D^{B_1}(C_2^{(0)}) = -D^{B_2}(C_2^{(0)}) = 1, \\
&D^{E}(C_4) = \begin{pmatrix} 0 & -1 \\ 1 & 0 \end{pmatrix}, && D^{E}(C_2^{(0)}) = \begin{pmatrix} 1 & 0 \\ 0 & -1 \end{pmatrix}.
\end{aligned} \tag{3.113}$$

The representation matrices of $D^E(\mathrm{D}_4)$ coincide with the matrices in coordinate transformation:

$$\begin{pmatrix} x' \\ y' \end{pmatrix} = D^E(R) \begin{pmatrix} x \\ y \end{pmatrix}, \qquad R \in \mathrm{D}_4.$$

Second, by separation of variables, $\psi(x, y) = X(x)Y(y)$, the Hamiltonian equation becomes

$$\begin{aligned}
&X'' + E_1 X = 0, \qquad Y'' + E_2 Y = 0, \qquad |x| \leqslant \pi, \qquad |y| \leqslant \pi, \\
&X(\pm\pi) = Y(\pm\pi) = 0, \qquad E = E_1 + E_2.
\end{aligned}$$

The solutions are

$$X(x) = \begin{cases} \sin(mx), & E_1 = m^2, \\ \cos\left[(2m-1)x/2\right], & E_1 = (2m-1)^2/4, \end{cases}$$

$$Y(y) = \begin{cases} \sin(ny), & E_2 = n^2, \\ \cos\left[(2n-1)y/2\right], & E_2 = (2n-1)^2/4, \end{cases}$$

where m and n are both positive integers. There are five kinds of energy levels with the following eigenfunctions ($m \neq n$):

(1) $E = (2m-1)^2/2$, $\psi = \cos\left[(2m-1)x/2\right] \cos\left[(2m-1)y/2\right]$.

(2) $E = 2m^2$, $\psi = \sin(mx)\sin(my)$. Usually, we do not study the state with $\psi = 0$.

(3) $E = m^2 + n^2$, $\psi_1 = \sin(mx)\sin(ny)$, $\psi_2 = \sin(nx)\sin(my)$.

(4) $E = (2m-1)^2/4 + (2n-1)^2/4$,

$$\begin{aligned}
\psi_1 &= \cos\left[(2m-1)x/2\right] \cos\left[(2n-1)y/2\right], \\
\psi_2 &= \cos\left[(2n-1)x/2\right] \cos\left[(2m-1)y/2\right].
\end{aligned}$$

(5) $E = m^2 + (2n-1)^2/4$,

$$\begin{aligned}
\psi_1 &= \sin(mx) \cos\left[(2n-1)y/2\right], \\
\psi_2 &= \cos\left[(2n-1)x/2\right] \sin(my).
\end{aligned}$$

In order to determine the representation corresponding to the energy in the above five cases, one calculates the matrices of C_4 and $C_2^{(0)}$ in those basis functions by Eq. (3.91).

(1) $P_{C_4}\psi = P_{C_2^{(0)}}\psi = \psi$, $D(C_4) = D(C_2^{(0)}) = 1$, and the representation is D^{A_1}.

(2) $P_{C_4}\psi = P_{C_2^{(0)}}\psi = -\psi$, $D(C_4) = D(C_2^{(0)}) = -1$, and the representation is D^{B_2}.

(3) $P_{C_4}\psi_1 = -\psi_2$, $P_{C_4}\psi_2 = -\psi_1$, $P_{C_2^{(0)}}\psi_1 = -\psi_1$, $P_{C_2^{(0)}}\psi_2 = -\psi_2$,

$$D(C_4) = \begin{pmatrix} 0 & -1 \\ -1 & 0 \end{pmatrix}, \qquad D(C_2^{(0)}) = \begin{pmatrix} -1 & 0 \\ 0 & -1 \end{pmatrix}.$$

Due to Eq. (3.93), the representation is equivalent to $D^{A_2} \oplus D^{B_2}$. $D(C_2^{(0)})$ is a constant matrix and $D(C_4)$ can be diagonalized as

$$X^{-1}D(C_4)X = \begin{pmatrix} 1 & 0 \\ 0 & -1 \end{pmatrix}, \qquad X = \frac{1}{\sqrt{2}}\begin{pmatrix} 1 & 1 \\ -1 & 1 \end{pmatrix},$$

$$\Phi_1 = (\psi_1 - \psi_2)/\sqrt{2}, \qquad \Phi_2 = (\psi_1 + \psi_2)/\sqrt{2}.$$

Φ_1 belongs to D^{A_2} and Φ_2 belongs to D^{B_2}.

(4) $P_{C_4}\psi_1 = \psi_2$, $P_{C_4}\psi_2 = \psi_1$, $P_{C_2^{(0)}}\psi_1 = \psi_1$, $P_{C_2^{(0)}}\psi_2 = \psi_2$,

$$D(C_4) = \begin{pmatrix} 0 & 1 \\ 1 & 0 \end{pmatrix}, \qquad D(C_2^{(0)}) = \begin{pmatrix} 1 & 0 \\ 0 & 1 \end{pmatrix}.$$

Due to Eq. (3.93), the representation is equivalent to $D^{A_1} \oplus D^{B_1}$.

$$Y^{-1}D(C_4)Y = \begin{pmatrix} 1 & 0 \\ 0 & -1 \end{pmatrix}, \qquad Y = \frac{1}{\sqrt{2}}\begin{pmatrix} 1 & 1 \\ 1 & -1 \end{pmatrix},$$

$$\Phi_1 = (\psi_1 + \psi_2)/\sqrt{2}, \qquad \Phi_2 = (\psi_1 - \psi_2)/\sqrt{2}.$$

Φ_1 belongs to D^{A_1} and Φ_2 belongs to D^{B_1}.

(5) $P_{C_4}\psi_1 = \psi_2$, $P_{C_4}\psi_2 = -\psi_1$, $P_{C_2^{(0)}}\psi_1 = \psi_1$, $P_{C_2^{(0)}}\psi_2 = -\psi_2$,

$$D(C_4) = \begin{pmatrix} 0 & -1 \\ 1 & 0 \end{pmatrix}, \qquad D(C_2^{(0)}) = \begin{pmatrix} 1 & 0 \\ 0 & -1 \end{pmatrix}.$$

This is the irreducible representation D^E.

Among five kinds of energy levels, the energy levels (1), (2), and (5) are normal degeneracies, and the energy levels (3) and (4) are accidental degeneracies. For some special energies the multiplicity of degeneracy may be higher. For example, the energy $E = 65$ in case (3) with $m = 1$, $n = 8$ is degeneracy with $m = 4$, $n = 7$ of case (3). $E = 50$ in case (3) with $m = 1$ and $n = 7$ is degeneracy with $m = 5$ of case (2). The corresponding representation is the direct sum of representations in two cases. There is no new difficulty in analyzing those energy levels with higher degeneracy.

At last, we discuss the shift and split of energy in the action of a symmetry perturbation Hamiltonian $H_1 = \varepsilon x^2y^2$. The group D_4 is the common symmetry group for both H_0 and H_1. Under the symmetry perturbation H_1, the energy of normal degeneracy makes a shift but does not split. For example, in case (5)

$$\begin{aligned}
\langle \psi_1|H_1|\psi_1\rangle &= \langle \psi_2|H_1|\psi_2\rangle = \varepsilon\left\{\frac{\pi^3}{3} - \frac{\pi}{2m^2}\right\}\left\{\frac{\pi^3}{3} - \frac{2\pi}{(2n+1)^2}\right\}, \\
\langle \psi_1|H_1|\psi_2\rangle &= \langle \psi_2|H_1|\psi_1\rangle = 0 .
\end{aligned}$$

There is only one matrix element that needs to be calculated.

The energy of accidental degeneracy will split under the symmetry perturbation H_1. For the cases (3) and (4), we have chosen the basis functions Φ_μ by group theory such that they belong to the irreducible representations. In those basis functions H_1 is diagonalized.

$$\begin{aligned}
\langle \Phi_1|H_1|\Phi_1\rangle &= \varepsilon\left\{\frac{\pi^3}{3} - \frac{\pi}{2m^2}\right\}\left\{\frac{\pi^3}{3} - \frac{\pi}{2n^2}\right\} \\
&\quad - \varepsilon(-1)^{m-n}\left\{\frac{2\pi}{(m-n)^2} - \frac{2\pi}{(m+n)^2}\right\}^2, \\
\langle \Phi_2|H_1|\Phi_2\rangle &= \varepsilon\left\{\frac{\pi^3}{3} - \frac{\pi}{2m^2}\right\}\left\{\frac{\pi^3}{3} - \frac{\pi}{2n^2}\right\} \\
&\quad + \varepsilon(-1)^{m-n}\left\{\frac{2\pi}{(m-n)^2} - \frac{2\pi}{(m+n)^2}\right\}^2, \\
\langle \Phi_1|H_1|\Phi_2\rangle &= \langle \Phi_2|H_1|\Phi_1\rangle = 0.
\end{aligned}$$

In comparison with the basis function ψ_μ, the calculation in the basis functions Φ_μ is simplified.

3.7 Irreducible Bases in Group Algebra

3.7.1 *Decomposition of Regular Representation*

The group algebra $\mathcal{L}$ of a finite group G of order g is left invariant under left- and right-multiplication with elements of G. Usually, one takes the elements of G to be the basis vectors in $\mathcal{L}$, called the natural bases. The representation in $\mathcal{L}$ with respect to the natural bases is the regular representation, shown in Eqs. (3.6) and (3.12). The regular representation of G is reducible, where the multiplicity of each irreducible representation $D^j(\mathrm{G})$ is its dimension m_j (see Eq. (3.61)).

The Schur theorem shows that the representation matrix elements $D^j_{\mu\nu}(R)$ of all inequivalent irreducible representations of G construct a complete set of orthogonal basis vectors in $\mathcal{L}$ (see Corollary 3.4.1). Define a new set of basis vectors in $\mathcal{L}$:

$$\phi^j_{\mu\nu} = \frac{m_j}{g} \sum_{R\in G} D^j_{\mu\nu}(R)^* R. \tag{3.114}$$

From Theorem 2.1 and Eq. (3.53) we have

$$\begin{aligned}
S\phi^j_{\mu\nu} &= \frac{m_j}{g} \sum_{R\in G} D^j_{\mu\nu}(R)^* SR = \frac{m_j}{g} \sum_{T\in G} D^j_{\mu\nu}(S^{-1}T)^* T \\
&= \frac{m_j}{g} \sum_{\rho} D^j_{\rho\mu}(S) \left[\sum_{T\in G} D^j_{\rho\nu}(T)^* T \right] = \sum_{\rho} \phi^j_{\rho\nu} D^j_{\rho\mu}(S), \\
\phi^j_{\mu\nu} S &= \frac{m_j}{g} \sum_{R\in G} D^j_{\mu\nu}(R)^* RS = \frac{m_j}{g} \sum_{T\in G} D^j_{\mu\nu}(TS^{-1})^* T \\
&= \frac{m_j}{g} \sum_{\rho} \left[\sum_{T\in G} D^j_{\mu\rho}(T)^* T \right] D^j_{\nu\rho}(S) = \sum_{\rho} D^j_{\nu\rho}(S) \phi^j_{\mu\rho}, \\
\phi^j_{\mu\rho}\phi^k_{\tau\nu} &= \frac{m_j m_k}{g^2} \sum_{R\in G} \sum_{S\in G} D^j_{\mu\rho}(R)^* D^j_{\tau\nu}(S)^* RS \\
&= \frac{m_j m_k}{g^2} \sum_{R\in G} \sum_{T\in G} D^j_{\mu\rho}(R)^* D^k_{\tau\nu}(R^{-1}T)^* T \\
&= \frac{m_j m_k}{g^2} \sum_{\lambda} \left[\sum_{R\in G} D^j_{\mu\rho}(R)^* D^k_{\lambda\tau}(R) \right] \sum_{T\in G} D^k_{\lambda\nu}(T)^* T \\
&= \delta_{jk}\delta_{\rho\tau}\phi^j_{\mu\nu}.
\end{aligned}$$

Thus, $\phi^j_{\mu\nu}$ **belongs to the irreducible representation** $D^j(\mathrm{G})$ for left- and right-multiplication with elements:

$$S\phi^j_{\mu\nu} = \sum_{\rho} \phi^j_{\rho\nu} D^j_{\rho\mu}(S), \qquad \phi^j_{\mu\nu} S = \sum_{\rho} D^j_{\nu\rho}(S) \phi^j_{\mu\rho}, \tag{3.115}$$

and satisfies the **recurrence relation**:

$$\phi^j_{\mu\rho}\phi^k_{\tau\nu} = \delta_{jk}\delta_{\rho\tau}\phi^j_{\mu\nu}. \tag{3.116}$$

It means that $\phi^j_{\mu\mu}$ is a projective operator:

$$\phi^j_{\mu\mu}\phi^k_{\nu\nu} = \delta_{jk}\delta_{\mu\nu}\phi^j_{\mu\mu}, \qquad \phi^j_{\mu\nu} = \phi^j_{\mu\mu}\phi^j_{\mu\nu} = \phi^j_{\mu\nu}\phi^j_{\nu\nu}. \tag{3.117}$$

Those basis vectors $\phi^j_{\mu\nu}$, satisfying Eqs. (3.115) and (3.116), are called **the irreducible bases in the group algebra $\mathcal{L}$**, where the regular representation is completely reduced. **The irreducible bases $\phi^j_{\mu\nu}$ can be calculated from $D^j(\mathrm{G})$** by Eq. (3.114). Conversely, **$D^j(\mathrm{G})$ can be calculated from $\phi^j_{\mu\nu}$**. The irreducible bases are determined by the irreducible representations up to a similarity transformation (see Eq. (4.90)).

The static wave function $\psi_\rho(x)$ of a quantum system can be combined into those belonging to the irreducible representations $D^j(\mathrm{G})$ of its symmetry group G. If $\phi^j_{\mu\nu}\psi_\rho(x)$ is non-vanishing, $\psi_\rho(x)$ is projected into the functions belonging to the irreducible representation $D^j(\mathrm{G})$:

$$P_S\left[\phi^j_{\mu\nu}\psi_\rho(x)\right] = \sum_\tau \left[\phi^j_{\tau\nu}\psi_\rho(x)\right] D^j_{\tau\mu}(S), \tag{3.118}$$

where the symmetric transformation R in Eq. (3.114) for $\phi^j_{\mu\nu}$ should be replaced by the symmetric operator P_R. The calculation for the decomposition of $\psi_\rho(x)$ is greatly simplified.

3.7.2 *Irreducible Bases of* C_N *and* D_N

The irreducible bases ϕ^j for the cyclic group C_N are calculated from its irreducible representations given in Eq. (3.70),

$$e^{(j)} = \phi^j = \frac{1}{N}\sum_{m=0}^{N-1} \mathrm{e}^{\mathrm{i}2mj\pi/N}\left(C_N\right)^m. \tag{3.119}$$

This $e^{(j)}$ is a projective operator such that $e^{(j)}\psi(x)$ is an eigenfunction of C_N with the eigenvalue $\mathrm{e}^{-\mathrm{i}2j\pi/N}$ (see Eqs. (3.129) and (5.99)).

The irreducible bases ϕ^A, ϕ^B, and $\phi^{E_j}_{\mu\nu}$, $1 \leqslant j \leqslant n$, for the group D_{2n+1} are calculated from Eqs. (3.85) and (3.86),

$$\begin{aligned}
\phi^A &= \frac{1}{4n+2}\sum_{m=0}^{2n}\left[\left(C_{2n+1}\right)^m + C_2^{(m)}\right], \\
\phi^B &= \frac{1}{4n+2}\sum_{m=0}^{2n}\left[\left(C_{2n+1}\right)^m - C_2^{(m)}\right], \\
\phi^{E_j}_{11} &= \frac{1}{2n+1}\sum_{m=0}^{2n}\cos\left(\frac{2mj\pi}{2n+1}\right)\left[\left(C_{2n+1}\right)^m + C_2^{(m)}\right],
\end{aligned}$$

$$\begin{aligned}
\phi_{22}^{E_j} &= \frac{1}{2n+1}\sum_{m=0}^{2n} \cos\left(\frac{2mj\pi}{2n+1}\right)\left[(C_{2n+1})^m - C_2^{(m)}\right], \\
\phi_{12}^{E_j} &= \frac{1}{2n+1}\sum_{m=0}^{2n} \sin\left(\frac{2mj\pi}{2n+1}\right)\left[-(C_{2n+1})^m + C_2^{(m)}\right], \\
\phi_{21}^{E_j} &= \frac{1}{2n+1}\sum_{m=0}^{2n} \sin\left(\frac{2mj\pi}{2n+1}\right)\left[(C_{2n+1})^m + C_2^{(m)}\right],
\end{aligned} \tag{3.120}$$

The irreducible bases ϕ^{A_a}, ϕ^{B_a}, $a = 1,\ 2$, and $\phi_{\mu\nu}^{E_j}$, $1 \leqslant j \leqslant n-1$, for the group D_{2n} are calculated from Eqs. (3.87) and (3.88):

$$\begin{aligned}
\phi^{A_1} &= \frac{1}{4n}\sum_{m=0}^{n-1}\left[(C_{2n})^{2m} + (C_{2n})^{2m+1} + C_2^{(2m)} + C_2^{(2m+1)}\right], \\
\phi^{A_2} &= \frac{1}{4n}\sum_{m=0}^{n-1}\left[(C_{2n})^{2m} + (C_{2n})^{2m+1} - C_2^{(2m)} - C_2^{(2m+1)}\right], \\
\phi^{B_1} &= \frac{1}{4n}\sum_{m=0}^{n-1}\left[(C_{2n})^{2m} - (C_{2n})^{2m+1} + C_2^{(2m)} - C_2^{(2m+1)}\right], \\
\phi^{B_2} &= \frac{1}{4n}\sum_{m=0}^{n-1}\left[(C_{2n})^{2m} - (C_{2n})^{2m+1} - C_2^{(2m)} + C_2^{(2m+1)}\right], \\
\phi_{11}^{E_j} &= \frac{1}{2n}\sum_{m=0}^{2n-1} \cos\left(\frac{mj\pi}{n}\right)\left[(C_{2n})^m + C_2^{(m)}\right], \\
\phi_{22}^{E_j} &= \frac{1}{2n}\sum_{m=0}^{2n-1} \cos\left(\frac{mj\pi}{n}\right)\left[(C_{2n})^m - C_2^{(m)}\right], \\
\phi_{12}^{E_j} &= \frac{1}{2n}\sum_{m=0}^{2n-1} \sin\left(\frac{mj\pi}{n}\right)\left[-(C_{2n})^m + C_2^{(m)}\right], \\
\phi_{21}^{E_j} &= \frac{1}{2n}\sum_{m=0}^{2n-1} \sin\left(\frac{mj\pi}{n}\right)\left[(C_{2n})^m + C_2^{(m)}\right].
\end{aligned} \tag{3.121}$$

3.7.3 *Irreducible Bases of* **T** *and* **O**

The proper symmetry groups of a cube and a tetrahedron is **O** and **T**, respectively. We have known in subsection 2.5.2 that the group **O** contains 24 elements and five self-reciprocal classes: $\mathcal{C}_1 = \{E\}$, $\mathcal{C}_2 = \{S_k; 1 \leqslant k \leqslant 6\}$, $\mathcal{C}_3 = \{R_j^{\pm 1}; 1 \leqslant j \leqslant 4\}$, $\mathcal{C}_4 = \{T_\mu^{\pm 1}; \mu = x,\ y,\ z\}$, and $\mathcal{C}_5 = \{T_\mu^2; \mu = x,\ y,\ z\}$. The class $\mathcal{C}_3$ in **O** is divided into two reciprocal classes of **T**: $\mathcal{C}_3 = \{R_j; 1 \leqslant j \leqslant 4\}$ and $\mathcal{C}_3^{(-1)} = \{R_j^{-1}; 1 \leqslant j \leqslant 4\}$, so that the group

T contains 12 elements and four classes: $\mathcal{C}_1$, $\mathcal{C}_5$, $\mathcal{C}_3$, and $\mathcal{C}_3^{(-1)}$. **T** is the invariant subgroup of **O** with index two. The set of two classes $\mathcal{C}_1$ and $\mathcal{C}_5$ forms an invariant subgroup D_2 of both **T** and **O**.

Since $1^2 + 1^2 + 1^2 + 3^2 = 12$, the group **T** has three inequivalent representations D^A, D^E, and $D^{E'}$ of one dimension and one irreducible representation D^T of three dimensions. Three representations of one dimension are calculated from the quotient group $\mathbf{T}/\mathrm{D}_2 \approx \mathrm{C}_3$ (see Table 3.2). Due to Eq. (3.68), $|\chi^T(R_j)|^2 = 12/4 - |\chi^A(R_j)|^2 - |\chi^E(R_j)|^2 - |\chi^{E'}(R_j)|^2 = 0$ and $\chi^T(R_j^{-1}) = 0$. Due to Eq. (3.55) $\chi^A(E)^*\chi^T(E) + 3\chi^A(T_z^2)^*\chi^T(T_z^2) = 1 \times 3 + 3 \times 1 \times \chi^T(T_z^2) = 0$, so that $\chi^T(T_z^2) = -1$. The character tables of **T** is listed in Table 3.17, where the convention for the representations in the theory of crystals is used.

Table 3.17 The character table of T ($\omega = \mathrm{e}^{-\mathrm{i}2\pi/3}$)

T	E	$3C_2$	$4C_3'$	$4C_3'^2$
A	1	1	1	1
E	1	1	ω	ω^2
E'	1	1	ω^2	ω
T	3	-1	0	0

Table 3.18 The character table of O

O	E	$3C_4^2$	$8C_3'$	$6C_4$	$6C_2''$
A	1	1	1	1	1
B	1	1	1	-1	-1
E	2	2	-1	0	0
T_1	3	-1	0	1	-1
T_2	3	-1	0	-1	1

Since $1^2+1^2+2^2+3^2+3^2 = 24$, the group **O** has two inequivalent representations D^A and D^B of one dimension, one irreducible representation D^E of two dimensions, and two inequivalent irreducible representation D^{T_1} and D^{T_2} of three dimensions. Two one-dimensional representations are those of the quotient group $\mathbf{O}/\mathbf{T} \approx \mathrm{C}_2$ (see Table 3.1). The two-dimensional representation is that of the quotient group $\mathbf{O}/\mathrm{D}_2 \approx \mathrm{D}_3$ (see Table 3.6). Equations (2.61) and (2.14) shows the representation matrices of $D^E(\mathbf{O})$. The subduced representation $D^{T_1}(\mathbf{T})$ of $D^{T_1}(\mathbf{O})$ cannot be reducible, otherwise, $\chi^{T_1}(T_z^2) = 3$ and violates Eq. (3.68): $3|\chi^{T_1}(T_z^2)|^2 = 27 > 24$. Thus, $D^{T_1}(\mathbf{T}) = D^T(\mathbf{T})$, and then, $\chi^{T_1}(T_z^2) = -1$, and $\chi^{T_1}(R_1) = 0$. Due to Eq. (3.55), $6|\chi^{T_1}(T_z)|^2 + 6|\chi^{T_1}(S_1)|^2 = 24 - (3)^2 - 3 \times (-1)^2 = 12 \neq 0$. Thus, $D^B(\mathbf{O}) \times D^{T_1}(\mathbf{O})$ is also an irreducible representation of three dimensions, inequivalent to $D^{T_1}(\mathbf{O})$. Namely, $D^{T_2}(\mathbf{O}) = D^B(\mathbf{O}) \times D^{T_1}(\mathbf{O})$, and $\chi^{T_1}(T_z) = -\chi^{T_2}(T_z)$, $\chi^{T_1}(S_1) = -\chi^{T_2}(S_1)$. Due to the orthogonality (3.55) between $\chi^B(\mathbf{O})$ and $\chi^{T_1}(\mathbf{O})$, $\chi^{T_1}(T_z) = -\chi^{T_1}(S_1)$. Without loss of generality, we choose $\chi^{T_1}(T_z) = 1$. The character table of $\mathbf{O}$ is listed in Table 3.18, where the convention in the theory of crystals is used.

The three-dimensional representations of $\mathbf{T}$ and $\mathbf{O}$ can be calculated by the method of coordinate transformation, similar to Eq. (2.12). The representation matrix $D^{T_1}(R)$ of each element R in $\mathbf{O}$ is denoted by a transformation from $(x,\ y,\ z)$ to $(x',\ y',\ z')$ in the three-dimensional coordinate system:

$$\begin{pmatrix} x' \\ y' \\ z' \end{pmatrix} = D^{T_1}(R) \begin{pmatrix} x \\ y \\ z \end{pmatrix}, \qquad \forall\, R \in \mathbf{O}. \tag{3.122}$$

According to the definitions (2.49) and (2.50) and by Fig. 2.2, one is able to determine the transformations of the three coordinate axes in the elements of $\mathbf{O}$ as shown in Table 3.19. From Table 3.19 the representations matrices $D^{T_1}(R)$ can be calculated. For example,

$$\begin{aligned}
&D^{T_1}(T_z) = \begin{pmatrix} 0 & -1 & 0 \\ 1 & 0 & 0 \\ 0 & 0 & 1 \end{pmatrix}, \qquad D^{T_1}(R_1) = \begin{pmatrix} 0 & 0 & 1 \\ 1 & 0 & 0 \\ 0 & 1 & 0 \end{pmatrix}, \\
&D^{T_1}(S_1) = \begin{pmatrix} 0 & 1 & 0 \\ 1 & 0 & 0 \\ 0 & 0 & -1 \end{pmatrix}, \qquad D^{T_1}(S_2) = \begin{pmatrix} 0 & -1 & 0 \\ -1 & 0 & 0 \\ 0 & 0 & -1 \end{pmatrix}, \\
&D^{T_2}(R) = D^B(R) \times D^{T_1}(R), \qquad \text{for} \ \ R \in \mathbf{O}.
\end{aligned} \tag{3.123}$$

The remaining are left as exercise. Note that $\chi^{T_1}(T_z) = 1$ and $\chi^{T_1}(T_z^2) = -1$. Then, the subduced representation of $D^{T_1}(\mathbf{O})$ is $D^T(\mathbf{T})$.

Table 3.19 Transformations of coordinate axes in elements of O

	T_x	T_y	T_z	R_1	R_1	R_1	R_4	S_1	S_2	S_3	S_4	S_5	S_6
X	X	$-Z$	Y	Y	$-Y$	Y	$-Y$	Y	$-Y$	$-X$	$-X$	Z	$-Z$
Y	Z	Y	$-X$	Z	Z	$-Z$	$-Z$	X	$-X$	Y	$-Y$	$-Y$	$-Y$
Z	$-Y$	X	Z	X	$-X$	$-X$	X	$-Z$	$-Z$	Z	$-Z$	X	$-X$

In terms of Eqs. (3.123) and (3.114) one is able to calculate the irreducible bases for $\mathbf{T}$ and $\mathbf{O}$. In the following we only list the results and left the calculating process as exercise (Note Eqs. (2.55) and (2.56)).

Define the brief symbols due to Eq. (2.61):

$$\begin{aligned} &\mathcal{D}_0 \equiv E + T_x^2 + T_y^2 + T_z^2, &&\mathcal{D}_1 \equiv R_1 + R_2 + R_3 + R_4, \\ &\mathcal{D}_2 \equiv R_1^2 + R_2^2 + R_3^2 + R_4^2, &&\mathcal{D}_3 \equiv S_1 + S_2 + T_z + T_z^3, \\ &\mathcal{D}_4 \equiv S_5 + S_6 + T_y + T_y^3, &&\mathcal{D}_5 \equiv S_3 + S_4 + T_x + T_x^3. \end{aligned} \tag{3.124}$$

The irreducible bases for the group **T** are

$$\begin{aligned} &\phi^A = [\mathcal{D}_0 + \mathcal{D}_1 + \mathcal{D}_2]/12, &&\phi^E = [\mathcal{D}_0 + \omega^2\mathcal{D}_1 + \omega\mathcal{D}_2]/12, \\ &\phi^{E'} = [\mathcal{D}_0 + \omega\mathcal{D}_1 + \omega^2\mathcal{D}_2]/12, &&\phi_{11}^T = [E + T_x^2 - T_y^2 - T_z^2]/4, \\ &\phi_{22}^T = [E - T_x^2 + T_y^2 - T_z^2]/4, &&\phi_{33}^T = [E - T_x^2 - T_y^2 + T_z^2]/4, \\ &\phi_{12}^T = [R_1^2 - R_2^2 + R_3^2 - R_4^2]/4, &&\phi_{21}^T = [R_1 - R_2 + R_3 - R_4]/4, \\ &\phi_{23}^T = [R_1^2 + R_2^2 - R_3^2 - R_4^2]/4, &&\phi_{32}^T = [R_1 + R_2 - R_3 - R_4]/4, \\ &\phi_{31}^T = [R_1^2 - R_2^2 - R_3^2 + R_4^2]/4, &&\phi_{13}^T = [R_1 - R_2 - R_3 + R_4]/4. \end{aligned} \tag{3.125}$$

The irreducible bases for the group **O** are

$$\begin{aligned} &\phi^A = [\mathcal{D}_0 + \mathcal{D}_1 + \mathcal{D}_2 + \mathcal{D}_3 + \mathcal{D}_4 + \mathcal{D}_5]/24, \\ &\phi^B = [\mathcal{D}_0 + \mathcal{D}_1 + \mathcal{D}_2 - \mathcal{D}_3 - \mathcal{D}_4 - \mathcal{D}_5]/24, \\ &\phi_{11}^E = [2\mathcal{D}_0 - \mathcal{D}_1 - \mathcal{D}_2 + 2\mathcal{D}_3 - \mathcal{D}_4 - \mathcal{D}_5]/24, \\ &\phi_{22}^E = [2\mathcal{D}_0 - \mathcal{D}_1 - \mathcal{D}_2 - 2\mathcal{D}_3 + \mathcal{D}_4 + \mathcal{D}_5]/24, \\ &\phi_{12}^E = [-\mathcal{D}_1 + \mathcal{D}_2 + \mathcal{D}_4 - \mathcal{D}_5]/(8\sqrt{3}), \\ &\phi_{21}^E = [\mathcal{D}_1 - \mathcal{D}_2 + \mathcal{D}_4 - \mathcal{D}_5]/(8\sqrt{3}), \\ &\phi_{11}^{T_a} = \left[(E + T_x^2 - T_y^2 - T_z^2) \pm (T_x + T_x^3 - S_3 - S_4)\right]/8, \\ &\phi_{22}^{T_a} = \left[(E - T_x^2 + T_y^2 - T_z^2) \pm (T_y + T_y^3 - S_5 - S_6)\right]/8, \\ &\phi_{33}^{T_a} = \left[(E - T_x^2 - T_y^2 + T_z^2) \pm (T_z + T_z^3 - S_1 - S_2)\right]/8, \\ &\phi_{12}^{T_a} = \left[(R_1^2 - R_2^2 + R_3^2 - R_4^2) \pm (-T_z + T_z^3 + S_1 - S_2)\right]/8, \\ &\phi_{21}^{T_a} = \left[(R_1 - R_2 + R_3 - R_4) \pm (T_z - T_z^3 + S_1 - S_2)\right]/8, \\ &\phi_{23}^{T_a} = \left[(R_1^2 + R_2^2 - R_3^2 - R_4^2) \pm (-T_x + T_x^3 + S_3 - S_4)\right]/8, \\ &\phi_{32}^{T_a} = \left[(R_1 + R_2 - R_3 - R_4) \pm (T_x - T_x^3 + S_3 - S_4)\right]/8, \\ &\phi_{31}^{T_a} = \left[(R_1^2 - R_2^2 - R_3^2 + R_4^2) \pm (-T_y + T_y^3 + S_5 - S_6)\right]/8, \\ &\phi_{13}^{T_a} = \left[(R_1 - R_2 - R_3 + R_4) \pm (T_y - T_y^3 + S_5 - S_6)\right]/8, \end{aligned} \tag{3.126}$$

where $a = 1, 2$, the upper sign is for the representation D^{T_1}, and the lower sign is for D^{T_2}.

The calculation of the irreducible bases by Eq. (3.114) needs the representation matrices of all elements in G, so that it becomes complicated for a group with high order, say **I**. In terms of the projective operator $e^{(j)}$

given in Eq. (3.119), the calculation can be greatly simplified (see subsection 5.4.6). In the following we demonstrate this method by calculating the irreducible bases for the group **O**.

The dimensions of the irreducible representations of **O** are not larger than three so that the row (column) can be enumerated by the exponent μ of the eigenvalue ω^μ of a 3-fold rotation R_1, where $\omega = \mathrm{e}^{-\mathrm{i}2\pi/3} = \overline{\omega}^{-1}$. Diagonalize $D^E(R_1)$ and $D^{T_1}(R_1)$ (see Eqs. (2.14), (2.61)) and (3.123)):

$$X = \frac{1}{\sqrt{2}}\begin{pmatrix} 1 & \mathrm{i} \\ \mathrm{i} & 1 \end{pmatrix}, \quad \overline{D}^E(R_1) = X^{-1}\frac{1}{2}\begin{pmatrix} -1 & -\sqrt{3} \\ \sqrt{3} & -1 \end{pmatrix} X = \begin{pmatrix} \omega & 0 \\ 0 & \overline{\omega} \end{pmatrix},$$
$$Y = \frac{1}{\sqrt{3}}\begin{pmatrix} \overline{\omega} & 1 & \omega \\ \omega & 1 & \overline{\omega} \\ 1 & 1 & 1 \end{pmatrix}, \quad \overline{D}^{T_1}(R_1) = Y^{-1}\begin{pmatrix} 0 & 0 & 1 \\ 1 & 0 & 0 \\ 0 & 1 & 0 \end{pmatrix} Y = \begin{pmatrix} \omega & 0 & 0 \\ 0 & 1 & 0 \\ 0 & 0 & \overline{\omega} \end{pmatrix}. \tag{3.127}$$

The row (column) index of $D^j(\mathbf{O})$ is taken as: 0 for $D^A(\mathbf{O})$ and $D^B(\mathbf{O})$; 1 and $\overline{1}$ for $D^E(\mathbf{O})$; 1, 0, and $\overline{1}$ for $D^{T_1}(\mathbf{O})$ and $D^{T_2}(\mathbf{O})$, where $\overline{\mu} = -\mu$.

The irreducible bases $\phi^j_{\mu\nu}$, defined in Eq. (3.114), satisfy

$$R_1\phi^j_{\mu\nu} = \omega^\mu \phi^j_{\mu\nu}, \qquad \phi^j_{\mu\nu}R_1 = \omega^\nu \phi^j_{\mu\nu}. \tag{3.128}$$

Introduce the projective operator $e^{(\mu)}$ of the cyclic group C_3 (see Eq. (3.119)),

$$e^{(\mu)} = \frac{1}{3}\sum_{\rho=0}^{2} \omega^{-\rho\mu} R_1^\rho, \qquad \phi^j_{\mu\nu} = e^{(\mu)}\phi^j_{\mu\nu} = \phi^j_{\mu\nu}e^{(\nu)}. \tag{3.129}$$

Following Eq. (3.128), we define the vectors $\Phi^{(t)}_{\mu\nu}$ in the group algebra of **O** by $\Phi^{(t)}_{\mu\nu} \propto e^{(\mu)}R^{(t)}e^{(\nu)}$, where $R^{(t)} \in \mathbf{O}$ are chosen to make $\Phi^{(t)}_{\mu\nu}$ linearly independent. In this way, $\phi^j_{\mu\nu}$ **are the combinations of** $\Phi^{(t)}_{\mu\nu}$ **with the same subscripts** μ **and** ν.

Letting $R^{(1)} = E$, one has

$$\begin{aligned}
\Phi^{(1)}_{\mu\nu} &= 3e^{(\mu)}Ee^{(\nu)} = \frac{1}{3}\sum_{\rho=0}^{2}\sum_{\lambda=0}^{2} \omega^{-\rho\mu-\lambda\nu} R_1^{\rho+\lambda} \\
&= \frac{1}{3}\sum_{\tau=0}^{2} \eta^{-\tau\mu} R_1^\tau \left[\sum_{\lambda=0}^{2} \omega^{\lambda(\mu-\nu)}\right] = \delta_{\mu\nu}\sum_{\tau=0}^{2} \omega^{-\tau\mu}R_1^\tau,
\end{aligned}$$
$$\Phi^{(1)}_{\mu\mu} = E + \omega^{-\mu}R_1 + \omega^{-2\mu}R_1^2. \tag{3.130}$$

Choosing $R^{(2)} = S_2$, one has

$$\begin{aligned}
\Phi_{\mu\nu}^{(2)} &= 3e^{(\mu)}S_2e^{(\nu)} = \frac{1}{3}\sum_{\rho=0}^{2}\sum_{\lambda=0}^{2}\omega^{-\rho\mu-\lambda\nu}R_1^{\rho-\lambda}S_2 \\
&= \frac{1}{3}\sum_{\tau=0}^{2}\omega^{-\tau\mu}R_1^{\tau}S_2\left[\sum_{\lambda=0}^{2}\omega^{-\lambda(\nu+\mu)}\right] = \delta_{\mu(-\nu)}\sum_{\tau=0}^{2}\omega^{-2\tau\mu}R_1^{\tau}S_2R_1^{-\tau}, \\
\Phi_{\mu\overline{\mu}}^{(2)} &= S_2+\omega^{-2\mu}S_4+\omega^{\mu}S_6,
\end{aligned} \tag{3.131}$$

where $R_1S_2 = S_2R_1^{-1}$ is used because S_2 is a 2-fold rotation about an axis perpendicular to the axis of R_1. Letting $R^{(3)} = S_1$, one has

$$\begin{aligned}
\Phi_{\mu\nu}^{(3)} &= 9e^{(\mu)}S_1e^{(\nu)} = \sum_{\rho=0}^{2}\sum_{\lambda=0}^{2}\omega^{-\rho\mu+\lambda\nu}R_1^{\rho}S_1R_1^{-\lambda} \\
&= \sum_{\lambda=0}^{2}\sum_{\tau=0}^{2}\omega^{-(\lambda+\tau)\mu+\lambda\nu}R_1^{\lambda+\tau}S_1R_1^{-\lambda}, \\
&= \sum_{\lambda=0}^{2}\omega^{\lambda(\nu-\mu)}R_1^{\lambda}\left(\sum_{\tau=0}^{2}\omega^{-\tau\mu}R_1^{\tau}S_1\right)R_1^{-\lambda} \\
&= S_1+\omega^{-\mu}T_y^3+\omega^{-2\mu}T_x \\
&\quad +\omega^{\nu-\mu}\left[S_3+\omega^{-\mu}T_z^3+\omega^{-2\mu}T_y\right] \\
&\quad +\omega^{2\nu-2\mu}\left[S_5+\omega^{-\mu}T_x^3+\omega^{-2\mu}T_z\right],
\end{aligned} \tag{3.132}$$

where

$$\begin{aligned}
R_1S_1 &\leftrightarrow (2\ 1\ 3)(2\ 4) = (1\ 3\ 2\ 4) \leftrightarrow T_y^3, \\
R_1^2S_1 &\leftrightarrow (1\ 2\ 3)(2\ 4) = (3\ 1\ 2\ 4) \leftrightarrow T_x.
\end{aligned}$$

Letting $R^{(4)} = T_z^2$, one has

$$\begin{aligned}
\Phi_{\mu\nu}^{(4)} &= 9e^{(\mu)}T_z^2e^{(\nu)} = \sum_{\rho=0}^{2}\sum_{\lambda=0}^{2}\omega^{-\rho\mu+\lambda\nu}R_1^{\rho}T_z^2R_1^{-\lambda} \\
&= \sum_{\lambda=0}^{2}\sum_{\tau=0}^{2}\omega^{-(\lambda+\tau)\mu+\lambda\nu}R_1^{\lambda+\tau}T_z^2R_1^{-\lambda} \\
&= \sum_{\lambda=0}^{2}\omega^{\lambda(\nu-\mu)}R_1^{\lambda}\left(\sum_{\tau=0}^{2}\omega^{-\tau\mu}R_1^{\tau}T_z^2\right)R_1^{-\lambda} \\
&= T_z^2+\omega^{-\mu}R_4+\omega^{-2\mu}R_2^2 \\
&\quad +\omega^{\nu-\mu}\left[T_x^2+\omega^{-\mu}R_3+\omega^{-2\mu}R_4^2\right] \\
&\quad +\omega^{\mu-\nu}\left[T_y^2+\omega^{-\mu}R_2+\omega^{-2\mu}R_3^2\right],
\end{aligned} \tag{3.133}$$

where

$$\begin{aligned}
R_1T_z^2 &\leftrightarrow (2\ 1\ 3)(1\ 3)(2\ 4) = (1\ 2\ 4) \leftrightarrow R_4,\\
R_1^2T_z^2 &\leftrightarrow (1\ 2\ 3)(1\ 3)(2\ 4) = (3\ 2\ 4) \leftrightarrow R_2^2,\\
R_1R_4R_1^{-1} &\leftrightarrow (2\ 1\ 3)(1\ 2\ 4)(1\ 2\ 3) = (3\ 1\ 4) \leftrightarrow R_3,\\
R_1R_3R_1^{-1} &\leftrightarrow (2\ 1\ 3)(3\ 1\ 4)(1\ 2\ 3) = (2\ 3\ 4) \leftrightarrow R_2.
\end{aligned}$$

Now,

$$\begin{aligned}
\phi^j_{\mu\nu} &= \frac{m_j}{g}\left[D^j_{\mu\nu}(E)^*E + D^j_{\mu\nu}(S_2)^*S_2 + D^j_{\mu\nu}(S_1)^*S_1 + D^j_{\mu\nu}(T_z^2)^*T_z^2 + \cdots\right]\\
&= \frac{m_j}{g}\left[\delta_{\mu\nu}\varPhi^{(1)}_{\mu\mu} + D^j_{\mu\nu}(S_2)^*\varPhi^{(2)}_{\mu\nu} + D^j_{\mu\nu}(S_1)^*\varPhi^{(3)}_{\mu\nu} + D^j_{\mu\nu}(T_z^2)^*\varPhi^{(4)}_{\mu\nu}\right].
\end{aligned}$$

Thus, the elements of **O** in Eq. (3.114) are combined into a few subsets $\varPhi^{(t)}_{\mu\nu}$ such that **the expression of $\phi^j_{\mu\nu}$ depends on the representation matrices of only a few elements.** It greatly simplifies the calculation.

Because $D^A(R) = 1$ for $R \in \mathbf{O}$, and $D^B(T_z^2) = -D^B(S_1) = -D^B(S_2) = 1$, one has

$$\begin{aligned}
\phi^A_{11} &= \frac{1}{24}\left[\varPhi^{(1)}_{11} + \varPhi^{(2)}_{11} + \varPhi^{(3)}_{11} + \varPhi^{(4)}_{11}\right],\\
\phi^B_{11} &= \frac{1}{24}\left[\varPhi^{(1)}_{11} - \varPhi^{(2)}_{11} - \varPhi^{(3)}_{11} + \varPhi^{(4)}_{11}\right].
\end{aligned} \tag{3.134}$$

From Eqs. (2.14), (2.61), and (3.127) one has

$$\overline{D}^E(E) = \overline{D}^E(T_z^2) = \mathbf{1},$$

$$\overline{D}^E(S_1) = \overline{D}^E(S_2) = X^{-1}\begin{pmatrix}1 & 0\\ 0 & -1\end{pmatrix}X = \begin{pmatrix}0 & \mathrm{i}\\ -\mathrm{i} & 0\end{pmatrix},$$

$$\begin{aligned}
\phi^E_{11} &= \frac{1}{12}\left[\varPhi^{(1)}_{11} + \varPhi^{(4)}_{11}\right], & \phi^E_{\overline{1}\,\overline{1}} &= \frac{1}{12}\left[\varPhi^{(1)}_{\overline{1}\,\overline{1}} + \varPhi^{(4)}_{\overline{1}\,\overline{1}}\right],\\
\phi^E_{1\overline{1}} &= \frac{-\mathrm{i}}{12}\left[\varPhi^{(2)}_{1\overline{1}} + \varPhi^{(3)}_{1\overline{1}}\right], & \phi^E_{\overline{1}1} &= \frac{\mathrm{i}}{12}\left[\varPhi^{(2)}_{\overline{1}1} + \varPhi^{(3)}_{\overline{1}1}\right].
\end{aligned} \tag{3.135}$$

From Eqs. (3.123) and (3.127) one has

$$\overline{D}^{T_1}(S_2) = Y^{-1}\begin{pmatrix}0 & -1 & 0\\ -1 & 0 & 0\\ 0 & 0 & -1\end{pmatrix}Y = \begin{pmatrix}0 & 0 & -1\\ 0 & -1 & 0\\ -1 & 0 & 0\end{pmatrix},$$

$$\overline{D}^{T_1}(S_1) = Y^{-1}\begin{pmatrix}0 & 1 & 0\\ 1 & 0 & 0\\ 0 & 0 & -1\end{pmatrix}Y = \frac{1}{3}\begin{pmatrix}-2 & -2 & 1\\ -2 & 1 & -2\\ 1 & -2 & -2\end{pmatrix},$$

$$\overline{D}^{T_1}(T_z^2) = Y^{-1}\begin{pmatrix}-1 & 0 & 0\\ 0 & -1 & 0\\ 0 & 0 & 1\end{pmatrix}Y = \frac{1}{3}\begin{pmatrix}-1 & 2 & 2\\ 2 & -1 & 2\\ 2 & 2 & -1\end{pmatrix},$$

$$\phi_{11}^{T_a} = \frac{1}{24}\left[3\Phi_{11}^{(1)} \mp 2\Phi_{11}^{(3)} - \Phi_{11}^{(4)}\right], \qquad \phi_{01}^{T_a} = \frac{1}{12}\left[\mp\Phi_{01}^{(3)} + \Phi_{01}^{(4)}\right],$$

$$\phi_{\overline{1}1}^{T_a} = \frac{1}{24}\left[\mp 3\Phi_{\overline{1}1}^{(2)} \pm \Phi_{\overline{1}1}^{(3)} + 2\Phi_{\overline{1}1}^{(4)}\right], \qquad \phi_{10}^{T_a} = \frac{1}{12}\left[\mp\Phi_{10}^{(3)} + \Phi_{10}^{(4)}\right],$$

$$\phi_{\overline{1}0}^{T_a} = \frac{1}{12}\left[\mp\Phi_{\overline{1}0}^{(3)} + \Phi_{\overline{1}0}^{(4)}\right], \qquad \phi_{1\overline{1}}^{T_a} = \frac{1}{24}\left[\mp 3\Phi_{1\overline{1}}^{(2)} \pm \Phi_{1\overline{1}}^{(3)} + 2\Phi_{1\overline{1}}^{(4)}\right],$$

$$\phi_{0\overline{1}}^{T_a} = \frac{1}{12}\left[\mp\Phi_{0\overline{1}}^{(3)} + \Phi_{0\overline{1}}^{(4)}\right], \qquad \phi_{\overline{1}\overline{1}}^{T_a} = \frac{1}{24}\left[3\Phi_{\overline{1}\overline{1}}^{(1)} \mp 2\Phi_{\overline{1}\overline{1}}^{(3)} - \Phi_{\overline{1}\overline{1}}^{(4)}\right],$$

$$\phi_{00}^{T_a} = \frac{1}{24}\left[3\Phi_{00}^{(1)} \mp 3\Phi_{00}^{(2)} \pm \Phi_{00}^{(3)} - \Phi_{00}^{(4)}\right], \tag{3.136}$$

where $a = 1$, 2, the upper sign is for the representation D^{T_1}, and the lower sign is for D^{T_2}.

3.8 Exercises

1. Let $D(\mathrm{G})$ denote a faithful representation of a non-Abelian group G, and $D(R)$ denote the representation matrix of the element $R \in \mathrm{G}$. Please judge whether the set $\overline{D}(\mathrm{G})$ forms a representation of G where $\overline{D}(R)$ is equal to the following matrix.

   (a) $\overline{D}(R) = D(R)^\dagger$; (b) $\overline{D}(R) = D(R)^T$;

   (c) $\overline{D}(R) = D(R^{-1})$; (d) $\overline{D}(R) = D(R)^*$;

   (e) $\overline{D}(R) = D(R^{-1})^\dagger$; (f) $\overline{D}(R) = \det D(R)$;

   (g) $\overline{D}(R) = \mathrm{Tr}\, D(R)$.

2. Prove that the module of every representation matrix in a one-dimensional representation of a finite group is equal to 1.
3. Prove that any irreducible representation of an infinite abelian group is one-dimensional.
4. Prove that the similarity transformation matrix between two equivalent irreducible unitary representations of a finite group, if restricting its determinant to be 1, has to be unitary.
5. Show that the sum of the characters of all elements in an irreducible representation of a finite group G, except for the identical representation, is equal to zero.

6. Calculate the similarity transformation matrix X between two equivalent regular representations $D(R)$ and $\overline{D}(R)$ of the group D_3, given in Eqs. (3.8) and (3.14). Prove the general result (3.40) for any finite group.

7. Calculate the combination coefficients $f(\alpha,\beta,\gamma)$ in the product $\mathsf{C}_\alpha \mathsf{C}_\beta$ of two class operators (see Eq. (3.20)) for the group D_3 from Table 2.5.

8. Let C_α denote the class operator (see Eq. (3.16)) of the class $\mathcal{C}_\alpha$, which contains $n(\alpha)$ elements, in a finite group G. Let $\chi^j(\mathcal{C}_\alpha)$ denote the character of $\mathcal{C}_\alpha$ in an irreducible representation $D^j(\mathrm{G})$ of dimension m_j. Prove that the representation matrix of C_α in $D^j(\mathrm{G})$ is a constant matrix, and calculate this constant.

9. If the group G is a direct product $\mathrm{H}_1 \otimes \mathrm{H}_2$ of two subgroups, show that the direct product $D^j(\mathrm{H}_1) \times D^k(\mathrm{H}_2)$ of two irreducible representations of the subgroups is an irreducible representation of G.

10. The irreducible representation $D^E(\mathrm{D}_3)$ of two dimensions is calculated by the method of coordinate transformation (see Eqs. (2.12) and (2.14)), where the representation matrices of the generators of D_3 are

$$D^E(C_3) = \frac{1}{2}\begin{pmatrix} -1 & -\sqrt{3} \\ \sqrt{3} & -1 \end{pmatrix}, \qquad D^E(C_2^{(0)}) = \begin{pmatrix} 1 & 0 \\ 0 & -1 \end{pmatrix}.$$

The four-dimensional functional space spanned by the following basis functions is left invariant under D_3 and corresponds to a representation $D(\mathrm{D}_3)$:

$$\psi_1(x,y) = x^3, \quad \psi_2(x,y) = x^2 y, \quad \psi_3(x,y) = xy^2, \quad \psi_4(x,y) = y^3.$$

Calculate the representation matrices $D(C_3)$ and $D(C_2^{(0)})$ of the generators of D_3. Then, decompose this representation $D(\mathrm{D}_3)$ into the direct sum of the irreducible representations of D_3, and calculate the new basis functions $\phi^j_\mu(x,y)$, belonging to the irreducible representation $D^j(\mathrm{D}_3)$, as the linear combinations of the original basis functions $\psi_\rho(x,y)$.

11. Please check the character tables of the groups C_N, D_N, **T**, and **O** whether Corollary 3.4.4 holds.

12. Calculate the character table of the group G given in Prob. 23 of Chap. 2.

13. Calculate the Clebsch–Gordan series and the Clebsch–Gordan coefficients for the direct product representation of each two irreducible representations of the group **O**.

14. Calculate the unitary similarity transformation matrix X for decomposing the self-direct product of the three-dimensional irreducible unitary representation D^T of the group **T**:

$$X^{-1}\left\{D^T(R)\times D^T(R)\right\}X=\sum_j a_j D^j(R),\qquad \forall R\in \mathbf{T}.$$

15. Please calculate the irreducible bases of the group **T** (see Eq. (3.125)).

16. Please calculate the irreducible bases of the group **O** (see Eq. (3.126)).

17. The multiplication table of the group G is given as follows.

left \ right	E	A	B	C	F	K	M	N
E	E	A	B	C	F	K	M	N
A	A	E	M	K	N	C	B	F
B	B	N	E	M	K	F	C	A
C	C	K	N	E	M	A	F	B
F	F	M	K	N	E	B	A	C
K	K	C	F	A	B	E	N	M
M	M	F	A	B	C	N	K	E
N	N	B	C	F	A	M	E	K

(a) Write the character table for the group G.
(b) If we know that the representation matrices of the generators A and B in a two-dimensional irreducible representation D(G) are

$$D(A)=\begin{pmatrix}1 & 0\\ 0 & -1\end{pmatrix},\qquad D(B)=\begin{pmatrix}0 & 1\\ 1 & 0\end{pmatrix},$$

and two sets of basis functions ψ_μ and ϕ_ν transform according to this two-dimensional representation of G, respectively:

$$P_R\psi_\mu=\sum_{\mu'}\psi_{\mu'}D_{\mu'\mu}(R),\qquad P_R\phi_\nu=\sum_{\nu'}\phi_{\nu'}D_{\nu'\nu}(R),$$

please combine the product functions $\psi_\mu\phi_\nu$ such that the basis functions belong to the irreducible representations of G.

Chapter 4

PERMUTATION GROUPS

The irreducible bases in the group algebra $\mathcal{L}$ of a finite group G can be calculated directly from Eq. (3.114) if all inequivalent irreducible representations of G have known. The order of the permutation group S_n is very large such that its irreducible representations are hard to calculate generally. In this chapter we will study an opposite problem that the representation theory of the permutation group is established from its irreducible bases. This method is so-called the **Young operator method** that is an elegant theory and is widely used in modern physics.

4.1 Ideals and Idempotents

In order to calculate the irreducible representations of the permutation group S_n from its irreducible bases, we have to introduce some new concepts in the group algebra $\mathcal{L}$. Those concepts are suitable in principle to other algebras, for example, to the Lie algebra.

4.1.1 *Ideals*

Let $\mathcal{L}$ be the group algebra of a finite group G. $\mathcal{L}_a$ is a subspace of $\mathcal{L}$ if $\mathcal{L}_a \subset \mathcal{L}$. A subspace $\mathcal{L}_a$ of $\mathcal{L}$ is a subalgebra of the group algebra $\mathcal{L}$ if

$$yx \in \mathcal{L}_a \qquad \forall\, x \in \mathcal{L}_a \text{ and } y \in \mathcal{L}_a. \tag{4.1}$$

A subalgebra $\mathcal{L}_a$ of $\mathcal{L}$ is called a **left ideal** of the group algebra $\mathcal{L}$ if

$$tx \in \mathcal{L}_a, \qquad \forall\, t \in \mathcal{L} \text{ and } x \in \mathcal{L}_a. \tag{4.2}$$

The whole algebra $\mathcal{L}$ and the null space $\emptyset$ are two **trivial** left ideals in $\mathcal{L}$. A subalgebra $\mathcal{L}_t \subset \mathcal{L}_a$ is called a **left sub-ideal** of a left ideal $\mathcal{L}_a$ of $\mathcal{L}$ if

$\mathcal{L}_t$ is a left ideal of $\mathcal{L}_a$.

Choosing a complete set of basis vectors x_μ in a left ideal $\mathcal{L}_a$, one can calculate the representation by left-multiplying with elements R of G,

$$Rx_\mu = \sum_{\nu=1}^{m} x_\nu D_{\nu\mu}(R) \in \mathcal{L}_a, \tag{4.3}$$

where m is the dimension of $\mathcal{L}_a$. $D(\mathrm{G})$ is called the **representation corresponding to the left ideal** $\mathcal{L}_a$. A left ideal is called **minimal** if its corresponding representation is irreducible. **A minimal left ideal does not contain nontrivial left sub-ideal.** Two left ideals are called **equivalent** if their corresponding representations are equivalent. One may choose the basis vectors in two equivalent left ideals $\mathcal{L}_a$ and $\mathcal{L}_b$ such that two corresponding representations are the same as each other:

$$Ry_\mu = \sum_{\nu=1}^{m} y_\nu D_{\nu\mu}(R) \in \mathcal{L}_b. \tag{4.4}$$

Define a one-to-one correspondence between two left ideals $\mathcal{L}_a$ and $\mathcal{L}_b$:

$$\sum_{\nu=1}^{m} x_\nu c_\nu \in \mathcal{L}_a \longleftrightarrow \sum_{\nu=1}^{m} y_\nu c_\nu \in \mathcal{L}_b. \tag{4.5}$$

This correspondence is left invariant under left-multiplying with elements of G:

$$Rx_\mu = \sum_{\nu=1}^{m} x_\nu D_{\nu\mu}(R) \longleftrightarrow Ry_\mu = \sum_{\nu=1}^{m} y_\nu D_{\nu\mu}(R). \tag{4.6}$$

Conversely, two left ideals $\mathcal{L}_a$ and $\mathcal{L}_b$ are equivalent if there is a one-to-one correspondence between $\mathcal{L}_a$ and $\mathcal{L}_b$ which is left invariant under left-multiplying with elements of G. This is another definition for two equivalent left ideals.

Similarly, a subalgebra $\mathcal{R}_a$ of $\mathcal{L}$ is called a right ideal of $\mathcal{L}$ if

$$xt \in \mathcal{R}_a, \qquad \forall\, t \in \mathcal{L} \text{ and } x \in \mathcal{R}_a. \tag{4.7}$$

A subalgebra $\mathcal{R}_t \subset \mathcal{R}_a$ is called a right sub-ideal of $\mathcal{R}_a$ if $\mathcal{R}_t$ is a right ideal of $\mathcal{R}_a$, too. Choosing a complete set of basis vectors x_μ in the right ideal $\mathcal{R}_a$, one can calculate the representation corresponding to $\mathcal{R}_a$:

$$y_\mu R = \sum_{\nu=1}^{m} D_{\mu\nu}(R) x_\nu \in \mathcal{R}_a, \qquad R \in \mathrm{G}, \tag{4.8}$$

where m is the dimension of $\mathcal{R}_a$. A right ideal is called minimal if its corresponding representation is irreducible. Two right ideals are equivalent if their corresponding representations are equivalent. There is a one-to-one correspondence between two equivalent right ideals and this correspondence is left invariant under right-multiplying with elements of G.

4.1.2 *Idempotents*

A projective operator $e_a \in \mathcal{L}$, satisfying $e_a^2 = e_a$, is called an **idempotent** of $\mathcal{L}$. n projective operators e_a, satisfying

$$e_a e_b = \delta_{ab} e_a, \qquad 1 \leqslant a \leqslant n, \qquad 1 \leqslant b \leqslant n, \tag{4.9}$$

are called the mutually **orthogonal idempotents** of $\mathcal{L}$. The set of $te_a, \forall t \in \mathcal{L}$, forms a left ideal $\mathcal{L}_a$ of $\mathcal{L}$: $\mathcal{L}e_a = \mathcal{L}_a$. $\mathcal{L}_a$ is called the **left ideal generated by** e_a. An important property of $t \in \mathcal{L}_a$ is

$$t = te_a, \quad \text{if} \quad t \in \mathcal{L}_a = \mathcal{L}e_a. \tag{4.10}$$

In fact, if $t \in \mathcal{L}_a$ and $t = re_a$, then, $te_a = re_ae_a = re_a = t$. Thus, there is no common vector in two left ideals $\mathcal{L}_a$ and $\mathcal{L}_b$, generated by two orthogonal idempotents e_a and e_b, respectively, because if $t \in \mathcal{L}_a$ and $t \in \mathcal{L}_b$, then, $t = te_a = (te_a)e_b = 0$. For n mutually orthogonal idempotents e_a, one has

$$\mathcal{L}e_a = \mathcal{L}_a, \qquad \mathcal{L}\sum_{a=1}^{n} e_a = \bigoplus_{a=1}^{n} \mathcal{L}_a. \tag{4.11}$$

The set $\mathcal{L}'_a$ of all $t \in \mathcal{L}$ satisfying $te_a = 0$ forms a left ideal of $\mathcal{L}$, too. In fact, the idempotent generating $\mathcal{L}'_a$ is $e'_a = E - e_a$, and

$$\left(e'_a\right)^2 = e'_a, \quad e_a e'_a = e'_a e_a = 0, \quad \mathcal{L} = \mathcal{L}e_a + \mathcal{L}e'_a = \mathcal{L}_a \oplus \mathcal{L}'_a. \tag{4.12}$$

$\mathcal{L}_a$ and $\mathcal{L}'_a$ are supplement. In fact, from any vector $z \in \mathcal{L}$, one is able to obtain two left ideals: $\mathcal{L}_z = \mathcal{L}z$ and $\mathcal{L}'_z$ is the set of all $t \in \mathcal{L}$ satisfying $tz = 0$. However, $\mathcal{L}_z$ and $\mathcal{L}'_z$ may not be supplement, and there may be common vectors belonging to both $\mathcal{L}_z$ and $\mathcal{L}'_z$. An extreme example is a vector z in the group algebra of the permutation group S_n,

$$z = (E + P_1)P_2(E - P_1) \neq 0, \qquad z^2 = 0, \tag{4.13}$$

where $P_1 = (1\ 2)$ and $P_2 = (2\ 3)$. Thus, $\mathcal{L}_z \subset \mathcal{L}'_z$, and there exist some vectors in $\mathcal{L}$ not belonging to $\mathcal{L}'_z$, e.g. the identity E.

An idempotent e_a is called **primitive** if the representation corresponding to the left ideal generated by e_a is irreducible. Two idempotents are **equivalent** if their corresponding representations are equivalent.

If an idempotent e_a is not primitive, the representation $D^a(\mathrm{G})$ corresponding to the left ideal $\mathcal{L}_a$ generated by e_a is reducible, say $D^a(\mathrm{G}) \simeq D^{(1)}(\mathrm{G}) \oplus D^{(2)}(\mathrm{G})$. Two representation spaces $\mathcal{L}_1$ of $D^{(1)}(\mathrm{G})$ and $\mathcal{L}_2$ of $D^{(2)}(\mathrm{G})$ both are the left sub-ideals of $\mathcal{L}_a$: $\mathcal{L}_a = \mathcal{L}_1 \bigoplus \mathcal{L}_2$. Any vector $t \in \mathcal{L}_a$ can be decomposed into a sum of two vectors, respectively belonging to $\mathcal{L}_1$ and $\mathcal{L}_2$, **uniquely**:

$$t = te_a = t_1 + t_2, \qquad t_1 \in \mathcal{L}_1, \qquad t_2 \in \mathcal{L}_2. \tag{4.14}$$

$e_a \in \mathcal{L}_a$ can also be decomposed uniquely: $e_a = e_1 + e_2$, where $e_1 \in \mathcal{L}_1$ and $e_2 \in \mathcal{L}_2$. Thus, $t = te_a = te_1 + te_2$, and then, $t_1 = te_1$ and $t_2 = te_2$. If $t \in \mathcal{L}_1$, $t = t_1 = te_1$ and $t_2 = te_2 = 0$. Replacing t by e_1 or e_2, one has

$$e_a = e_1 + e_2, \quad e_1^2 = e_1, \quad e_2^2 = e_2, \quad e_1e_2 = e_2e_1 = 0. \tag{4.15}$$

A non-primitive idempotent e_a can be decomposed into the sum of two orthogonal idempotents. The decomposition can be done continually until all orthogonal idempotents are primitive. A primitive idempotent cannot be made this decomposition further.

The representation corresponding to the group algebra $\mathcal{L}$ is the regular representation, which can be decomposed into a direct sum of all inequivalent irreducible representations $D^j(\mathrm{G})$ of G where the multiplicity of each $D^j(\mathrm{G})$ is its dimension m_j. Thus, $\mathcal{L}$ can be decomposed into a direct sum of the minimal left ideals $\mathcal{L}_\mu^{(j)}$, and the identity E can be decomposed into the sum of the orthogonal primitive idempotents $\mathrm{e}_\mu^{(j)}$:

$$\begin{aligned} &E = \sum_{j=1}^{g_c} \mathrm{e}^{(j)}, \qquad \mathrm{e}^{(j)} = \sum_{\mu=1}^{m_j} \mathrm{e}_\mu^{(j)}, \qquad \mathrm{e}_\mu^{(j)}\mathrm{e}_\nu^{(k)} = \delta_{jk}\delta_{\mu\nu}\mathrm{e}_\mu^{(j)}, \\ &\mathcal{L} = \bigoplus_{j=1}^{g_c} \mathcal{L}^{(j)} = \bigoplus_{j=1}^{g_c} \mathcal{L}\mathrm{e}^{(j)}, \qquad \mathcal{L}^{(j)} = \bigoplus_{\mu=1}^{m_j} \mathcal{L}_\mu^{(j)} = \bigoplus_{\mu=1}^{m_j} \mathcal{L}\mathrm{e}_\mu^{(j)}. \end{aligned} \tag{4.16}$$

Similarly, $\mathcal{R}_a = e_a\mathcal{L}$ is called the right ideal generated by e_a:

$$t = e_at, \qquad \text{if } \; t \in \mathcal{R}_a = e_a\mathcal{L}. \tag{4.17}$$

For the orthogonal idempotents e_a,

$$\left[\sum_{a=1}^{n} e_a\right] \mathcal{L} = \bigoplus_{a=1}^{n} \mathcal{R}_a, \qquad \mathcal{R}_a = e_a\mathcal{L}.$$

$\mathcal{L}$ can also be decomposed into the direct sum of the minimal right ideals $\mathcal{R}_\mu^{(j)}$ generated by the orthogonal idempotents $e_\mu^{(j)}$ given in Eq. (4.16):

$$\mathcal{L} = \bigoplus_{j=1}^{g_c} \mathcal{R}^{(j)} = \bigoplus_{j=1}^{g_c} e^{(j)}\mathcal{L}, \qquad \mathcal{R}^{(j)} = \bigoplus_{\mu=1}^{m_j} \mathcal{R}_\mu^{(j)} = \bigoplus_{\mu=1}^{m_j} e_\mu^{(j)}\mathcal{L}. \tag{4.18}$$

The orthogonal primitive idempotents $e_\mu^{(j)}$ of $\mathcal{L}$ are related to the irreducible bases $\phi_{\mu\nu}^j$ in $\mathcal{L}$ (see subsection 3.7.1),

$$\begin{aligned} e_\mu^{(j)} &= \phi_{\mu\mu}^j = \frac{m_j}{g} \sum_{R\in G} D_{\mu\mu}^j(R)^* R, \\ e^{(j)} &= \sum_{\mu=1}^{m_j} e_\mu^{(j)} = \frac{m_j}{g} \sum_{R\in G} \chi^j(R)^* R, \end{aligned} \tag{4.19}$$

even though the explicit forms of $\phi_{\mu\mu}^j$ for the permutation groups S_n are to be calculated. Because $m_j = \chi^j(E)$ and Eq. (3.68) one has

$$\sum_{j=1}^{g_c}\sum_{\mu=1}^{m_j} e_\mu^{(j)} = \sum_{j=1}^{g_c} e^{(j)} = \sum_{R\in G} \left[\frac{1}{g}\sum_{j=1}^{g_c} \chi^j(E)\chi^j(R)^*\right] R = E.$$

The representation corresponding to the left ideal $\mathcal{L}_\mu^{(j)} = \mathcal{L}e_\mu^{(j)}$ is equivalent to that corresponding to the right ideal $\mathcal{R}_\mu^{(j)} = e_\mu^{(j)}\mathcal{L}$. The m_j basis vectors in $\mathcal{L}_\mu^{(j)}$ are $\phi_{\nu\mu}^j$ with fixed j and μ, and the m_j basis vectors in $\mathcal{R}_\mu^{(j)}$ are $\phi_{\mu\nu}^j$ with fixed j and μ. Therefore, $\mathcal{L}_\mu^{(j)} \neq \mathcal{R}_\mu^{(j)}$. However, the m_j^2 basis vectors both in $\mathcal{L}^{(j)} = \mathcal{L}e^{(j)}$ and in $\mathcal{R}^{(j)} = e^{(j)}\mathcal{L}$ are $\phi_{\nu\mu}^j$ with fixed j, so that $\mathcal{L}^{(j)} = \mathcal{R}^{(j)}$.

Although the left ideal $\mathcal{L}_a = \mathcal{L}e_a$ and the right ideal $\mathcal{R}_a = e_a\mathcal{L}$, generated by one idempotent e_a, correspond the equivalent representations, but generally $\mathcal{L}_a \neq \mathcal{R}_a$. If $\mathcal{L}e_a = e_a\mathcal{L} = \mathcal{I}_a$, the ideal $\mathcal{I}_a$ is called a **two-side ideal** of $\mathcal{L}$. A two-side ideal is called **simple** if it does not contain nontrivial two-side sub-ideals: not whole and not the null space. $\mathcal{L}^{(j)}$ is not only a two-side ideal of $\mathcal{L}$, but also simple, which will be proved in Corollary 4.1.1.

4.1.3 *Primitive Idempotents*

In order to study the inequivalent irreducible representations of a finite group, one needs the criterions for equivalence, irreducibility and completeness of representations corresponding to the left (right) ideals.

Theorem 4.1 Two primitive idempotents e_a and e_b are equivalent if and only if there exists at least one element $S \in G$ satisfying

$$e_a S e_b \neq 0. \tag{4.20}$$

Proof The key in the proof is that both left ideals $\mathcal{L}_a$ and $\mathcal{L}_b$, respectively generated by e_a and e_b, are minimal such that the left sub-ideal contained in them must be trivial: the whole or the null space. If the condition (4.20) holds, there is a map from $\mathcal{L}_a$ onto $\mathcal{L}_b$:

$$x \in \mathcal{L}_a \quad \longrightarrow \quad y = x e_a S e_b \in \mathcal{L}_b. \tag{4.21}$$

Evidently, the map is invariant to left-multiplying with any element

$$Rx \in \mathcal{L}_a \quad \longrightarrow \quad Ry = Rx e_a S e_b \in \mathcal{L}_b. \tag{4.22}$$

We need to show that the map is bijection, namely there is a one-to-one correspondence between the vectors in two left ideals such that two minimal left ideals are equivalent.

First, we show that the map is surjective. If the set of y mapped from x is a subspace $\mathcal{L}_y$ of the minimal left ideal $\mathcal{L}_b$, due to Eq. (4.22), $\mathcal{L}_y$ is a left sub-ideal of $\mathcal{L}_b$. Since $\mathcal{L}_y$ contains $e_a S e_b \neq 0$, it is not the null space such that $\mathcal{L}_y = \mathcal{L}_b$.

Second, we show that $y \neq 0$ if $x \neq 0$. Otherwise, the set $\mathcal{L}_x$ of x which maps to $y = 0$ is a left sub-ideal of $\mathcal{L}_a$. Since $\mathcal{L}_x$ does not contain e_a, $\mathcal{L}_x$ is the null space.

Third, we show that the map is injective. If two vectors x and x' in $\mathcal{L}_a$ both map to the same y in $\mathcal{L}_b$, the difference $(x - x')$ maps to the null vector so that $x = x'$.

Conversely, if two minimal left ideals $\mathcal{L}_a$ and $\mathcal{L}_b$ are equivalent, the idempotent e_a in $\mathcal{L}_a$ maps to $t = t e_b \neq 0$ in $\mathcal{L}_b$, and $e_a e_a$ maps to $e_a t = e_a t e_b \neq 0$. Thus, equation (4.20) holds for at least one element S. □

Corollary 4.1.1 $\mathcal{L}^{(j)}$ given in Eq. (4.16) is the simple two-side ideal of the group algebra $\mathcal{L}$ of a finite group G.

Proof From Eq. (4.16), $e^{(j)} = \sum_\mu e_\mu^{(j)}$ is the sum of all mutually orthogonal primitive idempotents $e_\mu^{(j)}$, corresponding to the irreducible representation $D^j(\mathrm{G})$, and $\mathcal{L}^{(j)} = \mathcal{L}e^{(j)}$ is the left ideal of $\mathcal{L}$ generated by $e^{(j)}$. From Theorem 4.1, for any $t \in \mathcal{L}$ one has

$$e^{(j)}t = e^{(j)}tE = \sum_{k=1}^{g_c}\sum_{\mu,\nu=1}^{m_k} e_\mu^{(j)} t e_\nu^{(k)} = \left(e^{(j)}t\right)\sum_{\nu=1}^{m_j} e_\nu^{(j)} \in \mathcal{L}^{(j)}.$$

Thus, $\mathcal{L}^{(j)}$ is a two-side ideal.

If $\mathcal{I} \subset \mathcal{L}^{(j)}$ is a non-trivial two-side sub-ideal, one can choose at least one minimal left sub-ideal $\mathcal{L}_0 \subset \mathcal{I}$ generated by a primitive idempotent e_0. Since $e_0 \in \mathcal{I} \subset \mathcal{L}^{(j)}$, $e_0 = e_0 e^{(j)} \neq 0$, and there is at least one μ such that $e_0 e_\mu^{(j)} \neq 0$. From Theorem 4.1, e_0 corresponds to the irreducible representation $D^j(\mathrm{G})$, and there at least one element $R_\nu \in \mathrm{G}$ for every ν, satisfying $e_0 R_\nu e_\nu^{(j)} \neq 0$. $e_0 R_\nu e_\nu^{(j)}$ belongs to both $\mathcal{I}$ and $\mathcal{L}_\nu^{(j)}$ so that $\mathcal{L}_\nu^{(j)} \subset \mathcal{I}$, $\mathcal{L}^{(j)} \subset \mathcal{I}$ and $\mathcal{L}^{(j)} = \mathcal{I}$. □

Corollary 4.1.2 The representation corresponding to the left ideal $\mathcal{L}_0$ (or the right ideal $\mathcal{R}_0$) generated by a primitive idempotent e_0 is the irreducible representation $D^j(\mathrm{G})$ if and only if $\mathcal{L}_0 \subset \mathcal{L}^{(j)}$ (or $\mathcal{R}_0 \subset \mathcal{L}^{(j)}$).

Proof From Corollary 4.1.1, if $\mathcal{L}_0 \subset \mathcal{L}^{(j)}$, $e_0 = e_0 e^{(j)}$ and there is at least one μ such that $e_0 e_\mu^{(j)} \neq 0$. The representation corresponding to $\mathcal{L}_0$ is $D^j(\mathrm{G})$. Conversely, if the representation corresponding to $\mathcal{L}_0$ generated by e_0 is $D^j(\mathrm{G})$, $e_0 e_\mu^{(k)} = 0$ for $k \neq j$. Thus, $e_0 = e_0 E = e_0 e^{(j)}$, and $\mathcal{L}e_0 \subset \mathcal{L}^{(j)}$. The same arguments are applicable to $\mathcal{R}_0$. □

Theorem 4.2 An idempotent e_a is primitive if and only if

$$e_a t e_a = \lambda_t e_a, \qquad \forall\, t \in \mathcal{L}, \tag{4.23}$$

where λ_t is a constant depending on t and is allowed to be vanishing.

Proof First, prove the condition (4.23) to be sufficient by reduction to absurdity. If the condition (4.23) holds but e_a is not a primitive idempotent, e_a can be decomposed into a sum $e_a = e_1 + e_2$ given in Eq. (4.15). Thus,

$$e_1 = e_a e_1 e_a = \lambda_1 e_a, \qquad e_1 = e_1^2 = e_1\left(\lambda_1 e_a\right) = \lambda_1 e_1.$$

The solutions are $e_1 = 0$ $(e_a = e_2)$ or $\lambda_1 = 1$ $(e_a = e_1)$, both of which mean that e_a cannot be decomposed.

Second, prove the condition (4.23) to be necessary. If $e_a t e_a = 0$, equation (4.23) holds. If $e_a t e_a \neq 0$, $e_a t e_a$ provides a one-to-one correspondence from the minimal left ideal $\mathcal{L}_a$ onto itself, which is left invariant under left-multiplying with any element in G. The basis x_μ in $\mathcal{L}_a$ maps to $y_\mu = x_\mu e_a t e_a = \sum_\nu x_\nu M_{\nu\mu}$. The irreducible representations in both sets of basis vectors are the same, $M^{-1}D(R)M = D(R)$. From Corollary 3.3.1, M is a constant matrix, denoted by $\lambda_t \mathbf{1}$. Thus, $y_\mu = \lambda_t x_\mu = \lambda_t x_\mu e_a$. □

If $\lambda_t \neq 0$, another idempotent $e'_a = \lambda_t^{-1} t e_a$ generates the same minimal left ideal as that generated by e_a. In fact, since $e'_a e_a = e'_a$ and $e_a e'_a = e_a$, one has $\mathcal{L}e_a = \mathcal{L}e_a e'_a \subset \mathcal{L}e'_a$ and $\mathcal{L}e'_a = \mathcal{L}e'_a e_a \subset \mathcal{L}e_a$. Thus, $\mathcal{L}e_a = \mathcal{L}e'_a$. **The idempotent generating a minimal left ideal is not unique.** Similarly, both e_a and $e''_a = \lambda_t^{-1} e_a t$ generate the same minimal right ideal.

Theorem 4.3 The direct sum of n left ideals $\mathcal{L}_a$ respectively generated by the orthogonal idempotents e_a is equal to the group algebra $\mathcal{L}$ if and only if the sum of e_a is equal to the identity E:

$$\mathcal{L} = \bigoplus_{a=1}^{n} \mathcal{L}_a \iff E = \sum_{a=1}^{n} e_a. \tag{4.24}$$

Proof If E is equal to the sum of e_a, $t = tE = \sum_a te_a$. Thus, any vector t in $\mathcal{L}$ can be decomposed uniquely into a sum of vectors te_a belonging to $\mathcal{L}_a$, respectively. Conversely, if $\mathcal{L}$ is the direct sum of $\mathcal{L}_a$, E can be decomposed uniquely into a sum of vectors belonging to $\mathcal{L}_a$, respectively, where the vector belonging to $\mathcal{L}_a$ is $Ee_a = e_a$. □

4.2 Primitive Idempotents of S_n

The key in the Young operator method is to define the Young operators and to prove that they are proportional to the primitive idempotents of the permutation groups S_n.

4.2.1 *Young Patterns*

A class in the permutation group S_n is characterized by a partition of n, $(\ell) = (\ell_1, \ell_2, \ldots, \ell_m)$ with $\sum_j \ell_j = n$. Since the number of the inequivalent irreducible representations of a finite group is equal to the number of its classes, the irreducible representation of S_n can also be described by a partition of n, denoted by

$$\begin{aligned} &[\lambda] = [\lambda_1,\ \lambda_2,\ \ldots,\ \lambda_m], \\ &\lambda_1 \geqslant \lambda_2 \geqslant \ldots \geqslant \lambda_m > 0, \end{aligned} \qquad \sum_{j=1}^{m} \lambda_j = n. \tag{4.25}$$

Remind that the irreducible representation $[\lambda]$ has no relation with the class (ℓ), no matter whether the partitions are the same or not. We define a Young pattern $[\lambda]$ (or called a Young diagram) based on a partition $[\lambda]$, and will show later that a Young pattern $[\lambda]$ characterizes an irreducible representation of S_n. **A Young pattern $[\lambda]$ consists of n boxes lined up on the top and on the left, where the jth row contains λ_j boxes.** For example, the Young pattern $[3, 2]$ is

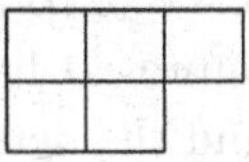

For a Young pattern, the number of boxes in the upper row is not less than that in the lower row, and the number of boxes in the left column is not less than that in the right column. In order to emphasize the above rules for a Young pattern, a Young pattern is sometimes called a regular Young pattern. In fact, we will not be interested in an irregular Young pattern.

A Young pattern $[\lambda]$ is said to be **larger** than a Young pattern $[\lambda']$ if

$$\lambda_i = \lambda_i', \qquad 1 \leqslant i < j, \qquad \text{and} \quad \lambda_j > \lambda_j'. \tag{4.26}$$

Although there is no analytic formula for the number of partitions of n, **one can list all Young patterns with n boxes in the decreasing order from the largest to the smallest.** First, λ_1 decreases from n one by one. Second, for a given λ_1, λ_2 decreases from the minimum of λ_1 and $(n - \lambda_1)$ one by one. Third, for the given λ_1 and λ_2, λ_3 decreases from the minimum of λ_2 and $(n - \lambda_1 - \lambda_2)$ one by one, and so on. For example, the Young patterns for $n = 7$ are listed as follows:

$$\begin{array}{llll}
[7], & [6,1], & [5,2], & [5,1,1], \\
[4,3], & [4,2,1], & [4,1,1,1], & [3,3,1], \\
[3,2,2], & [3,2,1,1], & [3,1,1,1,1], & [2,2,2,1], \\
[2,2,1,1,1], & [2,1,1,1,1,1], & [1,1,1,1,1,1,1]. &
\end{array}$$

4.2.2 *Young Tableaux*

Filling n digits 1, 2, ..., n arbitrarily into the Young pattern $[\lambda]$ with n boxes, one obtains a Young tableau. There are $n!$ different Young tableaux for a given Young pattern with n boxes. A Young tableau is called **standard** if the digits in each row increase from left to right, and the digits in each column increase from up to down. It is proved that the number $d_{[\lambda]}$ of the standard Young tableaux for the Young pattern $[\lambda]$ with m rows is

$$d_{[\lambda]}(S_n) = \frac{n!}{r_1! r_2! \ldots r_m!} \prod_{j<k}^{m} (r_j - r_k), \tag{4.27}$$

where $r_j = \lambda_j + m - j$. The square sum of the numbers $d_{[\lambda]}(S_n)$ is $n!$

$$\sum_{[\lambda]} d_{[\lambda]}(\mathrm{S}_n)^2 = n!. \tag{4.28}$$

The proof can be found in Chap. IV of [Boerner (1963)].

The formula (4.27) can be improved because there are some common factors between the numerator and the denominator. For example,

$$d_{[3,2,1,1]}(\mathrm{S}_7) = 7! \times \frac{5\times 4\times 2}{6!} \times \frac{3\times 2}{4!} \times \frac{1}{2!} \times \frac{1}{1!} = 35.$$

A simpler formula for calculating $d_{[\lambda]}(\mathrm{S}_n)$ is called the **hook rule**. The hook number h_{ij} of the box at the jth column of the ith row in a Young pattern $[\lambda]$ is the number of boxes at its right in the ith row, plus the number of boxes below it in the jth column, and plus 1. $Y_h^{[\lambda]}$ is a symbol to denote the product $\prod_{ij} h_{ij}$, and is drawn as a tableau of the Young pattern $[\lambda]$ where the box in the jth column of the ith row is filled with its hook number h_{ij}. The number $d_{[\lambda]}(\mathrm{S}_n)$ of standard Young tableaux for the Young pattern $[\lambda]$ is

$$d_{[\lambda]}(\mathrm{S}_n) = \frac{n!}{\prod_{ij} h_{ij}} = \frac{n!}{Y_h^{[\lambda]}}. \tag{4.29}$$

For comparison, $d_{[3,2,1,1]}$ is re-calculated from Eq. (4.29),

$$Y_h^{[3,2,1,1]} = \begin{array}{|c|c|c|}\hline 6 & 3 & 1 \\ \hline 4 & 1 & \multicolumn{1}{c}{} \\ \cline{1-2} 2 & \multicolumn{2}{c}{} \\ \cline{1-1} 1 & \multicolumn{2}{c}{} \\ \cline{1-1} \end{array}, \qquad d_{[3,2,1,1]}(\mathrm{S}_7) = \frac{7!}{6\times 3\times 4\times 2} = 35.$$

Two formulas are the same because for each row i,

$$\prod_{j=1}^{\lambda_i} h_{ij} = r_i! \left\{ \prod_{j=i+1}^{m} (r_i - r_j) \right\}^{-1}.$$

Compare the filled digits in two standard Young tableaux of a given Young pattern from left to right of the first row, and then those of the second row, and so on. For the first different filled digits, **a smaller filled digit corresponds to a smaller standard Young tableau**. Enumerate the standard Young tableaux of a given Young pattern $[\lambda]$ by an integer μ from 1 to $d_{[\lambda]}$. This increasing order is the so-called **dictionary order**. The readers are suggested to learn how to arrange the standard Young tableaux of a given Young pattern in the dictionary order. For example, the order of the standard Young tableaux of the Young pattern [3, 2] is

1	2	3
4	5	

1	2	4
3	5	

1	2	5
3	4	

1	3	4
2	5	

1	3	5
2	4	

4.2.3 *Young Operators*

A permutation of the digits in the jth row of a given Young tableau is called a horizontal permutation P_j of the Young tableau, and a permutation of the digits in the kth column is called its vertical permutation Q_k. The product of horizontal permutations is a horizontal permutation, denoted by P. The product of vertical permutations is a vertical permutation, denoted by Q. The sum of all horizontal permutations of a given Young tableau is called its **horizontal operator**, and the sum of all vertical permutations multiplied by their permutation parities is called its **vertical operator**. The Young operator of a given Young tableau is the product of $\mathcal{P}$ and $\mathcal{Q}$:

$$\mathcal{Y} = \mathcal{P}\mathcal{Q}, \qquad \mathcal{P} = \sum P = \prod_j \left(\sum P_j\right), \qquad \mathcal{Q} = \sum \delta(Q) Q = \prod_k \left(\sum \delta(Q_k) Q_k\right). \tag{4.30}$$

P and Q are also said to be the horizontal permutation and the vertical permutation of the Young operator $\mathcal{Y}$, respectively. A Young operator is said to be **standard** if its Young tableau is standard. For a given Young operator $\mathcal{Y}$, its Young tableau and its Young pattern are fixed. For convenience, they are usually called the Young pattern $\mathcal{Y}$ and the Young tableau $\mathcal{Y}$, respectively. The symbol $\mathcal{Y}$ alone denotes the Young operator itself. Except for the identity E, no horizontal permutation is also a vertical permutation of the same Young tableau, and vice versa. The readers are suggested to be familiar with writing the expansion of a Young operator of a given Young tableau. The following is an example:

1	2	3
4	5	

$$\begin{aligned} \mathcal{Y} = &\{E + (1\ 2) + (1\ 3) + (2\ 3) + (1\ 2\ 3) + (3\ 2\ 1)\} \\ &\cdot \{E + (4\ 5)\}\{E - (1\ 4)\}\{E - (2\ 5)\} \end{aligned}$$

$$\begin{aligned}
&= \{E + (1\ 2) + (1\ 3) + (2\ 3) + (1\ 2\ 3) + (3\ 2\ 1) + (4\ 5) + (1\ 2)(4\ 5) \\
&\quad + (1\ 3)(4\ 5) + (2\ 3)(4\ 5) + (1\ 2\ 3)(4\ 5) + (3\ 2\ 1)(4\ 5)\} \\
&\quad \cdot \{E - (1\ 4) - (2\ 5) + (1\ 4)(2\ 5)\} \\
&= \{E + (1\ 2) + (1\ 3) + (2\ 3) + (1\ 2\ 3) + (3\ 2\ 1) + (4\ 5) + (1\ 2)(4\ 5) \\
&\quad + (1\ 3)(4\ 5) + (2\ 3)(4\ 5) + (1\ 2\ 3)(4\ 5) + (3\ 2\ 1)(4\ 5)\} \\
&\quad - \{(1\ 4) + (2\ 1\ 4) + (3\ 1\ 4) + (2\ 3)(1\ 4) + (2\ 3\ 1\ 4) + (3\ 2\ 1\ 4) \\
&\quad + (5\ 4\ 1) + (2\ 1\ 5\ 4) + (3\ 1\ 5\ 4) + (2\ 3)(5\ 4\ 1) + (2\ 3\ 1\ 5\ 4) \\
&\quad + (3\ 2\ 1\ 5\ 4)\} - \{(2\ 5) + (1\ 2\ 5) + (1\ 3)(2\ 5) + (3\ 2\ 5) + (3\ 1\ 2\ 5) \\
&\quad + (1\ 3\ 2\ 5) + (4\ 5\ 2) + (1\ 2\ 4\ 5) + (1\ 3)(4\ 5\ 2) + (3\ 2\ 4\ 5) \\
&\quad + (3\ 1\ 2\ 4\ 5) + (1\ 3\ 2\ 4\ 5)\} + \{(1\ 4)(2\ 5) + (1\ 4\ 2\ 5) + (3\ 1\ 4)(2\ 5) \\
&\quad + (1\ 4)(3\ 2\ 5) + (3\ 1\ 4\ 2\ 5) + (1\ 4\ 3\ 2\ 5) + (4\ 1\ 5\ 2) + (4\ 2)(1\ 5) \\
&\quad + (4\ 3\ 1\ 5\ 2) + (3\ 2\ 4\ 1\ 5) + (4\ 2)(3\ 1\ 5) + (4\ 3\ 2)(1\ 5)\}.
\end{aligned}$$

From the definition of a Young operator or from the above example one sees that a Young operator is a vector in the group algebra $\mathcal{L}$ of the permutation group S_n,

$$\mathcal{Y} = \sum_{R \in S_n} F(R) R, \tag{4.31}$$

where the coefficients $F(R)$ are taken to be 1, -1, or 0:

$$F(E) = F(P) = \delta(Q)F(Q) = \delta(Q)F(PQ) = 1. \tag{4.32}$$

The permutation R with nonvanishing coefficient $F(R)$ in Eq. (4.31) is called a permutation belonging to the Young operator $\mathcal{Y}$ (and to the Young tableau $\mathcal{Y}$), and the remaining permutations do not belong to $\mathcal{Y}$. We will discuss later the criterion whether a permutation belongs to the Young operator $\mathcal{Y}$ or not (see Corollary 4.5.4).

It is easy to determine a permutation S **uniquely** which transforms a Young tableau $\mathcal{Y}$ to another Young tableau $\mathcal{Y}'$ with the same Young pattern. In fact, the first row of S is filled with the digits of the Young tableau $\mathcal{Y}$ and the second row of S is filled with the corresponding digits of the Young tableau $\mathcal{Y}'$ in the same order. For example,

Young tableau $\mathcal{Y} =$

1	3	5
2	4	

Young tableau $\mathcal{Y}' =$

1	2	3
4	5	

$$S = \begin{pmatrix} 1 & 3 & 5 & 2 & 4 \\ 1 & 2 & 3 & 4 & 5 \end{pmatrix}.$$

Then, one has from the interchanging rule (2.41)

$$\begin{aligned} &S\mathcal{Y}S^{-1} = \mathcal{Y}', \qquad S\mathcal{P}S^{-1} = \mathcal{P}', \qquad S\mathcal{Q}S^{-1} = \mathcal{Q}', \\ &\qquad SPS^{-1} = P', \qquad SQS^{-1} = Q'. \end{aligned} \tag{4.33}$$

4.2.4 *Symmetry of Young Operators*

For a given Young operator $\mathcal{Y}$, the set of its horizontal permutations forms a subgroup of S_n, and the horizontal operator $\mathcal{P}$ is the sum of elements in the subgroup. Similarly, the set of its vertical permutations forms a subgroup of S_n, and the vertical operator $\mathcal{Q}$ is the algebraic sum of elements in the subgroup with their permutation parities. From the rearrangement theorem (Theorem 2.1), one has

$$P\mathcal{P} = \mathcal{P}P = \mathcal{P}, \qquad Q\mathcal{Q} = \mathcal{Q}Q = \delta(Q)\mathcal{Q}. \tag{4.34}$$

The Young operator $\mathcal{Y}$ satisfies

$$P\mathcal{Y} = \mathcal{Y} = \delta(Q)\mathcal{Y}Q. \tag{4.35}$$

In the literature, $\mathcal{P}$ and $\mathcal{Q}$ are also called the symmetrizer and antisymmetrizer of a Young tableau $\mathcal{Y}$, respectively. It is the **fundamental property of a Young operator**. Except for the normalization condition $F(E) = 1$, equation (4.32) can be derived from this property. In fact, from Eq. (4.35) one has

$$\begin{aligned} P^{-1}\mathcal{Y} &= \sum_{R\in S_n} F(R)P^{-1}R = \sum_{S\in S_n} F(PS)S = \mathcal{Y} = \sum_{S\in S_n} F(S)S, \\ \mathcal{Y}Q^{-1} &= \sum_{R\in S_n} F(R)RQ^{-1} = \sum_{S\in S_n} F(SQ)S = \delta(Q)\mathcal{Y} = \sum_{S\in S_n} \delta(Q)F(S)S. \end{aligned}$$

Thus,

$$F(S) = F(PS) = \delta(Q)F(SQ) = \delta(Q)F(PSQ). \tag{4.36}$$

Letting $S = E$ and assuming $F(E) = 1$, one obtains Eq. (4.32).

Fock found another important property of a Young operator.

Theorem 4.4 (Fock conditions) Let λ and λ' be the numbers of boxes in the jth row and in the j'th row of a Young tableau $\mathcal{Y}$, respectively. We denote by a_μ the digits filled in the jth row and by b_ν those in the j'th row. If $\lambda \geqslant \lambda'$, then

$$\left\{ E + \sum_{\mu=1}^{\lambda} (a_\mu \; b_\nu) \right\} \mathcal{Y} = 0. \tag{4.37}$$

Let τ and τ' be the numbers of boxes in the kth column and in the k'th column of a Young tableau $\mathcal{Y}$, respectively. We denote by c_μ the digits filled in the kth column and by d_ν those in the k'th column. If $\tau \geqslant \tau'$, then

$$\mathcal{Y}\left\{E - \sum_{\mu=1}^{\tau} (c_\mu\ d_\nu)\right\} = 0. \tag{4.38}$$

The following example shows the filling positions of a_μ, b_ν, c_μ, and d_ν in a Young tableau, where $j = k = 2$ and $j' = k' = 3$:

a_1	a_2	a_3	a_4	a_5
	b_ν			

	c_1		
	c_2	d_ν	
	c_3		
	c_4		
	c_5		

Proof The proofs for two Fock conditions are similar. In the following we will prove the condition (4.37) as example. Let $(\sum P_j)$ and $(\sum P_{j'})$ denote the sums of horizontal permutations in the jth row and in the j'th row, respectively. $\mathcal{P}'$ is the sum of horizontal permutations not containing the digits in the jth row and in the j'th row. Hence,

$$\mathcal{Y} = \mathcal{P}\mathcal{Q} = \left(\sum P_j\right)\left(\sum P_{j'}\right)\mathcal{P}'\mathcal{Q}.$$

$(\sum P_j)$ is the sum of $\lambda!$ different permutations of the λ objects in the jth row, so that it is the totally symmetric operator of them. Left-multiplying it with $\left\{E + \sum_\mu (a_\mu\ b_\nu)\right\}$, one obtains a sum of $(\lambda+1)!$ different permutations of the object b_ν and λ objects in the jth row. Hence, the sum is the totally symmetric operator for those $(\lambda+1)$ objects, denoted by $\sum P_j(b_\nu)$. From interchanging rule (2.41), when each term $P_{j'}$ in the sum $(\sum P_{j'})$ moves from the right-side of $\sum P_j(b_\nu)$ to its left-side, the only change of $\sum P_j(b_\nu)$ is that b_ν may be replaced with another digit b_ρ in the j' row,

$$\left[\sum P_j(b_\nu)\right] P_{j'} = P_{j'}\left[\sum P_j(b_\rho)\right]. \tag{4.39}$$

$\sum P_j(b_\rho)$ commutes with $\mathcal{P}'$. Since $\lambda \geqslant \lambda'$, there exists an object a_ρ in the jth row which is located at the same column as b_ρ. The transposition $(a_\rho\ b_\rho)$ leaves $[\sum P_j(b_\rho)]$ invariant but changes the sign of $\mathcal{Q}$, so that

$$\left[\sum P_j(b_\rho)\right]\mathcal{Q} = \left[\sum P_j(b_\rho)\right](a_\rho\ b_\rho)\mathcal{Q} = -\left[\sum P_j(b_\rho)\right]\mathcal{Q} = 0. \tag{4.40}$$

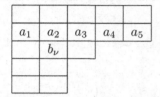

Thus, Eq. (4.37) is proved. □

The key in the proof is that if the box number λ in the jth row is not less than the box number λ' in the j'th row, $\mathcal{Y}$ is annihilated from the left by symmetrizing b_ν in the j'th row with all a_μ in the jth row. Note that Eq. (4.37) holds if $j < j'$, but when $j > j'$, it holds only if $\lambda = \lambda'$. Similarly, if the box number τ in the kth column is not less than the box number τ' in the k'th column, $\mathcal{Y}$ is annihilated from the right by antisymmetrizing d_ν in the k'th column with all c_μ in the kth column. Equation (4.38) holds if $k < k'$, but when $k > k'$, it holds only if $\tau = \tau'$.

4.2.5 *Products of Young Operators*

For two Young operators $\mathcal{Y}$ and $\mathcal{Y}'$, one cannot derive $\mathcal{Y}\mathcal{Y}' = 0$ from $\mathcal{Y}'\mathcal{Y} = 0$, and vice versa. Two Young operators are called orthogonal only when two products vanish simultaneously. The method used in Eq. (4.40) is helpful to prove that a product of two Young operators vanishes.

Theorem 4.5 If there exist two digits a and b which filled in one row of a Young tableau $\mathcal{Y}$ and in one column of a Young tableau $\mathcal{Y}'$ simultaneously, or equivalently, if $T_0 = (a\ b)$ is the horizontal transposition of a Young operator $\mathcal{Y} = \mathcal{P}\mathcal{Q}$ and the vertical transposition of a Young operator $\mathcal{Y}' = \mathcal{P}'\mathcal{Q}'$ simultaneously, then

$$\mathcal{Q}'\mathcal{P} = 0 \quad \text{and} \quad \mathcal{Y}'\mathcal{Y} = 0. \tag{4.41}$$

Proof $\mathcal{Q}'\mathcal{P} = \mathcal{Q}'T_0\mathcal{P} = -\mathcal{Q}'\mathcal{P} = 0$. □

The subscript 0 emphasizes T_0 to be a transposition. In proving the following corollaries, the key is to find the pair of digits.

Corollary 4.5.1 If a Young pattern $\mathcal{Y}'$ is less than a Young pattern $\mathcal{Y}$, then $\mathcal{Y}'\mathcal{Y} = 0$.

Proof Let $[\lambda']$ and $[\lambda]$ denote the Young patterns $\mathcal{Y}'$ and $\mathcal{Y}$, respectively. Since the Young pattern $\mathcal{Y}'$ is less than the Young pattern $\mathcal{Y}$, there is a row number j such that $\lambda'_i = \lambda_i$ when $i < j$ and $\lambda'_j < \lambda_j$.

Check the digits filled in the first $(j-1)$ rows of the Young tableau $\mathcal{Y}$ to see whether there exist two digits in one row of the Young tableau $\mathcal{Y}$ which also occur in one column of the Young tableau $\mathcal{Y}'$. If yes, from Theorem 4.5, $\mathcal{Y}'\mathcal{Y} = 0$. If no, there is a vertical permutation Q' of the Young tableau $\mathcal{Y}'$ which transforms the Young tableau $\mathcal{Y}'$ to the Young tableau $\mathcal{Y}''$ such that each row among the first $(j-1)$ rows of both the Young tableau $\mathcal{Y}$

and the Young tableau $\mathcal{Y}''$ contains the same digits. Note that

$$\mathcal{Y}'' = Q'\mathcal{Y}'Q'^{-1} = \delta(Q')Q'\mathcal{Y}', \qquad \mathcal{Y}' = \delta(Q')Q'^{-1}\mathcal{Y}''.$$

Since $\lambda'_j < \lambda_j$, there exist at least two digits in the jth row of the Young tableau $\mathcal{Y}$, which occur in one column of the Young tableau $\mathcal{Y}''$. Thus, $\mathcal{Y}''\mathcal{Y} = 0$, and then, $\mathcal{Y}'\mathcal{Y} = 0$. In this corollary the Young operators are not necessary to be standard. □

Corollary 4.5.2 For a given Young pattern, if a standard Young tableau $\mathcal{Y}'$ is larger than a standard Young tableau $\mathcal{Y}$, then $\mathcal{Y}'\mathcal{Y} = 0$.

Proof Compare the filled digits in two standard Young tableaux $\mathcal{Y}$ and $\mathcal{Y}'$ from left to right of the first row, and then those of the second row, and so on. If the first different filled digits occur at the jth column of the ith row where the digit in the Young tableau $\mathcal{Y}'$ is a and the digit in the Young tableau $\mathcal{Y}$ is b, then $a > b$. Thus, b has to be filled in the j'th column of the i'th row of the standard Young tableau $\mathcal{Y}'$, where $j' < j$ and $i' > i$. The digits filled in the j'th column of the ith row of both the Young tableaux $\mathcal{Y}$ and $\mathcal{Y}'$ are the same, say c. Thus, the pair of digits b and c occurs both in the same row of the Young tableau $\mathcal{Y}$ and in the same column of the Young tableau $\mathcal{Y}'$, so that from Theorem 4.5, $\mathcal{Y}'\mathcal{Y} = 0$. □

Corollary 4.5.3 For a given Young pattern, if any two digits in one column of the Young tableau $\mathcal{Y}'$ never occur in one row of the Young tableau $\mathcal{Y}$, the permutation R transforming the Young tableau $\mathcal{Y}$ to the Young tableau $\mathcal{Y}'$ belongs to both the Young tableaux $\mathcal{Y}$ and $\mathcal{Y}'$.

Proof Remind that the condition of Corollary 4.5.3 is equivalent to that any two digits in one row of the Young tableau $\mathcal{Y}$ never occur in one column of the Young tableau $\mathcal{Y}'$.

Let P denote the horizontal permutation of the Young tableau $\mathcal{Y}$ which transforms the Young tableau $\mathcal{Y}$ to the Young tableau $\mathcal{Y}''$ such that each column of both the Young tableau $\mathcal{Y}'$ and the Young tableau $\mathcal{Y}''$ contains the same digits. Thus, the permutation transforming the Young tableau $\mathcal{Y}''$ to the Young tableau $\mathcal{Y}'$ is a vertical permutation Q'' of the Young tableau $\mathcal{Y}''$, $Q'' = PQP^{-1}$, where Q is a vertical permutation of the Young tableau $\mathcal{Y}$. Therefore, $R = Q''P = PQ$ and $R = RRR^{-1} = (RPR^{-1})(RQR^{-1})$. □

The converse and negative theorem (?) of Corollary 4.5.3 says that if R transforming the Young tableau $\mathcal{Y}$ to the Young tableau $\mathcal{Y}'$ does not belong to $\mathcal{Y}$, there at least exists a horizontal transposition P_0 of $\mathcal{Y}$ which is also a vertical transposition of $\mathcal{Y}'$, $P_0 = Q'_0 = RQ_0R^{-1}$. Namely,

$$R = P_0 R Q_0. \tag{4.42}$$

Corollary 4.5.4 A permutation R does not belong to $\mathcal{Y}$ if and only if there are a horizontal transposition P_0 and a vertical transposition Q_0 of $\mathcal{Y}$ such that Eq. (4.42) holds.

Proof From the converse and negative theorem of Corollary 4.5.3, if R does not belong to the Young tableau $\mathcal{Y}$, Eq. (4.42) holds. Conversely, if $R = P_0 R Q_0$, from Eq. (4.32), $F(R) = F(P_0 R Q_0) = -F(R) = 0$. □

4.2.6 *Young Operators as Primitive Idempotents*

In this subsection we are going to show that the Young operator $\mathcal{Y}$ is proportional to the primitive idempotent of the permutation group S_n. Then, in the next section the irreducible representations of S_n will be calculated.

Theorem 4.6 If a vector $\mathcal{X}$ in the group algebra $\mathcal{L}$ of S_n satisfies

$$P\mathcal{X} = \delta(Q)\mathcal{X}Q = \mathcal{X} \tag{4.43}$$

for all horizontal permutations P and vertical permutations Q of $\mathcal{Y}$, then $\mathcal{X}$ is proportional to the Young operator $\mathcal{Y}$:

$$\mathcal{X} = \lambda\mathcal{Y}. \tag{4.44}$$

Proof Let $\mathcal{X} = \sum_{R \in S_n} F_1(R)R$. Similar to the proof of Eq. (4.36), from Eq. (4.43) one has

$$F_1(S) = F_1(PS) = \delta(Q)F_1(SQ) = \delta(Q)F_1(PSQ).$$

Taking $S = E$ and $F_1(E) = \lambda$, one obtains

$$\lambda = F_1(E) = F_1(P) = \delta(Q)F_1(Q) = \delta(Q)F_1(PQ).$$

If a permutation R does not belong to $\mathcal{Y}$, then from Corollary 4.5.4

$$F_1(R) = F_1(P_0 R Q_0) = -F_1(R) = 0.$$

In comparison with (4.32), Eq. (4.44) is proved. □

Since $\mathcal{Y}t\mathcal{Y}$ satisfies Eq. (4.43), Corollary 4.6.1 follows.

Corollary 4.6.1 For any vector t in the group algebra $\mathcal{L}$ of S_n, one has

$$\mathcal{Y}t\mathcal{Y} = \lambda_t\mathcal{Y}, \tag{4.45}$$

where λ_t is a constant depending upon t and is allowed to be zero.

Corollary 4.6.2 The square of a Young operator $\mathcal{Y}$ is not vanishing:

$$\mathcal{Y}\mathcal{Y} = \lambda\mathcal{Y} \neq 0. \tag{4.46}$$

Proof We are going to calculate the constant λ explicitly and to show that λ is non-vanishing. The right ideal generated by $\mathcal{Y}$ is $\mathcal{R} = \mathcal{Y}\mathcal{L}$. Let f denote the dimension of $\mathcal{R}$. $f \neq 0$ because $\mathcal{R}$ contains at least a nonvanishing vector $\mathcal{Y}$. Take a complete set of basis vectors x_μ in the group algebra $\mathcal{L}$ of S_n such that the first f basis vectors x_μ, $\mu \leqslant f$, constitute a set of bases in $\mathcal{R}$, and the last $(n! - f)$ basis vectors x_μ, $\mu > f$, do not belong to $\mathcal{R}$. Since $x_\mu \in \mathcal{R} = \mathcal{Y}\mathcal{L}$ for $\mu \leqslant f$, x_μ can be expressed as a product of $\mathcal{Y}$ and a vector $y_\mu \in \mathcal{L}$:

$$x_\mu = \mathcal{Y}y_\mu, \qquad 1 \leqslant \mu \leqslant f. \tag{4.47}$$

Now, calculate the product $\mathcal{Y}x_\mu$ from two viewpoints. On one hand, x_μ is a basis vector in $\mathcal{L}$, and $\mathcal{Y}$ is an operator acting on x_μ. The representation matrix $\overline{D}(\mathcal{Y})$ of $\mathcal{Y}$ with respect to the basis vectors x_μ is calculated by

$$\mathcal{Y}x_\mu = \sum_{\nu=1}^{n!} x_\nu \overline{D}_{\nu\mu}(\mathcal{Y}). \tag{4.48}$$

Since the representation $\overline{D}(\mathrm{S}_n)$ is equivalent to the regular one, one has

$$\mathrm{Tr}\ \overline{D}(\mathcal{Y}) = \mathrm{Tr}\ \overline{D}(E) = n!. \tag{4.49}$$

On the other hand, since $\mathcal{Y}x_\mu \in \mathcal{Y}\mathcal{L} = \mathcal{R}$, the summation in Eq. (4.48) only contains the terms with $\nu \leqslant f$,

$$\overline{D}_{\nu\mu}(\mathcal{Y}) = 0 \qquad \text{when} \ \ \nu > f. \tag{4.50}$$

When $\mu \leqslant f$, from Eq. (4.47) one has $\mathcal{Y}x_\mu = \mathcal{Y}\mathcal{Y}y_\mu = \lambda\mathcal{Y}y_\mu = \lambda x_\mu$. Thus, $\overline{D}_{\nu\mu}(\mathcal{Y}) = \delta_{\nu\mu}\lambda$ when $\mu \leqslant f$, and $\mathrm{Tr}\ \overline{D}(\mathcal{Y}) = f\lambda$. In comparison with Eq. (4.49), due to $f \neq 0$, one obtains

$$\lambda = n!/f \neq 0, \tag{4.51}$$

□

Corollary 4.6.3 $a = (f/n!)\mathcal{Y}$ is a primitive idempotent of the permutation group S_n.

Corollary 4.6.4 If any two digits in one column of the Young tableau $\mathcal{Y}'$ never occur in one row of the Young tableau $\mathcal{Y}$, then $\mathcal{Y}'\mathcal{Y} \neq 0$.

Proof From Corollary 4.5.3, $R = PQ$ belongs to $\mathcal{Y}$ if $\mathcal{Y}' = R\mathcal{Y}R^{-1}$. Then, $\mathcal{Y}'\mathcal{Y} = R\mathcal{Y}Q^{-1}P^{-1}\mathcal{Y} = \delta(Q)R\mathcal{Y}\mathcal{Y} = \delta(Q)\lambda R\mathcal{Y} \neq 0$. □

Corollary 4.6.5 Two minimal left ideals generated by the Young operators $\mathcal{Y}$ and $\mathcal{Y}'$, respectively, are equivalent if and only if their Young patterns are the same as each other.

Proof If the Young patterns $\mathcal{Y}$ and $\mathcal{Y}'$ are the same, there exist a permutation R transforming the Young tableau $\mathcal{Y}$ to the Young tableau $\mathcal{Y}'$ such that $\mathcal{Y}' = R\mathcal{Y}R^{-1}$, and $\mathcal{Y}'R\mathcal{Y} = R\mathcal{Y}\mathcal{Y} \neq 0$. From Theorem 4.1, two left ideals are equivalent.

Conversely, if two Young patterns $\mathcal{Y}$ and $\mathcal{Y}'$ are different, without loss of generality, the Young pattern $\mathcal{Y}$ is assumed to be larger than the Young pattern $\mathcal{Y}'$. For any permutation R, the Young pattern $\mathcal{Y}''$, where $\mathcal{Y}'' = R\mathcal{Y}R^{-1}$, is the same as the Young pattern $\mathcal{Y}$. Then, due to Corollary 4.5.1, $\mathcal{Y}'\mathcal{Y}'' = 0$ and $\mathcal{Y}'R\mathcal{Y} = \mathcal{Y}'\mathcal{Y}''R = 0$. □

Therefore, an irreducible representation of S_n can be characterized by a Young pattern $[\lambda]$. Two representations denoted by different Young patterns are inequivalent. Thus, Corollary 4.6.6 follows Theorem 4.1.

Corollary 4.6.6. Two Young operators corresponding to different Young patterns $\mathcal{Y}$ and $\mathcal{Y}'$ are orthogonal to each other, $\mathcal{Y}\mathcal{Y}' = \mathcal{Y}'\mathcal{Y} = 0$.

The number of all different Young patterns with n boxes is equal to the number of classes in S_n. Hence, the inequivalent irreducible representations denoted by all Young patterns with n boxes are complete for S_n.

4.3 Irreducible Representations of S_n

4.3.1 *Orthogonal Young Operators*

Two Young operators corresponding to different Young patterns are orthogonal to each other. However, two standard Young operators corresponding to the same Young pattern are not necessary to be orthogonal. The non-orthogonal standard Young operators occur only for S_n with $n \geqslant 5$. For $n = 5$ there are two Young patterns, $[3, 2]$ and $[2, 2, 1]$, where some standard Young operators are not orthogonal to each other. For example, list the standard Young tableaux for $[3, 2]$ from the smallest to the largest:

$\mathcal{Y}_1^{[3,2]}$

1	2	3
4	5	

$\mathcal{Y}_2^{[3,2]}$

1	2	4
3	5	

$\mathcal{Y}_3^{[3,2]}$

1	2	5
3	4	

$\mathcal{Y}_4^{[3,2]}$

1	3	4
2	5	

$\mathcal{Y}_5^{[3,2]}$

1	3	5
2	4	

Due to Corollary 4.5.2, $\mathcal{Y}_\mu^{[3,2]}\mathcal{Y}_\nu^{[3,2]} = 0$ when $\mu > \nu$. Check the product $\mathcal{Y}_\mu^{[3,2]}\mathcal{Y}_\nu^{[3,2]}$ with $\mu < \nu$ one by one whether two digits in one row of the Young tableau $\mathcal{Y}_\nu^{[3,2]}$ never occur in one column of the Young tableau $\mathcal{Y}_\mu^{[3,2]}$. The result is that only

$$\mathcal{Y}_1^{[3,2]}\mathcal{Y}_5^{[3,2]} \neq 0. \tag{4.52}$$

The permutation $R_{15} = P_5Q_5 = P_1Q_1$ transforming the Young tableau $\mathcal{Y}_5^{[3,2]}$ to the Young tableau $\mathcal{Y}_1^{[3,2]}$ is

$$\begin{aligned} R_{15} &= \begin{pmatrix} 1\,3\,5\,2\,4 \\ 1\,2\,3\,4\,5 \end{pmatrix} = (3\ 2\ 4\ 5) = (2\ 4)\,(4\ 5\ 3) \\ &= (2\ 4)\,(5\ 3)\ \ (3\ 4) = (4\ 5)\,(3\ 2)\ \ (2\ 5)\,, \end{aligned} \tag{4.53}$$

$$P_5 = (2\ 4)\,(5\ 3)\,, \quad Q_5 = (3\ 4)\,, \quad P_1 = (4\ 5)(3\ 2), \quad Q_1 = (2\ 5).$$

This is the general method to calculate the permutation between two non-orthogonal Young operators.

For a given Young pattern, we want to orthogonalize the standard Young operators by left-multiplying or right-multiplying them with some vectors y_μ in $\mathcal{L}$. In the above example, there are two sets of orthogonal Young operators. One set is

$$\mathcal{Y}_1' = \mathcal{Y}_1^{[3,2]}\,[E - P_5]\,, \qquad \mathcal{Y}_\nu' = \mathcal{Y}_\nu^{[3,2]}, \qquad \nu > 1. \tag{4.54}$$

Since $\mathcal{Y}_1^{[3,2]}P_5 = \mathcal{Y}_1^{[3,2]}R_{15}Q_5^{-1} = \delta(Q_5)R_{15}\mathcal{Y}_5^{[3,2]}$, $\mathcal{Y}_1'\mathcal{Y}_\mu^{[3,2]} = \mathcal{Y}_1^{[3,2]}\mathcal{Y}_\mu^{[3,2]}$ when $\mu < 5$. $\mathcal{Y}_1'\mathcal{Y}_5^{[3,2]} = \mathcal{Y}_1^{[3,2]}(E - P_5)\mathcal{Y}_5^{[3,2]} = 0$. The other set is

$$\mathcal{Y}_5'' = [E + Q_1]\,\mathcal{Y}_5^{[3,2]}, \qquad \mathcal{Y}_\mu'' = \mathcal{Y}_\mu^{[3,2]}, \qquad \mu < 5. \tag{4.55}$$

Since $Q_1\mathcal{Y}_5^{[3,2]} = P_1^{-1}R_{15}\mathcal{Y}_5^{[3,2]} = \mathcal{Y}_1^{[3,2]}R_{15}$, $\mathcal{Y}_\nu^{[3,2]}\mathcal{Y}_5'' = \mathcal{Y}_\nu^{[3,2]}\mathcal{Y}_5^{[3,2]}$ when $\nu > 1$. $\mathcal{Y}_1^{[3,2]}\mathcal{Y}_5'' = \mathcal{Y}_1^{[3,2]}(E + Q_1)\mathcal{Y}_5^{[3,2]} = 0$.

Generally, for a given Young pattern $[\lambda]$ where the standard Young operators $\mathcal{Y}_\mu^{[\lambda]}$ are not orthogonal completely, we want to choose some vectors $y_\mu^{[\lambda]}$ or $\overline{y}_\mu^{[\lambda]}$ such that the new set of $\mathcal{Y}_\mu^{[\lambda]}y_\mu^{[\lambda]}$ or the new set of $\overline{y}_\mu^{[\lambda]}\mathcal{Y}_\mu^{[\lambda]}$ are mutually orthogonal. We will discuss the first set in detail and give the result for the second set. Since the Young pattern $[\lambda]$ is fixed, in the following we will omit the superscript $[\lambda]$ for simplicity.

The problem is to find y_μ such that

$$\mathcal{Q}_\mu y_\mu \mathcal{P}_\nu = \delta_{\mu\nu}\mathcal{Q}_\mu\mathcal{P}_\mu, \qquad 1 \leqslant \mu \leqslant d, \qquad 1 \leqslant \nu \leqslant d. \tag{4.56}$$

Then, $\mathcal{Y}_\mu y_\mu \mathcal{Y}_\nu = \delta_{\mu\nu}\mathcal{Y}_\mu \mathcal{Y}_\mu$. Let $R_{\mu\nu}$ denote the permutation transforming the standard Young tableau $\mathcal{Y}_\nu$ to the standard Young tableau $\mathcal{Y}_\mu$,

$$\begin{aligned} &R_{\mu\nu}\mathcal{Y}_\nu = \mathcal{Y}_\mu R_{\mu\nu}, \qquad R_{\mu\nu}\mathcal{P}_\nu = \mathcal{P}_\mu R_{\mu\nu}, \qquad R_{\mu\nu}\mathcal{Q}_\nu = \mathcal{Q}_\mu R_{\mu\nu}, \\ &R_{\mu\rho}R_{\rho\nu} = R_{\mu\nu}, \qquad R_{\mu\mu} = E. \end{aligned} \tag{4.57}$$

Due to Corollary 4.5.3,

$$R_{\mu\nu} = P_\nu^{(\mu)} Q_\nu^{(\mu)} = \overline{P}_\mu^{(\nu)} \overline{Q}_\mu^{(\nu)} \qquad \text{when } \mathcal{Y}_\mu \mathcal{Y}_\nu \neq 0, \tag{4.58}$$

where $P_\nu^{(\mu)}$ and $Q_\nu^{(\mu)}$ are the horizontal permutation and the vertical permutation of $\mathcal{Y}_\nu$, respectively, and $\overline{P}_\mu^{(\nu)}$ and $\overline{Q}_\mu^{(\nu)}$ are those of $\mathcal{Y}_\mu$. Let

$$P_{\mu\nu} = \begin{cases} P_\nu^{(\mu)} & \text{when } \mathcal{Y}_\mu \mathcal{Y}_\nu \neq 0, \\ 0 & \text{when } \mathcal{Y}_\mu \mathcal{Y}_\nu = 0. \end{cases} \tag{4.59}$$

Obviously, when $\mathcal{Y}_\mu \mathcal{Y}_\nu \neq 0$,

$$\begin{aligned} &P_{\mu\nu}\mathcal{Q}_\nu = R_{\mu\nu}\mathcal{Q}_\nu \left(Q_\nu^{(\mu)}\right)^{-1} = \mathcal{Q}_\mu P_{\mu\nu}, \\ &\mathcal{P}_\nu P_{\mu\nu} = P_{\mu\nu}\mathcal{P}_\nu = \mathcal{P}_\nu, \\ &\mathcal{Y}_\mu P_{\mu\nu} = R_{\mu\nu}\mathcal{Y}_\nu \left(Q_\nu^{(\mu)}\right)^{-1} = \delta(Q_\nu^{(\mu)}) R_{\mu\nu}\mathcal{Y}_\nu. \end{aligned} \tag{4.60}$$

Define y_μ one by one from $\mu = d$ to $\mu = 1$,

$$y_\mu = E - \sum_{\rho=\mu+1}^{d} P_{\mu\rho} y_\rho, \qquad y_d = E, \qquad 1 \leqslant \mu \leqslant d, \tag{4.61}$$

where y_μ is the algebraic sum of elements of S_n with coefficients ± 1. Prove Eq. (4.56) by mathematical induction. First, equation (4.56) holds when $\mu = d$ owing to Corollary 4.5.2. Suppose that equation (4.56) holds for $\mu > \tau$. For $\mu = \tau$, Eq. (4.56) also holds because

$$\begin{aligned} \mathcal{Q}_\tau y_\tau \mathcal{P}_\nu &= \mathcal{Q}_\tau \mathcal{P}_\nu - \sum_{\rho=\tau+1}^{d} \mathcal{Q}_\tau P_{\tau\rho} y_\rho \mathcal{P}_\nu = \mathcal{Q}_\tau \mathcal{P}_\nu - \sum_{\rho=\tau+1}^{d} \delta_{\rho\nu} P_{\tau\rho} \mathcal{Q}_\rho \mathcal{P}_\rho \\ &= \begin{cases} 0 & \text{when } \nu < \tau, \\ \mathcal{Q}_\tau \mathcal{P}_\tau & \text{when } \nu = \tau, \\ \mathcal{Q}_\tau \mathcal{P}_\nu - P_{\tau\nu}\mathcal{Q}_\nu \mathcal{P}_\nu = 0 & \text{when } \nu > \tau. \end{cases} \end{aligned}$$

Note that due to Eq. (4.60),

$$\mathcal{Y}_\mu y_\mu = \mathcal{Y}_\mu - \sum_{\mathcal{Y}_\mu \mathcal{Y}_\nu \neq 0,\ \nu=\mu+1}^{d} \delta(Q_\nu^{(\mu)}) R_{\mu\nu} \mathcal{Y}_\nu y_\nu = \sum_{\rho=\mu}^{d} t_\rho \mathcal{Y}_\rho, \tag{4.62}$$

where the sum index ν in the middle expression runs over from $\mu+1$ to d and in the condition $\mathcal{Y}_\mu\mathcal{Y}_\nu \neq 0$, and t_ρ in the last expression is a vector in $\mathcal{L}$ which is allowed to be zero.

Similarly, let

$$\overline{Q}_{\mu\nu} = \begin{cases} \delta(\overline{Q}_\mu^{(\nu)})\overline{Q}_\mu^{(\nu)} & \text{when } \mathcal{Y}_\mu\mathcal{Y}_\nu \neq 0, \\ 0 & \text{when } \mathcal{Y}_\mu\mathcal{Y}_\nu = 0. \end{cases} \tag{4.63}$$

$\overline{y}_\nu$ are defined one by one from $\nu=1$ to $\nu=d$,

$$\overline{y}_\nu = E - \sum_{\rho=1}^{\nu-1} \overline{y}_\rho \overline{Q}_{\rho\nu}, \qquad \overline{y}_1 = E, \qquad 1 \leqslant \nu \leqslant d, \tag{4.64}$$

such that $\mathcal{Y}_\nu\overline{y}_\mu\mathcal{Y}_\mu = \delta_{\nu\mu}\mathcal{Y}_\nu\mathcal{Y}_\nu$.

Theorem 4.7 The following $e_\mu^{[\lambda]}$ constitute a complete set of orthogonal primitive idempotents:

$$e_\mu^{[\lambda]} = \frac{d_{[\lambda]}}{n!}\mathcal{Y}_\mu^{[\lambda]} y_\mu^{[\lambda]}, \qquad 1 \leqslant \mu \leqslant d_{[\lambda]}, \tag{4.65}$$

and the identical element E can be decomposed as

$$E = \frac{1}{n!}\sum_{[\lambda]} d_{[\lambda]} \sum_{\mu=1}^{d_{[\lambda]}} \mathcal{Y}_\mu^{[\lambda]} y_\mu^{[\lambda]}. \tag{4.66}$$

Proof For a given Young pattern $[\lambda]$, there are $d_{[\lambda]}$ orthogonal primitive idempotents $e_\mu^{[\lambda]}$, given in (4.65) but replacing $d_{[\lambda]}$ with $f_{[\lambda]}$, where $f_{[\lambda]}$ is the dimension of the right ideal $\mathcal{L}_\mu^{[\lambda]}$ generated by $e_\mu^{[\lambda]}$ (see Corollaries 4.6.3). Since the multiplicity of each irreducible representation in the decomposition of the regular representation is equal to the dimension of the representation, one has $d_{[\lambda]} \leqslant f_{[\lambda]}$. On the other hand, the square sum of the dimensions of the inequivalent irreducible representations of a finite group G is equal to its order,

$$\sum_{[\lambda]} f_{[\lambda]}^2 = n!. \tag{4.67}$$

In comparison with (4.21) one obtains

$$d_{[\lambda]} = f_{[\lambda]}. \tag{4.68}$$

Therefore, **the direct sum of the left ideals $\mathcal{L}_\mu^{[\lambda]} = \mathcal{L}e_\mu^{[\lambda]}$ is equal to the group algebra** $\mathcal{L}$, and Eq. (4.66) follows. The direct sum of the right ideals $\mathcal{R}_\mu^{[\lambda]} = e_\mu^{[\lambda]}\mathcal{L}$ is equal to the group algebra $\mathcal{L}$, too. □

In the same reason, $\overline{e}_\mu^{[\lambda]}$ also constitute a complete set of orthogonal primitive idempotents:

$$\overline{e}_\mu^{[\lambda]} = \frac{d_{[\lambda]}}{n!}\overline{y}_\mu^{[\lambda]}\mathcal{Y}_\mu^{[\lambda]}, \qquad 1 \leqslant \mu \leqslant d_{[\lambda]}, \tag{4.69}$$

and the identical element E can be decomposed as

$$E = \frac{1}{n!}\sum_{[\lambda]} d_{[\lambda]} \sum_{\mu=1}^{d_{[\lambda]}} \overline{y}_\mu^{[\lambda]}\mathcal{Y}_\mu^{[\lambda]}. \tag{4.70}$$

4.3.2 *Irreducible Bases*

In terms of the permutation $R_{\mu\nu}^{[\lambda]}$ transforming the standard Young tableau $\mathcal{Y}_\nu^{[\lambda]}$ to the standard Young tableau $\mathcal{Y}_\mu^{[\lambda]}$ (see Eq. (4.57)), we define

$$\begin{aligned}
\phi_{\mu\mu}^{[\lambda]} &= e_\mu^{[\lambda]} = \left(d_{[\lambda]}/n!\right)\mathcal{Y}_\mu^{[\lambda]}y_\mu^{[\lambda]}, \\
\phi_{\mu\nu}^{[\lambda]} &= e_\mu^{[\lambda]}R_{\mu\nu}^{[\lambda]}e_\nu^{[\lambda]} = \left(d_{[\lambda]}/n!\right)^2\mathcal{Y}_\mu^{[\lambda]}y_\mu^{[\lambda]}R_{\mu\nu}^{[\lambda]}\mathcal{Y}_\nu^{[\lambda]}y_\nu^{[\lambda]} \\
&= \left(d_{[\lambda]}/n!\right)^2\mathcal{Y}_\mu^{[\lambda]}y_\mu^{[\lambda]}\mathcal{Y}_\mu^{[\lambda]}R_{\mu\nu}^{[\lambda]}y_\nu^{[\lambda]} \\
&= \left(d_{[\lambda]}/n!\right)\mathcal{Y}_\mu^{[\lambda]}R_{\mu\nu}^{[\lambda]}y_\nu^{[\lambda]} \\
&= \left(d_{[\lambda]}/n!\right)R_{\mu\nu}^{[\lambda]}\mathcal{Y}_\nu^{[\lambda]}y_\nu^{[\lambda]} \\
&= R_{\mu\nu}^{[\lambda]}e_\nu^{[\lambda]}.
\end{aligned} \tag{4.71}$$

The idempotent $e_\mu^{[\lambda]}$ on the left of Eq. (4.71) can be removed. Thus, $\phi_{\mu\nu}^{[\lambda]}$ satisfy the recurrence relation (see Eq. (3.116)):

$$\phi_{\mu\rho}^{[\lambda]}\phi_{\tau\nu}^{[\lambda]} = \delta_{\rho\tau}e_\mu^{[\lambda]}R_{\mu\rho}^{[\lambda]}\left(e_\rho^{[\rho]}R_{\rho\nu}^{[\lambda]}e_\nu^{[\lambda]}\right) = \delta_{\rho\tau}e_\mu^{[\lambda]}R_{\mu\rho}^{[\lambda]}R_{\rho\nu}^{[\lambda]}e_\nu^{[\lambda]} = \delta_{\rho\tau}\phi_{\mu\nu}^{[\lambda]}. \tag{4.72}$$

Due to the orthogonality of $e_\mu^{[\lambda]}$, $\phi_{\mu\nu}^{[\lambda]}$ with $1 \leqslant \mu \leqslant d_{[\lambda]}$ **constitute the complete set of basis vectors of the left ideal $\mathcal{L}_\nu^{[\lambda]} = \mathcal{L}e_\nu^{[\lambda]}$** in $\mathcal{L}$, and $\phi_{\mu\nu}^{[\lambda]}$ with $1 \leqslant \nu \leqslant d_{[\lambda]}$ constitute the complete set of basis vectors of the right ideal $\mathcal{R}_\mu^{[\lambda]} = e_\mu^{[\lambda]}\mathcal{L}$ in $\mathcal{L}$ (see Eq. (3.115)):

$$S\phi_{\mu\nu}^{[\lambda]} = \sum_{\rho=1}^{d_{[\lambda]}} \phi_{\rho\nu}^{[\lambda]}D_{\rho\mu}^{[\lambda]}(S), \qquad \phi_{\mu\nu}^{[\lambda]}S = \sum_{\rho=1}^{d_{[\lambda]}} D_{\nu\rho}^{[\lambda]}(S)\phi_{\mu\rho}^{[\lambda]}. \tag{4.73}$$

$D^{[\lambda]}_{\rho\mu}(S)$ will be calculated from Eq. (4.73) in the next subsection. Although the irreducible bases in the group algebra $\mathcal{L}$ of the permutation group S_n depend on the chosen forms of the inequivalent irreducible representations of S_n, in this book we simply call $\phi^{[\lambda]}_{\mu\nu}$ **the irreducible bases in the group algebra** $\mathcal{L}$ of S_n for convenience. The irreducible bases will change when the representations make a similarity transformation (see Eq. (4.90)). Due to Eq. (4.71), Theorem 4.8 follows. The same arguments are applicable to Corollary 4.8.1 where $e^{[\lambda]}_{\mu}$ is replaced with $\overline{e}^{[\lambda]}_{\mu}$.

Theorem 4.8 $R^{[\lambda]}_{\mu\nu}\mathcal{Y}^{[\lambda]}_{\nu}$ with $1 \leqslant \mu \leqslant d_{[\lambda]}$ constitute the complete set of basis vectors of the left ideal $\mathcal{L}\mathcal{Y}^{[\lambda]}_{\nu}$ in $\mathcal{L}$.

Corollary 4.8.1 $\mathcal{Y}^{[\lambda]}_{\mu}R^{[\lambda]}_{\mu\nu}$ with $1 \leqslant \nu \leqslant d_{[\lambda]}$ constitute the complete set of basis vectors of the right ideal $\mathcal{Y}^{[\lambda]}_{\mu}\mathcal{L}$ in $\mathcal{L}$.

4.3.3 *Representation Matrices*

In this subsection we are going to study the calculation method for the elements $D^{[\lambda]}_{\nu\mu}(S)$ of the representation matrix. Since the Young pattern $[\lambda]$ is fixed, we omit the superscript $[\lambda]$ for simplicity. Replacing ν with τ in the first equality of Eq. (4.73) and left-multiplying it with $\phi_{\tau\nu}$, we obtain

$$D_{\nu\mu}(S)e_{\tau} = \phi_{\tau\nu}S\phi_{\mu\tau} = (d/n!)^2 \, R_{\tau\nu}\mathcal{Y}_{\nu}y_{\nu}S\mathcal{Y}_{\mu}R_{\mu\tau}y_{\tau}, \tag{4.74}$$

where τ is arbitrary, $1 \leqslant \tau \leqslant d$. Due to Theorem 4.2, the right-hand side of Eq. (4.74) is proportional to e_{τ}. The representation matrix of any element S of S_n in the representation $[\lambda]$ is calculated from Eq. (4.74).

In calculating $D_{\nu\mu}(S)$, one has to move two Young operators together in Eq. (4.74) such that the product of two Young operators becomes one Young operator. y_{ν} is an algebraic sum of group elements T_k with the coefficients ± 1 and can be expressed formally as follows,

$$y_{\nu} = \sum_k \delta_k T_k, \qquad \delta_k = \pm 1. \tag{4.75}$$

Let $\mathcal{Y}_{\nu k}$ denote the Young tableau transformed from the Young tableau $\mathcal{Y}_{\nu}$ by $(T_k)^{-1}$, and let $\mathcal{Y}_{\mu}(S)$ denote the Young tableau transformed from the Young tableau $\mathcal{Y}_{\mu}$ by S. We obtain

$$D_{\nu\mu}(S)e_{\tau} = \sum_k \delta_k \, (d/n!)^2 \, R_{\tau\nu}T_k\mathcal{Y}_{\nu k}\mathcal{Y}_{\mu}(S)SR_{\mu\tau}y_{\tau}. \tag{4.76}$$

Now, we calculate the product of two Young operators. If two digits in one row of the Young tableau $\mathcal{Y}_\mu(S)$ occur in one column of the Young tableau $\mathcal{Y}_{\nu k}$, the product $\mathcal{Y}_{\nu k}\mathcal{Y}_\mu(S)$ vanishes. Otherwise, from Corollary 4.5.3, the permutation transforming the Young tableau $\mathcal{Y}_\mu(S)$ to the Young tableau $\mathcal{Y}_{\nu k}$ can be expressed as

$$P_\mu(S)Q_\mu(S) = \left[P_\mu(S)Q_\mu(S)Q_\mu(S)Q_\mu(S)^{-1}P_\mu(S)^{-1}\right]P_\mu(S) = Q_{\nu k}P_\mu(S).$$

The product of permutations in the square bracket is a vertical permutation of the Young tableau $\mathcal{Y}_{\nu k}$, denoted by $Q_{\nu k}$. $(Q_{\nu k})^{-1}$ transforms the Young tableau $\mathcal{Y}_{\nu k}$ to the Young tableau $\mathcal{Y}'$ such that **every row of both the Young tableau $\mathcal{Y}'$ and the Young tableau $\mathcal{Y}_\mu(S)$ contains the same digits**. Hence,

$$\begin{aligned}(d/n!)\,\mathcal{Y}_{\nu k}\mathcal{Y}_\mu(S) &= (d/n!)\,\mathcal{Y}_{\nu k}\delta(Q_{\nu k})Q_{\nu k}P_\mu(S)\mathcal{Y}_\mu(S)\\ &= (d/n!)\,\delta(Q_{\nu k})Q_{\nu k}P_\mu(S)\mathcal{Y}_\mu(S)\mathcal{Y}_\mu(S)\\ &= \delta(Q_{\nu k})Q_{\nu k}P_\mu(S)\mathcal{Y}_\mu(S).\end{aligned}$$

Substituting it into Eq. (4.76), one obtains

$$D_{\nu\mu}(S)e_\tau = \sum_k \delta_k\delta(Q_{\nu k})\left\{R_{\tau\nu}T_kQ_{\nu k}P_\mu(S)SR_{\mu\tau}\right\}e_\tau. \tag{4.77}$$

The product of permutations in the curly bracket of Eq. (4.77) has to be equal to the identity E because the right-hand side of Eq. (4.77) is proportional to $e_\tau = (d/n!)\mathcal{Y}_\tau y_\tau$. In fact, $R_{\mu\tau}$ first transforms the Young tableau $\mathcal{Y}_\tau$ to the Young tableau $\mathcal{Y}_\mu$. Then, S transforms the Young tableau $\mathcal{Y}_\mu$ to the Young tableau $\mathcal{Y}_\mu(S)$. Third, $Q_{\nu k}P_\mu(S)$ transforms the Young tableau $\mathcal{Y}_\mu(S)$ to the Young tableau $\mathcal{Y}_{\nu k}$. Fourth, T_k transforms the Young tableau $\mathcal{Y}_{\nu k}$ to the Young tableau $\mathcal{Y}_\nu$. At last $R_{\tau\nu}$ transforms the Young tableau $\mathcal{Y}_\nu$ to the original Young tableau $\mathcal{Y}_\tau$. Namely, the product of permutations leaves the Young tableau $\mathcal{Y}_\tau$ invariant so that it is equal to the identity E. Hence,

$$D_{\nu\mu}(S) = \sum_k \delta_k\delta(Q_{\nu k}). \tag{4.78}$$

It means that the elements $D_{\nu\mu}(S)$ of the representation matrix are always integers. **Each irreducible representation of the permutation group S_n is real and its characters are integers.**

Equation (4.78) provides a method for calculating $D_{\nu\mu}(S)$. δ_k is known from Eq. (4.75). **$\delta(Q_{\nu k})$ can be calculated by comparing two Young tableaux $\mathcal{Y}_{\nu k}$** and $\mathcal{Y}_\mu(S)$. If two digits in one row of the Young tableau

$\mathcal{Y}_\mu(S)$ occur in one column of the Young tableau $\mathcal{Y}_{\nu k}$, $\delta(Q_{\nu k}) = 0$. Otherwise, from Corollary 4.4.3, there is a vertical permutation $Q_{\nu k}^{-1}$ of $\mathcal{Y}_{\nu k}$ that transforms the Young tableau $\mathcal{Y}_{\nu k}$ into the Young tableau $\mathcal{Y}'$ such that every row of both the Young tableau $\mathcal{Y}'$ and the Young tableau $\mathcal{Y}_\mu(S)$ contains the same digits. $\delta(Q_{\nu k})$ is the permutation parity of $Q_{\nu k}^{-1}$.

Table 4.1 Tabular method for calculating the irreducible representation matrix $D_{\nu\mu}^{[\lambda]}(S)$ of S_n

$\sum_k \delta_k \left\{\text{Young tableau } \mathcal{Y}_{\nu k}^{[3,2]}\right\}$	Young tableau $\mathcal{Y}_\mu^{[3,2]}(S)$				
	2 3 4 5 1	2 3 5 4 1	2 3 1 4 5	2 4 5 3 1	2 4 1 3 5
1 2 3 / 4 5 − 1 4 5 / 2 3	$-1-0$	$0-1$	$1-0$	$0+1$	$0-0$
1 2 4 3 5	-1	0	0	0	1
1 2 5 3 4	0	-1	0	0	0
1 3 4 2 5	-1	0	0	1	0
1 3 5 2 4	0	-1	0	1	0

$$[\lambda] = [3, 2], \qquad S = (1\ 2\ 3\ 4\ 5)\,, \qquad y_\nu^{[3,2]} = \sum_k \delta_k T_k,$$

$$T_k^{-1} \text{ transforms the Young tableau } \mathcal{Y}_\nu^{[3,2]} \text{ to the Young tableau } \mathcal{Y}_{\nu k}^{[3,2]},$$

$$S \text{ transforms the Young tableau } \mathcal{Y}_\mu^{[3,2]} \text{ to the Young tableau } \mathcal{Y}_\mu^{[3,2]}(S).$$

The matrix element $D_{\nu\mu}^{[\lambda]}(S)$ of S in the irreducible representation $[\lambda]$ of S_n can be calculated by the **tabular method** as follows. Let $\mathcal{Y}_\mu^{[\lambda]}(S)$ denote the Young tableau transformed from the standard Young tableau $\mathcal{Y}_\mu^{[\lambda]}$ by the permutation S. List $\mathcal{Y}_\mu^{[\lambda]}(S)$, $1 \leqslant \mu \leqslant d_{[\lambda]}$, on the first row of the table to designate its columns. Let $y_\nu^{[\lambda]} = \sum_k \delta_k T_k$ and let $\mathcal{Y}_{\nu k}^{[\lambda]}$ denote the Young tableau transformed from the standard Young tableau $\mathcal{Y}_\nu^{[\lambda]}$ by the permutation T_k^{-1}. List the sum of the Young tableaux $\sum_k \delta_k \mathcal{Y}_{\nu k}^{[\lambda]}$, $1 \leqslant \nu \leqslant d_{[\lambda]}$, on the first column of the table to designate its rows. The representation matrix element $D_{\nu\mu}^{[\lambda]}(S)$ is equal to $\sum_k \delta_k \delta(Q_{\nu k}^{[\lambda]})$, which is filled in the μth column of the νth row of the table. $\delta(Q_{\nu k}^{[\lambda]})$ is calculated by comparing two Young tableaux $\mathcal{Y}_{\nu k}^{[\lambda]}$ and $\mathcal{Y}_\mu^{[\lambda]}(S)$. $\delta(Q_{\nu k}^{[\lambda]}) = 0$ if there are two digits in one row of the Young tableau $\mathcal{Y}_\mu(S)$ which occur in one column of the Young tableau $\mathcal{Y}_{\nu k}^{[\lambda]}$. Otherwise, $\delta(Q_{\nu k}^{[\lambda]})$ is the permutation parity of the vertical permutation $Q_{\nu k}^{[\lambda]}$ of the Young tableau $\mathcal{Y}_{\nu k}^{[\lambda]}$, which transforms the Young tableau $\mathcal{Y}_{\nu k}^{[\lambda]}$ to the Young tableau $\mathcal{Y}'$ such that every row of both the Young tableau $\mathcal{Y}'$ and the Young tableau $\mathcal{Y}_{\mu}^{[\lambda]}(S)$ contains the same digits. The sum of the diagonal elements in the table is the character $\chi^{[\lambda]}(S)$. The calculated irreducible representation $D^{[\lambda]}(S)$ is generally not unitary.

As example, in Table 4.1 the representation matrix $D^{[3,2]}(S)$ of S_5, where $S = (1\ 2\ 3\ 4\ 5)$, is calculated by the tabular method. $D^{[3,2]}\ [(1\ 2)]$ can be calculated similarly.

$$D^{[3,2]}\ [(1\ 2\ 3\ 4\ 5)] = \begin{pmatrix} -1 & -1 & 1 & 1 & 0 \\ -1 & 0 & 0 & 0 & 1 \\ 0 & -1 & 0 & 0 & 0 \\ -1 & 0 & 0 & 1 & 0 \\ 0 & -1 & 0 & 1 & 0 \end{pmatrix}, \quad D^{[3,2]}\ [(1\ 2)] = \begin{pmatrix} 1 & 0 & 0 & -1 & -1 \\ 0 & 1 & 0 & -1 & 0 \\ 0 & 0 & 1 & 0 & -1 \\ 0 & 0 & 0 & -1 & 0 \\ 0 & 0 & 0 & 0 & -1 \end{pmatrix}.$$

Then, the representation matrix, as well as the character, of any element in S_5 can be calculated. The characters for the representation $[3,2]$ of S_5 are listed in Table 4.2.

Table 4.2. Character table for representation $[3,2]$ of S_5

Class (ℓ)	(1^5)	$(2,1^3)$	$(2^2,1)$	$(3,1^2)$	$(3,2)$	$(4,1)$	(5)
Character	5	1	1	-1	1	-1	0

4.3.4 *Characters by Graphic Method*

The representation matrices of S_n, as well as the characters in the representation, can be calculated by the tabular method. However, there is a graphic method for calculating the character of a class (ℓ) in the irreducible representation $[\lambda]$, which is much simpler than the tabular method.

The integers ℓ_j in the partition (ℓ) can be arranged in any order, but **the calculation will be simpler to arrange in the increasing order**, $\ell_1 \leqslant \ell_2 \leqslant \cdots \leqslant \ell_m$. One first fills ℓ_1 digits 1 into the Young pattern $[\lambda]$ according to the following rule, and then, fills ℓ_2 digits 2, and so on, until filling ℓ_m digits m. The filling rule is as follows:

(a) The boxes filled with each digit, say j, are connected such that from the lowest and the leftmost box one can go through all the boxes filled with j only upward and rightward.

(b) Each time when ℓ_j digits j are filled, all the boxes filled with digits $i \leqslant j$ form a regular Young pattern, namely, the boxes are lined up on the top and on the left such that the number of boxes in the upper row is not less than that in the lower row and there is no unfilled box embedding between two filled boxes in each row.

It is said to be one regular application if all digits are filled into the Young pattern according to the rule. The filling parity for the digit j is defined to be 1 if the number of rows of the boxes filled with j is odd, and to be -1 if it is even. **The filling parity of a regular application is defined to be the product of the filling parities of m digits.** The character $\chi^{[\lambda]}[(\ell)]$ of the class (ℓ) in the representation $[\lambda]$ is equal to **the sum of the filling parities of all regular applications.** $\chi^{[\lambda]}[(\ell)] = 0$ if there is no regular application, namely, if m digits cannot be filled in the Young pattern according to the above rule. For the class (1^n) composed of only the identity E, each regular application is just a standard Young tableau, so its character is nothing but the dimension of the representation. Usually, **the character of the class (1^n) is calculated by the hook rule instead of the graphic method.** In Table 4.3 all the regular applications of each class for the Young pattern $[3, 2]$ are listed, and their characters are calculated. The five regular applications of the class (1^5) are omitted.

Table 4.3 The character table for the representation [3,2] of S_5 calculated by the graphic method

Class	(1^5)	$(1^3, 2)$	$(1, 2^2)$	$(1^2, 3)$	$(2, 3)$	$(1, 4)$	(5)
Regular application		1 2 3 4 4	1 2 2 3 3	1 3 3 2 3	1 2 2 1 2	1 2 2 2 2	
Filling parity		1	1	-1	1	-1	
$\chi^{[3,2]}[(\ell)]$	5	1	1	-1	1	-1	0

If one changes the order of ℓ_j for (ℓ), there may be more regular applications, but the calculated results of the characters are the same. For example, if one changes the order for the class $(1, 2, 2)$ to be $(2, 2, 1)$, there are three regular applications in the Young pattern $[3, 2]$:

$$\begin{array}{lccc} & \begin{array}{l}1\,1\,3\\2\,2\end{array} & \begin{array}{l}1\,2\,2\\1\,3\end{array} & \begin{array}{l}1\,2\,3\\1\,2\end{array} \\ \text{Filling parity} = & 1, & -1, & 1 \end{array} \qquad \chi(1, 2^2) = 1 - 1 + 1 = 1.$$

The Young pattern $[n]$ with one row has one standard Young tableau, and the corresponding Young operator is the sum of all elements in S_n. $[n]$

characterizes the identical representation, where the representation matrix of each element in S_n is 1. The Young pattern $[1^n]$ with one column also has one standard Young tableau, and the corresponding Young operator is the sum of all elements in S_n, multiplied with their permutation parities. $[1^n]$ **characterizes the antisymmetric representation**, where the representation matrix of each element R is its permutation parity $\delta(R)$.

Two Young patterns related by a transpose are called the **associated Young patterns**. Let $[\widetilde{\lambda}]$ denote the associated Young pattern of $[\lambda]$. The numbers of the standard Young tableaux of two associated Young patterns are the same. The larger standard Young tableau of $[\lambda]$ becomes the smaller standard Young tableau of $[\widetilde{\lambda}]$. According to the graphic method, **the characters of a class (ℓ) in two associated Young patterns differ only with a permutation parity $\delta[(\ell)]$ of the class:**

$$\chi^{[\widetilde{\lambda}]}[(\ell)] = \delta[(\ell)]\chi^{[\lambda]}[(\ell)], \qquad D^{[\widetilde{\lambda}]}(S_n) \simeq D^{[1^n]}(S_n) \times D^{[\lambda]}(S_n). \tag{4.79}$$

In fact, the transpose of each regular application of $[\lambda]$ is a regular application of $[\widetilde{\lambda}]$, where the positions of each digit, say j, are the same as each other except for the transpose. The sum of the row number and the column number of boxes filled with the digits j in both $[\lambda]$ and $[\widetilde{\lambda}]$ is (ℓ_j+1), so that the product of two filling parities for the digit j in two associated Young patterns is equal to the permutation parity $(-1)^{\ell_j+1}$ of the cycle with length ℓ_j. If $[\widetilde{\lambda}] = [\lambda]$, the Young pattern $[\lambda]$ is called **self-associated**, and the character of a class (ℓ) with odd permutation parity is zero.

If $[\lambda] \neq [\widetilde{\lambda}]$, the subduced representation $D^{[\lambda]}(S'_n)$ of the irreducible representation $D^{[\lambda]}(S_n)$ with respect to the alternating subgroup S'_n of S_n is irreducible and is equivalent to $D^{[\widetilde{\lambda}]}(S'_n)$. For the self-associated Young pattern $[\lambda]$, $[\lambda] \simeq [\widetilde{\lambda}]$, **the subduced representation $D^{[\lambda]}(S'_n)$ is decomposed into the direct sum of two inequivalent irreducible representations with the same dimension.** (see Prob. 18.)

4.3.5 *The Permutation Group S_3*

As an example, we calculate the irreducible bases and the inequivalent irreducible representations of S_3 by the Young operator method. S_3 is isomorphic onto the symmetry group D_3 of a regular triangle. There are six elements and three classes in S_3. The class (1^3) contains only the identity E. The class $(2,1)$ contains three elements, $C_2^{(0)} = (2\ 3)$, $C_2^{(1)} = (3\ 1)$, and $C_2^{(2)} = (1\ 2)$. The class (3) contains two elements, $C_3 = (3\ 2\ 1)$ and

$C_3^2 = (1\ 2\ 3)$. There are three Young patterns for S_3.

The Young pattern [3] characterizes the identical representation, where the representation matrix of any group element is equal to 1.

$$\mathcal{Y}^{[3]} \qquad \begin{array}{|c|c|c|}\hline 1 & 2 & 3 \\ \hline \end{array}$$

$$\phi^{[3]} = e^{[3]} = \frac{1}{6}\mathcal{Y}^{[3]} = \frac{1}{6}\left\{E + (1\ 2) + (2\ 3) + (3\ 1) + (1\ 2\ 3) + (3\ 2\ 1)\right\}.$$

For the Young pattern $[2,1]$, there are two standard Young tableaux. The representation $[2,1]$ is two-dimensional.

$$\mathcal{Y}_1^{[2,1]} \quad \begin{array}{|c|c|}\hline 1 & 2 \\ \hline 3 \\ \cline{1-1} \end{array} \qquad \text{and} \qquad \mathcal{Y}_2^{[2,1]} \quad \begin{array}{|c|c|}\hline 1 & 3 \\ \hline 2 \\ \cline{1-1} \end{array}$$

The idempotents and the irreducible bases are

$$\begin{aligned}
\phi_{11}^{[2,1]} &= e_1^{[2,1]} = \mathcal{Y}_1^{[2,1]}/3 = \{E + (1\ 2) - (1\ 3) - (2\ 1\ 3)\}/3, \\
\phi_{22}^{[2,1]} &= e_2^{[2,1]} = \mathcal{Y}_2^{[2,1]}/3 = \{E + (1\ 3) - (1\ 2) - (3\ 1\ 2)\}/3, \\
\phi_{21}^{[2,1]} &= (2\ 3)e_1^{[2,1]} = \{(2\ 3) + (3\ 2\ 1) - (2\ 3\ 1) - (2\ 1)\}/3, \\
\phi_{12}^{[2,1]} &= (2\ 3)e_2^{[2,1]} = \{(2\ 3) + (2\ 3\ 1) - (3\ 2\ 1) - (3\ 1)\}/3.
\end{aligned} \tag{4.80}$$

The representation matrices of some elements in $[2,1]$ of S_3 are calculated by the tabular method (see Table 4.4).

Table 4.4. The representation matrices of some elements in $[2,1]$ of S_3

$\mathcal{Y}_\nu$	(1 2)		(2 3)		(1 2 3)	
	2 1 3	2 3 1	1 3 2	1 2 3	2 3 1	2 1 3
1 2 3	1	−1	0	1	−1	1
1 3 2	0	−1	1	0	−1	0

Since the irreducible bases are not orthonormal, this representation is not real orthogonal. The similarity transformation Z transforms this representation to the real orthogonal one given in Eq. (2.14):

$$Z = \frac{1}{2}\begin{pmatrix} 3 & -\sqrt{3} \\ 3 & \sqrt{3} \end{pmatrix}, \qquad Z^{-1}D^{[2,1]}(R)Z = D(R),$$

$$D\left[(2\ 3)\right] = \begin{pmatrix} 1 & 0 \\ 0 & -1 \end{pmatrix}, \qquad D\left[(1\ 2\ 3)\right] = \frac{1}{2}\begin{pmatrix} -1 & \sqrt{3} \\ -\sqrt{3} & -1 \end{pmatrix},$$

$$\begin{aligned} f_1 &= (3/2)\left(\phi_{11}^{[2,1]} + \phi_{21}^{[2,1]}\right) = \left\{E + (2\ 3) - (3\ 1) - (1\ 2\ 3)\right\}/2, \\ f_2 &= (\sqrt{3}/2)\left(-\phi_{11}^{[2,1]} + \phi_{21}^{[2,1]}\right) \\ &= \left\{-E + (2\ 3) + (3\ 1) - 2(1\ 2) - (1\ 2\ 3) + 2(3\ 2\ 1)\right\}/(2\sqrt{3}). \end{aligned}$$

The Young pattern $[1,1,1] = [1^3]$ characterizes the antisymmetric representation, where the representation matrix of each group element R is equal to its permutation parity $\delta(R)$.

$$\mathcal{Y}^{[1^3]} \quad \begin{array}{|c|} \hline 1 \\ \hline 2 \\ \hline 3 \\ \hline \end{array}$$

$$\phi^{[1^3]} = e^{[1^3]} = \frac{1}{6}\mathcal{Y}^{[1^3]} = \frac{1}{6}\left\{E - (1\ 2) - (2\ 3) - (3\ 1) + (1\ 2\ 3) + (3\ 2\ 1)\right\}.$$

4.3.6 *Inner Product of Irreducible Representations*

The direct product of two irreducible representations of S_n is specially called the **inner product**, because there is another product called the outer product (see section 4.5). The inner product can be decomposed into the Clebsch–Gordan series by the character formula (3.58).

$$\begin{aligned} &C^{-1}\left[D^{[\lambda]}(R) \times D^{[\mu]}(R)\right]C = \bigoplus_{\nu} a_{\lambda\mu\nu} D^{[\nu]}(R), \\ &\chi^{[\lambda]}(R)\chi^{[\mu]}(R) = \sum_{\nu} a_{\lambda\mu\nu}\chi^{[\nu]}(R), \\ &a_{\lambda\mu\nu} = \frac{1}{n!}\sum_{R\in S_n} \chi^{[\lambda]}(R)\chi^{[\mu]}(R)\chi^{[\nu]}(R). \end{aligned} \tag{4.81}$$

Since the characters in the irreducible representations of S_n are real, $a_{\lambda\mu\nu}$ **is totally symmetric with respect to three subscripts**. Noting Eq. (4.79) , one has

$$[n] \times [\lambda] \simeq [\lambda], \qquad [1^n] \times [\lambda] \simeq [\tilde{\lambda}], \qquad [\lambda] \times [\mu] \simeq [\tilde{\lambda}] \times [\tilde{\mu}]. \tag{4.82}$$

Due to Eq. (4.81) one concludes that there is one identical representation $[n]$ in the decomposition of $[\lambda] \times [\mu]$ if and only if $[\lambda] = [\mu]$, and there is one antisymmetric representation $[1^n]$ in the decomposition of $[\lambda] \times [\mu]$ if and only if $[\lambda] = [\tilde{\mu}]$. For the group S_3 one has

$$\begin{aligned} &[3] \times [3] \simeq [1^3] \times [1^3] \simeq [3], \qquad [3] \times [1^3] \simeq [1^3], \\ &[3] \times [2,1] \simeq [1^3] \times [2,1] \simeq [2,1], \\ &[2,1] \times [2,1] \simeq [3] \oplus [1^3] \oplus [2,1]. \end{aligned} \tag{4.83}$$

Some Clebsch–Gordan series are left as exercise (see Prob. 19).

4.3.7 *Character Table of the Group* **I**

The character table of the symmetry group **I** of an icosahedron can be calculated in terms of the isomorphic relation: **I**$\approx S_5'$. The class $\mathcal{C}_\alpha$ with the cycle structure (5) in S_5 is divided into two self-reciprocal classes for S_5', both containing 12 elements (see subsection 2.5.3). The subduced representations denoted by two associated Young patterns in S_5 with respect to S_5' are equivalent, but the subduced representation denoted by the self-associated Young patterns $[3,1^2]$ is a reducible representation of S_5', which contains two inequivalent irreducible representations with the same dimension due to Eqs. (3.62) and (3.65). The character table of S_5' is calculated by the graphic method and listed in Table 4.5, where both a and b are real and, due to Eq. (3.56) (or Eq. (3.68)), $a + b = 1$ and $ab = -1$. Then, a and b are two solutions of $x^2 - x - 1 = 0$. Without loss of generality, $a = (\sqrt{5}+1)/2 = 1 + 2\cos(2\pi/5) = p^{-1}$ and $b = (-\sqrt{5}+1)/2 = 1 + 2\cos(4\pi/5) = -p$. By the way, if the class $\mathcal{C}_\alpha$ in S_N is divided into a pair of reciprocal classes for S_N', $a = b^*$.

Table 4.5. The character table of S_5'

S_5'	E	$12(5)$	$12(5')$	$20(3,1^2)$	$15(2^2,1)$
$[5]$	1	1	1	1	1
$[4,1]$	4	-1	-1	1	0
$[3,2]$	5	0	0	-1	1
$[3,1^2]$	3	a	b	0	-1
$[3,1^2]'$	3	b	a	0	-1

$$a = (\sqrt{5}+1)/2, \quad b = (-\sqrt{5}+1)/2.$$

According to the convention in the theory of crystals, the inequivalent irreducible representations of the group **I** are denoted by $D^A(\mathbf{I})$ for one dimension, $D^{T_1}(\mathbf{I})$ and $D^{T_2}(\mathbf{I})$ for three dimensions, $D^G(\mathbf{I})$ for four dimensions, and $D^H(\mathbf{I})$ for five dimensions. In fact, $D^{T_1}(\mathbf{I})$ is the subduced representation of the self-representation of SO(3) such that $D^{T_1}(\hat{\boldsymbol{n}}, \omega) = 1 + 2\cos\omega$. This is another way to calculate $\chi^{[3,1^2]}(\mathrm{S}_5')$. The character table of $\mathbf{I}$ is listed in Table 4.6 in terms of $\mathbf{I} \approx \mathrm{S}_5'$. The character tables of the groups D_3, $\mathbf{O}$ and $\mathbf{T}$ can also be obtained in terms of the isomorphic relations: $\mathrm{D}_3 \approx \mathrm{S}_3$, $\mathbf{O} \approx \mathrm{S}_4$ and $\mathbf{T} \approx \mathrm{S}_4'$.

Table 4.6. The character table of I

$\mathbf{I}$	E	$12C_5$	$12C_5^2$	$20C_3$	$15C_2$
A	1	1	1	1	1
T_1	3	p^{-1}	$-p$	0	-1
T_2	3	$-p$	p^{-1}	0	-1
G	4	-1	-1	1	0
H	5	0	0	-1	1

$p = (\sqrt{5} - 1)/2, \quad p^{-1} = (\sqrt{5} + 1)/2.$

4.4 Real Orthogonal Representation of $\mathbf{S}_n$

The merit of the Young operator method for calculating the representations of S_n is that the basis vectors are well known and the representation matrices are related to Young operators. Its shortcoming is that the calculated representations are not real orthogonal. Although these representations are convenient in some cases even they are not real orthogonal (see Chap. 8), one may prefer the real orthogonal representations sometimes. In this section we are going to raise a method to combine the basis vectors such that the new representation is real orthogonal [Tong et al. (1992)].

With respect to the natural bases in the group algebra $\mathcal{L}$ of S_n, a transposition $(a\ d)$ is unitary and hermitian. Define a set of hermitian operators M_a in $\mathcal{L}$ by the sum of transpositions:

$$M_a = \sum_{d=1}^{a-1} (a\ d) = \sum_{d=1}^{a-1} P_{a-1}P_{a-2} \cdots P_{d+1}P_d P_{d+1} \cdots P_{a-2}P_{a-1}, \qquad (4.84)$$

where $2 \leqslant a \leqslant n$, $M_1 = 0$, and $P_a = (a\ a+1)$ is the transposition of two neighboring objects. From the definition (4.84) one has

$$M_{a+1} = P_a + P_a M_a P_a, \qquad P_a M_{a+1} = E + M_a P_a. \tag{4.85}$$

It is easy to show from (2.45) that

$$P_a M_b = M_b P_a, \qquad \text{if } \ b < a \ \text{ or } \ b > a + 1. \tag{4.86}$$

Then, the hermitian operators M_a are commutable with one another

$$[M_a,\ M_b] = 0. \tag{4.87}$$

Theorem 4.9 With respect to the irreducible bases $\phi_{\nu\rho}^{[\lambda]}$ (see Eq. (4.71)), the representation matrix $D^{[\lambda]}(M_a)$ of M_a is an upper triangular matrix with the known diagonal elements,

$$\begin{aligned} M_a \phi_{\nu\rho}^{[\lambda]} &= \sum_{\mu} \phi_{\mu\rho}^{[\lambda]} D_{\mu\nu}^{[\lambda]}(M_a), \\ D_{\nu\nu}^{[\lambda]}(M_a) = m_\nu(a), &\qquad D_{\mu\nu}^{[\lambda]}(M_a) = 0 \qquad \text{when } \mu > \nu, \end{aligned} \tag{4.88}$$

where the rows and columns are enumerated by the standard Young tableaux $\mathcal{Y}_\nu^{[\lambda]}$ in the increasing order of ν. If a is filled in the $c_\nu(a)$th column of the $r_\nu(a)$th row of the Young tableau $\mathcal{Y}_\nu^{[\lambda]}$, $m_\nu(a) = c_\nu(a) - r_\nu(a)$ is called the content of the digit a in the Young tableau $\mathcal{Y}_\nu^{[\lambda]}$.

Proof Decompose M_a as $M_a = M_a^{(1)} + M_a^{(2)} + M_a^{(3)} + M_a^{(4)}$,

$$\begin{aligned} M_a^{(1)} &= \sum_i \left(a \ \ a_i\right), & M_a^{(2)} &= \sum_j \left(a \ \ b_j\right) + \sum_k \left(a \ \ d_k\right), \\ M_a^{(3)} &= -\sum_k \left(a \ \ d_k\right), & M_a^{(4)} &= \sum_\ell \left(a \ \ t_\ell\right), \end{aligned}$$

where a_i denotes the digits filled in the boxes on the left of the box a at the $r_\nu(a)$th row of the Young tableau $\mathcal{Y}_\nu^{[\lambda]}$, b_j and d_k denote the digits, respectively smaller and larger than a, filled in the boxes at the first $[r_\nu(a) - 1]$ rows, and t_ℓ denote the digits less than a and filled in those r'th rows with $r' > r_\nu$. From the symmetric property (4.35) and the Fock condition (4.37) one has

$$M_a^{(1)} \phi_{\nu\rho} = \{c_\nu(a) - 1\}\, \phi_{\nu\rho}, \qquad M_a^{(2)} \phi_{\nu\rho} = \{1 - r_\nu(a)\}\, \phi_{\nu\rho}.$$

When applying each transposition in $M_a^{(3)}$ and $M_a^{(4)}$ to the Young tableau $\mathcal{Y}_\nu^{[\lambda]}$, a smaller digit in a lower row is interchanged with a larger digit in the upper row. Although the transformed Young tableau is generally not standard, it can be proved in terms of the similar method used in the proof for Corollary 4.5.2 that

$$\mathcal{Y}_\mu^{[\lambda]} M_a^{(3)} \mathcal{Y}_\nu^{[\lambda]} = \mathcal{Y}_\mu^{[\lambda]} M_a^{(4)} \mathcal{Y}_\nu^{[\lambda]} = 0, \qquad \text{when } \mu \geqslant \nu.$$

Due to Eq. (4.62) one has

$$\begin{aligned} &\mathcal{Y}_\mu^{[\lambda]} y_\mu^{[\lambda]} M_a^{(3)} \mathcal{Y}_\nu^{[\lambda]} = \mathcal{Y}_\mu^{[\lambda]} y_\mu^{[\lambda]} M_a^{(4)} \mathcal{Y}_\nu^{[\lambda]} = 0, \\ &\phi_{\rho\mu}^{[\lambda]} M_a^{(3)} \phi_{\nu\rho}^{[\lambda]} = \phi_{\rho\mu}^{[\lambda]} M_a^{(4)} \phi_{\nu\rho}^{[\lambda]} = 0, \end{aligned} \qquad \text{when } \mu \geqslant \nu. \qquad \square$$

Due to Corollary 3.1.1, there exists a real similarity transformation which transforms the representation $D^{[\lambda]}(S_n)$ to a real orthogonal representation $\overline{D}^{[\lambda]}(S_n)$ where $\overline{D}^{[\lambda]}(M_a)$ are real symmetric. Then, $\overline{D}^{[\lambda]}(M_a)$ can be diagonalized simultaneously by a real orthogonal similarity transformation. Altogether,

$$\begin{aligned} &\overline{D}^{[\lambda]}(P_a) = X_{[\lambda]}^{-1} D^{[\lambda]}(P_a) X_{[\lambda]} = \overline{D}^{[\lambda]}(P_a)^* = \overline{D}^{[\lambda]}(P_a)^T, \\ &\overline{D}_{\mu\nu}^{[\lambda]}(M_a) = \left[X_{[\lambda]}^{-1} D^{[\lambda]}(M_a) X_{[\lambda]}\right]_{\mu\nu} = \delta_{\mu\nu} m_\nu(a). \end{aligned} \tag{4.89}$$

Because there are no two different standard Young tableaux $\mathcal{Y}_\mu^{[\lambda]}$ and $\mathcal{Y}_\nu^{[\lambda]}$ satisfying $m_\mu(a) = m_\nu(a)$ for every a, the eigenvalues $m_\nu(a)$ of M_a are not degenerate. The new basis vectors $\psi_{\mu\nu}^{[\lambda]}$ after the similarity transformation $X_{[\lambda]}$ are called the **orthogonal bases** in the group algebra $\mathcal{L}$ of S_n:

$$\begin{aligned} &\psi_{\mu\nu}^{[\lambda]} = \sum_{\rho=1}^{d} \sum_{\tau=1}^{d} \left(X_{[\lambda]}^{-1}\right)_{\nu\tau} \phi_{\rho\tau}^{[\lambda]} \left(X_{[\lambda]}\right)_{\rho\mu}, \\ &R\psi_{\mu\nu}^{[\lambda]} = \sum_\rho \psi_{\rho\nu}^{[\lambda]} \overline{D}_{\rho\mu}^{[\lambda]}(R), \qquad \psi_{\mu\nu}^{[\lambda]} R = \sum_\rho \overline{D}_{\nu\rho}^{[\lambda]}(R) \psi_{\mu\rho}^{[\lambda]}. \end{aligned} \tag{4.90}$$

$\overline{D}^{[\lambda]}(P_a)$ is calculated by substituting Eq. (4.89) into Eq. (4.86):

$$\overline{D}_{\mu\nu}^{[\lambda]}(P_a)\left\{m_\mu(b) - m_\nu(b)\right\} = 0, \qquad \text{if } b < a \text{ or } b > a+1. \tag{4.91}$$

Namely, $\overline{D}_{\mu\nu}^{[\lambda]}(P_a) \neq 0$ only if $m_\mu(b) = m_\nu(b)$ for every b except for $b = a$ and $b = a+1$. Define the Young tableau $\mathcal{Y}_{\nu_a}^{[\lambda]}$ which is obtained from the Young tableau $\mathcal{Y}_\nu^{[\lambda]}$ by interchanging a and $(a+1)$. Thus, $\overline{D}^{[\lambda]}(P_a)$ contains a submatrix of two dimensions at ν- and ν_a-columns if $\mathcal{Y}_{\nu_a}^{[\lambda]}$ is standard, and a submatrix of one dimension at ν-column if $\mathcal{Y}_{\nu_a}^{[\lambda]}$ is not.

$\mathcal{Y}_{\nu_a}^{[\lambda]}$ is not standard, if a and $(a+1)$ occur in the **same row** in the Young tableau $\mathcal{Y}_\nu^{[\lambda]}$, $m = m_\nu(a) - m_\nu(a+1) = -1$, or in the **same column**, $m = m_\nu(a) - m_\nu(a+1) = 1$. From Eq. (4.85) one obtains

$$\overline{D}_{\nu\nu}(P_a) = \begin{cases} 1, & a \text{ and } (a+1) \text{ occur in the same row,} \\ -1, & a \text{ and } (a+1) \text{ occur in the same column.} \end{cases} \tag{4.92}$$

Otherwise, the Young tableau $\mathcal{Y}_{\nu_a}^{[\lambda]}$ is standard. Without loss of generality, we assume $\nu < \nu_a$, namely, a is located in the right of and upper than $(a+1)$ in the Young tableau $\mathcal{Y}_{\nu}^{[\lambda]}$:

$$\begin{aligned} m_\nu(a) = m_{\nu_a}(a+1) > m_\nu(a+1) = m_{\nu_a}(a), \\ m = m_\nu(a) - m_\nu(a+1) \geqslant 2. \end{aligned} \tag{4.93}$$

Substituting Eq. (4.89) into Eq. (4.85) and noting $P_a^2 = E$, one obtains

$$\begin{aligned} -\overline{D}_{\nu\nu}^{[\lambda]}(P_a) = \overline{D}_{\nu_a\nu_a}^{[\lambda]}(P_a) = m^{-1}, \\ \overline{D}_{\nu\nu_a}^{[\lambda]}(P_a) = \overline{D}_{\nu_a\nu}^{[\lambda]}(P_a) = \frac{\sqrt{m^2-1}}{m}. \end{aligned} \tag{4.94}$$

It is proved that the square root can be taken positive by choosing the phase angles of the basis vectors and noting the condition (2.45). Thus, the two-dimensional submatrix $\overline{D}_{\mu\nu}^{[\lambda]}(P_a)$ at ν and ν_a columns (rows) is

$$\frac{1}{m}\begin{pmatrix} -1 & \sqrt{m^2-1} \\ \sqrt{m^2-1} & 1 \end{pmatrix}, \qquad m = m_\nu(a) - m_\nu(a+1) \geqslant 2, \tag{4.95}$$

where m is equal to the steps of going from a to $(a+1)$ in the Young tableau $\mathcal{Y}_{\nu}^{[\lambda]}$ downward or leftward. The conclusion is that $\overline{D}^{[\lambda]}(P_a)$ is **a block matrix with submatrices of one or two dimensions, given in Eqs. (4.92) and (4.95)**. Since $D^{[\lambda]}(M_a)$ are upper triangular real matrices and $\overline{D}^{[\lambda]}(M_a)$ are real diagonal, $X_{[\lambda]}$ **has to be an upper triangular real matrix**, which can be calculated from Eq. (4.89).

As an example, we calculate the similarity transformation matrix $X_{[3,2]}$ for the representation $[3,2]$ of S_5, where we neglect the superscript $[3,2]$ for simplicity. The standard Young tableaux with the Young pattern $[3,2]$ are listed from $\mathcal{Y}_1$ to $\mathcal{Y}_5$ as follows:

1 2 3	1 2 4	1 2 5	1 3 4	1 3 5
4 5	3 5	3 4	2 5	2 4

The representation matrices $D(P_a)$ are calculated by the tabular method (see section 4.3.3)

$$D(P_1) = \begin{pmatrix} 1 & 0 & 0 & -1 & -1 \\ 0 & 1 & 0 & -1 & 0 \\ 0 & 0 & 1 & 0 & -1 \\ 0 & 0 & 0 & -1 & 0 \\ 0 & 0 & 0 & 0 & -1 \end{pmatrix}, \qquad D(P_2) = \begin{pmatrix} 1 & 0 & 0 & 0 & 0 \\ 0 & 0 & 0 & 1 & 0 \\ 0 & 0 & 0 & 0 & 1 \\ 0 & 1 & 0 & 0 & 0 \\ 0 & 0 & 1 & 0 & 0 \end{pmatrix},$$

$$D(P_3) = \begin{pmatrix} 0 & 1 & 0 & 0 & -1 \\ 1 & 0 & 0 & 0 & -1 \\ 0 & 0 & 1 & 0 & -1 \\ 0 & 0 & 0 & 1 & -1 \\ 0 & 0 & 0 & 0 & -1 \end{pmatrix}, \qquad D(P_4) = \begin{pmatrix} 1 & 0 & 0 & 0 & 0 \\ 0 & 0 & 1 & 0 & 0 \\ 0 & 1 & 0 & 0 & 0 \\ 0 & 0 & 0 & 0 & 1 \\ 0 & 0 & 0 & 1 & 0 \end{pmatrix}.$$

The real orthogonal representation matrices $\overline{D}(P_a)$ are calculated from Eqs. (4.92) and (4.95). For example, in the calculation of $\overline{D}(P_2)$ one needs to check the Young tableaux $\mathcal{Y}_\nu$ in the interchanging of 2 and 3: The Young tableau $\mathcal{Y}_2$ is changed to the Young tableau $\mathcal{Y}_4$ with $m = 2$, and the Young tableau $\mathcal{Y}_3$ is changed to the Young tableau $\mathcal{Y}_5$ with $m = 2$. So that $\overline{D}(P_2)$ is a block matrix with one 1×1 submatrix and two 2×2 submatrices.

$$\overline{D}(P_1) = \begin{pmatrix} 1 & 0 & 0 & 0 & 0 \\ 0 & 1 & 0 & 0 & 0 \\ 0 & 0 & 1 & 0 & 0 \\ 0 & 0 & 0 & -1 & 0 \\ 0 & 0 & 0 & 0 & -1 \end{pmatrix}, \qquad \overline{D}(P_2) = \frac{1}{2} \begin{pmatrix} 2 & 0 & 0 & 0 & 0 \\ 0 & -1 & 0 & \sqrt{3} & 0 \\ 0 & 0 & -1 & 0 & \sqrt{3} \\ 0 & \sqrt{3} & 0 & 1 & 0 \\ 0 & 0 & \sqrt{3} & 0 & 1 \end{pmatrix},$$

$$\overline{D}(P_3) = \frac{1}{3} \begin{pmatrix} -1 & \sqrt{8} & 0 & 0 & 0 \\ \sqrt{8} & 1 & 0 & 0 & 0 \\ 0 & 0 & 3 & 0 & 0 \\ 0 & 0 & 0 & 3 & 0 \\ 0 & 0 & 0 & 0 & -3 \end{pmatrix}, \qquad \overline{D}(P_4) = \frac{1}{2} \begin{pmatrix} 2 & 0 & 0 & 0 & 0 \\ 0 & -1 & \sqrt{3} & 0 & 0 \\ 0 & \sqrt{3} & 1 & 0 & 0 \\ 0 & 0 & 0 & -1 & \sqrt{3} \\ 0 & 0 & 0 & \sqrt{3} & 1 \end{pmatrix}.$$

The similarity transformation matrix X, $D(P_a)X = X\overline{D}(P_a)$, is an upper triangular one, whose column matrices X_μ are denoted by

$$X_1 = \begin{pmatrix} 1 \\ 0 \\ 0 \\ 0 \\ 0 \end{pmatrix}, \quad X_2 = \begin{pmatrix} a_1 \\ a_2 \\ 0 \\ 0 \\ 0 \end{pmatrix}, \quad X_3 = \begin{pmatrix} b_1 \\ b_2 \\ b_3 \\ 0 \\ 0 \end{pmatrix}, \quad X_4 = \begin{pmatrix} c_1 \\ c_2 \\ c_3 \\ c_4 \\ 0 \end{pmatrix}, \quad X_5 = \begin{pmatrix} d_1 \\ d_2 \\ d_3 \\ d_4 \\ d_5 \end{pmatrix}.$$

Neglecting the rows with all elements vanishing, from

$$D(P_3)X_1 = -\frac{1}{3}X_1 + \frac{\sqrt{8}}{3}X_2, \qquad \begin{pmatrix} 0 \\ 1 \end{pmatrix} = -\frac{1}{3}\begin{pmatrix} 1 \\ 0 \end{pmatrix} + \frac{\sqrt{8}}{3}\begin{pmatrix} a_1 \\ a_2 \end{pmatrix},$$

one obtains $a_1 = 1/\sqrt{8}$ and $a_2 = 3/\sqrt{8}$. From

$$D(P_4)X_2 = -\frac{1}{2}X_2 + \frac{\sqrt{3}}{2}X_3, \qquad \frac{1}{\sqrt{8}}\begin{pmatrix}1\\0\\3\end{pmatrix} = -\frac{1}{2\sqrt{8}}\begin{pmatrix}1\\3\\0\end{pmatrix} + \frac{\sqrt{3}}{2}\begin{pmatrix}b_1\\b_2\\b_3\end{pmatrix},$$

one obtains $b_1 = b_2 = \sqrt{3/8}$ and $b_3 = \sqrt{3/2}$. From

$$D(P_2)X_2 = -\frac{1}{2}X_2 + \frac{\sqrt{3}}{2}X_4, \qquad \frac{1}{\sqrt{8}}\begin{pmatrix}1\\0\\0\\3\end{pmatrix} = -\frac{1}{2\sqrt{8}}\begin{pmatrix}1\\3\\0\\0\end{pmatrix} + \frac{\sqrt{3}}{2}\begin{pmatrix}c_1\\c_2\\c_3\\c_4\end{pmatrix},$$

one obtains $c_1 = c_2 = \sqrt{3/8}$, $c_3 = 0$, and $c_4 = \sqrt{3/2}$. From

$$D(P_4)X_4 = -\frac{1}{2}X_4 + \frac{\sqrt{3}}{2}X_5, \qquad \sqrt{\frac{3}{8}}\begin{pmatrix}1\\0\\1\\0\\2\end{pmatrix} = -\frac{1}{2}\sqrt{\frac{3}{8}}\begin{pmatrix}1\\1\\0\\2\\0\end{pmatrix} + \frac{\sqrt{3}}{2}\begin{pmatrix}d_1\\d_2\\d_3\\d_4\\d_5\end{pmatrix},$$

one obtains $d_1 = 3/\sqrt{8}$, $d_2 = 1/\sqrt{8}$, $d_3 = d_4 = 1/\sqrt{2}$, and $d_5 = \sqrt{2}$. At last, the similarity transformation matrix X is

$$X_{[3,2]} = \frac{1}{\sqrt{8}}\begin{pmatrix}\sqrt{8} & 1 & \sqrt{3} & \sqrt{3} & 3\\ 0 & 3 & \sqrt{3} & \sqrt{3} & 1\\ 0 & 0 & 2\sqrt{3} & 0 & 2\\ 0 & 0 & 0 & 2\sqrt{3} & 2\\ 0 & 0 & 0 & 0 & 4\end{pmatrix}. \tag{4.96}$$

It is easy to check that $X_{[3,2]}$ satisfies (4.89). Removing a factor $1/\sqrt{8}$, one obtains the orthogonal bases $\psi_{\nu 1}$ from Eq. (4.90) as follows:

$$\begin{aligned}
\psi_{11} &= \sqrt{8}e_1,\\
\psi_{21} &= \{E + 3(3\ 4)\}\, e_1,\\
\psi_{31} &= \sqrt{3}\,\{E + (3\ 4) + 2(3\ 5\ 4)\}\, e_1,\\
\psi_{41} &= \sqrt{3}\,\{E + (3\ 4) + 2(2\ 3\ 4)\}\, e_1,\\
\psi_{51} &= \{3E + (3\ 4) + 2(3\ 5\ 4) + 2(2\ 3\ 4) + 4(2\ 3\ 5\ 4)\}\, e_1.
\end{aligned}$$

The similarity transformation matrices $X_{[\lambda]}$, $D^{[\lambda]}(P_a)X_{[\lambda]} = X_{[\lambda]}\overline{D}^{[\lambda]}(P_a)$, for some representations of S_n are calculated in Probs. 13 and 14:

$$X_{[2,1]} = X_{[2,2]} = \frac{1}{\sqrt{3}} \begin{pmatrix} \sqrt{3} & 1 \\ 0 & 2 \end{pmatrix},$$

$$X_{[3,1]} = \frac{1}{\sqrt{8}} \begin{pmatrix} \sqrt{8} & 1 & \sqrt{3} \\ 0 & 3 & \sqrt{3} \\ 0 & 0 & \sqrt{12} \end{pmatrix}, \quad X_{[2,1,1]} = \frac{1}{\sqrt{6}} \begin{pmatrix} \sqrt{6} & \sqrt{2} & -1 \\ 0 & \sqrt{8} & 1 \\ 0 & 0 & 3 \end{pmatrix}. \tag{4.97}$$

4.5 Outer Product of Irreducible Representations of S_n

The outer product of irreducible representations of the permutation groups is related to the induced representation of an irreducible representation of the subgroup $\mathrm{S}_n \otimes \mathrm{S}_m$ with respect to the group S_{n+m}. In this section the graphic method for calculating the decomposition of the outer product of irreducible representations will be explained.

4.5.1 *Group* S_{n+m} *and Its Subgroup* $\mathrm{S}_n \otimes \mathrm{S}_m$

First of all, we introduce our notations for the outer product of irreducible representations. The group algebra $\mathcal{L}$ of the permutation group S_{n+m} among $(n+m)$ objects is $(n+m)!$ dimensional. The primitive idempotent $e_{\mathcal{T}}^{[\omega]}$ in $\mathcal{L}$ generates the minimal left ideal $\mathcal{L}_{\mathcal{T}}^{[\omega]} = \mathcal{L}e_{\mathcal{T}}^{[\omega]}$, which corresponds to the $d_{[\omega]}$-dimensional irreducible representation $D^{[\omega]}(\mathrm{S}_{n+m})$ of S_{n+m}, where the Young pattern $[\omega]$ contains $(n+m)$ boxes.

Let S_n denote the permutation subgroup among the first n objects in the $(n+m)$ objects, and by S_m the subgroup among the last m objects. The product of elements $R \in \mathrm{S}_n$ and $S \in \mathrm{S}_m$ is commutable, $RS = SR$, because they permute different objects. The only common element in two subgroups is the identity E. Thus, the direct product $\mathrm{S}_n \otimes \mathrm{S}_m$ of two subgroups is a subgroup of S_{n+m} with index N,

$$N = \binom{n+m}{n} = \frac{(n+m)!}{n!m!}. \tag{4.98}$$

Its left-coset is denoted by $T_\alpha (\mathrm{S}_n \otimes \mathrm{S}_m)$, $2 \leqslant \alpha \leqslant N$. Although those permutations T_α are not unique, we define T_α as follows:

$$\begin{gathered} T_\alpha = \begin{pmatrix} 1 & 2 & \cdots & n & n+1 & n+2 & \cdots & n+m \\ a_1 & a_2 & \cdots & a_n & b_1 & b_2 & \cdots & b_m \end{pmatrix}, \\ a_j \neq b_k, \quad 1 \leqslant a_1 < a_2 < \cdots < a_n \leqslant n+m, \\ 1 \leqslant b_1 < b_2 < \cdots < b_m \leqslant n+m. \end{gathered} \tag{4.99}$$

The group algebra of the subgroup $S_n \otimes S_m$ is denoted by $\mathcal{L}_{nm}$, which is $n!m!$ dimensional. $\mathcal{L}$ is decomposed into the direct sum of $T_\alpha \mathcal{L}_{nm}$:

$$\mathcal{L} = \bigoplus_{\alpha=1}^{N} T_\alpha \mathcal{L}_{nm}, \qquad T_1 = E. \tag{4.100}$$

Let $e_\mu^{[\lambda_1]} e_\nu^{[\lambda_2]}$ denote the primitive idempotent of $\mathcal{L}_{nm}$. $e_\mu^{[\lambda_1]} e_\nu^{[\lambda_2]}$ generates the minimal left ideal $\mathcal{L}_{\mu\nu}^{[\lambda_1][\lambda_2]} = \mathcal{L}_{nm} e_\mu^{[\lambda_1]} e_\nu^{[\lambda_2]}$ of $\mathcal{L}_{nm}$, corresponding to the irreducible representation $D^{[\lambda_1]}(S_n) \times D^{[\lambda_2]}(S_m)$ of $S_n \otimes S_m$ with the dimension $d_{[\lambda_1]} d_{[\lambda_2]}$. The Young patterns $[\lambda_1]$ and $[\lambda_2]$ contain n and m boxes, respectively. Generally, $e_\mu^{[\lambda_1]} e_\nu^{[\lambda_2]}$ is an idempotent of $\mathcal{L}$ but not primitive, and $\mathcal{L}_{\mu\nu}^{[\lambda_1][\lambda_2]}$ is a subalgebra of $\mathcal{L}$ but not its left ideal.

Now, we discuss two related problems. One problem is the subduced representation $D^{[\omega]}(S_n \otimes S_m)$ of an irreducible representation $D^{[\omega]}(S_{n+m})$ of S_{n+m} with respect to the subgroup $S_n \otimes S_m$. This subduced representation $D^{[\omega]}(S_n \otimes S_m)$ is generally reducible and can be decomposed into a direct sum of $D^{[\lambda_1]}(S_n) \times D^{[\lambda_2]}(S_m)$ of $S_n \otimes S_m$:

$$\begin{aligned} &D^{[\omega]}(RS) \simeq \bigoplus_{\lambda_1,\lambda_2} a^\omega_{\lambda_1\lambda_2}\, D^{[\lambda_1]}(R) \times D^{[\lambda_2]}(S), \\ &R \in S_n, \quad S \in S_m, \qquad d_{[\omega]} = \sum_{\lambda_1,\lambda_2} a^\omega_{\lambda_1\lambda_2}\, d_{[\lambda_1]} d_{[\lambda_2]}. \end{aligned} \tag{4.101}$$

$a^\omega_{\lambda_1\lambda_2}$ is calculated by the character formula in the subgroup:

$$a^\omega_{\lambda_1\lambda_2} = \frac{1}{n!m!} \sum_{RS \in S_n \otimes S_m} \chi^{[\lambda_1]}(R) \chi^{[\lambda_2]}(S) \chi^{[\omega]}(RS). \tag{4.102}$$

where $\chi^{[\lambda_1]}(R)\chi^{[\lambda_2]}(S)$ and $\chi^{[\omega]}(RS)$ are the characters of the element $RS \in S_n \otimes S_m$ in the representations $D^{[\lambda_1]}(S_n) \times D^{[\lambda_2]}(S_m)$ and $D^{[\omega]}(S_n \otimes S_m)$, respectively.

From the viewpoint of the group algebra, if a vector t_j in $\mathcal{L}$ satisfies

$$e_\mu^{[\lambda_1]} e_\nu^{[\lambda_2]} t_j e_\tau^{[\omega]} \neq 0, \tag{4.103}$$

$e_\mu^{[\lambda_1]} e_\nu^{[\lambda_2]} t_j e_\tau^{[\omega]}$ provides a map from the minimal left ideal $\mathcal{L}_{\mu\nu}^{[\lambda_1][\lambda_2]}$ of $\mathcal{L}_{nm}$ onto the left ideal $\mathcal{L}_\tau^{[\omega]}$ of $\mathcal{L}$:

$$\mathcal{L}_{nm} e_\mu^{[\lambda_1]} e_\nu^{[\lambda_2]} t_j e_\tau^{[\omega]} \subset \mathcal{L} e_\tau^{[\omega]} = \mathcal{L}_\tau^{[\omega]}. \tag{4.104}$$

Namely, $D^{[\lambda_1]}(S_n) \times D^{[\lambda_2]}(S_m)$ is contained in the subduced representation $D^{[\omega]}(S_n \otimes S_m)$. Due to Eq. (4.101), there are $a^\omega_{\lambda_1\lambda_2}$ linearly independent vectors $e_\mu^{[\lambda_1]} e_\nu^{[\lambda_2]} t_j e_\tau^{[\omega]}$, $1 \leqslant j \leqslant a^\omega_{\lambda_1\lambda_2}$. One replaces $t_j e_\tau^{[\omega]}$ with the complete bases $R_{\tau'\tau}e_\tau^{[\omega]}$ in $\mathcal{L}_\tau^{[\omega]}$ one by one until $a_{\lambda\mu}^\omega$ independent $e_\mu^{[\lambda_1]}e_\nu^{[\lambda_2]}t_je_\tau^{[\omega]}$ have been found. The Clebsch–Gordan coefficients for the decomposition can also be calculated from the map (4.104) (see Ex. 4 in subsection 4.5.2.).

The other problem is the induced representation of an irreducible representation $D^{[\lambda_1]}(S_n) \times D^{[\lambda_2]}(S_m)$ of the subgroup $S_n \otimes S_m$ with respect to the group S_{n+m}. The induced representation is called the outer product of irreducible representations in the permutation group, denoted by $D^{[\lambda_1]\otimes[\lambda_2]}(S_{n+m})$. The outer product is generally reducible and can be decomposed into a direct sum of the irreducible representations $D^{[\omega]}(S_{n+m})$ of S_{n+m},

$$\begin{aligned} &D^{[\lambda_1]\otimes[\lambda_2]}(T) \simeq \bigoplus_\omega b_{\lambda_1\lambda_2}^\omega D^{[\omega]}(T), \qquad T \in S_{n+m}, \\ &\frac{(n+m)!}{n!m!}d_{[\lambda_1]}d_{[\lambda_2]} = \sum_\omega b_{\lambda_1\lambda_2}^\omega \ d_{[\omega]}. \end{aligned} \tag{4.105}$$

Due to the Frobenius theorem (Theorem 3.6),

$$b_{\lambda_1\lambda_2}^\omega = \frac{1}{(n+m)!}\sum_{T\in S_{n+m}} \chi^{[\lambda_1]\otimes[\lambda_2]}(T)\chi^{[\omega]}(T) = a_{\lambda_1\lambda_2}^\omega, \tag{4.106}$$

where $\chi^{[\lambda_1]\otimes[\lambda_2]}(T)$ is the character of the element $T \in S_{n+m}$ in the outer product representation $D^{[\lambda_1]\otimes[\lambda_2]}(T)$.

From the viewpoint of the group algebra, $e_\mu^{[\lambda_1]}e_\nu^{[\lambda_2]}$ is an idempotent of $\mathcal{L}$, but generally not primitive. $e_\mu^{[\lambda_1]}e_\nu^{[\lambda_2]}$ generates a left ideal $\mathcal{L}_{\mu\nu}^{[\lambda_1]\otimes[\lambda_2]}$ of $\mathcal{L}$, corresponding to a representation $[\lambda_1] \otimes [\lambda_2]$ of S_{n+m}.

$$\begin{aligned} \mathcal{L}_{\mu\nu}^{[\lambda_1]\otimes[\lambda_2]} &= \mathcal{L}e_\mu^{[\lambda_1]}e_\nu^{[\lambda_2]} = \bigoplus_{\alpha=1}^N T_\alpha \mathcal{L}_{nm}e_\mu^{[\lambda_1]}e_\nu^{[\lambda_2]} \\ &= \bigoplus_{\alpha=1}^N T_\alpha \mathcal{L}_{\mu\nu}^{[\lambda_1][\lambda_2]}, \qquad T_1 = E. \end{aligned} \tag{4.107}$$

Each term in the direct sum is a $d_{[\lambda_1]}d_{[\lambda_2]}$ dimensional subspace of $\mathcal{L}_{\mu\nu}^{[\lambda_1]\otimes[\lambda_2]}$, and there is no common vector between any two subspaces. Hence, the dimension of the representation $[\lambda_1] \otimes [\lambda_2]$ is

$$d_{[\lambda_1]\otimes[\lambda_2]} = \frac{(n+m)!}{n!m!}d_{[\lambda_1]}d_{[\lambda_2]}. \tag{4.108}$$

Since

$$\mathcal{L}e_\tau^{[\omega]}t_j'e_\mu^{[\lambda_1]}e_\nu^{[\lambda_2]} \subset \mathcal{L}e_\mu^{[\lambda_1]}e_\nu^{[\lambda_2]} = \mathcal{L}_{\mu\nu}^{[\lambda_1]\otimes[\lambda_2]}, \tag{4.109}$$

$e_\tau^{[\omega]} t'_j e_\mu^{[\lambda_1]} e_\nu^{[\lambda_2]}$ provides a map from the minimal left ideal $\mathcal{L}_\tau^{[\omega]}$ of $\mathcal{L}$ onto the left ideal $\mathcal{L}_{\mu\nu}^{[\lambda_1]\otimes[\lambda_2]}$ of $\mathcal{L}$.

The left ideal and the right ideal generated by one primitive idempotent of a group algebra correspond to the same irreducible representation. Therefore, the number of the linear independent vectors $e_\tau^{[\omega]} t'_j e_\mu^{[\lambda_1]} e_\nu^{[\lambda_2]}$ is equal to that of $e_\mu^{[\lambda_1]} e_\nu^{[\lambda_2]} t_j e_\tau^{[\omega]}$. It is just the Frobenius theorem (Theorem 3.6). One replaces $t'_j e_\mu^{[\lambda_1]} e_\nu^{[\lambda_2]}$ with the complete bases $R_{\mu'\mu} e_\mu^{[\lambda_1]} R_{\nu'\nu} e_\nu^{[\lambda_2]}$ in $\mathcal{L}_{\mu\nu}^{[\lambda_1][\lambda_2]}$ one by one until $b_{\lambda_1\lambda_2}^\omega$ independent $e_\tau^{[\omega]} t'_j e_\mu^{[\lambda_1]} e_\nu^{[\lambda_2]}$ have been found. The Clebsch–Gordan coefficients for the decomposition can also be calculated from the map (4.109) (see Ex. 2 in subsection 4.5.2.).

4.5.2 *Littlewood–Richardson Rule*

There is a simple graphic method for calculating $b_{\lambda_1\lambda_2}^\omega$, called the Littlewood–Richardson rule. Choose one representation between $[\lambda_1]$ and $[\lambda_2]$, say $[\lambda_2]$, with less box number for convenience. Fill the boxes in the jth row of the Young pattern $[\lambda_2]$ with digit j. Attach the boxes of the Young pattern $[\lambda_2]$ row by row, beginning from the first row, into the Young pattern $[\lambda_1]$ according to the following rule.

(a) Each time when the boxes in one row of the Young pattern $[\lambda_2]$ have been attached, the resultant diagram constitutes a regular Young pattern. Namely, it is lined up on the top and on the left, where the box number of the higher row is not less than that of the lower row.

(b) The boxes filled with the same digit never occur in the same column of the resultant diagram.

(c) After all the boxes of the Young pattern $[\lambda_2]$ are attached, read the digits filled in the boxes of the resultant diagram from right to left, row by row beginning from the first row, such that at every step of the reading process the number of boxes filled by a smaller digit is never less than that filled by a larger digit.

When all the boxes of the Young pattern $[\lambda_2]$ were attached according to the above rule, the representation $[\omega]$ of S_{n+m}, which is the resultant diagram, appears in the decomposition of the outer product $[\lambda_1] \otimes [\lambda_2]$. The times of appearing of a Young pattern $[\omega]$ in the different resultant diagrams is the multiplicity of $[\omega]$ in the decomposition. Let us give some examples to show how to use the Littlewood–Richardson rule.

Ex. 1. Calculate the decomposition of $[2,1] \otimes [2,1]$ of S_6.

$$\begin{array}{|c|c|}\hline \times & \times \\ \hline \times & \multicolumn{1}{c}{} \\ \cline{1-1} \end{array} \otimes \begin{array}{|c|c|}\hline 1 & 1 \\ \hline 2 & \multicolumn{1}{c}{} \\ \cline{1-1} \end{array}$$

Attaching the boxes filled by the digit 1 to the first Young pattern according to the Littlewood–Richardson rule, one obtains

$$\begin{array}{llll} \times & \times & 1 & 1 \\ \times & & & \end{array} \qquad \begin{array}{lll} \times & \times & 1 \\ \times & 1 & \end{array} \qquad \begin{array}{lll} \times & \times & 1 \\ \times & & \\ 1 & & \end{array} \qquad \begin{array}{ll} \times & \times \\ \times & 1 \\ 1 & \end{array}$$

Then, attach the box filled by the digit 2 to the above diagrams, respectively. Note that the box filled by the digit 2 cannot be attached in the first row due to the rule (c). It also cannot be attached to the second row of the fourth diagram. The allowed diagrams are as follows:

$$\begin{array}{llll} \times & \times & 1 & 1 \\ \times & 2 & & \end{array} \qquad \begin{array}{llll} \times & \times & 1 & 1 \\ \times & & & \\ 2 & & & \end{array} \qquad \begin{array}{lll} \times & \times & 1 \\ \times & 1 & 2 \end{array} \qquad \begin{array}{lll} \times & \times & 1 \\ \times & 1 & \\ 2 & & \end{array}$$

$$\begin{array}{lll} \times & \times & 1 \\ \times & 2 & \\ 1 & & \end{array} \qquad \begin{array}{lll} \times & \times & 1 \\ \times & & \\ 1 & & \\ 2 & & \end{array} \qquad \begin{array}{ll} \times & \times \\ \times & 1 \\ 1 & 2 \end{array} \qquad \begin{array}{ll} \times & \times \\ \times & 1 \\ 1 & \\ 2 & \end{array}$$

The decomposition of $[2,1] \otimes [2,1]$ and their dimensions are as follows

$$\begin{aligned} [2,1] \otimes [2,1] \simeq & [4,2] \oplus [4,1,1] \oplus [3,3] \oplus 2\,[3,2,1] \\ & \oplus [3,1,1,1] \oplus [2,2,2] \oplus [2,2,1,1], \end{aligned} \tag{4.110}$$

$$20 \times 2 \times 2 = 9 + 10 + 5 + 2 \times 16 + 10 + 5 + 9.$$

Ex. 2. Calculate the similarity transformation matrix X for the decomposition of $[2] \otimes [1,1]$ of S_4.

$$\begin{array}{|c|c|}\hline \times & \times \\ \hline \end{array} \otimes \begin{array}{|c|}\hline 1 \\ \hline 2 \\ \hline \end{array} \simeq \begin{array}{|c|c|c|}\hline \times & \times & 1 \\ \hline 2 & \multicolumn{2}{c}{} \\ \cline{1-1} \end{array} \oplus \begin{array}{|c|c|}\hline \times & \times \\ \hline 1 & \multicolumn{1}{c}{} \\ \cline{1-1} 2 & \multicolumn{1}{c}{} \\ \cline{1-1} \end{array}$$

The representation $[2] \times [1,1]$ of the subgroup $S_2 \otimes S_2$ is one-dimensional where the basis in the representation space of $\mathcal{L}_{22}$:

$$\phi_1 = \mathcal{Y}^{[2]} \mathcal{Y}^{[1,1]} = [E + (1\ 2)][E - (3\ 4)] = (1\ 2)\phi_1 = -(3\ 4)\phi_1.$$

Extending the space $\mathcal{L}_{22}\phi_1$ to $\mathcal{L}\phi_1$ (see Eq. (4.100)), where $\mathcal{L}_{22}$ and $\mathcal{L}$ are the group algebras of $S_2 \otimes S_2$ and S_4, respectively:

$$\begin{aligned}
&\phi_2 = (1\ 3)\phi_1, \qquad \phi_3 = (1\ 4)\phi_1, \qquad \phi_4 = (2\ 3)\phi_1, \\
&\phi_5 = (2\ 4)\phi_1, \qquad \phi_6 = (1\ 3)(2\ 4)\phi_1 = -(1\ 4)(2\ 3)\phi_1,
\end{aligned}$$

where $(1\ 3)(2\ 4)\phi_1 = -(1\ 3)(2\ 4)(3\ 4)\phi_1 = -(1\ 3\ 2\ 4)\phi_1 = -(4\ 1)(3\ 2)\phi_1$.

The representations $[3,1]$ and $[2,1^2]$ of S_4 both are three-dimensional. Taking the Young operators

$$\begin{aligned}
\mathcal{Y}_1^{[3,1]} &= [E+(1\ 2)+(1\ 3)+(2\ 3)+(1\ 2\ 3)+(3\ 2\ 1)][E-(1\ 4)] \\
&= [E+(1\ 3)+(2\ 3)][E+(1\ 2)] - [E+(1\ 2)+(2\ 3)][E+(1\ 3)](1\ 4) \\
&= [E+(1\ 3)+(2\ 3)][E+(1\ 2)] - [E+(1\ 2)+(2\ 3)](1\ 4)[E+(4\ 3)], \\
\mathcal{Y}_1^{[2,1^2]} &= [E+(1\ 2)][E-(1\ 3)-(1\ 4)-(3\ 4)+(1\ 3\ 4)+(4\ 3\ 1)] \\
&= [E+(1\ 2)][E-(1\ 3)-(1\ 4)][E-(3\ 4)] \\
&= [E+(1\ 2)-(1\ 3)-(2\ 3)(1\ 2)-(1\ 4)-(2\ 4)(1\ 2)][E-(3\ 4)].
\end{aligned}$$

One is able to expand the basis vectors in the left ideals $\mathcal{L}\mathcal{Y}_1^{[3,1]}$ and $\mathcal{L}\mathcal{Y}_1^{[2,1^2]}$ with respect to ϕ_j (see Eq. (4.109)):

$$\begin{aligned}
&\psi_1 = \mathcal{Y}_1^{[3,1]}\phi_1 = 2\phi_1 + 2\phi_2 + 2\phi_4, \\
&\psi_2 = (3\ 4)\psi_1 = -2\phi_1 - 2\phi_3 - 2\phi_5, \\
&\psi_3 = (2\ 3\ 4)\psi_1 = (4\ 2)\psi_1 = 2\phi_5 + 2\phi_6 - 2\phi_4, \\
&\psi_4 = \mathcal{Y}_1^{[2,1^2]}\phi_1 = 4\phi_1 - 2\phi_2 - 2\phi_4 - 2\phi_3 - 2\phi_5, \\
&\psi_5 = (2\ 3)\psi_4 = 4\phi_4 - 2\phi_2 - 2\phi_1 + 2\phi_6 + 2\phi_5, \\
&\psi_6 = (2\ 4\ 3)\psi_4 = -(2\ 4)\psi_4 = -4\phi_5 + 2\phi_6 - 2\phi_4 + 2\phi_3 + 2\phi_1, \\
&\psi_\mu = \sum_{\nu=1}^{6} \phi_\nu X_{\nu\mu}, \qquad \{[2]\otimes[1,1]\}X = X\{[3,1]\oplus[2,1^2]\}.
\end{aligned} \tag{4.111}$$

The detailed calculation is left as exercise (see Prob. 24).

The Littlewood–Richardson rule can also be used to calculate the decomposition of the subduced representation $[\omega]$ with respect to the subgroup $S_n \otimes S_m$ due to the Frobenius theorem (see Eq. (4.106)). In the decomposition, $[\lambda_1]$ and $[\lambda_2]$ run over all representations of S_n and S_m, respectively, but the resultant diagram obtained according to the Littlewood–Richardson rule has to be the given $[\omega]$.

Ex. 3. Calculate the decomposition of the subduced representation of $[3,2,1]$ of S_6 with respect to the subgroup $S_3 \otimes S_3$.

$$\begin{array}{llllll} \times\ \times\ \times & \times\ \times\ 1 & \times\ \times\ 1 & \times\ \times\ 1 & \times\ \times\ 1 & \times\ 1\ 1 \\ 1\ 1 & \oplus \times\ 1 & \oplus \times\ 1 & \oplus \times\ 2 & \oplus \times\ 2 & \oplus \times\ 2 \\ 2 & 1 & 2 & 1 & 3 & \times \end{array}$$

Hence,

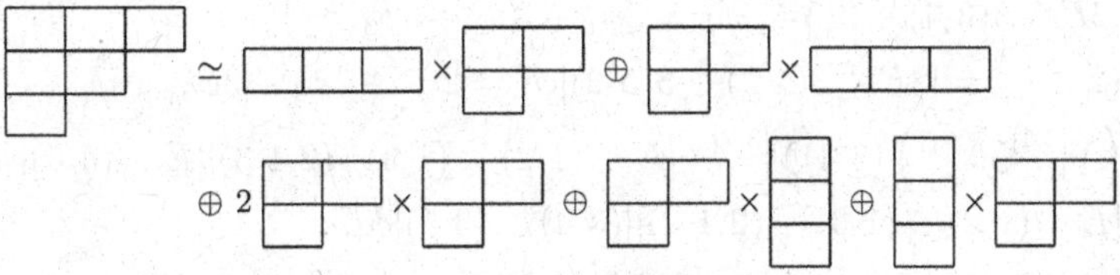

Check the dimensions in the decomposition,

$$16 = 1 \times 2 + 2 \times 1 + 2 \times 2 \times 2 + 2 \times 1 + 1 \times 2.$$

Ex. 4. Calculate the similarity transformation matrix for the decomposition of the subduced representation of $[3, 2]$ of S_6 with respect to the subgroup $S_3 \otimes S_2$.

$$\begin{array}{l} \times\ \times\ \times \\ 1\ 1 \end{array} \oplus \begin{array}{l} \times\ \times\ 1 \\ \times\ 1 \end{array} \oplus \begin{array}{l} \times\ \times\ 1 \\ \times\ 2 \end{array}$$

$$[3, 2] \simeq [3] \times [2] \ \oplus \ [2, 1] \times [2] \ \oplus \ [2, 1] \times [1, 1].$$

The basis vectors for the representation $[3, 2]$ of S_5 is

$$\begin{aligned} \phi_1 = \mathcal{Y}_1^{[3,2]} &= [E + (1\ 2) + (1\ 3) + (2\ 3) + (1\ 2\ 3) + (3\ 2\ 1)] \\ &\quad \cdot [E + (4\ 5)][E - (1\ 4)][E - (2\ 5)], \\ \phi_2 &= (3\ 4)\phi_1, \qquad\qquad \phi_3 = (3\ 5\ 4)\phi_1 = (3\ 5)\phi_1, \\ \phi_4 &= (2\ 3\ 4)\phi_1 = (4\ 2)\phi_1, \quad \phi_5 = (2\ 3\ 5\ 4)\phi_1 = (5\ 2)\phi_1. \end{aligned}$$

In later calculation the Fock condition (4.37) will be used:

$$\begin{aligned} [E + (1\ 4) + (2\ 4) + (3\ 4)]\phi_1 = 0, \\ [E + (1\ 5) + (2\ 5) + (3\ 5)]\phi_1 = 0. \end{aligned}$$

Because $\mathcal{Y}_1^{[2,1]}\mathcal{Y}_1^{[3,2]} = 0$ where the Young tableau $\mathcal{Y}_1^{[2,1]}$ is 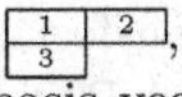, one has to choose, say $t = (3\ 4)$ in Eq. (4.103). Thus, the new basis vectors $\psi_\mu = \sum_{\nu=1}^5 \phi_\nu X_{\nu\mu}$ are

$$\begin{aligned} \psi_1 &= \left(\mathcal{Y}^{[3]}\mathcal{Y}^{[2]}\right) \mathcal{Y}_1^{[3,2]} \\ &= [E + (1\ 2) + (1\ 3) + (2\ 3) + (1\ 2\ 3) + (3\ 2\ 1)][E + (4\ 5)]\mathcal{Y}_1^{[3,2]} = 12\phi_1, \end{aligned}$$

$$
\begin{aligned}
\psi_2 &= \left(\mathcal{Y}_1^{[2,1]}\mathcal{Y}^{[2]}\right)(3\ 4)\mathcal{Y}_1^{[3,2]} = [E+(1\ 2)-(1\ 3)-(2\ 1\ 3)][E+(4\ 5)](3\ 4)\phi_1 \\
&= [E+(1\ 2)-(1\ 3)-(2\ 1\ 3)][(3\ 4)+(3\ 5)]\phi_1 \\
&= [2(3\ 4)-(4\ 1)-(4\ 2)+2(3\ 5)-(5\ 1)-(5\ 2)]\phi_1 \\
&= [2E+3(3\ 4)+3(3\ 5)]\phi_1 = 2\phi_1+3\phi_2+3\phi_3, \\
\psi_3 &= (2\ 3)\psi_2 = [2E+3(4\ 2)+3(5\ 2)]\phi_1 = 2\phi_1+3\phi_4+3\phi_5, \\
\psi_4 &= \left(\mathcal{Y}_1^{[2,1]}\mathcal{Y}^{[1,1]}\right)(3\ 4)\mathcal{Y}_1^{[3,2]} = [E+(1\ 2)-(1\ 3)-(2\ 1\ 3)][E-(4\ 5)](3\ 4)\phi_1 \\
&= [E+(1\ 2)-(1\ 3)-(2\ 1\ 3)][(3\ 4)-(3\ 5)]\phi_1 \\
&= [2(3\ 4)-(4\ 1)-(4\ 2)-2(3\ 5)+(5\ 1)+(5\ 2)]\phi_1 \\
&= [3(3\ 4)-3(3\ 5)]\phi_1 = 3\phi_2-3\phi_3, \\
\psi_5 &= (2\ 3)\psi_4 = [3(4\ 2)-3(5\ 2)]\phi_1 = 3\phi_4-3\phi_5,
\end{aligned}
$$

$$X^{-1}\{[2]\otimes[1,1]\}X = [3]\times[2] \ \oplus\ [2,1]\times[2] \ \oplus\ [2,1]\times[1,1].$$

As exercise, the readers are asked to write the similarity transformation matrix X explicitly, and check the similarity transformation relation for the generators (1 2), (2 3) and (4 5) of $S_3 \otimes S_2$.

4.6 Exercises

1. Calculate the number $d_{[\lambda]}(S_n)$ of the standard Young tableaux for each Young pattern $[\lambda]$ of the permutation groups S_5, S_6, and S_7.
2. Write the Young operators for the following Young tableaux:

   (a) $\begin{array}{|c|c|c|}\hline 1&2&3\\ \hline 4 \\ \cline{1-1}\end{array}$; (b) $\begin{array}{|c|c|}\hline 1&2\\ \hline 3&4\\ \hline\end{array}$; (c) $\begin{array}{|c|c|c|c|}\hline 1&2&3&4\\ \hline 5 \\ \cline{1-1}\end{array}$.
3. Write all standard Young tableaux $\mathcal{Y}_\mu$ for the Young pattern [4,2] of S_6 from the smallest to the largest, and calculate the permutations $R_{\mu 1}$ transforming the standard Young tableau $\mathcal{Y}_1$ to the remaining $\mathcal{Y}_\mu$.
4. Prove $\mathcal{Y}'\mathcal{Y} \neq 0$, and express the permutation $R = PQ = P'Q'$, which transforms the Young tableau $\mathcal{Y}$ to the Young tableau $\mathcal{Y}'$, where PQ and $P'Q'$ belong to the Young tableaux $\mathcal{Y}$ and $\mathcal{Y}'$, respectively:

   The Young tableau $\mathcal{Y}$ The Young tableau $\mathcal{Y}'$

   $\begin{array}{|c|c|c|c|}\hline 1&2&4&7\\ \hline 3&5&9 \\ \cline{1-3} 6&8 \\ \cline{1-2}\end{array}$ $\qquad$ $\begin{array}{|c|c|c|c|}\hline 1&2&3&4\\ \hline 5&6&7 \\ \cline{1-3} 8&9 \\ \cline{1-2}\end{array}$

5. Try to expand all nonstandard Young operators $\mathcal{Y}$ of the Young pattern [3,1] in S_4 with respect to the standard Young operators $\mathcal{Y}_\mu$, $\mathcal{Y} = \sum_\mu t_\mu \mathcal{Y}_\mu$, where t_μ is the vector in the group space of S_4.

6. Expand explicitly the identity of S_4 with respect to the Young operators in the group algebra of S_4.

7. Prove that $\overline{y}_\mu$ given in Eq. (4.64) satisfy $\mathcal{Y}_\nu \overline{y}_\mu \mathcal{Y}_\mu = \delta_{\nu\mu} \mathcal{Y}_\nu \mathcal{Y}_\nu$.

8. Calculate the orthogonal primitive idempotents for the Young pattern $[2, 2, 1]$ of the permutation group S_5.

9. Calculate the representation matrices of the generators (1 2) and (1 2 3 4 5) of S_5 in the irreducible representation $[2, 2, 1]$ by the tabular method.

10. Calculate the orthogonal primitive idempotents for the Young pattern $[4, 2]$ of the permutation group S_6.

11. Calculate the orthogonal primitive idempotents for the Young pattern $[3, 2, 1]$ of the permutation group S_6.

12. Calculate the Clebsch–Gordan coefficients for the decomposition of the self-product of the irreducible representation $[2, 1]$ of S_3 in the irreducible bases and in the orthogonal bases, respectively.

13. Calculate the irreducible bases for the irreducible representation $D^{[3,1]}(S_4)$ in the group algebra $\mathcal{L}$ of S_4, and calculate the representation matrix $D^{[3,1]}(P_a)$ of the transpositions of the neighboring objects by the tabular method.

14. Calculate the real orthogonal representation matrix $\overline{D}^{[3,1]}(P_a)$ of the transpositions of the neighboring objects in the irreducible representation $[3, 1]$ of S_4 by the formulas (4.92) and (4.95). Calculate the similarity transformation matrix $X_{[3,1]}$: $D^{[3,1]}(P_a)X_{[3,1]} = X_{[3,1]}\overline{D}^{[3,1]}(P_a)$. Calculate the orthogonal bases $\psi_{\mu\nu}^{[3,1]}$ in the group algebra of S_4.

15. Calculate the similarity transformation matrix $X_{[\lambda]}$: $D^{[\lambda]}(S_4)X_{[\lambda]} = X_{[\lambda]}\overline{D}^{[\lambda]}(S_4)$, where $D^{[\lambda]}(S_4)$ and $\overline{D}^{[\lambda]}(S_4)$ are calculated by the tabular method and by Eqs. (4.92) and (4.95) respectively. (1) $[\lambda] = [2, 1^2]$; (2) $[\lambda] = [2, 2]$.

16. Calculate the character table of each irreducible representation of S_5 by the graphic method (see Table 4.2).

17. Calculate the character table of each irreducible representation of S_6 by the graphic method.

18. Prove that the subduced representation of the irreducible representation $D^{[\lambda]}(S_n)$ with respect to the alternating subgroup S'_n of S_n, where $[\lambda]$ is a self-associated Young pattern of S_n, $[\tilde{\lambda}] = [\lambda]$, is decomposed into the direct sum of two inequivalent irreducible representations with the same dimension.
Hint: Use Eqs. (3.62), (3.65) and Prob. 18 in Chap. 2.

19. Calculate the character tables of the alternating subgroups S'_6 and S'_7.
Hint: See the calculation for Tables 4.5. Note that S'_7 contains a pair of reciprocal classes.

20. Calculate the Clebsch–Gordan series for the decomposition of the inner product of each pair of the irreducible representations for the permutation groups S_4 by the character method.

21. Calculate the Clebsch–Gordan series for the decomposition of the inner self-product of the irreducible representation $[3, 2]$ of S_5.

22. Calculate the decomposition of the following outer products of the representations in the permutation groups by the Littlewood–Richardson rule:
(1) $[3, 2, 1] \otimes [3]$, (2) $[3, 2] \otimes [2, 1]$, (3) $[2, 1] \otimes [4, 2^3]$.

23. Calculate the decomposition of the subduced representations of the following irreducible representations of S_6 with respect to the subgroup $S_3 \otimes S_3$:
(1) $[4, 2]$, (2) $[2, 2, 1, 1]$, (3) $[3, 3]$.

24. Please complete the calculation for Eq. (4.111), write the similarity transformation matrix X explicitly, and check the similarity transformation relation for the generators (1 2) and (1 2 3 4) of S_4.

25. Calculate the similarity transformation matrix in the decomposition the induced representation of two-dimensional irreducible representation $[2, 1]$ of S_3 with respect to S_4.

26. Calculate the similarity transformation matrix in the decomposition of the outer product representation $[2, 1] \otimes [2]$ with respect to S_5.

27. Calculate the similarity transformation matrix in the decomposition of the subduced representation of the irreducible representation $[3, 3]$ of S_6 with respect to the subgroup $S_3 \otimes S_3$.

Chapter 5

THREE-DIMENSIONAL ROTATION GROUP

The spherical symmetry is the most important symmetry both in the classical mechanics and in the quantum mechanics, because not only a system with the spherical symmetry is easier to deal with, but also many systems in physics have approximative spherical symmetry. A system with the spherical symmetry has a symmetry center, usually chosen as the origin, so that this system is isotropic with respect to the origin. This system is invariant under all rotations about any axis across the origin. The symmetry group of this system is the three-dimensional rotation group SO(3). The group SO(3) is the simplest Lie group. Therefore, the study on the group SO(3) is very important both in physics and in mathematics.

5.1 Rotations in Three-dimensional Space

There are two viewpoints in describing the transformation of a system, including the rotation. In this book we adopt the viewpoint where **the coordinate frame K is fixed with the laboratory and the system is transformed in K**. In another viewpoint the coordinate frame transforms but the system is fixed. **The transformation in one viewpoint are the inverse of that in the other viewpoint.**

Now, we discuss the rotation R in the three-dimensional real space. Let K be a perpendicular coordinate frame which is laboratory-fixed. The position of an arbitrary point P in K is described by a position vector $\boldsymbol{r}$ which is a vector pointing from the origin O to P. The coordinate basis vectors in K are denoted by $\boldsymbol{e}_a$, $a = 1,\ 2,\ 3$,

$$\boldsymbol{r} = \sum_{a=1}^{3} \boldsymbol{e}_a x_a \ . \tag{5.1}$$

There is a one-to-one correspondence between the position vector $\boldsymbol{r}$ and three coordinates x_a where $\boldsymbol{e}_a$ are fixed. In a space rotation R the point P of the system transforms to P' whose position vector is denoted by $\boldsymbol{r}' = \sum_a \boldsymbol{e}_a x'_a$. **Since the position of the origin and the distance of any two points are left invariant under** R, R can be denoted by a three-dimensional real orthogonal matrix:

$$\begin{pmatrix} x'_1 \\ x'_2 \\ x'_3 \end{pmatrix} = \begin{pmatrix} R_{11} & R_{12} & R_{13} \\ R_{21} & R_{22} & R_{23} \\ R_{31} & R_{32} & R_{33} \end{pmatrix} \begin{pmatrix} x_1 \\ x_2 \\ x_3 \end{pmatrix}, \qquad \underline{x}' = R\underline{x},$$
$$\left(\underline{x}'\right)^T \underline{x}' = \underline{x}^T \underline{x}, \qquad R^T R = \mathbf{1}, \qquad R^* = R. \tag{5.2}$$

The set of all three-dimensional real orthogonal matrices, in the multiplication rule of matrices, satisfies four axioms of a group and forms a group, denoted by O(3).

Let K' be a body-fixed frame with the coordinate basis vectors $\boldsymbol{e}'_a$. The vector $\boldsymbol{r}'$ is fixed with the frame K' under the rotation R,

$$\boldsymbol{r}' = \sum_{a=1}^{3} \boldsymbol{e}_a x'_a = \sum_{b=1}^{3} \boldsymbol{e}'_b x_b. \tag{5.3}$$

Substituting Eq. (5.2) into Eq. (5.3), one obtains

$$\boldsymbol{e}'_b = \sum_{a=1}^{3} \boldsymbol{e}_a R_{ab}. \tag{5.4}$$

The chirality of a coordinate frame is determined by the mixed product of coordinate basis vectors. Due to

$$\boldsymbol{e}_1 \cdot (\boldsymbol{e}_2 \times \boldsymbol{e}_3) = 1, \tag{5.5}$$

K is a right-handed coordinate frame. The chirality of K' depends on

$$\boldsymbol{e}'_1 \cdot (\boldsymbol{e}'_2 \times \boldsymbol{e}'_3) = \det R = \pm 1. \tag{5.6}$$

K' is right-handed if det $R = 1$ and R is a proper rotation. K' is left-handed if det $R = -1$ and R is an improper rotation. The set of all proper rotations forms an invariant subgroup of O(3) with index two, called the three-dimensional rotation group SO(3). The set of all improper rotations forms the only coset of SO(3) in O(3). The space inversion σ is the representative element in the coset. Every improper rotation S is equal to the product of σ and a proper rotation S', $S = \sigma S' = S'\sigma$.

A proper rotation about the Z-axis through an angle ω is denoted by $R(\boldsymbol{e}_3, \omega)$ (see Fig. 3.1):

$$\begin{aligned} x_1' &= x_1 \cos\omega - x_2 \sin\omega, \\ x_2' &= x_1 \sin\omega + x_2 \cos\omega, \\ x_3' &= x_3, \end{aligned} \qquad R(\boldsymbol{e}_3, \omega) = \begin{pmatrix} \cos\omega & -\sin\omega & 0 \\ \sin\omega & \cos\omega & 0 \\ 0 & 0 & 1 \end{pmatrix}. \tag{5.7}$$

In terms of the Pauli matrices,

$$\begin{gathered} \sigma_1 = \begin{pmatrix} 0 & 1 \\ 1 & 0 \end{pmatrix}, \qquad \sigma_2 = \begin{pmatrix} 0 & -\mathrm{i} \\ \mathrm{i} & 0 \end{pmatrix}, \qquad \sigma_3 = \begin{pmatrix} 1 & 0 \\ 0 & -1 \end{pmatrix}, \\ \sigma_a \sigma_b = \delta_{ab} \mathbf{1} + \mathrm{i} \sum_{c=1}^{3} \epsilon_{abc} \sigma_c, \qquad \mathrm{Tr}\, \sigma_a = 0, \qquad \mathrm{Tr}\,(\sigma_a \sigma_b) = 2\delta_{ab}, \end{gathered} \tag{5.8}$$

$$\begin{aligned} \exp\{-\mathrm{i}\omega\sigma_2\} &= \sum_n \frac{1}{n!} \omega^n \left(-\mathrm{i}\sigma_2\right)^n \\ &= \mathbf{1} \left(1 - \frac{1}{2!}\omega^2 + \frac{1}{4!}\omega^4 - \cdots \right) - \mathrm{i}\sigma_2 \left(\omega - \frac{1}{3!}\omega^3 + \frac{1}{5!}\omega^5 - \cdots \right) \\ &= \mathbf{1} \cos\omega - \mathrm{i}\sigma_2 \sin\omega = \begin{pmatrix} \cos\omega & -\sin\omega \\ \sin\omega & \cos\omega \end{pmatrix}, \end{aligned} \tag{5.9}$$

$R(\boldsymbol{e}_3, \omega)$ can also be expressed as an exponential function of matrix

$$R(\boldsymbol{e}_3, \omega) = \exp\{-\mathrm{i}\omega T_3\} = \begin{pmatrix} \cos\omega & -\sin\omega & 0 \\ \sin\omega & \cos\omega & 0 \\ 0 & 0 & 1 \end{pmatrix}, \quad T_3 = \begin{pmatrix} 0 & -\mathrm{i} & 0 \\ \mathrm{i} & 0 & 0 \\ 0 & 0 & 0 \end{pmatrix}.$$

From the cyclic combination of three axes, one has

$$R(\boldsymbol{e}_1, \omega) = \exp\{-\mathrm{i}\omega T_1\} = \begin{pmatrix} 1 & 0 & 0 \\ 0 & \cos\omega & -\sin\omega \\ 0 & \sin\omega & \cos\omega \end{pmatrix}, \quad T_1 = \begin{pmatrix} 0 & 0 & 0 \\ 0 & 0 & -\mathrm{i} \\ 0 & \mathrm{i} & 0 \end{pmatrix},$$

$$R(\boldsymbol{e}_2, \omega) = \exp\{-\mathrm{i}\omega T_2\} = \begin{pmatrix} \cos\omega & 0 & \sin\omega \\ 0 & 1 & 0 \\ -\sin\omega & 0 & \cos\omega \end{pmatrix}, \quad T_2 = \begin{pmatrix} 0 & 0 & \mathrm{i} \\ 0 & 0 & 0 \\ -\mathrm{i} & 0 & 0 \end{pmatrix}.$$

Write in the uniform form:

$$R(\boldsymbol{e}_a, \omega) = \exp\{-\mathrm{i}\omega T_a\}, \qquad (T_a)_{bc} = -\mathrm{i}\epsilon_{abc}. \tag{5.10}$$

Define $S(\varphi, \theta) = R(\boldsymbol{e}_3, \varphi) R(\boldsymbol{e}_2, \theta)$, which transforms $\boldsymbol{e}_3$ to the direction $\hat{\boldsymbol{n}}(\theta, \varphi)$, where θ and φ are the polar angle and the azimuthal angle of $\hat{\boldsymbol{n}}$,

$$S(\varphi,\theta)=R(\boldsymbol{e}_3,\varphi)R(\boldsymbol{e}_2,\theta)=\begin{pmatrix}\cos\varphi\cos\theta & -\sin\varphi & \cos\varphi\sin\theta\\ \sin\varphi\cos\theta & \cos\varphi & \sin\varphi\sin\theta\\ -\sin\theta & 0 & \cos\theta\end{pmatrix},$$
$$S(\varphi,\theta)\begin{pmatrix}0\\0\\1\end{pmatrix}=\begin{pmatrix}\cos\varphi\sin\theta\\ \sin\varphi\sin\theta\\ \cos\theta\end{pmatrix}=\begin{pmatrix}n_1\\n_2\\n_3\end{pmatrix}. \tag{5.11}$$

It is easy to check

$$S(\varphi,\theta)T_3S(\varphi,\theta)^{-1}=n_1T_1+n_2T_2+n_3T_3=\hat{\boldsymbol{n}}\cdot\boldsymbol{T},$$
$$\boldsymbol{T}=\boldsymbol{e}_1T_1+\boldsymbol{e}_2T_2+\boldsymbol{e}_3T_3. \tag{5.12}$$

Thus, the rotation $R(\hat{\boldsymbol{n}},\omega)$ about the direction $\hat{\boldsymbol{n}}$ through an angle ω can be expressed as an exponential function of matrix,

$$R(\hat{\boldsymbol{n}},\omega)=S(\varphi,\theta)R(\boldsymbol{e}_3,\omega)S(\varphi,\theta)^{-1}=\exp\{-\mathrm{i}\omega ST_3S^{-1}\}$$
$$=\exp\{-\mathrm{i}\omega\hat{\boldsymbol{n}}\cdot\boldsymbol{T}\}=\exp\left\{-\mathrm{i}\sum_{a=1}^{3}\omega_aT_a\right\}. \tag{5.13}$$

$$\omega_1=\omega\sin\theta\cos\varphi,\qquad \omega_2=\omega\sin\theta\sin\varphi,\qquad \omega_3=\omega\cos\theta. \tag{5.14}$$

Remind that $R(\hat{\boldsymbol{n}},\omega)\hat{\boldsymbol{n}}=\hat{\boldsymbol{n}}$, Tr $R(\hat{\boldsymbol{n}},\omega)=1+2\cos\omega$, and

$$R(\hat{\boldsymbol{n}},\omega+2\pi)=R(\hat{\boldsymbol{n}},\omega)=R(-\hat{\boldsymbol{n}},2\pi-\omega),\quad R(\hat{\boldsymbol{n}},\pi)=R(-\hat{\boldsymbol{n}},\pi). \tag{5.15}$$

Conversely, any element R in SO(3) is a rotation about a direction $\hat{\boldsymbol{n}}$ through an angle ω where $\hat{\boldsymbol{n}}$ is the eigenvector of R with the eigenvalue 1 and ω is determined by Tr $R=1+2\cos\omega$. Thus, the element $R(\hat{\boldsymbol{n}},\omega)\in$ SO(3) can be described by $\boldsymbol{\omega}$ through its spherical coordinates (ω,θ,φ) or its rectangular coordinates $(\omega_1,\omega_2,\omega_3)$. **The variation region of $\boldsymbol{\omega}$ is a spheroid with radius π, where two end points of a diameter on the sphere describe the same element (rotation).** Equation (5.13) shows that the elements $R(\hat{\boldsymbol{n}},\omega)$ with the same ω form a class in SO(3).

5.2 Fundamental Concept of a Lie Group

5.2.1 *The Composition Functions of a Lie Group*

A group G is called a **continuous group** if its element R can be characterized by g independent and continuous real parameters r_ρ, $1\leqslant\rho\leqslant g$, varying in a g-dimensional region. The region is called the **group space** of G and g is the **order** of G. It is required that there is a one-to-one correspondence between the group element R and the set of parameters r_ρ, at least in the region with non-vanishing measure. The "measure" of a region is a mathematical concept. "The region with non-vanishing measure" may be understood roughly as a region whose dimension is equal to that of the group space. For example, the group space of SO(3) is a spheroid with radius π. Each point inside the sphere corresponds to one and only one element of SO(3), but on the sphere where the measure is vanishing, two end points of a diameter correspond to one element. Hereafter, a point in the group space which corresponds to an element R of G will be simply called the element R of G for convenience.

The parameters t_ρ of the product $T = RS$ of two elements are the single-valued functions of the parameters of the factors R and S:

$$t_\rho = f_\rho(r_1 \ldots, r_g; s_1 \ldots, s_g) = f_\rho(r; s). \tag{5.16}$$

g functions $f_\rho(r;s)$ are called the **composition functions**, which describe the multiplication rule of elements of G completely. A continuous group is called a **Lie group** if its composition functions are the analytic functions or at least piecewisely. The mathematical analysis can be used in studying a Lie group so that the theory of Lie groups has been studied thoroughly.

The definition domain of $f_\rho(r;s)$ is square of the group space, and the value domain of $f_\rho(r;s)$ is the group space. In order to meet four axioms for a group, the composition functions have to satisfy the conditions:

$$\begin{aligned} &f_\rho(f(r;s);t) = f_\rho(r; f(s;t)), \\ &f_\rho(e;r) = r_\rho, \qquad f_\rho(\overline{r};r) = e_\rho, \end{aligned} \tag{5.17}$$

where $\overline{r}_\rho$ are the parameters of the inverse R^{-1}, and e_ρ are the parameters of the identity E. **e_ρ are usually taken to be zero for convenience**. As a matter of fact, the composition functions are mainly used in the theoretical analysis, but rarely in the practical calculations. Even for the simplest Lie group, such as SO(3), the composition functions are quite complicated and their forms do not write explicitly.

Many fundamental concepts of a group are also suitable for a Lie group, such as the concepts of an abelian group, a subgroup, a coset, the conjugate elements, a class, an invariant subgroup, the quotient group, the isomorphism, the homomorphism, a linear representation, the character, the equivalent representations, an irreducible representation, a self-conjugate representation, and so on. The matrix elements and the characters of a representation of a Lie group are the single-valued analytic functions of the group parameters in the group space, at least in the region where the measure is not vanishing.

5.2.2 *The Local Property of a Lie Group*

The group elements in a Lie group G are said to be **adjacent** if their parameters differ only "slightly" from one another. The parameters e_ρ of the identity E are chosen to be zero. Thus, the parameters α_ρ of the element $A(\alpha)$, which is adjacent to the identity E, are infinitesimal, so that $A(\alpha)$ is called the **infinitesimal element**. Evidently, RA and AR both are adjacent to the element R. The product of R^{-1} and an element adjacent to R is an infinitesimal element. To speak roughly, the point R moves continuously in the group space if R is multiplied by infinite number of infinitesimal elements continually. Conversely, if the points R and E can be connected by a continuous curve embedded completely in the group space, R is equal to the product of infinite number of infinitesimal elements. **The infinitesimal elements are related to the differential calculus of the group elements so that they describe the local property of a Lie group.** The precise version of this statement will be given in Theorem 5.1.

The product of two infinitesimal elements $A(\alpha)$ and $B(\beta)$ is an infinitesimal element. The parameters of AB can be expanded as a Taylor series with respect to the infinitesimal parameters α_λ and β_λ:

$$\begin{aligned} f_\rho(\alpha;\beta) &= f_\rho(0;0) + \sum_{\lambda=1}^{g} \left(\alpha_\lambda \left. \frac{\partial f_\rho(\alpha;0)}{\partial \alpha_\lambda} \right|_{\alpha=0} + \beta_\lambda \left. \frac{\partial f_\rho(0;\beta)}{\partial \beta_\lambda} \right|_{\beta=0} \right) \\ &= \alpha_\rho + \beta_\rho, \end{aligned}$$

where $e_\rho = 0$, $AE = A$, $EB = B$, and the infinitesimal quantities of the second order have been removed. Therefore, the product of two infinitesimal elements are commutable, and the parameters of the product are the sum of those of two elements. **It does not mean that the Lie group is abelian.** Denoting by $\overline{\alpha}_\rho$ the parameters of the inverse A^{-1} of an infinitesimal element $A(\alpha)$, one has

$$\overline{\alpha}_\rho = -\alpha_\rho. \tag{5.18}$$

5.2.3 Generators and Differential Operators

There are infinite number of infinitesimal elements in a Lie group G. How to study their property in the transformation group P_G and in a representation $D(G)$? P_R is the transformation operator of the scalar function $\psi(x)$ corresponding to the element R of G: $P_R\psi(x) = \psi(R^{-1}x)$, where x denotes the coordinates of all degrees of freedom in the system.

Letting $A(\alpha)$ denote an infinitesimal element, we have

$$\begin{aligned} P_A\psi(x) &= \psi(x) + \sum_{a\rho} \overline{\alpha}_\rho \left.\frac{\partial(A^{-1}x)_a}{\partial\overline{\alpha}_\rho}\right|_{\overline{\alpha}=0} \left.\frac{\partial\psi(A^{-1}x)}{\partial(A^{-1}x)_a}\right|_{\overline{\alpha}=0} \\ &= \psi(x) - \mathrm{i}\sum_{\rho=1}^{g} \alpha_\rho I_\rho^{(0)}\psi(x), \\ I_\rho^{(0)} &= -\mathrm{i}\sum_a \left.\frac{\partial(Ax)_a}{\partial\alpha_\rho}\right|_{\alpha=0} \frac{\partial}{\partial x_a}. \end{aligned} \tag{5.19}$$

The differential operators $I_\rho^{(0)}$, whose number is g, characterize the action of every infinitesimal element $A(\alpha)$ on a scalar function $\psi(x)$.

For the group SO(3), its element R is a rotation in a real three-dimensional space. If the system is a mass point and x denotes three coordinates of the mass point, we obtain from Eq. (5.10),

$$(Ax)_a = \sum_b \left\{\delta_{ab} - \mathrm{i}\sum_d \alpha_d (T_d)_{ab}\right\} x_b = x_a - \sum_{bd} \alpha_d \epsilon_{dab} x_b. \tag{5.20}$$

Substituting Eq. (5.20) into Eq. (5.19), we have

$$I_d^{(0)} = -\mathrm{i}\sum_{ab} \epsilon_{dba} x_b \frac{\partial}{\partial x_a} = L_d, \tag{5.21}$$

where the natural units are used, $\hbar = c = 1$. The differential operators $I_d^{(0)}$ of SO(3) are nothing but **the orbital angular momentum operators.** For an n-body system, $I_d^{(0)}$ is the total orbital angular momentum operators (see Eq. (5.231)).

If an m-dimensional functional space $\mathcal{L}$, spanned by the basis functions $\psi_\mu(x)$, is left invariant under the transformation group P_G and corresponds to a representation $D(G)$ of a Lie group G,

$$P_R\psi_\mu(x) = \sum_{\nu=1}^{m} \psi_\nu(x) D_{\nu\mu}(R), \tag{5.22}$$

we expand the representation matrix $D(A)$ of an infinitesimal element $A(\alpha)$:

$$D(A) = \mathbf{1} - \mathrm{i} \sum_{\rho=1}^{g} \alpha_\rho I_\rho, \qquad I_\rho = \mathrm{i} \left. \frac{\partial D(A)}{\partial \alpha_\rho} \right|_{\alpha=0} . \tag{5.23}$$

I_ρ is nothing but the matrix forms of the differential operators $I_\rho^{(0)}$ in the basis functions $\psi_\mu(x)$ of $\mathcal{L}$,

$$I_\rho^{(0)} \psi_\mu(x) = \sum_{\nu=1}^{m} \psi_\nu(x) \left(I_\rho\right)_{\nu\mu} . \tag{5.24}$$

I_ρ, $1 \leqslant \rho \leqslant g$, are called the **generators** in a representation $D(\mathrm{G})$ of G, which characterize the transformation of the basis functions $\psi_\mu(x)$, as well as all functions in $\mathcal{L}$, under the action of every infinitesimal element $A(\alpha)$. If $D(\mathrm{G})$ is a unitary representation, I_ρ are hermitian because we introduce a factor $-\mathrm{i}$ in front of I_ρ in Eq. (5.23).

5.2.4 *The Global Property of a Lie Group*

In fact, the global property of a Lie group is the **topological property of the group space** of the Lie group. We will explain the global property of a Lie group with the language familiar to physicists.

The first is the **continuity of the group space**. A Lie group G′ is called **mixed** if its group space falls into several disjoint pieces. The connected piece on which the identity E lies corresponds to a subset G of G′, $E \in G$, which forms an invariant Lie subgroup G of G′. In fact, each $R \in \mathrm{G}$, whose corresponding point in the group space connects the point E by a continuous curve embedded completely in the group space, is equal to the product of infinite number of infinitesimal elements. Thus, the inverse R^{-1} and the product RS where $R \in \mathrm{G}$ and $S \in \mathrm{G}$, as well as TRT^{-1} for each $T \in \mathrm{G}'$, all are equal to the products of infinite number of infinitesimal elements and belong to G. Therefore, a mixed Lie group G′ contains an invariant Lie subgroup G, which is called a **connected Lie group**. The set of elements related to the other connected piece is the coset of G. **The property of the mixed Lie group G′ can be characterized completely by G and the representative elements belonging to each coset, respectively**. For example, the group space of O(3) falls into two pieces where $\det R = 1$ and $\det R = -1$, respectively. SO(3) has a connected group space with $\det R = 1$ and is an invariant subgroup of O(3).

A representative element in the coset with det $R = -1$ is usually taken to be the space inversion σ. The SO(3) group and σ completely characterize the group O(3) because any improper rotation in O(3) is a product of σ and a proper rotation belonging to SO(3). Hereafter, we will mainly study the connected Lie groups.

The second is the **degree of continuity** of the group space. A Lie group G is called **multiply-connected** with the degree of continuity n if the connected curves of any two points in the group space are divided into n sets where any two curves in each set can be changed continuously from one to another completely in the group space, but two curves in different sets cannot, where "**continuous change of two curves in the group space**" means that there exist the variable curves $f(t)$ that depend on a continuous parameter t, $0 \leqslant t \leqslant 1$, such that each $f(t)$ embeds completely in the group space, and $f(0)$ and $f(1)$ coincide with the two curves, respectively. The Lie group with the degree of continuity $n = 1$ is called **simply-connected.** As proved in mathematics, for a connected Lie group G with the degree of continuity n, there exists a simply-connected Lie group G$'$ such that G$' \sim$ G with an n-to-one correspondence. G$'$ is called the **covering group** of G.

Consider the group SO(2) composed of all rotations $R(\boldsymbol{e}_3, \varphi)$ about the Z-axis through an angle φ. φ varies in a unit circumference, which is the group space of SO(2). SO(2) is an abelian Lie group, whose elements satisfy the condition $R(\boldsymbol{e}_3, \varphi + 2\pi) = R(\boldsymbol{e}_3, \varphi)$. Each irreducible representation of SO(2) is one-dimensional, denoted by an integer m,

$$D^m(\varphi) = \mathrm{e}^{-im\varphi} = D^m(\varphi + 2\pi). \tag{5.25}$$

Any two points in the group space of SO(2) (circumference) can be connected by a curve in the circumference through several loops around the circumference. The loops contained in the curve is enumerated by an integer n: -1 for each clockwise loop and $+1$ for each counter-clockwise loop. The curves connecting two points are divided into sets signed by n. The degree of continuity of SO(2) is infinite.

The covering group of SO(2) is the additive group G (see Prob. 2 in Chap. 2), which is composed of all real numbers φ where the multiplication law of elements is defined by the addition of numbers. Since the elements with the parameters $\varphi + 2\pi$ and φ in the additive group G are different, G $\sim$SO(2) with $\infty : 1$ correspondence. The irreducible representation of G is $\mathrm{e}^{-i\tau\varphi}$, where τ is any complex number.

The group space of SO(3) is a spheroid with radius π. Two end points of a diameter on the sphere correspond to the same element. A curve connecting two points in the group space of SO(3) may contain a jump between two end points of a diameter. When the curve changes in the group space continuously, two end points change in pair and cannot disappear. Discuss a curve connecting the origin O to P and containing two jumps at the sphere (see Fig. 5.1.). Moving two jumping points to coincide oppositely, one is able to shrunk the loop between two jumping points to a point continuously in the group space, such that two jumps disappear. Namely, the curve with two jumps becomes a curve without jump continuously in the group space. The connected curves from the origin O to an arbitrary point P are divided into two sets containing even or odd number of the jumps, respectively. Any two curves in each set can be changed continuously from each other completely in the group space, but two curves in different sets cannot. **The group SO(3) is doubly-connected**. We will show that the two-dimensional unimodular unitary group SU(2) is the covering group of SO(3).

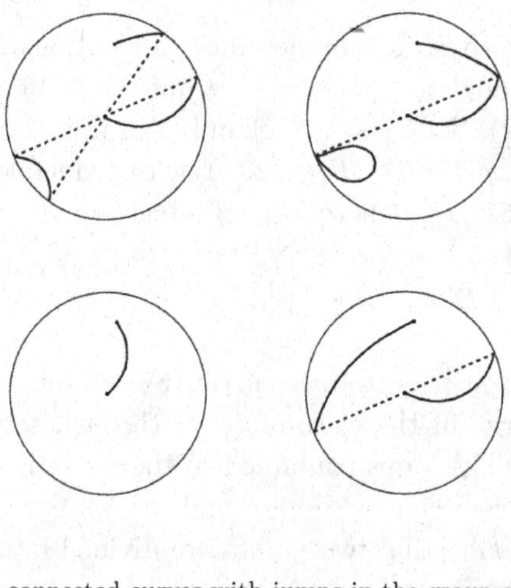

Fig. 5.1 The connected curves with jumps in the group space of SO(3)

The third is the **compactness** of the group space. In an Euclidean space, a closed finite region (including the boundary) is compact. An open finite region (without the boundary) or an infinite region is not compact. **A Lie group is compact if its group space is compact**. The SO(3) group is compact. The Lorentz group is not compact. If one chooses the relative velocity of two inertia systems to be a parameter of the Lorentz group, its variation region is open because the velocity can tend to the velocity c of light, but cannot equal to c. In order to generalize the formulas for the finite group given in Chap. 3 to a Lie group, **the key is to define the average of a group function over the group elements.** For a Lie group the average becomes an integral over the group parameters, instead of a sum over the group elements. As proved in mathematics, **the integral can be defined only for a compact Lie group.**

5.3 The Covering Group of SO(3)

5.3.1 *The Group SU(2)*

The set of all two-dimensional unimodular ($\det u = 1$) unitary matrices u, in the multiplication rule of matrices, forms a Lie group, called SU(2). An arbitrary element u in SU(2)

$$u = \begin{pmatrix} a & b \\ c & d \end{pmatrix} \in \mathrm{SU}(2), \qquad u^\dagger = u^{-1}, \qquad \det\ u = 1,$$

satisfies $aa^* + cc^* = bb^* + dd^* = ad - bc = 1$ and $ab^* + cd^* = 0$. The solution is $a = d^*$, $b = -c^*$, and $|c|^2 + |d|^2 = 1$. Letting $d = \cos(\omega/2) + \mathrm{i}\sin(\omega/2)\cos\theta$ and $c = \sin(\omega/2)\sin\theta\,(\sin\varphi - \mathrm{i}\cos\varphi)$, we rewrite the matrix u as

$$u(\hat{\boldsymbol{n}}, \omega) = \mathbf{1}\cos\left(\frac{\omega}{2}\right) - \mathrm{i}\left(\boldsymbol{\sigma}\cdot\hat{\boldsymbol{n}}\right)\sin\left(\frac{\omega}{2}\right) = \exp\left[-\mathrm{i}\omega\left(\boldsymbol{\sigma}\cdot\hat{\boldsymbol{n}}\right)/2\right], \tag{5.26}$$

where σ_a are the Pauli matrices given in Eq. (5.9), and $\boldsymbol{\sigma} = \sum_a \boldsymbol{e}_a\sigma_a$. It is straightforward to prove

$$\begin{aligned} \boldsymbol{\sigma}\cdot\hat{\boldsymbol{n}} &= \sum_{a=1}^{3}\sigma_a n_a = \begin{pmatrix} n_3 & n_1 - in_2 \\ n_1 + in_2 & -n_3 \end{pmatrix}, \\ \left(\boldsymbol{\sigma}\cdot\hat{\boldsymbol{m}}\right)\left(\boldsymbol{\sigma}\cdot\hat{\boldsymbol{n}}\right) &= \mathbf{1}\left(\hat{\boldsymbol{m}}\cdot\hat{\boldsymbol{n}}\right) + \mathrm{i}\boldsymbol{\sigma}\cdot\left(\hat{\boldsymbol{m}}\times\hat{\boldsymbol{n}}\right) \\ &= \mathbf{1}\sum_{a=1}^{3} m_a n_a + \mathrm{i}\sum_{abc}\epsilon_{abc}\sigma_a m_b n_c. \end{aligned} \tag{5.27}$$

Then,

$$\begin{aligned} & u(\hat{\boldsymbol{n}}, \omega_1)u(\hat{\boldsymbol{n}}, \omega_2) = u(\hat{\boldsymbol{n}}, \omega_1 + \omega_2), \\ & u(\hat{\boldsymbol{n}}, 4\pi) = \mathbf{1}, \qquad u(\hat{\boldsymbol{n}}, 2\pi) = -\mathbf{1}, \\ & u(\hat{\boldsymbol{n}}, \omega) = u(-\hat{\boldsymbol{n}}, 4\pi - \omega) = -u(-\hat{\boldsymbol{n}}, 2\pi - \omega). \end{aligned} \tag{5.28}$$

Thus, the element $u(\hat{\boldsymbol{n}},\omega) \in \mathrm{SU}(2)$ can be characterized by $\boldsymbol{\omega} = \omega\hat{\boldsymbol{n}}(\theta,\varphi)$ through its spherical coordinates (ω,θ,φ) or its rectangular coordinates $(\omega_1,\omega_2,\omega_3)$. The variation region of $\boldsymbol{\omega}$, which is the group space of SU(2), is a spheroid with radius 2π, where all points on the sphere ($\omega = 2\pi$) represent the same element $-\mathbf{1}$. Although the curve connecting any two points in the group space of SU(2) may contain a jump on the sphere, the jump can be looked as a continuous curve on the sphere and can be shrunk in the group space of SU(2). Thus, the group space of SU(2) is a simply-connected closed region, and the SU(2) **group is a simply-connected compact Lie group**.

5.3.2 *Homomorphism of* SU(2) *onto* SO(3)

Any real linear combination of three Pauli matrices is a traceless hermitian matrix. Conversely, any two-dimensional traceless hermitian matrix X can be expressed as a real linear combination of the Pauli matrices. Taking the components x_a of the position vector $\boldsymbol{r}$ of an arbitrary point P in the real three-dimensional space to be the coefficients in X, one obtains a one-to-one correspondence between X and $\boldsymbol{r}$:

$$X = \sum_{a=1}^{3} \sigma_a x_a = \boldsymbol{\sigma}\cdot\boldsymbol{r} = \begin{pmatrix} x_3 & x_1 - ix_2 \\ x_1 + ix_2 & -x_3 \end{pmatrix}, \qquad x_a = \frac{1}{2}\,\mathrm{Tr}\,(X\sigma_a), \qquad \det X = -\sum_{a=1}^{3} x_a^2 = -r^2. \tag{5.29}$$

Taking a similarity transformation $u(\hat{\boldsymbol{n}},\omega) \in \mathrm{SU}(2)$, one obtains another traceless hermitian matrix X' with the same determinant,

$$X' = u(\hat{\boldsymbol{n}},\omega)Xu(\hat{\boldsymbol{n}},\omega)^{-1}, \qquad \det\ X' = \det\ X. \tag{5.30}$$

X' corresponds to the position vector $\boldsymbol{r}'$ of another point P':

$$u(\hat{\boldsymbol{n}},\omega)\,(\boldsymbol{\sigma}\cdot\boldsymbol{r})\,u(\hat{\boldsymbol{n}},\omega)^{-1} = \boldsymbol{\sigma}\cdot\boldsymbol{r}', \qquad x'_b = \sum_a R_{ba}x_a, \tag{5.31}$$

where R is a real orthogonal matrix, $R \in \mathrm{O}(3)$. $u(\hat{\boldsymbol{n}},\omega)$ can be obtained from $\mathbf{1}$ by the continuous change, so can R. Thus, det $R = 1$ and $R \in \mathrm{SO}(3)$. Now, we calculate the R matrix explicitly. Decompose $\boldsymbol{r}$ into two vectors parallel and perpendicular to $\hat{\boldsymbol{n}}$, respectively:

$$\boldsymbol{r} = \hat{\boldsymbol{n}}a + \hat{\boldsymbol{m}}b, \qquad \hat{\boldsymbol{n}}\cdot\hat{\boldsymbol{m}} = 0, \qquad (\boldsymbol{\sigma}\cdot\boldsymbol{r}) = \boldsymbol{\sigma}\cdot\hat{\boldsymbol{n}}a + \boldsymbol{\sigma}\cdot\hat{\boldsymbol{m}}b.$$

From Eqs. (5.26) and (5.27) one has

$$
\begin{aligned}
&u(\hat{\boldsymbol{n}},\omega)\,(\boldsymbol{\sigma}\cdot\hat{\boldsymbol{n}}) = (\boldsymbol{\sigma}\cdot\hat{\boldsymbol{n}})\,u(\hat{\boldsymbol{n}},\omega),\\
&(\boldsymbol{\sigma}\cdot\hat{\boldsymbol{n}})\,(\boldsymbol{\sigma}\cdot\hat{\boldsymbol{m}}) = -\,(\boldsymbol{\sigma}\cdot\hat{\boldsymbol{m}})\,(\boldsymbol{\sigma}\cdot\hat{\boldsymbol{n}}) = \mathrm{i}\boldsymbol{\sigma}\cdot(\hat{\boldsymbol{n}}\times\hat{\boldsymbol{m}})\,,\\
&(\boldsymbol{\sigma}\cdot\hat{\boldsymbol{n}})\,(\boldsymbol{\sigma}\cdot\hat{\boldsymbol{m}})\,(\boldsymbol{\sigma}\cdot\hat{\boldsymbol{n}}) = \mathrm{i}\left\{\boldsymbol{\sigma}\cdot(\hat{\boldsymbol{n}}\times\hat{\boldsymbol{m}})\right\}(\boldsymbol{\sigma}\cdot\hat{\boldsymbol{n}})\\
&\qquad\qquad = -\boldsymbol{\sigma}\cdot\{(\hat{\boldsymbol{n}}\times\hat{\boldsymbol{m}})\times\hat{\boldsymbol{n}}\} = -\boldsymbol{\sigma}\cdot\hat{\boldsymbol{m}}.
\end{aligned}
$$

Thus,

$$
\begin{aligned}
&\left\{\mathbf{1}\cos\left(\frac{\omega}{2}\right) - \mathrm{i}\,(\boldsymbol{\sigma}\cdot\hat{\boldsymbol{n}})\sin\left(\frac{\omega}{2}\right)\right\}(\boldsymbol{\sigma}\cdot\hat{\boldsymbol{m}})\left\{\mathbf{1}\cos\left(\frac{\omega}{2}\right) + \mathrm{i}\,(\boldsymbol{\sigma}\cdot\hat{\boldsymbol{n}})\sin\left(\frac{\omega}{2}\right)\right\}\\
&\quad = (\boldsymbol{\sigma}\cdot\hat{\boldsymbol{m}})\left\{\cos\left(\frac{\omega}{2}\right)\right\}^2 - \frac{\mathrm{i}}{2}\sin\omega\left\{(\boldsymbol{\sigma}\cdot\hat{\boldsymbol{n}})\,(\boldsymbol{\sigma}\cdot\hat{\boldsymbol{m}}) - (\boldsymbol{\sigma}\cdot\hat{\boldsymbol{m}})\,(\boldsymbol{\sigma}\cdot\hat{\boldsymbol{n}})\right\}\\
&\qquad + (\boldsymbol{\sigma}\cdot\hat{\boldsymbol{n}})\,(\boldsymbol{\sigma}\cdot\hat{\boldsymbol{m}})\,(\boldsymbol{\sigma}\cdot\hat{\boldsymbol{n}})\left\{\sin\left(\frac{\omega}{2}\right)\right\}^2\\
&\quad = \boldsymbol{\sigma}\cdot\{\hat{\boldsymbol{m}}\cos\omega + (\hat{\boldsymbol{n}}\times\hat{\boldsymbol{m}})\sin\omega\}\,.
\end{aligned}
$$

Substituting back to Eq. (5.31), one has

$$
\boldsymbol{r}' = \hat{\boldsymbol{n}}a + \hat{\boldsymbol{m}}b\cos\omega + (\hat{\boldsymbol{n}}\times\hat{\boldsymbol{m}})\,b\sin\omega.
$$

From Fig. 5.2 one sees that $\boldsymbol{r}' = R(\hat{\boldsymbol{n}},\omega)\boldsymbol{r}$. Thus,

$$
\begin{aligned}
&u(\hat{\boldsymbol{n}},\omega)\,(\boldsymbol{\sigma}\cdot\boldsymbol{r})\,u(\hat{\boldsymbol{n}},\omega)^{-1} = \boldsymbol{\sigma}\cdot\boldsymbol{r}' = \boldsymbol{\sigma}\cdot[R(\hat{\boldsymbol{n}},\omega)\boldsymbol{r}]\,,\\
&u(\hat{\boldsymbol{n}},\omega)\sigma_a u(\hat{\boldsymbol{n}},\omega)^{-1} = \sum_{b=1}^{3}\sigma_b R_{ba}(\hat{\boldsymbol{n}},\omega).
\end{aligned}
\tag{5.32}
$$

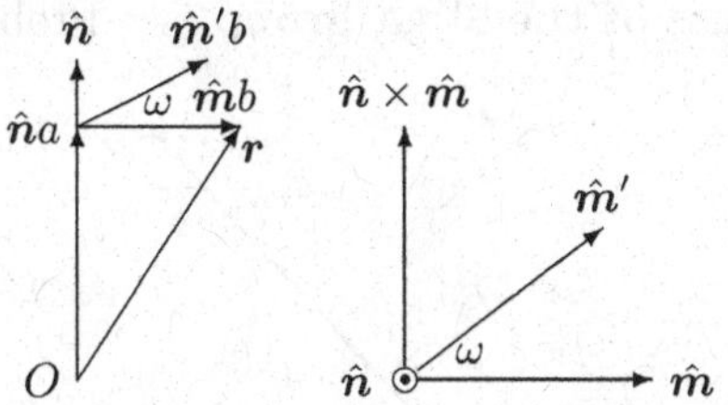

Fig. 5.2 Rotation of a vector $\boldsymbol{r}$ about $\hat{\boldsymbol{n}}$ through ω.

Conversely, for an arbitrary element $R \in \mathrm{SO}(3)$ which rotates $\boldsymbol{r}$ to $\boldsymbol{r}'$, two traceless Hermitian matrices $X = \boldsymbol{\sigma}\cdot\boldsymbol{r}$ and $X' = \boldsymbol{\sigma}\cdot\boldsymbol{r}'$ have the same determinant and can be related by a similarity transformation $u \in \mathrm{SU}(2)$ (see Eq. (5.30)). If $u_1Xu_1^{-1} = u_2Xu_2^{-1} = X'$, $u_2^{-1}u_1$ commutes with any matrix X, so it is a constant matrix, $u_1 = \lambda u_2$. Due to their determinants, $\lambda = \pm 1$. Thus, Eq. (5.32) gives a two-to-one correspondence between

$\pm u(\hat{\boldsymbol{n}},\omega) \in \mathrm{SU}(2)$ and $R(\hat{\boldsymbol{n}},\omega) \in \mathrm{SO}(3)$, and this correspondence is left invariant under the multiplication of group elements:

$$u_1\sigma_a u_1^{-1} = \sum_b \sigma_b \left(R_1\right)_{ba}, \qquad u_2\sigma_b u_2^{-1} = \sum_d \sigma_d \left(R_2\right)_{db},$$
$$u_2u_1\sigma_a u_1^{-1}u_2^{-1} = \sum_b u_2\sigma_b u_2^{-1} \left(R_1\right)_{ba} = \sum_d \sigma_d \sum_b \left\{\left(R_2\right)_{db}\left(R_1\right)_{ba}\right\}.$$

Hence, SU(2) is homomorphic onto SO(3):

$$\mathrm{SU}(2) \sim \mathrm{SO}(3). \tag{5.33}$$

The elements of SO(3) and SU(2) both are characterized by the parameter $\boldsymbol{\omega}$. The group space of SO(3) is the spheroid with radius π, which is doubly-connected. The group space of SU(2) is the spheroid with radius 2π, which is simply-connected. Inside the spheroid with radius π there is a one-to-one correspondence between elements of SU(2) and SO(3). $u(\hat{\boldsymbol{n}},\omega)$ in the ring with $\pi < \omega < 2\pi$ is equal to $-u(-\hat{\boldsymbol{n}},2\pi-\omega)$ owing to Eq. (5.28). The pair of $u(\hat{\boldsymbol{n}},\omega)$ and $u(-\hat{\boldsymbol{n}},2\pi-\omega)$ of SU(2) maps onto one element $R(\hat{\boldsymbol{n}},\omega)$ of SO(3) (see Fig. 5.3). **The group SU(2) is the covering group of** SO(3). A representation of SO(3), called a **single-valued** representation, is an unfaithful representation of SU(2). To speak strictly, a faithful representation of SU(2) is not a representation of SO(3). However, due to physical reason, it is often called a **double-valued** representation of SO(3). Similar to the classes in SO(3), **the elements** $u(\hat{\boldsymbol{n}},\omega)$ **with the same** ω **form a class of the** SU(2) **group** (see Prob. 3).

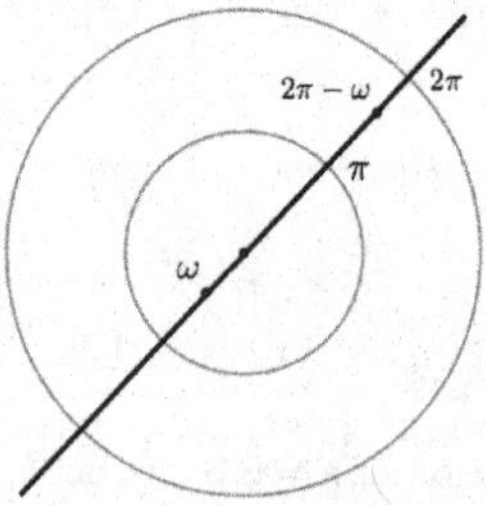

Fig. 5.3 Two elements of SU(2) correspond to one element of SO(3).

Due to the homomorphism of SU(2) onto SO(3), it is convenient to simply call $u(\hat{\boldsymbol{n}},\omega) \in \mathrm{SU}(2)$ a "rotation" about the direction $\boldsymbol{n}$ through an angle ω. It will be known later that the group SU(2) is related to the spinor. As usually say, **the period of a spinor in "rotation" is** 4π.

5.3.3 *Group Integral for Parameters* ω

Many properties of a finite group are based on the concept of the average of a group function which is left invariant under the left- and right-multiplication with any group element. For a Lie group, the average of a group function is generalized to an integral over the group space, called the group integral

$$\frac{1}{g}\sum_{R\in G} F(R) \longrightarrow \int dRF(R) = \int (dr)\, W(R)F(R). \tag{5.34}$$

The weight function $W(R)$ has to be introduced generally. Even though one may assume $W(R) = 1$ in one set of group parameters, $W(R)$ will appear after an integral transformation. **$W(R)$ can be understood as the relative density of elements in the neighborhood of R.**

The group integral is linear with respect to the group function:

$$\int dR\ [aF_1(R) + bF_2(R)] = a\int dR\ F_1(R) + b\int dR\ F_2(R). \tag{5.35}$$

If $F(R) \geqslant 0$ and $F(R)$ is not equal to zero at least in a region with non-vanishing measure, the average of $F(R)$ should be larger than zero. Thus, **$W(R)$ has to be single-valued, finite, integrable, and not equal to or less than zero in any region with non-vanishing measure.** $W(R)$ is normalized by

$$\frac{1}{g}\sum_{R\in G} 1 = 1 \longrightarrow \int dR = \int (dr)\, W(R) = 1. \tag{5.36}$$

Furthermore, the weight function $W(R)$ should be left invariant under the left- and right-multiplication with any group element:

$$\int dR\ F(R) = \int dR\ F(SR) = \int dR\ F(RS). \tag{5.37}$$

Letting $T = SR$, one obtains from Eq. (5.37)

$$\begin{aligned}\int (dt)W(T)F(T) &= \int dT\ F(T) = \int dR\ F(R)\\ &= \int dR\ F(SR) = \int dR\ F(T) = \int (dr)W(R)F(T).\end{aligned}$$

To speak roughly, the relative "number" of elements in the neighborhood of any group element is the same. In comparison with the density function $W(E) = W_0$ in the neighborhood of the identity E,

$$(dr)\, W(R) = (dt)\, W(T) = (d\alpha)\, W_0. \tag{5.38}$$

$W(R)$ can be calculated through the Jacobi determinant in the replacement of integral variables. Let A denote the infinitesimal element with the parameters α_μ, and by R' the element with the parameters r'_μ in the neighborhood of a given element R. The Jacobi determinants for $R' = AR$ and $A = R'R^{-1}$ are, respectively,

$$W_0 = W(R) \left| \det \left\{ \frac{\partial f_j(\alpha; r)}{\partial \alpha_k} \right\} \right|_{\alpha=0}, \tag{5.39}$$

$$W(R) = W_0 \left| \det \left\{ \frac{\partial f_j(r'; \overline{r})}{\partial r'_k} \right\} \right|_{r'=r}. \tag{5.40}$$

Two conditions for $W(R)$ are equivalent. From any of them $W(R)$ can be calculated where W_0 is determined by the normalization condition (5.36).

For the group SU(2) the weight function is $W(\hat{\boldsymbol{n}}, \omega)$,

$$(du) = W(\hat{\boldsymbol{n}}, \omega) d\omega_1 d\omega_2 d\omega_3 = W(\hat{\boldsymbol{n}}, \omega)\omega^2 \sin\theta d\omega d\theta d\varphi. \tag{5.41}$$

The elements $u(\hat{\boldsymbol{n}}, \omega)$ with the same ω form a class of SU(2). $u(\hat{\boldsymbol{n}}, \omega)$ and the elements in its neighborhood can be related by the same similarity transformation with those where $\hat{\boldsymbol{n}}$ is changed to another $\hat{\boldsymbol{n}}'$, such that the density function $W(\hat{\boldsymbol{n}}, \omega)$ should be independent of the direction $\hat{\boldsymbol{n}}$. In fact, the Jacobi determinant in the integral transformation between $u(\hat{\boldsymbol{n}}, \omega)$ and $u(\hat{\boldsymbol{n}'}, \omega)$ is the determinant of a rotation R which is det $R = 1$. Thus, $W(\hat{\boldsymbol{n}}, \omega) = W(\omega)$. Taking $\hat{\boldsymbol{n}} = \mathrm{e}_3$ in Eq. (5.39) for SU(2), one writes the element R as $u(\boldsymbol{e}_3, \omega)$ and the infinitesimal element A as $u(\boldsymbol{\alpha})$:

$$\begin{aligned}
u(\boldsymbol{e}_3, \omega) &= \mathbf{1}\cos(\omega/2) - \mathrm{i}\sigma_3 \sin(\omega/2), \\
u(\boldsymbol{\alpha}) &= \mathbf{1} - \mathrm{i}\left(\sigma_1\alpha_1 + \sigma_2\alpha_2 + \sigma_3\alpha_3\right)/2, \\
u(\boldsymbol{\alpha})u(\boldsymbol{e}_3, \omega) &= \mathbf{1}\cos\left(\omega'/2\right) - \mathrm{i}\left(\boldsymbol{\sigma}\cdot\hat{\boldsymbol{n}}'\right)\sin\left(\omega'/2\right) \\
&= \mathbf{1}\left\{\cos(\omega/2) - (\alpha_3/2)\sin(\omega/2)\right\} \\
&\quad - \mathrm{i}\sigma_1\left\{\alpha_1\cos(\omega/2) + \alpha_2\sin(\omega/2)\right\}/2 \\
&\quad - \mathrm{i}\sigma_2\left\{\alpha_2\cos(\omega/2) - \alpha_1\sin(\omega/2)\right\}/2 \\
&\quad - \mathrm{i}\sigma_3\left\{\alpha_3\cos(\omega/2) + 2\sin(\omega/2)\right\}/2,
\end{aligned}$$

where only the terms up to the first order of α_a is retained because $\boldsymbol{\alpha}$ will be vanishing after the derivative in Eq. (5.39). The parameters ω' and $\boldsymbol{n}'$ are calculated as follows:

$$\cos(\omega'/2) = \cos(\omega/2) - (\alpha_3/2)\sin(\omega/2) = \cos\{(\omega + \alpha_3)/2\},$$

$$\begin{aligned}
&\sin(\omega'/2) = \sin\{(\omega+\alpha_3)/2\} = \sin(\omega/2) + (\alpha_3/2)\cos(\omega/2),\\
&\omega' n_1' = \omega\{\sin(\omega/2)\}^{-1}\{\alpha_1\cos(\omega/2) + \alpha_2\sin(\omega/2)\}/2,\\
&\omega' n_2' = \omega\{\sin(\omega/2)\}^{-1}\{\alpha_2\cos(\omega/2) - \alpha_1\sin(\omega/2)\}/2,\\
&\omega' n_3' = \omega'\{\sin(\omega'/2)\}^{-1}\{\alpha_3\cos(\omega/2) + 2\sin(\omega/2)\}/2 = \omega' = \omega + \alpha_3.
\end{aligned}$$

Note that the curly bracket in the last equation contains a term independent of $\boldsymbol{\alpha}$ such that ω' in the front factor cannot be replaced by ω. Substituting them into Eq. (5.39), one has

$$\begin{aligned}
\frac{W_0}{W(\omega)} &= \left|\det\left\{\frac{\partial(\omega' n_a')}{\partial \alpha_b}\right\}\right|_{\alpha=0} = \begin{vmatrix} (\omega/2)\cot(\omega/2) & \omega/2 & 0 \\ -\omega/2 & (\omega/2)\cot(\omega/2) & 0 \\ 0 & 0 & 1 \end{vmatrix}\\
&= \omega^2\left\{4\sin^2(\omega/2)\right\}^{-1}.
\end{aligned}$$

Then, $W(\omega) = W_0 4\omega^{-2}\sin^2(\omega/2)$. W_0 is solved from the normalization condition (5.36):

$$1 = 4W_0 \int_0^{2\pi} \sin^2(\omega/2)d\omega \int_0^{\pi} \sin\theta d\theta \int_{-\pi}^{\pi} d\varphi = 16\pi^2 W_0.$$

Thus, the group integral of a function $F(u) = F(\boldsymbol{\omega})$ of SU(2) is

$$\int (du)F(u) = \frac{1}{4\pi^2}\int_{-\pi}^{\pi} d\varphi \int_0^{\pi} d\theta\ \sin\theta \int_0^{2\pi} d\omega\ \sin^2(\omega/2)F(\boldsymbol{\omega}). \tag{5.42}$$

For the group SO(3), the radius of the variation region of ω reduces to π,

$$\int (dR)F(R) = \frac{1}{2\pi^2}\int_{-\pi}^{\pi} d\varphi \int_0^{\pi} d\theta\ \sin\theta \int_0^{\pi} d\omega\ \sin^2(\omega/2)F(\boldsymbol{\omega}). \tag{5.43}$$

For a class function, $F(\boldsymbol{\omega}) = F(\omega)$, the integral over θ and φ can be done first, namely, $\int \sin\theta d\theta d\varphi = 4\pi$.

Now, for the compact Lie group SU(2), as well as SO(3), the average of a group function which is left invariant under the left- and right-multiplication with any group element has been defined. The following conclusions hold like those for a finite group.

(a) Any representation is equivalent to a unitary representation, and two equivalent unitary representations can be related through a unitary similarity transformation.

(b) Any real representation is equivalent to a real orthogonal representation, and two equivalent real orthogonal representations can be related through a real orthogonal similarity transformation.

(c) A representation is reducible if and only if there exists a nonconstant matrix commutable with its every representation matrix. Any reducible representation is completely reducible.

(d) For two inequivalent irreducible unitary representations, the representation matrix elements $D^j_{\mu\rho}(\boldsymbol{\omega})$ and the characters $\chi^j(\omega)$ satisfy the orthogonal relations:

$$\begin{aligned}
&\frac{1}{4\pi^2}\int_{-\pi}^{\pi} d\varphi \int_0^{\pi} d\theta \, \sin\theta \int_0^{2\pi} d\omega \, \sin^2(\omega/2) D^j_{\mu\rho}(\boldsymbol{\omega})^* D^k_{\nu\lambda}(\boldsymbol{\omega}) \\
&\quad = \frac{1}{m_j}\delta_{jk}\delta_{\mu\nu}\delta_{\rho\lambda},
\end{aligned} \tag{5.44}$$

$$\frac{1}{\pi}\int_0^{2\pi} d\omega \, \sin^2(\omega/2)\chi^j(\omega)^*\chi^k(\omega) = \delta_{jk}. \tag{5.45}$$

Any representation can be expanded with respect to the irreducible representations:

$$\begin{aligned}
&X^{-1}D(\boldsymbol{\omega})X = \bigoplus_j a_j D^j(\boldsymbol{\omega}), \qquad \chi(\omega) = \sum_j a_j \chi^j(\omega), \\
&a_j = \frac{1}{\pi}\int_0^{2\pi} d\omega \, \sin^2(\omega/2)\chi^j(\omega)^*\chi(\omega).
\end{aligned} \tag{5.46}$$

(e) Two representations $D(\boldsymbol{\omega})$ and $\overline{D}(\boldsymbol{\omega})$ are equivalent if and only if the characters of every element in them are equal to each other: $\chi(\omega) = \overline{\chi}(\omega)$. A representation $D(\boldsymbol{\omega})$ is irreducible if and only if the character $\chi(\omega)$ in the representation satisfies

$$\frac{1}{\pi}\int_0^{2\pi} d\omega \, \sin^2(\omega/2)\left|\chi(\omega)\right|^2 = 1. \tag{5.47}$$

(f) The similarity transformation matrix X, which relates the irreducible unitary self-conjugate representation and its conjugate one, is symmetric or antisymmetric. X is symmetric if and only if the representation is real.

(g) Although the number of the irreducible basis $\phi^j_{\nu\mu}$ for SU(2), satisfying Eq. (3.114), becomes infinite, the formulas can still be used:

$$\phi^j_{\nu\mu} = \frac{m_j}{4\pi^2}\int_{-\pi}^{\pi} d\varphi \int_0^{\pi} d\theta \, \sin\theta \int_0^{2\pi} d\omega \, \sin^2(\omega/2) D^j_{\nu\mu}(\hat{\boldsymbol{n}},\omega)^* u(\hat{\boldsymbol{n}},\omega),$$

$$e^{(j)}_\mu = \phi^j_{\mu\mu} = \frac{m_j}{4\pi^2}\int_{-\pi}^{\pi} d\varphi \int_0^{\pi} d\theta \, \sin\theta \int_0^{2\pi} d\omega \, \sin^2(\omega/2) D^j_{\mu\mu}(\hat{\boldsymbol{n}},\omega)^* u(\hat{\boldsymbol{n}},\omega),$$

$$\begin{aligned}
&\phi^j_{\nu\rho}\phi^j_{\tau\mu} = \delta_{\rho\tau}\phi^j_{\nu\mu}, \quad e^{(i)}_\nu e^{(j)}_\mu = \delta_{ij}\delta_{\nu\mu}e^{(j)}_\mu, \quad e^{(i)}_\nu t e^{(j)}_\mu = \delta_{ij}e^{(j)}_\nu t e^{(j)}_\mu, \\
&u(\hat{\boldsymbol{n}}',\omega')\,\phi^j_{\rho\mu} = \sum_\tau \phi^j_{\tau\mu}\, D^j_{\tau\rho}[\hat{\boldsymbol{n}}',\omega'], \\
&\phi^j_{\nu\rho}\, u(\hat{\boldsymbol{n}}',\omega') = \sum_\tau D^j_{\rho\tau}[\hat{\boldsymbol{n}}',\omega']\,\phi^j_{\nu\tau},
\end{aligned} \tag{5.48}$$

where the group element $u(\hat{\boldsymbol{n}},\omega)$ in $\phi^j_{\nu\mu}$ should be replaced by the transformation operator $P_{u(\hat{\boldsymbol{n}},\omega)}$ when $\phi^j_{\nu\mu}$ acts on the scalar wave functions. t is a linear combination of group elements.

It is proved that the group integral can be defined for a compact Lie group such that the average of a group function is left invariant under the left- and right-multiplication with any group element. Thus, the above conclusions can be generalized to any compact Lie group.

5.4 Irreducible Representations of SU(2)

5.4.1 *Euler Angles*

The geometric meaning of the parameters $\boldsymbol{\omega} = \hat{\boldsymbol{n}}\omega$ for characterizing an element $R(\hat{\boldsymbol{n}},\omega)$ in SO(3) is evident: $R(\hat{\boldsymbol{n}},\omega)$ is a rotation about the direction $\hat{\boldsymbol{n}}$ through an angle ω. The more important merit of the parameters $\boldsymbol{\omega}$ is that, at least in the neighborhood of the identity E, there is a one-to-one correspondence between the group element and the point in the group space. This merit makes the parameters $\boldsymbol{\omega}$ more suitable in theoretical study. However, the parameters $\boldsymbol{\omega}$ are not very convenient in the practical calculation, because the parameters $\boldsymbol{\omega}$ are not easy to be determined from the rotation matrix R or from the relative positions of two frames before and after the rotation R. On the other hand, there is another set of parameters, called the Euler angles, which are easy to be determined from the rotation matrix R and from the relative positions of two frames in the rotation R. The shortcoming of the Euler angles is that the Euler angles of an element in the neighborhood of the identity E are not unique. This shortcoming makes the Euler angles inconvenient in theoretical study.

For a given matrix $R \in$ SO(3), its third column is a unit vector $\hat{\boldsymbol{m}}(\beta,\alpha)$, where β and α are its polar angle and azimuthal angle, respectively. Since

$$R\boldsymbol{e}_3 = \hat{\boldsymbol{m}}(\beta,\alpha), \qquad S(\alpha,\beta)\boldsymbol{e}_3 = \hat{\boldsymbol{m}}(\beta,\alpha), \tag{5.49}$$

where $S(\alpha,\beta) = R(\boldsymbol{e}_3,\alpha)R(\boldsymbol{e}_2,\beta)$ is defined in Eq. (5.11), $S(\alpha,\beta)^{-1}R$

leaves the Z-axis invariant so that it is a rotation about the Z-axis, denoted by $R(\mathbf{e}_3, \gamma)$. Therefore, **every rotation in the three-dimensional space can be expressed as a product of three rotations about the coordinate axes:**

$$R(\alpha, \beta, \gamma) = R(\mathbf{e}_3, \alpha)R(\mathbf{e}_2, \beta)R(\mathbf{e}_3, \gamma)$$
$$= \begin{pmatrix} c_\alpha c_\beta c_\gamma - s_\alpha s_\gamma & -c_\alpha c_\beta s_\gamma - s_\alpha c_\gamma & c_\alpha s_\beta \\ s_\alpha c_\beta c_\gamma + c_\alpha s_\gamma & -s_\alpha c_\beta s_\gamma + c_\alpha c_\gamma & s_\alpha s_\beta \\ -s_\beta c_\gamma & s_\beta s_\gamma & c_\beta \end{pmatrix}, \tag{5.50}$$

where $c_\alpha = \cos\alpha$, $s_\alpha = \sin\alpha$, etc. Three angles α, β, and γ are called the **Euler angles**. When the matrix R is given, **the unit vector $\hat{\boldsymbol{m}}(\beta, \alpha)$ in the third column of R** has the polar angle β and the azimuthal angle α, and **the unit vector $\hat{\boldsymbol{m}}'(\beta, \pi - \gamma)$ in the third row of R** has the polar angle β and the azimuthal angle $(\pi - \gamma)$. If the rotation R transforms the coordinate frame K to K', then in the K coordinate frame, **the polar angle and the azimuthal angle of the Z'-axis of K' are β and α**, respectively. In the K' coordinate frame, **the polar angle and the azimuthal angle of the Z-axis of K are β and $(\pi - \gamma)$**. The definition domain of the Euler angles in SO(3) are

$$-\pi \leqslant \alpha \leqslant \pi, \qquad 0 \leqslant \beta \leqslant \pi, \qquad -\pi \leqslant \gamma \leqslant \pi. \tag{5.51}$$

When $\beta = 0$ or π,

$$R(\alpha, 0, \gamma) = R(\mathbf{e}_3, \alpha + \gamma), \quad R(\alpha, \pi, \gamma) = R(\mathbf{e}_3, \alpha - \gamma)R(\mathbf{e}_2, \pi). \tag{5.52}$$

In both cases, only one parameter in α and γ is independent. This is the shortcoming of the parameters of the Euler angles.

In Eq. (5.50) a rotation is expressed as a product of three rotations about the coordinate axes of the laboratory frame K. If one chooses the rotations about the coordinate axes of the body-fixed frame K', **the product order will be opposite.** (see Fig. 5.4 [Edmonds (1957)])

$$\begin{aligned} R(\alpha, \beta, \gamma) = &\left\{[R(\mathbf{e}_3, \alpha)R(\mathbf{e}_2, \beta)]\, R(\mathbf{e}_3, \gamma)\, [R(\mathbf{e}_3, \alpha)R(\mathbf{e}_2, \beta)]^{-1}\right\} \\ &\cdot \left\{R(\mathbf{e}_3, \alpha)R(\mathbf{e}_2, \beta)R(\mathbf{e}_3, \alpha)^{-1}\right\} R(\mathbf{e}_3, \alpha). \end{aligned} \tag{5.53}$$

R is a successive rotation which first rotates about the Z'-axis in K', which coincides with K, through the angle α, then rotates about the Y'-axis in the new K' frame through the angle β, and at last rotates about the Z'-axis in the newer K' frame through the angle γ.

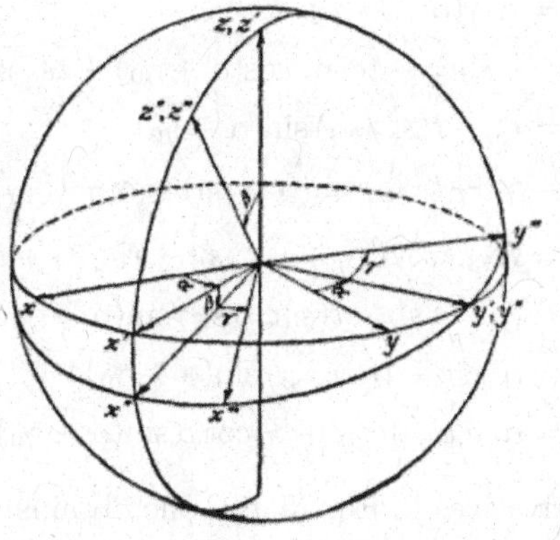

Fig. 5.4 The Euler angles of SO(3).

5.4.2 *Group Integral for the Euler Angles*

The group integral of SO(3) with the parameters of the Euler angles can also be calculated by Eq. (5.39). The difficult is that, in a region without measure near the identity E ($\beta = 0$), the Euler angles of an infinitesimal element $A(\alpha_0, \beta_0, \gamma_0)$ are multiple-valued (see Eq. (5.52)):

$$\alpha_0 = -\gamma_0 + r, \qquad \beta_0 = t, \tag{5.54}$$

where r and t are infinitesimal quantities, but α_0 and γ_0 are finite. Thus, W_0 is infinitesimal, and one has to preserve the terms with the first order of t or r in $W_0/W(\alpha, \beta, \gamma)$. Letting $c_0 = \cos\gamma_0$, $s_0 = \sin\gamma_0$, $c_\alpha = \cos\alpha$, $s_\alpha = \sin\alpha$, and so on, one calculates from Eq. (5.50):

$$A(\alpha_0, \beta_0, \gamma_0) = \frac{1}{2}\begin{pmatrix} 2 - r^2 - t^2 c_0^2 & -2r + t^2 s_0 c_0 & 2t(c_0 + r s_0) \\ 2r + t^2 s_0 c_0 & 2 - r^2 - t^2 s_0^2 & 2t(-s_0 + r c_0) \\ -2tc_0 & 2ts_0 & 2 - t^2 \end{pmatrix}.$$

Then, one is able to solve the functions α', β', and γ' on α_0, β_0, and γ_0:

$$R(\alpha', \beta', \gamma') = A(\alpha_0, \beta_0, \gamma_0) R(\alpha, \beta, \gamma),$$

$$\begin{aligned} \cos\beta' &= R_{33}(\alpha', \beta', \gamma') = c_\beta - t s_\beta \cos(\alpha + \gamma_0), \\ \beta' &= \beta + t\cos(\alpha + \gamma_0) = \beta + \beta_0 \cos(\alpha + \gamma_0), \\ \sin\beta' &= s_\beta + t c_\beta \cos(\alpha + \gamma_0), \end{aligned}$$

$$\begin{aligned}
-\sin\beta'\cos\gamma' &= R_{31}(\alpha',\beta',\gamma') \\
&= -s_\beta c_\gamma - t c_\beta c_\gamma \cos(\alpha+\gamma_0) + t s_\gamma \sin(\alpha+\gamma_0), \\
\cos\gamma' &= c_\gamma - t(s_\gamma/s_\beta)\sin(\alpha+\gamma_0), \\
\gamma' &= \gamma + (t/s_\beta)\sin(\alpha+\gamma_0) = \gamma + (\beta_0/s_\beta)\sin(\alpha+\gamma_0), \\
\sin\beta'\cos\alpha' &= R_{13}(\alpha',\beta',\gamma') = c_\alpha s_\beta - r s_\alpha s_\beta + t c_0 c_\beta, \\
\cos\alpha' &= c_\alpha - r s_\alpha + t(s_\alpha c_\beta/s_\beta)\sin(\alpha+\gamma_0), \\
\alpha' &= \alpha + r - t(c_\beta/s_\beta)\sin(\alpha+\gamma_0) \\
&= \alpha + \alpha_0 + \gamma_0 - \beta_0 \cot\beta \sin(\alpha+\gamma_0).
\end{aligned}$$

Calculating the derivatives in Eq. (5.39), one obtains

$$\frac{W_0}{W(R)} = \frac{W(\alpha_0,\beta_0,\gamma_0)}{W(\alpha,\beta,\gamma)} = \begin{vmatrix} \dfrac{\partial\alpha'}{\partial\alpha_0} & \dfrac{\partial\alpha'}{\partial\beta_0} & \dfrac{\partial\alpha'}{\partial\gamma_0} \\ \dfrac{\partial\beta'}{\partial\alpha_0} & \dfrac{\partial\beta'}{\partial\beta_0} & \dfrac{\partial\beta'}{\partial\gamma_0} \\ \dfrac{\partial\gamma'}{\partial\alpha_0} & \dfrac{\partial\gamma'}{\partial\beta_0} & \dfrac{\partial\gamma'}{\partial\gamma_0} \end{vmatrix}$$

$$= \begin{vmatrix} 1 & -\cot\beta\sin(\alpha+\gamma_0) & 1 - t\cot\beta\cos(\alpha+\gamma_0) \\ 0 & \cos(\alpha+\gamma_0) & -t\sin(\alpha+\gamma_0) \\ 0 & \sin(\alpha+\gamma_0)/\sin\beta & t\cos(\alpha+\gamma_0)/\sin\beta \end{vmatrix} = \frac{t}{\sin\beta}.$$

Thus, $W(R) = W(\alpha,\beta,\gamma) = (W_0/t)\sin\beta = W_0'\sin\beta$. From the normalization condition (5.36) one has

$$\int dR = W_0' \int_{-\pi}^{\pi} d\alpha \int_0^{\pi} \sin\beta d\beta \int_{-\pi}^{\pi} d\gamma = 8\pi^2 W_0' = 1.$$

Then, $W_0' = 1/(8\pi^2)$. The group integral of SO(3) with the parameters of the Euler angles is

$$\int F(R)dR = \frac{1}{8\pi^2} \int_{-\pi}^{\pi} d\alpha \int_0^{\pi} \sin\beta d\beta \int_{-\pi}^{\pi} F(\alpha,\beta,\gamma)d\gamma, \tag{5.55}$$

The result $W(R) = W_0'\sin\beta$ can be understood as follows. Let R rotate the K frame to the K' frame. Let P and Q denote the intersections of the Z'-axis and the X'-axis of K' with the unit sphere in K, respectively. The position of P on the unit sphere is characterized by the polar angle β and the azimuthal angle α of the direction OP. For a given P, the position of Q is characterized by the angle γ. For the elements in the neighborhood of R, P changes in the area ($\sin\beta d\alpha d\beta$), and Q changes in the arc ($d\gamma$)

when P is given. Since the unit sphere is isotropic, the relative "number" of elements in the neighborhood of R is proportional to the product of the area ($\sin\beta d\alpha d\beta$) and the arc ($d\gamma$).

For the SU(2) group, $u(\alpha,\beta,\gamma)=u(\boldsymbol{e}_3,\alpha)u(\boldsymbol{e}_2,\beta)u(\boldsymbol{e}_3,\gamma)$, and the definition domain of the Euler angles is enlarged,

$$-\pi\leqslant\alpha\leqslant\pi,\qquad 0\leqslant\beta\leqslant\pi,\qquad -2\pi\leqslant\gamma\leqslant 2\pi. \tag{5.56}$$

The group integral for SU(2) is

$$\int F(u)du=\frac{1}{16\pi^2}\int_{-\pi}^{\pi}d\alpha\int_0^{\pi}\sin\beta d\beta\int_{-2\pi}^{2\pi}F(\alpha,\beta,\gamma)d\gamma. \tag{5.57}$$

5.4.3 *Linear Representations of SU(2)*

The basic method for calculating the representation of a group G is to find a functional space $\mathcal{L}^{(j)}$ which is left invariant under G. Applying the transformation operator P_R to the basis functions ψ_μ^j in $\mathcal{L}^{(j)}$, one obtains a representation $D^j(R)$ of G:

$$P_R\psi_\mu^j(x)=\psi_\mu^j(R^{-1}x)=\sum_\nu \psi_\nu^j(x)\ D_{\nu\mu}^j(R).$$

The element $u\in$ SU(2) is a unitary transformation in a two-dimensional complex space,

$$\begin{pmatrix}\xi'\\ \eta'\end{pmatrix}=u\begin{pmatrix}\xi\\ \eta\end{pmatrix},\qquad \begin{pmatrix}\xi''\\ \eta''\end{pmatrix}=u^{-1}\begin{pmatrix}\xi\\ \eta\end{pmatrix},$$

where $u^{-1}=\mathbf{1}\cos(\omega/2)+\mathrm{i}\,(\boldsymbol{\sigma}\cdot\boldsymbol{n})\sin(\omega/2)$,

$$u^{-1}=\begin{pmatrix}\cos(\omega/2)+\mathrm{i}n_3\sin(\omega/2) & \sin(\omega/2)\,(n_2+\mathrm{i}n_1)\\ \sin(\omega/2)\,(-n_2+\mathrm{i}n_1) & \cos(\omega/2)-\mathrm{i}n_3\sin(\omega/2)\end{pmatrix}.$$

The homogeneous polynomials of order n with respect to ξ and η span an $(n+1)$-dimensional space $\mathcal{L}^{(n)}$ which is left invariant under the group SU(2). The basis functions $\xi^m\eta^{n-m}$ in $\mathcal{L}^{(n)}$, $m=0,\ 1,\ \ldots,\ n$, can be re-chosen for convenience in physics:

$$\begin{aligned}&\psi_\mu^j(\xi,\eta)=\frac{(-1)^{j-\mu}}{\sqrt{(j-\mu)!(j+\mu)!}}\xi^{j-\mu}\eta^{j+\mu},\\ &j=n/2=0,\ 1/2,\ 1,\ 3/2,\ \ldots,\\ &\mu=j-m=j,\ (j-1),\ \ldots,\ -(j-1),\ -j.\end{aligned} \tag{5.58}$$

Calculate the matrix form $D_{\nu\mu}^j(u)$ of P_u in the basis functions $\psi_\mu^j(\xi,\eta)$,

$$P_u\psi^j_\mu(\xi,\eta) = \psi^j_\mu(\xi'',\eta'') = \sum_\nu \psi^j_\nu(\xi,\eta)D^j_{\nu\mu}(u). \tag{5.59}$$

When $\hat{\boldsymbol{n}} = \boldsymbol{e}_3$, $\xi'' = \xi\exp(\mathrm{i}\omega/2)$ and $\eta'' = \eta\exp(-\mathrm{i}\omega/2)$. Then,

$$P_u\psi^j_\mu(\xi,\eta) = \frac{(-1)^{j-\mu}\left(\xi e^{\mathrm{i}\omega/2}\right)^{j-\mu}\left(\eta e^{-\mathrm{i}\omega/2}\right)^{j+\mu}}{\sqrt{(j-\mu)!(j+\mu)!}} = \psi^j_\mu(\xi,\eta)e^{-\mathrm{i}\mu\omega},$$

$$D^j_{\nu\mu}(\boldsymbol{e}_3,\omega) = \delta_{\nu\mu}e^{-\mathrm{i}\mu\omega}. \tag{5.60}$$

When $\hat{\boldsymbol{n}} = \boldsymbol{e}_2$, $\xi'' = \xi\cos(\omega/2)+\eta\sin(\omega/2)$ and $\eta'' = -\xi\sin(\omega/2)+\eta\cos(\omega/2)$. Then,

$$\begin{aligned} P_u\psi^j_\mu(\xi,\eta) &= \frac{(-1)^{j-\mu}\left[\xi\cos(\omega/2)+\eta\sin(\omega/2)\right]^{j-\mu}}{\sqrt{(j-\mu)!}\sqrt{(j+\mu)!}} \\ &\quad\times\left[-\xi\sin(\omega/2)+\eta\cos(\omega/2)\right]^{j+\mu} \\ &= (-1)^{j-\mu}\sum_{n=0}^{j-\mu}\frac{\sqrt{(j-\mu)!}\left\{\xi\cos(\omega/2)\right\}^{j-\mu-n}\left\{\eta\sin(\omega/2)\right\}^n}{(j-\mu-n)!n!} \\ &\quad\times\sum_{m=0}^{j+\mu}\frac{\sqrt{(j+\mu)!}\left\{-\xi\sin(\omega/2)\right\}^{j+\mu-m}\left\{\eta\cos(\omega/2)\right\}^m}{(j+\mu-m)!m!}. \end{aligned}$$

In order to express the right-hand side of the equality to be a linear combination of $\psi^j_\nu(\xi,\eta)$, one makes a replacement of the summation indices from m to ν, $\nu = n+m-j$,

$$P_u\psi^j_\mu(\xi,\eta) = \sum_{\nu=-j}^{j}\frac{(-1)^{j-\nu}\xi^{j-\nu}\eta^{j+\nu}}{\{(j-\nu)!(j+\nu)!\}}D^j_{\nu\mu}(\boldsymbol{e}_2,\omega),$$

$$\begin{aligned} d^j_{\nu\mu}(\omega) &\equiv D^j_{\nu\mu}(\boldsymbol{e}_2,\omega) \\ &= \sum_n \frac{(-1)^n\left\{(j+\nu)!(j-\nu)!(j+\mu)!(j-\mu)!\right\}^{1/2}}{(j+\nu-n)!(j-\mu-n)!n!(n-\nu+\mu)!} \\ &\quad\cdot\left\{\cos(\omega/2)\right\}^{2j+\nu-\mu-2n}\left\{\sin(\omega/2)\right\}^{2n-\nu+\mu}, \end{aligned} \tag{5.61}$$

where n runs from the maximum between 0 and $(\nu-\mu)$ to the minimum between $(j+\nu)$ and $(j-\mu)$. Hence, the representation D^j of SU(2) is

$$\begin{aligned} D^j_{\nu\mu}(\alpha,\beta,\gamma) &= \left\{D^j(\boldsymbol{e}_3,\alpha)D^j(\boldsymbol{e}_2,\beta)D^j(\boldsymbol{e}_3,\gamma)\right\}_{\nu\mu} \\ &= e^{-i\nu\alpha}d^j_{\nu\mu}(\beta)e^{-i\mu\gamma}, \end{aligned} \tag{5.62}$$

where $D^j(\boldsymbol{e}_3,\omega)$ is diagonal and $D^j(\boldsymbol{e}_2,\omega) = d^j(\omega)$ is real. As a convention, the arranging order of row (column) is j, $(j-1)$, ..., $-(j-1)$, $-j$.

The representation matrix d^j has some evident symmetry, which can be proved by replacement of summation indices:

$$\begin{aligned}
&d^j_{\nu\mu}(\pi) = (-1)^{j-\mu}\delta_{\nu(-\mu)}, \qquad d^j_{\nu\mu}(2\pi) = (-1)^{2j}\delta_{\nu\mu}, \\
&d^j_{\nu\mu}(\omega) = d^j_{(-\mu)(-\nu)}(\omega) = (-1)^{\mu-\nu}d^j_{\nu\mu}(-\omega) \\
&\quad = (-1)^{\mu-\nu}d^j_{\mu\nu}(\omega) = d^j_{\mu\nu}(-\omega) = (-1)^{\mu-\nu}d^j_{(-\nu)(-\mu)}(\omega) \\
&\quad = (-1)^{j+\nu}d^j_{\nu(-\mu)}(\pi-\omega) = (-1)^{j-\mu}d^j_{(-\nu)\mu}(\pi-\omega).
\end{aligned} \tag{5.63}$$

From Eq. (5.60), one has

$$\begin{aligned}
d^j_{\mu j}(\beta) &= d^j_{(-j)(-\mu)}(\beta) = (-1)^{j-\mu}d^j_{j\mu}(\beta) = (-1)^{j-\mu}d^j_{(-\mu)(-j)}(\beta) \\
&= \left\{\frac{(2j)!}{(j-\mu)!(j+\mu)!}\right\}^{1/2} \{\cos(\omega/2)\}^{j+\mu}\{\sin(\omega/2)\}^{j-\mu}, \\
d^\ell_{00}(\beta) &= \sum_{n=0}^{\ell}(-1)^n\left\{\frac{\ell!\,[\cos(\omega/2)]^{\ell-n}\,[\sin(\omega/2)]^n}{n!(\ell-n)!}\right\}^2 .
\end{aligned} \tag{5.64}$$

$d^j_{\nu\mu}(\omega)$ with $0 \leqslant j \leqslant 4$ are listed as follows where $c = \cos(\omega/2)$ and $s = \sin(\omega/2)$:

$$\begin{aligned}
&d^0_{0,0}(\omega) = 1, \qquad d^{1/2}_{1/2,1/2}(\omega) = c, \\
&d^1_{1,1}(\omega) = c^2, \qquad d^1_{0,1}(\omega) = \sqrt{2}cs, \qquad d^1_{0,0}(\omega) = c^2 - s^2, \\
&d^{3/2}_{3/2,3/2}(\omega) = c^3, \qquad d^{3/2}_{1/2,3/2}(\omega) = \sqrt{3}c^2s, \qquad d^{3/2}_{1/2,1/2}(\omega) = c^3 - 2cs^2, \\
&d^2_{2,2}(\omega) = c^4, \qquad d^2_{1,2}(\omega) = 2c^3s, \qquad d^2_{1,1}(\omega) = c^4 - 3c^2s^2, \\
&d^2_{0,2}(\omega) = \sqrt{6}c^2s^2, \quad d^2_{0,1}(\omega) = \sqrt{6}\left(c^3s - cs^3\right), \quad d^2_{0,0}(\omega) = c^4 - 4c^2s^2 + s^4,
\end{aligned}$$

$$\begin{aligned}
&d^{5/2}_{5/2,5/2}(\omega) = c^5, && d^{5/2}_{3/2,5/2}(\omega) = \sqrt{5}c^4s, \\
&d^{5/2}_{3/2,3/2}(\omega) = c^5 - 4c^3s^2, && d^{5/2}_{1/2,5/2}(\omega) = \sqrt{10}c^3s^2, \\
&d^{5/2}_{1/2,3/2}(\omega) = \sqrt{2}\left(2c^4s - 3c^2s^3\right), && d^{5/2}_{1/2,1/2}(\omega) = c^5 - 6c^3s^2 + 3cs^4, \\
&d^3_{3,3}(\omega) = c^6, && d^3_{2,3}(\omega) = \sqrt{6}c^5s, \\
&d^3_{2,2}(\omega) = c^6 - 5c^4s^2, && d^3_{1,3}(\omega) = \sqrt{15}c^4s^2, \\
&d^3_{1,2}(\omega) = \sqrt{10}\left(c^5s - 2c^3s^3\right), && d^3_{1,1}(\omega) = c^6 - 8c^4s^2 + 6c^2s^4, \\
&d^3_{0,3}(\omega) = 2\sqrt{5}c^3s^3, && d^3_{0,2}(\omega) = \sqrt{30}\left(c^4s^2 - c^2s^4\right), \\
&d^3_{0,1}(\omega) = 2\sqrt{3}\left(c^5s - 3c^3s^3 + cs^5\right), && d^3_{0,0}(\omega) = c^6 - 9c^4s^2 + 9c^2s^4 - s^6, \\
&d^{7/2}_{7/2,7/2}(\omega) = c^7, && d^{7/2}_{5/2,7/2}(\omega) = \sqrt{7}c^6s, \\
&d^{7/2}_{5/2,5/2}(\omega) = c^7 - 6c^5s^2, && d^{7/2}_{3/2,7/2}(\omega) = \sqrt{21}c^5s^2, \\
&d^{7/2}_{3/2,5/2}(\omega) = \sqrt{3}\left(2c^6s - 5c^4s^3\right), && d^{7/2}_{3/2,3/2}(\omega) = c^7 - 10c^5s^2 + 10c^3s^4,
\end{aligned}$$

$$d^{7/2}_{1/2,7/2}(\omega) = \sqrt{35}c^4 s^3, \qquad d^{7/2}_{1/2,5/2}(\omega) = \sqrt{5}\left(3c^5 s^2 - 4c^3 s^4\right),$$

$$\begin{aligned} d^{7/2}_{1/2,3/2}(\omega) &= \sqrt{15}\left(c^6 s - 4c^4 s^3 + 2c^2 s^5\right), \\ d^{7/2}_{1/2,1/2}(\omega) &= c^7 - 12c^5 s^2 + 18c^3 s^4 - 4cs^6, \end{aligned} \tag{5.65}$$

$$\begin{aligned} d^4_{4,4}(\omega) &= c^8, & d^4_{3,4}(\omega) &= 2\sqrt{2}c^7 s, \\ d^4_{3,3}(\omega) &= c^8 - 7c^6 s^2, & d^4_{2,4}(\omega) &= 2\sqrt{7}c^6 s^2, \\ d^4_{2,3}(\omega) &= \sqrt{14}\left(c^7 s - 3c^5 s^3\right), & d^4_{2,2}(\omega) &= c^8 - 12c^6 s^2 + 15c^4 s^4, \\ d^4_{1,4}(\omega) &= 2\sqrt{14}c^5 s^3, & d^4_{1,3}(\omega) &= \sqrt{7}\left(3c^6 s^2 - 5c^4 s^4\right), \end{aligned}$$

$$\begin{aligned} d^4_{1,2}(\omega) &= \sqrt{2}\left(3c^7 s - 15c^5 s^3 + 10c^3 s^5\right), \\ d^4_{1,1}(\omega) &= c^8 - 15c^6 s^2 + 30c^4 s^4 - 5c^2 s^6, \\ d^4_{0,4}(\omega) &= \sqrt{70}c^4 s^4, \\ d^4_{0,3}(\omega) &= 2\sqrt{35}\left(c^5 s^3 - c^3 s^5\right), \\ d^4_{0,2}(\omega) &= \sqrt{10}\left(3c^6 s^2 - 8c^4 s^4 + 3c^2 s^6\right), \\ d^4_{0,1}(\omega) &= 2\sqrt{5}\left(c^7 s - 6c^5 s^3 + 6c^3 s^5 - cs^7\right), \\ d^4_{0,0}(\omega) &= c^8 - 16c^6 s^2 + 36c^4 s^4 - 16c^2 s^6 + s^8. \end{aligned}$$

The remaining elements of $d^j(\omega)$ can be calculated by

$$d^j_{\nu\mu}(\omega) = (-1)^{\mu-\nu} d^j_{-\nu-\mu}(\omega) = (-1)^{\mu-\nu} d^j_{\mu\nu}(\omega) = (-1)^{j-\mu} d^j_{(-\nu)\mu}(\pi - \omega).$$

$d^j_{\nu\mu}(\pi - \omega)$ can be obtained from $d^j_{\nu\mu}(\omega)$ by interchanging s and c.

Now, we analyze the property of the representation D^j of SU(2).

(a) The dimension of D^j is $(2j+1)$, $j = 0,\ 1/2,\ 1,\ 3/2,\ \ldots$. $D^0(u) = 1$ and $D^{1/2}(u) = u$.

(b) When j is an integer, D^j is a single-valued representation of SO(3), but an unfaithful one of SU(2). When j is half of an odd integer, D^j is a double-valued representation of SO(3) and a faithful one of SU(2).

(c) d^j is a real orthogonal matrix and D^j is a unitary representation.

(d) The character of the class with angle ω can be calculated from the diagonal matrix $D^j(\boldsymbol{e}_3, \omega)$:

$$\chi^j(\omega) = \sum_{\mu=-j}^{j} \mathrm{e}^{-\mathrm{i}\mu\omega} = \sum_{\mu=-j}^{j} \mathrm{e}^{\mathrm{i}\mu\omega} = \frac{\sin\{(j+1/2)\omega\}}{\sin(\omega/2)}. \tag{5.66}$$

Since $\chi^j(\omega)$ satisfies the orthogonal relation (5.45), the representations D^j are inequivalent irreducible representations of SU(2).

We are going to show by reduction to absurdity that there is no other irreducible representation of SU(2) with a finite dimension which is inequivalent to all D^j. If this representation exists, its character $\chi(\omega)$ satisfies

$$\begin{aligned} 0 &= \frac{1}{\pi}\int_0^{2\pi} d\omega\ \sin^2(\omega/2)\chi^j(\omega)^*\chi(\omega) \\ &= \frac{1}{\pi}\int_0^{2\pi} d\omega\ \sin\left\{(j+1/2)\omega\right\}\left\{\chi(\omega)\sin(\omega/2)\right\}. \end{aligned}$$

From the theory of the Fourier series, the orthogonal basis functions $\sin\{(j+1/2)\omega\}$, where j are half of non-negative integers, are complete in the region $[0,\ 2\pi]$, namely, any function $\{\chi(\omega)\sin(\omega/2)\}$ orthogonal to all $\sin\{(j+1/2)\omega\}$ must vanish.

When the group parameters are taken to be the Euler angles, the orthogonal relation (5.44) becomes

$$\begin{aligned} &\frac{1}{16\pi^2}\int_{-\pi}^{\pi} d\alpha \int_0^{\pi} d\beta\ \sin\beta \int_{-2\pi}^{2\pi} d\gamma\ D^j_{\mu\rho}(\alpha,\beta,\gamma)^* D^k_{\nu\lambda}(\alpha,\beta,\gamma) \\ &\quad = \frac{1}{2}\delta_{\mu\nu}\delta_{\rho\lambda}\int_0^{\pi} d\beta\ \sin\beta d^j_{\mu\rho}(\beta) d^k_{\mu\rho}(\beta) = \frac{1}{2j+1}\delta_{jk}\delta_{\mu\nu}\delta_{\rho\lambda}, \end{aligned} \tag{5.67}$$

(e) The generators in the representation D^j are calculated by expanding the representation matrices of the rotations about the coordinate axes,

$$\begin{aligned} &D^j_{\nu\mu}(\boldsymbol{e}_3,\omega) = \delta_{\nu\mu}\left\{1 - \mathrm{i}\mu\omega + \ \cdots\ \right\}, \\ &d^j_{\nu\mu}(\omega) = \delta_{\nu\mu} + (\omega/2)\left\{\delta_{\nu(\mu-1)}\Gamma^j_\mu - \delta_{\nu(\mu+1)}\Gamma^j_\nu\right\}, \\ &D^j_{\nu\mu}(\boldsymbol{e}_1,\omega) = \left\{D^j(\boldsymbol{e}_3,-\pi/2)d^j(\omega)D^j(\boldsymbol{e}_3,\pi/2)\right\}_{\nu\mu} \\ &\quad = \delta_{\nu\mu} + (\omega/2)\left\{-\mathrm{i}\delta_{\nu(\mu-1)}\Gamma^j_\mu - \mathrm{i}\delta_{\nu(\mu+1)}\Gamma^j_\nu\right\}, \\ &\Gamma^j_\nu = \Gamma^j_{-\nu+1} = \left\{(j+\nu)(j-\nu+1)\right\}^{1/2}, \\ &\left(I^j_1\right)_{\nu\mu} = \frac{1}{2}\left[\delta_{\nu(\mu+1)}\Gamma^j_\nu + \delta_{\nu(\mu-1)}\Gamma^j_{-\nu}\right], \\ &\left(I^j_2\right)_{\nu\mu} = -\frac{\mathrm{i}}{2}\left[\delta_{\nu(\mu+1)}\Gamma^j_\nu - \delta_{\nu(\mu-1)}\Gamma^j_{-\nu}\right], \\ &\left(I^j_3\right)_{\nu\mu} = \mu\delta_{\nu\mu}, \qquad \left(I^j_\pm\right)_{\nu\mu} = \left(I^j_1 \pm \mathrm{i}I^j_2\right)_{\nu\mu} = \delta_{\nu(\mu\pm1)}\Gamma^j_{\pm\nu}. \end{aligned} \tag{5.68}$$

In comparison with T_a (see Eq. (5.11)), the representation D^1 is equivalent to the self-representation of SO(3) (see Prob. 8 of Chap. 1):

$$\begin{aligned} &M^{-1}T_aM = I^1_a, \qquad a = 1,\ 2,\ 3, \\ &M^{-1}R(\alpha,\beta,\gamma)M = D^1(\alpha,\beta,\gamma), \end{aligned} \qquad M = \frac{1}{\sqrt{2}}\begin{pmatrix} -1 & 0 & 1 \\ -\mathrm{i} & 0 & -\mathrm{i} \\ 0 & \sqrt{2} & 0 \end{pmatrix}. \tag{5.69}$$

The representation space of D^j contains 2j+1 basis functions $\psi^j_\mu(x)$:

$$\begin{aligned} & P_R\psi^j_\mu(x) = \sum_{\nu=-j}^{j} \psi^j_\nu(x) D^j_{\nu\mu}(R), \quad I^{(0)}_a \psi^j_\mu(x) = \sum_{\nu=-j}^{j} \psi^j_\nu(x) \left(I^j_a\right)_{\nu\mu}, \\ & I^{(0)}_3 \psi^j_\mu(x) = \mu \psi^j_\mu(x), \quad \left(I^{(0)}_1 \pm \mathrm{i} I^{(0)}_2\right) \psi^j_\mu(x) = \Gamma^j_{\mp\mu} \psi^j_{\mu\pm 1}(x), \\ & I^{(0)}_- \psi^j_{j-n}(x) = \Gamma^j_{j-n} \psi^j_{j-n-1}(x), \quad \Gamma^j_{j-n} = \sqrt{(2j-n)(n+1)}. \end{aligned} \tag{5.70}$$

Because the generators $I^{(0)}_a$ of SO(3) are related to the orbital angular momentum operators L_a (see Eq. (5.21)), we calculate the operator $\left(I^{(0)}\right)^2$ related to the square operator L^2 of angular momentum:

$$\begin{aligned} & \left[\sum_{a=1}^{3} \left(I^j_a\right)^2\right] \psi^j_\mu(x) = \frac{1}{2}\left[I^j_+ I^j_- + I^j_- I^j_+ + 2\left(I^j_3\right)^2\right] \psi^j_\mu(x) \\ & \quad = \frac{1}{2}\left[\left(\Gamma^j_\mu\right)^2 + \left(\Gamma^j_{\mu+1}\right)^2 + 2\mu^2\right] \psi^j_\mu(x) = j(j+1)\psi^j_\mu(x). \end{aligned} \tag{5.71}$$

(f) Since the character χ^j is real, D^j is self-conjugate: $X^{-1} D^j X = \left(D^j\right)^*$. X changes the sign of ω in $D^j(\boldsymbol{e}_3, \omega)$, but leaves $D^j(\boldsymbol{e}_2, \omega)$ invariant. Namely, X is $d^j(\pi)$:

$$d^j(\pi)^{-1} D^j(\alpha, \beta, \gamma) d^j(\pi) = D^j(\alpha, \beta, \gamma)^*. \tag{5.72}$$

Due to Eq. (5.63), when j is an integer ℓ, $d^\ell(\pi)$ is symmetric and D^ℓ is real. When j is half of an odd integer, $d^j(\pi)$ is antisymmetric and D^j is self-conjugate, but not real.

5.4.4 *Spherical Harmonic Functions*

Consider a single-body system moving in a spherically symmetric potential with respect to the origin. Its Hamiltonian $H(\boldsymbol{r})$ is left invariant under the rotation R about any axis through the origin, $H(\boldsymbol{r})P_R = P_R H(\boldsymbol{r})$. The symmetry group of the system is SO(3). If the energy level E is degenerate with multiplicity n, there are n linearly independent eigenfunctions $\psi_\mu(x)$, $H(\boldsymbol{r})\psi_\mu(\boldsymbol{r}) = E\psi_\mu(x)$, where $\mu = 1,\ 2,\ \cdots,\ n$. $P_R\psi_\mu(\boldsymbol{r})$ is also an eigenfunction of $H(\boldsymbol{r})$ with the same energy E so that $P_R\psi_\mu(\boldsymbol{r})$ can be expanded with respect to the basis functions $\psi_\nu(\boldsymbol{r})$:

$$P_R\psi_\mu(\boldsymbol{r}) = \psi_\mu(R^{-1}\boldsymbol{r}) = \sum_{\nu=1}^{n} \psi_\nu(\boldsymbol{r}) D_{\nu\mu}(R). \tag{5.73}$$

Since $\psi_\mu(\boldsymbol{r})$ is left invariant under a rotation through an angle 2π, the representation $D(R)$ of SO(3) corresponding to the energy E is single-valued. Generally, $D(R)$ is reducible,

$$\begin{aligned} X^{-1}D(R)X &= \bigoplus_\ell a_\ell D^\ell(R), \\ X^{-1}I_aX &= \bigoplus_\ell a_\ell I_a^\ell, \qquad \chi(R) = \sum_\ell a_\ell \chi^\ell(R), \end{aligned} \tag{5.74}$$

where I_a are the generators in the representation $D(R)$ and ℓ is an integer. a_ℓ can be calculated by the characters,

$$\begin{aligned} a_\ell &= \frac{2}{\pi}\int_0^\pi d\omega \ \sin^2(\omega/2)\chi^\ell(\omega)\chi(\omega) \\ &= \frac{2}{\pi}\int_0^\pi d\omega \ \sin(\omega/2)\sin\{(\ell+1/2)\omega\}\chi(\omega). \end{aligned}$$

Substituting a_ℓ into Eq. (5.74), where I_a is taken to be I_3 first, and then to be $I_\pm$, one obtains the similarity transformation matrix X. The row of X is enumerated by ρ, and the column by $\ell m\lambda$, where λ is needed when $a_\ell > 1$. The new basis function $\psi^\ell_{m\lambda}(\boldsymbol{r})$ is

$$\begin{aligned} \psi^\ell_{m\lambda}(\boldsymbol{r}) &= \sum_\mu \psi_\mu(\boldsymbol{r})X_{\mu,\ell m\lambda}\,, \\ P_R\psi^\ell_{m\lambda}(\boldsymbol{r}) &= \psi^\ell_{m\lambda}(R^{-1}\boldsymbol{r}) = \sum_{m'} \psi^\ell_{m'\lambda}(\boldsymbol{r})D^\ell_{m'm}(R). \end{aligned} \tag{5.75}$$

Hence, the static wave function of the single-body system with spherical symmetry can be chosen as $\psi^\ell_{m\lambda}(\boldsymbol{r})$ which belongs to the m-row of the irreducible representation D^ℓ of SO(3). $\psi^\ell_{m\lambda}(\boldsymbol{r})$ is the common eigenfunction of, in addition to the Hamiltonian $H(x)$, the orbital angular momentum L^2 and L_3 (see Eqs. (5.21), (5.70), and (5.71)):

$$\begin{aligned} L^2\psi^\ell_{m\lambda}(\boldsymbol{r}) &= \ell(\ell+1)\psi^\ell_{mr}(\boldsymbol{r}), \\ L_3\psi^\ell_{m\lambda}(\boldsymbol{r}) &= m\psi^\ell_{m\lambda}(\boldsymbol{r}), \\ L_\pm\psi^\ell_{m\lambda}(\boldsymbol{r}) &= \Gamma^\ell_{\mp m}\psi^\ell_{(m\pm 1)\lambda}(\boldsymbol{r}), \end{aligned} \tag{5.76}$$

where $L_\pm = L_1 \pm \mathrm{i}L_2$ and $L^2 = \sum_a L_a^2$.

Now, neglect the index λ in $\psi^\ell_{m\lambda}(\boldsymbol{r})$ for simplicity. Let $\boldsymbol{r} = (r,\ \theta,\ \varphi)$ and $\boldsymbol{r}_0 = (r,\ 0,\ 0)$. The rotation $T = R(\varphi,\ \theta,\ \gamma)$ changes the point $\boldsymbol{r}_0$ at the Z-axis to $\boldsymbol{r}$, $\boldsymbol{r} = T\boldsymbol{r}_0$. From Eq. (5.76), one has

$$\begin{aligned}\psi^\ell_m(\boldsymbol{r}) &= \psi^\ell_m(T\boldsymbol{r}_0) = P_{T^{-1}}\psi^\ell_m(\boldsymbol{r}_0) = \sum_{m'} \psi^\ell_{m'}(\boldsymbol{r}_0)D^\ell_{mm'}(T)^* \\ &= \sum_{m'} \psi^\ell_{m'}(\boldsymbol{r}_0)e^{im\varphi}d^\ell_{mm'}(\theta)e^{im'\gamma}.\end{aligned} \tag{5.77}$$

Since the left-hand side of Eq. (5.77) is independent of γ, the right-hand side of Eq. (5.77) has to be independent of γ. Because the exponent functions $e^{im'\gamma}$ are linearly independent, the terms with $m' \neq 0$ on the right-hand side of Eq. (5.77) have to vanish. Letting

$$\psi^\ell_{m'}(\boldsymbol{r}_0) = \delta_{m'0}\left(\frac{2\ell+1}{4\pi}\right)^{1/2}\phi_\ell(r), \tag{5.78}$$

one has

$$\psi^\ell_m(\boldsymbol{r}) = \phi_\ell(r)Y^\ell_m(\theta,\varphi), \tag{5.79}$$

$$\begin{aligned}Y^\ell_m(\theta,\varphi) &= \left(\frac{2\ell+1}{4\pi}\right)^{1/2} D^\ell_{m0}(\varphi,\theta,0)^* \\ &= \left(\frac{2\ell+1}{4\pi}\right)^{1/2} e^{im\varphi}d^\ell_{m0}(\theta).\end{aligned} \tag{5.80}$$

The static wave function $\psi^\ell_m(\boldsymbol{r})$ of the single-body system with the spherical symmetry can be decomposed as a product of a radial function $\phi_\ell(r)$ and an angular function $Y^\ell_m(\theta,\varphi)$. Since the radial function $\phi_\ell(r)$ is left invariant under any rotation, $Y^\ell_m(\theta,\varphi)$ **belongs to the** mth **row of the irreducible representation** D^ℓ of SO(3) so that it is the common eigenfunction of L^2 and L_3 as given in Eq. (5.76). In quantum mechanics, $Y^\ell_m(\theta,\varphi)$ is called the **spherical harmonic function.** Due to Eq. (5.67) $Y^\ell_m(\theta,\varphi)$ is orthonormal

$$\begin{aligned}&\int_{-\pi}^{\pi} d\varphi \int_0^{\pi} d\theta\ \sin\theta Y^\ell_m(\theta,\varphi)^* Y^{\ell'}_{m'}(\theta,\varphi) = \frac{\{(2\ell+1)(2\ell'+1)\}^{1/2}}{4\pi} \\ &\quad\times \int_{-\pi}^{\pi} d\varphi \int_0^{\pi} d\theta\ \sin\theta D^\ell_{m0}(\varphi,\theta,0)D^{\ell'}_{m'0}(\varphi,\theta,0)^* \\ &= \delta_{\ell\ell'}\delta_{mm'}.\end{aligned} \tag{5.81}$$

From the symmetry (5.64) of $d^\ell(\theta)$, one has

$$Y^\ell_m(\theta,\varphi)^* = \left(\frac{2\ell+1}{4\pi}\right)^{1/2} e^{-im\varphi}d^\ell_{m0}(\theta) = (-1)^m Y^\ell_{-m}(\theta,\varphi). \tag{5.82}$$

The Legendre function is defined as

$$P_\ell(\cos\theta) = \left(\frac{4\pi}{2\ell+1}\right)^{1/2} Y_0^\ell(\theta,0) = d_{00}^\ell(\theta), \tag{5.83}$$

which satisfies

$$\int_0^\pi d\theta \,\sin\theta P_\ell(\cos\theta)P_{\ell'}(\cos\theta) = \frac{2\delta_{\ell\ell'}}{2\ell+1}\,. \tag{5.84}$$

In the space inversion, $\boldsymbol{r} \longrightarrow -\boldsymbol{r}$, $\theta \longrightarrow \pi-\theta$, and $\varphi \longrightarrow \pi+\varphi$,

$$\begin{aligned} Y_m^\ell(\pi-\theta,\pi+\varphi) &= \left(\frac{2\ell+1}{4\pi}\right)^{1/2} \mathrm{e}^{-\mathrm{i}m(\pi+\varphi)} d_{m0}^\ell(\pi-\theta) \\ &= (-1)^\ell Y_m^\ell(\theta,\varphi). \end{aligned} \tag{5.85}$$

The parity of the spherical harmonic function is $(-1)^\ell$.

$Y_m^\ell(\theta,\varphi)$ **can be expressed by the Jacobi polynomials**, which are given as formula 8.960 in [Gradshteyn and Ryzhik (2007)]:

$$P_n^{(\alpha,\beta)}(x) = 2^{-n}\sum_{r=0}^{n}\binom{n+\alpha}{r}\binom{n+\beta}{n-r}(x-1)^{n-r}(1+x)^r.$$

Taking $\alpha=\beta=m$, $n=\ell-m$ and $r'=\ell-m-r$, one has

$$\begin{aligned} P_{\ell-m}^{(m,m)}(\cos\theta) &= \sum_{r=0}^{\ell-m}\binom{\ell}{r}\binom{\ell}{\ell-m-r}[-\sin^2(\theta/2)]^{\ell-m-r}\cos^{2r}(\theta/2) \\ &= \frac{2^m \ell! d_{-m0}^\ell(\theta)}{(\sin\theta)^m\sqrt{(\ell+m)!(\ell-m)!}} \\ &= \frac{2^m \ell!}{(-\sin\theta)^m}\sqrt{\frac{4\pi}{(2\ell+1)(\ell+m)!(\ell-m)!}}\mathrm{e}^{-\mathrm{i}m\varphi}Y_m^\ell(\theta,\varphi). \end{aligned} \tag{5.86}$$

5.4.5 *Harmonic Polynomials*

Multiplying $Y_\ell^\ell(\theta,\varphi)$ with r^ℓ, one obtains

$$\begin{aligned} \mathcal{Y}_\ell^\ell(\mathbf{r}) = r^\ell Y_\ell^\ell(\boldsymbol{r}) &= \sqrt{\frac{2\ell+1}{4\pi}}r^\ell \mathrm{e}^{\mathrm{i}\ell\varphi}d_{\ell 0}^\ell(\theta) \\ &= \frac{(-1)^\ell}{2^\ell \ell!}\sqrt{\frac{(2\ell+1)!}{4\pi}}(x_1+\mathrm{i}x_2)^\ell, \\ \nabla^2\mathcal{Y}_\ell^\ell(\mathbf{r}) &= \left[\sum_{a=1}^{3}\frac{\partial^2}{(\partial x_a)^2}\right]\mathcal{Y}_\ell^\ell(\mathbf{r}) = 0. \end{aligned} \tag{5.87}$$

$\mathcal{Y}^{\ell}_{\ell}(\mathbf{r})$ is a homogeneous polynomial of order ℓ with respect to the rectangular coordinates (x_1, x_2, x_3) and satisfies the Laplace equation. Since the Laplace operator $\bigtriangledown^2$ is left invariant under any rotation $R \in \mathrm{SO}(3)$, **$\mathcal{Y}^{\ell}_{m}(\boldsymbol{r})$ is also a homogeneous polynomial of order ℓ and satisfies the Laplace equation**, called the **harmonic polynomial**,

$$\mathcal{Y}^{\ell}_{m}(\boldsymbol{r}) = r^{\ell} Y^{\ell}_{m}(\boldsymbol{r}), \qquad \bigtriangledown^2 \mathcal{Y}^{\ell}_{m}(\boldsymbol{r}) = 0. \tag{5.88}$$

The explicit expression of $\mathcal{Y}^{\ell}_{m}(\boldsymbol{r})$ can be calculated by the action of the lowering operator L_- (see Eqs. (5.70) and (5.76)):

$$L_-(x_1 + \mathrm{i}x_2) = -2x_3, \quad L_-(x_3) = x_1 - \mathrm{i}x_2, \quad L_-(x_1 - \mathrm{i}x_2) = 0, \tag{5.89}$$

$$\begin{aligned} \mathcal{Y}^{\ell}_{m}(\mathbf{r}) = & \frac{(-1)^m}{2^m}\sqrt{\frac{(2\ell+1)(\ell+m)!(\ell-m)!}{4\pi}}(x_1+\mathrm{i}x_2)^m \\ & \times \left[\sum_{s=0}^{[(\ell-m)/2]} \frac{(-1)^s x_3^{\ell-m-2s}(x_1^2+x_2^2)^s}{4^s s!(m+s)!(\ell-m-2s)!}\right], \quad 0 \leqslant m \leqslant \ell. \end{aligned} \tag{5.90}$$

$\mathcal{Y}^{\ell}_{-|m|}(\mathbf{r}) = (-1)^m \mathcal{Y}^{\ell}_{|m|}(\mathbf{r})^*$. This expression (5.90) can be proved by mathematical induction (see Prob. 9). In fact, the first term ($s = 0$) in Eq. (5.90) is calculated by Eqs. (5.89) and (5.70), and the remaining terms are calculated by the Laplace equation:

$$\begin{aligned} & \bigtriangledown^2 \left[(x_1+\mathrm{i}x_2)^m x_3^a (x_1^2+x_2^2)^s\right] = (x_1+\mathrm{i}x_2)^m \\ & \quad \times \left[a(a-1)x_3^{a-2}(x_1^2+x_2^2)^s + 4s(m+s)x_3^a(x_1^2+x_2^2)^{s-1}\right]. \end{aligned} \tag{5.91}$$

$\mathcal{Y}^{\ell}_{m}(\boldsymbol{r})$ with $\ell \leqslant 3$ are listed as follows.

$$\begin{aligned}
\mathcal{Y}^0_0(\boldsymbol{r}) &= \sqrt{\frac{1}{4\pi}}, & \mathcal{Y}^2_0(\boldsymbol{r}) &= \sqrt{\frac{5}{16\pi}}\left(3x_3^2 - r^2\right), \\
\mathcal{Y}^1_{\pm 1}(\boldsymbol{r}) &= \mp\sqrt{\frac{3}{8\pi}}\left(x_1 \pm \mathrm{i}x_2\right), & \mathcal{Y}^3_{\pm 3}(\boldsymbol{r}) &= \mp\sqrt{\frac{35}{64\pi}}\left(x_1 \pm \mathrm{i}x_2\right)^3, \\
\mathcal{Y}^1_0(\boldsymbol{r}) &= \sqrt{\frac{3}{4\pi}}\, x_3, & \mathcal{Y}^3_{\pm 2}(\boldsymbol{r}) &= \sqrt{\frac{105}{32\pi}}\left(x_1 \pm \mathrm{i}x_2\right)^2 x_3, \\
\mathcal{Y}^2_{\pm 2}(\boldsymbol{r}) &= \sqrt{\frac{15}{32\pi}}\left(x_1 \pm \mathrm{i}x_2\right)^2, & \mathcal{Y}^3_{\pm 1}(\boldsymbol{r}) &= \mp\sqrt{\frac{21}{64\pi}}\left(x_1 \pm \mathrm{i}x_2\right)\left(5x_3^2 - r^2\right), \\
\mathcal{Y}^2_{\pm 1}(\boldsymbol{r}) &= \mp\sqrt{\frac{15}{8\pi}}\left(x_1 \pm \mathrm{i}x_2\right) x_3, & \mathcal{Y}^3_0(\boldsymbol{r}) &= \sqrt{\frac{7}{16\pi}}\left(5x_3^2 - 3r^2\right) x_3.
\end{aligned}$$

5.4.6 *Irreducible Bases of the Group* **I**

The explicit forms of the irreducible bases of **I** depend upon the irreducible representations of **I**, which are usually chosen as the subduced representation of the irreducible representation $D^{\ell}(\mathrm{SO}(3))$ with respect to **I**. Since $D^{\ell}_{\mu\nu}(T_0) = D^{\ell}_{\mu\nu}(\mathbf{e}_3, 2\pi/5) = \delta_{\mu\nu}\eta^{\mu}$, where $\eta = \mathrm{e}^{-\mathrm{i}2\pi/5}$, we obtain from Eq. (5.66) and Table 4.6,

$$\begin{aligned} &D^0(\mathbf{I}) = D^A(\mathbf{I}), \qquad D^1(\mathbf{I}) = D^{T_1}(\mathbf{I}),\\ &D^2(\mathbf{I}) = D^H(\mathbf{I}), \qquad X^{-1}D^3(\mathbf{I})X = D^{T_2}(\mathbf{I}) \oplus D^G(\mathbf{I}). \end{aligned} \tag{5.92}$$

Enumerating the row (column) index of $D^j(\mathrm{I})$ by the exponent μ of the diagonal element η^{μ} of $D^j(T_0)$:

$$\begin{array}{ll} & \text{row (column) index}: \\ D^A(T_0) = 1, & 0, \\ D^{T_1}(T_0) = \mathrm{diag}\{\eta,\ 1,\ \eta^{-1}\}, & 1,\ 0,\ \overline{1}, \\ D^{T_2}(T_0) = \mathrm{diag}\{\eta^2,\ 1,\ \eta^{-2}\}, & 2,\ 0,\ \overline{2}, \\ D^G(T_0) = \mathrm{diag}\{\eta^2,\ \eta,\ \eta^{-1},\ \eta^{-2}\}, & 2,\ 1,\ \overline{1},\ \overline{2}, \\ D^H(T_0) = \mathrm{diag}\{\eta^2,\ \eta,\ 1,\ \eta^{-1},\ \eta^{-2}\}, & 2,\ 1,\ 0,\ \overline{1},\ \overline{2}, \end{array} \tag{5.93}$$

where $\overline{\mu} = -\mu$. Thus, we have Table 5.1.

Table 5.1. The character tables of I and SO(3)

I	E	$12C_5$	$12C_5^2$	$20C_3$	$15C_2$	Subduced rep.
A	1	1	1	1	1	
T_1	3	p^{-1}	$-p$	0	-1	
T_2	3	$-p$	p^{-1}	0	-1	
G	4	-1	-1	1	0	
H	5	0	0	-1	1	
$D^0(\mathrm{SO}(3))$	1	1	1	1	1	$D^A(\mathrm{I})$
$D^1(\mathrm{SO}(3))$	3	p^{-1}	$-p$	0	-1	$D^{T_1}(\mathrm{I})$
$D^2(\mathrm{SO}(3))$	5	0	0	-1	1	$D^H(\mathrm{I})$
$D^3(\mathrm{SO}(3))$	7	$-p^{-1}$	$-p^{-1}$	1	-1	$D^{T_2}(\mathrm{I}) \oplus D^G(\mathrm{I})$

η satisfies the following formulas:

$$\begin{aligned} &\sum_{a=0}^{4} \eta^a = \sum_{a=0}^{4} \eta^{2a} = \sum_{a=0}^{4} \eta^{3a} = \sum_{a=0}^{4} \eta^{4a} = 0, \quad \sum_{a=0}^{4} \eta^{5a} = 5, \\ &p = \eta + \eta^{-1} = \frac{\sqrt{5}-1}{2}, \quad p^{-1} = -(\eta^2 + \eta^{-2}) = \frac{\sqrt{5}+1}{2}, \end{aligned} \tag{5.94}$$

There are two pairs of the equal diagonal elements in $D^3(T_0)$: $\eta^3 = \eta^{-2}$ and $\eta^2 = \eta^{-3}$, such that X in Eq. (5.92) is decomposed as YZ, where Y is a simple similarity transformation and Z is a block matrix containing two submatrices of two dimensions:

$$X = \begin{pmatrix} 0 & 0 & C & 0 & 0 & 0 & D^* \\ A & 0 & 0 & -B^* & 0 & 0 & 0 \\ 0 & 0 & 0 & 0 & 1 & 0 & 0 \\ 0 & 1 & 0 & 0 & 0 & 0 & 0 \\ 0 & 0 & 0 & 0 & 0 & 1 & 0 \\ 0 & 0 & D & 0 & 0 & 0 & -C^* \\ B & 0 & 0 & A^* & 0 & 0 & 0 \end{pmatrix}, \tag{5.95}$$

where $|A|^2 + |B|^2 = |C|^2 + |D|^2 = 1$. We enumerate the row index μ of $X_{\mu\nu}$ as that of $D^3(\mathrm{SO}(3))$, $\mu = 3, 2, 1, 0, \overline{1}, \overline{2}, \overline{3}$, and the column index ν as that of $D^{T_2}(\mathrm{I}) \oplus D^G(\mathrm{I})$, $\nu = T_22, T_20, T_2\overline{2}, G2, G1, G\overline{1}, G\overline{2}$.

Another generator of $\mathbf{I}$ is a 2-fold rotation $S_1 = R(0, 2\theta_4, \pi)$ (see Fig. 2.3 and Prob. 7). Since $X^{-1}D^3(S_1)X = D^{T_2}(S_1) \oplus D^G(S_1)$,

$$\begin{aligned} 0 &= \left[X^{-1}D^3(S_1)X\right]_{(T_20)(G1)} = X^*_{0(T_20)} D^3_{01}(S_1) X_{1(G1)} \\ &= -d^3_{01}(2\theta_4) = -\frac{\sqrt{3}}{4}\sin(2\theta_4)(5\cos^2(2\theta_4) - 1), \end{aligned}$$

where $d^3_{01}(2\theta_4)$ is given in Eq. (5.65). Due to $\sin(2\theta_4) \neq 0$ and $\theta_1 = 2\theta_4$, we obtain

$$\cos(2\theta_4) = \cos\theta_1 = \sqrt{1/5}, \quad \sin\theta_1 = 2/\sqrt{5}, \quad \tan\theta_1 = 2. \tag{5.96}$$

Defining the parameters $c = \cos\theta_4$ and $s = \sin\theta_4$, where $c^2 = p^{-1}/\sqrt{5}$, $s^2 = p/\sqrt{5}$, and $sc = 1/\sqrt{5}$, we have

$$\begin{aligned} &\left[X^{-1}D^3(S_1)X\right]_{(G1)(T_22)} = X^*_{1(G1)} D^3_{12}(S_1) X_{2(T_22)} + X^*_{1(G1)} D^3_{1\overline{3}}(S_1) X_{\overline{3}(T_22)} \\ &\quad = \sqrt{10}(c^5 s - 2c^3 s^3)A - \sqrt{15}(c^2 s^4)B = p(\sqrt{2}A - \sqrt{3}B)/5 = 0, \\ &\left[X^{-1}D^3(S_1)X\right]_{(G1)(T_2\overline{2})} = X^*_{1(G1)} D^3_{13}(S_1) X_{3(T_2\overline{2})} + X^*_{1(G1)} D^3_{1\overline{2}}(S_1) X_{\overline{2}(T_2\overline{2})} \\ &\quad = -\sqrt{15}(c^4 s^2)C + \sqrt{10}(-2c^3 s^3 + cs^5)D = -p^{-1}(\sqrt{3}C + \sqrt{2}D)/5 = 0. \end{aligned}$$

Taking $A = D = \sqrt{3/5}$, we have $B = -C = \sqrt{2/5}$:

$$\begin{aligned} &X_{1(G1)} = X_{0(T_20)} = X_{\overline{1}(G\overline{1})} = 1, \\ &X_{2(T_22)} = X_{\overline{2}(T_2\overline{2})} = X_{\overline{3}(G2)} = X_{3(G\overline{2})} = \sqrt{3/5}, \\ &X_{\overline{3}(T_22)} = -X_{3(T_2\overline{2})} = -X_{2(G2)} = X_{\overline{2}(G\overline{2})} = \sqrt{2/5}. \end{aligned} \tag{5.97}$$

The remaining of $X_{\mu\nu}$ vanish.

The irreducible bases $\phi^j_{\mu\nu}$ of **I**, defined in Eq. (3.114), satisfy

$$T_0\phi^j_{\mu\nu} = \eta^\mu \phi^j_{\mu\nu}, \qquad \phi^j_{\mu\nu}T_0 = \eta^\nu \phi^j_{\mu\nu}. \tag{5.98}$$

In terms of the projective operator $e^{(\mu)}$ of the cyclic group C_5 (see Eq. (3.119)),

$$e^{(\mu)} = \frac{1}{5}\sum_{\rho=0}^{4} \eta^{-\rho\mu}T_0^\rho, \qquad \phi^j_{\mu\nu} = e^{(\mu)}\phi^j_{\mu\nu} = \phi^j_{\mu\nu}e^{(\nu)}. \tag{5.99}$$

Following Eq. (5.99), we define the vectors $\Phi^{(t)}_{\mu\nu}$ in the group algebra of **I** by $\Phi^{(t)}_{\mu\nu} \propto e^{(\mu)}R^{(t)}e^{(\nu)}$, where $R^{(t)} \in \mathbf{I}$ are chosen to make $\Phi^{(t)}_{\mu\nu}$ linearly independent. Letting $R^{(1)} = E$, we have

$$\begin{aligned}
\Phi^{(1)}_{\mu\nu} &= 5e^{(\mu)}Ee^{(\nu)} = \frac{1}{5}\sum_{\rho=0}^{4}\sum_{\lambda=0}^{4} \eta^{-\rho\mu-\lambda\nu}T_0^{\rho+\lambda} \\
&= \frac{1}{5}\sum_{\tau=0}^{4}\eta^{-\tau\mu}T_0^\tau\left[\sum_{\lambda=0}^{4}\eta^{-\lambda(\nu-\mu)}\right] = \delta_{\mu\nu}\sum_{\tau=0}^{4}\eta^{-\tau\mu}T_0^\tau,
\end{aligned}$$

$$\Phi^{(1)}_{\mu\mu} = \sum_{\tau=0}^{4}\eta^{-\tau\mu}T_0^\tau = E + \eta^{-\mu}T_0 + \eta^{-2\mu}T_0^2 + \eta^{2\mu}T_0^3 + \eta^{\mu}T_0^4. \tag{5.100}$$

Choosing $R^{(2)} \neq T_0^\tau$, say $R^{(2)} = S_{12}$, which is the 2-fold rotation about the Y-axis, we have

$$\begin{aligned}
\Phi^{(2)}_{\mu\nu} &= 5e^{(\mu)}S_{12}e^{(\nu)} = \frac{1}{5}\sum_{\rho=0}^{4}\sum_{\lambda=0}^{4} \eta^{-\rho\mu-\lambda\nu}T_0^{\rho-\lambda}S_{12} \\
&= \frac{1}{5}\sum_{\tau=0}^{4}\eta^{-\tau\mu}T_0^\tau S_{12}\left[\sum_{\lambda=0}^{4}\eta^{-\lambda(\nu+\mu)}\right] = \delta_{\mu(-\nu)}\sum_{\tau=0}^{4}\eta^{-2\tau\mu}T_0^\tau S_{12}T_0^{-\tau},
\end{aligned}$$

where $T_0S_{12} = S_{12}T_0^{-1}$ is used. From Eqs. (2.64) and (2.65) we have

$$\Phi^{(2)}_{\mu\overline{\mu}} = S_{12} + \eta^{-2\mu}S_{13} + \eta^{\mu}S_{14} + \eta^{-\mu}S_{15} + \eta^{2\mu}S_{11}. \tag{5.101}$$

Letting $R^{(3)} = S_1$, we have

$$\begin{aligned}
\Phi^{(3)}_{\mu\nu} &= 25e^{(\mu)}S_1e^{(\nu)} = \sum_{\rho=0}^{4}\sum_{\lambda=0}^{4}\eta^{-\rho\mu+\lambda\nu}T_0^\rho S_1 T_0^{-\lambda} \\
&= \sum_{\lambda=0}^{4}\sum_{\tau=0}^{4}\eta^{-(\lambda+\tau)\mu+\lambda\nu}T_0^{\lambda+\tau}S_1T_0^{-\lambda},
\end{aligned}$$

$$\begin{aligned}
\Phi_{\mu\nu}^{(3)} &= \sum_{\lambda=0}^{4}\sum_{\tau=0}^{4} \eta^{\lambda(\nu-\mu)-\tau\mu} T_0^{\lambda}\left(T_0^{\tau} S_1\right) T_0^{-\lambda} \\
&= S_1 + \eta^{-\mu} R_1^2 + \eta^{-2\mu} T_2^4 + \eta^{2\mu} T_5 + \eta^{\mu} R_5 \\
&\quad + \eta^{\nu-\mu}\left[S_2 + \eta^{-\mu} R_2^2 + \eta^{-2\mu} T_3^4 + \eta^{2\mu} T_1 + \eta^{\mu} R_1\right] \\
&\quad + \eta^{2\nu-2\mu}\left[S_3 + \eta^{-\mu} R_3^2 + \eta^{-2\mu} T_4^4 + \eta^{2\mu} T_2 + \eta^{\mu} R_2\right] \\
&\quad + \eta^{2\mu-2\nu}\left[S_4 + \eta^{-\mu} R_4^2 + \eta^{-2\mu} T_5^4 + \eta^{2\mu} T_3 + \eta^{\mu} R_3\right] \\
&\quad + \eta^{\mu-\nu}\left[S_5 + \eta^{-\mu} R_5^2 + \eta^{-2\mu} T_1^4 + \eta^{2\mu} T_4 + \eta^{\mu} R_4\right],
\end{aligned} \tag{5.102}$$

where

$$\begin{aligned}
&T_0 S_1 \leftrightarrow (1\ 2\ 3\ 4\ 5)(2\ 3)(4\ 5) = (4\ 1\ 2) \leftrightarrow R_1^2, \\
&T_0^2 S_1 \leftrightarrow (1\ 2\ 3\ 4\ 5)(4\ 1\ 2) = (5\ 1\ 3\ 4\ 2) \leftrightarrow T_2^4, \\
&T_0^3 S_1 \leftrightarrow (1\ 2\ 3\ 4\ 5)(5\ 1\ 3\ 4\ 2) = (3\ 5\ 2\ 1\ 4) \leftrightarrow T_5, \\
&T_0^4 S_1 \leftrightarrow (1\ 2\ 3\ 4\ 5)(3\ 5\ 2\ 1\ 4) = (3\ 1\ 5) \leftrightarrow R_5.
\end{aligned}$$

Letting $R^{(4)} = S_6$, we have

$$\begin{aligned}
\Phi_{\mu\nu}^{(4)} &= 25 e^{(\mu)} S_6 e^{(\nu)} = \sum_{\rho=0}^{4}\sum_{\lambda=0}^{4} \eta^{-\rho\mu+\lambda\nu} T_0^{\rho} S_6 T_0^{-\lambda} \\
&= \sum_{\lambda=0}^{4}\sum_{\tau=0}^{4} \eta^{-(\lambda+\tau)\mu+\lambda\nu} T_0^{\lambda+\tau} S_6 T_0^{-\lambda},
\end{aligned}$$

$$\begin{aligned}
\Phi_{\mu\nu}^{(4)} &= \sum_{\lambda=0}^{4}\sum_{\tau=0}^{4} \eta^{\lambda(\nu-\mu)-\tau\mu} T_0^{\lambda}\left(T_0^{\tau} S_6\right) T_0^{-\lambda} \\
&= S_6 + \eta^{-\mu} T_2^3 + \eta^{-2\mu} R_7^2 + \eta^{2\mu} R_{10} + \eta^{\mu} T_1^2 \\
&\quad + \eta^{\nu-\mu}\left[S_7 + \eta^{-\mu} T_3^3 + \eta^{-2\mu} R_8^2 + \eta^{2\mu} R_6 + \eta^{\mu} T_2^2\right] \\
&\quad + \eta^{2\nu-2\mu}\left[S_8 + \eta^{-\mu} T_4^3 + \eta^{-2\mu} R_9^2 + \eta^{2\mu} R_7 + \eta^{\mu} T_3^2\right] \\
&\quad + \eta^{2\mu-2\nu}\left[S_9 + \eta^{-\mu} T_5^3 + \eta^{-2\mu} R_{10}^2 + \eta^{2\mu} R_8 + \eta^{\mu} T_4^2\right] \\
&\quad + \eta^{\mu-\nu}\left[S_{10} + \eta^{-\mu} T_1^3 + \eta^{-2\mu} R_6^2 + \eta^{2\mu} R_9 + \eta^{\mu} T_5^2\right],
\end{aligned} \tag{5.103}$$

where

$$\begin{aligned}
&T_0 S_6 \leftrightarrow (1\ 2\ 3\ 4\ 5)(1\ 3)(2\ 5) = (1\ 4\ 5\ 3\ 2) = T_2^3, \\
&T_0^2 S_6 \leftrightarrow (1\ 2\ 3\ 4\ 5)(1\ 4\ 5\ 3\ 2) = (1\ 5\ 4) = R_7^2, \\
&T_0^3 S_6 \leftrightarrow (1\ 2\ 3\ 4\ 5)(1\ 5\ 4) = (2\ 3\ 4) = R_{10}, \\
&T_0^4 S_6 \leftrightarrow (1\ 2\ 3\ 4\ 5)(2\ 3\ 4) = (5\ 1\ 2\ 4\ 3) = T_1^2.
\end{aligned}$$

Now, $\phi^j_{\mu\nu}$ in Eq. (3.114) is the combinations of $\Phi^{(t)}_{\mu\nu}$ with the same subscripts μ and ν such that the calculation of $\phi^j_{\mu\nu}$ depends upon the representation matrices of only four elements, including the identity:

$$\begin{aligned}
&E, \quad S_{12} = R(\boldsymbol{e}_y, \pi), \quad S_1 = R(0, 2\theta_4, \pi) = R(\boldsymbol{e}_y, \theta_1)R(\boldsymbol{e}_z, \pi),\\
&S_6 = R(\pi/5, 2\theta_5, 4\pi/5) = R(\boldsymbol{e}_z, \pi - 4\pi/5)R(\boldsymbol{e}_y, \pi - \theta_1)R(\boldsymbol{e}_z, 4\pi/5),\\
&\phi^j_{\mu\nu} = \frac{m_j}{g}\left[D^j_{\mu\nu}(E)^* E + D^j_{\mu\nu}(S_{12})^* S_{12} + D^j_{\mu\nu}(S_1)^* S_1 + D^j_{\mu\nu}(S_6)^* S_6 + \cdots\right]\\
&\quad = \frac{m_j}{g}\left[\delta_{\mu\nu}\Phi^{(1)}_{\mu\mu} + \delta_{\mu\overline{\nu}}D^j_{\mu\overline{\mu}}(S_{12})^* \Phi^{(2)}_{\mu\overline{\mu}} + D^j_{\mu\nu}(S_1)^* \Phi^{(3)}_{\mu\nu} + D^j_{\mu\nu}(S_6)^* \Phi^{(4)}_{\mu\nu}\right].
\end{aligned} \tag{5.104}$$

The calculation is greatly simplified.

For the representation $D^A(\mathbf{I})$, $D^A(R) = 1$ for each $R \in \mathbf{I}$ and

$$\phi^A_{00} = \frac{1}{60}\sum_{t=1}^{4} \Phi^{(t)}_{00} = \frac{1}{60}\sum_{R\in\mathbf{I}} R. \tag{5.105}$$

For the representation $D^{T_1}(\mathbf{I}) = D^1(\mathbf{I})$,

$$D^{T_1}_{\mu\nu}(S_{12}) = (-1)^{1+\mu}\delta_{\mu\overline{\nu}}, \qquad D^{T_1}(S_1) = \frac{1}{\sqrt{5}}\begin{pmatrix} -p^{-1} & -\sqrt{2} & -p \\ -\sqrt{2} & 1 & \sqrt{2} \\ -p & \sqrt{2} & -p^{-1} \end{pmatrix},$$

$$D^{T_1}(S_6) = \frac{1}{\sqrt{5}}\begin{pmatrix} -p & \sqrt{2}\eta^{-2} & -\eta p^{-1} \\ \sqrt{2}\eta^2 & -1 & -\sqrt{2}\eta^{-2} \\ -\eta^{-1}p^{-1} & -\sqrt{2}\eta^2 & -p \end{pmatrix}.$$

$$\begin{aligned}
\phi^{T_1}_{11} &= \frac{1}{20\sqrt{5}}\left[\sqrt{5}\Phi^{(1)}_{11} - p^{-1}\Phi^{(3)}_{11} - p\Phi^{(4)}_{11}\right],\\
\phi^{T_1}_{01} &= \frac{1}{20\sqrt{5}}\left[-\sqrt{2}\Phi^{(3)}_{01} + \sqrt{2}\eta^{-2}\Phi^{(4)}_{01}\right],\\
\phi^{T_1}_{\overline{1}1} &= \frac{1}{20\sqrt{5}}\left[\sqrt{5}\Phi^{(2)}_{\overline{1}1} - p\Phi^{(3)}_{\overline{1}1} - \eta p^{-1}\Phi^{(4)}_{\overline{1}1}\right],\\
\phi^{T_1}_{10} &= \frac{1}{20\sqrt{5}}\left[-\sqrt{2}\Phi^{(3)}_{10} + \sqrt{2}\eta^2\Phi^{(4)}_{10}\right],\\
\phi^{T_1}_{00} &= \frac{1}{20\sqrt{5}}\left[\sqrt{5}\Phi^{(1)}_{00} - \sqrt{5}\Phi^{(2)}_{00} + \Phi^{(3)}_{\overline{1}0} - \Phi^{(4)}_{00}\right],\\
\phi^{T_1}_{0\overline{1}} &= \frac{1}{20\sqrt{5}}\left[\sqrt{2}\Phi^{(3)}_{0\overline{1}} - \sqrt{2}\eta^{-2}\Phi^{(4)}_{0\overline{1}}\right],\\
\phi^{T_1}_{1\overline{1}} &= \frac{1}{20\sqrt{5}}\left[\sqrt{5}\Phi^{(2)}_{1\overline{1}} - p\Phi^{(3)}_{1\overline{1}} - \eta^{-1}p^{-1}\Phi^{(4)}_{1\overline{1}}\right],\\
\phi^{T_1}_{0\overline{1}} &= \frac{1}{20\sqrt{5}}\left[\sqrt{2}\Phi^{(3)}_{0\overline{1}} - \sqrt{2}\eta^2\Phi^{(4)}_{0\overline{1}}\right],\\
\phi^{T_1}_{\overline{1}\overline{1}} &= \frac{1}{20\sqrt{5}}\left[\sqrt{5}\Phi^{(1)}_{\overline{1}\overline{1}} - p^{-1}\Phi^{(3)}_{\overline{1}\overline{1}} - p\Phi^{(4)}_{\overline{1}\overline{1}}\right].
\end{aligned} \tag{5.106}$$

For the representation $D^H(\mathbf{I}) = D^2(\mathbf{I})$,

$$D^H_{\mu\nu}(S_{12}) = (-1)^\mu \delta_{\mu\overline{\nu}}, \qquad D^H(S_1) = \frac{1}{5}\begin{pmatrix} p^{-2} & 2p^{-1} & \sqrt{6} & 2p & p^2 \\ 2p^{-1} & p^2 & -\sqrt{6} & -p^{-2} & -2p \\ \sqrt{6} & -\sqrt{6} & -1 & \sqrt{6} & \sqrt{6} \\ 2p & -p^{-2} & \sqrt{6} & p^2 & -2p^{-1} \\ p^2 & -2p & \sqrt{6} & -2p^{-1} & p^{-2} \end{pmatrix},$$

$$D^H(S_6) = \frac{1}{5}\begin{pmatrix} p^2 & -2\eta^{-2}p & \eta\sqrt{6} & -2\eta^{-1}p^{-1} & \eta^2 p^{-2} \\ -2\eta^2 p & p^{-2} & -\sqrt{6}\eta^{-2} & -\eta p^2 & 2\eta^{-1}p^{-1} \\ \sqrt{6}\eta^{-1} & -\sqrt{6}\eta^2 & -1 & \sqrt{6}\eta^{-2} & \sqrt{6}\eta \\ -2\eta p^{-1} & -\eta^{-1}p^2 & \sqrt{6}\eta^2 & p^{-2} & 2\eta^{-2}p \\ \eta^{-2}p^{-2} & 2\eta p^{-1} & \sqrt{6}\eta^{-1} & 2\eta^2 p & p^2 \end{pmatrix},$$

one has

$$\begin{aligned}
\phi^H_{22} &= \frac{1}{60}\left[5\Phi^{(1)}_{22} + p^{-2}\Phi^{(3)}_{22} + p^2\Phi^{(4)}_{22}\right], \\
\phi^H_{12} &= \frac{1}{60}\left[2p^{-1}\Phi^{(3)}_{12} - 2\eta^{-2}p\Phi^{(4)}_{12}\right], \\
\phi^H_{02} &= \frac{1}{60}\left[\sqrt{6}\Phi^{(3)}_{02} + \sqrt{6}\eta\Phi^{(4)}_{02}\right], \\
\phi^H_{\overline{1}2} &= \frac{1}{60}\left[2p\Phi^{(3)}_{\overline{1}2} - 2\eta^{-1}p^{-1}\Phi^{(4)}_{\overline{1}2}\right], \\
\phi^H_{\overline{2}2} &= \frac{1}{60}\left[5\Phi^{(2)}_{\overline{2}2} + p^2\Phi^{(3)}_{\overline{2}2} + \eta^2p^{-2}\Phi^{(4)}_{\overline{2}2}\right], \\
\phi^H_{21} &= \frac{1}{60}\left[2p^{-1}\Phi^{(3)}_{21} - 2\eta^2p\Phi^{(4)}_{21}\right], \\
\phi^H_{11} &= \frac{1}{60}\left[5\Phi^{(5)}_{11} + p^2\Phi^{(3)}_{11} + p^{-2}\Phi^{(4)}_{11}\right], \\
\phi^H_{01} &= \frac{1}{60}\left[-\sqrt{6}\Phi^{(3)}_{01} - \sqrt{6}\eta^{-2}\Phi^{(4)}_{01}\right], \\
\phi^H_{\overline{1}1} &= \frac{1}{60}\left[-5\Phi^{(2)}_{\overline{1}1} - p^{-2}\Phi^{(3)}_{\overline{1}1} - \eta p^2\Phi^{(4)}_{\overline{1}1}\right], \\
\phi^H_{\overline{2}1} &= \frac{1}{60}\left[-2p\Phi^{(3)}_{\overline{2}1} + 2\eta^{-1}p^{-1}\Phi^{(4)}_{\overline{2}1}\right], \\
\phi^H_{20} &= \frac{1}{60}\left[\sqrt{6}\Phi^{(3)}_{20} + \sqrt{6}\eta^{-1}\Phi^{(4)}_{20}\right], \\
\phi^H_{10} &= \frac{1}{60}\left[-\sqrt{6}\Phi^{(3)}_{10} - \sqrt{6}\eta^2\Phi^{(4)}_{10}\right], \\
\phi^H_{00} &= \frac{1}{60}\left[5\Phi^{(1)}_{00} + 5\Phi^{(2)}_{00} - \Phi^{(3)}_{00} - \Phi^{(4)}_{00}\right], \\
\phi^H_{\overline{1}0} &= \frac{1}{60}\left[\sqrt{6}\Phi^{(3)}_{\overline{1}0} + \sqrt{6}\eta^{-2}\Phi^{(4)}_{\overline{1}0}\right], \\
\phi^H_{\overline{2}0} &= \frac{1}{60}\left[\sqrt{6}\Phi^{(3)}_{\overline{2}0} + \sqrt{6}\eta\Phi^{(4)}_{\overline{2}0}\right],
\end{aligned}$$

$$
\begin{aligned}
\phi^H_{2\bar{1}} &= \frac{1}{60}\left[2p\varPhi^{(3)}_{2\bar{1}} - 2\eta p^{-1}\varPhi^{(4)}_{2\bar{1}}\right], \\
\phi^H_{1\bar{1}} &= \frac{1}{60}\left[-5\varPhi^{(2)}_{1\bar{1}} - p^{-2}\varPhi^{(3)}_{1\bar{1}} - \eta^{-1}p^{2}\varPhi^{(4)}_{1\bar{1}}\right], \\
\phi^H_{0\bar{1}} &= \frac{1}{60}\left[\sqrt{6}\varPhi^{(3)}_{0\bar{1}} + \sqrt{6}\eta^{2}\varPhi^{(4)}_{0\bar{1}}\right], \\
\phi^H_{\overline{11}} &= \frac{1}{60}\left[5\varPhi^{(1)}_{\overline{11}} + p^{2}\varPhi^{(3)}_{\overline{11}} + p^{-2}\varPhi^{(4)}_{\overline{11}}\right], \\
\phi^H_{\overline{21}} &= \frac{1}{60}\left[-2p^{-1}\varPhi^{(3)}_{\overline{21}} + 2\eta^{-2}p\varPhi^{(4)}_{\overline{21}}\right], \\
\phi^H_{2\bar{2}} &= \frac{1}{60}\left[5\varPhi^{(2)}_{2\bar{2}} + p^{2}\varPhi^{(3)}_{2\bar{2}} + \eta^{-2}p^{-2}\varPhi^{(4)}_{2\bar{2}}\right], \\
\phi^H_{1\bar{2}} &= \frac{1}{60}\left[-2p\varPhi^{(3)}_{1\bar{2}} + 2\eta p^{-1}\varPhi^{(4)}_{1\bar{2}}\right], \\
\phi^H_{0\bar{2}} &= \frac{1}{60}\left[\sqrt{6}\varPhi^{(3)}_{0\bar{2}} + \sqrt{6}\eta^{-1}\varPhi^{(4)}_{0\bar{2}}\right]; \\
\phi^H_{\overline{12}} &= \frac{1}{60}\left[-2p^{-1}\varPhi^{(3)}_{\overline{12}} + 2\eta^{2}p\varPhi^{(4)}_{\overline{12}}\right], \\
\phi^H_{\overline{22}} &= \frac{1}{60}\left[5\varPhi^{(1)}_{\overline{22}} + p^{-2}\varPhi^{(3)}_{\overline{22}} + p^{2}\varPhi^{(4)}_{\overline{22}}\right],
\end{aligned}
\tag{5.107}
$$

The representation matrices $D^{T_2}(R)$ and $D^G(R)$ with $R \in \mathrm{I}$ are calculated by $X^{-1}D^3(R)X = D^{T_2}(R) \oplus D^G(R)$:

$$
\begin{aligned}
D^{T_2}_{22}(R) &= \sqrt{3/5}D^3_{22}(R)\sqrt{3/5} + \sqrt{2/5}D^3_{\bar{3}2}(R)\sqrt{3/5} \\
&\quad + \sqrt{3/5}D^3_{2\bar{3}}(R)\sqrt{2/5} + \sqrt{2/5}D^3_{\overline{33}}(R)\sqrt{2/5}, \\
D^{T_2}_{\overline{22}}(R) &= (-\sqrt{2/5})D^3_{33}(R)(-\sqrt{2/5}) + \sqrt{3/5}D^3_{\bar{2}3}(R)(-\sqrt{2/5}) \\
&\quad + (-\sqrt{2/5})D^3_{3\bar{2}}(R)\sqrt{3/5} + \sqrt{3/5}D^3_{\overline{22}}(R)\sqrt{3/5}, \\
D^{T_2}_{2\bar{2}}(R) &= \sqrt{3/5}D^3_{23}(R)(-\sqrt{2/5}) + \sqrt{2/5}D^3_{\bar{3}3}(R)(-\sqrt{2/5}) \\
&\quad + \sqrt{3/5}D^3_{2\bar{2}}(R)\sqrt{3/5} + \sqrt{2/5}D^3_{\overline{32}}(R)\sqrt{3/5}, \\
D^{T_2}_{\bar{2}2}(R) &= (-\sqrt{2/5})D^3_{32}(R)\sqrt{3/5} + \sqrt{3/5}D^3_{\bar{2}2}(R)\sqrt{3/5} \\
&\quad + (-\sqrt{2/5})D^3_{3\bar{3}}(R)\sqrt{2/5} + \sqrt{3/5}D^3_{\overline{23}}(R)\sqrt{2/5}, \\
D^{T_2}_{00}(R) &= D^3_{00}(R), \\
D^{T_2}_{02}(R) &= D^3_{02}(R)\sqrt{3/5} + D^3_{0\bar{3}}(R)\sqrt{2/5}, \\
D^{T_2}_{0\bar{2}}(R) &= D^3_{03}(R)(-\sqrt{2/5}) + D^3_{0\bar{2}}(R)\sqrt{3/5}, \\
D^{T_2}_{20}(R) &= \sqrt{3/5}D^3_{20}(R) + \sqrt{2/5}D^3_{\bar{3}0}(R), \\
D^{T_2}_{\bar{2}0}(R) &= (-\sqrt{2/5})D^3_{30}(R) + \sqrt{3/5}D^3_{\bar{2}0}(R).
\end{aligned}
\tag{5.108}
$$

Thus,

$$D^{T_2}_{\mu\nu}(S_{12}) = -\delta_{\mu\overline{\nu}}, \qquad D^{T_2}(S_1) = \sqrt{\frac{1}{5}}\begin{pmatrix} -p & \sqrt{2} & p^{-1} \\ \sqrt{2} & -1 & \sqrt{2} \\ p^{-1} & \sqrt{2} & -p \end{pmatrix},$$

$$D^{T_2}(S_6) = \sqrt{\frac{1}{5}}\begin{pmatrix} -p^{-1} & -\sqrt{2}\eta & \eta^2 p \\ -\sqrt{2}\eta^{-1} & 1 & -\sqrt{2}\eta \\ \eta^{-2}p & -\sqrt{2}\eta^{-1} & -p^{-1} \end{pmatrix},$$

$$\begin{aligned}
\phi^{T_2}_{22} &= \frac{1}{20\sqrt{5}}\left[\sqrt{5}\,\Phi^{(1)}_{22} - p\Phi^{(3)}_{22} - p^{-1}\,\Phi^{(4)}_{22}\right], \\
\phi^{T_2}_{0\overline{2}} &= \frac{1}{20\sqrt{5}}\left[\sqrt{2}\,\Phi^{(3)}_{02} - \sqrt{2}\eta\,\Phi^{(4)}_{02}\right], \\
\phi^{T_2}_{\overline{2}2} &= \frac{1}{20\sqrt{5}}\left[-\sqrt{5}\,\Phi^{(2)}_{\overline{2}2} + p^{-1}\,\Phi^{(3)}_{\overline{2}2} + \eta^2 p\Phi^{(4)}_{\overline{2}2}\right], \\
\phi^{T_2}_{20} &= \frac{1}{20\sqrt{5}}\left[\sqrt{2}\,\Phi^{(3)}_{20} - \sqrt{2}\eta^{-1}\,\Phi^{(4)}_{20}\right], \\
\phi^{T_2}_{00} &= \frac{1}{20\sqrt{5}}\left[\sqrt{5}\,\Phi^{(1)}_{00} - \sqrt{5}\,\Phi^{(2)}_{00} - \Phi^{(3)}_{00} + \Phi^{(4)}_{00}\right], \qquad (5.109) \\
\phi^{T_2}_{\overline{2}0} &= \frac{1}{20\sqrt{5}}\left[\sqrt{2}\,\Phi^{(3)}_{\overline{2}0} - \sqrt{2}\eta\,\Phi^{(4)}_{\overline{2}0}\right], \\
\phi^{T_2}_{2\overline{2}} &= \frac{1}{20\sqrt{5}}\left[-\sqrt{5}\,\Phi^{(2)}_{2\overline{2}} + p^{-1}\,\Phi^{(3)}_{2\overline{2}} + \eta^{-2}p\Phi^{(4)}_{2\overline{2}}\right], \\
\phi^{T_2}_{0\overline{2}} &= \frac{1}{20\sqrt{5}}\left[\sqrt{2}\,\Phi^{(3)}_{0\overline{2}} - \sqrt{2}\eta^{-1}\,\Phi^{(4)}_{0\overline{2}}\right], \\
\phi^{T_2}_{\overline{2}\,\overline{2}} &= \frac{1}{20\sqrt{5}}\left[\sqrt{5}\,\Phi^{(1)}_{\overline{2}\,\overline{2}} - p\Phi^{(3)}_{\overline{2}\,\overline{2}} - p^{-1}\,\Phi^{(4)}_{\overline{2}\,\overline{2}}\right].
\end{aligned}$$

$$\begin{aligned}
D^{G}_{22}(R) &= (-\sqrt{2/5})D^3_{22}(R)(-\sqrt{2/5}) + \sqrt{3/5}D^3_{\overline{3}2}(R)(-\sqrt{2/5}) \\
&\quad + (-\sqrt{2/5})D^3_{2\overline{3}}(R)\sqrt{3/5} + \sqrt{3/5}D^3_{\overline{3}\overline{3}}(R)\sqrt{3/5}, \\
D^{G}_{\overline{2}\overline{2}}(R) &= \sqrt{3/5}D^3_{33}(R)\sqrt{3/5} + \sqrt{2/5}D^3_{\overline{2}3}(R)\sqrt{3/5} \\
&\quad + \sqrt{3/5}D^3_{3\overline{2}}(R)\sqrt{2/5} + \sqrt{2/5}D^3_{\overline{2}\overline{2}}(R)\sqrt{2/5}, \\
D^{G}_{2\overline{2}}(R) &= (-\sqrt{2/5})D^3_{23}(R)\sqrt{3/5} + \sqrt{3/5}D^3_{\overline{3}3}(R)\sqrt{3/5} \\
&\quad + (-\sqrt{2/5})D^3_{2\overline{2}}(R)\sqrt{2/5} + \sqrt{3/5}D^3_{\overline{3}\overline{2}}(R)\sqrt{2/5}, \\
D^{G}_{\overline{2}2}(R) &= \sqrt{3/5}D^3_{32}(R)(-\sqrt{2/5}) + \sqrt{2/5}D^3_{\overline{2}2}(R)(-\sqrt{2/5}) \\
&\quad + \sqrt{3/5}D^3_{3\overline{3}}(R)\sqrt{3/5} + \sqrt{2/5}D^3_{\overline{2}\overline{3}}(R)\sqrt{3/5}, \\
D^{G}_{21}(R) &= (-\sqrt{2/5})D^3_{21}(R) + \sqrt{3/5}D^3_{\overline{3}1}(R), \\
D^{G}_{12}(R) &= D^3_{12}(R)(-\sqrt{2/5}) + D^3_{1\overline{3}}(R)\sqrt{3/5}, \\
D^{G}_{1\overline{2}}(R) &= D^3_{13}(R)\sqrt{3/5} + D^3_{1\overline{2}}(R)\sqrt{2/5}, \\
D^{G}_{\overline{2}1}(R) &= \sqrt{3/5}D^3_{31}(R) + \sqrt{2/5}D^3_{\overline{2}1}(R),
\end{aligned}$$

$$
\begin{aligned}
&D^G_{2\overline{1}}(R) = (-\sqrt{2/5})D^3_{2\overline{1}}(R) + \sqrt{3/5}D^3_{3\overline{1}}(R),\\
&D^G_{\overline{1}2}(R) = D^3_{\overline{1}2}(R)(-\sqrt{2/5}) + D^3_{\overline{1}3}(R)\sqrt{3/5},\\
&D^G_{\overline{1}\,\overline{2}}(R) = D^3_{\overline{1}3}(R)\sqrt{3/5} + D^3_{\overline{1}\,\overline{2}}(R)\sqrt{2/5},\\
&D^G_{\overline{2}\,\overline{1}}(R) = \sqrt{3/5}D^3_{3\overline{1}}(R) + \sqrt{2/5}D^3_{\overline{2}\,\overline{1}}(R),\\
&D^G_{11}(R) = D^3_{11}(R),\\
&D^G_{\overline{1}\,\overline{1}}(R) = D^3_{\overline{1}\overline{1}}(R),\\
&D^G_{1\overline{1}}(R) = D^3_{1\overline{1}}(R),\\
&D^G_{\overline{1}1}(R) = D^3_{\overline{1}1}(R).
\end{aligned}
\tag{5.110}
$$

Thus,

$$
D^G_{\mu\nu}(S_{12}) = \delta_{\mu\overline{\nu}}, \qquad D^G(S_1) = \sqrt{\frac{1}{5}}\begin{pmatrix} -1 & -p & -p^{-1} & 1 \\ -p & 1 & -1 & -p^{-1} \\ -p^{-1} & -1 & 1 & -p \\ 1 & -p^{-1} & -p & -1 \end{pmatrix},
$$

$$
D^G(S_6) = \sqrt{\frac{1}{5}}\begin{pmatrix} 1 & -\eta^{-2}p^{-1} & -\eta^{-1}p & -\eta^2 \\ -\eta^2 p^{-1} & -1 & \eta & -\eta^{-1}p \\ -\eta p & \eta^{-1} & -1 & -\eta^{-2}p^{-1} \\ -\eta^{-2} & -\eta p & -\eta^2 p^{-1} & 1 \end{pmatrix}.
$$

$$
\begin{aligned}
\phi^G_{22} &= \frac{1}{15\sqrt{5}}\left[\sqrt{5}\,\Phi^{(1)}_{22} - \Phi^{(3)}_{22} + \Phi^{(4)}_{22}\right],\\
\phi^G_{12} &= \frac{1}{15\sqrt{5}}\left[-p\Phi^{(3)}_{12} - \eta^{-2}p^{-1}\,\Phi^{(4)}_{12}\right],\\
\phi^G_{\overline{1}2} &= \frac{1}{15\sqrt{5}}\left[-p^{-1}\,\Phi^{(3)}_{\overline{1}2} - \eta^{-1}p\,\Phi^{(4)}_{\overline{1}2}\right],\\
\phi^G_{\overline{2}2} &= \frac{1}{15\sqrt{5}}\left[\sqrt{5}\,\Phi^{(2)}_{\overline{2}2} + \Phi^{(3)}_{\overline{2}2} - \eta^2\,\Phi^{(4)}_{\overline{2}2}\right],\\
\phi^G_{21} &= \frac{1}{15\sqrt{5}}\left[-p\Phi^{(3)}_{21} - \eta^2 p^{-1}\,\Phi^{(4)}_{21}\right],\\
\phi^G_{11} &= \frac{1}{15\sqrt{5}}\left[\sqrt{5}\,\Phi^{(1)}_{11} + \Phi^{(3)}_{11} - \Phi^{(4)}_{11}\right],\\
\phi^G_{\overline{1}1} &= \frac{1}{15\sqrt{5}}\left[\sqrt{5}\,\Phi^{(2)}_{\overline{1}1} - \Phi^{(3)}_{\overline{1}1} + \eta\,\Phi^{(4)}_{\overline{1}1}\right],\\
\phi^G_{\overline{2}1} &= \frac{1}{15\sqrt{5}}\left[-p^{-1}\,\Phi^{(3)}_{\overline{2}1} - \eta^{-1}p\,\Phi^{(4)}_{\overline{2}1}\right],
\end{aligned}
$$

$$\begin{aligned}
\phi^G_{2\overline{1}} &= \frac{1}{15\sqrt{5}}\left[-p^{-1}\varPhi^{(3)}_{2\overline{1}} - \eta p \varPhi^{(4)}_{2\overline{1}}\right], \\
\phi^G_{1\overline{1}} &= \frac{1}{15\sqrt{5}}\left[\sqrt{5}\varPhi^{(2)}_{11} - \varPhi^{(3)}_{1\overline{1}} + \eta^{-1}\varPhi^{(4)}_{1\overline{1}}\right], \\
\phi^G_{\overline{1}\,\overline{1}} &= \frac{1}{15\sqrt{5}}\left[\sqrt{5}\varPhi^{(1)}_{\overline{1}\,\overline{1}} + \varPhi^{(3)}_{\overline{1}\,\overline{1}} - \varPhi^{(4)}_{\overline{1}\,\overline{1}}\right], \\
\phi^G_{\overline{2}\,\overline{1}} &= \frac{1}{15\sqrt{5}}\left[-p\varPhi^{(3)}_{\overline{2}\,\overline{1}} - \eta^{-2}p^{-1}\varPhi^{(4)}_{\overline{2}\,\overline{1}}\right], \\
\phi^G_{2\overline{2}} &= \frac{1}{15\sqrt{5}}\left[\sqrt{5}\varPhi^{(2)}_{2\overline{2}} + \varPhi^{(3)}_{2\overline{2}} - \eta^{-2}\varPhi^{(4)}_{2\overline{2}}\right], \\
\phi^G_{1\overline{2}} &= \frac{1}{15\sqrt{5}}\left[-p^{-1}\varPhi^{(3)}_{1\overline{2}} - \eta p \varPhi^{(4)}_{1\overline{2}}\right], \\
\phi^G_{\overline{1}\,\overline{2}} &= \frac{1}{15\sqrt{5}}\left[-p\varPhi^{(3)}_{\overline{1}\,\overline{2}} - \eta^{2}p^{-1}\varPhi^{(4)}_{\overline{1}\,\overline{2}}\right], \\
\phi^G_{\overline{2}\,\overline{2}} &= \frac{1}{15\sqrt{5}}\left[\sqrt{5}\varPhi^{(1)}_{\overline{2}\,\overline{2}} - \varPhi^{(3)}_{\overline{2}\,\overline{2}} + \varPhi^{(4)}_{\overline{2}\,\overline{2}}\right].
\end{aligned} \tag{5.111}$$

Due to Eq. (3.114) **the representation matric $D^j(R)$ of any element $R \in \mathbf{I}$ can be found from the irreducible bases $\phi^j_{\mu\nu}$ of I.**

5.5 The Lie Theorems

The Lie theorems are the fundamental theorems in the theory of Lie groups which characterize the importance of generators.

5.5.1 *The First Lie Theorem*

Theorem 5.1 The representation of a connected Lie group G is completely determined by its generators.

Proof This theorem solves the problem that **the property of a connected Lie group G is determined completely by the infinitesimal elements.** We will prove that the representation matrix $D(R)$ of G can be calculated from the generators of the representation $D(\mathrm{G})$. It is the strict version for "any element R in a connected Lie group is equal to the product of infinite number of infinitesimal elements".

Let $RS = T$ and $t_\rho = f_\rho(r; s)$. In the representation $D(\mathrm{G})$ one has $D(R) = D(T)D(S^{-1})$. Make the partial derivative on both sides of the equation with respect to the parameters r_λ with S fixed,

$$\frac{\partial D(R)}{\partial r_\lambda} = \frac{\partial D(T)}{\partial r_\lambda} D(S^{-1}) = \sum_\rho \frac{\partial D(T)}{\partial t_\rho} \frac{\partial f_\rho(r;s)}{\partial r_\lambda} D(S^{-1}).$$

Then, taking $S = R^{-1}$ and $T = E$, one obtains from Eq. (5.23)

$$\frac{\partial D(R)}{\partial r_\lambda} = -\mathrm{i}\left\{\sum_\tau I_\tau S_{\tau\lambda}(r)\right\} D(R), \tag{5.112}$$

where

$$S_{\tau\lambda}(r) \equiv \left.\frac{\partial f_\tau(r;s)}{\partial r_\lambda}\right|_{s=\overline{r}} = \left.\frac{\partial f_\tau(r';\overline{r})}{\partial r'_\lambda}\right|_{r'=r}. \tag{5.113}$$

$S_{\tau\lambda}(r)$ is independent of the representation $D(\mathrm{G})$, but depends on the choice of parameters of the Lie group G. The determinant of $S(r)$ is nothing but the Jacobi determinant in the integral transformation (see Eq. (5.39)) so that $S(r)$ is non-singular. Let $\overline{S}(r)$ denote the inverse matrix of $S(r)$:

$$\sum_\rho \overline{S}_{\tau\rho}(r) S_{\rho\lambda}(r) = \delta_{\tau\lambda}. \tag{5.114}$$

Letting $R = E$ in Eq. (5.113), one has

$$S_{\tau\lambda}(E) = \delta_{\tau\lambda}, \tag{5.115}$$

and Eq. (5.112) reduces to Eq. (5.23).

Equation (5.112) is solved with the m^2 boundary conditions

$$D_{\mu\nu}(R)|_{R=E} = \delta_{\mu\nu}, \tag{5.116}$$

where m is the dimension of the representation. Since $D(R)$ exists, one may choose a convenient path to integrate Eq. (5.112) such that only one variable changes in each segment of the path. First, let all $r_\rho = 0$ except for r_1 which changes from 0. Equation (5.112) with the boundary condition (5.116) is a differential equation of first order and its solution $D(r_1,\ 0,\ \ldots,\ 0)$ is an exponential function. Then, fix r_1 and let the remaining $r_\rho = 0$ except for r_2 which changes from 0. Equation (5.112) with the boundary condition $D(r_1,\ 0,\ \ldots,\ 0)$ is a differential equation of first order again and its solution $D(r_1,\ r_2,\ 0,\ \ldots,\ 0)$ is the product of two exponential functions. In this way one can obtain the solution $D(R)$, which is a product of a few exponential functions. □

Corollary 5.1.1 Two representations of a connected Lie group are equivalent if and only if their generators are related by a common similarity transformation

$$I_\rho^{(1)} = X^{-1} I_\rho^{(2)} X.$$

Corollary 5.1.2 A representation of a connected Lie group is reducible if and only if its representation space contains a nontrivial invariant subspace with respect to its all generators.

Corollary 5.1.3 If $I_\rho^{(1)}$ and $I_\rho^{(2)}$ are the generators of two inequivalent irreducible representations $D^{(1)}(\mathrm{G})$ and $D^{(2)}(\mathrm{G})$ of a connected Lie group G and X satisfies

$$X I_\rho^{(1)} = I_\rho^{(2)} X$$

for all generators, then $X = 0$.

Corollary 5.1.4 If X commutes with all generators I_ρ of an irreducible representation $D(\mathrm{G})$ of a connected Lie group G, then X is a constant matrix.

Due to Theorem 5.1, the corollaries are obvious. For the mixed Lie group, one can choose one element in each connected piece in the group space. The corollaries hold if, in addition to the generators, the representation matrices of those chosen elements also satisfy the conditions.

5.5.2 *The Second Lie Theorem*

Theorem 5.2 The generators in any representation of a Lie group G with order g satisfy the common commutative relations

$$I_\rho I_\sigma - I_\sigma I_\rho = \mathrm{i} \sum_{\tau=1}^{g} C_{\rho\sigma}{}^{\tau} I_\tau, \tag{5.117}$$

where $C_{\rho\sigma}{}^{\tau}$ are called the **structure constants** of G. The structure constants are real and independent of the representations. Conversely, if g matrices satisfy the commutative relations (5.117), they are the generators of a representation of G.

Proof According to the theory of differential equations, two solutions in integrating Eq. (5.112) with the boundary condition (5.116) through two different paths, which can be deformed in the group space, are the same if and only if

$$\frac{\partial^2 D(R)}{\partial r_\omega \partial r_\lambda} = \frac{\partial^2 D(R)}{\partial r_\lambda \partial r_\omega}. \tag{5.118}$$

Differentiating Eq. (5.112) with respect to r_ω, one has

$$\frac{\partial^2 D(R)}{\partial r_\omega \partial r_\lambda} = -\mathrm{i}\sum_\tau I_\tau \frac{\partial S_{\tau\lambda}(r)}{\partial r_\omega} D(R) - \sum_{\tau\eta} I_\tau I_\eta S_{\tau\lambda}(r) S_{\eta\omega}(r) D(R).$$

By interchanging ω and λ, it becomes

$$\frac{\partial^2 D(R)}{\partial r_\lambda \partial r_\omega} = -\mathrm{i}\sum_\tau I_\tau \frac{\partial S_{\tau\omega}(r)}{\partial r_\lambda} D(R) - \sum_{\tau\eta} I_\tau I_\eta S_{\tau\omega}(r) S_{\eta\lambda}(r) D(R).$$

Then, Eq. (5.118) becomes

$$\sum_{\tau\omega} \{I_\tau I_\eta - I_\eta I_\tau\} S_{\tau\omega}(r) S_{\eta\lambda}(r) = \mathrm{i}\sum_\tau I_\tau \left\{ \frac{\partial S_{\tau\lambda}(r)}{\partial r_\omega} - \frac{\partial S_{\tau\omega}(r)}{\partial r_\lambda} \right\}.$$

Right-multiplying it with $\overline{S}(r)$ (see Eq. (5.114)), one has

$$I_\rho I_\sigma - I_\sigma I_\rho = \mathrm{i}\sum_\tau I_\tau \left\{ \sum_{\omega\eta} \left(\frac{\partial S_{\tau\lambda}(r)}{\partial r_\omega} - \frac{\partial S_{\tau\omega}(r)}{\partial r_\lambda} \right) \overline{S}_{\omega\rho}(r) \overline{S}_{\lambda\sigma}(r) \right\}. \tag{5.119}$$

The left-hand side of Eq. (5.119) is independent of R, so that the quantity in the curly bracket of Eq. (5.119) does not depend on R, either. Letting $R = E$, we obtain Eq. (5.117) where

$$C_{\rho\sigma}{}^\tau = \left. \left\{ \frac{\partial S_{\tau\sigma}(r)}{r_\rho} - \frac{\partial S_{\tau\rho}(r)}{r_\sigma} \right\} \right|_{r=0}. \tag{5.120}$$

Hence, the structure constants are **real and independent of the representations.**

Conversely, if g matrices I_ρ satisfy the commutative relations (5.117), Eqs. (5.119) and (5.118) also hold, so that the solution $D(R)$ is independent of the paths. Due to Eq. (5.115), I_ρ satisfy Eq. (5.25) obviously. The remaining problem is whether the solution $D(R)$ constitute a representation of G, namely, whether $D(R)D(S) = D(T)$ if $RS = T$.

Integrating Eq. (5.112) with the boundary condition (5.116), first from E to S, one obtains $D(S)$. Then, integrating Eq. (5.112) with the boundary condition $D(S)$ from S to T, one has

$$\frac{\partial D(T)}{\partial t_\tau} = -\mathrm{i}\sum_\rho I_\rho S_{\rho\tau}(t) D(T)\,, \qquad D(T)|_{T=S} = D(S).$$

Thus, for $T = RS$,

$$\begin{aligned}
\frac{\partial D(T)}{\partial r_\sigma} &= \sum_\tau \frac{\partial D(T)}{\partial t_\tau}\frac{\partial f_\tau(r;s)}{\partial r_\sigma} \\
&= -\mathrm{i}\sum_\rho I_\rho \left\{\sum_\tau S_{\rho\tau}(t)\frac{\partial f_\tau(r;s)}{\partial r_\sigma}\right\} D(T),
\end{aligned}$$

$$\begin{aligned}
\sum_\tau S_{\rho\tau}(t)\frac{\partial f_\tau(r;s)}{\partial r_\sigma} &= \sum_\tau \left.\frac{\partial f_\rho(t;u)}{\partial t_\tau}\right|_{u=\bar{t}} \frac{\partial f_\tau(r;s)}{\partial r_\sigma} \\
&= \left.\frac{\partial f_\rho\{f(r;s);u\}}{\partial r_\sigma}\right|_{u=\bar{t}} = \left.\frac{\partial f_\rho\{r;f(s;u)\}}{\partial r_\sigma}\right|_{f(s;u)=\bar{r}} = S_{\rho\sigma}(r).
\end{aligned}$$

On the other hand, right-multiplying Eqs. (5.112) and (5.116) with $D(S)$, one has

$$\begin{aligned}
&\frac{\partial D(R)D(S)}{\partial r_\sigma} = -\mathrm{i}\sum_\rho I_\rho S_{\rho\sigma}(r) D(R)D(S) , \\
&D(R)D(S)|_{R=E} = D(S).
\end{aligned}$$

Because both $D(T)$ and $D(R)D(S)$ satisfy the same equation with the same boundary condition, they equal to each other. □

Since the generator I_ρ is the matrix form of a differential operator $I_\rho^{(0)}$ in the representation space, $I_\rho^{(0)}$ satisfies the same commutative relation

$$\left[I_\rho^{(0)},\ I_\sigma^{(0)}\right] = \mathrm{i}\sum_\tau C_{\rho\sigma}{}^\tau I_\tau^{(0)}. \tag{5.121}$$

Usually, the structure constants $C_{\rho\sigma}{}^\tau$ are not calculated from Eq. (5.120). Instead, for a given Lie group and its parameters, $C_{\rho\sigma}{}^\tau$ are calculated from the commutative relations of the generators in a known faithful representation, say the self-representation. The differential operators of the SO(3) group are the orbital angular momentum operators,

$$[L_a,\ L_b] = \mathrm{i}\sum_d \epsilon_{abd} L_d, \qquad C_{ab}{}^d = \epsilon_{abd}. \tag{5.122}$$

From Eq. (5.122) one obtains that the structure constants $C_{ab}{}^d$ of both the SO(3) group and the SU(2) group are the totally antisymmetric tensor ϵ_{abd}. The generators of every representation of the SO(3) group and the SU(2) group, including their self-representations (see Eqs. (5.10) and (5.31)), have to satisfy the common commutative relations (5.122). In quantum mechanics, the matrix forms (5.68) of the angular momentum operators are calculated from the commutative relations (see Prob. 15). This method is widely used in semi-simple Lie algebras (see Chap. 7).

5.5.3 *The Third Lie Theorem*

What conditions should the structure constants $C_{\rho\sigma}{}^{\tau}$ of a Lie group satisfy? From Eq. (5.117) and the Jacobi identity:

$$\begin{aligned}
&[[I_\rho, I_\sigma], I_\tau] + [[I_\sigma, I_\tau], I_\rho] + [[I_\tau, I_\rho], I_\sigma] \\
&= I_\rho I_\sigma I_\tau - I_\sigma I_\rho I_\tau - I_\tau I_\rho I_\sigma + I_\tau I_\sigma I_\rho + I_\sigma I_\tau I_\rho - I_\tau I_\sigma I_\rho \\
&\quad - I_\rho I_\sigma I_\tau + I_\rho I_\tau I_\sigma + I_\tau I_\rho I_\sigma - I_\rho I_\tau I_\sigma - I_\sigma I_\tau I_\rho + I_\sigma I_\rho I_\tau = 0.
\end{aligned}$$

we obtain

$$\begin{aligned}
&C_{\rho\sigma}{}^{\tau} = -C_{\sigma\rho}{}^{\tau}, \\
&\sum_\omega \left\{ C_{\rho\sigma}{}^{\omega} C_{\omega\tau}{}^{\eta} + C_{\sigma\tau}{}^{\omega} C_{\omega\rho}{}^{\eta} + C_{\tau\rho}{}^{\omega} C_{\omega\sigma}{}^{\eta} \right\} = 0.
\end{aligned} \tag{5.123}$$

Theorem 5.3 A set of constants $C_{\rho\sigma}{}^{\tau}$ can be the structure constants of a Lie group if and only if they satisfy Eq. (5.123).

The Lie groups can be classified based on this theorem (see Chap. 7). Two Lie groups with the same structure constants are said to be **locally isomorphic**. Two locally isomorphic Lie groups are not isomorphic generally. There are two typical counter-examples. SU(2) and SO(3) have the same structure constants, but globally they are only homomorphic. The two-dimensional unitary group U(2) contains a subgroup SU(2) as well as a subgroup U(1) composed by constant matrices in U(2). However, two subgroups contain two common elements $\pm\mathbf{1}$. Thus, U(2) is not isomorphic onto the group SU(2)$\otimes$ U(1), but they are locally isomorphic.

5.5.4 *Adjoint Representation of a Lie Group*

Let $RSR^{-1} = T$, where the parameters of T are the functions of parameters of S and R,

$$t_\rho = \psi_\rho(s_1, s_2, \ldots; r_1, r_2, \ldots) \equiv \psi_\rho(s; r). \tag{5.124}$$

For a faithful representation $D(G)$, one calculates the derivative of $D(R)D(S)D(R)^{-1} = D(T)$ with respect to the parameter s_j, and then, takes $s_j = 0$,

$$\begin{aligned}
&D(R) \left. \frac{\partial D(S)}{\partial s_j} D(R)^{-1} \right|_{s=0} = \sum_k \left. \frac{\partial D(T)}{\partial t_k} \right|_{t=0} \left. \frac{\partial \psi_k(s;r)}{\partial s_j} \right|_{s=0}, \\
&D(R) I_j D(R)^{-1} = \sum_k I_k D^{\text{ad}}_{kj}(R), \qquad D^{\text{ad}}_{kj}(R) = \left. \frac{\partial \psi_k(s;r)}{\partial s_j} \right|_{s=0}.
\end{aligned} \tag{5.125}$$

Similarly,

$$P_R I_j^{(0)} P_R^{-1} = \sum_k I_k^{(0)} D_{kj}^{\rm ad}(R). \tag{5.126}$$

Equation (5.125) (or Eq. (5.126)) gives a one-to-one or many-to-one correspondence between the group element R and the matrix $D^{\rm ad}(R)$, and the correspondence is left invariant under the multiplication of group elements. Thus, $D^{\rm ad}(R)$ is a representation of a Lie group, called the **adjoint representation**. The dimension of the adjoint representation is the order g of G. Every Lie group has its adjoint representation.

When R is an infinitesimal element, from Eq. (5.125) one has

$$D(R) = \mathbf{1} - {\rm i}\sum_\tau r_\tau I_\tau, \qquad D_{\sigma\rho}^{\rm ad}(R) = \delta_{\sigma\rho} - {\rm i}\sum_\tau r_\tau \left(I_\tau^{\rm ad}\right)_{\sigma\rho},$$

$${\rm i}\sum_\sigma C_{\tau\rho}{}^\sigma I_\sigma = [I_\tau,\ I_\rho] = \sum_\sigma I_\sigma \left(I_\tau^{\rm ad}\right)_{\sigma\rho}.$$

Thus, the generators of the adjoint representation $D^{\rm ad}(R)$ of a Lie group are directly related with the structure constants,

$$\left(I_\rho^{\rm ad}\right)_{\sigma\tau} = {\rm i}C_{\rho\tau}{}^\sigma. \tag{5.127}$$

It is easy to check that $\left(I_\rho^{\rm ad}\right)_{\sigma\tau}$ satisfy the commutative relation (5.117),

$$\begin{aligned}
\left[I_\rho^{\rm ad},\ I_\sigma^{\rm ad}\right]_{\eta\tau} &= \sum_\omega \left\{\left(I_\rho^{\rm ad}\right)_{\eta\omega}\left(I_\sigma^{\rm ad}\right)_{\omega\tau} - \left(I_\sigma^{\rm ad}\right)_{\eta\omega}\left(I_\rho^{\rm ad}\right)_{\omega\tau}\right\} \\
&= -\sum_\omega \left\{C_{\rho\omega}{}^\eta C_{\sigma\tau}{}^\omega - C_{\sigma\omega}{}^\eta C_{\rho\tau}{}^\omega\right\} \\
&= -\sum_\omega \left\{C_{\rho\omega}{}^\eta C_{\sigma\tau}{}^\omega + C_{\sigma\omega}{}^\eta C_{\tau\rho}{}^\omega\right\} \\
&= \sum_\omega C_{\tau\omega}{}^\eta C_{\rho\sigma}{}^\omega = {\rm i}\sum_\omega C_{\rho\sigma}{}^\omega \left(I_\omega^{\rm ad}\right)_{\eta\tau}.
\end{aligned}$$

Usually, the adjoint representation of a Lie group is calculated neither by derivative in Eq. (5.125), nor by solving the differential equation (5.112) with the generators (5.127). In fact, the adjoint representation is determined by comparing the known representation and its generators with Eqs. (5.125) and (5.127). The group SU(2) and the group SO(3) have **the same adjoint representation**:

$$\left(I_a^{\rm ad}\right)_{bd} = {\rm i}C_{ad}{}^b = -{\rm i}\epsilon_{abd} = (T_a)_{bd}, \tag{5.128}$$

which is nothing but the **self-representation** of SO(3). This conclusion can also be seen from comparing Eq. (5.32) with Eq. (5.126). Thus, for the transformation operator P_R of SO(3), equation (5.126) becomes

$$P_R L_a P_R^{-1} = \sum_{b=1}^{3} L_b R_{ba}. \tag{5.129}$$

Let $R = R(\varphi, \theta, 0) = S(\varphi, \theta)$. Denoting by $\hat{\boldsymbol{n}}(\theta, \varphi)$ the unit vector $R_{\cdot 3}$, we obtain the angular momentum operator along the direction $\hat{\boldsymbol{n}}$,

$$\boldsymbol{L} \cdot \hat{\boldsymbol{n}}(\theta, \varphi) = P_R L_3 P_R^{-1}, \qquad R = R(\varphi, \theta, 0). \tag{5.130}$$

Since $P_{R(\boldsymbol{e}_3,\gamma)}$ commutes with L_3, R in Eq. (5.130) can be replaced with $R(\varphi, \theta, \gamma)$. But it is convenient to take $\gamma = 0$. If $\psi_m^\ell(x)$ belongs to the mth row of the irreducible representation $D^\ell(SO(3))$, $\psi_m^\ell(x)$ is the eigenfunction of L_3 with the eigenvalue m, and $P_R \psi_m^\ell(x)$ is the eigenfunction of $\boldsymbol{L} \cdot \hat{\boldsymbol{n}}$ with the eigenvalue m where $R = R(\varphi, \theta, \gamma)$ and $\hat{\boldsymbol{n}} = \hat{\boldsymbol{n}}(\theta, \varphi)$. $\psi_m^\ell(x)$ and $P_R \psi_m^\ell(x)$ both are the eigenfunctions of L^2 with the eigenvalue $\ell(\ell + 1)$.

5.6 Clebsch–Gordan Coefficients of SU(2)

5.6.1 *Direct Product of Representations*

For a compound system composed by two subsystems with the spherical symmetry, the wave functions of the subsystems belong to given irreducible representations of SU(2), respectively

$$\begin{aligned} P_u \psi_\mu^j(x^{(1)}) &= \sum_{\mu'} \psi_{\mu'}^j(x^{(1)}) D_{\mu'\mu}^j(u), \\ P_u \psi_\nu^k(x^{(2)}) &= \sum_{\nu'} \psi_{\nu'}^k(x^{(2)}) D_{\nu'\nu}^k(u). \end{aligned} \tag{5.131}$$

They are the eigenfunctions of the angular momentum operators, respectively. Here, we generalize the rotation of SO(3) to that of SU(2) in order to make the calculated formulas suitable for both SO(3) and SU(2). The wave function of the compound system is the product $\psi_\mu^j(x^{(1)})\psi_\nu^k(x^{(2)})$ which transforms under $u \in$ SU(2) according to the direct product of the representations:

$$\begin{aligned} &P_u \left\{\psi_\mu^j(x^{(1)})\psi_\nu^k(x^{(2)})\right\} \\ &\quad = \sum_{\mu'\nu'} \left\{\psi_{\mu'}^j(x^{(1)})\psi_{\nu'}^k(x^{(2)})\right\} \left[D^j(u) \times D^k(u)\right]_{\mu'\nu',\mu\nu}. \end{aligned} \tag{5.132}$$

Generally, the representation $D^j(u) \times D^k(u)$ is reducible and can be decomposed as a direct sum of irreducible representations D^J of SU(2) by a similarity transformation C^{jk},

$$\left(C^{jk}\right)^{-1}\left\{D^j(u)\times D^k(u)\right\}C^{jk}=\bigoplus_J a_J D^J(u),$$
$$\chi^j(\omega)\chi^k(\omega)=\sum_J a_J\chi^J(\omega). \tag{5.133}$$

The series in Eq. (5.133) is called the **Clebsch–Gordan series**, where the multiplicity a_J can be calculated by the formula (5.46) for characters. Here, we apply the formula (5.66) directly. When $j \geqslant k$,

$$\chi^j(\omega)=\sum_{\mu=-j}^{j} \mathrm{e}^{-\mathrm{i}\mu\omega}=\sum_{\mu=-j}^{j} \mathrm{e}^{\mathrm{i}\mu\omega}=\frac{\mathrm{e}^{\mathrm{i}(j+1)\omega}-\mathrm{e}^{-\mathrm{i}j\omega}}{\mathrm{e}^{\mathrm{i}\omega}-1},$$
$$\chi^j(\omega)\chi^k(\omega)=\sum_{\mu=-k}^{k}\frac{\mathrm{e}^{\mathrm{i}(j+\mu+1)\omega}-\mathrm{e}^{-\mathrm{i}(j+\mu)\omega}}{\mathrm{e}^{\mathrm{i}\omega}-1}$$
$$=\sum_{J=j-k}^{j+k}\frac{\mathrm{e}^{\mathrm{i}(J+1)\omega}-\mathrm{e}^{-\mathrm{i}J\omega}}{\mathrm{e}^{\mathrm{i}\omega}-1}=\sum_{J=j-k}^{j+k}\chi^J(\omega).$$

When $j<k$, $j-k$ is replaced with $k-j$. Generally, one has

$$a_J=\begin{cases}1, & J=j+k,\ j+k-1,\ \ldots,\ |j-k|,\\ 0, & \text{the remaining cases.}\end{cases} \tag{5.134}$$

In the coupling of the angular momentums j and k of two subsystems, the total angular momentum J of the compound system can take $(2k+1)$ values (when $j\geqslant k$) or $(2j+1)$ values (when $j\leqslant k$) given in Eq. (5.134), where **each value of J appears once**. Three angular momentums j, k, and J satisfy the rule that each one in three is not larger than the sum of the other two and not less than their difference. In addition, the three angular momentums j, k, and J are all integers, or one is integer and the other two are half of odd integers. This rule is called the **rule of a triangle**, denoted by $\Delta(j,k,J)$.

The row of the matrix C^{jk} is enumerated by $\mu\nu$, and its column by JM, The new basis functions are combined by the matrix elements of C^{jk},

$$\Psi^J_M(x^{(1)},x^{(2)})=\sum_{\mu\nu}\left\{\psi^j_\mu(x^{(1)})\psi^k_\nu(x^{(2)})\right\}C^{jk}_{\mu\nu,JM},$$
$$P_u\Psi^J_M(x^{(1)},x^{(2)})=\sum_{M'}\Psi^J_{M'}(x^{(1)},x^{(2)})D^J_{M'M}(u), \tag{5.135}$$

where $|\mu|\leqslant j$, $|\nu|\leqslant k$, and $|M|\leqslant J$. The matrix elements $C^{jk}_{\mu\nu,JM}$ are called the **Clebsch–Gordan coefficients**, or briefly, CG coefficients. Since the coefficients are related to the addition of angular momentums, the CG coefficients are also called the **vector coupling coefficients**. Sometimes, the CG coefficients $C^{jk}_{\mu\nu,JM}$ are denoted by the Dirac symbols $\langle jk\mu\nu|jkJM\rangle$ or $\langle j\mu,k\nu|JM\rangle$.

Now, let us discuss the properties of the CG coefficients. Rewrite Eq. (5.133) in the form of generators. Since the generators of the direct product $D^j(u)\times D^k(u)$ of two representations are $I^j_a\times \mathbf{1}_{2k+1}+\mathbf{1}_{2j+1}\times I^k_a$, one has

$$\sum_\rho \left(I^j_a\right)_{\mu\rho} C^{jk}_{\rho\nu,JM}+\sum_\lambda \left(I^k_a\right)_{\nu\lambda} C^{jk}_{\mu\lambda,JM}=\sum_N C^{jk}_{\mu\nu,JN}\left(I^J_a\right)_{NM}. \tag{5.136}$$

When $a=3$, I_3 is diagonal and Eq. (5.136) becomes $(\mu+\nu)C^{jk}_{\mu\nu,JM} = MC^{jk}_{\mu\nu,JM}$. Namely,

$$C^{jk}_{\mu\nu,JM}=0, \qquad \text{when} \quad M\neq\mu+\nu. \tag{5.137}$$

In the sum of two angular momentums, **the component along the Z-axis is summed like a scalar**. Replacing I_a in Eq. (5.136) with $I_\pm = I_1\pm \mathrm{i}I_2$, one has from Eq. (5.70)

$$\Gamma^j_{\pm\mu}C^{jk}_{(\mu\mp1)\nu,JM}+\Gamma^k_{\pm\nu}C^{jk}_{\mu(\nu\mp1),JM}=C^{jk}_{\mu\nu,J(M\pm1)}\Gamma^J_{\mp M}. \tag{5.138}$$

This is the recurrence relation for calculating the CG coefficients of SU(2).

Since SU(2) is a compact Lie group, D^j as well as C^{jk} is unitary. There is one undetermined phase angle for each J in the CG coefficients (see subsection 3.6.2). Choose the phase angle such that $C^{jk}_{j(-k),J(j-k)}$ is real and positive. Taking the lower signs in Eq. (5.138) with $\mu=j$ fixed and $\nu=-k$, $(-k+1)$, ..., $(k-1)$ one by one, one obtains that the first term on the left-hand side of Eq. (5.138) vanishes and $C^{jk}_{j\nu,J(j+\nu)}$ are all real positive. In the same way, taking the upper sign in Eq. (5.138) with $\nu=-k$ fixed and $\mu=j$, $(j-1)$, ..., $-j+1$ one by one, one obtains that the second term on the left-hand side of Eq. (5.138) vanishes and $C^{jk}_{\mu(-k),J(\mu-k)}$ are all real positive. Again from Eq. (5.138), the CG coefficients for SU(2) all are real and C^{jk} is a real orthogonal matrix:

$$\begin{aligned}\sum_{\mu\nu} C^{jk}_{\mu\nu JM}C^{jk}_{\mu\nu J'M'} &= \delta_{JJ'}\delta_{MM'},\\ \sum_{JM} C^{jk}_{\mu\nu JM}C^{jk}_{\mu'\nu' JM} &= \delta_{\mu\mu'}\delta_{\nu\nu'}.\end{aligned} \tag{5.139}$$

Due to Eq. (5.137) one has

$$\begin{aligned}
&\sum_{\mu} C^{jk}_{\mu(M-\mu),JM} C^{jk}_{\mu(M-\mu),J'M} = \delta_{JJ'}, \\
&\sum_{J} C^{jk}_{\mu(M-\mu),JM} C^{jk}_{\mu'(M-\mu'),JM} = \delta_{\mu\mu'}.
\end{aligned} \tag{5.140}$$

5.6.2 *Calculation of Clebsch–Gordan Coefficients*

There have been a few softwares to calculate the Clebsch–Gordan coefficients of SU(2), say in Mathematica one has

$$C^{jk}_{\mu\nu,J(\mu+\nu)} = \text{ClebschGordan}[\{j,\mu\},\{k,\nu\},\{J,\mu+\nu\}]. \tag{5.141}$$

Thus, there is no need to calculate the CG coefficients of SU(2) in detail. We only sketch the main ideals of two methods for calculation, and analysis the symmetry of the CG coefficients. One is based on the recurrence relation (5.138), and the other is based on the group integral for Eq. (5.133). In the first method, one takes the upper sign in Eq. (5.138) with $M = J$ fixed and $\mu = j$, $(j-1)$, ..., $-j+1$ one by one. Then, the term on the right-hand side of Eq. (5.138) vanishes such that $C^{jk}_{\mu(J-\mu),JJ}$ are expressed by $C^{jk}_{j(J-j),JJ}$, which is real positive and can be calculated through the normalized condition (5.140). Then, $(J-M)$ times application of Eq. (5.138) with the lower sign, $C^{jk}_{\mu(M-\mu),JM}$ is expressed as a series of $C^{jk}_{m(J-m),JJ}$. Making a replacement of summation index m with $n = m - \mu$, one obtains the Racah form of the CG coefficients of SU(2):

$$\begin{aligned}
&C^{jk}_{\mu(M-\mu),JM} \\
&\quad = \left\{ \frac{(2J+1)(j+k-J)!(J+M)!(J-M)!(j-\mu)!(k-M+\mu)!}{(J+j+k+1)!(J+j-k)!(J-j+k)!(j+\mu)!(k+M-\mu)!} \right\}^{1/2} \\
&\qquad \times \sum_{n} \frac{(-1)^{n+j-\mu}(J+k-\mu-n)!(j+\mu+n)!}{(j-\mu-n)!(J-M-n)!n!(n+\mu+k-J)!},
\end{aligned} \tag{5.142}$$

where n runs from the maximum between 0 and $(J-k-\mu)$ to the minimum between $(j-\mu)$ and $(J-M)$. The detailed calculation can be found in §5.8 of [Ma (1993)], where the quantum parameter q is taken to be 1.

In the second method (see Chap. 17 in [Wigner (1959)]), equation (5.133) is rewritten in the parameters of Euler angles. In terms of the orthogonal relation (5.67) on the D^J function, one has

$$\begin{aligned}
&(2J+1)^{-1}C^{jk}_{\mu\nu,J(\mu+\nu)}C^{jk}_{\rho\lambda,J(\rho+\lambda)} \\
&\quad = \int dR\, D^{J}_{(\mu+\nu)(\rho+\lambda)}(R)^* D^{j}_{\mu\rho}(R) D^{k}_{\nu\lambda}(R) \\
&\quad = \frac{1}{2}\int_0^{\pi} d\beta\ \sin\beta d^{J}_{(\mu+\nu)(\rho+\lambda)}(\beta) d^{j}_{\mu\rho}(\beta) d^{k}_{\nu\lambda}(\beta).
\end{aligned} \tag{5.143}$$

Letting $\mu = \rho = j$ and $\nu = \lambda = -k$, one obtains from Eqs. (5.61) and (5.63),

$$C^{jk}_{j(-k),J(j-k)} = \left\{\frac{(2J+1)(2j)!(2k)!}{(J+j+k+1)!(j+k-J)!}\right\}^{1/2}.$$

Substituting the result back to Eq. (5.143), one obtains the Wigner form of the CG coefficients of SU(2):

$$\begin{aligned}
C^{jk}_{\mu\nu,J(\mu+\nu)} &= \Delta(j,k,J)\left\{\frac{(2J+1)(J+\mu+\nu)!(J-\mu-\nu)!}{(j+\mu)!(j-\mu)!(k+\nu)!(k-\nu)!}\right\}^{1/2} \\
&\times \sum_n \frac{(-1)^n (J+\mu-\nu-n)!(n+j+k-\mu+\nu)!}{(J-j-\nu-n)!(J-k+\mu-n)!(n+k+\nu)!(n+j-\mu)!}, \\
\Delta(j,k,J) &= \left\{\frac{(j+k-J)!(k+J-j)!(J+j-k)!}{(j+k+J+1)!}\right\}^{1/2},
\end{aligned} \tag{5.144}$$

where n runs from the maximum between $(-k-\nu)$ and $(-j+\mu)$ to the minimum between $(J-j-\nu)$ and $(J-k+\mu)$. Two forms of the CG coefficients of SU(2) are equivalent. From them one obtains the following symmetry of the Clebsch–Gordan coefficients of SU(2) where $M = \mu+\nu$,

$$\begin{aligned}
C^{jk}_{\mu\nu,JM} &= C^{kj}_{(-\nu)(-\mu)J(-M)} = (-1)^{j+k-J}\, C^{kj}_{\nu\mu JM} \\
&= (-1)^{j+k-J}\, C^{jk}_{(-\mu)(-\nu)J(-M)} \\
&= (-1)^{k-J-\mu}\left(\frac{2J+1}{2k+1}\right)^{1/2} C^{Jj}_{(-M)\mu k(-\nu)} \\
&= (-1)^{j-J+\nu}\left(\frac{2J+1}{2j+1}\right)^{1/2} C^{kJ}_{\nu(-M)j(-\mu)}.
\end{aligned} \tag{5.145}$$

Wigner defined the $3j$-symbols with more symmetry:

$$\begin{pmatrix} j & k & \ell \\ \mu & \nu & \rho \end{pmatrix} = (-1)^{j-k-\rho}(2\ell+1)^{-1/2} C^{jk}_{\mu\nu\ell(-\rho)}, \tag{5.146}$$

where $|j-k| \leqslant \ell \leqslant j+k$ and $\mu+\nu+\rho = 0$. The $3j$-symbols satisfy

$$(-1)^{j+k+\ell}\begin{pmatrix} j & k & \ell \\ \mu & \nu & \rho \end{pmatrix} = \begin{pmatrix} k & j & \ell \\ \nu & \mu & \rho \end{pmatrix} = \begin{pmatrix} j & \ell & k \\ \mu & \rho & \nu \end{pmatrix} = \begin{pmatrix} j & k & \ell \\ -\mu & -\nu & -\rho \end{pmatrix},$$

$$\sum_{\ell}(2\ell+1)\begin{pmatrix} j & k & \ell \\ \mu & -\mu-\rho & \rho \end{pmatrix}\begin{pmatrix} j & k & \ell \\ \mu' & -\mu'-\rho & \rho \end{pmatrix} = \delta_{\mu\mu'}, \tag{5.147}$$

$$\sum_{\mu}\begin{pmatrix} j & k & \ell \\ \mu & -\mu-\rho & \rho \end{pmatrix}\begin{pmatrix} j & k & \ell' \\ \mu & -\mu-\rho & \rho \end{pmatrix} = (2\ell+1)^{-1}\,\delta_{\ell\ell'},$$

and can be calculated in Mathematica:

$$\begin{pmatrix} j & k & J \\ \mu & \nu & -\mu-\nu \end{pmatrix} = \text{ThreeJSymbol}[\{j,\mu\},\{k,\nu\},\{J,-\mu-\nu\}]. \tag{5.148}$$

Equation (5.143) can be rewritten as

$$\int dR\, D^{j}_{\mu\mu'}(R) D^{k}_{\nu\nu'}(R) D^{J}_{\rho\rho'}(R) = \begin{pmatrix} j & k & J \\ \mu & \nu & \rho \end{pmatrix}\begin{pmatrix} j & k & J \\ \mu' & \nu' & \rho' \end{pmatrix}. \tag{5.149}$$

5.6.3 *Applications*

(a) *The permutation symmetry of wave functions in a two-body system*

The wave function of a two-body system with a given angular momentum is

$$\Psi^L_M(\boldsymbol{x}^{(1)},\boldsymbol{x}^{(2)}) = \sum_{m_1 m_2} C^{\ell_1\ell_2}_{m_1 m_2,LM}\psi^{\ell_1}_{m_1}(\boldsymbol{x}^{(1)})\psi^{\ell_2}_{m_2}(\boldsymbol{x}^{(2)}).$$

In the permutation of two particles, one has

$$\begin{aligned}\Psi^L_M(\boldsymbol{x}^{(2)},\boldsymbol{x}^{(1)}) &= \sum_{m_1 m_2} C^{\ell_1\ell_2}_{m_1 m_2,LM}\psi^{\ell_1}_{m_1}(\boldsymbol{x}^{(2)})\psi^{\ell_2}_{m_2}(\boldsymbol{x}^{(1)}) \\ &= \sum_{m_2 m_1} (-1)^{L-\ell_1-\ell_2} C^{\ell_2\ell_1}_{m_2 m_1,LM}\psi^{\ell_2}_{m_2}(\boldsymbol{x}^{(1)})\psi^{\ell_1}_{m_1}(\boldsymbol{x}^{(2)}).\end{aligned}$$

If $\ell_1 = \ell_2 = \ell$, then

$$\Psi^L_M(\boldsymbol{x}^{(2)},\boldsymbol{x}^{(1)}) = (-1)^{L-2\ell}\,\Psi^L_M(\boldsymbol{x}^{(1)},\boldsymbol{x}^{(2)}). \tag{5.150}$$

For example, in a system composed of two electrons with P wave ($\ell = 1$), the wave function with the total angular momentum $L = 2$ and 0 is symmetric in permutation, and that with $L = 1$ is antisymmetric.

(b) *The expansion of the Legendre function*

The product of two spherical harmonic functions of two subsystems with the same ℓ can be combined to be left invariant under rotation ($L = 0$)

$$\begin{aligned}\Phi_0^0(\hat{\boldsymbol{n}}_1,\hat{\boldsymbol{n}}_2) &= \sum_m C_{(-m)m,00}^{\ell\ell} Y_{-m}^\ell(\hat{\boldsymbol{n}}_1) Y_m^\ell(\hat{\boldsymbol{n}}_2) \\ &= \sum_m \frac{(-1)^{\ell+m}}{(2\ell+1)^{1/2}} Y_{-m}^\ell(\hat{\boldsymbol{n}}_1) Y_m^\ell(\hat{\boldsymbol{n}}_2).\end{aligned}$$

Let R rotate $\hat{\boldsymbol{n}}_1$ to Z-axis and $\hat{\boldsymbol{n}}_2$ to the XZ plane with positive X component. Denoting by θ the angle between $\hat{\boldsymbol{n}}_1$ and $\hat{\boldsymbol{n}}_2$, one has

$$\begin{aligned}Y_{-m}^\ell(\boldsymbol{e}_3) &= Y_{-m}^\ell(0,0) = \delta_{m0}\left(\frac{2\ell+1}{4\pi}\right)^{1/2}, \\ Y_0^\ell(\theta,0) &= \left(\frac{2\ell+1}{4\pi}\right)^{1/2} P_\ell(\cos\theta), \\ Y_m^\ell(\hat{\boldsymbol{n}}_1)^* &= (-1)^m Y_{-m}^\ell(\hat{\boldsymbol{n}}_1).\end{aligned}$$

Hence,

$$\Phi_0^0(\hat{\boldsymbol{n}}_1,\hat{\boldsymbol{n}}_2) = \frac{(-1)^\ell(2\ell+1)^{1/2}}{4\pi} P_\ell(\cos\theta),$$

$$\begin{aligned}P_\ell(\hat{\boldsymbol{n}}_1\cdot\hat{\boldsymbol{n}}_2) = P_\ell(\cos\theta) &= \frac{4\pi}{2\ell+1}\sum_m (-1)^m Y_{-m}^\ell(\hat{\boldsymbol{n}}_1) Y_m^\ell(\hat{\boldsymbol{n}}_2) \\ &= \frac{4\pi}{2\ell+1}\sum_m Y_m^\ell(\hat{\boldsymbol{n}}_1)^* Y_m^\ell(\hat{\boldsymbol{n}}_2).\end{aligned} \tag{5.151}$$

In the method of partial waves in quantum mechanics, a plane wave $\exp(\mathrm{i}\boldsymbol{k}\cdot\boldsymbol{r})$, which is a solution of the d'Alembert equation, can be expanded with respect to the Legendre function. Now, the expansion is written in the spherical harmonic functions

$$\begin{aligned}\exp(\mathrm{i}\boldsymbol{k}\cdot\boldsymbol{r}) = \exp(\mathrm{i}kr\cos\theta) &= \sum_{\ell=0}^\infty \mathrm{i}^\ell(2\ell+1) j_\ell(kr) P_\ell(\cos\theta) \\ &= 4\pi\sum_{\ell=0}^\infty \mathrm{i}^\ell j_\ell(kr) \sum_{m=-\ell}^{\ell} Y_m^\ell(\hat{\boldsymbol{k}})^* Y_m^\ell(\hat{\boldsymbol{r}}),\end{aligned} \tag{5.152}$$

where j_ℓ is the spherical Bessel function, and $\hat{\boldsymbol{k}}$ and $\hat{\boldsymbol{r}}$ are the unit vectors along their directions, respectively.

(c) *The Expansion of the product of two spherical harmonic functions*

Replacing $D^j(u)$ in Eq. (5.133) with the spherical harmonic function (see Eq. (5.80)), one has

$$\begin{aligned}Y_{m_1}^{\ell_1}(\hat{\boldsymbol{n}}) Y_{m_2}^{\ell_2}(\hat{\boldsymbol{n}}) = \sum_L &\left\{\frac{(2\ell_1+1)(2\ell_2+1)}{4\pi(2L+1)}\right\}^{1/2} \\ &\times C_{00,L0}^{\ell_1\ell_2} C_{m_1m_2,L(m_1+m_2)}^{\ell_1\ell_2} Y_{m_1+m_2}^L(\hat{\boldsymbol{n}}).\end{aligned} \tag{5.153}$$

Since $C_{00L0}^{\ell_1\ell_2} = 0$ when $L - \ell_1 - \ell_2$ is an odd integer, the parities of the wave functions on the two sides of Eq. (5.153) are the same. In terms of the orthogonal relations (5.139) and (5.81), one obtains

$$\begin{aligned} C_{00,L0}^{\ell_1\ell_2} Y_M^L(\hat{\boldsymbol{n}}) = &\left\{\frac{4\pi(2L+1)}{(2\ell_1+1)(2\ell_2+1)}\right\}^{1/2} \\ &\times \sum_m C_{m(M-m),LM}^{\ell_1\ell_2} Y_m^{\ell_1}(\hat{\boldsymbol{n}}) Y_{M-m}^{\ell_2}(\hat{\boldsymbol{n}}), \end{aligned} \tag{5.154}$$

$$\begin{aligned} &\int Y_M^L(\theta,\varphi)^* Y_{m_1}^{\ell_1}(\theta,\varphi) Y_{m_2}^{\ell_2}(\theta,\varphi) \sin\theta d\theta d\varphi \\ &= \left\{\frac{(2\ell_1+1)(2\ell_2+1)}{4\pi(2L+1)}\right\}^{1/2} C_{00,L0}^{\ell_1\ell_2} C_{m_1m_2,LM}^{\ell_1\ell_2}. \end{aligned} \tag{5.155}$$

5.6.4 *Sum of Three Angular Momentums*

If a compound system consists of three subsystems with the spherical symmetry, the wave function of the compound system with a given angular momentum can be obtained by combining the wave functions of three subsystems. This is a decomposition problem of the direct product of three irreducible representations of SU(2). In the decomposition, the multiplicity a_J of an irreducible representation may be larger than one, and the result will depend upon the order of combinations. For example,

$$\begin{aligned} D^1 \times D^1 \times D^1 &\simeq \{D^2 \oplus D^1 \oplus D^0\} \times D^1 \\ &\simeq \{D^3 \oplus D^2 \oplus D^1\} \oplus \{D^2 \oplus D^1 \oplus D^0\} \oplus D^1 \\ &\simeq D^3 \oplus 2D^2 \oplus 3D^1 \oplus D^0, \end{aligned}$$

where $a_2 = 2$ and $a_1 = 3$. The functional spaces belonging to the same representation D^J can be combined arbitrarily, where the combination coefficients are independent of the row number M of the representation (the magnetic quantum number).

Discuss the wave functions with the same total angular momentum J but different order of combination. One is to combine the angular momentums of the first two subsystems into J_{12}, then to combine it with the angular momentum of the third one. The other is to combine the angular momentums of the last two subsystems into J_{23}, then to combine the angular momentum of the first one with J_{23}.

$$\Psi^J_M(J_{12}) = \sum_\rho C^{J_{12}\ell}_{(M-\rho)\rho,JM} \left\{ \sum_{\mu\nu} C^{jk}_{\mu\nu,J_{12}(M-\rho)} \psi^j_\mu(\boldsymbol{x}^{(1)})\psi^k_\nu(\boldsymbol{x}^{(2)}) \right\} \psi^\ell_\rho(\boldsymbol{x}^{(3)}),$$

$$\Psi^J_M(J_{23}) = \sum_\mu C^{jJ_{23}}_{\mu(M-\mu),JM} \psi^j_\mu(\boldsymbol{x}^{(1)}) \left\{ \sum_{\nu\rho} C^{k\ell}_{\nu\rho,J_{23}(M-\mu)} \psi^k_\nu(\boldsymbol{x}^{(2)})\psi^\ell_\rho(\boldsymbol{x}^{(3)}) \right\}. \tag{5.156}$$

Two wave functions can be related by a unitary transformation X, which depends on J_{12} and J_{23}, in addition to J, j, k, and ℓ, but is independent of the magnetic quantum numbers,

$$\Psi^J_M(J_{23}) = \sum_{J_{12}} \Psi^J_M(J_{12}) X_{J_{12}J_{23}}. \tag{5.157}$$

Extracting a factor from $X_{J_{12}J_{23}}$, one uses the more symmetric coefficients, called the Racah coefficients $W[jkJ\ell; J_{12}J_{23}]$, or the $6j$-symbols,

$$\begin{aligned} X_{J_{12}J_{23}} &= \{(2J_{12}+1)(2J_{23}+1)\}^{1/2} W[jkJ\ell; J_{12}J_{23}] \\ &= (-1)^{j+k+\ell+J} \{(2J_{12}+1)(2J_{23}+1)\}^{1/2} \begin{Bmatrix} j & k & J_{12} \\ \ell & J & J_{23} \end{Bmatrix}. \end{aligned} \tag{5.158}$$

Substituting Eq. (5.156) into Eq. (5.157), one obtains

$$\begin{aligned} C^{jJ_{23}}_{\mu(M-\mu),JM}\ C^{k\ell}_{\nu\rho,J_{23}(M-\mu)} &= \sum_{J_{12}} C^{J_{12}\ell}_{(M-\rho)\rho,JM}\ C^{jk}_{\mu\nu,J_{12}(M-\rho)} \\ &\times \{(2J_{12}+1)(2J_{23}+1)\}^{1/2} W[jkJ\ell; J_{12}J_{23}]. \end{aligned} \tag{5.159}$$

In terms of the orthogonal relations (5.139), the Racah coefficients can be expressed as a product of four CG coefficients:

$$\begin{aligned} & C^{J_{12}\ell}_{(M-\rho)\rho,JM}\ \{(2J_{12}+1)(2J_{23}+1)\}^{1/2} W[jkJ\ell; J_{12}J_{23}] \\ & \quad = \sum_{\mu\nu} C^{jJ_{23}}_{\mu(M-\mu),JM}\ C^{k\ell}_{\nu\rho,J_{23}(M-\mu)} C^{jk}_{\mu\nu,J_{12}(M-\rho)}, \\ & \{(2J_{12}+1)(2J_{23}+1)\}^{1/2} W[jkJ\ell; J_{12}J_{23}]\delta_{JJ'} \\ & \quad = \sum_{\mu\nu\rho} C^{jJ_{23}}_{\mu(M-\mu),JM}\ C^{k\ell}_{\nu\rho,J_{23}(M-\mu)} C^{J_{12}\ell}_{(M-\rho)\rho,J'M}\ C^{jk}_{\mu\nu,J_{12}(M-\rho)}. \end{aligned} \tag{5.160}$$

$$\begin{aligned} \begin{Bmatrix} j_1 & j_2 & j_3 \\ \ell_1 & \ell_2 & \ell_3 \end{Bmatrix} &= \sum_{\text{all } m \text{ and } \mu} (-1)^{\ell_1+\ell_2+\ell_3+m_1+m_2+m_3} \begin{pmatrix} j_1 & j_2 & j_3 \\ \mu_1 & \mu_2 & \mu_3 \end{pmatrix} \\ &\times \begin{pmatrix} j_1 & \ell_2 & \ell_3 \\ \mu_1 & m_2 & -m_3 \end{pmatrix} \begin{pmatrix} \ell_1 & j_2 & \ell_3 \\ -m_1 & \mu_2 & m_3 \end{pmatrix} \begin{pmatrix} \ell_1 & \ell_2 & j_3 \\ m_1 & -m_2 & \mu_3 \end{pmatrix}. \end{aligned} \tag{5.161}$$

Obviously, the Racah coefficients and the $6j$-symbols all are real, and X is a real orthogonal matrix. Hence, the Racah coefficients satisfy the orthogonal relations

$$\begin{aligned}
&\sum_{J_{12}} (2J_{12}+1)W[jkJ\ell;J_{12}J_{23}]W[jkJ\ell;J_{12}J_{23}'] = \frac{\delta_{J_{23}J_{23}'}}{2J_{23}+1}, \\
&\sum_{J_{23}} (2J_{23}+1)W[jkJ\ell;J_{12}J_{23}]W[jkJ\ell;J_{12}'J_{23}] = \frac{\delta_{J_{12}J_{12}'}}{2J_{12}+1}.
\end{aligned} \tag{5.162}$$

The analytic form of the Racah coefficients can be calculated from Eq. (5.160):

$$\begin{aligned}
W[abcd;ef] = &(-1)^{a+b+c+d}\ \Delta(a,b,e)\Delta(d,e,c)\Delta(b,d,f)\Delta(a,f,c) \\
&\cdot \sum_z (-1)^z(z+1)!\{(z-a-b-e)!(z-c-d-e)! \\
&\cdot (z-b-d-f)!(z-a-c-f)!(a+b+c+d-z)! \\
&\cdot (a+d+e+f-z)!(b+c+e+f-z)!\}^{-1},
\end{aligned} \tag{5.163}$$

$$z = \max\left\{\begin{matrix} a+b+e \\ c+d+e \\ a+c+f \\ b+d+f \end{matrix}\right\}, \quad \ldots, \quad \min\left\{\begin{matrix} a+b+c+d \\ a+d+e+f \\ b+c+e+f \end{matrix}\right\},$$

where a, b, c, d, e, and f have to satisfy four conditions of the triangle rules given in Fig. 5.5. The detailed calculation is given in §5.4.2 of [Ma (1993)], where the quantum parameter q is taken to be 1.

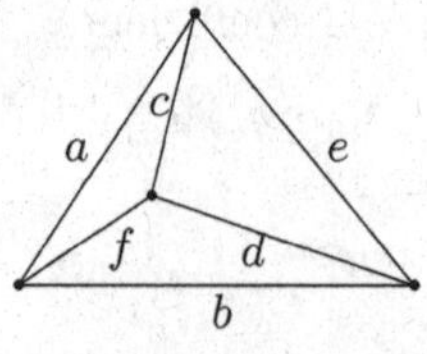

Fig. 5.5 Four triangle rules of the Racah coefficients

The formulas on the Racah coefficients are quite complicated. A graphic method may be helpful to understand them. Let three oriented lines intersected at one point, as shown in Fig. 5.6 (a), denote a CG coefficient. Remind that $\left(C^{jk}\right)^{-1}$ is the transpose of C^{jk}. When $J=0$ (or $j=0$, $k=0$) the corresponding oriented line can be omitted. **The magnetic quantum number in the solid line connecting two ends is summed, and that in the dotted line is equal to each other, but not summed.**

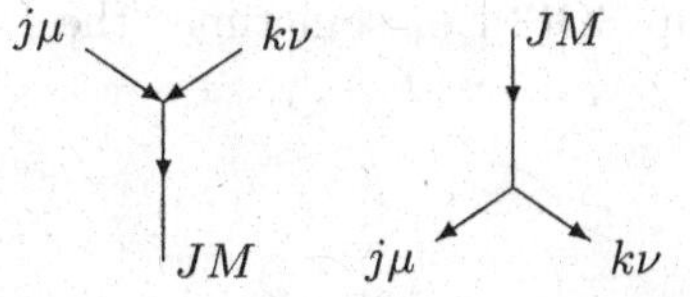

(a) $C^{jk}_{\mu\nu,JM}$

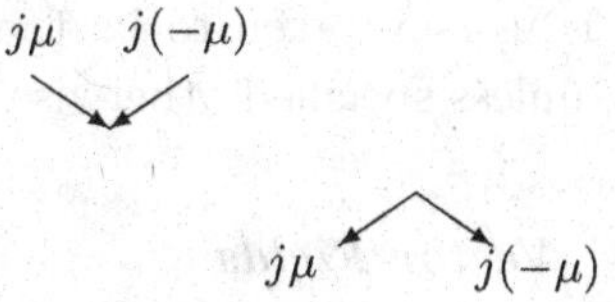

(b) $C^{jj}_{\mu(-\mu),00}$

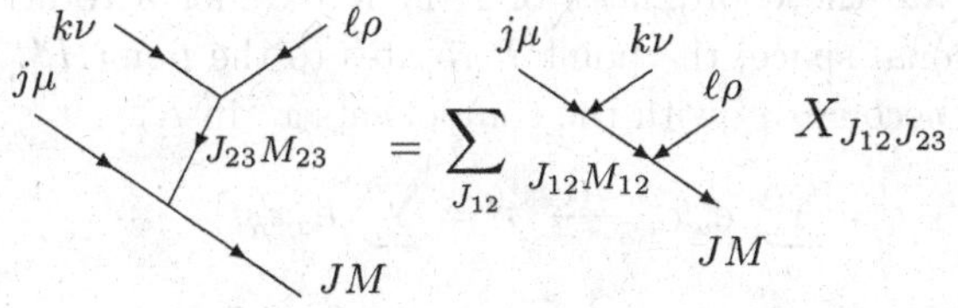

(c) The definition (5.159) for the Racah coefficients

$\{(2J_{12}+1)(2J_{23}+1)\}^{1/2}\, W[jkJ\ell; J_{12}J_{23}] \;=$

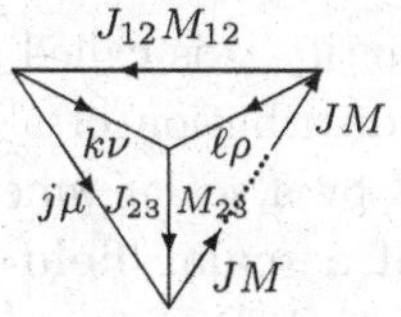

(d) The expansion (5.160) of the Racah coefficients.

Fig. 5.6 Diagram for the Racah coefficients.

The $6j$-symbols introduced by Wigner are more symmetric. There are 144 $6j$-symbols to be equal to each other:

$$\begin{aligned} &\begin{Bmatrix} a & b & e \\ d & c & f \end{Bmatrix} = \begin{Bmatrix} b & a & e \\ c & d & f \end{Bmatrix} = \begin{Bmatrix} a & e & b \\ d & f & c \end{Bmatrix} = \begin{Bmatrix} d & c & e \\ a & b & f \end{Bmatrix} \\ &= \begin{Bmatrix} (a+b-c+d)/2 & (a+b+c-d)/2 & e \\ (a-b+c+d)/2 & (-a+b+c+d)/2 & f \end{Bmatrix}. \end{aligned} \tag{5.164}$$

5.7 Tensors and Spinors

We have discussed the concepts of a scalar and a vector. Now, we are going to introduce the concepts of a tensor and a spinor. The definitions for a tensor and a spinor, as well as those for a scalar and a vector, are based on their transformation properties in a matrix group. In physics, the matrix group is usually taken to be the group SO(3), or sometimes the Lorentz group, unless specified otherwise.

5.7.1 *Vector Fields*

Let $\boldsymbol{r} = \sum_a \boldsymbol{e}_a x_a$ denote the position vector of an arbitrary point P in a real three-dimensional space, where $\boldsymbol{e}_a$ are the basis vectors in the laboratory frame K and x_a are the coordinates of P in K. Under a rotation R in the three-dimensional space, the point P rotates to the point P', and the position vector $\boldsymbol{r}$ becomes $\boldsymbol{r}'$ with the components x'_a in K,

$$\begin{aligned} \boldsymbol{r} = \sum_a \boldsymbol{e}_a x_a &\xrightarrow{R} \boldsymbol{r}' = \sum_a \boldsymbol{e}_a x'_a, \\ x_a &\xrightarrow{R} x'_a = \sum_{b=1}^{3} R_{ab} x_b. \end{aligned} \tag{5.165}$$

A quantity ψ is called a scalar if it is left invariant under the rotation R. The distribution of a scalar is called a scalar field. A scalar field is described by a scalar function $\psi(x)$. **After the rotation R, the value $\psi'(Rx)$ of a scalar field at the point P' is equal to the value $\psi(x)$ at the point P before the rotation.** The transformation of a scalar function in a rotation R is characterized by the operator P_R,

$$\psi(x) \xrightarrow{R} \psi'(Rx) = \psi(x), \qquad \psi'(x) \equiv P_R\psi(x) = \psi(R^{-1}x). \tag{5.166}$$

A quantity $\boldsymbol{V}$ is called a vector if its components $\boldsymbol{V}_a$ transform in a rotation R like x_a,

$$\begin{aligned} \boldsymbol{V} = \sum_a \boldsymbol{e}_a \boldsymbol{V}_a &\xrightarrow{R} \boldsymbol{V}' = \sum_a \boldsymbol{e}_a \boldsymbol{V}'_a, \\ \boldsymbol{V}_a &\xrightarrow{R} \boldsymbol{V}'_a = \sum_{b=1}^{3} R_{ab} \boldsymbol{V}_b. \end{aligned} \tag{5.167}$$

A vector $\boldsymbol{V}$, whose three components as a whole, characterizes the state of the system. Let Q_R denote the transformation operator for $\boldsymbol{V}$,

$$\boldsymbol{V}'_a \equiv (Q_R \boldsymbol{V})_a = \sum_{b=1}^{3} R_{ab} \boldsymbol{V}_b. \tag{5.168}$$

In fact, Q_R for a vector is the R matrix. The body-fixed frame K' and its basis vectors $\boldsymbol{e}'_a$ are fixed and rotate together with the system. The basis vector $\boldsymbol{e}_b$ is a special vector whose components all are zero except for one:

$(e_b)_a = \delta_{ab}$. The basis vector e'_b in K' transforms under the rotation R from e_b,

$$\begin{aligned}(e_b)_a \xrightarrow{R} (e'_b)_a &= (Q_R e_b)_a = \sum_{d=1}^{3} R_{ad} (e_b)_d \\ &= R_{ab} = \sum_{c=1}^{3} (e_c)_a R_{cb}.\end{aligned} \tag{5.169}$$

The distribution of a vector is called a vector field. A vector field is described by three functions $\boldsymbol{V}(x)_a$. After the rotation R, the vector $\boldsymbol{V}'(Rx)_a$ at the point P' is transformed according to Eq. (5.168) from the vector $\boldsymbol{V}(x)_a$ at the point P before the rotation: $\boldsymbol{V}'(Rx)_a = \sum_b R_{ab}\boldsymbol{V}(x)_b$. Introduce the transformation operator $O_R = P_R Q_R = Q_R P_R$ for the vector field in the rotation R,

$$\begin{aligned}\boldsymbol{V}(x) \xrightarrow{R} \boldsymbol{V}'(x) &\equiv O_R \boldsymbol{V}(x) = \sum_{a=1}^{3} e_a \left[O_R \boldsymbol{V}(x)\right]_a \\ &= \sum_{a=1}^{3} e_a \sum_{d=1}^{3} R_{ad} \boldsymbol{V}(R^{-1}x)_d = \sum_{d=1}^{3} \left[\sum_{a=1}^{3} e_a R_{ad}\right] \boldsymbol{V}(R^{-1}x)_d.\end{aligned} \tag{5.170}$$

$$\begin{aligned}&O_R \boldsymbol{V}(x) = R\boldsymbol{V}(R^{-1}x), && \left[O_R \boldsymbol{V}(x)\right]_a = \sum_{b=1}^{3} R_{ab} \boldsymbol{V}(R^{-1}x)_b, \\ &P_R \boldsymbol{V}(x) = \boldsymbol{V}(R^{-1}x), && \left[P_R \boldsymbol{V}(x)\right]_a = \boldsymbol{V}(R^{-1}x)_a, \\ &Q_R \boldsymbol{V}(x) = R\boldsymbol{V}(x), && \left[Q_R \boldsymbol{V}(x)\right]_a = \sum_{b=1}^{3} R_{ab} \boldsymbol{V}(x)_b, \\ &Q_R e_a = e'_a = \sum_{d=1}^{3} e_d R_{da}, && (Q_R e_a)_b = (e'_a)_b = \sum_{d=1}^{3} (e_d)_b R_{da}.\end{aligned} \tag{5.171}$$

Before the rotation R, two frames K and K' coincides with each other, $e'_a = e_a$, and $\boldsymbol{V}'(x)_a = \boldsymbol{V}(x)_a$. Expand the vector field $\boldsymbol{V}_a$ with respect to the basis vectors e_a:

$$[\boldsymbol{V}(x)]_b = \sum_{a=1}^{3} [e_a]_b V(x)_a = V(x)_b. \tag{5.172}$$

In the formula $\boldsymbol{V}(x) = \sum_a e_a \boldsymbol{V}(x)_a$, e_a is fixed in the rotation and the component $\boldsymbol{V}(x)_a$, denoted by the symbols in the bold form and called the vector field, transforms according to Eq. (5.170). In the formula $\boldsymbol{V}(x) = \sum_d e_d V(x)_d$, e_d are the basis vectors in K' which coincide with the basis vectors $\boldsymbol{e}_a$ in K before rotation and transforms according to Eq. (5.169), and $V(x)_d$ are only the coefficients and transform like the scalar.

The position vector field $\boldsymbol{r}(x)$ is a special vector field. In fact, $\boldsymbol{r}(x)_a = x_a$, and $\boldsymbol{r}(R^{-1}x)_a = (R^{-1}x)_a$,

$$[O_R\boldsymbol{r}(x)]_a = \sum_b R_{ab}\boldsymbol{r}(R^{-1}x)_b = x_a = \boldsymbol{r}(x)_a.$$

$$\begin{aligned} O_R\boldsymbol{r}(x) &= \boldsymbol{r}(x), \qquad [Q_R\boldsymbol{r}(x)]_a = \sum_b R_{ab}x_b, \\ [P_R\boldsymbol{r}(x)]_a &= \boldsymbol{r}(R^{-1}x)_a = \sum_b (R^{-1})_{ab}x_b. \end{aligned} \tag{5.173}$$

The operators O_R, Q_R and P_R all are the linear unitary operators. The operator $L(x)$, describing a mechanics quantity, transforms in a rotation R as follows:

$$L(x) \xrightarrow{R} O_R L(x) O_R^{-1}. \tag{5.174}$$

5.7.2 *Tensor Fields*

A tensor of rank n contains n subscripts and 3^n components, which as a whole characterize the state of the system. In the rotation, each subscript transforms like a vector subscript,

$$\boldsymbol{T}_{a_1a_2\ldots a_n} \xrightarrow{R} (O_R\boldsymbol{T})_{a_1a_2\ldots a_n} = \sum_{b_1b_2\ldots b_n} R_{a_1b_1}\ldots R_{a_nb_n}\boldsymbol{T}_{b_1b_2\ldots b_n}. \tag{5.175}$$

A distribution of a tensor is called a tensor field. A tensor field of rank n is described by 3^n functions $\boldsymbol{T}(x)_{a_1a_2\ldots a_n}$ and transforms in a rotation R as

$$\begin{aligned} [O_R\boldsymbol{T}(x)]_{a_1a_2\ldots a_n} &= \sum_{b_1b_2\ldots b_n} R_{a_1b_1}R_{a_2b_2}\ldots R_{a_nb_n}\boldsymbol{T}(R^{-1}x)_{b_1b_2\ldots b_n}, \\ O_R = Q_RP_R, &\qquad Q_R = R\times R\times\cdots\times R, \\ [P_R\boldsymbol{T}(x)]_{a_1a_2\ldots a_n} &= \boldsymbol{T}(R^{-1}x)_{a_1a_2\ldots a_n}, \\ [Q_R\boldsymbol{T}(x)]_{a_1a_2\ldots a_n} &= \sum_{b_1b_2\ldots b_n} R_{a_1b_1}R_{a_2b_2}\ldots R_{a_nb_n}\boldsymbol{T}(x)_{b_1b_2\ldots b_n}. \end{aligned} \tag{5.176}$$

The basis tensor $\boldsymbol{\theta}_{d_1d_2\ldots d_n}$ is a tensor containing only one nonvanishing component which is equal to 1

$$(\boldsymbol{\theta}_{d_1d_2\ldots d_n})_{a_1\ldots a_n} = \delta_{d_1a_1}\delta_{d_2a_2}\ \ldots\ \delta_{d_na_n}. \tag{5.177}$$

A tensor field can be expanded with respect to the basis tensor,

$$
\begin{aligned}
\boldsymbol{T}(x)_{a_1\ldots a_n} &= \sum_{d_1\ldots d_n} (\boldsymbol{\theta}_{d_1\ldots d_n})_{a_1\ldots a_n} T(x)_{d_1\ldots d_n} = T(x)_{a_1\ldots a_n}, \\
[O_R \boldsymbol{T}(x)]_{a_1 a_2\ldots a_n} &= \sum_{d_1\ldots d_n} (Q_R \boldsymbol{\theta}_{d_1\ldots d_n})_{a_1\ldots a_n} P_R T(x)_{d_1\ldots d_n}, \\
(Q_R \boldsymbol{\theta}_{d_1 d_2\ldots d_n})_{a_1 a_2\ldots a_n} &= \sum_{b_1\ldots b_n} R_{a_1 b_1} R_{a_2 b_2} \ldots R_{a_n b_n} (\boldsymbol{\theta}_{d_1 d_2\ldots d_n})_{b_1 b_2\ldots b_n} \\
&= R_{a_1 d_1} R_{a_2 d_2} \ldots R_{a_n d_n} = \sum_{b_1\ldots b_n} (\boldsymbol{\theta}_{b_1\ldots b_n})_{a_1\ldots a_n} R_{b_1 d_1} R_{b_2 d_2} \ldots R_{b_n d_n}, \\
P_R T(x)_{d_1\ldots d_n} &= T(R^{-1}x)_{d_1\ldots d_n}.
\end{aligned} \tag{5.178}
$$

A scalar field is the tensor field of rank 0, and a vector field is the tensor of rank 1.

5.7.3 *Spinor Fields*

A spinor of rank s contains $(2s+1)$ components which as a whole characterize the state of the system. The spinor transforms in a rotation R as follows:

$$
\Psi_\sigma^{(s)} \xrightarrow{R} \left(O_R \Psi^{(s)}\right)_\sigma = \sum_\lambda D^s_{\sigma\lambda}(R) \Psi_\lambda^{(s)}. \tag{5.179}
$$

A distribution of a spinor is called a spinor field. A spinor field of rank s is described by $(2s+1)$ functions $\Psi^{(s)}(x)_\sigma$ transforming in a rotation R as

$$
\begin{aligned}
&\left[O_R \Psi^{(s)}(x)\right]_\sigma = \sum_\lambda D^s_{\sigma\lambda}(R) \Psi^{(s)}(R^{-1}x)_\lambda, \\
&O_R = Q_R P_R, \qquad Q_R = D^s(R), \\
&\left[P_R \Psi^{(s)}(x)\right]_\sigma = \Psi^{(s)}(R^{-1}x)_\sigma, \\
&\left[Q_R \Psi^{(s)}(x)\right]_\sigma = \sum_\lambda D^s_{\sigma\lambda}(R) \Psi^{(s)}(x)_\lambda.
\end{aligned} \tag{5.180}
$$

Usually, a spinor is denoted by a $(2s+1)\times 1$ column matrix. The spinor of rank 1/2 is called a **fundamental spinor**, or briefly a spinor, where the superscript 1/2 is often neglected. A fundamental spinor for the group SO(3) has two components and is denoted by a 2×1 column matrix.

The basis spinor $\boldsymbol{e}^{(s)}(\rho)$ is a special spinor containing only one nonvanishing component which is equal to 1,

$$
\boldsymbol{e}^{(s)}(\rho)_\sigma = \delta_{\rho\sigma}, \tag{5.181}
$$

where the ordinal index ρ is indicated inside the round brackets. In a rotation R,

$$\begin{aligned}\left[O_R \mathbf{e}^{(s)}(\rho)\right]_\sigma = \left[Q_R \mathbf{e}^{(s)}(\rho)\right]_\sigma &= \sum_{\lambda=-s}^{s} D^s_{\sigma\lambda}(R)\mathbf{e}^{(s)}(\rho)_\lambda \\ = D^s_{\sigma\rho}(R) &= \sum_{\lambda=-s}^{s} \mathbf{e}^{(s)}(\lambda)_\sigma D^s_{\lambda\rho}(R).\end{aligned} \tag{5.182}$$

A spinor field can be expanded with respect to the basis spinor,

$$\begin{aligned}&\Psi^{(s)}(x) = \sum_{\rho=-s}^{s} \mathbf{e}^{(s)}(\rho)\psi^{(s)}(x)_\rho, \\ &O_R\Psi^{(s)}(x) = \sum_{\rho=-s}^{s} \left\{Q_R \mathbf{e}^{(s)}(\rho)\right\}\left\{P_R\psi^{(s)}(x)\right\}_\rho, \\ &Q_R \mathbf{e}^{(s)}(\rho) = \sum_{\lambda=-s}^{s} \mathbf{e}^{(s)}(\lambda)D^s_{\lambda\rho}(R), \qquad P_R\psi^{(s)}(x) = \psi^{(s)}(R^{-1}x).\end{aligned} \tag{5.183}$$

A scalar is the spinor of rank 0. A vector is the spinor of rank 1 because the self-representation of SO(3) is equivalent to D^1 (see Eq. (5.69)). The basis spinor $\boldsymbol{e}^{(1)}(\rho)$ of rank one is also called the **spherical harmonic basis** of vectors (see Eq. (5.80)),

$$\begin{aligned}&\boldsymbol{V}(x) = \sum_{a=1}^{3} \mathbf{e}_a V(x)_a = \sum_{\rho=-1}^{1} \mathbf{e}^{(1)}(\rho)\psi^{(1)}(x)_\rho, \\ &\mathbf{e}^{(1)}(\rho) = \sum_{a=1}^{3} \mathbf{e}_a M_{a\rho}, \qquad \psi^{(1)}(x)_\rho = \sum_{a=1}^{3} \left(M^{-1}\right)_{\rho a} V(x)_a, \\ &\begin{cases}\mathbf{e}^{(1)}(1) = -\left(\mathbf{e}_1 + \mathrm{i}\mathbf{e}_2\right)/\sqrt{2}, \\ \mathbf{e}^{(1)}(0) = \mathbf{e}_3, \\ \mathbf{e}^{(1)}(-1) = \left(\mathbf{e}_1 - \mathrm{i}\mathbf{e}_2\right)/\sqrt{2},\end{cases} \\ &\begin{cases}\psi^{(1)}(x)_1 = -\left[V(x)_1 - \mathrm{i}V(x)_2\right]/\sqrt{2}, \\ \psi^{(1)}(x)_0 = V(x)_3, \\ \psi^{(1)}(x)_{-1} = \left[V(x)_1 + \mathrm{i}V(x)_2\right]/\sqrt{2}.\end{cases}\end{aligned} \tag{5.184}$$

5.7.4 *Total Angular Momentum Operator*

Discuss a system characterized by a spinor field. The Hamiltonian of the system is isotropic so that the group SO(3) is the symmetry group of the system,

$$O_R H(x) = H(x) O_R, \tag{5.185}$$

where the transformation operator O_R for a spinor field is divided into two operators, $O_R = P_R Q_R$. For the infinitesimal elements,

$$\begin{aligned}
P_A &= 1 - \mathrm{i}\sum_{a=1}^{3} \alpha_a L_a, \\
Q_A &= \mathbf{1} - \mathrm{i}\sum_{a=1}^{3} \alpha_a S_a, \\
O_A &= \mathbf{1} - \mathrm{i}\sum_{a=1}^{3} \alpha_a \left(L_a + S_a\right) = \mathbf{1} - \mathrm{i}\sum_{a=1}^{3} \alpha_a J_a,
\end{aligned} \tag{5.186}$$

where S_a is the generator of D^s, and L_a is the differential operator of P_R, called the orbital angular momentum operator in physics. Their sum is denoted by J_a

$$J_a = L_a + S_a. \tag{5.187}$$

J_a, S_a, and L_a all satisfy the typical commutative relations of angular momentums.

The static wave functions with energy E span an invariant functional space. Its basis function $\Psi_\rho(x)$ is a spinor field, transforming in a rotation R according to Eq. (5.180). But, after the transformation, it has to be a combination of the basis functions,

$$O_R \Psi_\rho(x) = D^s(R)\, \Psi_\rho(R^{-1}x) = \sum_\lambda \Psi_\lambda(x) D_{\lambda\rho}(R). \tag{5.188}$$

The set of the combinative coefficients $D_{\lambda\rho}(R)$ forms a representation of SO(3). Decomposing the representation $D(R)$ into the direct sum of the irreducible representations of SO(3) by the method of group theory, the static wave function is combined to be $\Psi^j_\mu(x)$ belonging to the μth row of the irreducible representation D^j,

$$O_R \Psi^j_\mu(x) = D^s(R)\, \Psi^j_\mu(R^{-1}x) = \sum_\nu \Psi^j_\nu(x) D^j_{\nu\mu}(R). \tag{5.189}$$

Namely, $\Psi^j_\mu(x)$ is the common eigenfunction of the generators J^2 and J_3,

$$\begin{aligned}
&J^2\, \Psi^j_\mu(x) = j(j+1)\, \Psi^j_\mu(x), \\
&J_3\, \Psi^j_\mu(x) = \mu\, \Psi^j_\mu(x), \\
&J_\pm\, \Psi^j_\mu(x) = \Gamma^j_{\mp\mu}\, \Psi^j_{\mu\pm 1}(x), \\
&J^2 = J_1^2 + J_2^2 + J_3^2, \qquad J_\pm = J_1 \pm i J_2.
\end{aligned} \tag{5.190}$$

Now, for a isotropic system characterized by a spinor field, the conserved angular momentum is not the orbital angular momentum, but J^2 and J_3.

J_a is the sum of the orbital angular momentum L_a and another quantity S_a related to the spinor. Both J_a and S_a satisfy the typical commutative relations of angular momentums. S_a should be the mathematical description of the spinor angular momentum, discovered and measured in experiments. Therefore, S_a **is called the operator of the spinor angular momentum** and J_a is **the operator of the total angular momentum**. The total angular momentum is conserved in a spherically symmetric system characterized by a spinor field. From Eq. (5.182) the basis spinor $\boldsymbol{e}^{(s)}(\rho)$ is the common eigenfunction of the operators of the total angular momentum and the spinor angular momentum,

$$\begin{aligned} &J_3\boldsymbol{e}^{(s)}(\rho) = S_3\boldsymbol{e}^{(s)}(\rho) = \rho\boldsymbol{e}^{(s)}(\rho), \\ &J_\pm\boldsymbol{e}^{(s)}(\rho) = S_\pm\boldsymbol{e}^{(s)}(\rho) = \Gamma^s_{\mp\rho}\boldsymbol{e}^{(s)}(\rho\pm 1), \\ &J^2\boldsymbol{e}^{(s)}(\rho) = S^2\boldsymbol{e}^{(s)}(\rho) = s(s+1)\boldsymbol{e}^{(s)}(\rho), \\ &S^2 = S_1^2 + S_2^2 + S_3^2, \qquad S_\pm = S_1 \pm iS_2. \end{aligned} \tag{5.191}$$

There are three sets of the mutual commutable angular momentum operators, one set consists of L^2, L_3, S^2, and S_3, the other set consists of J^2, J_3, L^2, and S^2, and the third set consists of J^2, J_3, S^2, and $\boldsymbol{S}\cdot\hat{\boldsymbol{n}} = \sum_a S_a n_a$, where $\boldsymbol{n} = \boldsymbol{r}/r$. The common eigenfunctions of the first set are the product of the spherical harmonic functions $Y^\ell_m(\hat{\boldsymbol{n}})$ and the basis spinor $\boldsymbol{e}^{(s)}(\rho)$. Second, combining them by CG coefficients, one obtains the **spherical spinor function**, which is the common eigenfunction of the second set with the eigenvalues $j(j+1)$, μ, $\ell(\ell+1)$, and $s(s+1)$,

$$Y^{j\ell s}_\mu(\hat{\boldsymbol{n}}) = \sum_\rho C^{s\ell}_{\rho(\mu-\rho)j\mu}\boldsymbol{e}^{(s)}(\rho)Y^\ell_{\mu-\rho}(\hat{\boldsymbol{n}}). \tag{5.192}$$

Another conserved quantity $\hat{\kappa}$ is introduced in quantum mechanics

$$\hat{\kappa} = 2\boldsymbol{L}\cdot\boldsymbol{S} + 1 = J^2 - L^2 - S^2 + 1. \tag{5.193}$$

The spherical spinor function is also the eigenfunction of $\hat{\kappa}$. When $s = 1/2$, the eigenvalue of $\hat{\kappa}$ is

$$\kappa = j(j+1) - \ell(\ell+1) + \frac{1}{4} = \begin{cases} j+1/2 > 0, & \ell = j - 1/2, \\ -j-1/2 < 0, & \ell = j + 1/2. \end{cases} \tag{5.194}$$

The non-vanishing integer κ gives three quantities simultaneously: $j = |\kappa| - 1/2$, $\ell = j - \kappa/(2|\kappa|)$ and $s = 1/2$. Calculating the CG coefficients by Mathematica, one obtains the spherical spinor functions for $s = 1/2$:

$$Y_{|\kappa|,\mu}(\hat{\boldsymbol{n}}) = \begin{pmatrix} \left(\dfrac{j+\mu}{2j}\right)^{1/2} Y^{j-1/2}_{\mu-1/2}(\hat{\boldsymbol{n}}) \\ \left(\dfrac{j-\mu}{2j}\right)^{1/2} Y^{j-1/2}_{\mu+1/2}(\hat{\boldsymbol{n}}) \end{pmatrix},$$
$$Y_{-|\kappa|,\mu}(\hat{\boldsymbol{n}}) = \begin{pmatrix} -\left(\dfrac{j-\mu+1}{2j+2}\right)^{1/2} Y^{j+1/2}_{\mu-1/2}(\hat{\boldsymbol{n}}) \\ \left(\dfrac{j+\mu+1}{2j+2}\right)^{1/2} Y^{j+1/2}_{\mu+1/2}(\hat{\boldsymbol{n}}) \end{pmatrix}. \tag{5.195}$$

Since the self-inverse operator $\boldsymbol{\sigma}\cdot\hat{\boldsymbol{n}}$ commutes with J_3 and J^2,

$$(\boldsymbol{\sigma}\cdot\hat{\boldsymbol{n}})\, Y_{|\kappa|,\mu}(\hat{\boldsymbol{n}}) = C_1 Y_{|\kappa|,\mu}(\hat{\boldsymbol{n}}) + C_2 Y_{-|\kappa|,\mu}(\hat{\boldsymbol{n}}),$$

where the coefficients C_1 and C_2 are independent of μ. Letting $\mu = j$, we obtain from Eq. (5.90) and the first equality of Eq. (5.195):

$$\begin{aligned}
&n_3\left[\frac{(-1)^{j-1/2}}{2^{j-1/2}(j-1/2)!}\sqrt{\frac{(2j)!}{4\pi}}(n_1+\mathrm{i}n_2)^{j-1/2}\right] \\
&\quad = -\left[-\sqrt{\frac{1}{2j+2}}\frac{(-1)^{j-1/2}}{2^{j-1/2}(j-1/2)!}\sqrt{\frac{(2j+2)(2j)!}{4\pi}}(n_1+\mathrm{i}n_2)^{j-1/2}n_3\right], \\
&(n_1+\mathrm{i}n_2)\left[\frac{(-1)^{j-1/2}}{2^{j-1/2}(j-1/2)!}\sqrt{\frac{(2j)!}{4\pi}}(n_1+\mathrm{i}n_2)^{j-1/2}\right] \\
&\quad = -\left[\sqrt{\frac{2j+1}{2j+2}}\frac{(-1)^{j+1/2}}{2^{j+1/2}(j+1/2)!}\sqrt{\frac{(2j+2)!}{4\pi}}(n_1+\mathrm{i}n_2)^{j+1/2}\right].
\end{aligned}$$

Thus, $C_1 = 0$, $C_2 = -1$, and

$$(\boldsymbol{\sigma}\cdot\hat{\boldsymbol{n}})\, Y_{\kappa,\mu}(\hat{\boldsymbol{n}}) = -Y_{-\kappa,\mu}(\hat{\boldsymbol{n}}). \tag{5.196}$$

Third, from Eqs. (5.191) and (5.130), we have

$$\begin{aligned}
&(\boldsymbol{S}\cdot\hat{\boldsymbol{n}})\, Q_T \boldsymbol{e}^{(s)}(\rho) = \rho Q_T \boldsymbol{e}^{(s)}(\rho), \\
&\hat{\boldsymbol{n}} = \boldsymbol{r}/r = \boldsymbol{n}(\theta,\varphi), \quad T = R(\varphi,\theta,\gamma).
\end{aligned} \tag{5.197}$$

Letting the spinor $\Psi^j_\mu(\boldsymbol{r})$ belong to the μth row of the irreducible representation $D^j(\mathrm{SU}(2))$ and following Eq. (5.77), we obtain

$$\begin{aligned}
\left[\Psi^j_\mu(\boldsymbol{r})\right]_\sigma &= \left[\Psi^j_\mu(T\boldsymbol{r}_0)\right]_\sigma = Q_T O_{T^{-1}}\left[\Psi^j_\mu(\boldsymbol{r}_0)\right]_\sigma \\
&= \sum_{\nu\rho} D^s_{\sigma\rho}(T_0)\left[\Psi^j_\nu(\boldsymbol{r}_0)\right]_\rho \mathrm{e}^{\mathrm{i}\mu\varphi} d^j_{\mu\nu}(\theta)\mathrm{e}^{\mathrm{i}(\nu-\rho)\gamma},
\end{aligned} \tag{5.198}$$

where $\boldsymbol{r}_0 = (r, 0, 0)$ and $T_0 = R(\varphi, \theta, 0)$. The left-hand side of Eq. (5.198) is independent of γ, so its right-hand side. Because the exponent functions $\mathrm{e}^{\mathrm{i}(\nu-\rho)\gamma}$ are linearly independent, the terms with $\nu \neq \rho$ on the right-hand side of Eq. (5.198) have to vanish. Letting $\left[\Psi_\nu^j(\boldsymbol{r}_0)\right]_\rho = \delta_{\nu\rho}$, i.e., $\Psi_\nu^j(\boldsymbol{r}_0) = \boldsymbol{e}^{(s)}(\nu)$, we obtain from Eq. (5.183)

$$\Psi_\mu^j(\boldsymbol{r}) = Q_{T_0}\boldsymbol{e}^{(s)}(\nu)\mathrm{e}^{\mathrm{i}\mu\varphi}d_{\mu\nu}^j(\theta) = \sum_{\rho=-s}^{s} \boldsymbol{e}^{(s)}(\rho)D_{\rho\nu}^s(T_0)D_{\mu\nu}^j(T_0)^*. \qquad (5.199)$$

$\Psi_\mu^j(\boldsymbol{r})$ is the common eigenfunction of the third set of the angular momentum operators with the eigenvalues $j(j+1)$, μ, $s(s+1)$, and ν. When $s = 1/2$, $\mu = j$, and $\nu = 1/2$, equation (5.199) coincides with Eq. (5.196):

$$\begin{aligned}\Psi_j^j(\boldsymbol{r}) &= \begin{pmatrix} \dfrac{(-1)^{j-1/2}\mathrm{e}^{\mathrm{i}(j-1/2)\varphi}}{2^{j-1/2}(j-1/2)!}\sqrt{\dfrac{(2j)!}{j+1/2}}\cos^2(\theta/2)\sin^{j-1/2}(\theta) \\ \dfrac{(-1)^{j-1/2}\mathrm{e}^{\mathrm{i}(j+1/2)\varphi}}{2^{j+1/2}(j+1/2)!}\sqrt{\dfrac{(2j+1)!}{2}}\sin^{j+1/2}(\theta) \end{pmatrix} \\ &= \sqrt{\frac{2\pi}{2j+1}}\left[Y_{|\kappa|,j}(\hat{\boldsymbol{n}}) - Y_{-|\kappa|,j}(\hat{\boldsymbol{n}})\right].\end{aligned}$$

5.8 Irreducible Tensor Operators and Their Application

5.8.1 *Irreducible Tensor Operators*

In quantum mechanics a physical quantity is characterized by a linear operator $L(x)$ which consists of the coordinate operators x_a, the differential operators $\partial/\partial x_a$, and the matrix operators σ_a. In a rotation R, $L(x)$ transforms as

$$L(x) \xrightarrow{R} O_R L(x) O_R^{-1}.$$

A set of $(2k+1)$ **operators** $L_\rho^k(x)$, $-k \leqslant \rho \leqslant k$, **is called the irreducible tensor operators** of rank k if those operators transform in the rotation R of SO(3) as

$$O_R L_\rho^k(x) O_R^{-1} = \sum_{\lambda=-k}^{k} L_\lambda^k(x) D_{\lambda\rho}^k(R). \qquad (5.200)$$

Each operator in the set of the irreducible tensor operators characterizes a physical quantity independently, but in a rotation it relates to the other operators in the set as given in Eq. (5.200). Rewriting Eq. (5.200) in the form of generators, one has

$$\begin{aligned}
&\left[J_3,\ L^k_\rho(x)\right] = \rho L^k_\rho(x),\\
&\left[J_\pm,\ L^k_\rho(x)\right] = \{(k\mp\rho)(k\pm\rho+1)\}^{1/2}\, L^k_{\rho\pm1}(x),\\
&\sum_{a=1}^{3}\ \left[J_a,\ \left[J_a,\ L^k_\rho(x)\right]\right] = k(k+1)L^k_\rho(x).
\end{aligned} \tag{5.201}$$

This is an equivalent definition for the irreducible tensor operators. The irreducible tensor operator of rank 0 is called a symmetric operator or a scalar operator which is left invariant under any rotation R. The irreducible tensor operators of rank 1 are called the vector operators. $L^k_\rho(x)$ are called the irreducible tensor operators of rank k with respect to the orbital space or the spinor space if replacing O_R with P_R or Q_R, respectively.

A typical example for the irreducible tensor operators is the electric multipole operators which are proportional to $Y^k_\rho(\hat{\boldsymbol{n}})$,

$$O_R Y^k_\rho(\hat{\boldsymbol{n}})O_R^{-1} = P_R Y^k_\rho(\hat{\boldsymbol{n}})P_R^{-1} = \sum_\lambda\ Y^k_\lambda(\hat{\boldsymbol{n}})D^k_{\lambda\rho}(R). \tag{5.202}$$

They are the irreducible tensor operators of rank k with respect to the whole space and to the orbital space, but the scalar operators with respect to the spinor space. The electric dipole operators $Y^1_\mu(\hat{\boldsymbol{n}})$ $(k=1)$ are the combinations of the coordinate operators,

$$\begin{aligned}
&\left(\frac{4\pi}{3}\right)^{1/2} rY^1_1(\hat{\boldsymbol{n}}) = -\frac{1}{\sqrt{2}}(x_1+\mathrm{i}x_2),\\
&\left(\frac{4\pi}{3}\right)^{1/2} rY^1_0(\hat{\boldsymbol{n}}) = x_3,\\
&\left(\frac{4\pi}{3}\right)^{1/2} rY^1_{-1}(\hat{\boldsymbol{n}}) = \frac{1}{\sqrt{2}}(x_1-\mathrm{i}x_2),\\
&O_R Y^1_\rho(\hat{\boldsymbol{n}})O_R^{-1} = P_R Y^1_\rho(\hat{\boldsymbol{n}})P_R^{-1} = \sum_\lambda\ Y^1_\lambda(\hat{\boldsymbol{n}})D^1_{\lambda\rho}(R),\\
&O_R x_a O_R^{-1} = P_R x_a P_R^{-1} = \sum_b\ x_b R_{ba}, \qquad Q_R x_a Q_R^{-1} = x_a.
\end{aligned} \tag{5.203}$$

They are the vector operators with respect to the whole space and to the orbital space, and the scalar operators with respect to the spinor space. In the form of generators, one has

$$[J_a, x_b] = [L_a, x_b] = \mathrm{i}\sum_{d=1}^{3}\ \epsilon_{abd}x_d, \qquad [S_a, x_b] = 0. \tag{5.204}$$

In quantum mechanics there are some familiar operators having similar commutative relations,

$$
\begin{aligned}
&[J_a, p_b] = [L_a, p_b] = \mathrm{i}\sum_{d=1}^{3} \epsilon_{abd} p_d, \qquad [S_a, p_b] = 0, \\
&[J_a, L_b] = [L_a, L_b] = \mathrm{i}\sum_{d=1}^{3} \epsilon_{abd} L_d, \qquad [S_a, L_b] = 0, \\
&[J_a, J_b] = \mathrm{i}\sum_{d=1}^{3} \epsilon_{abd} J_d, \\
&[J_a, S_b] = [S_a, S_b] = \mathrm{i}\sum_{d=1}^{3} \epsilon_{abd} S_d, \qquad [L_a, S_b] = 0.
\end{aligned}
\tag{5.205}
$$

$$
\begin{aligned}
&O_R p_a O_R^{-1} = P_R p_a P_R^{-1} = \sum_b p_b R_{ba}, \qquad Q_R p_a Q_R^{-1} = p_a, \\
&O_R L_a O_R^{-1} = P_R L_a P_R^{-1} = \sum_b L_b R_{ba}, \qquad Q_R L_a Q_R^{-1} = L_a, \\
&O_R J_a O_R^{-1} = \sum_b J_b R_{ba}, \\
&O_R S_a O_R^{-1} = Q_R S_a Q_R^{-1} = \sum_b S_b R_{ba}, \qquad P_R S_a P_R^{-1} = S_a.
\end{aligned}
\tag{5.206}
$$

They all are the vector operators with respect to the whole space. For the vector operators, their components along a given direction $\hat{\boldsymbol{n}}$ are

$$
\begin{aligned}
&\hat{\boldsymbol{n}} \cdot \boldsymbol{r} = O_R x_3 O_R^{-1} = P_R x_3 P_R^{-1}, \qquad \hat{\boldsymbol{n}} \cdot \boldsymbol{p} = O_R p_3 O_R^{-1} = P_R p_3 P_R^{-1}, \\
&\hat{\boldsymbol{n}} \cdot \boldsymbol{L} = O_R L_3 O_R^{-1} = P_R L_3 P_R^{-1}, \qquad \hat{\boldsymbol{n}} \cdot \boldsymbol{J} = O_R J_3 O_R^{-1}, \\
&\hat{\boldsymbol{n}} \cdot \boldsymbol{S} = O_R S_3 O_R^{-1} = Q_R S_3 Q_R^{-1}, \qquad R = R(\varphi, \theta, 0).
\end{aligned}
\tag{5.207}
$$

5.8.2 *Wigner–Eckart Theorem*

The static wave functions of a spherically symmetric system belong to an irreducible representation of SO(3),

$$
\begin{aligned}
&O_R \Psi_\mu^j(x) = \sum_\nu \Psi_\nu^j(x) D_{\nu\mu}^j(R), \\
&O_R \Phi_{\mu'}^{j'}(x) = \sum_{\nu'} \Phi_{\nu'}^{j'}(x) D_{\nu'\mu'}^{j'}(R).
\end{aligned}
$$

The calculation on the expectation values of the irreducible tensor operators in the static wave functions can be greatly simplified by making use of the symmetry. Since

$$O_R\left\{L^k_\rho(x)\,\Psi^j_\mu(x)\right\} = \left\{O_R L^k_\rho(x) O_R^{-1}\right\}\left\{O_R\,\Psi^j_\mu(x)\right\}$$
$$= \sum_{\lambda\nu} \left\{L^k_\lambda(x)\,\Psi^j_\nu(x)\right\}\left\{D^k_{\lambda\rho}(R) D^j_{\nu\mu}(R)\right\},$$

$L^k_\rho(x)\,\Psi^j_\mu(x)$ transform according to the direct product of representations. Combining them by the Clebsch–Gordan coefficients, we obtain $F^J_M(x)$ belonging to the irreducible representation:

$$\begin{aligned} F^J_M(x) &= \sum_\rho L^k_\rho(x)\,\Psi^j_{M-\rho}(x) C^{kj}_{\rho(M-\rho)JM}, \\ L^k_\rho(x)\,\Psi^j_\mu(x) &= \sum_J F^J_{\rho+\mu}(x) C^{kj}_{\rho\mu J(\rho+\mu)}, \\ O_R F^J_M(x) &= \sum_{M'} F^J_{M'}(x) D^J_{M'M}(R). \end{aligned} \tag{5.208}$$

From the Wigner–Eckart theorem (see Theorem 3.7),

$$\langle \Phi^{j'}_{\mu'}(x)|F^J_M(x)\rangle = \delta_{j'J}\delta_{\mu'M}\langle \Phi^{j'}\|L^k\|\,\Psi^j\rangle, \tag{5.209}$$

where the constant $\langle \Phi^{j'}\|L^k\|\,\Psi^j\rangle$, called the reduced matrix element, is independent of the subscripts μ', ρ, μ, and M. It is related to the explicit forms of Φ, L, and Ψ and it depends on the indices j', k, and j. Hence,

$$\begin{aligned} \langle \Phi^{j'}_{\mu'}(x)|L^k_\rho(x)|\,\Psi^j_\mu(x)\rangle &= \sum_J C^{kj}_{\rho\mu J(\rho+\mu)}\langle \Phi^{j'}_{\mu'}(x)|F^J_{\rho+\mu}(x)\rangle \\ &= C^{kj}_{\rho\mu j'\mu'}\langle \Phi^{j'}\|L^k\|\,\Psi^j\rangle. \end{aligned} \tag{5.210}$$

In quantum mechanics, the observation quantities are usually expressed as the matric elements of the irreducible tensor operators in the static wave functions. There are $(2j'+1)(2k+1)(2j+1)$ matrix elements in the form of $\langle \Phi^{j'}_{\mu'}(x)|L^k_\rho(x)|\,\Psi^j_\mu(x)\rangle$. Equation (5.210) shows that **the rotational properties of the matrix elements are fully demonstrated in the CG coefficients**, and the detailed properties of the system and the operators are remained in the reduced matrix element. The Wigner–Eckart theorem simplifies the calculation of $(2j'+1)(2k+1)(2j+1)$ matrix elements to the calculation of only one parameter $\langle \Phi^{j'}\|L^k\|\,\Psi^j\rangle$. If there is one matrix element with given subscripts μ', ρ, and μ which can be calculated, the remaining matrix elements all are calculable. In most cases, all matrix elements are hard to be calculated. For example, the explicit forms of the wave functions $\Psi^j_\mu(x)$ and $\Phi^{j'}_{\mu'}(x)$ are unknown. However, through eliminating the parameter, Eq. (5.210) gives some relations among the matrix elements which sometimes can be observed in experiments. Furthermore, the CG coefficients in Eq. (5.210) have to vanish when the following conditions are not satisfied

$$\begin{aligned} &|\mu'| = |\rho + \mu| \leqslant j', \qquad |\rho| \leqslant k, \qquad |\mu| \leqslant j, \\ &|k - j| \leqslant j' \leqslant k + j. \end{aligned} \tag{5.211}$$

Those conditions are called the **selection rules** in quantum mechanics. The above method holds when O_R is replaced with P_R or Q_R.

5.8.3 *Selection Rule and Relative Intensity of Radiation*

An isolated atom is isotropic and its symmetry group is SO(3), no matter how complicated its internal construction is. Its static wave function belongs to the irreducible representation of SO(3). In the language of quantum mechanics, the static wave function $\Psi^J_M(x)$ is the common eigenfunction of $H(x)$, J^2, and J_3. The intensity of the electric dipole radiation between two states is proportional to the square of the matrix element of x_a

$$|\langle \Psi^{J'}_{M'}(x)|x_a|\Psi^J_M(x)\rangle|^2,$$

because the electric dipole operator is proportional to the distance of the pair of electric charges. From Eq. (5.203) the operators x_a are the combinations of the harmonic polynomials $\mathcal{Y}^1_\rho(\hat{\boldsymbol{n}}) = rY^1_\rho(\hat{\boldsymbol{n}})$, which are the vector operators with respect to the whole space and to the orbital space, and the scalar operators with respect to the spinor space. The square of the absolute value of the matrix element $|\langle \Psi^{J'}_{M'}|Y^1_\rho|\Psi^J_M\rangle|^2$ demonstrates the intensity of the electric dipole radiation, where the light is polarized in the Z-axis if $\rho = 0$, and is right- or left-circularly polarized about Z axis if $\rho = \pm 1$.

The Wigner–Eckart theorem shows

$$\langle \Psi^{J'}_{M'}(x)|Y^1_\rho(\hat{\boldsymbol{n}})|\Psi^J_M(x)\rangle = C^{1J}_{\rho M, J'M'}\langle \Psi^{J'}||Y^1||\Psi^J\rangle. \tag{5.212}$$

The selection rules are the conditions where the CG coefficients are not vanishing,

$$\Delta M = M' - M = \rho = 0 \ \text{ or } \ \pm 1, \qquad |J - 1| \leqslant J' \leqslant J + 1. \tag{5.213}$$

When $J = 0$, J' cannot be equal to 0, so the section rule for J' is usually written as

$$\Delta J = J' - J = \pm 1 \text{ or } 0, \qquad 0 \not\longrightarrow 0.$$

Read as "the transition from 0 to 0 is forbidden". Considering the space inversion, the parities of the initial state and the final state have to be opposite because x_a has odd parity.

If the spin-orbital interaction is weak, the wave functions are first coupled into the product of two parts, the orbital part with a total orbital angular momentum L and the spinor part with a total spinor angular momentum S. Then, the static wave function is the combination of those functions,

$$\Psi_M^{JLS} = \sum_\sigma C_{(M-\sigma)\sigma,JM}^{LS} \Psi_{M-\sigma}^L W_\sigma^S . \tag{5.214}$$

In physics, this case is called the LS **coupling**. Ψ_M^{JLS} is the common eigenfunction of J^2, J_3, L^2, and S^2. In this case, there are additional selection rules obtained from the rotational properties of the wave functions and the operators in the orbital space (the action of P_R) and the spinor space (the action of Q_R) [see Eq. (5.203)],

$$\begin{aligned} &\Delta L = L' - L = \pm 1 \text{ or } 0, \qquad 0 \not\longrightarrow 0, \\ &\Delta S = S' - S = 0. \end{aligned} \tag{5.215}$$

Remind that in the complicated system of n electrons, the parity is not equal to $(-1)^L$. The transition with $\Delta L = 0$ does not violate the conservation law of parity.

The relative intensity of the electric dipole radiation is proportional to the absolute square of the CG coefficients in Eq. (5.212),

$$\begin{aligned} &|\langle \Psi_{M+1}^{J'} | Y_1^1 | \Psi_M^J \rangle|^2 \ : \ |\langle \Psi_M^{J'} | Y_0^1 | \Psi_M^J \rangle|^2 \ : \ |\langle \Psi_{M-1}^{J'} | Y_{-1}^1 | \Psi_M^J \rangle|^2 \\ &= \left(C_{1M,J'(M+1)}^{1J} \right)^2 \ : \ \left(C_{0M,J'M}^{1J} \right)^2 \ : \ \left(C_{(-1)M,J'(M-1)}^{1J} \right)^2 . \end{aligned}$$

For instance, when $J' = J - 1$, the ratio of the relative intensity is (by Mathematica, see Eq. (5.141) or Prob. 20)

$$(J-M)(J-M-1) \ : \ 2(J-M)(J+M) \ : \ (J+M)(J+M-1).$$

The total intensity of radiation with the given J' and J

$$\sum_{\rho=-1}^{1} \left(C_{\rho M,J'(M+\rho)}^{1J} \right)^2 = \frac{2J'+1}{2J+1} \sum_{\rho=-1}^{1} \left(C_{\rho(-M-\rho),J(-M)}^{1J'} \right)^2 = \frac{2J'+1}{2J+1}.$$

The total intensity is independent of M. Namely, the total transition probability is independent of the direction of the total angular momentum.

5.8.4 *Landé Factor and Zeeman Effects*

The Stern–Gerlach experiment discovered that a hydrogen atom in S-wave deflects as it passes through an asymmetric magnetic field. The two spectral lines on the film shows that the atom has magnetic momentum. This is the earliest experiment observing the spin of an electron. It was measured that **the geromagnetic ratio for the spin angular momentum is double of that for the orbital angular momentum.** Namely, the operator of total magnetic momentum of an atom $\mathcal{M}_a$ is

$$\mathcal{M}_a = -\frac{e}{2m_e}\left(L_a + 2S_a\right) = -\frac{e}{2m_e}\left(J_a + S_a\right) = -\frac{egJ_a}{2m_e},$$

where m_e denotes the mass of the electron, g is called the Landé factor, and the natural units are used, $c = \hbar = 1$. The expectation value of $\mathcal{M}_3$ in the state Ψ^J_M is

$$\overline{\mathcal{M}_3} = \langle \Psi^J_M|\mathcal{M}_3|\Psi^J_M\rangle = \frac{-egM}{2m_e} = -\frac{eM}{2m_e} - \frac{e}{2m_e}\langle \Psi^J_M|S_3|\Psi^J_M\rangle,$$

$$(g-1)M = \langle \Psi^J_M|S_3|\Psi^J_M\rangle. \tag{5.216}$$

The matrix elements of two vector operators J_a and S_a in the same states are proportional,

$$\langle \Psi^J_M|S_a|\Psi^J_{M'}\rangle = A_J\langle \Psi^J_M|J_a|\Psi^J_{M'}\rangle, \tag{5.217}$$

where the proportional coefficient A_J is independent of the subscripts a, M, and M'. Since $J_a\Psi^J_M$ is a linear combination of $\Psi^J_{M'}$, Eq. (5.217) holds if $\Psi^J_{M'}$ is replaced with $J_a\Psi^J_M$,

$$\sum_{a=1}^{3} \langle \Psi^J_M|S_aJ_a|\Psi^J_M\rangle = A_J\sum_{a=1}^{3} \langle \Psi^J_M|J_aJ_a|\Psi^J_M\rangle = A_JJ(J+1).$$

If Ψ^J_M is a state by LS coupling (see Eq. (5.214)), it is the common eigenfunction of J^2, J_3, L^2, and S^2. Since

$$\sum_{a=1}^{3} L_a^2 = \sum_{a=1}^{3} \left(J_a - S_a\right)^2 = \sum_{a=1}^{3} J_a^2 + \sum_{a=1}^{3} S_a^2 - 2\sum_{a=1}^{3} S_aJ_a,$$

one has

$$\sum_{a=1}^{3} \langle \Psi^J_M|S_aJ_a|\Psi^J_M\rangle = \left\{J(J+1) + S(S+1) - L(L+1)\right\}/2,$$

and

$$A_J = \frac{J(J+1)+S(S+1)-L(L+1)}{2J(J+1)}.$$

Taking $a = 3$ and $M' = M$ in Eq. (5.217) and comparing it with Eq. (5.216), one obtains

$$g = 1 + A_J = \frac{3J(J+1)+S(S+1)-L(L+1)}{2J(J+1)}. \tag{5.218}$$

For the system of a single electron, $S = 1/2$ and $L = J \mp 1/2$. Thus,

$$g = \begin{cases} \dfrac{2J+1}{2J}, & L = J - \dfrac{1}{2}, \\[2ex] \dfrac{2J+1}{2(J+1)}, & L = J + \dfrac{1}{2}. \end{cases} \tag{5.219}$$

If the atom is in a strong magnetic field such that the spin-orbital interaction can be neglected, the wave function can be described by $\psi_m^L W_\sigma^S$, in which the energy E is

$$E = E_0 + \frac{e}{2m_e}(m + 2\sigma)H. \tag{5.220}$$

E_0 is the energy when the magnetic field is removed. Equation (5.220) shows that the change of energy of the system in a magnetic field is independent of L and S.

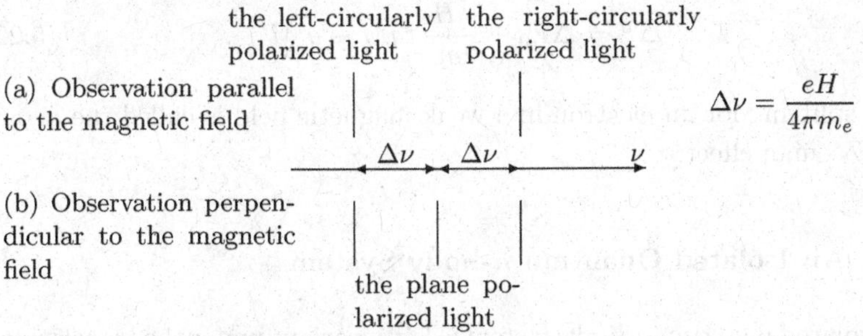

Fig. 5.7 The normal Zeeman effect

The selection rule for the electric dipole radiation is

$$\begin{aligned} &\Delta L = \pm 1 \text{ or } 0, \qquad 0 \not\longrightarrow 0. \\ &\Delta m = \pm 1,\ 0, \qquad \Delta S = 0, \qquad \Delta\sigma = 0. \end{aligned} \tag{5.221}$$

The energy of the radiative photon in the electric dipole transition is

$$\Delta E = \begin{cases} \Delta E_0, & \Delta m = 0, \\ \Delta E_0 \mp \dfrac{eH}{2m_e}, & \Delta m = \pm 1. \end{cases} \tag{5.222}$$

The radiative light is polarized in the Z-axis when $\Delta m = 0$ and is the circularly polarized about Z-axis when $\Delta m = \pm 1$. When the observation is made in the direction parallel to the magnetic field (the Z-axis), the plane polarized light cannot be seen. Two spectral lines of the circularly polarized light deflect from the original position (ΔE_0) in equidistance. The red-moved line is the left-circularly polarized light ($m - m' = -\Delta m = -1$), and the other is the right-circularly polarized light ($m - m' = 1$). When the observation is made in the direction perpendicular to the magnetic field, three lines can be seen where the middle line is the plane light polarized parallel to the magnetic field. This splitting of an electron in a strong magnetic field is called the normal Zeeman effect (see Fig. 5.7).

For an electron in a weak magnetic field, the spin-orbital interaction cannot be neglected. The wave function has to be characterized by Ψ_M^{JLS} given in Eq. (5.214). The energy of the electron in the magnetic field is

$$E = E_0 + \frac{e}{2m_e} gMH .$$

The observed splitting of spectrum is related with the quantum numbers J, L, S, and M of both the initial state and the final state:

$$\Delta E = \Delta E_0 + \frac{eH}{2m_e} \left(gM - g'M'\right) . \tag{5.223}$$

This splitting for an electron in a weak magnetic field is called the anomalous Zeeman effect.

5.9 An Isolated Quantum n-body System

An isolated quantum n-body system is left invariant under the translation of space-time and the spatial rotation so that the energy, the total momentum, and total orbital angular momentum of the system are conserved. The motion of the center-of-mass and the global rotation of the system can be separated from its internal motion, and its Schrödinger equation can be reduced to the radial equation, depending only on $(3n - 6)$ internal degrees of freedom. This separation is important essentially in theory, and simplifies the calculation explicitly at least for three- and four-body systems.

5.9.1 *Separation of the Motion of Center-of-Mass*

Let $\boldsymbol{r}_k$ and m_k, $k = 1, 2, \ldots, n$, denote the position vectors and the masses of n particles, respectively. The total mass of the n-body quantum system is $M = \sum_k m_k$ and its Schrödinger equation is

$$-\frac{1}{2}\sum_{k=1}^{n} m_k^{-1} \bigtriangledown_{\boldsymbol{r}_k}^2 \Psi + V\Psi = E\Psi, \tag{5.224}$$

where $\bigtriangledown_{\boldsymbol{r}_k}^2$ is the Laplace operator with respect to $\boldsymbol{r}_k$, and V is a pair potential, depending upon the distance $|\boldsymbol{r}_j - \boldsymbol{r}_k|$ of each pair of particles. The natural units are used for convenience, $\hbar = c = 1$.

Replace the position vectors $\boldsymbol{r}_k$ by the Jacobi coordinate vectors $\boldsymbol{R}_j$:

$$\begin{aligned} &\boldsymbol{R}_0 = M^{-1/2} \sum_{k=1}^{n} m_k \boldsymbol{r}_k, \qquad \boldsymbol{R}_j = \mu_j^{1/2} \left(\boldsymbol{r}_j - \sum_{k=j+1}^{n} \frac{m_k \boldsymbol{r}_k}{M_{j+1}} \right), \\ &M_j = \sum_{k=j}^{n} m_k, \qquad M_1 = M, \qquad 1 \leqslant j \leqslant (n-1), \end{aligned} \tag{5.225}$$

where $\boldsymbol{R}_0$ describes the position of the center of mass, $\boldsymbol{R}_1$ describes the mass-weighted separation from the first particle to the center of mass of the remaining particles, $\boldsymbol{R}_2$ describes the mass-weighted separation from the second particle to the center of mass of the remaining $(n-2)$ particles, and so on. An additional factor $\sqrt{M}$ is included in the Jacobi coordinate vectors for convenience. The mass-weighted factors μ_j in the formulas for $\boldsymbol{R}_j$ are determined by the condition

$$\sum_{k=1}^{n} m_k \boldsymbol{r}_k^2 = \sum_{j=0}^{n-1} \boldsymbol{R}_j^2. \tag{5.226}$$

μ_j can be calculated one by one from the following schemes. In the center-of-mass frame, let the first $(j-1)$ particles be located at the origin and the last $(n-j)$ particles coincide with each other,

$$\begin{aligned} &\boldsymbol{r}_k = 0, \qquad 1 \leqslant k < j, \\ &\boldsymbol{r}_{j+1} = \boldsymbol{r}_{j+2} = \cdots = \boldsymbol{r}_n = -m_j \boldsymbol{r}_j / M_{j+1}, \end{aligned}$$

then, $\boldsymbol{R}_k = 0$ for $k \neq j$,

$$\begin{aligned} \sum_{k=j}^{n} m_k \boldsymbol{r}_k^2 &= m_j \boldsymbol{r}_j^2 + M_{j+1} \left(-m_j \boldsymbol{r}_j / M_{j+1}\right)^2 = \left(m_j M_j / M_{j+1}\right) \boldsymbol{r}_j^2 \\ &= \boldsymbol{R}_j^2 = \mu_j \left[\boldsymbol{r}_j + m_j \boldsymbol{r}_j / M_{j+1}\right]^2 = \mu_j \left[M_j \boldsymbol{r}_j / M_{j+1}\right]^2, \end{aligned}$$

and

$$\mu_j = \frac{m_j M_{j+1}}{M_j}. \tag{5.227}$$

Rewrite the linear combinations between $\boldsymbol{r}_k$ and $\boldsymbol{R}_j$ as

$$\boldsymbol{R}_j = \sum_{k=1}^{n} X_{jk} m_k^{1/2} \boldsymbol{r}_k, \qquad \nabla_{\boldsymbol{r}_k} = m_k^{1/2} \sum_{j=0}^{n-1} X_{jk} \nabla_{\boldsymbol{R}_j}. \tag{5.228}$$

Substituting Eq. (5.228) into Eq. (5.226), one concludes that X is a real orthogonal matrix,

$$\sum_{j=0}^{n-1} \boldsymbol{R}_j^2 = \sum_{kt} (m_k m_t)^{1/2} \boldsymbol{r}_k \boldsymbol{r}_t \sum_{j=0}^{n-1} X_{jk} X_{jt} = \sum_{k=1}^{n} m_k \boldsymbol{r}_k^2.$$

Then, $\boldsymbol{r}_k = m_k^{-1/2} \sum_j \boldsymbol{R}_j X_{jk}$, and the Laplace operator and the orbital angular momentum operator $\boldsymbol{L}$ are directly expressed in $\boldsymbol{R}_j$:

$$\begin{aligned} &\sum_{k=1}^{n} m_k^{-1} \nabla_{\boldsymbol{r}_k}^2 = \sum_{j=0}^{n-1} \nabla_{\boldsymbol{R}_j}^2, \\ &\boldsymbol{L} = -\mathrm{i} \sum_{k=1}^{n} \boldsymbol{r}_k \times \nabla_{\boldsymbol{r}_k} = -\mathrm{i} \sum_{j=0}^{n-1} \boldsymbol{R}_j \times \nabla_{\boldsymbol{R}_j}, \end{aligned} \tag{5.229}$$

By the separation of variables, one has

$$\begin{aligned} &\boldsymbol{\Psi}(\boldsymbol{R}_0, \boldsymbol{R}_1, \ldots, \boldsymbol{R}_{n-1}) = \phi(\boldsymbol{R}_0)\, \Psi(\boldsymbol{R}_1, \ldots, \boldsymbol{R}_{n-1}), \\ &\phi(\boldsymbol{R}_0) = e^{\mathrm{i}\boldsymbol{P}_c \cdot \boldsymbol{R}_0/\sqrt{M}}, \qquad -\frac{1}{2} \nabla_{\boldsymbol{R}_0}^2 \phi(\boldsymbol{R}_0) = \frac{P_c^2}{2M} \phi(\boldsymbol{R}_0), \\ &-\frac{1}{2} \nabla^2 \Psi + V \Psi = \left(E - \frac{P_c^2}{2M}\right) \Psi, \qquad \nabla^2 = \sum_{j=1}^{n-1} \nabla_{\boldsymbol{R}_j}^2. \end{aligned} \tag{5.230}$$

The inverse transformation of Eq. (5.225) is $(k > j)$

$$\begin{aligned} \boldsymbol{r}_k &= \left[\frac{M_{k+1}}{m_k M_k}\right]^{1/2} \boldsymbol{R}_k - \sum_{j=1}^{k-1} \left[\frac{m_j}{M_j M_{j+1}}\right]^{1/2} \boldsymbol{R}_j + M^{-1/2} \boldsymbol{R}_0, \\ \boldsymbol{r}_k - \boldsymbol{r}_j &= \left[\frac{M_{k+1}}{m_k M_k}\right]^{1/2} \boldsymbol{R}_k - \sum_{i=j+1}^{k-1} \left[\frac{m_i}{M_i M_{i+1}}\right]^{1/2} \boldsymbol{R}_i - \left[\frac{M_j}{m_j M_{j+1}}\right]^{1/2} \boldsymbol{R}_j. \end{aligned} \tag{5.231}$$

$|\boldsymbol{r}_k - \boldsymbol{r}_j|^2$ are the functions of $\boldsymbol{R}_i \cdot \boldsymbol{R}_t$ which are rotational invariant and independent of $\boldsymbol{R}_0$. The last equality in Eq. (5.230) is the Schrödinger equation in the center-of-mass frame, which is independent of $\boldsymbol{R}_0$ and spherically symmetric. Hereafter, we will neglect the motion of center-of-mass by simply assuming $\boldsymbol{R}_0 = 0$ and $\boldsymbol{P}_c = 0$ in Eq. (5.230) for convenience.

5.9.2 *Separation of the Rotational Degree of Freedoms*

For a quantum n-body system, there are $(n-1)$ Jacobi coordinate vectors. Choose the body-fixed frame where $\boldsymbol{R}_1$ is parallel to its Z-axis and $\boldsymbol{R}_2$ is located in its XZ-plane with a non-negative X-component. Then, the Jacobi coordinate vectors $\boldsymbol{R}_j$ can be expressed by $\boldsymbol{R}_1$, $\boldsymbol{R}_2$ and $\boldsymbol{R}_1 \times \boldsymbol{R}_2$. The complete set of internal variables are

$$\begin{array}{lll} \xi_j = \boldsymbol{R}_j \cdot \boldsymbol{R}_1, & \eta_j = \boldsymbol{R}_j \cdot \boldsymbol{R}_2, & \zeta_j = \boldsymbol{R}_j \cdot (\boldsymbol{R}_1 \times \boldsymbol{R}_2), \\ 1 \leqslant j \leqslant (n-1), & \eta_1 = \xi_2, & \zeta_1 = \zeta_2 = 0, \end{array} \tag{5.232}$$

which are left invariant under the global rotation of the system. The number of the internal variables is $(3n-6)$, where ξ_j and η_j have even parity, but ζ_j have odd parity. Introduce a set of functions of internal variables,

$$\begin{aligned} \Omega_j &= (\boldsymbol{R}_1 \times \boldsymbol{R}_j) \cdot (\boldsymbol{R}_1 \times \boldsymbol{R}_2) = \xi_1\eta_j - \xi_2\xi_j, \\ \omega_j &= (\boldsymbol{R}_2 \times \boldsymbol{R}_j) \cdot (\boldsymbol{R}_1 \times \boldsymbol{R}_2) = \xi_2\eta_j - \eta_2\xi_j, \\ \Omega_1 &= \omega_2 = 0, \qquad \Omega_2 = -\omega_1 = (\boldsymbol{R}_1 \times \boldsymbol{R}_2)^2. \end{aligned} \tag{5.233}$$

Thus, the components R'_{jb} of $\boldsymbol{R}_j$, $1 \leqslant j \leqslant n-1$, in the body-fixed frame are

$$R'_{jx} = \Omega_j \left(\xi_1\Omega_2\right)^{-1/2}, \qquad R'_{jy} = \zeta_j\Omega_2^{-1/2}, \qquad R'_{jz} = \xi_j\xi_1^{-1/2}. \tag{5.234}$$

Especially, $R_{1x} = R_{1y} = R_{2y} = 0$, $R_{1z} = \sqrt{\xi_1}$, $R_{2x} = \sqrt{\Omega_2/\xi_1}$, and $R_{2z} = \xi_2/\sqrt{\xi_1}$. The volume element of the configuration space can be calculated from the Jacobi determinant by replacement of variables:

$$\prod_{j=1}^{n-1} dR_{jx}dR_{jy}dR_{jz} = \frac{1}{4}\Omega_2^{3-n}\sin\beta d\alpha d\beta d\gamma d\xi_1 d\xi_2 d\eta_2 \prod_{j=3}^{n-1} d\xi_j d\eta_j d\zeta_j. \tag{5.235}$$

The domains of definition of ξ_1 and η_2 are $(0,\infty)$ and the domains of definition of the remaining variables are $(-\infty,\infty)$. The potential V is a function of only the internal variables due to Eq. (5.231) and

$$\boldsymbol{R}_j \cdot \boldsymbol{R}_k = \Omega_2^{-1}\left(\Omega_j\eta_k - \omega_j\xi_k + \zeta_j\zeta_k\right), \tag{5.236}$$

Due to the spherical symmetry of the system, the static wave function $\Psi^\ell_\mu(x)$ must belong to the irreducible representation $D^\ell(\mathrm{SO}(3))$:

$$P_T \Psi^\ell_\mu(x) = \Psi^\ell_\mu(T^{-1}x) = \sum_\nu \Psi^\ell_\nu(x) D^\ell_{\nu\mu}(T). \tag{5.237}$$

Separate the coordinates x into two parts: $x = \{R, \xi\}$, where the Eular angles $R = R(\alpha, \beta, \gamma)$ denote the rotational variables, and ξ denotes the set of $(3n-6)$ internal variables. Letting $T = R^{-1}$, and replacing x by $x_0 = Tx = \{E, \xi\}$ in Eq. (5.237), we have (see Eq. (19.6) in [Wigner (1959)])

$$\Psi^\ell_\mu(x) = \sum_\nu \Psi^\ell_\nu(x_0) D^\ell_{\nu\mu}(R^{-1}) = \sum_{\nu=-\ell}^{\ell} \phi^\ell_\nu(\xi) D^\ell_{\mu\nu}(R)^*, \tag{5.238}$$

where $\phi^\ell_\nu(\xi) = \Psi^\ell_\nu(x_0)$. Therefore, the $(2\ell+1)$ sets of the basis functions $D^\ell_{\mu\nu}(R)^*$, $-\ell \leqslant \nu \leqslant \ell$, **constitute the independent and complete sets of eigenfunctions of the angular momentum L^2 and L_3 for the n-body system.** Unfortunately, $D^\ell_{\mu\nu}(R)^*$ depends on the Eular angles, and, in deriving the radial equations, the derivative of the Euler angles leads to singularity.

Due to spherical symmetry, we take $\mu = \ell$ in Eq. (5.238) for simplicity and try to find another complete sets of the basis functions, equivalent to $D^\ell_{\ell\nu}(R)^*$ and **depending on the rectangular coordinates such that the singularity disappears.** Those new sets of the basis functions can be combined from the harmonic polynomials by the CG coefficients:

$$\mathcal{Y}^{q(n-q)}_{\ell\ell}(\boldsymbol{R}_1, \boldsymbol{R}_2) = \sum_m \mathcal{Y}^q_m(\boldsymbol{R}_1) \mathcal{Y}^{n-q}_{\ell-m}(\boldsymbol{R}_2) C^{q(n-q)}_{m(\ell-m),\ell\ell}. \tag{5.239}$$

When $n = \ell$ and $n = \ell+1$, one has

$$\begin{aligned}
\mathcal{Y}^{q(\ell-q)}_{\ell\ell}(\boldsymbol{R}_1, \boldsymbol{R}_2) &= (-1)^\ell \left\{ \frac{[(2q+1)!(2\ell-2q+1)!]^{1/2}}{q!(\ell-q)!2^{\ell+2}\pi} \right\} \\
&\quad \cdot (R_{1x} + \mathrm{i}R_{1y})^q (R_{2x} + \mathrm{i}R_{2y})^{\ell-q}, \\
\mathcal{Y}^{q(\ell-q+1)}_{\ell\ell}(\boldsymbol{R}_1, \boldsymbol{R}_2) &= (-1)^\ell \left\{ \frac{(2q+1)!(2\ell-2q+3)!}{2q(\ell+1)(\ell-q+1)} \right\}^{1/2} \\
&\quad \cdot \left\{(q-1)!(\ell-q)!2^{\ell+2}\pi\right\}^{-1} (R_{1x} + \mathrm{i}R_{1y})^{q-1} (R_{2x} + \mathrm{i}R_{2y})^{\ell-q} \\
&\quad \cdot \left\{(R_{1x} + \mathrm{i}R_{1y})R_{2z} - R_{1z}(R_{2x} + \mathrm{i}R_{2y})\right\}.
\end{aligned} \tag{5.240}$$

It will be explained in subsection 10.4.4 that the remaining eigenfunctions contains the factor of the internal variables such that they are not independent. Removing the coefficients for convenience, we obtain $(2\ell+1)$

sets of the basis functions $Q_q^{\ell\lambda}(\boldsymbol{R}_1, \boldsymbol{R}_2)$, called the **generalized harmonic polynomials** in an n-body system: [Hsiang and Hsiang (1998)]

$$Q_q^{\ell\lambda}(\boldsymbol{R}_1, \boldsymbol{R}_2) = \frac{X^{q-\lambda}Y^{\ell-q}Z^{\lambda}}{(q-\lambda)!(\ell-q)!}, \qquad \lambda \leqslant q \leqslant \ell, \qquad \lambda = 0, 1,$$
$$X \equiv R_{1x} + \mathrm{i}R_{1y}, \qquad Y \equiv R_{2x} + \mathrm{i}R_{2y}, \qquad Z \equiv XR_{2z} - R_{1z}Y. \tag{5.241}$$

$Q_q^{\ell\lambda}(\boldsymbol{R}_1, \boldsymbol{R}_2)$ is a homogeneous polynomial of degree $(\ell + \lambda)$ in the components of the Jacobi coordinate vectors $\boldsymbol{R}_1$ and $\boldsymbol{R}_2$. It is the common eigenfunction of $\mathbf{L}^2$, L_z, $\mathbf{L}^2_{\mathbf{R}_1}$, $\mathbf{L}^2_{\mathbf{R}_2}$, $\triangle_{\mathbf{R}_1}$, $\triangle_{\mathbf{R}_2}$, $\nabla_{\mathbf{R}_1} \cdot \nabla_{\mathbf{R}_2}$, and the space inversion with the eigenvalues $\ell(\ell+1)$, ℓ, $q(q+1)$, $(\ell-q+\lambda)(\ell-q+\lambda+1)$, 0, 0, 0, and $(-1)^{\ell+\lambda}$, respectively, where $\mathbf{L}^2_{\mathbf{R}_1}$ $(\mathbf{L}^2_{\mathbf{R}_2})$ is the square of the partial angular momentum, and $\triangle_{\mathbf{R}_1}$ $(\triangle_{\mathbf{R}_2})$ is the Laplace operator with respect to the Jacobi coordinate vector $\mathbf{R}_1$ $(\mathbf{R}_2)$. Remind that

$$Q_q^{\ell 1}(\boldsymbol{R}_1, \boldsymbol{R}_2) = Q_{q-1}^{(\ell-1)0}(\boldsymbol{R}_1, \boldsymbol{R}_2)Z, \quad Z^2 = \eta_2 X^2 - \xi_2 XY + \xi_1 Y^2. \tag{5.242}$$

Thus, all the homogeneous polynomial of degree ℓ with respect to X, Y and Z are the combinations of $Q_q^{\ell\lambda}(\boldsymbol{R}_1, \boldsymbol{R}_2)$, where the coefficients are the functions of the internal variables. We are going to prove that any $D^{\ell}_{\ell\nu}(R)^*$ is the homogeneous polynomial of degree ℓ with respect to X, Y and Z where the coefficients are the function of the internal variables (see Eq. (5.244)). So that $Q_q^{\ell\lambda}(\boldsymbol{R}_1, \boldsymbol{R}_2)$ **are another complete sets of the basis functions of eigenfunctions of the angular momentum** L^2 and L_3 for the n-body system.

Let $R(\alpha, \beta, \gamma)$ denote the rotation of the system (see Eq. (5.50)). Two sets of the Jacobi coordinate vectors $\boldsymbol{R}_j$ and $\boldsymbol{R}'_j$, respectively in the center-of-mass frame and in the body-fixed frame, are related by $\boldsymbol{R}_j = R(\alpha, \beta, \gamma)\boldsymbol{R}'_j$ (see Eq. (5.234)). Through a straightforward calculation, we obtain

$$\begin{aligned}
&X = R_{1x} + \mathrm{i}R_{1y} = \xi_1^{1/2}\mathrm{e}^{\mathrm{i}\alpha}s_\beta, \\
&Y = R_{2x} + \mathrm{i}R_{2y} = (\Omega_2/\xi_1)^{1/2}\,\mathrm{e}^{\mathrm{i}\alpha}\left(c_\beta c_\gamma + \mathrm{i}s_\gamma\right) + \xi_2\xi_1^{-1/2}\mathrm{e}^{\mathrm{i}\alpha}s_\beta, \\
&R_{1z} = \xi_1^{1/2}c_\beta, \qquad R_{2z} = -\left(\Omega_2/\xi_1\right)^{1/2}s_\beta c_\gamma + \xi_2\xi_1^{-1/2}c_\beta, \\
&Z = \left(R_{1x} + \mathrm{i}R_{1y}\right)R_{2z} - R_{1z}\left(R_{2x} + \mathrm{i}R_{2y}\right) = -\Omega_2^{1/2}\mathrm{e}^{\mathrm{i}\alpha}\left(c_\gamma + \mathrm{i}c_\beta s_\gamma\right), \\
&R_{jx} + iR_{jy} = \Omega_2^{-1}\left\{-\omega_j X + \Omega_j Y - i\zeta_j Z\right\}, \\
&\left(R_{jx} + iR_{jy}\right)R_{kz} - R_{jz}\left(R_{kx} + iR_{ky}\right) = \Omega_2^{-1}\left\{i\left(\eta_j\zeta_k - \eta_k\zeta_j\right)X\right. \\
&\qquad \left.- i\left(\xi_j\zeta_k - \xi_k\zeta_j\right)Y + \left(\xi_j\eta_k - \xi_k\eta_j\right)Z\right\}.
\end{aligned} \tag{5.243}$$

From Eq. (5.64) one has

$$\begin{aligned} D^{\ell}_{\ell(\pm m)}(\alpha,\beta,\gamma)^* &= \frac{(-1)^{\ell-m}\sqrt{(2\ell)!}}{2^{\ell}\sqrt{(\ell+m)!(\ell-m)!}} e^{i(\ell\alpha\pm m\gamma)} s_{\beta}^{\ell-m}\left(1\pm c_{\beta}\right)^m \\ &= \frac{(-1)^{\ell-m}\sqrt{(2\ell)!}}{2^{\ell}\sqrt{(\ell+m)!(\ell-m)!}} \left(e^{i\alpha}s_{\beta}\right)^{\ell-m} \\ &\times \ \left[e^{i\alpha}\left(c_{\gamma}+ic_{\beta}s_{\gamma}\right)\pm e^{i\alpha}\left(c_{\beta}c_{\gamma}+is_{\gamma}\right)\right]^m, \end{aligned} \tag{5.244}$$

where

$$\begin{aligned} &e^{i\alpha}s_{\beta}=\xi_1^{-1/2}X, \qquad e^{i\alpha}\left(c_{\gamma}+ic_{\beta}s_{\gamma}\right)=-\Omega_2^{-1/2}Z, \\ &e^{i\alpha}\left(c_{\beta}c_{\gamma}+is_{\gamma}\right)=-\xi_2\left(\xi_1\Omega_2\right)^{-1/2}X+\left(\xi_1/\Omega_2\right)^{1/2}Y. \end{aligned} \tag{5.245}$$

As proved in Eqs. (5.244) and (5.245), the common eigenfunction $\Psi^{\ell\tau}_{\ell}(x)$ of L^2, L_z and space inversion P_σ with the eigenvalues $\ell(\ell+1)$, ℓ and $(-1)^{\tau}$ in an n-body system, given in Eq. (5.238), can be expanded with respect to $Q^{\ell\lambda}_q(\boldsymbol{R}_1,\boldsymbol{R}_2)$:

$$\begin{aligned} &\Psi^{\ell\tau}_{\ell}(x)=\sum_{q=\lambda}^{\ell}\sum_{\lambda=0}^{1}\phi^{\ell\tau}_{q\lambda}(\xi,\eta,\zeta)Q^{\ell\lambda}_q(\boldsymbol{R}_1,\boldsymbol{R}_2), \\ &\phi^{\ell\tau}_{q\lambda}(\xi,\eta,\zeta)=\phi^{\ell\tau}_{q\lambda}(\xi_1,\ldots,\xi_{n-1},\eta_2,\ldots,\eta_{n-1},\zeta_3,\ldots,\zeta_{n-1}), \\ &P_\sigma\phi^{\ell\tau}_{q\lambda}(\xi,\eta,\zeta)=(-1)^{\lambda-\tau}\phi^{\ell\tau}_{q\lambda}(\xi,\eta,\zeta). \end{aligned} \tag{5.246}$$

When substituting Eq. (5.246) into the Schrödinger equation (5.230), the action of the Laplace operator to the function $\Psi^{\ell\tau}_{\ell}(\boldsymbol{R}_1,\ldots,\boldsymbol{R}_{n-1})$ is separated into three parts. The first is its action on the radial functions $\phi^{\ell\tau}_{q\lambda}(\xi,\eta,\zeta)$ which can be calculated by the replacement of variables directly,

$$\begin{aligned} \nabla^2\phi^{\ell\lambda}_{q\tau}(\xi,\eta,\zeta) &= \Big\{4\xi_1\partial^2_{\xi_1}+4\eta_2\partial^2_{\eta_2}+\left(\xi_1+\eta_2\right)\partial^2_{\xi_2} \\ &+4\xi_2\left(\partial_{\xi_1}+\partial_{\eta_2}\right)\partial_{\xi_2}+6\left(\partial_{\xi_1}+\partial_{\eta_2}\right)+\sum_{j=3}^{n-1}\Big[\xi_1\partial^2_{\xi_j}+\eta_2\partial^2_{\eta_j} \\ &+\Omega_2\partial^2_{\zeta_j}+2\xi_2\partial_{\xi_j}\partial_{\eta_j}+4\left(\xi_j\partial_{\xi_j}+\zeta_j\partial_{\zeta_j}\right)\partial_{\xi_1} \\ &+4\left(\eta_j\partial_{\eta_j}+\zeta_j\partial_{\zeta_j}\right)\partial_{\eta_2}+2\left(\eta_j\partial_{\xi_j}+\xi_j\partial_{\eta_j}\right)\partial_{\xi_2}\Big] \\ &+\Omega_2^{-1}\sum_{j,k=3}^{n-1}\Big[\left(\Omega_j\eta_k-\omega_j\xi_k+\zeta_j\zeta_k\right)\left(\partial_{\xi_j}\partial_{\xi_k}+\partial_{\eta_j}\partial_{\eta_k}\right) \\ &-2\left(\omega_j\zeta_k-\omega_k\zeta_j\right)\partial_{\xi_j}\partial_{\zeta_k}+2\left(\Omega_j\zeta_k-\Omega_k\zeta_j\right)\partial_{\eta_j}\partial_{\zeta_k} \\ &+\left(\Omega_j\Omega_k+\omega_j\omega_k+\xi_1\zeta_j\zeta_k+\eta_2\zeta_j\zeta_k\right)\partial_{\zeta_j}\partial_{\zeta_k}\Big]\Big\}\,\phi^{\ell\lambda}_{q\tau}(\xi,\eta,\zeta). \end{aligned} \tag{5.247}$$

The second is its action on the generalized harmonic polynomials $Q_q^{\ell\tau}(\boldsymbol{R}_1, \boldsymbol{R}_2)$ which is vanishing because $Q_q^{\ell\tau}(\boldsymbol{R}_1, \boldsymbol{R}_2)$ satisfies the Laplace equation. The third is the mixed application

$$2\left\{\left(\partial_{\xi_1}\phi_{q\tau}^{\ell\lambda}\right)2\boldsymbol{R}_1 + \left(\partial_{\xi_2}\phi_{q\tau}^{\ell\lambda}\right)\boldsymbol{R}_2 + \sum_{j=3}^{n-1}\left[\left(\partial_{\xi_j}\phi_{q\tau}^{\ell\lambda}\right)\boldsymbol{R}_j \right.\right.$$
$$\left.\left. + \left(\partial_{\zeta_j}\phi_{q\tau}^{\ell\lambda}\right)(\boldsymbol{R}_2\times\boldsymbol{R}_j)\right]\right\}\cdot\nabla_{\boldsymbol{R}_1}Q_q^{\ell\tau} + 2\left\{\left(\partial_{\xi_2}\phi_{q\tau}^{\ell\lambda}\right)\boldsymbol{R}_1 + \left(\partial_{\eta_2}\phi_{q\tau}^{\ell\lambda}\right)2\boldsymbol{R}_2\right.$$
$$\left. + \sum_{j=3}^{n-1}\left[\left(\partial_{\eta_j}\phi_{q\tau}^{\ell\lambda}\right)\boldsymbol{R}_j + \left(\partial_{\zeta_j}\phi_{q\tau}^{\ell\lambda}\right)(\boldsymbol{R}_j\times\boldsymbol{R}_1)\right]\right\}\cdot\nabla_{\boldsymbol{R}_2}Q_q^{\ell\tau}.$$

In terms of Eqs. (5.241) and (5.243) one obtains

$$\begin{aligned}
&\boldsymbol{R}_1\cdot\nabla_{\boldsymbol{R}_1}Q_q^{\ell\tau} = qQ_q^{\ell\tau}, \qquad \boldsymbol{R}_2\cdot\nabla_{\boldsymbol{R}_2}Q_q^{\ell\tau} = (\ell-q+\tau)\,Q_q^{\ell\tau},\\
&\boldsymbol{R}_2\cdot\nabla_{\boldsymbol{R}_1}Q_q^{\ell\tau} = (\ell-q+1)\,Q_{q-1}^{\ell\tau}, \qquad \boldsymbol{R}_1\cdot\nabla_{\boldsymbol{R}_2}Q_q^{\ell\tau} = (q-\tau+1)\,Q_{q+1}^{\ell\tau},\\
&\boldsymbol{R}_j\cdot\nabla_{\boldsymbol{R}_1}Q_q^{\ell 0} = \Omega_2^{-1}\left\{-\omega_j qQ_q^{\ell 0} + \Omega_j(\ell-q+1)\,Q_{q-1}^{\ell 0} - i\zeta_j Q_q^{\ell 1}\right\},\\
&\boldsymbol{R}_j\cdot\nabla_{\boldsymbol{R}_2}Q_q^{\ell 0} = \Omega_2^{-1}\left\{-\omega_j(q+1)Q_{q+1}^{\ell 0} + \Omega_j(\ell-q)\,Q_q^{\ell 0} - i\zeta_j Q_{q+1}^{\ell 1}\right\},\\
&\boldsymbol{R}_j\cdot\nabla_{\boldsymbol{R}_1}Q_q^{\ell 1} = \Omega_2^{-1}\left\{-i\eta_2\zeta_j q^2 Q_q^{\ell 0} + i\xi_2\zeta_j(2q-1)(\ell-q+1)Q_{q-1}^{\ell 0}\right.\\
&\qquad \left. - i\xi_1\zeta_j(\ell-q+2)(\ell-q+1)Q_{q-2}^{\ell 0} - \omega_j qQ_q^{\ell 1} + \Omega_j(\ell-q+1)Q_{q-1}^{\ell 1}\right\},\\
&\boldsymbol{R}_j\cdot\nabla_{\boldsymbol{R}_2}Q_q^{\ell 1} = \Omega_2^{-1}\left\{-i\eta_2\zeta_j(q+1)qQ_{q+1}^{\ell 0} + i\xi_2\zeta_j q(2\ell-2q+1)Q_q^{\ell 0}\right.\\
&\qquad \left. - i\xi_1\zeta_j(\ell-q+1)^2 Q_{q-1}^{\ell 0} - \omega_j qQ_{q+1}^{\ell 1} + \Omega_j(\ell-q+1)Q_q^{\ell 1}\right\},\\
&(\boldsymbol{R}_2\times\boldsymbol{R}_j)\cdot\nabla_{\boldsymbol{R}_1}Q_q^{\ell 0} = \Omega_2^{-1}\left\{\eta_2\zeta_j qQ_q^{\ell 0} - \xi_2\zeta_j(\ell-q+1)\,Q_{q-1}^{\ell 0} - i\omega_j Q_q^{\ell 1}\right\},\\
&(\boldsymbol{R}_j\times\boldsymbol{R}_1)\cdot\nabla_{\boldsymbol{R}_2}Q_q^{\ell 0} = \Omega_2^{-1}\left\{-\xi_2\zeta_j(q+1)Q_{q+1}^{\ell 0} + \xi_1\zeta_j(\ell-q)\,Q_q^{\ell 0}\right.\\
&\qquad \left. + i\Omega_j Q_{q+1}^{\ell 1}\right\},\\
&(\boldsymbol{R}_2\times\boldsymbol{R}_j)\cdot\nabla_{\boldsymbol{R}_1}Q_q^{\ell 1} = \Omega_2^{-1}\left\{-i\eta_2\omega_j q^2 Q_q^{\ell 0} + i\xi_2\omega_j(2q-1)(\ell-q+1)\,Q_{q-1}^{\ell 0}\right.\\
&\qquad \left. - i\xi_1\omega_j(\ell-q+2)(\ell-q+1)Q_{q-2}^{\ell 0} + \eta_2\zeta_j qQ_q^{\ell 1} - \xi_2\zeta_j(\ell-q+1)\,Q_{q-1}^{\ell 1}\right\},\\
&(\boldsymbol{R}_j\times\boldsymbol{R}_1)\cdot\nabla_{\boldsymbol{R}_2}Q_q^{\ell 1} = \Omega_2^{-1}\left\{i\eta_2\Omega_j(q+1)qQ_{q+1}^{\ell 0} - i\xi_2\Omega_j q\,(2\ell-2q+1)\,Q_q^{\ell 0}\right.\\
&\qquad \left. + i\xi_1\Omega_j(\ell-q+1)^2 Q_{q-1}^{\ell 0} - \xi_2\zeta_j qQ_{q+1}^{\ell 1} + \xi_1\zeta_j(\ell-q+1)\,Q_q^{\ell 1}\right\}.
\end{aligned}$$

Now, the radial equations are

$$\triangle\phi_{q0}^{\ell\lambda} + 4\left\{q\partial_{\xi_1} + (\ell-q)\partial_{\eta_2}\right\}\phi_{q0}^{\ell\lambda} + 2q\partial_{\xi_2}\phi_{(q-1)0}^{\ell\lambda} + 2(\ell-q)\partial_{\xi_2}\phi_{(q+1)0}^{\ell\lambda}$$
$$+\sum_{j=3}^{n-1}2\Omega_2^{-1}\left\{\left[-\omega_j q\partial_{\xi_j} + \Omega_j(\ell-q)\partial_{\eta_j} + \eta_2\zeta_j q\partial_{\zeta_j} + \xi_1\zeta_j(\ell-q)\partial_{\zeta_j}\right]\phi_{q0}^{\ell\lambda}\right.$$

$$
\begin{aligned}
&- q\left[\omega_j\partial_{\eta_j} + \xi_2\zeta_j\partial_{\zeta_j}\right]\phi^{\ell\lambda}_{(q-1)0} + (\ell - q)\left[\Omega_j\partial_{\xi_j} - \xi_2\zeta_j\partial_{\zeta_j}\right]\phi^{\ell\lambda}_{(q+1)0} \\
&- i\eta_2 q(q-1)\left[\zeta_j\partial_{\eta_j} - \Omega_j\partial_{\zeta_j}\right]\phi^{\ell\lambda}_{(q-1)1} - iq\left[\eta_2\zeta_j q\partial_{\xi_j}\right. \\
&\left. - \xi_2\zeta_j(2\ell - 2q + 1)\partial_{\eta_j} + \eta_2\omega_j q\partial_{\zeta_j} + \xi_2\Omega_j(2\ell - 2q + 1)\partial_{\zeta_j}\right]\phi^{\ell\lambda}_{q1} \\
&+ \mathrm{i}(\ell - q)\left[\xi_2\zeta_j(2q+1)\partial_{\xi_j} - \xi_1\zeta_j(\ell - q)\partial_{\eta_j} + \xi_2\omega_j(2q+1)\partial_{\zeta_j}\right. \\
&\left. + \xi_1\Omega_j(\ell - q)\partial_{\zeta_j}\right]\phi^{\ell\lambda}_{(q+1)1} - i\xi_1(\ell - q)(\ell - q - 1)\left[\zeta_j\partial_{\xi_j} + \omega_j\partial_{\zeta_j}\right]\phi^{\ell\lambda}_{(q+2)1}\Big\} \\
&= -2\left[E - V\right]\phi^{\ell\lambda}_{q0},
\end{aligned}
$$

$$
\begin{aligned}
&\triangle\phi^{\ell\lambda}_{q1} + 4\left\{q\partial_{\xi_1} + (\ell - q + 1)\partial_{\eta_2}\right\}\phi^{\ell\lambda}_{q1} + 2(q-1)\partial_{\xi_2}\phi^{\ell\lambda}_{(q-1)1} \\
&+ 2(\ell - q)\partial_{\xi_2}\phi^{\ell\lambda}_{(q+1)1} + \sum_{j=3}^{n-1} 2\Omega_2^{-1}\left\{\left[-\omega_j q\partial_{\xi_j} + \Omega_j(\ell - q + 1)\partial_{\eta_j}\right.\right. \\
&\left. + \eta_2\zeta_j q\partial_{\zeta_j} + \xi_1\zeta_j(\ell - q + 1)\partial_{\zeta_j}\right]\phi^{\ell\lambda}_{q1} \\
&- (q-1)\left[\omega_j\partial_{\eta_j} + \xi_2\zeta_j\partial_{\zeta_j}\right]\phi^{\ell\lambda}_{(q-1)1} + (\ell - q)\left[\Omega_j\partial_{\xi_j} - \xi_2\zeta_j\partial_{\zeta_j}\right]\phi^{\ell\lambda}_{(q+1)1} \\
&\left. - i\left[\zeta_j\partial_{\eta_j} - \Omega_j\partial_{\zeta_j}\right]\phi^{\ell\lambda}_{(q-1)0} - i\left[\zeta_j\partial_{\xi_j} + \omega_j\partial_{\zeta_j}\right]\phi^{\ell\lambda}_{q0}\right\} \\
&= -2\left[E - V\right]\phi^{\ell\lambda}_{q1},
\end{aligned} \tag{5.248}
$$

where $\triangle\phi^{\ell\lambda}_{q\tau}$ was given in Eq. (5.247). When $n = 3$, there are only two Jacobi vectors, $\boldsymbol{R}_1$ and $\boldsymbol{R}_2$, and three internal variables, ξ_1, ξ_2 and η_2. Since all internal variables have even parity, Eq. (5.246) reduces to

$$
\Psi^{\ell\lambda}_{\ell}(x) = \sum_{q=\lambda}^{\ell} \phi^{\ell\lambda}_{q}(\xi_1, \xi_2, \eta_2) Q^{\ell\lambda}_{q}(\boldsymbol{R}_1, \boldsymbol{R}_2), \tag{5.249}
$$

The radial equation is [Schwartz (1961); Hsiang and Hsiang (1998)]

$$
\begin{aligned}
&\nabla^2\phi^{\ell\lambda}_{q} + 4q\partial_{\xi_1}\phi^{\ell\lambda}_{q} + 4(\ell - q + \lambda)\partial_{\eta_2}\phi^{\ell\lambda}_{q} + 2(q - \lambda)\partial_{\xi_2}\phi^{\ell\lambda}_{q-1} \\
&\qquad + 2(\ell - q)\partial_{\xi_2}\phi^{\ell\lambda}_{q+1} = -2\left(E - V\right)\phi^{\ell\lambda}_{q}, \\
&\nabla^2\phi^{\ell\lambda}_{q}(\xi_1, \xi_2, \eta_2) = \Big\{4\xi_1\partial^2_{\xi_1} + 4\eta_2\partial^2_{\eta_2} + 6\left(\partial_{\xi_1} + \partial_{\eta_2}\right) \\
&\qquad + \left(\xi_1 + \eta_2\right)\partial^2_{\xi_2} + 4\xi_2\left(\partial_{\xi_1} + \partial_{\eta_2}\right)\partial_{\xi_2}\Big\}\psi^{\ell\lambda}_{q}(\xi_1, \xi_2, \eta_2), \\
&\lambda \leqslant q \leqslant \ell, \qquad \lambda = 0, 1.
\end{aligned} \tag{5.250}
$$

In the calculation of the energy levels of the helium atom, a transformation of variables is useful for improving convergence of the series [Duan et al.

(2001a); Duan et al. (2002)]. For the calculation of four-body system the formula of the rotational invariant is helpful [Ma and Yan (2015)].

5.10 Exercises

1. Prove the preliminary formula by mathematical induction:
$$\begin{aligned} e^{\alpha}\beta e^{-\alpha} &= \beta + \frac{1}{1!}[\alpha,\ \beta] + \frac{1}{2!}[\alpha,\ [\alpha,\ \beta]] + \cdots \\ &= \sum_{n=0}^{\infty} \frac{1}{n!} \overbrace{[\alpha,\ [\alpha,\ \cdots [\alpha,}^{n}\ \beta] \cdots]\,], \end{aligned}$$
where α and β are two matrices with the same dimension. Please prove Eq. (5.32) in terms of this formula, where $u(\hat{\boldsymbol{n}}, \omega)$ and $R(\hat{\boldsymbol{n}}, \omega)$ are given in Eqs. (5.26) and (5.13). Thus, SU(2) is homomorphic onto SO(3).
2. Expand $R(\hat{\boldsymbol{n}}, \omega) = \exp(-i\omega\hat{\boldsymbol{n}} \cdot \boldsymbol{T})$ as a sum of matrices with the finite terms.
Hint: $(\hat{\boldsymbol{n}} \cdot \boldsymbol{T})^3 = \hat{\boldsymbol{n}} \cdot \boldsymbol{T}$.
3. Prove that the elements $u(\hat{\boldsymbol{n}}, \omega)$ with the same ω form a class of the SU(2) group.
4. Check the following multiplication formulas in the group **O** in terms of the homomorphism SU(2) $\sim$ SO(3):
$$T_z R_1 = S_3, \qquad T_z T_x = R_1, \qquad R_1 R_2 = R_3^2.$$
5. Calculate the Euler angles for the following transformation matrices $R(\alpha, \beta, \gamma)$, and write their representation matrices $D^j(\alpha, \beta, \gamma)$ of SO(3):
$$\frac{1}{4}\begin{pmatrix} -\sqrt{3}-2 & \sqrt{3}-2 & -\sqrt{2} \\ \sqrt{3}-2 & -\sqrt{3}-2 & \sqrt{2} \\ -\sqrt{2} & \sqrt{2} & 2\sqrt{3} \end{pmatrix}, \qquad \frac{1}{2}\begin{pmatrix} \sqrt{3} & -1 & 0 \\ -1 & -\sqrt{3} & 0 \\ 0 & 0 & -2 \end{pmatrix}$$
$$\frac{1}{8}\begin{pmatrix} \sqrt{6}+2\sqrt{3} & 3\sqrt{2}-2 & 2\sqrt{6} \\ \sqrt{2}-6 & \sqrt{6}+2\sqrt{3} & 2\sqrt{2} \\ -2\sqrt{2} & -2\sqrt{6} & 4\sqrt{2} \end{pmatrix}.$$
6. Calculate the Euler angles for the following rotations $R(\hat{\boldsymbol{n}}, \omega)$, and write their representation matrices $D^j(R)$ of SO(3).
   (a) $\hat{\boldsymbol{n}} = \boldsymbol{e}_1 \sin\theta + \boldsymbol{e}_3 \cos\theta$ where θ is an acute angle;

(b) $\hat{\boldsymbol{n}} = (\boldsymbol{e}_1 + \boldsymbol{e}_2 + \boldsymbol{e}_3)/\sqrt{3}$ and $\omega = 2\pi/3$;
(c) $\hat{\boldsymbol{n}} = (\boldsymbol{e}_1 + \boldsymbol{e}_2)/\sqrt{2}$ and $\omega = \pi$.

7. Calculate the Euler angles for the rotations T_0, T_2, R_1, R_2, R_6, S_1, S_2, S_6, S_{11}, and S_{12} in the icosahedron group **I**, where the definitions of the elements are given in subsection 2.5.3.

8. Decompose the subduced representation of the irreducible representation D^3 of SO(3) as the direct sum of the irreducible representations of the subgroup D_3, which are given in Eq. (2.14) and Table 3.5, and find the similarity transformation matrix.

9. Prove Eq. (5.90) by mathematical induction.

10. Calculate the decomposition of the subduced representation of each irreducible representation of the icosahedron group **I** with respect to the subgroups C_5, D_5, and **T**, respectively.

11. Decompose the subduced representations of the irreducible representations D^{20} and D^{18} of SO(3) with respect to the subgroup **I**, respectively.

12. Calculate the Clebsch–Gordan series in the decomposition of the direct product representation in terms of the character table of the group **I**:

(1) $D^{T_1} \times D^{T_1}$; (2) $D^{T_1} \times D^{T_2}$; (3) $D^{T_2} \times D^{T_2}$; (4) $D^{T_1} \times D^{G}$;
(5) $D^{T_2} \times D^{G}$; (6) $D^{T_1} \times D^{H}$; (7) $D^{T_2} \times D^{H}$; (8) $D^{G} \times D^{G}$;
(9) $D^{G} \times D^{H}$; (10) $D^{H} \times D^{H}$.

13. Calculate the representation matrices of the elements S_{14}, R_3 and R_7^2 of the group **I** in the irreducible representations $D^{T_1}(\mathbf{I})$ and $D^{T_2}(\mathbf{I})$ from its irreducible bases (5.106) and (5.109).

14. For any one-order Lie group with the composition function $f(r;s)$, please find a new parameter r' such that the new composition function is the additive function, $f'(r';s') = r' + s'$. The Lorentz transformation $A(v)$ for the boost along the Z-axis with the relative velocity v is taken in the following form. It forms a Lie group of one order:

$$A(v) = \begin{pmatrix} 1 & 0 & 0 & 0 \\ 0 & 1 & 0 & 0 \\ 0 & 0 & \gamma & -\mathrm{i}\gamma v/c \\ 0 & 0 & \mathrm{i}\gamma v/c & \gamma \end{pmatrix}, \qquad \begin{aligned} \gamma &= \left(1 - v^2/c^2\right)^{-1/2}, \\ f(v_1; v_2) &= \frac{v_1 + v_2}{1 + v_1 v_2/c^2}. \end{aligned}$$

Find the new parameter with the additive composition function.

15. Calculate the representation matrices of the generators in an irreducible representation of SU(2) with a finite dimension by the second Lie theorem.

16. Calculate the eigenfunction with the eigenvalue m of the orbital angular momentum operator $\boldsymbol{L} \cdot \hat{\boldsymbol{a}}$ in terms of the linear combination of the spherical harmonic functions $Y^{\ell}_m(\hat{\boldsymbol{n}})$, where $\hat{\boldsymbol{a}} = (\boldsymbol{e}_1 - \boldsymbol{e}_2)/\sqrt{2}$.

17. Let the function $\psi^{\ell}_m(x)$ belong to the mth row of the irreducible representation D^{ℓ} of SO(3). Calculate the eigenfunction with the eigenvalue m of the orbital angular momentum operator $\boldsymbol{L}\cdot\hat{\boldsymbol{b}}$ in terms of the linear combination of $\psi^{\ell}_m(x)^*$, where $\hat{\boldsymbol{b}} = \left(\sqrt{3}\boldsymbol{e}_2 + \boldsymbol{e}_3\right)/2$.

Hint: Use Eq. (5.72).

18. Calculate $\left\{d^{\ell}(\theta)\left(I^{\ell}_3\right)^2 d^{\ell}(\theta)^{-1}\right\}_{mm}$, where $d^{\ell}(\theta) = D^{\ell}(\boldsymbol{e}_2, \theta)$ and I^{ℓ}_3 is the third generator in the representation $D^{\ell}(\mathrm{SO}(3))$.

Hint: Use the property of the adjoint representation.

19. Directly calculate the Clebsch–Gordan coefficients for the direct product representation of two irreducible representations of SU(2) in terms of the raising and lowering operators $J_{\pm}$: (a) $D^{1/2} \times D^{1/2}$, (b) $D^{1/2} \times D^1$, (c) $D^1 \times D^1$, (d) $D^1 \times D^{3/2}$.

20. Prove the Clebsch–Gordan coefficients in the following formulas by the raising and lowering operators $J_{\pm}$:

(a) In the decomposition of $D^{1/2} \times D^j$,

$$\begin{aligned}
\|(j+1/2), M\rangle &= \left(\frac{j+M+1/2}{2j+1}\right)^{1/2} |1/2, 1/2\rangle|j, M-1/2\rangle \\
&\quad + \left(\frac{j-M+1/2}{2j+1}\right)^{1/2} |1/2, -1/2\rangle|j, M+1/2\rangle, \\
\|(j-1/2), M\rangle &= \left(\frac{j-M+1/2}{2j+1}\right)^{1/2} |1/2, 1/2\rangle|j, M-1/2\rangle \\
&\quad - \left(\frac{j+M+1/2}{2j+1}\right)^{1/2} |1/2, -1/2\rangle|j, M+1/2\rangle.
\end{aligned}$$

(b) In the decomposition of $D^1 \times D^j$,

$$\|(j+1), M\rangle = \left\{\frac{(j+M)(j+M+1)}{2(2j+1)(j+1)}\right\}^{1/2} |1, 1\rangle|j, M-1\rangle$$

$$+\left\{\frac{(j-M+1)(j+M+1)}{(2j+1)(j+1)}\right\}^{1/2}|1,0\rangle|j,M\rangle$$
$$+\left\{\frac{(j-M)(j-M+1)}{2(2j+1)(j+1)}\right\}^{1/2}|1,-1\rangle|j,M+1\rangle,$$
$$||j,M\rangle=\left\{\frac{(j+M)(j-M+1)}{2j(j+1)}\right\}^{1/2}|1,1\rangle|j,M-1\rangle$$
$$-\frac{M}{[j(j+1)]^{1/2}}|1,0\rangle|j,M\rangle$$
$$-\left\{\frac{(j-M)(j+M+1)}{2j(j+1)}\right\}^{1/2}|1,-1\rangle|j,M+1\rangle,$$
$$||(j-1),M\rangle=\left\{\frac{(j-M)(j-M+1)}{2j(2j+1)}\right\}^{1/2}|1,1\rangle|j,M-1\rangle$$
$$-\left\{\frac{(j-M)(j+M)}{j(2j+1)}\right\}^{1/2}|1,0\rangle|j,M\rangle$$
$$+\left\{\frac{(j+M)(j+M+1)}{2j(2j+1)}\right\}^{1/2}|1,-1\rangle|j,M+1\rangle.$$

21. Calculate the eigenfunctions of the total spinor angular momentum in a three-electron system.

22. Establish the differential equation satisfied by the matrix elements $D^j_{\nu\mu}(\alpha,\beta,\gamma)$ of the representation D^j of SO(3).

$$\left\{-\frac{1}{s_\beta}\frac{\partial}{\partial\beta}s_\beta\frac{\partial}{\partial\beta}-\frac{1}{s_\beta^2}\left(\frac{\partial^2}{\partial\gamma^2}+\frac{\partial^2}{\partial\alpha^2}\right)+\frac{2c_\beta}{s_\beta^2}\frac{\partial^2}{\partial\alpha\partial\gamma}\right\}D^j_{\nu\mu}(\alpha,\beta,\gamma)$$
$$=j(j+1)D^j_{\nu\mu}(\alpha,\beta,\gamma).$$

Hint: In terms of Eqs. (5.62) and (5.129).

23. Discuss all inequivalent and irreducible unitary representations of the SO(3) group and the SO(2,1) group.

Chapter 6

SYMMETRY OF CRYSTALS

The study on the symmetry of crystals is a typical example of application of group theory to physics. Due to the translation symmetry of crystals, through a systematic study by group theory, all the crystals are classified completely into 7 crystal systems, 14 Bravais lattices, and 230 distinct types of crystal structures. There are 32 crystallographic point groups (11 of which are proper point groups), and 230 space groups (73 of which are symmorphic space groups). In this chapter we will study the symmetry group of crystals, their representations, and the classification of crystals.

6.1 Symmetry Group of Crystals

The fundamental character of a crystal is the spatial periodic array of atoms or group of atoms composing the crystal. The abstraction of such periodic array of the atoms or group of atoms into array of points form the **crystal lattice structure**. By the periodic boundary condition, the crystal is left invariant under the following translation (see [Ren (2017)] for the crystals of finite size):

$$\boldsymbol{r} \longrightarrow T(\boldsymbol{\ell})\boldsymbol{r} = \boldsymbol{r} + \boldsymbol{\ell}, \tag{6.1}$$

where $\boldsymbol{\ell}$ is called the **vector of crystal lattice**. Three fundamental periods $\boldsymbol{a}_j$ of a crystal lattice, which are not coplanar, are taken to be the basis vectors of crystal lattice, or briefly called the **lattice bases**. The lattice bases are said to be **primitive** if each vector of crystal lattice is an integral linear combination of the lattice bases. For simplicity, we only use the primitive lattice bases unless specified otherwise. For a given set of lattice bases $\boldsymbol{a}_i$, a vector of crystal lattice is characterized by three integers ℓ_i:

$$\boldsymbol{\ell} = \boldsymbol{a}_1\ell_1 + \boldsymbol{a}_2\ell_2 + \boldsymbol{a}_3\ell_3 = \sum_{i=1}^{3} \boldsymbol{a}_i\ell_i, \qquad \ell_i \text{ are integers.} \tag{6.2}$$

The multiplication of two translations is defined to be a translation where two translation vectors are added. The set of all translations $T(\boldsymbol{\ell})$, which leaves the crystal invariant, forms an abelian group, called the **translation group** $\mathcal{T}$ of the crystal. The periodic boundary condition makes the translation group $\mathcal{T}$ a finite group even though its order is very large.

Usually, in addition to the translation symmetry, there are some other **symmetric operations**, which also leaves the crystal invariant. A general symmetric operation is composed of the rotation, the space inversion, and the translation, and is denoted by $g(R, \boldsymbol{\alpha})$,

$$\boldsymbol{r} \longrightarrow g(R, \boldsymbol{\alpha})\boldsymbol{r} = R\boldsymbol{r} + \boldsymbol{\alpha}, \tag{6.3}$$

where $R \in \mathrm{O}(3)$ is a proper or improper rotation, and $\boldsymbol{\alpha}$ is a translation vector, not necessary to be a vector of crystal lattice $\boldsymbol{\ell}$. If $\boldsymbol{\alpha} = 0$, $g(R, 0) = R$ is a proper or improper rotation which leaves the origin invariant. If $R = E$, $\boldsymbol{\alpha}$ has to be a vector of crystal lattice $\boldsymbol{\ell}$ and $g(E, \boldsymbol{\ell}) = T(\boldsymbol{\ell})$. The multiplication of two symmetric operations is defined as their successive applications,

$$g(R, \boldsymbol{\alpha})g(R', \boldsymbol{\beta})\boldsymbol{r} = g(R, \boldsymbol{\alpha})\left\{R'\boldsymbol{r} + \boldsymbol{\beta}\right\} = RR'\boldsymbol{r} + \boldsymbol{\alpha} + R\boldsymbol{\beta},$$

$$g(R, \boldsymbol{\alpha})g(R', \boldsymbol{\beta}) = g(RR', \boldsymbol{\alpha} + R\boldsymbol{\beta}). \tag{6.4}$$

The inverse of $g(R, \boldsymbol{\alpha})$ is

$$g(R, \boldsymbol{\alpha})^{-1} = g(R^{-1}, -R^{-1}\boldsymbol{\alpha}). \tag{6.5}$$

The set of all symmetric operations $g(R, \boldsymbol{\alpha})$ of a crystal, under the multiplication rule (6.4), forms a group $\mathcal{S}$, called the **space group** of the crystal. In the multiplication (6.4) of two symmetric operations, the rotational part obeys the multiplication rule of two rotations, but the translational part is affected by the rotational part. Therefore, the set of the rotational parts R in $g(R, \boldsymbol{\alpha})$ forms a group G, called the **crystallographic point group**.

The translation group $\mathcal{T}$ is an invariant subgroup of the space group $\mathcal{S}$ because the conjugate element of a translation $T(\boldsymbol{\ell})$ is still a translation,

$$g(R, \boldsymbol{\alpha})T(\boldsymbol{\ell})g(R, \boldsymbol{\alpha})^{-1} = g\left(E, \boldsymbol{\alpha} + R(\boldsymbol{\ell} - R^{-1}\boldsymbol{\alpha})\right) = T(R\boldsymbol{\ell}),$$

$$R\boldsymbol{\ell} = \boldsymbol{\ell}'. \tag{6.6}$$

The general symmetric operation $g(R, \boldsymbol{\alpha})$ can be decomposed by separating the vector of crystal lattice $\boldsymbol{\ell}$ from $\boldsymbol{\alpha}$,

$$\begin{aligned} g(R, \boldsymbol{\alpha}) &= T(\boldsymbol{\ell})g(R, \boldsymbol{t}), \qquad \boldsymbol{\alpha} = \boldsymbol{\ell} + \boldsymbol{t} \\ \boldsymbol{t} &= \sum_{j=1}^{3} \boldsymbol{a}_j t_j, \qquad 0 \leqslant t_j < 1. \end{aligned} \tag{6.7}$$

Any symmetric operation $g(R, \boldsymbol{\beta})$ in a crystal for a given R satisfies

$$g(R, \boldsymbol{t})^{-1} g(R, \boldsymbol{\beta}) = T(-R^{-1}\boldsymbol{t} + R^{-1}\boldsymbol{\beta}) = T(\boldsymbol{\ell}).$$

$\boldsymbol{\beta} = \boldsymbol{t} + R\boldsymbol{\ell} = \boldsymbol{t} + \boldsymbol{\ell}'$. Thus, for a given crystal, **the vector $\boldsymbol{t}$ given in** Eq. (6.7) **depends upon R uniquely**. Furthermore, equation (6.7) shows that the coset of $\mathcal{T}$ is completely determined by the rotation R. **The quotient group of $\mathcal{T}$ with respect to $\mathcal{S}$ is isomorphic to the crystallographic point group G**,

$$\mathrm{G} = \mathcal{T}/\mathcal{S}. \tag{6.8}$$

Generally, R is not an element of $\mathcal{S}$, and G is not a subgroup of $\mathcal{S}$. The space group $\mathcal{S}$ is called the **symmorphic space group** if G is the subgroup of $\mathcal{S}$. Namely, in a symmorphic space group, $\boldsymbol{t}$ in each $g(R, \boldsymbol{t})$ vanishes, and any element in $\mathcal{S}$ can be expressed as $g(R, \boldsymbol{\ell}) = T(\boldsymbol{\ell})R$.

It will be seen that **equation** (6.6) **is a fundamental constraint for the possible crystallographic point groups, the crystal systems, and the Bravais lattices.**

6.2 Crystallographic Point Groups

6.2.1 *Elements in a Crystallographic Point Group*

In the crystal theory, it is convenient to choose the lattice bases $\boldsymbol{a}_j$ to be the basis vectors. The merit for this choice is that, due to Eq. (6.6), **all elements** of the matrix $D_{ij}(R)$ of R in the bases are **integers**:

$$R\boldsymbol{a}_j = \sum_{i=1}^{3} \boldsymbol{a}_i D_{ij}(R) = \boldsymbol{\ell}. \tag{6.9}$$

The shortcoming for this choice is that $\mathbf{a}_j$ are generally not orthonormal and $D(R)$ is real but not orthogonal. Let $\boldsymbol{e}_a$ be the orthonormal basis vectors in the real three-dimensional space. The matrix of the rotation R in the basis vectors $\boldsymbol{e}_a$ is denoted by $\overline{D}(R)$,

$$\boldsymbol{e}_a \cdot \boldsymbol{e}_b = \delta_{ab}, \qquad R\boldsymbol{e}_a = \sum_{b=1}^{3} \boldsymbol{e}_b \overline{D}_{ba}(R). \tag{6.10}$$

$\overline{D}(R)$ is a real orthogonal matrix, related with $D(R)$ by a real similarity transformation X,

$$\boldsymbol{a}_i = \sum_{d=1}^{3} \boldsymbol{e}_d X_{di}, \qquad D(R) = X^{-1}\overline{D}(R)X. \tag{6.11}$$

We define another set of basis vectors $\mathbf{b}_j$ satisfying

$$\boldsymbol{b}_j = \sum_{a=1}^{3} \left(X^{-1}\right)_{ja} \boldsymbol{e}_a, \qquad \boldsymbol{b}_j \cdot \boldsymbol{a}_i = \delta_{ji}. \tag{6.12}$$

$\mathbf{b}_j$ are called the basis vectors of the reciprocal crystal lattice, or briefly called the **reciprocal lattice bases**. The matrix form of R in the basis vectors $\boldsymbol{b}_i$ is $\left[D(R)^{-1}\right]^T$:

$$\begin{aligned} R\boldsymbol{b}_i &= \sum_{d=1}^{3} \left(X^{-1}\right)_{id} R\boldsymbol{e}_d = \sum_{dc} \left(X^{-1}\right)_{id} \boldsymbol{e}_c \overline{D}_{cd}(R) \\ &= \sum_{j=1}^{3} \left\{ \sum_{dc} \left(X^{-1}\right)_{id} \left[\overline{D}(R)^{-1}\right]_{dc} X_{cj} \right\} \boldsymbol{b}_j, \\ &= \sum_{j=1}^{3} \left[D(R)^{-1}\right]_{ij} \boldsymbol{b}_j. \end{aligned} \tag{6.13}$$

A rotation R in the crystal theory is usually expressed in the form of a **double-vector**. A double-vector can be simply understood as an ordered pair of vectors, called left-vector and right-vector, respectively. Two vectors obey the usual vector calculation rules independently. The double-vector form of a rotation R is

$$\begin{aligned} \vec{\vec{R}} &= \sum_{ij} D_{ij}(R)\boldsymbol{a}_i\boldsymbol{b}_j = \sum_{ij} \left[D(R)^{-1}\right]_{ij} \boldsymbol{b}_j\boldsymbol{a}_i, \\ D_{ij}(R) &= \boldsymbol{b}_i \cdot \vec{\vec{R}} \cdot \boldsymbol{a}_j = \boldsymbol{a}_j \cdot \vec{\vec{R}}^{-1} \cdot \boldsymbol{b}_i. \end{aligned} \tag{6.14}$$

Equation (6.14) shows the method how to write the double-vector of R. $\vec{\vec{R}}$ is expanded by $\boldsymbol{a}_i\boldsymbol{b}_j$ (or by $\boldsymbol{b}_j\boldsymbol{a}_i$) with the coefficient $D_{ij}(R)$ (or $\left[D(R)^{-1}\right]_{ij}$). The double-vector of the identity E is

$$\vec{\vec{1}} = \sum_{j=1}^{3} \boldsymbol{a}_j\boldsymbol{b}_j = \sum_{j=1}^{3} \boldsymbol{b}_j\boldsymbol{a}_j. \tag{6.15}$$

The double-vector of the space inversion σ is $-\vec{\vec{1}}$, and that of a rotation $R(\hat{\boldsymbol{n}}, \omega)$ about the direction $\hat{\boldsymbol{n}}$ through an angle ω is (see Prob. 3)

$$\vec{\vec{R}}(\hat{\boldsymbol{n}}, \omega) = \hat{\boldsymbol{n}}\hat{\boldsymbol{n}} + \left(\vec{\vec{1}} - \hat{\boldsymbol{n}}\hat{\boldsymbol{n}}\right) \cos\omega + \left(\vec{\vec{1}} \times \hat{\boldsymbol{n}}\right) \sin\omega, \tag{6.16}$$

where $\hat{\boldsymbol{n}}\hat{\boldsymbol{n}}$ is the projective operator along the direction $\hat{\boldsymbol{n}}$, and $\left(\vec{\vec{1}} - \hat{\boldsymbol{n}}\hat{\boldsymbol{n}}\right)$ is that to the plane perpendicular to $\hat{\boldsymbol{n}}$.

From Eq. (6.9), $D_{ij}(R)$ are all integers, so the trace of $D(R)$:

$$\text{Tr } D(R) = \text{Tr } \overline{D}(R) = \pm(1 + 2\cos\omega) = \text{integer}, \tag{6.17}$$

where **the positive sign corresponds to the proper rotation, and the negative sign to the improper one.** Thus, $\cos\omega$ is a half integer,

$$\begin{aligned} &\cos\omega = 0, \quad \pm 1/2, \quad \text{and} \quad \pm 1, \\ &\omega = 2\pi m/N, \qquad N = 1,\ 2,\ 3,\ 4, \text{ or } 6, \qquad 0 \leqslant m < N. \end{aligned} \tag{6.18}$$

The element in a crystallographic point group has to be an N-fold proper or improper rotation, where N is 1, 2, 3, 4, or 6.

An integral combination $\boldsymbol{K}$ of the reciprocal lattice bases $\boldsymbol{b}_j$ is called the vector of the reciprocal crystal lattice. The inner product of $\boldsymbol{K}$ and $\boldsymbol{\ell}$ is an integer and is left invariant under the rotation,

$$\boldsymbol{K} \cdot \boldsymbol{\ell} = (R\boldsymbol{K}) \cdot (R\boldsymbol{\ell}) = \text{integer}, \tag{6.19}$$

Thus, $R\boldsymbol{K}$ is also a vector of the reciprocal crystal lattice

$$R\boldsymbol{K} = \boldsymbol{K}'. \tag{6.20}$$

Namely, **both the crystal lattice and the reciprocal crystal lattice have the same crystallographic point group.**

6.2.2 *Proper Crystallographic Point Groups*

First, if a proper crystallographic point group contains only one proper rotational axis, it is a cyclic group. It is denoted by C_N in the Schönflies notations, and by N in the international notations, where $N = 1$, 2, 3, 4, or 6. The generator of C_N is a rotation C_N about the rotational axis $\hat{\boldsymbol{n}}$ through an angle $2\pi/N$. The rotational axis $\hat{\boldsymbol{n}}$ of C_N is usually taken to be the Z-axis. The group C_N is an abelian group with order N. There are N inequivalent one-dimensional representations

$$D^m\left(C_N\right) = \exp\left(-\mathrm{i}2\pi m/N\right), \qquad 0 \leqslant m < N. \tag{6.21}$$

Second, if a proper crystallographic point group G contains more than one proper rotational axes, we are interested in the constraints on the number of axes. Let n_N denote the number of the N-fold proper axes contained in G, where $N = 2$, 3, 4, and 6. Except for the identity E, there is no common element contained in the cyclic groups with different rotational axes. Thus, n_N are related to the order g of G:

$$g = 1 + n_2 + 2n_3 + 3n_4 + 5n_6. \tag{6.22}$$

For each N-fold proper axis, let $\{C_N\}$ denote the sum of elements in the cyclic group C_N. The application of $\{C_N\}$ to $\hat{\boldsymbol{n}}$ is $N\hat{\boldsymbol{n}}$ and its application to a vector perpendicular to $\hat{\boldsymbol{n}}$ is zero. Thus, the double-vector of $\{C_N\}$ is

$$\{C_N\} = \sum_{m=0}^{N-1} (C_N)^m, \qquad \left\{\vec{\vec{C}}_N\right\} = N\hat{\boldsymbol{n}}\hat{\boldsymbol{n}}. \tag{6.23}$$

In a set of orthonormal bases including a basis vector $\hat{\boldsymbol{n}}$, Tr $D(\{C_N\}) = N$. The trace of the sum of all elements in C_N, except for the identity E, is $N - 3$. Thus, the trace of the sum of all elements in G is

$$3 + (2-3)n_2 + (3-3)n_3 + (4-3)n_4 + (6-3)n_6 = 3 - n_2 + n_4 + 3n_6.$$

On the other hand, from the rearrangement theorem (Theorem 2.1), for an arbitrary vector $\boldsymbol{r}$ one has

$$S\left\{\sum_{R\in G} R\boldsymbol{r}\right\} = \left\{\sum_{R\in G} R\boldsymbol{r}\right\}, \qquad S \in \mathrm{G}.$$

Since G contains at least two different rotation axes, the vector in the curly bracket has to be null. Thus, the sum of all elements in G vanishes

$$3 - n_2 + n_4 + 3n_6 = 0. \tag{6.24}$$

It is another constraint on the number of axes in G.

Since SU(2) $\sim$ SO(3), **the calculation of the product of two elements in** SO(3), i.e., the calculation of the product of two rotations about different axes in G, **can be replaced with that in** SU(2), **which is much simpler**, where the sign in $u(\hat{\boldsymbol{n}}, \omega)$ does not matter with the calculation. The fundamental calculation formulas are

$$\begin{aligned}
u(\hat{\boldsymbol{n}}, \omega) &= \mathbf{1}\cos(\omega/2) - \mathrm{i}\,(\boldsymbol{\sigma}\cdot\hat{\boldsymbol{n}})\sin(\omega/2), \\
\cos(\omega/2) &= \frac{1}{2}\mathrm{Tr}\,\{u(\hat{\boldsymbol{n}}, \omega)\}, \\
(\boldsymbol{\sigma}\cdot\hat{\boldsymbol{n}}_1)(\boldsymbol{\sigma}\cdot\hat{\boldsymbol{n}}_2) &= \mathbf{1}\,(\hat{\boldsymbol{n}}_1\cdot\hat{\boldsymbol{n}}_2) + \mathrm{i}\boldsymbol{\sigma}\cdot(\hat{\boldsymbol{n}}_1\times\hat{\boldsymbol{n}}_2),
\end{aligned} \tag{6.25}$$

where $\cos(\omega/2)$ is restricted to take only the following values

$$\cos(\omega/2) = 0, \quad \pm\frac{1}{2}, \quad \pm\frac{1}{\sqrt{2}}, \quad \pm\frac{\sqrt{3}}{2}, \quad \pm 1. \tag{6.26}$$

Third, if G contains only 2-fold proper axes, one obtains $n_2 = 3$ from Eq. (6.24). The matrix of a 2-fold proper rotation in SU(2) is

$$u(\hat{\boldsymbol{n}}, \pi) = -\mathrm{i}\,(\boldsymbol{\sigma}\cdot\hat{\boldsymbol{n}}).$$

The product of two 2-fold proper rotations about different axes is

$$u(\hat{\boldsymbol{n}}_1, \pi)u(\hat{\boldsymbol{n}}_2, \pi) = -\mathbf{1}\,(\hat{\boldsymbol{n}}_1\cdot\hat{\boldsymbol{n}}_2) - \mathrm{i}\vec{\sigma}\cdot(\hat{\boldsymbol{n}}_1\times\hat{\boldsymbol{n}}_2). \tag{6.27}$$

Thus, $\hat{\boldsymbol{n}}_1\cdot\hat{\boldsymbol{n}}_2 = 0$, and $\hat{\boldsymbol{n}}_3 = \hat{\boldsymbol{n}}_1\times\hat{\boldsymbol{n}}_2$. Three 2-fold axes are orthogonal to each other. The group G is D_2, which is an abelian group and isomorphic onto the inversion group V_4. Its group table is given in Table 2.4 and its character table is given in Table 3.5.

Fourth, if G contains, in addition to the 2-fold axes, only one N-fold proper axis where N is larger than 2, then the N-fold axis is called the **principal axis** and taken to be the Z-axis. **The remaining elements in G are 2-fold proper rotations with the axes perpendicular to the principal axis**. Due to the constraint (6.24), $n_2 = N$. The product of two 2-fold rotations has to be an N-fold rotation, $\omega = 2m\pi/N$, and due to Eq. (6.27), $\hat{\boldsymbol{n}}_1\cdot\hat{\boldsymbol{n}}_2 = \cos(m\pi/N)$, namely, N 2-fold axes uniformly-distributed in the XY plane, where the angle of two neighboring axes is π/N. This group G is just D_N, where $N = 3$, 4, or 6. The D_N group contains $2N$ elements whose multiplication rule is given in Eq. (2.11). The N-fold axis in D_N is non-polar. The character tables of D_3, D_4 and D_6 are given in Tables 3.6, 3.10 and 3.11, respectively. D_N can be expressed as a product of two subgroups, $C_N C_2'$, where C_2' is the cyclic group of 2-fold proper rotations about the X-axis, so that D_N is denoted by $N2'$ in the international notations.

Fifth, G contains only 2-fold and 3-fold proper axes. The matrix of 3-fold rotation in SU(2) is

$$u(\hat{\boldsymbol{n}}, \pm 2\pi/3) = \mathbf{1}/2 \mp \mathrm{i}\,(\boldsymbol{\sigma}\cdot\hat{\boldsymbol{n}})\,\sqrt{3}/2.$$

For the product of two 3-fold rotations about different axes, the cosine of half of its rotational angle is

$$\frac{1}{2}\mathrm{Tr}\,\{u(\hat{\boldsymbol{n}}_1, \pm 2\pi/3)u(\hat{\boldsymbol{n}}_2, 2\pi/3)\} = 1/4 \mp 3\,(\hat{\boldsymbol{n}}_1\cdot\hat{\boldsymbol{n}}_2)/4.$$

Due to Eq. (6.26), $\hat{\boldsymbol{n}}_1\cdot\hat{\boldsymbol{n}}_2 = \pm 1/3$. Without loss of generality, one has

$$\hat{\boldsymbol{n}}_1 \cdot \hat{\boldsymbol{n}}_2 = \cos\theta = -1/3, \qquad \theta = 109^\circ 28'. \tag{6.28}$$

Thus, from the products of 3-fold rotations one obtains four 3-fold axes uniformly-distributed in the three-dimensional space, $n_3 = 4$. The angle of two neighboring 3-fold axes is θ. Due to Eq. (6.24), $n_2 = 3$. Each 2-fold axis has to be along the angular bisector of two neighboring 3-fold axes. Since

$$u(\hat{\boldsymbol{n}}_1, -2\pi/3)u(\hat{\boldsymbol{n}}_2, 2\pi/3) = \mathrm{i}\boldsymbol{\sigma} \cdot \left(\sqrt{3}\hat{\boldsymbol{n}}_1 - \sqrt{3}\hat{\boldsymbol{n}}_2 + 3\hat{\boldsymbol{n}}_1 \times \hat{\boldsymbol{n}}_2\right)/4,$$
$$\hat{\boldsymbol{n}}_1 \cdot \left(\sqrt{3}\hat{\boldsymbol{n}}_1 - \sqrt{3}\hat{\boldsymbol{n}}_2 + 3\hat{\boldsymbol{n}}_1 \times \hat{\boldsymbol{n}}_2\right)/4 = 1/\sqrt{3},$$

the cosine of the angle between the neighboring 3-fold axis and 2-fold axis is $1/\sqrt{3}$. Three 2-fold axes and four 3-fold axes are equivalent to each other, respectively, but the 3-fold axes are polar. This group G is the proper symmetry group **T** of a tetrahedron. The product of two elements in **T** is calculated by the isomorphism $\mathbf{O} \approx \mathrm{S}_4$ (see Eq. (2.58)). The character table of **T** is given in Table 3.17. **T** is denoted by $3'22'$ in the international notations, because **T** can be expressed as a product of three subgroups, $\mathrm{C}_3'\mathrm{C}_2\mathrm{C}_2'$.

At last, if G contains more than one 6-fold axes, the angle between two 6-fold axes has to be $109^\circ 28'$, because a 6-fold axis is also a 3-fold axis. On the other hand, since a 6-fold axis is also a 2-fold axis, the product of two 2-fold rotations with this angle cannot satisfy the condition (6.26). Therefore, the only remaining case is that G contains more than one 4-fold proper axes. The matrices of 4-fold rotations in SU(2) are

$$u(\hat{\boldsymbol{n}}, \pm\pi/2) = \mathbf{1}/\sqrt{2} \mp \mathrm{i}\left(\boldsymbol{\sigma} \cdot \hat{\boldsymbol{n}}\right)/\sqrt{2}, \qquad u(\hat{\boldsymbol{n}}, \pi) = -\mathrm{i}\left(\boldsymbol{\sigma} \cdot \hat{\boldsymbol{n}}\right).$$

For a product of two 4-fold rotations about different axes, the cosine of half of its rotational angle is

$$\mathrm{Tr}\left\{u(\hat{\boldsymbol{n}}_1, \pm\pi/2)u(\hat{\boldsymbol{n}}_2, \pi/2)\right\}/2 = 1/2 \mp \left(\hat{\boldsymbol{n}}_1 \cdot \hat{\boldsymbol{n}}_2\right)/2,$$
$$\mathrm{Tr}\left\{u(\hat{\boldsymbol{n}}_1, \pi)u(\hat{\boldsymbol{n}}_2, \pi/2)\right\}/2 = -\left(\hat{\boldsymbol{n}}_1 \cdot \hat{\boldsymbol{n}}_2\right)/\sqrt{2}.$$

The only solution satisfying the condition (6.26) is $\hat{\mathbf{n}}_1 \cdot \hat{\boldsymbol{n}}_2 = 0$, namely, there are three 4-fold axes perpendicular to each other. They are non-polar and equivalent to each other. In the products of two 4-fold rotations, there are 3-fold rotations and 2-fold rotations,

$$u(\hat{\boldsymbol{n}}_1, \pi/2)u(\hat{\boldsymbol{n}}_2, \pi/2) = \mathbf{1}/2 - \mathrm{i}\boldsymbol{\sigma} \cdot \left(\hat{\boldsymbol{n}}_1 + \hat{\boldsymbol{n}}_2 + \hat{\boldsymbol{n}}_1 \times \hat{\boldsymbol{n}}_2\right)/2,$$
$$\hat{\boldsymbol{n}}_1 \cdot \left(\hat{\boldsymbol{n}}_1 + \hat{\boldsymbol{n}}_2 + \hat{\boldsymbol{n}}_1 \times \hat{\boldsymbol{n}}_2\right)/\sqrt{3} = 1/\sqrt{3},$$
$$u(\hat{\boldsymbol{n}}_1, \pi)u(\hat{\boldsymbol{n}}_2, \pi/2) = -\mathrm{i}\boldsymbol{\sigma} \cdot \left(\hat{\boldsymbol{n}}_1 + \hat{\boldsymbol{n}}_1 \times \hat{\boldsymbol{n}}_2\right)/\sqrt{2}.$$

The cosine of the angle between the neighboring 3-fold axis and 4-fold axis is $1/\sqrt{3}$. Four 3-fold axes are uniformly-distributed around a 4-fold axis. They are non-polar and equivalent to each other. Since $n_3 = 4$ and $n_4 = 3$, from Eq. (6.24), one has $n_2 = 6$. Each 2-fold axis is along the angular bisector of two neighboring 4-fold axes and perpendicular to the third 4-fold axis. Six 2-fold axes are equivalent to each other. This group G is the proper symmetry group **O** of a cube. The product of two elements in **O** is calculated by the isomorphism $\mathbf{O} \approx S_4$ (see Eq. (2.58)). The character table of **O** is given in Table 3.18. **O** is denoted by $3'42''$ in the international notations, because **O** can be expressed as a product of three subgroups $C_3'C_4C_2''$.

There are 11 proper crystallographic point groups G listed in Table 6.1, where g is the order of G, g_c is the number of classes in G, n_N is the number of N-fold proper axes contained in G, "No. of rep." denotes the number of the inequivalent irreducible representations with the given dimension, and G can be expressed as the product of subgroups, which is related to the symbol of G in the international notations. In the theory of crystals, an N-fold rotation about the Z-axis is denoted by C_N, the remaining N-fold rotation with $N > 2$ by C_N', a 2-fold rotation about the X-axis is denoted by C_2', and the remaining 2-fold rotation by C_N''.

Table 6.1 The proper crystallographic point group

G	g	g_c	n_N 2	3	4	6	No. of rep. 1d	2d	3d	Product of subgroups	Generators	Subgroup of index 2
C_1	1	1					1			C_1	C_1	
C_2	2	2	1				2			C_2	C_2	C_1
C_3	3	3		1			3			C_3	C_3	
C_4	4	4			1		4			C_4	C_4	C_2
C_6	6	6				1	6			C_6	C_6	C_3
D_2	4	4	3				4			C_2C_2'	C_2, C_2'	C_2
D_3	6	3	3	1			2	1		C_3C_2'	C_3, C_2'	C_3
D_4	8	5	4		1		4	1		C_4C_2'	C_4, C_2'	C_4,D_2
D_6	12	6	6			1	4	2		C_6C_2'	C_6, C_2'	C_6,D_3
T	12	4	3	4			3		1	$C_3'C_2C_2'$	C_3', C_2	
O	24	5	6	4	3		2	1	2	$C_3'C_4C_2''$	C_3', C_4	**T**

6.2.3 *Improper Crystallographic Point Group*

In subsection 2.6.2, we discussed the general method for finding the improper point groups from a proper point group, and introduce the meaning of the subscripts in the Schönflies notations. There are two types of improper point groups. An improper crystallographic point group G′ of

I-type is a direct product of a proper crystallographic point group G and the inversion group V_2, composed by the identity E and the space inversion σ. The irreducible representation of G′ is the direct product of two irreducible representations of G and V_2, respectively. There are 11 improper crystallographic point groups of I-type,

$$\begin{array}{lll} \mathrm{C}_i \approx V_2, & \mathrm{C}_{2h} = \mathrm{C}_2 \otimes \mathrm{C}_i, & \mathrm{C}_{3i} = \mathrm{C}_3 \otimes \mathrm{C}_i, \\ \mathrm{C}_{4h} = \mathrm{C}_4 \otimes \mathrm{C}_i, & \mathrm{C}_{6h} = \mathrm{C}_6 \otimes \mathrm{C}_i, & \mathrm{D}_{2h} = \mathrm{D}_2 \otimes \mathrm{C}_i, \\ \mathrm{D}_{3d} = \mathrm{D}_3 \otimes \mathrm{C}_i, & \mathrm{D}_{4h} = \mathrm{D}_4 \otimes \mathrm{C}_i, & \mathrm{D}_{6h} = \mathrm{D}_6 \otimes \mathrm{C}_i, \\ \mathrm{T}_h = \mathrm{T} \otimes \mathrm{C}_i, & \mathrm{O}_h = \mathrm{O} \otimes \mathrm{C}_i. & \end{array} \tag{6.29}$$

If a proper crystallographic point group G contains an invariant subgroup H of index 2, one can construct an improper crystallographic point group G′ of P-type by multiplying the elements in the coset of H in G with σ and leaving H invariant. G′ is isomorphic onto G. They have the same character table and the irreducible representations. From Table 6.1, there are 10 improper crystallographic point groups of P-type

$$\begin{array}{llll} \mathrm{C}_s \approx \mathrm{C}_2, & \mathrm{S}_4 \approx \mathrm{C}_4, & \mathrm{C}_{3h} \approx \mathrm{C}_6, & \mathrm{C}_{2v} \approx \mathrm{D}_2, \\ \mathrm{C}_{3v} \approx \mathrm{D}_3, & \mathrm{C}_{4v} \approx \mathrm{D}_{2d} \approx \mathrm{D}_4, & \mathrm{C}_{6v} \approx \mathrm{D}_{3h} \approx \mathrm{D}_6, & \mathrm{T}_d \approx \mathrm{O}. \end{array} \tag{6.30}$$

The property of 32 crystallographic point groups are listed in Table 6.2.

Table 6.2 Crystallographic point groups

Proper crystallographic point groups			Improper crystallographic point groups			
			P-type		I-type	
G	Product of subgroups	Subgroup of index 2	G′	Product	G′	Product
C_1	C_1				C_i	C_i
C_2	C_2	C_1	C_s	C_s	C_{2h}	$\mathrm{C}_i\mathrm{C}_2$
C_3	C_3				C_{3i}	C_{3i}
C_4	C_4	C_2	S_4	S_4	C_{4h}	$\mathrm{C}_i\mathrm{C}_4$
C_6	C_6	C_3	C_{3h}	C_{3h}	C_{6h}	$\mathrm{C}_i\mathrm{C}_6$
D_2	$\mathrm{C}_2\mathrm{C}_2'$	C_2	C_{2v}	$\mathrm{C}_2\mathrm{C}_s'$	D_{2h}	$\mathrm{C}_i\mathrm{C}_2\mathrm{C}_2'$
D_3	$\mathrm{C}_3\mathrm{C}_2'$	C_3	C_{3v}	$\mathrm{C}_3\mathrm{C}_s'$	D_{3d}	$\mathrm{C}_{3i}\mathrm{C}_2'$
D_4	$\mathrm{C}_4\mathrm{C}_2'$	C_4	C_{4v}	$\mathrm{C}_4\mathrm{C}_s'$	D_{4h}	$\mathrm{C}_i\mathrm{C}_4\mathrm{C}_2'$
		D_2	D_{2d}	$\mathrm{S}_4\mathrm{C}_2'$		
D_6	$\mathrm{C}_6\mathrm{C}_2'$	C_6	C_{6v}	$\mathrm{C}_6\mathrm{C}_s'$	D_{6h}	$\mathrm{C}_i\mathrm{C}_6\mathrm{C}_2'$
		D_3	D_{3h}	$\mathrm{C}_{3h}\mathrm{C}_2'$		
$\mathbf{T}$	$\mathrm{C}_3'\mathrm{C}_2\mathrm{C}_2'$				$\mathbf{T}_h$	$\mathrm{C}_{3i}'\mathrm{C}_2\mathrm{C}_2'$
$\mathbf{O}$	$\mathrm{C}_3'\mathrm{C}_4\mathrm{C}_2''$	$\mathbf{T}$	$\mathbf{T}_d$	$\mathrm{C}_3'\mathrm{S}_4\mathrm{C}_s''$	$\mathbf{O}_h$	$\mathrm{C}_{3i}'\mathrm{C}_4\mathrm{C}_2''$

6.2.4 *Double-Valued Representations of the Point Groups*

In the cases where relativistic effects are important, the spin degree of freedom of electrons needs to be considered in the Hamiltonian and the electron wave function becomes spinor. In these cases the element $R(\hat{\boldsymbol{n}},\omega)$ in a proper point group G is replaced with $u(\hat{\boldsymbol{n}},\omega)$, and G becomes so called a **double point group** $\tilde{\mathrm{G}}$. In fact, $\tilde{\mathrm{G}}$ is a subgroup of SU(2) and is homomorphic onto G with $2:1$ correspondence, $\tilde{\mathrm{G}} \sim \mathrm{G}$. Speaking strictly, a faithful representation of $\tilde{\mathrm{G}}$ is not a representation of G, but often called a double-valued representation of G. The single-valued representation of G is an unfaithful one of $\tilde{\mathrm{G}}$. For simplicity, we will also say **the single-valued or the double-valued representations of** $\tilde{\mathrm{G}}$. The double-valued representations of crystallographic point groups were first considered by [Bethe (1929)] and [Opechowski (1940)]. Let us study the properties of the double-valued representations of the proper point groups $\tilde{\mathrm{G}}$.

a) The element $\tilde{E} = u(\hat{\boldsymbol{n}},2\pi) = -\mathbf{1}$ in $\tilde{\mathrm{G}}$ is independent of the direction $\hat{\boldsymbol{n}}$ of its rotational axis. Just like the identity E, $\tilde{E}$ commutes with all elements in $\tilde{\mathrm{G}}$ and $\tilde{E}$ itself forms a class of $\tilde{\mathrm{G}}$. $D(\tilde{E}) = \mathbf{1}$ **for a single-valued representation and** $D(\tilde{E}) = -\mathbf{1}$ **for a double-valued one**. A subduced representation of $D^j(\mathrm{SU}(2))$ with respect to $\tilde{\mathrm{G}}$ only contains the single-valued representations when j is an integer, and the doubled-valued ones when j is half of an odd integer, respectively.

b) Since $u(\hat{\boldsymbol{n}},\omega)^{-1} = u(-\hat{\boldsymbol{n}},\omega) = u[\hat{\boldsymbol{n}},(4\pi-\omega)]$, not $u[\hat{\boldsymbol{n}},(2\pi-\omega)]$, in addition to the classes $\{E\}$ and $\{\tilde{E}\}$, the equivalent N-fold axes provide other $(N-1)$ **self-reciprocal classes** of $\tilde{\mathrm{G}}$ if the axes are non-polar, and $(N-1)$ **pairs of reciprocal classes** of $\tilde{\mathrm{G}}$ if the axes are polar.

The double point group $\tilde{\mathrm{C}}_N$ is isomorphic onto C_{2N}, whose irreducible representations are given in Eq. (3.70). For the double point group $\tilde{\mathrm{D}}_N$ with $N = 2$, 4, 6, all rotational axes are non-polar so that, in comparison with D_N, the number of classes in $\tilde{\mathrm{D}}_N$ increases by $N/2$ and its order increases by $2N$. From Eqs. (3.62), (3.65) and Corollary 3.4.4, $\tilde{\mathrm{D}}_N$ has $N/2$ inequivalent irreducible double-valued representations of two dimensions, all of which are self-conjugate. One of them is the subduced representation of $D^{1/2}(\mathrm{SU}(2))$. Another (when $N = 4$ and 6) is its direct product with an antisymmetric representation. The third representation of $\tilde{\mathrm{D}}_6$ can be calculated by Eq. (3.68) or by decomposition of the subduced representation of $D^{3/2}(\mathrm{SU}(2))$. Since the two-fold axes of $\tilde{\mathrm{D}}_3$ are polar, in comparison with D_3, the number of classes of $\tilde{\mathrm{D}}_3$ increases by 3 and its order increases by

6. From Eqs. (3.62), (3.65) and Corollary 3.4.4, $\tilde{\mathrm{D}}_3$ has a pair of conjugate double-valued one dimensional representations and one self-conjugate double-valued two dimensional irreducible representation. From Corollary 3.3.3 (see Eq. (3.60)), the subduced representation of $D^{1/2}(\mathrm{SU}(2))$ with respect to $\tilde{\mathrm{D}}_3$ is irreducible. The pair of one-dimensional representations of $\tilde{\mathrm{D}}_3$ can be calculated by Eq. (3.68) with noting $(C_2')^2 = \tilde{E}$.

Since the 3-fold axes of the double point group $\tilde{\mathbf{T}}$ are polar, in comparison with $\mathbf{T}$, the number of classes of $\tilde{\mathbf{T}}$ increases by 3 and its order increases by 12. From Eqs. (3.62) and (3.65) $\tilde{\mathbf{T}}$ has 3 inequivalent irreducible double-valued representations of two dimensions. One of them is the subduced representation of $D^{1/2}(\mathrm{SU}(2))$, and the other two are its direct products with two one-dimensional representations $D^{^1E}$ and $D^{^2E}$, respectively. Since all rotational axes of the double point group $\tilde{\mathbf{O}}$ are non-polar, in comparison with $\mathbf{O}$, the number of classes of $\tilde{\mathbf{O}}$ increases by 3 and its order increases by 24. From Eqs. (3.62) and (3.65) $\tilde{\mathbf{O}}$ has 3 self-conjugate inequivalent irreducible double-valued representations where two are two-dimensional, and one is four-dimensional. The subduced representations of $D^{1/2}(\mathrm{SU}(2))$ and $D^{3/2}(\mathrm{SU}(2))$ with respect to $\tilde{\mathbf{O}}$ both are irreducible, and the remaining double-valued two-dimensional representation is the direct products of one double-valued two-dimensional representation and a one-dimensional antisymmetric representation D^B.

Table 6.3 gives the character tables of the double point groups where the convention in the crystal theory is used: the self-conjugate representations of dimensions one, two, three, and four are denoted by A (B), E, T, and F, respectively, the pair of conjugate representations of one dimension is denoted by ${}^1E$ and ${}^2E$, and that of two dimensions by ${}^1F$ and ${}^2F$.

Table 6.3 Character Tables of the double point groups

$\tilde{\mathrm{D}}_2$	E	$\tilde{E}$	$2C_2$	$2C_2'$	$2C_2''$
A_1	1	1	1	1	1
A_2	1	1	1	−1	−1
B_1	1	1	−1	1	−1
B_2	1	1	−1	−1	1
$\tilde{E}$	2	−2	0	0	0

$\tilde{\mathrm{D}}_3$	E	$\tilde{E}$	$2C_3$	$2C_3^2$	$3C_2'$	$3C_2'{}^3$
A	1	1	1	1	1	1
B	1	1	1	1	−1	−1
E	2	2	−1	−1	0	0
${}^1\tilde{E}$	1	−1	−1	1	i	−i
${}^2\tilde{E}$	1	−1	−1	1	−i	i
$\tilde{E}_1$	2	−2	1	−1	0	0

(continued)

$\tilde{\mathrm{D}}_4$	E	$\tilde{E}$	$2C_4$	$2C_4^2$	$2C_4^3$	$4C_2'$	$4C_2''$
A_1	1	1	1	1	1	1	1
A_2	1	1	1	1	1	-1	-1
B_1	1	1	-1	1	-1	1	-1
B_2	1	1	-1	1	-1	-1	1
E	2	2	0	-2	0	0	0
$\tilde{E}_1$	2	-2	$\sqrt{2}$	0	$-\sqrt{2}$	0	0
$\tilde{E}_2$	2	-2	$-\sqrt{2}$	0	$\sqrt{2}$	0	0

$\tilde{\mathrm{D}}_6$	E	$\tilde{E}$	$2C_6$	$2C_6^2$	$2C_6^3$	$2C_6^4$	$2C_6^5$	$6C_2'$	$6C_2''$
A_1	1	1	1	1	1	1	1	1	1
A_2	1	1	1	1	1	1	1	-1	-1
B_1	1	1	-1	1	-1	1	-1	1	-1
B_2	1	1	-1	1	-1	1	-1	-1	1
E_1	2	2	1	-1	-2	-1	1	0	0
E_2	2	2	-1	-1	2	-1	-1	0	0
$\tilde{E}_1$	2	-2	$\sqrt{3}$	1	0	-1	$-\sqrt{3}$	0	0
$\tilde{E}_2$	2	-2	$-\sqrt{3}$	1	0	-1	$\sqrt{3}$	0	0
$\tilde{E}_3$	2	-2	0	-2	0	2	0	0	0

$\tilde{\mathrm{T}}$	E	$\tilde{E}$	$6C_2$	$4C_3'$	$4C_3'^{\,2}$	$4C_3'^{\,4}$	$4C_3'^{\,5}$
A	1	1	1	1	1	1	1
${}^1E$	1	1	1	ω	ω^*	ω	ω^*
${}^2E$	1	1	1	ω^*	ω	ω^*	ω
T	3	3	-1	0	0	0	0
E	2	-2	0	1	-1	-1	1
${}^1\tilde{F}$	2	-2	0	ω	$-\omega^*$	$-\omega$	ω^*
${}^2\tilde{F}$	2	-2	0	ω^*	$-\omega$	$-\omega^*$	ω

$\omega = e^{-i2\pi/3}$,

$\tilde{\mathrm{O}}$	E	$\tilde{E}$	$6C_4$	$6C_4^2$	$6C_4^3$	$8C_3'$	$8C_3'^{\,2}$	$12C_2''$
A	1	1	1	1	1	1	1	1
B	1	1	-1	1	-1	1	1	-1
E	2	2	0	2	0	-1	-1	0
T_1	3	3	1	-1	1	0	0	-1
T_2	3	3	-1	-1	-1	0	0	1
$\tilde{E}_1$	2	-2	$\sqrt{2}$	0	$-\sqrt{2}$	1	-1	0
$\tilde{E}_2$	2	-2	$-\sqrt{2}$	0	$\sqrt{2}$	1	-1	0
$\tilde{F}$	4	-4	0	0	0	-1	1	0

The representation matrices of the generators of the double point groups in the double-valued representations are listed as follows.

$$\tilde{\mathrm{D}}_2 : D^{1/2}(\tilde{\mathrm{D}}_2) = D^{\tilde{E}}(\tilde{\mathrm{D}}_2), \qquad D^{\tilde{E}}(C_2) = -i\sigma_3, \qquad D^{\tilde{E}}(C_2') = -i\sigma_1,$$

$$\tilde{\mathrm{D}}_3 : D^{1/2}(\tilde{\mathrm{D}}_3) = D^{\tilde{E}_1}(\tilde{\mathrm{D}}_3),$$

$$D^{\tilde{E}_1}(C_3) = \left(1 - i\sqrt{3}\sigma_3\right)/2, \qquad D^{\tilde{E}_1}(C_2') = -i\sigma_1,$$

$$D^{3/2}(\tilde{\mathrm{D}}_3) \approx D^{{}^1\tilde{E}}(\tilde{\mathrm{D}}_3) \oplus D^{\tilde{E}_1}(\tilde{\mathrm{D}}_3) \oplus D^{{}^2\tilde{E}}(\tilde{\mathrm{D}}_3),$$

$$D^{{}^2\tilde{E}}(\tilde{\mathrm{D}}_3) = D^{{}^1\tilde{E}}(\tilde{\mathrm{D}}_3)^* = D^{B}(\tilde{\mathrm{D}}_3) \times D^{{}^1\tilde{E}}(\tilde{\mathrm{D}}_3),$$

$$\begin{aligned}
\tilde{\mathrm{D}}_4: &\ D^{1/2}(\tilde{\mathrm{D}}_4) = D^{\tilde{E}_1}(\tilde{\mathrm{D}}_4), \qquad D^{\tilde{E}_2}(\tilde{\mathrm{D}}_4) = D^{B_1}(\tilde{\mathrm{D}}_4) \times D^{\tilde{E}_1}(\tilde{\mathrm{D}}_4), \\
&\ D^{\tilde{E}_1}(C_4) = (\mathbf{1} - \mathrm{i}\sigma_3)/\sqrt{2}, \qquad D^{\tilde{E}_1}(C_2') = -\mathrm{i}\sigma_1, \\
\tilde{\mathrm{D}}_6: &\ D^{1/2}(\tilde{\mathrm{D}}_6) = D^{\tilde{E}_1}(\tilde{\mathrm{D}}_6), \qquad D^{\tilde{E}_2}(\tilde{\mathrm{D}}_6) = D^{B_1}(\tilde{\mathrm{D}}_6) \times D^{\tilde{E}_1}(\tilde{\mathrm{D}}_6), \\
&\ X^{-1} D^{3/2}(\tilde{\mathrm{D}}_6) X = D^{\tilde{E}_1}(\tilde{\mathrm{D}}_6) \oplus D^{\tilde{E}_3}(\tilde{\mathrm{D}}_6), \quad X = (\sigma_3 \times \mathbf{1}) Y, \\
&\ Y \text{ is a simple similarity transformation matrix,} \\
&\ D^{\tilde{E}_1}(C_6) = -D^{\tilde{E}_2}(C_6) = \left(\mathbf{1}\sqrt{3} - \mathrm{i}\sigma_3\right)/2, \quad D^{\tilde{E}_3}(C_6) = -\mathrm{i}\sigma_3, \\
&\ D^{\tilde{E}_1}(C_2') = D^{\tilde{E}_2}(C_2') = D^{\tilde{E}_3}(C_2') = -\mathrm{i}\sigma_1, \\
\tilde{\mathbf{T}}: &\ D^{1/2}(\tilde{\mathbf{T}}) = D^{\tilde{E}}(\tilde{\mathbf{T}}), \\
&\ D^{\tilde{E}}(C_2) = -\mathrm{i}\sigma_3, \quad D^{\tilde{E}}(C_3') = \left[\mathbf{1} - \mathrm{i}\left(\sigma_1 + \sigma_2 + \sigma_3\right)\right]/2, \\
&\ D^{{}^1F}(\tilde{\mathbf{T}}) = D^{{}^1E}(\tilde{\mathbf{T}}) \times D^{\tilde{E}}(\tilde{\mathbf{T}}), \quad D^{{}^2F}(\tilde{\mathbf{T}}) = D^{{}^2E}(\tilde{\mathbf{T}}) \times D^{\tilde{E}}(\tilde{\mathbf{T}}), \\
\tilde{\mathbf{O}}: &\ D^{1/2}(\tilde{\mathbf{O}}) = D^{\tilde{E}_1}(\tilde{\mathbf{O}}), \qquad D^{\tilde{E}_2}(\tilde{\mathbf{O}}) = D^{B}(\tilde{\mathbf{O}}) \times D^{\tilde{E}_1}(\tilde{\mathbf{O}}), \\
&\ D^{\tilde{E}_1}(C_4) = -D^{\tilde{E}_2}(C_4) = (\mathbf{1} - \mathrm{i}\sigma_3)/\sqrt{2}, \\
&\ D^{\tilde{E}_1}(C_3') = D^{\tilde{E}_2}(C_3') = D^{1/2}(0, \tfrac{\pi}{2}, \tfrac{\pi}{2}) = \left[\mathbf{1} - \mathrm{i}\left(\sigma_1 + \sigma_2 + \sigma_3\right)\right]/2, \\
&\ D^{3/2}(\tilde{\mathbf{O}}) = D^{F}(\mathbf{O}), \qquad D^{F}(C_4) = \mathrm{diag}\{\mathrm{e}^{-\mathrm{i}3\pi/4}, \mathrm{e}^{-\mathrm{i}\pi/4}, \mathrm{e}^{\mathrm{i}\pi/4}, \mathrm{e}^{\mathrm{i}3\pi/4}\}, \\
&\ D^{F}(C_3') = \frac{1}{\sqrt{8}} \begin{pmatrix} \mathrm{e}^{-\mathrm{i}\frac{3\pi}{4}} & -\sqrt{3}\mathrm{e}^{-\mathrm{i}\frac{\pi}{4}} & \sqrt{3}\mathrm{e}^{\mathrm{i}\frac{\pi}{4}} & -\mathrm{e}^{\mathrm{i}\frac{3\pi}{4}} \\ \sqrt{3}\mathrm{e}^{-\mathrm{i}\frac{3\pi}{4}} & -\mathrm{e}^{-\mathrm{i}\frac{\pi}{4}} & -\mathrm{e}^{\mathrm{i}\frac{\pi}{4}} & \sqrt{3}\mathrm{e}^{\mathrm{i}\frac{3\pi}{4}} \\ \sqrt{3}\mathrm{e}^{-\mathrm{i}\frac{3\pi}{4}} & \mathrm{e}^{-\mathrm{i}\frac{\pi}{4}} & -\mathrm{e}^{\mathrm{i}\frac{\pi}{4}} & -\sqrt{3}\mathrm{e}^{\mathrm{i}\frac{3\pi}{4}} \\ \mathrm{e}^{-\mathrm{i}\frac{3\pi}{4}} & \sqrt{3}\mathrm{e}^{-\mathrm{i}\frac{\pi}{4}} & \sqrt{3}\mathrm{e}^{\mathrm{i}\frac{\pi}{4}} & \mathrm{e}^{\mathrm{i}\frac{3\pi}{4}} \end{pmatrix}.
\end{aligned} \tag{6.31}$$

The improper point group G′ of P-type is isomorphic onto the corresponding proper point groups G so that they have the same representations, both single-valued and double-valued. The improper point group G′ of I-type is equal to the direct product of a proper point group G and the inversion group $\mathrm{V}_2 = \{E, \sigma\}$ so that each irreducible representation of G′, single-valued or double-valued, is equal to the direct product of that of G and a one-dimensional representation of V_2.

In other cases where the Hamiltonian of the physical system contains the magnetic momentums of electrons, one has to study the magnetic point groups. As is well known, the magnetic momentums of electrons is left invariant under the space inversion σ, but changes sign under the time inversion τ, which is an anti-unitary operator. The readers who are interested in the magnetic point groups are suggested to read Chapter 7 in [Bradley and Cracknell (1972)].

6.3 Crystal Systems and Bravais Lattice

Because the translation group $\mathcal{T}$ is an invariant subgroup of the space group $\mathcal{S}$, we obtain the constraint (6.6) in the previous sections, which restricts the possible rotation R in the symmetric operation $g(R, \boldsymbol{\alpha})$. The conclusion is that there are only 32 crystallographic point groups. The constraint (6.6) also restricts the choice of the basis vectors $\boldsymbol{a}_i$ of crystal lattice for each given crystallographic point group. **In this section the lattice bases $\boldsymbol{a}_i$ may not be primitive.** The crystals are classified into seven crystal systems based on the restriction of the choice of $\boldsymbol{a}_i$. For each crystal system we will discuss the possibility of the vectors of crystal lattice $\boldsymbol{f}$ which are the fractional combinations of the lattice bases $\boldsymbol{a}_i$. Due to different $\boldsymbol{f}$, a crystal system is divided into different Bravais Lattices and the symmorphic space groups. There are 14 Bravais Lattices and 73 symmorphic space groups. The general space groups, where the crystallographic point group is not a subgroup of the space group, will be discussed in the next section.

6.3.1 *Restrictions on Vectors of Crystal Lattice*

The conditions (6.6) and (6.20) give the restriction on the lattice bases $\boldsymbol{a}_i$ and the reciprocal lattice bases $\boldsymbol{b}_j$ when the crystallographic point group G is given. The restriction for some crystallographic point groups may be the same. For instance, the space inversion σ makes no restriction on the lattice bases so that **the restriction for the improper crystallographic point groups, both P-type and I-type, are the same as that for the corresponding proper crystallographic point group.** It will be shown that except for $N = 2$, the restriction for C_N is the same as that for D_N, and the restriction for **T** is the same as that for **O**. Based on the restrictions, the crystals are classified into seven crystal systems listed as follows, where the crystallographic point groups are also listed.

(a) Triclinic crystal system, C_1 and C_i.
(b) Monoclinic crystal system, C_2, C_s, and C_{2h}.
(c) Orthorhombic crystal system, D_2, C_{2v}, and D_{2h}.
(d) Trigonal crystal system, C_3, C_{3i}, D_3, C_{3v}, and D_{3d}.
(e) Hexagonal crystal system, C_6, C_{3h}, C_{6h}, D_6, C_{6v}, D_{3h}, and D_{6h}.
(f) Tetragonal crystal system, C_4, S_4, C_{4h}, D_4, C_{4v}, D_{2d}, and D_{4h}.
(g) Cubic crystal system, T, T_h, O, T_d, and O_h.

In the trigonal crystal system, the restrictions on $\boldsymbol{a}_i$ in some cases are the same as those in the hexagonal crystal system. These cases can be merged into the hexagonal crystal system. But the remaining cases in the trigonal crystal system are called the **rhombohedral crystal system**.

For each crystal system, the lattice bases $\boldsymbol{a}_i$ are chosen in a given rule according to the restriction, and then, the possibility of the vectors of crystal lattice $\boldsymbol{f}$ which are the fractional combinations of $\boldsymbol{a}_i$ is studied,

$$\boldsymbol{f} = \boldsymbol{a}_1 f_1 + \boldsymbol{a}_2 f_2 + \boldsymbol{a}_3 f_3, \qquad 0 \leqslant f_i < 1. \tag{6.32}$$

Let $T(\boldsymbol{f}) \equiv T(f_1, f_2, f_3)$ denote the translation operator with $\boldsymbol{f}$. Later calculation will show that f_i can only take the value of 0 or 1/2. The set of the translations where the translation vectors are the integral combinations of $\boldsymbol{a}_i$ forms an invariant subgroup $\mathcal{T}_\ell$ of the translation group $\mathcal{T}$. The cosets of $\mathcal{T}_\ell$ are generated by $T(\boldsymbol{f})$ and its powers. The crystal system is divided into a distinct Bravais lattices, based on the cosets as follows:

(a) Primitive translation group, denoted by P:

$$\mathcal{T} = \mathcal{T}_\ell. \tag{6.33}$$

The primitive translation group for the rhombohedral crystal system is denoted by R.

(b) Body-centered translation group, denoted by I:

$$\mathcal{T} = \mathcal{T}_\ell \otimes \left\{E,\ T\left(\frac{1}{2}, \frac{1}{2}, \frac{1}{2}\right)\right\}. \tag{6.34}$$

(c) Base-centered translation group, denoted by A, B, and C:

$$\begin{aligned} A: &\quad \mathcal{T} = \mathcal{T}_\ell \otimes \left\{E,\ T\left(0, \frac{1}{2}, \frac{1}{2}\right)\right\}, \\ B: &\quad \mathcal{T} = \mathcal{T}_\ell \otimes \left\{E,\ T\left(\frac{1}{2}, 0, \frac{1}{2}\right)\right\}, \\ C: &\quad \mathcal{T} = \mathcal{T}_\ell \otimes \left\{E,\ T\left(\frac{1}{2}, \frac{1}{2}, 0\right)\right\}. \end{aligned} \tag{6.35}$$

(d) Face-centered translation group, denoted by F:

$$\mathcal{T} = \mathcal{T}_\ell \otimes \left\{E,\ T\left(0, \frac{1}{2}, \frac{1}{2}\right), T\left(\frac{1}{2}, 0, \frac{1}{2}\right), T\left(\frac{1}{2}, \frac{1}{2}, 0\right)\right\}. \tag{6.36}$$

Obviously, if two of A-, B- and C-type translation groups exist simultaneously for one crystal, its $\mathcal{T}$ is the face-centered translation group F. **Combining the translation group with the crystal system, one obtains 14 Bravais lattices.** A crystallographic point group headed by the symbol of the Bravais lattice denotes the symmorphic space group. In the following, we are going to count the 73 symmorphic space groups.

There are a few notation-systems for the space groups $\mathcal{S}$. The **Schönflies notations** are convenient for the crystallographic point groups, but very cumbersome for labeling the space groups. **The Mauguin–Hermann notations and the international notations are improved labeling schemes for the space groups.** Their comparison is listed in Table 6.4. We recommend the international notations for the space groups, where a proper cyclic subgroup C_N is denoted by a number N, an improper one by a number with a bar $\overline{N}$, and the inversion group V_2 (C_i) is denoted by $\pm$. N (or $\overline{N}$) stands for an N-fold cyclic subgroup along $\boldsymbol{a}_3$. N' (or $\overline{N}'$) with $N > 2$ stands for the remaining N-fold cyclic subgroup. $2'$ (or $\overline{2}'$) stands for a 2-fold cyclic subgroup along $\boldsymbol{a}_1$. $2''$ (or $\overline{2}''$) stands for the remaining 2-fold cyclic subgroup. We will explain the phenomena later that one crystallographic point group (e.g., D_3) corresponds to two notations in Table 6.4.

Table 6.4 Comparison between notations for space groups.

Sch. = the Schönflies notations,
INPG = the international notations for point groups,
MHN = the Mauguin–Hermann notations,
INSG = the international notations for space groups.

Sch.	INPG	MHN	INSG	Sch.	INPG	MHN	INSG
C_1	1	1	1	C_{3v}	$3m$	$31m$	$3\overline{2}''$
C_i	$\overline{1}$	$\overline{1}$	$\overline{1}$	D_{3d}	$\overline{3}m$	$\overline{3}\frac{2}{m}1$	$\overline{3}2'$
C_2	2	2	2	D_{3d}	$\overline{3}m$	$\overline{3}1\frac{2}{m}$	$\overline{3}2''$
C_s	m	m	$\overline{2}$	D_4	422	422	$42'$
C_{2h}	$2/m$	$2/m$	± 2	C_{4v}	$4mm$	$4mm$	$4\overline{2}'$
C_3	3	3	3	D_{2d}	$\overline{4}2m$	$\overline{4}2m$	$\overline{4}2'$
C_{3i}	$\overline{3}$	$\overline{3}$	$\overline{3}$	D_{2d}	$\overline{4}2m$	$\overline{4}m2$	$\overline{4}2''$
C_4	4	4	4	D_{4h}	$4/mmm$	$\frac{4}{m}\frac{2}{m}\frac{2}{m}$	$\pm 42'$
S_4	$\overline{4}$	$\overline{4}$	$\overline{4}$	D_6	622	622	$62'$
C_{4h}	$4/m$	$4/m$	± 4	C_{6v}	$6mm$	$6mm$	$6\overline{2}'$
C_6	6	6	6	D_{3h}	$\overline{6}m2$	$\overline{6}2m$	$\overline{6}2'$
C_{3h}	$\overline{6}$	$\overline{6}$	$\overline{6}$	D_{3h}	$\overline{6}m2$	$\overline{6}m2$	$\overline{6}2''$
C_{6h}	$6/m$	$6/m$	± 6	D_{6h}	$6/mmm$	$\frac{6}{m}\frac{2}{m}\frac{2}{m}$	$\pm 62'$
D_2	222	222	$22'$	T	23	23	$3'22'$
C_{2v}	$2mm$	$2mm$	$2\overline{2}'$	T_h	$m3$	$\frac{2}{m}3$	$\overline{3}'22'$
D_{2h}	mmm	$\frac{2}{m}\frac{2}{m}\frac{2}{m}$	$\pm 22'$	O	432	432	$3'42''$
D_3	32	321	$32'$	T_d	$\overline{4}3m$	$\overline{4}3m$	$3'\overline{4}\,\overline{2}''$
D_3	32	312	$32''$	O_h	$m3m$	$\frac{4}{m}3\frac{2}{m}$	$\overline{3}'42''$
C_{3v}	$3m$	$3m1$	$3\overline{2}'$				

We are going to show that, except for the trivial point group C_1, **there are vectors of both crystal lattice and reciprocal crystal lattice along each rotational axis**, and at least two non-collinear vectors of lattice bases and those of reciprocal crystal lattice at the plane perpendicular to a rotational axis. In fact, each C_N or S_N in a crystallographic point group G, except for C_1, contains C_2, C_s, or C_3 as a subgroup. If C_N contains C_2, $\boldsymbol{\ell}+C_2\boldsymbol{\ell}$ is parallel to the 2-fold axis, and $\boldsymbol{\ell}-C_2\boldsymbol{\ell}$ is perpendicular to it. Find another vector of crystal lattice $\boldsymbol{\ell}'$ outside of the plane spanned by $\boldsymbol{\ell}\pm C_2\boldsymbol{\ell}$. $\boldsymbol{\ell}'-C_2\boldsymbol{\ell}'$ is another vector of crystal lattice at the plane perpendicular to the 2-fold axis. The situation is similar when S_N contains C_s. If C_N or S_N contains C_3, $\boldsymbol{\ell}+C_3\boldsymbol{\ell}+C_3^2\boldsymbol{\ell}$ is along the 3-fold axis, and $\boldsymbol{\ell}-C_3\boldsymbol{\ell}$ and $\boldsymbol{\ell}-C_3^2\boldsymbol{\ell}$ are two noncollinear vectors of crystal lattice at the plane perpendicular to the 3-fold axis. Since the crystallographic point group for the reciprocal crystal lattice is the same as that for the crystal lattice, the same arguments are applicable.

In the following the length of $\boldsymbol{a}_i$ is denoted by a_i, the angle between $\boldsymbol{a}_1$ and $\boldsymbol{a}_2$ is denoted by α_3, and its cyclic combination.

6.3.2 *Triclinic Crystal System*

The crystallographic point groups in the triclinic crystal system are C_1 (1) and C_i ($\overline{1}$), where and later the number in the bracket is the symbol of international notations. The matrices of the identity E and the space inversion σ are constant ones, which make no restriction on the choice of the lattice bases. Three fundamental periods $\boldsymbol{a}_j$ of the crystal lattice are taken to be the **primitive** lattice bases. Thus, there are one Bravais lattice P and two symmorphic space groups $P1$ and $P\overline{1}$ in the triclinic crystal system.

6.3.3 *Monoclinic Crystal System*

The crystallographic point groups in the monoclinic crystal system are C_2 (2), C_s ($\overline{2}$), and C_{2h} (± 2). The **shortest vector** of crystal lattice along the 2-fold axis is chosen to be $\boldsymbol{a}_3$. In the plane perpendicular to $\boldsymbol{a}_3$, choose **two "primitive" vectors** of crystal lattice to be $\boldsymbol{a}_1$ and $\boldsymbol{a}_2$ such that $\boldsymbol{a}_1 \times \boldsymbol{a}_2$ is along the positive direction of $\boldsymbol{a}_3$, where "two primitive vectors of crystal lattice in a plane" means that each vector of crystal lattice in the plane is an integral combination of those two "primitive" vectors of crystal lattice. The length a_1 and a_2 may not be the shortest among the vectors of crystal lattice in the plane. Thus, in the monoclinic crystal system the lengths of the lattice bases have no restriction, and the restrictions on the angles are

$$\alpha_1 = \alpha_2 = \pi/2. \tag{6.37}$$

From Eq. (6.14) the double-vector and the matrix of the generator C_2 in the set of lattice bases are calculated as

$$\vec{\vec{2}} = -\boldsymbol{a}_1\boldsymbol{b}_1 - \boldsymbol{a}_2\boldsymbol{b}_2 + \boldsymbol{a}_3\boldsymbol{b}_3, \qquad D(C_2) = \text{diag}\{-1,\ -1,\ 1\}. \tag{6.38}$$

Discuss the possibility of the vector of crystal lattice $\boldsymbol{f}$ (6.32) with the fractional components. If $\boldsymbol{f}$ is a vector of crystal lattice, the following are also the vectors of crystal lattice

$$\boldsymbol{f} + \vec{\vec{2}} \cdot \boldsymbol{f} = \boldsymbol{a}_3\,(2f_3)\,, \qquad \boldsymbol{f} - \vec{\vec{2}} \cdot \boldsymbol{f} = \boldsymbol{a}_1\,(2f_1) + \boldsymbol{a}_2\,(2f_2)\,. \tag{6.39}$$

Hence, three f_j all can be 0 or $1/2$. However, $f_1 = f_2 = 0$ when $f_3 = 0$, because $\boldsymbol{a}_1$ and $\boldsymbol{a}_2$ are two "primitive" vectors of crystal lattice in the plane perpendicular to $\boldsymbol{a}_3$. $f_3 = 0$ when $f_1 = f_2 = 0$, because $\boldsymbol{a}_3$ is the shortest vector of crystal lattice along the 2-fold axis. Therefore, there are four possibilities for $\boldsymbol{f}$, $\boldsymbol{f} = 0$ (P-type), $\boldsymbol{f} = (\boldsymbol{a}_2 + \boldsymbol{a}_3)/2$ (A-type), $\boldsymbol{f} = (\boldsymbol{a}_1 + \boldsymbol{a}_3)/2$ (B-type), and $\boldsymbol{f} = (\boldsymbol{a}_1 + \boldsymbol{a}_2 + \boldsymbol{a}_3)/2$ (I-type). Since the C-type translation group does not exist in the crystal system, those of the A-type and the B-type cannot exist for one crystal, neither can the F-type one. The B-type translation group becomes the A-type one if $\boldsymbol{a}_2$ is chosen to be new $\boldsymbol{a}_1$ and $-\boldsymbol{a}_1$ to be new $\boldsymbol{a}_2$. Similarly, the I-type translation group becomes the A-type one if $\boldsymbol{a}_1 + \boldsymbol{a}_2$ is chosen to be new $\boldsymbol{a}_1$ and $\boldsymbol{a}_2$ is left the same. Thus, there are two Bravais lattices P and A, and six symmorphic space groups $P2$, $P\overline{2}$, $P \pm 2$, $A2$, $A\overline{2}$, and $A \pm 2$ in the monoclinic crystal system.

6.3.4 *Orthorhombic Crystal System*

The crystallographic point groups in the orthorhombic crystal system are D_2 $(22')$, C_{2v} $(2\overline{2}')$, and D_{2h} $(\pm 22')$. The point groups all contain three perpendicular 2-fold axes (proper or improper). The **shortest vector** of crystal lattice along one 2-fold proper axis is chosen to be $\boldsymbol{a}_3$, and those along the remaining two 2-fold axes to be $\boldsymbol{a}_1$ and $\boldsymbol{a}_2$ such that $\boldsymbol{a}_1 \times \boldsymbol{a}_2$ is along the positive direction of $\boldsymbol{a}_3$. Thus, in the orthorhombic crystal system the lengths of the lattice bases have no restriction, and the angles are

$$\alpha_1 = \alpha_2 = \alpha_3 = \pi/2. \tag{6.40}$$

In the set of lattice bases the double-vector and the matrix of the 2-fold rotation about $\boldsymbol{a}_3$ are given in Eq. (6.38), and those about $\boldsymbol{a}_1$ are calculated from Eq. (6.14)

$$\vec{\vec{2'}} = \boldsymbol{a}_1\boldsymbol{b}_1 - \boldsymbol{a}_2\boldsymbol{b}_2 - \boldsymbol{a}_3\boldsymbol{b}_3, \qquad D(C_2') = \text{diag}\{1, \ -1, \ -1\}. \qquad (6.41)$$

If $\boldsymbol{f}$ given in Eq. (6.32) is a vector of crystal lattice, $\boldsymbol{f} + \vec{\vec{R}} \cdot \boldsymbol{f}$ is a vector of crystal lattice, too, where $\vec{\vec{R}}$ is $\vec{\vec{2}}$, $\vec{\vec{2'}}$ or $\vec{\vec{2}} \cdot \vec{\vec{2'}}$. Hence, three f_j can be 0 or 1/2, respectively. Thus, for the orthorhombic crystal system, in addition to P-type, A-type, and B-type translation groups, there are C-type and F-type translation groups.

For the point groups D_2 and D_{2h}, A-, B-, and C-type translation groups are the same. For the point group C_{2v}, the 2-fold axis along $\boldsymbol{a}_3$ is proper, but two 2-fold axes along $\boldsymbol{a}_1$ and $\boldsymbol{a}_2$ are improper so that A- and B-type translation groups are the same, but C-type is different. Those different space groups are usually classified into the same Bravais lattices. Thus, there are four Bravais lattices P, C (A), I, and F, and 13 symmorphic space groups $P22'$, $P2\overline{2}'$, $P \pm 22'$, $C22'$, $C2\overline{2}'$, $A2\overline{2}'$, $C \pm 22'$, $I22'$, $I2\overline{2}'$, $I \pm 22'$, $F22'$, $F2\overline{2}'$, and $F \pm 22'$ in the orthorhombic crystal system.

6.3.5 *Trigonal and Hexagonal Crystal System*

The crystallographic point groups in the hexagonal crystal system are C_6 (6), C_{3h} $(\overline{6})$, C_{6h} (± 6), D_6 $(62')$, C_{6v} $(6\overline{2}')$, D_{3h} $(\overline{6}2')$, and D_{6h} $(\pm 62')$. The **shortest vector** of crystal lattice along the 6-fold axis is chosen to be $\boldsymbol{a}_3$. For the cases with the point groups C_6, C_{3h}, and C_{6h}, the **shortest vector** of crystal lattice in the plane perpendicular to $\boldsymbol{a}_3$ is chosen to be $\boldsymbol{a}_1$. For the cases with the point groups D_6, C_{6v}, D_{3h}, and D_{6h}, the **shortest vector** of crystal lattice along the 2-fold axes in the plane perpendicular to $\boldsymbol{a}_3$ is chosen to be $\boldsymbol{a}_1$. Then, $\boldsymbol{a}_2 = C_6^2\boldsymbol{a}_1$. Thus,

$$a_1 = a_2, \qquad \alpha_1 = \alpha_2 = \pi/2, \qquad \alpha_3 = 2\pi/3. \qquad (6.42)$$

For the cases with the point group D_{3h}, two space groups are **different** depending on whether $\boldsymbol{a}_1$ is along a proper or an improper 2-fold axis.

From Eq. (6.14), the double-vectors and the matrices of the generators C_6, C_2', and $C_2'' = C_2'C_6$ in the set of lattice bases are

$$\vec{\vec{6}} = (\boldsymbol{a}_1 + \boldsymbol{a}_2)\,\boldsymbol{b}_1 - \boldsymbol{a}_1\boldsymbol{b}_2 + \boldsymbol{a}_3\boldsymbol{b}_3, \quad D(C_6) = \begin{pmatrix} 1 & -1 & 0 \\ 1 & 0 & 0 \\ 0 & 0 & 1 \end{pmatrix},$$

$$\begin{aligned}
\overleftrightarrow{2'} &= \boldsymbol{a}_1\boldsymbol{b}_1 - (\boldsymbol{a}_1 + \boldsymbol{a}_2)\,\boldsymbol{b}_2 - \boldsymbol{a}_3\boldsymbol{b}_3, \quad & D(C_2') &= \begin{pmatrix} 1 & -1 & 0 \\ 0 & -1 & 0 \\ 0 & 0 & -1 \end{pmatrix}, \\
\overleftrightarrow{2''} &= -\boldsymbol{a}_2\boldsymbol{b}_1 - \boldsymbol{a}_1\boldsymbol{b}_2 - \boldsymbol{a}_3\boldsymbol{b}_3, & D(C_2') &= \begin{pmatrix} 0 & -1 & 0 \\ -1 & 0 & 0 \\ 0 & 0 & -1 \end{pmatrix}.
\end{aligned} \tag{6.43}$$

If $\boldsymbol{f}$ given in Eq. (6.32) is a vector of crystal lattice,

$$\begin{aligned}
&\boldsymbol{f} + \left(\overleftrightarrow{6}\right)^3 \cdot \boldsymbol{f} = 2f_3\boldsymbol{a}_3, \\
&\boldsymbol{f} + \left(\overleftrightarrow{6}\right)^2 \cdot \boldsymbol{f} + \left(\overleftrightarrow{6}\right)^4 \cdot \boldsymbol{f} = 3f_3\boldsymbol{a}_3, \\
&\boldsymbol{f} - \overleftrightarrow{6} \cdot \boldsymbol{f} = f_2\boldsymbol{a}_1 + (f_2 - f_1)\,\boldsymbol{a}_2,
\end{aligned} \tag{6.44}$$

are all vectors of crystal lattice. From the first two conditions in Eq. (6.44), $f_3 = 0$. For the cases with the point groups C_6, C_{3h}, and C_{6h}, the length of the vector in the third formula of Eq. (6.44) is less than a_1, hence $f_1 = f_2 = 0$. For the cases with the point groups D_6, C_{6v}, D_{3h}, and D_{6h},

$$\begin{aligned}
&\boldsymbol{f} + \overleftrightarrow{2'} \cdot \boldsymbol{f} = (2f_1 - f_2)\,\boldsymbol{a}_1, \\
&\boldsymbol{f} + \left(\overleftrightarrow{3} \cdot \overleftrightarrow{2'} \cdot \overleftrightarrow{3} \cdot \overleftrightarrow{3}\right) \cdot \boldsymbol{f} = (2f_2 - f_1)\,\boldsymbol{a}_2,
\end{aligned} \tag{6.45}$$

are also the vectors of crystal lattice. Thus, $2f_1 - f_2$ and $2f_2 - f_1$ are 0 or 1, respectively. They cannot be 1 at the same time because both f_1 and f_2 are less than 1. There are three solutions to Eq. (6.45):

$$f_1 = f_2 = 0, \qquad f_1 = 2f_2 = 2/3, \qquad 2f_1 = f_2 = 2/3. \tag{6.46}$$

However, both vectors $\boldsymbol{f} = (2\boldsymbol{a}_1 + \boldsymbol{a}_2)/3$ and $\boldsymbol{f}' = (\boldsymbol{a}_1 + 2\boldsymbol{a}_2)/3$ are along a 2-fold axis with the length less than a_1, so that they are ruled out. Thus, there is one Bravais lattice P and eight symmorphic space groups $P6$, $P\overline{6}$, $P{\pm}6$, $P62'$, $P6\overline{2}'$, $P\overline{6}2'$, $P\overline{6}2''$, and $P{\pm}62'$ in the hexagonal crystal system.

The crystallographic point groups in the trigonal crystal system are C_3 (3), C_{3i} ($\overline{3}$), D_3 ($32'$), C_{3v} ($3\overline{2}'$), and D_{3d} ($\overline{3}2'$). The **shortest vector** of crystal lattice along the 3-fold axis is chosen to be $\boldsymbol{a}_3$. For the cases with the point groups C_3 and C_{3i}, the **shortest vector** of crystal lattice in the plane perpendicular to $\boldsymbol{a}_3$ is chosen to be $\boldsymbol{a}_1$. For the cases with the point groups D_3, C_{3v}, and D_{3d}, the **shortest vector** of crystal lattice along the 2-fold axes is chosen to be $\boldsymbol{a}_1$. Then, $\boldsymbol{a}_2 = C_3\boldsymbol{a}_1$. Thus, the condition

(6.42) is satisfied. The double-vectors and the matrices of the generators $C_3 = C_6^2$ and C_2' in the set of lattice bases are given in Eq. (6.43).

If $\boldsymbol{f}$ given in Eq. (6.32) is a vector of crystal lattice, $\boldsymbol{f} + \overleftrightarrow{3} \cdot \boldsymbol{f} + \overleftrightarrow{3} \cdot \overleftrightarrow{3} \cdot \boldsymbol{f} = 3f_3\boldsymbol{a}_3$ is also a vector of crystal lattice. Thus,

$$f_3 = 0, \ 1/3, \quad \text{or} \ \ 2/3. \tag{6.47}$$

For the cases with the point groups D_3, C_{3v}, and D_{3d}, one also obtains Eqs. (6.45) and (6.46). When $f_1 = f_2 = 0$, f_3 has to be 0.

For the cases with the point groups C_3 and C_{3i}, the following vector is also a vector of crystal lattice

$$\boldsymbol{f} - \overleftrightarrow{3} \cdot \boldsymbol{f} = (f_1 + f_2)\,\boldsymbol{a}_1 + (2f_2 - f_1)\,\boldsymbol{a}_2.$$

After removing the integral part, if $(f_1 + f_2)$ and $(2f_2 - f_1)$ both are not equal to 0, the length of the vector is less than a_1, which is in conflict with the assumption. If $f_1 + f_2 = 0$, one has the first solution in Eq. (6.46), and if $f_1 + f_2 = 1$ and $(2f_2 - f_1) = 0$ or 1, one has the remaining solutions of Eq. (6.46).

From Eqs. (6.46) and (6.47), there are four possibilities for $\boldsymbol{f}$ in the trigonal crystal system, where $\boldsymbol{f}'$ is calculated from $2\boldsymbol{f}$,

$$\begin{aligned}
&\text{(a)} \ \ \boldsymbol{f} = 0, \\
&\text{(b)} \ \ \boldsymbol{f} = (2\boldsymbol{a}_1 + \boldsymbol{a}_2)/3, \ \ \text{and} \ \ \boldsymbol{f}' = (\boldsymbol{a}_1 + 2\boldsymbol{a}_2)/3, \\
&\text{(c)} \ \ \boldsymbol{f} = (2\boldsymbol{a}_1 + \boldsymbol{a}_2 + \boldsymbol{a}_3)/3, \ \ \text{and} \ \ \boldsymbol{f}' = (\boldsymbol{a}_1 + 2\boldsymbol{a}_2 + 2\boldsymbol{a}_3)/3, \\
&\text{(d)} \ \ \boldsymbol{f} = (\boldsymbol{a}_1 + 2\boldsymbol{a}_2 + \boldsymbol{a}_3)/3, \ \ \text{and} \ \ \boldsymbol{f}' = (2\boldsymbol{a}_1 + \boldsymbol{a}_2 + 2\boldsymbol{a}_3)/3.
\end{aligned} \tag{6.48}$$

The length of the vector $\boldsymbol{f}$ in (b) is less than a_1. It is not a vector of crystal lattice for the cases with the point groups C_3 and C_{3i}. For the cases with the point groups D_3, C_{3v}, and D_{3d}, a set of new lattice bases in (b) can be defined as Eq. (7.49) and Fig. 6.1 such that new lattice bases are primitive,

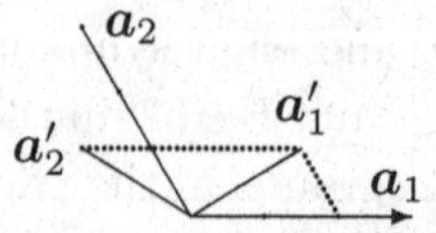

Fig. 6.1 New lattice bases in (b).

$$\begin{aligned}
&\boldsymbol{a}_1' = (2\boldsymbol{a}_1 + \boldsymbol{a}_2)/3 = \boldsymbol{f}, \qquad \boldsymbol{a}_2' = (-\boldsymbol{a}_1 + \boldsymbol{a}_2)/3 = \overleftrightarrow{3} \cdot \boldsymbol{a}_1', \\
&\boldsymbol{a}_1 = \boldsymbol{a}_1' - \boldsymbol{a}_2', \qquad \boldsymbol{a}_2 = \boldsymbol{a}_1' + 2\boldsymbol{a}_2', \qquad \boldsymbol{f}' = \boldsymbol{a}_1' + \boldsymbol{a}_2'.
\end{aligned} \tag{6.49}$$

However, in these cases new lattice bases $\boldsymbol{a}_1'$ and $\boldsymbol{a}_2'$ are not along a 2-fold axis, but along the angular bisector between two neighboring 2-fold axes (a direction with an angle $\pi/6$ to a 2-fold axis). Now, the 2-fold rotation is denoted by $\vec{\vec{2}}''$ given in Eq. (6.43) instead of $\vec{\vec{2}}'$. Two space groups are different depending on whether $\boldsymbol{a}_1$ is along a 2-fold axis or not. This Bravais lattice is merged into the P-type Bravais lattices in the hexagonal crystal system. In this case there are eight symmorphic space groups: $P3$, $P\overline{3}$, $P32'$, $P32''$, $P3\overline{2}'$, $P3\overline{2}''$, $P\overline{3}2'$, and $P\overline{3}2''$.

The length of $\boldsymbol{a}_1 + \boldsymbol{a}_2$ is the same as a_1. $\boldsymbol{a}_1 + \boldsymbol{a}_2$ is along a 2-fold axis for the cases with the point groups D_3, C_{3v}, and D_{3d}. Thus, for the cases of all five crystallographic point groups, $\boldsymbol{a}_1 + \boldsymbol{a}_2$ can also be chosen as a lattice basis instead of $\boldsymbol{a}_1$ such that the case (d) becomes the case (c). For the case (c), one defines new lattice bases, which are primitive

$$\begin{aligned}
\boldsymbol{a}_1' &= (2\boldsymbol{a}_1 + \boldsymbol{a}_2 + \boldsymbol{a}_3)/3 = \vec{\vec{3}} \cdot \boldsymbol{a}_3', & \boldsymbol{a}_1 &= \boldsymbol{a}_1' - \boldsymbol{a}_2', \\
\boldsymbol{a}_2' &= (-\boldsymbol{a}_1 + \boldsymbol{a}_2 + \boldsymbol{a}_3)/3 = \vec{\vec{3}} \cdot \boldsymbol{a}_1', & \boldsymbol{a}_2 &= \boldsymbol{a}_2' - \boldsymbol{a}_3', \\
\boldsymbol{a}_3' &= (-\boldsymbol{a}_1 - 2\boldsymbol{a}_2 + \boldsymbol{a}_3)/3 = \vec{\vec{3}} \cdot \boldsymbol{a}_2', & \boldsymbol{a}_3 &= \boldsymbol{a}_1' + \boldsymbol{a}_2' + \boldsymbol{a}_3'.
\end{aligned} \tag{6.50}$$

From Eq. (6.14), the double-vectors and the matrices of the generators C_3 and C_2' in the new lattice bases are

$$\begin{aligned}
&\vec{\vec{3}} = \boldsymbol{a}_2\boldsymbol{b}_1 + \boldsymbol{a}_3\boldsymbol{b}_2 + \boldsymbol{a}_1\boldsymbol{b}_3, \qquad && \vec{\vec{2}}' = -\boldsymbol{a}_2\boldsymbol{b}_1 - \boldsymbol{a}_1\boldsymbol{b}_2 - \boldsymbol{a}_3\boldsymbol{b}_3, \\
&D(C_3) = \begin{pmatrix} 0 & 0 & 1 \\ 1 & 0 & 0 \\ 0 & 1 & 0 \end{pmatrix}, && D(C_2') = \begin{pmatrix} 0 & -1 & 0 \\ -1 & 0 & 0 \\ 0 & 0 & -1 \end{pmatrix}.
\end{aligned} \tag{6.51}$$

It is called the R-type (rhombohedral) Bravais lattice where

$$a_1 = a_2 = a_3, \qquad \alpha_1 = \alpha_2 = \alpha_3. \tag{6.52}$$

There are five symmorphic space groups belonging to the R-type Bravais lattice: $R3$, $R\overline{3}$, $R32'$, $R3\overline{2}'$, and $R\overline{3}2'$.

6.3.6 *Tetragonal Crystal System*

The crystallographic point groups in the tetragonal crystal system are C_4 (4), S_4 ($\overline{4}$), C_{4h} (± 4), D_4 ($42'$), C_{4v} ($4\overline{2}'$), D_{2d} ($\overline{4}2'$), and D_{4h} ($\pm 42'$). The **shortest vector** of crystal lattice along the 4-fold axis is chosen to be $\boldsymbol{a}_3$. For the cases with the point groups C_4, S_4, and C_{4h}, the **shortest vector** of crystal lattice in the plane perpendicular to $\boldsymbol{a}_3$ is chosen to be $\boldsymbol{a}_1$. For the cases with the point groups D_4, C_{4v}, D_{2d}, and D_{4h}, the **shortest vector** of crystal lattice along the 2-fold axes in the plane perpendicular to $\boldsymbol{a}_3$ is chosen to be $\boldsymbol{a}_1$. Then, $\boldsymbol{a}_2 = C_4\boldsymbol{a}_1$. Thus,

$$a_1 = a_2, \qquad \alpha_1 = \alpha_2 = \alpha_3 = \pi/2. \tag{6.53}$$

For the cases with the point group D_{2d}, two space groups are different depending on whether $\boldsymbol{a}_1$ is along a proper or an improper 2-fold axis.

From Eq. (6.14), the double-vectors and the matrices of the generators C_4, C_2', and $C_2'' = C_4C_2'$ in the set of lattice bases are

$$\begin{aligned}
\vec{\vec{4}} &= \boldsymbol{a}_2\boldsymbol{b}_1 - \boldsymbol{a}_1\boldsymbol{b}_2 + \boldsymbol{a}_3\boldsymbol{b}_3, \qquad & D(C_4) &= \begin{pmatrix} 0 & -1 & 0 \\ 1 & 0 & 0 \\ 0 & 0 & 1 \end{pmatrix}, \\
\vec{\vec{2}}' &= \boldsymbol{a}_1\boldsymbol{b}_1 - \boldsymbol{a}_2\boldsymbol{b}_2 - \boldsymbol{a}_3\boldsymbol{b}_3, \qquad & D(C_2') &= \begin{pmatrix} 1 & 0 & 0 \\ 0 & -1 & 0 \\ 0 & 0 & -1 \end{pmatrix}, \\
\vec{\vec{2}}'' &= \boldsymbol{a}_2\boldsymbol{b}_1 + \boldsymbol{a}_1\boldsymbol{b}_2 - \boldsymbol{a}_3\boldsymbol{b}_3, \qquad & D(C_2'') &= \begin{pmatrix} 0 & 1 & 0 \\ 1 & 0 & 0 \\ 0 & 0 & -1 \end{pmatrix}.
\end{aligned} \tag{6.54}$$

If $\boldsymbol{f}$ given in Eq. (6.32) is a vector of crystal lattice, $\boldsymbol{f} + \vec{\vec{2}} \cdot \boldsymbol{f} = 2f_3\boldsymbol{a}_3$ is also a vector of crystal lattice, so that

$$f_3 = 0, \quad \text{or} \quad 1/2. \tag{6.55}$$

For the cases with the point groups D_4, C_{4v}, D_{2d}, and D_{4h},

$$\begin{aligned}
&\boldsymbol{f} + \vec{\vec{2}}' \cdot \boldsymbol{f} = 2f_1\boldsymbol{a}_1, \\
&\boldsymbol{f} + \left\{\vec{\vec{4}} \cdot \vec{\vec{2}}' \cdot \left(\vec{\vec{4}}\right)^3\right\} \cdot \boldsymbol{f} = 2f_2\boldsymbol{a}_2, \\
&\boldsymbol{f} + \vec{\vec{2}}'' \cdot \boldsymbol{f} = (f_1 + f_2)(\boldsymbol{a}_1 + \boldsymbol{a}_2),
\end{aligned}$$

are also the vectors of crystal lattice if $\boldsymbol{f}$ is. Thus, $f_1 = 0$ or $1/2$, $f_2 = 0$ or $1/2$, but $f_1 + f_2 \neq 1/2$ because the vector $(\boldsymbol{a}_1 + \boldsymbol{a}_2)/2$ is along a 2-fold axis and with the length less than a_1. Namely, $f_1 = f_2 = 0$ or $1/2$. Note that $f_3 = 0$ when $f_1 = f_2 = 0$. Conversely, $f_1 = f_2 = 0$ when $f_3 = 0$. Therefore, there are only P-type and I-type Bravais lattices.

For the cases with the point groups C_4, S_4, and C_{4h},

$$\begin{aligned}
\boldsymbol{f} - \vec{\vec{4}} \cdot \boldsymbol{f} &= \boldsymbol{a}_1(f_1 + f_2) + \boldsymbol{a}_2(f_2 - f_1) \equiv \boldsymbol{f}', \\
\boldsymbol{f}' - \vec{\vec{4}} \cdot \boldsymbol{f}' &= \boldsymbol{a}_1(2f_2) + \boldsymbol{a}_2(-2f_1) \equiv \boldsymbol{f}'',
\end{aligned} \tag{6.56}$$

are also the vectors of crystal lattice if $\boldsymbol{f}$ is. If $f_1 = f_2$, there are only two choices: a) $f_1 = f_2 = 0$, and then $f_3 = 0$; b) $f_1 = f_2 = 1/2$, and then $f_3 = 1/2$. There are only P-type and I-type Bravais lattices. We are going to show by Eq. (6.56) that, if $f_1 \neq f_2$, there exists a vector of crystal lattice with length less than a_1 in the plane perpendicular to $\boldsymbol{a}_3$, so that it is rule out. In fact, if one of f_1 and f_2 is zero, the other has to be 1/2 such that the length of $\boldsymbol{f}'$ is less than a_1. If $f_1 + f_2 \geqslant 1$,

$$a_1^2 \left\{ (f_1 + f_2 - 1)^2 + (f_2 - f_1)^2 \right\} = a_1^2 \left\{ 1 - 2\left(f_1 + f_2 - f_1^2 - f_2^2\right) \right\} < a_1^2.$$

The length of $\boldsymbol{f}' - \boldsymbol{a}_1$ is less than a_1. Thus, $f_1 + f_2 < 1$. In the same reason, $f_1' + f_2' = 2f_2$ has to be less than one, so does $2f_1 < 1$. Then,

$$|\vec{f}'|^2 = a_1^2 \left\{ (f_1 + f_2)^2 + (f_2 - f_1)^2 \right\} = a_1^2 \left\{ 2f_1^2 + 2f_2^2 \right\} < a_1^2.$$

Therefore, there are two Bravais lattices P and I, and 16 symmorphic space groups $P4$, $P\overline{4}$, $P\pm 4$, $P42'$, $P4\overline{2}'$, $P\overline{4}2'$, $P\overline{4}2''$, $P\pm 42'$, $I4$, $I\overline{4}$, $I\pm 4$, $I42'$, $I4\overline{2}'$, $I\overline{4}2'$, $I\overline{4}2''$, and $I \pm 42'$ in the tetragonal crystal system.

6.3.7 *Cubic Crystal System*

The crystallographic point groups in the cubic crystal system are $\mathbf{T}$ $(3'22')$, $\mathbf{T}_h$ $(\overline{3}'22')$, $\mathbf{O}$ $(3'42'')$, $\mathbf{T}_d$ $(3'\overline{4}\,\overline{2}'')$, and $\mathbf{O}_h$ $(\overline{3}42'')$. The **shortest vectors** of crystal lattice along the three 2-fold axes (for $\mathbf{T}$ and $\mathbf{T}_h$) or along the three 4-fold axes (for $\mathbf{O}$, $\mathbf{T}_d$, and $\mathbf{O}_h$) are chosen to be the lattice bases, where $\boldsymbol{a}_1 \times \boldsymbol{a}_2$ is along the positive direction of $\boldsymbol{a}_3$. Thus,

$$a_1 = a_2 = a_3, \qquad \alpha_1 = \alpha_2 = \alpha_3 = \pi/2. \tag{6.57}$$

In the set of lattice bases, the double-vectors and the matrices of C_4 about $\boldsymbol{a}_3$, C_2' about $\boldsymbol{a}_1$, and C_2'' about $(\boldsymbol{a}_1 + \boldsymbol{a}_2)$ are given in Eq. (6.54), and the matrix of C_3' about the direction $(\boldsymbol{a}_1 + \boldsymbol{a}_2 + \boldsymbol{a}_3)$ is given in Eq. (6.51).

If $\boldsymbol{f}$ given in Eq. (6.32) is a vector of crystal lattice, then, f_1, f_2 and f_3 all are 0 or 1/2, respectively, because there is a 2-fold axis along each direction of $\boldsymbol{a}_i$. Since three axes along $\boldsymbol{a}_i$ are equivalent, the Bravais lattices A, B, and C are merged to the F-type Bravais lattice. Thus, there are three Bravais lattices P, I, and F, and 15 symmorphic space groups $P3'22'$, $P\overline{3}'22'$, $P3'42''$, $P3'\overline{4}\,\overline{2}''$, $P\overline{3}'42''$, $I3'22'$, $I\overline{3}'22'$, $I3'42''$, $I3'\overline{4}\,\overline{2}''$, $I\overline{3}'42''$, $F3'22'$, $F\overline{3}'22'$, $F3'42''$, $F3'\overline{4}\,\overline{2}''$, and $F\overline{3}'42''$ in the cubic crystal system.

Seven crystal systems, 14 Bravais lattices and 73 symmorphic space groups as well as the relations between the lattice bases and the rotational axes are listed in Table 6.5. The matrices of the generator are listed in the note.

Table 6.5 The property of symmorphic space groups

Crystal system, Bravais	Point groups Sch.	Point groups INSG	Product of subgroups	Lattice bases	Direction of rotational axes		
Triclinic, P	C_1	1	C_1				
$P1$, $P\bar{1}$	C_i	$\bar{1}$	C_i				
Monoclinic,					$\boldsymbol{a}_3$		
P, A, *no.* = 6	C_2	2	C_2	$\alpha_1 = \alpha_2$	2		
$P2$, $P\bar{2}$, $P\pm 2$,	C_s	$\bar{2}$	C_s	$= \pi/2$	$\bar{2}$		
$A2$, $A\bar{2}$, $A\pm 2$	C_{2h}	± 2	C_iC_2		± 2		
Orthorhombic,					$\boldsymbol{a}_3$	$\boldsymbol{a}_1$	
P, $C(A)$, I, F	D_2	$22'$	C_2C_2'	$\alpha_1 = \alpha_2$	2	2	
no. = 13	C_{2v}	$2\bar{2}'$	C_2C_s'	$= \alpha_3$	± 2	± 2	
see Note (e)	D_{2h}	$\pm 22'$	$C_iC_2C_2'$	$= \pi/2$	2	$\bar{2}$	
Tetragonal,					$\boldsymbol{a}_3$	$\boldsymbol{a}_1$	$\boldsymbol{a}_1 + \boldsymbol{a}_2$
P, I *no.* = 16	C_4	4	C_4	$a_1 = a_2$	4		
$P4$, $P\bar{4}$, $P\pm 4$,	S_4	$\bar{4}$	S_4	$\alpha_1 = \alpha_2$	$\bar{4}$		
$P42'$, $P4\bar{2}'$,	C_{4h}	± 4	C_iC_4	$= \alpha_3$	± 4		
$P\bar{4}2'$, $P\bar{4}2''$,	D_4	$42'$	C_4C_2'	$= \pi/2$	4	2	2
$P\pm 42'$, $I4$,	C_{4v}	$4\bar{2}'$	C_4C_s'		4	$\bar{2}$	$\bar{2}$
$I\bar{4}$, $I\pm 4$, $I42'$,	D_{2d}	$\bar{4}2'$	S_4C_2'		$\bar{4}$	2	$\bar{2}$
$I4\bar{2}'$, $I\bar{4}2'$,	D_{2d}	$\bar{4}2''$	S_4C_2''		$\bar{4}$	$\bar{2}$	2
$I\bar{4}2''$, $I\pm 42'$	D_{4h}	$\pm 42'$	$C_iC_4C_2'$		± 4	± 2	± 2
Cubic,					$\boldsymbol{a}_3$	$\boldsymbol{a}'$	$\boldsymbol{a}_1 + \boldsymbol{a}_2$
	T	$3'22'$	$C_3'C_2C_2'$	$a_1 = a_2$	2	3	
P, I, F	T_h	$\bar{3}22'$	$C_{3i}'C_2C_2'$	$= a_3$	± 2	$\bar{3}$	
	O	$3'42''$	$C_3'C_4C_2''$	$\alpha_1 = \alpha_2$	4	3	2
no. = 15	T_d	$3'\bar{4}\,\bar{2}''$	$C_3S_4C_s''$	$= \alpha_3$	$\bar{4}$	3	$\bar{2}$
see Note (f)	O_h	$\bar{3}'42''$	$C_{3i}'C_4C_2''$	$= \pi/2$	± 4	$\bar{3}$	± 2
Rhombo-					$\boldsymbol{a}'$		$\boldsymbol{a}_1 - \boldsymbol{a}_2$
hedral,	C_3	3	C_3	$a_1 = a_2$	3		
R	C_{3i}	$\bar{3}$	C_{3i}	$= a_3$	$\bar{3}$		
no. = 5	D_3	$32'$	C_3C_2'	$\alpha_1 = \alpha_2$	3		2
$R3$, $R\bar{3}$, $R32'$,	C_{3v}	$3\bar{2}'$	C_3C_s'	$= \alpha_3$	3		$\bar{2}$
$R3\bar{2}'$, $R\bar{3}2'$	D_{3d}	$\bar{3}2'$	$C_{3i}C_2'$		$\bar{3}$		± 2
Hexagonal,					$\boldsymbol{a}_3$	$\boldsymbol{a}_1$	$\boldsymbol{a}_1 - \boldsymbol{a}_2$
	C_3	3	C_3	$a_1 = a_2$	3		
P	C_{3i}	$\bar{3}$	C_{3i}	$\alpha_1 = \alpha_2$	$\bar{3}$		
	D_3	$32'$	C_3C_2'	$= \pi/2$	3	2	
	D_3	$32''$	C_3C_2''	$\alpha_3 = 2\pi/3$	3		2
	C_{3v}	$3\bar{2}'$	C_3C_s'		3	$\bar{2}$	
	C_{3v}	$3\bar{2}''$	C_3C_s''		3		$\bar{2}$
no. = 16	D_{3d}	$\bar{3}2'$	$C_{3i}C_2'$		$\bar{3}$	± 2	
	D_{3d}	$\bar{3}2''$	$C_{3i}C_2''$		$\bar{3}$		± 2
$P3$, $P\bar{3}$,	C_6	6	C_6		6		
$P32'$, $P32''$,	C_{3h}	$\bar{6}$	C_{3h}		$\bar{6}$		
$P3\bar{2}'$, $P3\bar{2}''$,	C_{6h}	± 6	C_iC_6		± 6		
$P\bar{3}2'$, $P\bar{3}2''$	D_6	$62'$	C_6C_2'		6	2	2
$P6$, $P\bar{6}$,	C_{6v}	$6\bar{2}'$	C_6C_s'		6	$\bar{2}$	$\bar{2}$
$P\pm 6$, $P62'$,	D_{3h}	$\bar{6}2'$	$C_{3h}C_2'$		$\bar{6}$	2	$\bar{2}$
$P6\bar{2}'$, $P\bar{6}2'$,	D_{3h}	$\bar{6}2''$	$C_{3h}C_2''$		$\bar{6}$	$\bar{2}$	2
$P\bar{6}2''$, $P\pm 62'$	D_{6h}	$\pm 62'$	$C_iC_6C_2'$		± 6	± 2	± 2

Note. (a) Generators in the hexagonal crystal system:

$$C_6 = \begin{pmatrix} 1 & -1 & 0 \\ 1 & 0 & 0 \\ 0 & 0 & 1 \end{pmatrix}, \quad C_3 = \begin{pmatrix} 0 & -1 & 0 \\ 1 & -1 & 0 \\ 0 & 0 & 1 \end{pmatrix}, \quad C_2' = \begin{pmatrix} 1 & -1 & 0 \\ 0 & -1 & 0 \\ 0 & 0 & -1 \end{pmatrix}, \quad C_2'' = \begin{pmatrix} 0 & -1 & 0 \\ -1 & 0 & 0 \\ 0 & 0 & -1 \end{pmatrix}.$$

(b) Generators in the rhombohedral crystal system:

$$C_3 = \begin{pmatrix} 0 & 0 & 1 \\ 1 & 0 & 0 \\ 0 & 1 & 0 \end{pmatrix}, \qquad C_2' = \begin{pmatrix} 0 & -1 & 0 \\ -1 & 0 & 0 \\ 0 & 0 & -1 \end{pmatrix}.$$

(c) Generators of other cyclic subgroups:

$$C_2 = \begin{pmatrix} -1 & 0 & 0 \\ 0 & -1 & 0 \\ 0 & 0 & 1 \end{pmatrix}, \quad C_2' = \begin{pmatrix} 1 & 0 & 0 \\ 0 & -1 & 0 \\ 0 & 0 & -1 \end{pmatrix}, \quad C_2'' = \begin{pmatrix} 0 & 1 & 0 \\ 1 & 0 & 0 \\ 0 & 0 & -1 \end{pmatrix},$$

$$C_4 = \begin{pmatrix} 0 & -1 & 0 \\ 1 & 0 & 0 \\ 0 & 0 & 1 \end{pmatrix}, \quad C_3 = \begin{pmatrix} 0 & 0 & 1 \\ 1 & 0 & 0 \\ 0 & 1 & 0 \end{pmatrix}.$$

(d) $\boldsymbol{a}' = \boldsymbol{a}_1 + \boldsymbol{a}_2 + \boldsymbol{a}_3$.

(e) There are 13 symmorphic space groups $P22'$, $P2\overline{2}'$, $P\pm 22'$, $C22'$, $C2\overline{2}'$, $A2\overline{2}'$, $C\pm 22'$, $I22'$, $I2\overline{2}'$, $I\pm 22'$, $F22'$, $F2\overline{2}'$, and $F\pm 22'$ in the orthorhombic crystal system.

(f) There are 15 symmorphic space groups $P3'22'$, $P\overline{3}'22'$, $P3'42''$, $P3'\overline{4}\,\overline{2}''$, $P\overline{3}'42''$, $I3'22'$, $I\overline{3}'22'$, $I3'42''$, $I3'\overline{4}\,\overline{2}''$, $I\overline{3}'42''$, $F3'22'$, $F\overline{3}'22'$, $F3'42''$, $F3'\overline{4}\,\overline{2}''$, and $F\overline{3}'42''$ in the cubic crystal system.

6.4 Space Group

In this section we will study the space group $\mathcal{S}$ where at least one vector $\boldsymbol{t}$ in the symmetric operation $g(R, \boldsymbol{t})$ is not vanishing. We will first study the constraint on $\boldsymbol{t}$ from R and the dependence of $\boldsymbol{t}$ on the position of the origin. Then, we will briefly introduce the method of finding the inequivalent space groups. At last, we provide the method to analyze the symmetry property of a crystal based on the symbol of its space group in the international notations.

6.4.1 *Symmetry Elements*

For a given crystal with a crystallographic point group G, the symmetric operation in a coordinate frame is expressed generally as

$$\begin{aligned} &g(R, \boldsymbol{\alpha}) = T(\boldsymbol{\ell})g(R, \boldsymbol{t}), \qquad \boldsymbol{\alpha} = \boldsymbol{\ell} + \boldsymbol{t}, \\ &\boldsymbol{t} = \sum_{j=1}^{3} \boldsymbol{a}_j t_j, \qquad 0 \leqslant t_j < 1, \end{aligned} \tag{6.58}$$

where $\boldsymbol{t}$ depends on R uniquely. We first study the constraint on $\boldsymbol{t}$ from R.

Let $\left\{ \overleftrightarrow{R} \right\}$ denote the double-vector of the sum of the elements in the cyclic group generated by $R \in$ G. For an arbitrary vector $\boldsymbol{r}$, one has

$$\overleftrightarrow{R} \cdot \left\{ \overleftrightarrow{R} \right\} \cdot \boldsymbol{r} = \left\{ \overleftrightarrow{R} \right\} \cdot \boldsymbol{r}. \tag{6.59}$$

It means that the vector $\left\{ \overleftrightarrow{R} \right\} \cdot \boldsymbol{r}$ is left invariant under the rotation R. A proper rotation C_N ($N \neq 1$) leaves only the vector along its rotational axis $\hat{\boldsymbol{n}}$ invariant. The improper rotation S_2 is an inversion with respect to a plane perpendicular to its improper axis and leaves only the vector on the plane invariant. There is no vector which is left invariant under the improper rotation S_N ($N \neq 2$). Therefore,

$$\begin{aligned} &\left\{ \overleftrightarrow{C}_1 \right\} = \overleftrightarrow{1}, \qquad && \left\{ \overleftrightarrow{C}_N \right\} = N\hat{\boldsymbol{n}}\hat{\boldsymbol{n}}, \qquad && N \neq 1, \\ &\left\{ \overleftrightarrow{S}_2 \right\} = 2\left(\overleftrightarrow{1} - \hat{\boldsymbol{n}}\hat{\boldsymbol{n}}\right), \qquad && \left\{ \overleftrightarrow{S}_N \right\} = 0, \qquad && N \neq 2. \end{aligned} \tag{6.60}$$

Since $(R)^{N'} = E$, where $N' = 2N$ for $R = S_N$ with $N = 1$ or 3 and $N' = N$ for the remaining cases, the constraint

$$\{g(R, \boldsymbol{t})\}^{N'} = T\left(\left\{ \overleftrightarrow{R} \right\} \cdot \boldsymbol{t}\right) = T(\boldsymbol{\ell}), \tag{6.61}$$

gives a strong restriction on $\boldsymbol{t}$:

$$\begin{aligned} &\boldsymbol{t} = 0, && \text{when} \quad R = C_1 = E, \\ &N\left\{\hat{\boldsymbol{n}}\left(\hat{\boldsymbol{n}} \cdot \boldsymbol{t}\right)\right\} = N\boldsymbol{t}_{\parallel} = m\boldsymbol{a}_{\parallel}, && \text{when} \quad R = C_N, \quad N \neq 1, \\ &2\left\{\boldsymbol{t} - \hat{\boldsymbol{n}}\left(\hat{\boldsymbol{n}} \cdot \boldsymbol{t}\right)\right\} = 2\boldsymbol{t}_{\perp} = m\boldsymbol{a}_{\perp}, && \text{when} \quad R = S_2, \\ &\text{no restriction on} \quad \boldsymbol{t}, && \text{when} \quad R = S_N, \quad N \neq 2, \end{aligned} \tag{6.62}$$

where m is an integer, $\boldsymbol{t}_{\parallel} = \hat{\boldsymbol{n}}\left(\hat{\boldsymbol{n}} \cdot \boldsymbol{t}\right)$ is the parallel component of $\boldsymbol{t}$, and $\boldsymbol{t}_{\perp} = \boldsymbol{t} - \boldsymbol{t}_{\parallel}$ is the perpendicular component of $\boldsymbol{t}$. When $R = C_N$, $N \neq 1$, $\boldsymbol{a}_{\parallel}$ is the shortest vector of the crystal lattice along the rotational axis $\hat{\boldsymbol{n}}$ of C_N. When $R = S_2$, $\boldsymbol{a}_{\perp}$ is the shortest vector of the crystal lattice along the direction of $\boldsymbol{t}_{\perp}$. We are going to show that **the part of $\boldsymbol{t}$ restricted by the constraint (6.61) is independent of the choice of the origin**, and the remaining part can be removed by a suitable choice of the origin.

Let K and K_0 denote two laboratory frames with different origins O and O', respectively. If the position vector of O' in K is $\boldsymbol{r}_0$, any point

P with the position vector $\boldsymbol{r}$ in K has the position vector $\boldsymbol{r} - \boldsymbol{r}_0$ in K_0. Assume that a symmetric operation is described by $g(R, \boldsymbol{t})$ in K and by $g' = g(R, \boldsymbol{t}_0)$ in K_0, the symmetric operation moves O' to $R\boldsymbol{r}_0 + \boldsymbol{t}$ in K, and to $\boldsymbol{t}_0$ in K_0. Thus, $\boldsymbol{t}_0 = R\boldsymbol{r}_0 + \boldsymbol{t} - \boldsymbol{r}_0$,

$$g' = g\left[R, (R - E)\boldsymbol{r}_0 + \boldsymbol{t}\right] = T(-\boldsymbol{r}_0)g(R, \boldsymbol{t})T(\boldsymbol{r}_0). \qquad (6.63)$$

The operator form of a symmetric operation depends upon the choice of the origin of the coordinate frame. The suitable choice of the position of the origin may simplify the operator forms of the symmetric operations. For a given crystal **the origin has to be chosen uniformly.**

For $R = C_N$, $N \neq 1$, the shift $\boldsymbol{r}_0$ of the origin changes $\boldsymbol{t}$ with $(C_N - E)\boldsymbol{r}_0$, which is perpendicular to the rotational axis of C_N. In the perpendicular plane the eigenvalue of C_N is not equal to one such that the equation $(C_N - E)\boldsymbol{r}_0 = -\boldsymbol{t}_\perp$ has a unique solution $\boldsymbol{r}_0$. Thus, $\boldsymbol{t}_\perp = \left(\vec{\vec{1}} - \hat{\boldsymbol{n}}\hat{\boldsymbol{n}}\right) \cdot \boldsymbol{t}$ in $g(C_N, \boldsymbol{t})$ can be removed by the choice of the origin.

For $R = S_2$, the shift $\boldsymbol{r}_0$ of the origin changes $\boldsymbol{t}$ with $(S_2 - E)\boldsymbol{r}_0$. The equation $(S_2 - E)\boldsymbol{r}_0 = -\boldsymbol{t}_\parallel$ has a solution, $\boldsymbol{r}_0 = \boldsymbol{t}_\parallel/2$. Thus, $\boldsymbol{t}_\parallel = \hat{\boldsymbol{n}}(\hat{\boldsymbol{n}} \cdot \boldsymbol{t})$ in $g(S_2, \boldsymbol{t})$ can be removed by the choice of the origin.

For $R = S_N$, $N \neq 2$, the shift $\boldsymbol{r}_0$ of the origin changes $\boldsymbol{t}$ with $(S_N - E)\boldsymbol{r}_0$. Since the eigenvalue of S_N is not equal to one such that the equation $(S_N - E)\boldsymbol{r}_0 = -\boldsymbol{t}$ has a unique solution $\boldsymbol{r}_0$, the vector $\boldsymbol{t}$ in $g(S_N, \boldsymbol{t})$ can be removed by the choice of the origin.

In summary, the part of $\boldsymbol{t}$ which is not restricted by the constraint (6.61) can be removed by a suitable choice of the origin, but **the part of $\boldsymbol{t}$ restricted by the constraint (6.61) is independent of the choice of the origin.** Remind that the position of the origin can be further chosen in the condition that the simplified $\boldsymbol{t}$ does not change. We will discuss this with examples later.

A point is called the **symmetry center** of a symmetric operation $g(R, \boldsymbol{t})$ if the point is left invariant under $g(R, \boldsymbol{t})$. A symmetric operation $g(R, \boldsymbol{t})$ is called a **closed** one if it has a symmetry center, otherwise, it is an **open** one. The position vector $\boldsymbol{r}_0$ of a symmetry center of a symmetric operation $g(R, \boldsymbol{t})$ satisfies

$$\left[g(R, \boldsymbol{t}) - E\right]\boldsymbol{r}_0 = (R - E)\,\boldsymbol{r}_0 + \boldsymbol{t} = 0. \qquad (6.64)$$

In comparison with Eq. (6.63), **the vector $\boldsymbol{t}$ in a closed operation** $g(R, \boldsymbol{t})$ can be removed **if the symmetry center is chosen to be the origin of a new coordinate frame.** Since $R^{N'} = E$ one has

$$[g(R,t)]^{N'} = \begin{cases} E, & \text{closed operation,} \\ T(\ell) \neq E, & \text{open operation.} \end{cases} \tag{6.65}$$

Due to Eq. (6.61), the vector t in a closed operation $g(R,t)$ satisfies

$$\left\{\vec{\vec{R}}\right\} \cdot t = 0, \tag{6.66}$$

and m given in Eq. (6.62) has to be zero for a closed operation.

There are only two kinds of open operations where m in Eq. (6.62) does not vanish. In a suitable choice of the origin, two kinds of open operations are expressed as

$$\begin{aligned} & g\left(C_N, t_{\|}\right), \qquad t_{\|} = m a_{\|}/N \neq 0, \qquad 0 < m < N \neq 1, \\ & g\left(S_2, t_{\perp}\right), \qquad t_{\perp} = a_{\perp}/2 \neq 0. \end{aligned} \tag{6.67}$$

In the first kind of open operations, $t_{\|}$ is a multiple of the shortest vector of the crystal lattice along the rotational axis $\hat{n}$ of C_N divided by N. This axis is called a **screw axis, which is parallel to $\hat{n}$ and across the point** r_0, where r_0 is the solution of the following equation:

$$(E - C_N) r_0 = t - t_{\|} = t_{\perp}. \tag{6.68}$$

$t_{\|}$ is called **the gliding vector of the screw axis**. In the second kind of open operations, $t_{\perp}$ is on the inversion plane of S_2 and equal to half of the shortest vector of the crystal lattice along the direction of $t_{\perp}$. This inversion plane is called a **gliding plane, which is perpendicular to the improper rotational axis of S_2 and across the point r_0**, where r_0 satisfies

$$(E - S_2) r_0 = t - t_{\perp} = t_{\|}, \qquad r_0 = t_{\|}/2. \tag{6.69}$$

$t_{\perp}$ is called the gliding vector of the gliding plane. The gliding vectors of both the screw axis and the gliding plane have to satisfy the restriction (6.62).

A straight line is called a **symmetry straight line** of a symmetric operation $g(R,t)$ if the straight line is left invariant under $g(R,t)$. For a closed symmetric operation $g(C_N, t_{\perp})$, $N \neq 1$, a straight line, which is parallel to the rotational axis of C_N and across the symmetry center of $g(C_N, t_{\perp})$, is a symmetry straight line, and every point on this line is a symmetry center. A screw axis of an open operator $g(C_N, t_{\|})$ is its symmetry straight line although there is no symmetry center on it.

A plane is called a **symmetry plane** of a symmetric operation $g(R, \boldsymbol{t})$ if the plane is left invariant under $g(R, \boldsymbol{t})$. For a closed symmetric operation $g(S_2, \boldsymbol{t}_{\parallel})$, a plane, which is perpendicular to the improper rotational axis of S_2 and across the symmetry center of $g(S_2, \boldsymbol{t}_{\parallel})$, is a symmetry plane, and every point on this plane is a symmetry center. A gliding plane of an open operator $g(S_2, \boldsymbol{t}_{\perp})$ is its symmetry plane although there is no symmetry center on it.

6.4.2 *Symbols of a Space Group*

A symmetric operation of a crystal is generally written as

$$g(R, \boldsymbol{\alpha}) = T(\boldsymbol{L})g(R, \boldsymbol{t}), \qquad \boldsymbol{L} = \boldsymbol{\ell} \text{ or } \boldsymbol{\ell} + \boldsymbol{f}, \tag{6.70}$$

where $\boldsymbol{\ell}$ is an integral combination of $\boldsymbol{a}_j$. The possible vector of crystal lattice $\boldsymbol{f}$, which is a fractional combination of $\boldsymbol{a}_j$, are determined from its crystal system and its Bravais lattice. Conversely, from the symmetric operations $g(R, \boldsymbol{\alpha})$ of the crystal and the vectors of the crystal lattice we are able to determine its crystallographic point group, its crystal system, and its Bravais lattice.

When the crystallographic point group is C_1, there is only the symmorphic space group, which is the translation group: $\mathcal{S} = \mathcal{T}$. The remaining 31 crystallographic point groups G, from Table 6.2, can be expressed as a product of one, two, or three cyclic subgroups, respectively. For a given space group $\mathcal{S}$, each generator R of the cyclic subgroups has its corresponding translation vector $\boldsymbol{t}$, and the general form of the element in $\mathcal{S}$ is

$$T(\boldsymbol{L}) \left\{g(R, \boldsymbol{t})\right\}^n, \tag{6.71a}$$

$$T(\boldsymbol{L}) \left\{g(R, \boldsymbol{t})\right\}^n \left\{g(R_1, \boldsymbol{p})\right\}^{n_1}, \tag{6.71b}$$

$$T(\boldsymbol{L}) \left\{g(R, \boldsymbol{t})\right\}^n \left\{g(R_2, \boldsymbol{q})\right\}^{n_2} \left\{g(R_1, \boldsymbol{p})\right\}^{n_1}, \tag{6.71c}$$

where n, n_1, and n_2 are integers, and $\boldsymbol{t}$, $\boldsymbol{p}$, and $\boldsymbol{q}$ are the fractional combinations of $\boldsymbol{a}_j$ with the coefficients:

$$0 \leqslant t_j < 1, \qquad 0 \leqslant p_j < 1, \qquad 0 \leqslant q_j < 1. \tag{6.72}$$

In the international notations for the space groups (INSG), the symbol for a space group $\mathcal{S}$ is obtained from the symbol of its corresponding symmorphic space group by attaching the components of $\boldsymbol{t}$, $\boldsymbol{p}$, and $\boldsymbol{q}$ as subscripts to the numbers of the cyclic subgroups. If one of $\boldsymbol{t}$, $\boldsymbol{p}$, and $\boldsymbol{q}$ vanishes, the corresponding subscript zero can be omitted.

The translation vectors $\boldsymbol{t}$, $\boldsymbol{p}$, and $\boldsymbol{q}$ have to satisfy **three constraints.** The first constraint comes from Eq. (6.61) which was discussed in the preceding subsection. The second constraint comes from the group property. The product of two elements of $\mathcal{S}$ given in Eq. (6.71) has to be an element of $\mathcal{S}$ and can also be written in the form (6.71), where the translation vectors $\boldsymbol{t}$, $\boldsymbol{p}$, and $\boldsymbol{q}$ do not change in the product. The third constraint comes from the choice of the origin of the coordinate frame. **The choice of the origin of the coordinate frame affects the translation vectors $\boldsymbol{t}$, $\boldsymbol{p}$, and $\boldsymbol{q}$, as well as the symbol of the space group.** The same space group may be denoted by different symbols due to the different choices of the origin. They are called the **equivalent symbols.** For a given crystal, **the origin has to be chosen uniformly.**

The task to find out all **inequivalent symbols for the space groups** has been done completely and checked repeatedly. There are 230 space groups listed in Table 6.6, where the Schönflies notations (Sch.) and the international notations (INSG) for the space groups are given for comparison (see [Zachariasen (1951)]). The subscripts in the symbol for INSG are moved to a bracket for convenience. For example, the space group $D_{2h}^{24} = F \pm 2_{\frac{1}{4}\frac{1}{4}0} 2'_{0\frac{1}{4}\frac{1}{4}}$ with No. 70 is written as $F \pm 2(\frac{1}{4}\frac{1}{4}0)2'(0\frac{1}{4}\frac{1}{4})$.

Table 6.6 Space Group

1. Triclinic crystal system

Ordinal	Sch.	INSG	Ordinal	Sch.	INSG
*1	C_1^1	$P1$	*2	C_i^1	$P\overline{1}$

2. Monoclinic crystal system

Ordinal	Sch.	INSG	Ordinal	Sch.	INSG
*3	C_2^1	$P2$	4	C_2^2	$P2(0\frac{1}{2}0)$
*5	C_2^3	$A2$	*6	C_s^1	$P\overline{2}$
7	C_s^2	$P\overline{2}(\frac{1}{2}00)$	*8	C_s^3	$A\overline{2}$
9	C_s^4	$A\overline{2}(\frac{1}{2}00)$	*10	C_{2h}^1	$P \pm 2$
11	C_{2h}^2	$P \pm 2(0\frac{1}{2}0)$	*12	C_{2h}^3	$A \pm 2$
13	C_{2h}^4	$P \pm 2(\frac{1}{2}00)$	14	C_{2h}^5	$P \pm 2(\frac{1}{2}\frac{1}{2}0)$
15	C_{2h}^6	$A \pm 2(\frac{1}{2}00)$			

3. Orthorhombic crystal system

Ordinal	Sch.	INSG	Ordinal	Sch.	INSG
∗16	D_2^1	$P22'$	17	D_2^2	$P22'(0\frac{1}{2}0)$
18	D_2^3	$P22'(\frac{1}{2}\frac{1}{2}0)$	19	D_2^4	$P2(00\frac{1}{2})2'(\frac{1}{2}\frac{1}{2}0)$
20	D_2^5	$A22'(\frac{1}{2}00)$	∗21	D_2^6	$A22'$
∗22	D_2^7	$F22'$	∗23	D_2^8	$I22'$
24	D_2^9	$I2(00\frac{1}{2})2'(\frac{1}{2}\frac{1}{2}0)$	∗25	C_{2v}^1	$P2\overline{2}'$
26	C_{2v}^2	$P2(00\frac{1}{2})\overline{2}'$	27	C_{2v}^3	$P2\overline{2}'(00\frac{1}{2})$
28	C_{2v}^4	$P2\overline{2}'(\frac{1}{2}00)$	29	C_{2v}^5	$P2(00\frac{1}{2})\overline{2}'(\frac{1}{2}0\frac{1}{2})$
30	C_{2v}^6	$P2\overline{2}'(0\frac{1}{2}\frac{1}{2})$	31	C_{2v}^7	$P2(00\frac{1}{2})\overline{2}'(\frac{1}{2}00)$
32	C_{2v}^8	$P2\overline{2}'(\frac{1}{2}\frac{1}{2}0)$	33	C_{2v}^9	$P2(00\frac{1}{2})\overline{2}'(\frac{1}{2}\frac{1}{2}\frac{1}{2})$
34	C_{2v}^{10}	$P2\overline{2}'(\frac{1}{2}\frac{1}{2}\frac{1}{2})$	∗35	C_{2v}^{11}	$C2\overline{2}'$
36	C_{2v}^{12}	$C2(00\frac{1}{2})\overline{2}'$	37	C_{2v}^{13}	$C2\overline{2}'(00\frac{1}{2})$
∗38	C_{2v}^{14}	$A2\overline{2}'$	39	C_{2v}^{15}	$A2\overline{2}'(0\frac{1}{2}0)$
40	C_{2v}^{16}	$A2\overline{2}'(\frac{1}{2}00)$	41	C_{2v}^{17}	$A2\overline{2}'(\frac{1}{2}\frac{1}{2}0)$
∗42	C_{2v}^{18}	$F2\overline{2}'$	43	C_{2v}^{19}	$F2\overline{2}'(\frac{1}{4}\frac{1}{4}\frac{1}{4})$
∗44	C_{2v}^{20}	$I2\overline{2}'$	45	C_{2v}^{21}	$I2\overline{2}'(\frac{1}{2}\frac{1}{2}0)$
46	C_{2v}^{22}	$I2\overline{2}'(\frac{1}{2}00)$	∗47	D_{2h}^1	$P\pm22'$
48	D_{2h}^2	$P\pm2(\frac{1}{2}\frac{1}{2}0)2'(0\frac{1}{2}\frac{1}{2})$	49	D_{2h}^3	$P\pm22'(00\frac{1}{2})$
50	D_{2h}^4	$P\pm2(\frac{1}{2}\frac{1}{2}0)2'(0\frac{1}{2}0)$	51	D_{2h}^5	$P\pm22'(\frac{1}{2}00)$
52	D_{2h}^6	$P\pm2(\frac{1}{2}\frac{1}{2}0)2'(\frac{1}{2}\frac{1}{2}\frac{1}{2})$	53	D_{2h}^7	$P\pm22'(\frac{1}{2}0\frac{1}{2})$
54	D_{2h}^8	$P\pm2(\frac{1}{2}0\frac{1}{2})2'(00\frac{1}{2})$	55	D_{2h}^9	$P\pm22'(\frac{1}{2}\frac{1}{2}0)$
56	D_{2h}^{10}	$P\pm2(\frac{1}{2}\frac{1}{2}0)2'(\frac{1}{2}0\frac{1}{2})$	57	D_{2h}^{11}	$P\pm2(00\frac{1}{2})2'(0\frac{1}{2}0)$
58	D_{2h}^{12}	$P\pm22'(\frac{1}{2}\frac{1}{2}\frac{1}{2})$	59	D_{2h}^{13}	$P\pm2(00\frac{1}{2})2'(0\frac{1}{2}\frac{1}{2})$
60	D_{2h}^{14}	$P\pm2(\frac{1}{2}\frac{1}{2}0)2'(\frac{1}{2}\frac{1}{2}0)$	61	D_{2h}^{15}	$P\pm2(\frac{1}{2}0\frac{1}{2})2'(\frac{1}{2}\frac{1}{2}0)$
62	D_{2h}^{16}	$P\pm2(00\frac{1}{2})2'(\frac{1}{2}\frac{1}{2}\frac{1}{2})$	63	D_{2h}^{17}	$A\pm22'(\frac{1}{2}00)$
64	D_{2h}^{18}	$A\pm22'(\frac{1}{2}\frac{1}{2}0)$	∗65	D_{2h}^{19}	$A\pm22'$
66	D_{2h}^{20}	$A\pm2(\frac{1}{2}00)2'$	67	D_{2h}^{21}	$A\pm22'(0\frac{1}{2}0)$
68	D_{2h}^{22}	$A\pm2(\frac{1}{2}00)2'(0\frac{1}{2}0)$	∗69	D_{2h}^{23}	$F\pm22'$
70	D_{2h}^{24}	$F\pm2(\frac{1}{4}\frac{1}{4}0)2'(0\frac{1}{4}\frac{1}{4})$	∗71	D_{2h}^{25}	$I\pm22'$
72	D_{2h}^{26}	$I\pm22'(\frac{1}{2}\frac{1}{2}0)$	73	D_{2h}^{27}	$I\pm2(\frac{1}{2}0\frac{1}{2})2'(\frac{1}{2}\frac{1}{2}0)$
74	D_{2h}^{28}	$I\pm22'(\frac{1}{2}00)$			

4. Tetragonal crystal system

Ordinal	Sch.	INSG	Ordinal	Sch.	INSG
*75	C_4^1	$P4$	76	C_4^2	$P4(00\frac{1}{4})$
77	C_4^3	$P4(00\frac{1}{2})$	78	C_4^4	$P4(00\frac{3}{4})$
*79	C_4^5	$I4$	80	C_4^6	$I4(00\frac{1}{4})$
*81	S_4^1	$P\overline{4}$	*82	S_4^2	$I\overline{4}$
*83	C_{4h}^1	$P\pm 4$	84	C_{4h}^2	$P\pm 4(00\frac{1}{2})$
85	C_{4h}^3	$P\pm 4(\frac{1}{2}00)$	86	C_{4h}^4	$P\pm 4(0\frac{1}{2}\,\frac{1}{2})$
*87	C_{4h}^5	$I\pm 4$	88	C_{4h}^6	$I\pm 4(\frac{1}{4}\,\frac{1}{4}\,\frac{1}{4})$
*89	D_4^1	$P42'$	90	D_4^2	$P42'(\frac{1}{2}\,\frac{1}{2}0)$
91	D_4^3	$P4(00\frac{1}{4})2'$	92	D_4^4	$P4(00\frac{1}{4})2'(\frac{1}{2}\,\frac{1}{2}0)$
93	D_4^5	$P4(00\frac{1}{2})2'$	94	D_4^6	$P4(00\frac{1}{2})2'(\frac{1}{2}\,\frac{1}{2}0)$
95	D_4^7	$P4(00\frac{3}{4})2'$	96	D_4^8	$P4(00\frac{3}{4})2'(\frac{1}{2}\,\frac{1}{2}0)$
*97	D_4^9	$I42'$	98	D_4^{10}	$I4(00\frac{1}{4})2'$
*99	C_{4v}^1	$P4\overline{2}'$	100	C_{4v}^2	$P4\overline{2}'(\frac{1}{2}\,\frac{1}{2}0)$
101	C_{4v}^3	$P4(00\frac{1}{2})\overline{2}'(00\frac{1}{2})$	102	C_{4v}^4	$P4(00\frac{1}{2})\overline{2}'(\frac{1}{2}\,\frac{1}{2}\,\frac{1}{2})$
103	C_{4v}^5	$P4\overline{2}'(00\frac{1}{2})$	104	C_{4v}^6	$P4\overline{2}'(\frac{1}{2}\,\frac{1}{2}\,\frac{1}{2})$
105	C_{4v}^7	$P4(00\frac{1}{2})\overline{2}'$	106	C_{4v}^8	$P4(00\frac{1}{2})\overline{2}'(\frac{1}{2}\,\frac{1}{2}0)$
*107	C_{4v}^9	$I4\overline{2}'$	108	C_{4v}^{10}	$I4\overline{2}'(00\frac{1}{2})$
109	C_{4v}^{11}	$I4(00\frac{1}{4})\overline{2}'(\frac{1}{2}00)$	110	C_{4v}^{12}	$I4(00\frac{1}{4})\overline{2}'(\frac{1}{2}0\frac{1}{2})$
*111	D_{2d}^1	$P\overline{4}2'$	112	D_{2d}^2	$P\overline{4}2'(00\frac{1}{2})$
113	D_{2d}^3	$P\overline{4}2'(\frac{1}{2}\,\frac{1}{2}0)$	114	D_{2d}^4	$P\overline{4}2'(\frac{1}{2}\,\frac{1}{2}\,\frac{1}{2})$
*115	D_{2d}^5	$P\overline{4}2''$	116	D_{2d}^6	$P\overline{4}2''(00\frac{1}{2})$
117	D_{2d}^7	$P\overline{4}2''(\frac{1}{2}\,\frac{1}{2}0)$	118	D_{2d}^8	$P\overline{4}2''(\frac{1}{2}\,\frac{1}{2}\,\frac{1}{2})$
*119	D_{2d}^9	$I\overline{4}2''$	120	D_{2d}^{10}	$I\overline{4}2''(00\frac{1}{2})$
*121	D_{2d}^{11}	$I\overline{4}2'$	122	D_{2d}^{12}	$I\overline{4}2'(0\frac{1}{2}\,\frac{1}{4})$
*123	D_{4h}^1	$P\pm 42'$	124	D_{4h}^2	$P\pm 42'(00\frac{1}{2})$
125	D_{4h}^3	$P\pm 4(\frac{1}{2}00)2'(0\frac{1}{2}0)$	126	D_{4h}^4	$P\pm 4(\frac{1}{2}00)2'(0\frac{1}{2}\,\frac{1}{2})$
127	D_{4h}^5	$P\pm 42'(\frac{1}{2}\,\frac{1}{2}0)$	128	D_{4h}^6	$P\pm 42'(\frac{1}{2}\,\frac{1}{2}\,\frac{1}{2})$
129	D_{4h}^7	$P\pm 4(\frac{1}{2}00)2'(\frac{1}{2}00)$	130	D_{4h}^8	$P\pm 4(\frac{1}{2}00)2'(\frac{1}{2}0\frac{1}{2})$
131	D_{4h}^9	$P\pm 4(00\frac{1}{2})2'$	132	D_{4h}^{10}	$P\pm 4(00\frac{1}{2})2'(00\frac{1}{2})$
133	D_{4h}^{11}	$P\pm 4(0\frac{1}{2}\,\frac{1}{2})2'(0\frac{1}{2}0)$	134	D_{4h}^{12}	$P\pm 4(0\frac{1}{2}\,\frac{1}{2})2'(0\frac{1}{2}\,\frac{1}{2})$
135	D_{4h}^{13}	$P\pm 4(00\frac{1}{2})2'(\frac{1}{2}\,\frac{1}{2}0)$	136	D_{4h}^{14}	$P\pm 4(00\frac{1}{2})2'(\frac{1}{2}\,\frac{1}{2}\,\frac{1}{2})$
137	D_{4h}^{15}	$P\pm 4(0\frac{1}{2}\,\frac{1}{2})2'(\frac{1}{2}00)$	138	D_{4h}^{16}	$P\pm 4(0\frac{1}{2}\,\frac{1}{2})2'(\frac{1}{2}0\frac{1}{2})$
*139	D_{4h}^{17}	$I\pm 42'$	140	D_{4h}^{18}	$I\pm 42'(00\frac{1}{2})$
141	D_{4h}^{19}	$I\pm 4(\frac{1}{4}\,\frac{1}{4}\,\frac{1}{4})2'(\frac{1}{2}00)$	142	D_{4h}^{20}	$I\pm 4(\frac{1}{4}\,\frac{1}{4}\,\frac{1}{4})2'(\frac{1}{2}0\frac{1}{2})$

5. Trigonal crystal system

Ordinal	Sch.	INSG	Ordinal	Sch.	INSG
*143	C_3^1	$P3$	144	C_3^2	$P3(00\frac{1}{3})$
145	C_3^3	$P3(00\frac{2}{3})$	*146	C_3^4	$R3$
*147	C_{3i}^1	$P\overline{3}$	*148	C_{3i}^2	$R\overline{3}$
*149	D_3^1	$P32''$	*150	D_3^2	$P32'$
151	D_3^3	$P3(00\frac{1}{3})2''$	152	D_3^4	$P3(00\frac{1}{3})2'$
153	D_3^5	$P3(00\frac{2}{3})2''$	154	D_3^6	$P3(00\frac{2}{3})2'$
*155	D_3^7	$R32'$	*156	C_{3v}^1	$P3\overline{2}'$
*157	C_{3v}^2	$P3\overline{2}''$	158	C_{3v}^3	$P3\overline{2}'(00\frac{1}{2})$

5. Trigonal crystal system (continued)

Ordinal	Sch.	INSG	Ordinal	Sch.	INSG
159	C_{3v}^4	$P3\overline{2}''(00\frac{1}{2})$	∗160	C_{3v}^5	$R3\overline{2}'$
161	C_{3v}^6	$R3\overline{2}'(\frac{1}{2}\,\frac{1}{2}\,\frac{1}{2})$	∗162	D_{3d}^1	$P\overline{3}2''$
163	D_{3d}^2	$P\overline{3}2''(00\frac{1}{2})$	∗164	D_{3d}^3	$P\overline{3}2'$
165	D_{3d}^4	$P\overline{3}2'(00\frac{1}{2})$	∗166	D_{3d}^5	$R\overline{3}2'$
167	D_{3d}^6	$R\overline{3}2'(\frac{1}{2}\,\frac{1}{2}\,\frac{1}{2})$			

6. Hexagonal crystal system

Ordinal	Sch.	INSG	Ordinal	Sch.	INSG
∗168	C_6^1	$P6$	169	C_6^2	$P6(00\frac{1}{6})$
170	C_6^3	$P6(00\frac{5}{6})$	171	C_6^4	$P6(00\frac{1}{3})$
172	C_6^5	$P6(00\frac{2}{3})$	173	C_6^6	$P6(00\frac{1}{2})$
∗174	C_{3h}^1	$P\overline{6}$	∗175	C_{6h}^1	$P\pm 6$
176	C_{6h}^2	$P\pm 6(00\frac{1}{2})$	∗177	D_6^1	$P62'$
178	D_6^2	$P6(00\frac{1}{6})2'$	179	D_6^3	$P6(00\frac{5}{6})2'$
180	D_6^4	$P6(00\frac{1}{3})2'$	181	D_6^5	$P6(00\frac{2}{3})2'$
182	D_6^6	$P6(00\frac{1}{2})2'$	∗183	C_{6v}^1	$P6\overline{2}'$
184	C_{6v}^2	$P6\overline{2}'(00\frac{1}{2})$	185	C_{6v}^3	$P6(00\frac{1}{2})\overline{2}'(00\frac{1}{2})$
186	C_{6v}^4	$P6(00\frac{1}{2})\overline{2}'$	∗187	D_{3h}^1	$P\overline{6}2''$
188	D_{3h}^2	$P\overline{6}2''(00\frac{1}{2})$	∗189	D_{3h}^3	$P\overline{6}2'$
190	D_{3h}^4	$P\overline{6}2'(00\frac{1}{2})$	∗191	D_{6h}^1	$P\pm 62'$
192	D_{6h}^2	$P\pm 62'(00\frac{1}{2})$	193	D_{6h}^3	$P\pm 6(00\frac{1}{2})2'(00\frac{1}{2})$
194	D_{6h}^4	$P\pm 6(00\frac{1}{2})2'$			

7. Cubic crystal system

Ordinal	Sch.	INSG	Ordinal	Sch.	INSG
∗195	T^1	$P3'22'$	∗196	T^2	$F3'22'$
∗197	T^3	$I3'22'$	198	T^4	$P3'2(\frac{1}{2}0\frac{1}{2})2'(\frac{1}{2}\,\frac{1}{2}0)$
199	T^5	$I3'2(\frac{1}{2}0\frac{1}{2})2'(\frac{1}{2}\,\frac{1}{2}0)$	∗200	T_h^1	$P\overline{3}'22'$
201	T_h^2	$P\overline{3}'2(\frac{1}{2}\,\frac{1}{2}0)2'(0\frac{1}{2}\,\frac{1}{2})$	∗202	T_h^3	$F\overline{3}'22'$
203	T_h^4	$F\overline{3}'2(\frac{1}{4}\,\frac{1}{4}0)2'(0\frac{1}{4}\,\frac{1}{4})$	∗204	T_h^5	$I\overline{3}'22'$
205	T_h^6	$P\overline{3}'2(\frac{1}{2}0\frac{1}{2})2'(\frac{1}{2}\,\frac{1}{2}0)$	206	T_h^7	$I\overline{3}'2(0\frac{1}{2}0)2'(00\frac{1}{2})$
∗207	O^1	$P3'42''$	208	O^2	$P3'4(\frac{1}{2}\,\frac{1}{2}\,\frac{1}{2})2''(\frac{1}{2}\,\frac{1}{2}\,\frac{1}{2})$
∗209	O^3	$F3'42''$	210	O^4	$F3'4(\frac{1}{4}\,\frac{1}{4}\,\frac{1}{4})2''(\frac{1}{4}\,\frac{1}{4}\,\frac{1}{4})$
∗211	O^5	$I3'42''$	212	O^6	$P3'4(\frac{3}{4}\,\frac{1}{4}\,\frac{3}{4})2''(\frac{1}{4}\,\frac{3}{4}\,\frac{3}{4})$
213	O^7	$P3'4(\frac{1}{4}\,\frac{3}{4}\,\frac{1}{4})2''(\frac{3}{4}\,\frac{1}{4}\,\frac{1}{4})$	214	O^8	$I3'4(\frac{1}{4}\,\frac{3}{4}\,\frac{1}{4})2''(\frac{3}{4}\,\frac{1}{4}\,\frac{1}{4})$
∗215	T_d^1	$P3'\overline{4}\,\overline{2}''$	∗216	T_d^2	$F3'\overline{4}\,\overline{2}''$
∗217	T_d^3	$I3'\overline{4}\,\overline{2}''$	218	T_d^4	$P3'\overline{4}(\frac{1}{2}\,\frac{1}{2}\,\frac{1}{2})\overline{2}''(\frac{1}{2}\,\frac{1}{2}\,\frac{1}{2})$
219	T_d^5	$F3'\overline{4}(00\frac{1}{2})\overline{2}''(00\frac{1}{2})$	220	T_d^6	$I3'\overline{4}(\frac{1}{4}\,\frac{1}{4}\,\frac{1}{4})\overline{2}''(\frac{1}{4}\,\frac{1}{4}\,\frac{3}{4})$
∗221	O_h^1	$P\overline{3}'42''$	222	O_h^2	$P\overline{3}'4(\frac{1}{2}00)2''(00\frac{1}{2})$
223	O_h^3	$P\overline{3}'4(\frac{1}{2}\,\frac{1}{2}\,\frac{1}{2})2''(\frac{1}{2}\,\frac{1}{2}\,\frac{1}{2})$	224	O_h^4	$P\overline{3}'4(0\frac{1}{2}\,\frac{1}{2})2''(\frac{1}{2}\,\frac{1}{2}0)$
∗225	O_h^5	$F\overline{3}'42''$	226	O_h^6	$F\overline{3}'4(00\frac{1}{2})2''(00\frac{1}{2})$
227	O_h^7	$F\overline{3}'4(0\frac{1}{4}\,\frac{1}{4})2''(\frac{1}{4}\,\frac{1}{4}0)$	228	O_h^8	$F\overline{3}'4(\frac{1}{2}\,\frac{1}{4}\,\frac{1}{4})2''(\frac{1}{4}\,\frac{1}{4}\,\frac{1}{2})$
∗229	O_h^9	$I\overline{3}'42''$	230	O_h^{10}	$I\overline{3}'4(\frac{1}{4}\,\frac{3}{4}\,\frac{1}{4})2''(\frac{3}{4}\,\frac{1}{4}\,\frac{1}{4})$

6.4.3 *Method for Determining the Space Groups*

First, the origin of the coordinate frame is chosen to make the translation vector $\boldsymbol{t}$ in the factor $g(R,\boldsymbol{t})$ of Eq. (6.71) as simple as possible. Based on $g(R,\boldsymbol{t})$, the space groups are classified into three types.

Type A: $R = S_N$, $N \neq 2$. The origin of the coordinate frame is chosen at the symmetry center of $g(S_N,\boldsymbol{t})$ to make $\boldsymbol{t} = 0$. Under the condition $\boldsymbol{t} = 0$, the origin can be made further shift $\boldsymbol{r}_0$ satisfying

$$(S_N - E)\,\boldsymbol{r}_0 = \boldsymbol{L}. \tag{6.73}$$

This shift can be used to simplify $\boldsymbol{p}$ and $\boldsymbol{q}$. There are 15 crystallographic point groups belonging to type A: C_i, C_{3i}, S_4, C_{3h}, C_{2h}, C_{4h}, C_{6h}, D_{2d}, D_{3d}, D_{3h}, D_{2h}, D_{4h}, D_{6h}, T_h, and O_h.

Type B: $R = C_N$, $N \neq 1$. The origin of the coordinate frame is chosen at the symmetry straight line of $g(C_N,\boldsymbol{t})$ such that $\boldsymbol{t} = \boldsymbol{t}_{\parallel} = m\boldsymbol{a}_{\parallel}/N$, $0 \leqslant m < N$, where $\boldsymbol{a}_{\parallel}$ is the shortest vector of the crystal lattice along the rotational axis $\hat{\boldsymbol{n}}$ of C_N. The origin can be made further shift $\boldsymbol{r}_0$ satisfying

$$(C_N - E)\,\boldsymbol{r}_0 = \boldsymbol{L} + m'\boldsymbol{a}_{\parallel}/N, \qquad 0 \leqslant m' < N. \tag{6.74}$$

Note that the left-hand side of Eq. (6.74) is perpendicular to the rotational axis $\hat{\boldsymbol{n}}$. In the case where $\boldsymbol{L}$ contains a component along $\hat{\boldsymbol{n}}$, the second term with $m' \neq 0$ in the right-hand side of Eq. (6.74) occurs **to cancel** $\boldsymbol{t}_{\parallel}$ **or partly**. Furthermore, the origin can be made further shift $\boldsymbol{r}_0$ to simplify $\boldsymbol{p}$ and $\boldsymbol{q}$ under the condition that the simplified $\boldsymbol{t}$ does not change again. We will discuss this simplification with example later. There are 15 crystallographic point groups belonging to type B: C_2, C_3, C_4, C_6, D_2, C_{2v}, D_3, C_{3v}, D_4, C_{4v}, D_6, C_{6v}, T, O, and T_d.

Type C: $R = S_2$ and the crystallographic point group is C_s. The origin of the coordinate frame is chosen at the symmetry plane of $g(S_2,\boldsymbol{t})$ to make $\boldsymbol{t} = 0$ or $\boldsymbol{a}_{\perp}/2$. For the Bravais lattice P with $\boldsymbol{t} = \boldsymbol{a}_{\perp}/2$, we choose $\boldsymbol{a}_{\perp}$ to be the lattice basis $\boldsymbol{a}_1$ and obtain a space group $P\overline{2}_{\frac{1}{2}00}$. For the Bravais lattice A with $\boldsymbol{t} = \boldsymbol{a}_{\perp}/2$, the component $\boldsymbol{a}_2/2$ in $\boldsymbol{t}$, if exists, can be removed by a suitable choice of the origin because $\boldsymbol{a}_2/2 = \boldsymbol{f} - \boldsymbol{a}_3/2$. Therefore, in addition to the symmorphic space groups, there are two space groups in type C: $P\overline{2}_{\frac{1}{2}00}$ and $A\overline{2}_{\frac{1}{2}00}$.

For the space groups in type A and type B, the restrictions on the translation vectors $\boldsymbol{p}$ and $\boldsymbol{q}$ need to be studied further. It is found that the condition $R_1R^{-1} = RR_1$ holds for the space groups in the form of Eq. (6.71b) because in those space groups either $R = C_i$ or R_1 is a proper or improper 2-fold rotation with the rotational axis perpendicular to the rotational axis of R (see Table 6.2). Thus, $\boldsymbol{p}$ satisfies

$$g(R_1, \boldsymbol{p} + \boldsymbol{L}) = g(R, \boldsymbol{t})\, g(R_1, \boldsymbol{p})\, g(R, \boldsymbol{t}). \tag{6.75}$$

Under the restrictions (6.62) and (6.75), the origin of the coordinate frame is made further shift $\boldsymbol{r}_0$, satisfying Eq. (6.73) or Eq. (6.74), to simplify $\boldsymbol{p}$ as possible.

For the space groups in the form of Eq. (6.71c), the restriction like Eq. (6.75) becomes more complicated. A direct calculation shows

$$\begin{array}{lll} R_1R_2^{-1} = R_2R_1, & R_1R = R^iR_2^jR_1^k, & R_2R = R^{i'}R_2^{j'}R_1^{k'}, \\ i = k = i' = j' = 1, & j = k' = 0, & \text{for } D_{2h},\ D_{4h}, \text{ and } D_{6h}, \\ i = j = i' = j' = k' = 1, & k = 0, & \text{for } \mathbf{T} \text{ and } T_h, \\ i = i' = j' = -1, & j = 1, \quad k = k' = 0, & \text{for } \mathbf{O},\ \mathbf{T}_d, \text{ and } \mathbf{O}_h. \end{array} \tag{6.76}$$

Thus, $\boldsymbol{p}$ and $\boldsymbol{q}$ satisfy

$$\begin{aligned} g(R_1, \boldsymbol{p} + \boldsymbol{L}_1) &= g(R_2, \boldsymbol{q})\, g(R_1, \boldsymbol{p})\, g(R_2, \boldsymbol{q}), \\ g(R_1, \boldsymbol{p} + \boldsymbol{L}_2) &= g(R, \boldsymbol{t})^i\, g(R_2, \boldsymbol{q})^j\, g(R_1, \boldsymbol{p})^k\, g(R, \boldsymbol{t})^{-1}, \\ g(R_2, \boldsymbol{q} + \boldsymbol{L}_3) &= g(R, \boldsymbol{t})^{i'}\, g(R_2, \boldsymbol{q})^{j'}\, g(R_1, \boldsymbol{p})^{k'}\, g(R, \boldsymbol{t})^{-1}. \end{aligned} \tag{6.77}$$

Under the restrictions (6.62) and (6.77), the origin of the coordinate frame is made further shift $\boldsymbol{r}_0$, satisfying Eq. (6.73) or Eq. (6.74), to simplify $\boldsymbol{p}$ and $\boldsymbol{q}$ as possible.

6.4.4 *Example for the Space Groups in Type A*

Discuss a crystal with the crystallographic point group C_{4h} (± 4), which can be expressed as the product of subgroups C_iC_4. This crystal belongs to the tetragonal crystal system, $a_1 = a_2$ and $\alpha_1 = \alpha_2 = \alpha_3 = \pi/2$. The lattice basis $\boldsymbol{a}_3$ is along a 4-fold axis. The double-vector of C_4 is

$$\vec{\vec{4}} = \boldsymbol{a}_2\boldsymbol{b}_1 - \boldsymbol{a}_1\boldsymbol{b}_2 + \boldsymbol{a}_3\boldsymbol{b}_3.$$

There are two Bravais lattices P and I. $\boldsymbol{f} = 0$ for the Bravais lattices P, and $\boldsymbol{f} = (\boldsymbol{a}_1 + \boldsymbol{a}_2 + \boldsymbol{a}_3)/2$ for the Bravais lattices I.

The origin of the coordinate frame is taken at the symmetry center of $g(S_1, \boldsymbol{t})$, then, $\boldsymbol{t} = 0$. The origin is allowed to make further shift $\boldsymbol{r}_0$

satisfying $(S_1 - E)\,\boldsymbol{r}_0 = -2\boldsymbol{r}_0 = \boldsymbol{L}$.

$$-\boldsymbol{r}_0 = \frac{\boldsymbol{L}}{2} = \begin{cases} \boldsymbol{\ell}/2, & P \text{ lattice}, \\ \boldsymbol{\ell}/2 \text{ or } (\boldsymbol{a}_1 + \boldsymbol{a}_2 + \boldsymbol{a}_3)/4, & I \text{ lattice}. \end{cases}$$

Because

$$\begin{aligned} (C_4 - E)\boldsymbol{\ell}/2 &= -(\ell_2 + \ell_1)\,\boldsymbol{a}_1/2 + (\ell_1 - \ell_2)\,\boldsymbol{a}_2/2, \\ (C_4 - E)\,(\boldsymbol{a}_1 + \boldsymbol{a}_2 + \boldsymbol{a}_3)/4 &= -\boldsymbol{a}_1/2, \end{aligned}$$

$\boldsymbol{p}$ changes in the shift $\boldsymbol{r}_0$ of the origin

$$(C_4 - E)\boldsymbol{r}_0 = \begin{cases} (\boldsymbol{a}_1 - \boldsymbol{a}_2)/2 \text{ or } (\boldsymbol{a}_1 + \boldsymbol{a}_2)/2, & P \text{ lattice}, \\ (\boldsymbol{a}_1 - \boldsymbol{a}_2)/2,\ (\boldsymbol{a}_1 + \boldsymbol{a}_2)/2, \text{ or } \boldsymbol{a}_1/2, & I \text{ lattice}. \end{cases} \tag{6.78}$$

Due to Eq. (6.62), $4p_3$ has to be integer. From Eq. (6.75) we have

$$S_1 g(C_4, \boldsymbol{p}) S_1 = g(C_4, -\boldsymbol{p}) = g(C_4, \boldsymbol{p} + \boldsymbol{L}'),$$

$$\boldsymbol{p} = -\frac{\boldsymbol{L}'}{2} = \begin{cases} \boldsymbol{\ell}/2, & P \text{ lattice}, \\ \boldsymbol{\ell}/2 \text{ or } (\boldsymbol{a}_1 + \boldsymbol{a}_2 + \boldsymbol{a}_3)/4, & I \text{ lattice}. \end{cases}$$

Thus, for the Bravais lattice P, $\boldsymbol{p}$ can be

$$\begin{array}{llll} 0, & \boldsymbol{a}_1/2, & (\boldsymbol{a}_1 + \boldsymbol{a}_2)/2, & (\boldsymbol{a}_1 + \boldsymbol{a}_3)/2, \\ \boldsymbol{a}_2/2, & \boldsymbol{a}_3/2, & (\boldsymbol{a}_2 + \boldsymbol{a}_3)/2, & (\boldsymbol{a}_1 + \boldsymbol{a}_2 + \boldsymbol{a}_3)/2. \end{array}$$

Due to Eq. (6.78), $\boldsymbol{a}_1/2$ and $\boldsymbol{a}_2/2$ are equivalent, $(\boldsymbol{a}_1+\boldsymbol{a}_3)/2$ and $(\boldsymbol{a}_2+\boldsymbol{a}_3)/2$ are equivalent, $(\boldsymbol{a}_1 + \boldsymbol{a}_2 + \boldsymbol{a}_3)/2$ and $\boldsymbol{a}_3/2$ are equivalent, and $(\boldsymbol{a}_1 + \boldsymbol{a}_2)/2$ can be removed, namely, $\boldsymbol{p}$ can be 0, $\boldsymbol{a}_1/2$, $\boldsymbol{a}_3/2$, and $(\boldsymbol{a}_2 + \boldsymbol{a}_3)/2$ for the Bravais lattice P. For the Bravais lattice I, because $(\boldsymbol{a}_1 + \boldsymbol{a}_2 + \boldsymbol{a}_3)/2$ is a vector of crystal lattice, from Eq. (6.78) $\boldsymbol{a}_1/2$, $\boldsymbol{a}_3/2$, and $(\boldsymbol{a}_2 + \boldsymbol{a}_3)/2$ can be removed, but there is another $\boldsymbol{p} = (\boldsymbol{a}_1 + \boldsymbol{a}_2 + \boldsymbol{a}_3)/4$. In summary, the space groups with the crystallographic point group C_{4h} (± 4) are

$$P \pm 4, \quad P \pm 4_{\frac{1}{2}00}, \quad P \pm 4_{00\frac{1}{2}}, \quad P \pm 4_{0\frac{1}{2}\frac{1}{2}}, \quad I \pm 4, \quad I \pm 4_{\frac{1}{4}\frac{1}{4}\frac{1}{4}}. \tag{6.79}$$

6.4.5 *Example for the Space Groups in Type B*

Discuss a crystal with the crystallographic point group D_3, which can be expressed as the product of subgroups C_3C_2' (international symbol $32'$) or C_3C_2'' ($32''$), depending on whether the lattice basis $\boldsymbol{a}_1$ is along a 2-fold axis or not. The crystal belongs to the hexagonal crystal system or the rhombohedral crystal system, where $\boldsymbol{f} = 0$ in both cases.

(a) *The Bravais lattice P in the hexagonal system.*

$\boldsymbol{a}_3$ is along the 3-fold axis, $a_1 = a_2$, $\alpha_1 = \alpha_2 = \pi/2$, and $\alpha_3 = 2\pi/3$. The double-vectors of the generators are

$$\vec{\vec{3}} = \boldsymbol{a}_2\boldsymbol{b}_1 - (\boldsymbol{a}_1 + \boldsymbol{a}_2)\boldsymbol{b}_2 + \boldsymbol{a}_3\boldsymbol{b}_3, \qquad \text{for } 32' \text{ or } 32'',$$

$$\vec{\vec{2}}' = \boldsymbol{a}_1\boldsymbol{b}_1 - (\boldsymbol{a}_1 + \boldsymbol{a}_2)\boldsymbol{b}_2 - \boldsymbol{a}_3\boldsymbol{b}_3, \qquad \text{for } 32',$$

$$\vec{\vec{2}}'' = -\boldsymbol{a}_2\boldsymbol{b}_1 - \boldsymbol{a}_1\boldsymbol{b}_2 - \boldsymbol{a}_3\boldsymbol{b}_3, \qquad \text{for } 32''.$$

The origin of the coordinate frame is taken at the symmetry straight line of $g(C_3, \boldsymbol{t})$ such that $\boldsymbol{t} = 0$, $\boldsymbol{a}_3/3$, or $2\boldsymbol{a}_3/3$. The origin is allowed to make further shift $\boldsymbol{r}_0$ satisfying

$$(C_3 - E)\,\boldsymbol{r}_0 = -\boldsymbol{a}_1(r_{01} + r_{02}) + \boldsymbol{a}_2(r_{01} - 2r_{02}) = \boldsymbol{\ell},$$

$$r_{01} = (-2\ell_1 + \ell_2)/3, \qquad r_{02} = -(\ell_1 + \ell_2)/3, \qquad r_{03} \text{ arbitrary},$$

where ℓ_1 and ℓ_2 are integers. For the crystallographic point group $32'$, $\boldsymbol{p}$ changes in the shift $\boldsymbol{r}_0$ of the origin

$$\begin{aligned} (C_2' - E)\boldsymbol{r}_0 &= -\boldsymbol{a}_1 r_{02} - \boldsymbol{a}_2 2r_{02} - \boldsymbol{a}_3 2r_{03} \\ &= (\boldsymbol{a}_1 + 2\boldsymbol{a}_2)(\ell_1 + \ell_2)/3 - \boldsymbol{a}_3 2r_{03}. \end{aligned} \tag{6.80}$$

For the crystallographic point group $32''$, $\boldsymbol{p}$ changes in the shift $\boldsymbol{r}_0$ of the origin

$$\begin{aligned} (C_2'' - E)\boldsymbol{r}_0 &= -(\boldsymbol{a}_1 + \boldsymbol{a}_2)(r_{01} + r_{02}) - \boldsymbol{a}_3 2r_{03} \\ &= (\boldsymbol{a}_1 + \boldsymbol{a}_2)\ell_1 - \boldsymbol{a}_3 2r_{03}. \end{aligned} \tag{6.81}$$

On the other hand, for the crystallographic point group $32'$, from the constraint (6.61) the following vector is a vector of crystal lattice

$$(E + C_2')\boldsymbol{p} = \boldsymbol{a}_1(2p_1 - p_2).$$

Thus, $2p_1 - p_2$ has to be integer. Due to Eq. (6.75), $\boldsymbol{p}$ has to satisfy

$$\begin{aligned} g(C_2', \boldsymbol{p} + \boldsymbol{\ell}') &= g(C_3, \boldsymbol{t})g(C_2', \boldsymbol{p})g(C_3, \boldsymbol{t}) \\ &= g(C_2', \boldsymbol{t} + C_3\boldsymbol{p} + C_3C_2'\boldsymbol{t}) = g(C_2', C_3\boldsymbol{p}), \end{aligned}$$

where $\boldsymbol{t} + C_3C_2'\boldsymbol{t} = 0$ because $\boldsymbol{t}$ is along $\boldsymbol{a}_3$. Hence,

$$\boldsymbol{\ell}' = (C_3 - E)\boldsymbol{p} = -\boldsymbol{a}_1(p_1 + p_2) + \boldsymbol{a}_2(p_1 - 2p_2).$$

The solution is that p_3 is arbitrary, $p_1 = (-2\ell_1' + \ell_2')/3$ and $p_2 = -(\ell_1' + \ell_2')/3$. Then, $2p_1 - p_2 = \ell_2' - \ell_1'$ is integer. However, $\boldsymbol{p}$ can be removed by the shift $\boldsymbol{r}_0$ of the origin because

$$\boldsymbol{p} = (-2\ell_1' + \ell_2')(\boldsymbol{a}_1 + 2\boldsymbol{a}_2)/3 + (\ell_1' - \ell_2')\boldsymbol{a}_2 + p_3\boldsymbol{a}_3,$$

where the first and the third terms are removed by Eq. (6.80) and the second term is a vector of crystal lattice.

For the crystallographic point group 32″, the shortest vector of crystal lattice along the rotational axis of C_2'' is $(\boldsymbol{a}_1 - \boldsymbol{a}_2)$. From the constraint (6.61) the following vector is a vector of crystal lattice

$$(E + C_2'')\boldsymbol{p} = (\boldsymbol{a}_1 - \boldsymbol{a}_2)(p_1 - p_2).$$

Thus, $p_1 = p_2$. Due to Eq. (6.75), $\boldsymbol{p}$ has to satisfy

$$\begin{aligned} g(C_2'', \boldsymbol{p} + \boldsymbol{\ell}') &= g(C_3, \boldsymbol{t}) g(C_2'', \boldsymbol{p}) g(C_3, \boldsymbol{t}) \\ &= g(C_2'', \boldsymbol{t} + C_3\boldsymbol{p} + C_3C_2''\boldsymbol{t}) = g(C_2'', C_3\boldsymbol{p}), \end{aligned}$$

where $\boldsymbol{t} + C_3C_2''\boldsymbol{t} = 0$, because $\boldsymbol{t}$ is along $\boldsymbol{a}_3$. Hence,

$$\boldsymbol{\ell}' = (C_3 - E)\boldsymbol{p} = -\boldsymbol{a}_1(p_1 + p_2) + \boldsymbol{a}_2(p_1 - 2p_2) = -2p_1\boldsymbol{a}_1 - p_1\boldsymbol{a}_2,$$

namely, $p_1 = p_2 = 0$. p_3 can be removed by Eq. (6.81).

In summary, the space groups with the crystallographic point group D_3 in the hexagonal crystal system are

$$P32', \quad P3_{00\frac{1}{3}}2', \quad P3_{00\frac{2}{3}}2', \quad P32'', \quad P3_{00\frac{1}{3}}2'', \quad P3_{00\frac{2}{3}}2''. \tag{6.82}$$

(b) *The Bravais lattice R in the rhombohedral crystal system.*

Three lattice bases $\boldsymbol{a}_j$ are uniformly distributed around the 3-fold axis, $a_1 = a_2 = a_3$ and $\alpha_1 = \alpha_2 = \alpha_3$. The double-vectors of the generators are

$$\vec{\vec{C}}_3 = \boldsymbol{a}_2\boldsymbol{b}_1 + \boldsymbol{a}_3\boldsymbol{b}_2 + \boldsymbol{a}_1\boldsymbol{b}_3, \qquad \vec{\vec{C}}_2' = -\boldsymbol{a}_2\boldsymbol{b}_1 - \boldsymbol{a}_1\boldsymbol{b}_2 - \boldsymbol{a}_3\boldsymbol{b}_3.$$

The origin of the coordinate frame is taken at the symmetry straight line of $g(C_3, \boldsymbol{t})$, then,

$$\boldsymbol{t} = \boldsymbol{t}_{\|} = m\boldsymbol{a}_{\|}/3, \qquad \boldsymbol{a}_{\|} = \boldsymbol{a}_1 + \boldsymbol{a}_2 + \boldsymbol{a}_3, \qquad m = 0,\ 1,\ 2.$$

The origin is made a further shift $\boldsymbol{r}_0$ according to Eq. (6.74)

$$\begin{aligned} \boldsymbol{\ell} + (\boldsymbol{a}_1 + \boldsymbol{a}_2 + \boldsymbol{a}_3)m'/3 &= (C_3 - E)\,\boldsymbol{r}_0 \\ &= \boldsymbol{a}_1(-r_{01} + r_{03}) + \boldsymbol{a}_2(r_{01} - r_{02}) + \boldsymbol{a}_3(r_{02} - r_{03}). \end{aligned} \tag{6.83}$$

The sum of the three coefficients with respect to the lattice bases $\boldsymbol{a}_j$ on the right-hand side of Eq. (6.83) is 0, so the sum on the left-hand side: $m' = -\sum_j \ell_j$. Thus, $\boldsymbol{t}$ in $g(C_3, \boldsymbol{t})$ can be removed by a suitable choice of the origin. **The reason is that the component of $\boldsymbol{a}_j$ along the 3-fold axis is equal to $\boldsymbol{a}_{\|}/3$.**

Under the condition $\boldsymbol{t} = 0$, the origin is allowed to make further shift,

$$(C_3 - E)\,\boldsymbol{r}_0 = \boldsymbol{a}_1(-r_{01} + r_{03}) + \boldsymbol{a}_2(r_{01} - r_{02}) + \boldsymbol{a}_3(r_{02} - r_{03}) = \boldsymbol{\ell}.$$

The solution is that r_{03} is arbitrary, and both $r_{01} - r_{03}$ and $r_{02} - r_{03}$ are integers. $\boldsymbol{p}$ changes in the shift of the origin

$$\begin{aligned}(C_2' - E)\boldsymbol{r}_0 &= -(\boldsymbol{a}_1 + \boldsymbol{a}_2)(r_{01} + r_{02}) - 2\boldsymbol{a}_3 r_{03} \\ &= -(\boldsymbol{a}_1 + \boldsymbol{a}_2 + \boldsymbol{a}_3)2r_{03} - (\boldsymbol{a}_1 + \boldsymbol{a}_2)(r_{01} + r_{02} - 2r_{03}),\end{aligned} \tag{6.84}$$

namely, $\boldsymbol{p}$ changes an arbitrary multiple of $(\boldsymbol{a}_1 + \boldsymbol{a}_2 + \boldsymbol{a}_3)$.

On the other hand, from Eq. (6.61), $\boldsymbol{p}$ satisfies

$$(C_2' + E)\boldsymbol{p} = (\boldsymbol{a}_1 - \boldsymbol{a}_2)(p_1 - p_2) = \boldsymbol{\ell}.$$

Thus, $p_1 = p_2$. From Eq. (6.75) one has

$$g(C_2', \boldsymbol{p} + \boldsymbol{\ell}') = C_3 g(C_2', \boldsymbol{p}) C_3 = g(C_2', C_3\boldsymbol{p}).$$

Hence,

$$\boldsymbol{\ell}' = (C_3 - E)\boldsymbol{p} = \boldsymbol{a}_1(-p_1 + p_3) + \boldsymbol{a}_2(p_1 - p_2) + \boldsymbol{a}_3(p_2 - p_3).$$

The solution is $p_1 = p_2 = p_3$ which is just removed by a suitable choice of the origin [see Eq. (6.84)]. Therefore, the space groups with the crystallographic point group D_3 in the rhombohedral crystal system is only the symmorphic space group $R32$.

6.4.6 *Analysis of the Symmetry of a Crystal*

We have introduced the method for finding the space groups of crystals by group theory. However, the important problem for most readers is how **to analyze the symmetry of a crystal from its international symbol of the space group**. We will demonstrate this problem through an example.

Let us study a crystal with the space group

$$I \pm 4_{\frac{1}{4}\frac{1}{4}\frac{1}{4}} 2'_{\frac{1}{2}00}.$$

From the international symbol we know that the crystal belongs to the Bravais lattice I in the tetragonal crystal system with the crystallographic point group D_{4h}. $\boldsymbol{a}_3$ is along the 4-fold axis, $\boldsymbol{a}_1$ is along a 2-fold axis, and $\boldsymbol{a}_2 = C_4\boldsymbol{a}_1$. $\alpha_1 = \alpha_2 = \alpha_3 = \pi/2$ and $a_1 = a_2$. The shortest vector of crystal lattice along an inequivalent 2-fold axis is $\boldsymbol{a}_1 \pm \boldsymbol{a}_2$ with the length $\sqrt{2}a_1$. The lattice bases are not primitive. There is a vector of crystal lattice $\boldsymbol{f}$ which is a fractional combination of the lattice bases $\boldsymbol{a}_j$

$$\boldsymbol{L} = \boldsymbol{\ell} \text{ or } \boldsymbol{\ell} + \boldsymbol{f}, \qquad \boldsymbol{f} = (\boldsymbol{a}_1 + \boldsymbol{a}_2 + \boldsymbol{a}_3)/2.$$

The double-vectors of the generators in the space group are

$$\vec{\vec{C}}_4 = \boldsymbol{a}_2\boldsymbol{b}_1 - \boldsymbol{a}_1\boldsymbol{b}_2 + \boldsymbol{a}_3\boldsymbol{b}_3,$$
$$\vec{\vec{C}}_2' = \boldsymbol{a}_1\boldsymbol{b}_1 - \boldsymbol{a}_2\boldsymbol{b}_2 - \boldsymbol{a}_3\boldsymbol{b}_3.$$

The general form of an element in the space group is

$$T(\boldsymbol{\ell})T(\boldsymbol{f})^{n_1}S_1^{n_2}g\left[C_4, (\boldsymbol{a}_1+\boldsymbol{a}_2+\boldsymbol{a}_3)/4\right]^m g\left(C_2', \boldsymbol{a}_1/2\right)^{n_3},$$

where S_1 is the space inversion with respect to the origin of the coordinate system. n_1, n_2, and n_3 equal to 0 or 1, respectively. m is taken to be 0, 1, 2, or 3.

The rotational axis of $g(C_4, \boldsymbol{q})$ is parallel to $\boldsymbol{a}_3$, and $\boldsymbol{q}_\parallel = \boldsymbol{a}_3/4$, $\boldsymbol{q}_\perp = (\boldsymbol{a}_1+\boldsymbol{a}_2)/4$. Its screw axis is parallel to $\boldsymbol{a}_3$ and across the point $\boldsymbol{r}_0$

$$(C_4 - E)\,\boldsymbol{r}_0 + \boldsymbol{q}_\perp = 0, \qquad r_{01} + r_{02} = -r_{01} + r_{02} = 1/4.$$

The solution is $r_{01} = 0$, $r_{02} = 1/4$, namely, the screw axis of $g(C_4, \boldsymbol{q})$ is parallel to $\boldsymbol{a}_3$ and across the point $\boldsymbol{a}_2/4$ with the gliding vector $\boldsymbol{a}_3/4$.

The rotational axis of $g(C_2', \boldsymbol{p})$ is parallel to $\boldsymbol{a}_1$, and $\boldsymbol{p} = \boldsymbol{p}_\parallel = \boldsymbol{a}_1/2$. The screw axis of $g(C_2', \boldsymbol{p})$ is parallel to $\boldsymbol{a}_1$ and across the origin with the gliding vector $\boldsymbol{a}_1/2$. Since $S_1 g\left(C_2', \boldsymbol{a}_1/2\right) = T(-\boldsymbol{a}_1)g\left(S_2', \boldsymbol{a}_1/2\right)$, where S_2' is an improper 2-fold rotation about $\boldsymbol{a}_1$, $\boldsymbol{p} = \boldsymbol{p}_\parallel = \boldsymbol{a}_1/2$, and the symmetry plane of $g\left(S_2', \boldsymbol{a}_1/2\right)$ is perpendicular to $\boldsymbol{a}_1$ and across the point $\boldsymbol{a}_1/4$.

From an arbitrary point $\boldsymbol{r} = \boldsymbol{a}_1x_1 + \boldsymbol{a}_2x_2 + \boldsymbol{a}_3x_3$, one can obtain eight equivalent points in a crystal cell through the symmetric operations $g\left[C_4, (\boldsymbol{a}_1+\boldsymbol{a}_2+\boldsymbol{a}_3)/4\right]$ and $g\left(C_2', \boldsymbol{a}_1/2\right)$:

$$\begin{aligned}
\boldsymbol{r}_1 &= \boldsymbol{a}_1x_1 + \boldsymbol{a}_2x_2 + \boldsymbol{a}_3x_3,\\
\boldsymbol{r}_2 &= \boldsymbol{a}_1\left(-x_2+1/4\right) + \boldsymbol{a}_2\left(x_1+1/4\right) + \boldsymbol{a}_3\left(x_3+1/4\right),\\
\boldsymbol{r}_3 &= -\boldsymbol{a}_1x_1 + \boldsymbol{a}_2\left(-x_2+1/2\right) + \boldsymbol{a}_3\left(x_3+1/2\right),\\
\boldsymbol{r}_4 &= \boldsymbol{a}_1\left(x_2-1/4\right) + \boldsymbol{a}_2\left(-x_1+1/4\right) + \boldsymbol{a}_3\left(x_3+3/4\right),\\
\boldsymbol{r}_5 &= \boldsymbol{a}_1\left(x_1+1/2\right) - \boldsymbol{a}_2x_2 - \boldsymbol{a}_3x_3,\\
\boldsymbol{r}_6 &= \boldsymbol{a}_1\left(x_2+1/4\right) + \boldsymbol{a}_2\left(x_1+3/4\right) + \boldsymbol{a}_3\left(-x_3+1/4\right),\\
\boldsymbol{r}_7 &= \boldsymbol{a}_1\left(-x_1+1/2\right) + \boldsymbol{a}_2\left(x_2+1/2\right) + \boldsymbol{a}_3\left(-x_3+1/2\right),\\
\boldsymbol{r}_8 &= \boldsymbol{a}_1\left(-x_2-1/4\right) + \boldsymbol{a}_2\left(-x_1+3/4\right) + \boldsymbol{a}_3\left(-x_3+3/4\right).
\end{aligned}$$

Then, the remaining 24 equivalent points can be obtained through the symmetric operations $T(\boldsymbol{f})$ and S_1.

6.5 Linear Representations of Space Groups

The irreducible representations of $\mathcal{S}$, single-valued and double valued, will be studied in this section. The subduced representation of an irreducible representation of $\mathcal{S}$ with respect to the translation subgroup $\mathcal{T}$ is taken to be reduced such that it is diagonal, because $\mathcal{T}$ is abelian. This subduced representation is labeled by a **star of wave vectors**. The irreducible representation of $\mathcal{S}$ is associated with that of **the little group**. The single-valued and the double-valued irreducible representations of the little group will be studied in detail.

6.5.1 *Irreducible Representations of* $\mathcal{T}$

Let $\boldsymbol{a}_j$ denote the primitive basis vectors of crystal lattice and $\boldsymbol{\ell} = \sum_j \boldsymbol{a}_j \ell_j$ with integer coefficients ℓ_j denote a vector of crystal lattice. Let $\boldsymbol{b}_i$ denote the basis vectors of reciprocal crystal lattice and $\boldsymbol{K} = \sum_i \boldsymbol{b}_i K_i$ with integer coefficients K_i denote a vector of reciprocal crystal lattice. The inner product of $\boldsymbol{\ell}$ and $\boldsymbol{K}$ is an integer, invariant under the rotation $R \in \mathrm{G}$:

$$\boldsymbol{K} \cdot \boldsymbol{\ell} = \sum_j K_j \ell_j = (R\boldsymbol{K}) \cdot (R\boldsymbol{\ell}) = \text{integer}. \tag{6.85}$$

Although the actual crystal is finite (see [Ren (2017)]), the periodic boundary condition is usually assumed on the crystal to restore its translational symmetry such that the translation group $\mathcal{T}$ is **a finite abelian group**. A parallelepiped spanned by three lattice bases is called a crystal cell. Along three lattice bases $\boldsymbol{a}_j$ there are N_j crystal cells, respectively, where N_j are very large natural numbers. Assume that there are no common factor among three N_j for simplicity. Thus, the translation group $\mathcal{T}$ is a direct product of three translation subgroups with order $N_1 N_2 N_3$,

$$\mathcal{T} = \mathcal{T}^{(1)} \otimes \mathcal{T}^{(2)} \otimes \mathcal{T}^{(3)}. \tag{6.86}$$

When N_j goes to infinity, the finite crystal becomes an ideal crystal.

$\mathcal{T}^{(j)}$ is a cyclic group and has N_j one-dimensional inequivalent representations $D^{k_j}(\mathcal{T}_j)$, denoted by $k_j = p_j/N_j$, $0 \leqslant k_j < 1$,

$$D^{k_j}(\boldsymbol{a}_j \ell_j) = \exp(-\mathrm{i}2\pi p_j \ell_j / N_j) = \exp(-\mathrm{i}2\pi k_j \ell_j).$$

Thus, there are $(N_1 N_2 N_3)$ inequivalent one-dimensional representations $D^{\boldsymbol{k}}(\mathcal{T})$ denoted by a vector $\boldsymbol{k}$ in the space of the reciprocal crystal lattice,

$$\boldsymbol{k} = \sum_{j=1}^{3} \boldsymbol{b}_j k_j, \qquad k_j = p_j/N_j, \qquad 0 \leqslant k_j < 1. \tag{6.87}$$

The representation matrix of an element $T(\boldsymbol{\ell})$ of $\mathcal{T}$ in $D^{\boldsymbol{k}}(\mathcal{T})$ is

$$D^{\boldsymbol{k}}(\boldsymbol{\ell}) = e^{-i2\pi(\boldsymbol{k}\cdot\boldsymbol{\ell})} = \exp\left(-i2\pi \sum_{j=1}^{3} k_j \ell_j\right), \qquad 0 \leqslant k_j < 1. \tag{6.88}$$

$\boldsymbol{k}$ is called the **wave vector**, and the space of reciprocal crystal lattice is called the **space of wave vectors**, or $\boldsymbol{k}$ space. If the difference $\boldsymbol{k} - \boldsymbol{k}'$ is a vector of the reciprocal crystal lattice $\boldsymbol{K}$, $\boldsymbol{k}$ and $\boldsymbol{k}'$ both characterize the same representation of $\mathcal{T}$, and are called **the equivalent wave vectors.** In this meaning, there is the translation symmetry in the $\boldsymbol{k}$ space. In the parallelepiped of $\boldsymbol{k}$ space, $0 \leqslant k_j < 1$, there are $(N_1N_2N_3)$ inequivalent wave vectors $\boldsymbol{k}$ denoting the inequivalent irreducible representations of $\mathcal{T}$. When N_j goes to infinity, $\boldsymbol{k}$ varies continuously in that region.

6.5.2 *The Bloch Theorem*

Under the assumption of the periodic boundary condition, the Hamiltonian $H(x)$ of the crystal has the **translational symmetry** with the symmetry group $\mathcal{T}$. The static wave function $\psi_{\boldsymbol{k}}(\boldsymbol{r})$ of the crystal, scalar or spinor, belongs to the irreducible representation $D^{\boldsymbol{k}}(\mathcal{T})$ with the wave vector $\boldsymbol{k}$,

$$P_{T(\boldsymbol{\ell})}\psi_{\boldsymbol{k}}(\boldsymbol{r}) = \psi_{\boldsymbol{k}}(\boldsymbol{r} - \boldsymbol{\ell}) = e^{-i2\pi(\boldsymbol{k}\cdot\boldsymbol{\ell})}\psi_{\boldsymbol{k}}(\boldsymbol{r}). \tag{6.89}$$

$\psi_{\boldsymbol{k}}(\boldsymbol{r})$, called the **Bloch function**, can be written in the form

$$\psi_{\boldsymbol{k}}(\boldsymbol{r}) = e^{i2\pi\boldsymbol{k}\cdot\boldsymbol{r}}u(\boldsymbol{r}), \qquad u(\boldsymbol{r} - \boldsymbol{\ell}) = u(\boldsymbol{r}). \tag{6.90}$$

This is a fundamental theorem in solid physics [Bloch (1928)], called the Bloch theorem: **The wave function of a particle or quasi-particle moving in a periodical potential invariant under the translation group $\mathcal{T}$ can be written in the form of the Bloch function.**

6.5.3 *Star of Wave Vectors and the Little Group*

The region $0 \leqslant k_j < 1$ plays the role of a reciprocal crystal cell in the $\boldsymbol{k}$ space, just like a crystal cell in the crystal lattice. However, this region is not very convenient in the practical application because **the point in the region may move out of the region in the rotation** R of the crystallographic point group G. In solid-state physics the **Brillouin zone** is introduced as a reciprocal crystal cell. The perpendicular bisector plane of the line from the origin to a point of reciprocal crystal lattice is called the **Bragg plane**. The region surrounded the origin by the nearest Bragg planes is called **the first Brillouin zone**. The region from the first Brillouin zone to the next Bragg planes is called the second Brillouin zone, and so on. A Brillouin zone, except for the first one, falls into a few pieces, but each Brillouin zone has the same volume as the first one, and contains $(N_1N_2N_3)$ wave vectors equivalent to that in the first one, respectively. Under a rotation R of G, **a wave vector inside a Brillouin zone moves to that inside it, and a wave vector on its boundary moves to that on the boundary**. There is a one-to-one correspondence between a wave vector inside a Brillouin zone and an inequivalent irreducible representation of $\mathcal{T}$. But a few wave vectors on the boundary of a Brillouin zone may be equivalent, namely they may characterize the same representation.

Let $D(\mathcal{S})$ denote an m-dimensional irreducible unitary representation of the space group $\mathcal{S}$, single-valued or double-valued, and its subduced representation with respect to the translation group $\mathcal{T}$ is diagonal,

$$D(E, \boldsymbol{\ell}) = \text{diag}\left\{e^{-i2\pi \boldsymbol{k}_1\cdot\boldsymbol{\ell}},\ e^{-i2\pi \boldsymbol{k}_2\cdot\boldsymbol{\ell}},\ \cdots,\ e^{-i2\pi \boldsymbol{k}_m\cdot\boldsymbol{\ell}}\right\}, \tag{6.91}$$

where the subscript ρ in $\boldsymbol{k}_\rho$ is not the component index, but the row (column) index of the matrix. The wave vectors $\boldsymbol{k}_\rho$ with a different subscript ρ is not necessary to be different.

Because $\mathcal{T}$ is an invariant subgroup of $\mathcal{S}$ (see Eq. (6.6)),

$$\begin{aligned} T(\boldsymbol{\ell})g(R, \boldsymbol{\alpha}) &= g(R, \boldsymbol{\alpha})T(R^{-1}\boldsymbol{\ell}), \\ D(E, \boldsymbol{\ell})D(R, \boldsymbol{\alpha}) &= D(R, \boldsymbol{\alpha})D(E, R^{-1}\boldsymbol{\ell}). \end{aligned} \tag{6.92}$$

Since $D(E, \boldsymbol{\ell})$ is diagonal, one obtains

$$\exp\left\{-i2\pi \boldsymbol{k}_\tau \cdot \boldsymbol{\ell}\right\}\ D_{\tau\rho}(R, \boldsymbol{\alpha}) = D_{\tau\rho}(R, \boldsymbol{\alpha})\ \exp\left\{-i2\pi\left(R\boldsymbol{k}_\rho\right)\cdot \boldsymbol{\ell}\right\}.$$

Thus, $D_{\tau\rho}(R, \boldsymbol{\alpha}) \neq 0$ only if

$$R\boldsymbol{k}_\rho = \boldsymbol{k}_\tau \quad \text{or} \quad R\boldsymbol{k}_\rho = \boldsymbol{k}_\tau + \boldsymbol{K}, \tag{6.93}$$

$\boldsymbol{K} = 0$ unless $\boldsymbol{k}_\rho$ **is located on the boundary of a Brillouin zone.** The wave vectors in the first Brillouin zone is called mutual **conjugate** if they can be related by an element R in the crystallographic point group G through Eq. (6.93). The set of conjugate vectors is called a **star of wave vectors**. The number of inequivalent wave vectors in a star of wave vectors is called the **index of the star**, denoted by q.

Since the representation $D(\mathcal{S})$ is irreducible, any two wave vectors $\boldsymbol{k}_\rho$ and $\boldsymbol{k}_\tau$ in Eq. (6.91) are conjugate with each other, namely, there is an element R in G such that Eq. (6.93) holds. Otherwise, the wave vectors in Eq. (6.91) have to be divided into at least two sets, where $\boldsymbol{k}_\rho$ and $\boldsymbol{k}_\tau$ in different sets are not conjugate, so that element $D_{\tau\rho}(R, \boldsymbol{\alpha})$ of the representation matrix of every element $g(R, \boldsymbol{\alpha})$ in $\mathcal{S}$ vanishes and the representation is reducible. Since the eigenvalues of $D(E, \boldsymbol{\ell})$ does not change in the similarity transformation (6.92), the **multiplicity d of every $\boldsymbol{k}_\rho$ in $D(E, \boldsymbol{\ell})$ is the same** as each other. One is able to collect the same diagonal elements in Eq. (6.91) together by a simple similarity transformation,

$$D(E, \boldsymbol{\ell}) = \bigoplus_{\mu=1}^{q} \mathbf{1}_d \, \mathrm{e}^{-\mathrm{i}2\pi \boldsymbol{k}^{(\mu)} \cdot \boldsymbol{\ell}}, \qquad m = qd, \tag{6.94}$$

where $\boldsymbol{k}^{(\mu)} \neq \boldsymbol{k}^{(\nu)}$ if $\mu \neq \nu$. **Each irreducible representation $D(\mathcal{S})$ of $\mathcal{S}$ is associated with a star of wave vectors**, where the dimension m of the representation is a multiple of the index q of the star.

Arbitrarily choose one wave vector in the star, say $\boldsymbol{k}^{(1)}$. There are some elements Q in G, including the identity E, satisfying

$$Q\boldsymbol{k}^{(1)} = \boldsymbol{k}^{(1)} \quad \text{or} \quad \boldsymbol{k}^{(1)} + \boldsymbol{K}_Q. \tag{6.95}$$

It is said that Q leaves $\boldsymbol{k}^{(1)}$ invariant. The set of Q forms a subgroup $\mathrm{H}(\boldsymbol{k}^{(1)})$ of G, called the **little co-group**, or the symmetry group of $\boldsymbol{k}^{(1)}$. The index of $\mathrm{H}(\boldsymbol{k}^{(1)})$ in G is q. The left coset of $\mathrm{H}(\boldsymbol{k}^{(1)})$ is denoted by $R_\mu H(\boldsymbol{k}^{(1)})$,

$$R_\mu \boldsymbol{k}^{(1)} = \boldsymbol{k}^{(\mu)}, \quad \text{or} \quad \boldsymbol{k}^{(\mu)} + \boldsymbol{K}_\mu. \tag{6.96}$$

$\boldsymbol{K}_Q = 0$ and $\boldsymbol{K}_\mu = 0$ **unless the star of wave vectors is on the boundary of the first Brillouin zone.** The little co-group $\mathrm{H}(\boldsymbol{k}^{(\mu)})$ is conjugate with $H(\boldsymbol{k}^{(1)})$: $\mathrm{H}(\boldsymbol{k}^{(\mu)}) = R_\mu \mathrm{H}(\boldsymbol{k}^{(1)}) R_\mu^{-1}$. In this meaning, **the little co-group $\mathrm{H}(\boldsymbol{k}^{(1)})$ is associated with the star of wave vectors, not only with $\boldsymbol{k}^{(1)}$.** The crystallographic point group G can be decomposed as the union of the left cosets of $\mathrm{H}(\boldsymbol{k}^{(1)})$:

$$\mathrm{G} = \bigcup_{\mu=1}^{q} R_\mu \mathrm{H}(\boldsymbol{k}^{(1)}), \qquad R_1 = E. \tag{6.97}$$

Although the choices of R_μ are not unique, but as always, R_μ have been chosen. Any element in G can be expressed as $R_\mu Q$. For the spinor wave function, G is replaced by $\tilde{G}$. The actions of the elements E and $\tilde{E}$ in $\tilde{G}$ on the wave vectors $\boldsymbol{k}$ are the same.

The set of $g(Q, \boldsymbol{\alpha})$, where $Q \in H(\boldsymbol{k}^{(1)})$, forms a subgroup $\mathcal{S}(\boldsymbol{k}^{(1)})$ of $\mathcal{S}$, called the **little group**. The index of $\mathcal{S}(\boldsymbol{k}^{(1)})$ in $\mathcal{S}$ is q. The little group $\mathcal{S}(\boldsymbol{k}^{(1)})$ is associated with the star of wave vectors, not only with $\boldsymbol{k}^{(1)}$, and

$$\mathcal{S} = \bigcup_{\mu=1}^{q} g(R_\mu, \boldsymbol{t}_\mu)\mathcal{S}(\boldsymbol{k}^{(1)}), \qquad g(R_1, \boldsymbol{t}_1) = E, \tag{6.98}$$

where, as always, R_μ with their associated $\boldsymbol{t}_\mu$ have been chosen. $\mathcal{T}$ is an invariant subgroup of $\mathcal{S}(\boldsymbol{k}^{(1)})$, where the quotient group $\mathcal{S}(\boldsymbol{k}^{(1)})/\mathcal{T}$ is isomorphic onto $H(\boldsymbol{k}^{(1)})$. For any given element $g(R, \boldsymbol{t})$ in $\mathcal{S}$ and each R_ν, we are able to express the product $g(R, \boldsymbol{t})g(R_\nu, \boldsymbol{t}_\nu)$ as

$$g(R, \boldsymbol{t})g(R_\nu, \boldsymbol{t}_\nu) = g(R_\mu, \boldsymbol{t}_\mu)g(Q, \boldsymbol{\alpha}) = g(R_\mu, \boldsymbol{t}_\mu)T(\boldsymbol{\ell})g(Q, \boldsymbol{t}_Q), \tag{6.99}$$

where $\{\boldsymbol{t}, \boldsymbol{t}_\nu, R_\mu, \boldsymbol{t}_\mu, Q, \boldsymbol{t}_Q, \boldsymbol{\ell}\}$ **are completely determined from** R **and** R_ν.

6.5.4 *General Form of Irreducible Representation of* $\mathcal{S}$

For an m-dimensional unitary irreducible representation $D(\mathcal{S})$ of the space group $\mathcal{S}$, the representation matrix of $T(\boldsymbol{\ell})$ can be taken to be diagonal, as given in Eq. (6.94), which is **labeled by a star of wave vectors**. In the representation space $\mathcal{L}$ of $D(\mathcal{S})$ the basis functions are the Bloch functions.

The first d basis functions $\psi^{(a)}_{\boldsymbol{k}^{(1)}}(\boldsymbol{r})$ in $\mathcal{L}$, which are the eigenfunctions of $P_{T(\boldsymbol{\ell})}$ with the same eigenvalue $e^{-i2\pi\boldsymbol{k}^{(1)}\cdot\boldsymbol{\ell}}$, span a subspace $\mathcal{L}^{(1)}$ of $\mathcal{L}$:

$$\begin{aligned} &\psi^{(a)}_{\boldsymbol{k}^{(1)}}(\boldsymbol{r}) = e^{i2\pi\boldsymbol{k}^{(1)}\cdot\boldsymbol{r}}u_a(\boldsymbol{r}) \in \mathcal{L}^{(1)} \subset \mathcal{L}, \qquad u_a(\boldsymbol{r}-\boldsymbol{\ell}) = u_a(\boldsymbol{r}), \\ &P_{T(\boldsymbol{\ell})}\psi^{(a)}_{\boldsymbol{k}^{(1)}}(\boldsymbol{r}) = e^{-i2\pi\boldsymbol{k}^{(1)}\cdot\boldsymbol{\ell}}\psi^{(a)}_{\boldsymbol{k}^{(1)}}(\boldsymbol{r}), \qquad 1 \leqslant a \leqslant d. \end{aligned} \tag{6.100}$$

This subspace $\mathcal{L}^{(1)}$ is left invariant under the little group $\mathcal{S}(\mathrm{k}^{(1)})$ and corresponds to its d-dimensional unitary representation $\Gamma[\mathcal{S}(\boldsymbol{k}^{(1)})]$:

$$\begin{aligned} &P_{g(Q,\boldsymbol{t}_Q)}\psi^{(a)}_{\boldsymbol{k}^{(1)}}(\boldsymbol{r}) = \sum_{b=1}^{d} \psi^{(b)}_{\boldsymbol{k}^{(1)}}(\boldsymbol{r})\Gamma_{ba}(Q, \boldsymbol{t}_Q), \\ &\Gamma(E, \boldsymbol{\ell}) = \mathbf{1}_d e^{-i2\pi\boldsymbol{k}^{(1)}\cdot\boldsymbol{\ell}}. \end{aligned} \tag{6.101}$$

Remind that $Q \in \tilde{G}$ if $\psi^{(a)}_{\boldsymbol{k}^{(1)}}(\boldsymbol{r})$ is a spinor.

We define the remaining basis functions in $\mathcal{L}$ (see Eq. (6.92)) by

$$\begin{aligned}
&\psi_{\boldsymbol{k}^{(\mu)}}^{(a)}(\boldsymbol{r}) = P_{g(R_\mu,\boldsymbol{t}_\mu)}\psi_{\boldsymbol{k}^{(1)}}^{(a)}(\boldsymbol{r}), \qquad 1 < \mu \leqslant q, \\
&P_{T(\boldsymbol{\ell})}\psi_{\boldsymbol{k}^{(\mu)}}^{(a)}(\boldsymbol{r}) = P_{g(R_\mu,\boldsymbol{t}_\mu)}P_{T(R_\mu^{-1}\boldsymbol{\ell})}\psi_{\boldsymbol{k}^{(1)}}^{(a)}(\boldsymbol{r}) \\
&\quad = P_{g(R_\mu,\boldsymbol{t}_\mu)}\mathrm{e}^{-\mathrm{i}2\pi\boldsymbol{k}^{(1)}\cdot(R_\mu^{-1}\boldsymbol{\ell})}\psi_{\boldsymbol{k}^{(1)}}^{(a)}(\boldsymbol{r}) = \mathrm{e}^{-\mathrm{i}2\pi\boldsymbol{k}^{(\mu)}\cdot\boldsymbol{\ell}}\psi_{\boldsymbol{k}^{(\mu)}}^{(a)}(\boldsymbol{r}).
\end{aligned} \tag{6.102}$$

Thus, equation (6.94) holds for $T(\boldsymbol{\ell})$. The representation matrix of an element $g(R,\boldsymbol{t})$ of $\mathcal{S}$ is calculated from Eq. (6.99):

$$\begin{aligned}
P_{g(R,\boldsymbol{t})}\psi_{\boldsymbol{k}^{(\nu)}}^{(a)}(\boldsymbol{r}) &= P_{g(R,\boldsymbol{t})}P_{g(R_\nu,\boldsymbol{t}_\nu)}\psi_{\boldsymbol{k}^{(1)}}^{(a)}(\boldsymbol{r}) \\
&= P_{g(R_\mu,\boldsymbol{t}_\mu)}P_{T(\boldsymbol{\ell})}P_{g(Q,\boldsymbol{t}_Q)}\psi_{\boldsymbol{k}^{(1)}}^{(a)}(\boldsymbol{r}) \\
&= P_{g(R_\mu,\boldsymbol{t}_\mu)}\sum_{b=1}^{d}\psi_{\boldsymbol{k}^{(1)}}^{(b)}(\boldsymbol{r})\mathrm{e}^{-\mathrm{i}2\pi\boldsymbol{k}^{(1)}\cdot\boldsymbol{\ell}}\Gamma_{ba}(Q,\boldsymbol{t}_Q) \\
&= \sum_{b=1}^{d}\psi_{\boldsymbol{k}^{(\mu)}}^{(b)}(\boldsymbol{r})D_{\mu b,\nu a}(R,\boldsymbol{t}), \\
D_{\mu b,\nu a}(R,\boldsymbol{t}) &= \mathrm{e}^{-\mathrm{i}2\pi\boldsymbol{k}^{(1)}\cdot\boldsymbol{\ell}}\Gamma_{ba}(Q,\boldsymbol{t}_Q) = \Gamma_{ba}(Q,\boldsymbol{\ell}+\boldsymbol{t}_Q),
\end{aligned} \tag{6.103}$$

where $\{\mu, Q, \boldsymbol{t}_Q, \boldsymbol{\ell}\}$ are completely determined from R and ν. The row (column) index of $D(\mathcal{S})$ is denoted by two indices μ and b. **In the given ν-column of $D_{\mu b,\nu a}(R,\boldsymbol{t})$ there is only one nonvanishing d-dimensional submatrix at μ-row**, and vice versa. Especially, from Eq. (6.102), $D_{\nu b,1a}(R_\mu,\boldsymbol{t}_\mu) = \delta_{\nu\mu}\delta_{ba}$ and $D_{\mu b,\nu a}(R_\mu,\boldsymbol{t}_\mu) = \delta_{\nu 1}\delta_{ba}$.

The irreducible representation $D(\mathcal{S})$ of the space group $\mathcal{S}$ is determined completely by the irreducible representation $\Gamma(Q,\boldsymbol{\alpha})$ of the little group $\mathcal{S}(\boldsymbol{k}^{(1)})$ through Eqs. (6.100-103). The representation $D(\mathcal{S})$ is irreducible if and only if the representation $\Gamma[\mathcal{S}(\boldsymbol{k}^{(1)})]$ is irreducible. In fact, if $\Gamma[\mathcal{S}(\boldsymbol{k}^{(1)})]$ is reducible, there is a nonconstant matrix Y commutable with all matrices $\Gamma(Q,\boldsymbol{\alpha})$. Then, $X = \mathbf{1}_q \times Y$ commutes with all matrices $D(R,\boldsymbol{t})$ and $D(E,\boldsymbol{\ell})$, so that $D(\mathcal{S})$ is reducible. Conversely, if $D(\mathcal{S})$ is reducible, there is a nonconstant matrix X commutable with all representation matrices of $D(\mathcal{S})$. Since X commutes with $D(E,\boldsymbol{\ell})$, $X = \bigoplus_\mu Y_\mu$. Since X commutes with $D(R_\mu,\boldsymbol{t}_\mu)$, Y_μ is independent of μ, and then, Y_μ commutes with $\Gamma(Q,\boldsymbol{\alpha})$. Thus, $\Gamma(Q,\boldsymbol{\alpha})$ is reducible.

The static wave function of an electron moving in the crystal belongs to an irreducible representation of the space group $\mathcal{S}$ of the crystal. **All basis functions $\psi_{\boldsymbol{k}^{(\mu)}}^{(b)}(\boldsymbol{r})$, scalar or spinor, in an irreducible representation space of $\mathcal{S}$ are the degenerate eigenfunctions of the Hamiltonian with the same energy.**

6.5.5 *Irreducible Representation of A Little Group*

Now, we calculate the irreducible representation $\Gamma(Q, \boldsymbol{\alpha})$ of the little group $\mathcal{S}(\boldsymbol{k}^{(1)})$ from the definition (6.101). Define

$$\Gamma(Q, \boldsymbol{\alpha}) = \mathrm{e}^{-\mathrm{i}2\pi \boldsymbol{k}^{(1)}\cdot\boldsymbol{\alpha}}\overline{D}(Q), \qquad \overline{D}(E) = \mathbf{1}_d. \tag{6.104}$$

For the multiplication of the two elements of $\mathcal{S}(\boldsymbol{k}^{(1)})$,

$$g(Q, \boldsymbol{t})g(Q', \boldsymbol{t}') = g(QQ', \boldsymbol{t} + Q\boldsymbol{t}'), \tag{6.105}$$

we have

$$\begin{aligned}
\overline{D}(Q)\overline{D}(Q') &= \mathrm{e}^{\mathrm{i}2\pi\boldsymbol{k}^{(1)}\cdot(\boldsymbol{t}+\boldsymbol{t}')}\Gamma(Q, \boldsymbol{t})\Gamma(Q', \boldsymbol{t}') \\
&= \mathrm{e}^{\mathrm{i}2\pi\boldsymbol{k}^{(1)}\cdot(\boldsymbol{t}+\boldsymbol{t}')}\Gamma(QQ', \boldsymbol{t} + Q\boldsymbol{t}') \\
&= \mathrm{e}^{-\mathrm{i}2\pi\boldsymbol{k}^{(1)}\cdot(Q-E)\boldsymbol{t}'}\overline{D}(QQ') \\
&= \mathrm{e}^{-\mathrm{i}2\pi\boldsymbol{K}'_Q\cdot\boldsymbol{t}'}\overline{D}(QQ') \\
Q^{-1}\boldsymbol{k}^{(1)} &= \boldsymbol{k}^{(1)} + \boldsymbol{K}'_Q.
\end{aligned} \tag{6.106}$$

In the following two cases the exponent factor $\mathrm{e}^{-\mathrm{i}2\pi\boldsymbol{K}'_Q\cdot\boldsymbol{t}'}$ **becomes one** such that the set of $\overline{D}(Q)$ **forms an irreducible representation of the little co-group** $\mathrm{H}(\boldsymbol{k}^{(1)})$, which is well studied in subsection 6.2.4, including single-valued or double-valued one:

(a) **The space group $\mathcal{S}$ is symmorphic** so that all $\boldsymbol{t} = 0$ in $g(Q, \boldsymbol{t})$;

(b) **The star of wave vectors $\boldsymbol{k}^{(1)}$ is located inside the Brillouin zone** so that all $\boldsymbol{K}'_Q = 0$.

For those two cases, the irreducible representation $\Gamma(Q, \boldsymbol{\alpha})$ of the little group $\mathcal{S}(\boldsymbol{k}^{(1)})$ is directly given by the irreducible representation $\overline{D}(Q)$ of the little co-group $H(\boldsymbol{k}^{(1)})$ in Eq. (6.104). **Since the actions of both $R(\hat{\boldsymbol{n}}, \omega) \in \mathrm{SO}(3)$ and $u(\hat{\boldsymbol{n}}, \omega) \in \mathrm{SU}(2)$ on the coordinate vectors $\boldsymbol{r}$ and the wave vectors $\boldsymbol{k}$ are the same**, the single-valued and the double-valued irreducible representations of the space group $\mathcal{S}$ both can be calculated through those of the little co-group $\mathrm{H}(\boldsymbol{k}^{(1)})$.

For the non-symmorphic space group $\mathcal{S}$ where the star of wave vectors is on the boundary of the Brillouin zone, **the key for calculating its irreducible representations is to find a symmorphic space subgroup $\mathcal{S}'$ of $\mathcal{S}$** ([Bradley and Cracknell (1972)]). We need to show that the generators of $\mathcal{S}'$ can be expressed as products of elements of $\mathcal{S}$ and to choose the suitable representative elements $g\left(R_\rho, \boldsymbol{t}_\rho\right)$ in the cosets of $\mathcal{S}'$:

$$\mathcal{S} = \bigcup_{\mu} g\left(R_\rho, \boldsymbol{t}_\rho\right)\mathcal{S}', \quad g\left(R_1, \boldsymbol{t}_1\right) = E. \tag{6.107}$$

The irreducible representations $D(\mathcal{S}')$, **both single-valued and double-valued**, and their basis functions $\psi^{(a)}_{\boldsymbol{k}^{(\mu)}}(\boldsymbol{r})$ have been known. Extending the basis functions $\psi^{(a)}_{\boldsymbol{k}^{(\mu)}}(\boldsymbol{r})$ to $P_{g(R_\rho,\boldsymbol{t}_\rho)}\psi^{(a)}_{\boldsymbol{k}^{(\mu)}}(\boldsymbol{r})$, we are able to calculate **the irreducible representation of** $\mathcal{S}$ from **the induced representation of an irreducible representation of** $\mathcal{S}'$ or its decomposition. **The main calculation** is the product of the generators $g(R,\boldsymbol{t})$ of $\mathcal{S}$ and $g\left(R_\rho, \boldsymbol{t}_\rho\right)$:

$$g(R,\boldsymbol{t})g\left(R_\rho,\boldsymbol{t}_\rho\right) = g(R_\nu,\boldsymbol{t}_\nu)g(Q,\boldsymbol{\alpha}), \quad g(Q,\boldsymbol{\alpha}) \in \mathcal{S}'. \tag{6.108}$$

Two examples are given in the following, where we use the notations for the elements of the group $\mathbf{O}_h$ given in subsection 2.5.2 for convenience. Namely, the elements $3'$, $\overline{4}$ and $\overline{2}''$ in the group $\mathbf{O}_h$ are denoted by R_1, σT_z and σS_1, respectively. The readers are also suggested to look up the representations of the space groups on the web: http://www.cryst.ehu.es/rep/repres.html (provided by Prof. Hong-Jun Xiang of Fudan University).

Ex. 1. The irreducible representations of the space group $\mathbf{T}_d^5 = F3'\overline{4}_{0\,0\,\frac{1}{2}}\overline{2}''_{0\,0\,\frac{1}{2}}$ with No. 219.

From the international symbol, the crystal with the space group $\mathbf{T}_d^5$ belongs to the face-centred Bravais lattice F in the cubic crystal system with the point group $\mathbf{T}_d$. The lattice bases are perpendicular to each other: $\boldsymbol{a}_j = a\boldsymbol{e}_j$ (see subsection 6.3.7). The translation vector of crystal lattice $\boldsymbol{L} = \boldsymbol{\ell}$ or $\boldsymbol{\ell} + \boldsymbol{f}_j$, where $\boldsymbol{f}_1 = (\boldsymbol{a}_2 + \boldsymbol{a}_3)/2$ and its cyclic combination.

We are going to show that the symmorphic space group $\mathbf{T}^2 = F3'22'$ with No. 196 is the invariant subgroup of $\mathbf{T}_d^5$ of index two:

$$\mathbf{T}_d^5 = \mathbf{T}^2 \bigcup g\left(\sigma S_1, \{0, 0, \tfrac{1}{2}\}\right)\mathbf{T}^2. \tag{6.109}$$

In fact, both $\mathbf{T}^2 = F3'22'$ and $\mathbf{T}_d^5$ belong to the same Bravais lattice F and the same crystal system. In addition to the translation elements $T(\boldsymbol{L})$, the remaining generators of $\mathbf{T}^2$ are R_1 and T_z^2, which belong to $\mathbf{T}_d^5$ evidently

$$g\left(\sigma T_z, \left\{0, 0, \frac{1}{2}\right\}\right)^2 = T\left(0, 0, \frac{1}{2}\right)T\left(0, 0, -\frac{1}{2}\right)\sigma T_z \sigma T_z = T_z^2.$$

The first Brillouin zone for $\mathbf{T}_d^5$ and $\mathbf{T}^2$ are the same. We introduce the primitive lattice bases $\boldsymbol{a}'_j$ and the reciprocal ones $\boldsymbol{b}'_j$

$$\begin{array}{lll} \boldsymbol{a}_1' = (a/2)(0,1,1), & \boldsymbol{a}_2' = (a/2)(1,0,1), & \boldsymbol{a}_3' = (a/2)(1,1,0), \\ \boldsymbol{b}_1' = (1/a)(-1,1,1), & \boldsymbol{b}_2' = (1/a)(1,-1,1), & \boldsymbol{b}_3' = (1/a)(1,1,-1). \end{array}$$

The reciprocal vectors of crystal lattice along 4-fold axes, 3-fold axes, and 2-fold axes are $\boldsymbol{K}_1 = (2/a)(0,0,1)$, $\boldsymbol{K}_2 = (1/a)(1,1,1)$, $\boldsymbol{K}_3 = (2/a)(1,1,0)$, and so on, respectively. **The first Brillouin zone** is surrounded by the perpendicular bisector planes of $\boldsymbol{K}_1$ and $\boldsymbol{K}_2$ and so on. Namely, it is bounded by six squares perpendicular to the 4-fold axes and eight hexagons perpendicular to the 3-fold axes. The coordinates of 12 vertices of the first Brillouin zone in the rectangular coordinate frame are

$$\begin{array}{lll} \pm(1/2a)(2,1,0), & \pm(1/2a)(2,0,1), & \pm(1/2a)(2,-1,0), \\ \pm(1/2a)(2,0,-1), & \pm(1/2a)(0,2,1), & \pm(1/2a)(1,2,0), \\ \pm(1/2a)(0,2,-1), & \pm(1/2a)(-1,2,0), & \pm(1/2a)(1,0,2), \\ \pm(1/2a)(0,1,2), & \pm(1/2a)(-1,0,2), & \pm(1/2a)(0,-1,2). \end{array}$$

The irreducible representation $D(R,\boldsymbol{\alpha})$, single-valued or double-valued, and its basis functions $\psi^{(a)}_{\boldsymbol{k}^{(\mu)}}(\boldsymbol{r})$ of the symmorphic space group $\mathbf{T}^2$ associated with any star $\boldsymbol{k}^{(\mu)}$ of wave vectors, **including that at the boundary of the first Brillouin zone**, have been known. Extending the representation space by

$$\phi^{(a)}_{\boldsymbol{k}^{(\mu)}}(\boldsymbol{r}) = P_{g(\sigma S_1,\{0,0,\frac{1}{2}\})}\psi^{(a)}_{\boldsymbol{k}^{(\mu)}}(\boldsymbol{r}), \tag{6.110}$$

we obtain through careful calculation where $R(\hat{\boldsymbol{n}},\omega)$ is replaced by $u(\hat{\boldsymbol{n}},\omega)$ (see Prob. 7)

$$\begin{aligned} & g\left(\sigma S_1,\{0,0,\tfrac{1}{2}\}\right)^2 = T(0,0,1)\,\sigma S_1 \sigma S_1 = T(\boldsymbol{a}_3)\tilde{E}, \\ & T(\boldsymbol{L}) g\left[\sigma S_1,\{0,0,\tfrac{1}{2}\}\right] = g\left[\sigma S_1,\{0,0,\tfrac{1}{2}\}\right] T(S_1^{-1}\sigma \boldsymbol{L}), \\ & R_1 g\left[\sigma S_1,\{0,0,\tfrac{1}{2}\}\right] = g\left[\sigma S_1,\{0,0,\tfrac{1}{2}\}\right] T(-\boldsymbol{f}_1) R_3^5, \\ & T_z^2 g\left[\sigma S_1,\{0,0,\tfrac{1}{2}\}\right] = g\left[\sigma S_1,\{0,0,\tfrac{1}{2}\}\right] T_z^6. \end{aligned} \tag{6.111}$$

and $D(\tilde{E}) = \lambda \mathbf{1}$ where $\lambda = 1$ for the single-valued representation and $\lambda = -1$ for the double-valued one. Then, the irreducible representation $\Delta(\mathbf{T}_d^5)$ in the basis functions $\psi^{(a)}_{\boldsymbol{k}^{(\mu)}}(\boldsymbol{r})$ and $\phi^{(a)}_{\boldsymbol{k}^{(\mu)}}(\boldsymbol{r})$ is

$$\Delta_{\nu b,\mu a}\left[T(\boldsymbol{L})\right] = \delta_{\nu\mu}\delta_{ba}\begin{pmatrix} e^{-i2\pi \boldsymbol{k}^{(\mu)}\cdot \boldsymbol{L}} & 0 \\ 0 & e^{i2\pi\left(S_1 \boldsymbol{k}^{(\mu)}\right)\cdot \boldsymbol{L}} \end{pmatrix},$$

$$\begin{aligned}
&\Delta_{\nu b,\mu a}\left[\sigma S_1,\{0,0,\tfrac{1}{2}\}\right]=\delta_{\nu\mu}\delta_{ba}\begin{pmatrix}0 & \lambda e^{-i2\pi \boldsymbol{k}^{(\mu)}\cdot \boldsymbol{a}_3}\\ 1 & 0\end{pmatrix},\\
&\Delta_{\nu b,\mu a}(R_1)=\begin{pmatrix}D_{\nu b,\mu a}(R_1) & 0\\ 0 & e^{i2\pi \boldsymbol{k}^{(\nu)}\cdot \boldsymbol{f}_1}D_{\nu b,\mu a}(R_3^5)\end{pmatrix},\\
&\Delta_{\nu b,\mu a}(T_z^2)=\begin{pmatrix}D_{\nu b,\mu a}(T_z^2) & 0\\ 0 & D_{\nu b,\mu a}(T_z^6)\end{pmatrix},
\end{aligned}\tag{6.112}$$

where $\boldsymbol{L}=\boldsymbol{\ell}$ or $\boldsymbol{\ell}+\boldsymbol{f}_j$.

Ex. 2. The irreducible representation of the space group $\mathbf{O}_h^8=F\overline{3}'4_{\frac{1}{2}\frac{1}{4}\frac{1}{4}}2''_{\frac{1}{4}\frac{1}{4}\frac{1}{2}}$ with No. 228.

From the international symbol, the crystal with the space group $\mathbf{O}_h^8$ belongs to the face-centred Bravais lattice F in the cubic crystal system with the point group $\mathbf{O}_h$. The lattice bases $\boldsymbol{a}_j$ and their fractional combinations $\boldsymbol{f}_j$ are the same as those in $\mathbf{T}_d^5$ (see **Ex. 1**). The symbol $F\overline{3}'4_{\frac{1}{2}\frac{1}{4}\frac{1}{4}}2''_{\frac{1}{4}\frac{1}{4}\frac{1}{2}}$ is not convenient for finding its subgroup because

$$g\left(T_z,\left\{\frac{1}{2}\frac{1}{4}\frac{1}{4}\right\}\right)^2=T\left(\frac{1}{4},\frac{3}{4},\frac{1}{2}\right)T_z^2.$$

Making a shift $\boldsymbol{r}_0$ of the origin, one has (see Eq. (6.63))

$$\boldsymbol{r}_0=(\boldsymbol{a}_1+\boldsymbol{a}_2+\boldsymbol{a}_3)/8,\quad (T_z-E)\boldsymbol{r}_0=-\boldsymbol{a}_1/4,$$
$$(S_1-E)\boldsymbol{r}_0=-\boldsymbol{a}_3/4,\quad (\sigma R_1-E)\boldsymbol{r}_0=-\left(\boldsymbol{a}_1+\boldsymbol{a}_2+\boldsymbol{a}_3\right)/4.$$

Thus, an equivalent symbol of $\mathbf{O}_h^8$ is

$$\mathbf{O}_h^8=F\overline{3}'_{-\frac{1}{4}-\frac{1}{4}-\frac{1}{4}}4_{\frac{1}{4}\frac{1}{4}\frac{1}{4}}2''_{\frac{1}{4}\frac{1}{4}\frac{1}{4}}.\tag{6.113}$$

Replacing $R(\hat{\boldsymbol{n}},\omega)$ with $u(\hat{\boldsymbol{n}},\omega)$, we obtain

$$\begin{aligned}
&T(-\boldsymbol{f}_3)g\left[S_1,\{\tfrac{1}{4},\tfrac{1}{4},\tfrac{1}{4}\}\right]^2=\tilde{E},\qquad g\left[\sigma R_1,\{-\tfrac{1}{4},-\tfrac{1}{4},-\tfrac{1}{4}\}\right]^2=R_1^2,\\
&g\left[\sigma R_1,\{-\tfrac{1}{4},-\tfrac{1}{4},-\tfrac{1}{4}\}\right]\tilde{E}R_1^2=g\left[\sigma,\{-\tfrac{1}{4},-\tfrac{1}{4},-\tfrac{1}{4}\}\right],\\
&T\left(\boldsymbol{a}_3+\boldsymbol{f}_3\right)g\left[\sigma,\{-\tfrac{1}{4},-\tfrac{1}{4},-\tfrac{1}{4}\}\right]g\left[T_z,\{\tfrac{1}{4},\tfrac{1}{4},\tfrac{1}{4}\}\right]=g\left(\sigma T_z,\{0,0,\tfrac{1}{2}\}\right),\\
&T\left(\boldsymbol{a}_3+\boldsymbol{f}_3\right)g\left[\sigma,\{-\tfrac{1}{4},-\tfrac{1}{4},-\tfrac{1}{4}\}\right]g\left[S_1,\{\tfrac{1}{4},\tfrac{1}{4},\tfrac{1}{4}\}\right]=g\left(\sigma S_1,\{0,0,\tfrac{1}{2}\}\right).
\end{aligned}$$

Thus, the space group $\mathbf{T}_d^5=F3'\overline{4}_{00\frac{1}{2}}\overline{2}''_{00\frac{1}{2}}$ is the invariant subgroup of $\mathbf{O}_h^8$ with index two, so that $\mathbf{T}^5$ is a subgroup of $\mathbf{O}_h^8$ with index four:

$$\mathbf{O}_h^8=\mathbf{T}_d^5\bigcup g\left(\sigma,\{-\tfrac{1}{4},-\tfrac{1}{4},-\tfrac{1}{4}\}\right)\mathbf{T}_d^5.\tag{6.114}$$

The first Brilloiun zone for $\mathbf{O}_h^8$ is the same as that for $\mathbf{T}_d^5$ and that for $\mathbf{T}^2$. The irreducible representation $\Delta(R, \boldsymbol{\alpha})$ and its basis functions $\psi_{\boldsymbol{k}^{(\mu)}}^{(a)}(\boldsymbol{r})$ and $\phi_{\boldsymbol{k}^{(\mu)}}^{(a)}(\boldsymbol{r})$ of $\mathbf{T}_d^5$ associated with any star of wave vectors, including that at the boundary of the first Brillouin zone, have been obtained in **Ex.1**. Extending the representation space by

$$\begin{aligned} P_{g[\sigma,\{-\frac{1}{4},-\frac{1}{4},-\frac{1}{4}\}]}\psi_{\boldsymbol{k}^{(\mu)}}^{(a)}(\boldsymbol{r}) &= \overline{\psi}_{\boldsymbol{k}^{(\mu)}}^{(a)}(\boldsymbol{r}), \\ P_{g[\sigma,\{-\frac{1}{4},-\frac{1}{4},-\frac{1}{4}\}]}\phi_{\boldsymbol{k}^{(\mu)}}^{(a)}(\boldsymbol{r}) &= \overline{\phi}_{\boldsymbol{k}^{(\mu)}}^{(a)}(\boldsymbol{r}), \end{aligned} \tag{6.115}$$

we obtain the following formulas:

$$\begin{aligned} & g\left[\sigma, \{-\tfrac{1}{4}, -\tfrac{1}{4}, -\tfrac{1}{4}\}\right]^2 = E, \\ & T(\boldsymbol{L})g\left[\sigma, \{-\tfrac{1}{4}, -\tfrac{1}{4}, -\tfrac{1}{4}\}\right] = g\left[\sigma, \{-\tfrac{1}{4}, -\tfrac{1}{4}, -\tfrac{1}{4}\}\right]T(-\boldsymbol{L}), \\ & g\left[\sigma S_1, \{0, 0, \tfrac{1}{2}\}\right] g\left[\sigma, \{-\tfrac{1}{4}, -\tfrac{1}{4}, -\tfrac{1}{4}\}\right] \\ & \quad = g\left[\sigma, \{-\tfrac{1}{4}, -\tfrac{1}{4}, -\tfrac{1}{4}\}\right] T(-\boldsymbol{a}_3 - \boldsymbol{f}_3) g\left[\sigma S_1, \{0, 0, \tfrac{1}{2}\}\right], \\ & R_1 g\left[\sigma, \{-\tfrac{1}{4}, -\tfrac{1}{4}, -\tfrac{1}{4}\}\right] = g\left[\sigma, \{-\tfrac{1}{4}, -\tfrac{1}{4}, -\tfrac{1}{4}\}\right] R_1, \\ & T_z^2 g\left[\sigma, \{-\tfrac{1}{4}, -\tfrac{1}{4}, -\tfrac{1}{4}\}\right] = g\left[\sigma, \{-\tfrac{1}{4}, -\tfrac{1}{4}, -\tfrac{1}{4}\}\right] T(-\boldsymbol{f}_3) T_z^2. \end{aligned} \tag{6.116}$$

From Eqs. (6.115) and (6.116) the irreducible representation $\overline{D}(\mathbf{O}_h^8)$ of $\mathbf{O}_h^8$ in the basis functions $\psi_{\boldsymbol{k}^{(\mu)}}^{(a)}(\boldsymbol{r})$, $\phi_{\boldsymbol{k}^{(\mu)}}^{(a)}(\boldsymbol{r})$, $\overline{\psi}_{\boldsymbol{k}^{(\mu)}}^{(a)}(\boldsymbol{r})$, and $\overline{\phi}_{\boldsymbol{k}^{(\mu)}}^{(a)}(\boldsymbol{r})$ can be written directly in terms of $D(\mathbf{T}^5)$ (see Eq. (6.112)):

$$\begin{aligned} & \overline{D}_{\nu b,\mu a}\left[T(\boldsymbol{L})\right] = \delta_{\nu\mu}\delta_{ba} \\ & \quad \times \text{ diag}\{e^{-i2\pi \boldsymbol{k}^{(\mu)}\cdot \boldsymbol{L}},\ e^{i2\pi\left(S_1\boldsymbol{k}^{(\mu)}\right)\cdot \boldsymbol{L}},\ e^{i2\pi \boldsymbol{k}^{(\mu)}\cdot \boldsymbol{L}},\ e^{-i2\pi\left(S_1\boldsymbol{k}^{(\mu)}\right)\cdot \boldsymbol{L}}\}, \\ & \overline{D}_{\nu b,\mu a}\left[\sigma, \{-\tfrac{1}{4}, -\tfrac{1}{4}, -\tfrac{1}{4}\}\right] = \delta_{\nu\mu}\delta_{ba}\begin{pmatrix} 0 & 0 & 1 & 0 \\ 0 & 0 & 0 & 1 \\ 1 & 0 & 0 & 0 \\ 0 & 1 & 0 & 0 \end{pmatrix}, \\ & \overline{D}_{\nu b,\mu a}\left[\sigma S_1, \{0, 0, \tfrac{1}{2}\}\right] = \delta_{\nu\mu}\delta_{ab} \\ & \quad \times \begin{pmatrix} 0 & \lambda e^{-i2\pi \boldsymbol{k}^{(\mu)}\cdot \boldsymbol{a}_3} & 0 & 0 \\ 1 & 0 & 0 & 0 \\ 0 & 0 & 0 & \lambda e^{i2\pi \boldsymbol{k}^{(\mu)}\cdot \boldsymbol{f}_3} \\ 0 & 0 & e^{i2\pi \boldsymbol{k}^{(\mu)}\cdot(\boldsymbol{a}_3 - \boldsymbol{f}_3)} & 0 \end{pmatrix}, \\ & \overline{D}_{\nu b,\mu a}(R_1) = \text{diag}\{D_{\nu b,\mu a}(R_1), \\ & \quad e^{i2\pi \boldsymbol{k}^{(\nu)}\cdot \boldsymbol{f}_1} D_{\nu b,\mu a}(R_3^5),\ D_{\nu b,\mu a}(R_1),\ e^{i2\pi \boldsymbol{k}^{(\nu)}\cdot \boldsymbol{f}_1} D_{\nu b,\mu a}(R_3^5)\}, \end{aligned}$$

$$\overline{D}_{\nu b,\mu a}\left(T_z^2\right) = \text{diag}\{D_{\nu b,\mu a}\left(T_z^2\right),$$
$$D_{\nu b,\mu a}\left(T_z^6\right),\ \mathrm{e}^{\mathrm{i}2\pi \boldsymbol{k}^{(\nu)}\cdot \boldsymbol{f}_3} D_{\nu b,\mu a}\left(T_z^2\right),\ \mathrm{e}^{-\mathrm{i}2\pi \boldsymbol{k}^{(\nu)}\cdot \boldsymbol{f}_3} D_{\nu b,\mu a}\left(T_z^6\right)\}.$$

6.5.6 *Study on Graphene by Group Theory*

Recently, graphene is one focus of attention of theorists and experimentalists. The space group of graphene is $\mathrm{D}_{6h}^1 = P \pm 62'$ with No. 191, which belongs to the primitive Bravais lattice P in the hexagonal crystal system with the crystallographic point group D_{6h}.

Graphene is a planar crystal. The space of crystal lattice of graphene is given in Fig. 6.2. By making use of the convenient unit of length, the distance of two neighboring Carbon atoms is taken to be 2/3. The primitive lattice bases $\boldsymbol{a}_j$ and the reciprocal ones $\boldsymbol{b}_j$ are

$$\begin{aligned} \boldsymbol{a}_1 &= \frac{2}{\sqrt{3}}(1,\ 0), \qquad & \boldsymbol{a}_2 &= \frac{1}{\sqrt{3}}(-1,\ \sqrt{3}), \\ \boldsymbol{b}_1 &= \frac{1}{2}(\sqrt{3},\ 1), \qquad & \boldsymbol{b}_2 &= (0, 1). \end{aligned} \tag{6.117}$$

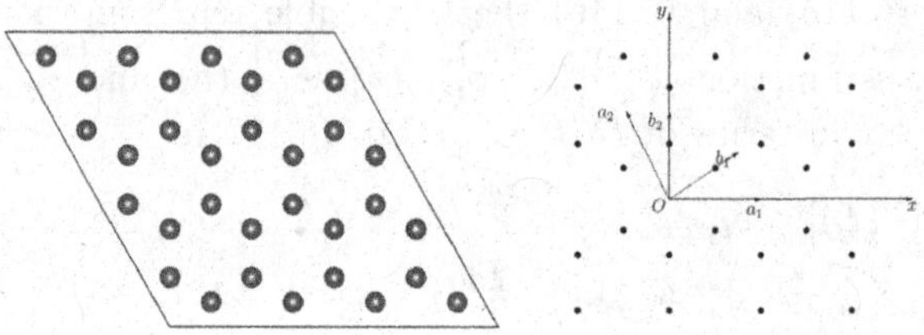

Fig. 6.2 The space of crystal lattice of graphene.

The irreducible representation of the space group $\mathcal{S}$ of graphene is associated with that of the little group $\mathcal{S}(\boldsymbol{k}^{(1)})$, as well as the little co-group $\mathrm{H}(\boldsymbol{k}^{(1)})$ through Eq. (6.104). A point $\boldsymbol{k}^{(1)}$ in the first Brillouin zone is called a **point of symmetry** if no point $\boldsymbol{k}$ in the neighborhood $\mathcal{N}$ of $\boldsymbol{k}^{(1)}$, except for $\boldsymbol{k}^{(1)}$, has the symmetry group $\mathrm{H}(\boldsymbol{k}^{(1)})$. A line in $\mathcal{N}$ passing through $\boldsymbol{k}^{(1)}$ is called a **line of symmetry** if all point $\boldsymbol{k}$ on the line has the symmetry group $\mathrm{H}(\boldsymbol{k}^{(1)})$, but any point out of the line does not. The first Brillouin zone of graphene is given in Fig. 6.3 where the points Γ, K_1, and M are the points of symmetry and the lines ΓK_1, ΓM, and $K_1 M$ are the lines of symmetry. Since graphene is a planar crystal, the little co-group $\mathrm{H}(\boldsymbol{k}^{(1)})$ for any $\boldsymbol{k}^{(1)}$ is the direct product of a subgroup and $\mathrm{C}_s = \{E, \sigma C_6^3\}$. Removing C_s, we list the little co-subgroup for some typical wave vectors $\boldsymbol{k}^{(1)}$ in Table 6.7, where $C_6^a C_2^{(b)} = C_2^{(a+b)}$ (see subsection 2.2.3).

Table 6.7 The little co-subgroups $\mathrm{H}(\boldsymbol{k}^{(1)})$

$\boldsymbol{k}^{(1)}$	Little co-subgroup	generators
Γ	$\mathrm{D}_6 = \{C_6^a, C_2^{(a)}, a \bmod 6\}$	C_6, $C_2^{(0)}$
ΓK_1	$\{E, C_2^{(2)}\} \approx \mathrm{C}_2$,	$C_2^{(2)}$
K_1	$\{E, C_6^2, C_6^4, C_2^{(0)}, C_2^{(2)}, C_2^{(4)}\} \approx \mathrm{D}_3$	C_6^2, $C_2^{(0)}$
$K_1 M$	$\{E, C_2^{(4)}\} \approx \mathrm{C}_2$,	$C_2^{(4)}$
M	$\{E, C_6^3, C_2^{(1)}, C_2^{(4)}\} \approx \mathrm{D}_2$	C_6^3, $C_2^{(1)}$
ΓM	$\{E, C_2^{(1)}\} \approx \mathrm{C}_2$,	$C_2^{(1)}$

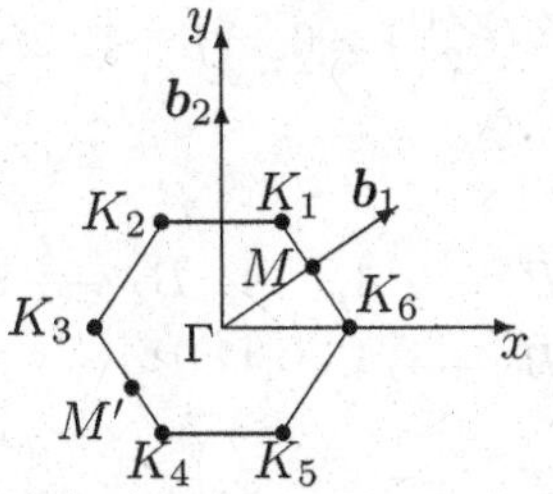

Fig. 6.3 The first Brillouin zone of graphene

The symmetry classification of nine lowest-lying valence was calculated by the first-principle calculations [Kogan and Nazarov 2012] (see Fig. 6.4). **The representation of the little co-subgroup on the line of symmetry should belong to the decomposition of the subduced representation of the irreducible representation of the little co-subgroup on the points of symmetry at its each end.** The representation matrices of elements of D_6, D_3, and D_2 (see Eq. (3.86)) are listed as follows:

$$\text{For } \mathrm{D}_6: \quad D^{A_1}(C_6) = D^{A_2}(C_6) = -D^{B_1}(C_6) = -D^{B_2}(C_6) = 1,$$
$$D^{A_1}(C_2^{(0)}) = -D^{A_2}(C_2^{(0)}) = D^{B_1}(C_2^{(0)}) = -D^{B_2}(C_2^{(0)}) = 1,$$
$$D^{E_j}(C_6) = \begin{pmatrix} \mathrm{e}^{-\mathrm{i}j\pi/3} & 0 \\ 0 & \mathrm{e}^{\mathrm{i}j\pi/3} \end{pmatrix}, \quad D^{E_j}(C_2^{(0)}) = \begin{pmatrix} 0 & 1 \\ 1 & 0 \end{pmatrix}, \quad j = 1, 2,$$

$$\begin{aligned}
\text{For } \mathrm{D}_3: \quad & D^A(C_6^2) = D^A(C_2^{(0)}) = D^B(C_6^2) = -D^B(C_2^{(0)}) = 1, \\
& D^E(C_6^2) = \begin{pmatrix} e^{-i2\pi/3} & 0 \\ 0 & e^{i2\pi/3} \end{pmatrix}, \quad D^E(C_2^{(0)}) = \begin{pmatrix} 0 & 1 \\ 1 & 0 \end{pmatrix}, \\
\text{For } \mathrm{D}_2: \quad & D^{A_1}(C_6^3) = D^{A_2}(C_6^3) = -D^{B_1}(C_6^3) = -D^{B_2}(C_6^3) = 1, \\
& D^{A_1}(C_2^{(1)}) = -D^{A_2}(C_2^{(1)}) = D^{B_1}(C_2^{(1)}) = -D^{B_2}(C_2^{(1)}) = 1.
\end{aligned}$$

There are only two inequivalent one-dimensional representations of the group C_2: the symmetric one A and the antisymmetric one B.

When the wave vector $\boldsymbol{k}^{(1)}$ goes from Γ to the lines ΓK_1 and ΓM, respectively, because

$$\begin{aligned}
& D^{A_1}(C_2^{(2)}) = -D^{A_2}(C_2^{(2)}) = D^{B_1}(C_2^{(2)}) = -D^{B_2}(C_2^{(2)}) = 1, \\
& D^{A_1}(C_2^{(1)}) = -D^{A_2}(C_2^{(1)}) = -D^{B_1}(C_2^{(1)}) = D^{B_2}(C_2^{(1)}) = 1, \\
& \chi^{E_j}(C_2^{(2)}) = \chi^{E_j}(C_2^{(1)}) = 0, \quad j = 1,\ 2,
\end{aligned}$$

we have

$$\begin{aligned}
\text{Line } \Gamma K_1: \quad & A_1, B_1 \longrightarrow A, \quad A_2, B_2 \longrightarrow B, \quad E_1, E_2 \longrightarrow A \oplus B, \\
\text{Line } \Gamma M: \quad & A_1, B_2 \longrightarrow A, \quad A_2, B_1 \longrightarrow B, \quad E_1, E_2 \longrightarrow A \oplus B.
\end{aligned}$$

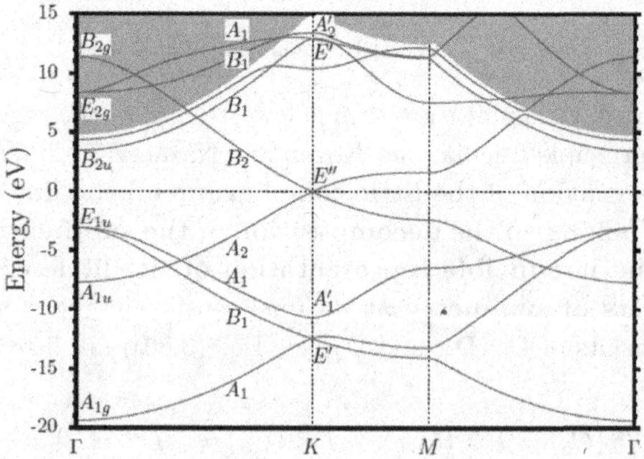

Fig. 6.4 Graphene band structure evaluated with use of the full potential linearized plane wave method. The solid lines are well-converged single graphene layer bands, while the gray background corresponds to continuous spectrum.

When the wave vector $\boldsymbol{k}^{(1)}$ goes from K_1 to the lines $K_1\Gamma$ and K_1M, respectively, because

$$D^A(C_2^{(2)}) = -D^B(C_2^{(2)}) = 1, \quad \chi^E(C_2^{(1)}) = 0,$$
$$D^A(C_2^{(4)}) = -D^B(C_2^{(4)}) = 1, \quad \chi^E(C_2^{(4)}) = 0,$$

we have

$$\text{Line } K_1\Gamma: \quad A \longrightarrow A, \quad B \longrightarrow B, \quad E \longrightarrow A \oplus B,$$
$$\text{Line } K_1M: \quad A \longrightarrow A, \quad B \longrightarrow B, \quad E \longrightarrow A \oplus B.$$

When the wave vector $\boldsymbol{k}^{(1)}$ goes from M to the lines MK_1 and $M\Gamma$, respectively, because

$$D^{A_1}(C_2^{(4)}) = -D^{A_2}(C_2^{(4)}) = -D^{B_1}(C_2^{(4)}) = D^{B_2}(C_2^{(4)}) = 1,$$
$$D^{A_1}(C_2^{(1)}) = -D^{A_2}(C_2^{(1)}) = D^{B_1}(C_2^{(1)}) = -D^{B_2}(C_2^{(1)}) = 1,$$

we have

$$\text{Line } MK_1: \quad A_1, B_2 \longrightarrow A, \quad A_2, B_1 \longrightarrow B,$$
$$\text{Line } M\Gamma: \quad A_1, B_1 \longrightarrow A, \quad A_2, B_2 \longrightarrow B.$$

Thus, the representations at different columns in Fig. 6.4 are given in Table 6.8, where the nomenclature for representations in Fig. 6.4 is changed in our notations:

$$A_{1g}, A_{1u} \rightarrow A_1, \quad B_{2u}, B_{2g} \rightarrow B_2, \quad E_{jg} \rightarrow E_j, \quad j = 1,\ 2, \qquad \text{Column } \Gamma,$$
$$A_1, A_2 \rightarrow A, \quad B_1, B_2 \rightarrow B, \qquad \text{Column } \Gamma K_1,$$
$$A_1' \rightarrow A, \quad A_2' \rightarrow B, \quad E', E'' \rightarrow E_1, \qquad \text{Column } K_1.$$

Table 6.8 The representations at columns in Fig.6.4

Column	Representations arranged upwards from bottom								
ΓM	A	A	B	A	A	A	B	A	A
Γ	A_1	A_1	E_1		B_2	B_2	E_2		B_2
ΓK_1	A	A	B	A	B	B	B	A	B
$K_1\Gamma$	A	B	A	A	B	B	B	A	B
K_1	E		A	E		B	E		B
K_1M	A	B	A	A	B	B	A	B	B
MK_1	A	B	A	A	B	A	B	B	B
M	A_1	A_2	A_1	A_1	B_1	B_2	B_1	B_1	B_1
$M\Gamma$	A	B	A	A	A	B	A	A	A

6.5.7 *Energy Band in a Crystal*

In this subsection we will discuss the dependence of the energy of an electron in a crystal on the wave vector $\boldsymbol{k}$ by the approximation of free electrons. An electron in a crystal is moving in a potential $V(\boldsymbol{r})$ with the symmetry of the space group $\mathcal{S}$. $V(\boldsymbol{r})$ varies near its average field V_0, and the difference $V_1(\boldsymbol{r}) = V(\boldsymbol{r}) - V_0$ plays the role of perturbation. The Hamiltonian of the electron is

$$H(\boldsymbol{r}) = H_0(\boldsymbol{r}) + V_1(\boldsymbol{r}), \qquad H_0(\boldsymbol{r}) = -\frac{\hbar^2}{2m_e}\nabla^2 + V_0, \tag{6.118}$$

where m_e is the mass of the electron. If the energy level is normal degeneracy, the eigenfunction of the energy belongs to an irreducible representation of the space group $\mathcal{S}$, namely, $\psi^{(b)}_{\boldsymbol{k}^{(\nu)}}(\boldsymbol{r})$ given in Eq. (6.102) has the same energy, independent of ν and b. For simplicity we denote $\boldsymbol{k}^{(\nu)}$ by $\boldsymbol{k}$ and omit the superscript and the subscript in the wave function,

$$\psi(\boldsymbol{r}) = \exp\left(\mathrm{i}2\pi\boldsymbol{k}\cdot\boldsymbol{r}\right)u(\boldsymbol{r}), \qquad u(\boldsymbol{r}-\boldsymbol{\ell}) = u(\boldsymbol{r}). \tag{6.119}$$

u can be made a Fourier expansion with respect to the vector of reciprocal crystal lattice $\boldsymbol{K}_n$

$$\begin{aligned} u(\boldsymbol{r}) &= \sum_n u_n \exp\left(-\mathrm{i}2\pi\boldsymbol{K}_n\cdot\boldsymbol{r}\right), \\ \psi(\boldsymbol{r}) &= \sum_n u_n \exp\left\{\mathrm{i}2\pi\left(\boldsymbol{k}-\boldsymbol{K}_n\right)\cdot\boldsymbol{r}\right\}, \end{aligned} \tag{6.120}$$

where u_n is a constant coefficient to be determined. Each term in the expansion (6.120) is an eigenfunction of H_0 with the eigenvalue E_n

$$E_n = \frac{h^2}{2m_e}\left(\boldsymbol{k}-\boldsymbol{K}_n\right)^2 + V_0. \tag{6.121}$$

E_n varies continuously as the wave vector $\boldsymbol{k}-\boldsymbol{K}_n$ changes.

Calculate the energy correction to the ground state with $n=0$. Since the first approximation vanishes, the energy correction comes from the second approximation of the energy. When $\boldsymbol{k}$ is near the center of the first Brillouin zone, E_0 depends on $\boldsymbol{k}^2$ continuously. $E_n - E_0$ is large and u_n with $n \neq 0$ is much smaller than u_0. When $\boldsymbol{k}$ goes to the boundary of the first Brillouin zone, $\left(\boldsymbol{k}-\boldsymbol{K}_n\right)^2$ may be closed to $\boldsymbol{k}^2$, and the energy interference occurs. The result is that one energy level enhances and the other depresses, so that an energy gap appears.

The condition for the boundary of the Brillouin zones is $\boldsymbol{k}^2 = (\boldsymbol{k} - \boldsymbol{K})^2$. The solution is $2\boldsymbol{k} \cdot \boldsymbol{K} = \boldsymbol{K}^2$ (see Fig. 6.5). It is the equation for the perpendicular bisector plane of the line from the origin to a point of reciprocal crystal lattice, called the Bragg plane in solid-state physics. The Brillouin zones are divided by the Bragg planes.

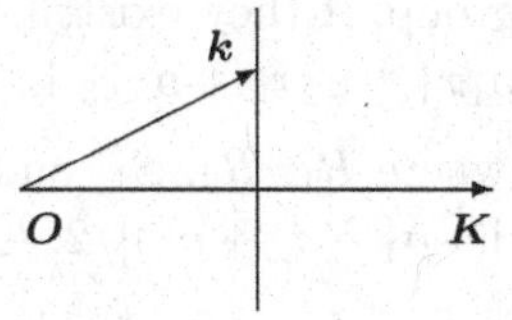

Fig. 6.5. The Bragg plane.

6.6 Exercises

1. In the rectangular coordinate frame, write the double-vector forms and the matrix forms of the proper and improper 6-fold rotations about the Z-axis.
2. Express the directions of the proper rotational axes in the group $\mathbf{O}_h$ by the basis vectors $\boldsymbol{e}_j$ of the rectangular coordinate frame. Express the double-vector forms of the generators of the proper rotational axes in $\mathbf{O}_h$ by $\boldsymbol{e}_j$.
3. Prove Eq. (6.16) for the double-vector form of a rotation $R(\hat{\boldsymbol{n}}, \omega)$: about the direction $\hat{\boldsymbol{n}}$ through the angle ω.
4. Let R denote a rotation about the direction $\hat{\boldsymbol{n}} = (\boldsymbol{e}_1 + \boldsymbol{e}_2)/\sqrt{2}$ through $2\pi/3$. Find the symmetry straight line for $g(R, \mathbf{t})$ where (a) $\mathbf{t} = \boldsymbol{e}_3$, (b) $\mathbf{t} = \boldsymbol{e}_1 + \boldsymbol{e}_3$. If the symmetry straight line is a screw line, find its gliding vector. Simplify $g(R, \mathbf{t})$ by moving the origin to the symmetry straight line for checking your result.
5. Let R denote an inversion with respect to the XY plane. Find the symmetry plane of $g(R, \mathbf{t})$ where (a) $\mathbf{t} = \boldsymbol{e}_3$, (b) $\mathbf{t} = \boldsymbol{e}_1 + \boldsymbol{e}_3$. If the symmetry plane is a gliding plane, find its gliding vector. Simplify $g(R, \mathbf{t})$ by moving the origin to the symmetry plane for checking your result.

6. Analyze the symmetric properties of the crystals with the following space groups: No. 52 [D_{2h}^6, $P \pm 2(\frac{1}{2}\frac{1}{2}0)2'(\frac{1}{2}\frac{1}{2}\frac{1}{2})$], No. 161 [$C_{3v}^6$, $R3\overline{2}'(\frac{1}{2}\frac{1}{2}\frac{1}{2})$], and No. 199 [$T^5$, $I3'2(\frac{1}{2}0\frac{1}{2})2'(\frac{1}{2}\frac{1}{2}0)$], respectively. Point (a) the general form of the symmetric operation; (b) the relations of directions and lengths among three lattice bases; (c) the symmetry straight line, symmetry plane, and the gliding vector of the generator in each cyclic subgroup of the space group, if they exist; (d) the equivalent point to an arbitrary point $\mathbf{r} = \mathbf{a}_1 x_1 + \mathbf{a}_2 x_2 + \mathbf{a}_3 x_3$ in the crystal cell.
7. Calculate Eq. (6.111), where R_1, R_3, S_1, and T_z^2 are replaced by $[\mathbf{1} - \mathrm{i}(\sigma_1 + \sigma_2 + \sigma_3)]/2$, $[\mathbf{1} - \mathrm{i}(-\sigma_1 - \sigma_2 + \sigma_3)]/2$, $-\mathrm{i}(\sigma_1 + \sigma_2)/\sqrt{2}$, and $-\mathrm{i}\sigma_3$, respectively.
8. Prove by Eq. (6.116) that the space group $\mathbf{T}^2$ is an invariant subgroup of $\mathbf{O}_h^8$ of index four with the quotient group $\mathbf{O}_h^8/\mathbf{T}^2 \approx \mathrm{V}_4$.
9. Find the symmorphic space subgroup of the non-symmorphic space group $\mathbf{O}_h^7$ with No. 227, and calculate the irreducible representation where the star of wave vectors is located on the boundary of the Brillouin zone.

Chapter 7

LIE GROUPS AND LIE ALGEBRAS

In Chap. 5 the fundamental concepts on Lie groups have been introduced through the SO(3) group and its covering group SU(2). In this chapter we will study the general theory on Lie groups and Lie algebras, such as the property of simple and semisimple Lie algebras, the regular commutative relations of the Cartan–Weyl bases of a semisimple Lie algebra, and the classification of the simple Lie algebras.

7.1 Lie Algebras and its Structure Constants

7.1.1 *Review on the Concepts of Lie Groups*

We will begin with a review on the concepts of Lie groups. An element R in a Lie group G can be characterized by g-independent real parameters r_A, $1 \leqslant A \leqslant g$, varying continuously in a g-dimensional region. g is called **the order** of the Lie group G and the region is called the **group space** of G. It is required that there is a **one-to-one correspondence between the set of parameters r_A and the group element R at least inside the region where the measure is not vanishing.** The parameters e_A of the identity E are usually taken to be zero for convenience. The parameters t_A of the product element are the real analytical functions of the parameters of the factor elements

$$T = RS, \qquad t_A = f_A(r_1 \cdots, r_g; s_1 \cdots, s_g) = f_A(r; s). \tag{7.1}$$

$f_A(r; s)$ are called the **composition functions**, characterizing the multiplication rule of elements in G completely. The composition functions have to satisfy some conditions to meet four axioms of a group.

The topological property of the group space describes the

global property of the Lie group G. If the group space of G falls into several disjoint pieces, G is called a **mixed Lie group**. G contains an invariant Lie subgroup H whose group space is a connected piece, to which the identity E belongs. H is called a **connected Lie group**. The set of elements belonging to the other connected piece is the coset of H in G. The property of the mixed Lie group G is characterized completely by H and the representative elements, respectively belonging to each coset. Hereafter, we will mainly study the connected Lie groups.

A connected Lie group G is called multiply connected with the **degree of continuity** n if the connected curves of any two points in the group space are separated into n sets where any two curves in each set can be changed continuously from one to another in the group space, but two curves in different sets cannot. The Lie group with the degree of continuity $n = 1$ is called simply-connected. For a Lie group G with the degree of continuity n, there exists a **covering group** G$'$ which is simply-connected and is homomorphic onto G with a n-to-one correspondence. A faithful representation of G$'$ is an n-valued representation of G.

A Lie group is **compact** if its group space is compact. If the group space is an Euclidean space, a closed finite region (including the boundary) is compact, and an open finite region (without the boundary) or an infinite region is not compact. **The group integral can be defined only for a compact Lie group so that the most properties of a finite group can be generalized to a compact Lie group**, such as Theorems 3.1, 3.3, 3.5, and 3.6. The representations of a noncompact Lie group G can be obtained through those of a compact Lie group which is "close" to G (see subsection 7.1.2).

A Lie group is called **simple** if it does not contain any nontrivial invariant Lie subgroup. A Lie group is called **semisimple** if it does not contain any abelian invariant Lie subgroup, including the whole Lie group. **A Lie group of order 1 is abelian so that it is simple but not semisimple.**

The infinitesimal elements describe the local property of a Lie group. An infinitesimal element R which is located at the neighborhood of the identity E in the group space is characterized by infinitesimal parameters r_A. In a representation D(G) of a Lie group G, the representation matrix of the identity E is the unit matrix, $D(E) = \mathbf{1}$, and that of an infinitesimal element $R(\alpha)$ can be expanded as a Taylor series. Up to the first order one has

$$D(R) = \mathbf{1} - \mathrm{i}\sum_A r_A I_A + \cdots, \qquad I_A = \mathrm{i}\left.\frac{\partial D(R)}{\partial r_A}\right|_{R=E}. \tag{7.2}$$

I_A are called the generators in a representation $D(\mathrm{G})$. **If the representation is faithful, the generators are linearly independent of each other.** The generators characterize the property of the infinitesimal elements in the representation of a Lie group G, so that, as shown by the Lie Theorems, they can characterize the local property of the Lie group.

The First Lie Theorem says that the representation of a Lie group G with a connected group space is completely determined by its generators. $D(R)$ satisfies a differential equation with a given boundary condition:

$$\frac{\partial D(R)}{\partial r_A} = -\mathrm{i}\sum_B I_B S_{BA}(r) D(R), \qquad D(R)|_{R=E} = \mathbf{1}, \tag{7.3}$$

where the real functions

$$S_{BA}(r) = \left.\frac{\partial f_B(r;s)}{\partial r_A}\right|_{S=R^{-1}} \tag{7.4}$$

are independent of the given representation. They depend upon the choice of the parameters of the Lie group.

The Second Lie Theorem says that the generators in any representation of a Lie group G satisfy the common commutative relations:

$$[I_A,\ I_B] = \mathrm{i}\sum_{D=1}^{g} C_{AB}{}^D I_D, \tag{7.5}$$

where the real numbers $C_{AB}{}^D$ are called the **structure constants** of G. Conversely, if g matrices satisfy the commutative relations (7.5), they are the generators in a representation of G with order g.

The structure constants are calculated from the generators in a known faithful representation. Although the structure constants are independent of the representation, they depend on the choice of the group parameters. **If the parameters are re-chosen, both the generators and the structure constants are changed,**

$$\begin{gathered} r'_A = \sum_B r_B \left(X^{-1}\right)_{BA}, \qquad I'_A = \sum_B X_{AB} I_B, \qquad \det X \neq 0, \\ C'_{AB}{}^D = \sum_{PQR} X_{AP} X_{BQ} C_{PQ}{}^R \left(X^{-1}\right)_{RD}. \end{gathered} \tag{7.6}$$

In fact,

$$[I'_A, I'_B] = \sum_{PQ} X_{AP} X_{BQ} [I_P, I_Q] = i \sum_{PQR} X_{AP} X_{BQ} C_{PQ}{}^R I_R$$
$$= i \sum_D \left\{ \sum_{PQR} X_{AP} X_{BQ} C_{PQ}{}^R \left(X^{-1}\right)_{RD} \right\} I'_D.$$

The Third Lie Theorem says that a set of constants $C_{AB}{}^D$ can be the structure constants of a Lie group if and only if they satisfy

$$C_{AB}{}^D = -C_{BA}{}^D,$$
$$\sum_P \left\{ C_{AB}{}^P C_{PD}{}^Q + C_{BD}{}^P C_{PA}{}^Q + C_{DA}{}^P C_{PB}{}^Q \right\} = 0. \qquad (7.7)$$

Based on this theorem the Lie groups can be classified by their structure constants. We will study the classification later in this chapter.

The **adjoint representation** is very important for a Lie group. If I_A are the generators in the representation $D(G)$ of a Lie group G, the adjoint representation $D^{ad}(G)$ satisfies

$$D(R) I_B D(R)^{-1} = \sum_D I_D D^{ad}_{DB}(R). \qquad (7.8)$$

When R is an infinitesimal element, one has

$$[I_A, I_B] = \sum_D I_D \left(I^{ad}_A\right)_{DB}, \qquad \left(I^{ad}_A\right)_{DB} = iC_{AB}{}^D. \qquad (7.9)$$

The structure constants characterize some important properties of a Lie group G. A Lie group is abelian if and only if its structure constants all vanish. A Lie group is simple if and only if its adjoint representation is irreducible. A Lie group is semisimple if and only if its adjoint representation is completely reducible and does not contain the identical representation.

We are going to show that if a Lie group G is compact, its parameters can be chosen such that **each structure constant of G is totally anti-symmetric** with respect to its three indices. In fact, the adjoint representation of a Lie group is real for any set of real group parameters owing to its definition (5.125). For a compact Lie group G, a real representation is equivalent to a real orthogonal one. Let Y^T denote the real similarity transformation changing the adjoint representation to be real orthogonal so that its generators $\overline{I}_A$ becomes pure imaginary and antisymmetric:

$$\begin{aligned}\left(\overline{I}_A\right)_{BD} = -\left(\overline{I}_A\right)_{DB} &= \left\{\left(Y^T\right)^{-1} I_A^{\text{ad}} Y^T\right\}_{BD} \\ &= \sum_{PQ} \left(Y^{-1}\right)_{PB} \left(I_A^{\text{ad}}\right)_{PQ} Y_{DQ}.\end{aligned} \tag{7.10}$$

Now, change the parameters in G such that the combination matrix X in Eq. (7.6) is taken to be Y in Eq. (7.10). Thus, each new structure constant of G is totally antisymmetric with respect to its three indices

$$\begin{aligned}C'^{\ D}_{AB} &= \sum_{PQR} Y_{AP} Y_{BQ} C_{PQ}^{\ \ R} \left(Y^{-1}\right)_{RD} \\ &= -\text{i} \sum_P Y_{AP} \left\{\sum_{QR} \left(Y^{-1}\right)_{RD} \left(I_P^{\text{ad}}\right)_{RQ} Y_{BQ}\right\} \\ &= -\text{i} \sum_P Y_{AP} \left(\overline{I}_P\right)_{DB} = -C'^{\ B}_{AD}.\end{aligned}$$

7.1.2 *The Lie Algebra*

Let I_A denote the generators in a faithful representation of a Lie group G. I_A satisfy the commutative relations (7.5), or equivalently,

$$\left[(-\text{i}I_A),\ (-\text{i}I_B)\right] = \sum_D C_{AB}^{\ \ D} (-\text{i}I_D). \tag{7.11}$$

For the given structure constants $C_{AB}^{\ \ D}$, the map (7.11) of two generators $(-\text{i}I_A)$ and $(-\text{i}I_B)$ onto a combination of generators $(-\text{i}I_D)$ is called the **Lie product**. The real linear space spanned by the basis vectors $(-\text{i}I_A)$ is closed for the Lie product of vectors in the space

$$X = \sum_A (-\text{i}I_A)\, x_A, \qquad Y = \sum_B (-\text{i}I_B)\, y_B,$$

$$[X,\ Y] = \sum_{AB} \left[(-\text{i}I_A),\ (-\text{i}I_B)\right] x_A y_B = \sum_D (-\text{i}I_D) \left\{\sum_{AB} x_A y_B C_{AB}^{\ \ D}\right\}.$$

This real space is called the **real Lie algebra** of a Lie group G, denoted by $\mathcal{L}_R$. Generalizing the real space $\mathcal{L}_R$ to a complex space $\mathcal{L}$ under the same product rule of vectors, one obtains a complex Lie algebra, briefly called a **Lie algebra**. Evidently, the Lie algebra is **not an associate algebra** because the Lie product satisfies the Jacobi identity

$$\left[I_A,\ \left[I_B,\ I_C\right]\right] - \left[\left[I_A,\ I_B\right],\ I_C\right] = \left[I_B,\ \left[I_A,\ I_C\right]\right]. \tag{7.12}$$

${C_{AB}}^D$ are the common structure constants of the Lie group G and the corresponding Lie algebra $\mathcal{L}$. A representation of a Lie group is also a representation (or called "module" in mathematics) of the corresponding Lie algebra. **The Lie algebra $\mathcal{L}$ is the complexification of the real Lie algebra $\mathcal{L}_R$,** and **$\mathcal{L}_R$ is a real form of $\mathcal{L}$.** In this textbook we only discuss those Lie algebras $\mathcal{L}$ which have their Lie groups G and their real forms $\mathcal{L}_R$.

The real Lie algebra is important for manifesting the compactness of a Lie group. For example, the SO(4) group is a real orthogonal group in the four-dimensional Euclidean space, which is compact. If the fourth coordinate is changed to be imaginary, $x_4 = \mathrm{i}ct$, the orthogonal group becomes the proper Lorentz group L_p, which is noncompact. Both SO(4) and L_p have the same Lie algebra, but different real Lie algebras. If the parameters of a group are allowed to be imaginary, the generators in the corresponding representations of SO(4) and L_p are the same but some parameters are different by a factor i. **A representation of L_p can be obtained from a representation of** SO(4) **by changing some real parameters of SO(4) to be imaginary**. This is the standard method to find the irreducible representation of a noncompact Lie group from that of a compact Lie group. We will discuss it in Chap. 11 in detail. The real Lie algebra of a compact Lie group is called the **compact real Lie algebra**.

Two Lie algebras are isomorphic if their structure constants are the same, but two Lie groups with the same structure constants are only **locally isomorphic** generally. There are two typical examples where two locally isomorphic Lie groups are not isomorphic. The SU(2) group is homomorphic onto the SO(3) group, but they have the same structure constants. The U(2) group contains two subgroups SU(2) and U(1), but does not contain a subgroup SU(2)$\otimes$U(1) because two subgroups SU(2) and U(1) have a common element $-\mathbf{1}$ other than the identity $\mathbf{1}$. Therefore, the U(2) group is not isomorphic onto the group SU(2)$\otimes$ U(1), although they are locally isomorphic onto each other.

The commonly used concepts for an algebra can also be used for a Lie algebra and for a real Lie algebra, such as subalgebras, ideals, abelian ideals, the direct sum, the semi-direct sum, the isomorphism, the homomorphism, and so on. For a Lie algebra, one also can introduce some concepts relating to a Lie group, such as a simple Lie algebra, a semisimple Lie algebra, representations, and so on.

A non-null subset $\mathcal{L}_1$ of a Lie algebra $\mathcal{L}$, $\mathcal{L}_1 \subset \mathcal{L}$, is a **Lie subalgebra** of

$\mathcal{L}$ if it is closed with respect to the sum and the Lie product of the vectors in the subset

$$c_1X + c_2Y \in \mathcal{L}_1 \qquad [X,\ Y] \in \mathcal{L}_1, \tag{7.13}$$

where $X \in \mathcal{L}_1$, $Y \in \mathcal{L}_1$, and c_1 and c_2 are two arbitrary complex numbers. Furthermore, a Lie subalgebra $\mathcal{L}_1$ of a Lie algebra $\mathcal{L}$ is **an ideal** of $\mathcal{L}$ if

$$[X,\ Y] \in \mathcal{L}_1, \qquad \forall\ X \in \mathcal{L}_1 \text{ and } Y \in \mathcal{L}. \tag{7.14}$$

There is no difference between a left-ideal and a right-ideal of a Lie algebra $\mathcal{L}$. A Lie algebra is called **simple** if it does not contain any ideal except for the whole algebra. A Lie algebra is called **semisimple** if it does not contain any abelian ideal. **A one-dimensional Lie algebra is abelian, so it is simple, but not semisimple.** A simple Lie algebra with dimension higher than 1 must be semisimple. The Lie algebra of a simple or a semisimple Lie group is simple or semisimple, respectively.

A Lie algebra $\mathcal{L}$ is called the direct sum of two Lie subalgebras, $\mathcal{L} = \mathcal{L}_1 \oplus \mathcal{L}_2$, if

$$\mathcal{L}_1 + \mathcal{L}_2 = \mathcal{L}, \qquad \mathcal{L}_1 \bigcap \mathcal{L}_2 = \emptyset, \qquad [\mathcal{L}_1,\ \mathcal{L}_2] = 0. \tag{7.15}$$

Evidently, both $\mathcal{L}_1$ and $\mathcal{L}_2$ are ideals of the direct sum $\mathcal{L}$. A Lie algebra $\mathcal{L}$ is called the semidirect sum of two Lie subalgebras, $\mathcal{L} = \mathcal{L}_1 \oplus_s \mathcal{L}_2$, if

$$\mathcal{L}_1 + \mathcal{L}_2 = \mathcal{L}, \qquad \mathcal{L}_1 \bigcap \mathcal{L}_2 = \emptyset, \qquad [\mathcal{L}_1,\ \mathcal{L}_2] \subset \mathcal{L}_1. \tag{7.16}$$

Now, $\mathcal{L}_1$ is an ideal of $\mathcal{L}$, but $\mathcal{L}_2$ is not.

If a Lie algebra $\mathcal{L}$ is homomorphic onto a Lie algebra $\mathcal{L}'$, $\mathcal{L}$ can be decomposed into a semidirect sum $\mathcal{L} = \mathcal{L}_1 \oplus_s \mathcal{L}_2$ such that $\mathcal{L}_2$ is isomorphic onto $\mathcal{L}'$. $\mathcal{L}_1$ is an ideal of $\mathcal{L}$ and maps to null vector in $\mathcal{L}'$. $\mathcal{L}_1$ is called the kernel of the homomorphism between $\mathcal{L}$ and $\mathcal{L}'$.

Define a series of Lie subalgebras $\mathcal{L}^{(n)}$ from a Lie algebra $\mathcal{L}$

$$\mathcal{L}^{(1)} = [\mathcal{L},\ \mathcal{L}], \qquad \mathcal{L}^{(2)} = \left[\mathcal{L}^{(1)},\ \mathcal{L}^{(1)}\right], \qquad \cdots . \tag{7.17}$$

A Lie algebra $\mathcal{L}$ is called solvable if there exists an integer m in the series of subalgebras $\mathcal{L}^{(n)}$ such that $\mathcal{L}^{(m)} = \emptyset$. It is proved that any Lie algebra can be decomposed into a semidirect sum of a solvable Lie algebra $\mathcal{L}_1$ and a semisimple Lie algebra $\mathcal{L}_2$, $\mathcal{L} = \mathcal{L}_1 \oplus_s \mathcal{L}_2$. Any irreducible representation with a finite dimension of a solvable Lie algebra is one-dimensional.

7.1.3 *The Killing Form and the Cartan Criteria*

Based on the Third Lie Theorem, the Lie groups as well as the Lie algebras can be classified by their structure constants. However, the structure constants depend on the choice of the group parameters. We hope to find the combination of the structure constants which reflects the essence of a Lie algebra.

The Killing form g_{AB} of a Lie algebra $\mathcal{L}$ is defined as a symmetric covariant tensor with respect to the parameter transformation (7.6) of $\mathcal{L}$:

$$\begin{aligned} g_{AB} &= \sum_{PQ} C_{AP}{}^{Q} C_{BQ}{}^{P} = -\text{Tr}\left(I_A^{\text{ad}} I_B^{\text{ad}}\right) = g_{BA}, \\ g'_{AB} &= \sum_{PQ} X_{AP} X_{BQ} g_{PQ}, \qquad g' = X g X^T. \end{aligned} \tag{7.18}$$

With respect to the transformation (7.6), the generators I_A is a covariant vector, the structure constant $C_{AB}{}^{D}$ is a mixed tensor of rank $(2,1)$, and the following covariant structure constant C_{ABD} is a totally antisymmetric tensor of rank three where g_{AB} is used as the metric tensor:

$$\begin{aligned} C_{ABD} &= \sum_{P} C_{AB}{}^{P} g_{PD} = \sum_{PQR} C_{AB}{}^{P} C_{PQ}{}^{R} C_{DR}{}^{Q} \\ &= -\sum_{PQR} \left\{C_{BQ}{}^{P} C_{PA}{}^{R} + C_{QA}{}^{P} C_{PB}{}^{R}\right\} C_{DR}{}^{Q} \\ &= \sum_{PQR} \left\{C_{BQ}{}^{P} C_{AP}{}^{R} C_{DR}{}^{Q} - C_{AQ}{}^{P} C_{BP}{}^{R} C_{DR}{}^{Q}\right\} \\ &= -C_{BAD} = -C_{ADB}, \\ C'_{ABD} &= \sum_{PQR} X_{AP} X_{BQ} X_{DR} C_{PQR}. \end{aligned} \tag{7.19}$$

Theorem 7.1 (The Cartan Criteria) A Lie algebra is semisimple if and only if its Killing form is nonsingular,

$$\det\ g \neq 0, \tag{7.20}$$

and a real semisimple Lie algebra is compact if and only if its Killing form is negative definite.

We will not prove this theorem here, but only give some explanation. The Killing form is a real symmetric matrix and can be diagonalized through a real orthogonal similarity transformation (see Eqs. (7.6) and (7.18)). Then, the diagonal elements of the Killing form can be changed to be ± 1 or 0 through a real scale transformation (7.18). For a Lie algebra, the sign of the eigenvalue of the Killing form g_{AB} can be changed in the parameter transformation, but the number of the zero eigenvalues of g_{AB} cannot be changed. For a real Lie algebra, the sign of the eigenvalue of g_{AB} cannot be changed in the parameter transformation.

Since the Killing form g_{AB} of a semisimple Lie algebra is nonsingular, g_{AB} has its inverse matrix g^{AB}

$$g^{AB} = g^{BA}, \qquad \sum_{D} g^{AD} g_{DB} = \delta^{A}_{B}. \tag{7.21}$$

g_{AB} and g^{AB} can be used as the metric tensors to raise the subscript of a tensor or to lower its superscript (see Eq. (7.19)). Define an operator C_n which is a homogeneous polynomial of order n with respect to the generators of the Lie algebra $\mathcal{L}$:

$$\begin{aligned} C_n = \sum_{(D)} \sum_{(A)} \sum_{(B)} & C_{A_1 D_1}^{\ \ D_2} C_{A_2 D_2}^{\ \ D_3} \cdots C_{A_n D_n}^{\ \ D_1} \\ & \times\ g^{A_1 B_1} \cdots g^{A_n B_n} I_{B_1} I_{B_2} \cdots I_{B_n}, \end{aligned} \tag{7.22}$$

which commutes with any generators I_A in $\mathcal{L}$,

$$[C_n,\ I_A] = 0, \qquad \forall\ I_A \in \mathcal{L}. \tag{7.23}$$

The readers are encouraged to prove Eq. (7.23) (see Prob. 1). C_n is called the Casimir operator of order n. The Casimir operator of order 2 is discussed in Chap. 5 (see Eq. (5.71)):

$$C_2 = \sum_{AB} g^{AB} I_A I_B. \tag{7.24}$$

7.2 The Regular Form of a Semisimple Lie Algebra

7.2.1 *The Inner Product in a Semisimple Lie Algebra*

Express the basis vector I_A in a semisimple Lie algebra $\mathcal{L}$ by the Dirac symbol which is commonly used in quantum mechanics [Georgi (1982)]

$$|I_A\rangle \equiv |A\rangle, \qquad 1 \leqslant A \leqslant g. \tag{7.25}$$

The inner product of two vectors in $\mathcal{L}$ is defined with its Killing form,

$$\langle A|B\rangle = -g_{AB} = \mathrm{Tr}\left(I_A^{\rm ad} I_B^{\rm ad}\right), \tag{7.26}$$

which is bilinear with respect to two vectors $\langle A|$ and $|B\rangle$

$$\begin{aligned}
&\langle A|(c_1B+c_2D)\rangle = c_1\langle A|B\rangle + c_2\langle A|D\rangle,\\
&\langle (c_1A+c_2D)|B\rangle = c_1\langle A|B\rangle + c_2\langle D|B\rangle,\\
&\langle A|B\rangle = \langle B|A\rangle,\\
&\langle A|D|B\rangle \equiv \langle A|\,[D,\ B]\rangle = \langle [A,\ D]\,|B\rangle.
\end{aligned} \tag{7.27}$$

When the group parameters are changed, the inner product will be related to the new Killing form in the same way,

$$\begin{aligned}
&|X_\mu\rangle = \left|\sum_A X_{\mu A} I_A\right\rangle = \sum_A X_{\mu A}|A\rangle,\\
&\langle X_\mu|X_\nu\rangle = -\sum_{AB} X_{\mu A} X_{\nu B} g_{AB} = -g_{\mu\nu}.
\end{aligned} \tag{7.28}$$

7.2.2 *The Cartan Subalgebra*

In a semisimple Lie algebra $\mathcal{L}$, any vector is also a linear operator which transforms a vector to another by the Lie product

$$X|Y\rangle = |\,[X,\ Y]\rangle. \tag{7.29}$$

One is able to calculate the eigenvalue and the eigenvector of a vector X in $\mathcal{L}$. Evidently, any operator is its own eigenvector with zero eigenvalue:

$$X|X\rangle = |\,[X,\ X]\rangle = 0.$$

Let ℓ_X denote the multiplicity of zero eigenvalue of X. Among all vectors in $\mathcal{L}$ the minimal ℓ_X is denoted by ℓ

$$\ell = \min \ell_X > 0. \tag{7.30}$$

ℓ is called the **rank** of the Lie algebra $\mathcal{L}$ as well as the rank of the Lie group G. A vector X in $\mathcal{L}$ is called **regular** if $\ell_X = \ell$.

Theorem 7.2 There are ℓ linearly independent eigenvectors H_j with zero eigenvalue for a regular vector X in a semisimple Lie algebra $\mathcal{L}$ with rank ℓ. H_j are commutable with each other

$$[H_j,\ H_k] = 0, \qquad 1 \leqslant j \leqslant \ell, \qquad 1 \leqslant k \leqslant \ell, \tag{7.31}$$

and X is a linear combination of H_j. In the remaining subspace of $\mathcal{L}$, the $(g-\ell)$ common eigenvectors $E_{\boldsymbol{\alpha}}$ of ℓ operators H_j are non-degenerate,

$$H_j|E_{\boldsymbol{\alpha}}\rangle = \alpha_j|E_{\boldsymbol{\alpha}}\rangle, \qquad [H_j,\ E_{\boldsymbol{\alpha}}] = \alpha_j E_{\boldsymbol{\alpha}}, \qquad \boldsymbol{\alpha} \neq 0. \tag{7.32}$$

The first sentence comes from the definition of a regular vector. We will not prove this theorem here. The abelian subalgebra $\mathcal{H}$ spanned by ℓ generators H_j is called the **Cartan subalgebra** of $\mathcal{L}$. $\mathcal{H}$ is not an ideal of $\mathcal{L}$. The set of H_j is the largest set of the mutually commutable generators in $\mathcal{L}$. The ℓ-dimensional vector $\boldsymbol{\alpha}$ is called a **root**, and the ℓ-dimensional space is called the **root space**. There is a one-to-one correspondence between the root $\boldsymbol{\alpha}$ and the generator $E_{\boldsymbol{\alpha}}$. Sometimes, H_j is also called the generator corresponding to the zero root, which is degeneracy with the multiplicity ℓ.

The choice of the Cartan subalgebra is not unique. In fact, for any group element $R \in G$, the set of $H'_j = RH_jR^{-1}$ also spans a Cartan subalgebra, conjugate to the original one.

7.2.3 *Regular Commutative Relations of Generators*

The basis vectors H_j and $E_{\boldsymbol{\alpha}}$ in a semisimple Lie algebra $\mathcal{L}$ are called the **Cartan–Weyl bases**, or the regular bases. The regular bases satisfy the commutative relations (7.31) and (7.32). We are going to study their remaining commutative relations.

From Eq. (7.27) one has

$$\langle E_{\boldsymbol{\beta}}|H_j|E_{\boldsymbol{\alpha}}\rangle = \alpha_j\langle E_{\boldsymbol{\beta}}|E_{\boldsymbol{\alpha}}\rangle = -\beta_j\langle E_{\boldsymbol{\beta}}|E_{\boldsymbol{\alpha}}\rangle ,$$

namely, $(\alpha_j + \beta_j)\langle E_{\boldsymbol{\beta}}|E_{\boldsymbol{\alpha}}\rangle = 0$,

$$\langle E_{\boldsymbol{\beta}}|E_{\boldsymbol{\alpha}}\rangle = 0 \qquad \text{if} \quad \boldsymbol{\alpha} \neq -\boldsymbol{\beta}. \tag{7.33}$$

For the same reason,

$$\langle H_j|E_{\boldsymbol{\alpha}}\rangle = -g_{j\boldsymbol{\alpha}} = 0. \tag{7.34}$$

For a given root $\boldsymbol{\alpha}$, if $-\boldsymbol{\alpha}$ is not a root, $g_{\boldsymbol{\alpha}B}$ all vanish, which conflicts to that $\mathcal{L}$ is semisimple. Thus, in a semisimple Lie algebra, **the roots $\pm\boldsymbol{\alpha}$ have to appear in pair**. Choose the factor $b_{\boldsymbol{\alpha}}$ in $E_{\boldsymbol{\alpha}}$ such that

$$\langle E_{-\boldsymbol{\alpha}}|E_{\boldsymbol{\alpha}}\rangle = -g_{(-\boldsymbol{\alpha})\boldsymbol{\alpha}} = 1. \tag{7.35}$$

The factor $b_{\boldsymbol{\alpha}}$ in $E_{\boldsymbol{\alpha}}$ can still be chosen in the condition $b_{\boldsymbol{\alpha}}b_{-\boldsymbol{\alpha}} = 1$.

The part g_{jk} of the Killing form related to the Cartan subalgebra $\mathcal{H}$ can be expressed by the components of roots:

$$\begin{gathered}\langle H_j|H_k\rangle = -g_{jk}, \\ g_{jk} = \sum_{AB} C_{jA}{}^{B} C_{kB}{}^{A} = \sum_{\boldsymbol{\alpha}\in\Delta} C_{j\boldsymbol{\alpha}}{}^{\boldsymbol{\alpha}} C_{k\boldsymbol{\alpha}}{}^{\boldsymbol{\alpha}} = -\sum_{\boldsymbol{\alpha}\in\Delta} \alpha_j\alpha_k,\end{gathered} \tag{7.36}$$

where Δ is the set of all roots in $\mathcal{L}$. Due to Eq. (7.34) g_{jk} is also nonsingular, det $(g_{jk}) \neq 0$. Define $-g_{jk}$ and its inverse $-g^{jk}$ to be the metric tensor in the root space such that they can be used to raise a subscript and to lower a superscript of the vector in the root space. The inner product of two vectors in the root space is defined as

$$\begin{aligned} \boldsymbol{V}\cdot\boldsymbol{U} &= -\sum_{jk} g^{jk}V_jU_k = \sum_k V^kU_k = -\sum_{jk} g_{jk}V^jU^k \\ &= \sum_{jk}\sum_{\alpha\in\Delta} (\alpha_jV^j)(\alpha_kU^k) = \sum_{\alpha\in\Delta} (\boldsymbol{\alpha}\cdot\boldsymbol{V})(\boldsymbol{\alpha}\cdot\boldsymbol{U}). \end{aligned} \tag{7.37}$$

From the Jacobi identity,

$$\begin{aligned} 0 = & \ [H_j,\ [E_{\boldsymbol{\alpha}},\ E_{\boldsymbol{\beta}}]] + [E_{\boldsymbol{\alpha}},\ [E_{\boldsymbol{\beta}},\ H_j]] + [E_{\boldsymbol{\beta}},\ [H_j,\ E_{\boldsymbol{\alpha}}]] \\ = & \ [H_j,\ [E_{\boldsymbol{\alpha}},\ E_{\boldsymbol{\beta}}]] - \beta_j [E_{\boldsymbol{\alpha}},\ E_{\boldsymbol{\beta}}] + \alpha_j [E_{\boldsymbol{\beta}},\ E_{\boldsymbol{\alpha}}], \end{aligned}$$

one has

$$\begin{aligned} [H_j,\ [E_{\boldsymbol{\alpha}},\ E_{\boldsymbol{\beta}}]] &= (\alpha_j+\beta_j)[E_{\boldsymbol{\alpha}},\ E_{\boldsymbol{\beta}}], \\ H_j\,|[E_{\boldsymbol{\alpha}},\ E_{\boldsymbol{\beta}}]\rangle &= (\alpha_j+\beta_j)\,|[E_{\boldsymbol{\alpha}},\ E_{\boldsymbol{\beta}}]\rangle. \end{aligned} \tag{7.38}$$

Since the non-zero root is non-degenerate, one concludes that $[E_{\boldsymbol{\alpha}},\ E_{\boldsymbol{\beta}}]$ is proportional to $E_{\boldsymbol{\alpha}+\boldsymbol{\beta}}$ if $\boldsymbol{\alpha}+\boldsymbol{\beta}$ is a root, and $[E_{\boldsymbol{\alpha}},\ E_{\boldsymbol{\beta}}] = 0$ if $\boldsymbol{\alpha}+\boldsymbol{\beta}$ is nonzero and not a root. When $\boldsymbol{\beta} = -\boldsymbol{\alpha}$, $[E_{\boldsymbol{\alpha}},\ E_{-\boldsymbol{\alpha}}]$ belongs to the Cartan subalgebra and can be expressed as $\sum_j \lambda^j H_j$, where

$$\begin{aligned} -\sum_j g_{kj}\lambda^j &= \sum_j \langle H_k|H_j\rangle\,\lambda^j = \langle H_k|\,[E_{\boldsymbol{\alpha}},\ E_{-\boldsymbol{\alpha}}]\rangle \\ &= \langle [H_k,\ E_{\boldsymbol{\alpha}}]\,|E_{-\boldsymbol{\alpha}}\rangle = \alpha_k\langle E_{\boldsymbol{\alpha}}|E_{-\boldsymbol{\alpha}}\rangle = \alpha_k. \\ \lambda^j &= -\sum_k g^{jk}\alpha_k = \alpha^j. \end{aligned} \tag{7.39}$$

λ^j is nothing but the contravariant component of the root $\boldsymbol{\alpha}$.

In summary, the regular commutative relations among the Cartan–Weyl bases in a semisimple Lie algebra $\mathcal{L}$ are

$$\begin{aligned} &[H_j,\ H_k] = 0, \qquad [H_j,\ E_{\boldsymbol{\alpha}}] = \alpha_j E_{\boldsymbol{\alpha}}, \\ &[E_{\boldsymbol{\alpha}},\ E_{\boldsymbol{\beta}}] = \begin{cases} N_{\boldsymbol{\alpha},\boldsymbol{\beta}}E_{\boldsymbol{\alpha}+\boldsymbol{\beta}}, & \text{when } \boldsymbol{\alpha}+\boldsymbol{\beta} \text{ is a root}, \\ \sum_j \alpha^jH_j = \boldsymbol{\alpha}\cdot\boldsymbol{H} \equiv H_{\boldsymbol{\alpha}}, & \text{when } \boldsymbol{\beta} = -\boldsymbol{\alpha}, \\ 0, & \text{the remaining cases}, \end{cases} \end{aligned} \tag{7.40}$$

where the antisymmetric coefficients $N_{\alpha,\beta} = -N_{\beta,\alpha}$ are to be determined. E_α can be multiplied with a factor b_α satisfying $b_\alpha b_{-\alpha} = 1$, namely, the ratio $b_\alpha/b_{-\alpha}$ can be chosen arbitrarily. H_j as well as the components α_j of a root α can be made an arbitrary nonsingular linear combination.

7.2.4 *The Inner Product of Roots*

Theorem 7.3 The inner product of any two non-zero roots α and β in a semisimple Lie algebra $\mathcal{L}$ satisfies

$$\Gamma(\alpha/\beta) \equiv \frac{2\alpha\cdot\beta}{\beta\cdot\beta} = \text{integer}, \tag{7.41}$$

and $\alpha - \Gamma(\alpha/\beta)\beta$ is a root in $\mathcal{L}$.

Proof We construct a root chain from a root α by adding and subtracting another root β successively

$$\cdots, \ (\alpha - 2\beta), \ (\alpha - \beta), \ \alpha, \ (\alpha + \beta), \ (\alpha + 2\beta), \ \cdots$$

Since the number of roots in $\mathcal{L}$ is finite, the root chain has to break off at two sides after a finite number of terms. Without loss of generality,

$$\begin{aligned} &\alpha + n\beta, && -q \leqslant n \leqslant p, && \text{all are the roots,} \\ &\alpha - (q+1)\beta \ \text{ and } \ \alpha + (p+1)\beta && && \text{are not the roots,} \end{aligned} \tag{7.42}$$

where both p and q are non-negative integers. From Eq. (7.40) we have

$$N_{\alpha,\beta} = -N_{\beta,\alpha}, \qquad N_{(\alpha+p\beta),\beta} = N_{(\alpha-q\beta),-\beta} = 0. \tag{7.43}$$

Letting

$$F_n = -N_{(\alpha+n\beta),\beta} N_{(\alpha+(n+1)\beta),-\beta}, \qquad F_p = F_{-q-1} = 0,$$

we obtain from the Jacobi identity

$$\begin{aligned} 0 &= [E_{\alpha+n\beta}, \ [E_\beta, \ E_{-\beta}]] + [E_\beta, \ [E_{-\beta}, \ E_{\alpha+n\beta}]] \\ &\quad + [E_{-\beta}, \ [E_{\alpha+n\beta}, \ E_\beta]] \\ &= \sum_j \beta^j \left[E_{\alpha+n\beta}, \ H_j\right] - N_{(\alpha+n\beta),-\beta} \left[E_\beta, \ E_{\alpha+(n-1)\beta}\right] \\ &\quad + N_{(\alpha+n\beta),\beta} \left[E_{-\beta}, \ E_{\alpha+(n+1)\beta}\right] \\ &= \left\{-\beta\cdot(\alpha+n\beta) - F_{n-1} + F_n\right\} E_{\alpha+n\beta}. \end{aligned}$$

Thus, F_n satisfies the recursive relation

$$
\begin{aligned}
F_n &= F_{n-1} + \boldsymbol{\beta} \cdot \{\boldsymbol{\alpha} + n\boldsymbol{\beta}\} \\
&= F_{n-2} + \boldsymbol{\beta} \cdot \{2\boldsymbol{\alpha} + (n+n-1)\,\boldsymbol{\beta}\} = \cdots \\
&= F_{n-(n+q+1)} + \boldsymbol{\beta} \cdot \left\{(n+q+1)\boldsymbol{\alpha} + \frac{1}{2}(n-q)(n+q+1)\,\boldsymbol{\beta}\right\} \\
&= \frac{1}{2}(n+q+1)\boldsymbol{\beta} \cdot \{2\boldsymbol{\alpha} + (n-q)\,\boldsymbol{\beta}\}.
\end{aligned}
\tag{7.44}
$$

When $n = p$, one has

$$2\boldsymbol{\alpha} \cdot \boldsymbol{\beta} = (q-p)\,(\boldsymbol{\beta} \cdot \boldsymbol{\beta}). \tag{7.45}$$

If $\boldsymbol{\beta} \cdot \boldsymbol{\beta} = 0$, $\boldsymbol{\beta}$ is orthogonal to each root $\boldsymbol{\alpha}$ in $\mathcal{L}$ so that $H_{\boldsymbol{\beta}}$ commutes with each generator in $\mathcal{L}$,

$$[H_{\boldsymbol{\beta}},\ H_k] = 0, \qquad [H_{\boldsymbol{\beta}},\ E_{\boldsymbol{\alpha}}] = (\boldsymbol{\beta} \cdot \boldsymbol{\alpha})\,E_{\boldsymbol{\alpha}} = 0. \tag{7.46}$$

Thus, $H_{\boldsymbol{\beta}}$ spans an abelian ideal in $\mathcal{L}$ which is in conflict with the fact that $\mathcal{L}$ is semisimple. Since $\boldsymbol{\beta} \cdot \boldsymbol{\beta} \neq 0$, from Eq. (7.45) we have

$$\Gamma\,(\boldsymbol{\alpha}/\boldsymbol{\beta}) = \frac{2\boldsymbol{\alpha} \cdot \boldsymbol{\beta}}{\boldsymbol{\beta} \cdot \boldsymbol{\beta}} = q - p. \tag{7.47}$$

It gives Eq. (7.41). Due to $-q \leqslant (p-q) \leqslant p$, $\boldsymbol{\alpha} - \Gamma(\boldsymbol{\alpha}/\boldsymbol{\beta})\boldsymbol{\beta}$ is a root.

At last, the theorem also holds if the root chain $\boldsymbol{\alpha} + n\boldsymbol{\beta}$ contains a zero root. Without loss of generality, let $\boldsymbol{\alpha} = m\boldsymbol{\beta}$. Defining the generator $E_{\boldsymbol{\alpha}-m\boldsymbol{\beta}} = E_0 = H_{\boldsymbol{\beta}}$, which corresponds to the zero root, one has

$$
\begin{aligned}
&\left[E_{\boldsymbol{\alpha}-(m-1)\boldsymbol{\beta}},\ E_{-\boldsymbol{\beta}}\right] = H_{\boldsymbol{\beta}}, && N_{(\boldsymbol{\alpha}-(m-1)\boldsymbol{\beta}),-\boldsymbol{\beta}} = 1, \\
&[E_{\boldsymbol{\alpha}-m\boldsymbol{\beta}},\ E_{\pm\boldsymbol{\beta}}] = \pm\,(\boldsymbol{\beta} \cdot \boldsymbol{\beta})\,E_{\pm\boldsymbol{\beta}}, && N_{(\boldsymbol{\alpha}-m\boldsymbol{\beta}),\pm\boldsymbol{\beta}} = \pm\,(\boldsymbol{\beta} \cdot \boldsymbol{\beta}), \\
&\left[E_{\boldsymbol{\alpha}-(m+1)\boldsymbol{\beta}},\ E_{\boldsymbol{\beta}}\right] = -H_{\boldsymbol{\beta}}, && N_{(\boldsymbol{\alpha}-(m+1)\boldsymbol{\beta}),\boldsymbol{\beta}} = -1.
\end{aligned}
$$

Thus, the above proof holds for this case. □

Corollary 7.3.1 The number of linearly independent roots in a semisimple Lie algebra $\mathcal{L}$ with rank ℓ is ℓ.

Proof Prove the corollary by reduction to absurdity. If the number is less than ℓ, there is at least a non-zero vector $\boldsymbol{V}$ orthogonal to each roots in $\mathcal{L}$, so that $\boldsymbol{V} \cdot \boldsymbol{H}$ commutes with each generator in $\mathcal{L}$. This contradicts to that $\mathcal{L}$ is semisimple. □

Corollary 7.3.2 In the root space of a semisimple Lie algebra, the inner product of any two vectors which are the real combinations of roots is real, and the inner self-product of a nonzero real combination of roots is positive.

Proof First, due to Eqs. (7.37) and (7.45) one has

$$\boldsymbol{\beta}\cdot\boldsymbol{\beta}=\sum_{\alpha\in\Delta}(\boldsymbol{\alpha}\cdot\boldsymbol{\beta})^2=\frac{1}{4}\left\{\sum_{\alpha\in\Delta}(q_\alpha-p_\alpha)^2\right\}(\boldsymbol{\beta}\cdot\boldsymbol{\beta})^2,$$

where q_α and p_α are the integral parameters in the root chain constructed from $\boldsymbol{\alpha}$ by adding and subtracting a root $\boldsymbol{\beta}$ successively. Since $\boldsymbol{\beta}\cdot\boldsymbol{\beta}\neq 0$,

$$\boldsymbol{\beta}\cdot\boldsymbol{\beta}=4\left\{\sum_{\alpha\in\Delta}(q_\alpha-p_\alpha)^2\right\}^{-1}=\text{real and positive.} \tag{7.48}$$

Then, due to Eq. (7.45), the inner product $\boldsymbol{\alpha}\cdot\boldsymbol{\beta}$ of any two roots is real.

Second, introducing two real combinations of roots

$$\boldsymbol{V}=\sum_{\beta\in\Delta}b_\beta\boldsymbol{\beta},\qquad \boldsymbol{U}=\sum_{\gamma\in\Delta}c_\gamma\boldsymbol{\gamma},$$

one has

$$\boldsymbol{V}\cdot\boldsymbol{U}=\sum_{\beta\in\Delta}\sum_{\gamma\in\Delta}b_\beta c_\gamma(\boldsymbol{\beta}\cdot\boldsymbol{\gamma})=\text{real}.$$

Due to Eq. (7.37),

$$\boldsymbol{V}\cdot\boldsymbol{V}=\sum_{\alpha\in\Delta}(\boldsymbol{\alpha}\cdot\boldsymbol{V})^2\geqslant 0.$$

$\boldsymbol{V}\cdot\boldsymbol{V}>0$ only if $\boldsymbol{V}\neq 0$. □

Corollary 7.3.3

$$N_{\alpha,\beta}N_{-\alpha,-\beta}=-\frac{1}{2}p(q+1)(\boldsymbol{\beta}\cdot\boldsymbol{\beta}). \tag{7.49}$$

Proof Letting $n=0$ in Eq. (7.44) we have

$$\begin{aligned}-N_{\alpha,\beta}N_{(\alpha+\beta),-\beta}=F_0&=\frac{1}{2}(q+1)\left(2\boldsymbol{\alpha}\cdot\boldsymbol{\beta}-q\boldsymbol{\beta}\cdot\boldsymbol{\beta}\right)\\&=-\frac{1}{2}p(q+1)\left(\boldsymbol{\beta}\cdot\boldsymbol{\beta}\right).\end{aligned} \tag{7.50}$$

If $\boldsymbol{\gamma}=\boldsymbol{\alpha}+\boldsymbol{\beta}$, and three roots are all non-zero,

$$\begin{aligned}\langle E_{-\alpha}|E_{-\beta}|E_\gamma\rangle&=\langle E_{-\alpha}\,|[E_{-\beta},\ E_\gamma]\rangle=N_{-\beta,\gamma}\langle E_{-\alpha}|E_\alpha\rangle=N_{-\beta,\gamma}\\&=\langle[E_{-\alpha},\ E_{-\beta}]|\,E_\gamma\rangle=N_{-\alpha,-\beta}\langle E_{-\gamma}|E_\gamma\rangle=N_{-\alpha,-\beta}.\end{aligned}$$

Due to Eq. (7.43),

$$N_{-\alpha,-\beta}=N_{-\beta,\gamma}=-N_{(\alpha+\beta),-\beta}. \tag{7.51}$$

Equation (7.49) follows Eq. (7.50). □

Corollary 7.3.4 $N_{\alpha,\beta} \neq 0$ if α, β, and $(\alpha+\beta)$ are all non-zero roots.

Corollary 7.3.5 Except for zero roots, there are only two roots $\pm\alpha$ along the direction of α.

Proof Let $t\alpha$ be a non-zero root, then $\Gamma(t\alpha/\alpha) = 2t$ and $\Gamma\left(\alpha/(t\alpha)\right) = 2/t$. From the conditions that $2t$ and $2/t$ are both integers, one obtains $t = \pm 1$, ± 2, or $\pm 1/2$. Since $N_{\alpha,\alpha} = 0$, $\pm 2\alpha$ is not a root. Then, $\pm\alpha/2$ is not a root, otherwise $\alpha = 2(\alpha/2)$ is not a root. □

7.2.5 *Positive Roots and Simple Roots*

Theorem 7.4 For a semisimple Lie algebra $\mathcal{L}$, the bases H_j in the Cartan subalgebra $\mathcal{H}$ of $\mathcal{L}$ can be chosen such that the roots are all real and the root space is real Euclidean.

Proof First, there exist ℓ linearly independent roots in $\mathcal{L}$, say $\beta^{(r)}$, $1 \leqslant r \leqslant \ell$, where ℓ is the rank of $\mathcal{L}$. Any root α can be expanded with respect to $\beta^{(r)}$, $\alpha = \sum_r x_r \beta^{(r)}$. Taking the inner product of each terms of the equation with $2\beta^{(s)}/\left(\beta^{(s)} \cdot \beta^{(s)}\right)$, one obtains

$$\Gamma\left(\alpha/\beta^{(s)}\right) = \sum_{r=1}^{\ell} x_r \Gamma\left(\beta^{(r)}/\beta^{(s)}\right).$$

They are the coupled linear algebraic equations with respect to the variables x_r. Since all the coefficients in the equations are integers and the solutions x_r do exist, each x_r has to be real. Namely, each root in $\mathcal{L}$ is a real combination of $\beta^{(r)}$.

Second, for the set of new basis vectors H_r in $\mathcal{H}$, $H_r = \beta^{(r)} \cdot \boldsymbol{H}$, g_{jk} becomes real,

$$g_{rs} = \sum_{jk} \beta^{(r)j}\beta^{(s)k} g_{jk} = -\sum_{\alpha\in\Delta} \left(\alpha\cdot\beta^{(r)}\right)\left(\alpha\cdot\beta^{(s)}\right) = \text{real}.$$

The new g_{rs} can be diagonalized through a real orthogonal transformation of H_r, and then become $-\delta_{rs}$ through a scale transformation.

Third, let H_j denote the transformed bases in $\mathcal{H}$, where $g_{jk} = -\delta_{jk}$. In the condition $g_{jk} = -\delta_{jk}$, H_r as well as the components of roots can still be made an orthogonal transformation. Due to Corollary 7.3.2, the root space is real Euclidean only if the basis roots $\beta^{(r)}$ can be transformed into real through the orthogonal transformation. Let $a_j^{(r)}$ and $b_j^{(r)}$ denote the real and imaginary parts of the jth component of $\boldsymbol{\beta}^{(r)}$, respectively, $\beta_j^{(r)} = a_j^{(r)} + ib_j^{(r)}$. Make a real orthogonal transformation on $a_j^{(1)}$ such that $a_1^{(1)}$ is non-negative and the remaining components vanish. Then, $a_1^{(1)} > 0$ and $b_1^{(1)} = 0$ because $\boldsymbol{\beta}^{(1)} \cdot \boldsymbol{\beta}^{(1)}$ is real and positive. Make a real orthogonal transformation on $b_j^{(1)}$ with $j > 1$ such that $b_2^{(1)}$ is non-negative and the remaining components vanish. The components of the transformed root $\boldsymbol{\beta}^{(1)}$ become $\beta_1^{(1)} = a_1^{(1)} = a$ and $\beta_2^{(1)} = ib_2^{(1)} = ib$, and the remaining components vanish. $a > b \geqslant 0$ due to $\boldsymbol{\beta}^{(1)} \cdot \boldsymbol{\beta}^{(1)} > 0$. Making an orthogonal transformation on the first two components,

$$\left(a^2 - b^2\right)^{-1/2} \begin{pmatrix} a & ib \\ -ib & a \end{pmatrix} \begin{pmatrix} a \\ ib \end{pmatrix} = \begin{pmatrix} \left(a^2 - b^2\right)^{1/2} \\ 0 \end{pmatrix},$$

one obtains the transformed root $\boldsymbol{\beta}^{(1)}$ such that its components all vanish except for its first component $\beta_1^{(1)}$ which is real and positive. Since $\boldsymbol{\beta}^{(1)} \cdot \boldsymbol{\beta}^{(r)}$ is real, the first component $\beta_1^{(r)}$ of each basis root $\boldsymbol{\beta}^{(r)}$ is real. Define new basis vectors in $\mathcal{L}$ by the real combinations, $\boldsymbol{\gamma}^{(r)} = \boldsymbol{\beta}^{(r)} - \left(\beta_1^{(r)}/\beta_1^{(1)}\right) \boldsymbol{\beta}^{(1)}$, where $r > 1$. Thus, the first components of $\boldsymbol{\gamma}^{(r)}$ with $r > 1$ vanish, and its remaining components are equal to those of $\boldsymbol{\beta}^{(r)}$. Each $\boldsymbol{\gamma}^{(r)}$ is not a null vector.

The rest can be deduced by analogy. Leaving the first one-dimensional subspace invariant, we make the orthogonal transformation on the remaining $(\ell - 1)$-dimensional subspace such that the first components $\beta_1^{(r)}$ are left invariant, the second component $a_2^{(2)}$ of the transformed root $\boldsymbol{\beta}^{(2)}$ is real and positive, and its remaining components all vanish. Corollary 7.3.2 is used in the proof. Thus, the first and the second components of all basis roots $\boldsymbol{\beta}^{(r)}$ are real. In the same way, one can prove that all the basis roots are real. Therefore, all roots in $\mathcal{L}$ are real, $g_{jk} = -\delta_{jk}$, and the root space is real Euclidean. □

For the given H_j in the Cartan subalgebra, a root $\boldsymbol{\alpha}$ is called positive if its first nonvanishing component is positive, and the root is negative if the component is negative. A positive root is called a **simple root** if it cannot be expressed as a non-negative integral combination of other positive roots. Therefore, **any positive (or negative) root is equal to a non-negative (or non-positive) integral combination of the simple roots**, and the sum of the coefficients in the combination is called the **level** of the positive root. Obviously, the number of the simple roots in a semisimple Lie algebra with rank ℓ is not less than ℓ. It is equal to ℓ if all simple roots are linearly independent of each other.

Theorem 7.5 The difference of two simple roots is not a root, the inner product of two simple roots is not larger than zero, and the number of the simple roots in a semisimple Lie algebra with rank ℓ is equal to ℓ.

Proof Let γ denote the difference of two simple roots $\boldsymbol{\alpha}$ and $\boldsymbol{\beta}$, $\boldsymbol{\gamma} = \boldsymbol{\alpha} - \boldsymbol{\beta}$. $\boldsymbol{\gamma}$ is not a positive root, otherwise $\boldsymbol{\alpha}$ is a sum of two positive roots $\boldsymbol{\beta}$ and $\boldsymbol{\gamma}$. $\boldsymbol{\gamma}$ is not a negative root, otherwise $\boldsymbol{\beta}$ is a sum of two positive roots $\boldsymbol{\alpha}$ and $-\boldsymbol{\gamma}$. Thus, the difference of two simple roots is not a root. Then, $q = 0$ in the root chain (7.42) where $\boldsymbol{\alpha}$ and $\boldsymbol{\beta}$ both are simple roots. From Eq. (7.47) we obtain that $2(\boldsymbol{\alpha} \cdot \boldsymbol{\beta}) = -p(\boldsymbol{\beta} \cdot \boldsymbol{\beta}) \leqslant 0$. At last, we prove by reduction to absurdity that simple roots are linearly independent. Assume that there is a real linear relation among simple roots,

$$\sum_j c_j \boldsymbol{\alpha}^{(j)} - \sum_k d_k \boldsymbol{\beta}^{(k)} = 0, \qquad c_j > 0, \qquad d_k > 0.$$

Since $\boldsymbol{\alpha}^{(j)} \neq \boldsymbol{\beta}^{(k)}$, $\boldsymbol{\alpha}^{(j)} \cdot \boldsymbol{\beta}^{(k)} \leqslant 0$. Then, it is in contradiction that the inner self-product of the nonzero vector $\boldsymbol{V}$ is not larger than zero,

$$\begin{aligned} \boldsymbol{V} &= \sum_j c_j \boldsymbol{\alpha}^{(j)} = \sum_k d_k \boldsymbol{\beta}^{(k)} \neq 0, \\ \boldsymbol{V} \cdot \boldsymbol{V} &= \sum_{jk} c_j d_k \boldsymbol{\alpha}^{(j)} \cdot \boldsymbol{\beta}^{(k)} \leqslant 0. \end{aligned}$$

□

Theorem 7.6 Up to isomorphism, any semisimple Lie algebra $\mathcal{L}$ has one and only one compact real form.

Proof We only sketch the proof. $E_{\boldsymbol{\alpha}}$ can still be multiplied with a factor $b_{\boldsymbol{\alpha}}$ satisfying $b_{\boldsymbol{\alpha}} b_{-\boldsymbol{\alpha}} = 1$ to preserve $g_{jk} = -\delta_{jk}$. From Eq. (7.40) one has

$$\frac{N'_{\boldsymbol{\alpha},\boldsymbol{\beta}}}{N'_{-\boldsymbol{\alpha},-\boldsymbol{\beta}}} = \frac{b_{\boldsymbol{\alpha}}}{b_{-\boldsymbol{\alpha}}} \frac{b_{\boldsymbol{\beta}}}{b_{-\boldsymbol{\beta}}} \frac{b_{-\boldsymbol{\alpha}-\boldsymbol{\beta}}}{b_{\boldsymbol{\alpha}+\boldsymbol{\beta}}} \frac{N_{\boldsymbol{\alpha},\boldsymbol{\beta}}}{N_{-\boldsymbol{\alpha},-\boldsymbol{\beta}}}.$$

Thus, one is able to make Eq. (7.52) holds for the positive root $\boldsymbol{\alpha}$ one by one as its level increases by choosing the ratios $b_{\boldsymbol{\alpha}}/b_{-\boldsymbol{\alpha}}$:

$$N_{-\boldsymbol{\alpha},-\boldsymbol{\beta}} = -N_{\boldsymbol{\alpha},\boldsymbol{\beta}}. \tag{7.52}$$

Thus, the Killing form g_{AB} becomes $-\delta_{AB}$ if $E_{\pm\boldsymbol{\alpha}}$ are replaced with

$$E_{\boldsymbol{\alpha}1} = \left(E_{\boldsymbol{\alpha}} + E_{-\boldsymbol{\alpha}}\right)/\sqrt{2}, \qquad E_{\boldsymbol{\alpha}2} = -\mathrm{i}\left(E_{\boldsymbol{\alpha}} - E_{-\boldsymbol{\alpha}}\right)/\sqrt{2}. \tag{7.53}$$

It can be shown straightforwardly that the structure constants $C_{AB}{}^{D}$ are all real. The real Lie algebra with the basis vectors $(-\mathrm{i}H_j)$, $(-\mathrm{i}E_{\alpha 1})$, and $(-\mathrm{i}E_{\alpha 2})$ is the compact real form of $\mathcal{L}$. □

Corollary 7.6.1 Any finite-dimensional representation of a semisimple Lie algebra is completely reducible.

Since the adjoint representation of a semisimple Lie algebra is completely reducible, the following Corollary follows.

Corollary 7.6.2 Any semisimple Lie algebra can be decomposed to a direct sum of some non-abelian simple Lie algebras .

Now, the classification problem of semisimple Lie algebras reduces to **the classification of simple Lie algebras with dimension larger than 1**. In the classification of simple Lie algebras one will find that in each simple Lie algebra $\mathcal{L}$, there is only one root ω whose level is the highest among all roots in $\mathcal{L}$. ω is called **the largest root** of $\mathcal{L}$, which is related to the adjoint representation of $\mathcal{L}$.

7.3 Classification of Simple Lie Algebras

A simple Lie algebra of one dimension is abelian. In this section we are going to study the classification of simple Lie algebras $\mathcal{L}$ with the dimension larger than one. In the compact real form of $\mathcal{L}$, the **root space is real Euclidean**, where the simple roots of $\mathcal{L}$ are denoted by $\boldsymbol{r}_\mu$.

7.3.1 *Angle between Two Simple Roots*

Table 7.1 The angles and the lengths of simple roots

θ	$\cos^2\theta$	$\Gamma(\boldsymbol{r}_\mu/\boldsymbol{r}_\nu)$	$\Gamma(\boldsymbol{r}_\nu/\boldsymbol{r}_\mu)$	$\lvert\boldsymbol{r}_\mu\rvert/\lvert\boldsymbol{r}_\nu\rvert$
$5\pi/6$ (150°)	3/4	−3	−1	$\sqrt{3}$
$3\pi/4$ (135°)	1/2	−2	−1	$\sqrt{2}$
$2\pi/3$ (120°)	1/4	−1	−1	1
$\pi/2$ (90°)	0	0	0	arbitrary

The inner product of two simple roots $\boldsymbol{r}_\mu$ is not larger than zero so that their angle θ is not less than $\pi/2$. The cosine square of θ is

$$\begin{aligned} 4\cos^2\theta &= 4\,\frac{(\boldsymbol{r}_\mu\cdot\boldsymbol{r}_\nu)^2}{\lvert\boldsymbol{r}_\mu\rvert^2\,\lvert\boldsymbol{r}_\nu\rvert^2} = \frac{2\boldsymbol{r}_\mu\cdot\boldsymbol{r}_\nu}{\lvert\boldsymbol{r}_\nu\rvert^2}\;\frac{2\boldsymbol{r}_\nu\cdot\boldsymbol{r}_\mu}{\lvert\boldsymbol{r}_\mu\rvert^2} \\ &= \Gamma(\boldsymbol{r}_\mu/\boldsymbol{r}_\nu)\,\Gamma(\boldsymbol{r}_\nu/\boldsymbol{r}_\mu) = \text{integer}. \end{aligned} \tag{7.54}$$

Without loss of generality, let $|\boldsymbol{r}_\mu|^2 \geqslant |\boldsymbol{r}_\nu|^2$. There are only four solutions for $4\cos^2\theta$ listed in Table 7.1.

7.3.2 *Dynkin Diagrams*

It will be shown later that there are only one or two different lengths among the simple roots in any simple Lie algebra. We draw the **Dynkin diagram** for a simple Lie algebra $\mathcal{L}$ where each longer simple root is denoted by **a white circle** and each shorter simple root, if it exists, by **a black circle**. Two circles, denoting two simple roots, are connected by a single link, a double link, or a triple link depending upon their angle to be $2\pi/3$, $3\pi/4$, or $5\pi/6$, respectively. The ratio of their square lengths is 1, 2, or 3, respectively. Two circles are not connected by any link if two simple roots are orthogonal and the ratio of their lengths is not restricted. Now, we are going to study what kinds of Dynkin diagrams of $\mathcal{L}$ are allowed based on the property of a simple Lie algebra.

1. *The Dynkin diagram of a simple Lie algebra is connected.*

If a Dynkin diagram is divided into two unconnected parts, two simple roots belonging to different parts, say $\boldsymbol{r}$ and $\boldsymbol{r}'$, are orthogonal to each other. Thus, the generators are divided into two sets, respectively related to the roots in two parts. The generators $E_{\boldsymbol{r}}$ and $\boldsymbol{r}\cdot\boldsymbol{H}$ belonging to the first set commute with the generators $E_{\boldsymbol{r}'}$ and $\boldsymbol{r}'\cdot\boldsymbol{H}$ belonging to the second one. Therefore, $\mathcal{L}$ is decomposed into the direct sum of two ideals, which contradicts to that $\mathcal{L}$ is simple.

2. *The Dynkin diagram contains no loop.*

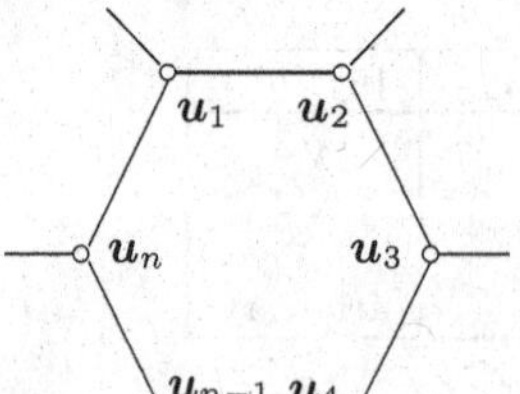

$$\boldsymbol{\alpha} = \sum_{j=1}^{n} \boldsymbol{u}_j \neq 0,$$

$$|\boldsymbol{\alpha}|^2 = \sum_{j=1}^{n} |\boldsymbol{u}_j|^2 + 2\sum_{j=1}^{n} \boldsymbol{u}_j \cdot \boldsymbol{u}_{j+1}$$

$$= \sum_{j=1}^{n} |\boldsymbol{u}_j|^2 \left\{1 + \Gamma(\boldsymbol{u}_{j+1}/\boldsymbol{u}_j)\right\} \leqslant 0.$$

If a Dynkin diagram contains a smallest loop composed of n circles, the neighboring circles denoting the simple roots $\boldsymbol{u}_j$ and $\boldsymbol{u}_{j+1}$ are connected by links and there is no link inside the loop. Assume $\boldsymbol{u}_{n+1} = \boldsymbol{u}_1$ for convenience. There is a contradiction that the sum $\boldsymbol{\alpha}$ of the simple roots is not vanishing but its square is not positive.

3. *The number of links fetching out from one circle is less than four.*

Let a simple root $\boldsymbol{r}$ connect with n simple roots $\boldsymbol{u}_j$, $1 \leqslant j \leqslant n$. From Eq. (7.54) the number of links fetching out from $\boldsymbol{r}$ is equal to $\sum_j \Gamma(\boldsymbol{r}/\boldsymbol{u}_j)\Gamma(\boldsymbol{u}_j/\boldsymbol{r})$. Since there is no loop, any two simple roots $\boldsymbol{u}_j$ are orthogonal to each other, and $\boldsymbol{r}$ is linearly independent of n simple roots $\boldsymbol{r}_j$. Thus,

$$|\boldsymbol{r}|^2 > \sum_{j=1}^{n} (\boldsymbol{r}\cdot\boldsymbol{u}_j)^2/|\boldsymbol{u}_j|^2 = \frac{|\boldsymbol{r}|^2}{4}\sum_{j=1}^{n}\Gamma(\boldsymbol{r}/\boldsymbol{u}_j)\Gamma(\boldsymbol{u}_j/\boldsymbol{r}). \tag{7.55}$$

The conclusion follows Eq. (7.55) by removing a factor $\boldsymbol{r}^2$.

From this property, one concludes that there is only one Dynkin diagram with a triple link:

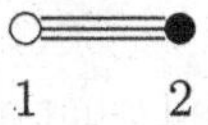

Its algebra is called the Lie algebra G_2. The following diagrams as well as those by interchanging the white and black circles are not allowed:

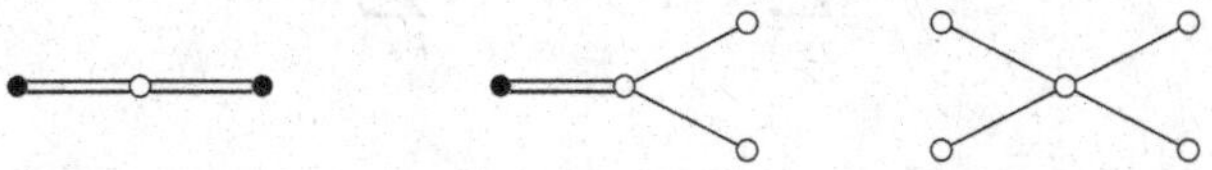

It will be shown that the diagrams by replacing the circle on the center of the above diagrams with a circle chain connected by single links are also not allowed:

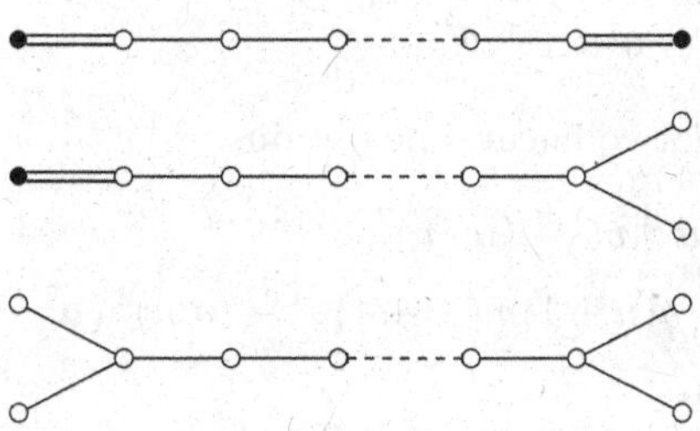

In fact, let the embedded circle chain contain m circles, and let $\boldsymbol{v}_j$ denote the simple root on the chain. Since $\boldsymbol{v}_j$, $1 \leqslant j \leqslant m$, all have the same length

v, the length of their sum, $\boldsymbol{v} = \sum_j \boldsymbol{v}_j$, is v, too:

$$|\boldsymbol{v}|^2 = \sum_{j=1}^{m} |\boldsymbol{v}_j|^2 + 2\sum_{j=1}^{m-1} \boldsymbol{v}_j \cdot \boldsymbol{v}_{j+1} = mv^2 - (m-1)v^2 = v^2.$$

Replacing $\boldsymbol{r}$ with $\boldsymbol{v}$ in the above proof, one also shows that the number of links fetching out from the circle chain is less than four. It also follows that any Dynkin diagram of a simple Lie algebra contains almost one double link or one triple link, and then, **there are only one or two different lengths among the simple roots in any simple Lie algebra.**

4. *The Dynkin diagrams with a double link*

The general form of the Dynkin diagrams with a double link is

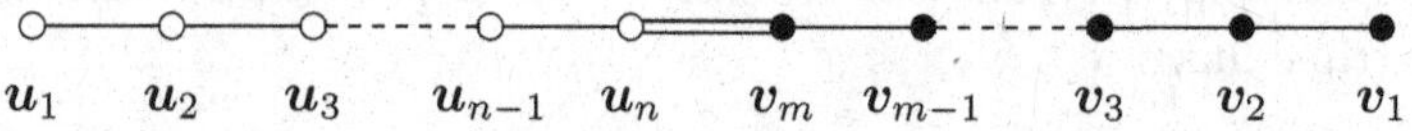

where the length of $\boldsymbol{v}_k$ is v and that of $\boldsymbol{u}_j$ is $\sqrt{2}v$, and $\boldsymbol{u}_n \cdot \boldsymbol{v}_m = (v^2/2)\Gamma(\boldsymbol{u}_n/\boldsymbol{v}_m) = -v^2$. Letting

$$\boldsymbol{u} = \sum_{j=1}^{n} j\boldsymbol{u}_j, \qquad \boldsymbol{v} = \sum_{k=1}^{m} k\boldsymbol{u}_k,$$

one has

$$\begin{aligned} |\boldsymbol{v}|^2 &= \sum_{k=1}^{m} k^2v^2 + \sum_{k=1}^{m-1} k(k+1)\left(-v^2\right) \\ &= v^2\left(m^2 - \sum_{k=1}^{m-1} k\right) = \frac{1}{2}m(m+1)v^2 . \\ |\boldsymbol{u}|^2 &= n(n+1)v^2. \end{aligned} \tag{7.56}$$

Since $\boldsymbol{u}$ and $\boldsymbol{v}$ are not collinear, one obtains

$$\begin{aligned} 0 < \ & |\boldsymbol{u}|^2|\boldsymbol{v}|^2 - (\boldsymbol{u}\cdot\boldsymbol{v})^2 \\ = \ & \frac{1}{2}n(n+1)m(m+1)v^4 - (mn)^2\left(\boldsymbol{u}_n \cdot \boldsymbol{v}_m\right)^2 \\ = \ & \frac{1}{2}nm\left(n+m+1-mn\right)v^4. \end{aligned}$$

Then, $(m-1)(n-1) < 2$. If $m = 1$, n is any positive integer, denoted by $\ell - 1$. The Dynkin diagram is

Its algebra is called the Lie algebra B_ℓ. If $n = 1$, m is an arbitrarily positive integer $\ell - 1$, and the Dynkin diagram is

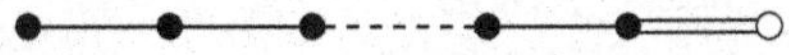

Its algebra is called the Lie algebra C_ℓ. If $n = m = 2$, the Dynkin diagram is

Its algebra is called the Lie algebra F_4.

5. *The Dynkin diagrams with a bifurcation*

The general form of the Dynkin diagrams with a bifurcation is

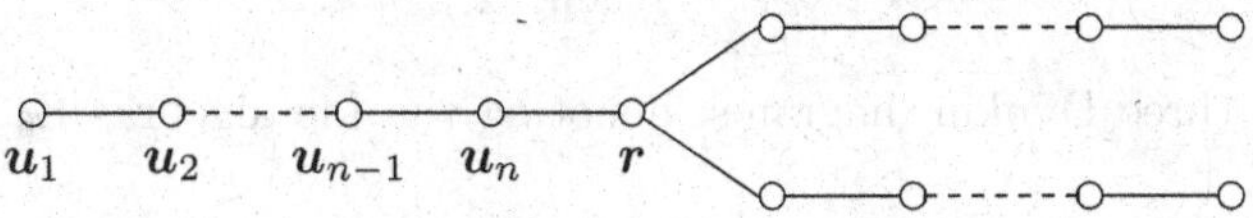

The lengths of all simple roots are the same and denoted by v. Let

$$\boldsymbol{u} = \sum_{j=1}^{n} j\boldsymbol{u}_j, \qquad \boldsymbol{v} = \sum_{k=1}^{m} k\boldsymbol{v}_k, \qquad \boldsymbol{w} = \sum_{s=1}^{p} s\boldsymbol{w}_s.$$

They are orthogonal to each other and linearly independent of $\boldsymbol{r}$, so that

$$v^2 = |\boldsymbol{r}|^2 > \frac{(\boldsymbol{r}\cdot\boldsymbol{u})^2}{|\boldsymbol{u}|^2} + \frac{(\boldsymbol{r}\cdot\boldsymbol{v})^2}{|\boldsymbol{v}|^2} + \frac{(\boldsymbol{r}\cdot\boldsymbol{w})^2}{|\boldsymbol{w}|^2}.$$

Due to Eq. (7.56),

$$\frac{(\boldsymbol{r}\cdot\boldsymbol{u})^2}{|\boldsymbol{u}|^2}=\frac{n^2v^4/4}{n(n+1)v^2/2}=\frac{nv^2}{2(n+1)}=\frac{v^2}{2}-\frac{v^2}{2(n+1)},$$
$$\frac{3}{2}-\frac{1}{2}\left(\frac{1}{n+1}+\frac{1}{m+1}+\frac{1}{p+1}\right)<1,$$
$$\frac{1}{n+1}+\frac{1}{m+1}+\frac{1}{p+1}>1.$$

Without loss of generality, letting $p\leqslant m\leqslant n$. Replacing n and m with p, we have

$$\frac{3}{p+1}>1, \qquad \text{then,}\ \ p=1.$$

Replacing n with m, we have

$$\frac{2}{m+1}>\frac{1}{2}, \qquad \text{then,}\ \ m=1\ \text{ or }\ 2.$$

If $m=p=1$, n is any positive integer, denoted by $\ell-3$. The Dynkin diagram is

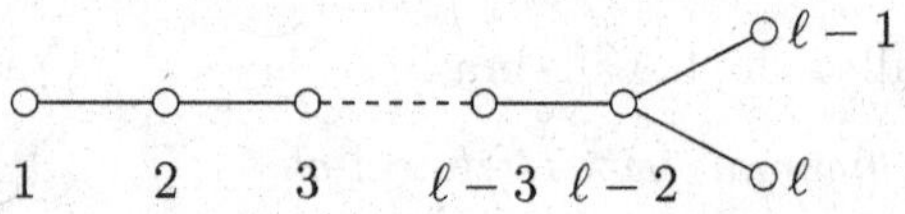

This algebra is called the Lie algebra D_ℓ. If $p=1$ and $m=2$, one has

$$\frac{1}{n+1}>\frac{1}{6}, \qquad \text{then,}\ \ 2\leqslant n\leqslant 4.$$

There are three Dynkin diagrams, denoting the Lie algebras E_6, E_7, and E_8:

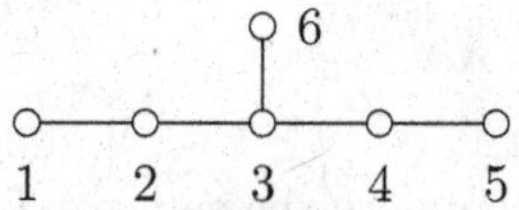

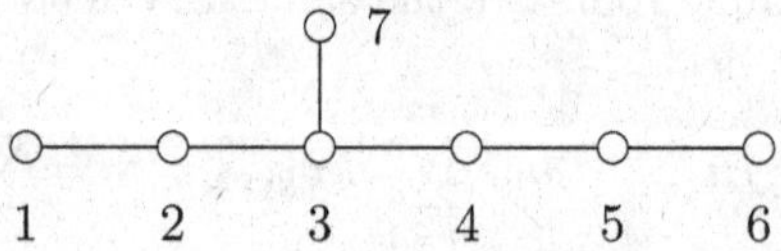

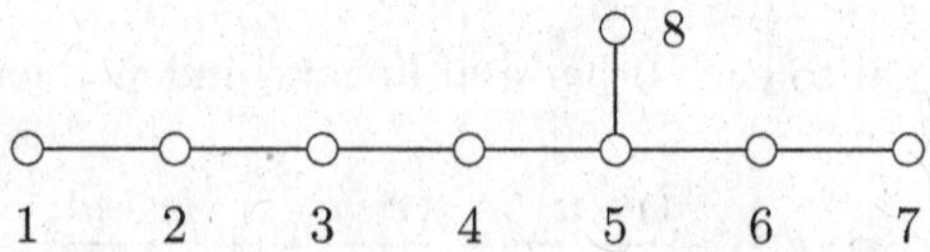

6. *The Dynkin diagram with only the single links denotes the Lie algebras* A_ℓ.

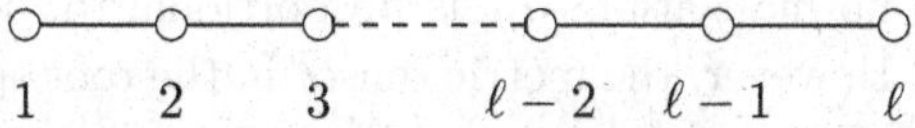

In summary, for the simple Lie algebras with dimension larger than 1, there are four sets of the **classical Lie algebras** A_ℓ, B_ℓ, C_ℓ, and D_ℓ and five **exceptional Lie algebras** G_2, F_4, E_6, E_7, and E_8. Their Dynkin diagrams are listed in Fig. 7.1. In the next section we are going to show that the Lie algebra of the $SU(\ell+1)$ group is A_ℓ, that of $SO(2\ell+1)$ is B_ℓ, that of $SO(2\ell)$ is D_ℓ, and that of $USp(2\ell)$ is C_ℓ. From Fig. 7.1 one finds that some Dynkin diagrams are the same such that the corresponding Lie groups are locally isomorphic:

$$\begin{array}{ll} A_1 \approx C_1 \approx B_1, & SU(2) \approx USp(2) \sim SO(3), \\ C_2 \approx B_2, & USp(4) \sim SO(5), \\ A_1 \oplus A_1 \approx D_2, & SU(2) \otimes SU(2)' \sim SO(4), \\ A_3 \approx D_3 & SU(4) \sim SO(6). \end{array} \tag{7.57}$$

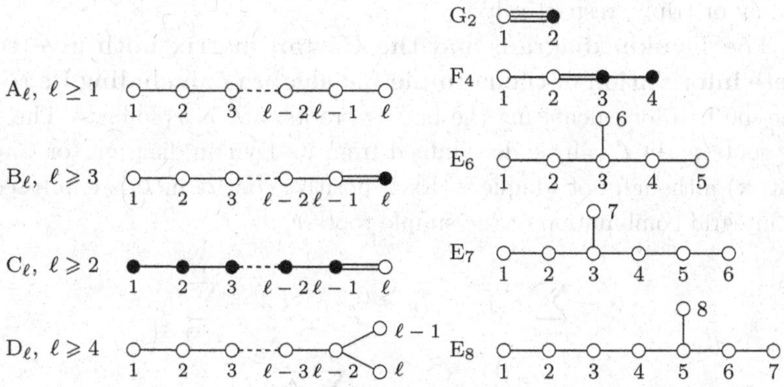

Fig. 7.1 The Dynkin diagrams of simple Lie algebras

If all generators in a given simple Lie algebra are multiplied with a common factor λ, the structure constants $C_{AB}{}^{D}$ are multiplied with λ and the Killing form g_{AB} is multiplied with λ^2. Let d_μ denote the half of square length of a simple root $\boldsymbol{r}_\mu$,

$$d_\mu = \frac{1}{2}\, \boldsymbol{r}_\mu \cdot \boldsymbol{r}_\mu. \tag{7.58}$$

Usually in mathematics, d_μ of the longer simple root $\boldsymbol{r}_\mu$ is normalized to be one. But in some old physics literature d_μ for Lie algebra A_ℓ is normalized to be 1/2. In the normalization, g_{jk} is proportional to, not equal to $-\delta_{jk}$ (see Eq. (7.136)). However, the metric tensor in the root space is re-defined to be δ_{jk}, instead of $-g_{jk}$ for preserving Eq. (7.40).

7.3.3 *The Cartan Matrix*

For a simple Lie algebra $\mathcal{L}$ of rank ℓ, there are ℓ simple roots $\boldsymbol{r}_\mu$. Define an ℓ-dimensional matrix A, called the Cartan matrix of $\mathcal{L}$,

$$A_{\mu\nu} = \Gamma\left(\boldsymbol{r}_\nu/\boldsymbol{r}_\mu\right) = \frac{2\boldsymbol{r}_\nu \cdot \boldsymbol{r}_\mu}{|\boldsymbol{r}_\mu|^2} = d_\mu^{-1}\left(\boldsymbol{r}_\nu \cdot \boldsymbol{r}_\mu\right). \tag{7.59}$$

For the given order of simple roots in $\mathcal{L}$, **the Dynkin diagram of $\mathcal{L}$ completely determines its Cartan matrix, and vice versa.** The diagonal element in the Cartan matrix A is always 2. When $\mu \neq \nu$, $A_{\mu\nu} = A_{\nu\mu} = 0$ if two simple roots $\boldsymbol{r}_\mu$ and $\boldsymbol{r}_\nu$ are disconnected in the Dynkin diagram. Otherwise, $A_{\mu\nu} = -1$, if the length of $\boldsymbol{r}_\mu$ is not less than that of $\boldsymbol{r}_\nu$. $A_{\nu\mu} = -2$, or -3, if the square length of $\boldsymbol{r}_\mu$ is longer than that of $\boldsymbol{r}_\nu$ doubly or triply, respectively.

The Dynkin diagram and the Cartan matrix both give complete information of their simple Lie algebra $\mathcal{L}$, including its roots. The method for calculating the positive roots of $\mathcal{L}$ is as follows. The simple roots $\boldsymbol{r}_\mu$ in $\mathcal{L}$ can be determined from its Dynkin diagram (or Cartan matrix) although not uniquely. Each positive root $\boldsymbol{\alpha}$ in $\mathcal{L}$ is expressed as an integral combination of the simple roots $\boldsymbol{r}_\mu$:

$$\begin{aligned} \boldsymbol{\alpha} &= \sum_{\mu=1}^{\ell} \boldsymbol{r}_\mu f_\mu, \quad f_\mu \geqslant 0, \qquad L_{\boldsymbol{\alpha}} = \sum_{\mu=1}^{\ell} f_\mu, \\ p &= q - \Gamma\left(\boldsymbol{\alpha}/\boldsymbol{r}_\nu\right) = q - \sum_{\mu=1}^{\ell} A_{\nu\mu} f_\mu. \end{aligned} \tag{7.60}$$

$L_{\boldsymbol{\alpha}}$ is called the level of a positive root $\boldsymbol{\alpha}$. The level of a simple root $\boldsymbol{r}_\mu$ is one. The positive roots in $\mathcal{L}$ can be calculated from Eq. (7.60) one by one as their levels increase.

As example, we calculate the positive roots of the Lie algebra G_2. The Cartan matrix of G_2 is $A_{11} = A_{22} = 2$, $A_{12} = \Gamma\left(\boldsymbol{r}_2/\boldsymbol{r}_1\right) = -1$ and $A_{21} = \Gamma\left(\boldsymbol{r}_1/\boldsymbol{r}_2\right) = -3$. The roots of the first level are two simple roots $\boldsymbol{r}_1 = \boldsymbol{e}_2\sqrt{2}$ and $\boldsymbol{r}_2 = \boldsymbol{e}_1\sqrt{1/6} - \boldsymbol{e}_2\sqrt{1/2}$. Since the difference of two simple roots is not a root, we have

$$p_1 = q_1 - \Gamma\left(\boldsymbol{r}_1/\boldsymbol{r}_2\right) = 0 - \left(A_{21} \times f_1 + A_{22} \times f_2\right) = -(-3 \times 1 + 2 \times 0) = 3.$$

Hence, $\boldsymbol{r}_1+\boldsymbol{r}_2$, $\boldsymbol{r}_1+2\boldsymbol{r}_2$, and $\boldsymbol{r}_1+3\boldsymbol{r}_2$ are roots, but $\boldsymbol{r}_1+4\boldsymbol{r}_2$ is not a root. On the other hand,

$$p_2 = q_2 - \Gamma(\boldsymbol{r}_2/\boldsymbol{r}_1) = 0 - (A_{11}\times f_1 + A_{12}\times f_2) = -(2\times 0 - 1\times 1) = 1,$$

$\boldsymbol{r}_2+\boldsymbol{r}_1$ is a root, $\boldsymbol{r}_2+2\boldsymbol{r}_1$ is not a root. Thus, $\boldsymbol{r}_1+\boldsymbol{r}_2$ is the only root of level two. $\boldsymbol{r}_1+2\boldsymbol{r}_2$ is the only root of level three. Because $2\boldsymbol{r}_1+2\boldsymbol{r}_2 = 2(\boldsymbol{r}_1+\boldsymbol{r}_2)$ is not a root, $\boldsymbol{\alpha} = \boldsymbol{r}_1+3\boldsymbol{r}_2$ is the only root of level four. Since $\boldsymbol{\alpha}+\boldsymbol{r}_2$ and $\boldsymbol{\alpha}-\boldsymbol{r}_1$ are not roots,

$$p = q - \Gamma(\boldsymbol{\alpha}/\boldsymbol{r}_1) = 0 - (A_{11}\times f_1 + A_{12}\times f_2) = -(2\times 1 - 1\times 3) = 1.$$

$\boldsymbol{\alpha}+\boldsymbol{r}_1 = 2\boldsymbol{r}_1+3\boldsymbol{r}_2$ is the only root of level five, and $\boldsymbol{\alpha}+2\boldsymbol{r}_1$ is not a root. There is no root of level six because $(2\boldsymbol{r}_1+3\boldsymbol{r}_2)+\boldsymbol{r}_2 = 2(\boldsymbol{r}_1+2\boldsymbol{r}_2)$. In summary, G_2 contains six positive roots including two simple roots.

7.4 Classical Simple Lie Algebras

7.4.1 *The SU(N) Group and its Lie Algebra*

The set of all $N\times N$ unimodular unitary matrices u,

$$u^\dagger u = \mathbf{1}, \qquad \det u = 1, \tag{7.61}$$

in the multiplication rule of matrices, constitutes a group, called the N-dimensional **special unitary matrix group**, denoted by SU(N). An N-dimensional complex matrix contains $2N^2$ real parameters. The column matrices of a unitary matrix are orthonormal to each other. There are N real constraints for the normalization and $N(N-1)$ real constraints for the orthogonality. One constraint comes from the determinant. Thus, the number of independent real parameters needed for characterizing the elements of SU(N) is $g = 2N^2 - N - N(N-1) - 1 = N^2 - 1$. The group space is a connected closed region so that SU(N) is a **simply-connected compact Lie group with order** $g = N^2 - 1$.

Any element u of SU(N) can be diagonalized through a unitary similarity transformation X,

$$\begin{aligned} &X^{-1}uX = \exp\{-i\Phi\}, \qquad &&\Phi = \text{diag}\{\varphi_1,\ \varphi_2,\ \cdots,\ \varphi_N\}, \\ &u = \exp(-iH), &&H = X\Phi X^{-1}. \end{aligned} \tag{7.62}$$

The phase angle φ_a is determined up to a multiple of 2π,

$$-\pi \leqslant \varphi_a \leqslant \pi, \qquad 1 \leqslant a \leqslant (N-1), \qquad \varphi_N = -\sum_{a=1}^{N-1} \varphi_a, \tag{7.63}$$

where the sum of the phase angles is chosen to be 0 owing to det $u = 1$.

Due to $X \in \mathrm{SU}(N)$, Eq. (7.62) shows that u is conjugate to a diagonal matrix $\exp(-\mathrm{i}\Phi)$. Thus, the classes of SU(N) is characterized by $(N-1)$ parameters φ_a given in Eq. (7.63). The integral on the classes of SU(N) is proved to be

$$\begin{aligned}
&\int (d\varphi) W(\varphi) F(\varphi) = \int_{-\pi}^{\pi} d\varphi_1 \cdots \int_{-\pi}^{\pi} d\varphi_{N-1} \int d\varphi_N \delta\left(\sum_{a=1}^{N} \varphi_a\right) W(\varphi) F(\varphi), \\
&W(\varphi) = \frac{1}{\Omega} \prod_{a<b}^{N} \sin^2\left(\frac{\varphi_a - \varphi_b}{2}\right), \\
&\Omega = \int_{-\pi}^{\pi} d\varphi_1 \cdots \int_{-\pi}^{\pi} d\varphi_{N-1} \int d\varphi_N \delta\left(\sum_{a=1}^{N} \varphi_a\right) W(\varphi).
\end{aligned} \tag{7.64}$$

The hermitian traceless matrix H in Eq. (7.62) can be expanded with respect to the hermitian traceless basis matrices of N dimensions, which are divided into three types, generalized from $\sigma_a/2$, respectively,

$$\begin{aligned}
&\left(T_{ab}^{(1)}\right)_{cd} = \left(T_{ba}^{(1)}\right)_{cd} = \left(T_{ab}^{(1)}\right)_{dc} = \frac{1}{2}\left(\delta_{ac}\delta_{bd} + \delta_{ad}\delta_{bc}\right), \\
&\left(T_{ab}^{(2)}\right)_{cd} = -\left(T_{ba}^{(2)}\right)_{cd} = -\left(T_{ab}^{(2)}\right)_{dc} = -\frac{\mathrm{i}}{2}\left(\delta_{ac}\delta_{bd} - \delta_{ad}\delta_{bc}\right), \\
&T_r^{(3)} = \left(\frac{1}{2r(r-1)}\right)^{1/2} \left\{\sum_{b=1}^{r-1} T_{bb}^{(1)} - (r-1)T_{rr}^{(1)}\right\}, \\
&T_{11}^{(1)} - T_{rr}^{(1)} = \sum_{s=2}^{r-1} \left(\frac{2}{s(s-1)}\right)^{1/2} T_s^{(3)} + \left(\frac{2r}{r-1}\right)^{1/2} T_r^{(3)},
\end{aligned} \tag{7.65}$$

where $1 \leqslant a < b \leqslant N$ and $2 \leqslant r \leqslant N$. The subscripts a and b are the ordinal indices of the matrix, while c and d are its row and column indices. The matrix $T_{ab}^{(1)}$ is symmetric with respect to both ab and cd, but $T_{ab}^{(2)}$ is antisymmetric. The symbol $T_{aa}^{(1)}$ is defined as

$$\left(T_{aa}^{(1)}\right)_{cd} = \delta_{ac}\delta_{ad}. \tag{7.66}$$

There are $N(N-1)/2$ basis matrices $T_{ab}^{(1)}$, $N(N-1)/2$ basis matrices $T_{ab}^{(2)}$, and $(N-1)$ basis matrices $T_a^{(3)}$. Altogether, the number of basis matrices is (N^2-1). The expanded coefficients of H with respect to three types of basis matrices are real:

$$u = e^{-iH}, \quad H = \sum_{a<b} \left\{ \omega_{ab}^{(1)} T_{ab}^{(1)} + \omega_{ab}^{(2)} T_{ab}^{(2)} \right\} + \sum_{a=2}^{N} \omega_a^{(3)} T_a^{(3)}. \tag{7.67}$$

Three types of basis matrices $T_{ab}^{(1)}$, $T_{ab}^{(2)}$, and $T_a^{(3)}$ are the generators in the self-representation of SU(N), and $\omega_{ab}^{(1)}$, $\omega_{ab}^{(2)}$ and $\omega_a^{(3)}$ are the parameters of the SU(N) group. Usually, the generators in three types are enumerated uniformly in the following order:

$$\begin{array}{llll} T_1 = T_{12}^{(1)}, & T_2 = T_{12}^{(2)}, & T_3 = T_2^{(3)}, & T_4 = T_{13}^{(1)}, \\ T_5 = T_{13}^{(2)}, & T_6 = T_{23}^{(1)}, & T_7 = T_{23}^{(2)}, & T_8 = T_3^{(3)}, \\ T_9 = T_{14}^{(1)}, & T_{10} = T_{14}^{(2)}, & T_{11} = T_{24}^{(1)}, & T_{12} = T_{24}^{(2)}, \\ T_{13} = T_{34}^{(1)}, & T_{14} = T_{34}^{(2)}, & T_{15} = T_4^{(3)}, & \cdots \end{array} \tag{7.68}$$

The generators T_A satisfy the orthonormal condition:

$$\text{Tr}\,(T_A T_B) = \frac{1}{2}\,\delta_{AB}, \qquad A,\ B = 1,\ 2,\ \cdots,\ (N^2-1). \tag{7.69}$$

The orthonormal condition (7.69) guarantees that the structure constant $C_{AB}{}^{D}$ is totally antisymmetric with respect to its three indices. In fact, multiplying the commutative relation (7.5) with T_P and taking the trace, one obtains

$$C_{AB}{}^{P} = -2i\text{Tr}\left\{T_A T_B T_P - T_B T_A T_P\right\}. \tag{7.70}$$

It shows that the SU(N) group is a compact Lie group. In physical literatures, the antisymmetric structure constants $C_{AB}{}^{D}$ are usually denoted by f_{ABD}.

Through a straightforward calculation, one obtains the commutative relations of generators in the self-representation as follows,

$$\begin{aligned} \left[T_{ab}^{(1)},\ T_{cd}^{(1)}\right] &= \frac{i}{2}\left(\delta_{bc}T_{ad}^{(2)} + \delta_{ad}T_{bc}^{(2)} + \delta_{ac}T_{bd}^{(2)} + \delta_{bd}T_{ac}^{(2)}\right), \\ \left[T_{ab}^{(2)},\ T_{cd}^{(2)}\right] &= \frac{-i}{2}\left(\delta_{bc}T_{ad}^{(2)} + \delta_{ad}T_{bc}^{(2)} - \delta_{ac}T_{bd}^{(2)} - \delta_{bd}T_{ac}^{(2)}\right), \\ \left[T_{ab}^{(1)},\ T_{cd}^{(2)}\right] &= \frac{-i}{2}\left(\delta_{bc}T_{ad}^{(1)} - \delta_{ad}T_{bc}^{(1)} + \delta_{ac}T_{bd}^{(1)} - \delta_{bd}T_{ac}^{(1)}\right), \\ \left[T_a^{(3)},\ T_{ab}^{(1)}\right] &= -i\left\{(a-1)/(2a)\right\}^{1/2} T_{ab}^{(2)}, \\ \left[T_a^{(3)},\ T_{ab}^{(2)}\right] &= i\left\{(a-1)/(2a)\right\}^{1/2} T_{ab}^{(1)}, \end{aligned}$$

$$\begin{aligned}
\left[T_r^{(3)},\, T_{ab}^{(1)}\right] &= \mathrm{i}\,\{2r(r-1)\}^{-1/2}\, T_{ab}^{(2)}, \\
\left[T_r^{(3)},\, T_{ab}^{(2)}\right] &= -\mathrm{i}\,\{2r(r-1)\}^{-1/2}\, T_{ab}^{(1)}, \\
\left[T_b^{(3)},\, T_{ab}^{(1)}\right] &= \mathrm{i}\,\{b/(2b-2)\}^{1/2}\, T_{ab}^{(2)}, \\
\left[T_b^{(3)},\, T_{ab}^{(2)}\right] &= -\mathrm{i}\,\{b/(2b-2)\}^{1/2}\, T_{ab}^{(1)}, \\
\left[T_a^{(3)},\, T_b^{(3)}\right] &= \left[T_b^{(3)},\, T_{ar}^{(1)}\right] = \left[T_a^{(3)},\, T_{rb}^{(1)}\right] \\
&= \left[T_b^{(3)},\, T_{ar}^{(2)}\right] = \left[T_a^{(3)},\, T_{rb}^{(2)}\right] = 0,
\end{aligned} \tag{7.71}$$

where $a < r < b$. Remind that from the second formula in Eq. (7.71), **the set of generators $T_{ab}^{(2)}$ spans a Lie subalgebra in** SU(N).

There are N constant matrices in the SU(N) group,

$$T_m = \omega^m \mathbf{1}, \qquad \omega = \exp\{-\mathrm{i}2\pi/N\}, \qquad 0 \leqslant m \leqslant (N-1). \tag{7.72}$$

They commute with any element in SU(N) and forms the **center** of SU(N), denoted by Z_N. Z_N is an abelian invariant subgroup of SU(N). The group space of the quotient group SU(N)/Z_N is connected with the degree of continuity N, and SU(N) is its covering group.

In order to combine the generators in the self-representation of SU(N) to be the Cartan–Weyl bases, one chooses $(N-1)$ commutable generators $T_a^{(3)}$ to span the Cartan subalgebra. In the mathematical convention,

$$H_j = \sqrt{2}\, T_{N-j+1}^{(3)}, \qquad j = 1,\, 2,\, \cdots,\, \ell, \qquad \ell \equiv N-1. \tag{7.73}$$

The factor $\sqrt{2}$ is introduced to make $d_\mu = 1$ for the simple roots. The normalized common eigenvectors of H_j are $E_{\pm\boldsymbol{\alpha}_{ab}} = T_{ab}^{(1)} \pm \mathrm{i}T_{ab}^{(2)}$ with eigenvalues $\pm(\boldsymbol{\alpha}_{ab})_j = \pm\sqrt{2}\,(\boldsymbol{V}_a - \boldsymbol{V}_b)_j$:

$$\begin{aligned}
&[H_j,\, E_{\pm\boldsymbol{\alpha}_{ab}}] = \pm\sqrt{2}\,(\boldsymbol{V}_a - \boldsymbol{V}_b)_j\ E_{\boldsymbol{\alpha}_{ab}}, && a < b, \\
&\sqrt{2}\,(\boldsymbol{V}_a)_j = \sqrt{2}\left(T_{N-j+1}^{(3)}\right)_{aa} = (H_j)_{aa}, && 1 \leqslant j \leqslant \ell, \\
&\sum_{a=1}^{\ell+1} \boldsymbol{V}_a = 0, \qquad \boldsymbol{V}_a \cdot \boldsymbol{V}_b = \frac{\delta_{ab}}{2} - \frac{1}{2(\ell+1)}, && 1 \leqslant a \leqslant \ell+1.
\end{aligned} \tag{7.74}$$

$\boldsymbol{V}_a$ can be written explicitly

$$\begin{aligned}
\sqrt{2}\boldsymbol{V}_1 &= \left\{\sqrt{\frac{1}{(\ell+1)\ell}}, \sqrt{\frac{1}{\ell(\ell-1)}}, \cdots, \sqrt{\frac{1}{6}}, \sqrt{\frac{1}{2}}\right\}, \\
\sqrt{2}\boldsymbol{V}_a &= \left\{\sqrt{\frac{1}{(\ell+1)\ell}}, \sqrt{\frac{1}{\ell(\ell-1)}}, \cdots, \sqrt{\frac{1}{(a+1)a}}, -\sqrt{\frac{a-1}{a}}, 0, \cdots, 0\right\}, \\
\sqrt{2}\left(\boldsymbol{V}_a\right)_j &= \begin{cases} [(N-j+1)(N-j)]^{-1/2}, & a < N-j+1, \\ -[(N-j)/(N-j+1)]^{1/2}, & a = N-j+1, \\ 0, & a > N-j+1. \end{cases}
\end{aligned} \tag{7.75}$$

Thus, the simple roots $\boldsymbol{r}_\mu$ of SU(N) are

$$\begin{aligned}
\boldsymbol{r}_\mu &= \sqrt{2}\left[\boldsymbol{V}_\mu - \boldsymbol{V}_{\mu+1}\right], & 1 \leqslant \mu \leqslant \ell, \\
\boldsymbol{\alpha}_{ab} &= \sqrt{2}\left[\boldsymbol{V}_a - \boldsymbol{V}_b\right] = \sum_{\mu=a}^{b-1} \boldsymbol{r}_\mu, & a < b.
\end{aligned} \tag{7.76}$$

From Eqs. (7.74) and (7.76), the inner product of two simple roots is

$$\boldsymbol{r}_\mu \cdot \boldsymbol{r}_\nu = 2\delta_{\mu\nu} - \delta_{\mu(\nu-1)} - \delta_{\mu(\nu+1)}, \qquad d_\mu = 1. \tag{7.77}$$

Therefore, the lengths of the simple roots of SU(N) are the same and the angle of two neighboring simple roots is $2\pi/3$, namely, the Lie algebra of SU($\ell+1$) is A_ℓ. The largest root in A_ℓ is

$$\boldsymbol{\omega} = \sqrt{2}\left[\boldsymbol{V}_1 - \boldsymbol{V}_{\ell+1}\right] = \sum_{\mu=1}^{\ell} \boldsymbol{r}_\mu. \tag{7.78}$$

7.4.2 *The SO(N) Group and its Lie Algebra*

The set of all $N \times N$ real orthogonal matrices R,

$$R^T R = \mathbf{1}, \qquad R^* = R, \tag{7.79}$$

in the multiplication rule of matrices, constitutes a group, called the N-dimensional **real orthogonal matrix group**, denoted by O(N). From Eq. (7.79) one has det $R = \pm 1$. Thus, O(N) is a mixed Lie group. The subset of elements with det $R = 1$ forms an invariant subgroup of O(N), denoted by SO(N). An N-dimensional real matrix contains N^2 real parameters. The column matrices of a real orthogonal matrix are orthonormal to each other. There are N real constraints for the normalization and $N(N-1)/2$ real constraints for the orthogonality. The constraint on the determinant is a discontinuous condition, which does not decrease the parameters. Thus, the number of independent real parameters needed for characterizing the elements of SO(N) is $g = N^2 - N - N(N-1)/2 = N(N-1)/2$. The group space is a doubly-connected closed region so that SO(N) is **a doubly-connected compact Lie group** with order $g = N(N-1)/2$.

SO(N) is a subgroup of SU(N). The elements of the subgroup SO(N) can be obtained by taking the parameters $\omega_{ab}^{(1)}$ and $\omega_a^{(3)}$ of SU(N) to be zero. In the usual convention generalized from SO(3), one takes $\omega_{ab} = \omega_{ab}^{(2)}/2$ and $T_{ab} = 2T_{ab}^{(2)}$ for SO(N),

$$\begin{aligned} &R = \exp\left\{-\mathrm{i}\sum_{a<b=2}^{N} \omega_{ab}T_{ab}\right\}, \quad (T_{ab})_{cd} = -\mathrm{i}\left(\delta_{ac}\delta_{bd} - \delta_{ad}\delta_{bc}\right), \\ &[T_{ab},\, T_{cd}] = -\mathrm{i}\left\{\delta_{bc}T_{ad} + \delta_{ad}T_{bc} - \delta_{bd}T_{ac} - \delta_{ac}T_{bd}\right\}. \end{aligned} \tag{7.80}$$

The generators in the self-representation of SO(N) satisfy the orthonormal condition:

$$\mathrm{Tr}\left(T_{ab}T_{cd}\right) = 2\left(\delta_{ac}\delta_{bd} - \delta_{ad}\delta_{bc}\right). \tag{7.81}$$

So that the structure constant is totally antisymmetric with respect to three indices. Two generators T_{ab} and T_{cd} of SO(N) are commutable with each other if their subscripts are all different. The commutable generators

$$H_j = T_{(2j-1)(2j)}, \qquad 1 \leqslant j \leqslant \ell. \tag{7.82}$$

span the Cartan subalgebra of both the groups SO(2ℓ) and SO($2\ell+1$). Combining the remaining generators to be the common eigenvectors $E_{\boldsymbol{\alpha}}$ of H_j, $[H_j,\, E_{\boldsymbol{\alpha}}] = \alpha_j E_{\boldsymbol{\alpha}}$, one obtains four types of generators $E_{\boldsymbol{\alpha}}$ for SO(2ℓ)

$$\begin{aligned} E_{ab}^{(1)} &= \frac{1}{2}\left[T_{(2a)(2b-1)} - \mathrm{i}T_{(2a-1)(2b-1)} - \mathrm{i}T_{(2a)(2b)} - T_{(2a-1)(2b)}\right], \\ E_{ab}^{(2)} &= \frac{1}{2}\left[T_{(2a)(2b-1)} + \mathrm{i}T_{(2a-1)(2b-1)} + \mathrm{i}T_{(2a)(2b)} - T_{(2a-1)(2b)}\right], \qquad a < b, \\ E_{ab}^{(3)} &= \frac{1}{2}\left[T_{(2a)(2b-1)} - \mathrm{i}T_{(2a-1)(2b-1)} + \mathrm{i}T_{(2a)(2b)} + T_{(2a-1)(2b)}\right], \\ E_{ab}^{(4)} &= \frac{1}{2}\left[T_{(2a)(2b-1)} + \mathrm{i}T_{(2a-1)(2b-1)} - \mathrm{i}T_{(2a)(2b)} + T_{(2a-1)(2b)}\right], \end{aligned} \tag{7.83}$$

with the eigenvalues (roots) $\{\boldsymbol{e}_a - \boldsymbol{e}_b\}_j$, $\{-\boldsymbol{e}_a + \boldsymbol{e}_b\}_j$, $\{\boldsymbol{e}_a + \boldsymbol{e}_b\}_j$, and $\{-\boldsymbol{e}_a - \boldsymbol{e}_b\}_j$, respectively, where $\{\boldsymbol{e}_a\}_j = \delta_{aj}$. Two more types of generators for SO($2\ell+1$) are

$$
\begin{aligned}
E_a^{(5)} &= \sqrt{\frac{1}{2}}\left[T_{(2a)(2\ell+1)} - \mathrm{i}T_{(2a-1)(2\ell+1)}\right], \\
E_a^{(6)} &= \sqrt{\frac{1}{2}}\left[T_{(2a)(2\ell+1)} + \mathrm{i}T_{(2a-1)(2\ell+1)}\right],
\end{aligned}
\tag{7.84}
$$

with the eigenvalues $\{\boldsymbol{e}_a\}_j$ and $-\{\boldsymbol{e}_a\}_j$. The generators labeled by the superscripts (1), (3), and (5) correspond to the positive roots.

The simple roots $\boldsymbol{r}_\mu$ and the largest root $\boldsymbol{\omega}$ for SO(2ℓ) are

$$
\begin{aligned}
&\boldsymbol{r}_\mu = \boldsymbol{e}_\mu - \boldsymbol{e}_{\mu+1}, \qquad \boldsymbol{r}_\ell = \boldsymbol{e}_{\ell-1} + \boldsymbol{e}_\ell, \qquad 1 \leqslant \mu \leqslant \ell - 1, \\
&\boldsymbol{e}_a - \boldsymbol{e}_b = \sum_{\mu=a}^{b-1} \boldsymbol{r}_\mu, \qquad \boldsymbol{e}_a + \boldsymbol{e}_b = \sum_{\mu=a}^{b-1} \boldsymbol{r}_\mu + 2\sum_{\nu=b}^{\ell-2} \boldsymbol{r}_\nu + \boldsymbol{r}_{\ell-1} + \boldsymbol{r}_\ell,
\end{aligned}
\tag{7.85}
$$

$$
\boldsymbol{\omega} = \boldsymbol{e}_1 + \boldsymbol{e}_2 = \boldsymbol{r}_1 + 2\sum_{\nu=2}^{\ell-2} \boldsymbol{r}_\nu + \boldsymbol{r}_{\ell-1} + \boldsymbol{r}_\ell. \tag{7.86}
$$

The angle of two neighboring roots is $2\pi/3$ except for $\boldsymbol{r}_\ell$, which is orthogonal to $\boldsymbol{r}_{\ell-1}$, but with an angle $2\pi/3$ to $\boldsymbol{r}_{\ell-2}$. $d_\mu = 1$ for all simple roots. Thus, the algebra of SO(2ℓ) is D_ℓ.

The simple roots $\boldsymbol{r}_\mu$ and the largest root $\boldsymbol{\omega}$ for SO($2\ell+1$) are

$$
\begin{aligned}
&\boldsymbol{r}_\mu = \boldsymbol{e}_\mu - \boldsymbol{e}_{\mu+1}, \qquad \boldsymbol{r}_\ell = \boldsymbol{e}_\ell, \qquad 1 \leqslant \mu \leqslant \ell - 1, \\
&\boldsymbol{e}_a - \boldsymbol{e}_b = \sum_{\mu=a}^{b-1} \boldsymbol{r}_\mu, \quad \boldsymbol{e}_a + \boldsymbol{e}_b = \sum_{\mu=a}^{b-1} \boldsymbol{r}_\mu + 2\sum_{\nu=b}^{\ell} \boldsymbol{r}_\nu, \quad \boldsymbol{e}_a = \sum_{\mu=a}^{\ell} \boldsymbol{r}_\mu,
\end{aligned}
\tag{7.87}
$$

$$
\boldsymbol{\omega} = \boldsymbol{e}_1 + \boldsymbol{e}_2 = \boldsymbol{r}_1 + 2\sum_{\nu=2}^{\ell} \boldsymbol{r}_\nu. \tag{7.88}
$$

Except for $\boldsymbol{r}_\ell$, the angle of two neighboring roots is $2\pi/3$ and $d_\mu = 1$. But, $d_\ell = 1/2$ and the angle between $\boldsymbol{r}_\ell$ and $\boldsymbol{r}_{\ell-1}$ is $3\pi/4$. Thus, the algebra of SO($2\ell+1$) is B_ℓ.

7.4.3 *The* USp(2ℓ) *Group and its Lie Algebra*

In a (2ℓ)-dimensional space, the vector index a is taken to be j or $\overline{j}$, $1 \leqslant j \leqslant \ell$, in the following order:

$$
a = 1,\ \overline{1},\ 2,\ \overline{2},\ \cdots,\ \ell,\ \overline{\ell}. \tag{7.89}
$$

Generalizing the real orthogonal matrix R in Eq. (7.79) by replacing the unit matrix $\mathbf{1}$ with a real antisymmetric matrix J, we obtain the real pseudo-orthogonal matrix R,

$$R^T JR = J, \qquad R^* = R, \tag{7.90}$$

where

$$J = \mathbf{1}_\ell \times (i\sigma_2) = -J^{-1} = -J^T, \qquad \det\, J = 1,$$

$$J_{ab} = \begin{cases} 1 & \text{when } a = j, \quad b = \overline{j}, \\ -1 & \text{when } a = \overline{j}, \quad b = j, \\ 0 & \text{the remaining cases.} \end{cases} \tag{7.91}$$

In the multiplication rule of matrices, the product of two real pseudo-orthogonal matrices is a real pseudo-orthogonal matrix. The matrix product satisfies the associative law. The unit matrix is also a real pseudo-orthogonal matrix and plays the role of the identity. The inverse $R^{-1} = -JR^TJ$ and the transpose R^T of R are also real pseudo-orthogonal,

$$RJR^T = J, \qquad \left(R^{-1}\right)^T JR^{-1} = (-JRJ)JR^{-1} = J. \tag{7.92}$$

Thus, the set of all (2ℓ)-dimensional real pseudo-orthogonal matrices R, in the multiplication rule of matrices, constitutes a group, called the (2ℓ)-dimensional **real symplectic group**, denoted by $\mathrm{Sp}(2\ell, R)$. We are going to show that $\det R = 1$ so that $\mathrm{Sp}(2\ell, R)$ is a **connected Lie group**.

Let R be a transformation matrix in a (2ℓ)-dimensional real space

$$x_a \xrightarrow{R} x'_a = \sum_b R_{ab}x_b, \qquad R \in \mathrm{Sp}(2\ell, R). \tag{7.93}$$

The pseudo-inner product of two vectors x and y is defined as

$$\{x,\ y\}_J \equiv \sum_{ab} x_a J_{ab} y_b = \sum_{j=1}^{\ell} \left(x_j y_{\overline{j}} - x_{\overline{j}} y_j\right), \tag{7.94}$$

which is left invariant under a real pseudo-orthogonal transformation R:

$$\{x,\ y\}_J = \{Rx,\ Ry\}_J. \tag{7.95}$$

The pseudo-inner self-product of a vector vanishes.

In terms of the totally antisymmetric tensor one has

$$\sum_{a_1\cdots a_{2\ell}} \epsilon_{a_1\cdots a_{2\ell}} J_{a_1a_2} J_{a_3a_4} \cdots J_{a_{2\ell-1}a_{2\ell}} = 2^\ell \ell!. \tag{7.96}$$

In fact, in the sum there is only $2^\ell \ell!$ non-vanishing terms, which come from the transpose of each J matrix and the permutations among ℓ different J. Let $\left(X^b_{\cdot}\right)_a$ denote the bth column matrix of a 2ℓ dimensional matrix X. The determinant of X is

$$\epsilon_{a_1\cdots a_{2\ell}} \det X = \sum_{b_1\cdots b_{2\ell}} \epsilon_{b_1\cdots b_{2\ell}} X_{a_1}^{b_1} \cdots X_{a_{2\ell}}^{b_{2\ell}}. \tag{7.97}$$

Thus,

$$\begin{aligned}
\det X &= \left(2^\ell \ell!\right)^{-1} \sum_{a_1\cdots a_{2\ell}} (\det X)\, \epsilon_{a_1\cdots a_{2\ell}} J_{a_1 a_2} \cdots J_{a_{2\ell-1}a_{2\ell}} \\
&= \left(2^\ell \ell!\right)^{-1} \sum_{a_1\cdots a_{2\ell}} J_{a_1 a_2} \cdots J_{a_{2\ell-1}a_{2\ell}} \sum_{b_1\cdots b_{2\ell}} \epsilon_{b_1\cdots b_{2\ell}} X_{a_1}^{b_1} \cdots X_{a_{2\ell}}^{b_{2\ell}} \\
&= \left(2^\ell \ell!\right)^{-1} \sum_{b_1\cdots b_{2\ell}} \epsilon_{b_1\cdots b_{2\ell}} \left\{X^{b_1},\, X^{b_2}\right\}_J \cdots \left\{X^{b_{2\ell-1}},\, X^{b_{2\ell}}\right\}_J \\
&= \left(2^\ell \ell!\right)^{-1} \sum_{b_1\cdots b_{2\ell}} \epsilon_{b_1\cdots b_{2\ell}} \left\{RX^{b_1},\, RX^{b_2}\right\}_J \cdots \left\{RX^{b_{2\ell-1}},\, RX^{b_{2\ell}}\right\}_J \\
&= \det\ (RX), \qquad R \in \mathrm{Sp}(2\ell, R).
\end{aligned}$$

Thus, $\det R = 1$ and the group space of $\mathrm{Sp}(2\ell, R)$ is simply-connected.

From the definition (7.90), a diagonal matrix R in $\mathrm{Sp}(2\ell, R)$ is given as

$$R = \mathrm{diag}\left\{\mathrm{e}^{\omega_1},\ \mathrm{e}^{-\omega_1},\ \mathrm{e}^{\omega_2},\ \mathrm{e}^{-\omega_2},\ \cdots,\ \mathrm{e}^{\omega_\ell},\ \mathrm{e}^{-\omega_\ell}\right\}, \tag{7.98}$$

where the real parameters ω_j have no boundary, $-\infty < \omega_j < \infty$, such that $\mathrm{Sp}(2\ell, R)$ is **not a compact Lie group**.

Replace the real matrix R in Eq. (7.90) with the unitary matrix u,

$$u^T J u = J, \qquad u^\dagger = u^{-1}. \tag{7.99}$$

The set of all (2ℓ)-dimensional unitary pseudo-orthogonal matrices u, in the multiplication rule of matrices, forms a group, called the (2ℓ)-dimensional **unitary symplectic group**, denoted by $\mathrm{USp}(2\ell)$. It can be proven similarly that $\det u = 1$. Since the diagonal matrix $u \in \mathrm{USp}(2\ell)$ has the diagonal elements $\mathrm{e}^{\pm \mathrm{i}\varphi_j}$, whose module is one, $\mathrm{USp}(2\ell)$ is a **connected compact Lie group**. By making complex conjugate on Eq. (7.99), we obtain that

$$u^* = J^{-1} u J. \tag{7.100}$$

The self-representation of $\mathrm{USp}(2\ell)$ is self-conjugate, which leads to the conclusion that **each irreducible representation of** $\mathrm{USp}(2\ell)$ **is self-conjugate**. When $\ell = 1$, $J = \mathrm{i}\sigma_2$ and $(\boldsymbol{\sigma}\cdot\hat{\boldsymbol{n}})^T J = -J(\boldsymbol{\sigma}\cdot\hat{\boldsymbol{n}})$, so that any element $u(\hat{\boldsymbol{n}}, \omega) = \mathbf{1}\cos(\omega/2) - \mathrm{i}(\boldsymbol{\sigma}\cdot\hat{\boldsymbol{n}})\sin(\omega/2) \in \mathrm{SU}(2)$ satisfies Eq. (7.99). Namely, $\mathrm{SU}(2) = \mathrm{USp}(2)$.

Discuss the real parameters in the self-representation of $\mathrm{Sp}(2\ell, R)$ and $\mathrm{USp}(2\ell)$. For their infinitesimal elements

$$\begin{array}{lll} R = \mathbf{1} - \mathrm{i}\alpha X \in \mathrm{Sp}(2\ell, R), & X^* = -X, & X^T = JXJ, \\ u = \mathbf{1} - \mathrm{i}\beta Y \in \mathrm{USp}(2\ell), & Y^\dagger = Y, & Y^T = JYJ. \end{array} \tag{7.101}$$

Y is a hermitian matrix and the component form of $Y^T = JYJ$ is:

$$Y_{kj} = -Y_{\overline{j}\,\overline{k}}, \qquad Y_{\overline{k}j} = Y_{\overline{j}k}.$$

$Y^T = JYJ$ gives $\ell^2 + \ell(\ell-1) = \ell(2\ell-1)$ real constraints. Thus, the order of USp(2ℓ) is $g = 4\ell^2 - \ell(2\ell-1) = \ell(2\ell+1)$. In the self-representation of USp(2ℓ), due to Eq. (7.101), the generators T_A can be expressed in terms of the matrices $T^{(r)}_{jk}$ $(r = 1, 2)$ in SU(ℓ) and the Pauli matrices σ_a:

$$\begin{array}{c} T^{(2)}_{jk} \times \mathbf{1}_2, \qquad T^{(1)}_{jk} \times \sigma_d, \qquad T^{(1)}_{jj} \times \sigma_d/\sqrt{2}, \\ 1 \leqslant d \leqslant 3, \qquad 1 \leqslant j < k \leqslant \ell. \end{array} \tag{7.102}$$

They satisfied the orthonormal condition $\mathrm{Tr}(T_A T_B) = \delta_{AB}$ so that the structure constant $C_{AB}{}^D$ of USp(2ℓ) is totally antisymmetric and the Killing form g_{AB} is a negative constant matrix. These coincide with the fact that USp(2ℓ) is compact. The generators in the self-representation of Sp$(2\ell, R)$ are pure imaginary and can be obtained from Eq. (7.102) by replacing σ_1 and σ_3 with $\tau_1 = \mathrm{i}\sigma_1$ and $\tau_3 = \mathrm{i}\sigma_3$, respectively. If one moves the additional factor i from the generators of Sp$(2\ell, R)$ to its parameters, the generators in the corresponding representations of Sp$(2\ell, R)$ and USp(2ℓ) are the same, but their parameters are different. This is the **standard method for calculating the irreducible representations of a noncompact Lie group**. We will calculate the representations of the Lorentz group by this method in Chap. 11.

The diagonal matrices in the generators (7.102) span the Cartan subalgebra of the USp(2ℓ) group,

$$H_j = T^{(1)}_{jj} \times \sigma_3/\sqrt{2}, \qquad 1 \leqslant j \leqslant \ell. \tag{7.103}$$

The remaining generators are combined to be the eigenvectors of H_j, $[H_j,\ E_{\boldsymbol{\alpha}}] = \alpha_j E_{\boldsymbol{\alpha}}$:

$$\begin{aligned} E^{(1)}_{ab} &= \left\{T^{(1)}_{ab} \times \sigma_3 + \mathrm{i}T^{(2)}_{ab} \times \mathbf{1}_2\right\}/\sqrt{2}, \\ E^{(2)}_{ab} &= \left\{T^{(1)}_{ab} \times \sigma_3 - \mathrm{i}T^{(2)}_{ab} \times \mathbf{1}_2\right\}/\sqrt{2}, \\ E^{(3)}_{ab} &= T^{(1)}_{ab} \times (\sigma_1 + \mathrm{i}\sigma_2)/\sqrt{2}, \\ E^{(4)}_{ab} &= T^{(1)}_{ab} \times (\sigma_1 - \mathrm{i}\sigma_2)/\sqrt{2}, \end{aligned}$$

$$\begin{aligned} E_a^{(5)} &= T_{aa}^{(1)} \times (\sigma_1 + \mathrm{i}\sigma_2)/2, \\ E_a^{(6)} &= T_{aa}^{(1)} \times (\sigma_1 - \mathrm{i}\sigma_2)/2, \end{aligned} \tag{7.104}$$

with the eigenvalues $\sqrt{1/2}\,(\boldsymbol{e}_a - \boldsymbol{e}_b)$, $-\sqrt{1/2}\,(\boldsymbol{e}_a - \boldsymbol{e}_b)$, $\sqrt{1/2}\,(\boldsymbol{e}_a + \boldsymbol{e}_b)$, $-\sqrt{1/2}(\boldsymbol{e}_a + \boldsymbol{e}_b)$, $\sqrt{2}\boldsymbol{e}_a$, and $-\sqrt{2}\boldsymbol{e}_a$, respectively, where $1 \leqslant a < b \leqslant \ell$. The generators labeled by the superscripts (1), (3), and (5) correspond to the positive roots. The simple roots $\boldsymbol{r}_\mu$ are

$$\begin{aligned} &\boldsymbol{r}_\mu = \sqrt{1/2}\,(\boldsymbol{e}_\mu - \boldsymbol{e}_{\mu+1}), \qquad d_\mu = 1/2, \quad 1 \leqslant \mu \leqslant \ell - 1, \\ &\boldsymbol{r}_\ell = \sqrt{2}\boldsymbol{e}_\ell, \qquad\qquad\qquad\quad d_\ell = 1. \end{aligned} \tag{7.105}$$

$$\begin{aligned} \sqrt{1/2}\,(\boldsymbol{e}_a - \boldsymbol{e}_b) &= \sum_{\mu=a}^{b-1} \boldsymbol{r}_\mu, \qquad \sqrt{2}\boldsymbol{e}_a = 2\sum_{\mu=a}^{\ell-1} \boldsymbol{r}_\mu + \boldsymbol{r}_\ell, \\ \sqrt{1/2}\,(\boldsymbol{e}_a + \boldsymbol{e}_b) &= \sum_{\mu=a}^{b-1} \boldsymbol{r}_\mu + 2\sum_{\mu=b}^{\ell-1} \boldsymbol{r}_\mu + \boldsymbol{r}_\ell. \end{aligned}$$

The angle of two neighboring simple roots is $2\pi/3$ except for the angle of $\boldsymbol{r}_{\ell-1}$ and $\boldsymbol{r}_\ell$ which is $3\pi/4$. Thus, the algebra of USp(2ℓ) is C_ℓ. The largest root $\boldsymbol{\omega}$ in C_ℓ is

$$\boldsymbol{\omega} = \sqrt{2}\boldsymbol{e}_1 = 2\sum_{\mu=1}^{\ell-1} \boldsymbol{r}_\mu + \boldsymbol{r}_\ell. \tag{7.106}$$

7.5 Representations of a Simple Lie Algebra

A representation of a Lie group is also a representation of its Lie algebra. Let $\mathcal{L}$ denote a simple Lie algebra with dimension larger than one in this section. $\mathcal{L}$ has its compact real form where its representation is equivalent to an unitary one. We are going to study the representation matrices of the generators in the unitary representation of $\mathcal{L}$.

7.5.1 *Representations and Weights*

In a unitary representation of a simple Lie algebra, the Cartan–Weyl bases of generators, denoted by $D(H_j)$ and $D(E_{\boldsymbol{\alpha}})$ for convenience, satisfy

$$D(H_j)^\dagger = D(H_j), \qquad D(E_{\boldsymbol{\alpha}})^\dagger = D(E_{-\boldsymbol{\alpha}}).$$

The basis vectors $|\boldsymbol{m}\rangle$ in the representation space, called the **basis states**, are usually chosen to be the common eigenvectors of the mutually commutable hermitian operators H_j,

$$H_j\, |\boldsymbol{m}\rangle = m_j\, |\boldsymbol{m}\rangle. \tag{7.107}$$

The ℓ-dimensional vector $\boldsymbol{m} = (m_1, \cdots, m_\ell)$, whose components are the eigenvalues m_j, is called the **weight** of the basis state $|\boldsymbol{m}\rangle$. The ℓ-dimensional space is called the **weight space.** Let n denote the number of the linearly independent basis states $|\boldsymbol{m}\rangle$ with the weight $\boldsymbol{m}$ in the representation space. n is called the multiplicity of the weight $\boldsymbol{m}$. A weight $\boldsymbol{m}$ is single if $n = 1$, and is **multiple** if $n > 1$. In the adjoint representation, the representation space is the Lie algebra itself, a weight is a root, and the weight space coincides with the root space. **The roots in $\mathcal{L}$ can be calculated from the weights in its adjoint representation.**

Except for the identical representation, the generators of any irreducible representation of $\mathcal{L}$ are **linearly independent of each other**, and, in this meaning, the representation of $\mathcal{L}$ is faithful. Thus, there are ℓ linearly independent weights in a representation of $\mathcal{L}$ with rank ℓ. From the regular commutative relations (7.40), we have

$$\text{Tr}\ D(E_{\boldsymbol{\alpha}}) = 0, \qquad \text{Tr}\ D(H_{\boldsymbol{\alpha}}) = 0, \qquad \text{Tr}\ D(H_j) = 0. \tag{7.108}$$

Namely, the sum of weights of all basis states in an irreducible representation of $\mathcal{L}$ vanishes,

$$\sum \boldsymbol{m} = 0. \tag{7.109}$$

A weight $\boldsymbol{m}$ is said to be higher than a weight $\boldsymbol{m}'$ if the first nonzero component of $\boldsymbol{m} - \boldsymbol{m}'$ is positive. For a positive root $\boldsymbol{\alpha}$, one has

$$\begin{aligned} H_j\left(E_{\pm\boldsymbol{\alpha}}\, |\boldsymbol{m}\rangle\right) = &\ [H_j,\ E_{\pm\boldsymbol{\alpha}}]\ |\boldsymbol{m}\rangle + E_{\pm\boldsymbol{\alpha}} H_j\, |\boldsymbol{m}\rangle \\ = &\ (m_j \pm \alpha_j)\ \left(E_{\pm\boldsymbol{\alpha}}\, |\boldsymbol{m}\rangle\right). \end{aligned} \tag{7.110}$$

Thus, the action of $E_{\boldsymbol{\alpha}}$ on $|\boldsymbol{m}\rangle$ raises its weight $\boldsymbol{m}$ by a positive root $\boldsymbol{\alpha}$, and that of $E_{-\boldsymbol{\alpha}}$ lowers $\boldsymbol{m}$ by $\boldsymbol{\alpha}$, so that $E_{\boldsymbol{\alpha}}$ is called the **raising operator** and $E_{-\boldsymbol{\alpha}}$ the **lowering operator**. In a representation space with a finite dimension, there is a basis state $|\boldsymbol{M}\rangle$ whose weight $\boldsymbol{M}$ is higher than the weight of any other basis state $|\boldsymbol{m}\rangle$. $\boldsymbol{M}$ is called the **highest weight** of the representation and $|\boldsymbol{M}\rangle$ is the highest weight state. Any generator $E_{\boldsymbol{\alpha}}$ with the positive root $\boldsymbol{\alpha}$ annihilates the highest weight state $|\boldsymbol{M}\rangle$,

$$E_{\boldsymbol{\alpha}}\, |\boldsymbol{M}\rangle = 0, \qquad \forall\ \text{positive root}\ \ \boldsymbol{\alpha}. \tag{7.111}$$

This is the main method to calculate the highest weight in a representation.

Theorem 7.7 The highest weight $\boldsymbol{M}$ in an irreducible representation of a simple Lie algebra $\mathcal{L}$ is single. Two irreducible representations of $\mathcal{L}$ are equivalent if and only if their highest weights are the same.

Proof Let $|\boldsymbol{M}\rangle$ be the highest weight state. We are going to prove that if there is another state $|\boldsymbol{M}\rangle'$ with the highest weight $\boldsymbol{M}$ in the representation space, it has to be proportional to $|\boldsymbol{M}\rangle$. Since the representation is irreducible, $|\boldsymbol{M}\rangle'$ can be expressed as a linear combination of the following states each of which has the highest weight $\boldsymbol{M}$,

$$E_{\boldsymbol{\lambda}} \cdots E_{\boldsymbol{\beta}} E_{\boldsymbol{\alpha}} \, |\boldsymbol{M}\rangle. \tag{7.112}$$

If a raising operator, say $E_{\boldsymbol{\rho}}$, is contained in Eq. (7.112), move $E_{\boldsymbol{\rho}}$ rightward according to $E_{\boldsymbol{\rho}} E_{\boldsymbol{\tau}} = E_{\boldsymbol{\tau}} E_{\boldsymbol{\rho}} + [E_{\boldsymbol{\rho}}, \ E_{\boldsymbol{\tau}}]$. The additional term in the move is $N_{\boldsymbol{\rho},\boldsymbol{\tau}} E_{\boldsymbol{\rho}+\boldsymbol{\tau}}$ if $\boldsymbol{\rho}+\boldsymbol{\tau}$ is a root, $\boldsymbol{\rho} \cdot \boldsymbol{H}$ if $\boldsymbol{\rho}+\boldsymbol{\tau} = 0$, and 0 if $\boldsymbol{\rho}+\boldsymbol{\tau}$ is not a root. In each case, the additional term contains less operators than the original one. When the raising operator $E_{\boldsymbol{\rho}}$ moves to the rightmost position, the term is annihilated owing to Eq. (7.111), and the number of operators contained in each remaining term decreases. We make those moves of the raising operators in each term until it does not contain any raising operator. In the same time the lowering operators also disappear in the result terms because the weight of each term is $\boldsymbol{M}$. Namely, each term is proportional to $|\boldsymbol{M}\rangle$.

The highest weights of two equivalent representations are obviously the same. Conversely, if the highest weights of two representations are the same, we define a correspondence between the basis states of two representations

$$|\boldsymbol{M}\rangle \longleftrightarrow |\boldsymbol{M}\rangle'$$
$$|j\rangle \equiv E_{\boldsymbol{\lambda}} \cdots E_{\boldsymbol{\beta}} E_{\boldsymbol{\alpha}} \, |\boldsymbol{M}\rangle \longleftrightarrow |j\rangle' \equiv E_{\boldsymbol{\lambda}} \cdots E_{\boldsymbol{\beta}} E_{\boldsymbol{\alpha}} \, |\boldsymbol{M}\rangle'.$$

In the following we show by reduction to absurdity that the correspondence is one-to-one, namely, any linear relation which holds in one representation space also holds in another representation space, and vice versa,

$$\forall \sum_j c_j \, |j\rangle = 0, \qquad \exists \sum_j c_j \, |j\rangle' \equiv |w\rangle' = 0.$$

In fact, if $|w\rangle' \neq 0$, it corresponds to the null state in the first representation space, so does the state $E_{\alpha} \, |w\rangle'$ obtained from $|w\rangle'$ by action of any generator E_{α}. Then, the set of those states constitutes an invariant subspace in the second representation space, which is not equal to the whole space because it does not contain $|\boldsymbol{M}\rangle'$ at least. It is in contradiction to that the representation is irreducible. □

Hereafter, **an irreducible representation of a simple Lie algebra $\mathcal{L}$ is also called the highest weight representation.** The highest weight of the adjoint representation is the largest root $\boldsymbol{\omega}$ of $\mathcal{L}$.

7.5.2 *Weyl Reflection and Equivalent Weights*

Theorem 7.8 If $\boldsymbol{\alpha}$ is a nonzero root in a simple Lie algebra $\mathcal{L}$ and $\boldsymbol{m}$ is a weight in an irreducible representation space of $\mathcal{L}$, then,

$$\frac{2\boldsymbol{m}\cdot\boldsymbol{\alpha}}{\boldsymbol{\alpha}\cdot\boldsymbol{\alpha}} \equiv \Gamma\left(\boldsymbol{m}/\boldsymbol{\alpha}\right) = \text{ integer,} \tag{7.113}$$

and $\boldsymbol{m} - \Gamma\left(\boldsymbol{m}/\boldsymbol{\alpha}\right)\boldsymbol{\alpha}$ is also a weight with the same multiplicity as $\boldsymbol{m}$.

Proof Although Theorem 7.8 looks similar to Theorem 7.3, there is an important difference between them that the multiplicity of a weight $\boldsymbol{m}$, which appears on the numerator of Eq. (7.113), may be larger than 1. For a multiple weight, **any combination of the basis states with the same weight is also a state in the representation space.** Before proving the theorem we need to make a rule for choosing the basis states.

Without loss of generality, we make a convention by choosing the sign of $\boldsymbol{\alpha}$ that

$$\boldsymbol{m}\cdot\boldsymbol{\alpha} \geqslant 0. \tag{7.114}$$

Arbitrarily choose a basis state $|\boldsymbol{m}\rangle$ with the weight $\boldsymbol{m}$ and apply to it with $E_{\boldsymbol{\alpha}}$ successively until the state is annihilated,

$$E_{\boldsymbol{\alpha}}^{n}\,|\boldsymbol{m}\rangle = |\boldsymbol{m}+n\boldsymbol{\alpha}\rangle \neq 0, \qquad 0 \leqslant n \leqslant p, \qquad E_{\boldsymbol{\alpha}}^{p+1}\,|\boldsymbol{m}\rangle = 0.$$

Forgetting the choice of $|\boldsymbol{m}\rangle$, we define a chain of basis states by applying $E_{-\boldsymbol{\alpha}}$ to $|\boldsymbol{m}+p\boldsymbol{\alpha}\rangle$ successively,

$$\begin{aligned} &|\boldsymbol{m}+n\boldsymbol{\alpha}\rangle \equiv E_{-\boldsymbol{\alpha}}^{p-n}\,|\boldsymbol{m}+p\boldsymbol{\alpha}\rangle \neq 0, \qquad && -q \leqslant n \leqslant p, \\ &E_{-\boldsymbol{\alpha}}^{p+q+1}\,|\boldsymbol{m}+p\boldsymbol{\alpha}\rangle = 0, && p+q+1 > 0. \end{aligned} \tag{7.115}$$

We do not care whether or not the state $E_{-\boldsymbol{\alpha}}^{p}\,|\boldsymbol{m}+p\boldsymbol{\alpha}\rangle$ coincides with the original state $|\boldsymbol{m}\rangle$. We are going to prove by mathematical induction whether the subspace spanned by the basis states $|\boldsymbol{m}+n\boldsymbol{\alpha}\rangle$ is closed in the application of $E_{\pm\boldsymbol{\alpha}}$, namely whether

$$E_{\boldsymbol{\alpha}}\ |\boldsymbol{m}+n\boldsymbol{\alpha}\rangle = B_n\ |\boldsymbol{m}+(n+1)\boldsymbol{\alpha}\rangle, \qquad -q \leqslant n \leqslant p. \tag{7.116}$$

Equation (7.116) holds for $n = p$ and $n = p-1$,

$$\begin{aligned}
&E_{\boldsymbol{\alpha}}\ |\boldsymbol{m}+p\boldsymbol{\alpha}\rangle = 0, \\
&E_{\boldsymbol{\alpha}}\ |\boldsymbol{m}+(p-1)\boldsymbol{\alpha}\rangle = [E_{\boldsymbol{\alpha}},\ E_{-\boldsymbol{\alpha}}]\ |\boldsymbol{m}+p\boldsymbol{\alpha}\rangle \\
&\quad = (\boldsymbol{\alpha}\cdot\boldsymbol{H})\ |\boldsymbol{m}+p\boldsymbol{\alpha}\rangle = (\boldsymbol{m}\cdot\boldsymbol{\alpha}+p|\boldsymbol{\alpha}|^2)\ |\boldsymbol{m}+p\boldsymbol{\alpha}\rangle, \\
&B_p = 0, \qquad B_{p-1} = \boldsymbol{m}\cdot\boldsymbol{\alpha}+p|\boldsymbol{\alpha}|^2.
\end{aligned} \tag{7.117}$$

If equation (7.116) holds for $n > k$, when $n = k$ we have

$$\begin{aligned}
E_{\boldsymbol{\alpha}}\ |\boldsymbol{m}+k\boldsymbol{\alpha}\rangle &= E_{\boldsymbol{\alpha}}E_{-\boldsymbol{\alpha}}\ |\boldsymbol{m}+(k+1)\boldsymbol{\alpha}\rangle \\
&= [E_{\boldsymbol{\alpha}},\ E_{-\boldsymbol{\alpha}}]\ |\boldsymbol{m}+(k+1)\boldsymbol{\alpha}\rangle + E_{-\boldsymbol{\alpha}}E_{\boldsymbol{\alpha}}\ |\boldsymbol{m}+(k+1)\boldsymbol{\alpha}\rangle \\
&= \left\{\boldsymbol{m}\cdot\boldsymbol{\alpha}+(k+1)|\boldsymbol{\alpha}|^2+B_{k+1}\right\}\ |\boldsymbol{m}+(k+1)\boldsymbol{\alpha}\rangle.
\end{aligned}$$

Equation (7.116) is proved, and B_n satisfies the recursive relation

$$\begin{aligned}
B_n &= B_{n+1}+\boldsymbol{m}\cdot\boldsymbol{\alpha}+(n+1)|\boldsymbol{\alpha}|^2 \\
&= B_{n+2}+2\boldsymbol{m}\cdot\boldsymbol{\alpha}+\{(n+1)+(n+2)\}\,|\boldsymbol{\alpha}|^2 \\
&= \cdots \\
&= B_{n+(p-n)}+(p-n)\boldsymbol{m}\cdot\boldsymbol{\alpha}+\frac{1}{2}(p-n)(n+p+1)|\boldsymbol{\alpha}|^2 \\
&= \frac{1}{2}(p-n)\left\{2\boldsymbol{m}\cdot\boldsymbol{\alpha}+(n+p+1)|\boldsymbol{\alpha}|^2\right\}, \\
B_{-q} &= \frac{1}{2}(p+q)\left\{2\boldsymbol{m}\cdot\boldsymbol{\alpha}+(-q+p+1)|\boldsymbol{\alpha}|^2\right\}.
\end{aligned}$$

On the other hand,

$$\begin{aligned}
0 &= E_{\boldsymbol{\alpha}}E_{-\boldsymbol{\alpha}}\ |\boldsymbol{m}-q\boldsymbol{\alpha}\rangle \\
&= \{[E_{\boldsymbol{\alpha}},\ E_{-\boldsymbol{\alpha}}]+E_{-\boldsymbol{\alpha}}E_{\boldsymbol{\alpha}}\}\ |\boldsymbol{m}-q\boldsymbol{\alpha}\rangle \\
&= \left\{\boldsymbol{m}\cdot\boldsymbol{\alpha}-q|\boldsymbol{\alpha}|^2+B_{-q}\right\}\ |\boldsymbol{m}-q\boldsymbol{\alpha}\rangle, \\
B_{-q} &= -\boldsymbol{m}\cdot\boldsymbol{\alpha}+q|\boldsymbol{\alpha}|^2.
\end{aligned}$$

In comparison one obtains

$$\frac{1}{2}(p+q+1)\left\{2\boldsymbol{m}\cdot\boldsymbol{\alpha}-(q-p)|\boldsymbol{\alpha}|^2\right\} = 0.$$

Since $(p+q+1) > 0$ and $|\boldsymbol{\alpha}|^2 > 0$, equation (7.113) is proved where the integer is $q-p$,

$$\Gamma\left(\boldsymbol{m}/\boldsymbol{\alpha}\right) = \frac{2\boldsymbol{m}\cdot\boldsymbol{\alpha}}{|\boldsymbol{\alpha}|^2} = q-p = \text{ integer}. \tag{7.118}$$

Due to our convention (7.114), $q \geqslant p \geqslant 0$. The state chain $|\boldsymbol{m}+n\boldsymbol{\alpha}\rangle$ given in Eq. (7.115) contains both states $|\boldsymbol{m}\rangle$ and $|\boldsymbol{m}'\rangle$ with the weight $\boldsymbol{m}$ and $\boldsymbol{m}'$, respectively, where

$$\boldsymbol{m}' = \boldsymbol{m} - \Gamma(\boldsymbol{m}/\boldsymbol{\alpha})\,\boldsymbol{\alpha} = \boldsymbol{m} - (q-p)\boldsymbol{\alpha}. \tag{7.119}$$

If there is another state with the weight $\boldsymbol{m}$ in the subspace orthogonal to the state chain $|\boldsymbol{m}+n\boldsymbol{\alpha}\rangle$ given in Eq. (7.115), one is able to find another state chain $|\boldsymbol{m}+n\boldsymbol{\alpha}\rangle$ by repeating the above steps in the subspace. Even though the length $(p+q+1)$ of the new state chain may be different to the first one, the difference $(q-p)$ is same and the new state chain must contain two states with the weight $\boldsymbol{m}$ and $\boldsymbol{m}'$, respectively. If the multiplicity of the weight $\boldsymbol{m}$ is d, there are d state chains so that the multiplicity d' of the weight $\boldsymbol{m}'$ is not less than d. Conversely, if the state chain is calculated from the state $|\boldsymbol{m}'\rangle$, one obtains $d \geqslant d'$. Thus, two weights $\boldsymbol{m}$ and $\boldsymbol{m}'$ have the **same multiplicity** and are called the **equivalent weights**. □

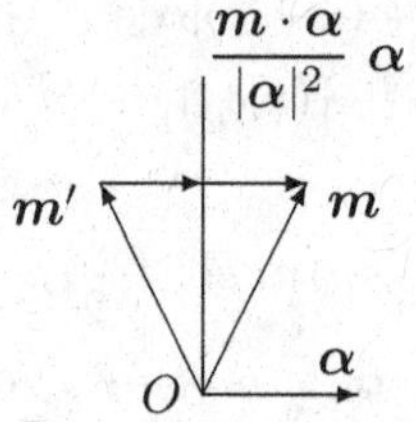

Fig. 7.2 A Weyl reflection

Two weights $\boldsymbol{m}$ and $\boldsymbol{m}'$ are the mirror images with respect to the plane which is perpendicular to $\boldsymbol{\alpha}$ and across the origin. This reflection is called a **Weyl reflection** in the weight space (see Fig. 7.2). The product of two Weyl reflections is defined as their successive applications. The set of all Weyl reflections and their products for a representation forms its **Weyl group** W. The weights related by any element of the Weyl group are equivalent, and the number of equivalent weights is called the **size of Weyl orbit**.

A weight $\boldsymbol{M}$ satisfying

$$\Gamma(\boldsymbol{M}/\boldsymbol{r}_\mu) = \text{ non-negative integer}, \qquad \forall \text{ simple root } \boldsymbol{r}_\mu, \tag{7.120}$$

is called a **dominant weight**. In an irreducible representation of a simple Lie algebra $\mathcal{L}$ there are a few dominant weights, single or multiple. **Each weight $\boldsymbol{m}$ in the representation is equivalent to one dominant**

weight. The dimension of the representation is equal to the sum of products of the multiplicity of each dominant weight and its size of Weyl orbit. **The highest weight of an irreducible representation is a dominant weight and simple** because of Eqs. (7.118), (7.111), and Theorem 7.7. Dynkin proved that a space constructed by applying the lowering operators $E_{-\boldsymbol{r}_\mu}$ successively to a state $|\boldsymbol{M}\rangle$ with a dominant weight $\boldsymbol{M}$ corresponds to an irreducible representation of $\mathcal{L}$ with a finite dimension. This is the foundation for the method of the **block weight diagram** which will be discussed in the next chapter.

Theorem 7.9 Any dominant weight $\boldsymbol{M}$ is the highest weight of one irreducible representation of a simple Lie algebra $\mathcal{L}$ with finite dimension.

7.5.3 *Mathematical Property of Representations*

The highest weight $\boldsymbol{M}$ gives the full property of an irreducible representation of a simple Lie algebra $\mathcal{L}$. In this subsection we only quote some mathematical results of the highest weight representation.

Let G be a compact simple Lie group with the Lie algebra $\mathcal{L}$, and H be the abelian Lie subgroup produced from the Cartan subalgebra $\mathcal{H}$. The elements in H are characterized by ℓ parameters $\boldsymbol{\varphi} \equiv (\varphi_1, \varphi_2, \cdots, \varphi_\ell)$,

$$R = \exp\{-\mathrm{i}\boldsymbol{\varphi}\cdot\boldsymbol{H}\} = \exp\left\{-\mathrm{i}\sum_{j=1}^{\ell}\varphi_j H_j\right\} \in \mathrm{H}. \tag{7.121}$$

Each element in G is conjugate to an element in H so that the class in G is characterized by the parameters $\boldsymbol{\varphi}$. In an irreducible representation $D^{\boldsymbol{M}}$ the character $\chi(\boldsymbol{M},\boldsymbol{\varphi})$ of the class $\boldsymbol{\varphi}$ is

$$\begin{aligned}\chi(\boldsymbol{M},\boldsymbol{\varphi}) &= \mathrm{Tr}\ \exp\left\{-\mathrm{i}\sum_{j=1}^{\ell}\varphi_j D(H_j)\right\} \\ &= \sum_{\boldsymbol{m}} b(\boldsymbol{m})\exp\{-\mathrm{i}\boldsymbol{\varphi}\cdot\boldsymbol{m}\}.\end{aligned} \tag{7.122}$$

$b(\boldsymbol{m})$ is the multiplicity of the weight $\boldsymbol{m}$, and in fact, the sum in Eq. (7.122) runs over all basis states in the representation space. The character $\chi(\boldsymbol{M},\boldsymbol{\varphi})$ can be calculated by the girdle $\xi(\boldsymbol{K},\boldsymbol{\varphi})$ introduced by Weyl,

$$\xi(\boldsymbol{K},\boldsymbol{\varphi}) = \sum_{S\in\mathrm{W}}\delta_S\exp\left\{-\mathrm{i}\sum_{j=1}^{\ell}(S\boldsymbol{K})_j\varphi_j\right\}, \tag{7.123}$$

where S is the element in the Weyl group W and the reflection parity δ_S is 1 if S is a product of even Weyl reflections, and -1 if S is a product of odd Weyl reflections. $\boldsymbol{K} = \boldsymbol{M} + \boldsymbol{\rho}$, where $2\boldsymbol{\rho}$ is the sum of all positive roots in $\mathcal{L}$. Then,

$$\chi(\boldsymbol{M}, \varphi) = \frac{\xi(\boldsymbol{K}, \varphi)}{\xi(\boldsymbol{\rho}, \varphi)}. \tag{7.124}$$

$\xi(\boldsymbol{\rho}, \varphi)$ is independent of the representation and is related to the class integral of the Lie group. The orthonormal relations of the characters of two irreducible representations are expressed as

$$\begin{aligned} &\int \chi(\boldsymbol{M}, \varphi)^* \chi(\boldsymbol{M}', \varphi) |\xi(\boldsymbol{\rho}, \varphi)|^2 (d\varphi) \\ &\quad = \int \xi(\boldsymbol{K}, \varphi)^* \xi(\boldsymbol{K}', \varphi)(d\varphi) = \delta_{\boldsymbol{K}\boldsymbol{K}'}. \end{aligned} \tag{7.125}$$

When φ goes to zero, $\chi(\boldsymbol{M}, \varphi)$ is an indefinite form and tends to the dimension $d(\boldsymbol{M})$,

$$d(\boldsymbol{M}) = \prod_{\alpha \in \Delta_+} \left\{ 1 + \frac{\boldsymbol{M} \cdot \boldsymbol{\alpha}}{\boldsymbol{\rho} \cdot \boldsymbol{\alpha}} \right\}, \tag{7.126}$$

where Δ_+ is the set of the positive roots in $\mathcal{L}$.

7.5.4 *Fundamental Dominant Weights*

In the weight space of a simple Lie algebra $\mathcal{L}$ with rank ℓ, there are ℓ dominant weights $\boldsymbol{w}_\mu$ satisfying

$$\Gamma\left(\boldsymbol{w}_\mu / \boldsymbol{r}_\nu\right) = d_\nu^{-1} \left(\boldsymbol{w}_\mu \cdot \boldsymbol{r}_\nu\right) = \delta_{\mu\nu}, \qquad d_\nu = \boldsymbol{r}_\nu \cdot \boldsymbol{r}_\nu / 2. \tag{7.127}$$

$\boldsymbol{w}_\mu$ is called the **fundamental dominant weight**. An irreducible representation of $\mathcal{L}$ is called the **basic representation** of $\mathcal{L}$ if its highest weight is a fundamental dominant weight $\boldsymbol{w}_\mu$. From Theorem 7.8 and Eq. (7.120) any weight $\boldsymbol{m}$ is an integral linear combination of $\boldsymbol{w}_\mu$:

$$\begin{aligned} \boldsymbol{M} &= \sum_{\mu=1}^{\ell} \boldsymbol{w}_\mu M_\mu, \quad M_\mu = \Gamma\left(\boldsymbol{M} / \boldsymbol{r}_\mu\right) = \text{non-negative integer}, \\ \boldsymbol{m} &= \sum_{\mu=1}^{\ell} \boldsymbol{w}_\mu m_\mu, \qquad m_\mu = \Gamma\left(\boldsymbol{m} / \boldsymbol{r}_\mu\right) = \text{integer}. \end{aligned} \tag{7.128}$$

The Weyl reflection relation of two equivalent weights also becomes simpler,

$$\boldsymbol{m}' \xrightarrow{\boldsymbol{r}_\mu} \boldsymbol{m} - m_\mu \boldsymbol{r}_\mu, \qquad \boldsymbol{m} = \sum_\mu m_\mu \boldsymbol{w}_\mu. \tag{7.129}$$

Comparing Eq. (7.127) with Eq. (7.59), one has

$$\boldsymbol{r}_\mu = \sum_{\nu=1}^{\ell} \boldsymbol{w}_\nu A_{\nu\mu}, \qquad \boldsymbol{w}_\nu = \sum_{\mu=1}^{\ell} \boldsymbol{r}_\mu \left(A^{-1}\right)_{\mu\nu}. \tag{7.130}$$

$A_{\nu\mu}$ is the component of the simple root $\boldsymbol{r}_\mu$ with respect to $\boldsymbol{w}_\nu$. Hereafter, we will usually choose the fundamental dominant weight $\boldsymbol{w}_\mu$ to be the basis vectors in the weight space and in the root space where the components of both the weight and the root are integers, and

$$\begin{aligned} &(\boldsymbol{r}_\mu \cdot \boldsymbol{r}_\nu) = d_\mu A_{\mu\nu}, && (\boldsymbol{r}_\mu \cdot \boldsymbol{w}_\nu) = d_\mu \delta_{\mu\nu}, \\ &(\boldsymbol{w}_\mu \cdot \boldsymbol{w}_\nu) = d_\mu \left(A^{-1}\right)_{\mu\nu}, && d_\mu = (\boldsymbol{r}_\mu \cdot \boldsymbol{r}_\mu)/2. \end{aligned} \tag{7.131}$$

$d_\mu A_{\mu\nu}$ is called the symmetrized Cartan matrix. The shortage of the fundamental dominant weights is that they are not orthonormal.

7.5.5 *The Casimir Operator of Order 2*

The Casimir operator C_2 of order 2 of a simple Lie algebra $\mathcal{L}$ is given in Eq. (7.24). In the mathematical convention, d_μ for the longer simple root is normalized to be one (see subsection 7.3.2), the Killing form is a constant matrix, and the structure constant $C_{AB}{}^D$ is totally antisymmetric with respect to its three indices. The Casimir operator C_2 is defined as

$$C_2 = \sum_A I_A I_A = \sum_{j=1}^{\ell} H_j H_j + \sum_{\alpha \in \Delta_+} \{E_{\boldsymbol{\alpha}} E_{-\boldsymbol{\alpha}} + E_{-\boldsymbol{\alpha}} E_{\boldsymbol{\alpha}}\}. \tag{7.132}$$

Since C_2 commutes with each generator I_B, $D^{\boldsymbol{M}}(C_2)$ in an irreducible representation $D^{\boldsymbol{M}}$ of $\mathcal{L}$ is a constant matrix $\mathbf{1}C_2(\boldsymbol{M})$ owing to the Schur theorem. The constant $C_2(\boldsymbol{M})$ is called the Casimir invariant of order 2 in the representation $D^{\boldsymbol{M}}$, or briefly the **Casimir invariant**. Applying C_2 to the highest weight state $|\boldsymbol{M}\rangle$, one obtains from Eq. (7.111),

$$\begin{aligned} C_2 |\boldsymbol{M}\rangle &= \sum_{j=1}^{\ell} H_j H_j |\boldsymbol{M}\rangle + \sum_{\alpha \in \Delta_+} [E_{\boldsymbol{\alpha}}, E_{-\boldsymbol{\alpha}}] \, |\boldsymbol{M}\rangle, \\ C_2(\boldsymbol{M}) &= \boldsymbol{M} \cdot (\boldsymbol{M} + 2\boldsymbol{\rho}) = \sum_{\mu,\nu=1}^{\ell} M_\mu d_\mu \left(A^{-1}\right)_{\mu\nu} (M_\nu + 2), \end{aligned} \tag{7.133}$$

where $2\boldsymbol{\rho}$ is the sum of the positive roots in $\mathcal{L}$. It can be shown as follows that $\boldsymbol{\rho}$ is also equal to the sum of the fundamental dominant weights

$$2\rho = \sum_{\alpha \in \Delta_+} \boldsymbol{\alpha} = 2\sum_{\mu=1}^{\ell} \boldsymbol{w}_\mu. \tag{7.134}$$

In fact, for a given simple root $\boldsymbol{r}_\mu$, one is able to construct a root chain in Δ_+ from any positive root $\boldsymbol{\alpha}$, except for $\boldsymbol{r}_\mu$,

$$(\boldsymbol{\alpha} - q\boldsymbol{r}_\mu), \ \cdots, \ (\boldsymbol{\alpha} - \boldsymbol{r}_\mu), \ \boldsymbol{\alpha}, \ (\boldsymbol{\alpha} + \boldsymbol{r}_\mu), \ \cdots, \ (\boldsymbol{\alpha} + p\boldsymbol{r}_\mu).$$

The sum of the roots in the root chain is $(p+q+1)\left[\boldsymbol{\alpha} + \boldsymbol{r}_\mu(p-q)/2\right]$, which is orthogonal to $\boldsymbol{r}_\mu$ due to Eq. (7.45). Thus, $(2\boldsymbol{\rho})\cdot\boldsymbol{r}_\mu = \boldsymbol{r}_\mu\cdot\boldsymbol{r}_\mu = 2d_\mu$ holds for each simple root $\boldsymbol{r}_\mu$. Then, equation (7.134) follows Eq. (7.127).

There is another way to calculate the Casimir invariant. Let I_A^M denote the matrix form of the generator I_A in the representation D^M of $\mathcal{L}$. Define a g-dimensional matrix $T(\boldsymbol{M})$, whose element is $T_{AB}(\boldsymbol{M}) = \text{Tr}\left[I_A^M I_B^M\right]$. $T(\boldsymbol{M})$ commutes with each generator I_A^{ad} in the adjoint representation due to Eq. (7.9),

$$\begin{aligned}
\left[T(\boldsymbol{M}),\ I_A^{\text{ad}}\right]_{BD} &= \sum_P \left\{\text{Tr}\left(I_B^M I_P^M\right)\left(I_A^{\text{ad}}\right)_{PD} - \left(I_A^{\text{ad}}\right)_{BP}\text{Tr}\left(I_P^M I_D^M\right)\right\} \\
&= \text{iTr}\sum_P \left\{I_B^M I_P^M C_{AD}{}^P + C_{AB}{}^P I_P^M I_D^M\right\} \\
&= \text{Tr}\left\{I_B^M\left(I_A^M I_D^M - I_D^M I_A^M\right) + \left(I_A^M I_B^M - I_B^M I_A^M\right) I_D^M\right\} = 0.
\end{aligned}$$

Since the adjoint representation of $\mathcal{L}$ is irreducible, $T(\boldsymbol{M})$ is a constant matrix: $T_{AB}(\boldsymbol{M}) = \delta_{AB}T_2(\boldsymbol{M})$. Taking the trace of $T(\boldsymbol{M})$, one has

$$g\ T_2(\boldsymbol{M}) = \sum_A T_{AA}(\boldsymbol{M}) = \text{Tr}\left\{\sum_A I_A^M I_A^M\right\} = d(\boldsymbol{M})C_2(\boldsymbol{M}), \tag{7.135}$$

where $d(\boldsymbol{M})$ is the dimension of the representation D^M. The highest weight $\boldsymbol{M}$ of the adjoint representation is the largest root $\boldsymbol{\omega}$,

$$g_{AB} = -\text{Tr}\left(I_A^{\text{ad}} I_B^{\text{ad}}\right) = -\delta_{AB}T_2(\boldsymbol{\omega}) = -\delta_{AB}C_2(\boldsymbol{\omega}). \tag{7.136}$$

7.5.6 *Main Data of Simple Lie Algebras*

Now, we collect the main data of the simple Lie algebras, such as the Dynkin diagram, the Cartan matrix A and its inverse A^{-1}, the simple roots $\boldsymbol{r}_\mu$, the positive roots, the largest root $\boldsymbol{\omega}$, the fundamental dominant weights $\boldsymbol{w}_\mu$, some Casimir invariants $C_2(\boldsymbol{M})$, and the formulas for the dimensions $d(\boldsymbol{M})$.

In the data, $\boldsymbol{M} = \sum_\mu \boldsymbol{w}_\mu M_\mu$ and the Killing form is $g_{AB} = -\delta_{AB} C_2(\boldsymbol{\omega})$. $\boldsymbol{e}_\mu$ are the Euclidean basis vectors. The basis vectors $\boldsymbol{V}_a$ are given in Eq. (7.74). $D^{\boldsymbol{M}_0}$ is the self-representation. $D^{\boldsymbol{M}_s}$ is the spinor representation of the SO(N) group. In the mathematical conventions, the half of square length of a longer simple root is $d_\mu = (\boldsymbol{r}_\mu \cdot \boldsymbol{r}_\mu)/2 = 1$.

1. Lie Algebra A_ℓ and Lie Group SU$(\ell+1)$, $\ell \geqslant 1$.

The order of A_ℓ is $\ell(\ell+2)$. There are $\ell(\ell+1)$ roots, denoted by $\sqrt{2}\,(\boldsymbol{V}_a - \boldsymbol{V}_b)$. The Dynkin diagram of A_ℓ is

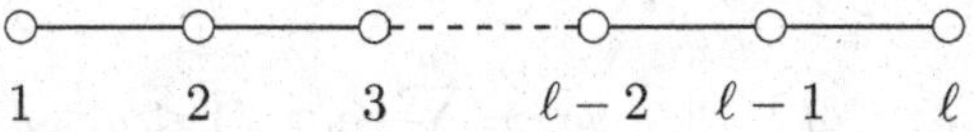

$$A = \begin{pmatrix} 2 & -1 & 0 & 0 & \cdots & 0 & 0 \\ -1 & 2 & -1 & 0 & \cdots & 0 & 0 \\ 0 & -1 & 2 & -1 & \cdots & 0 & 0 \\ \cdot & \cdot & \cdot & \cdot & \cdots & \cdot & \cdot \\ 0 & 0 & 0 & 0 & \cdots & 2 & -1 \\ 0 & 0 & 0 & 0 & \cdots & -1 & 2 \end{pmatrix},$$

$$A^{-1} = \frac{1}{\ell+1} \begin{pmatrix} 1\cdot\ell & 1\cdot(\ell-1) & 1\cdot(\ell-2) & \cdots & 1\cdot 2 & 1\cdot 1 \\ 1\cdot(\ell-1) & 2\cdot(\ell-1) & 2\cdot(\ell-2) & \cdots & 2\cdot 2 & 2\cdot 1 \\ 1\cdot(\ell-2) & 2\cdot(\ell-2) & 3\cdot(\ell-2) & \cdots & 3\cdot 2 & 3\cdot 1 \\ \cdot & \cdot & \cdot & \cdots & \cdot & \cdot \\ 1\cdot 2 & 2\cdot 2 & 3\cdot 2 & \cdots & (\ell-1)\cdot 2 & (\ell-1)\cdot 1 \\ 1\cdot 1 & 2\cdot 1 & 3\cdot 1 & \cdots & (\ell-1)\cdot 1 & \ell\cdot 1 \end{pmatrix}$$

$$\boldsymbol{r}_\mu = \sqrt{2}\,(\boldsymbol{V}_\mu - \boldsymbol{V}_{\mu+1}), \quad d_\mu = 1, \quad \boldsymbol{w}_\mu = \sqrt{2}\sum_{\nu=1}^{\mu} \boldsymbol{V}_\nu, \quad 1 \leqslant \mu \leqslant \ell,$$

$$\boldsymbol{M}_0 = \boldsymbol{w}_1, \qquad \boldsymbol{M}_{\text{adj}} = \boldsymbol{\omega} = \boldsymbol{w}_1 + \boldsymbol{w}_\ell,$$

$$d(\boldsymbol{M}_0) = \ell + 1, \qquad d(\boldsymbol{M}_{\text{adj}}) = \ell(\ell+2),$$

$$C_2(\boldsymbol{M}_0) = \frac{\ell^2 + 2\ell}{\ell+1}, \qquad C_2(\boldsymbol{M}_{\text{adj}}) = 2(\ell+1),$$

$$d(\boldsymbol{M}) = \prod_{a<b}^{\ell+1} \left\{1 + \sum_{\mu=a}^{b-1} \frac{M_\mu}{b-a}\right\}.$$

2. Lie Algebra B_ℓ and Lie Group SO$(2\ell+1)$, $\ell \geqslant 3$.

Being Lie algebras, $\mathrm{B}_1 \approx \mathrm{A}_1$ and $\mathrm{B}_2 \approx \mathrm{C}_2$. The order of B_ℓ is $\ell(2\ell+1)$. The Dynkin diagram of B_ℓ is

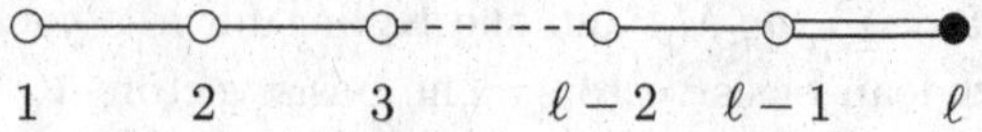

$$A = \begin{pmatrix} 2 & -1 & 0 & \cdots & 0 & 0 \\ -1 & 2 & -1 & \cdots & 0 & 0 \\ 0 & -1 & 2 & \cdots & 0 & 0 \\ \cdot & \cdot & \cdot & \cdots & \cdot & \cdot \\ 0 & 0 & 0 & \cdots & 2 & -1 \\ 0 & 0 & 0 & \cdots & -2 & 2 \end{pmatrix}, \quad A^{-1} = \begin{pmatrix} 1 & 1 & 1 & \cdots & 1 & 1/2 \\ 1 & 2 & 2 & \cdots & 2 & 1 \\ 1 & 2 & 3 & \cdots & 3 & 3/2 \\ \cdot & \cdot & \cdot & \cdots & \cdot & \cdot \\ 1 & 2 & 3 & \cdots & \ell-1 & (\ell-1)/2 \\ 1 & 2 & 3 & \cdots & \ell-1 & \ell/2 \end{pmatrix},$$

$$\boldsymbol{r}_\mu = \boldsymbol{e}_\mu - \boldsymbol{e}_{\mu+1}, \qquad d_\mu = 1, \quad \boldsymbol{w}_\mu = \sum_{\nu=1}^{\mu} \boldsymbol{e}_\nu, \quad 1 \leqslant \mu \leqslant \ell - 1,$$

$$\boldsymbol{r}_\ell = \boldsymbol{e}_\ell, \qquad d_\ell = 1/2, \qquad \boldsymbol{w}_\ell = \frac{1}{2}\sum_{\nu=1}^{\ell} \boldsymbol{e}_\nu,$$

$$\begin{array}{lll} \boldsymbol{M}_0 = \boldsymbol{w}_1\ , & \boldsymbol{M}_{\rm adj} = \boldsymbol{\omega} = \boldsymbol{w}_2, & \boldsymbol{M}_S = \boldsymbol{w}_\ell, \\ d(\boldsymbol{M}_0) = 2\ell + 1, & d(\boldsymbol{M}_{\rm adj}) = \ell(2\ell+1), & d(\boldsymbol{M}_S) = 2^\ell, \\ C_2(\boldsymbol{M}_0) = 2\ell, & C_2(\boldsymbol{M}_{\rm adj}) = 2(2\ell-1), & C_2(\boldsymbol{M}_S) = \ell(2\ell+1)/4. \end{array}$$

There are ℓ^2 positive roots, denoted by $\boldsymbol{e}_\mu$ and $\boldsymbol{e}_\mu \pm \boldsymbol{e}_\nu$.

$$d(\boldsymbol{M}) = \prod_{\lambda=1}^{\ell} \left\{ 1 + \frac{M_\ell + 2\sum_{\rho=\lambda}^{\ell-1} M_\rho}{1 + 2(\ell - \lambda)} \right\}$$

$$\cdot \prod_{\mu<\nu}^{\ell} \left\{ \left(1 + \frac{M_\ell + \sum_{\rho=\mu}^{\nu-1} M_\rho + \sum_{\rho=\nu}^{\ell-1} 2M_\rho}{1 + 2\ell - \mu - \nu} \right) \left(1 + \frac{\sum_{\rho=\mu}^{\nu-1} M_\rho}{\nu - \mu} \right) \right\}.$$

3. Lie Algebra C_ℓ and Lie Group USp(2ℓ), $\ell \geqslant 2$.

USp(2) = SU(2) and $C_1 = A_1$. The order of C_ℓ is $\ell(2\ell+1)$. The Dynkin diagram of C_ℓ is

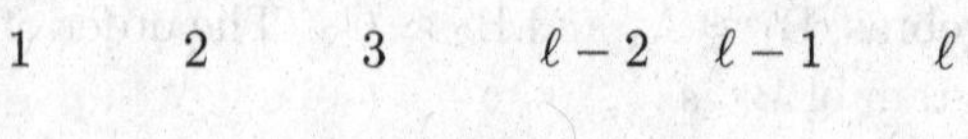

$$A = \begin{pmatrix} 2 & -1 & 0 & \cdots & 0 & 0 \\ -1 & 2 & -1 & \cdots & 0 & 0 \\ 0 & -1 & 2 & \cdots & 0 & 0 \\ \cdot & \cdot & \cdot & \cdots & \cdot & \cdot \\ 0 & 0 & 0 & \cdots & 2 & -2 \\ 0 & 0 & 0 & \cdots & -1 & 2 \end{pmatrix}, \quad A^{-1} = \begin{pmatrix} 1 & 1 & 1 & \cdots & 1 & 1 \\ 1 & 2 & 2 & \cdots & 2 & 2 \\ 1 & 2 & 3 & \cdots & 3 & 3 \\ \cdot & \cdot & \cdot & \cdots & \cdot & \cdot \\ 1 & 2 & 3 & \cdots & \ell-1 & \ell-1 \\ 1/2 & 1 & 3/2 & \cdots & (\ell-1)/2 & \ell/2 \end{pmatrix},$$

$$\begin{aligned} &\boldsymbol{r}_\mu = \sqrt{1/2}\,\{\boldsymbol{e}_\mu - \boldsymbol{e}_{\mu+1}\}, \quad d_\mu = 1/2, \quad 1 \leqslant \mu \leqslant (\ell-1), \\ &\boldsymbol{r}_\ell = \sqrt{2}\boldsymbol{e}_\ell, \qquad d_\ell = 1, \qquad \boldsymbol{w}_\mu = \sqrt{1/2}\sum_{\nu=1}^{\mu} \boldsymbol{e}_\nu, \quad 1 \leqslant \mu \leqslant \ell, \\ &\boldsymbol{M}_0 = \boldsymbol{w}_1, \qquad \boldsymbol{M}_{\text{adj}} = \boldsymbol{\omega} = 2\boldsymbol{w}_1, \\ &d(\boldsymbol{M}_0) = 2\ell, \qquad d(\boldsymbol{M}_{\text{adj}}) = \ell(2\ell+1), \\ &C_2(\boldsymbol{M}_0) = \ell + 1/2, \qquad C_2(\boldsymbol{M}_{\text{adj}}) = 2(\ell+1). \end{aligned}$$

There are ℓ^2 positive roots, denoted by $\sqrt{1/2}\,(\boldsymbol{e}_\mu \pm \boldsymbol{e}_\nu)$ and $\sqrt{2}\boldsymbol{e}_\mu$.

$$\begin{aligned} d(\boldsymbol{M}) = &\prod_{\lambda=1}^{\ell} \left\{ 1 + \frac{\sum_{\rho=\lambda}^{\ell} M_\rho}{\ell+1-\lambda} \right\} \\ &\cdot \prod_{\mu<\nu}^{\ell} \left\{ \left(1 + \frac{\sum_{\rho=\mu}^{\nu-1} M_\rho + \sum_{\rho=\nu}^{\ell} 2M_\rho}{2\ell+2-\mu-\nu} \right) \left(1 + \frac{\sum_{\rho=\mu}^{\nu-1} M_\rho}{\nu-\mu} \right) \right\}. \end{aligned}$$

4. Lie Algebra D_ℓ and Lie Group $SO(2\ell)$, $\ell \geqslant 4$.

Being Lie algebras, $D_2 \approx A_1 \oplus A_1$ and $D_3 \approx A_3$. The order of D_ℓ is $\ell(2\ell-1)$. The Dynkin diagram of D_ℓ is

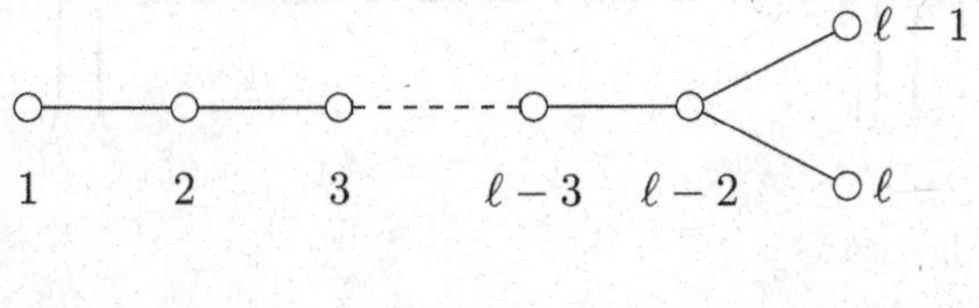

$$A = \begin{pmatrix} 2 & -1 & 0 & 0 & \cdots & 0 & 0 & 0 \\ -1 & 2 & -1 & 0 & \cdots & 0 & 0 & 0 \\ 0 & -1 & 2 & -1 & \cdots & 0 & 0 & 0 \\ \cdot & \cdot & \cdot & \cdot & \cdots & \cdot & \cdot & \cdot \\ 0 & 0 & 0 & 0 & \cdots & 2 & -1 & -1 \\ 0 & 0 & 0 & 0 & \cdots & -1 & 2 & 0 \\ 0 & 0 & 0 & 0 & \cdots & -1 & 0 & 2 \end{pmatrix},$$

$$A^{-1} = \frac{1}{2} \begin{pmatrix} 2 & 2 & 2 & \cdots & 2 & 1 & 1 \\ 2 & 4 & 4 & \cdots & 4 & 2 & 2 \\ 2 & 4 & 6 & \cdots & 6 & 3 & 3 \\ \cdot & \cdot & \cdot & \cdots & \cdot & \cdot & \cdot \\ 2 & 4 & 6 & \cdots & 2(\ell-2) & \ell-2 & \ell-2 \\ 1 & 2 & 3 & \cdots & \ell-2 & \ell/2 & (\ell-2)/2 \\ 1 & 2 & 3 & \cdots & \ell-2 & (\ell-2)/2 & \ell/2 \end{pmatrix},$$

$$\begin{aligned}
&\boldsymbol{r}_\mu = \boldsymbol{e}_\mu - \boldsymbol{e}_{\mu+1}, && 1 \leqslant \mu \leqslant (\ell-1), \\
&\boldsymbol{r}_\ell = \boldsymbol{e}_{\ell-1} + \boldsymbol{e}_\ell, && d_\mu = d_\ell = 1, \\
&\boldsymbol{w}_\nu = \sum_{\rho=1}^{\nu} \boldsymbol{e}_\rho, && 1 \leqslant \nu \leqslant \ell-2, \\
&\boldsymbol{w}_{\ell-1} = \frac{1}{2}\sum_{\rho=1}^{\ell-1} \boldsymbol{e}_\rho - \frac{1}{2}\boldsymbol{e}_\ell, && \boldsymbol{w}_\ell = \frac{1}{2}\sum_{\rho=1}^{\ell} \boldsymbol{e}_\rho, \\
&\boldsymbol{M}_0 = \boldsymbol{w}_1, && \boldsymbol{M}_{\text{adj}} = \boldsymbol{\omega} = \boldsymbol{w}_2, \\
&\boldsymbol{M}_{S1} = \boldsymbol{w}_{\ell-1}, && \boldsymbol{M}_{S2} = \boldsymbol{w}_\ell, \\
&d(\boldsymbol{M}_0) = 2\ell, && d(\boldsymbol{M}_{\text{adj}}) = \ell(2\ell-1), \\
&d(\boldsymbol{M}_{S1}) = d(\boldsymbol{M}_{S2}) = 2^{\ell-1}, && C_2(\boldsymbol{M}_0) = 2\ell-1, \\
&C_2(\boldsymbol{M}_{\text{adj}}) = 4(\ell-1), && C_2(\boldsymbol{M}_{S1}) = C_2(\boldsymbol{M}_{S2}) = \ell(2\ell-1)/4.
\end{aligned}$$

There are $\ell(\ell-1)$ positive roots, denoted by $\boldsymbol{e}_\mu \pm \boldsymbol{e}_\nu$.

$$d(\boldsymbol{M}) = \prod_{\mu<\nu}^{\ell} \left\{ \left(1 + \frac{\sum_{\rho=\mu}^{\nu-1} M_\rho + \sum_{\rho=\nu}^{\ell} 2M_\rho - M_{\ell-1} - M_\ell}{2\ell-\mu-\nu} \right) \left(1 + \frac{\sum_{\rho=\mu}^{\nu-1} M_\rho}{\nu-\mu} \right) \right\}.$$

5. Lie Algebra G_2.

The order of G_2 is 14. The Dynkin diagram of G_2 is

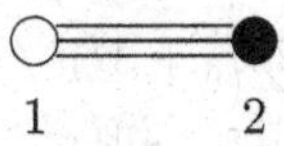

$$A = \begin{pmatrix} 2 & -1 \\ -3 & 2 \end{pmatrix}, \qquad A^{-1} = \begin{pmatrix} 2 & 1 \\ 3 & 2 \end{pmatrix},$$

$$\begin{aligned}
&\boldsymbol{r}_1 = \boldsymbol{e}_2\sqrt{2}, \quad d_1 = 1, && \boldsymbol{r}_2 = \boldsymbol{e}_1\sqrt{1/6} - \boldsymbol{e}_2\sqrt{1/2}, \quad d_2 = 1/3, \\
&\boldsymbol{M}_0 = \boldsymbol{w}_2 = \boldsymbol{e}_1\sqrt{2/3}, && \boldsymbol{M}_{\text{adj}} = \boldsymbol{\omega} = \boldsymbol{w}_1 = \boldsymbol{e}_1\sqrt{3/2} + \boldsymbol{e}_2\sqrt{1/2}, \\
&d(\boldsymbol{M}_0) = 7, && d(\boldsymbol{M}_{\text{adj}}) = 14, \\
&C_2(\boldsymbol{M}_0) = 4, && C_2(\boldsymbol{M}_{\text{adj}}) = 8.
\end{aligned}$$

In addition to the simple roots, there are four positive roots: $\boldsymbol{r}_1 + \boldsymbol{r}_2$, $\boldsymbol{r}_1 + 2\boldsymbol{r}_2$, $\boldsymbol{r}_1 + 3\boldsymbol{r}_2$, and $2\boldsymbol{r}_1 + 3\boldsymbol{r}_2$.

$$\begin{aligned}
d(\boldsymbol{M}) = \;& (1 + M_1)(1 + M_2)(4 + 3M_1 + M_2)(5 + 3M_1 + 2M_2) \\
& \cdot (2 + M_1 + M_2)(3 + 2M_1 + M_2)/120 \,.
\end{aligned}$$

6. Lie Algebra F_4.

The order of F_4 is 52. The Dynkin diagram of F_4 is

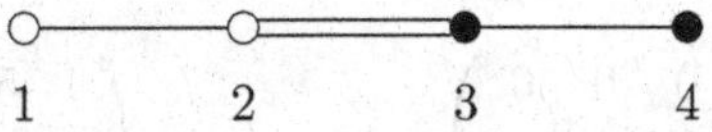

$$A = \begin{pmatrix} 2 & -1 & 0 & 0 \\ -1 & 2 & -1 & 0 \\ 0 & -2 & 2 & -1 \\ 0 & 0 & -1 & 2 \end{pmatrix}, \qquad A^{-1} = \begin{pmatrix} 2 & 3 & 2 & 1 \\ 3 & 6 & 4 & 2 \\ 4 & 8 & 6 & 3 \\ 2 & 4 & 3 & 2 \end{pmatrix},$$

$$\begin{aligned}
&\boldsymbol{r}_1 = \boldsymbol{e}_2 - \boldsymbol{e}_3, \qquad \boldsymbol{r}_2 = \boldsymbol{e}_3 - \boldsymbol{e}_4, \qquad d_1 = d_2 = 1, \\
&\boldsymbol{r}_3 = \boldsymbol{e}_4, \qquad \boldsymbol{r}_4 = (\boldsymbol{e}_1 - \boldsymbol{e}_2 - \boldsymbol{e}_3 - \boldsymbol{e}_4)/2, \qquad d_3 = d_4 = 1/2, \\
&\boldsymbol{w}_1 = \boldsymbol{e}_1 + \boldsymbol{e}_2, \qquad \boldsymbol{w}_2 = 2\boldsymbol{e}_1 + \boldsymbol{e}_2 + \boldsymbol{e}_3, \\
&\boldsymbol{w}_3 = (3\boldsymbol{e}_1 + \boldsymbol{e}_2 + \boldsymbol{e}_3 + \boldsymbol{e}_4)/2, \qquad \boldsymbol{w}_4 = \boldsymbol{e}_1, \\
&\boldsymbol{M}_0 = \boldsymbol{w}_4, \qquad d(\boldsymbol{M}_0) = 26, \qquad C_2(\boldsymbol{M}_0) = 12, \\
&\boldsymbol{M}_{\text{adj}} = \boldsymbol{\omega} = \boldsymbol{w}_1, \qquad d(\boldsymbol{M}_{\text{adj}}) = 52, \qquad C_2(\boldsymbol{M}_{\text{adj}}) = 18.
\end{aligned}$$

It can be shown (see Prob. 9) that $\boldsymbol{\alpha} = \boldsymbol{r}_2 + 2\boldsymbol{r}_3 + 2\boldsymbol{r}_4 = \boldsymbol{e}_1 - \boldsymbol{e}_2$ is a positive root of F_4 so that B_4 is a subalgebra of F_4. Thus, F_4 contains 48 roots, including 32 roots in B_4 and 16 other roots denoted by $a\boldsymbol{e}_1 + b\boldsymbol{e}_2 + c\boldsymbol{e}_3 + d\boldsymbol{e}_4$, where four parameters are taken to be ± 1 independently.

$$\begin{aligned}
d(\boldsymbol{M}) = &\{1+(2M_1+4M_2+3M_3+2M_4)/11\}\{1+M_1\}\{1+M_2\} \\
\cdot\ &\{1+M_3\}\{1+M_4\}\{1+(2M_1+2M_2+M_3)/5\} \\
\cdot\ &\{1+(2M_2+M_3)/3\}\{1+(2M_1+3M_2+2M_3+M_4)/8\} \\
\cdot\ &\{1+(M_2+M_3+M_4)/3\}\{1+(M_1+3M_2+2M_3+M_4)/7\} \\
\cdot\ &\{1+(M_1+M_2+M_3+M_4)/4\}\{1+(M_1+2M_2+2M_3+M_4)/6\} \\
\cdot\ &\{1+(M_1+2M_2+M_3+M_4)/5\}\{1+(M_1+2M_2+M_3)/4\} \\
\cdot\ &\{1+(M_1+M_2+M_3)/3\}\{1+(M_1+M_2)/2\}\{1+(M_2+M_3)/2\} \\
\cdot\ &\{1+(2M_1+4M_2+3M_3+M_4)/10\} \\
\cdot\ &\{1+(2M_2+2M_3+M_4)/5\}\{1+(2M_1+2M_2+2M_3+M_4)/7\} \\
\cdot\ &\{1+(2M_1+4M_2+2M_3+M_4)/9\}\{1+(M_3+M_4)/2\} \\
\cdot\ &\{1+(2M_2+M_3+M_4)/4\}\{1+(2M_1+2M_2+M_3+M_4)/6\}\ .
\end{aligned}$$

7. Lie Algebra E_6.

The order of E_6 is 78. The Dynkin diagram of E_6 is

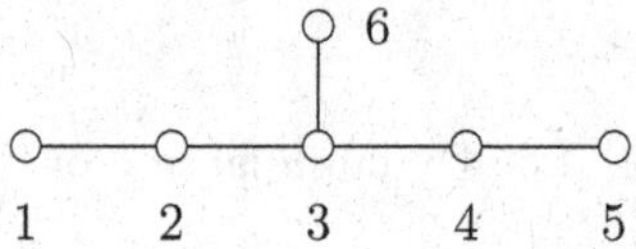

$$A = \begin{pmatrix} 2 & -1 & 0 & 0 & 0 & 0 \\ -1 & 2 & -1 & 0 & 0 & 0 \\ 0 & -1 & 2 & -1 & 0 & -1 \\ 0 & 0 & -1 & 2 & -1 & 0 \\ 0 & 0 & 0 & -1 & 2 & 0 \\ 0 & 0 & -1 & 0 & 0 & 2 \end{pmatrix}, \qquad A^{-1} = \frac{1}{3}\begin{pmatrix} 4 & 5 & 6 & 4 & 2 & 3 \\ 5 & 10 & 12 & 8 & 4 & 6 \\ 6 & 12 & 18 & 12 & 6 & 9 \\ 4 & 8 & 12 & 10 & 5 & 6 \\ 2 & 4 & 6 & 5 & 4 & 3 \\ 3 & 6 & 9 & 6 & 3 & 6 \end{pmatrix},$$

Let $\boldsymbol{V}_a$ denote the basis vectors for A_5 (see Eq. (7.74)). The first five components of $\boldsymbol{U}_a$ are the components of $\boldsymbol{V}_a$ for A_5, but the sixth component is 0. Each simple root in E_6 satisfies $d_\mu = 1$, $1 \leqslant \mu \leqslant 6$.

$$\begin{aligned}
&\boldsymbol{r}_1 = \sqrt{2}\boldsymbol{e}_6, && \boldsymbol{r}_2 = \sqrt{2}\,(\boldsymbol{U}_3+\boldsymbol{U}_4+\boldsymbol{U}_5) - \boldsymbol{e}_6/\sqrt{2}, \\
&\boldsymbol{r}_3 = \sqrt{2}\,(\boldsymbol{U}_2-\boldsymbol{U}_3), && \boldsymbol{r}_4 = \sqrt{2}\,(\boldsymbol{U}_3-\boldsymbol{U}_4), \\
&\boldsymbol{r}_5 = \sqrt{2}\,(\boldsymbol{U}_4-\boldsymbol{U}_5), && \boldsymbol{r}_6 = \sqrt{2}\,(\boldsymbol{U}_1-\boldsymbol{U}_2), \\
&\boldsymbol{w}_1 = -\sqrt{2}\boldsymbol{U}_6 + \boldsymbol{e}_6/\sqrt{2}, && \boldsymbol{w}_2 = -2\sqrt{2}\boldsymbol{U}_6, \\
&\boldsymbol{w}_3 = \sqrt{2}\,(\boldsymbol{U}_1+\boldsymbol{U}_2-2\boldsymbol{U}_6), && \boldsymbol{w}_4 = \sqrt{2}\,(\boldsymbol{U}_1+\boldsymbol{U}_2+\boldsymbol{U}_3-\boldsymbol{U}_6), \\
&\boldsymbol{w}_5 = -\sqrt{2}\,(\boldsymbol{U}_5+\boldsymbol{U}_6), && \boldsymbol{w}_6 = \sqrt{2}\,(\boldsymbol{U}_1-\boldsymbol{U}_6),
\end{aligned}$$

$$\begin{aligned} \boldsymbol{M}_0 &= \boldsymbol{w}_1, & \boldsymbol{M}_{\text{adj}} &= \boldsymbol{\omega} = \boldsymbol{w}_6, \\ d(\boldsymbol{M}_0) &= 27, & d(\boldsymbol{M}_{\text{adj}}) &= 78, \\ C_2(\boldsymbol{M}_0) &= 52/3, & C_2(\boldsymbol{M}_{\text{adj}}) &= 24. \end{aligned}$$

It can be shown (see Prob. 10) that $\boldsymbol{\alpha} = \boldsymbol{r}_1 + 2\boldsymbol{r}_2 + 2\boldsymbol{r}_3 + \boldsymbol{r}_4 + \boldsymbol{r}_6 = \sqrt{2}\,(\boldsymbol{U}_5 - \boldsymbol{U}_6)$ is a positive root of E_6 so that A_5 is a subalgebra of E_6. E_6 contains 72 roots, including all 30 roots $\sqrt{2}\,(\boldsymbol{U}_a - \boldsymbol{U}_b)$ in A_5, and 42 other roots denoted by $\pm\sqrt{2}\boldsymbol{e}_6$ and $\sqrt{2}\,(\boldsymbol{U}_a + \boldsymbol{U}_b + \boldsymbol{U}_c) \pm \boldsymbol{e}_6/\sqrt{2}$, where $1 \leqslant a < b < c \leqslant 6$.

8. Lie Algebra E_7

The order of E_7 is 133. The Dynkin diagram of E_7 is

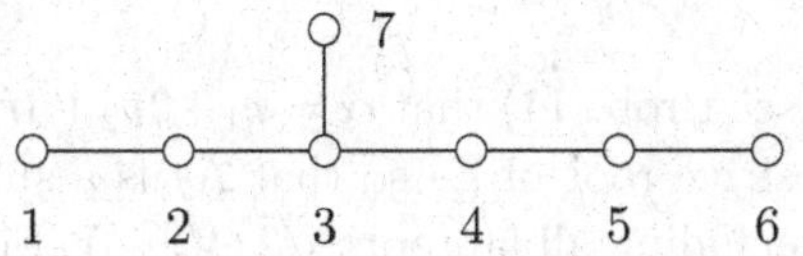

$$A = \begin{pmatrix} 2 & -1 & 0 & 0 & 0 & 0 & 0 \\ -1 & 2 & -1 & 0 & 0 & 0 & 0 \\ 0 & -1 & 2 & -1 & 0 & 0 & -1 \\ 0 & 0 & -1 & 2 & -1 & 0 & 0 \\ 0 & 0 & 0 & -1 & 2 & -1 & 0 \\ 0 & 0 & 0 & 0 & -1 & 2 & 0 \\ 0 & 0 & -1 & 0 & 0 & 0 & 2 \end{pmatrix},$$

$$A^{-1} = \frac{1}{2}\begin{pmatrix} 4 & 6 & 8 & 6 & 4 & 2 & 4 \\ 6 & 12 & 16 & 12 & 8 & 4 & 8 \\ 8 & 16 & 24 & 18 & 12 & 6 & 12 \\ 6 & 12 & 18 & 15 & 10 & 5 & 9 \\ 4 & 8 & 12 & 10 & 8 & 4 & 6 \\ 2 & 4 & 6 & 5 & 4 & 3 & 3 \\ 4 & 8 & 12 & 9 & 6 & 3 & 7 \end{pmatrix},$$

Let $\boldsymbol{V}_a$ denote the basis vectors for A_7 (see Eq. (7.74)). Each simple root in E_7 satisfies $d_\mu = 1$, $1 \leqslant \mu \leqslant 7$.

$$\begin{aligned}
&\boldsymbol{r}_1=\sqrt{2}\,(\boldsymbol{V}_1-\boldsymbol{V}_2)\;, && \boldsymbol{r}_2=\sqrt{2}\,(\boldsymbol{V}_2-\boldsymbol{V}_3)\,,\\
&\boldsymbol{r}_3=\sqrt{2}\,(\boldsymbol{V}_3-\boldsymbol{V}_4)\,, && \boldsymbol{r}_4=\sqrt{2}\,(\boldsymbol{V}_4-\boldsymbol{V}_5)\,,\\
&\boldsymbol{r}_5=\sqrt{2}\,(\boldsymbol{V}_5-\boldsymbol{V}_6)\,, && \boldsymbol{r}_6=\sqrt{2}\,(\boldsymbol{V}_6-\boldsymbol{V}_7)\,,\\
&\boldsymbol{r}_7=\sqrt{2}\,(\boldsymbol{V}_4+\boldsymbol{V}_5+\boldsymbol{V}_6+\boldsymbol{V}_7)\,, && \boldsymbol{w}_1=\sqrt{2}\,(\boldsymbol{V}_1-\boldsymbol{V}_8)\,,\\
&\boldsymbol{w}_2=\sqrt{2}\,(\boldsymbol{V}_1+\boldsymbol{V}_2-2\boldsymbol{V}_8)\,, && \boldsymbol{w}_3=\sqrt{2}\,(\boldsymbol{V}_1+\boldsymbol{V}_2+\boldsymbol{V}_3-3\boldsymbol{V}_8)\,,\\
&\boldsymbol{w}_4=-\sqrt{2}\,(\boldsymbol{V}_5+\boldsymbol{V}_6+\boldsymbol{V}_7+3\boldsymbol{V}_8)\,, && \boldsymbol{w}_5=-\sqrt{2}\,(\boldsymbol{V}_6+\boldsymbol{V}_7+2\boldsymbol{V}_8)\,,\\
&\boldsymbol{w}_6=-\sqrt{2}\,(\boldsymbol{V}_7+\boldsymbol{V}_8)\,, && \boldsymbol{w}_7=-2\sqrt{2}\boldsymbol{V}_8,\\
&\boldsymbol{M}_0=\boldsymbol{w}_6, && \boldsymbol{M}_{\rm adj}=\boldsymbol{\omega}=\boldsymbol{w}_1,\\
&d(\boldsymbol{M}_0)=56, && d(\boldsymbol{M}_{\rm adj})=133,\\
&C_2(\boldsymbol{M}_0)=57/2, && C_2(\boldsymbol{M}_{\rm adj})=36,
\end{aligned}$$

It can be shown (see Prob. 11) that $\boldsymbol{\alpha}=\boldsymbol{r}_1+2\boldsymbol{r}_2+3\boldsymbol{r}_3+2\boldsymbol{r}_4+\boldsymbol{r}_5+2\boldsymbol{r}_7=\sqrt{2}(\boldsymbol{V}_7-\boldsymbol{V}_8)$ is a positive root of E_7 so that A_7 is a subalgebra of E_7. E_7 contains 126 roots, including all 56 roots $\sqrt{2}\,(\boldsymbol{V}_a-\boldsymbol{V}_b)$ in A_7, and 70 other roots denoted by $\sqrt{2}\,(\boldsymbol{V}_a+\boldsymbol{V}_b+\boldsymbol{V}_c+\boldsymbol{V}_d)$, where $1\leqslant a<b<c<d\leqslant 8$.

9. Lie Algebra E_8

The order of E_8 is 248. The Dynkin diagram of E_8 is

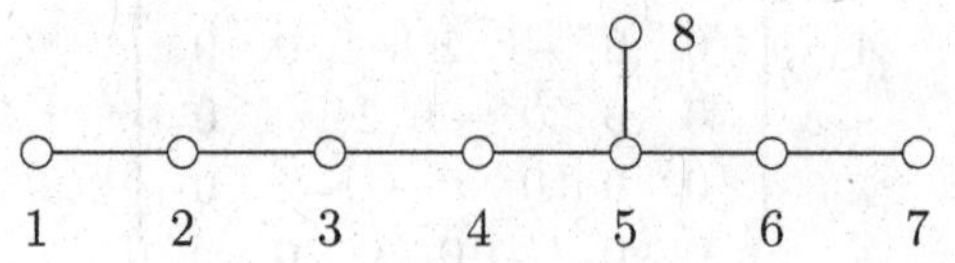

$$A=\begin{pmatrix}
2&-1&0&0&0&0&0&0\\
-1&2&-1&0&0&0&0&0\\
0&-1&2&-1&0&0&0&0\\
0&0&-1&2&-1&0&0&0\\
0&0&0&-1&2&-1&0&-1\\
0&0&0&0&-1&2&-1&0\\
0&0&0&0&0&-1&2&0\\
0&0&0&0&-1&0&0&2
\end{pmatrix},$$

$$A^{-1} = \begin{pmatrix} 2 & 3 & 4 & 5 & 6 & 4 & 2 & 3 \\ 3 & 6 & 8 & 10 & 12 & 8 & 4 & 6 \\ 4 & 8 & 12 & 15 & 18 & 12 & 6 & 9 \\ 5 & 10 & 15 & 20 & 24 & 16 & 8 & 12 \\ 6 & 12 & 18 & 24 & 30 & 20 & 10 & 15 \\ 4 & 8 & 12 & 16 & 20 & 14 & 7 & 10 \\ 2 & 4 & 6 & 8 & 10 & 7 & 4 & 5 \\ 3 & 6 & 9 & 12 & 15 & 10 & 5 & 8 \end{pmatrix},$$

Each simple root in E_8 satisfies $d_\mu = 1$, $1 \leqslant \mu \leqslant 8$,

$$\boldsymbol{r}_1 = \boldsymbol{e}_2 - \boldsymbol{e}_3, \qquad \boldsymbol{r}_2 = \boldsymbol{e}_3 - \boldsymbol{e}_4, \qquad \boldsymbol{r}_3 = \boldsymbol{e}_4 - \boldsymbol{e}_5,$$
$$\boldsymbol{r}_4 = \boldsymbol{e}_5 - \boldsymbol{e}_6, \qquad \boldsymbol{r}_5 = \boldsymbol{e}_6 - \boldsymbol{e}_7, \qquad \boldsymbol{r}_6 = \boldsymbol{e}_7 - \boldsymbol{e}_8,$$
$$\boldsymbol{r}_7 = 2^{-1}\left\{\boldsymbol{e}_1 + \boldsymbol{e}_8 - \sum_{j=2}^{7} \boldsymbol{e}_j\right\}, \qquad \boldsymbol{r}_8 = \boldsymbol{e}_7 + \boldsymbol{e}_8,$$
$$\boldsymbol{w}_1 = \boldsymbol{e}_1 + \boldsymbol{e}_2, \qquad \boldsymbol{w}_2 = 2\boldsymbol{e}_1 + \boldsymbol{e}_2 + \boldsymbol{e}_3,$$
$$\boldsymbol{w}_3 = 3\boldsymbol{e}_1 + \sum_{j=2}^{4} \boldsymbol{e}_j, \qquad \boldsymbol{w}_4 = 4\boldsymbol{e}_1 + \sum_{j=2}^{5} \boldsymbol{e}_j,$$
$$\boldsymbol{w}_5 = 5\boldsymbol{e}_1 + \sum_{j=2}^{6} \boldsymbol{e}_j, \qquad \boldsymbol{w}_6 = \frac{7}{2}\boldsymbol{e}_1 + \frac{1}{2}\sum_{j=2}^{7} \boldsymbol{e}_j - \frac{1}{2}\boldsymbol{e}_8,$$
$$\boldsymbol{w}_7 = 2\boldsymbol{e}_1, \qquad \boldsymbol{w}_8 = \frac{5}{2}\boldsymbol{e}_1 + \frac{1}{2}\sum_{j=2}^{8} \boldsymbol{e}_j,$$
$$\boldsymbol{M}_0 = \boldsymbol{M}_{\text{adj}} = \boldsymbol{\omega} = \boldsymbol{w}_1, \qquad d(\boldsymbol{M}_0) = 248, \quad C_2(\boldsymbol{M}_{\text{adj}}) = 60.$$

It can be shown (see Prob. 12) that $\boldsymbol{\alpha} = \boldsymbol{r}_2 + 2\boldsymbol{r}_3 + 3\boldsymbol{r}_4 + 4\boldsymbol{r}_5 + 3\boldsymbol{r}_6 + 2\boldsymbol{r}_7 + 2\boldsymbol{r}_8 = \boldsymbol{e}_1 - \boldsymbol{e}_2$ is a positive root of E_8 so that D_8 is a subalgebra of E_8. E_8 contains 240 roots, including all 112 roots $\pm(\boldsymbol{e}_\mu + \boldsymbol{e}_\nu)$ and $(\boldsymbol{e}_\mu - \boldsymbol{e}_\nu)$ in D_8, and 128 other roots denoted by $(\pm\boldsymbol{e}_1 \pm \boldsymbol{e}_2 \pm \boldsymbol{e}_3 \pm \boldsymbol{e}_4 \pm \boldsymbol{e}_5 \pm \boldsymbol{e}_6 \pm \boldsymbol{e}_7 \pm \boldsymbol{e}_8)/2$, where even signs are plus and the remaining signs are minus.

7.6 Exercises

1. Prove that the Casimir operator C_n given in Eq. (7.22) commutes with any generators I_A in a simple Lie algebra $\mathcal{L}$ as shown in Eq. (7.23).

2. Prove that the number $p+q+1$ in the root chain (7.42) for any two roots in a simple Lie algebra $\mathcal{L}$ is less than five.

3. Write the Cartan matrix of the Lie algebra E_6.

4. Draw the Dynkin diagram of a simple Lie algebra where its Cartan matrix is as follows, and indicate the enumeration for the simple roots.

$$A = \begin{pmatrix} 2 & -1 & 0 & 0 \\ -1 & 2 & -1 & 0 \\ 0 & -2 & 2 & -1 \\ 0 & 0 & -1 & 2 \end{pmatrix},$$

$$A_{\mu\nu} = \Gamma\left[\boldsymbol{r}_\nu/\boldsymbol{r}_\mu\right] = \frac{2\boldsymbol{r}_\nu \cdot \boldsymbol{r}_\mu}{\boldsymbol{r}_\mu \cdot \boldsymbol{r}_\mu}.$$

5. Calculate all simple roots and positive roots in the C_2 Lie algebra.

6. Calculate all simple roots and positive roots in the B_3 Lie algebra.

7. Calculate all simple roots and positive roots in the D_4 Lie algebra.

8. Calculate all simple roots and positive roots in the F_4 Lie algebra.

9. Please show that B_4 is a subalgebra of F_4. Namely, please point out by Eq. (7.60) that $\boldsymbol{\alpha} = \boldsymbol{r}_2 + 2\boldsymbol{r}_3 + 2\boldsymbol{r}_4$ is a positive root of F_4 (see 6. in subsection 7.5.6).

10. Please show that A_5 is a subalgebra of E_6. Namely, please point out by Eq. (7.60) that $\boldsymbol{\alpha} = \boldsymbol{r}_1 + 2\boldsymbol{r}_2 + 2\boldsymbol{r}_3 + \boldsymbol{r}_4 + \boldsymbol{r}_6$ is a positive root of E_6 (see 7. in subsection 7.5.6).

11. Please show that A_7 is a subalgebra of E_7. Namely, please point out by Eq. (7.60) that $\boldsymbol{\alpha} = \boldsymbol{r}_1 + 2\boldsymbol{r}_2 + 3\boldsymbol{r}_3 + 2\boldsymbol{r}_4 + \boldsymbol{r}_5 + 2\boldsymbol{r}_7$ is a positive root of E_7 (see 8. in subsection 7.5.6).

12. Please show that B_8 is a subalgebra of E_8. Namely, please point out by Eq. (7.60) that $\boldsymbol{\alpha} = \boldsymbol{r}_2 + 2\boldsymbol{r}_3 + 3\boldsymbol{r}_4 + 4\boldsymbol{r}_5 + 3\boldsymbol{r}_6 + 2\boldsymbol{r}_7 + 2\boldsymbol{r}_8$ is a positive root of E_8 (see 9. in subsection 7.5.6).

Chapter 8

GEL'FAND'S METHOD AND ITS GENERALIZATION

In this chapter we will explain the methods of calculating the basis states and the matrix elements of the generators in an irreducible representation of a simple Lie algebra $\mathcal{L}$ with dimension larger than one. Three Chevalley bases of generators, E_μ, F_μ and H_μ with the same subscript μ, span an A_1 subalgebra. The multiplets of those A_1 subalgebras embed in the representation space of $\mathcal{L}$ by a complicated way. The method of the block weight diagram analyzes the complicated overlapping of those multiplets and calculate the representation matrices of the generators. The Gel'fand's method and its generalization gives a certain rule to define the orthonormal basis states in the representation space of $\mathcal{L}$. The study on the physical background of the basis states is left to the subsequent chapters.

The direct product of two irreducible representations of $\mathcal{L}$ is decomposed by the method of dominant weight diagram.

8.1 Method of Block weight diagram

In this section we will calculate the representation matrices of generators in the highest weight representation of a simple Lie algebra $\mathcal{L}$.

8.1.1 *Chevalley Bases*

Due to Eq. (7.40), $E_{\alpha+\beta}$ can be expressed as the commutator of E_α and E_β, so that the representation matrices of only those generators related to the simple roots are needed to be calculated. Chevalley introduced 3ℓ bases E_μ, F_μ, and H_μ for those generators, $1 \leqslant \mu \leqslant \ell$, called the **Chevalley bases**,

$$\begin{aligned} &d_\mu^{-1/2} E_{\boldsymbol{r}_\mu} \longrightarrow E_\mu, \qquad d_\mu^{-1/2} E_{-\boldsymbol{r}_\mu} \longrightarrow F_\mu, \\ &d_\mu^{-1} \sum_{j=1}^{\ell} (\boldsymbol{r}_\mu)_j H_j = d_\mu^{-1} \boldsymbol{r}_\mu \cdot \boldsymbol{H} \longrightarrow H_\mu. \end{aligned} \tag{8.1}$$

Thus, the regular commutative relations (7.40) become

$$\begin{aligned} &[H_\mu, H_\nu] = 0, && [H_\mu, E_\nu] = A_{\mu\nu} E_\nu, \\ &[H_\mu, F_\nu] = -A_{\mu\nu} F_\nu, && [E_\mu, F_\nu] = \delta_{\mu\nu} H_\mu, \end{aligned} \tag{8.2}$$

where $A_{\mu\nu}$ is the Cartan matrix. Since $A_{\mu\mu} = 2$, three generators H_μ, E_μ and F_μ with the same subscript μ satisfy the same commutative relations as H, E and F in the Lie algebra A_1 (the SU(2) group),

$$\begin{aligned} &H = 2J_3, && E = J_1 + \mathrm{i}J_2, && F = J_1 - \mathrm{i}J_2, \\ &[H, E] = 2E, && [H, F] = -2F, && [E, F] = H, \\ &[H_\mu, E_\mu] = 2E_\mu, && [H_\mu, F_\mu] = -2F_\mu, && [E_\mu, F_\mu] = H_\mu. \end{aligned} \tag{8.3}$$

Thus, in the simple Lie algebra $\mathcal{L}$, **three generators** E_μ, F_μ, **and** H_μ **span an** A_1 **subalgebra, denoted by** $\mathcal{A}_\mu$. The set of the basis states in an irreducible representation of $\mathcal{A}_\mu$ is called an $\mathcal{A}_\mu$-multiplet. We will analyze the complicated overlapping of the $\mathcal{A}_\mu$-multiplets in the representation space of $D^{\boldsymbol{M}}$ of $\mathcal{L}$, and calculate the representation matrices of the generators.

Due to the root chain $\boldsymbol{r}_\mu + n\boldsymbol{r}_\nu$ with $0 \leqslant n \leqslant -A_{\nu\mu}$, the Chevalley bases satisfy the Serre relations:

$$\begin{aligned} &\underbrace{[E_\nu, [E_\nu, \cdots [E_\nu,}_{n} E_\mu] \cdots]] \neq 0, \qquad \underbrace{[E_\nu, [E_\nu, \cdots [E_\nu,}_{1-A_{\nu\mu}} E_\mu] \cdots]] = 0, \\ &\underbrace{[F_\nu, [F_\nu, \cdots [F_\nu,}_{n} F_\mu] \cdots]] \neq 0, \qquad \underbrace{[F_\nu, [F_\nu, \cdots [F_\nu,}_{1-A_{\nu\mu}} F_\mu] \cdots]] = 0. \end{aligned} \tag{8.4}$$

8.1.2 *Block Weight Diagram*

In an irreducible representation space of a simple Lie algebra $\mathcal{L}$, the common eigenstates of H_μ are chosen to be the basis states:

$$H_\mu \, |\boldsymbol{M}, \boldsymbol{m}\rangle = m_\mu \, |\boldsymbol{M}, \boldsymbol{m}\rangle. \tag{8.5}$$

As usually, the basis vectors in the weight space are the fundamental dominant weight $\boldsymbol{w}_\mu$, so that the components m_μ **all are integers.** Due to Eq.

(8.2) the action of E_ν raises the weight $\boldsymbol{m}$ of the basis state $|\boldsymbol{M},\boldsymbol{m}\rangle$ by a simple root $\boldsymbol{r}_\nu$ and the action of F_ν lowers $\boldsymbol{m}$ by $\boldsymbol{r}_\nu$:

$$H_\mu \{E_\nu |\boldsymbol{M},\boldsymbol{m}\rangle\} = (\boldsymbol{m}+\boldsymbol{r}_\nu)_\mu \{E_\nu |\boldsymbol{M},\boldsymbol{m}\rangle\},$$
$$H_\mu \{F_\nu |\boldsymbol{M},\boldsymbol{m}\rangle\} = (\boldsymbol{m}-\boldsymbol{r}_\nu)_\mu \{F_\nu |\boldsymbol{M},\boldsymbol{m}\rangle\},$$

where $(\boldsymbol{r}_\nu)_\mu = A_{\mu\nu} = \Gamma(\boldsymbol{r}_\nu/\boldsymbol{r}_\mu)$ only takes the value of 2, 0, -1, -2, or -3.

For a unitary representation, **the basis states are orthonormal.** When the weight is n-multiple, the n orthonormal basis states can be made a further unitary transformation. When $n=1$, the unitary transformation reduces to a phase transformation. We will show that **all matrix elements of generators become real** in terms of those unitary transformations.

A basis state $|\boldsymbol{M},\boldsymbol{m}\rangle$ with a given weight $\boldsymbol{m}$ in an highest weight representation $D^{\boldsymbol{M}}$ can be obtained by applications of a few lowering operators F_μ to the highest weight state $|\boldsymbol{M},\boldsymbol{M}\rangle$, so that the difference $\boldsymbol{M}-\boldsymbol{m}$ is equal to an integral sum of the simple roots $\boldsymbol{r}_\mu$:

$$\boldsymbol{M}-\boldsymbol{m} = \sum_{\mu=1}^{\ell} C_\mu \boldsymbol{r}_\mu, \qquad L(\boldsymbol{M},\boldsymbol{m}) = 1 + \sum_{\mu=1}^{\ell} C_\mu. \tag{8.6}$$

The integer $L(\boldsymbol{M},\boldsymbol{m})$ is called the **level** of the weight $\boldsymbol{m}$ in the representation $D^{\boldsymbol{M}}$. **The level $L(\boldsymbol{M},\boldsymbol{M})$ of the highest weight is 1.**

In the block weight diagram of a highest weight representation of $\mathcal{L}$, each orthonormal basis state with weight $\boldsymbol{m}$ is denoted by a block filled with its weight $\boldsymbol{m}=(m_1,\cdots)$. When a component m_μ is negative, it is denoted by a positive number with a bar. For example, a basis state with a single weight $\boldsymbol{m}=-2\boldsymbol{w}_1+\boldsymbol{w}_2$ is denoted by $\boxed{\overline{2},1}$. When the weight is multiple, an ordinal number is added as a subscript such as $\boxed{(\overline{2},1)_2}$. **The block for the weight $\boldsymbol{m}$ with the level $L(\boldsymbol{M},\boldsymbol{m})$ is located at the $L(\boldsymbol{M},\boldsymbol{m})$th row** in the block weight diagram. In the first row there is only one block for the highest weight state. Two blocks, whose basis states are related by a lowering operator F_μ, are located in the neighboring rows and connected by a special line (solid line, dotted line, double-line, and so on), corresponding to the index μ. **The line is attached with the matrix element of F_μ** between two states. The line is removed if the matrix element is zero. Usually, the rows of blocks in a block weight diagram are arranged **downwards** as their level increases. The order can also be arranged **upwards** sometimes (see Fig. 8.2 for the conjugate representation).

The main problem in the calculations of the basis states and the generators in D^{M} is how to pick up the $\mathcal{A}_\mu$-multiplets embedding in the representation space. We check each basis state $|\boldsymbol{M},\boldsymbol{m}\rangle$, one by one as the level increases from one, **whether the weight $\boldsymbol{m}$ has a positive component** m_μ and **whether the state is annihilated by** E_μ,

$$E_\mu\,|\boldsymbol{M},\boldsymbol{m}\rangle = 0, \qquad m_\mu > 0. \tag{8.7}$$

If yes, we construct an $\mathcal{A}_\mu$-multiplet from $|\boldsymbol{M},\boldsymbol{m}\rangle$ by applications of F_μ successively. The $\mathcal{A}_\mu$-multiplet contains $(m_\mu+1)$ orthonormal basis states, one by one located in the neighboring rows of the block weight diagram (see Eq. (5.70)):

$$\begin{aligned} &|\boldsymbol{M},\boldsymbol{m}-n\boldsymbol{r}_\mu\rangle, \quad 0\leqslant n\leqslant m_\mu, \qquad (F_\mu)^{m_\mu+1}\,|\boldsymbol{M},\boldsymbol{m}\rangle = 0, \\ &F_\mu\,|\boldsymbol{M},\boldsymbol{m}-n\boldsymbol{r}_\mu\rangle = \sqrt{(m_\mu-n)(n+1)}\,|\boldsymbol{M},\boldsymbol{m}-(n+1)\boldsymbol{r}_\mu\rangle. \end{aligned} \tag{8.8}$$

Three formulas in Eq. (8.2) have been used, and the remaining one

$$[E_\mu,\ F_\nu] = \delta_{\mu\nu}H_\mu, \tag{8.9}$$

is used to calculate the matrix elements of generators, including that given in Eq. (8.8).

The highest weight state $|\boldsymbol{M},\boldsymbol{M}\rangle$ satisfies the condition (8.7) and some $\mathcal{A}_\mu$-multiplets can be constructed from the highest weight state by Eq. (8.8). If there is a basis state in an $\mathcal{A}_\mu$-multiplet again satisfying the condition (8.7) but with another positive component, say m_ν, another $\mathcal{A}_\nu$-multiplet can be constructed by Eq. (8.8). This constructing process goes on sequentially. **The series of connected multiplets beginning from the highest weight state are called a path**. Each state in the connected multiplets is said to have a path connecting the highest weight state.

If a basis state $|\boldsymbol{M},\boldsymbol{m}\rangle$ belongs to one $\mathcal{A}_\mu$-multiplet with the highest weight state, or it has only one path connecting the highest weight state, the weight $\boldsymbol{m}$ is single. If all weights $\boldsymbol{m}$ in one $\mathcal{A}_\mu$-multiplet are single, the state $|\boldsymbol{M},\boldsymbol{m}-n\boldsymbol{r}_\mu\rangle$ given in Eq. (8.8), **including their phases**, is directly chosen to be the orthonormal basis state in the representation.

If a basis state $|\boldsymbol{M},\boldsymbol{m}\rangle$ has n different paths connecting the highest weight, **the multiplicity of the weight $\boldsymbol{m}$ is not larger than the path number of the weight $\boldsymbol{m}$.** Among equivalent weights, there must be one and only one dominant weight, which has the least level. **Equivalent weights have the same multiplicity.**

Check each basis state one by one as the level $L(\boldsymbol{M}, \boldsymbol{m})$ increases from one (from the highest weight state) whether or not it satisfies the condition (8.7). If yes, we construct the $\mathcal{A}_\mu$-multiplet by Eq. (8.8). This constructing process goes on sequentially until a basis state, which is annihilated by each lowering operator F_μ, appears (see Theorem 7.9). This basis state is called the **lowest weight state** in the representation $D^{\boldsymbol{M}}$. The weight $-\boldsymbol{N}$ of the lowest weight state has no positive component. **The representation $D^{\boldsymbol{N}}$ is the complex conjugate one of $D^{\boldsymbol{M}}$:**

$$\begin{aligned}
D^{\boldsymbol{M}}(R) &= \left[D^{\boldsymbol{N}}(R)\right]^* = \mathbf{1} - \mathrm{i}\sum_A \omega_A D^{\boldsymbol{M}}(I_A), \\
D^{\boldsymbol{N}}(R) &= \mathbf{1} - \mathrm{i}\sum_A \omega_A D^{\boldsymbol{N}}(I_A) = \mathbf{1} - \mathrm{i}\sum_A \omega_A \left\{-D^{\boldsymbol{M}}(I_A)^*\right\}, \\
D^{\boldsymbol{N}}(I_A) &= -D^{\boldsymbol{M}}(I_A)^* = -D^{\boldsymbol{M}}(I_A)^T.
\end{aligned}$$

The minus sign can be attracted into the basis states in the conjugate representation $D^{\boldsymbol{N}}$

$$|\boldsymbol{N}, -\boldsymbol{m}\rangle = (-1)^{L(\boldsymbol{M},\boldsymbol{m})-1}\ |\boldsymbol{M}, \boldsymbol{m}\rangle^*. \tag{8.10}$$

$$\begin{aligned}
D^{\boldsymbol{N}}_{-\boldsymbol{m}+\boldsymbol{r}_\mu,-\boldsymbol{m}}(E_\mu) &= D^{\boldsymbol{M}}_{\boldsymbol{m}-\boldsymbol{r}_\mu,\boldsymbol{m}}(F_\mu), \\
D^{\boldsymbol{N}}_{-\boldsymbol{m}-\boldsymbol{r}_\mu,-\boldsymbol{m}}(F_\mu) &= D^{\boldsymbol{M}}_{\boldsymbol{m}+\boldsymbol{r}_\mu,\boldsymbol{m}}(E_\mu).
\end{aligned} \tag{8.11}$$

The block weight diagrams of two conjugate representations are upside down with each other. If we change the arranging order of the blocks upwards in the block weight diagram for the conjugate representation, two block weight diagrams are the same except for $\boldsymbol{m}$ replaced by $-\boldsymbol{m}$ (see Fig. 8.2). **The representation $D^{\boldsymbol{M}}$ is self-conjugate if $\boldsymbol{N} = \boldsymbol{M}$.**

In drawing a block weight diagram one should check step by step. Check whether the difference of two weights related by F_μ is $\boldsymbol{r}_\mu$, whether Eq. (8.9) holds when it applies to each basis state, and whether the multiplicities of the equivalent weights are the same. In the completed block weight diagram, the number of blocks with the same level first increases and then decreases as the level increases, **symmetric up and down like a spindle.**

The method of the block weight diagram does not give a rule for choosing the orthonormal basis states with the multiple weights. **The calculation quantity depends greatly on how to choose the orthonormal basis states, especially when the multiplicity of the weight is high.** The Gel'fand's method and its generalization give a certain rule to choose the orthonormal basis states.

The formula for the multiplicity of a weight was given by [Kastant 1979]. An excellent table book [Bremner et al. (1985)] gave the useful data of the highest weight representations of all simple Lie algebras with the rank less than 13, such as their dimensions, the number of different levels, and the multiplicities and orbit sizes of the dominant weights in the representation.

8.2 Gel'fand's Method

The advantage of the method of the block weight diagram is intuitive and evident, but its disadvantage is without a rule to choose the basis states when the weight $\boldsymbol{m}$ is multiple. Gel'fand and Tsetlin (1950) gave a certain rule to define the orthonormal basis states for the Lie algebra A_ℓ.

8.2.1 *The Gel'fand's Bases*

Gel'fand and Tsetlin solved completely the problem on calculating the highest weight representation of the SU(N) group (the Lie algebra A_ℓ with $\ell = N-1$), including the basis states and the matrix elements of generators. The Dynkin diagram and simple roots of A_ℓ are (see subsection 7.5.6.)

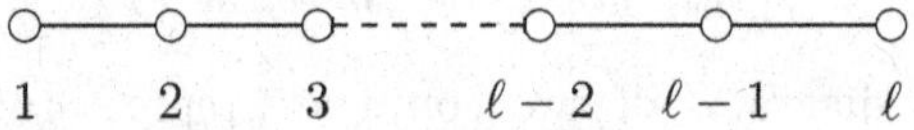

$$\mathrm{r}_\mu = -\boldsymbol{w}_{\mu-1} + 2\boldsymbol{w}_\mu - \boldsymbol{w}_{\mu+1}, \quad \boldsymbol{w}_0 = \boldsymbol{w}_{\ell+1} = 0, \quad 1 \leqslant \mu \leqslant \ell. \tag{8.12}$$

The element u of SU(N) is a unitary transformation in an N-dimensional complex space. If u is restricted to the transformation in the first n-dimensional complex space, SU(N) reduces to its subgroup SU(n). Those subgroups construct **a chain of Lie subgroups** of SU(N). Correspondingly, the Lie subalgebras A_μ construct **a chain of Lie subalgebras**.

$$\begin{gathered} \mathrm{SU}(2) \subset \mathrm{SU}(3) \subset \cdots \subset \mathrm{SU}(N-1) \subset \mathrm{SU}(N), \\ A_1 \subset A_2 \subset \cdots \subset A_\mu \subset \cdots \subset A_{\ell-1} \subset A_\ell, \quad \ell = N-1. \end{gathered} \tag{8.13}$$

For definiteness, we make a convention in this chapter that A_μ **denotes the subalgebra in the chain** (8.13), and $\mathcal{A}_\nu$ **is an** A_1 **Lie algebra, spanned by** H_ν, E_ν and F_ν. The Dynkin diagram of A_μ becomes that of $A_{\mu-1}$ by removing the last simple root $\boldsymbol{r}_\mu$:

$$A_1 = \mathcal{A}_1, \qquad A_\mu \supset A_{\mu-1} \oplus \mathcal{A}_\mu, \qquad 2 \leqslant \mu \leqslant \ell. \tag{8.14}$$

The key idea of Gel'fand and Tsetlin is to choose the orthonormal basis states in the highest weight representation $D^{\boldsymbol{M}}(\mathrm{A}_\ell)$ such that **the subduced representation of $D^{\boldsymbol{M}}(\mathrm{A}_\ell)$ with respect to each subalgebra A_μ is a reduced one.** Namely, each orthonormal basis state in $D^{\boldsymbol{M}}(\mathrm{A}_\ell)$, called the Gel'fand's basis, belongs to a highest weight representation of each subalgebra A_μ. The Gel'fand's basis is described by $N(N+1)/2$ parameters ω_{ab}, $1 \leqslant a \leqslant b \leqslant N$, satisfying the constraint:

$$\omega_{ab} \geqslant \omega_{a(b-1)} \geqslant \omega_{(a+1)b}. \tag{8.15}$$

The parameters are arranged as a triangle upside down:

$$|\omega_{ab}\rangle = \left|\begin{array}{ccccccc} \omega_{1N} & & \omega_{2N} & \cdots & \omega_{(N-1)N} & & \omega_{NN} \\ & \omega_{1(N-1)} & & \cdots & & \omega_{(N-1)(N-1)} & \\ & & \cdots & \cdots & \cdots & & \\ & & \omega_{12} & & \omega_{22} & & \\ & & & \omega_{11} & & & \end{array}\right\rangle. \tag{8.16}$$

The Gel'fand's basis $|\omega_{ab}\rangle$ belongs to the highest weight representation $D^{\boldsymbol{M}^{(\mu)}}$ of the subalgebra A_μ (see subsection 9.1.7):

$$\left(\boldsymbol{M}^{(\mu)}\right)_a = \omega_{a(\mu+1)} - \omega_{(a+1)(\mu+1)}, \quad 1 \leqslant a \leqslant \mu, \quad 1 \leqslant \mu \leqslant \ell, \tag{8.17}$$

where $\boldsymbol{M} = \boldsymbol{M}^{(\ell)}$. The components m_μ of the weight $\boldsymbol{m}$ are

$$\begin{aligned} & m_\mu = 2\Omega_\mu - \Omega_{\mu+1} - \Omega_{\mu-1}, \qquad 1 \leqslant \mu \leqslant \ell, \\ & \Omega_0 = 0, \quad \Omega_b = \sum_{a=1}^{b} \omega_{ab}, \qquad 1 \leqslant b \leqslant \ell + 1. \end{aligned} \tag{8.18}$$

The multiplicity of a weight $\boldsymbol{m}$ in a given representation $D^{\boldsymbol{M}}$ is the number of the possible Gel'fand's bases with different values of ω_{ab} satisfying the constraint (8.15) and the condition (8.18). The basis states with a multiple weight are described by a weight $(\boldsymbol{m})_r$ with an ordinal number r in the block weight diagram. **The rule of the correspondence between the Gel'fand's basis $|\omega_{ab}\rangle$ and $(\boldsymbol{m})_r$ can be known from examples.**

In the action of the raising operator E_μ, a Gel'fand's basis $|\omega_{ab}\rangle$ transforms to the linear combination of all possible Gel'fand's bases $|\omega'_{ab}\rangle$ where only one of $\omega_{a\mu}$ at the μth row increases by 1 under the constraint (8.15):

$$E_\mu\ |\omega_{ab}\rangle = \sum_{\nu=1}^{\mu} A_{\nu\mu}(\omega_{ab})\ |\omega_{ab} + \delta_{a\nu}\delta_{b\mu}\rangle. \tag{8.19}$$

Due to Eq. (8.18), the action of E_μ increases m_μ by 2 and decreases $m_{\mu\pm 1}$ by 1, so that it increases the weight $\boldsymbol{m}$ of $|\omega_{ab}\rangle$ by a simple root $\boldsymbol{r}_\mu$. The matrix element $A_{\nu\mu}(\omega_{ab})$ of E_μ is

$$\begin{aligned}
&A_{\nu\mu}(\omega_{ab}) \\
&= \left\{-\prod_{\tau=1}^{\mu-1}\left(\omega_{\tau(\mu-1)} - \omega_{\nu\mu} - \tau + \nu - 1\right)\prod_{\rho=1}^{\mu+1}\left(\omega_{\rho(\mu+1)} - \omega_{\nu\mu} - \rho + \nu\right)\right\}^{1/2} \\
&\times \left\{\prod_{\lambda\neq\nu,\lambda=1}^{\mu}\left(\omega_{\lambda\mu} - \omega_{\nu\mu} - \lambda + \nu\right)\left(\omega_{\lambda\mu} - \omega_{\nu\mu} - \lambda + \nu - 1\right)\right\}^{-1/2}.
\end{aligned} \tag{8.20}$$

The action of the lowering operator F_μ decreases one of $\omega_{a\mu}$ by 1, so that it decreases the weight $\boldsymbol{m}$ by $\boldsymbol{r}_\mu$. **The matrix form of F_μ is the transpose of that of E_μ.** Since all formulas in the Gel'fand's method depend only on the differences of the parameters ω_{ab}, **two Gel'fand's bases $|\omega_{ab}\rangle$ and $|\omega'_{ab}\rangle$ describe the same basis state in the same representation of A_ℓ if all parameters satisfy $\omega'_{ab} = \omega_{ab} - 1$.** ω_{NN} is usually taken to be zero in the Gel'fand's method.

When the parameters ω_{ab} in the basis state $|\omega_{ab}\rangle$ satisfy

$$\omega_{ab} = \omega_{a(b+1)}, \qquad 1 \leqslant a \leqslant b \leqslant \ell, \tag{8.21}$$

due to the constraint (8.15), any ω_{ab} cannot increases, so that the basis state $|\omega_{ab}\rangle$ with the parameters ω_{ab} satisfying Eq. (8.21) is annihilated by each raising operator E_μ and describes **the highest weight state**. In the same reason, the basis state with ω_{ab} satisfying $\omega_{a(b-1)} = \omega_{(a+1)b}$ is the lowest weight state. The components of the lowest weight $-\boldsymbol{N}$ satisfy

$$-N_\mu = \omega_{1\mu} - \omega_{1(\mu+1)} = \omega_{(N-\mu+1)N} - \omega_{(N-\mu)N} = -M_{N-\mu}. \tag{8.22}$$

Gel'fand and Tsetlin did not publish the proof for the formula (8.20). In fact, $A_{\nu\mu}(\omega_{ab})$ should satisfy Eq. (8.9). I have proved Eq. (8.20) by mathematical induction for the chain (8.13) of subalgebras (unpublished). In the following we are going to explain **how to calculate $A_{\nu\mu}(\omega_{ab})$ directly from Eq. (8.9) in corporation with the method of the block weight diagram.** The formula (8.20) is certainly helpful for calculating the matrix elements $A_{\nu\mu}(\omega_{ab})$ by computer. However, **by the following method it is able to compile the programs for this calculation, too.**

The Gel'fand's method is a recursive way. For the $\mathcal{A}_1$ subalgebra, we know from Eq. (8.8) where $m_\mu = \omega_{12} - \omega_{22}$ and $n = \omega_{12} - \omega_{11}$,

$$F_1\,|\omega_{ab}\rangle = \sqrt{(\omega_{11} - \omega_{22})(\omega_{12} - \omega_{11} + 1)}\;|\omega'_{ab}\rangle, \tag{8.23}$$

where ω'_{ab} satisfy the constraint (8.15) and $\omega'_{ab} = \omega_{ab}$ except for $\omega'_{11} = \omega_{11} - 1$. It is easy to check that Eq. (8.20) coincides with Eq. (8.23).

The set of the basis states in the highest weight representation of the subalgebra A_μ will be called the A_μ-multiplet for simplicity. The parameters ω_{ab} of the highest weight state in the A_μ-multiplet satisfy

$$\omega_{ab} = \omega_{a(b+1)}, \qquad 1 \leqslant a \leqslant b \leqslant \mu. \tag{8.24}$$

Equation (8.24) is back to Eq. (8.21) when $\mu = \ell$. Due to Eq. (8.14), **the representation space of A_ℓ is spanned by a few $A_{\ell-1}$-multiplets, and the generator F_ℓ relates the states belonging to different $A_{\ell-1}$-multiplets.** As a recursive way, the $A_{\ell-1}$-multiplet is assumed to be known and the calculation of A_ℓ-multiplet reduces to that of the $\mathcal{A}_\ell$-multiplets. Since $\mathcal{A}_\ell$ is only an A_1 Lie algebra, to calculate an $\mathcal{A}_\ell$-multiplet (see Eq. (8.7) and (8.8)) is much easier than to calculate a representation of A_ℓ.

The highest weight state $|\boldsymbol{M}, \boldsymbol{M}\rangle$ is also that of an $A_{\ell-1}$-multiplet. If $M_\ell > 0$ ($\omega_{\ell N} = \omega_{\ell\ell} > \omega_{NN} = 0$), $|\boldsymbol{M}, \boldsymbol{M}\rangle$ is the highest weight state of an $\mathcal{A}_\ell$-multiplet, where the action of F_ℓ on each basis state in this $\mathcal{A}_\ell$-multiplet only decreases $\omega_{\ell\ell}$ such that **each basis state in this $\mathcal{A}_\ell$-multiplet is also the highest weight state of an $A_{\ell-1}$-multiplet.**

Since each $A_{\ell-1}$-multiplet contains a few $A_{\ell-2}$-multiplets wholly. As the level $L(\boldsymbol{M}, \boldsymbol{m})$ increases from one (the state $|\boldsymbol{M}, \boldsymbol{M}\rangle$), **check the highest weight state of each $A_{\ell-2}$-multiplets in the $A_{\ell-1}$-multiplet one by one**, whether or not it satisfies the condition (8.7) with $\mu = \ell$. If yes, we construct an $\mathcal{A}_\ell$-multiplet by successive application of F_ℓ:

$$F_\ell \ |\omega_{ab}\rangle = \sum_{d=1}^{\ell} A_d \ |\omega_{ab}^{(d)}\rangle, \tag{8.25}$$

where $\omega_{ab}^{(d)} = \omega_{ab}$ except for $\omega_{d\ell}^{(d)} = \omega_{d\ell} - 1$. $A_d = 0$ if $\omega_{d\ell} = \omega_{d(\ell-1)}$ or $\omega_{d\ell} = \omega_{(d+1)(\ell+1)}$. **If** $E_{\ell-1} \ |\omega_{ab}^{(d)}\rangle \neq 0$,

$$E_{\ell-1} \ |\omega_{ab}^{(d)}\rangle = \sum_{c=1}^{\ell-1} B_c \ |\omega_{ab}^{(dc)}\rangle, \tag{8.26}$$

where $\omega_{ab}^{(dc)} = \omega_{ab}^{(d)}$ except for $\omega_{c(\ell-1)}^{(dc)} = \omega_{c(\ell-1)}^{(d)} + 1$. B_c is known because it is the matrix element of $E_{\ell-1}$ in an $A_{\ell-1}$-multiplet. On the other hand, we have

$$F_\ell E_{\ell-1} \ |\omega_{ab}\rangle = \sum_{r=1}^{\ell-1} C_r F_\ell \ |\overline{\omega}_{ab}^{(r)}\rangle = \sum_{r=1}^{\ell-1} \sum_{s=1}^{\ell} C_r D_s \ |\overline{\omega}_{ab}^{(rs)}\rangle, \tag{8.27}$$

where $\overline{\omega}_{ab}^{(r)} = \omega_{ab}$ except for $\overline{\omega}_{r(\ell-1)}^{(r)} = \omega_{r(\ell-1)} + 1$, and $\overline{\omega}_{ab}^{(rs)} = \overline{\omega}_{ab}^{(r)}$ except for $\overline{\omega}_{s\ell}^{(rs)} = \overline{\omega}_{s\ell}^{(r)} - 1$. C_r is known because it is the matrix element of $E_{\ell-1}$ in an $\mathrm{A}_{\ell-1}$-multiplet. D_s have been calculated because the level of the basis state $|\overline{\omega}_{ab}^{(r)}\rangle$ is less than that of the basis state $|\omega_{ab}\rangle$ by one. Due to $E_{\ell-1}F_\ell = F_\ell E_{\ell-1}$ (see Eq. (8.9)), we have

$$\sum_{r=1}^{\ell-1}\sum_{s=1}^{\ell} C_r D_s \, |\overline{\omega}_{ab}^{(rs)}\rangle = \sum_{d=1}^{\ell}\sum_{c=1}^{\ell-1} A_d B_c \, |\omega_{ab}^{(dc)}\rangle = E_{\ell-1}F_\ell \, |\omega_{ab}\rangle. \tag{8.28}$$

Comparing the coefficients of the same basis state, $|\overline{\omega}_{ab}^{(rs)}\rangle = |\omega_{ab}^{(dc)}\rangle$, we are able to obtain the matrix elements A_d of F_ℓ.

There is at most one basis state $|\omega_{ab}^{(d)}\rangle$ in Eq. (8.25) where $|\omega_{ab}^{(d)}\rangle$ is the highest weight state of an $\mathrm{A}_{\ell-1}$-multiplet. We choose the phase of $|\omega_{ab}^{(d)}\rangle$ such that its coefficient A_d is real and positive. This coefficient A_d is able to calculate by applying $E_\ell F_\ell = F_\ell E_\ell + H_\ell$ (see Eq. (8.9)) to $|\omega_{ab}\rangle$. **All basis states and the matrix elements of F_ℓ in the $\mathcal{A}_\ell$-multiplet can be calculated by successive application of** Eqs. (8.25-28).

When the $\mathcal{A}_\ell$-multiplet from the highest weight state of an $\mathrm{A}_{\ell-2}$-multiplet, which satisfies the condition (8.7) with $\mu = \ell$, is calculated, we can **apply the lowering operators F_μ in $\mathrm{A}_{\ell-2}$ to these states in the $\mathcal{A}_\ell$-multiplet appearing in Eq. (8.25) and obtain the similar equations to** Eq. (8.25) **and the similar $\mathcal{A}_\ell$-multiplets from the remaining basis states in the $\mathrm{A}_{\ell-2}$-multiplet**, because F_μ in $\mathrm{A}_{\ell-2}$ commute with F_ℓ. This property is called the "**parallel principle**" of $\mathcal{A}_\ell$-multiplets. The parallel principle greatly reduces the calculation quantity for $\mathcal{A}_\ell$-multiplets. It is also the reason why the matrix elements $A_{\nu\mu}(\omega_{ab})$ given in the formula (8.20) depend only upon the parameters $\omega_{b\mu}$ and $\omega_{b(\mu\pm1)}$.

In this method all $\mathrm{A}_{\ell-1}$-multiplets and $\mathcal{A}_\ell$-multiplets contained in the highest weight representation D^M of A_ℓ are calculated. The block weight diagram of the highest weight representation of A_ℓ is decomposed as **the block weight diagrams of $\mathrm{A}_{\ell-1}$-multiplets and of $\mathcal{A}_\ell$-multiplets.**

8.2.2 *Some Representations of* $\mathbf{A_2}$

The matrix elements of F_1 in an $\mathcal{A}_1$-multiplet are given in Eq. (8.23), which are the foundation of calculating the representations of A_2. The calculated results of the representations of A_2 (the SU(3) group), which are widely used in particle physics, are the foundation of those of A_3 and so on.

The key to calculate the highest weight representations of A_2 is to construct the $\mathcal{A}_2$-multiplet in the representation. The Gel'fand's bases in the representation $D^{\boldsymbol{M}}$ of A_2 is

$$\left|\begin{matrix} \omega_{13} & & \omega_{23} & & \omega_{33} \\ & \omega_{12} & & \omega_{22} & \\ & & \omega_{11} & & \end{matrix}\right\rangle^{L}, \qquad \begin{aligned} M_1 &= \omega_{13} - \omega_{23}, \\ M_2 &= \omega_{23} - \omega_{33}. \end{aligned} \tag{8.29}$$

It is convenient to add a superscript $L(\boldsymbol{M}, \boldsymbol{m})$ on the Gel'fand's bases.

For the representation with the highest weight $\boldsymbol{M} = M\boldsymbol{w}_1 = (M, 0)$, the parameters ω_{ab} in the highest weight state are $\omega_{13} = \omega_{12} = \omega_{11} = M$ and $\omega_{23} = \omega_{33} = \omega_{22} = 0$. **All weights in this representation are single.** The A_1-multiplet is constructed from the highest weight state by Eq. (8.23) where $(M+1)$ basis states are obtained by decreasing ω_{11} from M to 0. Those basis states all satisfy Eq. (8.7) with $\mu = 2$, so that an $\mathcal{A}_2$-multiplet is constructed from each of them where the highest weight state $|\boldsymbol{M}, \boldsymbol{M}\rangle$ spans the $\mathcal{A}_2$-singlet. Then, from each basis state in those $\mathcal{A}_2$-multiplets, if it satisfies Eq. (8.7) with $\mu = 1$, an A_1-multiplet is constructed by Eq. (8.23). The block weight diagram and the Gel'fand's bases for the representation $D^{\boldsymbol{M}}$ with $\boldsymbol{M} = M\boldsymbol{w}_1$ are given in Fig. 8.1. Its complex conjugate representation has the highest weight $\boldsymbol{M}' = M\boldsymbol{w}_2 = (0, M)$. The block weight diagram of $\boldsymbol{M}'$ is upside down of that of $\boldsymbol{M}$.

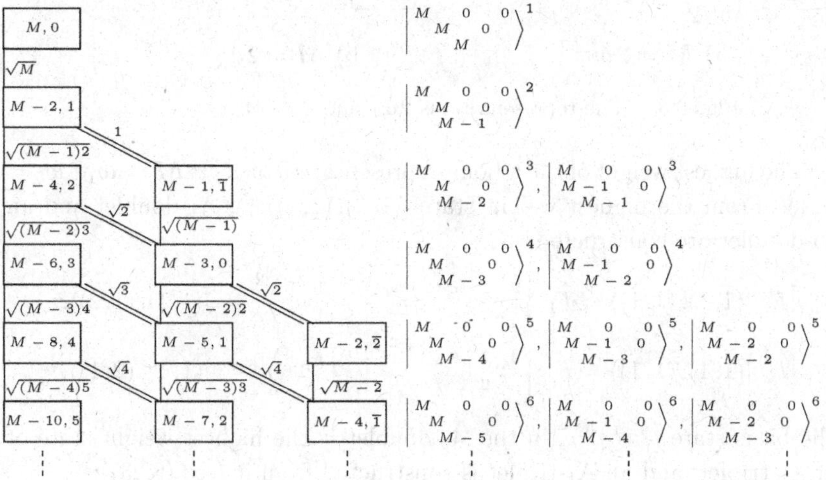

Fig. 8.1 The highest weight representation $D^{\boldsymbol{M}}$ of A_2 with $\boldsymbol{M} = M\boldsymbol{w}_1$.

Some examples for the representation $\boldsymbol{M} = M\boldsymbol{w}_1$ are given in Figs. 8.2,

8.3, and 8.4 with M = 1, 2, and 3, respectively. The representation $D^{\boldsymbol{w}_1}$ contains only one single dominant weight $\boldsymbol{w}_1$. The representation $D^{2\boldsymbol{w}_1}$ contains two single dominant weights: $2\boldsymbol{w}_1$ and $\boldsymbol{w}_2$. The representation $D^{3\boldsymbol{w}_1}$ contains three single dominant weights: $3\boldsymbol{w}_1$, $\boldsymbol{w}_1 + \boldsymbol{w}_2$, and 0.

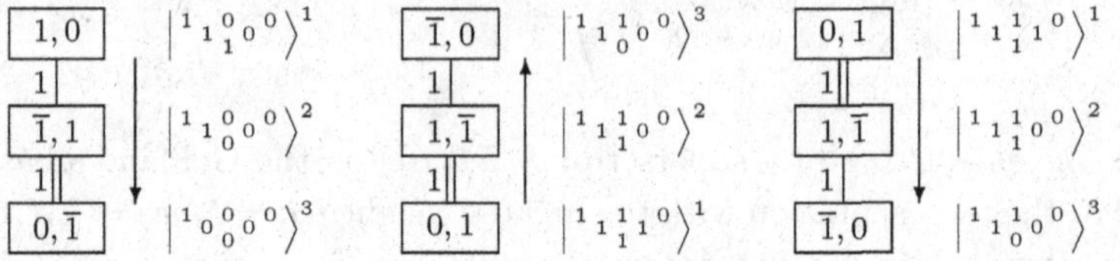

a) $\boldsymbol{M} = \boldsymbol{w}_1$　　　　b) $\boldsymbol{M} = \boldsymbol{w}_2$

Fig. 8.2　The basic representations of A_2.

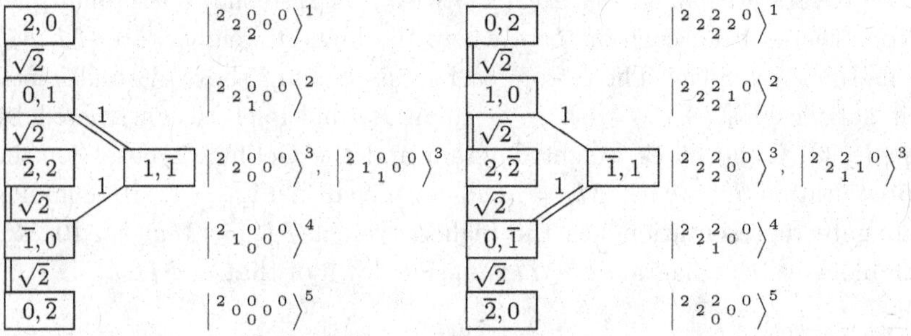

a) $\boldsymbol{M} = 2\boldsymbol{w}_1$　　　　b) $\boldsymbol{M} = 2\boldsymbol{w}_2$

Fig. 8.3　The representations $2\boldsymbol{w}_1$ and $2\boldsymbol{w}_2$ of A_2.

The highest weight of the adjoint representation of A_2 is $\boldsymbol{M} = \boldsymbol{w}_1 + \boldsymbol{w}_2 = (1,1)$. From the highest weight state $|(1,1),(1,1)\rangle$ an A_1-doublet and an $\mathcal{A}_2$-doublet are constructed:

$$F_1\, |(1,1),(1,1)\rangle = F_1 \left|\begin{smallmatrix} 2 & & 1 & & 0 \\ & 2 & & 1 & \\ & & 2 & & \end{smallmatrix}\right\rangle^1 = \left|\begin{smallmatrix} 2 & & 1 & & 0 \\ & 2 & & 1 & \\ & & 1 & & \end{smallmatrix}\right\rangle^2 = |(1,1),(\overline{1},2)\rangle,$$

$$F_2\, |(1,1),(1,1)\rangle = F_2 \left|\begin{smallmatrix} 2 & & 1 & & 0 \\ & 2 & & 1 & \\ & & 2 & & \end{smallmatrix}\right\rangle^1 = \left|\begin{smallmatrix} 2 & & 1 & & 0 \\ & 2 & & 0 & \\ & & 2 & & \end{smallmatrix}\right\rangle^2 = |(1,1),(2,\overline{1})\rangle.$$

The basis state $\left|\begin{smallmatrix} 2 & & 1 & & 0 \\ & 2 & & 0 & \\ & & 2 & & \end{smallmatrix}\right\rangle^2$ in the $\mathcal{A}_2$-doublet is the highest weight state of an A_1-triplet, and an A_1-triplet is constructed from it (see (8.23)):

$$F_1\, |(1,1),(2,\overline{1})\rangle = F_1 \left|\begin{smallmatrix} 2 & & 1 & & 0 \\ & 2 & & 0 & \\ & & 2 & & \end{smallmatrix}\right\rangle^2 = \sqrt{2} \left|\begin{smallmatrix} 2 & & 1 & & 0 \\ & 2 & & 0 & \\ & & 1 & & \end{smallmatrix}\right\rangle^3 = \sqrt{2}\, |(1,1),(0,0)_1\rangle,$$

$$F_1\,|(1,1),(0,0)_1\rangle = F_1\left|\begin{smallmatrix}2&&1&&0\\&2&&0&\\&&1&&\end{smallmatrix}\right\rangle^3 = \sqrt{2}\left|\begin{smallmatrix}2&&1&&0\\&2&&0&\\&&0&&\end{smallmatrix}\right\rangle^4 = \sqrt{2}\,|(1,1),(\overline{2},1)\rangle.$$

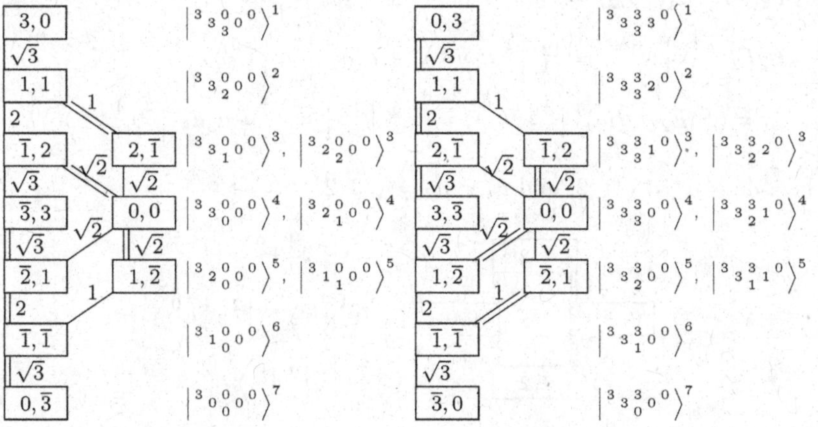

a) $\boldsymbol{M} = 3\boldsymbol{w}_1$ b) $\boldsymbol{M} = 3\boldsymbol{w}_2$

Fig. 8.4 The representations $3\boldsymbol{w}_1$ and $3\boldsymbol{w}_2$ of A_2.

The basis state $\left|\begin{smallmatrix}2&&1&&0\\&2&&1&\\&&1&&\end{smallmatrix}\right\rangle^2$ in the A_1-doublet satisfies the condition (8.7) with $\mu = 2$, and an $\mathcal{A}_2$-triplet is constructed from it (see Eq. (8.25)): .

$$F_2\,|(1,1),(\overline{1},2)\rangle = F_2\left|\begin{smallmatrix}2&&1&&0\\&2&&1&\\&&1&&\end{smallmatrix}\right\rangle^2 = a_1\left|\begin{smallmatrix}2&&1&&0\\&2&&0&\\&&1&&\end{smallmatrix}\right\rangle^3 + a_2\left|\begin{smallmatrix}2&&1&&0\\&1&&1&\\&&1&&\end{smallmatrix}\right\rangle^3.$$

The basis state $\left|\begin{smallmatrix}2&&1&&0\\&2&&0&\\&&1&&\end{smallmatrix}\right\rangle^3$ belongs to the A_1-triplet. The basis state $\left|\begin{smallmatrix}2&&1&&0\\&1&&1&\\&&1&&\end{smallmatrix}\right\rangle^3 = |(1,1),(0,0)_2\rangle$ spans an A_1-singlet, and its phase is chosen such that a_2 is real and positive. Following Eqs. (8.26-28), we have

$$\begin{aligned}
E_1F_2\left|\begin{smallmatrix}2&&1&&0\\&2&&1&\\&&1&&\end{smallmatrix}\right\rangle^2 &= \sqrt{2}a_1\left|\begin{smallmatrix}2&&1&&0\\&2&&0&\\&&2&&\end{smallmatrix}\right\rangle^2 \\
&= F_2E_1\left|\begin{smallmatrix}2&&1&&0\\&2&&1&\\&&1&&\end{smallmatrix}\right\rangle^2 = F_2\left|\begin{smallmatrix}2&&1&&0\\&2&&1&\\&&2&&\end{smallmatrix}\right\rangle^1 = \left|\begin{smallmatrix}2&&1&&0\\&2&&0&\\&&2&&\end{smallmatrix}\right\rangle^2, \\
E_2F_2\left|\begin{smallmatrix}2&&1&&0\\&2&&1&\\&&1&&\end{smallmatrix}\right\rangle^2 &= \left(a_1^2 + a_2^2\right)\left|\begin{smallmatrix}2&&1&&0\\&2&&1&\\&&1&&\end{smallmatrix}\right\rangle^2 \\
&= (F_2E_2 + H_2)\left|\begin{smallmatrix}2&&1&&0\\&2&&1&\\&&1&&\end{smallmatrix}\right\rangle^2 = 2\left|\begin{smallmatrix}2&&1&&0\\&2&&1&\\&&1&&\end{smallmatrix}\right\rangle^2.
\end{aligned}$$

Then, $a_1 = \sqrt{1/2}$, and $a_2 = \sqrt{2 - 1/2} = \sqrt{3/2}$. Using the method of Eqs. (8.25-28) to $\left|\begin{smallmatrix}2&&1&&0\\&2&&0&\\&&1&&\end{smallmatrix}\right\rangle^3$ and $\left|\begin{smallmatrix}2&&1&&0\\&1&&1&\\&&1&&\end{smallmatrix}\right\rangle^3$ further, and choosing the phase of $\left|\begin{smallmatrix}2&&1&&0\\&1&&0&\\&&1&&\end{smallmatrix}\right\rangle^4$ to make a_3 real and positive, we obtain

$$F_2\,|(1,1),(0,0)_1\rangle = F_2\left|\begin{matrix}2&&1&&0\\&2&&0&\\&&1&&\end{matrix}\right\rangle^3 = a_3\left|\begin{matrix}2&&1&&0\\&1&&0&\\&&1&&\end{matrix}\right\rangle^4,$$

$$F_2\,|(1,1),(0,0)_2\rangle = F_2\left|\begin{matrix}2&&1&&0\\&1&&1&\\&&1&&\end{matrix}\right\rangle^3 = a_4\left|\begin{matrix}2&&1&&0\\&1&&0&\\&&1&&\end{matrix}\right\rangle^4,$$

$$E_2F_2\left|\begin{matrix}2&&1&&0\\&2&&0&\\&&1&&\end{matrix}\right\rangle^3 = a_1^2\left|\begin{matrix}2&&1&&0\\&2&&0&\\&&1&&\end{matrix}\right\rangle^3 + a_1a_2\left|\begin{matrix}2&&1&&0\\&1&&1&\\&&1&&\end{matrix}\right\rangle^3$$

$$= (F_2E_2+H_2)\left|\begin{matrix}2&&1&&0\\&2&&0&\\&&1&&\end{matrix}\right\rangle^3 = a_3^2\left|\begin{matrix}2&&1&&0\\&2&&0&\\&&1&&\end{matrix}\right\rangle^3 + a_3a_4\left|\begin{matrix}2&&1&&0\\&1&&1&\\&&1&&\end{matrix}\right\rangle^3.$$

Then, $a_3 = a_1 = \sqrt{1/2}$, and $a_4 = a_2 = \sqrt{3/2}$.

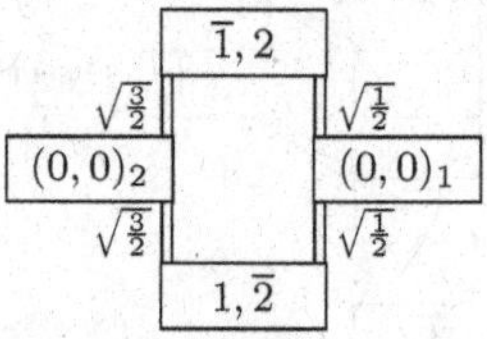

$$\left|\begin{matrix}2&&1&&0\\&2&&1&\\&&1&&\end{matrix}\right\rangle^2$$

$$\left|\begin{matrix}2&&1&&0\\&1&&1&\\&&1&&\end{matrix}\right\rangle^3,\quad \left|\begin{matrix}2&&1&&0\\&2&&0&\\&&1&&\end{matrix}\right\rangle^3$$

$$\left|\begin{matrix}2&&1&&0\\&1&&0&\\&&1&&\end{matrix}\right\rangle^4$$

The basis state $\left|\begin{smallmatrix}2&&1&&0\\&1&&0&\\&&1&&\end{smallmatrix}\right\rangle^4$ is the highest weight state of an A_1-doublet. The basis state $\left|\begin{smallmatrix}2&&1&&0\\&2&&0&\\&&0&&\end{smallmatrix}\right\rangle^4$ in the A_1-triplet satisfies the condition (8.7) with $\mu = 2$, and an $\mathcal{A}_2$-doublet is constructed from it. The produced basis state in both A_1-doublet and $\mathcal{A}_2$-doublet is the lowest weight state $\left|\begin{smallmatrix}2&&1&&0\\&1&&0&\\&&0&&\end{smallmatrix}\right\rangle^5 = |(1,1),(\overline{1},\overline{1})\rangle$ in this representation.

The block weight diagram and the Gel'fand's basis states of the adjoint representation are given in Fig. 8.5. The adjoint representation with $\boldsymbol{M} = (1,1)$ is 8-dimensional and contains a simple dominant weight $\boldsymbol{w}_1 + \boldsymbol{w}_2$ and a double dominant weight 0.

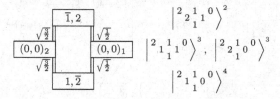

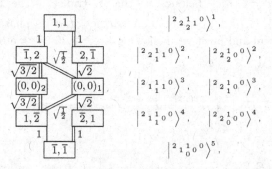

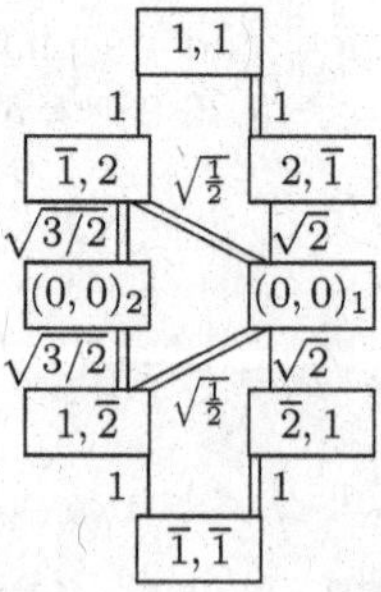

$$\left|\begin{matrix}2&&1&&0\\&2&&1&\\&&2&&\end{matrix}\right\rangle^1,$$

$$\left|\begin{matrix}2&&1&&0\\&2&&1&\\&&1&&\end{matrix}\right\rangle^2,\quad \left|\begin{matrix}2&&1&&0\\&2&&0&\\&&2&&\end{matrix}\right\rangle^2,$$

$$\left|\begin{matrix}2&&1&&0\\&1&&1&\\&&1&&\end{matrix}\right\rangle^3,\quad \left|\begin{matrix}2&&1&&0\\&2&&0&\\&&1&&\end{matrix}\right\rangle^3,$$

$$\left|\begin{matrix}2&&1&&0\\&1&&0&\\&&1&&\end{matrix}\right\rangle^4,\quad \left|\begin{matrix}2&&1&&0\\&2&&0&\\&&0&&\end{matrix}\right\rangle^4,$$

$$\left|\begin{matrix}2&&1&&0\\&1&&0&\\&&0&&\end{matrix}\right\rangle^5,$$

Fig. 8.5 The adjoint representation $\boldsymbol{w}_1 + \boldsymbol{w}_2$ of A_2.

The block weight diagrams and the Gel'fand's bases of the representation $\boldsymbol{M} = 2\boldsymbol{w}_1 + \boldsymbol{w}_2 = (2,1)$ and its conjugate one $\boldsymbol{M}' = \boldsymbol{w}_1 + 2\boldsymbol{w}_2 = (1,2)$ are given in Fig. 8.6 (see Prob. 1). They are 15 dimensional. The representation $M = (2,1)$ contains two simple dominant weights $2w_1 + w_2$ and $2w_2$ and one double dominant weight w_1.

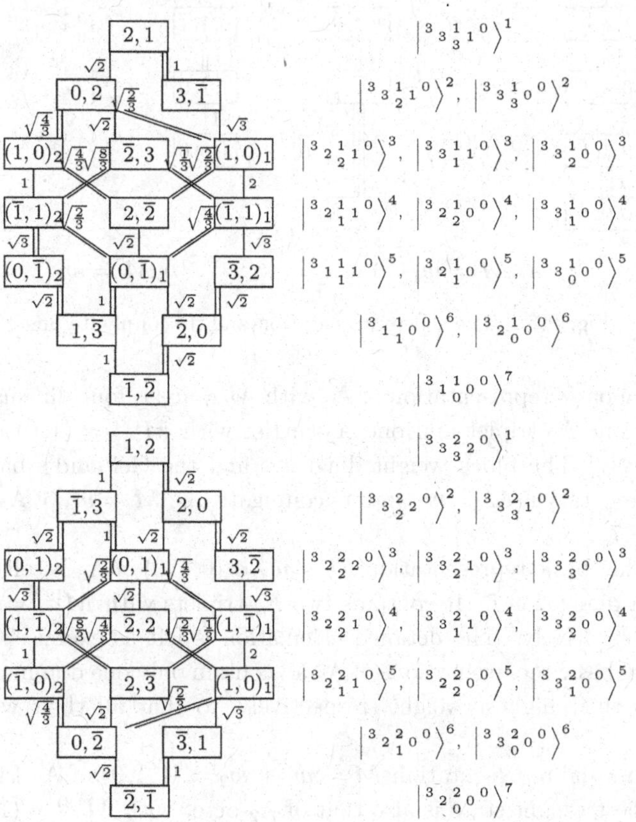

Fig. 8.6 The representation $2w_1 + w_2$ of A_2 and its conjugate one.

8.2.3 *Some Representations of* A_3

Based on the knowledge of representations of A_2, we calculate some representations of A_3 Lie algebra. The block weight diagram of A_3 is decomposed into the block weight diagrams of A_2-multiplets and those of $\mathcal{A}_3$-multiplets. We neglect the symbol M in the basis state $|M, m\rangle$ for simplicity.

The basic representation with $M = w_1 = (1,0,0)$ of A_3 is the self-representation of SU(4). An A_2-triplet with $M^{(2)} = (1,0)$ is constructed from the highest weight state $|(1,0,0)\rangle$. The last basis state satisfies the condition (8.7) with $\mu = 3$, and an $\mathcal{A}_3$-doublet is constructed from it.

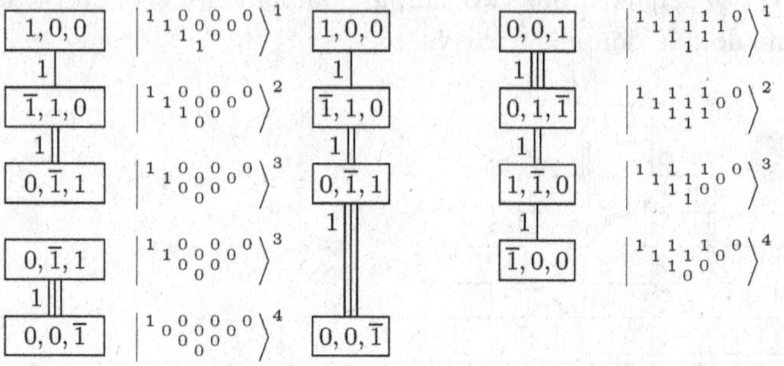

a) $M = w_1$ b) $M = w_3$

Fig. 8.7 The self-representation and its conjugate one of A_3.

Thus, the basic representation of A_3 with $M = w_1$ is four-dimensional and contains one A_2-triplet and one A_2-singlet with $M^{(2)} = (1, 0)$ and $(0, 0)$, respectively. The block weight diagrams and the Gel'fand's bases of the self-representation $M = w_1$ and its conjugate one $M = w_3$ of A_3 are given in Fig. 8.7.

Another basic representation $M = w_2 = (0, 1, 0)$ of A_3 is self-conjugate one with dimension 6. It contains two A_2-triplets with $M^{(2)} = (0, 1)$ and $(1, 0)$, respectively. The detailed calculation is left as exercise (see Prob. 2). Three basic representations of A_3 all contain only one dominant weight, which is their highest weight, respectively, so that all their weights are simple.

Discuss the representation $M = w_1 + w_2 = (1, 1, 0)$ of A_3 Lie algebra. The highest weight state is also that of A_2-octet with $M^{(2)} = (1, 1)$:

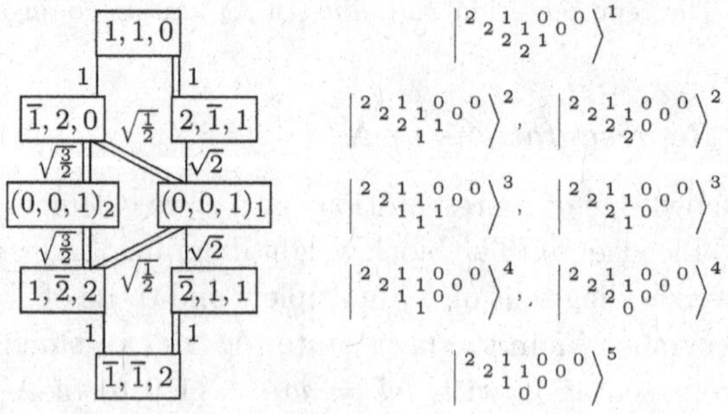

There are three A_1-multiplets whose highest weight states satisfy the condition (8.7) with $\mu = 3$. From $\left|\begin{smallmatrix}2 & & 1 & & 0 & & 0\\ & 2 & & 1 & & 0 & \\ & & 2 & & 0 & & \\ & & & 2 & & & \end{smallmatrix}\right\rangle^2$ we have an $\mathcal{A}_3$-doublet:

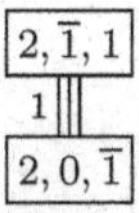

$$\left|\begin{smallmatrix}2&&1&&0&&0\\&2&&1&&0&\\&&2&&0&&\\&&&2&&&\end{smallmatrix}\right\rangle^{2} \qquad \left|\begin{smallmatrix}2&&1&&0&&0\\&2&&0&&0&\\&&2&&0&&\\&&&2&&&\end{smallmatrix}\right\rangle^{3}$$

The produced basis state is the highest weight state of an A_2-sixlet with $\boldsymbol{M}^{(2)} = (2,0)$ (see Fig. 8.3):

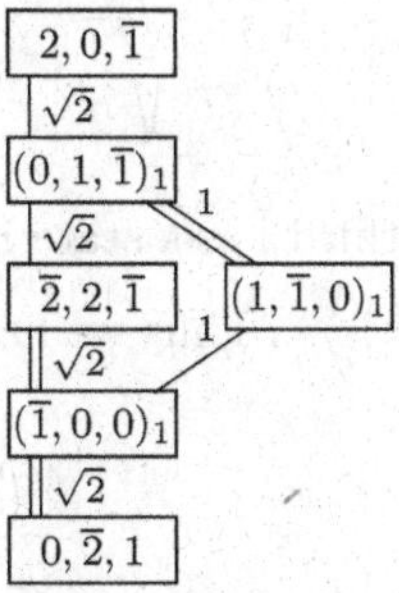

$$\left|\begin{smallmatrix}2&&1&&0&&0\\&2&&0&&0&\\&&2&&0&&\\&&&2&&&\end{smallmatrix}\right\rangle^{3}$$

$$\left|\begin{smallmatrix}2&&1&&0&&0\\&2&&0&&0&\\&&2&&0&&\\&&&1&&&\end{smallmatrix}\right\rangle^{4}$$

$$\left|\begin{smallmatrix}2&&1&&0&&0\\&2&&0&&0&\\&&2&&0&&\\&&&0&&&\end{smallmatrix}\right\rangle^{5}, \quad \left|\begin{smallmatrix}2&&1&&0&&0\\&2&&0&&0&\\&&1&&0&&\\&&&1&&&\end{smallmatrix}\right\rangle^{5}$$

$$\left|\begin{smallmatrix}2&&1&&0&&0\\&2&&0&&0&\\&&1&&0&&\\&&&0&&&\end{smallmatrix}\right\rangle^{6}$$

$$\left|\begin{smallmatrix}2&&1&&0&&0\\&2&&0&&0&\\&&0&&0&&\\&&&0&&&\end{smallmatrix}\right\rangle^{7}$$

where the superscript denotes the level in the representation of A_3, not in that of A_2 (see Fig. 8.3). Due to parallel principle we have

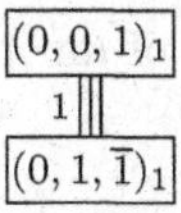

$$\left|\begin{smallmatrix}2&&1&&0&&0\\&2&&1&&0&\\&&2&&0&&\\&&&1&&&\end{smallmatrix}\right\rangle^{3} \qquad \left|\begin{smallmatrix}2&&1&&0&&0\\&2&&0&&0&\\&&2&&0&&\\&&&1&&&\end{smallmatrix}\right\rangle^{4}$$

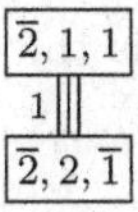

$$\left|\begin{smallmatrix}2&&1&&0&&0\\&2&&1&&0&\\&&2&&0&&\\&&&0&&&\end{smallmatrix}\right\rangle^{4} \qquad \left|\begin{smallmatrix}2&&1&&0&&0\\&2&&0&&0&\\&&2&&0&&\\&&&0&&&\end{smallmatrix}\right\rangle^{5}$$

From $\left|\begin{smallmatrix}2&&1&&0&&0\\&2&&1&&0&\\&&1&&1&&\\&&&1&&&\end{smallmatrix}\right\rangle^{3}$ we have an $\mathcal{A}_3$-doublet due to $m_3 = 1$:

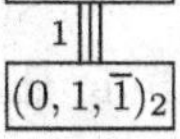

$$\left|\begin{smallmatrix}2&&1&&0&&0\\&2&&1&&0&\\&&1&&1&&\\&&&1&&&\end{smallmatrix}\right\rangle^{3} \qquad \left|\begin{smallmatrix}2&&1&&0&&0\\&1&&1&&0&\\&&1&&1&&\\&&&1&&&\end{smallmatrix}\right\rangle^{4}$$

The produced basis state is the highest weight state of an A_2-triplet with $\boldsymbol{M}^{(2)} = (0,1)$ (see Fig. 8.2):

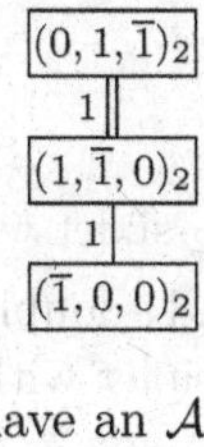

$$\left|\begin{smallmatrix}2&&1&&0&&0\\&1&&1&&0&\\&&1&&1&&\\&&&1&&&\end{smallmatrix}\right\rangle^{4}$$

$$\left|\begin{smallmatrix}2&&1&&0&&0\\&1&&1&&0&\\&&1&&0&&\\&&&1&&&\end{smallmatrix}\right\rangle^{5}$$

$$\left|\begin{smallmatrix}2&&1&&0&&0\\&1&&1&&0&\\&&1&&0&&\\&&&0&&&\end{smallmatrix}\right\rangle^{6}$$

From $\left|\begin{smallmatrix}2&&1&&0&&0\\&2&&1&&0&\\&&1&&0&&\\&&&1&&&\end{smallmatrix}\right\rangle^{4}$ we have an $\mathcal{A}_3$-triplet due to $m_3 = 2$:

$$F_3 \left|\begin{matrix}2&&1&&0&&0\\&2&&1&&0&\\&&1&&0&&\\&&&1&&&\end{matrix}\right\rangle^{4} = a\left|\begin{matrix}2&&1&&0&&0\\&2&&0&&0&\\&&1&&0&&\\&&&1&&&\end{matrix}\right\rangle^{5} + b\left|\begin{matrix}2&&1&&0&&0\\&1&&1&&0&\\&&1&&0&&\\&&&1&&&\end{matrix}\right\rangle^{5},$$

From Eq. (8.8) we have

$$
\begin{aligned}
E_2F_3 \left|\begin{matrix}2&&1&&0&&0\\&2&&1&&0&\\&&1&&0&&\\&&&1&&&\end{matrix}\right\rangle^4 &= a\left|\begin{matrix}2&&1&&0&&0\\&2&&0&&0&\\&&2&&0&&\\&&&1&&&\end{matrix}\right\rangle^4 + b\left|\begin{matrix}2&&1&&0&&0\\&1&&1&&0&\\&&1&&1&&\\&&&1&&&\end{matrix}\right\rangle^4 \\
= F_3E_2 \left|\begin{matrix}2&&1&&0&&0\\&2&&1&&0&\\&&1&&0&&\\&&&1&&&\end{matrix}\right\rangle^4 &= \sqrt{\frac{1}{2}}F_3\left|\begin{matrix}2&&1&&0&&0\\&2&&1&&0&\\&&2&&0&&\\&&&1&&&\end{matrix}\right\rangle^3 + \sqrt{\frac{3}{2}}F_3\left|\begin{matrix}2&&1&&0&&0\\&2&&1&&0&\\&&1&&1&&\\&&&1&&&\end{matrix}\right\rangle^3 \\
&= \sqrt{\frac{1}{2}}\left|\begin{matrix}2&&1&&0&&0\\&2&&0&&0&\\&&2&&0&&\\&&&1&&&\end{matrix}\right\rangle^4 + \sqrt{\frac{3}{2}}\left|\begin{matrix}2&&1&&0&&0\\&1&&1&&0&\\&&1&&1&&\\&&&1&&&\end{matrix}\right\rangle^4 .
\end{aligned}
$$

Then, $a = \sqrt{1/2}$ and $b = \sqrt{3/2}$. The third basis state in the $\mathcal{A}_3$-triplet is defined by $F_3^2 \left|\begin{smallmatrix}2&&1&&0&&0\\&2&&1&&0&\\&&1&&0&&\\&&&1&&&\end{smallmatrix}\right\rangle^4 = 2\left|\begin{smallmatrix}2&&1&&0&&0\\&1&&0&&0&\\&&1&&0&&\\&&&1&&&\end{smallmatrix}\right\rangle^6$. Thus we have from Eq. (8.9)

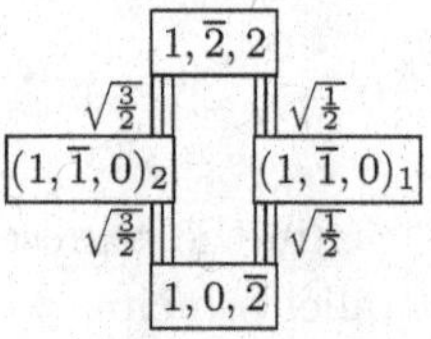

$$
\left|\begin{smallmatrix}2&&1&&0&&0\\&2&&1&&0&\\&&1&&0&&\\&&&1&&&\end{smallmatrix}\right\rangle^4
$$

$$
\left|\begin{smallmatrix}2&&1&&0&&0\\&1&&1&&0&\\&&1&&0&&\\&&&1&&&\end{smallmatrix}\right\rangle^5, \quad \left|\begin{smallmatrix}2&&1&&0&&0\\&2&&0&&0&\\&&1&&0&&\\&&&1&&&\end{smallmatrix}\right\rangle^5
$$

$$
\left|\begin{smallmatrix}2&&1&&0&&0\\&1&&0&&0&\\&&1&&0&&\\&&&1&&&\end{smallmatrix}\right\rangle^6
$$

The produced basis state is the highest weight state of an A$_2$-triplet with $\boldsymbol{M}^{(2)} = (1,0)$ (see Fig. 8.2):

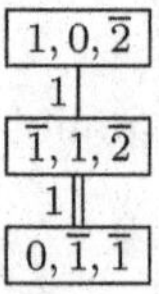

$$
\left|\begin{smallmatrix}2&&1&&0&&0\\&1&&0&&0&\\&&1&&0&&\\&&&1&&&\end{smallmatrix}\right\rangle^6
$$

$$
\left|\begin{smallmatrix}2&&1&&0&&0\\&1&&0&&0&\\&&1&&0&&\\&&&0&&&\end{smallmatrix}\right\rangle^7
$$

$$
\left|\begin{smallmatrix}2&&1&&0&&0\\&1&&0&&0&\\&&0&&0&&\\&&&0&&&\end{smallmatrix}\right\rangle^8
$$

The third state is the lowest weight state in this representation. From the parallel principle we have

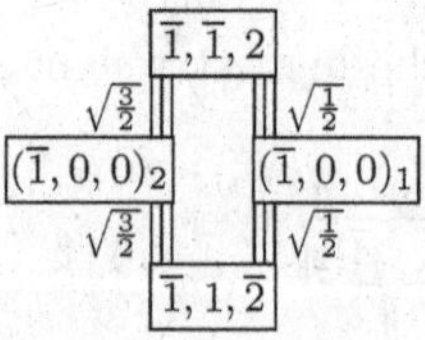

$$
\left|\begin{smallmatrix}2&&1&&0&&0\\&2&&1&&0&\\&&1&&0&&\\&&&0&&&\end{smallmatrix}\right\rangle^5
$$

$$
\left|\begin{smallmatrix}2&&1&&0&&0\\&1&&1&&0&\\&&1&&0&&\\&&&0&&&\end{smallmatrix}\right\rangle^6, \quad \left|\begin{smallmatrix}2&&1&&0&&0\\&2&&0&&0&\\&&1&&0&&\\&&&0&&&\end{smallmatrix}\right\rangle^6
$$

$$
\left|\begin{smallmatrix}2&&1&&0&&0\\&1&&0&&0&\\&&1&&0&&\\&&&0&&&\end{smallmatrix}\right\rangle^7
$$

The basis state $\left|\begin{smallmatrix}2&&1&&0&&0\\&2&&0&&0&\\&&0&&0&&\\&&&0&&&\end{smallmatrix}\right\rangle^7$ in the A$_2$-sixlet with $\boldsymbol{M}^{(2)} = (2,0)$ satisfies the condition (8.7) with $\mu = 3$, and an $\mathcal{A}_3$-doublet is constructed from it. The produced state belongs to the A$_2$-triplet with $\boldsymbol{M}^{(2)} = (1,0)$:

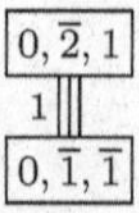

$$
\left|\begin{smallmatrix}2&&1&&0&&0\\&2&&0&&0&\\&&0&&0&&\\&&&0&&&\end{smallmatrix}\right\rangle^7
$$

$$
\left|\begin{smallmatrix}2&&1&&0&&0\\&1&&0&&0&\\&&0&&0&&\\&&&0&&&\end{smallmatrix}\right\rangle^8
$$

The highest weight representation $\boldsymbol{M} = \boldsymbol{w}_1 + \boldsymbol{w}_2 = (1,1,0)$ of A_3 is 20-dimensional and contains four A_2-multiplets with the highest weights $(1,1)$, $(2,0)$, $(0,1)$, and $(1,0)$, respectively. The basis states in this representation are related by two $\mathcal{A}_3$-triplets and five $\mathcal{A}_3$-doublets. This representation contains one single dominant weight $\boldsymbol{M}$ and one double dominant weight $\boldsymbol{m} = \boldsymbol{w}_3$. The block weight diagram of this representation of A_3 is given in Fig. 8.8. Its conjugate representation has the highest weight $\boldsymbol{M}' = (0,1,1)$, whose block weight diagram can be obtained from Fig. 8.8 by replacing $\boldsymbol{m}$ with $-\boldsymbol{m}$ and upside down. We suggest the readers to compare Fig. 8.8 with the block weight diagrams of A_2-multiplets and $\mathcal{A}_3$-multiplets.

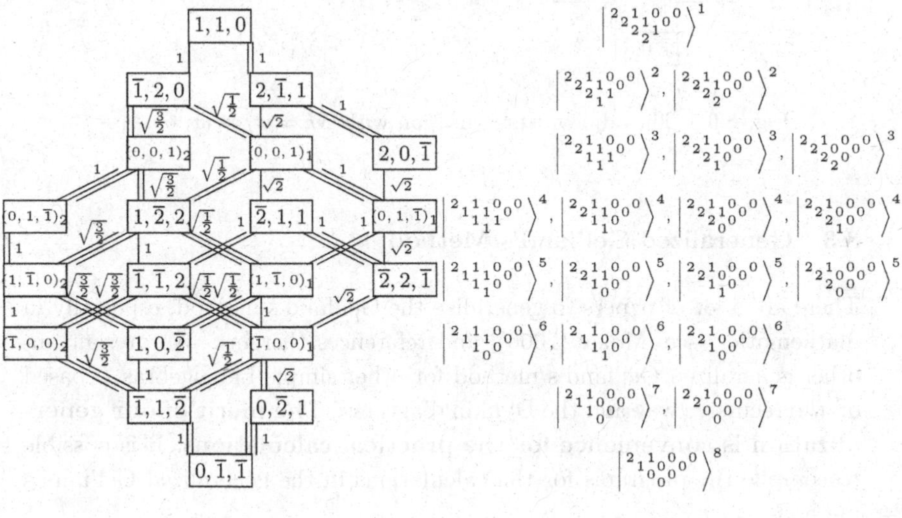

Fig. 8.8 The representation $\boldsymbol{M} = \boldsymbol{w}_1 + \boldsymbol{w}_2$ of A_3.

The representation $\boldsymbol{M} = \boldsymbol{w}_1 + \boldsymbol{w}_3 = (1,0,1)$ of A_3 is its adjoint representation. This representation of A_3 is 15-dimensional and contains two A_2-triplets with $\boldsymbol{M}^{(2)} = (1,0)$ and $(0,1)$, one A_2-octet with $\boldsymbol{M}^{(2)} = (1,1)$, and one A_2-singlet with $\boldsymbol{M}^{(2)} = (0,0)$. Their highest weight states are

$$\left|\begin{array}{cccc} 2 & 1 & 1 & 0 \\ & 2 \; 1 & 1 & \\ & 2 & 1 & \\ & & 2 & \end{array}\right\rangle^1, \quad \left|\begin{array}{cccc} 2 & 1 & 1 & 0 \\ & 1 \; 1 & 0 & \\ & 1 & 1 & \\ & & 1 & \end{array}\right\rangle^5, \quad \left|\begin{array}{cccc} 2 & 1 & 1 & 0 \\ & 2 \; 1 & 0 & \\ & 2 & 1 & \\ & & 2 & \end{array}\right\rangle^2, \quad \left|\begin{array}{cccc} 2 & 1 & 1 & 0 \\ & 1 \; 1 & 1 & \\ & 1 & 1 & \\ & & 1 & \end{array}\right\rangle^4 .$$

This representation contains one single dominant weight $\boldsymbol{M}$ and one triple dominant weight $\boldsymbol{m} = 0$. The block weight diagram and the Gel'fand's basis states of this representation is given in Fig. 8.9. The detailed calculation for this representation is left as exercise (see Prob. 2).

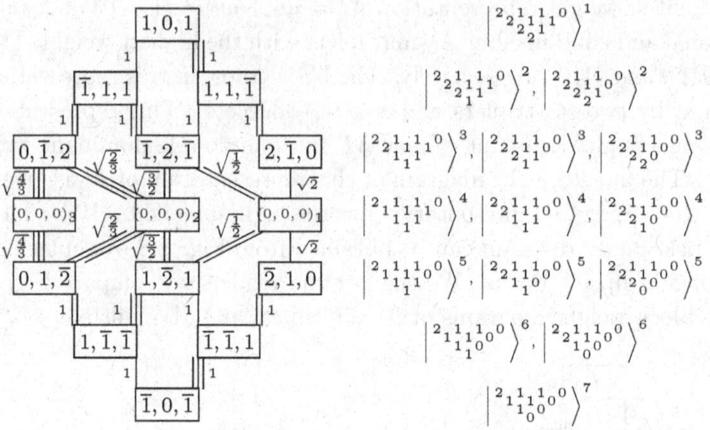

Fig. 8.9 The adjoint representation with $\boldsymbol{M} = \boldsymbol{w}_1 + \boldsymbol{w}_3$ of A_3.

8.3 Generalized Gel'fand's Method

There are a lot of papers to generalize the Gel'fand's method, especially in mathematics (see [Molev (2006)] and references therein). We present another generalized Gel'fand's method for other simple Lie algebras $\mathcal{L}$ based on the recursive way and the Dynkin diagrams. **The merit of our generalization is convenience for the practical calculations.** It is possible to compile the programs for the calculations in the generalized Gel'fand's method.

8.3.1 *Generalized Gel'fand's Bases*

The Gel'fand's method is a recursive way. By removing the last simple root $\boldsymbol{r}_\ell$, the Dynkin diagram of A_ℓ becomes that of $A_{\ell-1}$, and $A_\ell \supset A_{\ell-1} \oplus \mathcal{A}_\ell$. Each basis state in D^M of A_ℓ belongs to an $A_{\ell-1}$-multiplet, and F_ℓ relates some basis states belonging to different $A_{\ell-1}$-multiplets. However, from Fig. 7.1, **by removing the last simple root $\boldsymbol{r}_\ell$, the Dynkin diagram of each simple Lie algebra $\mathcal{L}$, except for F_4, becomes that of $A_{\ell-1}$, and $\mathcal{L} \supset A_{\ell-1} \oplus \mathcal{A}_\ell$.** Then, each basis state in D^M of $\mathcal{L}$ also belongs to an $A_{\ell-1}$-multiplet, and F_ℓ relates some basis states belonging to different $A_{\ell-1}$-multiplets. **The difference between A_ℓ and $\mathcal{L}$ is only the connecting way between the last simple root $\boldsymbol{r}_\ell$ and those simple roots in**

$A_{\ell-1}$. This is the foundation of our generalized Gel'fand's method.

We are going to generalize the Gel'fand's method to $\mathcal{L}$, where $\mathcal{L}$ **denotes a simple Lie algebra with rank** ℓ, **except for** F_4, **in this section**. The fundamental content in the generalized Gel'fand's method is as follows.

1). The basis states of the highest weight representation D^M of $\mathcal{L}$.

Each basis state in the highest weight representation D^M of $\mathcal{L}$ is described by the Gel'fand's basis of $A_{\ell-1}$ with a subscript m_ℓ, which is called the **generalized Gel'fand's basis**

$$|\omega_{ab}\rangle_{m_\ell}^L = \left| \begin{array}{ccccccc} \omega_{1\ell} & & \omega_{2\ell} \cdots & \omega_{(\ell-1)\ell} & & \omega_{\ell\ell} \\ & \omega_{1(\ell-1)} & \cdots & & \omega_{(\ell-1)(\ell-1)} & \\ & & \cdots \cdots & \cdots & & \\ & & \omega_{12} & \omega_{22} & & \\ & & & \omega_{11} & & \end{array} \right\rangle_{m_\ell}^{L}, \tag{8.30}$$

where the superscript L is the level $L(\boldsymbol{M}, \boldsymbol{m})$ in D^M of $\mathcal{L}$ and the subscript m_ℓ is the ℓth component of the weight $\boldsymbol{m}$. The constraint (8.15) and formulas (8.17-20) with $\mu < \ell$ still hold. Now, $\omega_{\ell\ell}$ **is allowed to be negative**. We neglect $\boldsymbol{M}$ in the expression (8.30) for simplicity.

2). The actions of the lowering operators F_ℓ in $\mathcal{A}_\ell$ and F_μ in $A_{\ell-1}$.

F_ℓ in $\mathcal{A}_\ell$ and F_μ in $A_{\ell-1}$ both increase the superscript L in Eq. (8.30) by one. F_ℓ in $\mathcal{A}_\ell$ decreases the subscript m_ℓ in Eq. (8.30) by two. F_μ in $A_{\ell-1}$ decreases one of $\omega_{a\mu}$, as well as Ω_μ, in Eq. (8.30) by one. In addition, F_ℓ and F_μ make the following actions for different $\mathcal{L}$.

For B_ℓ Lie algebra, $\boldsymbol{r}_\ell = 2\boldsymbol{w}_\ell - \boldsymbol{w}_{\ell-1}$ and $\boldsymbol{r}_{\ell-1} = -\boldsymbol{w}_{\ell-2} + 2\boldsymbol{w}_{\ell-1} - 2\boldsymbol{w}_\ell$. The action of F_ℓ decreases Ω_ℓ by one (increases $m_{\ell-1}$ by one). The action of $F_{\ell-1}$ increases the subscript m_ℓ by two.

For C_ℓ Lie algebra, $\boldsymbol{r}_\ell = 2\boldsymbol{w}_\ell - 2\boldsymbol{w}_{\ell-1}$ and $\boldsymbol{r}_{\ell-1} = -\boldsymbol{w}_{\ell-2} + 2\boldsymbol{w}_{\ell-1} - \boldsymbol{w}_\ell$. The action of F_ℓ decreases Ω_ℓ by two (increases $m_{\ell-1}$ by two). The action of $F_{\ell-1}$ increases the subscript m_ℓ by one.

For D_ℓ Lie algebra, $\boldsymbol{r}_\ell = 2\boldsymbol{w}_\ell - \boldsymbol{w}_{\ell-2}$ and $\boldsymbol{r}_{\ell-2} = -\boldsymbol{w}_{\ell-3} + 2\boldsymbol{w}_{\ell-2} - \boldsymbol{w}_{\ell-1} - \boldsymbol{w}_\ell$. The action of F_ℓ decreases $\Omega_{\ell-1}$ by one and decreases Ω_ℓ by two (increases $m_{\ell-2}$ by one and leaves $m_{\ell-1}$ invariant). The action of $F_{\ell-2}$ increases the subscript m_ℓ by one.

For G_2 Lie algebra, $\boldsymbol{r}_1 = 2\boldsymbol{w}_1 - 3\boldsymbol{w}_2$ and $\boldsymbol{r}_2 = 2\boldsymbol{w}_2 - \boldsymbol{w}_1$. If the basis state (8.30) denotes that of A_1-multiplet and the subscript denotes m_2, the action of F_2 decreases Ω_2 by one. The action of F_1 increases the subscript m_2 by three.

For E_6 Lie algebra, $\boldsymbol{r}_6 = 2\boldsymbol{w}_6 - \boldsymbol{w}_3$ and $\boldsymbol{r}_3 = -\boldsymbol{w}_2 + 2\boldsymbol{w}_3 - \boldsymbol{w}_4 - \boldsymbol{w}_6$. The action of F_6 decreases Ω_4 by one, Ω_5 by two, and Ω_6 by three (increases

m_3 by one and leaves m_4 and m_5 invariant). The action of F_3 increases the subscript m_6 by one. Similar actions occur for E_7 and E_8 Lie algebras.

3). $\mathcal{A}_\ell$-multiplets and the parallel principle.

The block weight diagram of $\mathcal{L}$ is decomposed to the block weight diagrams of $A_{\ell-1}$-multiplets and $\mathcal{A}_\ell$-multiplets. The **$A_{\ell-1}$-multiplets are calculated by the Gel'fand's method.** From the basis state satisfying the condition (8.7) with $\mu = \ell$, we construct an $\mathcal{A}_\ell$-multiplet, which is composed of some basis states belonging to different $A_{\ell-1}$-multiplets.

The parameters for the highest weight state $|\omega_{ab}\rangle^L_{m_\ell}$ of $D^{\boldsymbol{M}}$ of $\mathcal{L}$ satisfy $\omega_{\mu\nu} = \omega_{\mu(\nu+1)}$, $M_\mu = \omega_{\mu\ell} - \omega_{(\mu+1)\ell}$, and $M_\ell = m_\ell = \omega_{\ell\ell}$, $1 \leqslant \mu \leqslant \nu < \ell$. This highest weight state of $D^{\boldsymbol{M}}$ is also that of an $\mathcal{A}_\ell$-multiplet and that of an $A_{\ell-1}$-multiplet with $\boldsymbol{M}^{(\ell-1)}$ where $M^{(\ell-1)}_\mu = M_\mu$, $\mu < \ell$.

It is the key step in the generalized Gel'fand's method how to deal with Eq. (8.25). The condition $\omega_{d(\ell+1)} \leqslant \omega_{d\ell} \leqslant \omega_{(d+1)(\ell+1)}$ for A_ℓ restricts which $A_{\ell-1}$-multiplets occur in the representation space of $D^{\boldsymbol{M}}$ of A_ℓ. However, for the Lie algebra $\mathcal{L}$ the parameters $\omega_{a(\ell+1)}$ do not exist, and this condition is ineffective. We have to find the restriction condition in another form suitable for both A_ℓ and other $\mathcal{L}$ (see below). The decomposition of the subduced representation of an highest weight representation of $D^{\boldsymbol{M}}$ of $\mathcal{L}$ with respect to the subalgebra $A_{\ell-1}$ will be studied again in subsections 9.1.7, 10.1.2, 10.1.3, 10.3.4, and 12.1.2.

We check the highest weight state of the $A_{\ell-2}$-multiplet contained in all $A_{\ell-1}$-multiplets of $D^{\boldsymbol{M}}(A_\ell)$ one by one as the level $L(\boldsymbol{M}, \boldsymbol{m})$ increases from one, whether or not it satisfies the condition (8.7) with $\mu = \ell$, because the simple root $\boldsymbol{r}_\ell$ connects with the simple root $\boldsymbol{r}_{\ell-1}$ in the Dynkin diagram of A_ℓ. Now, $\boldsymbol{r}_\ell$ connects with the simple root $\boldsymbol{r}_\rho$ in the Dynkin diagram of other Lie algebra $\mathcal{L}$, where $\rho = \ell - 1$ for B_ℓ, C_ℓ, and G_2 ($\ell = 2$), $\rho = \ell - 2$ for D_ℓ, and $\rho = 3$ for E_6. For E_7 and E_8, ρ depends on the enumeration of the simple roots. Then, we have to **check the highest weight state of the $A_{\rho-1}$-multiplet** contained in all $A_{\ell-1}$-multiplet of $D^{\boldsymbol{M}}(\mathcal{L})$ one by one as the level $L(\boldsymbol{M}, \boldsymbol{m})$ increases from one, whether or not it satisfies the condition (8.7) with $\mu = \ell$. If yes, we construct an $\mathcal{A}_\ell$-multiplet by successive application of F_ℓ on $|\omega^{(d)}_{ab}\rangle$ through Eq. (8.25)

We define as **a certain rule** that there are only two cases for the basis states $|\omega^{(d)}_{ab}\rangle$ in Eq. (8.25). $|\omega^{(d)}_{ab}\rangle$ **belongs to an $A_{\ell-1}$-multiplet whose highest weight state occurs in the previous calculation**, namely, it has lower level L than $|\omega^{(d)}_{ab}\rangle$. Otherwise, $|\omega^{(d)}_{ab}\rangle$ **is the highest weight state of an $A_{\ell-1}$-multiplet.** If a basis state $|\omega^{(d)}_{ab}\rangle$ in Eq. (8.25) does not belong to these two cases, its coefficient A_d is zero. This is the restrict condition for which $A_{\ell-1}$-multiplets occur in the representation space of $D^M(\mathcal{L})$. **This restrict condition is also suitable for A_ℓ.**

In those two cases we are able to calculate the coefficients A_d (see Eqs. (8.25-28)) by the commutators $[E_\tau, F_\ell] = \delta_{\tau\ell} H_\ell$ with $\rho \leqslant \tau \leqslant \ell$, where the matrix elements of E_τ with $\tau \neq \ell$ are known because the $A_{\ell-1}$-multiplet have been known by the Gel'fand's method. The matrix element of E_ℓ is also known because it occurs in the previous calculation.

The problem occurs whether the weight of the basis state $|\omega_{ab}^{(d)}\rangle$ in Eq. (8.25), which is the highest weight state of an $A_{\ell-1}$-multiplet, is simple or not. The calculation shows that only for B_ℓ Lie algebra, the highest weight state $|\omega_{ab}^{(d)}\rangle$ of a new $A_{\ell-1}$-multiplet in Eq. (8.25) may be double degeneracy. **Two orthonormal basis states for the double degeneracy can be chosen arbitrarily, but it does not make essential difficult in the calculation.** The examples and the explanation will be given in Fig. 8.10 and in subsection 10.1.2.

When the $\mathcal{A}_\ell$-multiplet is obtained from the highest weight state $|\omega_{ab}\rangle$ of an $A_{\rho-1}$-multiplet, which satisfies the condition (8.7) with $\mu = \ell$, we can **obtain the similar equations to** Eq. (8.25) **and the similar $\mathcal{A}_\ell$-multiplets for the remaining basis states in the $A_{\rho-1}$-multiplet, by applying the lowering operators** F_μ in the $A_{\rho-1}$, because F_μ commute with F_ℓ. Namely, the parallel principle of $\mathcal{A}_\ell$-multiplets still hold for $\mathcal{L}$. Applying the lowering operators F_τ with $\rho < \tau < \ell$ to Eq. (8.25), we may obtain new equations similar to Eq. (8.25) in some simple cases, where there is no overlapping in the $\mathcal{A}_\ell$-multiplets (see pp. 565-568 in subsection 10.3.4 and Prob. 8 in Chap. 10). The parallel principle reduces the calculation quantity for calculating the $\mathcal{A}_\ell$-multiplets.

Generalized Gel'fand's method gives a certain rule to define the orthonormal basis states in the representation space of $\mathcal{L}$, where attention should be paid to **the correspondence between the generalized Gel'fand's basis $|\omega_{ab}\rangle^L_{m_\ell}$ and the basis states $(\boldsymbol{m})_r$ in the block weight diagram.** The readers are encouraged to draw the rule of correspondence from examples. In the subsequent subsections we are going to calculate some representations of C_ℓ, B_ℓ, D_ℓ, and G_2 as examples. **The symbol M in the basis state $|M, m\rangle$ is neglected for simplicity.** The representations of F_4 Lie algebra can be calculated by $F_4 \supset B_3 \oplus \mathcal{A}_4$ or $F_4 \supset \mathcal{A}_1 \oplus C_3$.

8.3.2 *Some Representations of* C_ℓ

The Lie algebra of the group USp(2ℓ) is C_ℓ. The Dynkin diagram is given in Fig. 7.1. The basis state in the highest weight representation D^M of C_ℓ is described by the Gel'fand's basis of $A_{\ell-1}$ (see Eq. (8.30)). The calculation of $D^M(C_\ell)$ depends on the knowledge of the representations of $A_{\ell-1}$. For definiteness, we discuss some representations of C_3. The basis state in the highest weight representation D^M of C_3 is described by the Gel'fand's basis of A_2 with a subscript m_3,

$$|\omega_{ab}\rangle^L_{m_3} = \left|\begin{array}{ccccc} \omega_{13} & & \omega_{23} & & \omega_{33} \\ & \omega_{12} & & \omega_{22} & \\ & & \omega_{11} & & \end{array}\right\rangle^L_{m_3}. \tag{8.31}$$

The action of F_3 increases the superscript L by one and decreases the subscript m_3 and Ω_3 both by two. The action of F_2 increases the superscript L and the subscript m_3 both by one and decreases Ω_2 by one. Each basis state in D^M of C_3 belongs to an A_2-multiplet, and F_3 relates some basis states belonging to different A_2-multiplets.

The parameters of the highest weight state of D^M of C_3 satisfy

$$\begin{aligned} &\omega_{33} = m_3 = M_3, \quad \omega_{23} = \omega_{22} = M_2 + \omega_{33}, \\ &\omega_{13} = \omega_{12} = \omega_{11} = M_1 + \omega_{23}. \end{aligned} \tag{8.32}$$

The parameters of the highest weight state of A_2-multiplet with the highest weight $\boldsymbol{M}^{(2)}$ in the representation D^M of C_3 satisfy

$$\begin{aligned} &\omega_{22} = \omega_{23}, \quad \omega_{11} = \omega_{12} = \omega_{13}, \\ &M_1^{(2)} = \omega_{13} - \omega_{23}, \quad M_2^{(2)} = \omega_{23} - \omega_{33}. \end{aligned} \tag{8.33}$$

Check each highest weight state ($\omega_{12} = \omega_{11}$) of A_1-multiplet contained in the A_2-multiplet, as the level $L(\boldsymbol{M}, \boldsymbol{m})$ increases, whether or not it satisfies the condition (8.7) with $\mu = 3$. If yes, we construct an $\mathcal{A}_3$-multiplet by successive application of F_3 (see Eqs. (8.25-28)). The "parallel principle" holds for each basis state in the A_1-multiplet.

The basic representation with $\boldsymbol{M} = \boldsymbol{w}_1 = (1, 0, 0)$ of C_3 is the self-representation of USp(6). From the highest weight state $|(1, 0, 0)\rangle$, we construct an A_2-triplet with $\boldsymbol{M}^{(2)} = (1, 0)$ (see Fig. 8.2):

$$\begin{array}{cc}
\boxed{1,0,0} & \left|\begin{smallmatrix} 1 & & 0 & & 0 \\ & 1 & & 0 & \\ & & 1 & & \end{smallmatrix}\right\rangle_0^1 \\
1\,\vert & \\
\boxed{\bar{1},1,0} & \left|\begin{smallmatrix} 1 & & 0 & & 0 \\ & 1 & & 0 & \\ & & 0 & & \end{smallmatrix}\right\rangle_0^2 \\
1\,\Vert & \\
\boxed{0,\bar{1},1} & \left|\begin{smallmatrix} 1 & & 0 & & 0 \\ & 0 & & 0 & \\ & & 0 & & \end{smallmatrix}\right\rangle_1^3
\end{array}$$

The last basis state $\left|\begin{smallmatrix} 1 & & 0 & & 0 \\ & 0 & & 0 & \\ & & 0 & & \end{smallmatrix}\right\rangle_1^3$ satisfies the condition (8.7) with $\mu = 3$ such that an $\mathcal{A}_3$-doublet is constructed:

$$\begin{array}{cc}
\boxed{0,\bar{1};1} & \left|\begin{smallmatrix} 1 & & 0 & & 0 \\ & 0 & & 0 & \\ & & 0 & & \end{smallmatrix}\right\rangle_1^3 \\
1\,\vert\vert\vert & \\
\boxed{0,1,\bar{1}} & \left|\begin{smallmatrix} 0 & & 0 & & \bar{1} \\ & 0 & & 0 & \\ & & 0 & & \end{smallmatrix}\right\rangle_{\bar{1}}^4
\end{array}$$

Why does the basis state $\left|\begin{smallmatrix} 1 & & 0 & & \bar{2} \\ & 0 & & 0 & \\ & & 0 & & \end{smallmatrix}\right\rangle_{\bar{1}}^4$ not appear in the action of F_3? Because the state is not the highest weight state of the A_2-multiplet with $\boldsymbol{M}^{(2)} = (1,2)$. This A_2-multiplet does not exist in the representation $D^{\boldsymbol{w}_1}$ of C_3. In other words, its coefficient is calculated to be zero when one applies $E_2F_3 = F_3E_2$ to $\left|\begin{smallmatrix} 1 & & 0 & & 0 \\ & 0 & & 0 & \\ & & 0 & & \end{smallmatrix}\right\rangle_1^3$. The basis state $\left|\begin{smallmatrix} 0 & & 0 & & \bar{1} \\ & 0 & & 0 & \\ & & 0 & & \end{smallmatrix}\right\rangle_{\bar{1}}^4$ is the highest weight state of the A_2-triplet with $\boldsymbol{M}^{(2)} = (0,1)$ and the A_2-triplet is constructed from it (see Fig. 8.2):

$$\begin{array}{cc}
\boxed{0,1,\bar{1}} & \left|\begin{smallmatrix} 0 & & 0 & & \bar{1} \\ & 0 & & 0 & \\ & & 0 & & \end{smallmatrix}\right\rangle_{\bar{1}}^4 \\
1\,\Vert & \\
\boxed{1,\bar{1},0} & \left|\begin{smallmatrix} 0 & & 0 & & \bar{1} \\ & 0 & & \bar{1} & \\ & & 0 & & \end{smallmatrix}\right\rangle_0^5 \\
1\,\vert & \\
\boxed{\bar{1},0,0} & \left|\begin{smallmatrix} 0 & & 0 & & \bar{1} \\ & 0 & & \bar{1} & \\ & & \bar{1} & & \end{smallmatrix}\right\rangle_0^6
\end{array}$$

The last basis state $\left|\begin{smallmatrix} 0 & & 0 & & \bar{1} \\ & 0 & & \bar{1} & \\ & & \bar{1} & & \end{smallmatrix}\right\rangle_0^6$ is the lowest weight state of this representation. In summary, the basic representation with $\boldsymbol{M} = \boldsymbol{w}_1 = (1,0,0)$ of C_3 is 6-dimensional and self-conjugate. It contains two A_2-triplets with $\boldsymbol{M}^{(2)} = (1,0)$, and $(0,1)$, respectively. There is only one dominant weight $\boldsymbol{M} = \boldsymbol{w}_1$ is this representation. Because the self-representation of C_3 is self-conjugate, **all representations** of C_3 **are self-conjugate**.

Another basic representation with $\boldsymbol{M} = \boldsymbol{w}_2 = (0,1,0)$ of C_3 is 14-dimensional. It contains one single dominant weights $\boldsymbol{w}_2$ and one double dominant weight 0. It contains two A_2-triplets and one A_2-octet with $\boldsymbol{M}^{(2)} = (0,1)$, $(1,0)$ and $(1,1)$, respectively. The highest weight states of those A_2-multiplets are

$$\begin{array}{ccc}
\left|\begin{smallmatrix} 1 & & 1 & & 0 \\ & 1 & & 1 & \\ & & 1 & & \end{smallmatrix}\right\rangle_0^1, & \left|\begin{smallmatrix} 0 & & \bar{1} & & \bar{1} \\ & 0 & & \bar{1} & \\ & & 0 & & \end{smallmatrix}\right\rangle_{\bar{1}}^7, & \left|\begin{smallmatrix} 1 & & 0 & & \bar{1} \\ & 1 & & 0 & \\ & & 1 & & \end{smallmatrix}\right\rangle_{\bar{1}}^3, \\
\vert(0,1,0)\rangle & \vert(1,0,\bar{1})\rangle & \vert(1,1,\bar{1})\rangle.
\end{array}$$

The detailed calculation is left as exercise (see Prob. 3).

The third basic representation of C_3 has the highest weight $\boldsymbol{M} = \boldsymbol{w}_3 = (0, 0, 1)$. Its highest weight state spans the A_2-singlet and satisfies the condition (8.7) with $\mu = 3$, from which an $\mathcal{A}_3$-doublet is constructed:

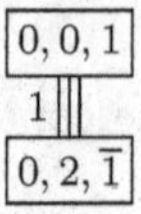

$$\left|\begin{smallmatrix} 1 & & 1 & & 1 \\ & 1 & & 1 & \\ & & 1 & & \end{smallmatrix}\right\rangle^{1}_{1}$$

$$\left|\begin{smallmatrix} 1 & & 1 & & \bar{1} \\ & 1 & & 1 & \\ & & 1 & & \end{smallmatrix}\right\rangle^{2}_{\bar{1}}$$

The basis state $\left|\begin{smallmatrix} 1 & & 1 & & \bar{1} \\ & 1 & & 1 & \\ & & 1 & & \end{smallmatrix}\right\rangle^{2}_{\bar{1}}$ is the highest weight state of an A_2-sixlet with $\boldsymbol{M}^{(2)} = (0, 2)$, and the A_2-sixlet is constructed from it (see Fig. 8.3):

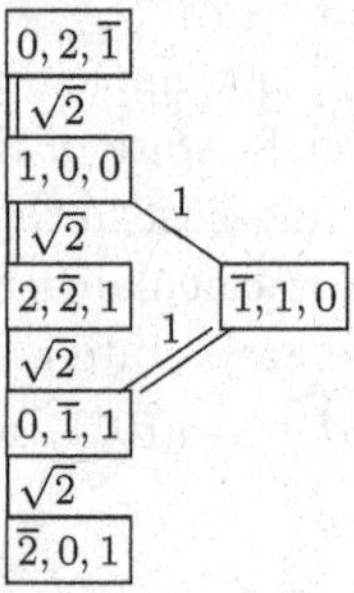

$$\left|\begin{smallmatrix} 1 & & 1 & & \bar{1} \\ & 1 & & 1 & \\ & & 1 & & \end{smallmatrix}\right\rangle^{2}_{\bar{1}}$$

$$\left|\begin{smallmatrix} 1 & & 1 & & \bar{1} \\ & 1 & & 0 & \\ & & 1 & & \end{smallmatrix}\right\rangle^{3}_{0}$$

$$\left|\begin{smallmatrix} 1 & & 1 & & \bar{1} \\ & 1 & & \bar{1} & \\ & & 1 & & \end{smallmatrix}\right\rangle^{4}_{1}, \quad \left|\begin{smallmatrix} 1 & & 1 & & \bar{1} \\ & 1 & & 0 & \\ & & 0 & & \end{smallmatrix}\right\rangle^{4}_{0} \quad .$$

$$\left|\begin{smallmatrix} 1 & & 1 & & \bar{1} \\ & 1 & & \bar{1} & \\ & & 0 & & \end{smallmatrix}\right\rangle^{5}_{1}$$

$$\left|\begin{smallmatrix} 1 & & 1 & & \bar{1} \\ & 1 & & \bar{1} & \\ & & \bar{1} & & \end{smallmatrix}\right\rangle^{6}_{1}$$

The highest weight state $\left|\begin{smallmatrix} 1 & & 1 & & \bar{1} \\ & 1 & & \bar{1} & \\ & & 1 & & \end{smallmatrix}\right\rangle^{4}_{1}$ of an A_1-triplet in the A_2-sixlet satisfies the condition (8.7) with $\mu = 3$, and an $\mathcal{A}_3$-doublet is constructed from it:

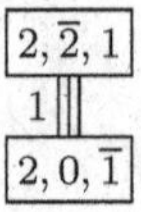

$$\left|\begin{smallmatrix} 1 & & 1 & & \bar{1} \\ & 1 & & \bar{1} & \\ & & 1 & & \end{smallmatrix}\right\rangle^{4}_{1}$$

$$\left|\begin{smallmatrix} 1 & & \bar{1} & & \bar{1} \\ & 1 & & \bar{1} & \\ & & 1 & & \end{smallmatrix}\right\rangle^{5}_{\bar{1}}$$

The basis state $\left|\begin{smallmatrix} 1 & & \bar{1} & & \bar{1} \\ & 1 & & \bar{1} & \\ & & 1 & & \end{smallmatrix}\right\rangle^{5}_{\bar{1}}$ is the highest weight state of an A_2-sixlet with $\boldsymbol{M}^{(2)} = (2, 0)$, and the A_2-sixlet is constructed from it (see Fig. 8.3):

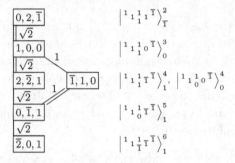

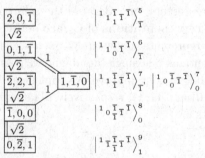

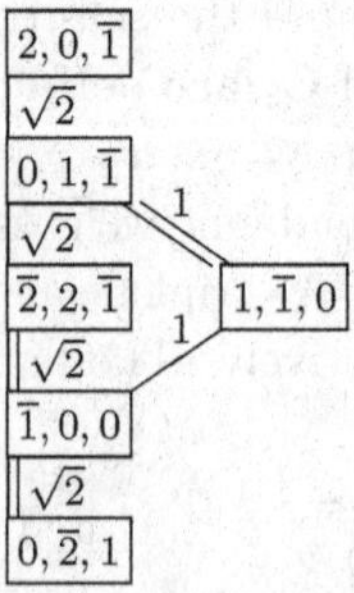

$$\left|\begin{smallmatrix} 1 & & \bar{1} & & \bar{1} \\ & 1 & & \bar{1} & \\ & & 1 & & \end{smallmatrix}\right\rangle^{5}_{\bar{1}}$$

$$\left|\begin{smallmatrix} 1 & & \bar{1} & & \bar{1} \\ & 1 & & \bar{1} & \\ & & 0 & & \end{smallmatrix}\right\rangle^{6}_{\bar{1}}$$

$$\left|\begin{smallmatrix} 1 & & \bar{1} & & \bar{1} \\ & 1 & & \bar{1} & \\ & & \bar{1} & & \end{smallmatrix}\right\rangle^{7}_{\bar{1}}, \quad \left|\begin{smallmatrix} 1 & & \bar{1} & & \bar{1} \\ & 0 & & \bar{1} & \\ & & 0 & & \end{smallmatrix}\right\rangle^{7}_{0}$$

$$\left|\begin{smallmatrix} 1 & & \bar{1} & & \bar{1} \\ & 0 & & \bar{1} & \\ & & \bar{1} & & \end{smallmatrix}\right\rangle^{8}_{0}$$

$$\left|\begin{smallmatrix} 1 & & \bar{1} & & \bar{1} \\ & \bar{1} & & \bar{1} & \\ & & \bar{1} & & \end{smallmatrix}\right\rangle^{9}_{1}$$

From the parallel principle we have

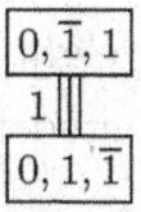

$\left|{}^{1}{}_{1}{}^{1}_{0}{}_{\bar{1}}{}^{\bar{1}}\right\rangle^{5}_{1}$
$\left|{}^{1}{}_{1}{}^{\bar{1}}_{0}{}_{\bar{1}}{}^{\bar{1}}\right\rangle^{6}_{\bar{1}}$

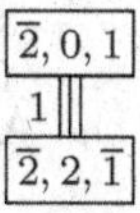

$\left|{}^{1}{}_{1}{}^{1}_{\bar{1}}{}_{\bar{1}}{}^{\bar{1}}\right\rangle^{6}_{1}$
$\left|{}^{1}{}_{1}{}^{\bar{1}}_{\bar{1}}{}_{\bar{1}}{}^{\bar{1}}\right\rangle^{7}_{\bar{1}}$

The basis state $\left|{}^{1}{}_{\bar{1}}{}^{\bar{1}}_{\bar{1}}{}_{\bar{1}}{}^{\bar{1}}\right\rangle^{9}_{1}$ in the A_2-sixlet with $\boldsymbol{M}^{(2)} = (2, 0)$ satisfies the condition (8.7) with $\mu = 3$, and an $\mathcal{A}_3$-doublet is constructed from it:

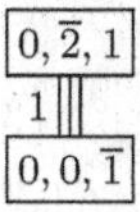

$\left|{}^{1}{}_{\bar{1}}{}^{\bar{1}}_{\bar{1}}{}_{\bar{1}}{}^{\bar{1}}\right\rangle^{9}_{1}$
$\left|{}^{\bar{1}}{}_{\bar{1}}{}^{\bar{1}}_{\bar{1}}{}_{\bar{1}}{}^{\bar{1}}\right\rangle^{10}_{\bar{1}}$

The basis state $\left|{}^{\bar{1}}{}_{\bar{1}}{}^{\bar{1}}_{\bar{1}}{}_{\bar{1}}{}^{\bar{1}}\right\rangle^{10}_{\bar{1}}$ spans an A_2-singlet and it is the lowest weight state in this representation. In summary, the basic representation with $\boldsymbol{M} = \boldsymbol{w}_3$ of C_3 is 14-dimensional. It contains two A_2-singlets and two A_2-sixlets with $\boldsymbol{M}^{(2)} = (0, 0)$, $(0, 0)$, $(0, 2)$, and $(2, 0)$, respectively. There are five $\mathcal{A}_3$-doublets in this representation. This representation contains two single dominant weights $\boldsymbol{w}_3$ and $\boldsymbol{w}_1$.

Now, we calculate the representation of C_3 with the highest weight $\boldsymbol{M} = \boldsymbol{w}_1 + \boldsymbol{w}_2 = (1, 1, 0)$. Please compare two representations $\boldsymbol{M} = (1, 1, 0)$ of A_3 and C_3. From the highest weight state $|(1, 1, 0)\rangle$ we construct an A_2-octet with $\boldsymbol{M}^{(2)} = (1, 1)$ (see Fig. 8.5):

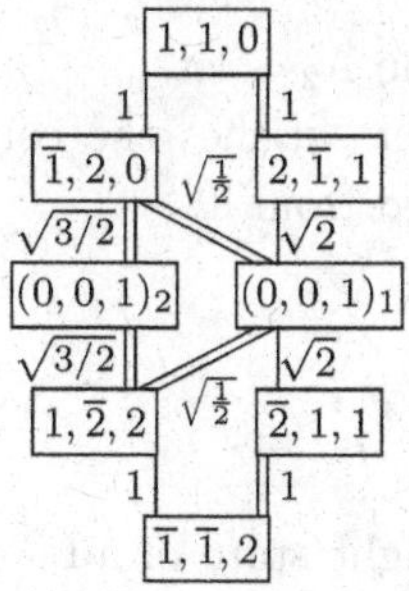

$\left|{}^{2}{}_{2}{}^{1}_{2}{}_{1}{}^{0}\right\rangle^{1}_{0}$

$\left|{}^{2}{}_{2}{}^{1}_{1}{}_{1}{}^{0}\right\rangle^{2}_{0}$, $\left|{}^{2}{}_{2}{}^{1}_{2}{}_{0}{}^{0}\right\rangle^{2}_{1}$

$\left|{}^{2}{}_{1}{}^{1}_{1}{}_{1}{}^{0}\right\rangle^{3}_{1}$, $\left|{}^{2}{}_{2}{}^{1}_{1}{}_{0}{}^{0}\right\rangle^{3}_{1}$

$\left|{}^{2}{}_{1}{}^{1}_{1}{}_{0}{}^{0}\right\rangle^{4}_{2}$, $\left|{}^{2}{}_{2}{}^{1}_{0}{}_{0}{}^{0}\right\rangle^{4}_{1}$

$\left|{}^{2}{}_{1}{}^{1}_{0}{}_{0}{}^{0}\right\rangle^{5}_{2}$

There are three highest weight states of A_1-multiplets in the A_2-octet which satisfy the condition (8.7) with $\mu = 3$. From them we construct $\mathcal{A}_3$-multiplets one by one.

The basis state $\left|{}^{2}{}_{2}{}^{1}_{2}{}_{0}{}^{0}\right\rangle^{2}_{1}$ in the A_2-octet satisfies the condition (8.7) with $\mu = 3$, and an $\mathcal{A}_3$-doublet is constructed from it:

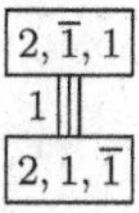

$\left|{}^{2}{}_{2}{}^{1}_{2}{}_{0}{}^{0}\right\rangle^{2}_{1}$
$\left|{}^{2}{}_{2}{}^{0}_{2}{}_{0}{}^{\bar{1}}\right\rangle^{3}_{\bar{1}}$

The produced basis state $\left|{}^{2}{}_{2}{}^{0}_{2}{}_{0}{}^{\bar 1}\right\rangle^{3}_{\bar 1}$ is the highest weight state of an A_2-15let with $\boldsymbol{M}^{(2)} = (2,1)$ and the A_2-15let is constructed from it (see Fig. 8.6):

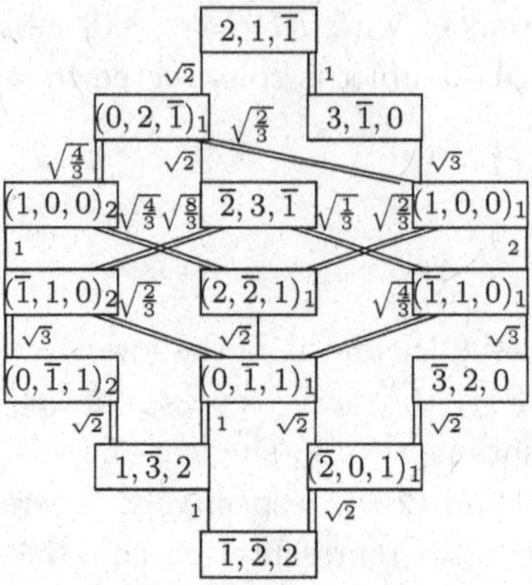

$\left|{}^{2}{}_{2}{}^{0}_{2}{}_{0}{}^{\bar 1}\right\rangle^{3}_{\bar 1}$

$\left|{}^{2}{}_{2}{}^{0}_{1}{}_{0}{}^{\bar 1}\right\rangle^{4}_{\bar 1}$, $\left|{}^{2}{}_{2}{}^{0}_{2}{}_{\bar 1}{}^{\bar 1}\right\rangle^{4}_{0}$

$\left|{}^{2}{}_{1}{}^{0}_{1}{}_{0}{}^{\bar 1}\right\rangle^{5}_{0}$, $\left|{}^{2}{}_{2}{}^{0}_{0}{}_{0}{}^{\bar 1}\right\rangle^{5}_{\bar 1}$, $\left|{}^{2}{}_{2}{}^{0}_{1}{}_{\bar 1}{}^{\bar 1}\right\rangle^{5}_{0}$

$\left|{}^{2}{}_{1}{}^{0}_{0}{}_{0}{}^{\bar 1}\right\rangle^{6}_{0}$, $\left|{}^{2}{}_{1}{}^{0}_{1}{}_{\bar 1}{}^{\bar 1}\right\rangle^{6}_{1}$, $\left|{}^{2}{}_{2}{}^{0}_{0}{}_{\bar 1}{}^{\bar 1}\right\rangle^{6}_{0}$

$\left|{}^{2}{}_{0}{}^{0}_{0}{}_{0}{}^{\bar 1}\right\rangle^{7}_{1}$, $\left|{}^{2}{}_{1}{}^{0}_{0}{}_{\bar 1}{}^{\bar 1}\right\rangle^{7}_{1}$, $\left|{}^{2}{}_{2}{}^{0}_{\bar 1}{}_{\bar 1}{}^{\bar 1}\right\rangle^{7}_{0}$

$\left|{}^{2}{}_{0}{}^{0}_{0}{}_{\bar 1}{}^{\bar 1}\right\rangle^{8}_{2}$, $\left|{}^{2}{}_{1}{}^{0}_{\bar 1}{}_{\bar 1}{}^{\bar 1}\right\rangle^{8}_{1}$

$\left|{}^{2}{}_{0}{}^{0}_{\bar 1}{}_{\bar 1}{}^{\bar 1}\right\rangle^{9}_{2}$

From the parallel principle we have

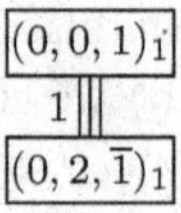

$\left|{}^{2}{}_{2}{}^{1}_{1}{}_{0}{}^{0}\right\rangle^{3}_{1}$

$\left|{}^{2}{}_{2}{}^{0}_{1}{}_{0}{}^{\bar 1}\right\rangle^{4}_{\bar 1}$

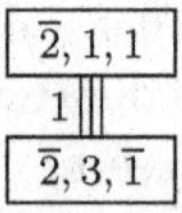

$\left|{}^{2}{}_{2}{}^{1}_{0}{}_{0}{}^{0}\right\rangle^{4}_{1}$

$\left|{}^{2}{}_{2}{}^{0}_{0}{}_{0}{}^{\bar 1}\right\rangle^{5}_{\bar 1}$

The produced basis states both belong to the A_2-15let.

The basis state $\left|{}^{2}{}_{1}{}^{1}_{1}{}_{1}{}^{0}\right\rangle^{3}_{1}$ in the A_2-octet satisfies the condition (8.7) with $\mu = 3$, and an $\mathcal{A}_3$-doublet is constructed from it:

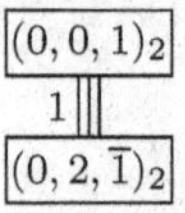

$\left|{}^{2}{}_{1}{}^{1}_{1}{}_{1}{}^{0}\right\rangle^{3}_{1}$

$\left|{}^{1}{}_{1}{}^{1}_{1}{}_{1}{}^{\bar 1}\right\rangle^{4}_{\bar 1}$

The produced basis state is the highest weight state of an A_2-sixlet with $\boldsymbol{M}^{(2)} = (0,2)$ and the A_2-sixlet is constructed from it (see Fig. 8.3):

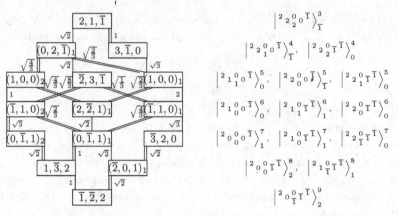

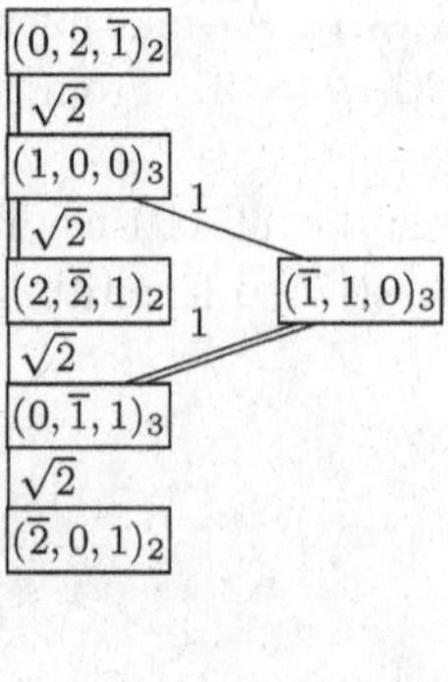

$\left|{}^{1}{}_{1}{}^{1}_{1}{}_{1}{}^{\bar 1}\right\rangle^{4}_{\bar 1}$

$\left|{}^{1}{}_{1}{}^{1}_{1}{}_{0}{}^{\bar 1}\right\rangle^{5}_{0}$

$\left|{}^{1}{}_{1}{}^{1}_{1}{}_{\bar 1}{}^{\bar 1}\right\rangle^{6}_{1}$, $\left|{}^{1}{}_{1}{}^{1}_{0}{}_{0}{}^{\bar 1}\right\rangle^{6}_{0}$

$\left|{}^{1}{}_{1}{}^{1}_{0}{}_{\bar 1}{}^{\bar 1}\right\rangle^{7}_{1}$

$\left|{}^{1}{}_{1}{}^{1}_{\bar 1}{}_{\bar 1}{}^{\bar 1}\right\rangle^{8}_{1}$

The basis state $\left|\begin{smallmatrix}2&&1&&0\\&1&&0&\\&&1&&\end{smallmatrix}\right\rangle^4_2$ in the A_2-octet satisfies the condition (8.7) with $\mu = 3$, and an $\mathcal{A}_3$-triplet is constructed from it:

$$F_3\left|\begin{matrix}2&&1&&0\\&1&&0&\\&&1&&\end{matrix}\right\rangle^4_2 = a_1\left|\begin{matrix}2&&0&&\bar{1}\\&1&&0&\\&&1&&\end{matrix}\right\rangle^5_0 + a_2\left|\begin{matrix}1&&1&&\bar{1}\\&1&&0&\\&&1&&\end{matrix}\right\rangle^5_0 + a_3\left|\begin{matrix}1&&0&&0\\&1&&0&\\&&1&&\end{matrix}\right\rangle^5_0,$$

The basis state $\left|\begin{smallmatrix}2&&0&&\bar{1}\\&1&&0&\\&&1&&\end{smallmatrix}\right\rangle^5_0$ belongs to the A_2-15let with $\boldsymbol{M}^{(2)} = (2,1)$, the basis state $\left|\begin{smallmatrix}1&&1&&\bar{1}\\&1&&0&\\&&1&&\end{smallmatrix}\right\rangle^5_0$ belongs to the A_2-sixlet with $\boldsymbol{M}^{(2)} = (0,2)$, and the basis state $\left|\begin{smallmatrix}1&&0&&0\\&1&&0&\\&&1&&\end{smallmatrix}\right\rangle^5_0$ is the highest weight state of an A_2-triplet with $\boldsymbol{M}^{(2)} = (1,0)$. The coefficients a_j can be calculated by applying Eq. (8.9) to $\left|\begin{smallmatrix}2&&1&&0\\&1&&0&\\&&1&&\end{smallmatrix}\right\rangle^4_2$ (see Eqs. (8.25-28)).

$$\begin{aligned}
E_2F_3\left|\begin{matrix}2&&1&&0\\&1&&0&\\&&1&&\end{matrix}\right\rangle^4_2 &= a_1\sqrt{\frac{4}{3}}\left|\begin{matrix}2&&0&&\bar{1}\\&2&&0&\\&&1&&\end{matrix}\right\rangle^4_{\bar{1}} + a_2\sqrt{2}\left|\begin{matrix}1&&1&&\bar{1}\\&1&&1&\\&&1&&\end{matrix}\right\rangle^4_{\bar{1}}\\
= F_3E_2\left|\begin{matrix}2&&1&&0\\&1&&0&\\&&1&&\end{matrix}\right\rangle^4_2 &= \sqrt{\frac{1}{2}}F_3\left|\begin{matrix}2&&1&&0\\&2&&0&\\&&1&&\end{matrix}\right\rangle^3_1 + \sqrt{\frac{3}{2}}F_3\left|\begin{matrix}2&&1&&0\\&1&&1&\\&&1&&\end{matrix}\right\rangle^3_1\\
&= \sqrt{\frac{1}{2}}\left|\begin{matrix}2&&0&&\bar{1}\\&2&&0&\\&&1&&\end{matrix}\right\rangle^4_{\bar{1}} + \sqrt{\frac{3}{2}}\left|\begin{matrix}1&&1&&\bar{1}\\&1&&1&\\&&1&&\end{matrix}\right\rangle^4_{\bar{1}}.
\end{aligned}$$

Thus, $a_1 = \sqrt{3/8}$ and $a_2 = \sqrt{3/4}$. By choosing the phase of $\left|\begin{smallmatrix}1&&0&&0\\&1&&0&\\&&1&&\end{smallmatrix}\right\rangle^5_0$, a_3 is real and positive: $a_3 = \sqrt{2 - 3/8 - 3/4} = \sqrt{7/8}$. From Fig. 8.2 we list three basis states of of the A_2-triplet with $\boldsymbol{M}^{(2)} = \boldsymbol{w}_1 = (1,0)$:

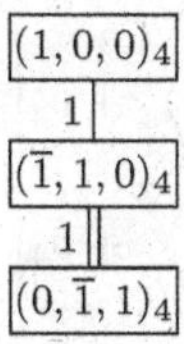

$$\left|\begin{matrix}1&&0&&0\\&1&&0&\\&&1&&\end{matrix}\right\rangle^5_0 \qquad \left|\begin{matrix}1&&0&&0\\&1&&0&\\&&0&&\end{matrix}\right\rangle^6_0 \qquad \left|\begin{matrix}1&&0&&0\\&0&&0&\\&&0&&\end{matrix}\right\rangle^7_1$$

By applying $E_3F_3 = F_3E_3 + H_3$ to $\left|\begin{smallmatrix}2&&0&&\bar{1}\\&1&&0&\\&&1&&\end{smallmatrix}\right\rangle^5_0$ and choosing the phase of $\left|\begin{smallmatrix}1&&0&&\bar{2}\\&1&&0&\\&&1&&\end{smallmatrix}\right\rangle^6_{\bar{2}}$, we obtain

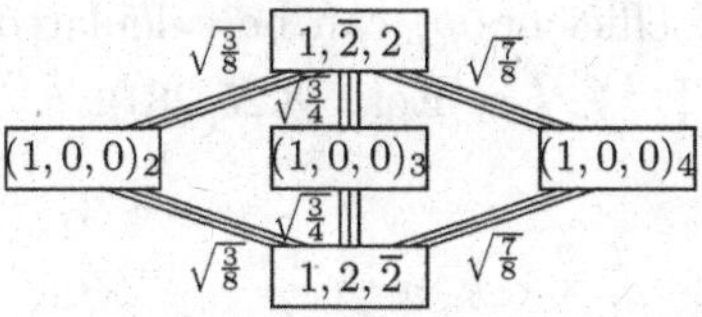

$$\left|\begin{matrix}2&&1&&0\\&1&&0&\\&&1&&\end{matrix}\right\rangle^4_2$$

$$\left|\begin{matrix}2&&0&&\bar{1}\\&1&&0&\\&&1&&\end{matrix}\right\rangle^5_0,\quad \left|\begin{matrix}1&&1&&\bar{1}\\&1&&0&\\&&1&&\end{matrix}\right\rangle^5_0,\quad \left|\begin{matrix}1&&0&&0\\&1&&0&\\&&1&&\end{matrix}\right\rangle^5_0$$

$$\left|\begin{matrix}1&&0&&\bar{2}\\&1&&0&\\&&1&&\end{matrix}\right\rangle^6_{\bar{2}}$$

The basis state $\left|\begin{smallmatrix}1&&0&&\bar{2}\\&1&&0&\\&&1&&\end{smallmatrix}\right\rangle^6_{\bar{2}}$ is the highest weight state of an A_2-15let with $\boldsymbol{M}^{(2)} = (1,2)$ and the A_2-15let is constructed from it (see Fig. 8.6):

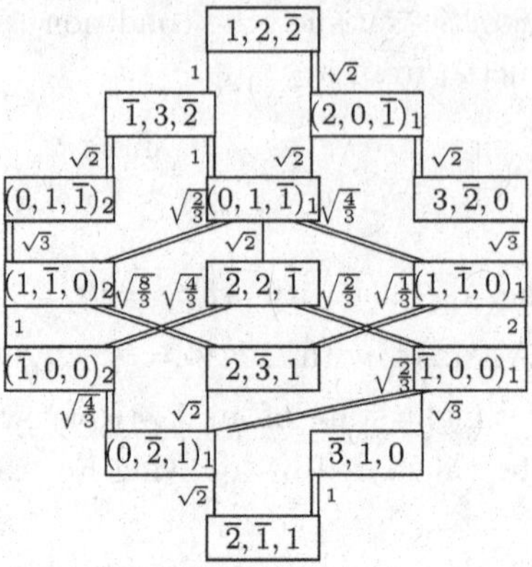

$$\left|\begin{smallmatrix}1&&0&&\overline{2}\\&1&&0&\\&&1&&\end{smallmatrix}\right\rangle^{6}_{\overline{2}}$$

$$\left|\begin{smallmatrix}1&&0&&\overline{2}\\&1&&0&\\&&0&&\end{smallmatrix}\right\rangle^{7}_{\overline{2}}, \quad \left|\begin{smallmatrix}1&&0&&\overline{2}\\&1&&\overline{1}&\\&&1&&\end{smallmatrix}\right\rangle^{7}_{\overline{1}}$$

$$\left|\begin{smallmatrix}1&&0&&\overline{2}\\&0&&0&\\&&0&&\end{smallmatrix}\right\rangle^{8}_{\overline{1}}, \quad \left|\begin{smallmatrix}1&&0&&\overline{2}\\&1&&\overline{1}&\\&&0&&\end{smallmatrix}\right\rangle^{8}_{\overline{1}}, \quad \left|\begin{smallmatrix}1&&0&&\overline{2}\\&1&&\overline{2}&\\&&1&&\end{smallmatrix}\right\rangle^{8}_{0}$$

$$\left|\begin{smallmatrix}1&&0&&\overline{2}\\&0&&\overline{1}&\\&&0&&\end{smallmatrix}\right\rangle^{9}_{0}, \quad \left|\begin{smallmatrix}1&&0&&\overline{2}\\&1&&\overline{1}&\\&&\overline{1}&&\end{smallmatrix}\right\rangle^{9}_{\overline{1}}, \quad \left|\begin{smallmatrix}1&&0&&\overline{2}\\&1&&\overline{2}&\\&&0&&\end{smallmatrix}\right\rangle^{9}_{0}$$

$$\left|\begin{smallmatrix}1&&0&&\overline{2}\\&0&&\overline{1}&\\&&\overline{1}&&\end{smallmatrix}\right\rangle^{10}_{0}, \quad \left|\begin{smallmatrix}1&&0&&\overline{2}\\&0&&\overline{2}&\\&&0&&\end{smallmatrix}\right\rangle^{10}_{1}, \quad \left|\begin{smallmatrix}1&&0&&\overline{2}\\&1&&\overline{2}&\\&&\overline{1}&&\end{smallmatrix}\right\rangle^{10}_{0}$$

$$\left|\begin{smallmatrix}1&&0&&\overline{2}\\&0&&\overline{2}&\\&&\overline{1}&&\end{smallmatrix}\right\rangle^{11}_{1}, \quad \left|\begin{smallmatrix}1&&0&&\overline{2}\\&1&&\overline{2}&\\&&\overline{2}&&\end{smallmatrix}\right\rangle^{11}_{0}$$

$$\left|\begin{smallmatrix}1&&0&&\overline{2}\\&0&&\overline{2}&\\&&\overline{2}&&\end{smallmatrix}\right\rangle^{12}_{1}$$

Due to the parallel principle we obtain

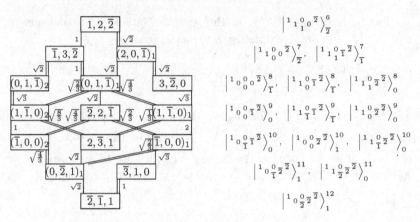

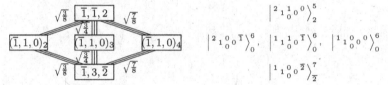

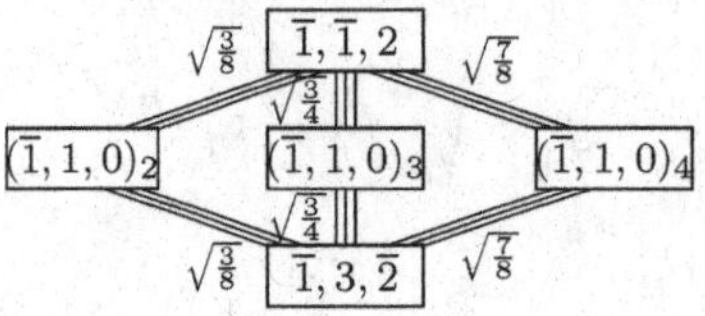

$$\left|\begin{smallmatrix}2&&1&&0\\&1&&0&\\&&0&&\end{smallmatrix}\right\rangle^{5}_{2}$$

$$\left|\begin{smallmatrix}2&&0&&\overline{1}\\&1&&0&\\&&0&&\end{smallmatrix}\right\rangle^{6}_{0}, \quad \left|\begin{smallmatrix}1&&1&&\overline{1}\\&1&&0&\\&&0&&\end{smallmatrix}\right\rangle^{6}_{0}, \quad \left|\begin{smallmatrix}1&&0&&0\\&1&&0&\\&&0&&\end{smallmatrix}\right\rangle^{6}_{0}$$

$$\left|\begin{smallmatrix}1&&0&&\overline{2}\\&1&&0&\\&&0&&\end{smallmatrix}\right\rangle^{7}_{\overline{2}}$$

The basis state $\left|\begin{smallmatrix}2&&0&&\overline{1}\\&1&&\overline{1}&\\&&1&&\end{smallmatrix}\right\rangle^{6}_{1}$ in the A_2-15let with $\boldsymbol{M}^{(2)} = (2,1)$ and the basis state $\left|\begin{smallmatrix}1&&1&&\overline{1}\\&1&&\overline{1}&\\&&1&&\end{smallmatrix}\right\rangle^{6}_{1}$ in the A_2-sixlet with $\boldsymbol{M}^{(2)} = (0,2)$ both are the highest weight states of the A_1-triplet with $\boldsymbol{M}^{(1)} = 2\boldsymbol{w}_1$ and both satisfy the condition (8.7) with $\mu = 3$. Then, applying F_3 to them, we obtain

$$F_3\left|\begin{matrix}2&&0&&\overline{1}\\&1&&\overline{1}&\\&&1&&\end{matrix}\right\rangle^{6}_{1} = b_1\left|\begin{matrix}1&&0&&\overline{2}\\&1&&\overline{1}&\\&&1&&\end{matrix}\right\rangle^{7}_{\overline{1}} + b_2\left|\begin{matrix}1&&\overline{1}&&\overline{1}\\&1&&\overline{1}&\\&&1&&\end{matrix}\right\rangle^{7}_{\overline{1}},$$

$$F_3\left|\begin{matrix}1&&1&&\overline{1}\\&1&&\overline{1}&\\&&1&&\end{matrix}\right\rangle^{6}_{1} = b_3\left|\begin{matrix}1&&0&&\overline{2}\\&1&&\overline{1}&\\&&1&&\end{matrix}\right\rangle^{7}_{\overline{1}} + b_4\left|\begin{matrix}1&&\overline{1}&&\overline{1}\\&1&&\overline{1}&\\&&1&&\end{matrix}\right\rangle^{7}_{\overline{1}}.$$

The first produced basis state $\left|\begin{smallmatrix}1&&0&&\overline{2}\\&1&&\overline{1}&\\&&1&&\end{smallmatrix}\right\rangle^{7}_{\overline{1}}$ belongs to the A_2-15let with $\boldsymbol{M}^{(2)} = (1,2)$, and the second one $\left|\begin{smallmatrix}1&&\overline{1}&&\overline{1}\\&1&&\overline{1}&\\&&1&&\end{smallmatrix}\right\rangle^{7}_{\overline{1}}$ is the highest weight state of an A_2-sixlet with $\boldsymbol{M}^{(2)} = (2,0)$. b_2 is real and positive by choosing the phase of the second basis state. The coefficients b_j can be calculated by applying Eq. (8.9) to $\left|\begin{smallmatrix}2&&0&&\overline{1}\\&1&&\overline{1}&\\&&1&&\end{smallmatrix}\right\rangle^{6}_{1}$ and $\left|\begin{smallmatrix}1&&1&&\overline{1}\\&1&&\overline{1}&\\&&1&&\end{smallmatrix}\right\rangle^{6}_{1}$ (see Eqs. (8.25-28)).

$$E_2F_3\left|\begin{matrix}2&&0&&\overline{1}\\&1&&\overline{1}&\\&&1&&\end{matrix}\right\rangle^{6}_{1} = b_1\sqrt{2}\left|\begin{matrix}1&&0&&\overline{2}\\&1&&0&\\&&1&&\end{matrix}\right\rangle^{6}_{\overline{2}}$$

$$= F_3E_2\left|\begin{matrix}2&&0&&\overline{1}\\&1&&\overline{1}&\\&&1&&\end{matrix}\right\rangle^{6}_{1} = F_3\left[\sqrt{\frac{2}{3}}\left|\begin{matrix}2&&0&&\overline{1}\\&2&&\overline{1}&\\&&1&&\end{matrix}\right\rangle^{5}_{0} + \sqrt{\frac{4}{3}}\left|\begin{matrix}2&&0&&\overline{1}\\&1&&0&\\&&1&&\end{matrix}\right\rangle^{5}_{0}\right] = \sqrt{\frac{1}{2}}\left|\begin{matrix}1&&0&&\overline{2}\\&1&&0&\\&&1&&\end{matrix}\right\rangle^{6}_{\overline{2}},$$

$$E_2F_3\left|{}^{1}{}_{1}{}^{1}_{1}{}_{\bar{1}}{}^{\bar{1}}\right\rangle^{6}_{1}=b_3\sqrt{2}\left|{}^{1}{}_{1}{}^{0}_{1}{}_{0}{}^{\bar{2}}\right\rangle^{6}_{\bar{2}}$$

$$=F_3E_2\left|{}^{1}{}_{1}{}^{1}_{1}{}_{\bar{1}}{}^{\bar{1}}\right\rangle^{6}_{2}=F_3\sqrt{2}\left|{}^{1}{}_{1}{}^{1}_{1}{}_{0}{}^{\bar{1}}\right\rangle^{5}_{0}=\sqrt{\frac{3}{2}}\left|{}^{1}{}_{1}{}^{0}_{1}{}_{0}{}^{\bar{2}}\right\rangle^{6}_{\bar{2}},$$

$$E_3F_3\left|{}^{2}{}_{1}{}^{0}_{1}{}_{\bar{1}}{}^{\bar{1}}\right\rangle^{6}_{1}=(b_1^2+b_2^2)\left|{}^{2}{}_{1}{}^{0}_{1}{}_{\bar{1}}{}^{\bar{1}}\right\rangle^{6}_{1}+(b_1b_3+b_2b_4)\left|{}^{1}{}_{1}{}^{1}_{1}{}_{\bar{1}}{}^{\bar{1}}\right\rangle^{6}_{1}$$

$$=(F_3E_3+H_3)\left|{}^{2}{}_{1}{}^{0}_{1}{}_{\bar{1}}{}^{\bar{1}}\right\rangle^{6}_{1}=\left|{}^{2}{}_{1}{}^{0}_{1}{}_{\bar{1}}{}^{\bar{1}}\right\rangle^{6}_{1}.$$

Thus, $b_1 = 1/2$, $b_3 = \sqrt{3}/2$, $b_2 = \sqrt{1-b_1^2} = \sqrt{3}/2$, and $b_4 = -b_1b_3/b_2 = -1/2$. The produced basis state $\left|{}^{1}{}_{1}{}^{\bar{1}}_{1}{}_{\bar{1}}{}^{\bar{1}}\right\rangle^{7}_{\bar{1}}$ is the the highest weight state of an A$_2$-sixlet with $\boldsymbol{M}^{(2)} = (2, 0)$:

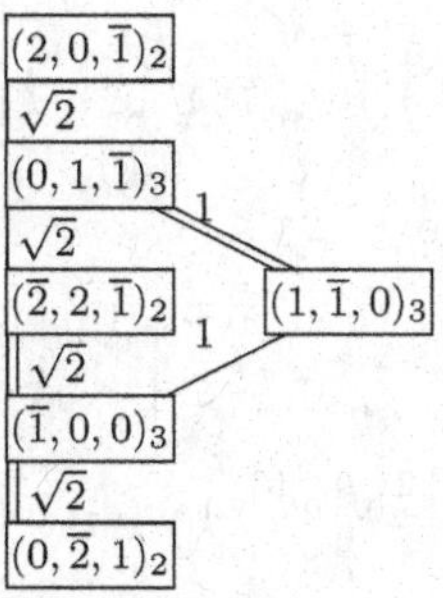

$\left|{}^{1}{}_{1}{}^{\bar{1}}_{1}{}_{\bar{1}}{}^{\bar{1}}\right\rangle^{7}_{\bar{1}}$

$\left|{}^{1}{}_{1}{}^{\bar{1}}_{0}{}_{\bar{1}}{}^{\bar{1}}\right\rangle^{8}_{\bar{1}}$

$\left|{}^{1}{}_{1}{}^{\bar{1}}_{\bar{1}}{}_{\bar{1}}{}^{\bar{1}}\right\rangle^{9}_{\bar{1}}$, $\left|{}^{1}{}_{0}{}^{\bar{1}}_{0}{}_{\bar{1}}{}^{\bar{1}}\right\rangle^{9}_{0}$

$\left|{}^{1}{}_{0}{}^{\bar{1}}_{\bar{1}}{}_{\bar{1}}{}^{\bar{1}}\right\rangle^{10}_{0}$

$\left|{}^{1}{}_{\bar{1}}{}^{\bar{1}}_{\bar{1}}{}_{\bar{1}}{}^{\bar{1}}\right\rangle^{11}_{1}$

Due to the parallel principle we have

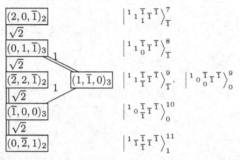

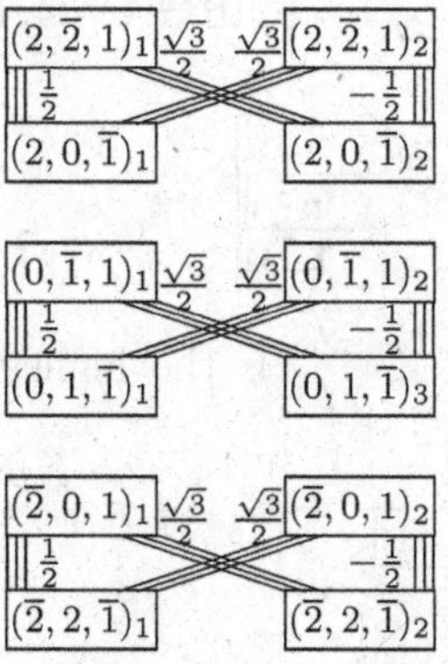

$\left|{}^{2}{}_{1}{}^{0}_{1}{}_{\bar{1}}{}^{\bar{1}}\right\rangle^{6}_{1}$, $\left|{}^{1}{}_{1}{}^{1}_{1}{}_{\bar{1}}{}^{\bar{1}}\right\rangle^{6}_{1}$

$\left|{}^{1}{}_{1}{}^{0}_{1}{}_{\bar{1}}{}^{\bar{2}}\right\rangle^{7}_{\bar{1}}$, $\left|{}^{1}{}_{1}{}^{\bar{1}}_{1}{}_{\bar{1}}{}^{\bar{1}}\right\rangle^{7}_{\bar{1}}$

$\left|{}^{2}{}_{1}{}^{0}_{0}{}_{\bar{1}}{}^{\bar{1}}\right\rangle^{7}_{1}$, $\left|{}^{1}{}_{1}{}^{1}_{0}{}_{\bar{1}}{}^{\bar{1}}\right\rangle^{7}_{1}$

$\left|{}^{1}{}_{1}{}^{0}_{0}{}_{\bar{1}}{}^{\bar{2}}\right\rangle^{8}_{\bar{1}}$, $\left|{}^{1}{}_{1}{}^{\bar{1}}_{0}{}_{\bar{1}}{}^{\bar{1}}\right\rangle^{8}_{\bar{1}}$

$\left|{}^{2}{}_{1}{}^{0}_{\bar{1}}{}_{\bar{1}}{}^{\bar{1}}\right\rangle^{8}_{1}$, $\left|{}^{1}{}_{1}{}^{1}_{\bar{1}}{}_{\bar{1}}{}^{\bar{1}}\right\rangle^{8}_{1}$

$\left|{}^{1}{}_{1}{}^{0}_{\bar{1}}{}_{\bar{1}}{}^{\bar{2}}\right\rangle^{9}_{\bar{1}}$, $\left|{}^{1}{}_{1}{}^{\bar{1}}_{\bar{1}}{}_{\bar{1}}{}^{\bar{1}}\right\rangle^{9}_{\bar{1}}$

The basis state $\left|{}^{2}{}_{0}{}^{0}_{0}{}_{0}{}^{\bar{1}}\right\rangle^{7}_{1}$ in the A$_2$-15let with $\boldsymbol{M}^{(2)} = (2, 1)$ and the basis state $\left|{}^{1}{}_{0}{}^{0}_{0}{}_{0}{}^{0}\right\rangle^{7}_{1}$ in the A$_2$-triplet with $\boldsymbol{M}^{(2)} = (1, 0)$ both span an A$_1$-singlet and both satisfy the condition (8.7) with $\mu = 3$. Then, applying F_3 to them, we obtain

$$F_3\left|\begin{smallmatrix}2&&0&&\bar{1}\\&0&&0&\\&&0&&\end{smallmatrix}\right\rangle^{7}_{1}=c_1\left|\begin{smallmatrix}1&&0&&\bar{2}\\&0&&0&\\&&0&&\end{smallmatrix}\right\rangle^{8}_{\bar{1}}+c_2\left|\begin{smallmatrix}0&&0&&\bar{1}\\&0&&0&\\&&0&&\end{smallmatrix}\right\rangle^{8}_{\bar{1}},$$

$$F_3\left|\begin{smallmatrix}1&&0&&0\\&0&&0&\\&&0&&\end{smallmatrix}\right\rangle^{7}_{1}=c_3\left|\begin{smallmatrix}1&&0&&\bar{2}\\&0&&0&\\&&0&&\end{smallmatrix}\right\rangle^{8}_{\bar{1}}+c_4\left|\begin{smallmatrix}0&&0&&\bar{1}\\&0&&0&\\&&0&&\end{smallmatrix}\right\rangle^{8}_{\bar{1}}.$$

The first basis state $\left|\begin{smallmatrix}1&&0&&\bar{2}\\&0&&0&\\&&0&&\end{smallmatrix}\right\rangle^{8}_{\bar{1}}$ belongs to the A_2-15let with $\boldsymbol{M}^{(2)}=(1,2)$, and the second one $\left|\begin{smallmatrix}0&&0&&\bar{1}\\&0&&0&\\&&0&&\end{smallmatrix}\right\rangle^{8}_{\bar{1}}$ is the highest weight state of an A_2-triplet with $\boldsymbol{M}^{(2)}=(0,1)$. c_2 is made to be real and positive by choosing the phase of the second basis state. The coefficients c_j can be calculated by applying Eq. (8.9) to $\left|\begin{smallmatrix}2&&0&&\bar{1}\\&0&&0&\\&&0&&\end{smallmatrix}\right\rangle^{7}_{1}$ and $\left|\begin{smallmatrix}1&&0&&0\\&0&&0&\\&&0&&\end{smallmatrix}\right\rangle^{7}_{1}$ (see Eqs. (8.25-28)).

$$\begin{aligned}
&E_2F_3\left|\begin{smallmatrix}2&&0&&\bar{1}\\&0&&0&\\&&0&&\end{smallmatrix}\right\rangle^{7}_{1}=c_1\sqrt{2}\left|\begin{smallmatrix}1&&0&&\bar{2}\\&1&&0&\\&&0&&\end{smallmatrix}\right\rangle^{7}_{\bar{2}}\\
&=F_3E_2\left|\begin{smallmatrix}2&&0&&\bar{1}\\&0&&0&\\&&0&&\end{smallmatrix}\right\rangle^{7}_{1}=F_3\sqrt{3}\left|\begin{smallmatrix}2&&0&&\bar{1}\\&1&&0&\\&&0&&\end{smallmatrix}\right\rangle^{6}_{0}=\sqrt{\frac{9}{8}}\left|\begin{smallmatrix}1&&0&&\bar{2}\\&1&&0&\\&&0&&\end{smallmatrix}\right\rangle^{7}_{\bar{2}},\\
&E_2F_3\left|\begin{smallmatrix}1&&0&&0\\&0&&0&\\&&0&&\end{smallmatrix}\right\rangle^{7}_{1}=c_3\sqrt{2}\left|\begin{smallmatrix}1&&0&&\bar{2}\\&1&&0&\\&&0&&\end{smallmatrix}\right\rangle^{7}_{\bar{2}}\\
&=F_3E_2\left|\begin{smallmatrix}1&&0&&0\\&0&&0&\\&&0&&\end{smallmatrix}\right\rangle^{7}_{1}=F_3\left|\begin{smallmatrix}1&&0&&0\\&1&&0&\\&&1&&\end{smallmatrix}\right\rangle^{6}_{0}=\sqrt{\frac{7}{8}}\left|\begin{smallmatrix}1&&0&&\bar{2}\\&1&&0&\\&&0&&\end{smallmatrix}\right\rangle^{7}_{\bar{2}},\\
&E_3F_3\left|\begin{smallmatrix}2&&0&&\bar{1}\\&0&&0&\\&&0&&\end{smallmatrix}\right\rangle^{7}_{1}=(c_1^2+c_2^2)\left|\begin{smallmatrix}2&&0&&\bar{1}\\&0&&0&\\&&0&&\end{smallmatrix}\right\rangle^{7}_{1}+(c_1c_3+c_2c_4)\left|\begin{smallmatrix}1&&0&&0\\&0&&0&\\&&0&&\end{smallmatrix}\right\rangle^{7}_{1}\\
&\qquad=(F_3E_3+H_3)\left|\begin{smallmatrix}2&&0&&\bar{1}\\&0&&0&\\&&0&&\end{smallmatrix}\right\rangle^{7}_{1}=\left|\begin{smallmatrix}2&&0&&\bar{1}\\&0&&0&\\&&0&&\end{smallmatrix}\right\rangle^{7}_{1}.
\end{aligned}$$

Thus, $c_1=3/4$, $c_3=\sqrt{7}/4$, $c_2=\sqrt{1-9/16}=\sqrt{7}/4$, and $c_4=-c_1c_3/c_2=-3/4$.

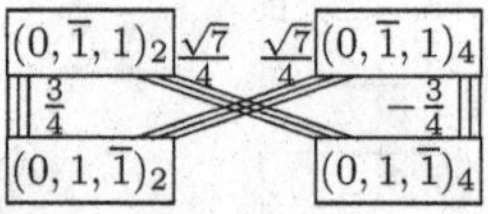

$\left|\begin{smallmatrix}2&&0&&\bar{1}\\&0&&0&\\&&0&&\end{smallmatrix}\right\rangle^{7}_{1}$, $\left|\begin{smallmatrix}1&&0&&0\\&0&&0&\\&&0&&\end{smallmatrix}\right\rangle^{7}_{1}$

$\left|\begin{smallmatrix}1&&0&&\bar{2}\\&0&&0&\\&&0&&\end{smallmatrix}\right\rangle^{8}_{\bar{1}}$, $\left|\begin{smallmatrix}0&&0&&\bar{1}\\&0&&0&\\&&0&&\end{smallmatrix}\right\rangle^{8}_{\bar{1}}$

The produced basis state $\left|\begin{smallmatrix}0&&0&&\bar{1}\\&0&&0&\\&&0&&\end{smallmatrix}\right\rangle^{8}_{\bar{1}}$ is the highest weight state of an A_2-triplet with $\boldsymbol{M}^{(2)}=(0,1)$:

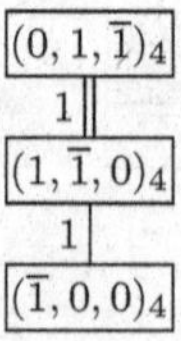

$\left|\begin{smallmatrix}0&&0&&\bar{1}\\&0&&0&\\&&0&&\end{smallmatrix}\right\rangle^{8}_{\bar{1}}$

$\left|\begin{smallmatrix}0&&0&&\bar{1}\\&0&&\bar{1}&\\&&0&&\end{smallmatrix}\right\rangle^{9}_{0}$

$\left|\begin{smallmatrix}0&&0&&\bar{1}\\&0&&\bar{1}&\\&&\bar{1}&&\end{smallmatrix}\right\rangle^{10}_{0}$

The basis state $\left|\begin{smallmatrix}2&&0&&\bar{1}\\&0&&\bar{1}&\\&&0&&\end{smallmatrix}\right\rangle^{8}_{2}$ in the A_2-15let with $\boldsymbol{M}^{(2)}=(2,1)$ satisfies the condition (8.7) with $\mu=3$ and an $\mathcal{A}_3$-triplet is constructed from it:

$$F_3\left|\begin{smallmatrix} 2 & & 0 & & \bar{1} \\ & 0 & & \bar{1} & \\ & & 0 & & \end{smallmatrix}\right\rangle^{8}_{2} = d_1\left|\begin{smallmatrix} 1 & & 0 & & \bar{2} \\ & 0 & & \bar{1} & \\ & & 0 & & \end{smallmatrix}\right\rangle^{9}_{0} + d_2\left|\begin{smallmatrix} 1 & & \bar{1} & & \bar{1} \\ & 0 & & \bar{1} & \\ & & 0 & & \end{smallmatrix}\right\rangle^{9}_{0} + d_3\left|\begin{smallmatrix} 0 & & 0 & & \bar{1} \\ & 0 & & \bar{1} & \\ & & 0 & & \end{smallmatrix}\right\rangle^{9}_{0},$$

The produced basis states belong to the A_2-15let with $\boldsymbol{M}^{(2)} = (1,2)$, the A_2-sixlet with $\boldsymbol{M}^{(2)} = (2,0)$, and the A_2-triplet with $\boldsymbol{M}^{(2)} = (0,1)$, respectively. The coefficients d_j can be calculated by applying Eq. (8.9) to $\left|\begin{smallmatrix} 2 & & 0 & & \bar{1} \\ & 0 & & \bar{1} & \\ & & 0 & & \end{smallmatrix}\right\rangle^{8}_{2}$ (see Eqs. (8.25-28)).

$$\begin{aligned}
E_2F_3\left|\begin{smallmatrix} 2 & & 0 & & \bar{1} \\ & 0 & & \bar{1} & \\ & & 0 & & \end{smallmatrix}\right\rangle^{8}_{2} &= d_1\sqrt{\frac{2}{3}}\left|\begin{smallmatrix} 1 & & 0 & & \bar{2} \\ & 1 & & \bar{1} & \\ & & 0 & & \end{smallmatrix}\right\rangle^{8}_{\bar{1}} + d_1\sqrt{3}\left|\begin{smallmatrix} 1 & & 0 & & \bar{2} \\ & 0 & & 0 & \\ & & 0 & & \end{smallmatrix}\right\rangle^{8}_{\bar{1}} \\
&\quad + d_2\left|\begin{smallmatrix} 1 & & \bar{1} & & \bar{1} \\ & 1 & & \bar{1} & \\ & & 0 & & \end{smallmatrix}\right\rangle^{8}_{\bar{1}} + d_3\left|\begin{smallmatrix} 0 & & 0 & & \bar{1} \\ & 0 & & 0 & \\ & & 0 & & \end{smallmatrix}\right\rangle^{8}_{\bar{1}} \\
= F_3E_2\left|\begin{smallmatrix} 2 & & 0 & & \bar{1} \\ & 0 & & \bar{1} & \\ & & 0 & & \end{smallmatrix}\right\rangle^{8}_{2} &= F_3\left[\left|\begin{smallmatrix} 2 & & 0 & & \bar{1} \\ & 1 & & \bar{1} & \\ & & 0 & & \end{smallmatrix}\right\rangle^{7}_{1} + \sqrt{2}\left|\begin{smallmatrix} 2 & & 0 & & \bar{1} \\ & 0 & & 0 & \\ & & 0 & & \end{smallmatrix}\right\rangle^{7}_{1}\right] \\
&= \frac{1}{2}\left|\begin{smallmatrix} 1 & & 0 & & \bar{2} \\ & 1 & & \bar{1} & \\ & & 0 & & \end{smallmatrix}\right\rangle^{8}_{\bar{1}} + \frac{\sqrt{3}}{2}\left|\begin{smallmatrix} 1 & & \bar{1} & & \bar{1} \\ & 1 & & \bar{1} & \\ & & 0 & & \end{smallmatrix}\right\rangle^{8}_{\bar{1}} \\
&\quad + \sqrt{\frac{9}{8}}\left|\begin{smallmatrix} 1 & & 0 & & \bar{2} \\ & 0 & & 0 & \\ & & 0 & & \end{smallmatrix}\right\rangle^{8}_{\bar{1}} + \sqrt{\frac{7}{8}}\left|\begin{smallmatrix} 0 & & 0 & & \bar{1} \\ & 0 & & 0 & \\ & & 0 & & \end{smallmatrix}\right\rangle^{8}_{\bar{1}}.
\end{aligned}$$

Thus, $d_1 = \sqrt{3/8}$, $d_2 = \sqrt{3}/2$, and $d_3 = \sqrt{7/8}$. By applying $E_3F_3 = F_3E_3 + H_3$ to $\left|\begin{smallmatrix} 1 & & 0 & & \bar{2} \\ & 0 & & \bar{1} & \\ & & 0 & & \end{smallmatrix}\right\rangle^{9}_{0}$ and choosing the phase of $\left|\begin{smallmatrix} 0 & & \bar{1} & & \bar{2} \\ & 0 & & \bar{1} & \\ & & 0 & & \end{smallmatrix}\right\rangle^{10}_{\bar{2}}$, we obtain

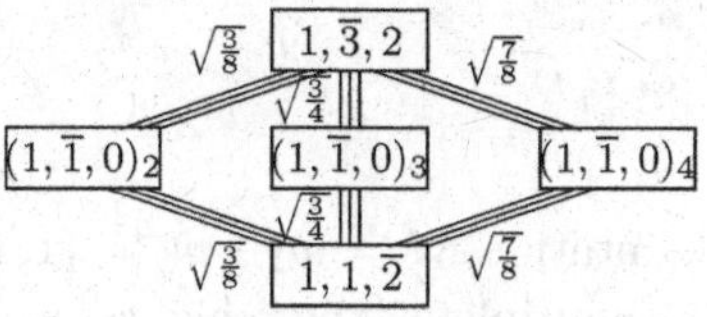

$$\left|\begin{smallmatrix} 2 & & 0 & & \bar{1} \\ & 0 & & \bar{1} & \\ & & 0 & & \end{smallmatrix}\right\rangle^{8}_{2}$$

$$\left|\begin{smallmatrix} 1 & & 0 & & \bar{2} \\ & 0 & & \bar{1} & \\ & & 0 & & \end{smallmatrix}\right\rangle^{9}_{0}, \quad \left|\begin{smallmatrix} 1 & & \bar{1} & & \bar{1} \\ & 0 & & \bar{1} & \\ & & 0 & & \end{smallmatrix}\right\rangle^{9}_{0}, \quad \left|\begin{smallmatrix} 0 & & 0 & & \bar{1} \\ & 0 & & \bar{1} & \\ & & 0 & & \end{smallmatrix}\right\rangle^{9}_{0}$$

$$\left|\begin{smallmatrix} 0 & & \bar{1} & & \bar{2} \\ & 0 & & \bar{1} & \\ & & 0 & & \end{smallmatrix}\right\rangle^{10}_{\bar{2}}$$

The produced basis state $\left|\begin{smallmatrix} 0 & & \bar{1} & & \bar{2} \\ & 0 & & \bar{1} & \\ & & 0 & & \end{smallmatrix}\right\rangle^{10}_{\bar{2}}$ is the highest weight state of an A_2-octet with $\boldsymbol{M}^{(2)} = (1,1)$ (see Fig. 8.5):

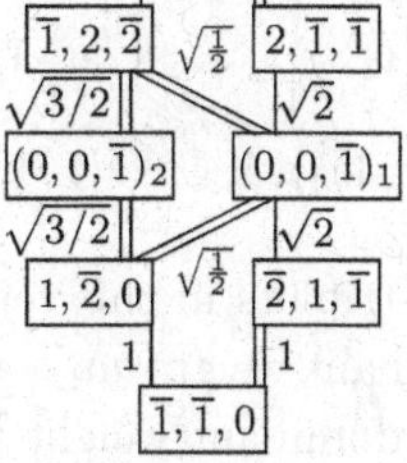

$$\left|\begin{smallmatrix} 0 & & \bar{1} & & \bar{2} \\ & 0 & & \bar{1} & \\ & & 0 & & \end{smallmatrix}\right\rangle^{10}_{\bar{2}}$$

$$\left|\begin{smallmatrix} 0 & & \bar{1} & & \bar{2} \\ & 0 & & \bar{1} & \\ & & \bar{1} & & \end{smallmatrix}\right\rangle^{11}_{\bar{2}}, \quad \left|\begin{smallmatrix} 0 & & \bar{1} & & \bar{2} \\ & 0 & & \bar{2} & \\ & & 0 & & \end{smallmatrix}\right\rangle^{11}_{\bar{1}}$$

$$\left|\begin{smallmatrix} 0 & & \bar{1} & & \bar{2} \\ & \bar{1} & & \bar{1} & \\ & & \bar{1} & & \end{smallmatrix}\right\rangle^{12}_{\bar{1}}, \quad \left|\begin{smallmatrix} 0 & & \bar{1} & & \bar{2} \\ & 0 & & \bar{2} & \\ & & \bar{1} & & \end{smallmatrix}\right\rangle^{12}_{\bar{1}}$$

$$\left|\begin{smallmatrix} 0 & & \bar{1} & & \bar{2} \\ & \bar{1} & & \bar{2} & \\ & & \bar{1} & & \end{smallmatrix}\right\rangle^{13}_{0}, \quad \left|\begin{smallmatrix} 0 & & \bar{1} & & \bar{2} \\ & 0 & & \bar{2} & \\ & & \bar{2} & & \end{smallmatrix}\right\rangle^{13}_{\bar{1}}$$

$$\left|\begin{smallmatrix} 0 & & \bar{1} & & \bar{2} \\ & \bar{1} & & \bar{2} & \\ & & \bar{2} & & \end{smallmatrix}\right\rangle^{14}_{0}$$

The basis state $\left|\begin{smallmatrix} 0 & & \bar1 & & \bar2 \\ & \bar1 & & \bar2 & \\ & & \bar2 & & \end{smallmatrix}\right\rangle^{14}_{0}$ is the lowest weight state in this representation of C_3 with $\boldsymbol{M} = (1, 1, 0)$. From the parallel principle we have

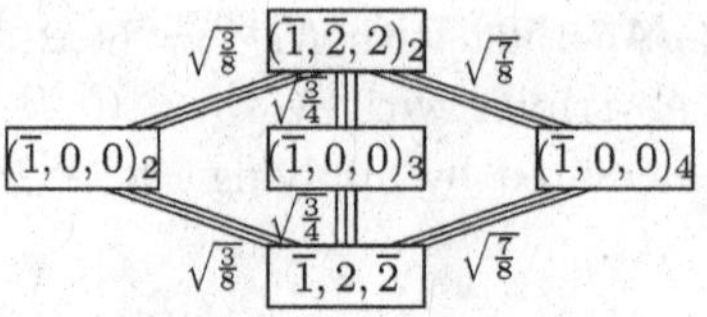

$$\left|\begin{smallmatrix} 2 & & 0 & & \bar1 \\ & 0 & & \bar1 & \\ & & \bar1 & & \end{smallmatrix}\right\rangle^{9}_{2}$$

$$\left|\begin{smallmatrix} 1 & & 0 & & \bar2 \\ & 0 & & \bar1 & \\ & & \bar1 & & \end{smallmatrix}\right\rangle^{10}_{0}, \quad \left|\begin{smallmatrix} 1 & & \bar1 & & \bar1 \\ & 0 & & \bar1 & \\ & & \bar1 & & \end{smallmatrix}\right\rangle^{10}_{0}, \quad \left|\begin{smallmatrix} 0 & & 0 & & \bar1 \\ & 0 & & \bar1 & \\ & & \bar1 & & \end{smallmatrix}\right\rangle^{10}_{0}$$

$$\left|\begin{smallmatrix} 0 & & \bar1 & & \bar2 \\ & 0 & & \bar1 & \\ & & \bar1 & & \end{smallmatrix}\right\rangle^{11}_{\bar2}$$

The basis state $\left|\begin{smallmatrix} 1 & & 0 & & \bar2 \\ & 0 & & \bar2 & \\ & & 0 & & \end{smallmatrix}\right\rangle^{10}_{1}$ is the highest weight state of an A_1-triplet in the A_2-15let with $\boldsymbol{M}^{(2)} = (1, 2)$ and satisfies the condition (8.7) with $\mu = 3$. Three $\mathcal{A}_3$-doublets can be constructed between two A_1-triplets in the A_2-15let and in the A_2-octet due to the parallel principle:

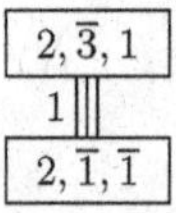

$$\left|\begin{smallmatrix} 1 & & 0 & & \bar2 \\ & 0 & & \bar2 & \\ & & 0 & & \end{smallmatrix}\right\rangle^{10}_{1}, \qquad \left|\begin{smallmatrix} 0 & & \bar1 & & \bar2 \\ & 0 & & \bar2 & \\ & & 0 & & \end{smallmatrix}\right\rangle^{11}_{\bar1}$$

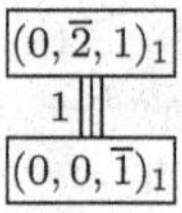

$$\left|\begin{smallmatrix} 1 & & 0 & & \bar2 \\ & 0 & & \bar2 & \\ & & \bar1 & & \end{smallmatrix}\right\rangle^{11}_{1}, \qquad \left|\begin{smallmatrix} 0 & & \bar1 & & \bar2 \\ & 0 & & \bar2 & \\ & & \bar1 & & \end{smallmatrix}\right\rangle^{12}_{\bar1}$$

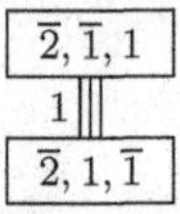

$$\left|\begin{smallmatrix} 1 & & 0 & & \bar2 \\ & 0 & & \bar2 & \\ & & \bar2 & & \end{smallmatrix}\right\rangle^{12}_{1}, \qquad \left|\begin{smallmatrix} 0 & & \bar1 & & \bar2 \\ & 0 & & \bar2 & \\ & & \bar2 & & \end{smallmatrix}\right\rangle^{13}_{\bar1}$$

The basis state $\left|\begin{smallmatrix} 1 & & \bar1 & & \bar1 \\ & \bar1 & & \bar1 & \\ & & \bar1 & & \end{smallmatrix}\right\rangle^{11}_{1}$ in the A_2-sixlet with $\boldsymbol{M}^{(2)} = (2, 0)$ satisfies the condition (8.7) with $\mu = 3$ and an $\mathcal{A}_3$-doublet can be constructed between two A_1-singlets in the A_2-sixlet and in the A_2-octet:

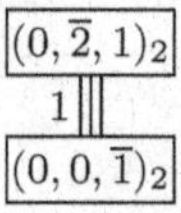

$$\left|\begin{smallmatrix} 1 & & \bar1 & & \bar1 \\ & \bar1 & & \bar1 & \\ & & \bar1 & & \end{smallmatrix}\right\rangle^{11}_{1}, \qquad \left|\begin{smallmatrix} 0 & & \bar1 & & \bar2 \\ & \bar1 & & \bar1 & \\ & & \bar1 & & \end{smallmatrix}\right\rangle^{12}_{\bar1}$$

In summary, the highest weight representation $\boldsymbol{M} = \boldsymbol{w}_1 + \boldsymbol{w}_2 = (1, 1, 0)$ of C_3 is 64-dimensional, and contains 8 A_2-multiplets, whose highest weight states are

$$\left|\begin{smallmatrix} 2 & & 1 & & 0 \\ & 2 & & 1 & \\ & & 2 & & \end{smallmatrix}\right\rangle^{1}_{0}, \quad \left|\begin{smallmatrix} 2 & & 0 & & \bar1 \\ & 2 & & 0 & \\ & & 2 & & \end{smallmatrix}\right\rangle^{3}_{\bar1}, \quad \left|\begin{smallmatrix} 1 & & 1 & & \bar1 \\ & 1 & & 1 & \\ & & 1 & & \end{smallmatrix}\right\rangle^{4}_{\bar1}, \quad \left|\begin{smallmatrix} 1 & & 0 & & 0 \\ & 1 & & 0 & \\ & & 1 & & \end{smallmatrix}\right\rangle^{5}_{0},$$

$$|(1, 1, 0)\rangle, \quad |(2, 1, \bar1)\rangle, \quad |(0, 2, \bar1)_2\rangle, \quad |(1, 0, 0)_4\rangle,$$

$$\left|\begin{smallmatrix} 1 & & 0 & & \bar2 \\ & 1 & & 0 & \\ & & 1 & & \end{smallmatrix}\right\rangle^{6}_{\bar2}, \quad \left|\begin{smallmatrix} 1 & & \bar1 & & \bar1 \\ & 1 & & \bar1 & \\ & & 1 & & \end{smallmatrix}\right\rangle^{7}_{\bar1}, \quad \left|\begin{smallmatrix} 0 & & 0 & & \bar1 \\ & 0 & & 0 & \\ & & 0 & & \end{smallmatrix}\right\rangle^{8}_{\bar1}, \quad \left|\begin{smallmatrix} 0 & & \bar1 & & \bar2 \\ & 0 & & \bar1 & \\ & & 0 & & \end{smallmatrix}\right\rangle^{10}_{\bar2},$$

$$|(1, 2, \bar2)\rangle, \quad |(2, 0, \bar1)_2\rangle, \quad |(0, 1, \bar1)_4\rangle, \quad |(1, 1, \bar2)\rangle.$$

There are sixteen $\mathcal{A}_3$-doublets, and four $\mathcal{A}_3$-triplets in this representation. This representation contains one single dominant weight $\boldsymbol{w}_1 + \boldsymbol{w}_2$, one double dominant weight $\boldsymbol{w}_3$ and one quadruple dominant weight $\boldsymbol{w}_1$:

$$\left|\begin{array}{ccccc} 2 & & 1 & & 0 \\ & 2 & & 1 & \\ & & 2 & & \end{array}\right\rangle_0^1, \qquad \left|\begin{array}{ccccc} 2 & & 1 & & 0 \\ & 2 & & 0 & \\ & & 1 & & \end{array}\right\rangle_1^3, \qquad \left|\begin{array}{ccccc} 2 & & 1 & & 0 \\ & 1 & & 1 & \\ & & 1 & & \end{array}\right\rangle_1^3,$$

$$|(1,1,0)\rangle, \qquad |(0,0,1)_1\rangle, \qquad |(0,0,1)_2\rangle,$$

$$\left|\begin{array}{ccccc} 2 & & 0 & & \bar{1} \\ & 2 & & \bar{1} & \\ & & 1 & & \end{array}\right\rangle_0^5, \quad \left|\begin{array}{ccccc} 2 & & 0 & & \bar{1} \\ & 1 & & 0 & \\ & & 1 & & \end{array}\right\rangle_0^5, \quad \left|\begin{array}{ccccc} 1 & & 1 & & \bar{1} \\ & 1 & & 0 & \\ & & 1 & & \end{array}\right\rangle_0^5, \quad \left|\begin{array}{ccccc} 1 & & 0 & & 0 \\ & 1 & & 0 & \\ & & 1 & & \end{array}\right\rangle_0^5,$$

$$|(1,0,0)_1\rangle, \qquad |(1,0,0)_2\rangle, \qquad |(1,0,0)_3\rangle, \qquad |(1,0,0)_4\rangle.$$

From the calculation results we find that the parameters of the highest weight states of A_2-multiplet in the representation D^M of C_3 satisfy two additional relations: $m_3 = \omega_{33}$ and $\omega_{33} \geqslant -M_1 - M_2 = -2$. The generalized relations, $m_\ell = \omega_{\ell\ell}$ and $\omega_{\ell\ell} \geqslant -\sum_\mu M_\mu$, hold for C_ℓ Lie algebra. We suggest readers to think over these relations.

8.3.3 *Some Representations of* B_ℓ

The Lie algebra of the group $SO(2\ell+1)$ is B_ℓ. The Dynkin diagram is given in Fig. 7.1. The basis state in the highest weight representation D^M of B_ℓ is described by the Gel'fand's basis of $A_{\ell-1}$ (see Eq. (8.30)). The calculation of $D^M(B_\ell)$ depends on the knowledge of the representations of $A_{\ell-1}$. For definiteness, we discuss some representations of B_3. The basis state in the highest weight representation D^M of B_3 is described by the Gel'fand's basis of A_2 with a subscript m_3 (see Eq. (8.30)),

$$|\omega_{ab}\rangle_{m_3}^L = \left|\begin{array}{ccccc} \omega_{13} & & \omega_{23} & & \omega_{33} \\ & \omega_{12} & & \omega_{22} & \\ & & \omega_{11} & & \end{array}\right\rangle_{m_3}^L. \tag{8.34}$$

The action of F_3 increases the superscript L by one, decreases the subscript m_3 by two, and decreases Ω_3 by one. The action of F_2 increases L by one, increases m_3 by two and decreases Ω_2 by one. Each basis state in $D^M(B_3)$ belongs to an A_2-multiplet, and F_3 relates some basis states belonging to different A_2-multiplets.

The parameters of the highest weight state of A_2-multiplet with the highest weight $M^{(2)}$ in the representation D^M of B_3 satisfy

$$\omega_{22} = \omega_{23}, \quad \omega_{11} = \omega_{12} = \omega_{13}, \tag{8.35}$$

where $M_1^{(2)} = \omega_{13} - \omega_{23}$ and $M_2^{(2)} = \omega_{23} - \omega_{33}$. The parameters of the highest weight state of D^M of B_3 also satisfy Eq. (8.35), but $\omega_{33} = m_3 = M_3$, $\omega_{23} = M_2 + \omega_{33}$, and $\omega_{13} = M_1 + \omega_{23}$.

Check each highest weight state ($\omega_{12} = \omega_{11}$) of the A_1-multiplet contained in the A_2-multiplet, as the level $L(\boldsymbol{M}, \boldsymbol{m})$ increases, whether or not it satisfies the condition (8.7) with $\mu = 3$. If yes, we construct an $\mathcal{A}_3$-multiplet by successive application of F_3 (see Eqs. (8.25-28)). The "parallel principle" holds for each basis state in the A_1-multiplet.

The basic representation $\boldsymbol{M} = \boldsymbol{w}_1 = (1, 0, 0)$ of B_3 is the self-representation of SO(7). From the highest weight state $|\boldsymbol{M}\rangle$, an A_2-triplet with $\boldsymbol{M}^{(2)} = (1, 0)$ is constructed:

$$\begin{array}{cc}
\boxed{1,0,0} & \left|\,{}^{1}\,{}_{1}\,{}^{0}_{1}\,{}_{0}\,{}^{0}\right\rangle^{1}_{0} \\
1\,| & \\
\boxed{\overline{1},1,0} & \left|\,{}^{1}\,{}_{1}\,{}^{0}_{0}\,{}_{0}\,{}^{0}\right\rangle^{2}_{0} \\
1\,\| & \\
\boxed{0,\overline{1},2} & \left|\,{}^{1}\,{}_{0}\,{}^{0}_{0}\,{}_{0}\,{}^{0}\right\rangle^{3}_{2}
\end{array}$$

The last basis state satisfies the condition (8.7) with $\mu = 3$ and an $\mathcal{A}_3$-triplet is constructed from it:

$$\begin{array}{cc}
\boxed{0,\overline{1},2} & \left|\,{}^{1}\,{}_{0}\,{}^{0}_{0}\,{}_{0}\,{}^{0}\right\rangle^{3}_{2} \\
\sqrt{2}\,||| & \\
\boxed{0,0,0} & \left|\,{}^{0}\,{}_{0}\,{}^{0}_{0}\,{}_{0}\,{}^{0}\right\rangle^{4}_{0} \\
\sqrt{2}\,||| & \\
\boxed{0,1,\overline{2}} & \left|\,{}^{0}\,{}_{0}\,{}^{0}_{0}\,{}_{0}\,{}^{\overline{1}}\right\rangle^{5}_{\overline{2}}
\end{array}$$

The basis state $\left|\,{}^{0}\,{}_{0}\,{}^{0}_{0}\,{}_{0}\,{}^{0}\right\rangle^{4}_{0}$ spans an A_2-singlet, and the basis state $\left|\,{}^{0}\,{}_{0}\,{}^{0}_{0}\,{}_{0}\,{}^{\overline{1}}\right\rangle^{5}_{\overline{2}}$ is the highest weight state of an A_2-triplet with $\boldsymbol{M}^{(2)} = (0, 1)$:

$$\begin{array}{cc}
\boxed{0,1,\overline{2}} & \left|\,{}^{0}\,{}_{0}\,{}^{0}_{0}\,{}_{0}\,{}^{\overline{1}}\right\rangle^{5}_{\overline{2}} \\
1\,\| & \\
\boxed{1,\overline{1},0} & \left|\,{}^{0}\,{}_{0}\,{}^{0}_{0}\,{}_{\overline{1}}\,{}^{\overline{1}}\right\rangle^{6}_{0} \\
1\,| & \\
\boxed{\overline{1},0,0} & \left|\,{}^{0}\,{}_{0}\,{}^{0}_{\overline{1}}\,{}_{\overline{1}}\,{}^{\overline{1}}\right\rangle^{7}_{0}
\end{array}$$

The last basis state is the lowest weight state of this representation. In summary, the self-representation with $\boldsymbol{M} = \boldsymbol{w}_1$ of SO(7) is real and 7-dimensional. It contains two A_2-triplets and one A_2-singlet with $\boldsymbol{M}^{(2)} = (1, 0)$, $(0, 1)$, and $(0, 0)$, respectively. There are two single dominant weights $\boldsymbol{w}_1$ and 0 in this representation.

The second basic representation $\boldsymbol{M} = \boldsymbol{w}_2 = (0, 1, 0)$ of B_3 is the adjoint representation of SO(7) which is real and 21-dimensional. It contains four A_2-triplets, one A_2-octet, and one A_2-singlet whose highest weight states are

$$\left|\begin{smallmatrix}1&&1&&0\\&1&&1&\\&&1&&\end{smallmatrix}\right\rangle^{1}_{0},\qquad \left|\begin{smallmatrix}1&&0&&0\\&1&&0&\\&&1&&\end{smallmatrix}\right\rangle^{3}_{0},\qquad \left|\begin{smallmatrix}1&&0&&\bar{1}\\&1&&0&\\&&1&&\end{smallmatrix}\right\rangle^{4}_{\bar{2}},$$

$$\left|\begin{smallmatrix}0&&0&&0\\&0&&0&\\&&0&&\end{smallmatrix}\right\rangle^{6}_{0},\qquad \left|\begin{smallmatrix}0&&0&&\bar{1}\\&0&&0&\\&&0&&\end{smallmatrix}\right\rangle^{7}_{\bar{2}},\qquad \left|\begin{smallmatrix}0&&\bar{1}&&\bar{1}\\&0&&\bar{1}&\\&&0&&\end{smallmatrix}\right\rangle^{9}_{\bar{2}}.$$

There are two single dominant weights $\boldsymbol{w}_2$, $\boldsymbol{w}_1$ and one triple dominant weight 0 in this representation:

$$\left|\begin{smallmatrix}1&&1&&0\\&1&&1&\\&&1&&\end{smallmatrix}\right\rangle^{1}_{0},\quad \left|\begin{smallmatrix}1&&0&&0\\&1&&0&\\&&1&&\end{smallmatrix}\right\rangle^{3}_{0},\quad \left|\begin{smallmatrix}1&&0&&\bar{1}\\&1&&\bar{1}&\\&&0&&\end{smallmatrix}\right\rangle^{6}_{0},\quad \left|\begin{smallmatrix}1&&0&&\bar{1}\\&0&&0&\\&&0&&\end{smallmatrix}\right\rangle^{6}_{0},\quad \left|\begin{smallmatrix}0&&0&&0\\&0&&0&\\&&0&&\end{smallmatrix}\right\rangle^{6}_{0}.$$

The detailed calculation is left as exercise (see Prob. 4).

The third basic representation $\boldsymbol{M} = \boldsymbol{w}_3 = (0, 0, 1)$ of B_3 is the spinor representation of SO(7). An A_2-singlet and an $\mathcal{A}_3$-doublet can be constructed from the highest weight state:

$$\begin{array}{cl} \boxed{0,0,1} & \left|\begin{smallmatrix}1&&1&&1\\&1&&1&\\&&1&&\end{smallmatrix}\right\rangle^{1}_{1} \\ 1\,\Vert & \\ \boxed{0,1,\bar{1}} & \left|\begin{smallmatrix}1&&1&&0\\&1&&1&\\&&1&&\end{smallmatrix}\right\rangle^{2}_{\bar{1}} \end{array}$$

The basis state $\left|\begin{smallmatrix}1&&1&&0\\&1&&1&\\&&1&&\end{smallmatrix}\right\rangle^{2}_{\bar{1}}$ is the highest weight state of an A_2-triplet with $\boldsymbol{M}^{(2)} = (0, 1)$ (see Fig. 8.2):

$$\begin{array}{cl} \boxed{0,1,\bar{1}} & \left|\begin{smallmatrix}1&&1&&0\\&1&&1&\\&&1&&\end{smallmatrix}\right\rangle^{2}_{\bar{1}} \\ 1\,\Vert & \\ \boxed{1,\bar{1},1} & \left|\begin{smallmatrix}1&&1&&0\\&1&&0&\\&&1&&\end{smallmatrix}\right\rangle^{3}_{1} \\ 1\,| & \\ \boxed{\bar{1},0,1} & \left|\begin{smallmatrix}1&&1&&0\\&1&&0&\\&&0&&\end{smallmatrix}\right\rangle^{4}_{1} \end{array}$$

The last two basis states span an A_1-doublet and both satisfy the condition (8.7) with $\mu = 3$. From the parallel principle we obtain two $\mathcal{A}_3$-doublets:

$$\begin{array}{clccl} \boxed{1,\bar{1},1} & \left|\begin{smallmatrix}1&&1&&0\\&1&&0&\\&&1&&\end{smallmatrix}\right\rangle^{3}_{1} & \qquad & \boxed{\bar{1},0,1} & \left|\begin{smallmatrix}1&&1&&0\\&1&&0&\\&&0&&\end{smallmatrix}\right\rangle^{4}_{1} \\ 1\,\Vert & & & 1\,\Vert & \\ \boxed{1,0,\bar{1}} & \left|\begin{smallmatrix}1&&0&&0\\&1&&0&\\&&1&&\end{smallmatrix}\right\rangle^{4}_{\bar{1}} & & \boxed{\bar{1},1,\bar{1}} & \left|\begin{smallmatrix}1&&0&&0\\&1&&0&\\&&0&&\end{smallmatrix}\right\rangle^{5}_{\bar{1}} \end{array}$$

The basis state $\left|\begin{smallmatrix}1&&0&&0\\&1&&0&\\&&1&&\end{smallmatrix}\right\rangle^{4}_{\bar{1}}$ is the highest weight state of an A_2-triplet with $\boldsymbol{M}^{(2)} = (1, 0)$, and the basis state $\left|\begin{smallmatrix}1&&0&&0\\&1&&0&\\&&0&&\end{smallmatrix}\right\rangle^{5}_{\bar{1}}$ belongs to the A_2-triplet (see Fig. 8.2):

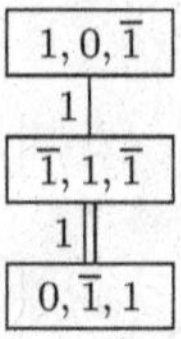

$$\left|\begin{smallmatrix}1 & & 0 & & 0\\ & 1 & & 0 & \\ & & 1 & & \end{smallmatrix}\right\rangle^{4}_{\bar{1}} \qquad \left|\begin{smallmatrix}1 & & 0 & & 0\\ & 1 & & 0 & \\ & & 0 & & \end{smallmatrix}\right\rangle^{5}_{\bar{1}} \qquad \left|\begin{smallmatrix}1 & & 0 & & 0\\ & 0 & & 0 & \\ & & 0 & & \end{smallmatrix}\right\rangle^{6}_{1}$$

The last basis state satisfies the condition (8.7) with $\mu = 3$ and an $\mathcal{A}_3$-doublet can be constructed from it:

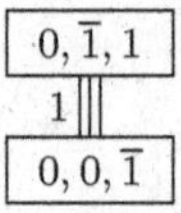

$$\left|\begin{smallmatrix}1 & & 0 & & 0\\ & 0 & & 0 & \\ & & 0 & & \end{smallmatrix}\right\rangle^{6}_{1} \qquad \left|\begin{smallmatrix}0 & & 0 & & 0\\ & 0 & & 0 & \\ & & 0 & & \end{smallmatrix}\right\rangle^{7}_{\bar{1}}$$

In summary, the spinor representation $\boldsymbol{M} = \boldsymbol{w}_3$ of SO(7) is self-conjugate and 8-dimension. It contains two A_2-singlets and two A_2-triplets with $\boldsymbol{M}^{(2)} = (0,0)$, $(0,0)$, $(0,1)$, and $(1,0)$, respectively. There is only one single dominant weight $\boldsymbol{w}_3$ in this representation.

There is an **important property** for some highest weight representations of B_ℓ: the $A_{\ell-1}$-multiplet produced by the action of the lowering operator F_ℓ may be **double degeneracy**. Its example is given as follows, and the explanation will be given in subsection 10.1.2. The dimension of the representation of B_3 with $\boldsymbol{M} = 2\boldsymbol{w}_2 + 2\boldsymbol{w}_3 = (0,2,2)$ is 2079. We calculate only part of the basis states of this representation and give part of the block weight diagram in Fig. 8.10.

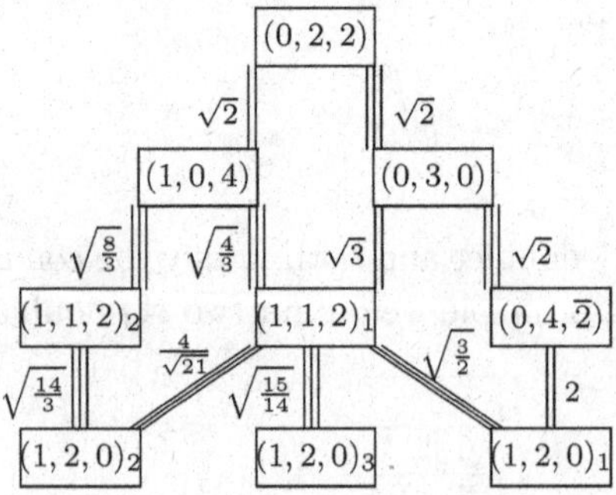

$$\left|\begin{smallmatrix}4 & & 4 & & 2\\ & 4 & & 4 & \\ & & 4 & & \end{smallmatrix}\right\rangle^{1}_{2}$$

$$\left|\begin{smallmatrix}4 & & 4 & & 2\\ & 4 & & 3 & \\ & & 4 & & \end{smallmatrix}\right\rangle^{2}_{4}, \left|\begin{smallmatrix}4 & & 4 & & 1\\ & 4 & & 4 & \\ & & 4 & & \end{smallmatrix}\right\rangle^{2}_{0}$$

$$\left|\begin{smallmatrix}4 & & 3 & & 2\\ & 4 & & 3 & \\ & & 4 & & \end{smallmatrix}\right\rangle^{3}_{2}, \left|\begin{smallmatrix}4 & & 4 & & 1\\ & 4 & & 3 & \\ & & 4 & & \end{smallmatrix}\right\rangle^{3}_{2}, \left|\begin{smallmatrix}4 & & 4 & & 0\\ & 4 & & 4 & \\ & & 4 & & \end{smallmatrix}\right\rangle^{3}_{\bar{2}}$$

$$\left|\begin{smallmatrix}4 & & 3 & & 1\\ & 4 & & 3 & \\ & & 4 & & \end{smallmatrix}\right\rangle^{4}_{0}, \left|\begin{smallmatrix}4 & & 3 & & 1\\ & 4 & & 3 & \\ & & 4 & & \end{smallmatrix}\right\rangle'^{4}_{0}, \left|\begin{smallmatrix}4 & & 4 & & 0\\ & 4 & & 3 & \\ & & 4 & & \end{smallmatrix}\right\rangle^{4}_{0}$$

Fig. 8.10. Part of basis states in the representation (0, 2, 2) of B_3.

The highest weight state $|(0,2,2)\rangle$ is that of an A_2-sixlet and that of $\mathcal{A}_3$-triplet:

$$\begin{aligned}
F_2\left|(0,\dot{2},2)\right\rangle &= F_2\left|\begin{smallmatrix} 4 & & 4 & & 2 \\ & 4 & & 4 & \\ & & 4 & & \end{smallmatrix}\right\rangle_2^1 = \sqrt{2}\left|\begin{smallmatrix} 4 & & 4 & & 2 \\ & 4 & & 3 & \\ & & 4 & & \end{smallmatrix}\right\rangle_4^2 = \sqrt{2}\left|(1,0,4)\right\rangle, \\
F_3\left|(0,2,2)\right\rangle &= F_3\left|\begin{smallmatrix} 4 & & 4 & & 2 \\ & 4 & & 4 & \\ & & 4 & & \end{smallmatrix}\right\rangle_2^1 = \sqrt{2}\left|\begin{smallmatrix} 4 & & 4 & & 1 \\ & 4 & & 4 & \\ & & 4 & & \end{smallmatrix}\right\rangle_0^2 = \sqrt{2}\left|(0,3,0)\right\rangle, \\
F_3\left|(0,3,0)\right\rangle &= F_3\left|\begin{smallmatrix} 4 & & 4 & & 1 \\ & 4 & & 4 & \\ & & 4 & & \end{smallmatrix}\right\rangle_0^2 = \sqrt{2}\left|\begin{smallmatrix} 4 & & 4 & & 0 \\ & 4 & & 4 & \\ & & 4 & & \end{smallmatrix}\right\rangle_{\overline{2}}^3 = \sqrt{2}\left|(0,4,\overline{2})\right\rangle.
\end{aligned}$$

The states $\left|\begin{smallmatrix} 4 & & 4 & & 1 \\ & 4 & & 4 & \\ & & 4 & & \end{smallmatrix}\right\rangle_0^2$ and $\left|\begin{smallmatrix} 4 & & 4 & & 0 \\ & 4 & & 4 & \\ & & 4 & & \end{smallmatrix}\right\rangle_{\overline{2}}^3$ are the highest weight states of an A_2-decuplet with $\boldsymbol{M}^{(2)} = (0,3)$ and an A_2-15let with $\boldsymbol{M}^{(2)} = (0,4)$, respectively.

$$\begin{aligned}
F_2\left|(0,3,0)\right\rangle &= F_2\left|\begin{smallmatrix} 4 & & 4 & & 1 \\ & 4 & & 4 & \\ & & 4 & & \end{smallmatrix}\right\rangle_0^2 = \sqrt{3}\left|\begin{smallmatrix} 4 & & 4 & & 1 \\ & 4 & & 3 & \\ & & 4 & & \end{smallmatrix}\right\rangle_2^3 = \sqrt{3}\left|(1,1,2)_1\right\rangle, \\
F_2\left|(0,4,\overline{2})\right\rangle &= F_2\left|\begin{smallmatrix} 4 & & 4 & & 0 \\ & 4 & & 4 & \\ & & 4 & & \end{smallmatrix}\right\rangle_{\overline{2}}^3 = 2\left|\begin{smallmatrix} 4 & & 4 & & 0 \\ & 4 & & 3 & \\ & & 4 & & \end{smallmatrix}\right\rangle_0^4 = 2\left|(1,2,0)_1\right\rangle.
\end{aligned}$$

The basis state $\left|\begin{smallmatrix} 4 & & 4 & & 2 \\ & 4 & & 3 & \\ & & 4 & & \end{smallmatrix}\right\rangle_4^2 = \left|(1,0,4)\right\rangle$ is the highest weight state of an $\mathcal{A}_3$-quintuplet. By the action of F_3 we obtain

$$\begin{aligned}
F_3\left|(1,0,4)\right\rangle = F_3\left|\begin{smallmatrix} 4 & & 4 & & 2 \\ & 4 & & 3 & \\ & & 4 & & \end{smallmatrix}\right\rangle_4^2 &= a_1\left|\begin{smallmatrix} 4 & & 4 & & 1 \\ & 4 & & 3 & \\ & & 4 & & \end{smallmatrix}\right\rangle_2^3 + a_2\left|\begin{smallmatrix} 4 & & 3 & & 2 \\ & 4 & & 3 & \\ & & 4 & & \end{smallmatrix}\right\rangle_2^3 \\
&= a_1\left|(1,1,2)_1\right\rangle + a_2\left|(1,1,2)_2\right\rangle.
\end{aligned}$$

The basis state $\left|\begin{smallmatrix} 4 & & 4 & & 1 \\ & 4 & & 3 & \\ & & 4 & & \end{smallmatrix}\right\rangle_2^3$ belongs to the A_2-decuplet with $\boldsymbol{M}^{(2)} = (0,3)$ and the basis state $\left|\begin{smallmatrix} 4 & & 3 & & 2 \\ & 4 & & 3 & \\ & & 4 & & \end{smallmatrix}\right\rangle_2^3$ is the highest weight state of an A_2-octet with $\boldsymbol{M}^{(2)} = (1,1)$. a_2 is taken to be real and positive by choosing the phase of $\left|\begin{smallmatrix} 4 & & 3 & & 2 \\ & 4 & & 3 & \\ & & 4 & & \end{smallmatrix}\right\rangle_2^3$. Then, the coefficients a_j can be calculated by applying Eq. (8.9) to $\left|\begin{smallmatrix} 4 & & 4 & & 2 \\ & 4 & & 3 & \\ & & 4 & & \end{smallmatrix}\right\rangle_4^2$ (see Eqs. (8.25-28)).

$$\begin{aligned}
& E_2F_3\left|\begin{smallmatrix} 4 & & 4 & & 2 \\ & 4 & & 3 & \\ & & 4 & & \end{smallmatrix}\right\rangle_4^2 = \sqrt{3}a_1\left|\begin{smallmatrix} 4 & & 4 & & 1 \\ & 4 & & 4 & \\ & & 4 & & \end{smallmatrix}\right\rangle_0^2 \\
& \quad = F_3E_2\left|\begin{smallmatrix} 4 & & 4 & & 2 \\ & 4 & & 3 & \\ & & 4 & & \end{smallmatrix}\right\rangle_4^2 = \sqrt{2}F_3\left|\begin{smallmatrix} 4 & & 4 & & 2 \\ & 4 & & 4 & \\ & & 4 & & \end{smallmatrix}\right\rangle_2^1 = 2\left|\begin{smallmatrix} 4 & & 4 & & 1 \\ & 4 & & 4 & \\ & & 4 & & \end{smallmatrix}\right\rangle_0^2, \\
& E_3F_3\left|\begin{smallmatrix} 4 & & 4 & & 2 \\ & 4 & & 3 & \\ & & 4 & & \end{smallmatrix}\right\rangle_4^2 = (a_1^2 + a_2^2)\left|\begin{smallmatrix} 4 & & 4 & & 2 \\ & 4 & & 3 & \\ & & 4 & & \end{smallmatrix}\right\rangle_0^2 \\
& \quad = (F_3E_3 + H_3)\left|\begin{smallmatrix} 4 & & 4 & & 2 \\ & 4 & & 3 & \\ & & 4 & & \end{smallmatrix}\right\rangle_4^2 = 4\left|\begin{smallmatrix} 4 & & 4 & & 2 \\ & 4 & & 3 & \\ & & 4 & & \end{smallmatrix}\right\rangle_0^2.
\end{aligned}$$

Thus, $a_1 = \sqrt{4/3}$ and $a_2 = \sqrt{4 - 4/3} = \sqrt{8/3}$. Assume that

$$\begin{aligned}
F_3\left|(1,1,2)_1\right\rangle &= F_3\left|\begin{smallmatrix} 4 & & 4 & & 1 \\ & 4 & & 3 & \\ & & 4 & & \end{smallmatrix}\right\rangle^{3}_{2} = b_1\left|\begin{smallmatrix} 4 & & 4 & & 0 \\ & 4 & & 3 & \\ & & 4 & & \end{smallmatrix}\right\rangle^{4}_{0} + b_2\left|\begin{smallmatrix} 4 & & 3 & & 1 \\ & 4 & & 3 & \\ & & 4 & & \end{smallmatrix}\right\rangle^{4}_{0} \\
&= b_1\left|(1,2,0)_1\right\rangle + b_2\left|(1,2,0)_2\right\rangle, \\
F_3\left|(1,1,2)_2\right\rangle &= F_3\left|\begin{smallmatrix} 4 & & 3 & & 2 \\ & 4 & & 3 & \\ & & 4 & & \end{smallmatrix}\right\rangle^{3}_{2} = b_3\left|\begin{smallmatrix} 4 & & 4 & & 0 \\ & 4 & & 3 & \\ & & 4 & & \end{smallmatrix}\right\rangle^{4}_{0} + b_4\left|\begin{smallmatrix} 4 & & 3 & & 1 \\ & 4 & & 3 & \\ & & 4 & & \end{smallmatrix}\right\rangle^{4}_{0} \\
&= b_3\left|(1,2,0)_1\right\rangle + b_4\left|(1,2,0)_2\right\rangle.
\end{aligned}$$

The basis state $\left|\begin{smallmatrix} 4 & & 4 & & 0 \\ & 4 & & 3 & \\ & & 4 & & \end{smallmatrix}\right\rangle^{4}_{0}$ belongs to the A_2-15let with $\boldsymbol{M}^{(2)} = (0,4)$ and the basis state $\left|\begin{smallmatrix} 4 & & 3 & & 1 \\ & 4 & & 3 & \\ & & 4 & & \end{smallmatrix}\right\rangle^{4}_{0}$ is the highest weight state of an A_2-15let with $\boldsymbol{M}^{(2)} = (1,2)$. b_4 is taken to be real and positive by choosing the phase of $\left|\begin{smallmatrix} 4 & & 3 & & 1 \\ & 4 & & 3 & \\ & & 4 & & \end{smallmatrix}\right\rangle^{4}_{0}$. The coefficients b_j are calculated in terms of Eqs. (8.25-28):

$$\begin{aligned}
&E_2F_3\left|\begin{smallmatrix} 4 & & 4 & & 1 \\ & 4 & & 3 & \\ & & 4 & & \end{smallmatrix}\right\rangle^{3}_{2} = 2b_1\left|\begin{smallmatrix} 4 & & 4 & & 0 \\ & 4 & & 4 & \\ & & 4 & & \end{smallmatrix}\right\rangle^{3}_{\overline{2}} \\
&\quad = F_3E_2\left|\begin{smallmatrix} 4 & & 4 & & 1 \\ & 4 & & 3 & \\ & & 4 & & \end{smallmatrix}\right\rangle^{3}_{2} = \sqrt{3}F_3\left|\begin{smallmatrix} 4 & & 4 & & 1 \\ & 4 & & 4 & \\ & & 4 & & \end{smallmatrix}\right\rangle^{2}_{0} = \sqrt{6}\left|\begin{smallmatrix} 4 & & 4 & & 0 \\ & 4 & & 4 & \\ & & 4 & & \end{smallmatrix}\right\rangle^{3}_{\overline{2}}, \\
&E_2F_3\left|\begin{smallmatrix} 4 & & 3 & & 2 \\ & 4 & & 3 & \\ & & 4 & & \end{smallmatrix}\right\rangle^{3}_{2} = 2b_3\left|\begin{smallmatrix} 4 & & 4 & & 0 \\ & 4 & & 4 & \\ & & 4 & & \end{smallmatrix}\right\rangle^{3}_{\overline{2}} = F_3E_2\left|\begin{smallmatrix} 4 & & 3 & & 2 \\ & 4 & & 3 & \\ & & 4 & & \end{smallmatrix}\right\rangle^{3}_{2} = 0, \\
&E_3F_3\left|\begin{smallmatrix} 4 & & 3 & & 2 \\ & 4 & & 3 & \\ & & 4 & & \end{smallmatrix}\right\rangle^{3}_{2} = (b_3^2 + b_4^2)\left|\begin{smallmatrix} 4 & & 3 & & 2 \\ & 4 & & 3 & \\ & & 4 & & \end{smallmatrix}\right\rangle^{3}_{2} + (b_3b_1 + b_4b_2)\left|\begin{smallmatrix} 4 & & 4 & & 1 \\ & 4 & & 3 & \\ & & 4 & & \end{smallmatrix}\right\rangle^{3}_{2} \\
&\quad = (F_3E_3 + H_3)\left|\begin{smallmatrix} 4 & & 3 & & 2 \\ & 4 & & 3 & \\ & & 4 & & \end{smallmatrix}\right\rangle^{3}_{2} \\
&\quad = \left(\frac{8}{3} + 2\right)\left|\begin{smallmatrix} 4 & & 3 & & 2 \\ & 4 & & 3 & \\ & & 4 & & \end{smallmatrix}\right\rangle^{3}_{2} + \sqrt{\frac{8}{3}} \cdot \sqrt{\frac{4}{3}}\left|\begin{smallmatrix} 4 & & 4 & & 1 \\ & 4 & & 3 & \\ & & 4 & & \end{smallmatrix}\right\rangle^{3}_{2}.
\end{aligned}$$

Thus, $b_1 = \sqrt{3/2}$, $b_3 = 0$, $b_4 = \sqrt{14/3}$, and $b_2 = \sqrt{32/9}/\sqrt{14/3} = \sqrt{16/21}$. However, a contradiction occurs in applying $E_3F_3 = F_3E_3 + H_3$ to $\left|\begin{smallmatrix} 4 & & 4 & & 1 \\ & 4 & & 3 & \\ & & 4 & & \end{smallmatrix}\right\rangle^{3}_{2}$:

$$\begin{aligned}
E_3F_3\left|\begin{smallmatrix} 4 & & 4 & & 1 \\ & 4 & & 3 & \\ & & 4 & & \end{smallmatrix}\right\rangle^{3}_{2} &= (b_1^2 + b_2^2)\left|\begin{smallmatrix} 4 & & 4 & & 1 \\ & 4 & & 3 & \\ & & 4 & & \end{smallmatrix}\right\rangle^{3}_{2} + (b_1b_3 + b_2b_4)\left|\begin{smallmatrix} 4 & & 3 & & 2 \\ & 4 & & 3 & \\ & & 4 & & \end{smallmatrix}\right\rangle^{3}_{2} \\
&= (F_3E_3 + H_3)\left|\begin{smallmatrix} 4 & & 4 & & 1 \\ & 4 & & 3 & \\ & & 4 & & \end{smallmatrix}\right\rangle^{3}_{2} \\
&= \left(\frac{4}{3} + 2\right)\left|\begin{smallmatrix} 4 & & 4 & & 1 \\ & 4 & & 3 & \\ & & 4 & & \end{smallmatrix}\right\rangle^{3}_{2} + \sqrt{\frac{4}{3}} \cdot \sqrt{\frac{8}{3}}\left|\begin{smallmatrix} 4 & & 3 & & 2 \\ & 4 & & 3 & \\ & & 4 & & \end{smallmatrix}\right\rangle^{3}_{2}, \\
b_1^2 + b_2^2 &= \frac{3}{2} + \frac{16}{21} = \frac{10}{3} - \frac{15}{14} \neq \frac{10}{3}.
\end{aligned}$$

We have to introduce another basis state $\left|(1,2,0)_3\right\rangle$ such that

$$
\begin{aligned}
F_3\,|(1,1,2)_1\rangle &= F_3 \left|\begin{smallmatrix} 4 & & 4 & & 1 \\ & 4 & & 3 & \\ & & 4 & & \end{smallmatrix}\right\rangle^{3}_{2} \\
&= \sqrt{\frac{3}{2}} \left|\begin{smallmatrix} 4 & & 4 & & 0 \\ & 4 & & 3 & \\ & & 4 & & \end{smallmatrix}\right\rangle^{4}_{0} + \sqrt{\frac{16}{21}} \left|\begin{smallmatrix} 4 & & 3 & & 1 \\ & 4 & & 3 & \\ & & 4 & & \end{smallmatrix}\right\rangle^{4}_{0} + \sqrt{\frac{15}{14}} \left|\begin{smallmatrix} 4 & & 3 & & 1 \\ & 4 & & 3 & \\ & & 4 & & \end{smallmatrix}\right\rangle'^{4}_{0} \\
&= \sqrt{\frac{3}{2}}\,|(1,2,0)_1\rangle + \sqrt{\frac{16}{21}}\,|(1,2,0)_2\rangle + \sqrt{\frac{15}{14}}\,|(1,2,0)_3\rangle, \\
F_3\,|(1,1,2)_2\rangle &= F_3 \left|\begin{smallmatrix} 4 & & 3 & & 2 \\ & 4 & & 3 & \\ & & 4 & & \end{smallmatrix}\right\rangle^{3}_{2} = \sqrt{\frac{14}{3}} \left|\begin{smallmatrix} 4 & & 3 & & 1 \\ & 4 & & 3 & \\ & & 4 & & \end{smallmatrix}\right\rangle^{4}_{0} = \sqrt{\frac{14}{3}}\,|(1,2,0)_2\rangle.
\end{aligned}
$$

In the action of F_3 to the states $\left|\begin{smallmatrix} 4 & & 4 & & 1 \\ & 4 & & 3 & \\ & & 4 & & \end{smallmatrix}\right\rangle^{3}_{2}$ and $\left|\begin{smallmatrix} 4 & & 3 & & 2 \\ & 4 & & 3 & \\ & & 4 & & \end{smallmatrix}\right\rangle^{3}_{2}$, the double degeneracy of the highest weight state $\left|\begin{smallmatrix} 4 & & 3 & & 1 \\ & 4 & & 3 & \\ & & 4 & & \end{smallmatrix}\right\rangle^{4}_{0}$ of the A_2-15let with $\boldsymbol{M}^{(2)} = (1,2)$ occurs in the representation $D^{\boldsymbol{M}}$ of B_3 with $\boldsymbol{M} = (0,2,2)$.

8.3.4 *Some Representations of* D_ℓ

The Lie algebra of the group $SO(2\ell)$ is D_ℓ. The Dynkin diagram is given in Fig. 7.1. The basis state in the highest weight representation $D^{\boldsymbol{M}}$ of D_ℓ is described by the Gel'fand's basis of $A_{\ell-1}$ (see Eq. (8.30)). The calculation of $D^{\boldsymbol{M}}$ of D_ℓ depends on the knowledge of the representations of $A_{\ell-1}$. For definiteness, we discuss some representations of D_4. The basis state in the highest weight representation $D^{\boldsymbol{M}}$ of D_4 is described by the Gel'fand's basis of A_3 with a subscript m_4:

$$
\left|\begin{smallmatrix} \omega_{14} & & \omega_{24} & & \omega_{34} & & \omega_{44} \\ & \omega_{13} & & \omega_{23} & & \omega_{33} & \\ & & \omega_{12} & & \omega_{22} & & \\ & & & \omega_{11} & & & \end{smallmatrix}\right\rangle^{L}_{m_4}. \qquad (8.36)
$$

The action of F_4 increases the superscript L by one, and decreases the subscript m_4, Ω_3, and Ω_4 by two, one, and two, respectively. The action of F_2 increases L and m_4 both by one and decreases Ω_2 by one (see Eq. (8.18)). The parameters of the highest weight state of A_3-multiplet with the highest weight $\boldsymbol{M}^{(3)}$ satisfy

$$
\omega_{11} = \omega_{12} = \omega_{13} = \omega_{14}, \quad \omega_{22} = \omega_{23} = \omega_{24}, \quad \omega_{33} = \omega_{34}, \qquad (8.37)
$$

where $M_1^{(3)} = \omega_{14} - \omega_{24}$, $M_2^{(3)} = \omega_{24} - \omega_{34}$, and $M_3^{(3)} = \omega_{34} - \omega_{44}$. The parameters of the highest weight state in $D^{\boldsymbol{M}}$ of D_4 also satisfy Eq. (8.37), but $\omega_{44} = m_4 = M_4$ and $\omega_{\mu 4} = M_\mu + \omega_{(\mu+1)4}$, $1 \leqslant \mu \leqslant 3$.

Check each highest weight state of A_1-multiplet ($\omega_{12} = \omega_{11}$) contained in the A_3-multiplet, as the level $L(\boldsymbol{M}, \boldsymbol{m})$ increases, whether or not it satisfies the condition (8.7) with $\mu = 4$. If yes, we construct an $\mathcal{A}_4$-multiplet by successive application of F_4 (see Eqs. (8.25-28)). **The "parallel principle" holds when F_1 applies to the $\mathcal{A}_4$-multiplet** because F_1 commutes with F_4. Sometimes, the "parallel principle" also holds for the application of F_3 (see pp. 565-568 in subsection 10.3.4 and Prob. 8 in Chap. 10).

Discuss the basic representation $M = w_1 = (1,0,0,0)$ of D_4. An A_3-quadruplet with $M^{(3)} = (1,0,0)$ is constructed from the highest weight state (see Fig. 8.7):

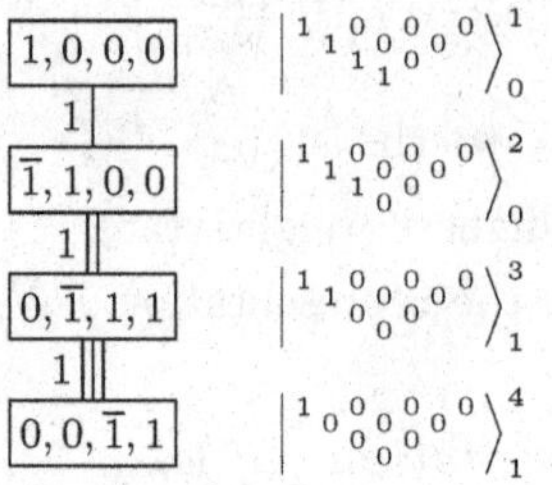

$$\boxed{1,0,0,0} \qquad \left|\,{}^1\,{}_1\,{}^0_1\,{}^0_0\,{}^0\,{}^0\,\right\rangle_0^1$$

$$\Big\|\,1$$

$$\boxed{\bar{1},1,0,0} \qquad \left|\,{}^1\,{}_1\,{}^0_1\,{}^0_0\,{}^0\,{}^0\,\right\rangle_0^2$$

$$\Big\|\,1$$

$$\boxed{0,\bar{1},1,1} \qquad \left|\,{}^1\,{}_1\,{}^0_0\,{}^0_0\,{}^0\,{}^0\,\right\rangle_1^3$$

$$\Big\|\,1$$

$$\boxed{0,0,\bar{1},1} \qquad \left|\,{}^1\,{}_0\,{}^0_0\,{}^0_0\,{}^0\,{}^0\,\right\rangle_1^4$$

The basis state $\left|\,{}^1\,{}_1\,{}^0_0\,{}^0_0\,{}^0\,{}^0\,\right\rangle_1^3$ satisfy the condition (8.7) with $\mu = 4$, and an $\mathcal{A}_4$-doublet is constructed from it. Another $\mathcal{A}_4$-doublet is obtained by the action of F_3 due to the parallel principle.

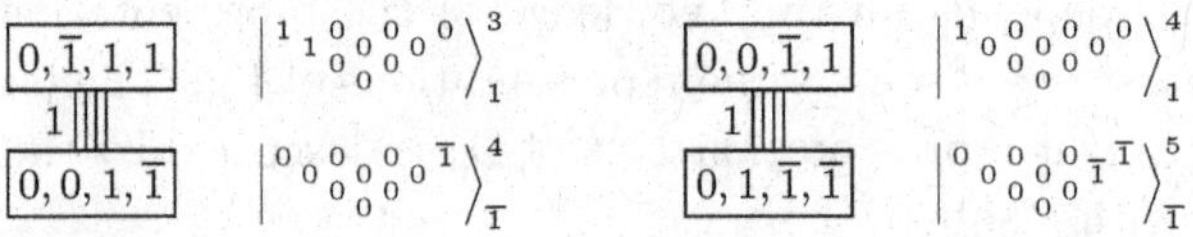

$$\boxed{0,\bar{1},1,1} \quad \left|\,{}^1\,{}_1\,{}^0_0\,{}^0_0\,{}^0\,{}^0\,\right\rangle_1^3 \qquad\qquad \boxed{0,0,\bar{1},1} \quad \left|\,{}^1\,{}_0\,{}^0_0\,{}^0_0\,{}^0\,{}^0\,\right\rangle_1^4$$

$$\Big\|\,1 \qquad\qquad\qquad\qquad\qquad\qquad\qquad \Big\|\,1$$

$$\boxed{0,0,1,\bar{1}} \quad \left|\,{}^0\,{}_0\,{}^0_0\,{}^0_0\,{}^{\bar1}\,{}^0\,\right\rangle_{\bar1}^4 \qquad\qquad \boxed{0,1,\bar{1},\bar{1}} \quad \left|\,{}^0\,{}_0\,{}^0_0\,{}^0_0\,{}^{\bar1}\,{}^{\bar1}\,\right\rangle_{\bar1}^5$$

The basis state $\left|\,{}^0\,{}_0\,{}^0_0\,{}^0_0\,{}^{\bar1}\,{}^0\,\right\rangle_0^4$ is the highest weight state of an A_3-quadruplet with $M^{(3)} = (0,0,1)$. The A_3-quadruplet contains four basis states (see Fig. 8.7), including two basis states produced in above two $\mathcal{A}_4$-doublet:

$$\boxed{0,0,1,\bar{1}} \qquad \left|\,{}^0\,{}_0\,{}^0_0\,{}^0_0\,{}^{\bar1}\,{}^0\,\right\rangle_{\bar1}^4$$

$$\Big\|\,1$$

$$\boxed{0,1,\bar{1},\bar{1}} \qquad \left|\,{}^0\,{}_0\,{}^0_0\,{}^0_0\,{}^{\bar1}\,{}^{\bar1}\,\right\rangle_{\bar1}^5$$

$$\Big\|\,1$$

$$\boxed{1,\bar{1},0,0} \qquad \left|\,{}^0\,{}_0\,{}^0_0\,{}^0_{\bar1}\,{}^{\bar1}\,{}^{\bar1}\,\right\rangle_0^6$$

$$\Big\|\,1$$

$$\boxed{\bar{1},0,0,0} \qquad \left|\,{}^0\,{}_0\,{}^0_0\,{}^0_{\bar1}\,{}^{\bar1}\,{}^{\bar1}\,\right\rangle_0^7$$

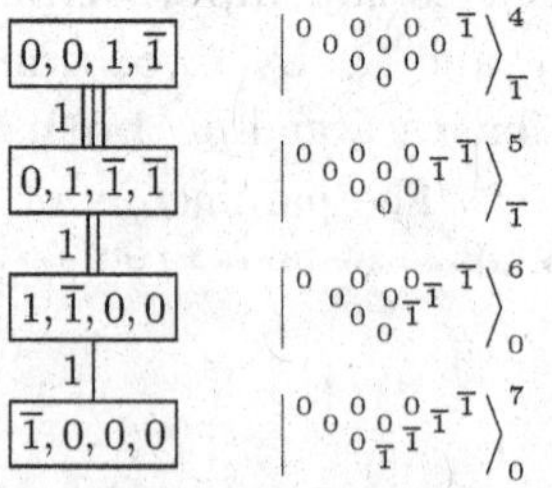

The last basis state is the lowest weight state in this representation of D_4. The representation of D_4 with $M = w_1$ is the self-representation of the group SO(8), which contains two A_3-quadruplets with $M^{(3)} = w_1$ and w_3. Due to the symmetry of the Dynkin diagram of D_4, three basic representations with $M = w_1$, w_3, and w_4 all are 8-dimensional real representations.

Two basic representations with $\boldsymbol{M} = \boldsymbol{w}_3$ and $\boldsymbol{w}_4$ are called the spinor representations of SO(8) (see section 10.3.). Because the A_3 subalgebra has been chosen, the spinor representation with $\boldsymbol{M} = \boldsymbol{w}_3$ also contains two A_3-quadruplets with $\boldsymbol{M}^{(3)} = \boldsymbol{w}_1$ and $\boldsymbol{w}_3$, but the spinor representation with $\boldsymbol{M} = \boldsymbol{w}_4$ contains two A_3-singlets with $\boldsymbol{M}^{(3)} = 0$ and one A_3-sixlet with $\boldsymbol{M}^{(3)} = (0, 1, 0)$ (see Prob. 5). These three basic representations all contain only one single dominant weight, which is their highest weight, respectively.

The remaining basic representation of D_4 with $\boldsymbol{M} = \boldsymbol{w}_2 = (0, 1, 0, 0)$ is the adjoint representation of SO(8). It is 28-dimensional and contains two A_3-sixlets with $A^{(3)} = \boldsymbol{w}_2$, one A_3-15let with $A^{(3)} = \boldsymbol{w}_1 + \boldsymbol{w}_3$, and one A_3-singlet, whose highest weight states respectively are

$$\left|\begin{array}{ccccccc} 1 & & 1 & & 0 & & 0 \\ & 1 & & 1 & & 0 & \\ & & 1 & & 1 & & \\ & & & 1 & & & \end{array}\right\rangle^{1}_{0}, \quad \left|\begin{array}{ccccccc} 0 & & 0 & & \bar{1} & & \bar{1} \\ & 0 & & 0 & & \bar{1} & \\ & & 0 & & 0 & & \\ & & & 0 & & & \end{array}\right\rangle^{7}_{\bar{2}}, \quad \left|\begin{array}{ccccccc} 1 & & 0 & & 0 & & \bar{1} \\ & 1 & & 0 & & 0 & \\ & & 1 & & 0 & & \\ & & & 1 & & & \end{array}\right\rangle^{3}_{\bar{1}}, \quad \left|\begin{array}{ccccccc} 0 & & 0 & & 0 & & 0 \\ & 0 & & 0 & & 0 & \\ & & 0 & & 0 & & \\ & & & 0 & & & \end{array}\right\rangle^{6}_{0}.$$

It contains one single dominant weight $\boldsymbol{w}_2$ and one quadruple dominant weight 0:

$$\left|\begin{array}{ccccccc} 1 & & 1 & & 0 & & 0 \\ & 1 & & 1 & & 0 & \\ & & 1 & & 1 & & \\ & & & 1 & & & \end{array}\right\rangle^{1}_{0},$$

$$\left|\begin{array}{ccccccc} 1 & & 0 & & 0 & & \bar{1} \\ & 1 & & 0 & & \bar{1} & \\ & & 1 & & \bar{1} & & \\ & & & 0 & & & \end{array}\right\rangle^{6}_{0}, \quad \left|\begin{array}{ccccccc} 1 & & 0 & & 0 & & \bar{1} \\ & 1 & & 0 & & \bar{1} & \\ & & 0 & & 0 & & \\ & & & 0 & & & \end{array}\right\rangle^{6}_{0}, \quad \left|\begin{array}{ccccccc} 1 & & 0 & & 0 & & \bar{1} \\ & 0 & & 0 & & 0 & \\ & & 0 & & 0 & & \\ & & & 0 & & & \end{array}\right\rangle^{6}_{0}, \quad \left|\begin{array}{ccccccc} 0 & & 0 & & 0 & & 0 \\ & 0 & & 0 & & 0 & \\ & & 0 & & 0 & & \\ & & & 0 & & & \end{array}\right\rangle^{6}_{0}.$$

The detailed calculation is left as exercise (see Prob. 5).

8.3.5 *Some Representations of* G_2

The Dynkin diagram and the simple roots of G_2 are

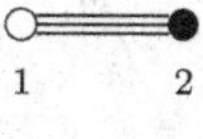

$$\boldsymbol{r}_1 = 2\boldsymbol{w}_1 - 3\boldsymbol{w}_2, \qquad \boldsymbol{r}_2 = -\boldsymbol{w}_1 + 2\boldsymbol{w}_2.$$

Each basis state in the highest weight representation $D^{\boldsymbol{M}}$ of G_2 is described by the Gel'fand's basis of A_1 with a superscript L and a subscript m_2:

$$\left|\begin{array}{ccc} \omega_{12} & & \omega_{22} \\ & \omega_{11} & \end{array}\right\rangle^{L}_{m_2}, \qquad M^{(1)} = \omega_{12} - \omega_{22}. \tag{8.38}$$

The action of F_2 increases L by one, decreases m_2 by two, and decreases $\Omega_2 = \omega_{12} + \omega_{22}$ by one. The action of F_1 increases L by one, increases m_2 by three, and decreases ω_{11} by one. The parameters of the highest weight state in $D^{\boldsymbol{M}}(G_2)$ satisfy $\omega_{22} = m_2 = M_2$ and $\omega_{11} = \omega_{12} = M_1 + \omega_{22}$.

The parameters of the highest weight state of an A_1-multiplet with $M^{(1)}$ in $D^{M}(G_2)$ satisfy $\omega_{11} = \omega_{12} = M^{(1)} + \omega_{22}$.

Check each basis state in the A_1-multiplet, as the level $L(\boldsymbol{M}, \boldsymbol{m})$ increases, whether or not it satisfies the condition (8.7) with $\mu = 2$. If yes, we construct an $\mathcal{A}_2$-multiplet by successive application of F_2 (see Eqs. (8.25-28)). The highest weight state in D^{M} of G_2 is also that of an A_1-multiplet with $M^{(1)} = M_1$ and that of an $\mathcal{A}_2$-multiplet if $M_2 > 0$. Since the rank of G_2 is two, the direct calculation of its representations by the method of the block weight diagram is also simple. In Fig. 8.11 the calculation results of two basic representations of G_2 are listed (see the detail in Prob. 6).

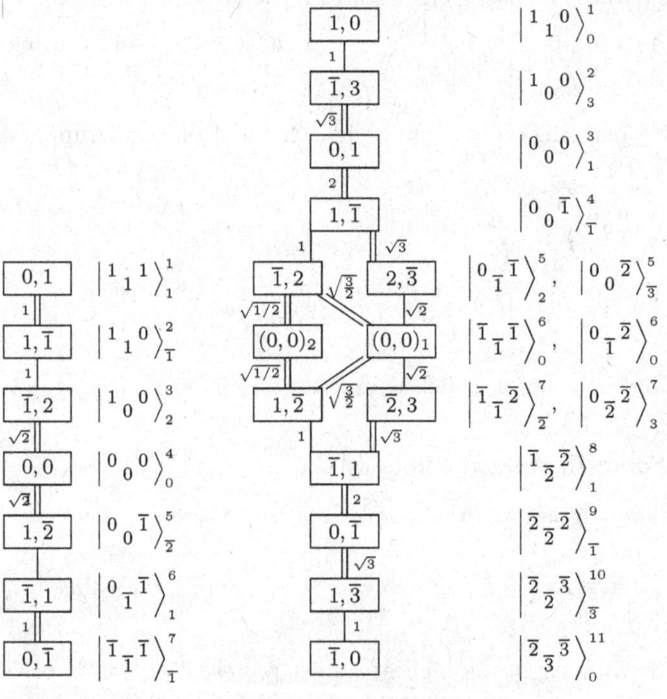
(a) $\boldsymbol{M} = \boldsymbol{w}_2$ (b) $\boldsymbol{M} = \boldsymbol{w}_1$

Fig. 8.11. The block weight diagram and the generalized Gel'fand's bases of the basic representations of G_2.

The basic representation with $\boldsymbol{M} = \boldsymbol{w}_2 = (0, 1)$ is 7-dimensional and self-conjugate. It contains three A_1-singlets and two A_1-doublets. There are two single dominant weights $(0, 1)$ and $(0, 0)$ in this representation. Another basic representation with $\boldsymbol{M} = \boldsymbol{w}_1 = (1, 0)$ is the adjoint representation of

G_2. It is 14-dimensional and self-conjugate. It contains three A_1-singlets, four A_1-doublets, and one A_1-triplet. There are one single dominant weight $(1,0)$ and one double dominant weight $(0,0)$ in this representation.

8.4 Planar Weight Diagrams

For a two-rank Lie algebra, each weight has two components so that the weights and their multiplicities in a representation $D^{\boldsymbol{M}}$ can be drawn evidently in a planar diagram, called the **planar weight diagram**. Since the fundamental dominant weights $\boldsymbol{w}_\mu$ are not orthonormal generally, it is more convenient to express the weights and roots in the unit vectors $\boldsymbol{e}_a$ along the coordinate axes. **The planar weight diagrams of two conjugate representations are related to each other by the inversion with respect to the origin.** The planar weight diagram of a self-conjugate representation is symmetric with respect to the inversion. There are only three rank-two simple Lie algebras: A_2, $B_2 \approx C_2$, and G_2.

The Lie group of A_2 is SU(3). The generators in its self-representation are (see Eqs. (7.73-76)):

$$\begin{aligned}
H_1 &= \sqrt{2}T_3^{(3)} = \sqrt{2}T_8 = 6^{-1/2}\text{diag}\{1,\ 1,\ -2\},\\
H_2 &= \sqrt{2}T_2^{(3)} = \sqrt{2}T_3 = 2^{-1/2}\text{diag}\{1,\ -1,\ 0\},\\
E_{\boldsymbol{r}_1} &= T_{12}^{(1)} + \text{i}T_{12}^{(2)} = T_1 + \text{i}T_2 = \begin{pmatrix} 0 & 1 & 0\\ 0 & 0 & 0\\ 0 & 0 & 0\end{pmatrix},\\
E_{\boldsymbol{r}_2} &= T_{23}^{(1)} + \text{i}T_{23}^{(2)} = T_6 + \text{i}T_7 = \begin{pmatrix} 0 & 0 & 0\\ 0 & 0 & 1\\ 0 & 0 & 0\end{pmatrix}.
\end{aligned} \tag{8.39}$$

According to $[H_\mu,\ E_{\boldsymbol{r}_\nu}] = (\boldsymbol{r}_\nu)_\mu E_{\boldsymbol{r}_\nu}$, we obtain

$$\begin{aligned}
&\boldsymbol{r}_1 = \sqrt{2}\,\boldsymbol{e}_2, \quad \boldsymbol{r}_2 = \sqrt{\frac{3}{2}}\,\boldsymbol{e}_1 - \frac{1}{\sqrt{2}}\,\boldsymbol{e}_2,\\
&\boldsymbol{w}_1 = \frac{1}{3}\,(2\boldsymbol{r}_1 + \boldsymbol{r}_2) = \frac{1}{\sqrt{6}}\,\boldsymbol{e}_1 + \frac{1}{\sqrt{2}}\,\boldsymbol{e}_2,\\
&\boldsymbol{w}_2 = \frac{1}{3}\,(\boldsymbol{r}_1 + 2\boldsymbol{r}_2) = \sqrt{\frac{2}{3}}\,\boldsymbol{e}_1.
\end{aligned} \tag{8.40}$$

In the physical convention, $\boldsymbol{e}_1$ is along the vertical axis and $\boldsymbol{e}_2$ is along the abscissa axis. The planar weight diagram of the representation with $\boldsymbol{M} = M\boldsymbol{w}_2$ is a regular triangle, that with $\boldsymbol{M} = M\boldsymbol{w}_1$ is a regular triangle upside down. The planar weight diagram of the remaining representation is a hexagon where the multiplicities of the weights on the boundary is one and the multiplicity increases one by one as the point of the weight goes from the boundary line to the center until the hexagon becomes a regular triangle or the origin, where the multiplicities of weights are the same. The planar weight diagrams of some representations of A_2 are given in Fig. 8.12.

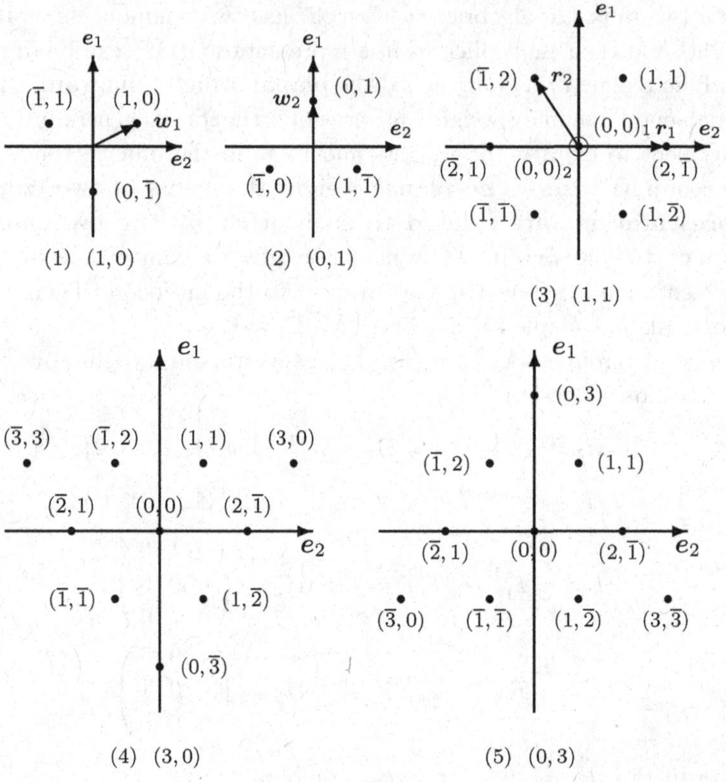

Fig. 8.12. The planar weight diagrams of some representations of SU(3)

The Lie group of B_2 is SO(5). The generators in its self-representation are (see Eqs. (7.82-85)):

$$
\begin{aligned}
&H_1 = T_{12}, && H_2 = T_{34}, \\
&E_{\mathbf{r}_1} = \frac{1}{2}\left\{T_{23} - \mathrm{i}T_{13} - \mathrm{i}T_{24} - T_{14}\right\}, && E_{\mathbf{r}_2} = \sqrt{\frac{1}{2}}\left\{T_{45} - \mathrm{i}T_{35}\right\}.
\end{aligned}
\tag{8.41}
$$

The simple roots and the dominant weights are

$$
\begin{aligned}
&r_1 = e_1 - e_2, && r_2 = e_2, \\
&w_1 = r_1 + r_2 = e_1, && w_2 = \frac{1}{2}r_1 + r_2 = \frac{1}{2}(e_1 + e_2) .
\end{aligned} \tag{8.42}
$$

The planar weight diagrams of the basic representations and the adjoint representation with $M = 2w_2$ are listed in Fig. 8.13. The Lie group of C_2 is USp(4). The C_2 is locally isomorphic onto B_2, but with different simple roots and dominant weights:

$$
\begin{aligned}
&r_1 = \sqrt{1/2}(e_1 - e_2), && r_2 = \sqrt{2}e_2, \\
&w_1 = r_1 + \frac{1}{2}r_2 = \frac{1}{\sqrt{2}}e_1, && w_2 = r_1 + r_2 = \frac{1}{\sqrt{2}}(e_1 + e_2) .
\end{aligned} \tag{8.43}
$$

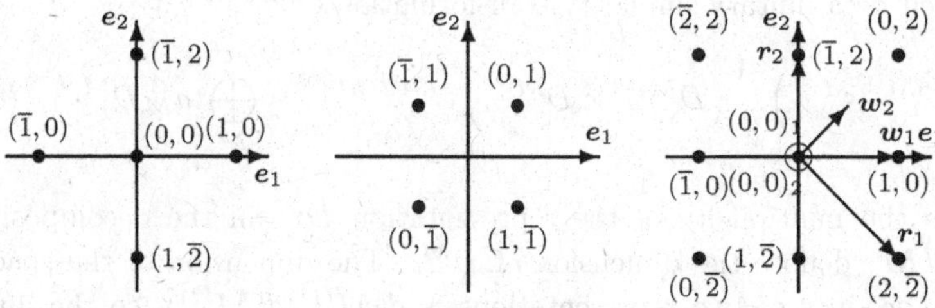

(a) (1, 0) (b) (0, 1) (c) (0, 2)

Fig. 8.13 The planar weight diagrams of some representations of SO(5).

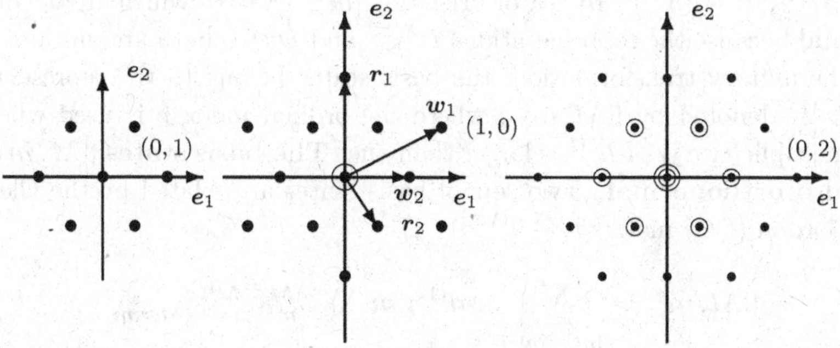

(a) (0,1) (b) (1,0) (c) (0,2)

Fig. 8.14 The planar weight diagrams of some representations of G_2

The simple roots and the dominant weights of G_2 are

$$\begin{aligned} &r_1 = \sqrt{2}\, e_2, & &r_2 = \frac{1}{\sqrt{6}}\, e_1 - \frac{1}{\sqrt{2}}\, e_2, \\ &w_1 = 2r_1 + 3r_2 = \sqrt{\frac{3}{2}}\, e_1 + \frac{1}{\sqrt{2}}\, e_2, & &w_2 = r_1 + 2r_2 = \sqrt{\frac{2}{3}}\, e_1. \end{aligned} \tag{8.44}$$

The planar weight diagrams of some representations of G_2 are listed in Fig. 8.14.

8.5 Clebsch–Gordan Coefficients

The direct product $D^{M^{(1)}} \times D^{M^{(2)}}$ of two irreducible representations $D^{M^{(1)}}$ and $D^{M^{(2)}}$ of a simple Lie algebra $\mathcal{L}$ is generally reducible and can be reduced by a unitary similarity transformation $C^{M^{(1)}M^{(2)}}$,

$$\left(C^{M^{(1)}M^{(2)}}\right)^{-1} \left(D^{M^{(1)}} \times D^{M^{(2)}}\right) C^{M^{(1)}M^{(2)}} = \bigoplus_{M} a_M D^M. \tag{8.45}$$

a_M is the multiplicity of the representation D^M in the decomposition. Let $d(M)$ denote the dimension of D^M. The dimension of the space of direct product of two representations is $d(M^{(1)})d(M^{(2)})$, so the dimension of $C^{M^{(1)}M^{(2)}}$. There are two sets of basis vectors in the product space. Before the unitary transformation, the basis states are the products of two basis states of the representations $D^{M^{(1)}}$ and $D^{M^{(2)}}$, denoted by $|M^{(1)}, m^{(1)}\rangle|M^{(2)}, m^{(2)}\rangle$, or briefly by $|m^{(1)}\rangle|m^{(2)}\rangle$, which are **orthonormal** because the representations $D^{M^{(1)}}$ and $D^{M^{(2)}}$ both are unitary. After the unitary transformation, the basis states belong to the representation D^M, denoted by $||M, m\rangle_r$, where the ordinal index r is used when the multiplicity a_M of D^M is larger than one. **The basis states $||M, m\rangle_r$ are also orthonormal**. Two sets of basis states are related by the Clebsch–Gordan (CG) matrix $C^{M^{(1)}M^{(2)}}$,

$$||M, m\rangle_r = \sum_{m^{(1)}, m^{(2)}} |m^{(1)}\rangle|m^{(2)}\rangle\, C^{M^{(1)}M^{(2)}}_{m^{(1)}, m^{(2)}; M, r, m}\,, \tag{8.46}$$

where the sum runs over all basis states, including those with the multiple weights $m^{(1)}$ and $m^{(2)}$. The series on the right-hand side in Eq. (8.45) is called the **CG series**, and the matrix elements of $C^{M^{(1)}M^{(2)}}$ are called the **CG coefficients**. Applying the generator I_A to two sides of Eq. (8.46), we have

$$I_A \| \boldsymbol{M}, \boldsymbol{m} \rangle_r = \sum_{\boldsymbol{m}^{(1)}, \boldsymbol{m}^{(2)}} \left\{ |\boldsymbol{m}^{(1)}\rangle \left(I_A |\boldsymbol{m}^{(2)}\rangle \right) + \left(I_A |\boldsymbol{m}^{(1)}\rangle \right) |\boldsymbol{m}^{(2)}\rangle \right\} C^{\boldsymbol{M}^{(1)}\boldsymbol{M}^{(2)}}_{\boldsymbol{m}^{(1)},\boldsymbol{m}^{(2)};\boldsymbol{M},r,\boldsymbol{m}} . \tag{8.47}$$

When I_A is taken to be H_μ, we obtain

$$\boldsymbol{m}\, C^{\boldsymbol{M}^{(1)}\boldsymbol{M}^{(2)}}_{\boldsymbol{m}^{(1)},\boldsymbol{m}^{(2)};\boldsymbol{M},r,\boldsymbol{m}} = \left(\boldsymbol{m}^{(1)} + \boldsymbol{m}^{(2)} \right) C^{\boldsymbol{M}^{(1)}\boldsymbol{M}^{(2)}}_{\boldsymbol{m}^{(1)},\boldsymbol{m}^{(2)};\boldsymbol{M},r,\boldsymbol{m}} ,$$

$$C^{\boldsymbol{M}^{(1)}\boldsymbol{M}^{(2)}}_{\boldsymbol{m}^{(1)},\boldsymbol{m}^{(2)};\boldsymbol{M},r,\boldsymbol{m}} = 0, \qquad \text{when } \boldsymbol{m} \neq \boldsymbol{m}^{(1)} + \boldsymbol{m}^{(2)}. \tag{8.48}$$

The weights of the basis states before and after the transformation are the same.

There are two main tasks in reducing the direct product of two representations. One is **to determine the CG series**. The other is **to find the expansions (8.49) for the highest weight states** $\| \boldsymbol{M}, \boldsymbol{M} \rangle_r$

$$\| \boldsymbol{M}, \boldsymbol{M} \rangle_r = \sum_{\boldsymbol{m}^{(1)}} |\boldsymbol{m}^{(1)}\rangle |\boldsymbol{M} - \boldsymbol{m}^{(1)}\rangle \, C^{\boldsymbol{M}^{(1)}\boldsymbol{M}^{(2)}}_{\boldsymbol{m}^{(1)},(\boldsymbol{M}-\boldsymbol{m}^{(1)});\boldsymbol{M},r,\boldsymbol{M}} . \tag{8.49}$$

Applying the lowering operators F_μ to Eq. (8.49), we are able to calculate the expansions (8.49) for the remaining basis states in terms of the block weight diagram. Namely, the Clebsch–Gordan coefficients can be calculated. In this section the method of the dominant weight diagrams will be introduced to calculate the CG series and CG coefficients. **The generalized Gel'fand method plays an important pole to simplify the calculation.**

For simplicity, **we denote the representation $D^{\boldsymbol{M}}$ by its highest weight $\boldsymbol{M}$ directly in this section**, if without confusion.

8.5.1 *Representations in the CG Series*

Let $\boldsymbol{M}_\alpha = \boldsymbol{m}^{(1)} + \boldsymbol{m}^{(2)}$ denote the dominant weight in the product space of $\boldsymbol{M}^{(1)} \times \boldsymbol{M}^{(2)}$. The highest weight $\boldsymbol{M}_0$ in the product space is single and its basis state is

$$\boldsymbol{M}_0 = \boldsymbol{M}^{(1)} + \boldsymbol{M}^{(2)}, \qquad \| \boldsymbol{M}_0, \boldsymbol{M}_0 \rangle = |\boldsymbol{M}^{(1)}\rangle |\boldsymbol{M}^{(2)}\rangle. \tag{8.50}$$

The remaining dominant weights $\boldsymbol{M}_\alpha$ in the product space satisfies

$$\boldsymbol{M}_\alpha = \boldsymbol{M}_0 - \sum_\mu \alpha_\mu \boldsymbol{r}_\mu . \tag{8.51}$$

$\alpha = \sum_\mu \alpha_\mu$ is called the level of $\boldsymbol{M}_\alpha$. If there are a few highest weights with the same level α, they can be denoted by $\boldsymbol{M}_\alpha$, $\boldsymbol{M}_{\alpha'}$, and so on.

Let $OS(\boldsymbol{M}_\alpha)$ denote the size of the Weyl orbit of $\boldsymbol{M}_\alpha$. The dimension of the product space is

$$d(\boldsymbol{M}^{(1)})d(\boldsymbol{M}^{(2)}) = \sum_\alpha OS(\boldsymbol{M}_\alpha)n_\alpha, \tag{8.52}$$

where n_α denotes the multiplicity of $\boldsymbol{M}_\alpha$ in the product space, which can be counted in terms of two block weight diagrams of $\boldsymbol{M}^{(1)}$ and $\boldsymbol{M}^{(2)}$. In fact, $n_\alpha = \sum N_\alpha^{(1,2)}$, where the sum runs over all weights $\boldsymbol{m}^{(1)}$ and $\boldsymbol{m}^{(2)}$ satisfying the condition $\boldsymbol{m}^{(1)} + \boldsymbol{m}^{(2)} = \boldsymbol{M}_\alpha$ and $N_\alpha^{(1,2)}$ is the product of the multiplicity of $\boldsymbol{m}^{(1)}$ in the representation $\boldsymbol{M}^{(1)}$ and the multiplicity of $\boldsymbol{m}^{(2)}$ in $\boldsymbol{M}^{(2)}$.

What kind of representations $\boldsymbol{M}$ will appear in the CG series? **$\boldsymbol{M}$ is certainly one of the dominant weights $\boldsymbol{M}_\alpha$ in the product space, but $\boldsymbol{M}_\alpha$ is not necessary to appear in the CG series.** If $\boldsymbol{M}$ appears in the CG series, its highest weight state $||\boldsymbol{M}, \boldsymbol{M}\rangle_r$, given in Eq. (8.49), is annihilated by each E_μ. If the $|\boldsymbol{m}^{(1)}\rangle$ in one term of $||\boldsymbol{M}, \boldsymbol{M}\rangle_r$ is not equal to $\boldsymbol{M}^{(1)}$, there must exist a raising operator E_μ such that

$$E_\mu \; |\boldsymbol{m}^{(1)}\rangle = a \; |\boldsymbol{m}^{(1)} + \boldsymbol{r}_\mu\rangle + \cdots \; , \qquad a \neq 0.$$

The suspension points denote the possible states due to the multiple weight. Since E_μ annihilates the highest weight state $||\boldsymbol{M}, \boldsymbol{M}\rangle_r$ (see Eq. (7.111)), there must be another term $|\boldsymbol{m}^{(1)} + \boldsymbol{r}_\mu\rangle|\boldsymbol{M} - \boldsymbol{m}^{(1)} - \boldsymbol{r}_\mu\rangle$ in $||\boldsymbol{M}, \boldsymbol{M}\rangle_r$ such that $E_\mu|\boldsymbol{M} - \boldsymbol{m}^{(1)} - \boldsymbol{r}_\mu\rangle = b|\boldsymbol{M} - \boldsymbol{m}^{(1)}\rangle + \cdots$, and

$$aC^{\boldsymbol{M}^{(1)}\boldsymbol{M}^{(2)}}_{\boldsymbol{m}^{(1)},(\boldsymbol{M}-\boldsymbol{m}^{(1)});\boldsymbol{M},r,\boldsymbol{M}} + bC^{\boldsymbol{M}^{(1)}\boldsymbol{M}^{(2)}}_{(\boldsymbol{m}^{(1)}+\boldsymbol{r}_\mu),(\boldsymbol{M}-\boldsymbol{m}^{(1)}-\boldsymbol{r}_\mu);\boldsymbol{M},r,\boldsymbol{M}} = 0.$$

Thus, if $||\boldsymbol{M}, \boldsymbol{M}\rangle_r$ contains a term with $\boldsymbol{m}^{(1)} \neq \boldsymbol{M}^{(1)}$, it has to contain another term with a higher weight, say $\boldsymbol{m}^{(1)} + \boldsymbol{r}_\mu$. Furthermore, the highest weight state $||\boldsymbol{M}, \boldsymbol{M}\rangle_r$ has to contain a term $|\boldsymbol{M}^{(1)}\rangle|\boldsymbol{M} - \boldsymbol{M}^{(1)}\rangle$ and, for the same reason, to contain another term $|\boldsymbol{M} - \boldsymbol{M}^{(2)}\rangle|\boldsymbol{M}^{(2)}\rangle$. Namely, we obtain **another necessary condition for a highest weight representation $\boldsymbol{M}$ appearing in the CG series (8.45)** ($a_{\boldsymbol{M}} > 0$):

$$\begin{aligned} &\boldsymbol{M} - \boldsymbol{M}^{(1)} \;\text{ is a weight } \boldsymbol{m}^{(2)} \text{ in } \;\boldsymbol{M}^{(2)}, \\ &\boldsymbol{M} - \boldsymbol{M}^{(2)} \;\text{ is a weight } \boldsymbol{m}^{(1)} \text{ in } \;\boldsymbol{M}^{(1)}. \end{aligned} \tag{8.53}$$

Those $\boldsymbol{M}$ satisfying the condition (8.53) can be calculated in terms of the block weight diagrams of $\boldsymbol{M}^{(1)}$ and $\boldsymbol{M}^{(2)}$, although the coefficient $a_{\boldsymbol{M}}$ in the CG series is to be determined by the method of the dominant weight diagram given in the next subsection.

Theorem 8.1 The Clebsch–Gordan series (8.45) for the direct product $\boldsymbol{M}^{(1)} \times \boldsymbol{M}^{(2)}$ of two highest weight representations of a simple Lie algebra $\mathcal{L}$, whose dimension is more than one, contains the identical representation $\boldsymbol{M} = 0$ if and only if $\boldsymbol{M}^{(1)}$ is equivalent to the conjugate representation of $\boldsymbol{M}^{(2)}$. In this case the Clebsch–Gordan series contains one and only one identical representation.

This theorem is just the application of Theorem 3.8 to the simple Lie algebra $\mathcal{L}$. Due to Eq. (8.11), the basis state $||0,0\rangle$ of the identical representation is

$$||0,0\rangle = d(\boldsymbol{M}^{(1)})^{-1} \sum_{\boldsymbol{m}^{(1)}} (-1)^{L(\boldsymbol{M}^{(1)},\boldsymbol{m}^{(1)})-1} \, |\boldsymbol{m}^{(1)}\rangle| - \boldsymbol{m}^{(1)}\rangle, \qquad (8.54)$$

where the sum runs over all states $\boldsymbol{m}^{(1)}$ in the representation $\boldsymbol{M}^{(1)}$ with the level $L(\boldsymbol{M}^{(1)}, \boldsymbol{m}^{(1)})$.

8.5.2 *Method of Dominant Weight Diagram*

We are going to calculate the multiplicity a_α of $\boldsymbol{M}_\alpha$ in the CG series defined in Eq. (8.45) and the expansion of the highest weight state $||\boldsymbol{M}_\alpha, \boldsymbol{M}_\alpha\rangle_r$ of the representation $\boldsymbol{M}_\alpha$ with respect to $|\boldsymbol{m}^{(1)}\rangle|\boldsymbol{m}^{(2)}\rangle$ defined in Eq. (8.49).

Let $n_\alpha^{(1)}$, $n_\alpha^{(2)}$ and $n_{\beta\alpha}$ denote the multiplicities of the dominant weight $\boldsymbol{M}_\alpha$ in the representations $\boldsymbol{M}^{(1)}$, $\boldsymbol{M}^{(2)}$ and $\boldsymbol{M}_\beta$, respectively. From the definition we have $n_{\beta\alpha} = 0$ if $\beta > \alpha$ and $n_{\alpha\alpha} = 1$. If there are two representations $\boldsymbol{M}_\alpha$ and $\boldsymbol{M}_{\alpha'}$ with the same level in the CG series, $n_{\alpha'\alpha} = n_{\alpha\alpha'} = 0$. Thus,

$$\begin{aligned} a_\alpha &= n_\alpha - \sum_{\beta=0}^{\alpha-1} a_\beta n_{\beta\alpha}, \qquad \alpha > 0, \qquad a_0 = 1 \\ d(\boldsymbol{M}^{(1)})d(\boldsymbol{M}^{(2)}) &= \sum_\alpha a_\alpha d(\boldsymbol{M}_\alpha). \end{aligned} \qquad (8.55)$$

The multiplicity a_α of the representation $\boldsymbol{M}_\alpha$ in the CG series (8.45) can be calculated successively as α increases from 1. Remind that $a_\alpha = 0$ if $\boldsymbol{M}_\alpha$ does not satisfy the condition (8.53).

The dominant weight diagram consists of a few columns. In the leftmost column the products $OS(\boldsymbol{M}_\beta) \times n_\beta$ are listed and their sum is the dimension $d(\boldsymbol{M}^{(1)})d(\boldsymbol{M}^{(2)})$ of the product space. In the second column the boxes with $\boldsymbol{M}_\beta$ are demonstrated one by one at each row. On the right of the diagram, each representation $\boldsymbol{M}_\alpha$ with $a_\alpha > 0$ corresponds to one column and is arranged column by column in the increasing order of α. The column for the representation $\boldsymbol{M}_\alpha$ is demonstrated by a box with $\boldsymbol{M}_\alpha$ on the top, and its multiplicity a_α in the CG series, if $a_\alpha > 1$, is denoted above the box. The multiplicities $n_{\alpha\beta}$ of $\boldsymbol{M}_\beta$ in $\boldsymbol{M}_\alpha$ is listed at the row for $\boldsymbol{M}_\beta$ in this column. In the last row of the table, the dimension of the product space is calculated by the sum of $a_\alpha d(\boldsymbol{M}_\alpha)$, and the Clebsch–Gordan series is listed.

The general method for calculating the expansion of the highest weight state $||\boldsymbol{M}_\alpha, \boldsymbol{M}_\alpha\rangle_r$ of $\boldsymbol{M}_\alpha$ in the decomposition of $\boldsymbol{M}^{(1)} \times \boldsymbol{M}^{(2)}$ is based on the property (7.111) that every raising operator E_μ annihilates the highest weight state. As the recursive way in the generalized Gel'fand's method, **the highest weight state $||\boldsymbol{M}_\alpha, \boldsymbol{M}_\alpha\rangle_r$ is first expressed as a linear combination of the normalized highest weight states of the subalgebra** $\mathrm{A}_{\ell-1}$, and then, **the coefficients of the combinations are determined by the annihilation condition of F_ℓ.**

Now, we summarize the main steps for calculating the CG series (8.45) of the direct product $\boldsymbol{M}^{(1)} \times \boldsymbol{M}^{(2)}$ in a simple Lie algebra $\mathcal{L} \supset \mathrm{A}_{\ell-1} \oplus \mathcal{A}_\ell$ $(\ell > 1)$, including $\mathcal{L} = \mathrm{A}_\ell$, and the expansions (8.49) of $\boldsymbol{M}_\alpha$.

a) Find the document weights $\boldsymbol{M}_\alpha$ and their multiplicities n_α in the product space in terms of the block weight diagrams of $\boldsymbol{M}^{(1)}$ and $\boldsymbol{M}^{(2)}$.

b) Draw the block weight diagram of $\boldsymbol{M}_0$ at least partly. List the multiplicities $n_{0\beta}$ and the generalized Gel'fand basis states $|\omega_{ab}\rangle^L_{m_\ell}$ with the weights $\boldsymbol{M}_\beta$ in the representation $\boldsymbol{M}_0$, where $\Omega_b = \sum_a \omega_{ab}$ are calculated.

c) In the same way, calculate the multiplicities $n_{\beta\alpha}$ of the weights $\boldsymbol{M}_\alpha$ in the representation $\boldsymbol{M}_\beta$. Then, calculate a_α successively by Eq. (8.55) and draw the dominant weight diagram for $\boldsymbol{M}^{(1)} \times \boldsymbol{M}^{(2)}$ in $\mathcal{L}$.

d) We expand the highest weight state (8.49) of $\boldsymbol{M}_\alpha$ as a linear combination of the normalized highest weight states of $\mathrm{A}_{\ell-1}$-multiplets, and then, determine the coefficients by the annihilation condition of E_ℓ. Let $|\omega_{ab}\rangle_{m_\ell}|\omega'_{ab}\rangle_{m'_\ell}$ denote the basis state of arbitrary term in the expansion (8.49). In all terms of one highest weight state of $\mathrm{A}_{\ell-1}$-multiplet, $\omega_{a\ell}$ and $\omega'_{a\ell}$ are the same, respectively. In addition, $\sum_{a=1}^{b}(\omega_{ab} + \omega'_{ab}) = \Omega_b$ should satisfy the calculated values (see **b)**). The expansion (8.49) for $\boldsymbol{M}_0$ is given in Eq. (8.50). The expansion (8.49) for $\boldsymbol{M}_\alpha = 0$ is given in Eq. (8.54), if $\boldsymbol{M}^{(1)}$ and $\left(\boldsymbol{M}^{(2)}\right)^*$ are equivalent to each other.

e) When $\boldsymbol{M}^{(1)} = \boldsymbol{M}^{(2)}$, the expansion (8.49) of the basis state in each representation $\boldsymbol{M}_\alpha$ can be combined to be symmetric or antisymmetric with respect to the interchange between $|\boldsymbol{m}^{(1)}\rangle$ and $|\boldsymbol{m}^{(2)}\rangle$, called the **symmet-**

ric or antisymmetric representations. The sum of dimensions of the symmetric representations is $d(\boldsymbol{M}^{(1)})\left[d(\boldsymbol{M}^{(1)})+1\right]/2$ and that of the antisymmetric ones is $d(\boldsymbol{M}^{(1)})\left[d(\boldsymbol{M}^{(1)})-1\right]/2$.

The method will be explained in detail by some examples in the subsequent subsections.

8.5.3 *The CG series for* A_ℓ

The block weight diagrams of some representations of A_ℓ are given in subsections 8.2.2 and 8.2.3. The planar weight diagrams of some representations of A_2 are given in Fig. 8.12. The sizes of Weyl orbit are $OS(0,0) = 1$, $OS(M_1,0) = OS(0,M_2) = 3$ and $OS(M_1,M_2) = 6$ for A_2, and $OS(0,0,0) = 1$, $OS(M_1,0,0) = OS(0,0,M_3) = 4$, $OS(0,M_2,0) = 6$, $OS(M_1,M_2,0) = OS(M_2,0,M_3) = OS(0,M_2,M_3) = 12$ and $OS(M_1,M_2,M_3) = 24$ for A_3, where M_1, M_2 and M_3 are non-vanishing.

Ex. 1. The decomposition of $(1,0)\times(0,1)$ in A_2.

From Fig. 8.2, there are two dominant weights in the product space: $\boldsymbol{M}_0 = (1,1)$ with a state $|(1,0)\rangle|(0,1)\rangle$ and $\boldsymbol{M}_1 = (0,0)$ with three states $|(1,0)\rangle|(\overline{1},0)\rangle$, $|(\overline{1},1)\rangle|(1,\overline{1})\rangle$ and $|(0,\overline{1})\rangle|(0,1)\rangle$. Since the representation $(1,1)$ contains a single dominant weight $(1,1)$ and a double dominant weight $(0,0)$ (see Fig. 8.5), from Eq. (8.55) we obtain $a_1 = n_1 - n_{01} = 3 - 2 = 1$. The dominant weight diagram for $(1,0)\times(0,1)$ in A_2 is given in Fig. 8.15.

$OS(\boldsymbol{M}_\alpha)\times n_\alpha$		$\boxed{1,1}$	$\boxed{0,0}$
6×1	$\boxed{1,1}$	1	
1×3	$\boxed{0,0}$	2	1
$3\times 3=9$	$=$	$8\ +$	1

$$(1,0)\times(0,1) \simeq (1,1) + (0,0)$$

Fig. 8.15 The dominant weight diagram for $(1,0)\times(0,1)$ in A_2.

The expansions for the highest weight states of $\boldsymbol{M}_0$ and $\boldsymbol{M}_1$ are given in Eqs. (8.50) and (8.54):

$$\|(1,1),(1,1)\rangle = |(1,0)\rangle|(0,1)\rangle = \left|\begin{smallmatrix}1 & & 0 & & 0\\ & 1 & & 0 & \\ & & 1 & & \end{smallmatrix}\right\rangle^1 \left|\begin{smallmatrix}1 & & 1 & & 0\\ & 1 & & 1 & \\ & & 1 & & \end{smallmatrix}\right\rangle^1,$$

$$\begin{aligned}\|(0,0),(0,0)\rangle &= \sqrt{1/3}\left[\,|(1,0)\rangle|(\overline{1},0)\rangle - |(\overline{1},1)\rangle|(1,\overline{1})\rangle + |(0,\overline{1})\rangle|(0,1)\rangle\right] \\ &= \sqrt{\frac{1}{3}}\left\{\left|\begin{array}{ccccc}1&&0&&0\\&1&&0&\\&&1&&\end{array}\right\rangle^{1}\left|\begin{array}{ccccc}1&&1&&0\\&1&&0&\\&&0&&\end{array}\right\rangle^{3}-\left|\begin{array}{ccccc}1&&0&&0\\&1&&0&\\&&0&&\end{array}\right\rangle^{2}\left|\begin{array}{ccccc}1&&1&&0\\&1&&0&\\&&1&&\end{array}\right\rangle^{2}\right. \\ &\qquad\left.+\left|\begin{array}{ccccc}1&&0&&0\\&0&&0&\\&&0&&\end{array}\right\rangle^{3}\left|\begin{array}{ccccc}1&&1&&0\\&1&&1&\\&&1&&\end{array}\right\rangle^{1}\right\}.\end{aligned}$$

The expansions of the remaining basis states in the representation $(1,1)$ can be calculated by the action F_μ (see Fig. 8.5 and Eq. (8.47)):

$$\begin{aligned}
&\|(1,1),(\overline{1},2)\rangle = F_1\|(1,1),(1,1)\rangle = |(\overline{1},1)\rangle|(0,1)\rangle,\\
&\|(1,1),(2,\overline{1})\rangle = F_2\|(1,1),(1,1)\rangle = |(1,0)\rangle|(1,\overline{1})\rangle,\\
&\|(1,1),(0,0)_1\rangle = \sqrt{1/2}F_1\|(1,1),(2,\overline{1})\rangle\\
&\quad = \sqrt{1/2}\left\{|(1,0)\rangle|(\overline{1},0)\rangle + |(\overline{1},1)\rangle|(1,\overline{1})\rangle\right\},\\
&\|(1,1),(0,0)_2\rangle = \sqrt{2/3}\left\{F_2\|(1,1),(\overline{1},2)\rangle - \sqrt{1/2}\|(1,1),(0,0)_1\rangle\right\}\\
&\quad = \sqrt{1/6}\left\{-|(1,0)\rangle|(\overline{1},0)\rangle + |(\overline{1},1)\rangle|(1,\overline{1})\rangle + 2\,|(0,\overline{1})\rangle|(0,1)\rangle\right\}\\
&\|(1,1),(\overline{2},1)\rangle = \sqrt{1/2}F_1\|(1,1),(0,0)_1\rangle = |(\overline{1},1)\rangle|(\overline{1},0)\rangle,\\
&\|(1,1),(1,\overline{2})\rangle = \sqrt{2}F_2\|(1,1),(0,0)_1\rangle = |(0,\overline{1})\rangle|(1,\overline{1})\rangle,\\
&\|(1,1),(\overline{1},\overline{1})\rangle = F_2\|(1,1),(\overline{2},1)\rangle = |(0,\overline{1})\rangle|(\overline{1},0)\rangle.
\end{aligned}$$

Ex. 2. The decomposition of $(1,1)\times(1,1)$ in A_2.

From Fig. 8.5 there are five dominant weights, (2,2), (3,0), (0,3), (1,1), and (0,0) in the product space. Neglecting the states by interchanging $|\boldsymbol{m}^{(1)}\rangle$ and $|\boldsymbol{m}^{(2)}\rangle$, we list the basis states $|\boldsymbol{m}^{(1)}\rangle|\boldsymbol{m}^{(2)}\rangle$ with the dominant weights $\boldsymbol{M}_\alpha$ in the direct space as follows.

α	$\boldsymbol{M}_\alpha$	n_α	Independent basis states $\lvert\boldsymbol{m}^{(1)}\rangle\lvert\boldsymbol{m}^{(2)}\rangle$
0	$(2,2)$	1	$\lvert(1,1)\rangle\lvert(1,1)\rangle$,
1	$(3,0)$	2	$\lvert(1,1)\rangle\lvert(2,\overline{1})\rangle$,
1′	$(0,3)$	2	$\lvert(1,1)\rangle\lvert(\overline{1},2)\rangle$,
2	$(1,1)$	6	$\lvert(1,1)\rangle\lvert(0,0)_1\rangle$, $\lvert(1,1)\rangle\lvert(0,0)_2\rangle$, $\lvert(\overline{1},2)\rangle\lvert(2,\overline{1})\rangle$,
3	$(0,0)$	10	$\lvert(1,1)\rangle\lvert(\overline{1},\overline{1})\rangle$, $\lvert(2,\overline{1})\rangle\lvert(\overline{2},1)\rangle$, $\lvert(\overline{1},2)\rangle\lvert(1,\overline{2})\rangle$, $\lvert(0,0)_a\rangle\lvert(0,0)_b\rangle$, $a,b=1,2$.

The multiplicity $n_{0\alpha}$ of the dominant weight $\boldsymbol{M}_\alpha$ in the representation $\boldsymbol{M}_0=(2,2)$ is calculated as follows.

$$M_0 = (2,2)$$

α	M_α	$\Omega_3,\ \Omega_2,\ \Omega_1$	$n_{0\alpha}$	the basis states
0	$(2,2)$	$6,6,4$	1	$\left\vert\begin{smallmatrix}4 & & 2 & & 0\\ & 4 & & 2 & \\ & & 4 & & \end{smallmatrix}\right\rangle^1$
1	$(3,0)$	$6,5,4$	1	$\left\vert\begin{smallmatrix}4 & & 2 & & 0\\ & 4 & & 1 & \\ & & 4 & & \end{smallmatrix}\right\rangle^2$
$1'$	$(0,3)$	$6,6,3$	1	$\left\vert\begin{smallmatrix}4 & & 2 & & 0\\ & 4 & & 2 & \\ & & 3 & & \end{smallmatrix}\right\rangle^2$
2	$(1,1)$	$6,5,3$	2	$\left\vert\begin{smallmatrix}4 & & 2 & & 0\\ & 4 & & 1 & \\ & & 3 & & \end{smallmatrix}\right\rangle^3$, $\left\vert\begin{smallmatrix}4 & & 2 & & 0\\ & 3 & & 2 & \\ & & 3 & & \end{smallmatrix}\right\rangle^3$
3	$(0,0)$	$6,4,2$	3	$\left\vert\begin{smallmatrix}4 & & 2 & & 0\\ & 4 & & 0 & \\ & & 2 & & \end{smallmatrix}\right\rangle^5$, $\left\vert\begin{smallmatrix}4 & & 2 & & 0\\ & 3 & & 1 & \\ & & 2 & & \end{smallmatrix}\right\rangle^5$, $\left\vert\begin{smallmatrix}4 & & 2 & & 0\\ & 2 & & 2 & \\ & & 2 & & \end{smallmatrix}\right\rangle^5$

Due to Figs. 8.4 and 8.5, we obtain $n_{12} = n_{13} = n_{1'2} = n_{1'3} = 1$ and $n_{23} = 2$. Thus, the multiplicity a_M in the CG series (8.45) can be calculated:

$$\begin{aligned}
&a_1 = n_1 - n_{01} = 2 - 1 = 1, \quad a_{1'} = n_{1'} - n_{01'} = 2 - 1 = 1,\\
&a_2 = n_2 - n_{02} - n_{12} - n_{1'2} = 6 - 2 - 1 - 1 = 2,\\
&a_3 = n_3 - n_{03} - n_{13} - n_{1'3} - a_2 \times n_{23} = 10 - 3 - 1 - 1 - 2 \times 2 = 1.
\end{aligned}$$

The dominant weight diagram for $(1,1)\times(1,1)$ in A_2 is given in Fig. 8.16.

O.S. × multiplicity		[2,2]	[3,0]	[0,3]	2 [1,1]	[0,0]
3×1	[2,2]	1				
6×2	[3,0]	1	1			
6×2	[0,3]	1	0	1		
3×6	[1,1]	2	1	1	1	
3×10	[0,0]	3	1	1	2	1

$$8\times 8 = 64 = 27 \quad +10 \quad +10 + 2\times 8 \quad +1$$

$$(1,1)\times(1,1) \simeq (2,2)\oplus(3,0)\oplus(0,3)\oplus 2\times(1,1)\oplus(0,0)$$

Fig. 8.16 The dominant weight diagram for $(1,1)\times(1,1)$ in A_2.

The expansions (8.49) for M_0 and M_3 are given in Eqs. (8.50) and (8.54), and those for M_1 and $M_{1'}$ are easy to calculated by the property (7.111).

$$\|(2,2),(2,2)\rangle_S = |(1,1)\rangle|(1,1)\rangle = \left\vert\begin{smallmatrix}2 & & 1 & & 0\\ & 2 & & 1 & \\ & & 2 & & \end{smallmatrix}\right\rangle^1 \left\vert\begin{smallmatrix}2 & & 1 & & 0\\ & 2 & & 1 & \\ & & 2 & & \end{smallmatrix}\right\rangle^1,$$

$$
\begin{aligned}
||(3,0),(3,0)\rangle_A &= \sqrt{\frac{1}{2}}\,\left[\,|(1,1)\rangle|(2,\overline{1})\rangle - |(2,\overline{1})\rangle|(1,1)\rangle\,\right] \\
&= \sqrt{\frac{1}{2}}\left[\left|\begin{smallmatrix}2&&1&&0\\&2&&1&\\&&2&&\end{smallmatrix}\right\rangle^1 \left|\begin{smallmatrix}2&&1&&0\\&2&&0&\\&&2&&\end{smallmatrix}\right\rangle^2 - \left|\begin{smallmatrix}2&&1&&0\\&2&&0&\\&&2&&\end{smallmatrix}\right\rangle^2 \left|\begin{smallmatrix}2&&1&&0\\&2&&1&\\&&2&&\end{smallmatrix}\right\rangle^1\right], \\
||(0,3),(0,3)\rangle_A &= \sqrt{\frac{1}{2}}\,\left[\,|(1,1)\rangle|(\overline{1},2)\rangle - |(\overline{1},2)\rangle|(1,1)\rangle\,\right] \\
&= \sqrt{\frac{1}{2}}\left[\left|\begin{smallmatrix}2&&1&&0\\&2&&1&\\&&2&&\end{smallmatrix}\right\rangle^1 \left|\begin{smallmatrix}2&&1&&0\\&2&&1&\\&&1&&\end{smallmatrix}\right\rangle^2 - \left|\begin{smallmatrix}2&&1&&0\\&2&&1&\\&&1&&\end{smallmatrix}\right\rangle^2 \left|\begin{smallmatrix}2&&1&&0\\&2&&1&\\&&2&&\end{smallmatrix}\right\rangle^1\right], \\
||(0,0),(0,0)\rangle_S & \\
&= \sqrt{1/8}\,\{\,|(1,1)\rangle|(\overline{1},\overline{1})\rangle - |(2,\overline{1})\rangle|(\overline{2},1)\rangle - |(\overline{1},2)\rangle|(1,\overline{2})\rangle \\
&\quad + |(0,0)_1\rangle|(0,0)_1\rangle + |(0,0)_2\rangle|(0,0)_2\rangle - |(1,\overline{2})\rangle|(\overline{1},2)\rangle \\
&\quad - |(\overline{2},1)\rangle|(2,\overline{1})\rangle + |(\overline{1},\overline{1})\rangle|(1,1)\rangle\} \\
&= \sqrt{\frac{1}{8}}\left\{\left|\begin{smallmatrix}2&&1&&0\\&2&&1&\\&&2&&\end{smallmatrix}\right\rangle^1 \left|\begin{smallmatrix}2&&1&&0\\&1&&0&\\&&0&&\end{smallmatrix}\right\rangle^5 - \left|\begin{smallmatrix}2&&1&&0\\&2&&0&\\&&2&&\end{smallmatrix}\right\rangle^2 \left|\begin{smallmatrix}2&&1&&0\\&2&&0&\\&&0&&\end{smallmatrix}\right\rangle^4\right. \\
&\quad - \left|\begin{smallmatrix}2&&1&&0\\&2&&1&\\&&1&&\end{smallmatrix}\right\rangle^2 \left|\begin{smallmatrix}2&&1&&0\\&1&&0&\\&&1&&\end{smallmatrix}\right\rangle^4 + \left|\begin{smallmatrix}2&&1&&0\\&2&&0&\\&&1&&\end{smallmatrix}\right\rangle^3 \left|\begin{smallmatrix}2&&1&&0\\&2&&0&\\&&1&&\end{smallmatrix}\right\rangle^3 \\
&\quad + \left|\begin{smallmatrix}2&&1&&0\\&1&&1&\\&&1&&\end{smallmatrix}\right\rangle^3 \left|\begin{smallmatrix}2&&1&&0\\&1&&1&\\&&1&&\end{smallmatrix}\right\rangle^3 - \left|\begin{smallmatrix}2&&1&&0\\&1&&0&\\&&1&&\end{smallmatrix}\right\rangle^4 \left|\begin{smallmatrix}2&&1&&0\\&2&&1&\\&&1&&\end{smallmatrix}\right\rangle^2 \\
&\quad \left. - \left|\begin{smallmatrix}2&&1&&0\\&2&&0&\\&&0&&\end{smallmatrix}\right\rangle^4 \left|\begin{smallmatrix}2&&1&&0\\&2&&0&\\&&2&&\end{smallmatrix}\right\rangle^2 + \left|\begin{smallmatrix}2&&1&&0\\&1&&0&\\&&0&&\end{smallmatrix}\right\rangle^5 \left|\begin{smallmatrix}2&&1&&0\\&2&&1&\\&&2&&\end{smallmatrix}\right\rangle^1\right\}.
\end{aligned}
$$

We express the highest weight state (8.49) for $M_2 = (1,1)$ as a linear combination of the highest weight states of A_1-triplets.

$$
\begin{aligned}
||(1,1),(1,1)\rangle &= \frac{a_1}{\sqrt{3}}\left[\left|\begin{smallmatrix}2&&1&&0\\&2&&1&\\&&2&&\end{smallmatrix}\right\rangle^1 \left|\begin{smallmatrix}2&&1&&0\\&2&&0&\\&&1&&\end{smallmatrix}\right\rangle^3 - \sqrt{2}\left|\begin{smallmatrix}2&&1&&0\\&2&&1&\\&&1&&\end{smallmatrix}\right\rangle^2 \left|\begin{smallmatrix}2&&1&&0\\&2&&0&\\&&2&&\end{smallmatrix}\right\rangle^2\right] \\
&\quad + a_2\left|\begin{smallmatrix}2&&1&&0\\&2&&1&\\&&2&&\end{smallmatrix}\right\rangle^1 \left|\begin{smallmatrix}2&&1&&0\\&1&&1&\\&&1&&\end{smallmatrix}\right\rangle^3 + b_2\left|\begin{smallmatrix}2&&1&&0\\&1&&1&\\&&1&&\end{smallmatrix}\right\rangle^3 \left|\begin{smallmatrix}2&&1&&0\\&2&&1&\\&&2&&\end{smallmatrix}\right\rangle^1 \\
&\quad + \frac{b_1}{\sqrt{3}}\left[\left|\begin{smallmatrix}2&&1&&0\\&2&&0&\\&&1&&\end{smallmatrix}\right\rangle^3 \left|\begin{smallmatrix}2&&1&&0\\&2&&1&\\&&2&&\end{smallmatrix}\right\rangle^1 - \sqrt{2}\left|\begin{smallmatrix}2&&1&&0\\&2&&0&\\&&2&&\end{smallmatrix}\right\rangle^2 \left|\begin{smallmatrix}2&&1&&0\\&2&&1&\\&&1&&\end{smallmatrix}\right\rangle^2\right].
\end{aligned}
$$

Then, from the annihilation condition of E_2 we obtain

$$
\begin{aligned}
E_2\,||(1,1),(1,1)\rangle &= \frac{a_1}{\sqrt{3}}\left[\sqrt{\frac{1}{2}}\left|\begin{smallmatrix}2&&1&&0\\&2&&1&\\&&2&&\end{smallmatrix}\right\rangle^1 \left|\begin{smallmatrix}2&&1&&0\\&2&&1&\\&&1&&\end{smallmatrix}\right\rangle^2 - \sqrt{2}\left|\begin{smallmatrix}2&&1&&0\\&2&&1&\\&&1&&\end{smallmatrix}\right\rangle^2 \left|\begin{smallmatrix}2&&1&&0\\&2&&1&\\&&2&&\end{smallmatrix}\right\rangle^1\right] \\
&\quad + a_2\sqrt{\tfrac{3}{2}}\left|\begin{smallmatrix}2&&1&&0\\&2&&1&\\&&2&&\end{smallmatrix}\right\rangle^1 \left|\begin{smallmatrix}2&&1&&0\\&2&&1&\\&&1&&\end{smallmatrix}\right\rangle^2 + b_2\sqrt{\tfrac{3}{2}}\left|\begin{smallmatrix}2&&1&&0\\&2&&1&\\&&1&&\end{smallmatrix}\right\rangle^2 \left|\begin{smallmatrix}2&&1&&0\\&2&&1&\\&&2&&\end{smallmatrix}\right\rangle^1
\end{aligned}
$$

$$+\ \frac{b_1}{\sqrt{3}}\left[\sqrt{\frac{1}{2}}\left|\begin{smallmatrix}2&&1&&0\\&2&&1&\\&&1&&\end{smallmatrix}\right\rangle^{2}\left|\begin{smallmatrix}2&&1&&0\\&2&&1&\\&&2&&\end{smallmatrix}\right\rangle^{1}-\sqrt{2}\left|\begin{smallmatrix}2&&1&&0\\&2&&1&\\&&2&&\end{smallmatrix}\right\rangle^{1}\left|\begin{smallmatrix}2&&1&&0\\&2&&1&\\&&1&&\end{smallmatrix}\right\rangle^{2}\right]=0.$$

Thus, $a_1+3a_2=2b_1$ and $b_1+3b_2=2a_1$. For the symmetric representation $(1,1)_S$, we have $a_1=b_1=3a_2=3b_2$. For the antisymmetric representation $(1,1)_A$, we have $a_1'=-b_1'=-a_2'=b_1'$. After normalization, $(a_2)^{-2}=2(9+1)=20$, $(a_1')^{-2}=4$, and

$$\begin{aligned}
&||(1,1),(1,1)\rangle_S\\
&\quad=\sqrt{1/20}\,\{\sqrt{3}\,|(1,1)\rangle|(0,0)_1\rangle+\sqrt{3}\,|(0,0)_1\rangle|(1,1)\rangle+|(1,1)\rangle|(0,0)_2\rangle\\
&\qquad+|(0,0)_2\rangle|(1,1)\rangle-\sqrt{6}\,|(\overline{1},2)\rangle|(2,\overline{1})\rangle-\sqrt{6}\,|(2,\overline{1})\rangle|(\overline{1},2)\rangle\}\\
&\quad=\sqrt{\frac{3}{20}}\left[\left|\begin{smallmatrix}2&&1&&0\\&2&&1&\\&&2&&\end{smallmatrix}\right\rangle^{1}\left|\begin{smallmatrix}2&&1&&0\\&2&&0&\\&&1&&\end{smallmatrix}\right\rangle^{3}+\left|\begin{smallmatrix}2&&1&&0\\&2&&0&\\&&1&&\end{smallmatrix}\right\rangle^{3}\left|\begin{smallmatrix}2&&1&&0\\&2&&1&\\&&2&&\end{smallmatrix}\right\rangle^{1}\right]\\
&\qquad+\sqrt{\frac{1}{20}}\left[\left|\begin{smallmatrix}2&&1&&0\\&2&&1&\\&&2&&\end{smallmatrix}\right\rangle^{1}\left|\begin{smallmatrix}2&&1&&0\\&1&&1&\\&&1&&\end{smallmatrix}\right\rangle^{3}+\left|\begin{smallmatrix}2&&1&&0\\&1&&1&\\&&1&&\end{smallmatrix}\right\rangle^{3}\left|\begin{smallmatrix}2&&1&&0\\&2&&1&\\&&2&&\end{smallmatrix}\right\rangle^{1}\right]\\
&\qquad-\sqrt{\frac{3}{10}}\left[\left|\begin{smallmatrix}2&&1&&0\\&2&&1&\\&&1&&\end{smallmatrix}\right\rangle^{2}\left|\begin{smallmatrix}2&&1&&0\\&2&&0&\\&&2&&\end{smallmatrix}\right\rangle^{2}+\left|\begin{smallmatrix}2&&1&&0\\&2&&0&\\&&2&&\end{smallmatrix}\right\rangle^{2}\left|\begin{smallmatrix}2&&1&&0\\&2&&1&\\&&1&&\end{smallmatrix}\right\rangle^{2}\right],\\
&||(1,1),(1,1)\rangle_A\\
&\quad=\sqrt{1/12}\,\{\,|(1,1)\rangle|(0,0)_1\rangle-|(0,0)_1\rangle|(1,1)\rangle-\sqrt{3}\,|(1,1)\rangle|(0,0)_2\rangle\\
&\qquad+\sqrt{3}\,|(0,0)_2\rangle|(1,1)\rangle-\sqrt{2}\,|(\overline{1},2)\rangle|(2,\overline{1})\rangle+\sqrt{2}\,|(2,\overline{1})\rangle|(\overline{1},2)\rangle\}\\
&\quad=\sqrt{\frac{1}{12}}\left[\left|\begin{smallmatrix}2&&1&&0\\&2&&1&\\&&2&&\end{smallmatrix}\right\rangle^{1}\left|\begin{smallmatrix}2&&1&&0\\&2&&0&\\&&1&&\end{smallmatrix}\right\rangle^{3}-\left|\begin{smallmatrix}2&&1&&0\\&2&&0&\\&&1&&\end{smallmatrix}\right\rangle^{3}\left|\begin{smallmatrix}2&&1&&0\\&2&&1&\\&&2&&\end{smallmatrix}\right\rangle^{1}\right]\\
&\qquad-\frac{1}{2}\left[\left|\begin{smallmatrix}2&&1&&0\\&2&&1&\\&&2&&\end{smallmatrix}\right\rangle^{1}\left|\begin{smallmatrix}2&&1&&0\\&1&&1&\\&&1&&\end{smallmatrix}\right\rangle^{3}-\left|\begin{smallmatrix}2&&1&&0\\&1&&1&\\&&1&&\end{smallmatrix}\right\rangle^{3}\left|\begin{smallmatrix}2&&1&&0\\&2&&1&\\&&2&&\end{smallmatrix}\right\rangle^{1}\right]\\
&\qquad-\sqrt{\frac{1}{6}}\left[\left|\begin{smallmatrix}2&&1&&0\\&2&&1&\\&&1&&\end{smallmatrix}\right\rangle^{2}\left|\begin{smallmatrix}2&&1&&0\\&2&&0&\\&&2&&\end{smallmatrix}\right\rangle^{2}-\left|\begin{smallmatrix}2&&1&&0\\&2&&0&\\&&2&&\end{smallmatrix}\right\rangle^{2}\left|\begin{smallmatrix}2&&1&&0\\&2&&1&\\&&1&&\end{smallmatrix}\right\rangle^{2}\right].
\end{aligned}\tag{8.56}$$

Ex. 3. The decomposition of $(3,0)\times(0,3)$ in A_2.

From Fig. 8.4, there are eight dominant weights in the product space as given in Eq. (8.57). Since $\boldsymbol{M}_1$, $\boldsymbol{M}_{1'}$, $\boldsymbol{M}_3$, and $\boldsymbol{M}_{3'}$ do not satisfy the condition (8.53), they do not appear in the CG series for $(3,0)\times(0,3)$: $a_1=a_{1'}=a_3=a_{3'}=0$. The multiplicities $n_{\beta\alpha}$ can be counted by the Gel'fand's bases. The results are $n_{01}=n_{01'}=n_{23}=n_{23'}=n_{34}=n_{3'4}=n_{3'5}=n_{34}=n_{35}=1$, $n_{02}=n_{03}=n_{03'}=n_{24}=n_{45}=2$, $n_{04}=n_{25}=3$

and $n_{05} = 4$. Then, the multiplicities a_M in the CG series (8.45) can be calculated: $a_2 = n_2 - n_{02} = 2 - 1 = 1$, $a_4 = n_4 - n_{04} - n_{24} = 6 - 3 - 2 = 1$, $a_5 = n_5 - n_{05} - n_{25} - n_{45} = 10 - 4 - 3 - 2 = 1$. The dominant weight diagram for $(3,0) \times (0,3)$ of A_2 is calculated in Fig. 8.17.

α	$\boldsymbol{M}_\alpha$	n_α	Independent basis states $\vert\boldsymbol{m}^{(1)}\rangle\vert\boldsymbol{m}^{(2)}\rangle$
0	$(3,3)$	1	$\vert(3,0)\rangle\vert(0,3)\rangle$
1	$(4,1)$	2	$\vert(3,0)\rangle\vert(1,1)\rangle$
$1'$	$(1,4)$	2	$\vert(1,1)\rangle\vert(0,3)\rangle$
2	$(2,2)$	3	$\vert(3,0)\rangle\vert(\overline{1},2)\rangle$, $\vert(1,1)\rangle\vert(1,1)\rangle$, $\vert(2,\overline{1})\rangle\vert(0,3)\rangle$
3	$(3,0)$	3	$\vert(3,0)\rangle\vert(0,0)\rangle$, $\vert(1,1)\rangle\vert(2,\overline{1})\rangle$, $\vert(2,\overline{1})\rangle\vert(1,1)\rangle$
$3'$	$(0,3)$	3	$\vert(1,1)\rangle\vert(\overline{1},2)\rangle$, $\vert(\overline{1},2)\rangle\vert(1,1)\rangle$, $\vert(0,0)\rangle\vert(0,3)\rangle$
4	$(1,1)$	6	$\vert(3,0)\rangle\vert(\overline{2},1)\rangle$, $\vert(1,1)\rangle\vert(0,0)\rangle$, $\vert(\overline{1},2)\rangle\vert(2,\overline{1})\rangle$,
			$\vert(2,\overline{1})\rangle\vert(\overline{1},2)\rangle$, $\vert(0,0)\rangle\vert(1,1)\rangle$, $\vert(1,\overline{2})\rangle\vert(0,3)\rangle$
5	$(0,0)$	10	$\vert\boldsymbol{m}\rangle\vert-\boldsymbol{m}\rangle$, where $\boldsymbol{m} \in D^{(3,0)}(A_2)$.

(8.57)

$$\boldsymbol{M}_0 = (3,3)$$

α	$\boldsymbol{M}_\alpha$	Ω_3, Ω_2, Ω_1	$n_{0\alpha}$	the basis states
0	$(3,3)$	$9,9,6$	1	$\left\vert\begin{smallmatrix}6&&3&&0\\&6&&3&\\&&6&&\end{smallmatrix}\right\rangle^1$
1	$(4,1)$	$9,9,5$	1	$\left\vert\begin{smallmatrix}6&&3&&0\\&6&&3&\\&&5&&\end{smallmatrix}\right\rangle^2$
$1'$	$(1,4)$	$9,8,6$	1	$\left\vert\begin{smallmatrix}6&&3&&0\\&6&&2&\\&&6&&\end{smallmatrix}\right\rangle^2$
2	$(2,2)$	$9,8,5$	2	$\left\vert\begin{smallmatrix}6&&3&&0\\&6&&2&\\&&5&&\end{smallmatrix}\right\rangle^3$, $\left\vert\begin{smallmatrix}6&&3&&0\\&5&&3&\\&&5&&\end{smallmatrix}\right\rangle^3$
3	$(3,0)$	$9,8,4$	2	$\left\vert\begin{smallmatrix}6&&3&&0\\&6&&2&\\&&4&&\end{smallmatrix}\right\rangle^4$, $\left\vert\begin{smallmatrix}6&&3&&0\\&5&&3&\\&&4&&\end{smallmatrix}\right\rangle^4$
$3'$	$(0,3)$	$9,7,5$	2	$\left\vert\begin{smallmatrix}6&&3&&0\\&6&&1&\\&&5&&\end{smallmatrix}\right\rangle^4$, $\left\vert\begin{smallmatrix}6&&3&&0\\&5&&2&\\&&5&&\end{smallmatrix}\right\rangle^4$
4	$(1,1)$	$9,7,4$	3	$\left\vert\begin{smallmatrix}6&&3&&0\\&6&&1&\\&&4&&\end{smallmatrix}\right\rangle^5$, $\left\vert\begin{smallmatrix}6&&3&&0\\&5&&2&\\&&4&&\end{smallmatrix}\right\rangle^5$, $\left\vert\begin{smallmatrix}6&&3&&0\\&4&&3&\\&&4&&\end{smallmatrix}\right\rangle^5$
5	$(0,0)$	$9,6,3$	4	$\left\vert\begin{smallmatrix}6&&3&&0\\&6&&0&\\&&3&&\end{smallmatrix}\right\rangle^7$, $\left\vert\begin{smallmatrix}6&&3&&0\\&5&&1&\\&&3&&\end{smallmatrix}\right\rangle^7$, $\left\vert\begin{smallmatrix}6&&3&&0\\&4&&2&\\&&3&&\end{smallmatrix}\right\rangle^7$, $\left\vert\begin{smallmatrix}6&&3&&0\\&3&&3&\\&&3&&\end{smallmatrix}\right\rangle^7$

The expansions (8.49) for $\boldsymbol{M}_0$ and $\boldsymbol{M}_5$ are given in Eqs. (8.50) and (8.54).

$$||(3,3),(3,3)\rangle = |(3,0)\rangle|(0,3)\rangle = \left|\begin{smallmatrix}3&&0&&0\\&3&&0&\\&&3&&\end{smallmatrix}\right\rangle^1 \left|\begin{smallmatrix}3&&3&&0\\&3&&3&\\&&3&&\end{smallmatrix}\right\rangle^1,$$

$$\begin{aligned}||(0,0),(0,0)\rangle = \sqrt{1/10}\,\{\ & |(3,0)\rangle|(\overline{3},0)\rangle - |(1,1)\rangle|(\overline{1},\overline{1})\rangle \\ & - |(\overline{1},2)\rangle|(1,\overline{2})\rangle - |(2,\overline{1})\rangle|(\overline{2},1)\rangle - |(\overline{3},3)\rangle|(3,\overline{3})\rangle \\ & - |(0,0)\rangle|(0,0)\rangle - |(\overline{2},1)\rangle|(2,\overline{1})\rangle - |(1,\overline{2})\rangle|(\overline{1},2)\rangle \\ & - |(\overline{1},\overline{1})\rangle|(1,1)\rangle - |(0,\overline{3})\rangle|(0,3)\rangle\}\end{aligned}$$

$$
\begin{aligned}
= \sqrt{\frac{1}{10}} \Bigg\{ & \left|\begin{smallmatrix}3&&0&&0\\&3&&0&\\&&3&&\end{smallmatrix}\right\rangle^{1} \left|\begin{smallmatrix}3&&3&&0\\&3&&0&\\&&0&&\end{smallmatrix}\right\rangle^{7} - \left|\begin{smallmatrix}3&&0&&0\\&3&&0&\\&&2&&\end{smallmatrix}\right\rangle^{2} \left|\begin{smallmatrix}3&&3&&0\\&3&&0&\\&&1&&\end{smallmatrix}\right\rangle^{6} \\
+ & \left|\begin{smallmatrix}3&&0&&0\\&3&&0&\\&&1&&\end{smallmatrix}\right\rangle^{3} \left|\begin{smallmatrix}3&&3&&0\\&3&&0&\\&&2&&\end{smallmatrix}\right\rangle^{5} + \left|\begin{smallmatrix}3&&0&&0\\&2&&0&\\&&2&&\end{smallmatrix}\right\rangle^{3} \left|\begin{smallmatrix}3&&3&&0\\&3&&1&\\&&1&&\end{smallmatrix}\right\rangle^{5} \\
- & \left|\begin{smallmatrix}3&&0&&0\\&3&&0&\\&&0&&\end{smallmatrix}\right\rangle^{4} \left|\begin{smallmatrix}3&&3&&0\\&3&&0&\\&&3&&\end{smallmatrix}\right\rangle^{4} - \left|\begin{smallmatrix}3&&0&&0\\&2&&0&\\&&1&&\end{smallmatrix}\right\rangle^{4} \left|\begin{smallmatrix}3&&3&&0\\&3&&1&\\&&2&&\end{smallmatrix}\right\rangle^{4} \\
+ & \left|\begin{smallmatrix}3&&0&&0\\&2&&0&\\&&0&&\end{smallmatrix}\right\rangle^{5} \left|\begin{smallmatrix}3&&3&&0\\&3&&1&\\&&3&&\end{smallmatrix}\right\rangle^{3} + \left|\begin{smallmatrix}3&&0&&0\\&1&&0&\\&&1&&\end{smallmatrix}\right\rangle^{5} \left|\begin{smallmatrix}3&&3&&0\\&3&&2&\\&&2&&\end{smallmatrix}\right\rangle^{3} \\
- & \left|\begin{smallmatrix}3&&0&&0\\&1&&0&\\&&0&&\end{smallmatrix}\right\rangle^{6} \left|\begin{smallmatrix}3&&3&&0\\&3&&2&\\&&3&&\end{smallmatrix}\right\rangle^{2} + \left|\begin{smallmatrix}3&&0&&0\\&0&&0&\\&&0&&\end{smallmatrix}\right\rangle^{7} \left|\begin{smallmatrix}3&&3&&0\\&3&&3&\\&&3&&\end{smallmatrix}\right\rangle^{1} \Bigg\}.
\end{aligned}
$$

O.S. × multiplicity		[3,3]	[2,2]	[1,1]	[0,0]
6 × 1	[3,3]	1			
6 × 1	[4,1]	1			
6 × 1	[1,4]	1			
6 × 3	[2,2]	2	1		
3 × 3	[3,0]	2	1		
3 × 3	[0,3]	2	1		
6 × 6	[1,1]	3	2	1	
1 × 10	[0,0]	4	3	2	1

$10 \times 10 = 100 = 64 + 27 + 8 + 1$

$(3,0) \times (0,3) \simeq (3,3) \oplus (2,2) \oplus (1,1) \oplus (0,0).$

Fig. 8.17 The dominant weight diagram for $(3,0) \times (0,3)$ in A_2.

We express the expansions (8.49) for $\boldsymbol{M}_2 = (2,2)$ and $\boldsymbol{M}_4 = (1,1)$ as the linear combinations of the highest weight states of A_1-triplets and A_1-doublets, respectively, and then, calculate the coefficients in the expansions by the annihilation condition of E_2.

$$
\begin{aligned}
\|(2,2),(2,2)\rangle = & \frac{a_1}{2}\left[\sqrt{3} \left|\begin{smallmatrix}3&&0&&0\\&3&&0&\\&&3&&\end{smallmatrix}\right\rangle^{1} \left|\begin{smallmatrix}3&&3&&0\\&3&&2&\\&&2&&\end{smallmatrix}\right\rangle^{3} - \left|\begin{smallmatrix}3&&0&&0\\&3&&0&\\&&2&&\end{smallmatrix}\right\rangle^{2} \left|\begin{smallmatrix}3&&3&&0\\&3&&2&\\&&3&&\end{smallmatrix}\right\rangle^{2}\right] \\
& + a_2 \left|\begin{smallmatrix}3&&0&&0\\&2&&0&\\&&2&&\end{smallmatrix}\right\rangle^{3} \left|\begin{smallmatrix}3&&3&&0\\&3&&3&\\&&3&&\end{smallmatrix}\right\rangle^{1},
\end{aligned}
$$

$$E_2 \,||(2,2),(2,2)\rangle = -\frac{\sqrt{3}a_1}{2}\left|\begin{smallmatrix}3&&0&&0\\&3&&0&\\&&2&&\end{smallmatrix}\right\rangle^{2}\left|\begin{smallmatrix}3&&3&&0\\&3&&3&\\&&3&&\end{smallmatrix}\right\rangle^{1}$$
$$+\; a_2\left|\begin{smallmatrix}3&&0&&0\\&3&&0&\\&&2&&\end{smallmatrix}\right\rangle^{2}\left|\begin{smallmatrix}3&&3&&0\\&3&&3&\\&&3&&\end{smallmatrix}\right\rangle^{1} = 0.$$

Thus, $a_2 = a_1\sqrt{3}/2$. After normalization, $(a_1)^{-2} = 1 + 3/4 = 7/4$,

$$||(2,2),(2,2)\rangle = \frac{1}{\sqrt{7}}\left\{\sqrt{3}|(3,0)\rangle|(\overline{1},2)\rangle - |(1,1)\rangle|(1,1)\rangle + \sqrt{3}|(2,\overline{1})\rangle|(0,3)\rangle\right\}$$
$$= \sqrt{\frac{1}{7}}\left[\sqrt{3}\left|\begin{smallmatrix}3&&0&&0\\&3&&0&\\&&3&&\end{smallmatrix}\right\rangle^{1}\left|\begin{smallmatrix}3&&3&&0\\&3&&2&\\&&2&&\end{smallmatrix}\right\rangle^{3} - \left|\begin{smallmatrix}3&&0&&0\\&3&&0&\\&&2&&\end{smallmatrix}\right\rangle^{2}\left|\begin{smallmatrix}3&&3&&0\\&3&&2&\\&&3&&\end{smallmatrix}\right\rangle^{2}\right.$$
$$\left.+\sqrt{3}\left|\begin{smallmatrix}3&&0&&0\\&2&&0&\\&&2&&\end{smallmatrix}\right\rangle^{3}\left|\begin{smallmatrix}3&&3&&0\\&3&&3&\\&&3&&\end{smallmatrix}\right\rangle^{1}\right\},$$

$$||(1,1),(1,1)\rangle = \frac{b_1}{\sqrt{6}}\left[\sqrt{3}\left|\begin{smallmatrix}3&&0&&0\\&3&&0&\\&&3&&\end{smallmatrix}\right\rangle^{1}\left|\begin{smallmatrix}3&&3&&0\\&3&&1&\\&&1&&\end{smallmatrix}\right\rangle^{5} - \sqrt{2}\left|\begin{smallmatrix}3&&0&&0\\&3&&0&\\&&2&&\end{smallmatrix}\right\rangle^{2}\left|\begin{smallmatrix}3&&3&&0\\&3&&1&\\&&2&&\end{smallmatrix}\right\rangle^{4}\right.$$
$$\left.+\left|\begin{smallmatrix}3&&0&&0\\&3&&0&\\&&1&&\end{smallmatrix}\right\rangle^{3}\left|\begin{smallmatrix}3&&3&&0\\&3&&1&\\&&3&&\end{smallmatrix}\right\rangle^{3}\right] + \frac{b_2}{\sqrt{3}}\left[\sqrt{2}\left|\begin{smallmatrix}3&&0&&0\\&2&&0&\\&&2&&\end{smallmatrix}\right\rangle^{3}\left|\begin{smallmatrix}3&&3&&0\\&3&&2&\\&&2&&\end{smallmatrix}\right\rangle^{3}\right.$$
$$\left.-\left|\begin{smallmatrix}3&&0&&0\\&2&&0&\\&&1&&\end{smallmatrix}\right\rangle^{4}\left|\begin{smallmatrix}3&&3&&0\\&3&&2&\\&&3&&\end{smallmatrix}\right\rangle^{2}\right] + b_3\left|\begin{smallmatrix}3&&0&&0\\&1&&0&\\&&1&&\end{smallmatrix}\right\rangle^{5}\left|\begin{smallmatrix}3&&3&&0\\&3&&3&\\&&3&&\end{smallmatrix}\right\rangle^{1},$$

$$E_2\,||(1,1),(1,1)\rangle = \frac{2b_1}{\sqrt{6}}\left[-\left|\begin{smallmatrix}3&&0&&0\\&3&&0&\\&&2&&\end{smallmatrix}\right\rangle^{2}\left|\begin{smallmatrix}3&&3&&0\\&3&&2&\\&&2&&\end{smallmatrix}\right\rangle^{3} + \left|\begin{smallmatrix}3&&0&&0\\&3&&0&\\&&1&&\end{smallmatrix}\right\rangle^{3}\left|\begin{smallmatrix}3&&3&&0\\&3&&2&\\&&3&&\end{smallmatrix}\right\rangle^{2}\right]$$
$$+\;\frac{b_2}{\sqrt{3}}\left[\sqrt{2}\left|\begin{smallmatrix}3&&0&&0\\&3&&0&\\&&2&&\end{smallmatrix}\right\rangle^{2}\left|\begin{smallmatrix}3&&3&&0\\&3&&2&\\&&2&&\end{smallmatrix}\right\rangle^{3} - \sqrt{3}\left|\begin{smallmatrix}3&&0&&0\\&2&&0&\\&&1&&\end{smallmatrix}\right\rangle^{4}\left|\begin{smallmatrix}3&&3&&0\\&3&&3&\\&&3&&\end{smallmatrix}\right\rangle^{1}\right.$$
$$\left.-\sqrt{2}\left|\begin{smallmatrix}3&&0&&0\\&3&&0&\\&&1&&\end{smallmatrix}\right\rangle^{3}\left|\begin{smallmatrix}3&&3&&0\\&3&&2&\\&&3&&\end{smallmatrix}\right\rangle^{2}\right] + b_3\sqrt{2}\left|\begin{smallmatrix}3&&0&&0\\&2&&0&\\&&1&&\end{smallmatrix}\right\rangle^{4}\left|\begin{smallmatrix}3&&3&&0\\&3&&3&\\&&3&&\end{smallmatrix}\right\rangle^{1} = 0.$$

Thus, $b_1 = b_2 = \sqrt{2}b_3$. After normalization, $(b_3)^{-2} = 2 + 2 + 1 = 5$,

$$||(1,1),(1,1)\rangle = \sqrt{\frac{1}{15}}\left\{\sqrt{3}\,|(3,0)\rangle|(\overline{2},1)\rangle - \sqrt{2}\,|(1,1)\rangle|(0,0)\rangle\right.$$
$$+\;|(\overline{1},2)\rangle|(2,\overline{1})\rangle + 2\,|(2,\overline{1})\rangle|(\overline{1},2)\rangle$$
$$\left.-\sqrt{2}\,|(0,0)\rangle|(1,1)\rangle + \sqrt{3}\,|(1,\overline{1})\rangle|(0,1)\rangle\right\}$$
$$= \sqrt{\frac{1}{15}}\left[\sqrt{3}\left|\begin{smallmatrix}3&&0&&0\\&3&&0&\\&&3&&\end{smallmatrix}\right\rangle^{1}\left|\begin{smallmatrix}3&&3&&0\\&3&&1&\\&&1&&\end{smallmatrix}\right\rangle^{5} - \sqrt{2}\left|\begin{smallmatrix}3&&0&&0\\&3&&0&\\&&2&&\end{smallmatrix}\right\rangle^{2}\left|\begin{smallmatrix}3&&3&&0\\&3&&1&\\&&2&&\end{smallmatrix}\right\rangle^{4}\right.$$
$$+\left|\begin{smallmatrix}3&&0&&0\\&3&&0&\\&&1&&\end{smallmatrix}\right\rangle^{3}\left|\begin{smallmatrix}3&&3&&0\\&3&&1&\\&&3&&\end{smallmatrix}\right\rangle^{3} + 2\left|\begin{smallmatrix}3&&0&&0\\&2&&0&\\&&2&&\end{smallmatrix}\right\rangle^{3}\left|\begin{smallmatrix}3&&3&&0\\&3&&2&\\&&2&&\end{smallmatrix}\right\rangle^{3}$$
$$\left.-\sqrt{2}\left|\begin{smallmatrix}3&&0&&0\\&2&&0&\\&&1&&\end{smallmatrix}\right\rangle^{4}\left|\begin{smallmatrix}3&&3&&0\\&3&&2&\\&&3&&\end{smallmatrix}\right\rangle^{2} + \sqrt{3}\left|\begin{smallmatrix}3&&0&&0\\&1&&0&\\&&1&&\end{smallmatrix}\right\rangle^{5}\left|\begin{smallmatrix}3&&3&&0\\&3&&3&\\&&3&&\end{smallmatrix}\right\rangle^{1}\right\}.$$

Ex. 4. The decomposition of $(1,1) \times (2,1)$ in A_2.

From Figs. 8.5 and 8.6 there are six dominant weights $\boldsymbol{M}_\alpha$ in the direct product space of $(1,1) \times (2,1)$ in A_2. The multiplicities n_α of $\boldsymbol{M}_\alpha$ in the direct product space can be counted in Figs. 8.5 and 8.6. The multiplicities $n_{\alpha\beta}$ of the dominant weight $\boldsymbol{M}_\beta$ in the representation $\boldsymbol{M}_\alpha$ can be counted by the Gel'fand's bases (see the statement following Eq. (8.18)), where $n_{0\alpha}$ are calculated as example.

$$\boldsymbol{M}_0 = (3,2)$$

α	$\boldsymbol{M}_\alpha$	$\Omega_3,\ \Omega_2,\ \Omega_1$	$n_{0\alpha}$	the basis states
0	$(3,2)$	7, 7, 5	1	$\left\lvert\begin{smallmatrix}5&&2&&0\\&5&&2&\\&&5&&\end{smallmatrix}\right\rangle^1$
1	$(4,0)$	7, 6, 5	1	$\left\lvert\begin{smallmatrix}5&&2&&0\\&5&&1&\\&&5&&\end{smallmatrix}\right\rangle^2$
$1'$	$(1,3)$	7, 7, 4	1	$\left\lvert\begin{smallmatrix}5&&2&&0\\&5&&2&\\&&4&&\end{smallmatrix}\right\rangle^2$
2	$(2,1)$	7, 6, 4	2	$\left\lvert\begin{smallmatrix}5&&2&&0\\&5&&1&\\&&4&&\end{smallmatrix}\right\rangle^3$, $\left\lvert\begin{smallmatrix}5&&2&&0\\&4&&2&\\&&4&&\end{smallmatrix}\right\rangle^3$
3	$(0,2)$	7, 6, 3	3	$\left\lvert\begin{smallmatrix}5&&2&&0\\&5&&1&\\&&3&&\end{smallmatrix}\right\rangle^4$, $\left\lvert\begin{smallmatrix}5&&2&&0\\&4&&2&\\&&3&&\end{smallmatrix}\right\rangle^4$, $\left\lvert\begin{smallmatrix}5&&2&&0\\&3&&3&\\&&3&&\end{smallmatrix}\right\rangle^4$
4	$(1,0)$	7, 5, 3	3	$\left\lvert\begin{smallmatrix}5&&2&&0\\&5&&0&\\&&3&&\end{smallmatrix}\right\rangle^5$, $\left\lvert\begin{smallmatrix}5&&2&&0\\&4&&1&\\&&3&&\end{smallmatrix}\right\rangle^5$, $\left\lvert\begin{smallmatrix}5&&2&&0\\&3&&2&\\&&3&&\end{smallmatrix}\right\rangle^5$

Then, the dominant weight diagram for $(1,1) \times (2,1)$ in A_2 is calculated in Fig. 8.18.

O.S. × multiplicity		[3,2]	[4,0]	[1,3]	[2,1]	[0,2]	[1,0]
6 × 1	[3,2]	1					
3 × 2	[4,0]	1	1				
6 × 2	[1,3]	1	0	1	2		
6 × 6	[2,1]	2	1	1	1		
3 × 8	[0,2]	3	1	2	1	1	
3 × 12	[1,0]	3	1	2	2	1	1

$$8 \times 15 = 120 = \quad 42 + 15 + 24 + 2 \times 15 + 6 + 1$$

$$(1,1) \times (2,1) \simeq (3,2) \oplus (4,0) \oplus (1,3) \oplus 2 \times (2,1) \oplus (0,2) \oplus (1,0).$$

Fig. 8.18 The dominant weight diagram for $(1,1) \times (2,1)$ in A_2.

The highest weight states of the first three representations in the Clebsch–Gordan series are easy to calculate.

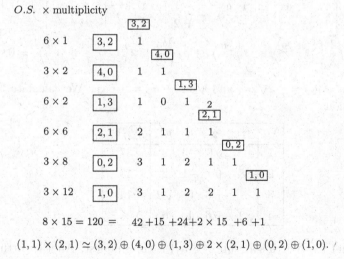

$$\|(3,2),(3,2)\rangle = \left|\begin{smallmatrix} 2 & & 1 & & 0 \\ & 2 & & 1 & \\ & & 2 & & \end{smallmatrix}\right\rangle^1 \left|\begin{smallmatrix} 3 & & 1 & & 0 \\ & 3 & & 1 & \\ & & 3 & & \end{smallmatrix}\right\rangle^1,$$

$$\|(4,0),(4,0)\rangle = \sqrt{\frac{1}{2}}\left[\left|\begin{smallmatrix} 2 & & 1 & & 0 \\ & 2 & & 1 & \\ & & 2 & & \end{smallmatrix}\right\rangle^1 \left|\begin{smallmatrix} 3 & & 1 & & 0 \\ & 3 & & 0 & \\ & & 3 & & \end{smallmatrix}\right\rangle^2 - \left|\begin{smallmatrix} 2 & & 1 & & 0 \\ & 2 & & 0 & \\ & & 2 & & \end{smallmatrix}\right\rangle^2 \left|\begin{smallmatrix} 3 & & 1 & & 0 \\ & 3 & & 1 & \\ & & 3 & & \end{smallmatrix}\right\rangle^1\right],$$

$$\|(1,3),(1,3)\rangle = \sqrt{\frac{1}{3}}\left[\left|\begin{smallmatrix} 2 & & 1 & & 0 \\ & 2 & & 1 & \\ & & 2 & & \end{smallmatrix}\right\rangle^1 \left|\begin{smallmatrix} 3 & & 1 & & 0 \\ & 3 & & 1 & \\ & & 2 & & \end{smallmatrix}\right\rangle^2 - \sqrt{2}\left|\begin{smallmatrix} 2 & & 1 & & 0 \\ & 2 & & 1 & \\ & & 1 & & \end{smallmatrix}\right\rangle^2 \left|\begin{smallmatrix} 3 & & 1 & & 0 \\ & 3 & & 1 & \\ & & 3 & & \end{smallmatrix}\right\rangle^1\right].$$

For the highest weight state of $\boldsymbol{M}_2 = (2,1)$, we express it as a linear combination of the normalized highest weight states of A_1-multiplets.

$$\begin{aligned}\|(2,1),(2,1)\rangle = {} & \frac{a_1}{2}\left[\left|\begin{smallmatrix} 2 & & 1 & & 0 \\ & 2 & & 1 & \\ & & 2 & & \end{smallmatrix}\right\rangle^1 \left|\begin{smallmatrix} 3 & & 1 & & 0 \\ & 3 & & 0 & \\ & & 2 & & \end{smallmatrix}\right\rangle^3 - \sqrt{3}\left|\begin{smallmatrix} 2 & & 1 & & 0 \\ & 2 & & 1 & \\ & & 1 & & \end{smallmatrix}\right\rangle^2 \left|\begin{smallmatrix} 3 & & 1 & & 0 \\ & 3 & & 0 & \\ & & 3 & & \end{smallmatrix}\right\rangle^2\right] \\ & + \frac{a_2}{\sqrt{2}}\left[\left|\begin{smallmatrix} 2 & & 1 & & 0 \\ & 2 & & 0 & \\ & & 2 & & \end{smallmatrix}\right\rangle^2 \left|\begin{smallmatrix} 3 & & 1 & & 0 \\ & 3 & & 1 & \\ & & 2 & & \end{smallmatrix}\right\rangle^2 - \left|\begin{smallmatrix} 2 & & 1 & & 0 \\ & 2 & & 0 & \\ & & 1 & & \end{smallmatrix}\right\rangle^3 \left|\begin{smallmatrix} 3 & & 1 & & 0 \\ & 3 & & 1 & \\ & & 3 & & \end{smallmatrix}\right\rangle^1\right] \\ & + a_3\left|\begin{smallmatrix} 2 & & 1 & & 0 \\ & 2 & & 1 & \\ & & 2 & & \end{smallmatrix}\right\rangle^1 \left|\begin{smallmatrix} 3 & & 1 & & 0 \\ & 2 & & 1 & \\ & & 2 & & \end{smallmatrix}\right\rangle^3 + a_4\left|\begin{smallmatrix} 2 & & 1 & & 0 \\ & 1 & & 1 & \\ & & 1 & & \end{smallmatrix}\right\rangle^3 \left|\begin{smallmatrix} 3 & & 1 & & 0 \\ & 3 & & 1 & \\ & & 3 & & \end{smallmatrix}\right\rangle^1.\end{aligned}$$

Then, from the annihilation condition of E_2 we obtain

$$\begin{aligned}E_2\,\|(2,1),(2,1)\rangle = {} & \frac{a_1}{2}\left[\sqrt{\frac{2}{3}}\left|\begin{smallmatrix} 2 & & 1 & & 0 \\ & 2 & & 1 & \\ & & 2 & & \end{smallmatrix}\right\rangle^1 \left|\begin{smallmatrix} 3 & & 1 & & 0 \\ & 3 & & 1 & \\ & & 2 & & \end{smallmatrix}\right\rangle^2 - \sqrt{3}\left|\begin{smallmatrix} 2 & & 1 & & 0 \\ & 2 & & 1 & \\ & & 1 & & \end{smallmatrix}\right\rangle^2 \left|\begin{smallmatrix} 3 & & 1 & & 0 \\ & 3 & & 1 & \\ & & 3 & & \end{smallmatrix}\right\rangle^1\right] \\ & + \frac{a_2}{\sqrt{2}}\left[\left|\begin{smallmatrix} 2 & & 1 & & 0 \\ & 2 & & 1 & \\ & & 2 & & \end{smallmatrix}\right\rangle^1 \left|\begin{smallmatrix} 3 & & 1 & & 0 \\ & 3 & & 1 & \\ & & 2 & & \end{smallmatrix}\right\rangle^2 - \sqrt{\frac{1}{2}}\left|\begin{smallmatrix} 2 & & 1 & & 0 \\ & 2 & & 1 & \\ & & 1 & & \end{smallmatrix}\right\rangle^2 \left|\begin{smallmatrix} 3 & & 1 & & 0 \\ & 3 & & 1 & \\ & & 3 & & \end{smallmatrix}\right\rangle^1\right] \\ & + a_3\sqrt{\frac{4}{3}}\left|\begin{smallmatrix} 2 & & 1 & & 0 \\ & 2 & & 1 & \\ & & 2 & & \end{smallmatrix}\right\rangle^1 \left|\begin{smallmatrix} 3 & & 1 & & 0 \\ & 3 & & 1 & \\ & & 2 & & \end{smallmatrix}\right\rangle^2 + a_4\sqrt{\frac{3}{2}}\left|\begin{smallmatrix} 2 & & 1 & & 0 \\ & 2 & & 1 & \\ & & 1 & & \end{smallmatrix}\right\rangle^2 \left|\begin{smallmatrix} 3 & & 1 & & 0 \\ & 3 & & 1 & \\ & & 3 & & \end{smallmatrix}\right\rangle^1 = 0.\end{aligned}$$

Thus, $a_1 + \sqrt{3}a_2 = -2\sqrt{2}a_3$ and $\sqrt{3}a_1 + a_2 = \sqrt{6}a_4$. We calculate four basic solutions:

No.	a_1	a_2	a_3	a_4	$\sum_j a_j^2$
1	0	$2\sqrt{6}$	-3	2	37
2	$\sqrt{8}$	0	-1	2	13
3	3	$-\sqrt{3}$	0	$\sqrt{2}$	14
4	$\sqrt{2}$	$-\sqrt{6}$	1	0	9

We find that the fourth solution $\psi^{(4)}(a_j)$ is the simplest. Another orthogonal solution can be calculated uniquely up to a constant factor. Since $\sum_j a_j^{(2)}a_j^{(4)} = 3$, $[3\psi^{(2)}(a_j) - \psi^{(4)}(a_j)]/\sqrt{2}$, where $a_1 = 5$, $a_2 = \sqrt{3}$, $a_3 = -2\sqrt{2}$, $a_4 = 3\sqrt{2}$, is orthogonal to $\psi^{(4)}(a_j)$. After normalization, we obtain two orthogonal solutions as follows.

$$
\begin{aligned}
\|(2,1),(2,1)\rangle_1 = \frac{1}{3\sqrt{2}} \Bigg\{ & \left|\begin{smallmatrix}2&&1&&0\\&2&&1&\\&&2&&\end{smallmatrix}\right\rangle^{1} \left|\begin{smallmatrix}3&&1&&0\\&3&&0&\\&&2&&\end{smallmatrix}\right\rangle^{3} - \sqrt{3} \left|\begin{smallmatrix}2&&1&&0\\&2&&1&\\&&1&&\end{smallmatrix}\right\rangle^{2} \left|\begin{smallmatrix}3&&1&&0\\&3&&0&\\&&3&&\end{smallmatrix}\right\rangle^{2} \\
& - \sqrt{6} \left|\begin{smallmatrix}2&&1&&0\\&2&&0&\\&&2&&\end{smallmatrix}\right\rangle^{2} \left|\begin{smallmatrix}3&&1&&0\\&3&&1&\\&&2&&\end{smallmatrix}\right\rangle^{2} + \sqrt{6} \left|\begin{smallmatrix}2&&1&&0\\&2&&0&\\&&1&&\end{smallmatrix}\right\rangle^{3} \left|\begin{smallmatrix}3&&1&&0\\&3&&1&\\&&3&&\end{smallmatrix}\right\rangle^{1} \\
& + \sqrt{2} \left|\begin{smallmatrix}2&&1&&0\\&2&&1&\\&&2&&\end{smallmatrix}\right\rangle^{1} \left|\begin{smallmatrix}3&&1&&0\\&2&&1&\\&&2&&\end{smallmatrix}\right\rangle^{3} \Bigg\}, \\
\|(2,1),(2,1)\rangle_2 = \frac{1}{6\sqrt{6}} \Bigg\{ & 5 \left|\begin{smallmatrix}2&&1&&0\\&2&&1&\\&&2&&\end{smallmatrix}\right\rangle^{1} \left|\begin{smallmatrix}3&&1&&0\\&3&&0&\\&&2&&\end{smallmatrix}\right\rangle^{3} - 5\sqrt{3} \left|\begin{smallmatrix}2&&1&&0\\&2&&1&\\&&1&&\end{smallmatrix}\right\rangle^{2} \left|\begin{smallmatrix}3&&1&&0\\&3&&0&\\&&3&&\end{smallmatrix}\right\rangle^{2} \\
& + \sqrt{6} \left|\begin{smallmatrix}2&&1&&0\\&2&&0&\\&&2&&\end{smallmatrix}\right\rangle^{2} \left|\begin{smallmatrix}3&&1&&0\\&3&&1&\\&&2&&\end{smallmatrix}\right\rangle^{2} - \sqrt{6} \left|\begin{smallmatrix}2&&1&&0\\&2&&0&\\&&1&&\end{smallmatrix}\right\rangle^{3} \left|\begin{smallmatrix}3&&1&&0\\&3&&1&\\&&3&&\end{smallmatrix}\right\rangle^{1} \\
& - 4\sqrt{2} \left|\begin{smallmatrix}2&&1&&0\\&2&&1&\\&&2&&\end{smallmatrix}\right\rangle^{1} \left|\begin{smallmatrix}3&&1&&0\\&2&&1&\\&&2&&\end{smallmatrix}\right\rangle^{3} + 6\sqrt{2} \left|\begin{smallmatrix}2&&1&&0\\&1&&1&\\&&1&&\end{smallmatrix}\right\rangle^{3} \left|\begin{smallmatrix}3&&1&&0\\&3&&1&\\&&3&&\end{smallmatrix}\right\rangle^{1} \Bigg\}.
\end{aligned} \tag{8.58}
$$

For the highest weight state of $M_3 = (0,2)$, we express it as a linear combination of the normalized highest weight states of A_1-multiplets.

$$
\begin{aligned}
\|(0,2),(0,2)\rangle = & \frac{b_1}{\sqrt{2}} \left[\left|\begin{smallmatrix}2&&1&&0\\&2&&1&\\&&2&&\end{smallmatrix}\right\rangle^{1} \left|\begin{smallmatrix}3&&1&&0\\&2&&1&\\&&1&&\end{smallmatrix}\right\rangle^{4} - \left|\begin{smallmatrix}2&&1&&0\\&2&&1&\\&&1&&\end{smallmatrix}\right\rangle^{2} \left|\begin{smallmatrix}3&&1&&0\\&2&&1&\\&&2&&\end{smallmatrix}\right\rangle^{3} \right] \\
& + \frac{b_2}{\sqrt{3}} \left[\left|\begin{smallmatrix}2&&1&&0\\&2&&0&\\&&2&&\end{smallmatrix}\right\rangle^{2} \left|\begin{smallmatrix}3&&1&&0\\&3&&1&\\&&1&&\end{smallmatrix}\right\rangle^{3} - \left|\begin{smallmatrix}2&&1&&0\\&2&&0&\\&&1&&\end{smallmatrix}\right\rangle^{3} \left|\begin{smallmatrix}3&&1&&0\\&3&&1&\\&&2&&\end{smallmatrix}\right\rangle^{2} \right. \\
& \left. + \left|\begin{smallmatrix}2&&1&&0\\&2&&0&\\&&0&&\end{smallmatrix}\right\rangle^{4} \left|\begin{smallmatrix}3&&1&&0\\&3&&1&\\&&3&&\end{smallmatrix}\right\rangle^{1} \right].
\end{aligned}
$$

Then, from the annihilation condition of E_2 we obtain

$$
\begin{aligned}
E_2 \|(0,2),(0,2)\rangle = & \frac{b_1}{\sqrt{2}} \left[\sqrt{\frac{8}{3}} \left|\begin{smallmatrix}2&&1&&0\\&2&&1&\\&&2&&\end{smallmatrix}\right\rangle^{1} \left|\begin{smallmatrix}3&&1&&0\\&3&&1&\\&&1&&\end{smallmatrix}\right\rangle^{3} - \sqrt{\frac{4}{3}} \left|\begin{smallmatrix}2&&1&&0\\&2&&1&\\&&1&&\end{smallmatrix}\right\rangle^{2} \left|\begin{smallmatrix}3&&1&&0\\&3&&1&\\&&2&&\end{smallmatrix}\right\rangle^{2} \right] \\
& + \frac{b_2}{\sqrt{3}} \left[\left|\begin{smallmatrix}2&&1&&0\\&2&&1&\\&&2&&\end{smallmatrix}\right\rangle^{1} \left|\begin{smallmatrix}3&&1&&0\\&3&&1&\\&&1&&\end{smallmatrix}\right\rangle^{3} - \sqrt{\frac{1}{2}} \left|\begin{smallmatrix}2&&1&&0\\&2&&1&\\&&1&&\end{smallmatrix}\right\rangle^{2} \left|\begin{smallmatrix}3&&1&&0\\&3&&1&\\&&2&&\end{smallmatrix}\right\rangle^{2} \right] = 0.
\end{aligned}
$$

Thus, $b_2 = -2b_1$. After normalization we obtain $b_1^{-2} = 1 + 2^2 = 5$ and

$$
\begin{aligned}
\|(0,2),(0,2)\rangle = \frac{1}{\sqrt{30}} \Bigg\{ & \sqrt{3} \left|\begin{smallmatrix}2&&1&&0\\&2&&1&\\&&2&&\end{smallmatrix}\right\rangle^{1} \left|\begin{smallmatrix}3&&1&&0\\&2&&1&\\&&1&&\end{smallmatrix}\right\rangle^{4} - \sqrt{3} \left|\begin{smallmatrix}2&&1&&0\\&2&&1&\\&&1&&\end{smallmatrix}\right\rangle^{2} \left|\begin{smallmatrix}3&&1&&0\\&2&&1&\\&&2&&\end{smallmatrix}\right\rangle^{3} \\
& - 2\sqrt{2} \left|\begin{smallmatrix}2&&1&&0\\&2&&0&\\&&2&&\end{smallmatrix}\right\rangle^{2} \left|\begin{smallmatrix}3&&1&&0\\&3&&1&\\&&1&&\end{smallmatrix}\right\rangle^{3} + 2\sqrt{2} \left|\begin{smallmatrix}2&&1&&0\\&2&&0&\\&&1&&\end{smallmatrix}\right\rangle^{3} \left|\begin{smallmatrix}3&&1&&0\\&3&&1&\\&&2&&\end{smallmatrix}\right\rangle^{2} \\
& - 2\sqrt{2} \left|\begin{smallmatrix}2&&1&&0\\&2&&0&\\&&0&&\end{smallmatrix}\right\rangle^{4} \left|\begin{smallmatrix}3&&1&&0\\&3&&1&\\&&3&&\end{smallmatrix}\right\rangle^{1} \Bigg\}.
\end{aligned} \tag{8.59}
$$

For the highest weight state of $M_4 = (1,0)$, we express it as a linear combination of the normalized highest weight states of A_1-multiplets.

$$
\begin{aligned}
\|(1,0),(1,0)\rangle = {} & \frac{c_1}{\sqrt{3}}\left[\left|\begin{smallmatrix}2&&1&&0\\&2&&1&\\&&2&&\end{smallmatrix}\right\rangle^{1}\left|\begin{smallmatrix}3&&1&&0\\&2&&0&\\&&1&&\end{smallmatrix}\right\rangle^{5}-\sqrt{2}\left|\begin{smallmatrix}2&&1&&0\\&2&&1&\\&&1&&\end{smallmatrix}\right\rangle^{2}\left|\begin{smallmatrix}3&&1&&0\\&2&&0&\\&&2&&\end{smallmatrix}\right\rangle^{4}\right] \\
& + c_2\left|\begin{smallmatrix}2&&1&&0\\&2&&1&\\&&2&&\end{smallmatrix}\right\rangle^{1}\left|\begin{smallmatrix}3&&1&&0\\&1&&1&\\&&1&&\end{smallmatrix}\right\rangle^{5}+\frac{c_3}{\sqrt{6}}\left[\left|\begin{smallmatrix}2&&1&&0\\&2&&0&\\&&2&&\end{smallmatrix}\right\rangle^{2}\left|\begin{smallmatrix}3&&1&&0\\&3&&0&\\&&1&&\end{smallmatrix}\right\rangle^{4}\right. \\
& \left.-\sqrt{2}\left|\begin{smallmatrix}2&&1&&0\\&2&&0&\\&&1&&\end{smallmatrix}\right\rangle^{3}\left|\begin{smallmatrix}3&&1&&0\\&3&&0&\\&&2&&\end{smallmatrix}\right\rangle^{3}+\sqrt{3}\left|\begin{smallmatrix}2&&1&&0\\&2&&0&\\&&0&&\end{smallmatrix}\right\rangle^{4}\left|\begin{smallmatrix}3&&1&&0\\&3&&0&\\&&3&&\end{smallmatrix}\right\rangle^{2}\right] \\
& + \frac{c_4}{\sqrt{3}}\left[\sqrt{2}\left|\begin{smallmatrix}2&&1&&0\\&2&&0&\\&&2&&\end{smallmatrix}\right\rangle^{2}\left|\begin{smallmatrix}3&&1&&0\\&2&&1&\\&&1&&\end{smallmatrix}\right\rangle^{4}-\left|\begin{smallmatrix}2&&1&&0\\&2&&0&\\&&1&&\end{smallmatrix}\right\rangle^{3}\left|\begin{smallmatrix}3&&1&&0\\&2&&1&\\&&2&&\end{smallmatrix}\right\rangle^{3}\right] \\
& + c_5\left|\begin{smallmatrix}2&&1&&0\\&1&&1&\\&&1&&\end{smallmatrix}\right\rangle^{3}\left|\begin{smallmatrix}3&&1&&0\\&2&&1&\\&&2&&\end{smallmatrix}\right\rangle^{3} \\
& + \frac{c_6}{\sqrt{3}}\left[\left|\begin{smallmatrix}2&&1&&0\\&1&&0&\\&&1&&\end{smallmatrix}\right\rangle^{4}\left|\begin{smallmatrix}3&&1&&0\\&3&&1&\\&&2&&\end{smallmatrix}\right\rangle^{2}-\sqrt{2}\left|\begin{smallmatrix}2&&1&&0\\&1&&0&\\&&0&&\end{smallmatrix}\right\rangle^{5}\left|\begin{smallmatrix}3&&1&&0\\&3&&1&\\&&3&&\end{smallmatrix}\right\rangle^{1}\right].
\end{aligned}
$$

Then, from the annihilation condition of E_2 we obtain

$$
\begin{aligned}
E_2\,\|(1,0),(1,0)\rangle = {} & \frac{c_1}{\sqrt{3}}\left[\left(\sqrt{\frac{4}{3}}\left|\begin{smallmatrix}2&&1&&0\\&2&&1&\\&&2&&\end{smallmatrix}\right\rangle^{1}\left|\begin{smallmatrix}3&&1&&0\\&3&&0&\\&&1&&\end{smallmatrix}\right\rangle^{4}\right.\right. \\
& \left.+\sqrt{\frac{2}{3}}\left|\begin{smallmatrix}2&&1&&0\\&2&&1&\\&&2&&\end{smallmatrix}\right\rangle^{1}\left|\begin{smallmatrix}3&&1&&0\\&2&&1&\\&&1&&\end{smallmatrix}\right\rangle^{4}\right)-\sqrt{2}\left(\sqrt{\frac{2}{3}}\left|\begin{smallmatrix}2&&1&&0\\&2&&1&\\&&1&&\end{smallmatrix}\right\rangle^{2}\left|\begin{smallmatrix}3&&1&&0\\&3&&0&\\&&2&&\end{smallmatrix}\right\rangle^{3}\right. \\
& \left.\left.+\sqrt{\frac{4}{3}}\left|\begin{smallmatrix}2&&1&&0\\&2&&1&\\&&1&&\end{smallmatrix}\right\rangle^{2}\left|\begin{smallmatrix}3&&1&&0\\&2&&1&\\&&2&&\end{smallmatrix}\right\rangle^{3}\right)\right]+c_2\sqrt{3}\left|\begin{smallmatrix}2&&1&&0\\&2&&1&\\&&2&&\end{smallmatrix}\right\rangle^{1}\left|\begin{smallmatrix}3&&1&&0\\&2&&1&\\&&1&&\end{smallmatrix}\right\rangle^{4} \\
& + \frac{c_3}{\sqrt{6}}\left[\sqrt{\frac{1}{3}}\left|\begin{smallmatrix}2&&1&&0\\&2&&0&\\&&2&&\end{smallmatrix}\right\rangle^{2}\left|\begin{smallmatrix}3&&1&&0\\&3&&1&\\&&1&&\end{smallmatrix}\right\rangle^{3}+\left|\begin{smallmatrix}2&&1&&0\\&2&&1&\\&&2&&\end{smallmatrix}\right\rangle^{1}\left|\begin{smallmatrix}3&&1&&0\\&3&&0&\\&&1&&\end{smallmatrix}\right\rangle^{4}\right. \\
& -\sqrt{\frac{4}{3}}\left|\begin{smallmatrix}2&&1&&0\\&2&&0&\\&&1&&\end{smallmatrix}\right\rangle^{3}\left|\begin{smallmatrix}3&&1&&0\\&3&&1&\\&&2&&\end{smallmatrix}\right\rangle^{2}-\left|\begin{smallmatrix}2&&1&&0\\&2&&1&\\&&1&&\end{smallmatrix}\right\rangle^{2}\left|\begin{smallmatrix}3&&1&&0\\&3&&0&\\&&2&&\end{smallmatrix}\right\rangle^{3} \\
& \left.+\sqrt{3}\left|\begin{smallmatrix}2&&1&&0\\&2&&0&\\&&0&&\end{smallmatrix}\right\rangle^{4}\left|\begin{smallmatrix}3&&1&&0\\&3&&1&\\&&3&&\end{smallmatrix}\right\rangle^{1}\right]+\frac{c_4}{\sqrt{3}}\left[\sqrt{\frac{16}{3}}\left|\begin{smallmatrix}2&&1&&0\\&2&&0&\\&&2&&\end{smallmatrix}\right\rangle^{2}\left|\begin{smallmatrix}3&&1&&0\\&3&&1&\\&&1&&\end{smallmatrix}\right\rangle^{3}\right. \\
& +\sqrt{2}\left|\begin{smallmatrix}2&&1&&0\\&2&&1&\\&&2&&\end{smallmatrix}\right\rangle^{1}\left|\begin{smallmatrix}3&&1&&0\\&2&&1&\\&&1&&\end{smallmatrix}\right\rangle^{4}-\sqrt{\frac{1}{2}}\left|\begin{smallmatrix}2&&1&&0\\&2&&1&\\&&1&&\end{smallmatrix}\right\rangle^{2}\left|\begin{smallmatrix}3&&1&&0\\&2&&1&\\&&2&&\end{smallmatrix}\right\rangle^{3} \\
& \left.-\sqrt{\frac{4}{3}}\left|\begin{smallmatrix}2&&1&&0\\&2&&0&\\&&1&&\end{smallmatrix}\right\rangle^{3}\left|\begin{smallmatrix}3&&1&&0\\&3&&1&\\&&2&&\end{smallmatrix}\right\rangle^{2}\right]+c_5\left(\sqrt{\frac{4}{3}}\left|\begin{smallmatrix}2&&1&&0\\&1&&1&\\&&1&&\end{smallmatrix}\right\rangle^{3}\left|\begin{smallmatrix}3&&1&&0\\&3&&1&\\&&2&&\end{smallmatrix}\right\rangle^{2}\right. \\
& \left.+\sqrt{\frac{3}{2}}\left|\begin{smallmatrix}2&&1&&0\\&2&&1&\\&&1&&\end{smallmatrix}\right\rangle^{2}\left|\begin{smallmatrix}3&&1&&0\\&2&&1&\\&&2&&\end{smallmatrix}\right\rangle^{3}\right)+\frac{c_6}{\sqrt{3}}\left[\sqrt{\frac{1}{2}}\left|\begin{smallmatrix}2&&1&&0\\&2&&0&\\&&1&&\end{smallmatrix}\right\rangle^{3}\left|\begin{smallmatrix}3&&1&&0\\&3&&1&\\&&2&&\end{smallmatrix}\right\rangle^{2}\right. \\
& \left.+\sqrt{\frac{3}{2}}\left|\begin{smallmatrix}2&&1&&0\\&1&&1&\\&&1&&\end{smallmatrix}\right\rangle^{3}\left|\begin{smallmatrix}3&&1&&0\\&3&&1&\\&&2&&\end{smallmatrix}\right\rangle^{2}-\sqrt{2}\left|\begin{smallmatrix}2&&1&&0\\&2&&0&\\&&0&&\end{smallmatrix}\right\rangle^{4}\left|\begin{smallmatrix}3&&1&&0\\&3&&1&\\&&3&&\end{smallmatrix}\right\rangle^{1}\right]=0.
\end{aligned}
$$

Collecting the same states, we have

$$
\begin{aligned}
&\left|\begin{smallmatrix}2&&1&&0\\&2&&1&\\&&2&&\end{smallmatrix}\right\rangle^{1}\left|\begin{smallmatrix}3&&1&&0\\&3&&0&\\&&1&&\end{smallmatrix}\right\rangle^{4}: && c_1(2/3)+c_3\sqrt{1/6}=0,\\
&\left|\begin{smallmatrix}2&&1&&0\\&2&&1&\\&&2&&\end{smallmatrix}\right\rangle^{1}\left|\begin{smallmatrix}3&&1&&0\\&2&&1&\\&&1&&\end{smallmatrix}\right\rangle^{4}: && c_1\sqrt{2/9}+c_2\sqrt{3}+c_4\sqrt{2/3}=0,\\
&\left|\begin{smallmatrix}2&&1&&0\\&2&&1&\\&&1&&\end{smallmatrix}\right\rangle^{2}\left|\begin{smallmatrix}3&&1&&0\\&3&&0&\\&&2&&\end{smallmatrix}\right\rangle^{3}: && -c_1(2/3)-c_3\sqrt{1/6}=0,\\
&\left|\begin{smallmatrix}2&&1&&0\\&2&&1&\\&&1&&\end{smallmatrix}\right\rangle^{2}\left|\begin{smallmatrix}3&&1&&0\\&2&&1&\\&&2&&\end{smallmatrix}\right\rangle^{3}: && -c_1\sqrt{8/9}-c_4\sqrt{1/6}+c_5\sqrt{3/2}=0,\\
&\left|\begin{smallmatrix}2&&1&&0\\&2&&0&\\&&2&&\end{smallmatrix}\right\rangle^{2}\left|\begin{smallmatrix}3&&1&&0\\&3&&1&\\&&1&&\end{smallmatrix}\right\rangle^{3}: && c_3\sqrt{1/18}+c_4(4/3)=0,\\
&\left|\begin{smallmatrix}2&&1&&0\\&2&&0&\\&&1&&\end{smallmatrix}\right\rangle^{3}\left|\begin{smallmatrix}3&&1&&0\\&3&&1&\\&&2&&\end{smallmatrix}\right\rangle^{2}: && -c_3\sqrt{2/9}-c_4(2/3)+c_6\sqrt{1/6}=0,\\
&\left|\begin{smallmatrix}2&&1&&0\\&1&&1&\\&&1&&\end{smallmatrix}\right\rangle^{3}\left|\begin{smallmatrix}3&&1&&0\\&3&&1&\\&&2&&\end{smallmatrix}\right\rangle^{2}: && c_5\sqrt{4/3}+c_6\sqrt{1/2}=0,\\
&\left|\begin{smallmatrix}2&&1&&0\\&2&&0&\\&&0&&\end{smallmatrix}\right\rangle^{4}\left|\begin{smallmatrix}3&&1&&0\\&3&&1&\\&&3&&\end{smallmatrix}\right\rangle^{1}: && c_3\sqrt{1/2}-c_6\sqrt{2/3}=0.
\end{aligned}
$$

The solution is $c_1 = 2\sqrt{3}$, $c_2 = -\sqrt{2}$, $c_3 = -4\sqrt{2}$, $c_4 = 1$, $c_5 = 3$, $c_6 = -2\sqrt{6}$. After normalization we have

$$
\begin{aligned}
\|(1,0),(1,0)\rangle = &\sqrt{\frac{1}{240}}\left\{2\sqrt{3}\left|\begin{smallmatrix}2&&1&&0\\&2&&1&\\&&2&&\end{smallmatrix}\right\rangle^{1}\left|\begin{smallmatrix}3&&1&&0\\&2&&0&\\&&1&&\end{smallmatrix}\right\rangle^{5}\right.\\
&-2\sqrt{6}\left|\begin{smallmatrix}2&&1&&0\\&2&&1&\\&&1&&\end{smallmatrix}\right\rangle^{2}\left|\begin{smallmatrix}3&&1&&0\\&2&&0&\\&&2&&\end{smallmatrix}\right\rangle^{4}-\sqrt{6}\left|\begin{smallmatrix}2&&1&&0\\&2&&1&\\&&2&&\end{smallmatrix}\right\rangle^{1}\left|\begin{smallmatrix}3&&1&&0\\&1&&1&\\&&1&&\end{smallmatrix}\right\rangle^{5}\\
&-4\left|\begin{smallmatrix}2&&1&&0\\&2&&0&\\&&2&&\end{smallmatrix}\right\rangle^{2}\left|\begin{smallmatrix}3&&1&&0\\&3&&0&\\&&1&&\end{smallmatrix}\right\rangle^{4}+4\sqrt{2}\left|\begin{smallmatrix}2&&1&&0\\&2&&0&\\&&1&&\end{smallmatrix}\right\rangle^{3}\left|\begin{smallmatrix}3&&1&&0\\&3&&0&\\&&2&&\end{smallmatrix}\right\rangle^{3}\\
&-4\sqrt{3}\left|\begin{smallmatrix}2&&1&&0\\&2&&0&\\&&0&&\end{smallmatrix}\right\rangle^{4}\left|\begin{smallmatrix}3&&1&&0\\&3&&0&\\&&3&&\end{smallmatrix}\right\rangle^{2}+\sqrt{2}\left|\begin{smallmatrix}2&&1&&0\\&2&&0&\\&&2&&\end{smallmatrix}\right\rangle^{2}\left|\begin{smallmatrix}3&&1&&0\\&2&&1&\\&&1&&\end{smallmatrix}\right\rangle^{4}\\
&-\left|\begin{smallmatrix}2&&1&&0\\&2&&0&\\&&1&&\end{smallmatrix}\right\rangle^{3}\left|\begin{smallmatrix}3&&1&&0\\&2&&1&\\&&2&&\end{smallmatrix}\right\rangle^{3}+3\sqrt{3}\left|\begin{smallmatrix}2&&1&&0\\&1&&1&\\&&1&&\end{smallmatrix}\right\rangle^{3}\left|\begin{smallmatrix}3&&1&&0\\&2&&1&\\&&2&&\end{smallmatrix}\right\rangle^{3}\\
&\left.-2\sqrt{6}\left|\begin{smallmatrix}2&&1&&0\\&1&&0&\\&&1&&\end{smallmatrix}\right\rangle^{4}\left|\begin{smallmatrix}3&&1&&0\\&3&&1&\\&&2&&\end{smallmatrix}\right\rangle^{2}+4\sqrt{3}\left|\begin{smallmatrix}2&&1&&0\\&1&&0&\\&&0&&\end{smallmatrix}\right\rangle^{5}\left|\begin{smallmatrix}3&&1&&0\\&3&&1&\\&&3&&\end{smallmatrix}\right\rangle^{1}\right\}.
\end{aligned}
\tag{8.60}
$$

Ex. 5. The decomposition of $(2,2)\times(2,2)$ in A_2.

From Fig. 8.5 there are thirteen dominant weights $\boldsymbol{M}_\alpha$ in the direct product space of $(2,2)\times(2,2)$ in A_2. The multiplicities n_α of $\boldsymbol{M}_\alpha$ in the direct product space can be counted in Fig. 8.5. The multiplicities $n_{\alpha\beta}$ of the dominant weight $\boldsymbol{M}_\beta$ in the representation $\boldsymbol{M}_\alpha$ can be counted by the Gel'fand's bases (see the statement following Eq. (8.18)). Then, the dominant weight diagram for $(2,2) \times (2,2)$ of A_2 is calculated in Fig 8.19.

$O.S.$ × multiplicity		4,4	5,2	2,5	2 3,3	6,0	0,6	2 4,1	2 1,4	3 2,2	3,0	0,3	2 1,1	0,0
6 × 1	4,4	1												
6 × 2	5,2	1	1											
6 × 2	2,5	1	0	1										
6 × 6	3,3	2	1	1	1									
3 × 3	6,0	1	1	0	0	1								
3 × 3	0,6	1	0	1	0	0	1							
6 × 10	4,1	2	2	1	1	1	0	1						
6 × 10	1,4	2	1	2	1	0	1	0	1					
6 × 20	2,2	3	2	2	2	1	1	1	1	1				
3 × 24	3,0	3	3	2	2	1	1	2	1	1	1			
3 × 24	0,3	3	2	3	2	1	1	1	2	1	0	1		
6 × 36	1,1	4	3	3	3	1	1	2	2	2	1	1	1	
1 × 45	0,0	5	3	3	4	1	1	2	2	3	1	1	2	1

$$27 \times 27 = 729 = 125+81+81+2\times 64+28+28+2\times 35+2\times 35+3\times 27+10+10+2\times 8+1$$

$$(2,2)\times(2,2) \simeq (4,4)\oplus(5,2)\oplus(2,5)\oplus 2\times(3,3)\oplus(6,0)\oplus 2\times(4,1)\oplus 2\times(1,4)$$
$$\oplus 3\times(2,2)\oplus(3,0)\oplus(0,3)\oplus 2\times(1,1)\oplus(0,0).$$

Fig. 8.19 The dominant weight diagram for $(2,2)\times(2,2)$ in A_2.

We are going to calculate only the highest weight states of $\boldsymbol{M}_4 = (2,2)$ in detail because the multiplicity of the representation $\boldsymbol{M}_4 = (2,2)$ in the decomposition of $(2,2)\times(2,2)$ in A_2 is three. The remaining calculations are left as exercise.

First we express the highest weight state of $\boldsymbol{M}_4 = (2,2)$ as a linear combination of the normalized highest weight states of A_1-multiplets.

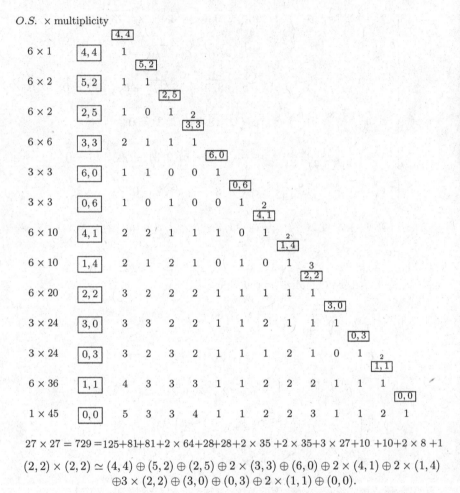

$$
\begin{aligned}
\|(2,2),(2,2)\rangle =\ & \frac{a_1}{\sqrt{10}}\left[\left|\begin{smallmatrix}4\;2\;0\\4\;2\\4\end{smallmatrix}\right\rangle^{1}\left|\begin{smallmatrix}4\;2\;0\\4\;0\\2\end{smallmatrix}\right\rangle^{5}-\sqrt{3}\left|\begin{smallmatrix}4\;2\;0\\4\;2\\3\end{smallmatrix}\right\rangle^{2}\left|\begin{smallmatrix}4\;2\;0\\4\;0\\3\end{smallmatrix}\right\rangle^{4}\right.\\
&\left.+\sqrt{6}\left|\begin{smallmatrix}4\;2\;0\\4\;2\\2\end{smallmatrix}\right\rangle^{3}\left|\begin{smallmatrix}4\;2\;0\\4\;0\\4\end{smallmatrix}\right\rangle^{3}\right]+\frac{a_2}{\sqrt{2}}\left[\left|\begin{smallmatrix}4\;2\;0\\4\;2\\4\end{smallmatrix}\right\rangle^{1}\left|\begin{smallmatrix}4\;2\;0\\3\;1\\2\end{smallmatrix}\right\rangle^{5}\right.\\
&\left.-\left|\begin{smallmatrix}4\;2\;0\\4\;2\\3\end{smallmatrix}\right\rangle^{2}\left|\begin{smallmatrix}4\;2\;0\\3\;1\\3\end{smallmatrix}\right\rangle^{4}\right]+a_3\left|\begin{smallmatrix}4\;2\;0\\4\;2\\4\end{smallmatrix}\right\rangle^{1}\left|\begin{smallmatrix}4\;2\;0\\2\;2\\2\end{smallmatrix}\right\rangle^{5}\\
&+\frac{a_4}{2}\left[\sqrt{3}\left|\begin{smallmatrix}4\;2\;0\\4\;1\\4\end{smallmatrix}\right\rangle^{2}\left|\begin{smallmatrix}4\;2\;0\\3\;2\\2\end{smallmatrix}\right\rangle^{4}-\left|\begin{smallmatrix}4\;2\;0\\4\;1\\3\end{smallmatrix}\right\rangle^{3}\left|\begin{smallmatrix}4\;2\;0\\3\;2\\3\end{smallmatrix}\right\rangle^{3}\right]\\
&+\text{ the terms } b_j \text{ obtained by interchanging } |\boldsymbol{m}^{(1)}\rangle \text{ and } |\boldsymbol{m}^{(2)}\rangle\\
&+\frac{a_5}{\sqrt{10}}\left[\sqrt{3}\left|\begin{smallmatrix}4\;2\;0\\4\;1\\4\end{smallmatrix}\right\rangle^{2}\left|\begin{smallmatrix}4\;2\;0\\4\;1\\2\end{smallmatrix}\right\rangle^{4}-2\left|\begin{smallmatrix}4\;2\;0\\4\;1\\3\end{smallmatrix}\right\rangle^{3}\left|\begin{smallmatrix}4\;2\;0\\4\;1\\3\end{smallmatrix}\right\rangle^{3}\right.\\
&\left.+\sqrt{3}\left|\begin{smallmatrix}4\;2\;0\\4\;1\\2\end{smallmatrix}\right\rangle^{4}\left|\begin{smallmatrix}4\;2\;0\\4\;1\\4\end{smallmatrix}\right\rangle^{2}\right]+a_6\left|\begin{smallmatrix}4\;2\;0\\3\;2\\3\end{smallmatrix}\right\rangle^{3}\left|\begin{smallmatrix}4\;2\;0\\3\;2\\3\end{smallmatrix}\right\rangle^{3}
\end{aligned}
$$

The coefficients are determined by the annihilation condition of E_2.

$$
\begin{aligned}
E_2\,\|(2,2),(2,2)\rangle =\ & \frac{a_1}{\sqrt{10}}\left[\left|\begin{smallmatrix}4\;2\;0\\4\;2\\4\end{smallmatrix}\right\rangle^{1}\left|\begin{smallmatrix}4\;2\;0\\4\;1\\2\end{smallmatrix}\right\rangle^{4}\right.\\
&\left.-\sqrt{3}\times\sqrt{\frac{3}{2}}\left|\begin{smallmatrix}4\;2\;0\\4\;2\\3\end{smallmatrix}\right\rangle^{2}\left|\begin{smallmatrix}4\;2\;0\\4\;1\\3\end{smallmatrix}\right\rangle^{3}+\sqrt{6}\times\sqrt{2}\left|\begin{smallmatrix}4\;2\;0\\4\;2\\2\end{smallmatrix}\right\rangle^{3}\left|\begin{smallmatrix}4\;2\;0\\4\;1\\4\end{smallmatrix}\right\rangle^{2}\right]\\
&+\frac{a_2}{\sqrt{2}}\left[\left|\begin{smallmatrix}4\;2\;0\\4\;2\\4\end{smallmatrix}\right\rangle^{1}\left(\sqrt{\frac{5}{3}}\left|\begin{smallmatrix}4\;2\;0\\4\;1\\2\end{smallmatrix}\right\rangle^{4}+\sqrt{\frac{4}{3}}\left|\begin{smallmatrix}4\;2\;0\\3\;2\\2\end{smallmatrix}\right\rangle^{4}\right)\right.\\
&\left.-\left|\begin{smallmatrix}4\;2\;0\\4\;2\\3\end{smallmatrix}\right\rangle^{2}\left(\sqrt{\frac{5}{6}}\left|\begin{smallmatrix}4\;2\;0\\4\;1\\3\end{smallmatrix}\right\rangle^{3}+\sqrt{\frac{8}{3}}\left|\begin{smallmatrix}4\;2\;0\\3\;2\\3\end{smallmatrix}\right\rangle^{3}\right)\right]+a_3 2\left|\begin{smallmatrix}4\;2\;0\\4\;2\\4\end{smallmatrix}\right\rangle^{1}\left|\begin{smallmatrix}4\;2\;0\\3\;2\\2\end{smallmatrix}\right\rangle^{4}\\
&+\frac{a_4}{2}\left[\sqrt{3}\left(\sqrt{\frac{10}{3}}\left|\begin{smallmatrix}4\;2\;0\\4\;1\\4\end{smallmatrix}\right\rangle^{2}\left|\begin{smallmatrix}4\;2\;0\\4\;2\\2\end{smallmatrix}\right\rangle^{3}+\sqrt{2}\left|\begin{smallmatrix}4\;2\;0\\4\;2\\4\end{smallmatrix}\right\rangle^{1}\left|\begin{smallmatrix}4\;2\;0\\3\;2\\2\end{smallmatrix}\right\rangle^{4}\right)\right.\\
&\left.-\left(\sqrt{\frac{5}{3}}\left|\begin{smallmatrix}4\;2\;0\\4\;1\\3\end{smallmatrix}\right\rangle^{3}\left|\begin{smallmatrix}4\;2\;0\\4\;2\\3\end{smallmatrix}\right\rangle^{2}+\sqrt{\frac{4}{3}}\left|\begin{smallmatrix}4\;2\;0\\4\;2\\3\end{smallmatrix}\right\rangle^{2}\left|\begin{smallmatrix}4\;2\;0\\3\;2\\3\end{smallmatrix}\right\rangle^{3}\right)\right]\\
&+\text{ the terms } b_j \text{ obtained by interchanging } |\boldsymbol{m}^{(1)}\rangle \text{ and } |\boldsymbol{m}^{(2)}\rangle\\
&+\frac{a_5}{\sqrt{10}}\left[\sqrt{3}\left(\sqrt{\frac{2}{3}}\left|\begin{smallmatrix}4\;2\;0\\4\;1\\4\end{smallmatrix}\right\rangle^{2}\left|\begin{smallmatrix}4\;2\;0\\4\;2\\2\end{smallmatrix}\right\rangle^{3}+\sqrt{2}\left|\begin{smallmatrix}4\;2\;0\\4\;2\\4\end{smallmatrix}\right\rangle^{1}\left|\begin{smallmatrix}4\;2\;0\\4\;1\\2\end{smallmatrix}\right\rangle^{4}\right)\right.\\
&-2\times\sqrt{\frac{4}{3}}\left(\left|\begin{smallmatrix}4\;2\;0\\4\;1\\3\end{smallmatrix}\right\rangle^{3}\left|\begin{smallmatrix}4\;2\;0\\4\;2\\3\end{smallmatrix}\right\rangle^{2}+\left|\begin{smallmatrix}4\;2\;0\\4\;2\\3\end{smallmatrix}\right\rangle^{2}\left|\begin{smallmatrix}4\;2\;0\\4\;1\\3\end{smallmatrix}\right\rangle^{3}\right)\\
&\left.+\sqrt{3}\left(\sqrt{2}\left|\begin{smallmatrix}4\;2\;0\\4\;1\\2\end{smallmatrix}\right\rangle^{4}\left|\begin{smallmatrix}4\;2\;0\\4\;2\\4\end{smallmatrix}\right\rangle^{1}+\sqrt{\frac{2}{3}}\left|\begin{smallmatrix}4\;2\;0\\4\;2\\2\end{smallmatrix}\right\rangle^{3}\left|\begin{smallmatrix}4\;2\;0\\4\;1\\4\end{smallmatrix}\right\rangle^{2}\right)\right]
\end{aligned}
$$

$$+\,a_6\sqrt{\frac{5}{3}}\left(\left|\begin{smallmatrix}4&&2&&0\\&3&&2&\\&&3&&\end{smallmatrix}\right\rangle^{3}\left|\begin{smallmatrix}4&&2&&0\\&4&&2&\\&&3&&\end{smallmatrix}\right\rangle^{2}+\left|\begin{smallmatrix}4&&2&&0\\&4&&2&\\&&3&&\end{smallmatrix}\right\rangle^{2}\left|\begin{smallmatrix}4&&2&&0\\&3&&2&\\&&3&&\end{smallmatrix}\right\rangle^{3}\right)=0.$$

Collecting the same basis states, we obtain the equations for the coefficients.

$$\left|\begin{smallmatrix}4&&2&&0\\&4&&2&\\&&4&&\end{smallmatrix}\right\rangle^{1}\left|\begin{smallmatrix}4&&2&&0\\&4&&1&\\&&2&&\end{smallmatrix}\right\rangle^{4}:\quad a_1\sqrt{\tfrac{1}{10}}+a_2\sqrt{\tfrac{5}{6}}+a_5\sqrt{\tfrac{3}{5}}=0$$

$$\left|\begin{smallmatrix}4&&2&&0\\&4&&2&\\&&4&&\end{smallmatrix}\right\rangle^{1}\left|\begin{smallmatrix}4&&2&&0\\&3&&2&\\&&2&&\end{smallmatrix}\right\rangle^{4}:\quad a_2\sqrt{\tfrac{2}{3}}+2a_3+a_4\sqrt{\tfrac{3}{2}}=0$$

$$\left|\begin{smallmatrix}4&&2&&0\\&4&&2&\\&&3&&\end{smallmatrix}\right\rangle^{2}\left|\begin{smallmatrix}4&&2&&0\\&4&&1&\\&&3&&\end{smallmatrix}\right\rangle^{3}:\quad -\tfrac{3}{2\sqrt{5}}a_1-a_2\sqrt{\tfrac{5}{12}}-b_4\sqrt{\tfrac{5}{12}}-a_5\sqrt{\tfrac{8}{15}}=0$$

$$\left|\begin{smallmatrix}4&&2&&0\\&4&&2&\\&&3&&\end{smallmatrix}\right\rangle^{2}\left|\begin{smallmatrix}4&&2&&0\\&3&&2&\\&&3&&\end{smallmatrix}\right\rangle^{3}:\quad -a_2\sqrt{\tfrac{4}{3}}-a_4\sqrt{\tfrac{1}{3}}+a_6\sqrt{\tfrac{5}{3}}=0$$

$$\left|\begin{smallmatrix}4&&2&&0\\&4&&2&\\&&2&&\end{smallmatrix}\right\rangle^{3}\left|\begin{smallmatrix}4&&2&&0\\&4&&1&\\&&4&&\end{smallmatrix}\right\rangle^{2}:\quad a_1\sqrt{\tfrac{6}{5}}+b_4\sqrt{\tfrac{5}{2}}+a_5\sqrt{\tfrac{1}{5}}=0$$

For the antisymmetric representation, $a_j=-b_j$ and $a_5=a_6=0$, we solve the equations and normalize the solution:

$$a_1=5N,\quad a_2=-\sqrt{3}N,\quad a_3=\sqrt{2}N,\quad a_4=2\sqrt{3}N,$$

where the normalization factor is $N^{-2}=2(25+3+2+12)=84$.

For the symmetric representation, $b_j=a_j$, we obtain the equations for the coefficients:

$$a_3=a_1/\sqrt{8}+a_2/(2\sqrt{6}),\qquad a_4=-a_1/\sqrt{3}+a_2/3,$$
$$a_5=-a_1\sqrt{1/6}-a_2(5/6)\sqrt{2},\qquad a_6=-a_1/\sqrt{15}+a_2(7/3\sqrt{5}).$$

The six basic solutions $\psi^{(i)}(a_j)$, $1\leqslant i\leqslant 6$, are

No.	a_1	a_2	a_3	a_4	a_5	a_6	N'
1	0	$6\sqrt{10}$	$\sqrt{15}$	$2\sqrt{10}$	$-10\sqrt{5}$	$14\sqrt{2}$	1622
2	$2\sqrt{30}$	0	$\sqrt{15}$	$-2\sqrt{10}$	$-2\sqrt{5}$	$-2\sqrt{2}$	378
3	$\sqrt{15}$	$-3\sqrt{5}$	0	$-2\sqrt{5}$	$2\sqrt{10}$	-8	344
4	$\sqrt{30}$	$3\sqrt{10}$	$\sqrt{15}$	0	$-6\sqrt{5}$	$6\sqrt{2}$	522
5	$5\sqrt{5}$	$-\sqrt{15}$	$\sqrt{10}$	$-2\sqrt{15}$	0	$-4\sqrt{3}$	468
6	7	$\sqrt{3}$	$\sqrt{8}$	$-2\sqrt{3}$	$-2\sqrt{6}$	0	168

where $N'=N^{-2}=2(a_1^2+a_2^2+a_3^2+a_4^2)+a_5+a_6$ and N is the normalization factor. It seems that the solution $\psi^{(6)}(a_j)$ is the simplest. The solution orthogonal to $\psi^{(6)}(a_j)$ is unique up to a constant factor. Since $2(a_1^{(3)}a_1^{(6)} + a_2^{(3)}a_2^{(6)} + a_3^{(3)}a_3^{(6)} + a_4^{(3)}a_4^{(6)}) + a_5^{(3)}a_5^{(6)} + a_6^{(3)}a_6^{(6)} = 8\sqrt{15}$, $[21\psi^{(3)}(a_j) - \sqrt{15}\psi^{(6)}(a_j)]/2$, where $a_1 = 7\sqrt{15}$, $a_2 = -33\sqrt{5}$, $a_3 = -\sqrt{30}$, $a_4 = -18\sqrt{5}$, $a_5 = 24\sqrt{10}$ and $a_6 = -84$, is orthogonal to $\psi^{(6)}(a_j)$. Thus, we obtain three orthogonal highest weight states of the representation $(2,2)$ in the decomposition of $(2,2) \times (2,2)$ in A_2.

$$\begin{aligned}
||(2,2),(2,2)\rangle_A = \sqrt{\frac{1}{168}}\Bigg\{ & \sqrt{5}\left|\begin{smallmatrix}4&&2&&0\\&4&&2&\\&&4&&\end{smallmatrix}\right\rangle^{1}\left|\begin{smallmatrix}4&&2&&0\\&4&&0&\\&&2&&\end{smallmatrix}\right\rangle^{5} - \sqrt{15}\left|\begin{smallmatrix}4&&2&&0\\&4&&2&\\&&3&&\end{smallmatrix}\right\rangle^{2}\left|\begin{smallmatrix}4&&2&&0\\&4&&0&\\&&3&&\end{smallmatrix}\right\rangle^{4} \\
& + \sqrt{30}\left|\begin{smallmatrix}4&&2&&0\\&4&&2&\\&&2&&\end{smallmatrix}\right\rangle^{3}\left|\begin{smallmatrix}4&&2&&0\\&4&&0&\\&&4&&\end{smallmatrix}\right\rangle^{3} - \sqrt{3}\left|\begin{smallmatrix}4&&2&&0\\&4&&2&\\&&4&&\end{smallmatrix}\right\rangle^{1}\left|\begin{smallmatrix}4&&2&&0\\&3&&1&\\&&2&&\end{smallmatrix}\right\rangle^{5} \\
& + \sqrt{3}\left|\begin{smallmatrix}4&&2&&0\\&4&&2&\\&&3&&\end{smallmatrix}\right\rangle^{2}\left|\begin{smallmatrix}4&&2&&0\\&3&&1&\\&&3&&\end{smallmatrix}\right\rangle^{4} + 2\left|\begin{smallmatrix}4&&2&&0\\&4&&2&\\&&4&&\end{smallmatrix}\right\rangle^{1}\left|\begin{smallmatrix}4&&2&&0\\&2&&2&\\&&2&&\end{smallmatrix}\right\rangle^{5} \\
& + 3\sqrt{2}\left|\begin{smallmatrix}4&&2&&0\\&4&&1&\\&&4&&\end{smallmatrix}\right\rangle^{2}\left|\begin{smallmatrix}4&&2&&0\\&3&&2&\\&&2&&\end{smallmatrix}\right\rangle^{4} - \sqrt{6}\left|\begin{smallmatrix}4&&2&&0\\&4&&1&\\&&3&&\end{smallmatrix}\right\rangle^{3}\left|\begin{smallmatrix}4&&2&&0\\&3&&2&\\&&3&&\end{smallmatrix}\right\rangle^{3} \\
& - \text{ the terms obtained by interchanging } |\boldsymbol{m}^{(1)}\rangle \text{ and } |\boldsymbol{m}^{(2)}\rangle \Bigg\},
\end{aligned}$$

$$\begin{aligned}
||(2,2),(2,2)\rangle_{S1} = \sqrt{\frac{1}{1680}}\Bigg\{ & 7\left|\begin{smallmatrix}4&&2&&0\\&4&&2&\\&&4&&\end{smallmatrix}\right\rangle^{1}\left|\begin{smallmatrix}4&&2&&0\\&4&&0&\\&&2&&\end{smallmatrix}\right\rangle^{5} - 7\sqrt{3}\left|\begin{smallmatrix}4&&2&&0\\&4&&2&\\&&3&&\end{smallmatrix}\right\rangle^{2}\left|\begin{smallmatrix}4&&2&&0\\&4&&0&\\&&3&&\end{smallmatrix}\right\rangle^{4} \\
& + 7\sqrt{6}\left|\begin{smallmatrix}4&&2&&0\\&4&&2&\\&&2&&\end{smallmatrix}\right\rangle^{3}\left|\begin{smallmatrix}4&&2&&0\\&4&&0&\\&&4&&\end{smallmatrix}\right\rangle^{3} + \sqrt{15}\left|\begin{smallmatrix}4&&2&&0\\&4&&2&\\&&4&&\end{smallmatrix}\right\rangle^{1}\left|\begin{smallmatrix}4&&2&&0\\&3&&1&\\&&2&&\end{smallmatrix}\right\rangle^{5} \\
& - \sqrt{15}\left|\begin{smallmatrix}4&&2&&0\\&4&&2&\\&&3&&\end{smallmatrix}\right\rangle^{2}\left|\begin{smallmatrix}4&&2&&0\\&3&&1&\\&&3&&\end{smallmatrix}\right\rangle^{4} + 4\sqrt{5}\left|\begin{smallmatrix}4&&2&&0\\&4&&2&\\&&4&&\end{smallmatrix}\right\rangle^{1}\left|\begin{smallmatrix}4&&2&&0\\&2&&2&\\&&2&&\end{smallmatrix}\right\rangle^{5} \\
& - 3\sqrt{10}\left|\begin{smallmatrix}4&&2&&0\\&4&&1&\\&&4&&\end{smallmatrix}\right\rangle^{2}\left|\begin{smallmatrix}4&&2&&0\\&3&&2&\\&&2&&\end{smallmatrix}\right\rangle^{4} + \sqrt{30}\left|\begin{smallmatrix}4&&2&&0\\&4&&1&\\&&3&&\end{smallmatrix}\right\rangle^{3}\left|\begin{smallmatrix}4&&2&&0\\&3&&2&\\&&3&&\end{smallmatrix}\right\rangle^{3} \\
& + \text{ the terms obtained by interchanging } |\boldsymbol{m}^{(1)}\rangle \text{ and } |\boldsymbol{m}^{(2)}\rangle \\
& - 6\sqrt{2}\left|\begin{smallmatrix}4&&2&&0\\&4&&1&\\&&4&&\end{smallmatrix}\right\rangle^{2}\left|\begin{smallmatrix}4&&2&&0\\&4&&1&\\&&2&&\end{smallmatrix}\right\rangle^{4} + 4\sqrt{6}\left|\begin{smallmatrix}4&&2&&0\\&4&&1&\\&&3&&\end{smallmatrix}\right\rangle^{3}\left|\begin{smallmatrix}4&&2&&0\\&4&&1&\\&&3&&\end{smallmatrix}\right\rangle^{3} \\
& - 6\sqrt{2}\left|\begin{smallmatrix}4&&2&&0\\&4&&1&\\&&2&&\end{smallmatrix}\right\rangle^{4}\left|\begin{smallmatrix}4&&2&&0\\&4&&1&\\&&4&&\end{smallmatrix}\right\rangle^{2}\Bigg\},
\end{aligned}$$

$$\begin{aligned}
||(2,2),(2,2)\rangle_{S2} = \sqrt{\frac{1}{56952}}\Bigg\{ & 7\sqrt{3}\left|\begin{smallmatrix}4&&2&&0\\&4&&2&\\&&4&&\end{smallmatrix}\right\rangle^{1}\left|\begin{smallmatrix}4&&2&&0\\&4&&0&\\&&2&&\end{smallmatrix}\right\rangle^{5} - 21\left|\begin{smallmatrix}4&&2&&0\\&4&&2&\\&&3&&\end{smallmatrix}\right\rangle^{2}\left|\begin{smallmatrix}4&&2&&0\\&4&&0&\\&&3&&\end{smallmatrix}\right\rangle^{4} \\
& + 21\sqrt{2}\left|\begin{smallmatrix}4&&2&&0\\&4&&2&\\&&2&&\end{smallmatrix}\right\rangle^{3}\left|\begin{smallmatrix}4&&2&&0\\&4&&0&\\&&4&&\end{smallmatrix}\right\rangle^{3} - 33\sqrt{5}\left|\begin{smallmatrix}4&&2&&0\\&4&&2&\\&&4&&\end{smallmatrix}\right\rangle^{1}\left|\begin{smallmatrix}4&&2&&0\\&3&&1&\\&&2&&\end{smallmatrix}\right\rangle^{5} \\
& + 33\sqrt{5}\left|\begin{smallmatrix}4&&2&&0\\&4&&2&\\&&3&&\end{smallmatrix}\right\rangle^{2}\left|\begin{smallmatrix}4&&2&&0\\&3&&1&\\&&3&&\end{smallmatrix}\right\rangle^{4} - 2\sqrt{15}\left|\begin{smallmatrix}4&&2&&0\\&4&&2&\\&&4&&\end{smallmatrix}\right\rangle^{1}\left|\begin{smallmatrix}4&&2&&0\\&2&&2&\\&&2&&\end{smallmatrix}\right\rangle^{5}
\end{aligned}$$

$$
\begin{aligned}
&-9\sqrt{30}\left|\begin{smallmatrix}4&&2&&0\\&4&&1&\\&&4&&\end{smallmatrix}\right\rangle^{2}\left|\begin{smallmatrix}4&&2&&0\\&3&&2&\\&&2&&\end{smallmatrix}\right\rangle^{4}+9\sqrt{10}\left|\begin{smallmatrix}4&&2&&0\\&4&&1&\\&&3&&\end{smallmatrix}\right\rangle^{3}\left|\begin{smallmatrix}4&&2&&0\\&3&&2&\\&&3&&\end{smallmatrix}\right\rangle^{3}\\
&+\ \text{the terms obtained by interchanging } |\boldsymbol{m}^{(1)}\rangle \text{ and } |\boldsymbol{m}^{(2)}\rangle\\
&+24\sqrt{6}\left|\begin{smallmatrix}4&&2&&0\\&4&&1&\\&&4&&\end{smallmatrix}\right\rangle^{2}\left|\begin{smallmatrix}4&&2&&0\\&4&&1&\\&&2&&\end{smallmatrix}\right\rangle^{4}-48\sqrt{2}\left|\begin{smallmatrix}4&&2&&0\\&4&&1&\\&&3&&\end{smallmatrix}\right\rangle^{3}\left|\begin{smallmatrix}4&&2&&0\\&4&&1&\\&&3&&\end{smallmatrix}\right\rangle^{3}\\
&+24\sqrt{6}\left|\begin{smallmatrix}4&&2&&0\\&4&&1&\\&&2&&\end{smallmatrix}\right\rangle^{4}\left|\begin{smallmatrix}4&&2&&0\\&4&&1&\\&&4&&\end{smallmatrix}\right\rangle^{2}-84\sqrt{2}\left|\begin{smallmatrix}4&&2&&0\\&3&&2&\\&&3&&\end{smallmatrix}\right\rangle^{3}\left|\begin{smallmatrix}4&&2&&0\\&3&&2&\\&&3&&\end{smallmatrix}\right\rangle^{3}\Bigg\}.
\end{aligned}
$$

Ex. 6. The decomposition of $(1,0,1)\times(1,0,1)$ in A_3.

From Fig. 8.9 there are six dominant weights $\boldsymbol{M}_\alpha$ in the product space of $(1,0,1)\times(1,0,1)$ in A_3. The multiplicities n_α of $\boldsymbol{M}_\alpha$ in the direct product space can be counted in Fig. 8.9. The results are $n_0=1$, $n_1=n_{1'}=2$, $n_2=4$, $n_3=10$, and $n_4=21$. The multiplicities $n_{\alpha\beta}$ of the dominant weight $\boldsymbol{M}_\beta$ in the representation $\boldsymbol{M}_\alpha$ can be counted by the Gel'fand's bases (see the statement following Eq. (8.18)), where $n_{0\alpha}$ are counted as example.

$\boldsymbol{M}_0=(2,0,2)$

α	$\boldsymbol{M}_\alpha$	$\Omega_4\Omega_3\Omega_2\Omega_1$	$n_{0\alpha}$	the basis states
0	$(2,0,2)$	$8,8,6,4$	1	$\left\vert\begin{smallmatrix}4&&2&&2&&0\\&4&&2&&2&\\&&4&&2&&\\&&&4&&&\end{smallmatrix}\right\rangle^{1}$
1	$(2,1,0)$	$8,8,6,3$	1	$\left\vert\begin{smallmatrix}4&&2&&2&&0\\&4&&2&&1&\\&&4&&2&&\\&&&4&&&\end{smallmatrix}\right\rangle^{2}$
1′	$(0,1,2)$	$8,7,6,4$	1	$\left\vert\begin{smallmatrix}4&&2&&2&&0\\&4&&2&&2&\\&&4&&2&&\\&&&3&&&\end{smallmatrix}\right\rangle^{2}$
2	$(0,2,0)$	$8,7,6,3$	1	$\left\vert\begin{smallmatrix}4&&2&&2&&0\\&4&&2&&1&\\&&4&&2&&\\&&&3&&&\end{smallmatrix}\right\rangle^{3}$
3	$(1,0,1)$	$8,7,5,3$	3	$\left\vert\begin{smallmatrix}4&&2&&2&&0\\&4&&2&&1&\\&&4&&1&&\\&&&3&&&\end{smallmatrix}\right\rangle^{4}$, $\left\vert\begin{smallmatrix}4&&2&&2&&0\\&4&&2&&1&\\&&3&&2&&\\&&&3&&&\end{smallmatrix}\right\rangle^{4}$, $\left\vert\begin{smallmatrix}4&&2&&2&&0\\&3&&2&&2&\\&&3&&2&&\\&&&3&&&\end{smallmatrix}\right\rangle^{4}$
4	$(0,0,0)$	$8,6,4,2$	6	$\left\vert\begin{smallmatrix}4&&2&&2&&0\\&4&&2&&0&\\&&4&&0&&\\&&&2&&&\end{smallmatrix}\right\rangle^{7}$, $\left\vert\begin{smallmatrix}4&&2&&2&&0\\&4&&2&&0&\\&&3&&1&&\\&&&2&&&\end{smallmatrix}\right\rangle^{7}$, $\left\vert\begin{smallmatrix}4&&2&&2&&0\\&3&&2&&1&\\&&2&&2&&\\&&&2&&&\end{smallmatrix}\right\rangle^{7}$,
				$\left\vert\begin{smallmatrix}4&&2&&2&&0\\&3&&2&&1&\\&&3&&1&&\\&&&2&&&\end{smallmatrix}\right\rangle^{7}$, $\left\vert\begin{smallmatrix}4&&2&&2&&0\\&3&&2&&1&\\&&2&&2&&\\&&&2&&&\end{smallmatrix}\right\rangle^{7}$, $\left\vert\begin{smallmatrix}4&&2&&2&&0\\&2&&2&&2&\\&&2&&2&&\\&&&2&&&\end{smallmatrix}\right\rangle^{7}$

Thus, the dominant weight diagram of $(1,0,1)\times(1,0,1)$ of A_3 is calculated in Fig. 8.20.

$OS(\boldsymbol{M}_\alpha)\times n_\alpha$							
		[2,0,2]					
12×1	[2,0,2]	1	[2,1,0]				
12×2	[2,1,0]	1	1	[0,1,2]			
12×2	[0,1,2]	1	0	1	[0,2,0]		
6×4	[0,2,0]	1	1	1	1	2 [1,0,1]	
12×10	[1,0,1]	3	2	2	1	1	[0,0,0]
1×21	[0,0,0]	6	3	3	2	3	1
$15\times 15=225$	$=$	84 +	45 +	45 +	20	$+2\times 15$ +	1

$$(1,0,1)\times(1,0,1)\simeq(2,0,2)\oplus(2,1,0)\oplus(0,1,2)\oplus(0,2,0)\oplus 2\times(1,0,1)+(0,0,0)$$

Fig. 8.20 The dominant weight diagram for $(1,0,1)\times(1,0,1)$ in A_3

The expansions (8.49) for the highest weight states of $\boldsymbol{M}_0$ and $\boldsymbol{M}_4$ are given in Eqs. (8.50) and (8.54). The expansions (8.49) for the highest weight states of $\boldsymbol{M}_1$, $\boldsymbol{M}_{1'}$ and $\boldsymbol{M}_2$ can be calculated easily.

$$||(2,0,2),(2,0,2)\rangle_S=|(1,0,1)\rangle|(1,0,1)\rangle=\left|\begin{smallmatrix}2&&1&&1&&0\\&2&&1&&1&\\&&2&&1&&\\&&&2&&&\end{smallmatrix}\right\rangle^{1}\left|\begin{smallmatrix}2&&1&&1&&0\\&2&&1&&1&\\&&2&&1&&\\&&&2&&&\end{smallmatrix}\right\rangle^{1},$$

$$\begin{aligned}||(2,1,0),(2,1,0)\rangle_A&=\sqrt{\frac{1}{2}}\ \left\{\ |(1,0,1)\rangle|(1,1,\overline{1})\rangle-|(1,1,\overline{1})\rangle|(1,0,1)\rangle\ \right\}\\&=\sqrt{\frac{1}{2}}\left\{\left|\begin{smallmatrix}2&&1&&1&&0\\&2&&1&&1&\\&&2&&1&&\\&&&2&&&\end{smallmatrix}\right\rangle^{1}\left|\begin{smallmatrix}2&&1&&1&&0\\&2&&1&&0&\\&&2&&1&&\\&&&2&&&\end{smallmatrix}\right\rangle^{2}-\left|\begin{smallmatrix}2&&1&&1&&0\\&2&&1&&0&\\&&2&&1&&\\&&&2&&&\end{smallmatrix}\right\rangle^{2}\left|\begin{smallmatrix}2&&1&&1&&0\\&2&&1&&1&\\&&2&&1&&\\&&&2&&&\end{smallmatrix}\right\rangle^{1}\right\},\end{aligned}$$

$$\begin{aligned}||(0,1,2),(0,1,2)\rangle_A&=\sqrt{\frac{1}{2}}\left\{\ |(1,0,1)\rangle|(\overline{1},1,1)\rangle-|(\overline{1},1,1)\rangle|(1,0,1)\rangle\ \right\}\\&=\sqrt{\frac{1}{2}}\left\{\left|\begin{smallmatrix}2&&1&&1&&0\\&2&&1&&1&\\&&2&&1&&\\&&&2&&&\end{smallmatrix}\right\rangle^{1}\left|\begin{smallmatrix}2&&1&&1&&0\\&2&&1&&1&\\&&2&&1&&\\&&&1&&&\end{smallmatrix}\right\rangle^{2}-\left|\begin{smallmatrix}2&&1&&1&&0\\&2&&1&&1&\\&&2&&1&&\\&&&1&&&\end{smallmatrix}\right\rangle^{2}\left|\begin{smallmatrix}2&&1&&1&&0\\&2&&1&&1&\\&&2&&1&&\\&&&2&&&\end{smallmatrix}\right\rangle^{1}\right\},\end{aligned}$$

$$\begin{aligned}||(0,2,0),(0,2,0)\rangle_S&=\frac{1}{2}\left\{\ |(1,0,1)\rangle|(\overline{1},2,\overline{1})\rangle-\ |(1,1,\overline{1})\rangle|(\overline{1},1,1)\rangle\right.\\&\quad\left.-\ |(\overline{1},1,1)\rangle|(1,1,\overline{1})\rangle+|(\overline{1},2,\overline{1})\rangle|(1,0,1)\rangle\ \right\}\\&=\frac{1}{2}\left\{\left|\begin{smallmatrix}2&&1&&1&&0\\&2&&1&&1&\\&&2&&1&&\\&&&2&&&\end{smallmatrix}\right\rangle^{1}\left|\begin{smallmatrix}2&&1&&1&&0\\&2&&1&&0&\\&&2&&1&&\\&&&1&&&\end{smallmatrix}\right\rangle^{3}-\left|\begin{smallmatrix}2&&1&&1&&0\\&2&&1&&1&\\&&2&&1&&\\&&&1&&&\end{smallmatrix}\right\rangle^{2}\left|\begin{smallmatrix}2&&1&&1&&0\\&2&&1&&0&\\&&2&&1&&\\&&&2&&&\end{smallmatrix}\right\rangle^{2}\right.\\&\quad\left.-\left|\begin{smallmatrix}2&&1&&1&&0\\&2&&1&&0&\\&&2&&1&&\\&&&2&&&\end{smallmatrix}\right\rangle^{2}\left|\begin{smallmatrix}2&&1&&1&&0\\&2&&1&&1&\\&&2&&1&&\\&&&1&&&\end{smallmatrix}\right\rangle^{2}+\left|\begin{smallmatrix}2&&1&&1&&0\\&2&&1&&0&\\&&2&&1&&\\&&&1&&&\end{smallmatrix}\right\rangle^{3}\left|\begin{smallmatrix}2&&1&&1&&0\\&2&&1&&1&\\&&2&&1&&\\&&&2&&&\end{smallmatrix}\right\rangle^{1}\right\},\end{aligned}$$

$$\begin{aligned}
|(0,0,0),(0,0,0)\rangle_S = \sqrt{1/15}\,\{ & \,|(1,0,1)\rangle|(\bar{1},0,\bar{1})\rangle - |(\bar{1},1,1)\rangle|(1,\bar{1},\bar{1})\rangle \\
& - |(1,1,\bar{1})\rangle|(\bar{1},\bar{1},1)\rangle + |(0,\bar{1},2)\rangle|(0,1,\bar{2})\rangle + |(0,2,0)\rangle|(0,\bar{2},0)\rangle \\
& + |(2,\bar{1},0)\rangle|(\bar{2},1,0)\rangle - |(0,0,0)_1\rangle|(0,0,0)_1\rangle - |(0,0,0)_2\rangle|(0,0,0)_2\rangle \\
& - |(0,0,0)_3\rangle|(0,0,0)_3\rangle + |(\bar{2},1,0)\rangle|(2,\bar{1},0)\rangle + |(0,\bar{2},0)\rangle|(0,2,0)\rangle \\
& + |(0,1,\bar{2})\rangle|(0,\bar{1},2)\rangle - |(\bar{1},\bar{1},1)\rangle|(1,1,\bar{1})\rangle - |(1,\bar{1},\bar{1})\rangle|(\bar{1},1,1)\rangle \\
& + |(\bar{1},0,\bar{1})\rangle|(1,0,1)\rangle\}
\end{aligned}$$

$$\begin{aligned}
= \sqrt{\frac{1}{15}}\Bigg\{ & \left|\begin{array}{ccccccc}2&&1&&1&&0\\&2&&1&&1&\\&&2&&1&&\\&&&2&&&\end{array}\right\rangle^{1}\left|\begin{array}{ccccccc}2&&1&&1&&0\\&1&&1&&0&\\&&1&&0&&\\&&&0&&&\end{array}\right\rangle^{7} - \left|\begin{array}{ccccccc}2&&1&&1&&0\\&2&&1&&1&\\&&2&&1&&\\&&&1&&&\end{array}\right\rangle^{2}\left|\begin{array}{ccccccc}2&&1&&1&&0\\&1&&1&&0&\\&&1&&0&&\\&&&1&&&\end{array}\right\rangle^{6} \\
& - \left|\begin{array}{ccccccc}2&&1&&1&&0\\&2&&1&&0&\\&&2&&1&&\\&&&2&&&\end{array}\right\rangle^{2}\left|\begin{array}{ccccccc}2&&1&&1&&0\\&2&&1&&0&\\&&1&&0&&\\&&&0&&&\end{array}\right\rangle^{6} + \left|\begin{array}{ccccccc}2&&1&&1&&0\\&2&&1&&1&\\&&1&&1&&\\&&&1&&&\end{array}\right\rangle^{3}\left|\begin{array}{ccccccc}2&&1&&1&&0\\&1&&1&&0&\\&&1&&1&&\\&&&1&&&\end{array}\right\rangle^{5} \\
& + \left|\begin{array}{ccccccc}2&&1&&1&&0\\&2&&1&&0&\\&&2&&1&&\\&&&1&&&\end{array}\right\rangle^{3}\left|\begin{array}{ccccccc}2&&1&&1&&0\\&2&&1&&0&\\&&1&&0&&\\&&&1&&&\end{array}\right\rangle^{5} + \left|\begin{array}{ccccccc}2&&1&&1&&0\\&2&&1&&0&\\&&2&&0&&\\&&&2&&&\end{array}\right\rangle^{3}\left|\begin{array}{ccccccc}2&&1&&1&&0\\&2&&1&&0&\\&&2&&0&&\\&&&0&&&\end{array}\right\rangle^{5} \\
& + \text{ the terms } b_j \text{ obtained by interchanging } |\boldsymbol{m}^{(1)}\rangle \text{ and } |\boldsymbol{m}^{(2)}\rangle \\
& - \left|\begin{array}{ccccccc}2&&1&&1&&0\\&2&&1&&0&\\&&2&&0&&\\&&&1&&&\end{array}\right\rangle^{4}\left|\begin{array}{ccccccc}2&&1&&1&&0\\&2&&1&&0&\\&&2&&0&&\\&&&1&&&\end{array}\right\rangle^{4} - \left|\begin{array}{ccccccc}2&&1&&1&&0\\&2&&1&&0&\\&&1&&1&&\\&&&1&&&\end{array}\right\rangle^{4}\left|\begin{array}{ccccccc}2&&1&&1&&0\\&2&&1&&0&\\&&1&&1&&\\&&&1&&&\end{array}\right\rangle^{4} \\
& - \left|\begin{array}{ccccccc}2&&1&&1&&0\\&1&&1&&1&\\&&1&&1&&\\&&&1&&&\end{array}\right\rangle^{4}\left|\begin{array}{ccccccc}2&&1&&1&&0\\&1&&1&&1&\\&&1&&1&&\\&&&1&&&\end{array}\right\rangle^{4}\Bigg\}.
\end{aligned}$$

For the representation $\boldsymbol{M}_3 = (1,0,1)$, we express its highest weight state as a linear combination of the normalized highest weight states of A_2-multiplets (see Prob. 8), and then, calculate the coefficients in the expansion by the annihilation condition of E_3.

$$\begin{aligned}
||(1,0,1),(1,0,1)\rangle = {} & a_1 \left|\begin{array}{ccccccc}2&&1&&1&&0\\&2&&1&&1&\\&&2&&1&&\\&&&2&&&\end{array}\right\rangle^{1}\left|\begin{array}{ccccccc}2&&1&&1&&0\\&1&&1&&1&\\&&1&&1&&\\&&&1&&&\end{array}\right\rangle^{4} \\
& + \frac{a_2}{4}\Bigg[\sqrt{3}\left|\begin{array}{ccccccc}2&&1&&1&&0\\&2&&1&&1&\\&&2&&1&&\\&&&2&&&\end{array}\right\rangle^{1}\left|\begin{array}{ccccccc}2&&1&&1&&0\\&2&&1&&0&\\&&2&&0&&\\&&&1&&&\end{array}\right\rangle^{4} - \left|\begin{array}{ccccccc}2&&1&&1&&0\\&2&&1&&1&\\&&2&&1&&\\&&&2&&&\end{array}\right\rangle^{1}\left|\begin{array}{ccccccc}2&&1&&1&&0\\&2&&1&&0&\\&&1&&1&&\\&&&1&&&\end{array}\right\rangle^{4} \\
& -\sqrt{6}\left|\begin{array}{ccccccc}2&&1&&1&&0\\&2&&1&&1&\\&&2&&1&&\\&&&1&&&\end{array}\right\rangle^{2}\left|\begin{array}{ccccccc}2&&1&&1&&0\\&2&&1&&0&\\&&2&&0&&\\&&&2&&&\end{array}\right\rangle^{3} + \sqrt{6}\left|\begin{array}{ccccccc}2&&1&&1&&0\\&2&&1&&1&\\&&1&&1&&\\&&&1&&&\end{array}\right\rangle^{3}\left|\begin{array}{ccccccc}2&&1&&1&&0\\&2&&1&&0&\\&&2&&1&&\\&&&2&&&\end{array}\right\rangle^{2}\Bigg] \\
& + \text{ the terms } b_j \text{ obtained by interchanging } |\boldsymbol{m}^{(1)}\rangle \text{ and } |\boldsymbol{m}^{(2)}\rangle.
\end{aligned}$$

$$E_3\,||(1,0,1),(1,0,1)\rangle = a_1\sqrt{\frac{4}{3}}\left|\begin{array}{ccccccc}2&&1&&1&&0\\&2&&1&&1&\\&&2&&1&&\\&&&2&&&\end{array}\right\rangle^{1}\left|\begin{array}{ccccccc}2&&1&&1&&0\\&2&&1&&1&\\&&1&&1&&\\&&&1&&&\end{array}\right\rangle^{3}$$

$$+\frac{a_2}{4}\left[-\sqrt{\frac{2}{3}}\left|\begin{smallmatrix}2&&1&&1&&0\\&2&&1&&1&\\&&2&&1&&\\&&&2&&&\end{smallmatrix}\right\rangle^1\left|\begin{smallmatrix}2&&1&&1&&0\\&2&&1&&1&\\&&1&&1&&\\&&&1&&&\end{smallmatrix}\right\rangle^3+\sqrt{6}\left|\begin{smallmatrix}2&&1&&1&&0\\&2&&1&&1&\\&&1&&1&&\\&&&1&&&\end{smallmatrix}\right\rangle^3\left|\begin{smallmatrix}2&&1&&1&&0\\&2&&1&&1&\\&&2&&1&&\\&&&2&&&\end{smallmatrix}\right\rangle^1\right]$$

+ the terms b_j obtained by interchanging $|\boldsymbol{m}^{(1)}\rangle$ and $|\boldsymbol{m}^{(2)}\rangle = 0$.

Thus, $4\sqrt{2}a_1 = a_2 - 3b_2$. For the symmetric representation $(1,0,1)_S$, $a_j = b_j$, $a_2 = -2\sqrt{2}a_1$, and $a_1 = -[2(1+8)]^{-1/2} = -1/(3\sqrt{2})$. For the antisymmetric representation $(1,0,1)_A$, $a_j = -b_j$, $a_2 = \sqrt{2}a_1$, and $a_1^{-2} = 2(1+2) = 6$.

$$\begin{aligned}
||(1,0,1),(1,0,1)\rangle_S = (1/6)\, \{&\sqrt{3}\,|(1,0,1)\rangle|(0,0,0)_1\rangle - |(1,0,1)\rangle|(0,0,0)_2\rangle \\
&- \sqrt{2}\,|(1,0,1)\rangle|(0,0,0)_3\rangle - \sqrt{6}\,|(\overline{1},1,1)\rangle|(2,\overline{1},0)\rangle + \sqrt{6}\,|(0,\overline{1},2)\rangle|(1,1,\overline{1})\rangle \\
&+ \sqrt{6}\,|(1,1,\overline{1})\rangle|(0,\overline{1},2)\rangle - \sqrt{6}\,|(2,\overline{1},0)\rangle|(\overline{1},1,1)\rangle - \sqrt{2}\,|(0,0,0)_3\rangle|(1,0,1)\rangle \\
&- |(0,0,0)_2\rangle|(1,0,1)\rangle + \sqrt{3}\,|(0,0,0)_1\rangle|(1,0,1)\rangle\}
\end{aligned}$$

$$\begin{aligned}
=\frac{1}{6}\Bigg\{&\sqrt{3}\left|\begin{smallmatrix}2&&1&&1&&0\\&2&&1&&1&\\&&2&&1&&\\&&&2&&&\end{smallmatrix}\right\rangle^1\left|\begin{smallmatrix}2&&1&&1&&0\\&2&&1&&0&\\&&2&&0&&\\&&&1&&&\end{smallmatrix}\right\rangle^4 \\
&-\left|\begin{smallmatrix}2&&1&&1&&0\\&2&&1&&1&\\&&2&&1&&\\&&&2&&&\end{smallmatrix}\right\rangle^1\left|\begin{smallmatrix}2&&1&&1&&0\\&2&&1&&0&\\&&1&&1&&\\&&&1&&&\end{smallmatrix}\right\rangle^4-\sqrt{2}\left|\begin{smallmatrix}2&&1&&1&&0\\&2&&1&&1&\\&&2&&1&&\\&&&2&&&\end{smallmatrix}\right\rangle^1\left|\begin{smallmatrix}2&&1&&1&&0\\&1&&1&&1&\\&&1&&1&&\\&&&1&&&\end{smallmatrix}\right\rangle^4 \\
&-\sqrt{6}\left|\begin{smallmatrix}2&&1&&1&&0\\&2&&1&&1&\\&&2&&1&&\\&&&1&&&\end{smallmatrix}\right\rangle^2\left|\begin{smallmatrix}2&&1&&1&&0\\&2&&1&&0&\\&&2&&0&&\\&&&2&&&\end{smallmatrix}\right\rangle^3+\sqrt{6}\left|\begin{smallmatrix}2&&1&&1&&0\\&2&&1&&1&\\&&1&&1&&\\&&&1&&&\end{smallmatrix}\right\rangle^3\left|\begin{smallmatrix}2&&1&&1&&0\\&2&&1&&0&\\&&2&&1&&\\&&&2&&&\end{smallmatrix}\right\rangle^2\Bigg\}
\end{aligned}$$

+ the terms obtained by interchanging $|\boldsymbol{m}^{(1)}\rangle$ and $|\boldsymbol{m}^{(2)}\rangle$,

$$\begin{aligned}
||(1,0,1),(1,0,1)\rangle_A = \sqrt{1/48}\, \{&\sqrt{3}\,|(1,0,1)\rangle|(0,0,0)_1\rangle - |(1,0,1)\rangle|(0,0,0)_2\rangle \\
&+ \sqrt{8}\,|(1,0,1)\rangle|(0,0,0)_3\rangle - \sqrt{6}\,|(\overline{1},1,1)\rangle|(2,\overline{1},0)\rangle + \sqrt{6}\,|(0,\overline{1},2)\rangle|(1,1,\overline{1})\rangle \\
&- \sqrt{6}\,|(1,1,\overline{1})\rangle|(0,\overline{1},2)\rangle + \sqrt{6}\,|(2,\overline{1},0)\rangle|(\overline{1},1,1)\rangle - \sqrt{8}\,|(0,0,0)_3\rangle|(1,0,1)\rangle \\
&+ |(0,0,0)_2\rangle|(1,0,1)\rangle - \sqrt{3}\,|(0,0,0)_1\rangle|(1,0,1)\rangle\}
\end{aligned}$$

$$\begin{aligned}
=\sqrt{\frac{1}{48}}\Bigg\{&\sqrt{3}\left|\begin{smallmatrix}2&&1&&1&&0\\&2&&1&&1&\\&&2&&1&&\\&&&2&&&\end{smallmatrix}\right\rangle^1\left|\begin{smallmatrix}2&&1&&1&&0\\&2&&1&&0&\\&&2&&0&&\\&&&1&&&\end{smallmatrix}\right\rangle^4 \\
&-\left|\begin{smallmatrix}2&&1&&1&&0\\&2&&1&&1&\\&&2&&1&&\\&&&2&&&\end{smallmatrix}\right\rangle^1\left|\begin{smallmatrix}2&&1&&1&&0\\&2&&1&&0&\\&&1&&1&&\\&&&1&&&\end{smallmatrix}\right\rangle^4+\sqrt{8}\left|\begin{smallmatrix}2&&1&&1&&0\\&2&&1&&1&\\&&2&&1&&\\&&&2&&&\end{smallmatrix}\right\rangle^1\left|\begin{smallmatrix}2&&1&&1&&0\\&1&&1&&1&\\&&1&&1&&\\&&&1&&&\end{smallmatrix}\right\rangle^4 \\
&-\sqrt{6}\left|\begin{smallmatrix}2&&1&&1&&0\\&2&&1&&1&\\&&2&&1&&\\&&&1&&&\end{smallmatrix}\right\rangle^2\left|\begin{smallmatrix}2&&1&&1&&0\\&2&&1&&0&\\&&2&&0&&\\&&&2&&&\end{smallmatrix}\right\rangle^3+\sqrt{6}\left|\begin{smallmatrix}2&&1&&1&&0\\&2&&1&&1&\\&&1&&1&&\\&&&1&&&\end{smallmatrix}\right\rangle^3\left|\begin{smallmatrix}2&&1&&1&&0\\&2&&1&&0&\\&&2&&1&&\\&&&2&&&\end{smallmatrix}\right\rangle^2\Bigg\}
\end{aligned}$$

− the terms obtained by interchanging $|\boldsymbol{m}^{(1)}\rangle$ and $|\boldsymbol{m}^{(2)}\rangle$.

8.5.4 *The CG series for* C_3

The block weight diagrams of some representations of C_3 are given in subsection 8.3.2. The size of Weyl orbit is $OS(0,0,0) = 1$, $OS(M_1,0,0) = 6$, $OS(0,M_2,0) = 12$, $OS(0,0,M_3) = 8$, $OS(M_1,M_2,0) = OS(M_1,0,M_3) = OS(0,M_2,M_3) = 24$, and $OS(M_1,M_2,M_3) = 48$, where M_1, M_2 and M_3 all are non-vanishing.

Ex. 1. The decomposition of $(1,0,0) \times (0,1,0)$ in C_3.

From the block weight diagrams of $\boldsymbol{M}^{(1)} = (1,0,0)$ and $\boldsymbol{M}^{(2)} = (0,1,0)$ (see subsection 8.3.2), there are three dominant weights in the product space, whose basis states $|\boldsymbol{m}^{(1)}\rangle|\boldsymbol{m}^{(2)}\rangle$ are listed as follows.

α	$\boldsymbol{M}_\alpha$	n_α	Independent basis states $\lvert\boldsymbol{m}^{(1)}\rangle\lvert\boldsymbol{m}^{(2)}\rangle$
0	$(1,1,0)$	1	$\lvert(1,0,0)\rangle\lvert(0,1,0)\rangle$
1	$(0,0,1)$	3	$\lvert(1,0,0)\rangle\lvert(\overline{1},0,1)\rangle$, $\lvert(\overline{1},1,0)\rangle\lvert(1,\overline{1},1)\rangle$, $\lvert(0,\overline{1},1)\rangle\lvert(0,1,0)\rangle$
2	$(1,0,0)$	6	$\lvert(1,0,0)\rangle\lvert(0,0,0)_1\rangle, \lvert(1,0,0)\rangle\lvert(0,0,0)_2\rangle, \lvert(\overline{1},1,0)\rangle\lvert(2,\overline{1},0)\rangle,$ $\lvert(0,\overline{1},1)\rangle\lvert(1,1,\overline{1})\rangle, \lvert(0,1,\overline{1})\rangle\lvert(1,\overline{1},1)\rangle, \lvert(1,\overline{1},0)\rangle\lvert(0,1,0)\rangle$

The multiplicities $n_{\alpha\beta}$ are given in subsection 8.3.2. The generalized Gel'fand's bases of $\boldsymbol{M}_\alpha$ in the representation $\boldsymbol{M}_0 = (1,1,0)$ are listed as follows. Then, the dominant weight diagram for $(1,0,0) \times (0,1,0)$ of C_3 is calculated in Fig. 8.21.

$$\boldsymbol{M}_0 = (1,1,0)$$

α	$\boldsymbol{M}_\alpha$	$\Omega_3\Omega_2\Omega_1$	$n_{0\alpha}$	the basis states
0	$(1,1,0)$	3, 3, 2	1	$\left\lvert\begin{smallmatrix}2&&1&&0\\&2&&1&\\&&2&&\end{smallmatrix}\right\rangle_0^1$
1	$(0,0,1)$	3, 2, 1	2	$\left\lvert\begin{smallmatrix}2&&1&&0\\&2&&0&\\&&1&&\end{smallmatrix}\right\rangle_1^3$, $\left\lvert\begin{smallmatrix}2&&1&&0\\&1&&1&\\&&1&&\end{smallmatrix}\right\rangle_1^3$
2	$(1,0,0)$	1, 1, 1	4	$\left\lvert\begin{smallmatrix}2&&0&&\overline{1}\\&2&&\overline{1}&\\&&1&&\end{smallmatrix}\right\rangle_0^5$, $\left\lvert\begin{smallmatrix}2&&0&&\overline{1}\\&1&&0&\\&&1&&\end{smallmatrix}\right\rangle_0^5$, $\left\lvert\begin{smallmatrix}1&&1&&\overline{1}\\&1&&0&\\&&1&&\end{smallmatrix}\right\rangle_0^5$, $\left\lvert\begin{smallmatrix}1&&0&&0\\&1&&0&\\&&1&&\end{smallmatrix}\right\rangle_0^5$

$OS(\boldsymbol{M}_c) \times n_c$		$\boxed{1,1,0}$		
24×1	$\boxed{1,1,0}$	1	$\boxed{0,0,1}$	
8×3	$\boxed{0,0,1}$	2	1	$\boxed{1,0,0}$
6×6	$\boxed{1,0,0}$	4	1	1
$6 \times 14 = 84$	$=$	64	$+$ 14	$+$ 6

$$(1,0,0) \times (0,1,0) \simeq (1,1,0) + (0,0,1) + (1,0,0)$$

Fig. 8.21 The dominant weight diagram for $(1,0,0) \times (0,1,0)$ in C_3.

The expansions (8.49) for the highest weight states of $\boldsymbol{M}_0$ is given in Eq. (8.50), and that for $\boldsymbol{M}_1$ is easy to calculate (see Eq. (8.54)).

$$\begin{aligned}
\|(1,1,0),(1,1,0)\rangle &= \left|\begin{smallmatrix} 1 & & 0 & & 0 \\ & 1 & & 0 & \\ & & 1 & & \end{smallmatrix}\right\rangle^{1}_{0} \left|\begin{smallmatrix} 1 & & 1 & & 0 \\ & 1 & & 1 & \\ & & 1 & & \end{smallmatrix}\right\rangle^{1}_{0}, \\
\|(0,0,1),(0,0,1)\rangle &= \sqrt{\frac{1}{3}}\left[\left|\begin{smallmatrix} 1 & & 0 & & 0 \\ & 1 & & 0 & \\ & & 1 & & \end{smallmatrix}\right\rangle^{1}_{0} \left|\begin{smallmatrix} 1 & & 1 & & 0 \\ & 1 & & 0 & \\ & & 0 & & \end{smallmatrix}\right\rangle^{3}_{1} - \left|\begin{smallmatrix} 1 & & 0 & & 0 \\ & 1 & & 0 & \\ & & 0 & & \end{smallmatrix}\right\rangle^{2}_{0} \left|\begin{smallmatrix} 1 & & 1 & & 0 \\ & 1 & & 0 & \\ & & 1 & & \end{smallmatrix}\right\rangle^{2}_{1}\right. \\
&\quad \left. + \left|\begin{smallmatrix} 1 & & 0 & & 0 \\ & 0 & & 0 & \\ & & 0 & & \end{smallmatrix}\right\rangle^{3}_{1} \left|\begin{smallmatrix} 1 & & 1 & & 0 \\ & 1 & & 1 & \\ & & 1 & & \end{smallmatrix}\right\rangle^{1}_{0}\right].
\end{aligned}$$

We express the highest weight state of $\boldsymbol{M}_2 = (1,0,0)$ as a linear combination of two highest weight states of A$_2$-triplets with $\boldsymbol{M}^{(2)} = (1,0)$ (see Prob. 8 for $(1,0)$ in the decomposition of $(1,0)\times(1,1)$).

$$\begin{aligned}
\|(1,0,0),(1,0,0)\rangle &= \frac{a_1}{4}\left[\sqrt{3}\left|\begin{smallmatrix} 1 & & 0 & & 0 \\ & 1 & & 0 & \\ & & 1 & & \end{smallmatrix}\right\rangle^{1}_{0} \left|\begin{smallmatrix} 1 & & 0 & & \bar{1} \\ & 1 & & \bar{1} & \\ & & 0 & & \end{smallmatrix}\right\rangle^{5}_{0} - \left|\begin{smallmatrix} 1 & & 0 & & 0 \\ & 1 & & 0 & \\ & & 1 & & \end{smallmatrix}\right\rangle^{1}_{0} \left|\begin{smallmatrix} 1 & & 0 & & \bar{1} \\ & 0 & & 0 & \\ & & 0 & & \end{smallmatrix}\right\rangle^{5}_{0}\right. \\
&\quad \left. -\sqrt{6}\left|\begin{smallmatrix} 1 & & 0 & & 0 \\ & 1 & & 0 & \\ & & 0 & & \end{smallmatrix}\right\rangle^{2}_{0} \left|\begin{smallmatrix} 1 & & 0 & & \bar{1} \\ & 1 & & \bar{1} & \\ & & 1 & & \end{smallmatrix}\right\rangle^{4}_{0} + \sqrt{6}\left|\begin{smallmatrix} 1 & & 0 & & 0 \\ & 0 & & 0 & \\ & & 0 & & \end{smallmatrix}\right\rangle^{3}_{1} \left|\begin{smallmatrix} 1 & & 0 & & \bar{1} \\ & 1 & & 0 & \\ & & 1 & & \end{smallmatrix}\right\rangle^{3}_{\bar{1}}\right] \\
&\quad + \frac{a_2}{\sqrt{2}}\left[\left|\begin{smallmatrix} 0 & & 0 & & \bar{1} \\ & 0 & & 0 & \\ & & 0 & & \end{smallmatrix}\right\rangle^{4}_{\bar{1}} \left|\begin{smallmatrix} 1 & & 1 & & 0 \\ & 1 & & 0 & \\ & & 1 & & \end{smallmatrix}\right\rangle^{2}_{1} - \left|\begin{smallmatrix} 0 & & 0 & & \bar{1} \\ & 0 & & \bar{1} & \\ & & 0 & & \end{smallmatrix}\right\rangle^{5}_{0} \left|\begin{smallmatrix} 1 & & 1 & & 0 \\ & 1 & & 1 & \\ & & 1 & & \end{smallmatrix}\right\rangle^{1}_{0}\right].
\end{aligned}$$

The coefficients are calculated by the annihilation condition of E_3:

$$\begin{aligned}
E_3\,\|(1,0,0),(1,0,0)\rangle &= \frac{a_1\sqrt{6}}{4}\left|\begin{smallmatrix} 1 & & 0 & & 0 \\ & 0 & & 0 & \\ & & 0 & & \end{smallmatrix}\right\rangle^{3}_{1} \left|\begin{smallmatrix} 1 & & 1 & & 0 \\ & 1 & & 0 & \\ & & 1 & & \end{smallmatrix}\right\rangle^{2}_{1} \\
&\quad + \frac{a_2}{\sqrt{2}}\left|\begin{smallmatrix} 1 & & 0 & & 0 \\ & 0 & & 0 & \\ & & 0 & & \end{smallmatrix}\right\rangle^{3}_{1} \left|\begin{smallmatrix} 1 & & 1 & & 0 \\ & 1 & & 0 & \\ & & 1 & & \end{smallmatrix}\right\rangle^{2}_{1} = 0.
\end{aligned}$$

Thus, $a_2 = -a_1\sqrt{3}/2$, and $a_1 = (1+3/4)^{-1/2} = \sqrt{4/7}$.

$$\begin{aligned}
\|(1,0,0),(1,0,0)\rangle &= \frac{1}{2\sqrt{7}}\left\{\sqrt{3}\left|\begin{smallmatrix} 1 & & 0 & & 0 \\ & 1 & & 0 & \\ & & 1 & & \end{smallmatrix}\right\rangle^{1}_{0} \left|\begin{smallmatrix} 1 & & 0 & & \bar{1} \\ & 1 & & \bar{1} & \\ & & 0 & & \end{smallmatrix}\right\rangle^{5}_{0} - \left|\begin{smallmatrix} 1 & & 0 & & 0 \\ & 1 & & 0 & \\ & & 1 & & \end{smallmatrix}\right\rangle^{1}_{0} \left|\begin{smallmatrix} 1 & & 0 & & \bar{1} \\ & 0 & & 0 & \\ & & 0 & & \end{smallmatrix}\right\rangle^{5}_{0}\right. \\
&\quad - \sqrt{6}\left|\begin{smallmatrix} 1 & & 0 & & 0 \\ & 1 & & 0 & \\ & & 0 & & \end{smallmatrix}\right\rangle^{2}_{0} \left|\begin{smallmatrix} 1 & & 0 & & \bar{1} \\ & 1 & & \bar{1} & \\ & & 1 & & \end{smallmatrix}\right\rangle^{4}_{0} + \sqrt{6}\left|\begin{smallmatrix} 1 & & 0 & & 0 \\ & 0 & & 0 & \\ & & 0 & & \end{smallmatrix}\right\rangle^{3}_{1} \left|\begin{smallmatrix} 1 & & 0 & & \bar{1} \\ & 1 & & 0 & \\ & & 1 & & \end{smallmatrix}\right\rangle^{3}_{\bar{1}} \\
&\quad \left. - \sqrt{6}\left|\begin{smallmatrix} 0 & & 0 & & \bar{1} \\ & 0 & & 0 & \\ & & 0 & & \end{smallmatrix}\right\rangle^{4}_{\bar{1}} \left|\begin{smallmatrix} 1 & & 1 & & 0 \\ & 1 & & 0 & \\ & & 1 & & \end{smallmatrix}\right\rangle^{2}_{1} + \sqrt{6}\left|\begin{smallmatrix} 0 & & 0 & & \bar{1} \\ & 0 & & \bar{1} & \\ & & 0 & & \end{smallmatrix}\right\rangle^{5}_{0} \left|\begin{smallmatrix} 1 & & 1 & & 0 \\ & 1 & & 1 & \\ & & 1 & & \end{smallmatrix}\right\rangle^{1}_{0}\right\}.
\end{aligned}$$

Ex. 2. The decomposition of $(1,1,0)\times(1,1,0)$ in C$_3$.

From the block weight diagram of $\boldsymbol{M} = (1,1,0)$ given in subsection 8.3.2, there are twelve dominant weights $\boldsymbol{M}_\alpha$ in the direct product space of $(1,1,0)\times(1,1,0)$ in C$_3$. The multiplicities n_α of $\boldsymbol{M}_\alpha$ in the direct product space can be counted from the block weight diagram of $\boldsymbol{M} = (1,1,0)$. The multiplicities $n_{\alpha\beta}$ of the dominant weight $\boldsymbol{M}_\beta$ in the representation $\boldsymbol{M}_\alpha$, which are left as exercise, can be counted from the block weight diagram of M_α or looked over the table book [Bremner et al. (1985)]. Then, the dominant weight diagram for $(1,1,0) \times (1,1,0)$ is calculated in Fig. 8.22.

$O.S.$ × multiplicity		2, 2, 0	3, 0, 1	0, 3, 0	1, 1, 1	4, 0, 0	0, 0, 2	2, 1, 0	0, 2, 0	1, 0, 1	2, 0, 0	0, 1, 0	0, 0, 0
24 × 1	2, 2, 0	1											
24 × 2	3, 0, 1	1	1										
12 × 2	0, 3, 0	1	0	1	2								
48 × 6	1, 1, 1	2	1	1	1								
6 × 4	4, 0, 0	2	1	0	0	1							
8 × 10	0, 0, 2	3	1	1	2	0	1	3					
24 × 18	2, 1, 0	5	3	2	2	1	0	1	2				
12 × 28	0, 2, 0	7	3	3	4	1	1	1	1	3			
24 × 44	1, 0, 1	9	6	4	6	1	1	2	1	1	2		
6 × 68	2, 0, 0	14	7	5	8	3	2	4	2	1	1	2	
12 × 102	0, 1, 0	17	11	8	12	3	2	6	3	3	1	1	
1 × 152	0, 0, 0	24	13	11	16	6	4	9	6	4	3	2	1
64 × 64 = 4096		924	525	385	256	512	84	189	90	70	21	14	1

Fig. 8.22 The dominant weight diagram for $(1,1,0) \times (1,1,0)$ in C_3.

The Clebsch–Gordan series for $(1,1,0) \times (1,1,0)$ in C_3 is

$$\begin{aligned}(1,1,0) \otimes (1,1,0) \simeq\ & (2,2,0) \oplus (3,0,1) \oplus (0,3,0) \oplus 2 \times (1,1,1) \oplus (4,0,0) \\ & \oplus (0,0,2) \oplus 3 \times (2,1,0) \oplus 2 \times (0,2,0) \oplus 3 \times (1,0,1) \\ & \oplus 2 \times (2,0,0) \oplus 2 \times (0,1,0) \oplus (0,0,0), \\ 64 \times 64 = 4096 =\ & 924 + 525 + 385 + 2 \times 512 + 126 + 84 + 3 \times 189 \\ & + 2 \times 90 + 3 \times 70 + 2 \times 21 + 2 \times 14 + 1.\end{aligned}$$

The highest weight states of the first six representations in the CG series are easy to calculate (see **Ex. 2** in subsection 8.5.3). The highest weight state of $M = 0$ is given in Eq. (8.54), which is neglected.

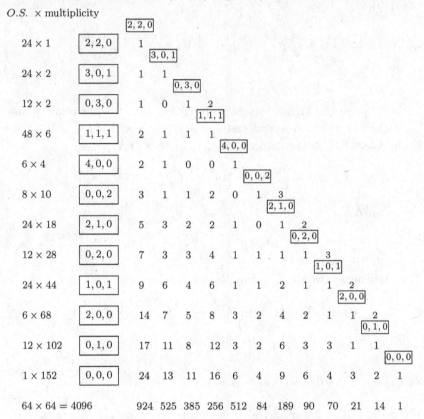

$$\|(2,2,0),(2,2,0)\rangle_S = \left|\begin{smallmatrix}2&&1&&0\\&2&&1&\\&&2&&\end{smallmatrix}\right\rangle^{1}_{0}\left|\begin{smallmatrix}2&&1&&0\\&2&&1&\\&&2&&\end{smallmatrix}\right\rangle^{1}_{0},$$

$$\|(3,0,1),(3,0,1)\rangle_A = \frac{1}{\sqrt{2}}\left\{\left|\begin{smallmatrix}2&&1&&0\\&2&&1&\\&&2&&\end{smallmatrix}\right\rangle^{1}_{0}\left|\begin{smallmatrix}2&&1&&0\\&2&&0&\\&&2&&\end{smallmatrix}\right\rangle^{2}_{1} - \left|\begin{smallmatrix}2&&1&&0\\&2&&0&\\&&2&&\end{smallmatrix}\right\rangle^{2}_{1}\left|\begin{smallmatrix}2&&1&&0\\&2&&1&\\&&2&&\end{smallmatrix}\right\rangle^{1}_{0}\right\},$$

$$\|(0,3,0),(0,3,0)\rangle_A = \frac{1}{\sqrt{2}}\left\{\left|\begin{smallmatrix}2&&1&&0\\&2&&1&\\&&2&&\end{smallmatrix}\right\rangle^{1}_{0}\left|\begin{smallmatrix}2&&1&&0\\&2&&1&\\&&1&&\end{smallmatrix}\right\rangle^{2}_{0} - \left|\begin{smallmatrix}2&&1&&0\\&2&&1&\\&&1&&\end{smallmatrix}\right\rangle^{2}_{0}\left|\begin{smallmatrix}2&&1&&0\\&2&&1&\\&&2&&\end{smallmatrix}\right\rangle^{1}_{0}\right\},$$

$$\|(1,1,0),(1,1,0)\rangle_S = \sqrt{\frac{1}{20}}\left\{\sqrt{3}\left|\begin{smallmatrix}2&&1&&0\\&2&&1&\\&&2&&\end{smallmatrix}\right\rangle^{1}_{0}\left|\begin{smallmatrix}2&&1&&0\\&2&&0&\\&&1&&\end{smallmatrix}\right\rangle^{3}_{1}\right.$$
$$+\left|\begin{smallmatrix}2&&1&&0\\&2&&1&\\&&1&&\end{smallmatrix}\right\rangle^{1}_{0}\left|\begin{smallmatrix}2&&1&&0\\&1&&1&\\&&1&&\end{smallmatrix}\right\rangle^{3}_{1} - \sqrt{6}\left|\begin{smallmatrix}2&&1&&0\\&2&&1&\\&&1&&\end{smallmatrix}\right\rangle^{2}_{0}\left|\begin{smallmatrix}2&&1&&0\\&2&&0&\\&&2&&\end{smallmatrix}\right\rangle^{2}_{1}$$

+terms obtained by interchanging $|\boldsymbol{m}^{(1)}\rangle$ and $|\boldsymbol{m}^{(2)}\rangle$.}

$$\|(1,1,0),(1,1,0)\rangle_A = \sqrt{\frac{1}{12}}\left\{\left|\begin{smallmatrix}2&&1&&0\\&2&&1&\\&&2&&\end{smallmatrix}\right\rangle^{1}_{0}\left|\begin{smallmatrix}2&&1&&0\\&2&&0&\\&&1&&\end{smallmatrix}\right\rangle^{3}_{1}\right.$$
$$-\sqrt{3}\left|\begin{smallmatrix}2&&1&&0\\&2&&1&\\&&1&&\end{smallmatrix}\right\rangle^{1}_{0}\left|\begin{smallmatrix}2&&1&&0\\&1&&1&\\&&1&&\end{smallmatrix}\right\rangle^{3}_{1} - \sqrt{2}\left|\begin{smallmatrix}2&&1&&0\\&2&&1&\\&&1&&\end{smallmatrix}\right\rangle^{2}_{0}\left|\begin{smallmatrix}2&&1&&0\\&2&&0&\\&&2&&\end{smallmatrix}\right\rangle^{2}_{1}$$

−terms obtained by interchanging $|\boldsymbol{m}^{(1)}\rangle$ and $|\boldsymbol{m}^{(2)}\rangle$.},

$$\|(4,0,0),(4,0,0)\rangle_A = \frac{1}{2}\left\{\left|\begin{smallmatrix}2&&1&&0\\&2&&1&\\&&2&&\end{smallmatrix}\right\rangle^{1}_{0}\left|\begin{smallmatrix}2&&0&&\bar{1}\\&2&&\bar{1}&\\&&2&&\end{smallmatrix}\right\rangle^{4}_{0} - \left|\begin{smallmatrix}2&&1&&0\\&2&&0&\\&&2&&\end{smallmatrix}\right\rangle^{2}_{1}\left|\begin{smallmatrix}2&&0&&\bar{1}\\&2&&0&\\&&2&&\end{smallmatrix}\right\rangle^{3}_{\bar{1}}\right.$$
$$\left.+\left|\begin{smallmatrix}2&&0&&\bar{1}\\&2&&0&\\&&2&&\end{smallmatrix}\right\rangle^{3}_{\bar{1}}\left|\begin{smallmatrix}2&&1&&0\\&2&&0&\\&&2&&\end{smallmatrix}\right\rangle^{2}_{1} - \left|\begin{smallmatrix}2&&0&&\bar{1}\\&2&&\bar{1}&\\&&2&&\end{smallmatrix}\right\rangle^{4}_{0}\left|\begin{smallmatrix}2&&1&&0\\&2&&1&\\&&2&&\end{smallmatrix}\right\rangle^{1}_{0}\right\}.$$

The calculation for the highest weight states of $\boldsymbol{M}_\alpha = (2,1,0)$, which is triple-multiplicity in the decomposition of $(1,1,0)\times(1,1,0)$ in C_3, is typical and general. We are going to calculate them in detail. The calculation for the highest weight states of the remaining four representations in the CG series is left as exercise (see Prob. 11).

We express the highest weight state of $(2,1,0)$ as a linear combination of the normalized highest weight states of A_2-multiplets (see **Ex. 4** in subsection 8.5.3).

$$\|(2,1,0),(2,1,0)\rangle = \frac{a_1}{3\sqrt{2}}\left[\left|\begin{smallmatrix}2&&1&&0\\&2&&1&\\&&2&&\end{smallmatrix}\right\rangle^{1}_{0}\left|\begin{smallmatrix}2&&0&&\bar{1}\\&2&&\bar{1}&\\&&1&&\end{smallmatrix}\right\rangle^{5}_{0} - \sqrt{3}\left|\begin{smallmatrix}2&&1&&0\\&2&&1&\\&&1&&\end{smallmatrix}\right\rangle^{2}_{0}\left|\begin{smallmatrix}2&&0&&\bar{1}\\&2&&\bar{1}&\\&&2&&\end{smallmatrix}\right\rangle^{4}_{0}\right.$$
$$-\sqrt{6}\left|\begin{smallmatrix}2&&1&&0\\&2&&0&\\&&2&&\end{smallmatrix}\right\rangle^{2}_{1}\left|\begin{smallmatrix}2&&0&&\bar{1}\\&2&&0&\\&&1&&\end{smallmatrix}\right\rangle^{4}_{\bar{1}} + \sqrt{6}\left|\begin{smallmatrix}2&&1&&0\\&2&&0&\\&&1&&\end{smallmatrix}\right\rangle^{3}_{1}\left|\begin{smallmatrix}2&&0&&\bar{1}\\&2&&0&\\&&2&&\end{smallmatrix}\right\rangle^{3}_{\bar{1}}$$
$$\left.+\sqrt{2}\left|\begin{smallmatrix}2&&1&&0\\&2&&1&\\&&2&&\end{smallmatrix}\right\rangle^{1}_{0}\left|\begin{smallmatrix}2&&0&&\bar{1}\\&1&&0&\\&&1&&\end{smallmatrix}\right\rangle^{5}_{0}\right] + \frac{a_2}{6\sqrt{6}}\left[5\left|\begin{smallmatrix}2&&1&&0\\&2&&1&\\&&2&&\end{smallmatrix}\right\rangle^{1}_{0}\left|\begin{smallmatrix}2&&0&&\bar{1}\\&2&&\bar{1}&\\&&1&&\end{smallmatrix}\right\rangle^{5}_{0}\right.$$
$$-5\sqrt{3}\left|\begin{smallmatrix}2&&1&&0\\&2&&1&\\&&1&&\end{smallmatrix}\right\rangle^{2}_{0}\left|\begin{smallmatrix}2&&0&&\bar{1}\\&2&&\bar{1}&\\&&2&&\end{smallmatrix}\right\rangle^{4}_{0} + \sqrt{6}\left|\begin{smallmatrix}2&&1&&0\\&2&&0&\\&&2&&\end{smallmatrix}\right\rangle^{2}_{1}\left|\begin{smallmatrix}2&&0&&\bar{1}\\&2&&0&\\&&1&&\end{smallmatrix}\right\rangle^{4}_{\bar{1}}$$

$$
\begin{aligned}
&- \sqrt{6}\left|\begin{smallmatrix}2&&1&&0\\&2&&0&\\&&1&&\end{smallmatrix}\right\rangle^{3}_{1}\left|\begin{smallmatrix}2&&0&&\bar{1}\\&2&&0&\\&&2&&\end{smallmatrix}\right\rangle^{3}_{\bar{1}} - 4\sqrt{2}\left|\begin{smallmatrix}2&&1&&0\\&2&&1&\\&&2&&\end{smallmatrix}\right\rangle^{1}_{0}\left|\begin{smallmatrix}2&&0&&\bar{1}\\&1&&0&\\&&1&&\end{smallmatrix}\right\rangle^{5}_{0}\\
&+ 6\sqrt{2}\left|\begin{smallmatrix}2&&1&&0\\&1&&1&\\&&1&&\end{smallmatrix}\right\rangle^{3}_{1}\left|\begin{smallmatrix}2&&0&&\bar{1}\\&2&&0&\\&&2&&\end{smallmatrix}\right\rangle^{3}_{\bar{1}}\Bigg] + \frac{a_3}{\sqrt{3}}\Bigg[\left|\begin{smallmatrix}2&&1&&0\\&2&&1&\\&&2&&\end{smallmatrix}\right\rangle^{1}_{0}\left|\begin{smallmatrix}1&&1&&\bar{1}\\&1&&0&\\&&1&&\end{smallmatrix}\right\rangle^{5}_{0}\\
&- \sqrt{2}\left|\begin{smallmatrix}2&&1&&0\\&2&&0&\\&&2&&\end{smallmatrix}\right\rangle^{2}_{1}\left|\begin{smallmatrix}1&&1&&\bar{1}\\&1&&1&\\&&1&&\end{smallmatrix}\right\rangle^{4}_{\bar{1}}\Bigg] + a_4\left|\begin{smallmatrix}2&&1&&0\\&2&&1&\\&&2&&\end{smallmatrix}\right\rangle^{1}_{0}\left|\begin{smallmatrix}1&&0&&0\\&1&&0&\\&&1&&\end{smallmatrix}\right\rangle^{5}_{0}\\
&+ \text{terms } b_j \text{ obtained by interchanging } |\boldsymbol{m}^{(1)}\rangle \text{ and } |\boldsymbol{m}^{(2)}\rangle\\
&= \frac{\sqrt{12}a_1 + 5a_2}{6\sqrt{6}}\left|\begin{smallmatrix}2&&1&&0\\&2&&1&\\&&2&&\end{smallmatrix}\right\rangle^{1}_{0}\left|\begin{smallmatrix}2&&0&&\bar{1}\\&2&&\bar{1}&\\&&1&&\end{smallmatrix}\right\rangle^{5}_{0} - \frac{\sqrt{12}a_1 + 5a_2}{6\sqrt{2}}\left|\begin{smallmatrix}2&&1&&0\\&2&&1&\\&&1&&\end{smallmatrix}\right\rangle^{2}_{0}\left|\begin{smallmatrix}2&&0&&\bar{1}\\&2&&\bar{1}&\\&&2&&\end{smallmatrix}\right\rangle^{4}_{0}\\
&- \frac{\sqrt{12}a_1 - a_2}{6}\left|\begin{smallmatrix}2&&1&&0\\&2&&0&\\&&2&&\end{smallmatrix}\right\rangle^{2}_{1}\left|\begin{smallmatrix}2&&0&&\bar{1}\\&2&&0&\\&&1&&\end{smallmatrix}\right\rangle^{4}_{\bar{1}} + \frac{\sqrt{12}a_1 - a_2}{6}\left|\begin{smallmatrix}2&&1&&0\\&2&&0&\\&&1&&\end{smallmatrix}\right\rangle^{3}_{1}\left|\begin{smallmatrix}2&&0&&\bar{1}\\&2&&0&\\&&2&&\end{smallmatrix}\right\rangle^{3}_{\bar{1}}\\
&+ \frac{\sqrt{3}a_1 - 2a_2}{3\sqrt{3}}\left|\begin{smallmatrix}2&&1&&0\\&2&&1&\\&&2&&\end{smallmatrix}\right\rangle^{1}_{0}\left|\begin{smallmatrix}2&&0&&\bar{1}\\&1&&0&\\&&1&&\end{smallmatrix}\right\rangle^{5}_{0} + \frac{a_2}{\sqrt{3}}\left|\begin{smallmatrix}2&&1&&0\\&1&&1&\\&&1&&\end{smallmatrix}\right\rangle^{3}_{1}\left|\begin{smallmatrix}2&&0&&\bar{1}\\&2&&0&\\&&2&&\end{smallmatrix}\right\rangle^{3}_{\bar{1}}\\
&+ \frac{a_3}{\sqrt{3}}\Bigg[\left|\begin{smallmatrix}2&&1&&0\\&2&&1&\\&&2&&\end{smallmatrix}\right\rangle^{1}_{0}\left|\begin{smallmatrix}1&&1&&\bar{1}\\&1&&0&\\&&1&&\end{smallmatrix}\right\rangle^{5}_{0} - \sqrt{2}\left|\begin{smallmatrix}2&&1&&0\\&2&&0&\\&&2&&\end{smallmatrix}\right\rangle^{2}_{1}\left|\begin{smallmatrix}1&&1&&\bar{1}\\&1&&1&\\&&1&&\end{smallmatrix}\right\rangle^{4}_{\bar{1}}\Bigg]\\
&+ a_4\left|\begin{smallmatrix}2&&1&&0\\&2&&1&\\&&2&&\end{smallmatrix}\right\rangle^{1}_{0}\left|\begin{smallmatrix}1&&0&&0\\&1&&0&\\&&1&&\end{smallmatrix}\right\rangle^{5}_{0}\\
&+ \text{terms } b_j \text{ obtained by interchanging } |\boldsymbol{m}^{(1)}\rangle \text{ and } |\boldsymbol{m}^{(2)}\rangle.
\end{aligned}
\tag{8.61}
$$

The coefficients are calculated by the annihilation condition of E_3.

$$
\begin{aligned}
E_3\,||(2,1,0),(2,1,0)\rangle &= -\frac{\sqrt{12}a_1 - a_2}{6}\left|\begin{smallmatrix}2&&1&&0\\&2&&0&\\&&2&&\end{smallmatrix}\right\rangle^{2}_{1}\left|\begin{smallmatrix}2&&1&&0\\&2&&0&\\&&1&&\end{smallmatrix}\right\rangle^{3}_{1}\\
&+ \frac{\sqrt{12}a_1 - a_2}{6}\left|\begin{smallmatrix}2&&1&&0\\&2&&0&\\&&1&&\end{smallmatrix}\right\rangle^{3}_{1}\left|\begin{smallmatrix}2&&1&&0\\&2&&0&\\&&2&&\end{smallmatrix}\right\rangle^{2}_{1} + \frac{\sqrt{3}a_1 - 2a_2}{3\sqrt{8}}\left|\begin{smallmatrix}2&&1&&0\\&2&&1&\\&&2&&\end{smallmatrix}\right\rangle^{1}_{0}\left|\begin{smallmatrix}2&&1&&0\\&1&&0&\\&&1&&\end{smallmatrix}\right\rangle^{4}_{2}\\
&+ \frac{a_2}{\sqrt{3}}\left|\begin{smallmatrix}2&&1&&0\\&1&&1&\\&&1&&\end{smallmatrix}\right\rangle^{3}_{1}\left|\begin{smallmatrix}2&&1&&0\\&2&&0&\\&&2&&\end{smallmatrix}\right\rangle^{2}_{1} + \frac{a_3}{2}\left|\begin{smallmatrix}2&&1&&0\\&2&&1&\\&&2&&\end{smallmatrix}\right\rangle^{1}_{0}\left|\begin{smallmatrix}1&&1&&\bar{1}\\&1&&0&\\&&1&&\end{smallmatrix}\right\rangle^{4}_{2}\\
&- a_3\sqrt{\frac{2}{3}}\left|\begin{smallmatrix}2&&1&&0\\&2&&0&\\&&2&&\end{smallmatrix}\right\rangle^{2}_{1}\left|\begin{smallmatrix}2&&1&&0\\&1&&1&\\&&1&&\end{smallmatrix}\right\rangle^{3}_{1} + a_4\sqrt{\frac{7}{8}}\left|\begin{smallmatrix}2&&1&&0\\&2&&1&\\&&2&&\end{smallmatrix}\right\rangle^{1}_{0}\left|\begin{smallmatrix}2&&1&&0\\&1&&0&\\&&1&&\end{smallmatrix}\right\rangle^{4}_{2}\\
&+ \text{terms } b_j \text{ obtained by interchanging } |\boldsymbol{m}^{(1)}\rangle \text{ and } |\boldsymbol{m}^{(2)}\rangle.
\end{aligned}
$$

Collecting the same terms, we are able to find the equations which the coefficients a_j and b_j satisfy.

$$
\left|\begin{smallmatrix}2&&1&&0\\&2&&1&\\&&2&&\end{smallmatrix}\right\rangle^{1}_{0}\left|\begin{smallmatrix}2&&1&&0\\&1&&0&\\&&1&&\end{smallmatrix}\right\rangle^{4}_{2}:\quad a_1\frac{1}{2\sqrt{6}} - a_2\frac{1}{3\sqrt{2}} + a_3\frac{1}{2} + a_4\sqrt{\frac{7}{8}} = 0,
$$

$$\left|\begin{smallmatrix}2&&1&&0\\&2&&0&\\&&2&&\end{smallmatrix}\right\rangle_{1}^{2}\left|\begin{smallmatrix}2&&1&&0\\&2&&0&\\&&1&&\end{smallmatrix}\right\rangle_{1}^{3}: \quad -a_1\frac{1}{\sqrt{3}}+b_1\frac{1}{\sqrt{3}}+a_2\frac{1}{6}-b_2\frac{1}{6}=0,$$

$$\left|\begin{smallmatrix}2&&1&&0\\&1&&1&\\&&1&&\end{smallmatrix}\right\rangle_{1}^{3}\left|\begin{smallmatrix}2&&1&&0\\&2&&0&\\&&2&&\end{smallmatrix}\right\rangle_{1}^{2}: \quad a_2\frac{1}{\sqrt{3}}-b_3\sqrt{\frac{2}{3}}=0.$$

For the antisymmetric representation, $a_j = -b_j$. Thus, $a_1 = \sqrt{7}N$, $a_2 = 2\sqrt{21}N$, $a_3 = -\sqrt{42}N$, $a_4 = 3\sqrt{3}N$, and $N^{-2} = 2(7+84+42+27) = 320$. The highest weight state of the anti-symmetric representation is

$$\begin{aligned}
||(2,1,0),(2,1,0)\rangle_A = \frac{1}{\sqrt{320}}\Bigg\{&\sqrt{14}\left|\begin{smallmatrix}2&&1&&0\\&2&&1&\\&&2&&\end{smallmatrix}\right\rangle_{0}^{1}\left|\begin{smallmatrix}2&&0&&\bar{1}\\&2&&\bar{1}&\\&&1&&\end{smallmatrix}\right\rangle_{0}^{5}\\
&-\sqrt{42}\left|\begin{smallmatrix}2&&1&&0\\&2&&1&\\&&1&&\end{smallmatrix}\right\rangle_{0}^{2}\left|\begin{smallmatrix}2&&0&&\bar{1}\\&2&&\bar{1}&\\&&2&&\end{smallmatrix}\right\rangle_{0}^{4}-\sqrt{7}\left|\begin{smallmatrix}2&&1&&0\\&2&&1&\\&&2&&\end{smallmatrix}\right\rangle_{0}^{1}\left|\begin{smallmatrix}2&&0&&\bar{1}\\&1&&0&\\&&1&&\end{smallmatrix}\right\rangle_{0}^{5}\\
&+2\sqrt{7}\left|\begin{smallmatrix}2&&1&&0\\&1&&1&\\&&1&&\end{smallmatrix}\right\rangle_{1}^{3}\left|\begin{smallmatrix}2&&0&&\bar{1}\\&2&&0&\\&&2&&\end{smallmatrix}\right\rangle_{\bar{1}}^{3}-\sqrt{14}\left|\begin{smallmatrix}2&&1&&0\\&2&&1&\\&&2&&\end{smallmatrix}\right\rangle_{0}^{1}\left|\begin{smallmatrix}1&&1&&\bar{1}\\&1&&0&\\&&1&&\end{smallmatrix}\right\rangle_{0}^{5}\\
&+2\sqrt{7}\left|\begin{smallmatrix}2&&1&&0\\&2&&0&\\&&2&&\end{smallmatrix}\right\rangle_{1}^{2}\left|\begin{smallmatrix}1&&1&&\bar{1}\\&1&&1&\\&&1&&\end{smallmatrix}\right\rangle_{\bar{1}}^{4}+3\sqrt{3}\left|\begin{smallmatrix}2&&1&&0\\&2&&1&\\&&2&&\end{smallmatrix}\right\rangle_{0}^{1}\left|\begin{smallmatrix}1&&0&&0\\&1&&0&\\&&1&&\end{smallmatrix}\right\rangle_{0}^{5}\\
&-\text{terms obtained by interchanging } |\boldsymbol{m}^{(1)}\rangle \text{ and } |\boldsymbol{m}^{(2)}\rangle\Bigg\}.
\end{aligned}$$

For the symmetric representations, $a_j = b_j$. We obtain that the equations for the coefficients are $a_2 = \sqrt{2}a_3$ and $a_4 = -[\sqrt{3}a_1 + \sqrt{2}a_3]/(3\sqrt{7})$. We calculate three basic solutions:

No.	a_1	a_2	a_3	a_4	$\sum_j a_j^2$
1	0	$3\sqrt{14}$	$3\sqrt{7}$	$-\sqrt{2}$	5492
2	$\sqrt{21}$	0	0	-1	22
3	$\sqrt{2}$	$-\sqrt{6}$	$-\sqrt{3}$	0	11

There are two sets of orthogonal solutions based on $\psi^{(2)}(a_j)$ or $\psi^{(3)}(a_j)$, respectively. Since $\sum_j a_j^{(2)}a_j^{(3)} = \sqrt{42}$, $\sqrt{21}\psi^{(2)}(a_j) - 11\sqrt{2}\psi^{(3)}(a_j)$, where $a_1 = -1$, $a_2 = 22\sqrt{3}$, $a_3 = 11\sqrt{6}$, and $a_4 = -\sqrt{21}$, is orthogonal to $\psi^{(2)}(a_j)$, and $11\psi^{(2)}(a_j) - \sqrt{42}\psi^{(3)}(a_j)$, where $a_1 = 9\sqrt{21}$, $a_2 = 6\sqrt{7}$, $a_3 = 3\sqrt{14}$, and $a_4 = -11$, is orthogonal to $\psi^{(3)}(a_j)$. Substituting a_j into Eq. (8.61) and making normalization, we obtain two sets of orthogonal solutions. The first set is

$$||(2,1,0),(2,1,0)\rangle_{S1} = \sqrt{\frac{1}{264}}\Bigg\{\sqrt{7}\left|\begin{smallmatrix}2&&1&&0\\&2&&1&\\&&2&&\end{smallmatrix}\right\rangle_{0}^{1}\left|\begin{smallmatrix}2&&0&&\bar{1}\\&2&&\bar{1}&\\&&1&&\end{smallmatrix}\right\rangle_{0}^{5}$$

$$
- \sqrt{21} \left|\begin{smallmatrix} 2 & & 1 & & 0 \\ & 2 & & 1 & \\ & & 1 & & \end{smallmatrix}\right\rangle^{2}_{0} \left|\begin{smallmatrix} 2 & & 0 & & \bar{1} \\ & 2 & & \bar{1} & \\ & & 2 & & \end{smallmatrix}\right\rangle^{4}_{0} - \sqrt{42} \left|\begin{smallmatrix} 2 & & 1 & & 0 \\ & 2 & & 0 & \\ & & 2 & & \end{smallmatrix}\right\rangle^{2}_{1} \left|\begin{smallmatrix} 2 & & 0 & & \bar{1} \\ & 2 & & 0 & \\ & & 1 & & \end{smallmatrix}\right\rangle^{4}_{\bar{1}}
$$

$$
+ \sqrt{42} \left|\begin{smallmatrix} 2 & & 1 & & 0 \\ & 2 & & 0 & \\ & & 1 & & \end{smallmatrix}\right\rangle^{3}_{1} \left|\begin{smallmatrix} 2 & & 0 & & \bar{1} \\ & 2 & & 0 & \\ & & 2 & & \end{smallmatrix}\right\rangle^{3}_{\bar{1}} + \sqrt{14} \left|\begin{smallmatrix} 2 & & 1 & & 0 \\ & 2 & & 1 & \\ & & 2 & & \end{smallmatrix}\right\rangle^{1}_{0} \left|\begin{smallmatrix} 2 & & 0 & & \bar{1} \\ & 1 & & 0 & \\ & & 1 & & \end{smallmatrix}\right\rangle^{5}_{0}
$$

$$
- \sqrt{6} \left|\begin{smallmatrix} 2 & & 1 & & 0 \\ & 2 & & 1 & \\ & & 2 & & \end{smallmatrix}\right\rangle^{1}_{0} \left|\begin{smallmatrix} 1 & & 0 & & 0 \\ & 1 & & 0 & \\ & & 1 & & \end{smallmatrix}\right\rangle^{5}_{0}
$$

$$
+ \text{terms obtained by interchanging } |\boldsymbol{m}^{(1)}\rangle \text{ and } |\boldsymbol{m}^{(2)}\rangle\Big\} ,
$$

$$
||(2,1,0),(2,1,0)\rangle_{S2} = \sqrt{\frac{1}{4400}} \Big\{ 9\sqrt{2} \left|\begin{smallmatrix} 2 & & 1 & & 0 \\ & 2 & & 1 & \\ & & 2 & & \end{smallmatrix}\right\rangle^{1}_{0} \left|\begin{smallmatrix} 2 & & 0 & & \bar{1} \\ & 2 & & \bar{1} & \\ & & 1 & & \end{smallmatrix}\right\rangle^{5}_{0}
$$

$$
- 9\sqrt{6} \left|\begin{smallmatrix} 2 & & 1 & & 0 \\ & 2 & & 1 & \\ & & 1 & & \end{smallmatrix}\right\rangle^{2}_{0} \left|\begin{smallmatrix} 2 & & 0 & & \bar{1} \\ & 2 & & \bar{1} & \\ & & 2 & & \end{smallmatrix}\right\rangle^{4}_{0} + 4\sqrt{3} \left|\begin{smallmatrix} 2 & & 1 & & 0 \\ & 2 & & 0 & \\ & & 2 & & \end{smallmatrix}\right\rangle^{2}_{1} \left|\begin{smallmatrix} 2 & & 0 & & \bar{1} \\ & 2 & & 0 & \\ & & 1 & & \end{smallmatrix}\right\rangle^{4}_{\bar{1}}
$$

$$
- 4\sqrt{3} \left|\begin{smallmatrix} 2 & & 1 & & 0 \\ & 2 & & 0 & \\ & & 1 & & \end{smallmatrix}\right\rangle^{3}_{1} \left|\begin{smallmatrix} 2 & & 0 & & \bar{1} \\ & 2 & & 0 & \\ & & 2 & & \end{smallmatrix}\right\rangle^{3}_{\bar{1}} - 15 \left|\begin{smallmatrix} 2 & & 1 & & 0 \\ & 2 & & 1 & \\ & & 2 & & \end{smallmatrix}\right\rangle^{1}_{0} \left|\begin{smallmatrix} 2 & & 0 & & \bar{1} \\ & 1 & & 0 & \\ & & 1 & & \end{smallmatrix}\right\rangle^{5}_{0}
$$

$$
+ 22 \left|\begin{smallmatrix} 2 & & 1 & & 0 \\ & 1 & & 1 & \\ & & 1 & & \end{smallmatrix}\right\rangle^{3}_{1} \left|\begin{smallmatrix} 2 & & 0 & & \bar{1} \\ & 2 & & 0 & \\ & & 2 & & \end{smallmatrix}\right\rangle^{3}_{\bar{1}} + 11\sqrt{2} \left|\begin{smallmatrix} 2 & & 1 & & 0 \\ & 2 & & 1 & \\ & & 2 & & \end{smallmatrix}\right\rangle^{1}_{0} \left|\begin{smallmatrix} 1 & & 1 & & \bar{1} \\ & 1 & & 0 & \\ & & 1 & & \end{smallmatrix}\right\rangle^{5}_{0}
$$

$$
- 22 \left|\begin{smallmatrix} 2 & & 1 & & 0 \\ & 2 & & 0 & \\ & & 2 & & \end{smallmatrix}\right\rangle^{2}_{1} \left|\begin{smallmatrix} 1 & & 1 & & \bar{1} \\ & 1 & & 1 & \\ & & 1 & & \end{smallmatrix}\right\rangle^{4}_{\bar{1}} - \sqrt{21} \left|\begin{smallmatrix} 2 & & 1 & & 0 \\ & 2 & & 1 & \\ & & 2 & & \end{smallmatrix}\right\rangle^{1}_{0} \left|\begin{smallmatrix} 1 & & 0 & & 0 \\ & 1 & & 0 & \\ & & 1 & & \end{smallmatrix}\right\rangle^{5}_{0}
$$

$$
+ \text{terms } b_j \text{ obtained by interchanging } |\boldsymbol{m}^{(1)}\rangle \text{ and } |\boldsymbol{m}^{(2)}\rangle\Big\} .
$$

The second set is

$$
||(2,1,0),(2,1,0)\rangle_{S1} = \sqrt{\frac{1}{88}} \Big\{ - \left|\begin{smallmatrix} 2 & & 1 & & 0 \\ & 2 & & 1 & \\ & & 2 & & \end{smallmatrix}\right\rangle^{1}_{0} \left|\begin{smallmatrix} 2 & & 0 & & \bar{1} \\ & 2 & & \bar{1} & \\ & & 1 & & \end{smallmatrix}\right\rangle^{5}_{0}
$$

$$
+ \sqrt{3} \left|\begin{smallmatrix} 2 & & 1 & & 0 \\ & 2 & & 1 & \\ & & 1 & & \end{smallmatrix}\right\rangle^{2}_{0} \left|\begin{smallmatrix} 2 & & 0 & & \bar{1} \\ & 2 & & \bar{1} & \\ & & 2 & & \end{smallmatrix}\right\rangle^{4}_{0} - \sqrt{6} \left|\begin{smallmatrix} 2 & & 1 & & 0 \\ & 2 & & 0 & \\ & & 2 & & \end{smallmatrix}\right\rangle^{2}_{1} \left|\begin{smallmatrix} 2 & & 0 & & \bar{1} \\ & 2 & & 0 & \\ & & 1 & & \end{smallmatrix}\right\rangle^{4}_{\bar{1}}
$$

$$
+ \sqrt{6} \left|\begin{smallmatrix} 2 & & 1 & & 0 \\ & 2 & & 0 & \\ & & 1 & & \end{smallmatrix}\right\rangle^{3}_{1} \left|\begin{smallmatrix} 2 & & 0 & & \bar{1} \\ & 2 & & 0 & \\ & & 2 & & \end{smallmatrix}\right\rangle^{3}_{\bar{1}} + 2\sqrt{2} \left|\begin{smallmatrix} 2 & & 1 & & 0 \\ & 2 & & 1 & \\ & & 2 & & \end{smallmatrix}\right\rangle^{1}_{0} \left|\begin{smallmatrix} 2 & & 0 & & \bar{1} \\ & 1 & & 0 & \\ & & 1 & & \end{smallmatrix}\right\rangle^{5}_{0}
$$

$$
- 2\sqrt{2} \left|\begin{smallmatrix} 2 & & 1 & & 0 \\ & 1 & & 1 & \\ & & 1 & & \end{smallmatrix}\right\rangle^{3}_{1} \left|\begin{smallmatrix} 2 & & 0 & & \bar{1} \\ & 2 & & 0 & \\ & & 2 & & \end{smallmatrix}\right\rangle^{3}_{\bar{1}} - 2 \left|\begin{smallmatrix} 2 & & 1 & & 0 \\ & 2 & & 1 & \\ & & 2 & & \end{smallmatrix}\right\rangle^{1}_{0} \left|\begin{smallmatrix} 1 & & 1 & & \bar{1} \\ & 1 & & 0 & \\ & & 1 & & \end{smallmatrix}\right\rangle^{5}_{0}
$$

$$
+ 2\sqrt{2} \left|\begin{smallmatrix} 2 & & 1 & & 0 \\ & 2 & & 0 & \\ & & 2 & & \end{smallmatrix}\right\rangle^{2}_{1} \left|\begin{smallmatrix} 1 & & 1 & & \bar{1} \\ & 1 & & 1 & \\ & & 1 & & \end{smallmatrix}\right\rangle^{4}_{\bar{1}}
$$

$$
+ \text{terms } b_j \text{ obtained by interchanging } |\boldsymbol{m}^{(1)}\rangle \text{ and } |\boldsymbol{m}^{(2)}\rangle\Big\} ,
$$

$$
||(2,1,0),(2,1,0)\rangle_{S2} = \sqrt{\frac{1}{13200}} \Big\{ 7\sqrt{14} \left|\begin{smallmatrix} 2 & & 1 & & 0 \\ & 2 & & 1 & \\ & & 2 & & \end{smallmatrix}\right\rangle^{1}_{0} \left|\begin{smallmatrix} 2 & & 0 & & \bar{1} \\ & 2 & & \bar{1} & \\ & & 1 & & \end{smallmatrix}\right\rangle^{5}_{0}
$$

$$
- 7\sqrt{42} \left|\begin{smallmatrix} 2 & & 1 & & 0 \\ & 2 & & 1 & \\ & & 1 & & \end{smallmatrix}\right\rangle^{2}_{0} \left|\begin{smallmatrix} 2 & & 0 & & \bar{1} \\ & 2 & & \bar{1} & \\ & & 2 & & \end{smallmatrix}\right\rangle^{4}_{0} - 8\sqrt{21} \left|\begin{smallmatrix} 2 & & 1 & & 0 \\ & 2 & & 0 & \\ & & 2 & & \end{smallmatrix}\right\rangle^{2}_{1} \left|\begin{smallmatrix} 2 & & 0 & & \bar{1} \\ & 2 & & 0 & \\ & & 1 & & \end{smallmatrix}\right\rangle^{4}_{\bar{1}}
$$

$$+\,8\sqrt{21}\left|\begin{array}{ccccc}2&&1&&0\\&2&&0&\\&&1&&\end{array}\right\rangle^{3}_{1}\left|\begin{array}{ccccc}2&&0&&\bar{1}\\&2&&0&\\&&2&&\end{array}\right\rangle^{3}_{\bar{1}}+5\sqrt{7}\left|\begin{array}{ccccc}2&&1&&0\\&2&&1&\\&&2&&\end{array}\right\rangle^{1}_{0}\left|\begin{array}{ccccc}2&&0&&\bar{1}\\&1&&0&\\&&1&&\end{array}\right\rangle^{5}_{0}$$

$$+\,6\sqrt{7}\left|\begin{array}{ccccc}2&&1&&0\\&1&&1&\\&&1&&\end{array}\right\rangle^{3}_{1}\left|\begin{array}{ccccc}2&&0&&\bar{1}\\&2&&0&\\&&2&&\end{array}\right\rangle^{3}_{\bar{1}}+3\sqrt{14}\left|\begin{array}{ccccc}2&&1&&0\\&2&&1&\\&&2&&\end{array}\right\rangle^{1}_{0}\left|\begin{array}{ccccc}1&&1&&\bar{1}\\&1&&0&\\&&1&&\end{array}\right\rangle^{5}_{0}$$

$$-\,6\sqrt{7}\left|\begin{array}{ccccc}2&&1&&0\\&2&&0&\\&&2&&\end{array}\right\rangle^{2}_{1}\left|\begin{array}{ccccc}1&&1&&\bar{1}\\&1&&1&\\&&1&&\end{array}\right\rangle^{4}_{\bar{1}}-11\sqrt{3}\left|\begin{array}{ccccc}2&&1&&0\\&2&&1&\\&&2&&\end{array}\right\rangle^{1}_{0}\left|\begin{array}{ccccc}1&&0&&0\\&1&&0&\\&&1&&\end{array}\right\rangle^{5}_{0}$$

$$+\,\text{terms } b_j \text{ obtained by interchanging } |\boldsymbol{m}^{(1)}\rangle \text{ and } |\boldsymbol{m}^{(2)}\rangle\Big\}\,.$$

8.6 Exercises

1. Calculate the block weight diagram and the Gel'fand's bases for the representation $D^{\boldsymbol{M}}$ of A_2 with $\boldsymbol{M} = 2\boldsymbol{w}_1 + \boldsymbol{w}_2$.
2. Calculate the block weight diagram of A_2-multiplets and $\mathcal{A}_3$-multiplets and the Gel'fand's bases for the following representations $D^{\boldsymbol{M}}$ of A_3: (a) $\boldsymbol{M} = \boldsymbol{w}_2$; (b) $\boldsymbol{M} = \boldsymbol{w}_1 + \boldsymbol{w}_2$; (c) $\boldsymbol{M} = \boldsymbol{w}_1 + \boldsymbol{w}_3$.
3. Calculate the block weight diagrams of A_2-multiplets and $\mathcal{A}_3$-multiplets and the generalized Gel'fand's bases for the following representations $D^{\boldsymbol{M}}$ of C_3: (a) $\boldsymbol{M} = \boldsymbol{w}_2$; (b) $\boldsymbol{M} = 2\boldsymbol{w}_1$; (c) $\boldsymbol{M} = 2\boldsymbol{w}_2$; (d) $\boldsymbol{M} = 2\boldsymbol{w}_3$; (e) $\boldsymbol{M} = \boldsymbol{w}_1 + \boldsymbol{w}_3$. List the dominant weights in $D^{\boldsymbol{M}}$.
4. Calculate the block weight diagrams of A_2-multiplets and $\mathcal{A}_3$-multiplets and the generalized Gel'fand's bases for the following representations $D^{\boldsymbol{M}}$ of B_3: (a) $\boldsymbol{M} = \boldsymbol{w}_2$; (b) $\boldsymbol{M} = 2\boldsymbol{w}_1$; (c) $\boldsymbol{M} = 2\boldsymbol{w}_2$; (d) $\boldsymbol{M} = 2\boldsymbol{w}_3$. List the dominant weights in $D^{\boldsymbol{M}}$.
5. Calculate the block weight diagrams of A_3-multiplets and $\mathcal{A}_4$-multiplets and the generalized Gel'fand's bases for the following representations $D^{\boldsymbol{M}}$ of D_4: (a) $\boldsymbol{M} = \boldsymbol{w}_2$; (b) $\boldsymbol{M} = \boldsymbol{w}_3$; (c) $\boldsymbol{M} = \boldsymbol{w}_4$. Please calculate all positive roots of D_4 in terms of the block weight diagram of the adjoint representation $\boldsymbol{M} = \boldsymbol{w}_2$.
6. Calculate the block weight diagrams of A_1-multiplets and $\mathcal{A}_2$-multiplets and the generalized Gel'fand's bases for the following representations of G_2: (a) $\boldsymbol{M} = \boldsymbol{w}_2$; (b) $\boldsymbol{M} = \boldsymbol{w}_1$; (c) $\boldsymbol{M} = 2\boldsymbol{w}_2$; (d) $\boldsymbol{M} = \boldsymbol{w}_1 + \boldsymbol{w}_2$; (e) $\boldsymbol{M} = 3\boldsymbol{w}_2$; (f) $\boldsymbol{M} = 2\boldsymbol{w}_1$.
7. Draw up-half of the block weight diagram of the adjoint representation $D^{\boldsymbol{M}}$ with $\boldsymbol{M} = \boldsymbol{w}_1 = (1, 0, 0, 0)$ of the Lie algebra F_4 until the states with zero weight appear. Then, list all positive root $\boldsymbol{\alpha}$ of F_4 by the combinations of the basis vectors $\boldsymbol{e}_a$ in the rectangular coordinate frame, and show that B_4 is a subalgebra of F_4.

8. Calculate the Clebsch–Gordan series and the Clebsch–Gordan coefficients for the following direct product representations in A_2: (a) $(1,0)\times(1,0)$; (b) $(0,1)\times(0,1)$; (c) $(1,0)\times(1,1)$; (d) $(1,1)\times(2,0)$; (e) $(1,1)\times(2,2)$; (f) $(1,1)\times(2,1)$.
9. Calculate the Clebsch–Gordan series for the direct product representation $(1,0,0)\times(0,1,0)$ in the A_3 Lie algebra and the expansion for the highest weight state of each irreducible representation in the Clebsch–Gordan series.
10. Calculate the Clebsch–Gordan series for the decomposition of the direct product representation $(1,0,0)\times(0,0,1)$ in the C_3 Lie algebra and the expansion for the highest weight states for the representations contained in the series.
11. Please complete the calculation for the decomposition of the direct product representation $(1,1,0)\times(1,1,0)$ in C_3.
12. Calculate the Clebsch–Gordan series for the following direct product representations in the G_2 Lie algebra and the expansion for the highest weight state of each irreducible representation in the CG series (see Prob. 6): (a) $(0,1)\times(0,1)$; (b) $(0,1)\times(1,0)$; (c) $(1,0)\times(1,0)$.
13. Calculate the Clebsch–Gordan series for the direct product representation $(0,0,0,1)\times(0,0,0,1)$ in the F_4 Lie algebra and the expansion for the highest weight state of each irreducible representation in the Clebsch–Gordan series, where the dimensions $d(\boldsymbol{M})$ of some representations $D^{\boldsymbol{M}}$ of F_4, the Weyl orbital sizes $OS(\boldsymbol{M})$, and the multiplicities of the dominant weights in the representation $D^{\boldsymbol{M}}$ are listed in the following table (see table book [Bremner et al. (1985)].

$\boldsymbol{M}$	$d(\boldsymbol{M})$	OS	The multiplicity of the dominant weight				
			$(0,0,0,0)$	$(0,0,0,1)$	$(1,0,0,0)$	$(0,0,1,0)$	$(0,0,0,2)$
$(0,0,0,0)$	1	1	1				
$(0,0,0,1)$	26	24	2	1			
$(1,0,0,0)$	52	24	4	1	1		
$(0,0,1,0)$	273	96	9	5	2	1	
$(0,0,0,2)$	324	24	12	5	3	1	1

Chapter 9

UNITARY GROUPS

The generalized Gel'fand's method gives a certain rule to define the orthonormal basis states in an irreducible representation of a simple Lie algebra $\mathcal{L}$, but does not give the tensor structure of the basis states, **which are important for calculating wave functions in physics**. In the present chapter we study the tensor structure and the wave functions of the orthonormal basis states belonging to the given irreducible representation of the unitary group SU(N) by combining with the method of block weight diagrams and the Gel'fand's method, and briefly introduce some applications to the particle physics. The tensor structure of the orthonormal basis tensors for the irreducible representations of the SO(N) group and the USp(2ℓ) group will be studied in the subsequent chapters.

9.1 Irreducible Representations of SU(N)

The concept of tensor is related to the transformation group. In this chapter, we study the tensor with respect to the SU(N) group. The element of SU(N) is an $N \times N$ unitary matrix, which transforms a vector $\boldsymbol{V}$ in a complex space of dimension N as follows:

$$\boldsymbol{V}_a \stackrel{u}{\longrightarrow} \boldsymbol{V}_a' \equiv (O_u \boldsymbol{V})_a = \sum_{b=1}^{N} u_{ab} \boldsymbol{V}_b, \qquad 1 \leqslant a \leqslant N. \tag{9.1}$$

A tensor $\boldsymbol{T}_{a_1,\ldots,a_n}$ of rank n contains n indices and N^n components. In the SU(N) transformation u, each index plays the role of a vector index,

$$\boldsymbol{T}_{a_1\ldots a_n} \stackrel{u}{\longrightarrow} (O_u \boldsymbol{T})_{a_1\ldots a_n} = \sum_{b_1\ldots b_n} u_{a_1 b_1} \ldots u_{a_n b_n} \boldsymbol{T}_{b_1\ldots b_n}. \tag{9.2}$$

Namely, the transformation matrix of a tensor of rank n is the direct product of n matrices u. The direct product of n self-representations of SU(N) is generally reducible. For example, the direct product of self-representations of SU(2) are decomposed as follows:

$$D^{1/2} \times D^{1/2} \times D^{1/2} \simeq D^{3/2} \oplus 2\ D^{1/2},$$

$$D^{1/2} \times D^{1/2} \times D^{1/2} \times D^{1/2} \simeq D^{2} \oplus 3\ D^{1} \oplus 2\ D^{0}.$$

We are going to decompose the direct product representation of SU(N) and find the irreducible basis tensors.

9.1.1 *Symmetry of the Tensor Space of* SU*(N)*

The tensor space $\mathcal{T}$ of rank n is an N^n-dimensional linear space which is invariant in SU(N) transformations, namely, the transformed tensor still belongs to $\mathcal{T}$. The u matrices appear in the transformation (9.2) as the product of the matrix elements u_{ab}, which is commutable. Therefore, **the permutation symmetry of the tensor indices is left invariant in the** SU(N) **transformation**. Namely, the subset of tensors with the same permutation symmetry of indices constitutes an invariant subspace of $\mathcal{T}$ in O_u. **How to describe a subspace with given permutation symmetry**? It is easy to describe the subspace of totally symmetric tensors and that of totally antisymmetric tensors. However, the subspace with mixed symmetry is hard to describe. One cannot define a tensor of rank 3 whose first two indices are symmetric and the last two indices are antisymmetric.

We begin with the definition of a permutation R acting on a tensor, which has to **satisfy the group property**. The definition for a transposition $(j\ k)$ contains no confusion because it is self-inverse,

$$[(j\ k)\boldsymbol{T}]_{a_1 \ldots a_j \ldots a_k \ldots a_n} = \boldsymbol{T}'_{a_1 \ldots a_j \ldots a_k \ldots a_n} = \boldsymbol{T}_{a_1 \ldots a_k \ldots a_j \ldots a_n}. \tag{9.3}$$

Since **a permutation R can be decomposed into a product of transpositions**, the action of a permutation R on a tensor, $R\boldsymbol{T} = \boldsymbol{T}_R$, is defined based on Eq. (9.3). For a simple example, we have

$$R = \begin{pmatrix} 1 & 2 & 3 \\ 2 & 3 & 1 \end{pmatrix} = \begin{pmatrix} 3 & 1 & 2 \\ 1 & 2 & 3 \end{pmatrix} = (1\ 2\ 3) = (1\ 2)(2\ 3)\ ,$$

$$[(2\ 3)\boldsymbol{T}]_{a_1 a_2 a_3} = \boldsymbol{T}'_{a_1 a_2 a_3} = \boldsymbol{T}_{a_1 a_3 a_2},$$

$$(R\boldsymbol{T})_{a_1 a_2 a_3} = [(1\ 2)\boldsymbol{T}']_{a_1 a_2 a_3} = \boldsymbol{T}'_{a_2 a_1 a_3} = \boldsymbol{T}_{a_2 a_3 a_1} \neq \boldsymbol{T}_{a_3 a_1 a_2}.$$

Generally,

$$(R\boldsymbol{T})_{a_1 \ldots a_n} \equiv (\boldsymbol{T}_R)_{a_1 \ldots a_n} = \boldsymbol{T}_{a_{r_1} \ldots a_{r_n}} \neq \boldsymbol{T}_{a_{\overline{r}_1} \ldots a_{\overline{r}_n}}. \tag{9.4}$$

where

$$R = \begin{pmatrix} 1 & 2 & \dots & n \\ r_1 & r_2 & \dots & r_n \end{pmatrix} = \begin{pmatrix} \overline{r}_1 & \overline{r}_2 & \dots & \overline{r}_n \\ 1 & 2 & \dots & n \end{pmatrix}, \tag{9.5}$$

The permutation R **moves the r_jth index a_{r_j}, NOT the $\overline{r}_j$th index $a_{\overline{r}_j}$, to the jth position.** Equivalently, R moves the jth index a_j to the $\overline{r}_j$th position, NOT to the r_jth position. The tensor space $\mathcal{T}$ is left invariant in the permutation group S_n.

It is well known that a tensor of rank 2 can be decomposed into the sum of symmetric and antisymmetric tensors,

$$\boldsymbol{T}_{ab} = \frac{1}{2}\{\boldsymbol{T}_{ab} + \boldsymbol{T}_{ba}\} + \frac{1}{2}\{\boldsymbol{T}_{ab} - \boldsymbol{T}_{ba}\}. \tag{9.6}$$

The decomposition can be written in terms of the Young operators,

$$\begin{aligned} \boldsymbol{T}_{ab} &= \frac{1}{2}\{E + (1\ 2)\}\boldsymbol{T}_{ab} + \frac{1}{2}\{E - (1\ 2)\}\boldsymbol{T}_{ab} \\ &= \frac{1}{2}\left\{\mathcal{Y}^{[2]} + \mathcal{Y}^{[1,1]}\right\}\boldsymbol{T}_{ab} = E\boldsymbol{T}_{ab}. \end{aligned} \tag{9.7}$$

Namely, the decomposition can be achieved by the expansion (4.66) of the identity E with respect to the Young operators,

$$\boldsymbol{T}_{a_1 \dots a_n} = E\boldsymbol{T}_{a_1 \dots a_n} = \frac{1}{n!}\sum_{[\lambda]} d_{[\lambda]} \sum_{\mu} \mathcal{Y}_\mu^{[\lambda]} y_\mu^{[\lambda]} \boldsymbol{T}_{a_1 \dots a_n}. \tag{9.8}$$

For example, a tensor of rank 3 is decomposed as

$$\boldsymbol{T}_{abc} = \frac{1}{6}\mathcal{Y}^{[3]}\boldsymbol{T}_{abc} + \frac{1}{3}\mathcal{Y}_1^{[2,1]}\boldsymbol{T}_{abc} + \frac{1}{3}\mathcal{Y}_2^{[2,1]}\boldsymbol{T}_{abc} + \frac{1}{6}\mathcal{Y}^{[1,1,1]}\boldsymbol{T}_{abc}. \tag{9.9}$$

The first term is a totally symmetric tensor, the last term is a totally antisymmetric tensor, and the remaining terms are tensors with mixed symmetry. **The Young operators**, which are the projective operators, **decompose the tensor space $\mathcal{T}$ into the sum of tensor subspaces**

$$\mathcal{T} = E\mathcal{T} = \frac{1}{n!}\bigoplus_{[\lambda]} d_{[\lambda]} \bigoplus_{\mu} \mathcal{Y}_\mu^{[\lambda]} y_\mu^{[\lambda]} \mathcal{T} = \bigoplus_{[\lambda]}\bigoplus_{\mu} \mathcal{T}_\mu^{[\lambda]}. \tag{9.10}$$

Let us study the property of the tensor subspace $\mathcal{T}_\mu^{[\lambda]}$. First, there is no common tensor between two subspaces $\mathcal{T}_\mu^{[\lambda]}$ and $\mathcal{T}_\nu^{[\omega]}$ because the Young operators $\mathcal{Y}_\mu^{[\lambda]} y_\mu^{[\lambda]}$ are orthogonal to each other. Thus, **the decomposition (9.10) is in the form of direct sum**. Second, the constant factor $d_{[\lambda]}/n!$ and the operator $y_\mu^{[\lambda]}$ do not make any change with the subspace $\mathcal{T}_\mu^{[\lambda]}$. In fact, due to $y_\mu^{[\lambda]}\mathcal{T} \subset \mathcal{T}$ and $\mathcal{Y}_\mu^{[\lambda]}\mathcal{T} \subset \mathcal{T}$, one has

$$\mathcal{Y}_\mu^{[\lambda]}\left\{y_\mu^{[\lambda]}\mathcal{T}\right\} \subset \mathcal{Y}_\mu^{[\lambda]}\mathcal{T}, \qquad \mathcal{Y}_\mu^{[\lambda]}\mathcal{T} = \mathcal{Y}_\mu^{[\lambda]}y_\mu^{[\lambda]}\left\{\frac{d_{[\lambda]}}{n!}\mathcal{Y}_\mu^{[\lambda]}\mathcal{T}\right\} \subset \mathcal{Y}_\mu^{[\lambda]}y_\mu^{[\lambda]}\mathcal{T}.$$

Thus,

$$\mathcal{T}_\mu^{[\lambda]} = \frac{d_{[\lambda]}}{n!}\,\mathcal{Y}_\mu^{[\lambda]}y_\mu^{[\lambda]}\mathcal{T} = \mathcal{Y}_\mu^{[\lambda]}\mathcal{T}. \tag{9.11}$$

For the same reason,

$$R\mathcal{T} = \mathcal{T}, \qquad \mathcal{Y}_\mu^{[\lambda]}R\mathcal{T} = \mathcal{T}_\mu^{[\lambda]}, \qquad \forall\, R \in \mathrm{S}_n. \tag{9.12}$$

Third, as shown in Theorem 9.1, the subspace $\mathcal{T}_\mu^{[\lambda]}$ is invariant in O_u.

Theorem 9.1 (Weyl reciprocity) Any permutation R in S_n for a tensor of rank n commutes with each transformation O_u in $\mathrm{SU}(N)$.

Proof The key of the proof is that the product of matrix elements u_{ab} in Eq. (9.2) are commutable:

$$\begin{aligned}
\left(O_uR\boldsymbol{T}\right)_{a_1\ldots a_n} &= \left(O_u\boldsymbol{T}_R\right)_{a_1\ldots a_n} = \sum_{b_1\ldots b_n} u_{a_1b_1}\ldots u_{a_nb_n}\left(\boldsymbol{T}_R\right)_{b_1\ldots b_n} \\
&= \sum_{b_1\ldots b_n} u_{a_{r_1}b_{r_1}}\ldots u_{a_{r_n}b_{r_n}}\boldsymbol{T}_{b_{r_1}\ldots b_{r_n}} = \left(O_u\boldsymbol{T}\right)_{a_{r_1}\ldots a_{r_n}} \\
&= \left(RO_u\boldsymbol{T}\right)_{a_1\ldots a_n}.
\end{aligned} \tag{9.13}$$

□

Thus, $O_u\left\{\mathcal{Y}_\mu^{[\lambda]}\boldsymbol{T}\right\} = \mathcal{Y}_\mu^{[\lambda]}\left\{O_u\boldsymbol{T}\right\} \subset \mathcal{Y}_\mu^{[\lambda]}\mathcal{T} = \mathcal{T}_\mu^{[\lambda]}$.

Corollary 9.1.1 The $\mathrm{SU}(N)$ transformation O_u leaves the subspace $\mathcal{T}_\mu^{[\lambda]}$ invariant.

Remind that R on the left of $\mathcal{Y}_\mu^{[\lambda]}\mathcal{T}$ may change the subspace

$$R_{\nu\mu}\mathcal{Y}_\mu^{[\lambda]}\mathcal{T} = \mathcal{Y}_\nu^{[\lambda]}R_{\nu\mu}\mathcal{T} = \mathcal{T}_\nu^{[\lambda]}, \tag{9.14}$$

where $R_{\nu\mu}$ is the permutation transforming the standard Young tableau $\mathcal{Y}_\mu^{[\lambda]}$ to the standard Young tableau $\mathcal{Y}_\nu^{[\lambda]}$.

9.1.2 *Basis Tensors in the Tensor Subspace*

First of all, we review the property of the basis vectors (see section 5.7). A basis vector $\boldsymbol{\theta}_d$ is a special vector with only one nonvanishing component, which is equal to 1: $\left(\boldsymbol{\theta}_d\right)_a = \delta_{da}$. Any vector $\boldsymbol{V}$ can be expanded with respect to the basis vectors:

$$\boldsymbol{V} = \sum_{d=1}^{N}\boldsymbol{\theta}_dV_d, \qquad \left(\boldsymbol{V}\right)_a = \sum_{d=1}^{N}\left(\boldsymbol{\theta}_d\right)_aV_d = V_a. \tag{9.15}$$

Remind that $(\boldsymbol{V})_a$ and V_a are different in the SU(N) transformation although they are equal in value.

A basis tensor $\boldsymbol{\theta}_{d_1\ldots d_n}$ is a special tensor with only one non-vanishing component which is equal to 1,

$$(\boldsymbol{\theta}_{d_1\ldots d_n})_{a_1\ldots a_n} = (\boldsymbol{\theta}_{d_1} \times \cdots \times \boldsymbol{\theta}_{d_n})_{a_1\ldots a_n} = \delta_{d_1a_1}\delta_{d_2a_2}\ldots\delta_{d_na_n}. \tag{9.16}$$

Being a tensor, the basis tensor transforms in O_u and in R as follows:

$$\begin{aligned}
(O_u\boldsymbol{\theta}_{d_1\ldots d_n})_{a_1\ldots a_n} &= \sum_{b_1\ldots b_n} u_{a_1b_1}\ldots u_{a_nb_n}\,(\boldsymbol{\theta}_{d_1\ldots d_n})_{b_1\ldots b_n} \\
&= u_{a_1d_1}\ldots u_{a_nd_n} = \sum_{b_1\ldots b_n} (\boldsymbol{\theta}_{b_1\ldots b_n})_{a_1\ldots a_n}\, u_{b_1d_1}\ldots u_{b_nd_n}, \\
(R\boldsymbol{\theta}_{d_1\ldots d_n})_{a_1\ldots a_n} &= (\boldsymbol{\theta}_{d_1\ldots d_n})_{a_{r_1}\ldots a_{r_n}} = \delta_{d_1a_{r_1}}\delta_{d_2a_{r_2}}\ldots\delta_{d_na_{r_n}} \\
&= \delta_{d_{\overline{r}_1}a_1}\delta_{d_{\overline{r}_2}a_2}\ldots\delta_{d_{\overline{r}_n}a_n} = \left(\boldsymbol{\theta}_{d_{\overline{r}_1}\ldots d_{\overline{r}_n}}\right)_{a_1\ldots a_n}.
\end{aligned}$$

Namely,

$$O_u\boldsymbol{\theta}_{d_1\ldots d_n} = \sum_{b_1\ldots b_n} \boldsymbol{\theta}_{b_1\ldots b_n} u_{b_1d_1}\ldots u_{b_nd_n}, \tag{9.17}$$

$$R\boldsymbol{\theta}_{d_1\ldots d_n} = \boldsymbol{\theta}_{d_{\overline{r}_1}\ldots d_{\overline{r}_n}} \neq \boldsymbol{\theta}_{d_{r_1}\ldots d_{r_n}}. \tag{9.18}$$

R transforms a basis tensor $\boldsymbol{\theta}_{d_1\ldots d_n}$ to another basis tensor $\boldsymbol{\theta}_{d_{\overline{r}_1}\ldots d_{\overline{r}_n}}$, where **the jth index d_j moves to the r_jth position, NOT to the $\overline{r}_j$th position**. Equivalently, the $\overline{r}_j$th index $d_{\overline{r}_j}$, NOT the r_jth index d_{r_j}, moves to the jth position. In a simple example $R = (1\ 2\ 3) = (1\ 2)(2\ 3)$,

$$(2\ 3)\boldsymbol{\theta}_{d_1d_2d_3} = \boldsymbol{\theta}_{d_1d_3d_2}, \qquad R\boldsymbol{\theta}_{d_1d_2d_3} = (1\ 2)\boldsymbol{\theta}_{d_1d_3d_2} = \boldsymbol{\theta}_{d_3d_1d_2} \neq \boldsymbol{\theta}_{d_2d_3d_1}.$$

Any tensor $\boldsymbol{T}$ can be expanded with respect to the basis tensors

$$\boldsymbol{T}_{a_1\ldots a_n} = \sum_{d_1\ldots d_n} (\boldsymbol{\theta}_{d_1\ldots d_n})_{a_1\ldots a_n} T_{d_1\ldots d_n} = T_{a_1\ldots a_n}. \tag{9.19}$$

In the SU(N) transformation O_u, $\boldsymbol{T}_{a_1\ldots a_n}$ is a tensor, but $T_{a_1\ldots a_n}$ is a scalar.

$\boldsymbol{\theta}_{d_1\ldots d_n}$ is the basis tensor of the tensor space $\mathcal{T}$, and $\mathcal{Y}_\mu^{[\lambda]}\boldsymbol{\theta}_{d_1\ldots d_n} \in \mathcal{T}_\mu^{[\lambda]}$ is a linear combination of the basis tensors $\boldsymbol{\theta}_{d_1\ldots d_n}$. Since the dimension of $\mathcal{T}_\mu^{[\lambda]}$ is less than that of $\mathcal{T}$, some $\mathcal{Y}_\mu^{[\lambda]}\boldsymbol{\theta}_{d_1\ldots d_n}$ may be vanishing or linearly dependent. **Our task is to find the complete set of the basis tensors in $\mathcal{T}_\mu^{[\lambda]}$, which are linearly independent.**

Usually, the expansion of $\mathcal{Y}_\mu^{[\lambda]}\boldsymbol{\theta}_{d_1\ldots d_n} \in \mathcal{T}_\mu^{[\lambda]}$ is quite long. Let us first to simplify the notations. For example, letting the Young tableau $\mathcal{Y}_2^{[3,2]} =$ $\begin{array}{|c|c|c|}\hline 1 & 2 & 4 \\ \hline 3 & 5 \\ \cline{1-2}\end{array}$, we have

$$\begin{aligned}
\mathcal{Y}_2^{[3,2]}\boldsymbol{\theta}_{11233} &= [E + (1\ 2) + (1\ 4) + (2\ 4) + (1\ 2\ 4) + (2\ 1\ 4)] \\
&\quad \times\ [E + (3\ 5)]\,[E - (1\ 3)]\,[E - (2\ 5)]\,\boldsymbol{\theta}_{11233} \\
&= [E + (3\ 5)]\,[E + (1\ 2) + (1\ 4) + (2\ 4) + (1\ 2\ 4) + (2\ 1\ 4)] \\
&\quad \times\ [\boldsymbol{\theta}_{11233} - \boldsymbol{\theta}_{13231} - \boldsymbol{\theta}_{21133} + \boldsymbol{\theta}_{23131}] \\
&= 2\,[\boldsymbol{\theta}_{11233} + \boldsymbol{\theta}_{31213} + \boldsymbol{\theta}_{13213}] + 2\,[\boldsymbol{\theta}_{11332} + \boldsymbol{\theta}_{31312} + \boldsymbol{\theta}_{13312}] \\
&\quad - 2\,[\boldsymbol{\theta}_{13231} + \boldsymbol{\theta}_{31231} + \boldsymbol{\theta}_{33211}] - 2\,[\boldsymbol{\theta}_{13132} + \boldsymbol{\theta}_{31132} + \boldsymbol{\theta}_{33112}] \\
&\quad - [\boldsymbol{\theta}_{21133} + \boldsymbol{\theta}_{12133} + \boldsymbol{\theta}_{13123} + \boldsymbol{\theta}_{31123} + \boldsymbol{\theta}_{23113} + \boldsymbol{\theta}_{32113}] \\
&\quad - [\boldsymbol{\theta}_{21331} + \boldsymbol{\theta}_{12331} + \boldsymbol{\theta}_{13321} + \boldsymbol{\theta}_{31321} + \boldsymbol{\theta}_{23311} + \boldsymbol{\theta}_{32311}] \\
&\quad + 4\,[\boldsymbol{\theta}_{23131} + \boldsymbol{\theta}_{32131} + \boldsymbol{\theta}_{33121}]\,.
\end{aligned} \tag{9.20}$$

Introduce a simplified notation to **denote the tensor expansion of $\mathcal{Y}\boldsymbol{\theta}_{a_1\ldots a_n}$ by a tensor Young tableau** where the box filled with j in the Young tableau $\mathcal{Y}$ is now filled with the subscript a_j. For example, the tensor in Eq. (9.20) is denoted by a tensor Young tableau $\begin{array}{|c|c|c|}\hline 1&1&3\\\hline 2&3\\\cline{1-2}\end{array}$. Another example is that if the Young tableau $\mathcal{Y}$ is

$$\begin{array}{|c|c|c|}\hline 1&2&6\\\hline 3&5&7\\\hline 4\\\cline{1-1}\end{array}\,,$$

the tensor $\mathcal{Y}\boldsymbol{\theta}_{a_1\ldots a_7}$ is denoted by a tensor Young tableau

$$\mathcal{Y}\boldsymbol{\theta}_{a_1\ldots a_7} = \begin{array}{|c|c|c|}\hline a_1&a_2&a_6\\\hline a_3&a_5&a_7\\\hline a_4\\\cline{1-1}\end{array}\,.$$

In different tensor subspaces, the same tensor Young tableau describes different tensors. For example, letting the Young tableaux $\mathcal{Y}_1^{[2,1]}$ and $\mathcal{Y}_2^{[2,1]}$ be $\begin{array}{|c|c|}\hline 1&2\\\hline 3\\\cline{1-1}\end{array}$ and $\begin{array}{|c|c|}\hline 1&3\\\hline 2\\\cline{1-1}\end{array}$, respectively, we have

$$\mathcal{Y}_1^{[2,1]}\boldsymbol{\theta}_{123} = \begin{array}{|c|c|}\hline 1&2\\\hline 3\\\cline{1-1}\end{array} = \boldsymbol{\theta}_{123} + \boldsymbol{\theta}_{213} - \boldsymbol{\theta}_{321} - \boldsymbol{\theta}_{231} \in \mathcal{Y}_1^{[2,1]}\mathcal{T},$$

$$\mathcal{Y}_2^{[2,1]}\boldsymbol{\theta}_{132} = \begin{array}{|c|c|}\hline 1&2\\\hline 3\\\cline{1-1}\end{array} = \boldsymbol{\theta}_{132} + \boldsymbol{\theta}_{231} - \boldsymbol{\theta}_{312} - \boldsymbol{\theta}_{213} \in \mathcal{Y}_2^{[2,1]}\mathcal{T}.$$

The tensor Young tableau $\begin{array}{|c|c|}\hline 1&2\\\hline 3\\\cline{1-1}\end{array}$ in two tensor subspaces $\mathcal{Y}_1^{[2,1]}\mathcal{T}$ and $\mathcal{Y}_2^{[2,1]}\mathcal{T}$ for SU(3) describes two different tensors. They are related by a permutation (2 3),

$$(2\ 3)\mathcal{Y}_1^{[2,1]}\boldsymbol{\theta}_{123} = \mathcal{Y}_2^{[2,1]}(2\ 3)\boldsymbol{\theta}_{123} = \mathcal{Y}_2^{[2,1]}\boldsymbol{\theta}_{132}\,. \tag{9.21}$$

Please do not get confused between **a tensor Young tableau** and **a Young tableau.**

Study the symmetry of the tensor Young tableaux in terms of the symmetry of a Young operator $\mathcal{Y}_\mu^{[\lambda]}$ and the Fock condition (see subsection 4.2.4)

$$\mathcal{Y}_\mu^{[\lambda]} Q = \delta(Q) \mathcal{Y}_\mu^{[\lambda]}, \qquad \mathcal{Y}_\mu^{[\lambda]} \left\{ E - \sum_\mu (c_\mu \ d_\nu) \right\} = 0.$$

Let Q_0 denote a vertical transposition belonging to the Young tableau $\mathcal{Y}_\mu^{[\lambda]}$. Say $Q_0 = (i \ j)$, Q_0 interchanges two digits i and j in the same column of the Young tableau $\mathcal{Y}_\mu^{[\lambda]}$. A minus sign occurs when right-multiplying Q_0 on $\mathcal{Y}_\mu^{[\lambda]}$, and two subscripts a_i and a_j are interchanged when left-multiplying Q_0 on the basis tensor $\boldsymbol{\theta}_{a_1 \ldots a_n}$. a_i and a_j are located in the same column of the tensor Young tableau $\mathcal{Y}_\mu^{[\lambda]} \boldsymbol{\theta}_{a_1 \ldots a_n}$. Thus, **two digits in the same column of a tensor Young tableau are antisymmetric**:

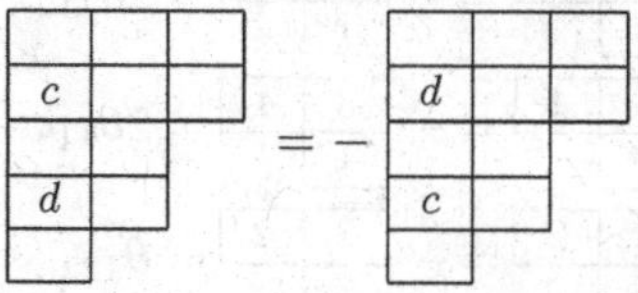

(9.22)

A tensor Young tableau with the repetitive digits in the same column vanishes. **The row number of a non-vanishing tensor Young tableau for SU(N) is not larger than N.** The Fock condition also gives some relations between the tensor Young tableaux in a tensor subspace $\mathcal{T}_\mu^{[\lambda]}$:

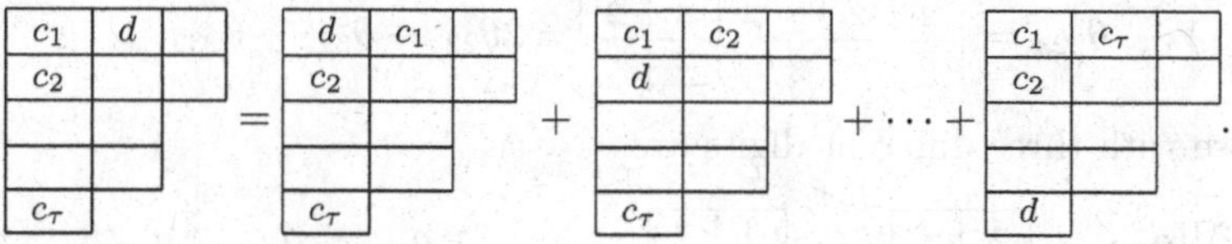

(9.23)

For SU(3) we have

$$\mathcal{Y}_1^{[2,1]} \boldsymbol{\theta}_{abc} = -\mathcal{Y}_1^{[2,1]} (1 \ 3) \boldsymbol{\theta}_{abc}, \qquad \mathcal{Y}_1^{[2,1]} \boldsymbol{\theta}_{abc} = \mathcal{Y}_1^{[2,1]} \left[(2 \ 1) + (2 \ 3) \right] \boldsymbol{\theta}_{abc},$$

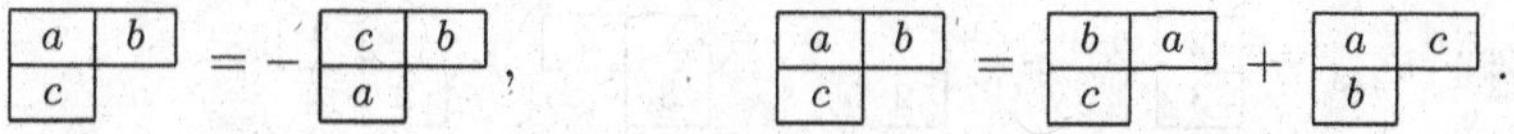

Remind that although left-multiplying with a horizontal permutation P of a Young operator $\mathcal{Y}$ does not change the Young operator $\mathcal{Y}$ as well as the tensor Young tableau $\mathcal{Y}\boldsymbol{\theta}_{a_1 \ldots a_n}$, it **does not mean a symmetry of digits** in the tensor Young tableau. For example, the expansion of $\mathcal{Y}_1^{[2,1]} \boldsymbol{\theta}_{abc}$ is **symmetric between the first two indices**, but NOT between a and b,

$$\begin{aligned}(1\ 2)\mathcal{Y}_1^{[2,1]}\boldsymbol{\theta}_{abc} &= (1\ 2)\left\{\boldsymbol{\theta}_{abc}+\boldsymbol{\theta}_{bac}-\boldsymbol{\theta}_{cba}-\boldsymbol{\theta}_{bca}\right\}\\ &= \boldsymbol{\theta}_{bac}+\boldsymbol{\theta}_{abc}-\boldsymbol{\theta}_{bca}-\boldsymbol{\theta}_{cba}=\mathcal{Y}_1^{[2,1]}\boldsymbol{\theta}_{abc}\,.\end{aligned}$$

Now we return to the problem how to find the complete set of the linearly independent basis tensors in the subspace $\mathcal{T}_\mu^{[\lambda]} = \mathcal{Y}_\mu^{[\lambda]}\mathcal{T}$, where $\mathcal{Y}_\mu^{[\lambda]}$ is a standard Young operator. Discuss a useful example first. The general form of a tensor Young tableau in the tensor subspace $\mathcal{T}_1^{[2,1]}$ of the tensor space $\mathcal{T}$ of rank 3 of SU(3) is

$$\begin{aligned}\begin{array}{|c|c|}\hline a & b\\\hline c\\\cline{1-1}\end{array} &= \mathcal{Y}_1^{[2,1]}\boldsymbol{\theta}_{abc}=\{E+(1\ 2)-(1\ 3)-(2\ 1)(1\ 3)\}\,\boldsymbol{\theta}_{abc}\\ &= \boldsymbol{\theta}_{abc}+\boldsymbol{\theta}_{bac}-\boldsymbol{\theta}_{cba}-\boldsymbol{\theta}_{bca}.\end{aligned} \tag{9.24}$$

The tensor Young tableau is vanishing when $a = c$. The tensor Young tableaux with a pair of same digits are

$$\begin{aligned}
\mathcal{Y}_1^{[2,1]}\boldsymbol{\theta}_{112} &= \begin{array}{|c|c|}\hline 1 & 1\\\hline 2\\\cline{1-1}\end{array} = -\begin{array}{|c|c|}\hline 2 & 1\\\hline 1\\\cline{1-1}\end{array} = 2\boldsymbol{\theta}_{112}-\boldsymbol{\theta}_{211}-\boldsymbol{\theta}_{121},\\
\mathcal{Y}_1^{[2,1]}\boldsymbol{\theta}_{113} &= \begin{array}{|c|c|}\hline 1 & 1\\\hline 3\\\cline{1-1}\end{array} = -\begin{array}{|c|c|}\hline 3 & 1\\\hline 1\\\cline{1-1}\end{array} = 2\boldsymbol{\theta}_{113}-\boldsymbol{\theta}_{311}-\boldsymbol{\theta}_{131},\\
\mathcal{Y}_1^{[2,1]}\boldsymbol{\theta}_{122} &= \begin{array}{|c|c|}\hline 1 & 2\\\hline 2\\\cline{1-1}\end{array} = -\begin{array}{|c|c|}\hline 2 & 2\\\hline 1\\\cline{1-1}\end{array} = \boldsymbol{\theta}_{122}+\boldsymbol{\theta}_{212}-2\boldsymbol{\theta}_{221},\\
\mathcal{Y}_1^{[2,1]}\boldsymbol{\theta}_{133} &= \begin{array}{|c|c|}\hline 1 & 3\\\hline 3\\\cline{1-1}\end{array} = -\begin{array}{|c|c|}\hline 3 & 3\\\hline 1\\\cline{1-1}\end{array} = \boldsymbol{\theta}_{133}+\boldsymbol{\theta}_{313}-2\boldsymbol{\theta}_{331},\\
\mathcal{Y}_1^{[2,1]}\boldsymbol{\theta}_{233} &= \begin{array}{|c|c|}\hline 2 & 3\\\hline 3\\\cline{1-1}\end{array} = -\begin{array}{|c|c|}\hline 3 & 3\\\hline 2\\\cline{1-1}\end{array} = \boldsymbol{\theta}_{233}+\boldsymbol{\theta}_{323}-2\boldsymbol{\theta}_{332},\\
\mathcal{Y}_1^{[2,1]}\boldsymbol{\theta}_{223} &= \begin{array}{|c|c|}\hline 2 & 2\\\hline 3\\\cline{1-1}\end{array} = -\begin{array}{|c|c|}\hline 3 & 2\\\hline 2\\\cline{1-1}\end{array} = 2\boldsymbol{\theta}_{223}-\boldsymbol{\theta}_{322}-\boldsymbol{\theta}_{232},
\end{aligned}$$

and those with three different digits are

$$\begin{aligned}
\mathcal{Y}_1^{[2,1]}\boldsymbol{\theta}_{123} &= \begin{array}{|c|c|}\hline 1 & 2\\\hline 3\\\cline{1-1}\end{array} = -\begin{array}{|c|c|}\hline 3 & 2\\\hline 1\\\cline{1-1}\end{array} = \boldsymbol{\theta}_{123}+\boldsymbol{\theta}_{213}-\boldsymbol{\theta}_{321}-\boldsymbol{\theta}_{231},\\
\mathcal{Y}_1^{[2,1]}\boldsymbol{\theta}_{132} &= \begin{array}{|c|c|}\hline 1 & 3\\\hline 2\\\cline{1-1}\end{array} = -\begin{array}{|c|c|}\hline 2 & 3\\\hline 1\\\cline{1-1}\end{array} = \boldsymbol{\theta}_{132}+\boldsymbol{\theta}_{312}-\boldsymbol{\theta}_{231}-\boldsymbol{\theta}_{321},\\
\mathcal{Y}_1^{[2,1]}\boldsymbol{\theta}_{213} &= \begin{array}{|c|c|}\hline 2 & 1\\\hline 3\\\cline{1-1}\end{array} = -\begin{array}{|c|c|}\hline 3 & 1\\\hline 2\\\cline{1-1}\end{array} = \begin{array}{|c|c|}\hline 1 & 2\\\hline 3\\\cline{1-1}\end{array} - \begin{array}{|c|c|}\hline 1 & 3\\\hline 2\\\cline{1-1}\end{array}\\
&= \boldsymbol{\theta}_{213}+\boldsymbol{\theta}_{123}-\boldsymbol{\theta}_{312}-\boldsymbol{\theta}_{132}.
\end{aligned}$$

The dimension of the tensor subspace $\mathcal{T}_1^{[2,1]}$ is 8. The eight linearly independent tensor Young tableaux have a common characteristic: **The digit in each column of the tableau increases downward and the digit**

in each row does not decrease rightward. A tensor Young tableau with this characteristic is called **standard**.

Theorem 9.2 The standard tensor Young tableaux constitute a complete set of basis tensors in the tensor subspace $\mathcal{T}_\mu^{[\lambda]}$.

Proof For any basis tensor $\boldsymbol{\theta}_{b_1 \ldots b_n}$, there is a permutation S to arrange its subscripts in the increasing order:

$$\mathcal{Y}_\mu^{[\lambda]} \boldsymbol{\theta}_{b_1 \ldots b_n} = \mathcal{Y}_\mu^{[\lambda]} S \boldsymbol{\theta}_{a_1 \ldots a_n}, \qquad a_1 \leqslant a_2 \leqslant \ldots \leqslant a_n.$$

On the other hand, $\mathcal{Y}_\mu^{[\lambda]} S$ belongs to the right-ideal $\mathcal{R}_\mu^{[\lambda]} = \mathcal{Y}_\mu^{[\lambda]} \mathcal{L}$ in the group algebra of S_n. The basis vectors in $\mathcal{R}_\mu^{[\lambda]}$ are $\mathcal{Y}_\mu^{[\lambda]} R_{\mu\nu}$. Thus, any tensor in the tensor subspace $\mathcal{T}_\mu^{[\lambda]}$ can be expressed as a linear combination of the following tensor Young tableaux:

$$\mathcal{Y}_\mu^{[\lambda]} R_{\mu\nu} \boldsymbol{\theta}_{a_1 \ldots a_n}, \qquad a_1 \leqslant a_2 \leqslant \ldots \leqslant a_n. \tag{9.25}$$

Namely, the tensor Young tableaux given in Eq. (9.25) constitute a complete set of basis tensors in the tensor subspace $\mathcal{T}_\mu^{[\lambda]}$. We are going to show that $\mathcal{Y}_\mu^{[\lambda]} R_{\mu\nu} \boldsymbol{\theta}_{a_1 \ldots a_n}$ **are standard if it is not vanishing.**

$R_{\mu\nu}$ is a permutation transforming the standard Young tableau $\mathcal{Y}_\nu^{[\lambda]}$ to the standard Young tableau $\mathcal{Y}_\mu^{[\lambda]}$ so that

$$\mathcal{Y}_\mu^{[\lambda]} R_{\mu\nu} \boldsymbol{\theta}_{a_1 \ldots a_n} = R_{\mu\nu} \left\{ \mathcal{Y}_\nu^{[\lambda]} \boldsymbol{\theta}_{a_1 \ldots a_n} \right\}. \tag{9.26}$$

It can be proven that **two tensors** $\mathcal{Y}_\mu^{[\lambda]} R_{\mu\nu} \boldsymbol{\theta}_{a_1 \ldots a_n}$ and $\mathcal{Y}_\nu^{[\lambda]} \boldsymbol{\theta}_{a_1 \ldots a_n}$ given in Eq. (8.26) **are described by the same tensor Young tableau in two different tensor subspaces** $\mathcal{T}_\mu^{[\lambda]}$ and $\mathcal{T}_\nu^{[\lambda]}$. Let

$$R_{\mu\nu} = \begin{pmatrix} \overline{r}_1 \, \overline{r}_2 \ldots \overline{r}_n \\ 1 \; 2 \; \ldots \; n \end{pmatrix}, \qquad R_{\mu\nu} \boldsymbol{\theta}_{a_1 \ldots a_n} = \boldsymbol{\theta}_{b_1 \ldots b_n}, \qquad b_j = a_{\overline{r}_j}.$$

Arbitrarily choose a box in the Young pattern $[\lambda]$, say the box filled with the digit j in the standard Young tableau $\mathcal{Y}_\mu^{[\lambda]}$. This box is filled with $\overline{r}_j$ in the standard Young tableau $\mathcal{Y}_\nu^{[\lambda]}$, is filled with $a_{\overline{r}_j}$ in the tensor Young tableau $\mathcal{Y}_\nu^{[\lambda]} \boldsymbol{\theta}_{a_1 \ldots a_n}$, and is filled with $b_j = a_{\overline{r}_j}$ in the tensor Young tableau $\mathcal{Y}_\mu^{[\lambda]} R_{\mu\nu} \boldsymbol{\theta}_{a_1 \ldots a_n}$. Therefore, two tensor Young tableaux are the same. Equation (9.21) is a special example of Eq. (9.26).

Since $\mathcal{Y}_\nu^{[\lambda]}$ is a standard Young tableau and $a_1 \leqslant a_2 \leqslant \ldots \leqslant a_n$, the tensor Young tableau $\mathcal{Y}_\nu^{[\lambda]} \boldsymbol{\theta}_{a_1 \ldots a_n}$ satisfies that the digit in each column of $\mathcal{Y}_\nu^{[\lambda]} \boldsymbol{\theta}_{a_1 \ldots a_n}$ does not decrease downward and the digit in each row does not decrease rightward. Thus, $\mathcal{Y}_\nu^{[\lambda]}\boldsymbol{\theta}_{a_1\ldots a_n}$ as well as $\mathcal{Y}_\mu^{[\lambda]}R_{\mu\nu}\boldsymbol{\theta}_{a_1\ldots a_n}$ is a standard tensor Young tableau or vanishing, depending on whether there are repetitive digits in a column of the tableau. **Different standard tensor Young tableaux are linearly independent.** □

9.1.3 *Chevalley Bases of Generators in* SU*(N)*

From Eqs. (7.75) and (7.76) one has

$$(\boldsymbol{r}_\mu)_j = \sqrt{2}\left[\boldsymbol{V}_\mu - \boldsymbol{V}_{\mu+1}\right]_j = \sqrt{\frac{\mu+1}{\mu}}\delta_{\mu(N-j)} - \sqrt{\frac{\mu-1}{\mu}}\delta_{\mu(N-j+1)}.$$

Substituting $(\boldsymbol{r}_\mu)_j$ into Eq. (8.1), from Eq. (7.73) we obtain the Chevalley bases of generators in the self-representation of SU(N) as follows:

$$\begin{aligned} H_\mu &= \sum_{j=1}^{\ell}(\boldsymbol{r}_\mu)_j H_j = \sqrt{\frac{2(\mu+1)}{\mu}}T_{\mu+1}^{(3)} - \sqrt{\frac{2(\mu-1)}{\mu}}T_\mu^{(3)} \\ &= T_{\mu\mu}^{(1)} - T_{(\mu+1)(\mu+1)}^{(1)}, \\ E_\mu &= E_{\boldsymbol{\alpha}_{\mu(\mu+1)}} = T_{\mu(\mu+1)}^{(1)} + \mathrm{i}T_{\mu(\mu+1)}^{(2)}, \quad F_\mu = T_{\mu(\mu+1)}^{(1)} - \mathrm{i}T_{\mu(\mu+1)}^{(2)}. \end{aligned} \tag{9.27}$$

Namely, each of H_μ, E_μ, and F_μ contains only a nonvanishing two-dimensional submatrix in the μth and $(\mu+1)$th columns and rows, which **is the same as that in the self-representation of** SU(2) (see Eq. (8.3)). Their actions on the basis vectors are

$$\begin{aligned} H_\mu\boldsymbol{\theta}_\nu &= \boldsymbol{\theta}_\mu\delta_{\nu\mu} - \boldsymbol{\theta}_{\mu+1}\delta_{\nu(\mu+1)}, \\ E_\mu\boldsymbol{\theta}_\nu &= \boldsymbol{\theta}_\mu\delta_{\nu(\mu+1)}, \qquad F_\mu\boldsymbol{\theta}_\nu = \boldsymbol{\theta}_{\mu+1}\delta_{\nu\mu}. \end{aligned} \tag{9.28}$$

In the SU(N) transformation, each index in a tensor plays the role of a vector index. A basis tensor in the action of a generator becomes the sum of the basis tensors, each of which is obtained by the action of the generator on one tensor index according to Eq. (9.28). Thus, the standard tensor Young tableau is the common eigenstate of H_μ where the eigenvalue is **the number of the digit μ filled in the tableau, subtracting the number of the digit** $(\mu+1)$. The eigenvalues constitutes the weight $\boldsymbol{m}$ of the standard tensor Young tableau. **Two tensor Young tableaux have the same weight if their sets of the filled digits are the same.** The standard tensor Young tableau in the action of F_μ becomes the sum of all possible tensor Young tableaux, each of which is obtained from the original one **by replacing one filled digit μ with the digit** $(\mu+1)$, and in the action of E_μ the sum of those is obtained by replacing $(\mu+1)$

with μ. The obtained Young tableaux may not be standard, but they can be transformed to the sum of the standard tensor Young tableaux by the symmetries (9.22) and (9.23). For example,

$$H_1 \begin{array}{|c|c|}\hline 1 & 1 \\ \hline 3 \\ \cline{1-1}\end{array} = 2 \begin{array}{|c|c|}\hline 1 & 1 \\ \hline 3 \\ \cline{1-1}\end{array}, \qquad H_2 \begin{array}{|c|c|}\hline 1 & 1 \\ \hline 3 \\ \cline{1-1}\end{array} = - \begin{array}{|c|c|}\hline 1 & 1 \\ \hline 3 \\ \cline{1-1}\end{array},$$

$$E_2 \begin{array}{|c|c|}\hline 1 & 1 \\ \hline 3 \\ \cline{1-1}\end{array} = \begin{array}{|c|c|}\hline 1 & 1 \\ \hline 2 \\ \cline{1-1}\end{array}, \qquad E_1 \begin{array}{|c|c|}\hline 1 & 1 \\ \hline 3 \\ \cline{1-1}\end{array} = F_2 \begin{array}{|c|c|}\hline 1 & 1 \\ \hline 3 \\ \cline{1-1}\end{array} = 0,$$

$$F_1 \begin{array}{|c|c|}\hline 1 & 1 \\ \hline 3 \\ \cline{1-1}\end{array} = \begin{array}{|c|c|}\hline 1 & 2 \\ \hline 3 \\ \cline{1-1}\end{array} + \begin{array}{|c|c|}\hline 2 & 1 \\ \hline 3 \\ \cline{1-1}\end{array} = 2 \begin{array}{|c|c|}\hline 1 & 2 \\ \hline 3 \\ \cline{1-1}\end{array} - \begin{array}{|c|c|}\hline 1 & 3 \\ \hline 2 \\ \cline{1-1}\end{array}.$$

9.1.4 *Irreducible Representations of* SU(N)

Theorem 9.3 The tensor subspace $\mathcal{T}_\mu^{[\lambda]} = \mathcal{Y}_\mu^{[\lambda]} \mathcal{T}$ corresponds to an irreducible representation of SU(N) with the highest weight $\boldsymbol{M}$,

$$\boldsymbol{M} = \sum_{\nu=1}^{N-1} \boldsymbol{w}_\nu M_\nu, \qquad M_\nu = \lambda_\nu - \lambda_{\nu+1}, \tag{9.29}$$

where λ_ν is the box number in the νth row of the Young pattern $[\lambda]$ and $\boldsymbol{w}_\nu$ are the fundamental dominant weights of SU(N).

Proof Let ϕ_0 denote the standard tensor Young tableau in $\mathcal{T}_\mu^{[\lambda]}$, where each box located in the αth row of the tableau is filled with digit α. Since a standard tensor Young tableau in the action of the raising operator E_ν becomes the sum of all possible tensor Young tableaux, each of which is obtained from the original one by replacing one filled digit ν with the digit $(\nu + 1)$, we have from Eq. (9.22) that $E_\nu \phi_0 = 0$ and ϕ_0 is the highest weight state in $\mathcal{T}_\mu^{[\lambda]}$ (see Eq. (7.111)) with the highest weight $\boldsymbol{M}$ given in Eq. (9.29).

Let ϕ denote any standard tensor Young tableau in $\mathcal{T}_\mu^{[\lambda]}$. Without loss of generality, we assume that each box located in the αth row of ϕ with $\alpha < \beta$ is filled with digit α and at least one box in the βth row of ϕ is filled with digit $\tau > \beta$. Since the tensor Young tableau ϕ is standard, $E_{\tau-1}\phi \neq 0$ and ϕ is not the highest weight state in $\mathcal{T}_\mu^{[\lambda]}$.

Because there is one and only one highest weight state in $\mathcal{T}_\mu^{[\lambda]}$, the subspace $\mathcal{T}_\mu^{[\lambda]}$ corresponds to the highest weight representation with the highest weight $\boldsymbol{M}$ given in Eq. (9.29). □

Corollary 9.3.1 Two tensor subspaces $\mathcal{T}_\mu^{[\lambda]}$ and $\mathcal{T}_\nu^{[\lambda]}$ with the same Young pattern $[\lambda]$ are equivalent.

Corollary 9.3.2 Two tensor subspaces $\mathcal{T}_\mu^{[\lambda]}$ and $\mathcal{T}_\nu^{[\lambda']}$ of SU(N) are equivalent if $\lambda_j = \lambda'_j + 1$, $1 \leqslant j \leqslant N$. The representation described by a Young pattern $[\lambda]$ with N row is equivalent to the representation described by a Young pattern $[\lambda'']$ with $(N-1)$ row:

$$[\lambda''] = [(\lambda_1 - \lambda_N), (\lambda_2 - \lambda_N), \ldots, (\lambda_{N-1} - \lambda_N), 0] .$$

Corollary 9.3.3 The basis tensors of SU(N) with the same tensor Young tableau in all tensor subspaces $\mathcal{T}_\mu^{[\lambda]}$ with the same Young pattern $[\lambda]$ constitute a complete set of basis tensors of the irreducible representation $[\lambda]$ of the permutation group S_n.

This Corollary is proved by Eq. (9.26) directly.

9.1.5 *Dimension of the Representation of* SU(N)

The dimension $d_{[\lambda]}(\text{SU}(N))$ of the irreducible representation $[\lambda]$ of SU(N) is equal to the number of the standard tensor Young tableaux in the tensor subspace $\mathcal{T}_\mu^{[\lambda]}$. There is a simple way, called the **hook rule**, to calculate $d_{[\lambda]}(\text{SU}(N))$. Please first review the hook rule for calculating the dimension of the irreducible representation of S_n (see subsection 4.2.2).

For a box at the jth column of the ith row in a Young pattern $[\lambda]$, we define its content $m_{ij} = j - i$ and its hook number h_{ij} which is the number of the boxes on its right in the ith row of the Young pattern, plus the number of the boxes below it in the jth column, and plus 1. The dimension $d_{[\lambda]}(\text{SU}(N))$ of the representation $[\lambda]$ of SU(N) is expressed by a quotient,

$$d_{[\lambda]}(\text{SU}(N)) = \prod_{ij} \frac{N + m_{ij}}{h_{ij}} = \frac{Y_A^{[\lambda]}}{Y_h^{[\lambda]}} . \tag{9.30}$$

$Y_A^{[\lambda]}$ is a tableau obtained from the Young pattern $[\lambda]$ by filling $(N + m_{ij})$ into the box located at the jth column of the ith row of the pattern, and $Y_h^{[\lambda]}$ is a tableau obtained from $[\lambda]$ by filling h_{ij} into that box. The symbol $Y_A^{[\lambda]}$ means the product of the filled digits in it, so does the symbol $Y_h^{[\lambda]}$.

For a one-row Young pattern, $[\lambda] = [n]$, $\mathcal{T}^{[n]}$ is the set of the totally symmetric tensors and its dimension is

$$d_{[n]}(\text{SU}(N)) = \prod_{j=0}^{n-1} \frac{N + j}{n - j} = \frac{(N + n - 1)!}{n!(N-1)!} = \begin{pmatrix} n + N - 1 \\ n \end{pmatrix} . \tag{9.31}$$

When $N = 2$ we have $d_{[n]}(\mathrm{SU}(2)) = n + 1$. The representation $[n]$ of SU(2) is denoted by D^j with $j = n/2$ in Chap. 5.

For a two-row Young pattern, $[\lambda] = [\mu, \nu]$,

$$\begin{aligned} d_{[\mu,\nu]}(\mathrm{SU}(N)) &= \frac{\begin{array}{|c|c|c|c|c|c|c|}\hline N & N+1 & \cdots & N+\nu-1 & N+\nu & \cdots & N+\mu-1 \\ \hline N-1 & N & \cdots & N+\nu-2 & \multicolumn{3}{c}{} \\ \cline{1-4}\end{array}}{\begin{array}{|c|c|c|c|c|c|c|}\hline \mu+1 & \mu & \cdots & \mu-\nu+2 & \mu-\nu & \cdots & 1 \\ \hline \nu & \nu-1 & \cdots & 1 & \multicolumn{3}{c}{} \\ \cline{1-4}\end{array}} \\ &= \frac{(N+\mu-1)!(N+\nu-2)!(\mu-\nu+1)}{(N-1)!(N-2)!(\mu+1)!\nu!}. \end{aligned} \tag{9.32}$$

When $N = 3$ we have $d_{[\mu,\nu]}(\mathrm{SU}(3)) = (\mu + 2)(\nu + 1)(\mu - \nu + 1)/2$. For example, $d_{[1]}(\mathrm{SU}(3)) = d_{[1,1]}(\mathrm{SU}(3)) = 3$, $d_{[2,1]}(\mathrm{SU}(3)) = 8$, $d_{[3]}(\mathrm{SU}(3)) = d_{[3,3]}(\mathrm{SU}(3)) = 10$, and $d_{[4,2]}(\mathrm{SU}(3)) = 27$.

For a one-column Young pattern $[1^n]$, $n \leqslant N$, the standard tensor Young tableau is a tableau filled with n digits downward in the increasing order so that its number is the combinatorial number of n among N,

$$d_{[1^n]}(\mathrm{SU}(N)) = \prod_{j=0}^{n-1} \frac{N-j}{n-j} = \frac{N!}{n!(N-n)!} = \binom{N}{n}. \tag{9.33}$$

When $n = N$, there is only one standard tensor Young tableau. $\mathcal{Y}^{[1^N]} = \sum_R \delta(R) R$ is an antisymmetrized operator. Letting $\boldsymbol{\phi} = \mathcal{Y}^{[1^N]}\boldsymbol{\theta}_{12\ldots N}$, we have

$$\boldsymbol{\phi}_{a_1\ldots a_N} = \epsilon_{a_1\ldots a_N}, \qquad \mathcal{Y}^{[1^N]}\boldsymbol{\theta}_{a_1\ldots a_N} = \epsilon_{a_1\ldots a_N}\boldsymbol{\phi}. \tag{9.34}$$

Due to the Weyl reciprocity, $\boldsymbol{\phi}$ is an invariant tensor in SU(N):

$$\begin{aligned} O_u\boldsymbol{\phi} &= \mathcal{Y}^{[1^N]} O_u \boldsymbol{\theta}_{12\ldots N} = \sum_{a_1\ldots a_N} \left(\mathcal{Y}^{[1^N]}\boldsymbol{\theta}_{a_1\ldots a_N}\right) u_{a_1 1} \cdots u_{a_N N} \\ &= \boldsymbol{\phi} \sum_{a_1\ldots a_N} \epsilon_{a_1\ldots a_N} u_{a_1 1} \cdots u_{a_N N} = \boldsymbol{\phi}\ \det u = \boldsymbol{\phi}. \end{aligned} \tag{9.35}$$

Thus, $[1^N]$ **describes the identical representation of** SU(N).

9.1.6 *Antisymmetric Wave Functions of Fermions*

An electron is a fermion with spin 1/2. The spinor wave function of an electron is described by a vector of SU(2) where the basis vector is denoted by $\boldsymbol{\theta}_\sigma$ with $\sigma = 1$ or 2:

$$\boldsymbol{\theta}_1 = \begin{pmatrix} 1 \\ 0 \end{pmatrix}, \qquad \boldsymbol{\theta}_2 = \begin{pmatrix} 0 \\ 1 \end{pmatrix}. \tag{9.36}$$

The spinor wave function of the system of n electrons is described by a tensor $\boldsymbol{T}$ of rank n of SU(2) with the basis tensors (see Eq. (9.16))

$$\boldsymbol{\theta}_{\sigma_1,\cdots,\sigma_n} = \boldsymbol{\theta}_{\sigma_1}(1) \times \boldsymbol{\theta}_{\sigma_2}(2) \times \cdots \times \boldsymbol{\theta}_{\sigma_n}(n). \tag{9.37}$$

In the physical convention, an ordinal number in the bracket is added to the basis vector for convenience and the sign "×" is omitted sometimes.

Due to Eq. (9.22), the row number of the non-vanishing tensor Young tableau for SU(2) is not larger than two. Let n denote the total number of electrons in the system. The tensor subspace $\mathcal{Y}_\mu^{[n-m,m]}\mathcal{T} = \mathcal{T}_\mu^{[n-m,m]}$ of SU(2) corresponds to the irreducible representation $[n-m,m] \simeq [n-2m]$, which is denoted by $D^j(\text{SU}(2))$ with $j = n/2 - m$ in Chap. 5. Namely, the tensor $\boldsymbol{T} \in \mathcal{T}_\mu^{[n-m,m]}$ is the spinor wave function of this system with the total spin $S = n/2 - m$. The standard tensor Young tableau in $\mathcal{T}_\mu^{[n-m,m]}$ is the basis spinor, where $2S_z$ is equal to the number of boxes filled by 1 (upspin electrons), subtracted by that filled by 2 (downspin electrons). The highest weight state is described by the standard tensor Young tableau where the boxes in the first row are filled by 1 and those in the second row filled by 2. The standard Young tableau $\mathcal{Y}_1^{[n-m,m]}$ is

1	2	$\cdots$	m	$m+1$	$\cdots$	$n-m$
$n-m+1$	$n-m+2$	$\cdots$	n			

(9.38)

The highest weight state $\mathcal{Y}_1^{[n-m,m]}Z$ in the subspace $\mathcal{Y}_1^{[n-m,m]}\mathcal{T}$ of SU(2) has the spin $S = S_z = n/2 - m$:

$$\begin{aligned} Z = \boldsymbol{\theta}_{\underbrace{11\cdots1}_{n-m}\underbrace{22\cdots2}_{m}} &= \boldsymbol{\theta}_1(1) \times \boldsymbol{\theta}_1(2) \times \cdots \times \boldsymbol{\theta}_1(n-m) \\ &\quad \times \boldsymbol{\theta}_2(n-m+1) \times \cdots \times \boldsymbol{\theta}_2(n-1) \times \boldsymbol{\theta}_2(n). \end{aligned} \tag{9.39}$$

Letting $R_{\mu 1} \in \text{S}_n$ denote the permutation transforming the standard Young tableau $\mathcal{Y}_1^{[n-m,m]}$ to the standard Young tableau $\mathcal{Y}_\mu^{[n-m,m]}$. $R_{\mu 1}\mathcal{Y}_1^{[n-m,m]}Z$ **is the basis spinors in the complete set** of the spinor wave functions of the system of n electrons with spin $S = S_z = n/2 - m$.

Let H= $\text{S}_{n-m} \otimes \text{S}_m$ denote the subgroup of S_n, where S_{n-m} is the permutation group for the first $n-m$ electrons and S_m is that for the last m electrons. The index κ of H in S_n is the combinatorial number of m among n. The left coset of H in S_n is denoted by P_αH, where, as always, P_α have been chosen and P_1 is the identity:

$$P_\alpha = \begin{pmatrix} 1 & 2 & \cdots & n-m & n-m+1 & n-m+2 & \cdots & n \\ a_1 & a_2 & \cdots & a_{n-m} & b_1 & b_2 & \cdots & b_m \end{pmatrix},$$
$$a_i \neq b_j, \quad \begin{array}{l} 1 \leqslant a_1 < a_2 < \cdots < a_{n-m} \leqslant n, \\ 1 \leqslant b_1 < b_2 \cdots < b_m \leqslant n, \end{array} \tag{9.40}$$

From Eqs. (9.38) and (9.39), any element T in H **leaves the spinor wave function $\mathcal{Y}_1^{[n-m,m]}Z$ invariant:**

$$T\left[\mathcal{Y}_1^{[n-m,m]}Z\right] = \mathcal{Y}_1^{[n-m,m]}Z, \qquad T \in \mathrm{H}. \tag{9.41}$$

Now, we study the totally antisymmetric wave function of the ground state with given total spin $S = S_z = n/2 - m$ in a system of n identical electrons moving in a one-dimensional harmonic oscillator potential. The hamiltonian equation of the system is

$$\sum_{a=1}^{n}\left[-\frac{\hbar^2}{2M}\frac{\partial^2}{\partial x_a^2} + \frac{M\omega^2 x_a^2}{2}\right]\Psi(x_1 \cdots x_n) = E\Psi(x_1 \cdots x_n), \tag{9.42}$$

where M denotes the mass of electron. As is well known in quantum mechanics, the static wave function $u_r(x)$ of the rth excited state for one electron moving in a one-dimensional harmonic oscillator potential is

$$\begin{aligned} &\left[-\frac{\hbar^2}{2M}\frac{d^2}{dx^2} + \frac{M\omega^2 x^2}{2}\right] u_r(x) = E_r u_r(x), \quad E_r = (r+1/2)\hbar\omega, \\ &u_r(x) = \left(\frac{\alpha}{2^r r!\sqrt{\pi}}\right)^{1/2} H_r(\alpha x)e^{-\alpha^2 x^2}, \quad \alpha = \sqrt{\frac{M\omega}{\hbar}}, \end{aligned} \tag{9.43}$$

where $H_r(x)$ is the Hermite polynomial of order r (see 8.95 in [Gradshteyn and Ryzhik (2007)]).

The spinor part of wave function of this system with total spin $S = S_z = n/2-m$ must be a combination of $R_{\mu 1}\mathcal{Y}_1^{[n-m,m]}Z$. $R_{\mu 1}\mathcal{Y}_1^{[n-m,m]}Z$ span the invariant space with respect to S_n, **corresponding to the irreducible representation of S_n denoted by the same Young pattern** $[n-m, m]$ (see Corollary 9.3.3). Due to Eq. (4.82) the orbital part of wave function of this system has to satisfy Eq. (9.42) and to belong to the representation $[2^m, 1^{n-2m}]$ of S_n which is the associated Young pattern of $[n-m, m]$.

Let $\psi(x_1 \cdots x_t)$ denote the totally anti-symmetric wave function where t electrons run in the different states from the ground state to the $(t-1)$th excited state, respectively:

$$\begin{aligned}\psi(x_1 \cdots x_t) &= \frac{1}{\sqrt{t!}} \det \left\{u_i(x_j)\right\}_{j=1,2,\cdots,t}^{i=0,1,\cdots,(t-1)} \\ &= \frac{1}{\sqrt{t!}} \begin{vmatrix} u_0(x_1) & u_1(x_1) & \cdots & u_{t-1}(x_1) \\ u_0(x_2) & u_1(x_2) & \cdots & u_{t-1}(x_2) \\ \cdots & \cdots & \cdots & \cdots \\ u_0(x_t) & u_1(x_t) & \cdots & u_{t-1}(x_t) \end{vmatrix},\end{aligned} \tag{9.44}$$

Thus, the orbital wave function $\psi(x_1 \cdots x_{n-m})\psi(x_{n-m+1} \cdots x_n)$ belongs to the anti-symmetric representation $[1^{n-m}] \times [1^m]$ of H:

$$\begin{aligned}&\delta(T)T\left[\psi(x_1 \cdots x_{n-m})\psi(x_{n-m+1} \cdots x_n)\right] \\ &= \psi(x_1 \cdots x_{n-m})\psi(x_{n-m+1} \cdots x_n), \qquad T \in \mathrm{H},\end{aligned} \tag{9.45}$$

and satisfies Eq. (9.42) with the energy E

$$\begin{aligned}E &= \sum_{r=0}^{n-m-1} \left(r + \frac{1}{2}\right)\hbar\omega + \sum_{s=0}^{m-1} \left(s + \frac{1}{2}\right)\hbar\omega \\ &= \left[\frac{(n-m)^2}{2} + \frac{m^2}{2}\right]\hbar\omega.\end{aligned} \tag{9.46}$$

The Young operator $\mathcal{Y}^{[1^n]}$ of S_n

$$\mathcal{Y}^{[1^n]} = \sum_{R \in \mathrm{S}_n} \delta(R)R = \sum_{\alpha=1}^{\kappa} \delta(P_\alpha)P_\alpha \sum_{T \in \mathrm{H}} \delta(T)T, \tag{9.47}$$

is the projective operator to the totally anti-symmetric wave function, where $\delta(R)$ is the permutation parity of R. Thus, the totally antisymmetric wave function of the ground state of the system of n identical electrons moving in the one-dimensional harmonic oscillator potential is

$$\begin{aligned}\Psi(x_1 \cdots x_n) &= \frac{\mathcal{Y}^{[1^n]}}{n!(n-m)!}\left[\psi(x_1 \cdots x_{n-m})\psi(x_{n-m+1} \cdots x_n)\mathcal{Y}_1^{[n-m,m]}Z\right] \\ &= \sum_{\alpha=1}^{\kappa} \delta(P_\alpha)P_\alpha \left[\psi(x_1 \cdots x_{n-m})\psi(x_{n-m+1} \cdots x_n)\mathcal{Y}_1^{[n-m,m]}Z\right],\end{aligned} \tag{9.48}$$

where the total spin is $S = S_z = n/2 - m$ and the energy is E given in Eq. (9.46). $\Psi(x_1 \cdots x_n)$ **is the unique anti-symmetric solution with this total spin and this energy** owing to the following reason. From the Littlewood–Richardson rule (see subsection 4.5.2), there is one and only one representation $[2^m, 1^{n-2m}]$ contained in the induced representation of $[1^{n-m}] \times [1^m]$ of H with respect to S_n. From Eq. (4.82) there is no anti-symmetric representation $[1^n]$ contained in the decomposition of $[\omega] \times [\lambda]$

except for $[\omega]$ is the associated Young pattern $[\tilde{\lambda}]$ of $[\lambda]$, and there is one and only one representation $[1^n]$ contained in the decomposition of $[\tilde{\lambda}] \times [\lambda]$. Now, since $[\psi(x_1 \cdots x_{n-m})\psi(x_{n-m+1} \cdots x_n)]$ belongs to the totally anti-symmetric representation $[1^{n-m}] \times [1^m]$ of H, there is one and only one solution which is totally anti-symmetric.

This method for finding anti-symmetric wave function of n identical electrons can be generalized to the problems for finding totally anti-symmetric or symmetric wave functions of the system of identical particles. As a similar problem, we calculate the wave function of the ground state for the system of n electrons in a one-dimensional harmonic trap with infinitely strongly delta function interaction [Guan et al. (2009)]. The hamiltonian equation of the system is

$$\left\{\sum_{a=1}^{n}\left[-\frac{\hbar^2}{2M}\frac{\partial^2}{\partial x_a^2}+\frac{M\omega^2 x_a^2}{2}\right]+g\sum_{i<j}\delta(x_i-x_j)\right\}\Psi(x_1\cdots x_n) = E\Psi(x_1\cdots x_n). \tag{9.49}$$

In the limit for very strong repulsive coupling $g \longrightarrow +\infty$, $\Psi(x_i = x_j) = 0$. Following the idea of [Girardeau (1960)] for bosons, we assume

$$\Psi(x_1 \cdots x_n) = \psi(x_1 \cdots x_n)\Phi(x_1 \cdots x_n), \tag{9.50}$$

where $\psi(x_1 \cdots x_n)$ given in Eq. (9.44) is totally anti-symmetric in S_n. So that $\Phi(x_1 \cdots x_n)$ has to be totally symmetric in S_n. We choose $\Phi(x_1 \cdots x_n)$ as the linear combination of $R\left[Q(x_1 \cdots x_n)\mathcal{Y}_1^{[n-m,m]}Z\right]$, where $R \in S_n$:

$$\begin{aligned}
Q(x_1\cdots x_n) &= \prod_{i=1}^{n-m}\prod_{j=n-m+1}^{n} \text{sgn}(x_i-x_j),\\
\text{sgn}(x_i-x_j) &= \frac{x_i-x_j}{|x_i-x_j|},\\
T\left[Q(x_1\cdots x_n)\mathcal{Y}_1^{[n-m,m]}Z\right] &= Q(x_1\cdots x_n)\mathcal{Y}_1^{[n-m,m]}Z, \qquad T\in \text{H}.
\end{aligned} \tag{9.51}$$

In the action of the sum of Laplace operators on $\Psi(x_1 \cdots x_n)$, we obtain some terms in the following forms.

$$\sum_{a=1}^{n}\left[-\frac{\hbar^2}{2M}\frac{\partial^2}{\partial x_a^2}+\frac{M\omega^2 x_a^2}{2}\right]\psi(x_1\cdots x_n)=E\psi(x_1\cdots x_n),\quad E=n^2\hbar\omega/2,$$

$$2\left[\frac{\partial}{\partial x_i}\psi(x_1\cdots x_n)\right]\left[\frac{\partial}{\partial x_i}\text{sgn}(x_i-x_j)\right]\propto\delta(x_i-x_j),$$

$$\left[\frac{\partial^2}{\partial x_i^2}+\frac{\partial^2}{\partial x_j^2}+2\frac{\partial^2}{\partial x_i\partial x_j}\right]\text{sgn}(x_i-x_j)=0.$$

The spinor part of wave function of this system with total spin $S=S_z=n/2-m$ must be a combination of $R_{\mu 1}\mathcal{Y}_1^{[n-m,m]}Z$, which belong to the representation $[n-m,m]$ of S_n. $RQ(x_1\cdots x_n)$ belongs to the induced representation of $[n-m]\times[m]$ of H with respect to S_n, in whose decomposition there is one and only one representation $[n-m,m]$ of S_n. There is only one symmetric solution for S_n from the combination of the product of $RQ(x_1\cdots x_n)$ and $R_{\mu 1}\mathcal{Y}_1^{[n-m,m]}Z$. In terms of the method like Eq. (9.48), we obtain the totally antisymmetric wave function of the ground state of the system of n identical electrons moving in a one-dimensional harmonic trap with infinitely strongly repulsive delta function interaction:

$$\begin{aligned}\Psi(x_1\cdots x_n)&=\psi(x_1\cdots x_n)\frac{\mathcal{Y}^{[n]}}{n!(n-m)!}\left[Q(x_1\cdots x_n)\mathcal{Y}_1^{[n-m,m]}Z\right]\\&=\psi(x_1\cdots x_n)\sum_{\alpha=1}^{K}P_\alpha\left[Q(x_1\cdots x_n)\mathcal{Y}_1^{[n-m,m]}Z\right],\end{aligned}\tag{9.52}$$

$$\mathcal{Y}^{[n]}=\sum_{R\in \mathrm{S}_n}R=\sum_{\alpha=1}^{K}P_\alpha\sum_{T\in \mathrm{H}}T,$$

where the total spin is $S=S_z=n/2-m$ and the energy is $E=n^2\hbar\omega/2$. The energy of the ground state of this system when the coupling g is finite can be calculated by the Thomas-Fermi method (see [Ma and Yang (2010)] and [Wang and Ma (2017)]).

9.1.7 *Subduced Representations*

The basis tensor in the representation $[\lambda]$ of SU(N) is given as the standard tensor Young tableau where N has to be filled only in the lowest boxes of some columns. Removing the boxes filled with N, one obtains a standard tensor Young tableau belongs to an irreducible representation $[\mu]$ of SU($N-1$), which is contained in the decomposition of the subduced representation of $[\lambda]$ of SU(N) with respect to SU($N-1$) (see the chain of subgroups given in Eq. (8.13)). For example,

$$\begin{array}{|c|c|c|c|c|}\hline \ & \ & \ & N & N \\ \hline \ & \ \\ \cline{1-2} \ & N \\ \cline{1-2} N \\ \cline{1-1}\end{array} \qquad \begin{array}{l}[\lambda]=[5,2,2,1], \\ [\mu]=[3,2,1].\end{array}$$

Removing the boxes filled with N from all standard tensor Young tableaux in the representation $[\lambda]$ of SU(N), we obtain all standard tensor Young tableaux in its subduced representation with respect to SU($N-1$). This is the method of decomposing the subduced representation of $[\lambda]$ of SU(N) with respect to its subgroup S($N-1$):

$$\begin{aligned} &[\lambda] \longrightarrow \bigoplus [\mu], \qquad d_{[\lambda]}(\mathrm{SU}(N)) = \sum_{[\mu]} d_{[\mu]}(\mathrm{SU}(N-1)), \\ &\lambda_1 \geqslant \mu_1 \geqslant \lambda_2 \geqslant \mu_2 \geqslant \cdots \geqslant \mu_{N-2} \geqslant \lambda_{N-1} \geqslant \mu_{N-1} \geqslant \lambda_N. \end{aligned} \tag{9.53}$$

In this decomposition of the subduced representation of $[\lambda]$, the multiplicity of each representation $[\mu]$ satisfies Eq. (9.53) **is one.** The parameters λ_a and μ_b in Eq. (9.53) are denoted by ω_{aN} and $\omega_{b(N-1)}$ in the Gel'fand's method (see Eq. (8.17)), respectively. Thus, the condition (9.53) is nothing but the constraint (8.15). By the successive applications of this method an irreducible representation of SU(N) is decomposed with respect to the subgroup chain (8.13). **The decomposition (9.53) is the foundation of the Gel'fand's method.**

9.2 Orthonormal Irreducible Basis Tensors

Define the inner product in the tensor space $\mathcal{T}$ such that the basis tensors $\boldsymbol{\theta}_{a_1\dots a_n}$ are orthonormal to each other. The tensor representation of SU(N) with the basis tensors $\boldsymbol{\theta}_{a_1\dots a_n}$ is the direct product of self-representations, which is unitary. Through the projection of a standard Young operator $\mathcal{Y}_\mu^{[\lambda]}$, the tensor space reduces to its subspace $\mathcal{T}_\mu^{[\lambda]} = \mathcal{Y}_\mu^{[\lambda]}\mathcal{T}$, whose basis tensors are the standard tensor Young tableaux. **The standard tensor Young tableaux are the integral combinations of $\boldsymbol{\theta}_{a_1\dots a_n}$, but they are generally not orthonormal**, although the standard tensor Young tableaux with different weights in the same tensor subspace $\mathcal{T}_\mu^{[\lambda]}$ are orthogonal. For example, in the tensor subspace $\mathcal{T}_1^{[2,1]}$ of SU(3), $\mathcal{Y}_1^{[2,1]}\boldsymbol{\theta}_{112}$ and $\mathcal{Y}_1^{[2,1]}\boldsymbol{\theta}_{123}$ are orthogonal, but normalized to 6 and 4, respectively. $\mathcal{Y}_1^{[2,1]}\boldsymbol{\theta}_{123}$ and $\mathcal{Y}_1^{[2,1]}\boldsymbol{\theta}_{132}$ are not orthogonal to each other (see Eq. (9.24)). The highest weight states in $\mathcal{T}_1^{[2,1]}$ and in $\mathcal{T}_2^{[2,1]}$ of SU(3) are not orthogonal.

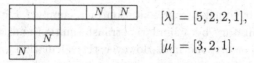

9.2.1 *Orthonormal Basis Tensors in $\mathcal{T}_\mu^{[\lambda]}$ of* A_2

A non-unitary irreducible representation of a compact Lie group can be changed to be unitary by a similarity transformation and the basis states are combined to be orthonormal. However, the calculation for the similarity transformation is quite complicated. On the other hand, the Gel'fand's bases of SU(N) are orthonormal, and the highest weight is single. **By applying the lowering operators F_μ successively to the standard tensor Young tableau for the highest weight state**, we are able to calculate the orthonormal basis tensors through the block weight diagram. Each orthonormal basis tensor is a combination of the standard tensor Young tableaux with the same weight in the same tensor subspace $\mathcal{T}_\mu^{[\lambda]}$ of SU(N), where the set of filled digits in each tensor Young tableau is the same. **The modules of the orthonormal basis states in this way are normalized to the module of the highest weight state, instead of 1.** This method will be explained through some examples in SU(3) as follows, and be generalized to the SO(N) groups and the USp(2ℓ) groups in the subsequent chapters.

The standard tensor Young tableaux in two basic representations (1, 0) and (0, 1) of SU(3) all are orthonormal to each other:

$$\begin{aligned}
F_1 \left|(1,0),(1,0)\right\rangle &= F_1 \left|\begin{array}{ccccc} 1 & & 0 & & 0 \\ & 1 & & 0 & \\ & & 1 & & \end{array}\right\rangle^1 = F_1 \ \boxed{1} \\
&= \left|(1,0),(\bar{1},1)\right\rangle = \left|\begin{array}{ccccc} 1 & & 0 & & 0 \\ & 1 & & 0 & \\ & & 0 & & \end{array}\right\rangle^2 = \boxed{2}\,, \\
F_2 \left|(1,0),(\bar{1},1)\right\rangle &= F_2 \left|\begin{array}{ccccc} 1 & & 0 & & 0 \\ & 1 & & 0 & \\ & & 0 & & \end{array}\right\rangle^2 = F_2 \ \boxed{2} \\
&= \left|(1,0),(0,\bar{1})\right\rangle = \left|\begin{array}{ccccc} 1 & & 0 & & 0 \\ & 0 & & 0 & \\ & & 0 & & \end{array}\right\rangle^3 = \boxed{3}\,, \\
F_2 \left|(0,1),(0,1)\right\rangle &= F_2 \left|\begin{array}{ccccc} 1 & & 1 & & 0 \\ & 1 & & 1 & \\ & & 1 & & \end{array}\right\rangle^1 = F_2 \begin{array}{|c|}\hline 1 \\ \hline 2 \\ \hline\end{array} \\
&= \left|(0,1),(1,\bar{1})\right\rangle = \left|\begin{array}{ccccc} 1 & & 1 & & 0 \\ & 1 & & 0 & \\ & & 1 & & \end{array}\right\rangle^2 = \begin{array}{|c|}\hline 1 \\ \hline 3 \\ \hline\end{array}\,, \\
F_1 \left|(0,1),(1,\bar{1})\right\rangle &= F_1 \left|\begin{array}{ccccc} 1 & & 1 & & 0 \\ & 1 & & 0 & \\ & & 1 & & \end{array}\right\rangle^2 = F_1 \begin{array}{|c|}\hline 1 \\ \hline 3 \\ \hline\end{array} \\
&= \left|(0,1),(\bar{1},0)\right\rangle = \left|\begin{array}{ccccc} 1 & & 1 & & 0 \\ & 1 & & 0 & \\ & & 0 & & \end{array}\right\rangle^3 = \begin{array}{|c|}\hline 2 \\ \hline 3 \\ \hline\end{array}\,.
\end{aligned}$$

The Young pattern of the representation of symmetric tensors of rank 3 of SU(3) is $[\lambda] = [3, 0]$. Its highest weight is $\boldsymbol{M} = (3, 0)$. There are three typical standard tensor Young tableaux which are normalized to 6, $2\sqrt{3}$, and $\sqrt{6}$, respectively,

$$\begin{array}{|c|c|c|}\hline \mathrm{a} & \mathrm{a} & \mathrm{a} \\ \hline\end{array} = 6\left\{\boldsymbol{\theta}_{aaa}\right\},$$

$$\begin{array}{|c|c|c|}\hline \mathrm{a} & \mathrm{b} & \mathrm{b} \\ \hline\end{array} = 2\sqrt{3}\left\{3^{-1/2}\left(\boldsymbol{\theta}_{abb}+\boldsymbol{\theta}_{bab}+\boldsymbol{\theta}_{bba}\right)\right\},$$

$$\begin{array}{|c|c|c|}\hline \mathrm{a} & \mathrm{b} & \mathrm{c} \\ \hline\end{array} = \sqrt{6}\left\{6^{-1/2}\left(\boldsymbol{\theta}_{abc}+\boldsymbol{\theta}_{acb}+\boldsymbol{\theta}_{bac}+\boldsymbol{\theta}_{bca}+\boldsymbol{\theta}_{cab}+\boldsymbol{\theta}_{cba}\right)\right\},$$

where a, b, and c are three different digits. In this representation, all weights are simple such that all standard tensor Young tableaux are orthogonal to each other. In the action of the lowering operators F_μ, the obtained tensor Young tableaux should be changed to be standard by the Eq. (9.23). Some calculations are as follows.

$$F_1\begin{array}{|c|c|c|}\hline 1 & 1 & 1 \\ \hline\end{array} = \begin{array}{|c|c|c|}\hline 1 & 1 & 2 \\ \hline\end{array} + \begin{array}{|c|c|c|}\hline 1 & 2 & 1 \\ \hline\end{array} + \begin{array}{|c|c|c|}\hline 2 & 1 & 1 \\ \hline\end{array}$$
$$= 3\begin{array}{|c|c|c|}\hline 1 & 1 & 2 \\ \hline\end{array},$$

$$F_1\begin{array}{|c|c|c|}\hline 1 & 1 & 3 \\ \hline\end{array} = \begin{array}{|c|c|c|}\hline 1 & 2 & 3 \\ \hline\end{array} + \begin{array}{|c|c|c|}\hline 2 & 1 & 3 \\ \hline\end{array} = 2\begin{array}{|c|c|c|}\hline 1 & 2 & 3 \\ \hline\end{array},$$

The orthogonal basis tensors of this representation $[3,0]$ of SU(3) are listed in Fig. 9.1.

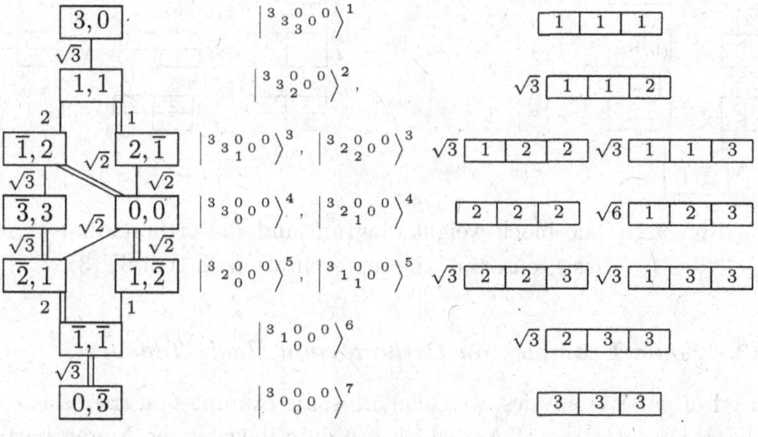

Fig. 9.1 The block weight diagram and the orthonormal basis tensors of the representation $[3,0]$ of SU(3).

The Young pattern for the adjoint representation of SU(3) is $[\lambda] = [2,1]$, where $\boldsymbol{M} = (1,1)$. The general form of the expansion of the tensor Young tableau in $\mathcal{T}_1^{[2,1]}$ is given in Eq. (9.24). There are two typical standard tensor Young tableaux which are normalized to $\sqrt{6}$ and 2, respectively,

$$\begin{array}{|c|c|}\hline a & a \\ \hline b \\ \cline{1-1}\end{array} = 2\theta_{aab} - \theta_{baa} - \theta_{aba},$$

$$\begin{array}{|c|c|}\hline a & b \\ \hline c \\ \cline{1-1}\end{array} = \theta_{abc} + \theta_{bac} - \theta_{cba} - \theta_{bca}\ ,$$

where a, b, and c are all different. The block weight diagrams and the standard tensor Young tableaux of [2, 1] of SU(3) are listed in Fig. 9.2. The calculations related to the multiple weight (0, 0) are as follows.

$$\begin{aligned}
|(1,1),(0,0)_1\rangle &= \sqrt{\frac{1}{2}}\ F_1\ |(1,1),(2,\overline{1})\rangle = \sqrt{\frac{1}{2}}\ F_1 \begin{array}{|c|c|}\hline 1 & 1 \\ \hline 3 \\ \cline{1-1}\end{array} \\
&= \sqrt{\frac{1}{2}}\left\{\begin{array}{|c|c|}\hline 1 & 2 \\ \hline 3 \\ \cline{1-1}\end{array} + \begin{array}{|c|c|}\hline 2 & 1 \\ \hline 3 \\ \cline{1-1}\end{array}\right\} = \sqrt{2}\begin{array}{|c|c|}\hline 1 & 2 \\ \hline 3 \\ \cline{1-1}\end{array} - \sqrt{\frac{1}{2}}\begin{array}{|c|c|}\hline 1 & 3 \\ \hline 2 \\ \cline{1-1}\end{array}, \\
|(1,1),(0,0)_2\rangle &= \sqrt{\frac{2}{3}}\left\{F_2\ |(1,1),(\overline{1},2)\rangle - \sqrt{\frac{1}{2}}\ |(1,1),(0,0)_1\rangle\right\} \\
&= \sqrt{\frac{2}{3}}\left\{F_2 \begin{array}{|c|c|}\hline 1 & 2 \\ \hline 2 \\ \cline{1-1}\end{array} - \begin{array}{|c|c|}\hline 1 & 2 \\ \hline 3 \\ \cline{1-1}\end{array} + \frac{1}{2}\begin{array}{|c|c|}\hline 1 & 3 \\ \hline 2 \\ \cline{1-1}\end{array}\right\} = \sqrt{\frac{3}{2}}\begin{array}{|c|c|}\hline 1 & 3 \\ \hline 2 \\ \cline{1-1}\end{array}.
\end{aligned}$$

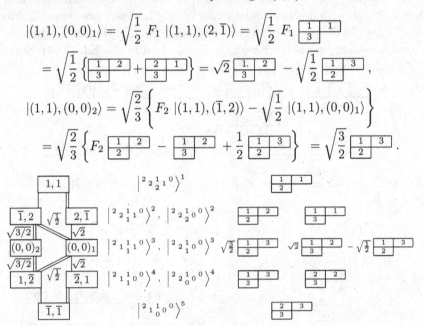

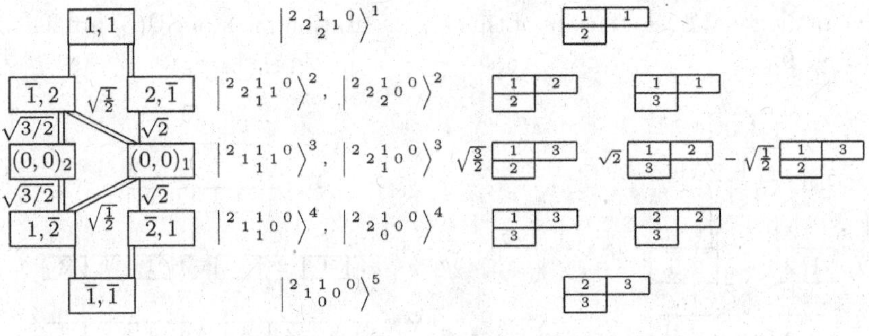

Fig. 9.2 The block weight diagram and the orthonormal basis tensors of the representation [2, 1] of SU(3).

9.2.2 *Some Examples on Orthonormal Basis Tensors*

In need of particle physics, we calculate some examples on the basis states in the representations of A_2, which are described by the Young patterns $[\lambda]$ with six boxes. We will also calculate the basis states in the adjoint representation of A_3. As studied in Chap. 8, each basis state in the representation $[\lambda]$ of A_ℓ belongs to an $A_{\ell-1}$-multiplet, and F_ℓ relates some basis states belonging to different $A_{\ell-1}$-multiplets.

Ex. 1. The representation [6, 0] of A_2.

Due to Eq. (9.31), the dimension of the representation [6, 0] of A_2 is

$8!/(2!6!) = 28$. All weights in $[6,0]$ are single. The highest weight state in $[6,0]$ of A_2 is also that of an A_1-sevenlet with $\boldsymbol{M}^{(1)} = 6\boldsymbol{w}_1$:

$$\begin{aligned}
|(6,0)\rangle &= \left|\begin{smallmatrix}6&&0&&0\\&6&&0&\\&&6&&\end{smallmatrix}\right\rangle^{1} = \begin{array}{|c|c|c|c|c|c|}\hline 1&1&1&1&1&1\\\hline\end{array}\,,\\
|(4,1)\rangle &= \left|\begin{smallmatrix}6&&0&&0\\&6&&0&\\&&5&&\end{smallmatrix}\right\rangle^{2} = \sqrt{\tfrac{1}{6}}F_1\,|(6,0)\rangle = \sqrt{6}\begin{array}{|c|c|c|c|c|c|}\hline 1&1&1&1&1&2\\\hline\end{array}\,,\\
|(2,2)\rangle &= \left|\begin{smallmatrix}6&&0&&0\\&6&&0&\\&&4&&\end{smallmatrix}\right\rangle^{3} = \sqrt{\tfrac{1}{10}}F_1\,|(4,1)\rangle = \sqrt{15}\begin{array}{|c|c|c|c|c|c|}\hline 1&1&1&1&2&2\\\hline\end{array}\,,\\
|(0,3)\rangle &= \left|\begin{smallmatrix}6&&0&&0\\&6&&0&\\&&3&&\end{smallmatrix}\right\rangle^{4} = \sqrt{\tfrac{1}{12}}F_1\,|(2,2)\rangle = \sqrt{20}\begin{array}{|c|c|c|c|c|c|}\hline 1&1&1&2&2&2\\\hline\end{array}\,,\\
|(\overline{2},4)\rangle &= \left|\begin{smallmatrix}6&&0&&0\\&6&&0&\\&&2&&\end{smallmatrix}\right\rangle^{5} = \sqrt{\tfrac{1}{12}}F_1\,|(0,3)\rangle = \sqrt{15}\begin{array}{|c|c|c|c|c|c|}\hline 1&1&2&2&2&2\\\hline\end{array}\,,\\
|(\overline{4},5)\rangle &= \left|\begin{smallmatrix}6&&0&&0\\&6&&0&\\&&1&&\end{smallmatrix}\right\rangle^{6} = \sqrt{\tfrac{1}{10}}F_1\,|(\overline{2},4)\rangle = \sqrt{6}\begin{array}{|c|c|c|c|c|c|}\hline 1&2&2&2&2&2\\\hline\end{array}\,,\\
|(\overline{6},6)\rangle &= \left|\begin{smallmatrix}6&&0&&0\\&6&&0&\\&&0&&\end{smallmatrix}\right\rangle^{7} = \sqrt{\tfrac{1}{6}}F_1\,|(\overline{4},5)\rangle = \begin{array}{|c|c|c|c|c|c|}\hline 2&2&2&2&2&2\\\hline\end{array}\,,
\end{aligned}$$

Seven basis states in the A_1-sevenlet all are the highest weight states of the $\mathcal{A}_2$-multiplets, which will be calculated one by one. The basis state $|(6,0)\rangle$ spans an $\mathcal{A}_2$-singlet. The basis state $|(4,1)\rangle$ is the highest weight state of an $\mathcal{A}_2$-doublet. The basis state $|(5,\overline{1})\rangle$ in the $\mathcal{A}_2$-doublet is the highest weight state of an A_1-sixlet with $\boldsymbol{M}^{(1)} = 5\boldsymbol{w}_1$.

$$\begin{aligned}
|(5,\overline{1})\rangle &= \left|\begin{smallmatrix}6&&0&&0\\&5&&0&\\&&5&&\end{smallmatrix}\right\rangle^{3} = F_2\,|(4,1)\rangle = \sqrt{6}\begin{array}{|c|c|c|c|c|c|}\hline 1&1&1&1&1&3\\\hline\end{array}\,;\\
|(3,0)\rangle &= \left|\begin{smallmatrix}6&&0&&0\\&5&&0&\\&&4&&\end{smallmatrix}\right\rangle^{4} = \sqrt{\tfrac{1}{5}}F_1\,|(5,\overline{1})\rangle = \sqrt{30}\begin{array}{|c|c|c|c|c|c|}\hline 1&1&1&1&2&3\\\hline\end{array}\,,\\
|(1,1)\rangle &= \left|\begin{smallmatrix}6&&0&&0\\&5&&0&\\&&3&&\end{smallmatrix}\right\rangle^{5} = \sqrt{\tfrac{1}{8}}F_1\,|(3,0)\rangle = \sqrt{60}\begin{array}{|c|c|c|c|c|c|}\hline 1&1&1&2&2&3\\\hline\end{array}\,,\\
|(\overline{1},2)\rangle &= \left|\begin{smallmatrix}6&&0&&0\\&5&&0&\\&&2&&\end{smallmatrix}\right\rangle^{6} = \tfrac{1}{3}F_1\,|(1,1)\rangle = \sqrt{60}\begin{array}{|c|c|c|c|c|c|}\hline 1&1&2&2&2&3\\\hline\end{array}\,,\\
|(\overline{3},3)\rangle &= \left|\begin{smallmatrix}6&&0&&0\\&5&&0&\\&&1&&\end{smallmatrix}\right\rangle^{7} = \sqrt{\tfrac{1}{8}}F_1\,|(\overline{1},2)\rangle = \sqrt{30}\begin{array}{|c|c|c|c|c|c|}\hline 1&2&2&2&2&3\\\hline\end{array}\,,\\
|(\overline{5},4)\rangle &= \left|\begin{smallmatrix}6&&0&&0\\&5&&0&\\&&0&&\end{smallmatrix}\right\rangle^{8} = \sqrt{\tfrac{1}{5}}F_1\,|(\overline{3},3)\rangle = \sqrt{6}\begin{array}{|c|c|c|c|c|c|}\hline 2&2&2&2&2&3\\\hline\end{array}\,.
\end{aligned}$$

The basis state $|(2,2)\rangle$ is the highest weight state of an $\mathcal{A}_2$-triplet. The basis state $|(3,0)\rangle$ belongs to the A_1-sixlet and the basis state $|(4,\overline{2})\rangle$ is the highest weight state of an A_1-quintuplet with $\boldsymbol{M}^{(1)} = 4\boldsymbol{w}_1$.

$$\begin{aligned}
&F_2\,|(2,2)\rangle = \sqrt{2}\,|(3,0)\rangle\,,\\
&|(4,\overline{2})\rangle = \left|\begin{smallmatrix}6&&0&&0\\&4&&0&\\&&4&&\end{smallmatrix}\right\rangle^{5} = \sqrt{\tfrac{1}{2}}F_2\,|(3,0)\rangle = \sqrt{15}\begin{array}{|c|c|c|c|c|c|}\hline 1&1&1&1&3&3\\\hline\end{array}\,;
\end{aligned}$$

$$|(2,\overline{1})\rangle = \left|\begin{smallmatrix}6&&0&&0\\&4&&0&\\&&3&&\end{smallmatrix}\right\rangle^6 = \sqrt{\tfrac{1}{4}}F_1\,|(4,\overline{2})\rangle = \sqrt{60}\,\begin{array}{|c|c|c|c|c|c|}\hline 1&1&1&2&3&3\\\hline\end{array}\,,$$

$$|(0,0)\rangle = \left|\begin{smallmatrix}6&&0&&0\\&4&&0&\\&&2&&\end{smallmatrix}\right\rangle^7 = \sqrt{\tfrac{1}{6}}F_1\,|(2,\overline{1})\rangle = \sqrt{90}\,\begin{array}{|c|c|c|c|c|c|}\hline 1&1&2&2&3&3\\\hline\end{array}\,,$$

$$|(\overline{2},1)\rangle = \left|\begin{smallmatrix}6&&0&&0\\&4&&0&\\&&1&&\end{smallmatrix}\right\rangle^8 = \sqrt{\tfrac{1}{6}}F_1\,|(0,0)\rangle = \sqrt{60}\,\begin{array}{|c|c|c|c|c|c|}\hline 1&2&2&2&3&3\\\hline\end{array}\,,$$

$$|(\overline{4},2)\rangle = \left|\begin{smallmatrix}6&&0&&0\\&4&&0&\\&&0&&\end{smallmatrix}\right\rangle^9 = \sqrt{\tfrac{1}{4}}F_1\,|(\overline{2},1)\rangle = \sqrt{15}\,\begin{array}{|c|c|c|c|c|c|}\hline 2&2&2&2&3&3\\\hline\end{array}\,.$$

The basis state $|(0,3)\rangle$ is the highest weight state of an $\mathcal{A}_2$-quadruplet. The basis state $|(3,\overline{3})\rangle$ in the $\mathcal{A}_2$-quadruplet is the highest weight state of A_1 with $\boldsymbol{M}^{(1)} = 3\boldsymbol{w}_1$.

$$F_2\,|(0,3)\rangle = \sqrt{3}\,|(1,1)\rangle\,, \quad F_2\,|(1,1)\rangle = 2\,|(2,\overline{1})\rangle\,,$$

$$|(3,\overline{3})\rangle = \left|\begin{smallmatrix}6&&0&&0\\&3&&0&\\&&3&&\end{smallmatrix}\right\rangle^7 = \sqrt{\tfrac{1}{3}}F_2\,|(2,\overline{1})\rangle = \sqrt{20}\,\begin{array}{|c|c|c|c|c|c|}\hline 1&1&1&3&3&3\\\hline\end{array}\,;$$

$$|(1,\overline{2})\rangle = \left|\begin{smallmatrix}6&&0&&0\\&3&&0&\\&&2&&\end{smallmatrix}\right\rangle^8 = \sqrt{\tfrac{1}{3}}F_1\,|(3,\overline{3})\rangle = \sqrt{60}\,\begin{array}{|c|c|c|c|c|c|}\hline 1&1&2&3&3&3\\\hline\end{array}\,,$$

$$|(\overline{1},\overline{1})\rangle = \left|\begin{smallmatrix}6&&0&&0\\&3&&0&\\&&1&&\end{smallmatrix}\right\rangle^9 = \sqrt{\tfrac{1}{4}}F_1\,|(1,\overline{2})\rangle = \sqrt{60}\,\begin{array}{|c|c|c|c|c|c|}\hline 1&2&2&3&3&3\\\hline\end{array}\,,$$

$$|(\overline{3},0)\rangle = \left|\begin{smallmatrix}6&&0&&0\\&3&&0&\\&&0&&\end{smallmatrix}\right\rangle^{10} = \sqrt{\tfrac{1}{3}}F_1\,|(\overline{1},\overline{1})\rangle = \sqrt{20}\,\begin{array}{|c|c|c|c|c|c|}\hline 2&2&2&3&3&3\\\hline\end{array}\,.$$

The basis state $|(\overline{2},4)\rangle$ is the highest weight state of an $\mathcal{A}_2$-quintuplet. The basis state $|(2,\overline{4})\rangle$ is the highest weight state of an A_1-triplet with $\boldsymbol{M}^{(1)} = 2\boldsymbol{w}_1$.

$$F_2\,|(\overline{2},4)\rangle = 2\,|(\overline{1},2)\rangle, \quad F_2\,|(\overline{1},2)\rangle = \sqrt{6}\,|(0,0)\rangle, \quad F_2\,|(0,0)\rangle = \sqrt{6}\,|(1,\overline{2})\rangle,$$

$$|(2,\overline{4})\rangle = \left|\begin{smallmatrix}6&&0&&0\\&2&&0&\\&&2&&\end{smallmatrix}\right\rangle^9 = \sqrt{\tfrac{1}{4}}F_2\,|(1,\overline{2})\rangle = \sqrt{15}\,\begin{array}{|c|c|c|c|c|c|}\hline 1&1&3&3&3&3\\\hline\end{array}\,;$$

$$|(0,\overline{3})\rangle = \left|\begin{smallmatrix}6&&0&&0\\&2&&0&\\&&1&&\end{smallmatrix}\right\rangle^{10} = \sqrt{\tfrac{1}{2}}F_1\,|(2,\overline{4})\rangle = \sqrt{30}\,\begin{array}{|c|c|c|c|c|c|}\hline 1&2&3&3&3&3\\\hline\end{array}\,,$$

$$|(\overline{2},\overline{2})\rangle = \left|\begin{smallmatrix}6&&0&&0\\&2&&0&\\&&0&&\end{smallmatrix}\right\rangle^{11} = \sqrt{\tfrac{1}{2}}F_1\,|(0,\overline{3})\rangle = \sqrt{15}\,\begin{array}{|c|c|c|c|c|c|}\hline 2&2&3&3&3&3\\\hline\end{array}\,.$$

The basis state $|(\overline{4},5)\rangle$ is the highest weight state of an $\mathcal{A}_2$-sixlet. The basis state $|(1,\overline{5})\rangle$ in the $\mathcal{A}_2$-sixlet is the highest weight state of an A_1-doublet with $\boldsymbol{M}^{(1)} = \boldsymbol{w}_1$.

$$F_2\,|(\overline{4},5)\rangle = \sqrt{5}\,|(\overline{3},3)\rangle\,, \quad F_2\,|(\overline{3},3)\rangle = \sqrt{8}\,|(\overline{2},1)\rangle\,,$$

$$F_2\,|(\overline{2},1)\rangle = 3\,|(\overline{1},\overline{1})\rangle\,, \quad F_2\,|(\overline{1},\overline{1})\rangle = \sqrt{8}\,|(0,\overline{3})\rangle\,,$$

$$|(1,\overline{5})\rangle = \left|\begin{smallmatrix}6&&0&&0\\&1&&0&\\&&1&&\end{smallmatrix}\right\rangle^{11} = \sqrt{\tfrac{1}{5}}F_2\,|(0,\overline{3})\rangle = \sqrt{6}\,\begin{array}{|c|c|c|c|c|c|}\hline 1&3&3&3&3&3\\\hline\end{array}\,;$$

$$|(\overline{1},\overline{4})\rangle = \left|\begin{smallmatrix}6&&0&&0\\&1&&0&\\&&0&&\end{smallmatrix}\right\rangle^{12} = F_1\,|(1,\overline{5})\rangle = \sqrt{6}\,\begin{array}{|c|c|c|c|c|c|}\hline 2&3&3&3&3&3\\\hline\end{array}\,.$$

The basis state $|(\overline{6},6)\rangle$ is the highest weight state of an $\mathcal{A}_2$-sevenlet. The lowest weight state of $[6,0]$ of A_2 belongs to this $\mathcal{A}_2$-sevenlet.

$$\begin{aligned}
&F_2\,|(\overline{6},6)\rangle = \sqrt{6}\,|(\overline{5},4)\rangle\,, \quad F_2\,|(\overline{5},4)\rangle = \sqrt{10}\,|(\overline{4},2)\rangle\,,\\
&F_2\,|(\overline{4},2)\rangle = \sqrt{12}\,|(\overline{3},0)\rangle\,, \quad F_2\,|(\overline{3},0)\rangle = \sqrt{12}\,|(\overline{2},\overline{2})\rangle\,,\\
&F_2\,|(\overline{2},\overline{2})\rangle = \sqrt{10}\,|(\overline{1},\overline{4})\rangle\,,\\
&|(0,\overline{6})\rangle = \left|\begin{smallmatrix}6&&0&&0\\&0&&0&\\&&0&&\end{smallmatrix}\right\rangle^{13} = \sqrt{\tfrac{1}{6}}F_2\,|(\overline{1},\overline{4})\rangle = \begin{array}{|c|c|c|c|c|c|}\hline 3&3&3&3&3&3\\\hline\end{array}\,.
\end{aligned}$$

Ex. 2. The representation $[5,1]$ of A_2.

Due to Eq. (9.32), the dimension of the representation $[5,1]$ of A_2 is $(7\times 2\times 5)/2 = 35$. The highest weight state in $[5,1]$ of A_2 is also that of an A_1-quintuplet with $\boldsymbol{M}^{(1)} = 4\boldsymbol{w}_1$.

$$\begin{aligned}
|(4,1)\rangle &= \left|\begin{smallmatrix}5&&1&&0\\&5&&1&\\&&5&&\end{smallmatrix}\right\rangle^{1} = \begin{array}{|c|c|c|c|c|}\hline 1&1&1&1&1\\\hline 2\\\cline{1-1}\end{array}\,,\\
|(2,2)\rangle &= \left|\begin{smallmatrix}5&&1&&0\\&5&&1&\\&&4&&\end{smallmatrix}\right\rangle^{2} = \sqrt{\tfrac{1}{4}}F_1\,|(4,1)\rangle = 2\,\begin{array}{|c|c|c|c|c|}\hline 1&1&1&1&2\\\hline 2\\\cline{1-1}\end{array}\,,\\
|(0,3)\rangle &= \left|\begin{smallmatrix}5&&1&&0\\&5&&1&\\&&3&&\end{smallmatrix}\right\rangle^{3} = \sqrt{\tfrac{1}{6}}F_1\,|(2,2)\rangle = \sqrt{6}\,\begin{array}{|c|c|c|c|c|}\hline 1&1&1&2&2\\\hline 2\\\cline{1-1}\end{array}\,,\\
|(\overline{2},4)\rangle &= \left|\begin{smallmatrix}5&&1&&0\\&5&&1&\\&&2&&\end{smallmatrix}\right\rangle^{4} = \sqrt{\tfrac{1}{6}}F_1\,|(0,3)\rangle = 2\,\begin{array}{|c|c|c|c|c|}\hline 1&1&2&2&2\\\hline 2\\\cline{1-1}\end{array}\,,\\
|(\overline{4},5)\rangle &= \left|\begin{smallmatrix}5&&1&&0\\&5&&1&\\&&1&&\end{smallmatrix}\right\rangle^{5} = \sqrt{\tfrac{1}{4}}F_1\,|(\overline{2},4)\rangle = \begin{array}{|c|c|c|c|c|}\hline 1&2&2&2&2\\\hline 2\\\cline{1-1}\end{array}\,.
\end{aligned}$$

Five basis states all are the highest weight states of the $\mathcal{A}_2$-multiplets, which will be calculated one by one. First, the basis state $|(4,1)\rangle$ is the highest weight states of an $\mathcal{A}_2$-doublet and its partner $|(5,\overline{1})\rangle$ is the highest weight states of an A_1-sixlet (see Eq. (9.23)).

$$\begin{aligned}
|(5,\overline{1})\rangle &= \left|\begin{smallmatrix}5&&1&&0\\&5&&0&\\&&5&&\end{smallmatrix}\right\rangle^{2} = F_2\,|(4,1)\rangle = \begin{array}{|c|c|c|c|c|}\hline 1&1&1&1&1\\\hline 3\\\cline{1-1}\end{array}\,;\\
|(3,0)_1\rangle &= \left|\begin{smallmatrix}5&&1&&0\\&5&&0&\\&&4&&\end{smallmatrix}\right\rangle^{3} = \sqrt{\tfrac{1}{5}}F_1\,|(5,\overline{1})\rangle\\
&= \sqrt{5}\,\begin{array}{|c|c|c|c|c|}\hline 1&1&1&1&2\\\hline 3\\\cline{1-1}\end{array} - \frac{1}{\sqrt{5}}\,\begin{array}{|c|c|c|c|c|}\hline 1&1&1&1&3\\\hline 2\\\cline{1-1}\end{array}\,,
\end{aligned}$$

$$
\begin{aligned}
|(1,1)_1\rangle &= \left|\begin{smallmatrix}5&&1&&0\\&5&&0&\\&&3&&\end{smallmatrix}\right\rangle^4 = \sqrt{\tfrac{1}{8}}F_1\,|(3,0)_1\rangle \\
&= \sqrt{10}\,\begin{array}{|c|c|c|c|c|}\hline 1&1&1&2&2\\\hline 3\\\cline{1-1}\end{array} - \sqrt{\tfrac{5}{8}}\,\begin{array}{|c|c|c|c|c|}\hline 1&1&1&2&3\\\hline 2\\\cline{1-1}\end{array} \\
&\quad - \frac{3}{\sqrt{40}}\,\begin{array}{|c|c|c|c|c|}\hline 1&1&1&2&3\\\hline 2\\\cline{1-1}\end{array} \\
&= \sqrt{10}\,\begin{array}{|c|c|c|c|c|}\hline 1&1&1&2&2\\\hline 3\\\cline{1-1}\end{array} - \frac{4}{\sqrt{10}}\,\begin{array}{|c|c|c|c|c|}\hline 1&1&1&2&3\\\hline 2\\\cline{1-1}\end{array}\,, \\
|(\overline{1},2)_1\rangle &= \left|\begin{smallmatrix}5&&1&&0\\&5&&0&\\&&2&&\end{smallmatrix}\right\rangle^5 = \tfrac{1}{3}F_1\,|(1,1)_1\rangle \\
&= \sqrt{10}\,\begin{array}{|c|c|c|c|c|}\hline 1&1&2&2&2\\\hline 3\\\cline{1-1}\end{array} - \frac{\sqrt{10}}{3}\,\begin{array}{|c|c|c|c|c|}\hline 1&1&2&2&3\\\hline 2\\\cline{1-1}\end{array} \\
&\quad - \frac{8}{3\sqrt{10}}\,\begin{array}{|c|c|c|c|c|}\hline 1&1&2&2&3\\\hline 2\\\cline{1-1}\end{array} \\
&= \sqrt{10}\,\begin{array}{|c|c|c|c|c|}\hline 1&1&2&2&2\\\hline 3\\\cline{1-1}\end{array} - \frac{6}{\sqrt{10}}\,\begin{array}{|c|c|c|c|c|}\hline 1&1&2&2&3\\\hline 2\\\cline{1-1}\end{array}\,, \\
|(\overline{3},3)_1\rangle &= \left|\begin{smallmatrix}5&&1&&0\\&5&&0&\\&&1&&\end{smallmatrix}\right\rangle^6 = \sqrt{\tfrac{1}{8}}F_1\,|(\overline{1},2)_1\rangle \\
&= \sqrt{5}\,\begin{array}{|c|c|c|c|c|}\hline 1&2&2&2&2\\\hline 3\\\cline{1-1}\end{array} - \frac{\sqrt{5}}{2}\,\begin{array}{|c|c|c|c|c|}\hline 1&2&2&2&3\\\hline 2\\\cline{1-1}\end{array} \\
&\quad - \frac{3}{2\sqrt{5}}\,\begin{array}{|c|c|c|c|c|}\hline 1&2&2&2&3\\\hline 2\\\cline{1-1}\end{array} \\
&= \sqrt{5}\,\begin{array}{|c|c|c|c|c|}\hline 1&2&2&2&2\\\hline 3\\\cline{1-1}\end{array} - \frac{4}{\sqrt{5}}\,\begin{array}{|c|c|c|c|c|}\hline 1&2&2&2&3\\\hline 2\\\cline{1-1}\end{array}\,, \\
|(\overline{5},4)\rangle &= \left|\begin{smallmatrix}5&&1&&0\\&5&&0&\\&&0&&\end{smallmatrix}\right\rangle^7 = \sqrt{\tfrac{1}{5}}F_1\,|(\overline{3},3)_1\rangle = \begin{array}{|c|c|c|c|c|}\hline 2&2&2&2&2\\\hline 3\\\cline{1-1}\end{array}\,.
\end{aligned}
$$

Second, $|(2,2)\rangle$ is the highest weight state of an $\mathcal{A}_2$-triplet. We calculate the basis states in the $\mathcal{A}_2$-triplet by Eqs. (8.25-28). Let

$$
\begin{aligned}
F_2\,|(2,2)\rangle &= F_2\left|\begin{smallmatrix}5&&1&&0\\&5&&1&\\&&4&&\end{smallmatrix}\right\rangle^2 = b_1\left|\begin{smallmatrix}5&&1&&0\\&5&&0&\\&&4&&\end{smallmatrix}\right\rangle^3 + c_1\left|\begin{smallmatrix}5&&1&&0\\&4&&1&\\&&4&&\end{smallmatrix}\right\rangle^3 \\
&= b_1\,|(3,0)_1\rangle + c_1\,|(3,0)_2\rangle = 2F_2\,\begin{array}{|c|c|c|c|c|}\hline 1&1&1&1&2\\\hline 2\\\cline{1-1}\end{array}\,, \\
E_1F_2\,|(2,2)\rangle &= \sqrt{5}b_1\,|(5,\overline{1})\rangle \\
&= F_2E_1\,|(2,2)\rangle = 2F_2\,|(4,1)\rangle = 2\,|(5,\overline{1})\rangle\,, \\
E_2F_2\,|(2,2)\rangle &= (b_1^2+c_1^2)\,|(2,2)\rangle = (F_2E_2+H_2)\,|(2,2)\rangle = 2\,|(2,2)\rangle\,.
\end{aligned}
$$

Choosing the phase of $|(3,0)_2\rangle$ such that c_1 is real and positive, we have $b_1=\sqrt{4/5}$, $c_1=\sqrt{2-4/5}=\sqrt{6/5}$, and

$$
|(3,0)_2\rangle = \left|\begin{smallmatrix}5&&1&&0\\&4&&1&\\&&4&&\end{smallmatrix}\right\rangle^3 = \sqrt{\tfrac{5}{6}}\left\{F_2\,|(2,2)\rangle - \sqrt{\tfrac{4}{5}}\,|(3,0)_1\rangle\right\}
$$

$$
\begin{aligned}
&= \sqrt{\tfrac{5}{6}} \times 2 \left\{ \begin{array}{|c|c|c|c|c|}\hline 1&1&1&1&2\\\hline 3\\\cline{1-1}\end{array} + \begin{array}{|c|c|c|c|c|}\hline 1&1&1&1&3\\\hline 2\\\cline{1-1}\end{array} \right\} \\
&\quad - \sqrt{\tfrac{2}{3}} \left\{ \sqrt{5} \begin{array}{|c|c|c|c|c|}\hline 1&1&1&1&2\\\hline 3\\\cline{1-1}\end{array} - \frac{1}{\sqrt{5}} \begin{array}{|c|c|c|c|c|}\hline 1&1&1&1&3\\\hline 2\\\cline{1-1}\end{array} \right\} \\
&= 2\sqrt{\tfrac{6}{5}} \begin{array}{|c|c|c|c|c|}\hline 1&1&1&1&3\\\hline 2\\\cline{1-1}\end{array} .
\end{aligned}
$$

By the action of $[E_2,\ F_2] = H_2$ on $|(3,0)_1\rangle$ we have

$$
\begin{aligned}
|(4,\overline{2})\rangle &= \left| \begin{smallmatrix} 5&&1&&0\\ &4&&0&\\ &&4&& \end{smallmatrix} \right\rangle^4 = \sqrt{\tfrac{5}{4}} F_2\ |(3,0)_1\rangle = \sqrt{\tfrac{5}{6}} F_2\ |(3,0)_2\rangle \\
&= 2 \begin{array}{|c|c|c|c|c|}\hline 1&1&1&1&3\\\hline 3\\\cline{1-1}\end{array} .
\end{aligned}
$$

The basis states $|(3,0)_2\rangle$ and $|(4,\overline{2})\rangle$ are the highest weight states of an A_1-quadruplet and an A_1-quintuplet, respectively.

$$
\begin{aligned}
|(1,1)_2\rangle &= \left| \begin{smallmatrix} 5&&1&&0\\ &4&&1&\\ &&3&& \end{smallmatrix} \right\rangle^4 = \sqrt{\tfrac{1}{3}} F_1\ |(3,0)_2\rangle = 6\sqrt{\tfrac{2}{5}} \begin{array}{|c|c|c|c|c|}\hline 1&1&1&2&3\\\hline 2\\\cline{1-1}\end{array} , \\
|(\overline{1},2)_2\rangle &= \left| \begin{smallmatrix} 5&&1&&0\\ &4&&1&\\ &&2&& \end{smallmatrix} \right\rangle^5 = \sqrt{\tfrac{1}{4}} F_1\ |(1,1)_2\rangle = 6\sqrt{\tfrac{2}{5}} \begin{array}{|c|c|c|c|c|}\hline 1&1&2&2&3\\\hline 2\\\cline{1-1}\end{array} , \\
|(\overline{3},3)_2\rangle &= \left| \begin{smallmatrix} 5&&1&&0\\ &4&&1&\\ &&1&& \end{smallmatrix} \right\rangle^6 = \sqrt{\tfrac{1}{3}} F_1\ |(\overline{1},2)_2\rangle = 2\sqrt{\tfrac{6}{5}} \begin{array}{|c|c|c|c|c|}\hline 1&2&2&2&3\\\hline 2\\\cline{1-1}\end{array} ; \\
|(2,\overline{1})_1\rangle &= \left| \begin{smallmatrix} 5&&1&&0\\ &4&&0&\\ &&3&& \end{smallmatrix} \right\rangle^5 = \sqrt{\tfrac{1}{4}} F_1\ |(4,\overline{2})\rangle \\
&= 4 \begin{array}{|c|c|c|c|c|}\hline 1&1&1&2&3\\\hline 3\\\cline{1-1}\end{array} - \begin{array}{|c|c|c|c|c|}\hline 1&1&1&3&3\\\hline 2\\\cline{1-1}\end{array} , \\
|(0,0)_1\rangle &= \left| \begin{smallmatrix} 5&&1&&0\\ &4&&0&\\ &&2&& \end{smallmatrix} \right\rangle^6 = \sqrt{\tfrac{1}{6}} F_1\ |(2,\overline{1})_1\rangle \\
&= 2\sqrt{6} \begin{array}{|c|c|c|c|c|}\hline 1&1&2&2&3\\\hline 3\\\cline{1-1}\end{array} - \sqrt{6} \begin{array}{|c|c|c|c|c|}\hline 1&1&2&3&3\\\hline 2\\\cline{1-1}\end{array} , \\
|(\overline{2},1)_1\rangle &= \left| \begin{smallmatrix} 5&&1&&0\\ &4&&0&\\ &&1&& \end{smallmatrix} \right\rangle^7 = \sqrt{\tfrac{1}{6}} F_1\ |(0,0)_1\rangle \\
&= 4 \begin{array}{|c|c|c|c|c|}\hline 1&2&2&2&3\\\hline 3\\\cline{1-1}\end{array} - 3 \begin{array}{|c|c|c|c|c|}\hline 1&2&2&3&3\\\hline 2\\\cline{1-1}\end{array} , \\
|(\overline{4},2)_1\rangle &= \left| \begin{smallmatrix} 5&&1&&0\\ &4&&0&\\ &&0&& \end{smallmatrix} \right\rangle^8 = \tfrac{1}{2} F_1\ |(\overline{2},1)_1\rangle = 2 \begin{array}{|c|c|c|c|c|}\hline 2&2&2&2&3\\\hline 3\\\cline{1-1}\end{array} .
\end{aligned}
$$

Third, $|(0,3)\rangle$ is the highest weight state of an $\mathcal{A}_2$-quadruplet. We calculate the basis states in the $\mathcal{A}_2$-quadruplet by Eqs. (8.25-28). Let

$$
\begin{aligned}
F_2\ |(0,3)\rangle &= F_2 \left| \begin{smallmatrix} 5&&1&&0\\ &5&&1&\\ &&3&& \end{smallmatrix} \right\rangle^3 = b_2 \left| \begin{smallmatrix} 5&&1&&0\\ &5&&0&\\ &&3&& \end{smallmatrix} \right\rangle^4 + c_2 \left| \begin{smallmatrix} 5&&1&&0\\ &4&&1&\\ &&3&& \end{smallmatrix} \right\rangle^4 \\
&= b_2\ |(1,1)_1\rangle + c_2\ |(1,1)_2\rangle ,
\end{aligned}
$$

$$E_1F_2\,|(0,3)\rangle = \sqrt{8}b_2\,|(3,0)_1\rangle + \sqrt{3}c_2\,|(3,0)_2\rangle$$
$$= F_2E_1\,|(0,3)\rangle = \sqrt{6}F_2\,|(2,2)\rangle = \sqrt{\tfrac{24}{5}}b_2\,|(3,0)_1\rangle + \sqrt{\tfrac{36}{5}}c_2\,|(3,0)_2\rangle\ .$$

Thus, $b_2 = \sqrt{3/5}$, $c_2 = \sqrt{12/5}$, and

$$F_2\,|(0,3)\rangle = \sqrt{6}\begin{array}{|c|c|c|c|c|}\hline 1&1&1&2&2\\\hline \multicolumn{1}{|c|}{3}\\\cline{1-1}\end{array} + 2\sqrt{6}\begin{array}{|c|c|c|c|c|}\hline 1&1&1&2&3\\\hline \multicolumn{1}{|c|}{2}\\\cline{1-1}\end{array}$$
$$= \sqrt{\tfrac{3}{5}}\,|(1,1)_1\rangle + \sqrt{\tfrac{12}{5}}\,|(1,1)_2\rangle$$
$$= \sqrt{\tfrac{3}{5}}\left\{\sqrt{10}\begin{array}{|c|c|c|c|c|}\hline 1&1&1&2&2\\\hline \multicolumn{1}{|c|}{3}\\\cline{1-1}\end{array} - \frac{4}{\sqrt{10}}\begin{array}{|c|c|c|c|c|}\hline 1&1&1&2&3\\\hline \multicolumn{1}{|c|}{2}\\\cline{1-1}\end{array}\right\}$$
$$+ \sqrt{\tfrac{12}{5}} \times 6\sqrt{\tfrac{2}{5}}\begin{array}{|c|c|c|c|c|}\hline 1&1&1&2&3\\\hline \multicolumn{1}{|c|}{2}\\\cline{1-1}\end{array}\ .$$

Then,

$$F_2\,|(1,1)_1\rangle = F_2\left|\begin{smallmatrix}5&&1&&0\\&5&&0&\\&&3&&\end{smallmatrix}\right\rangle^{4} = a_3\left|\begin{smallmatrix}5&&1&&0\\&4&&0&\\&&3&&\end{smallmatrix}\right\rangle^{5} = a_3\,|(2,\overline{1})_1\rangle\ ,$$
$$F_2\,|(1,1)_2\rangle = F_2\left|\begin{smallmatrix}5&&1&&0\\&4&&1&\\&&3&&\end{smallmatrix}\right\rangle^{4} = b_3\left|\begin{smallmatrix}5&&1&&0\\&4&&0&\\&&3&&\end{smallmatrix}\right\rangle^{5} + c_3\left|\begin{smallmatrix}5&&1&&0\\&3&&1&\\&&3&&\end{smallmatrix}\right\rangle^{5}$$
$$= b_3\,|(2,\overline{1})_1\rangle + c_3\,|(2,\overline{1})_2\rangle\ ,$$
$$E_1F_2\,|(1,1)_1\rangle = 2a_3\,|(4,\overline{2})\rangle$$
$$= F_2E_1\,|(1,1)_1\rangle = \sqrt{8}F_2\,|(3,0)_1\rangle = \sqrt{8}\times\sqrt{\tfrac{4}{5}}\,|(4,\overline{2})\rangle\ ,$$
$$E_1F_2\,|(1,1)_2\rangle = 2b_3\,|(4,\overline{2})\rangle$$
$$= F_2E_1\,|(1,1)_2\rangle = \sqrt{3}F_2\,|(3,0)_2\rangle = \sqrt{3}\times\sqrt{\tfrac{6}{5}}\,|(4,\overline{2})\rangle\ ,$$
$$E_2F_2\,|(1,1)_2\rangle = b_3a_3\,|(1,1)_1\rangle + (b_3^2+c_3^2)\,|(1,1)_2\rangle$$
$$= (F_2E_2 + H_2)\,|(1,1)_2\rangle = c_2b_2\,|(1,1)_1\rangle + (c_2^2+1)\,|(1,1)_2\rangle\ .$$

Choosing the phase of $|(2,\overline{1})_2\rangle$ such that c_3 is real and positive, we obtain $a_3 = \sqrt{8/5}$, $b_3 = \sqrt{9/10}$, and $c_3 = \sqrt{12/5+1-9/10} = \sqrt{5/2}$.

$$F_2\,|(1,1)_1\rangle = \frac{16}{\sqrt{10}}\begin{array}{|c|c|c|c|c|}\hline 1&1&1&2&3\\\hline \multicolumn{1}{|c|}{3}\\\cline{1-1}\end{array} - \frac{4}{\sqrt{10}}\begin{array}{|c|c|c|c|c|}\hline 1&1&1&3&3\\\hline \multicolumn{1}{|c|}{2}\\\cline{1-1}\end{array}$$
$$= \sqrt{\tfrac{8}{5}}\,|(2,\overline{1})_1\rangle = \sqrt{\tfrac{8}{5}}\left\{4\begin{array}{|c|c|c|c|c|}\hline 1&1&1&2&3\\\hline \multicolumn{1}{|c|}{3}\\\cline{1-1}\end{array} - \begin{array}{|c|c|c|c|c|}\hline 1&1&1&3&3\\\hline \multicolumn{1}{|c|}{2}\\\cline{1-1}\end{array}\right\}\ ,$$
$$|(2,\overline{1})_2\rangle = \left|\begin{smallmatrix}5&&1&&0\\&3&&1&\\&&3&&\end{smallmatrix}\right\rangle^{5} = \sqrt{\tfrac{2}{5}}\left\{F_2\,|(1,1)_2\rangle - \frac{3}{\sqrt{10}}\,|(2,\overline{1})_1\rangle\right\}$$
$$= \sqrt{\tfrac{2}{5}}\times 6\sqrt{\tfrac{2}{5}}\left\{\begin{array}{|c|c|c|c|c|}\hline 1&1&1&2&3\\\hline \multicolumn{1}{|c|}{3}\\\cline{1-1}\end{array} + \begin{array}{|c|c|c|c|c|}\hline 1&1&1&3&3\\\hline \multicolumn{1}{|c|}{2}\\\cline{1-1}\end{array}\right\}$$
$$-\frac{3}{5}\left\{4\begin{array}{|c|c|c|c|c|}\hline 1&1&1&2&3\\\hline \multicolumn{1}{|c|}{3}\\\cline{1-1}\end{array} - \begin{array}{|c|c|c|c|c|}\hline 1&1&1&3&3\\\hline \multicolumn{1}{|c|}{2}\\\cline{1-1}\end{array}\right\}$$
$$= 3\begin{array}{|c|c|c|c|c|}\hline 1&1&1&3&3\\\hline \multicolumn{1}{|c|}{2}\\\cline{1-1}\end{array}\ .$$

By the action of $[E_2, F_2] = H_2$ on $|(2,\overline{1})_2\rangle$, we obtain

$$\begin{aligned}|(3,\overline{3})\rangle &= \left|\begin{matrix}5&&1&&0\\&3&&0&\\&&3&&\end{matrix}\right\rangle^{6} = \sqrt{\tfrac{2}{3}}F_2\ |(2,\overline{1})_1\rangle = \sqrt{\tfrac{2}{3}}F_2\ |(2,\overline{1})_2\rangle \\ &= \sqrt{6}\ \begin{array}{|c|c|c|c|c|}\hline 1&1&1&3&3\\\hline 3\\\cline{1-1}\end{array}\ .\end{aligned}$$

The basis states $|(2,\overline{1})_2\rangle$ and $|(3,\overline{3})\rangle$ are the highest weight states of an A_1-triplet and an A_1-quadruplet, respectively.

$$\begin{aligned}|(0,0)_2\rangle &= \left|\begin{matrix}5&&1&&0\\&3&&1&\\&&2&&\end{matrix}\right\rangle^{6} = \sqrt{\tfrac{1}{2}}F_1\ |(2,\overline{1})_2\rangle = 3\sqrt{2}\ \begin{array}{|c|c|c|c|c|}\hline 1&1&2&3&3\\\hline 2\\\cline{1-1}\end{array}\ , \\ |(\overline{2},1)_2\rangle &= \left|\begin{matrix}5&&1&&0\\&3&&1&\\&&1&&\end{matrix}\right\rangle^{7} = \sqrt{\tfrac{1}{2}}F_1\ |(0,0)_2\rangle = 3\ \begin{array}{|c|c|c|c|c|}\hline 1&2&2&3&3\\\hline 2\\\cline{1-1}\end{array}\ ; \\ |(1,\overline{2})_1\rangle &= \left|\begin{matrix}5&&1&&0\\&3&&0&\\&&2&&\end{matrix}\right\rangle^{7} = \sqrt{\tfrac{1}{3}}F_1\ |(3,\overline{3})\rangle \\ &= 3\sqrt{2}\ \begin{array}{|c|c|c|c|c|}\hline 1&1&2&3&3\\\hline 3\\\cline{1-1}\end{array} - \sqrt{2}\ \begin{array}{|c|c|c|c|c|}\hline 1&1&3&3&3\\\hline 2\\\cline{1-1}\end{array}\ , \\ |(\overline{1},\overline{1})_1\rangle &= \left|\begin{matrix}5&&1&&0\\&3&&0&\\&&1&&\end{matrix}\right\rangle^{8} = \tfrac{1}{2}F_1\ |(1,\overline{2})_1\rangle \\ &= 3\sqrt{2}\ \begin{array}{|c|c|c|c|c|}\hline 1&2&2&3&3\\\hline 3\\\cline{1-1}\end{array} - 2\sqrt{2}\ \begin{array}{|c|c|c|c|c|}\hline 1&2&3&3&3\\\hline 2\\\cline{1-1}\end{array}\ , \\ |(\overline{3},0)\rangle &= \left|\begin{matrix}5&&1&&0\\&3&&0&\\&&0&&\end{matrix}\right\rangle^{9} = \sqrt{\tfrac{1}{3}}F_1\ |(\overline{1},\overline{1})_1\rangle = \sqrt{6}\ \begin{array}{|c|c|c|c|c|}\hline 2&2&2&3&3\\\hline 3\\\cline{1-1}\end{array}\ .\end{aligned}$$

The method in the remaining calculation is the same, so that we only list the results. $|(\overline{2},4)\rangle$ is the highest weight state of an $\mathcal{A}_2$-quintuplet.

$$\begin{aligned}F_2\ |(\overline{2},4)\rangle &= 2\ \begin{array}{|c|c|c|c|c|}\hline 1&1&2&2&2\\\hline 3\\\cline{1-1}\end{array} + 6\ \begin{array}{|c|c|c|c|c|}\hline 1&1&2&2&3\\\hline 2\\\cline{1-1}\end{array} \\ &= \sqrt{\tfrac{2}{5}}\ |(\overline{1},2)_1\rangle + \sqrt{\tfrac{18}{5}}\ |(\overline{1},2)_2\rangle \\ &= \sqrt{\tfrac{2}{5}}\left\{\sqrt{10}\ \begin{array}{|c|c|c|c|c|}\hline 1&1&2&2&2\\\hline 3\\\cline{1-1}\end{array} - \frac{6}{\sqrt{10}}\ \begin{array}{|c|c|c|c|c|}\hline 1&1&2&2&3\\\hline 2\\\cline{1-1}\end{array}\right\} \\ &\quad + \sqrt{\tfrac{18}{5}} \times 6\sqrt{\tfrac{2}{5}}\ \begin{array}{|c|c|c|c|c|}\hline 1&1&2&2&3\\\hline 2\\\cline{1-1}\end{array}\ , \\ F_2\ |(\overline{1},2)_1\rangle &= \frac{24}{\sqrt{10}}\ \begin{array}{|c|c|c|c|c|}\hline 1&1&2&2&3\\\hline 3\\\cline{1-1}\end{array} - \frac{12}{\sqrt{10}}\ \begin{array}{|c|c|c|c|c|}\hline 1&1&1&3&3\\\hline 2\\\cline{1-1}\end{array} \\ &= \sqrt{\tfrac{12}{5}}\ |(0,0)_1\rangle = \sqrt{\tfrac{12}{5}} \times \sqrt{6}\left\{2\ \begin{array}{|c|c|c|c|c|}\hline 1&1&2&2&3\\\hline 3\\\cline{1-1}\end{array} - \begin{array}{|c|c|c|c|c|}\hline 1&1&2&3&3\\\hline 2\\\cline{1-1}\end{array}\right\}, \\ F_2\ |(\overline{1},2)_2\rangle &= 6\sqrt{\tfrac{2}{5}}\left\{\begin{array}{|c|c|c|c|c|}\hline 1&1&2&2&3\\\hline 3\\\cline{1-1}\end{array} - 2\ \begin{array}{|c|c|c|c|c|}\hline 1&1&2&3&3\\\hline 2\\\cline{1-1}\end{array}\right\} \\ &= \sqrt{\tfrac{3}{5}}\ |(0,0)_1\rangle + \sqrt{5}\ |(0,0)_2\rangle \\ &= \sqrt{\tfrac{3}{5}} \times \sqrt{6}\left\{2\ \begin{array}{|c|c|c|c|c|}\hline 1&1&2&2&3\\\hline 3\\\cline{1-1}\end{array} - \begin{array}{|c|c|c|c|c|}\hline 1&1&2&3&3\\\hline 2\\\cline{1-1}\end{array}\right\}\end{aligned}$$

$$+\sqrt{5}\times 3\sqrt{2}\begin{array}{|c|c|c|c|c|}\hline 1&1&2&3&3\\\hline 2\\\cline{1-1}\end{array},$$

$$F_2\,|(0,0)_1\rangle=\sqrt{6}\left\{3\begin{array}{|c|c|c|c|c|}\hline 1&1&2&3&3\\\hline 3\\\cline{1-1}\end{array}-\begin{array}{|c|c|c|c|c|}\hline 1&1&3&3&3\\\hline 2\\\cline{1-1}\end{array}\right\}$$

$$=\sqrt{3}\,|(1,\overline{2})_1\rangle$$

$$=\sqrt{3}\times\sqrt{2}\left\{3\begin{array}{|c|c|c|c|c|}\hline 1&1&2&3&3\\\hline 3\\\cline{1-1}\end{array}-\begin{array}{|c|c|c|c|c|}\hline 1&1&3&3&3\\\hline 2\\\cline{1-1}\end{array}\right\},$$

$$|(1,\overline{2})_2\rangle=\left|\begin{array}{ccccc}5&&1&&0\\&2&&1&\\&&2&&\end{array}\right\rangle^{7}=\tfrac{1}{2}\left\{F_2\,|(0,0)_2\rangle-|(1,\overline{2})_1\rangle\right\}$$

$$=\tfrac{1}{2}\times 3\sqrt{2}\left\{\begin{array}{|c|c|c|c|c|}\hline 1&1&2&3&3\\\hline 3\\\cline{1-1}\end{array}+\begin{array}{|c|c|c|c|c|}\hline 1&1&3&3&3\\\hline 2\\\cline{1-1}\end{array}\right\}$$

$$-\tfrac{1}{2}\times\sqrt{2}\left\{3\begin{array}{|c|c|c|c|c|}\hline 1&1&2&3&3\\\hline 3\\\cline{1-1}\end{array}-\begin{array}{|c|c|c|c|c|}\hline 1&1&3&3&3\\\hline 2\\\cline{1-1}\end{array}\right\}$$

$$=2\sqrt{2}\begin{array}{|c|c|c|c|c|}\hline 1&1&3&3&3\\\hline 2\\\cline{1-1}\end{array}.$$

By the action of $[E_2,\ F_2]=H_2$ on $|(1,\overline{2})_2\rangle$, we obtain

$$|(2,\overline{4})\rangle=\left|\begin{array}{ccccc}5&&1&&0\\&2&&0&\\&&2&&\end{array}\right\rangle^{8}=\sqrt{\tfrac{1}{2}}F_2\,|(1,\overline{2})_1\rangle=\sqrt{\tfrac{1}{2}}F_2\,|(1,\overline{2})_2\rangle$$

$$=2\begin{array}{|c|c|c|c|c|}\hline 1&1&3&3&3\\\hline 3\\\cline{1-1}\end{array}.$$

The basis states $|(1,\overline{2})_2\rangle$ and $|(2,\overline{4})\rangle$ are the highest weight states of an A_1-doublet and an A_1-triplet, respectively.

$$|(\overline{1},\overline{1})_2\rangle=\left|\begin{array}{ccccc}5&&1&&0\\&2&&1&\\&&1&&\end{array}\right\rangle^{8}=F_1\,|(1,\overline{2})_2\rangle=2\sqrt{2}\begin{array}{|c|c|c|c|c|}\hline 1&2&3&3&3\\\hline 2\\\cline{1-1}\end{array};$$

$$|(0,\overline{3})_1\rangle=\left|\begin{array}{ccccc}5&&1&&0\\&2&&0&\\&&1&&\end{array}\right\rangle^{9}=\sqrt{\tfrac{1}{2}}F_1\,|(2,\overline{4})\rangle$$

$$=2\sqrt{2}\begin{array}{|c|c|c|c|c|}\hline 1&2&3&3&3\\\hline 3\\\cline{1-1}\end{array}-\sqrt{2}\begin{array}{|c|c|c|c|c|}\hline 1&3&3&3&3\\\hline 2\\\cline{1-1}\end{array},$$

$$|(\overline{2},\overline{2})\rangle=\left|\begin{array}{ccccc}5&&1&&0\\&2&&0&\\&&0&&\end{array}\right\rangle^{10}=\sqrt{\tfrac{1}{2}}F_1\,|(0,\overline{3})_1\rangle=2\begin{array}{|c|c|c|c|c|}\hline 2&2&3&3&3\\\hline 3\\\cline{1-1}\end{array}.$$

$|(\overline{4},5)\rangle$ is the highest weight state of an $\mathcal{A}_2$-sixlet.

$$F_2\,|(\overline{4},5)\rangle=\begin{array}{|c|c|c|c|c|}\hline 1&2&2&2&2\\\hline 3\\\cline{1-1}\end{array}+4\begin{array}{|c|c|c|c|c|}\hline 1&2&2&2&3\\\hline 2\\\cline{1-1}\end{array}$$

$$=\sqrt{\tfrac{1}{5}}\,|(\overline{3},3)_1\rangle+\sqrt{\tfrac{24}{5}}\,|(\overline{3},3)_2\rangle$$

$$=\sqrt{\tfrac{1}{5}}\left\{\sqrt{5}\begin{array}{|c|c|c|c|c|}\hline 1&2&2&2&2\\\hline 3\\\cline{1-1}\end{array}-\frac{4}{\sqrt{5}}\begin{array}{|c|c|c|c|c|}\hline 1&2&2&2&3\\\hline 2\\\cline{1-1}\end{array}\right\}$$

$$+\sqrt{\tfrac{24}{5}}\times\sqrt{\tfrac{24}{5}}\begin{array}{|c|c|c|c|c|}\hline 1&2&2&2&3\\\hline 2\\\cline{1-1}\end{array},$$

$$\begin{aligned}
F_2\,|(\overline{3},3)_1\rangle &= \tfrac{16}{\sqrt{5}}\,\ytableaushort{12223,3} - \tfrac{12}{\sqrt{5}}\,\ytableaushort{12223,3} \\
&= \tfrac{4}{\sqrt{5}}\,|(\overline{2},1)_1\rangle = \tfrac{4}{\sqrt{5}}\left\{4\,\ytableaushort{12223,3} - 3\,\ytableaushort{12233,2}\right\}, \\
F_2\,|(\overline{3},3)_2\rangle &= \sqrt{\tfrac{24}{5}}\left\{\ytableaushort{12223,3} + 3\,\ytableaushort{12223,3}\right\} \\
&= \sqrt{\tfrac{3}{10}}\,|(\overline{2},1)_1\rangle + \sqrt{\tfrac{15}{2}}\,|(\overline{2},1)_2\rangle \\
&= \sqrt{\tfrac{3}{10}}\left\{4\,\ytableaushort{12223,3} - 3\,\ytableaushort{12233,2}\right\} \\
&\quad + \sqrt{\tfrac{15}{2}} \times 3\,\ytableaushort{12233,2}, \\
F_2\,|(\overline{2},1)_1\rangle &= 9\,\ytableaushort{12233,3} - 6\,\ytableaushort{12333,2} = \tfrac{3}{\sqrt{2}}\,|(\overline{1},\overline{1})_1\rangle \\
&= \tfrac{3}{\sqrt{2}} \times \sqrt{2}\left\{3\,\ytableaushort{12233,3} - 2\,\ytableaushort{12333,2}\right\}, \\
F_2\,|(\overline{2},1)_2\rangle &= 3\,\ytableaushort{12233,3} + 6\,\ytableaushort{12333,2} \\
&= \sqrt{\tfrac{1}{2}}\,|(\overline{1},\overline{1})_1\rangle + \sqrt{8}\,|(\overline{1},\overline{1})_2\rangle \\
&= \sqrt{\tfrac{1}{2}} \times \sqrt{2}\left\{3\,\ytableaushort{12223,3} - 2\,\ytableaushort{12233,2}\right\} \\
&\quad + \sqrt{8} \times 2\sqrt{2}\,\ytableaushort{12333,2}, \\
F_2\,|(\overline{1},\overline{1})_1\rangle &= \sqrt{2}\left\{4\,\ytableaushort{12333,3} - 2\,\ytableaushort{13333,2}\right\} \\
&= 2\,|(0,\overline{3})_1\rangle = 2 \times \sqrt{2}\left\{2\,\ytableaushort{12333,3} - \ytableaushort{13333,2}\right\}, \\
|(0,\overline{3})_2\rangle &= \left|\begin{array}{ccccc} 5 & & 1 & & 0 \\ & 1 & & 1 & \\ & & 1 & & \end{array}\right\rangle^{9} = \sqrt{\tfrac{1}{6}}\left\{F_2\,|(\overline{1},\overline{1})_2\rangle - |(0,\overline{3})_1\rangle\right\} \\
&= \sqrt{\tfrac{1}{6}} \times 2\sqrt{2}\left\{\ytableaushort{12333,3} + \ytableaushort{13333,2}\right\} \\
&\quad - \sqrt{\tfrac{1}{6}} \times \sqrt{2}\left\{2\,\ytableaushort{12333,3} - \ytableaushort{13333,2}\right\} \\
&= \sqrt{3}\,\ytableaushort{13333,2}\,.
\end{aligned}$$

By the action of $[E_2,\ F_2] = H_2$ on $|(0,\overline{3})_2\rangle$, we obtain

$$\begin{aligned}
|(1,\overline{5})\rangle &= \left|\begin{array}{ccccc} 5 & & 1 & & 0 \\ & 1 & & 0 & \\ & & 1 & & \end{array}\right\rangle^{10} = \sqrt{\tfrac{1}{2}}F_2\,|(0,\overline{3})_1\rangle = \sqrt{\tfrac{1}{3}}F_2\,|(0,\overline{3})_2\rangle \\
&= \ytableaushort{13333,3}\,,
\end{aligned}$$

The basis states $|(0,\overline{3})_2\rangle$ is the A_1-singlet and $|(1,\overline{5})\rangle$ is the highest weight state of an A_1-doublet.

$$|(\overline{1},\overline{4})\rangle = \left|\begin{matrix} 5 & 1 & 0 \\ 1 & 0 & \\ 0 & & \end{matrix}\right\rangle^{11} = F_1\,|(1,\overline{5})\rangle = \begin{array}{|c|c|c|c|c|}\hline 2 & 3 & 3 & 3 & 3 \\ \hline 3 \\ \cline{1-1}\end{array}\,.$$

$|(\overline{5},4)\rangle$ is the highest weight state of an $\mathcal{A}_2$-sixlet.

$$F_2\,|(\overline{5},4)\rangle = 4\,\begin{array}{|c|c|c|c|c|}\hline 2 & 2 & 2 & 2 & 3 \\ \hline 3 \\ \cline{1-1}\end{array} = 2\,|(\overline{4},2)\rangle\,,$$

$$F_2\,|(\overline{4},2)\rangle = 6\,\begin{array}{|c|c|c|c|c|}\hline 2 & 2 & 2 & 3 & 3 \\ \hline 3 \\ \cline{1-1}\end{array} = \sqrt{6}\,|(\overline{3},0)\rangle\,,$$

$$F_2\,|(\overline{3},0)\rangle = 2\sqrt{6}\,\begin{array}{|c|c|c|c|c|}\hline 2 & 2 & 3 & 3 & 3 \\ \hline 3 \\ \cline{1-1}\end{array} = \sqrt{6}\,|(\overline{2},\overline{2})\rangle\,,$$

$$F_2\,|(\overline{2},\overline{2})\rangle = 2\,\begin{array}{|c|c|c|c|c|}\hline 2 & 3 & 3 & 3 & 3 \\ \hline 3 \\ \cline{1-1}\end{array} = 2\,|(\overline{1},\overline{4})\rangle\,.$$

Ex. 3. The representation $[4,2]$ of A_2.

Due to Eq. (9.32), the dimension of the representation $[4,2]$ of A_2 is $(6\times 3\times 3)/2 = 27$. The highest weight state in $[4,2]$ of A_2 is also that of an A_1-triplet with $\boldsymbol{M}^{(1)} = 2\boldsymbol{w}_1$.

$$|(2,2)\rangle = \left|\begin{matrix} 4 & 2 & 0 \\ 4 & 2 & \\ 4 & & \end{matrix}\right\rangle^{1} = \begin{array}{|c|c|c|c|}\hline 1 & 1 & 1 & 1 \\ \hline 2 & 2 \\ \cline{1-2}\end{array}\,,$$

$$|(0,3)\rangle = \left|\begin{matrix} 4 & 2 & 0 \\ 4 & 2 & \\ 3 & & \end{matrix}\right\rangle^{2} = \sqrt{\tfrac{1}{2}}F_1\,|(2,2)\rangle = \sqrt{2}\,\begin{array}{|c|c|c|c|}\hline 1 & 1 & 1 & 2 \\ \hline 2 & 2 \\ \cline{1-2}\end{array}\,,$$

$$|(\overline{2},4)\rangle = \left|\begin{matrix} 4 & 2 & 0 \\ 4 & 2 & \\ 2 & & \end{matrix}\right\rangle^{3} = \sqrt{\tfrac{1}{2}}F_1\,|(0,3)\rangle = \begin{array}{|c|c|c|c|}\hline 1 & 1 & 2 & 2 \\ \hline 2 & 2 \\ \cline{1-2}\end{array}\,.$$

Three basis states in the A_1-triplet all are the highest weight states of the $\mathcal{A}_2$-multiplets, respectively. First, we calculate the $\mathcal{A}_2$-triplet from its the highest weight state $|(2,2)\rangle$.

$$|(3,0)\rangle = \left|\begin{matrix} 4 & 2 & 0 \\ 4 & 1 & \\ 4 & & \end{matrix}\right\rangle^{2} = \sqrt{\tfrac{1}{2}}F_2\,|(2,2)\rangle = \sqrt{2}\,\begin{array}{|c|c|c|c|}\hline 1 & 1 & 1 & 1 \\ \hline 2 & 3 \\ \cline{1-2}\end{array}\,,$$

$$|(4,\overline{2})\rangle = \left|\begin{matrix} 4 & 2 & 0 \\ 4 & 0 & \\ 4 & & \end{matrix}\right\rangle^{3} = \sqrt{\tfrac{1}{2}}F_2\,|(3,0)\rangle = \begin{array}{|c|c|c|c|}\hline 1 & 1 & 1 & 1 \\ \hline 3 & 3 \\ \cline{1-2}\end{array}\,.$$

Two A_1-multiplets from their highest weight states $|(3,0)\rangle$ and $|(4,\overline{2})\rangle$ are calculated.

$$\begin{aligned}|(1,1)_1\rangle &= \left|\begin{matrix} 4 & 2 & 0 \\ 4 & 1 & \\ 3 & & \end{matrix}\right\rangle^{3} = \sqrt{\tfrac{1}{3}}F_1\,|(3,0)\rangle \\ &= \sqrt{6}\,\begin{array}{|c|c|c|c|}\hline 1 & 1 & 1 & 2 \\ \hline 2 & 3 \\ \cline{1-2}\end{array} - \sqrt{\tfrac{2}{3}}\,\begin{array}{|c|c|c|c|}\hline 1 & 1 & 1 & 3 \\ \hline 2 & 2 \\ \cline{1-2}\end{array}\,,\\ |(\overline{1},2)_1\rangle &= \left|\begin{matrix} 4 & 2 & 0 \\ 4 & 1 & \\ 2 & & \end{matrix}\right\rangle^{4} = \tfrac{1}{2}F_1\,|(1,1)_1\rangle \\ &= \sqrt{6}\,\begin{array}{|c|c|c|c|}\hline 1 & 1 & 2 & 2 \\ \hline 2 & 3 \\ \cline{1-2}\end{array} - \frac{4}{\sqrt{6}}\,\begin{array}{|c|c|c|c|}\hline 1 & 1 & 2 & 3 \\ \hline 2 & 2 \\ \cline{1-2}\end{array}\,,\\ |(\overline{3},3)\rangle &= \left|\begin{matrix} 4 & 2 & 0 \\ 4 & 1 & \\ 1 & & \end{matrix}\right\rangle^{5} = \sqrt{\tfrac{1}{3}}F_1\,|(\overline{1},2)_1\rangle = \sqrt{2}\,\begin{array}{|c|c|c|c|}\hline 1 & 2 & 2 & 2 \\ \hline 2 & 3 \\ \cline{1-2}\end{array}\,;\end{aligned}$$

$$
\begin{aligned}
|(2,\overline{1})_1\rangle &= \left|\begin{array}{ccccc}4&&2&&0\\&4&&0&\\&&3&&\end{array}\right\rangle^{4} = \tfrac{1}{2}F_1\ |(4,\overline{2})\rangle \\
&= 2\,\begin{array}{|c|c|c|c|}\hline 1&1&1&2\\\hline 3&3\\\cline{1-2}\end{array} - \begin{array}{|c|c|c|c|}\hline 1&1&1&3\\\hline 2&3\\\cline{1-2}\end{array}\ , \\
|(0,0)_1\rangle &= \left|\begin{array}{ccccc}4&&2&&0\\&4&&0&\\&&2&&\end{array}\right\rangle^{5} = \sqrt{\tfrac{1}{6}}F_1\ |(2,\overline{1})_1\rangle \\
&= \sqrt{6}\,\begin{array}{|c|c|c|c|}\hline 1&1&2&2\\\hline 3&3\\\cline{1-2}\end{array} - \sqrt{6}\,\begin{array}{|c|c|c|c|}\hline 1&1&2&3\\\hline 2&3\\\cline{1-2}\end{array} + \sqrt{\tfrac{1}{6}}\,\begin{array}{|c|c|c|c|}\hline 1&1&3&3\\\hline 2&2\\\cline{1-2}\end{array}\ , \\
|(\overline{2},1)_1\rangle &= \left|\begin{array}{ccccc}4&&2&&0\\&4&&0&\\&&1&&\end{array}\right\rangle^{6} = \sqrt{\tfrac{1}{6}}F_1\ |(0,0)_1\rangle \\
&= 2\,\begin{array}{|c|c|c|c|}\hline 1&2&2&2\\\hline 3&3\\\cline{1-2}\end{array} - \begin{array}{|c|c|c|c|}\hline 1&2&2&3\\\hline 2&3\\\cline{1-2}\end{array}\ , \\
|(\overline{4},2)\rangle &= \left|\begin{array}{ccccc}4&&2&&0\\&4&&0&\\&&0&&\end{array}\right\rangle^{7} = \tfrac{1}{2}F_1\ |(\overline{2},1)_1\rangle = \begin{array}{|c|c|c|c|}\hline 2&2&2&2\\\hline 3&3\\\cline{1-2}\end{array}\ .
\end{aligned}
$$

Second, we calculate the $\mathcal{A}_2$-quadruplet from its highest weight state $|(0,3)\rangle$ by Eqs. (8.25-28). Let

$$
\begin{aligned}
F_2\ |(0,3)\rangle &= F_2\left|\begin{array}{ccccc}4&&2&&0\\&4&&2&\\&&3&&\end{array}\right\rangle^{2} = a_1\left|\begin{array}{ccccc}4&&2&&0\\&4&&1&\\&&3&&\end{array}\right\rangle^{3} + b_1\left|\begin{array}{ccccc}4&&2&&0\\&3&&2&\\&&3&&\end{array}\right\rangle^{3} \\
&= a_1\ |(1,1)_1\rangle + b_1\ |(1,1)_2\rangle\ , \\
E_1F_2\ |(0,3)\rangle &= \sqrt{3}a_1\ |(3,0)\rangle \\
&= F_2E_1\ |(0,3)\rangle = \sqrt{2}F_2\ |(2,2)\rangle = \sqrt{2}\times\sqrt{2}\ |(3,0)\rangle\ , \\
E_2F_2\ |(0,3)\rangle &= (a_1^2+b_1^2)\ |(0,3)\rangle = (F_2E_2+H_2)\ |(0,3)\rangle = 3\ |(0,3)\rangle\ .
\end{aligned}
$$

Choosing the phase of $|(1,1)_2\rangle$ such that b_1 is real and positive, we obtain $a_1=\sqrt{4/3}$, $b_1=\sqrt{3-4/3}=\sqrt{5/3}$, and

$$
\begin{aligned}
|(1,1)_2\rangle &= \left|\begin{array}{ccccc}4&&2&&0\\&3&&2&\\&&3&&\end{array}\right\rangle^{3} = \sqrt{\tfrac{3}{5}}\left\{F_2\ |(0,3)\rangle - \sqrt{\tfrac{4}{3}}\ |(1,1)_1\rangle\right\} \\
&= \sqrt{\tfrac{3}{5}}\times\sqrt{2}\left\{2\,\begin{array}{|c|c|c|c|}\hline 1&1&1&2\\\hline 2&3\\\cline{1-2}\end{array} + \begin{array}{|c|c|c|c|}\hline 1&1&1&3\\\hline 2&2\\\cline{1-2}\end{array}\right\} \\
&\quad - \sqrt{\tfrac{4}{5}}\times\sqrt{\tfrac{2}{3}}\left\{3\,\begin{array}{|c|c|c|c|}\hline 1&1&1&2\\\hline 2&3\\\cline{1-2}\end{array} - \begin{array}{|c|c|c|c|}\hline 1&1&1&3\\\hline 2&2\\\cline{1-2}\end{array}\right\} \\
&= \sqrt{\tfrac{10}{3}}\,\begin{array}{|c|c|c|c|}\hline 1&1&1&3\\\hline 2&2\\\cline{1-2}\end{array}\ .
\end{aligned}
$$

The basis state $|(1,1)_2\rangle$ is the highest weight state of an A_1-doublet.

$$
|(\overline{1},2)_2\rangle = \left|\begin{array}{ccccc}4&&2&&0\\&3&&2&\\&&2&&\end{array}\right\rangle^{4} = F_1\ |(1,1)_2\rangle = \sqrt{\tfrac{10}{3}}\,\begin{array}{|c|c|c|c|}\hline 1&1&2&3\\\hline 2&2\\\cline{1-2}\end{array}\ .
$$

Let

$$\begin{aligned}
&F_2\,|(1,1)_1\rangle = F_2\left|\begin{smallmatrix}4&&2&&0\\&4&&1&\\&&3&&\end{smallmatrix}\right\rangle^{3} = a_2\left|\begin{smallmatrix}4&&2&&0\\&4&&0&\\&&3&&\end{smallmatrix}\right\rangle^{4} + b_2\left|\begin{smallmatrix}4&&2&&0\\&3&&1&\\&&3&&\end{smallmatrix}\right\rangle^{4}\\
&\quad = a_2\,|(2,\overline{1})_1\rangle + b_2\,|(2,\overline{1})_2\rangle\,,\\
&F_2\,|(1,1)_2\rangle = F_2\left|\begin{smallmatrix}4&&2&&0\\&3&&2&\\&&3&&\end{smallmatrix}\right\rangle^{3} = c_2\left|\begin{smallmatrix}4&&2&&0\\&3&&1&\\&&3&&\end{smallmatrix}\right\rangle^{4} = c_2\,|(2,\overline{1})_2\rangle\,,\\
&E_1F_2\,|(1,1)_1\rangle = 2a_2\,|(4,\overline{2})\rangle\\
&\quad = F_2E_1\,|(1,1)_1\rangle = \sqrt{3}F_2\,|(3,0)\rangle = \sqrt{3}\times\sqrt{2}\,|(4,\overline{2})\rangle\,,\\
&E_2F_2\,|(1,1)_1\rangle = (a_2^2+b_2^2)\,|(1,1)_1\rangle \;+\; b_2c_2\,|(1,1)_2\rangle\\
&\quad = (F_2E_2+H_2)\,|(1,1)_1\rangle = (a_1^2+1)\,|(1,1)_1\rangle \;+\; a_1b_1\,|(1,1)_2\rangle\,.
\end{aligned}$$

Choosing the phase of $|(2,\overline{1})_2\rangle$ such that b_2 is real and positive, we obtain $a_2=\sqrt{3/2}$, $b_2=\sqrt{4/3+1-3/2}=\sqrt{5/6}$, $c_2=\sqrt{4/3}\times\sqrt{5/3}/\sqrt{5/6}=\sqrt{8/3}$, and

$$\begin{aligned}
&|(2,\overline{1})_2\rangle = \left|\begin{smallmatrix}4&&2&&0\\&3&&1&\\&&3&&\end{smallmatrix}\right\rangle^{4} = \sqrt{\tfrac{3}{8}}\;F_2|(1,1)_2\rangle\\
&\quad = \sqrt{\tfrac{3}{8}}\times\sqrt{\tfrac{10}{3}}\times 2\,\begin{array}{|c|c|c|c|}\hline 1&1&1&3\\\hline 2&3\\\cline{1-2}\end{array} = \sqrt{5}\,\begin{array}{|c|c|c|c|}\hline 1&1&1&3\\\hline 2&3\\\cline{1-2}\end{array}\,,\\
&F_2|(1,1)_1\rangle = \sqrt{\tfrac{2}{3}}\left\{3\,\begin{array}{|c|c|c|c|}\hline 1&1&1&2\\\hline 3&3\\\cline{1-2}\end{array} + \begin{array}{|c|c|c|c|}\hline 1&1&1&3\\\hline 2&3\\\cline{1-2}\end{array}\right\}\\
&\quad = \sqrt{\tfrac{3}{2}}\;|(2,\overline{1})_1\rangle \;+\; \sqrt{\tfrac{5}{6}}\;|(2,\overline{1})_2\rangle\\
&\quad = \sqrt{\tfrac{3}{2}}\left\{2\,\begin{array}{|c|c|c|c|}\hline 1&1&1&2\\\hline 3&3\\\cline{1-2}\end{array} - \begin{array}{|c|c|c|c|}\hline 1&1&1&3\\\hline 2&3\\\cline{1-2}\end{array}\right\}\\
&\qquad + \sqrt{\tfrac{5}{6}}\times\sqrt{5}\,\begin{array}{|c|c|c|c|}\hline 1&1&1&3\\\hline 2&3\\\cline{1-2}\end{array}\,.
\end{aligned}$$

By the action of $[E_2,\;F_2]=H_2$ on $|(2,\overline{1})_2\rangle$, we obtain

$$|(3,\overline{3})\rangle = \left|\begin{smallmatrix}4&&2&&0\\&3&&0&\\&&3&&\end{smallmatrix}\right\rangle^{5} = \sqrt{2}\;|(2,\overline{1})_1\rangle = \sqrt{\tfrac{2}{5}}\;|(2,\overline{1})_2\rangle = \sqrt{2}\,\begin{array}{|c|c|c|c|}\hline 1&1&1&3\\\hline 3&3\\\cline{1-2}\end{array}\,.$$

The basis states $|(2,\overline{1})_2\rangle$ and $|(3,\overline{3})\rangle$ are the highest weight states of an A_1-triplet and an A_1-quadruplet, respectively.

$$\begin{aligned}
&|(0,0)_2\rangle = \left|\begin{smallmatrix}4&&2&&0\\&3&&1&\\&&2&&\end{smallmatrix}\right\rangle^{5} = \sqrt{\tfrac{1}{2}}F_1\;|(2,\overline{1})_2\rangle\\
&\quad = \sqrt{10}\,\begin{array}{|c|c|c|c|}\hline 1&1&2&3\\\hline 2&3\\\cline{1-2}\end{array} - \sqrt{\tfrac{5}{2}}\,\begin{array}{|c|c|c|c|}\hline 1&1&3&3\\\hline 2&2\\\cline{1-2}\end{array}\,,\\
&|(\overline{2},1)_2\rangle = \left|\begin{smallmatrix}4&&2&&0\\&3&&1&\\&&1&&\end{smallmatrix}\right\rangle^{6} = \sqrt{\tfrac{1}{2}}F_1\;|(0,0)_2\rangle = \sqrt{5}\,\begin{array}{|c|c|c|c|}\hline 1&2&2&3\\\hline 2&3\\\cline{1-2}\end{array}\,;\\
&|(1,\overline{2})_1\rangle = \left|\begin{smallmatrix}5&&1&&0\\&3&&0&\\&&2&&\end{smallmatrix}\right\rangle^{6} = \sqrt{\tfrac{1}{3}}F_1\;|(3,\overline{3})\rangle\\
&\quad = \sqrt{\tfrac{2}{3}}\left\{3\,\begin{array}{|c|c|c|c|}\hline 1&1&2&3\\\hline 3&3\\\cline{1-2}\end{array} - 2\,\begin{array}{|c|c|c|c|}\hline 1&1&3&3\\\hline 2&3\\\cline{1-2}\end{array}\right\}\,,
\end{aligned}$$

$$\begin{aligned}
\vert(\overline{1},\overline{1})_1\rangle &= \left\vert\begin{smallmatrix} 5 && 1 && 0 \\ & 3 && 0 & \\ && 1 && \end{smallmatrix}\right\rangle^{7} = \tfrac{1}{2}F_1\ \vert(1,\overline{2})_1\rangle \\
&= \sqrt{\tfrac{2}{3}}\left\{3\ \begin{array}{|c|c|c|c|}\hline 1&2&2&3\\ \hline 3&3\\ \cline{1-2}\end{array} - \begin{array}{|c|c|c|c|}\hline 1&2&3&3\\ \hline 2&3\\ \cline{1-2}\end{array}\right\}, \\
\vert(\overline{3},0)\rangle &= \left\vert\begin{smallmatrix} 5 && 1 && 0 \\ & 3 && 0 & \\ && 0 && \end{smallmatrix}\right\rangle^{9} = \sqrt{\tfrac{1}{3}}F_1\ \vert(\overline{1},\overline{1})_1\rangle \\
&= \sqrt{2}\ \begin{array}{|c|c|c|c|}\hline 2&2&2&3\\ \hline 3&3\\ \cline{1-2}\end{array}\ .
\end{aligned}$$

Third, we calculate the $\mathcal{A}_2$-quintuplet from its highest weight state $\vert(\overline{2},4)\rangle$ by Eqs. (8.25-28). Let

$$\begin{aligned}
&F_2\ \vert(\overline{2},4)\rangle = F_2 \left\vert\begin{smallmatrix} 4 && 2 && 0 \\ & 4 && 2 & \\ && 2 && \end{smallmatrix}\right\rangle^{3} = a_3 \left\vert\begin{smallmatrix} 4 && 2 && 0 \\ & 4 && 1 & \\ && 2 && \end{smallmatrix}\right\rangle^{4} + b_3 \left\vert\begin{smallmatrix} 4 && 2 && 0 \\ & 3 && 2 & \\ && 2 && \end{smallmatrix}\right\rangle^{4} \\
&\quad = a_3\ \vert(\overline{1},2)_1\rangle + b_3\ \vert(\overline{1},2)_2\rangle\ , \\
&E_1F_2\ \vert(\overline{2},4)\rangle = 2a_3\ \vert(1,1)_1\rangle\ +\ b_3\ \vert(1,1)_2\rangle \\
&\quad = F_2E_1\ \vert(\overline{2},4)\rangle = \sqrt{2}F_2\ \vert(0,3)\rangle = \sqrt{\tfrac{8}{3}}\ \vert(1,1)_1\rangle\ +\ \sqrt{\tfrac{10}{3}}\ \vert(1,1)_2\rangle\ .
\end{aligned}$$

Thus, $a_3 = \sqrt{2/3}$, $b_3 = \sqrt{10/3}$, and

$$\begin{aligned}
&F_2\ \vert(\overline{2},4)\rangle = F_2\ \begin{array}{|c|c|c|c|}\hline 1&1&2&2\\ \hline 2&2\\ \cline{1-2}\end{array} = 2\ \begin{array}{|c|c|c|c|}\hline 1&1&2&2\\ \hline 2&3\\ \cline{1-2}\end{array} + 2\ \begin{array}{|c|c|c|c|}\hline 1&1&2&3\\ \hline 2&2\\ \cline{1-2}\end{array} \\
&\quad = \sqrt{\tfrac{2}{3}}\ \vert(\overline{1},2)_1\rangle\ +\ \sqrt{\tfrac{10}{3}}\ \vert(\overline{1},2)_2\rangle \\
&\quad = \sqrt{\tfrac{2}{3}} \times \sqrt{\tfrac{2}{3}}\left\{3\ \begin{array}{|c|c|c|c|}\hline 1&1&2&2\\ \hline 2&3\\ \cline{1-2}\end{array} - 2\ \begin{array}{|c|c|c|c|}\hline 1&1&2&3\\ \hline 2&2\\ \cline{1-2}\end{array}\right\} \\
&\qquad + \sqrt{\tfrac{10}{3}} \times \sqrt{\tfrac{10}{3}}\ \begin{array}{|c|c|c|c|}\hline 1&1&2&3\\ \hline 2&2\\ \cline{1-2}\end{array}\ .
\end{aligned}$$

Let

$$\begin{aligned}
&F_2\ \vert(\overline{1},2)_1\rangle = F_2 \left\vert\begin{smallmatrix} 4 && 2 && 0 \\ & 4 && 1 & \\ && 2 && \end{smallmatrix}\right\rangle^{4} = a_4 \left\vert\begin{smallmatrix} 4 && 2 && 0 \\ & 4 && 0 & \\ && 2 && \end{smallmatrix}\right\rangle^{5} + b_4 \left\vert\begin{smallmatrix} 4 && 2 && 0 \\ & 3 && 1 & \\ && 2 && \end{smallmatrix}\right\rangle^{5} \\
&\quad = a_4\ \vert(0,0)_1\rangle + b_4\ \vert(0,0)_2\rangle\ , \\
&F_2\ \vert(\overline{1},2)_2\rangle = F_2 \left\vert\begin{smallmatrix} 4 && 2 && 0 \\ & 3 && 2 & \\ && 2 && \end{smallmatrix}\right\rangle^{4} = c_4 \left\vert\begin{smallmatrix} 4 && 2 && 0 \\ & 3 && 1 & \\ && 2 && \end{smallmatrix}\right\rangle^{5} + d_4 \left\vert\begin{smallmatrix} 4 && 2 && 0 \\ & 2 && 2 & \\ && 2 && \end{smallmatrix}\right\rangle^{5} \\
&\quad = c_4\ \vert(0,0)_2\rangle + d_4\ \vert(0,0)_3\rangle\ , \\
&E_1F_2\ \vert(\overline{1},2)_1\rangle = \sqrt{6}a_4\ \vert(2,\overline{1})_1\rangle\ +\ \sqrt{2}b_4\ \vert(2,\overline{1})_2\rangle \\
&\quad = F_2E_1\ \vert(\overline{1},2)_1\rangle = 2F_2\ \vert(1,1)_1\rangle = \sqrt{6}\ \vert(2,\overline{1})_1\rangle\ +\ \sqrt{\tfrac{10}{3}}\ \vert(2,\overline{1})_2\rangle\ , \\
&E_1F_2\ \vert(\overline{1},2)_2\rangle = \sqrt{2}c_4\ \vert(2,\overline{1})_2\rangle \\
&\quad = F_2E_1\ \vert(\overline{1},2)_2\rangle = F_2\ \vert(1,1)_2\rangle = \sqrt{\tfrac{8}{3}}\ \vert(2,\overline{1})_2\rangle\ ,
\end{aligned}$$

$$\begin{aligned}
&E_2F_2\ |(\overline{1},2)_2\rangle = c_4b_4\ |(\overline{1},2)_1\rangle + (c_4^2+d_4^2)\ |(\overline{1},2)_2\rangle \\
&\quad = (F_2E_2+H_2)\ |(\overline{1},2)_2\rangle = b_3a_3\ |(\overline{1},2)_1\rangle + (b_3^2+2)\ |(\overline{1},2)_2\rangle\ .
\end{aligned}$$

Choosing the phase of $|(0,0)_3\rangle$ such that d_4 is real and positive, we have $a_4 = 1$, $b_4 = \sqrt{5/3}$, $c_4 = \sqrt{4/3}$, $d_4 = \sqrt{10/3+2-4/3} = 2$, and

$$\begin{aligned}
F_2\ |(\overline{1},2)_1\rangle &= F_2\ \sqrt{\tfrac{2}{3}}\left\{3\ \begin{array}{|c|c|c|c|}\hline 1&1&2&2\\\hline 2&3\\\cline{1-2}\end{array} - 2\ \begin{array}{|c|c|c|c|}\hline 1&1&2&3\\\hline 2&2\\\cline{1-2}\end{array}\right\} \\
&= \sqrt{\tfrac{2}{3}}\left\{3\ \begin{array}{|c|c|c|c|}\hline 1&1&2&2\\\hline 3&3\\\cline{1-2}\end{array} + 2\ \begin{array}{|c|c|c|c|}\hline 1&1&2&3\\\hline 2&3\\\cline{1-2}\end{array} - 2\ \begin{array}{|c|c|c|c|}\hline 1&1&3&3\\\hline 2&2\\\cline{1-2}\end{array}\right\} \\
&= |(0,0)_1\rangle\ +\ \sqrt{\tfrac{5}{3}}\ |(0,0)_2\rangle \\
&= \sqrt{\tfrac{1}{6}}\left\{6\ \begin{array}{|c|c|c|c|}\hline 1&1&2&2\\\hline 3&3\\\cline{1-2}\end{array} - 6\ \begin{array}{|c|c|c|c|}\hline 1&1&2&3\\\hline 2&3\\\cline{1-2}\end{array} + \begin{array}{|c|c|c|c|}\hline 1&1&3&3\\\hline 2&2\\\cline{1-2}\end{array}\right\} \\
&\quad + \sqrt{\tfrac{5}{3}} \times \sqrt{\tfrac{5}{2}}\left\{2\ \begin{array}{|c|c|c|c|}\hline 1&1&2&3\\\hline 2&3\\\cline{1-2}\end{array} - \begin{array}{|c|c|c|c|}\hline 1&1&3&3\\\hline 2&2\\\cline{1-2}\end{array}\right\}, \\
|(0,0)_3\rangle &= \left|\begin{smallmatrix}4&&2&&0\\&2&&2&\\&&2&&\end{smallmatrix}\right\rangle^5 = \tfrac{1}{2}\left\{F_2\ |(\overline{1},2)_2\rangle - \sqrt{\tfrac{4}{3}}\ |(0,0)_2\rangle\right\} \\
&= \tfrac{1}{2} \times \sqrt{\tfrac{10}{3}}\left\{2\ \begin{array}{|c|c|c|c|}\hline 1&1&2&3\\\hline 2&3\\\cline{1-2}\end{array} + \begin{array}{|c|c|c|c|}\hline 1&1&3&3\\\hline 2&2\\\cline{1-2}\end{array}\right\} \\
&\quad - \sqrt{\tfrac{1}{3}} \times \sqrt{\tfrac{5}{2}}\left\{2\ \begin{array}{|c|c|c|c|}\hline 1&1&2&3\\\hline 2&3\\\cline{1-2}\end{array} - \begin{array}{|c|c|c|c|}\hline 1&1&3&3\\\hline 2&2\\\cline{1-2}\end{array}\right\} \\
&= \sqrt{\tfrac{10}{3}}\ \begin{array}{|c|c|c|c|}\hline 1&1&3&3\\\hline 2&2\\\cline{1-2}\end{array}\ .
\end{aligned}$$

Let

$$\begin{aligned}
&F_2\ |(0,0)_1\rangle = F_2 \left|\begin{smallmatrix}4&&2&&0\\&4&&0&\\&&2&&\end{smallmatrix}\right\rangle^5 = a_5 \left|\begin{smallmatrix}4&&2&&0\\&3&&0&\\&&2&&\end{smallmatrix}\right\rangle^6 = a_5\ |(1,\overline{2})_1\rangle\ , \\
&F_2\ |(0,0)_2\rangle = F_2 \left|\begin{smallmatrix}4&&2&&0\\&3&&1&\\&&2&&\end{smallmatrix}\right\rangle^5 = b_5 \left|\begin{smallmatrix}4&&2&&0\\&3&&0&\\&&2&&\end{smallmatrix}\right\rangle^6 + c_5 \left|\begin{smallmatrix}4&&2&&0\\&2&&1&\\&&2&&\end{smallmatrix}\right\rangle^6 \\
&\quad = b_5\ |(1,\overline{2})_1\rangle +\ c_5\ |(1,\overline{2})_2\rangle\ , \\
&F_2\ |(0,0)_3\rangle = F_2 \left|\begin{smallmatrix}4&&2&&0\\&2&&2&\\&&2&&\end{smallmatrix}\right\rangle^5 = d_5 \left|\begin{smallmatrix}4&&2&&0\\&2&&1&\\&&2&&\end{smallmatrix}\right\rangle^6 = d_5\ |(1,\overline{2})_2\rangle\ , \\
&E_1F_2\ |(0,0)_1\rangle = \sqrt{3}a_5\ |(3,\overline{3})\rangle \\
&\quad = F_2E_1\ |(0,0)_1\rangle = \sqrt{6}F_2\ |(2,\overline{1})_1\rangle = \sqrt{6} \times \sqrt{\tfrac{1}{2}}\ |(3,\overline{3})\rangle\ , \\
&E_1F_2\ |(0,0)_2\rangle = \sqrt{3}b_5\ |(3,\overline{3})\rangle \\
&\quad = F_2E_1\ |(0,0)_2\rangle = \sqrt{2}F_2\ |(2,\overline{1})_2\rangle = \sqrt{2} \times \sqrt{\tfrac{5}{2}}\ |(3,\overline{3})\rangle\ , \\
&E_2F_2\ |(0,0)_2\rangle = b_5a_5\ |(0,0)_1\rangle + (b_5^2+c_5^2)\ |(0,0)_2\rangle + c_5d_5\ |(0,0)_3\rangle \\
&\quad = (F_2E_2+H_2)\ |(0,0)_2\rangle \\
&\quad = b_4a_4\ |(0,0)_1\rangle + (b_4^2+c_4^2)\ |(0,0)_2\rangle + c_4d_4\ |(0,0)_3\rangle\ .
\end{aligned}$$

Choosing the phase of $|(1,\overline{2})_2\rangle$ such that c_5 is real and positive, we have $a_5 = 1$, $b_5 = \sqrt{5/3}$, $c_5 = \sqrt{5/3 + 4/3 - 5/3} = \sqrt{4/3}$, $d_5 = 2$, and

$$|(1,\overline{2})_2\rangle = \left|\begin{smallmatrix}4&&2&&0\\&2&&1&\\&&2&&\end{smallmatrix}\right\rangle^6 = \tfrac{1}{2}F_2\,|(0,0)_3\rangle = \sqrt{\tfrac{10}{3}}\,\young(1133,23)\,.$$

By the action of $[E_2,\ F_2] = H_2$ on $|(1,\overline{2})_2\rangle$, we obtain

$$\begin{aligned}|(2,\overline{4})\rangle &= \left|\begin{smallmatrix}4&&2&&0\\&2&&0&\\&&2&&\end{smallmatrix}\right\rangle^7 = \sqrt{\tfrac{3}{2}}F_2\,|(1,\overline{2})_1\rangle = \sqrt{\tfrac{3}{10}}F_2\,|(1,\overline{2})_2\rangle \\ &= \young(1133,33)\,.\end{aligned}$$

$|(1,\overline{2})_2\rangle$ and $|(2,\overline{4})\rangle$ are the highest weight states of an A$_1$-doublet and A$_1$-triplet, respectively.

$$\begin{aligned}|(\overline{1},\overline{1})_2\rangle &= \left|\begin{smallmatrix}4&&2&&0\\&2&&1&\\&&1&&\end{smallmatrix}\right\rangle^7 = F_1\,|(1,\overline{2})_2\rangle = \sqrt{\tfrac{10}{3}}\,\young(1233,23)\,,\\ |(0,\overline{3})\rangle &= \left|\begin{smallmatrix}4&&2&&0\\&2&&0&\\&&1&&\end{smallmatrix}\right\rangle^8 = \sqrt{\tfrac{1}{2}}F_1\,|(2,\overline{4})\rangle = \sqrt{2}\,\young(1233,33)\,,\\ |(\overline{2},\overline{2})\rangle &= \left|\begin{smallmatrix}4&&2&&0\\&2&&0&\\&&0&&\end{smallmatrix}\right\rangle^9 = \sqrt{\tfrac{1}{2}}F_1\,|(0,\overline{3})\rangle = \young(2233,33)\,.\end{aligned}$$

The remaining calculations are similar. We only list the results.

$$\begin{aligned}F_2\,|(\overline{3},3)\rangle =\ & \sqrt{2}\,\young(1222,33) + 2\sqrt{2}\,\young(1223,23)\\ =\ & \sqrt{\tfrac{1}{2}}\,|(\overline{2},1)_1\rangle\ +\ \sqrt{\tfrac{5}{2}}\,|(\overline{2},1)_2\rangle\\ =\ & \sqrt{\tfrac{1}{2}}\times\left\{2\,\young(1222,33) - \young(1223,23)\right\}\\ & + \sqrt{\tfrac{5}{2}}\times\sqrt{5}\,\young(1223,23)\,,\\ F_2\,|(\overline{2},1)_1\rangle =\ & 3\,\young(1223,33) - \young(1233,23) = \sqrt{\tfrac{3}{2}}\,|(\overline{1},\overline{1})_1\rangle\\ =\ & \sqrt{\tfrac{3}{2}}\times\sqrt{\tfrac{2}{3}}\left\{3\,\young(1223,33) - \young(1233,23)\right\},\\ F_2\,|(\overline{2},1)_2\rangle =\ & \sqrt{5}\left\{\young(1223,33) + \young(1233,23)\right\}\\ =\ & \sqrt{\tfrac{5}{6}}\,|(\overline{1},\overline{1})_1\rangle + \sqrt{\tfrac{8}{3}}\,|(\overline{1},\overline{1})_2\rangle\\ =\ & \sqrt{\tfrac{5}{6}}\times\sqrt{\tfrac{2}{3}}\left\{3\,\young(1223,33) - \young(1233,23)\right\}\\ =\ & \sqrt{\tfrac{8}{3}}\times\sqrt{\tfrac{10}{3}}\,\young(1233,23)\,,\\ \sqrt{\tfrac{3}{4}}F_2\,|(\overline{1},\overline{1})_1\rangle =\ & \sqrt{\tfrac{3}{5}}F_2\,|(\overline{1},\overline{1})_2\rangle = |(0,\overline{3})\rangle = \sqrt{2}\,\young(1233,33)\,;\\ |(\overline{3},0)\rangle =\ & \sqrt{\tfrac{1}{2}}F_2\,|(\overline{4},2)\rangle = \sqrt{\tfrac{1}{2}}F_2\,\young(2222,33) = \sqrt{2}\,\young(2223,33)\,,\end{aligned}$$

$$|(\overline{2},\overline{2})\rangle = \sqrt{\tfrac{1}{2}}F_2\ |(\overline{3},0)\rangle = \begin{array}{|c|c|c|c|}\hline 2&2&2&3\\\hline 3&3\\\cline{1-2}\end{array}\ .$$

Ex. 4. The representation $[2,1,1]$ of A_3.

The block weight diagram of the adjoint representation $[2,1,1]$ of A_3 is given in Fig. 8.9. The dimension of this representation is 15. The highest weight state in $[2,1,1]$ of A_3 is also the highest weight states of an A_2-triplet with $\boldsymbol{M}^{(2)} = (1,0)$ and an $\mathcal{A}_3$-doublet:

$$|(1,0,1)\rangle = \left|\begin{smallmatrix}2&&1&&1&&0\\&2&&1&&1&\\&&2&&1&&\\&&&2&&&\end{smallmatrix}\right\rangle^{1} = \begin{array}{|c|c|}\hline 1&1\\\hline 2\\\cline{1-1}3\\\cline{1-1}\end{array}\ ,$$

$$|(\overline{1},1,1)\rangle = F_1\ |(1,0,1)\rangle = \left|\begin{smallmatrix}2&&1&&1&&0\\&2&&1&&1&\\&&2&&1&&\\&&&1&&&\end{smallmatrix}\right\rangle^{2} = \begin{array}{|c|c|}\hline 1&2\\\hline 2\\\cline{1-1}3\\\cline{1-1}\end{array}\ ,$$

$$|(0,\overline{1},2)\rangle = F_2\ |(\overline{1},1,1)\rangle = \left|\begin{smallmatrix}2&&1&&1&&0\\&2&&1&&1&\\&&1&&1&&\\&&&1&&&\end{smallmatrix}\right\rangle^{3} = \begin{array}{|c|c|}\hline 1&3\\\hline 2\\\cline{1-1}3\\\cline{1-1}\end{array}\ ;$$

$$|(1,1,\overline{1})\rangle = F_3\ |(1,0,1)\rangle = \left|\begin{smallmatrix}2&&1&&1&&0\\&2&&1&&0&\\&&2&&1&&\\&&&2&&&\end{smallmatrix}\right\rangle^{2} = \begin{array}{|c|c|}\hline 1&1\\\hline 2\\\cline{1-1}4\\\cline{1-1}\end{array}\ .$$

The basis states $|(1,1,\overline{1})\rangle$ is the highest weight state of an A_2-octet with $\boldsymbol{M}^{(2)} = (1,1)$, from which we constructed the A_2-octet (see Fig. 8.5):

$$|(\overline{1},2,\overline{1})\rangle = F_1\ |(1,1,\overline{1})\rangle = \left|\begin{smallmatrix}2&&1&&1&&0\\&2&&1&&0&\\&&2&&1&&\\&&&1&&&\end{smallmatrix}\right\rangle^{3} = \begin{array}{|c|c|}\hline 1&2\\\hline 2\\\cline{1-1}4\\\cline{1-1}\end{array}\ ,$$

$$|(2,\overline{1},0)\rangle = F_2\ |(1,1,\overline{1})\rangle = \left|\begin{smallmatrix}2&&1&&1&&0\\&2&&1&&0&\\&&2&&0&&\\&&&2&&&\end{smallmatrix}\right\rangle^{3} = \begin{array}{|c|c|}\hline 1&1\\\hline 3\\\cline{1-1}4\\\cline{1-1}\end{array}\ ,$$

$$\begin{aligned}
|(0,0,0)_1\rangle &= \sqrt{\tfrac{1}{2}}F_1\ |(2,\overline{1},0)\rangle = \left|\begin{smallmatrix}2&&1&&1&&0\\&2&&1&&0&\\&&2&&0&&\\&&&1&&&\end{smallmatrix}\right\rangle^{4} \\
&= \sqrt{2}\begin{array}{|c|c|}\hline 1&2\\\hline 3\\\cline{1-1}4\\\cline{1-1}\end{array} - \sqrt{\tfrac{1}{2}}\begin{array}{|c|c|}\hline 1&3\\\hline 2\\\cline{1-1}4\\\cline{1-1}\end{array} + \sqrt{\tfrac{1}{2}}\begin{array}{|c|c|}\hline 1&4\\\hline 2\\\cline{1-1}3\\\cline{1-1}\end{array}\ ,
\end{aligned}$$

$$\begin{aligned}
|(0,0,0)_2\rangle &= \sqrt{\tfrac{2}{3}}\left\{F_2\ |(\overline{1},2,\overline{1})\rangle - \sqrt{\tfrac{1}{2}}\ |(0,0,0)_1\rangle\right\} = \left|\begin{smallmatrix}2&&1&&1&&0\\&2&&1&&0&\\&&1&&1&&\\&&&1&&&\end{smallmatrix}\right\rangle^{4} \\
&= \sqrt{\tfrac{2}{3}}\left\{\begin{array}{|c|c|}\hline 1&2\\\hline 3\\\cline{1-1}4\\\cline{1-1}\end{array} + \begin{array}{|c|c|}\hline 1&3\\\hline 2\\\cline{1-1}4\\\cline{1-1}\end{array}\right\} - \sqrt{\tfrac{1}{6}}\left\{2\begin{array}{|c|c|}\hline 1&2\\\hline 3\\\cline{1-1}4\\\cline{1-1}\end{array} - \begin{array}{|c|c|}\hline 1&3\\\hline 2\\\cline{1-1}4\\\cline{1-1}\end{array} + \begin{array}{|c|c|}\hline 1&4\\\hline 2\\\cline{1-1}3\\\cline{1-1}\end{array}\right\} \\
&= \sqrt{\tfrac{3}{2}}\begin{array}{|c|c|}\hline 1&3\\\hline 2\\\cline{1-1}4\\\cline{1-1}\end{array} - \sqrt{\tfrac{1}{6}}\begin{array}{|c|c|}\hline 1&4\\\hline 2\\\cline{1-1}3\\\cline{1-1}\end{array}\ ,
\end{aligned}$$

$$|(\overline{2},1,0)\rangle = \sqrt{\tfrac{1}{2}}F_1\ |(0,0,0)_1\rangle = \left|\begin{smallmatrix}2&&1&&1&&0\\&2&&1&&0&\\&&2&&0&&\\&&&0&&&\end{smallmatrix}\right\rangle^{5} = \begin{array}{|c|c|}\hline 2&2\\\hline 3\\\cline{1-1}4\\\cline{1-1}\end{array}\ ,$$

$$|(1,\overline{2},1)\rangle = \sqrt{2}F_2\ |(0,0,0)_1\rangle = \sqrt{\tfrac{2}{3}}F_2\ |(0,0,0)_2\rangle = \left|\begin{array}{ccccccc} 2 && 1 && 1 && 0 \\ & 2 && 1 && 0 & \\ && 1 && 0 && \\ &&& 1 &&& \end{array}\right\rangle^{5} = \begin{array}{|c|c|}\hline 1 & 3 \\ \hline 3 \\ \cline{1-1} 4 \\ \cline{1-1}\end{array}\ ,$$

$$|(\overline{1},\overline{1},1)\rangle = F_1\ |(1,\overline{2},1)\rangle = F_2\ |(\overline{2},1,0)\rangle = \left|\begin{array}{ccccccc} 2 && 1 && 1 && 0 \\ & 2 && 1 && 0 & \\ && 1 && 0 && \\ &&& 0 &&& \end{array}\right\rangle^{6} = \begin{array}{|c|c|}\hline 2 & 3 \\ \hline 3 \\ \cline{1-1} 4 \\ \cline{1-1}\end{array}\ .$$

Due to the parallel principle, the basis state $|(\overline{1},1,1)\rangle$ in the A_2-triplet with $\boldsymbol{M}^{(2)} = (1,0)$ satisfies

$$F_3\ |(\overline{1},1,1)\rangle = |(\overline{1},2,\overline{1})\rangle = \left|\begin{array}{ccccccc} 2 && 1 && 1 && 0 \\ & 2 && 1 && 0 & \\ && 2 && 1 && \\ &&& 1 &&& \end{array}\right\rangle^{3} = \begin{array}{|c|c|}\hline 1 & 2 \\ \hline 2 \\ \cline{1-1} 4 \\ \cline{1-1}\end{array}\ .$$

The basis state $|(0,\overline{1},2)\rangle$ in the A_2-triplet with $\boldsymbol{M}^{(2)} = (1,0)$ is the highest weight state of an $\mathcal{A}_3$-triplet. Let

$$\begin{aligned} F_3\ |(0,\overline{1},2)\rangle &= F_3\left|\begin{array}{ccccccc} 2 && 1 && 1 && 0 \\ & 2 && 1 && 1 & \\ && 1 && 1 && \\ &&& 1 &&& \end{array}\right\rangle^{3} = a\left|\begin{array}{ccccccc} 2 && 1 && 1 && 0 \\ & 2 && 1 && 0 & \\ && 1 && 1 && \\ &&& 1 &&& \end{array}\right\rangle^{4} + b\left|\begin{array}{ccccccc} 2 && 1 && 1 && 0 \\ & 1 && 1 && 1 & \\ && 1 && 1 && \\ &&& 1 &&& \end{array}\right\rangle^{4} \\ &= a\ |(0,0,0)_2\rangle + b\ |(0,0,0)_3\rangle\ . \end{aligned}$$

Choosing the phase of $|(0,0,0)_3\rangle$ such that b is real and positive, we obtain in terms of Eqs. (8.25-28):

$$\begin{aligned} E_2F_3\ |(0,\overline{1},2)\rangle &= \sqrt{\tfrac{3}{2}}\ a\ |(\overline{1},2,\overline{1})\rangle \\ &= F_3E_2\ |(0,\overline{1},2)\rangle = F_3\ |(\overline{1},1,1)\rangle = |(\overline{1},2,\overline{1})\rangle, \\ E_3F_3\ |(0,\overline{1},2)\rangle &= (a^2+b^2)\ |(0,\overline{1},2)\rangle \\ &= (F_3E_3 + H_3)\ |(0,\overline{1},2)\rangle = 2\ |(0,\overline{1},2)\rangle . \end{aligned}$$

Thus, $a = \sqrt{2/3}$, $b = \sqrt{2-2/3} = \sqrt{4/3}$, and

$$\begin{aligned} |(0,0,0)_3\rangle &= \left|\begin{array}{ccccccc} 2 && 1 && 1 && 0 \\ & 1 && 1 && 1 & \\ && 1 && 1 && \\ &&& 1 &&& \end{array}\right\rangle^{4} = \sqrt{\tfrac{3}{4}}\left\{F_3\ |(0,\overline{1},2)\rangle - \sqrt{\tfrac{2}{3}}\ |(0,0,0)_2\rangle\right\} \\ &= \sqrt{\tfrac{3}{4}}\left\{\begin{array}{|c|c|}\hline 1 & 3 \\ \hline 2 \\ \cline{1-1} 4 \\ \cline{1-1}\end{array} + \begin{array}{|c|c|}\hline 1 & 4 \\ \hline 2 \\ \cline{1-1} 3 \\ \cline{1-1}\end{array}\right\} - \sqrt{\tfrac{1}{2}}\left\{\sqrt{\tfrac{3}{2}}\begin{array}{|c|c|}\hline 1 & 3 \\ \hline 2 \\ \cline{1-1} 4 \\ \cline{1-1}\end{array} - \sqrt{\tfrac{1}{6}}\begin{array}{|c|c|}\hline 1 & 4 \\ \hline 2 \\ \cline{1-1} 3 \\ \cline{1-1}\end{array}\right\} \\ &= \sqrt{\tfrac{4}{3}}\begin{array}{|c|c|}\hline 1 & 4 \\ \hline 2 \\ \cline{1-1} 3 \\ \cline{1-1}\end{array}\ . \end{aligned}$$

By the action of F_3 on $|(0,0,0)_3\rangle$, we obtain

$$\begin{aligned} |(0,1,\overline{2})\rangle &= \left|\begin{array}{ccccccc} 2 && 1 && 1 && 0 \\ & 1 && 1 && 0 & \\ && 1 && 1 && \\ &&& 1 &&& \end{array}\right\rangle^{5} = \sqrt{\tfrac{3}{2}}F_3\ |(0,0,0)_2\rangle = \sqrt{\tfrac{3}{4}}F_3\ |(0,0,0)_3\rangle \\ &= \begin{array}{|c|c|}\hline 1 & 4 \\ \hline 2 \\ \cline{1-1} 4 \\ \cline{1-1}\end{array}\ . \end{aligned}$$

The basis state $|(0,1,\overline{2})\rangle$ is the highest weight state of an A_2-triplet, from which we construct the A_2-triplet.

$$
\begin{aligned}
|(1,\overline{1},\overline{1})\rangle &= \left|\begin{array}{ccccccc} 2 & & 1 & & 1 & & 0 \\ & 1 & & 1 & & 0 & \\ & & 1 & & 0 & & \\ & & & 1 & & & \end{array}\right\rangle^{6} = F_2\,|(0,1,\overline{2})\rangle = \begin{array}{|c|c|}\hline 1 & 4 \\ \hline 3 \\ \cline{1-1} 4 \\ \cline{1-1}\end{array}\,, \\
|(\overline{1},0,\overline{1})\rangle &= \left|\begin{array}{ccccccc} 2 & & 1 & & 1 & & 0 \\ & 1 & & 1 & & 0 & \\ & & 1 & & 0 & & \\ & & & 0 & & & \end{array}\right\rangle^{7} = F_1\,|(1,\overline{1},\overline{1})\rangle = \begin{array}{|c|c|}\hline 2 & 4 \\ \hline 3 \\ \cline{1-1} 4 \\ \cline{1-1}\end{array}\,.
\end{aligned}
$$

The basis states $|(1,\overline{2},1)\rangle$ and $|(\overline{1},\overline{1},1)\rangle$ in the A_2-octet are two highest weight states of the $\mathcal{A}_3$-doublets (the parallel principle).

$$
F_3\,|(1,\overline{2},1)\rangle = |(1,\overline{1},\overline{1})\rangle\,, \qquad F_3\,|(\overline{1},\overline{1},1)\rangle = |(\overline{1},0,\overline{1})\rangle\,.
$$

9.2.3 *Orthonormal Basis Tensors in* S_n

The basis tensors with same tensor Young tableau in all $\mathcal{T}_\mu^{[\lambda]}$ of SU(N) with given $[\lambda]$ constitute a complete set of the basis tensors in the representation $D^{[\lambda]}(\mathrm{S}_n)$, which is not unitary generally. Let $\phi_\mu^{[\lambda]} = \mathcal{Y}_\mu^{[\lambda]}\boldsymbol{\theta}_{b_1\ldots b_n}$ denote the highest weight state in $\mathcal{T}_\mu^{[\lambda]}$. Then, due to Eq. (9.26), the highest weight state in $\mathcal{T}_\nu^{[\lambda]}$ is

$$
\phi_\nu^{[\lambda]} = R_{\nu\mu}\phi_\mu^{[\lambda]} = \mathcal{Y}_\nu^{[\lambda]} R_{\nu\mu}\boldsymbol{\theta}_{b_1\ldots b_n} \in \mathcal{T}_\nu^{[\lambda]}. \tag{9.54}
$$

Letting $X_{[\lambda]}$ be the similarity transformation which changes $D^{[\lambda]}(\mathrm{S}_n)$ to the real orthogonl representation $\overline{D}^{[\lambda]}(\mathrm{S}_n)$, $\overline{D}^{[\lambda]}(R) = \left(X_{[\lambda]}\right)^{-1} D^{[\lambda]}(R) X_{[\lambda]}$, we obtain the orthonormal basis tensors $\psi_\nu^{[\lambda]}$ for S_n:

$$
\psi_\rho^{[\lambda]} = \sum_{\nu=1}^{d} \phi_\nu^{[\lambda]}\left(X_{[\lambda]}\right)_{\nu\rho}, \quad R\psi_\rho^{[\lambda]} = \sum_{\tau=1}^{d} \psi_\tau^{[\lambda]}\overline{D}_{\tau\rho}^{[\lambda]}(R), \quad R \in \mathrm{S}_n. \tag{9.55}
$$

For the adjoint representation $[2,1]$ of SU(3), $X_{[2,1]} = \begin{pmatrix} 1 & \sqrt{1/3} \\ 0 & \sqrt{4/3}\end{pmatrix}$, (see Eq. (4.97)),

$$
\begin{aligned}
\psi_1^{[2,1]} &= \phi_1 = \mathcal{Y}_1^{[2,1]}\boldsymbol{\theta}_{112} = 2\boldsymbol{\theta}_{112} - \boldsymbol{\theta}_{211} - \boldsymbol{\theta}_{121}, \\
\psi_2^{[2,1]} &= \sqrt{1/3}\,\{E + 2(2\ 3)\}\,\phi_1 = \sqrt{3}\,\{\boldsymbol{\theta}_{121} - \boldsymbol{\theta}_{211}\}\,, \\
\overline{D}^{[2,1]}(1\ 2) &= \begin{pmatrix} 1 & 0 \\ 0 & -1 \end{pmatrix}, \quad \overline{D}^{[2,1]}(2\ 3) = \frac{1}{2}\begin{pmatrix} -1 & \sqrt{3} \\ \sqrt{3} & 1 \end{pmatrix}.
\end{aligned} \tag{9.56}
$$

9.3 Direct Product of Tensor Representations

9.3.1 *Direct Product of Tensors*

Let $\boldsymbol{T}^{(1)}_{a_1\ldots a_n} \in \mathcal{T}^{(1)}$ and $\boldsymbol{T}^{(2)}_{b_1\ldots b_m} \in \mathcal{T}^{(2)}$ be two tensors of rank n and of rank m of SU(N), respectively. The ordered pair of two tensors, $\boldsymbol{T}^{(1)}_{a_1\ldots a_n}\boldsymbol{T}^{(2)}_{b_1\ldots b_m}$, constitutes a tensor of rank $(n+m)$, called the direct product of two tensors. Its product space is denoted by $\mathcal{T} = \mathcal{T}^{(1)}\mathcal{T}^{(2)}$. After the projections of two Young operators acting on two tensors, respectively, $\mathcal{T}$ maps into its subspace $\mathcal{T}^{[\lambda][\tau]}_{\mu\nu}$:

$$\mathcal{Y}^{[\lambda]}_{\mu}\mathcal{T}^{(1)}\mathcal{Y}^{[\tau]}_{\nu}\mathcal{T}^{(2)} = \mathcal{Y}^{[\lambda]}_{\mu}\mathcal{Y}^{[\tau]}_{\nu}\mathcal{T}^{(1)}\mathcal{T}^{(2)} = \mathcal{Y}^{[\lambda]}_{\mu}\mathcal{Y}^{[\tau]}_{\nu}\mathcal{T} = \mathcal{T}^{[\lambda][\tau]}_{\mu\nu},$$

where the Young patterns $[\lambda]$ and $[\tau]$ contain n and m boxes, respectively, and their row numbers are not larger than N. **The tensor subspace $\mathcal{T}^{[\lambda][\tau]}_{\mu\nu} \subset \mathcal{T}$ is left invariant in the SU(N) transformation and corresponds to the representation** $[\lambda] \times [\tau]$ with the dimension $d_{[\lambda]}(\mathrm{SU}(N))d_{[\tau]}(\mathrm{SU}(N))$, where we denote the representation directly by its Young pattern for convenience. Generally, the direct product representation is reducible. It can be decomposed as follows. Applying a Young operator $\mathcal{Y}^{[\omega]}_{\rho}$ to $\mathcal{T}$, where $[\omega]$ contains $(n+m)$ boxes and its row number is not larger than N, we have

$$\mathcal{Y}^{[\omega]}_{\rho}\mathcal{T}^{(1)}\mathcal{T}^{(2)} = \mathcal{Y}^{[\omega]}_{\rho}\mathcal{T} = \mathcal{T}^{[\omega]}_{\rho} \subset \mathcal{T}.$$

$\mathcal{T}^{[\omega]}_{\rho}$ is left invariant in the SU(N) transformation and corresponds to the representation $[\omega]$. If

$$\mathcal{Y}^{[\lambda]}_{\mu}\mathcal{Y}^{[\tau]}_{\nu}t_{\alpha}\mathcal{Y}^{[\omega]}_{\rho} \neq 0,$$

where t_α is a vector in the group algebra of the permutation group S_{n+m}, there is a subspace of $\mathcal{T}^{[\lambda][\tau]}_{\mu\nu}$, corresponding to the representation $[\omega]$,

$$\mathcal{Y}^{[\lambda]}_{\mu}\mathcal{Y}^{[\tau]}_{\nu}\left\{t_{\alpha}\mathcal{Y}^{[\omega]}_{\rho}\mathcal{T}^{(1)}\mathcal{T}^{(2)}\right\} \subset \mathcal{Y}^{[\lambda]}_{\mu}\mathcal{Y}^{[\tau]}_{\nu}\mathcal{T} = \mathcal{T}^{[\lambda][\tau]}_{\mu\nu}. \tag{9.57}$$

Remind that the leftmost operator $\mathcal{Y}^{[\lambda]}_{\mu}\mathcal{Y}^{[\tau]}_{\nu}$ determines that **the tensor subspace after the projection belongs to** $\mathcal{T}^{[\lambda][\tau]}_{\mu\nu}$ because the tensor in the curly bracket belongs to $\mathcal{T}$, and the rightmost operator $\mathcal{Y}^{[\omega]}_{\rho}$ determines **the property of the tensor subspace in the SU(N) transformations** owing to the Weyl reciprocity.

In comparison with Eq. (4.103), one can borrow the technique of the Littlewood–Richardson rule to calculate the decomposition of the direct product representation of SU(N)

$$[\lambda] \times [\tau] \simeq \bigoplus_{[\omega]} a^{\omega}_{\lambda\tau}[\omega]. \tag{9.58}$$

However, equation (9.58) is different from Eq. (4.105) because the representations in Eq. (9.58) are those of the SU(N) group, not those of the permutation group. If the row number of a Young pattern $[\omega]$ calculated by the Littlewood–Richardson rule is larger than N, $[\omega]$ should be removed from the Clebsch–Gordan series for SU(N), given in Eq. (9.58). The dimension formula of the decomposition (9.58) becomes

$$d_{[\lambda]}(\mathrm{SU}(N))d_{[\mu]}(\mathrm{SU}(N)) = \sum_{[\omega]} a^{\omega}_{\lambda\mu} d_{[\omega]}(\mathrm{SU}(N)). \tag{9.59}$$

For example, the direct product of two adjoint representations $[2,1] \times [2,1]$ of SU(3) and their dimension formula are

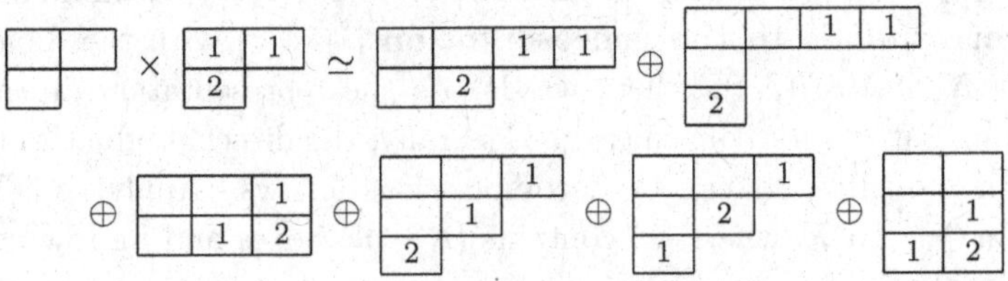

$$8 \times 8 = 27 + 10 + 10^* + 2 \times 8 + 1, \tag{9.60}$$

where 10^* denotes the representation $[3,3]$ which is conjugate with $[3,0]$. Please compare the decomposition with Example 1 in subsection 4.5.2.

An important example for the decomposition is the direct product of an arbitrary representation $[\lambda]$ and the totally antisymmetric tensor representation $[1^N]$ of rank N in SU(N). In the CG series calculated by the Littlewood–Richardson rule, there is only one representation with the row number not larger than N, which is the direct adhibition of two Young patterns

$$[\lambda] \times [1^N] \simeq [\lambda'], \qquad \lambda'_j = \lambda_j + 1. \tag{9.61}$$

Since $[1^N]$ is the identical representation (see Eq. 9.35)), $[\lambda']$ is equivalent to $[\lambda]$ (see Corollary 9.3.2). Therefore, the irreducible representations of SU(N) can be characterized by a Young pattern with the row number less than N, namely by $(N-1)$ parameters where $(N-1)$ is the rank to SU(N).

In order to calculate the Clebsch–Gordan coefficients, we need to write the expansions of the standard tensor Young tableaux with the highest weights of the representations appearing in the Clebsch–Gordan series. In writing an expansion for the highest weight state $||\boldsymbol{M}, \boldsymbol{M}\rangle$, we first find out all possible products of two standard tensor Young tableaux in two tensor subspaces where the sum of two weights is $\boldsymbol{M}$. **The coefficient of each term can be determined by the condition (7.111)**: The expansion is annihilated by each raising operators E_μ. In the following the expansions for the highest weight state $||\boldsymbol{M}, \boldsymbol{M}\rangle$ appearing in the decomposition of $[2,1] \times [2,1]$ of SU(3) are listed as examples, both by the standard tensor Young tableaux and by the basis states.

$$\begin{array}{|c|c|c|c|}\hline 1&1&1&1\\\hline 2&2\\\cline{1-2}\end{array} \sim \begin{array}{|c|c|}\hline 1&1\\\hline 2\\\cline{1-1}\end{array} \times \begin{array}{|c|c|}\hline 1&1\\\hline 2\\\cline{1-1}\end{array},$$

$$||(2,2),(2,2)\rangle = |(1,1)\rangle|(1,1)\rangle,$$

$$\begin{array}{|c|c|c|c|}\hline 1&1&1&1\\\hline 2\\\cline{1-1} 3\\\cline{1-1}\end{array} \sim \begin{array}{|c|c|}\hline 1&1\\\hline 2\\\cline{1-1}\end{array} \times \begin{array}{|c|c|}\hline 1&1\\\hline 3\\\cline{1-1}\end{array} - \begin{array}{|c|c|}\hline 1&1\\\hline 3\\\cline{1-1}\end{array} \times \begin{array}{|c|c|}\hline 1&1\\\hline 2\\\cline{1-1}\end{array},$$

$$||(3,0),(3,0)\rangle = \sqrt{1/2}\,\{|(1,1)\rangle|(2,\overline{1})\rangle - |(2,\overline{1})\rangle|(1,1)\rangle\},$$

$$\begin{array}{|c|c|c|}\hline 1&1&1\\\hline 2&2&2\\\hline\end{array} \sim \begin{array}{|c|c|}\hline 1&1\\\hline 2\\\cline{1-1}\end{array} \times \begin{array}{|c|c|}\hline 1&2\\\hline 2\\\cline{1-1}\end{array} - \begin{array}{|c|c|}\hline 1&2\\\hline 2\\\cline{1-1}\end{array} \times \begin{array}{|c|c|}\hline 1&1\\\hline 2\\\cline{1-1}\end{array},$$

$$||(0,3),(0,3)\rangle = \sqrt{1/2}\,\{|(1,1)\rangle|(\overline{1},2)\rangle - |(\overline{1},2)\rangle|(1,1)\rangle\},$$

$$\begin{array}{|c|c|c|}\hline 1&1&1\\\hline 2&2\\\cline{1-2} 3\\\cline{1-1}\end{array}_{\,S} \sim \begin{array}{|c|c|}\hline 1&1\\\hline 2\\\cline{1-1}\end{array} \times \begin{array}{|c|c|}\hline 1&2\\\hline 3\\\cline{1-1}\end{array} + \begin{array}{|c|c|}\hline 1&2\\\hline 3\\\cline{1-1}\end{array} \times \begin{array}{|c|c|}\hline 1&1\\\hline 2\\\cline{1-1}\end{array}$$
$$- \begin{array}{|c|c|}\hline 1&2\\\hline 2\\\cline{1-1}\end{array} \times \begin{array}{|c|c|}\hline 1&1\\\hline 3\\\cline{1-1}\end{array} - \begin{array}{|c|c|}\hline 1&1\\\hline 3\\\cline{1-1}\end{array} \times \begin{array}{|c|c|}\hline 1&2\\\hline 2\\\cline{1-1}\end{array},$$

$$\begin{aligned} ||(1,1),(1,1)\rangle_S = \sqrt{1/20}\,\{ & |(1,1)\rangle \left[\sqrt{3}\,|(0,0)_1\rangle + |(0,0)_2\rangle\right] \\ & + \left[\sqrt{3}\,|(0,0)_1\rangle + |(0,0)_2\rangle\right] |(1,1)\rangle \,\} \\ & - \sqrt{3/10}\,[\, |(\overline{1},2)\rangle\, |(2,\overline{1})\rangle + |(2,\overline{1})\rangle|(\overline{1},2)\rangle \,], \end{aligned}$$

$$\begin{array}{|c|c|c|}\hline 1&1&1\\\hline 2&2\\\cline{1-2} 3\\\cline{1-1}\end{array}_{\,A} \sim \begin{array}{|c|c|}\hline 1&1\\\hline 2\\\cline{1-1}\end{array} \times \begin{array}{|c|c|}\hline 1&2\\\hline 3\\\cline{1-1}\end{array} - \begin{array}{|c|c|}\hline 1&2\\\hline 3\\\cline{1-1}\end{array} \times \begin{array}{|c|c|}\hline 1&1\\\hline 2\\\cline{1-1}\end{array}$$
$$- 2 \begin{array}{|c|c|}\hline 1&1\\\hline 2\\\cline{1-1}\end{array} \times \begin{array}{|c|c|}\hline 1&3\\\hline 2\\\cline{1-1}\end{array} + 2 \begin{array}{|c|c|}\hline 1&3\\\hline 2\\\cline{1-1}\end{array} \times \begin{array}{|c|c|}\hline 1&1\\\hline 2\\\cline{1-1}\end{array}$$
$$- \begin{array}{|c|c|}\hline 1&2\\\hline 2\\\cline{1-1}\end{array} \times \begin{array}{|c|c|}\hline 1&1\\\hline 3\\\cline{1-1}\end{array} + \begin{array}{|c|c|}\hline 1&1\\\hline 3\\\cline{1-1}\end{array} \times \begin{array}{|c|c|}\hline 1&2\\\hline 2\\\cline{1-1}\end{array},$$

$$\begin{aligned} ||(1,1),(1,1)\rangle_A = \sqrt{1/12}\, \{ & |(1,1)\rangle\, |(0,0)_1\rangle - |(0,0)_1\rangle|(1,1)\rangle \\ & - \sqrt{3}[\, |(1,1)\rangle\, |(0,0)_2\rangle - |(0,0)_2\rangle|1,1\rangle \,] \end{aligned}$$

$$- \sqrt{2}\left[\, |(\overline{1},2)\rangle|(2,\overline{1})\rangle - |(2,\overline{1})\rangle|(\overline{1},2)\rangle \,\right]\Big\},$$

$$\begin{array}{|c|c|}\hline 1&1\\ \hline 2&2\\ \hline 3&3\\ \hline\end{array} \sim \begin{array}{|c|c|}\hline 1&1\\ \hline 2\\ \cline{1-1}\end{array} \times \begin{array}{|c|c|}\hline 2&3\\ \hline 3\\ \cline{1-1}\end{array} + \begin{array}{|c|c|}\hline 2&3\\ \hline 3\\ \cline{1-1}\end{array} \times \begin{array}{|c|c|}\hline 1&1\\ \hline 2\\ \cline{1-1}\end{array} - \begin{array}{|c|c|}\hline 1&1\\ \hline 3\\ \cline{1-1}\end{array} \times \begin{array}{|c|c|}\hline 2&2\\ \hline 3\\ \cline{1-1}\end{array}$$

$$- \begin{array}{|c|c|}\hline 2&2\\ \hline 3\\ \cline{1-1}\end{array} \times \begin{array}{|c|c|}\hline 1&1\\ \hline 3\\ \cline{1-1}\end{array} - \begin{array}{|c|c|}\hline 1&2\\ \hline 2\\ \cline{1-1}\end{array} \times \begin{array}{|c|c|}\hline 1&3\\ \hline 3\\ \cline{1-1}\end{array} - \begin{array}{|c|c|}\hline 1&3\\ \hline 3\\ \cline{1-1}\end{array} \times \begin{array}{|c|c|}\hline 1&2\\ \hline 2\\ \cline{1-1}\end{array}$$

$$+2 \begin{array}{|c|c|}\hline 1&2\\ \hline 3\\ \cline{1-1}\end{array} \times \begin{array}{|c|c|}\hline 1&2\\ \hline 3\\ \cline{1-1}\end{array} + 2 \begin{array}{|c|c|}\hline 1&3\\ \hline 2\\ \cline{1-1}\end{array} \times \begin{array}{|c|c|}\hline 1&3\\ \hline 2\\ \cline{1-1}\end{array}$$

$$- \begin{array}{|c|c|}\hline 1&2\\ \hline 3\\ \cline{1-1}\end{array} \times \begin{array}{|c|c|}\hline 1&3\\ \hline 2\\ \cline{1-1}\end{array} - \begin{array}{|c|c|}\hline 1&3\\ \hline 2\\ \cline{1-1}\end{array} \times \begin{array}{|c|c|}\hline 1&2\\ \hline 3\\ \cline{1-1}\end{array},$$

$$\begin{aligned} ||(0,0),(0,0)\rangle = \sqrt{1/8}\,\{ & |(1,1)\rangle|(\overline{1},\overline{1})\rangle - |(2,\overline{1})\rangle|(\overline{2},1)\rangle - |(\overline{1},2)\rangle|(1,\overline{2})\rangle \\ & + |(0,0)_1\rangle|(0,0)_1\rangle + |(0,0)_2\rangle|(0,0)_2\rangle - |(1,\overline{2})\rangle|(\overline{1},2)\rangle \\ & - |(\overline{2},1)\rangle|(2,\overline{1})\rangle + |(\overline{1},\overline{1})\rangle|(1,1)\rangle \} . \end{aligned}$$

9.3.2 *Covariant and Contravariant Tensors*

The self-representation of SU(N) is not equivalent to its conjugate representation. The conjugate of a covariant tensor is called a contravariant tensor. A contravariant tensor $\boldsymbol{T}^{a_1\ldots a_m}$ of rank m of SU(N) and its basis tensors $\boldsymbol{\theta}^{d_1\ldots d_m}$ transform in SU(N) as

$$\begin{aligned} (O_u\boldsymbol{T})^{a_1\ldots a_m} &= \sum_{b_1\ldots b_m} \boldsymbol{T}^{b_1\ldots b_m} \left(u^{-1}\right)_{b_1a_1} \cdots \left(u^{-1}\right)_{b_ma_m} \\ &= \sum_{b_1\ldots b_m} u^*_{a_1b_1} \cdots u^*_{a_mb_m} \boldsymbol{T}^{b_1\ldots b_m}, \\ O_u\boldsymbol{\theta}^{d_1\ldots d_m} &= \sum_{b_1\ldots b_m} \left(u^{-1}\right)_{d_1b_1} \cdots \left(u^{-1}\right)_{d_mb_m} \boldsymbol{\theta}^{b_1\ldots b_m} \\ &= \sum_{b_1\ldots b_m} \boldsymbol{\theta}^{b_1\ldots b_m} u^*_{b_1d_1} \cdots u^*_{b_md_m}. \end{aligned} \tag{9.62}$$

The contravariant tensor space is denoted by $\mathcal{T}^*$. The Weyl reciprocity holds for the contravariant tensors so that $\mathcal{T}^*$ can also be decomposed by the projection of the Young operators. The tensor subspace $\mathcal{Y}_\nu^{[\tau]}\mathcal{T}^* = \mathcal{T}_\nu^{[\tau]^*}$ corresponds to the irreducible representation denoted by $[\tau]^*$ which is the conjugate one of the representation $[\tau]$. The basis tensors in $\mathcal{T}_\nu^{[\tau]^*}$ are the standard tensor Young tableaux $\mathcal{Y}_\nu^{[\tau]}\boldsymbol{\theta}^{d_1\ldots d_m}$.

A tensor $\boldsymbol{T}^{b_1\ldots b_m}_{a_1\ldots a_n}$ is called a mixed tensor of rank (n,m) if it contains n

subscripts and m superscripts and transforms in SU(N) as

$$\begin{aligned}(O_u \boldsymbol{T})^{b_1 \ldots b_m}_{a_1 \ldots a_n} &= \sum_{(a')(b')} u_{a_1 a'_1} \cdots u_{a_n a'_n} \boldsymbol{T}^{b'_1 \ldots b'_m}_{a'_1 \ldots a'_n} \left(u^{-1}\right)_{b'_1 b_1} \cdots \left(u^{-1}\right)_{b'_m b_m} \\ &= \sum_{(a')(b')} u_{a_1 a'_1} \cdots u_{a_n a'_n} u^*_{b_1 b'_1} \cdots u^*_{b_m b'_m} \boldsymbol{T}^{b'_1 \ldots b'_m}_{a'_1 \ldots a'_n} .\end{aligned} \tag{9.63}$$

The trace tensor of a mixed tensor is the sum of components of the mixed tensor where one covariant index and one contravariant index are taken to be equal and run over from 1 to N, and is called the **contraction of a tensor**. The trace tensor subspace is left invariant in SU(N),

$$\begin{aligned}O_u \left(\sum_{c=1}^{N} \boldsymbol{T}^{cb_1 \ldots}_{ca_1 \ldots} \right) &= \sum_{cdd'} u_{cd} u^*_{cd'} \sum_{(a')(b')} u_{a_1 a'_1} \cdots u^*_{b_1 b'_1} \cdots \boldsymbol{T}^{d' b'_1 \ldots}_{d a'_1 \ldots} \\ &= \sum_{(a')(b')} u_{a_1 a'_1} \cdots u^*_{b_1 b'_1} \cdots \left(\sum_d \boldsymbol{T}^{d b'_1 \ldots}_{d a'_1 \ldots} \right) .\end{aligned} \tag{9.64}$$

The trace tensor of a mixed tensor of rank (n, m) is a mixed tensor of rank $(n-1, m-1)$. There is a special mixed tensor $\boldsymbol{D}^b_a$ of rank $(1,1)$ of SU(N) whose component is the Kronecker δ function

$$\begin{aligned}&\boldsymbol{D}^b_a = \delta^b_a = \begin{cases} 1 & \text{when } \ a = b, \\ 0 & \text{when } \ a \neq b, \end{cases} \\ &(O_u \boldsymbol{D})^b_a = \sum_{a'b'} u_{aa'} u^*_{bb'} \boldsymbol{D}^{b'}_{a'} = \sum_{a'b'} u_{aa'} u^*_{bb'} \delta^{b'}_{a'} = \delta^b_a = \boldsymbol{D}^b_a .\end{aligned} \tag{9.65}$$

The one-dimensional tensor subspace composed of $\boldsymbol{D}$ is left invariant in SU(N) and corresponds to the identical representation of SU(N). A mixed tensor can be decomposed into the sum of a series of traceless tensors with different ranks in terms of the invariant tensor $\boldsymbol{D}^b_a$. For example,

$$\boldsymbol{T}^b_a = \left\{ \boldsymbol{T}^b_a - \boldsymbol{D}^b_a \left(\frac{1}{N} \sum_c \boldsymbol{T}^c_c \right) \right\} + \boldsymbol{D}^b_a \left(\frac{1}{N} \sum_c \boldsymbol{T}^c_c \right) . \tag{9.66}$$

The first term is a traceless mixed tensor of rank $(1,1)$ and the trace tensor in the bracket of the second term is a scalar. **The decomposition of a mixed tensor into the sum of traceless tensors is straightforward, but tedious.** One has to write all possible terms and calculate the coefficients by the traceless conditions. For example,

$$\Phi^d_{ab} = T^d_{ab} + D^d_a \sum_{p=1}^N \left\{ c_1 T^p_{bp} + c_2 T^p_{pb} \right\} + D^d_b \sum_{p=1}^N \left\{ c_3 T^p_{ap} + c_4 T^p_{pa} \right\}.$$

Solving the traceless conditions $\sum_a \Phi^a_{ab} = 0$ and $\sum_b \Phi^b_{ab} = 0$, one obtains $c_2 = c_3 = -N/(N^2-1)$ and $c_1 = c_4 = 1/(N^2-1)$.

9.3.3 *Traceless Mixed Tensors*

Let $\mathcal{L}$ denote a traceless mixed tensor space of rank (n, m). The subspace $\mathcal{Y}^{[\lambda]}_\mu \mathcal{Y}^{[\tau]}_\nu \mathcal{L}$, where the covariant and contravariant parts of $\mathcal{L}$ are respectively projected by two Young operators $\mathcal{Y}^{[\lambda]}_\mu$ and $\mathcal{Y}^{[\tau]}_\nu$, corresponds a representation of SU(N), denoted by a pair of Young patterns $[\lambda]\backslash[\tau]^*$. The numbers of boxes contained in the Young patterns $[\lambda]$ and $[\tau]$ are n and m, respectively, and the numbers of rows in $[\lambda]$ and $[\tau]$ are r and s.

First, we study the condition whether the traceless mixed tensor subspace denoted by $[\lambda]\backslash[\tau]^*$ is vanishing or not. **It is a nonvanishing space if the number of constraints from the traceless conditions is less than the number of independent tensors.** Discuss a pair of tensor Young tableaux in the subspace where there are ℓ pairs of repetitive digits and $(r+s-2\ell)$ different digits in the first columns of two Young tableaux. We fix the $(r+s-2\ell)$ different digits and the digits in the remaining columns in the pair of tensor Young tableaux. Under these conditions, those pairs of tensor Young tableaux with different repetitive digits (ℓ pairs) are called "independent tensors". The **number of independent tensors** is

$$\binom{N-(r+s-2\ell)}{\ell}.$$

The traceless condition is written as a sum of tensors to be zero where the $(\ell-1)$ pairs of digits are fixed and only one pair of digits runs over from 1 to N. The **number of the traceless conditions** is the number of the possible values of the $(\ell-1)$ pairs of digits, that is,

$$\binom{N-(r+s-2\ell)}{\ell-1}.$$

The condition that the number of traceless conditions is less than the number of independent tensors becomes $[N-(r+s-2\ell)]/2 \geqslant \ell$. Namely, the condition for non-vanishing traceless mixed tensor space is

$$r+s \leqslant N. \tag{9.67}$$

Second, discuss a totally antisymmetric contravariant tensor subspace $\mathcal{T}^{[1^m]^*}$ of rank m, whose basis tensors are standard tensor Young tableaux $\mathcal{Y}^{[1^m]}\boldsymbol{\theta}^{b_1\ldots b_m}$. Multiplying the basis tensors with a totally antisymmetric tensor of rank N, we obtain

$$\phi_{a_1\ldots a_{N-m}} = \frac{1}{m!}\sum_{b_1\ldots b_m} \epsilon_{a_1\ldots a_{N-m}b_1\ldots b_m}\mathcal{Y}^{[1^m]}\boldsymbol{\theta}^{b_1\ldots b_m}. \tag{9.68}$$

In fact, there is only one term in the sum at the right-hand side of Eq. (9.68). The number of the basis tensors of two sets are the same

$$\binom{N}{m} = \binom{N}{N-m}.$$

The correspondence of two sets of basis tensors is one-to-one. The difference of two sets of basis tensors is only in the arranged order. **The representations with respect to two sets of basis tensors are equivalent.** The basis tensors $\phi_{a_1\ldots a_{N-m}}$ transform in SU(N) as follows:

$$\begin{aligned}
O_u\phi_{a_1\ldots a_{N-m}} &= \frac{1}{m!}\sum_{b_1\ldots b_m} \epsilon_{a_1\ldots a_{N-m}b_1\ldots b_m}\left(\mathcal{Y}^{[1^m]}O_u\boldsymbol{\theta}^{b_1\ldots b_m}\right)\\
&= \frac{1}{m!}\sum_{b_1\ldots b_m}\sum_{d_1\ldots d_{N-m}}\sum_{c_1\ldots c_{N-m}} \epsilon_{d_1\ldots d_{N-m}b_1\ldots b_m}\left(u^*_{c_1d_1}u_{c_1a_1}\right)\cdots\\
&\quad\times\left(u^*_{c_{N-m}d_{N-m}}u_{c_{N-m}a_{N-m}}\right)\sum_{t_1\ldots t_m}\mathcal{Y}^{[1^m]}\boldsymbol{\theta}^{t_1\ldots t_m}u^*_{t_1b_1}\cdots u^*_{t_mb_m}\\
&= \frac{1}{m!}\sum_{c_1\ldots c_{N-m}}\sum_{t_1\ldots t_m}\mathcal{Y}^{[1^m]}\boldsymbol{\theta}^{t_1\ldots t_m}u_{c_1a_1}\cdots u_{c_{N-m}a_{N-m}}\\
&\quad\times\left\{\sum_{d_1\ldots d_{N-m}}\sum_{b_1\ldots b_m}u^*_{c_1d_1}\cdots u^*_{c_{N-m}d_{N-m}}u^*_{t_1b_1}\cdots u^*_{t_mb_m}\epsilon_{d_1\ldots d_{N-m}b_1\ldots b_m}\right\}\\
&= \frac{1}{m!}\sum_{c_1\ldots c_{N-m}}\sum_{t_1\ldots t_m}\epsilon_{c_1\ldots c_{N-m}t_1\ldots t_m}\mathcal{Y}^{[1^m]}\boldsymbol{\theta}^{t_1\ldots t_m}u_{c_1a_1}\cdots u_{c_{N-m}a_{N-m}}\\
&= \sum_{c_1\ldots c_{N-m}}\phi_{c_1\ldots c_{N-m}}u_{c_1a_1}\cdots u_{c_{N-m}a_{N-m}},
\end{aligned} \tag{9.69}$$

where the formula for $\epsilon_{a_1\ldots a_N}$ is used,

$$\begin{aligned}
\sum_{d_1\ldots d_N} u_{a_1d_1}\cdots u_{a_Nd_N}\epsilon_{d_1\ldots d_N} &= \epsilon_{a_1\ldots a_N}\sum_{d_1\ldots d_N}u_{1d_1}\cdots u_{Nd_N}\epsilon_{d_1\ldots d_N}\\
&= (\det u)\epsilon_{a_1\ldots a_N} = \epsilon_{a_1\ldots a_N}.
\end{aligned} \tag{9.70}$$

Equation (9.69) shows that $\phi_{a_1\ldots a_{N-m}}$ is proportional to the basis tensor $\mathcal{Y}^{[1^{N-m}]}\theta_{a_1\ldots a_{N-m}}$ in the tensor subspace $\mathcal{T}^{[1^{N-m}]}$. Namely, the representations of two tensor subspaces $\mathcal{T}^{[1^m]^*}$ and $\mathcal{T}^{[1^{N-m}]}$ are equivalent,

$$[1^m]^* \simeq [1^{N-m}]. \tag{9.71}$$

Third, generalize $\mathcal{T}^{[1^m]^*}$ to the traceless tensor subspace $\mathcal{T}^{[1^m]^*}_{[\lambda]}$, corresponding to a pair of Young patterns $[\lambda]\backslash[1^m]^*$ where the row number of $[\lambda]$ is not larger than $N-m$. Let $\Omega^{b_1\ldots b_m}_{c\ldots}$, which are traceless between any pair of a covariant index c and a contravariant index b_j, denote the basis tensors in $\mathcal{T}^{[1^m]^*}_{[\lambda]}$

$$\Omega^{b_1\ldots b_j\ldots b_k\ldots b_m}_{c\ldots} = -\Omega^{b_1\ldots b_k\ldots b_j\ldots b_m}_{c\ldots}, \qquad \sum_{c=1}^{N}\sum_{b_j=1}^{N} \delta^c_{b_j}\Omega^{b_1\ldots b_j\ldots b_m}_{c\ldots} = 0. \tag{9.72}$$

Multiplying the basis tensors with a totally antisymmetric tensor of rank N, we obtain

$$\phi_{a_1\ldots a_{N-m}c\ldots} = \frac{1}{m!}\sum_{b_1\ldots b_m} \epsilon_{a_1\ldots a_{N-m}b_1\ldots b_m}\Omega^{b_1\ldots b_m}_{c\ldots}. \tag{9.73}$$

$\phi_{a_1\ldots a_{N-m}c\ldots}$ belongs to a covariant tensor subspace corresponding to a direct product representation, $[1^{N-m}]\times[\lambda]$, which is calculated by the Littlewood–Richardson rule. From the traceless condition (9.72) we have

$$\begin{aligned}&\sum_{a_1\ldots a_{N-m}c} \epsilon^{a_1\ldots a_{N-m}cd_2\ldots d_m}\phi_{a_1\ldots a_{N-m}c\ldots}\\ &\quad = \frac{1}{m!}\sum_{a_1\ldots a_{N-m}cb_1\ldots b_m} \epsilon^{a_1\ldots a_{N-m}cd_2\ldots d_m}\epsilon_{a_1\ldots a_{N-m}b_1\ldots b_m}\Omega^{b_1\ldots b_m}_{c\ldots} = 0,\end{aligned}$$

because each term in the sum contains m factors of δ functions including a factor related with c, say $\delta^c_{b_j}$. Thus, in the decomposition of $[1^{N-m}]\times[\lambda]$, there is only one nonvanishing term whose Young pattern $[\lambda']$ is obtained by adhibiting $[1^{N-m}]$ and $[\lambda]$ directly,

$$[\lambda]\backslash[1^m]^* \simeq [\lambda'], \qquad \lambda'_k = \lambda_k + 1, \qquad 1 \leqslant k \leqslant N-m. \tag{9.74}$$

Similarly, we have $[1^m]\backslash[\lambda]^* \simeq [\lambda']^*$, or equivalently,

$$[\tau]^* \simeq [1^{N-s}]\backslash[\tau']^*, \qquad \tau'_k = \tau_k - 1, \qquad 1 \leqslant k \leqslant s, \tag{9.75}$$

where the row number of $[\tau]$ is s, and $[\tau']$ is obtained from $[\tau]$ by removing its first column. The replacement of basis tensors for the equivalent tensor subspaces in Eq. (9.75) is

$$\phi^{c\ldots}_{a_1\ldots a_{N-s}} = \frac{1}{s!}\sum_{b_1\ldots b_s} \epsilon_{a_1\ldots a_{N-s}b_1\ldots b_s}\mathcal{Y}^{[\tau]}\boldsymbol{\theta}^{b_1\ldots b_s c\ldots}, \tag{9.76}$$

where $\phi^{c\ldots}_{a_1\ldots a_{N-s}}$ is traceless and the first column in the tensor Young tableau $\mathcal{Y}^{[\tau]}\boldsymbol{\theta}^{b_1\ldots b_s c\ldots}$ is filled by digits $b_1\ldots b_s$.

At last, discuss the general case where $\boldsymbol{\Omega}^{b_1\ldots b_s c\ldots}_{d\ldots}$ is the traceless basis tensor in the representation $[\lambda]\backslash[\tau]^*$. The row number of $[\tau]$ is s and that of $[\lambda]$ is not larger than $(N-s)$. The digits b_j in the traceless basis tensor $\boldsymbol{\Omega}^{b_1\ldots b_s c\ldots}_{d\ldots}$ are filled in the first column of the tensor Young tableau for the contravariant part. The tensor is traceless between each pair of one covariant and one contravariant indices, say d and b_j or d and c. Let

$$\phi^{c\ldots}_{a_1\ldots a_{N-s}d\ldots} = \frac{1}{s!}\sum_{b_1\ldots b_s} \epsilon_{a_1\ldots a_{N-s}b_1\ldots b_s}\boldsymbol{\Omega}^{b_1\ldots b_s c\ldots}_{d\ldots}. \tag{9.77}$$

The covariant part of $\phi^{c\ldots}_{a_1\ldots a_{N-s}d\ldots}$ corresponds to the representation $[1^{N-s}]\times[\lambda]$. In its decomposition by the Littlewood–Richardson rule, there is only one Young pattern which is adhibited by $[1^{N-s}]$ and $[\lambda]$ directly. Each of the remaining Young patterns in the Clebsch–Gordan series contains an operation of antisymmetrizing one covariant index, say d, and all new covariant indices a_j, which annihilates the traceless tensor $\boldsymbol{\Omega}^{b_1\ldots b_s c\ldots}_{d\ldots}$ because the factor $\delta^d_{b_j}$ in the product:

$$\sum_{a_1\cdots a_{N-s}} \epsilon^{a_1\ldots a_{N-s}db'_2\ldots b'_s}\epsilon_{a_1\ldots a_{N-s}b_1\ldots b_s}$$

Thus, the following two representations are equivalent

$$[\lambda]\backslash[\tau]^* \simeq [\lambda']\backslash[\tau']^*, \qquad \begin{array}{ll} \tau'_j = \tau_j - 1, & 1 \leqslant j \leqslant s, \\ \lambda'_k = \lambda_k + 1, & 1 \leqslant k \leqslant N-s, \end{array} \tag{9.78}$$

where $[\lambda']$ **is the Young pattern obtained by adhibiting** $[1^{N-s}]$ **and** $[\lambda]$ **directly, and** $[\tau']$ **is obtained from** $[\tau]$ **by removing its first column.** Successively applying the replacement (9.78), one is able to transform any traceless mixed tensor subspace denoted by $[\mu]\backslash[\tau]^*$ into a covariant tensor subspace denoted by a Young pattern $[\lambda]$ so that $[\mu]\backslash[\tau]^*$ is irreducible. Furthermore, a contravariant tensor subspace denoted by a Young pattern $[\tau]^*$ is equivalent to a covariant tensor subspace denoted by a Young pattern $[\lambda]$, where

$$[\tau]^* \simeq [\lambda], \qquad \lambda_j = \tau_1 - \tau_{N-j+1}, \qquad 1 \leqslant j \leqslant N. \tag{9.79}$$

Adhibiting the Young pattern $[\tau]$ upside down to the Young pattern $[\lambda]$, one obtains an $N \times \tau_1$ rectangle.

9.3.4 *Adjoint Representation of* SU(N)

As shown in subsection 7.5.6, the highest weight of the adjoint representation of SU(N) is $\boldsymbol{M} = \boldsymbol{w}_1 + \boldsymbol{w}_{N-1}$, corresponding to the Young pattern $[2, 1^{N-2}] \simeq [1]\backslash[1]^*$. In this subsection we are going to discuss the adjoint representation of SU(N) in another way.

From the definition (7.8), the adjoint representation of SU(N) satisfies

$$uT_Au^{-1} = \sum_{B=1}^{N^2-1} T_B D^{\text{ad}}_{BA}(u), \tag{9.80}$$

where T_A is the generator in the self-representation of SU(N). T_A is an $N \times N$ traceless hermitian matrix. A traceless mixed tensor $\boldsymbol{T}_a^b$ of rank (1,1) has a similar transformation,

$$\left(O_u\boldsymbol{T}\right)_a^b = \sum_{a'b'} u_{aa'}\boldsymbol{T}_{a'}^{b'}\left(u^{-1}\right)_{b'b}. \tag{9.81}$$

$\boldsymbol{T}_a^b$ can be looked as the matrix element of an $N \times N$ traceless hermitian matrix at the ath row and the bth column so that it can be expanded with respect to the generators $(T_A)_{ab}$ where the tensor $\sqrt{2}\boldsymbol{F}_A$ are the coefficients,

$$\boldsymbol{T}_a^b = \sqrt{2}\sum_{A=1}^{N^2-1} (T_A)_{ab}\,\boldsymbol{F}_A, \qquad \boldsymbol{F}_A = \sqrt{2}\sum_{ab} (T_A)_{ba}\,\boldsymbol{T}_a^b. \tag{9.82}$$

From the viewpoint of replacement of tensors, $\boldsymbol{T}_a^b$ and $\boldsymbol{F}_A$ are two tensors in the traceless tensor subspace of rank $(1,1)$ so that they correspond to the equivalent representation $[1]\backslash[1]^*$. Calculate the transformation of $\boldsymbol{F}_A$ in SU(N),

$$\begin{aligned}
O_u\boldsymbol{T} = u\boldsymbol{T}u^{-1} &= \sqrt{2}\sum_{A=1}^{N^2-1} uT_Au^{-1}\boldsymbol{F}_A \\
&= \sqrt{2}\sum_{B=1}^{N^2-1} T_B\left(\sum_{A=1}^{N^2-1} D^{\text{ad}}_{BA}(u)\boldsymbol{F}_A\right), \\
O_u\boldsymbol{T} &= \sqrt{2}\sum_{B=1}^{N^2-1} T_B\left(O_u\boldsymbol{F}\right)_B.
\end{aligned}$$

Thus,

$$(O_u \boldsymbol{F})_B = \sum_{A=1}^{N^2-1} D^{\text{ad}}_{BA}(u) \boldsymbol{F}_A. \tag{9.83}$$

$\boldsymbol{F}_A$ corresponds to the adjoint representation. It shows that the adjoint representation of SU(N) is equivalent to the representation $[1]\backslash[1]^* \simeq [2, 1^{N-2}]$. Both $\boldsymbol{T}^b_a$ and $\boldsymbol{F}_A$ are the tensors transformed according to the adjoint representation of SU(N). Two forms of $\boldsymbol{T}^b_a$ and $\boldsymbol{F}_A$ are commonly used in particle physics.

9.4 SU(3) Symmetry and Wave Functions of Hadrons

As an example of physical applications of group theory, we are going to study the flavor SU(3) symmetry in particle physics. The rank of SU(3) is 2 so that the planar weight diagram is more convenient to demonstrate the basis states in an irreducible representation of SU(3) than the block weight diagram. We will derive the mass relations of hadrons and calculate the wave functions of hadrons in the multiplets of the flavor SU(3).

9.4.1 *Quantum Numbers of Quarks*

In the theory of modern particle physics, the "elementary" particles are divided into four classes. The particles participating in the strong interaction are called the hadrons. Those fermions with spin 1/2, which do not participate the strong interaction are called the leptons. Those bosons mediating the interactions are called the gauge particles. The particles in the fourth class, called the Higgs bosons, are introduced in the theory for providing the static masses of other particles. There are three generations of leptons as well as their anti-particles. They are the charged leptons e, μ, and τ and their neutrinos ν_e, ν_μ, and ν_τ. The gauge particles include the photon mediating the electro-magnetic interaction, the neutral boson Z^0 and charged bosons $W^{\pm}$ mediating the weak interaction, and the gluons mediating the strong interaction. The graviton mediating the gravitational force is being studied in theory and in experiment. Some other particles presented in the supersymmetric theory are still not observed in experiments.

We focus our attention on the hadrons. Among hadrons, the fermions are called the baryons and the bosons are called the mesons. All hadrons are constructed by the more elementary particles called the quarks and antiquarks. In the modern theory there are 18 quarks and 18 antiquarks. Usually, the quarks are described by a visual language, "color" and "flavor" which are not in the common sense on color and flavor. There are three color quantum numbers, say red, yellow, and blue. The quarks with three colors are the bases of the color $\mathrm{SU}(3)_c$ group in the theory of quantum chromodynamics. They participate in the $\mathrm{SU}(3)_c$ gauge interaction mediated by the gluons. The theoretical and experimental researchers expect the so-called color confinement that a state with color cannot be observed in the recent experimental energy level. Thus, the quarks in the low energy have to appear in the colorless states, or called the color singlet of $\mathrm{SU}(3)_c$. Namely, three quarks construct the basis states of totally antisymmetric tensors of rank three, and a quark and an antiquark construct the trace state of the mixed tensor of rank $(1,1)$,

$$\sum_{abc} \epsilon^{abc}\, q_a q_b q_c, \qquad \text{and} \qquad \sum_{a=1}^{3} \overline{q}^a q_a. \tag{9.84}$$

Both states correspond to the identical representation of $\mathrm{SU}(3)_c$. The state with three quarks is a baryon state with baryon number 1. The pair of quark and antiquark is a meson state without the baryon number. Therefore, a quark brings the baryon number $1/3$ and an antiquark brings the baryon number $-1/3$. Some composite colorless states composed of the states (9.84) are being studied recently. The quark and antiquark states appearing in the following are understood to be the colorless states.

There are six flavor quantum numbers for quarks. They are divided into three generations in weak interaction. Each generation contains two quarks: the up quark u and the down quark d; the charm quark c and the strange quark s; and the top quark t and the bottom quark b. The first quark brings $2/3$ electric charge unit and the second $-1/3$ unit. In each generation, their left-hand states constitute a doublet with respect to the weak isospin, and the right-hand states are singlets. The u quark and the d quark are very light, and the s quark is a little bit heavier. They are called the light quarks. The c, b, and t quarks are heavier one by one, and are called the heavy quarks.

The up quark u and the down quark d constitute a doublet in the isospin $\mathrm{SU}(2)$ group. Although they have different electric charges, the isospin is conserved approximately in the strong interaction and plays an important role in particle physics. Generalizing the isospin symmetry, one assumes that three light quarks constitute a triplet of the flavor $\mathrm{SU}(3)$ group, which is a broken symmetry because the s quark is heavier than u and d quarks. However, the flavor SU(3) symmetry made some historical contributions in discovering new hadrons and predicting their properties. Even recently, the flavor SU(3) symmetry helps the research of hadron physics in some respects. In this section we pay attention to the application of the flavor SU(3) group to the hadron physics only from the viewpoint of group theory. The isospin SU(2) is a subgroup of the flavor SU(3) group. T_3 and the supercharge Y span the Cartan subalgebra of the flavor SU(3) group,

$$T_3 = \frac{1}{2}\begin{pmatrix} 1 & 0 & 0 \\ 0 & -1 & 0 \\ 0 & 0 & 0 \end{pmatrix}, \qquad Y = \frac{2}{\sqrt{3}}T_8 = \frac{1}{3}\begin{pmatrix} 1 & 0 & 0 \\ 0 & 1 & 0 \\ 0 & 0 & -2 \end{pmatrix}. \tag{9.85}$$

In addition to the color, the flavor quantum numbers (T, T_3, Y), and some spatial quantum numbers (spin and parity), there are a few inner quantum numbers for quarks such as the baryon number B, the electric charge Q, and the strange number S. They are related by

$$Y = B + S, \qquad Q = T_3 + Y/2. \tag{9.86}$$

The quantum numbers of quarks are listed in Table 9.1. The quantum numbers of antiquarks are changed in sign except for the isospin T.

Table 9.1 Quantum numbers of light quarks

Quark	B	T	T_3	S	Y	Q
u	1/3	1/2	1/2	0	1/3	2/3
d	1/3	1/2	− 1/2	0	1/3	− 1/3
s	1/3	0	0	− 1	− 2/3	− 1/3

9.4.2 *Planar Weight Diagrams of Mesons and Baryons*

The flavor SU(3) group contains two commonly used SU(2) ⊗ U(1) subgroups. One SU(2) is called T-spin where the generators in the Chevalley bases are $H_1 = 2T_3$, $E_1 = T_+ = T_1 + \mathrm{i}T_2$ and $F_1 = T_- = T_1 - \mathrm{i}T_2$. Y is the corresponding generator of U(1) which commutes with H_1, E_1 and F_1 and takes constant in a multiplet of T-spin. The other SU(2) is called U-spin where the generators are $H_2 = 3Y/2 - T_3$, $E_2 = U_+ = T_6 + \mathrm{i}T_7$, and $F_2 = U_- = T_6 - \mathrm{i}T_7$. Q is the corresponding generator of U(1) which commutes with H_2, E_2 and F_2 and takes constant in a multiplet of U-spin.

As the basis states of the flavor SU(3), three quarks satisfy

$$\begin{aligned}
&T_+\mathrm{d} = \mathrm{u}, && T_+\mathrm{u} = T_+\mathrm{s} = 0, && T_-\mathrm{u} = \mathrm{d}, && T_-\mathrm{d} = T_-\mathrm{s} = 0, \\
&U_+\mathrm{s} = \mathrm{d}, && U_+\mathrm{u} = U_+\mathrm{d} = 0, && U_-\mathrm{d} = \mathrm{s}, && U_-\mathrm{u} = U_-\mathrm{s} = 0; \\
&T_+\overline{\mathrm{u}} = -\overline{\mathrm{d}}, && T_+\overline{\mathrm{d}} = T_+\overline{\mathrm{s}} = 0, && T_-\overline{\mathrm{d}} = -\overline{\mathrm{u}}, && T_-\overline{\mathrm{u}} = T_-\overline{\mathrm{s}} = 0, \\
&U_+\overline{\mathrm{d}} = -\overline{\mathrm{s}}, && U_+\overline{\mathrm{u}} = U_+\overline{\mathrm{s}} = 0, && U_-\overline{\mathrm{s}} = -\overline{\mathrm{d}}, && U_-\overline{\mathrm{u}} = U_-\overline{\mathrm{d}} = 0.
\end{aligned} \tag{9.87}$$

In the planar weight diagram for the flavor SU(3), the abscissa axis and the ordinate axis are the eigenvalues of T_3 and T_8, respectively. The simple roots and the fundamental dominant weights are given in Eq. (8.40). $T_\pm$ moves the basis state along the abscissa axis such that **$T_\pm$ changes T_3 and Q both by ± 1**, but **leaves Y invariant**. $U_\pm$ moves the basis state along the direction with an angle $2\pi/3$ to the abscissa axis such that **$U_\pm$ changes T_3 by $\mp 1/2$, changes Y by ± 1**, but **leaves Q invariant**.

The hadrons are composed of three quarks or a pair of quark and antiquark, denoted by a suitable standard tensor Young tableaux. The filled digits 1, 2, and 3 in the covariant tensor Young tableaux are replaced with u, d, and s, respectively. $\overline{3}$, $-\overline{2}$, and $\overline{1}$ are replaced with $\overline{\mathrm{s}}$, $-\overline{\mathrm{d}}$, and $\overline{\mathrm{u}}$ in the contravariant tensor Young tableaux. In drawing the planar weight diagram of a given representation $[\lambda]$, one **first determines the position of the highest weight state**, where the box in the first row is filled with u and the box in the second row with d. Then, applying the lowering operators T_- and U_-, one calculates the positions of the other basis states. The basis states for a multiple weight have the same quantum numbers T, T_3, and Y, and are located in the same place of the planar weight diagram. **All basis states in an irreducible representation have the same spin and parity**. There are three types of the planar weight diagrams for the flavor SU(3).

(a) For a one-row Young pattern $[\lambda, 0]$, the planar weight diagram is a regular triangle upside down. The length of the edge of the triangle is λ. All weights are single. The highest weight state in $[\lambda, 0]$ is described by a standard tensor Young tableau where each box is filled by u so that

$$Y = \lambda/3, \qquad T_3 = \lambda/2. \tag{9.88}$$

The conjugate representation of $[\lambda, 0]$ is $[\lambda, \lambda] \simeq [\lambda, 0]^*$ whose planar weight diagram is a regular triangle.

From the highest weight state one constructs a T-multiplet by applications of the lowering operator T_-, where u **quark is replaced with** d

quark one by one. In the planar weight diagram, the states in the T-multiplet are located in a horizontal line with $Y = \lambda/3$. Recall that the one-row tensor Young tableau describes a totally symmetric tensor so that the quarks in the tableau can be interchanged symmetrically. From each state in the T-multiplet one constructs the U-multiplets by applications of the lowering operator U_-, where d **quark is replaced with s quark one by one.** In the planar weight diagram, the states in the U-multiplet are along a line with the angle $2\pi/3$ to the abscissa axis where Q is fixed. All weights in the representation are single. The planar weight diagrams of the representations $[1,0]$, $[1,1] \simeq [1,0]^*$, and $[3,0]$ are listed in Fig. 9.3.

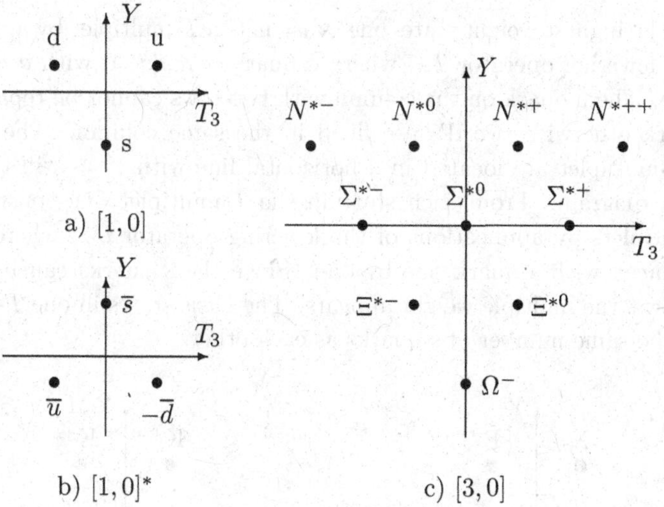

Fig. 9.3 The planar weight diagrams of $[\lambda, 0]$ of SU(3).

$[1,0]$ and $[1,0]^*$ describe the quarks and antiquarks, respectively. $[3,0]$ describe the baryon decuplet, observed in experiment with spin $3/2$ and the positive parity. The baryons in the decuplet are denoted by N^*, Σ^*, Ξ^*, and Ω,

$$\begin{aligned}
&N^{*++} = \begin{array}{|c|c|c|}\hline \text{u} & \text{u} & \text{u} \\ \hline\end{array}\,, && N^{*+} = \sqrt{3}\begin{array}{|c|c|c|}\hline \text{u} & \text{u} & \text{d} \\ \hline\end{array}\,, \\
&N^{*0} = \sqrt{3}\begin{array}{|c|c|c|}\hline \text{u} & \text{d} & \text{d} \\ \hline\end{array}\,, && N^{*-} = \begin{array}{|c|c|c|}\hline \text{d} & \text{d} & \text{d} \\ \hline\end{array}\,, \\
&\Sigma^{*+} = \sqrt{3}\begin{array}{|c|c|c|}\hline \text{u} & \text{u} & \text{s} \\ \hline\end{array}\,, && \Sigma^{*0} = \sqrt{6}\begin{array}{|c|c|c|}\hline \text{u} & \text{d} & \text{s} \\ \hline\end{array}\,, \\
&\Sigma^{*-} = \sqrt{3}\begin{array}{|c|c|c|}\hline \text{d} & \text{d} & \text{s} \\ \hline\end{array}\,, && \Xi^{*0} = \sqrt{3}\begin{array}{|c|c|c|}\hline \text{u} & \text{s} & \text{s} \\ \hline\end{array}\,, \\
&\Xi^{*-} = \sqrt{3}\begin{array}{|c|c|c|}\hline \text{d} & \text{s} & \text{s} \\ \hline\end{array}\,, && \Omega^{-} = \begin{array}{|c|c|c|}\hline \text{s} & \text{s} & \text{s} \\ \hline\end{array}\,.
\end{aligned} \tag{9.89}$$

(b) For a Young pattern $[2\lambda, \lambda] \simeq [\lambda, 0]\backslash[\lambda, 0]^*$, the planar weight diagram is a regular hexagon with the edge length λ. The weights on the edge are single. The multiplicities of the weights increase one by one as their positions go from the edge toward the origin, where the multiplicities of the weight is $\lambda+1$. The representation of $[2\lambda, \lambda]$ is self-conjugate and its planar weight diagram is symmetric in the inversion with respect to the origin.

The highest weight state in $[2\lambda, \lambda]$ is described by a standard tensor Young tableau where each box in the first row is filled by u and each box in the second row is filled by d so that

$$Y = \lambda, \qquad T_3 = \lambda/2. \tag{9.90}$$

From the highest weight state, one constructs a T-multiplet by applications of the lowering operator T_-, where u quark is replaced with d quark one by one. The u quark on the column with two rows cannot be replaced with d quark otherwise two d's are filled in the same column. The states in the T-multiplet are located in a horizontal line with $Y = \lambda$ in the planar weight diagram. From each state in the T-multiplet one constructs the U-multiplets by applications of the lowering operator U_-, where d quark is replaced with s quark one by one. Since the s quarks can be filled in two rows, the multiple weight appears. The basis states in one T-multiplet have the same number of s quarks as each other.

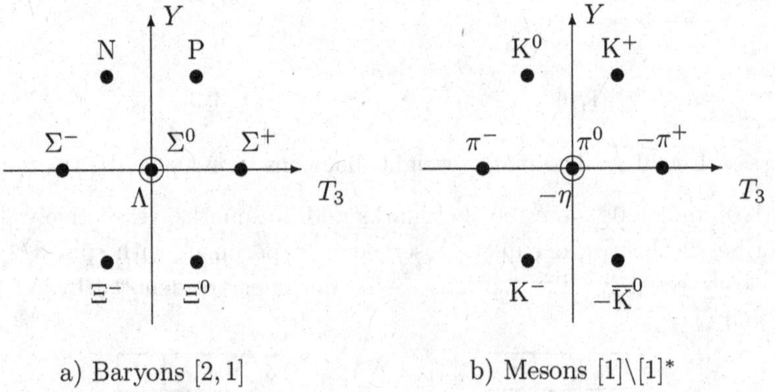

Fig. 9.4 The planar weight diagrams of adjoint representation of SU(3).

The planar weight diagrams of the representation $[2, 1]$ for the baryons and the representation $[1]\backslash[1]^*$ for the mesons are listed in Fig. 9.4. In experiments, the observed baryons in the octet $[2, 1]$, called P, N, Σ, Λ, and Ξ, have spin 1/2 and positive parity. The observed mesons in the octet $[1]\backslash[1]^*$ have negative parity. When the spin is 0, they are the scalar mesons, called K, π, η, and $\overline{\text{K}}$. When the spin is 1, they are the vector mesons, called K*, ρ, ϕ, and $\overline{K}^*$.

The tensor Young tableaux of the $(1/2)^+$ baryon octet and the 0^- meson octet are listed as follows. They can be calculated from the highest weight state by the lowering operators (see Fig. 9.2, and Eq. (9.87)).

$$
\begin{array}{ll}
\text{Baryon} & \text{Scalar meson} \\
\text{P} = \begin{array}{l}\text{uu}\\ \text{d}\end{array} & \text{K}^+ = \text{u}\overline{\text{s}} \\
\text{N} = \begin{array}{l}\text{ud}\\ \text{d}\end{array} & \text{K}^0 = \text{d}\overline{\text{s}} \\
\Sigma^+ = \begin{array}{l}\text{uu}\\ \text{s}\end{array} & -\pi^+ = -\,\text{u}\overline{\text{d}} \\
\Sigma^0 = \sqrt{1/2}\left\{2\begin{array}{l}\text{ud}\\ \text{s}\end{array} - \begin{array}{l}\text{us}\\ \text{d}\end{array}\right\} & \pi^0 = \sqrt{1/2}\left\{\text{u}\overline{\text{u}} - \text{d}\overline{\text{d}}\right\} \\
\Sigma^- = \begin{array}{l}\text{dd}\\ \text{s}\end{array} & \pi^- = \text{d}\overline{\text{u}} \\
\Lambda = \sqrt{3/2}\begin{array}{l}\text{us}\\ \text{d}\end{array} & -\eta = -\sqrt{1/6}\left\{\text{u}\overline{\text{u}} + \text{d}\overline{\text{d}} - 2\text{s}\overline{\text{s}}\right\} \\
\Xi^0 = \begin{array}{l}\text{us}\\ \text{s}\end{array} & -\overline{\text{K}^0} = -\,\text{s}\overline{\text{d}} \\
\Xi^- = \begin{array}{l}\text{ds}\\ \text{s}\end{array} & \text{K}^- = \text{s}\overline{\text{u}}
\end{array}
\tag{9.91}
$$

$$
\begin{aligned}
K^0 &= T_-\,K^+ = T_-\,\boxed{\text{u}}\backslash\boxed{\text{s}}^{\,*} = \boxed{\text{d}}\backslash\boxed{\text{s}}^{\,*}, \\
-\pi^+ &= U_-\,K^+ = U_-\,\boxed{\text{u}}\backslash\boxed{\text{s}}^{\,*} = -\,\boxed{\text{u}}\backslash\boxed{\text{d}}^{\,*}, \\
\pi^0 &= \sqrt{1/2}\,T_-\,(-\pi^+) = -\sqrt{1/2}\,T_-\,\boxed{\text{u}}\backslash\boxed{\text{d}}^{\,*} \\
&= \sqrt{1/2}\left\{\boxed{\text{u}}\backslash\boxed{\text{u}}^{\,*} - \boxed{\text{d}}\backslash\boxed{\text{d}}^{\,*}\right\}, \\
\pi^- &= \sqrt{1/2}\,T_-\,\pi^0 = (1/2)\,T_-\left\{\boxed{\text{u}}\backslash\boxed{\text{u}}^{\,*} - \boxed{\text{d}}\backslash\boxed{\text{d}}^{\,*}\right\} \\
&= \boxed{\text{d}}\backslash\boxed{\text{u}}^{\,*}, \\
-\eta &= \sqrt{2/3}\left\{U_-\,\text{K}^0 - \sqrt{1/2}\,\pi^0\right\} \\
&= \sqrt{1/6}\left\{2\,U_-\,\boxed{\text{d}}\backslash\boxed{\text{s}}^{\,*} - \boxed{\text{u}}\backslash\boxed{\text{u}}^{\,*} + \boxed{\text{d}}\backslash\boxed{\text{d}}^{\,*}\right\} \\
&= \sqrt{1/6}\left\{2\,\boxed{\text{s}}\backslash\boxed{\text{s}}^{\,*} - \boxed{\text{u}}\backslash\boxed{\text{u}}^{\,*} - \boxed{\text{d}}\backslash\boxed{\text{d}}^{\,*}\right\}, \\
-\overline{\text{K}^0} &= \sqrt{2}\,U_-\,\pi^0 = U_-\left\{\boxed{\text{u}}\backslash\boxed{\text{u}}^{\,*} - \boxed{\text{d}}\backslash\boxed{\text{d}}^{\,*}\right\} = -\,\boxed{\text{s}}\backslash\boxed{\text{d}}^{\,*}, \\
\overline{\text{K}^-} &= T_-\left(-\overline{\text{K}^0}\right) = -\,T_-\,\boxed{\text{s}}\backslash\boxed{\text{d}}^{\,*} = \boxed{\text{s}}\backslash\boxed{\text{u}}^{\,*}.
\end{aligned}
$$

Remind that the tensor Young tableaux calculated by Eq. (9.73) are traceless:

$$\begin{aligned}
3\,\begin{array}{|c|c|}\hline u & d \\ \hline s \\ \cline{1-1}\end{array} &= 2\,\begin{array}{|c|c|}\hline u & d \\ \hline s \\ \cline{1-1}\end{array} + \begin{array}{|c|c|}\hline d & u \\ \hline s \\ \cline{1-1}\end{array} + \begin{array}{|c|c|}\hline u & s \\ \hline d \\ \cline{1-1}\end{array} \\
&= -2\,\boxed{d}\backslash\boxed{d}^{\,*} + \boxed{u}\backslash\boxed{u}^{\,*} + \boxed{s}\backslash\boxed{s}^{\,*}, \\
3\,\begin{array}{|c|c|}\hline u & s \\ \hline d \\ \cline{1-1}\end{array} &= 2\,\begin{array}{|c|c|}\hline u & s \\ \hline d \\ \cline{1-1}\end{array} + \begin{array}{|c|c|}\hline s & u \\ \hline d \\ \cline{1-1}\end{array} + \begin{array}{|c|c|}\hline u & d \\ \hline s \\ \cline{1-1}\end{array} \\
&= 2\,\boxed{s}\backslash\boxed{s}^{\,*} - \boxed{u}\backslash\boxed{u}^{\,*} - \boxed{d}\backslash\boxed{d}^{\,*}.
\end{aligned} \tag{9.92}$$

(c) For a Young pattern $[\lambda_1, \lambda_2]$, $\lambda_1 > 2\lambda_2 > 0$, the planar weight diagram is a hexagon where the length of the top edge is $\lambda_1 - \lambda_2$, the length of the bottom edge is λ_2, and the lengths of two un-neighboring edges are the same. The weights on the edge are single. The multiplicities of the weights increase one by one as their positions go inside until the hexagon becomes a triangle upside down. The multiplicities of the weights inside the triangle are $\lambda_2 + 1$. The highest weight state has

$$Y = (\lambda_1 + \lambda_2)/3, \qquad T_3 = (\lambda_1 - \lambda_2)/2. \tag{9.93}$$

The conjugate representation of $[\lambda_1, \lambda_2]$ is $[\lambda_1, \lambda_1 - \lambda_2]$. Their planar weight diagrams are in the invertion with respect to the origin.

9.4.3 *Mass Formulas*

There is a naive model to study the mass formula of the hadrons in a multiplet of the flavor SU(3), which are made by quarks and antiquarks. Assume that the binding energy $-V$ for the hadrons in one multiplet of SU(3) are the same, and the difference of the masses of hadrons comes from the different quarks. The masses of u and d quarks and their antiquarks are m_1 and the masses of s quark and its antiquark are m_2.

The masses in the baryon decuplet are

$$\begin{aligned}
M(N^*) &= 3m_1 - V, & M(\Sigma^*) &= 2m_1 + m_2 - V, \\
M(\Xi^*) &= m_1 + 2m_2 - V, & M(\Omega) &= 3m_2 - V.
\end{aligned}$$

Then, the mass formula is obtained

$$M(\Omega) - M(\Xi^*) = M(\Xi^*) - M(\Sigma^*) = M(\Sigma^*) - M(N^*). \tag{9.94}$$

From the experiments, the observed average masses for the isospin multiplets are

$$M_{N^*} = 1232 \text{ MeV}, \qquad M_{\Sigma^*} = 1384.6 \text{ MeV},$$
$$M_{\Xi^*} = 1531.8 \text{ MeV}, \qquad M_{\Omega} = 1672.5 \text{ MeV}.$$
$$M(\Omega) - M(\Xi^*) = 140.7 \text{ MeV},$$
$$M(\Xi^*) - M(\Sigma^*) = 147.2 \text{ MeV},$$
$$M(\Sigma^*) - M(N^*) = 152.6 \text{ MeV}.$$

In the beginning of sixties of the last Century the simple model predicted that a baryon Ω with spin 3/2, positive parity, supercharge $Y = -2$, and electric charge $Q = -1$ should exist at the mass near 1680 MeV. It was found in 1962 as expected.

This model is too simple to explain the masses of the baryon octet, because both the baryons Σ and Λ are composed of one s quark and two quarks of u and d, but have different masses in experiment. Further analysis shows that the different masses of s quark and the quark of u or d may be demonstrated by a broken mass matrix $\hat{M}$. In addition to the symmetric mass $\hat{M}_0$, $\hat{M} - \hat{M}_0$ has the transformation property like the supercharge Y, called the "33" **symmetry broken**. Namely, the Hamiltonian contains a mass term $\overline{\psi}\hat{M}\psi$ which is left invariant in SU(3). How many parameters appear in the mass term $\overline{\psi}\hat{M}\psi$? Since $\hat{M} - \hat{M}_0$ belongs to the adjoint representation, the parameters come from the decomposition of $[2,1] \times [\lambda]$ to $[\lambda]$. When $[\lambda]$ is a one-row Young pattern, $[2,1] \times [\lambda]$ contains one $[\lambda]$ so that there is two mass parameters ($3m_1 - V$ and $m_2 - m_1$) as shown in Eq. (9.94). When $[\lambda]$ is a two-row Young pattern, $[2,1] \times [\lambda]$ contains two $[\lambda]$ so that a new mass parameter appears.

Because there are only three mass parameters, Gell-Mann, Nishijima, and Okubo expressed the mass operator $\hat{M}$ in the sum of generators: The constant term, the linear term and the square term:

$$\hat{M} = M_0 + M_1 Y + M_2 \left[Y^2 + cT(T+1)\right], \tag{9.95}$$

The square term includes the symmetric combination of generators and the anti-symmetric one. The antisymmetric combination of generators is the commutator which is proportional to the linear term of generators (see Eq. (7.5)). The symmetric combination, as shown in the last formula of Fig. 8.16, contains the representations $[4,2]$, $[2,1]_S$, and $[0,0]$, where we have to exclude $[4,2]$ now. For the baryon decuplet we have

$$Y^2 + cT(T+1) = a + bY.$$

For Ω in the baryon decuplet, $Y = -2$ and $T = 0$, one has $4 = a - 2b$. For Ξ^*, $Y = -1$ and $T = 1/2$, one has $1 + 3c/4 = a - b$. For Σ^*, $Y = 0$

and $T = 1$, one has $2c = a$. The solution is $a = -8$, $b = -6$, and $c = -4$. The solution meets the condition from the baryon N^*, where $Y = 1$ and $T = 3/2$. Thus, the Gell-Mann–Nishijima–Okubo mass formula is

$$M(T, Y) = M_0 + M_1 Y + M_2 \left\{ Y^2 - 4T(T+1) \right\}.$$

For the baryon octet, $M(N) = M_0 + M_1 - 2M_2$, $M(\Sigma) = M_0 - 8M_2$, $M(\Lambda) = M_0$, and $M(\Xi) = M_0 - M_1 - 2M_2$. Then,

$$\frac{M(N) + M(\Xi)}{2} = \frac{M(\Sigma) + 3M(\Lambda)}{4}. \tag{9.96}$$

The prediction fits the experiment data:

$$\begin{aligned} M(N) &= 938.9 \text{ MeV}, \qquad & M(\Sigma) &= 1193.1 \text{ MeV}, \\ M(\Lambda) &= 1115.7 \text{ MeV}, \qquad & M(\Xi) &= 1318.1 \text{ MeV}. \end{aligned}$$

The left-hand side of Eq. (9.96) is 1128.5 MeV, and the right-hand side is 1135.1 MeV. The formula (9.96) holds approximately for the mass square of the scalar meson octet. The experiment data are

$$m(\pi) = 138.0 \text{ MeV}, \qquad m(K) = 495.7 \text{ MeV}, \qquad m(\eta) = 547.5 \text{ MeV}.$$

The left-hand side of Eq. (9.96) for the mass square is 0.2457 GeV^2, and the right-hand side is 0.2296 GeV^2. The reason to meet the experiment data of mass square instead of mass is as follows. The wave functions of baryons satisfy the Dirac equation where the mass appears as linear term, but the wave functions of mesons satisfy the d'Alembert equation where the mass appears as square term. The formula is not in good agreement with the vector meson octet because there is a mixture between the meson octet and the meson singlet ω.

9.4.4 *Wave Functions of Mesons*

A meson in low energy is composed of a quark and an antiquark with zero orbit angular momentum. The wave function of a meson is a product of the color, the flavor, and the spinor wave functions. It is not needed to consider the permutation symmetry because the quark and the antiquark are not the identical particles. The mixed tensor of rank $(1, 1)$ is decomposed into a traceless tensor and a trace tensor (scalar),

$$\square \times \square^{*} = \square\!\!\diagdown\!\!\square^{*} \oplus \mathbf{1}.$$

Due to color confinement, the color wave function of a meson has to be in the colorless state, namely in the singlet of $\mathrm{SU}(3)_c$. The flavor wave function of a meson can be in the octet (traceless tensor) or singlet (trace tensor). For the spinor wave functions, the traceless tensor describes the vector mesons and the trace tensor the scalar mesons. These mesons meet the observation in experiments: the vector meson octet and singlet (ω) with the negative parity and the scalar meson octet and singlet (η') with the negative parity.

Let (ψ_+, ψ_-) denote the spinor wave functions for a quark and let $(-\overline{\psi_-}, \overline{\psi_+})$ denote that for an antiquark. The spinor wave functions for the scalar meson and for the vector mesons are

$$\begin{aligned} S=0, \quad & S_3=0: \quad && \sqrt{1/2}\left(\overline{\psi_+}\psi_+ + \overline{\psi_-}\psi_-\right), \\ S=1, \quad & S_3=1: && -\overline{\psi_-}\psi_+, \\ & S_3=0: && \sqrt{1/2}\left(\overline{\psi_+}\psi_+ - \overline{\psi_-}\psi_-\right), \\ & S_3=-1: && \overline{\psi_+}\psi_-. \end{aligned} \tag{9.97}$$

The flavor wave function of the singlet meson is the trace tensor,

$$\sqrt{1/3}\left\{\boxed{\text{u}}\backslash\boxed{\text{u}}^{\,*} + \boxed{\text{d}}\backslash\boxed{\text{d}}^{\,*} + \boxed{\text{s}}\backslash\boxed{\text{s}}^{\,*}\right\}. \tag{9.98}$$

The flavor wave functions of the octet mesons are given in Eq. (9.91). But, in the particle physics, the wave functions are preferred to be expressed in a matrix of three dimensions, where the row index denotes the covariant one and the column index denotes the contravariant one. The basis tensors of the scalar mesons are, for example,

$$\pi^+ = \begin{pmatrix} 0 & 1 & 0 \\ 0 & 0 & 0 \\ 0 & 0 & 0 \end{pmatrix}, \qquad \eta = \sqrt{\frac{1}{6}}\begin{pmatrix} 1 & 0 & 0 \\ 0 & 1 & 0 \\ 0 & 0 & -2 \end{pmatrix}.$$

The traceless tensor is expanded with respect to the basis tensors where the coefficients are written by the names of the mesons,

$$M = \begin{pmatrix} \dfrac{\pi^0}{\sqrt{2}} + \dfrac{\eta}{\sqrt{6}} & \pi^+ & \mathrm{K}^+ \\ \pi^- & -\dfrac{\pi^0}{\sqrt{2}} + \dfrac{\eta}{\sqrt{6}} & \mathrm{K}^0 \\ \mathrm{K}^- & \overline{\mathrm{K}^0} & -\dfrac{2\eta}{\sqrt{6}} \end{pmatrix}, \tag{9.99}$$

M transforms in the flavor SU(3) as follows:

$$M \xrightarrow{u} uMu^{-1}. \tag{9.100}$$

Through Eq. (9.82), the flavor wave functions can be expressed in those in the real orthogonal representation of eight dimensions,

$$\begin{aligned} &(\boldsymbol{M}_1 + \mathrm{i}\boldsymbol{M}_2)/\sqrt{2} = \pi^-, &&(\boldsymbol{M}_1 - \mathrm{i}\boldsymbol{M}_2)/\sqrt{2} = \pi^+, \\ &(\boldsymbol{M}_4 + \mathrm{i}\boldsymbol{M}_5)/\sqrt{2} = \mathrm{K}^-, &&(\boldsymbol{M}_4 - \mathrm{i}\boldsymbol{M}_5)/\sqrt{2} = \mathrm{K}^+, \\ &(\boldsymbol{M}_6 + \mathrm{i}\boldsymbol{M}_7)/\sqrt{2} = \overline{\mathrm{K}^0}, &&(\boldsymbol{M}_6 - \mathrm{i}\boldsymbol{M}_7)/\sqrt{2} = \mathrm{K}^0, \\ &\boldsymbol{M}_3 = \pi^0, &&\boldsymbol{M}_8 = \eta. \end{aligned} \tag{9.101}$$

Similarly, the flavor wave functions of the baryon octet are also expressed in a matrix of three dimensions,

$$\boldsymbol{B} = \begin{pmatrix} \dfrac{\Sigma^0}{\sqrt{2}} - \dfrac{\Lambda}{\sqrt{6}} & -\Sigma^+ & \mathrm{P} \\ \Sigma^- & -\dfrac{\Sigma^0}{\sqrt{2}} - \dfrac{\Lambda}{\sqrt{6}} & \mathrm{N} \\ \Xi^- & -\Xi^0 & \dfrac{2\Lambda}{\sqrt{6}} \end{pmatrix}, \tag{9.102}$$

where the minus sign comes from the definition of the particles as shown in Eq. (9.91).

9.4.5 *Wave Functions of Baryons*

A baryon in low energy is composed of three quarks which are identical particles satisfying the Fermi statistics. **Its total wave function has to be antisymmetric in the transposition between the quarks.** Assume that the orbital angular momentum of the low energy baryon is vanishing such that its total wave function is a product of the color, the flavor, and the spinor wave functions. Three quarks are described by a tensor of rank 3, which is decomposed by the Young operators,

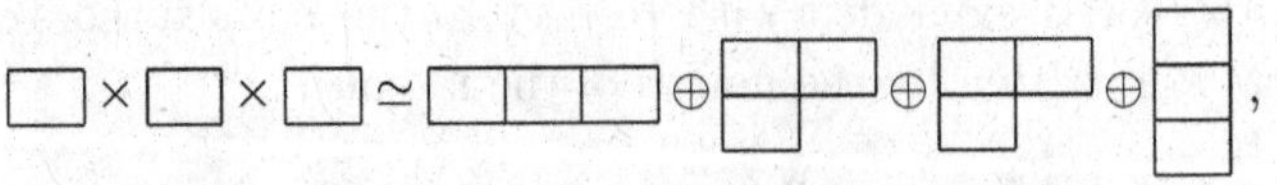

(9.103)

$$[1] \times [1] \times [1] \simeq [3] \oplus [2,1] \oplus [2,1] \oplus [1^3].$$

Those wave functions belong to the representations of the permutation group denoted by the same Young patterns. Due to the color confinement, the color wave function is in the color singlet $[1^3]$ which is totally antisymmetric in the quark transposition. The product of the flavor and the spinor wave functions has to be totally symmetric.

There are three choices for the flavor wave functions. The representation [3] of the flavor SU(3) corresponds to the decuplet which describes the totally symmetric states in the permutations. The representation [2,1] corresponds to the octet which describes the mixed symmetric states. The representation $[1^3]$ corresponds to the singlet which describes the totally antisymmetric state. However, there are only two choices for the spinor wave functions because the representation $[1^3]$ of SU(2) corresponds to the null space. The representation [3] of the spinor SU(2) corresponds to the quadruplet ($S = 3/2$) which describes the totally symmetric states in the permutations. The representation [2,1] of the spinor SU(2) corresponds to the doublet ($S = 1/2$) which describes the mixed symmetric states. Since the product of the flavor and the spinor wave functions are totally symmetric, the wave function of flavor decuplet has to multiply that of spinor quadruplet, and the wave function of flavor octet has to multiply that of spinor doublet where a suitable combination is needed such that the multiplied wave functions are combined to be totally symmetric with respect to the permutations. This coincides with the experimental data that the observed low-energy baryons are the baryon decuplet with spin-parity $(3/2)^+$ and the baryon octet with spin-parity $(1/2)^+$.

(a) **The $(3/2)^+$ baryon decuplet.**

The wave function is a product of the flavor and the spinor wave functions. Two examples are given: $N^{*+}_{1/2}$ with $T = 3/2$, $T_3 = 1/2$, $Y = 1$, and $S_3 = 1/2$, and $\Sigma^{*0}_{-1/2}$ with $T = 1$, $T_3 = 0$, $Y = 0$, and $S_3 = -1/2$.

$$\begin{array}{|c|c|c|}\hline u & u & d \\ \hline\end{array} \cdot \begin{array}{|c|c|c|}\hline + & + & - \\ \hline\end{array}$$

$$\begin{aligned} N^{*+}_{1/2} = \ & \frac{1}{3}\left\{u_+u_+d_- + u_+d_+u_- + d_+u_+u_- + u_+u_-d_+ + u_+d_-u_+ \right. \\ & \left. + \ d_+u_-u_+ + u_-u_+d_+ + u_-d_+u_+ + d_-u_+u_+\right\}. \end{aligned}$$

$$\begin{array}{|c|c|c|}\hline u & d & s \\ \hline\end{array} \cdot \begin{array}{|c|c|c|}\hline + & - & - \\ \hline\end{array}$$

$$\begin{aligned} \Sigma^{*0}_{-1/2} = \ & \frac{1}{3\sqrt{2}}\left\{u_+d_-s_- + u_+s_-d_- + d_+u_-s_- + d_+s_-u_- \right. \\ & + \ s_+u_-d_- + s_+d_-u_- + u_-d_+s_- + u_-s_+d_- + d_-u_+s_- \\ & + \ d_-s_+u_- + s_-u_+d_- + s_-d_+u_- + u_-d_-s_+ + u_-s_-d_+ \\ & \left. + \ d_-u_-s_+ + d_-s_-u_+ + s_-u_-d_+ + s_-d_-u_+\right\}. \end{aligned}$$

(b) **The $(1/2)^+$ baryon octet.**

Both the flavor and the spinor wave functions are in the mixed symmetry of the permutations. Their product has to be combined as the totally symmetric wave function. The representation matrices of generators of the permutation group S_3 are calculated in Table 4.4: $D^{[2,1]}[(1\ 2)] = \begin{pmatrix} 1 & -1 \\ 0 & -1 \end{pmatrix}$ and $D^{[2,1]}[(1\ 2\ 3)] = \begin{pmatrix} 1 & -1 \\ 1 & 0 \end{pmatrix}$. The common eigenvector of their direct products with eigenvalue 1 is

$$\begin{pmatrix} 2 \\ 1 \\ 1 \\ 2 \end{pmatrix} \quad \text{for} \quad \begin{pmatrix} 1 & -1 & -1 & 1 \\ 0 & -1 & 0 & 1 \\ 0 & 0 & -1 & 1 \\ 0 & 0 & 0 & 1 \end{pmatrix} \quad \text{and} \quad \begin{pmatrix} 1 & -1 & -1 & 1 \\ 1 & 0 & -1 & 0 \\ 1 & -1 & 0 & 0 \\ 1 & 0 & 0 & 0 \end{pmatrix}.$$

Write the wave function $P_{-1/2}$ of a proton ($T = T_3 = 1/2$ and $Y = 1$) with $S_3 = -1/2$ as example. Take the Young tableau $\mathcal{Y} = \begin{array}{|c|c|}\hline 1 & 2 \\ \hline 3 \\ \cline{1-1}\end{array}$. The basis tensors of the flavor wave functions are

$$\begin{array}{|c|c|}\hline u & u \\ \hline d \\ \cline{1-1}\end{array} = 2uud - duu - udu, \qquad (23)\begin{array}{|c|c|}\hline u & u \\ \hline d \\ \cline{1-1}\end{array} = 2udu - duu - uud.$$

Similarly, the basis tensors of the spinor wave functions are

$$\begin{array}{|c|c|}\hline + & - \\ \hline - \\ \cline{1-1}\end{array} = (+--) + (-+-) - 2(--+),$$

$$(23)\begin{array}{|c|c|}\hline + & - \\ \hline - \\ \cline{1-1}\end{array} = (+--) + (--+) - 2(-+-).$$

Thus, the wave function of a proton with $S_3 = -1/2$ is

$$\begin{aligned}
P_{-1/2} = & \begin{array}{|c|c|}\hline u & u \\ \hline d \\ \cline{1-1}\end{array} \left\{ 2 \begin{array}{|c|c|}\hline + & - \\ \hline - \\ \cline{1-1}\end{array} + \left[(2\ 3) \begin{array}{|c|c|}\hline + & - \\ \hline - \\ \cline{1-1}\end{array} \right] \right\} \\
& + \left[(2\ 3) \begin{array}{|c|c|}\hline u & u \\ \hline d \\ \cline{1-1}\end{array} \right] \left\{ \begin{array}{|c|c|}\hline + & - \\ \hline - \\ \cline{1-1}\end{array} + 2 \left[(2\ 3) \begin{array}{|c|c|}\hline + & - \\ \hline - \\ \cline{1-1}\end{array} \right] \right\} \\
= & \ \{2uud - duu - udu\} \cdot 3\{(+--) - (--+)\} \\
& + \{2udu - duu - uud\} \cdot 3\{(+--) - (-+-)\} \\
= & \ 3\{u_+u_-d_- - 2d_+u_-u_- + u_+d_-u_- - 2u_-u_-d_+ + d_-u_-u_+ \\
& + u_-d_-u_+ + u_-u_+d_- + d_-u_+u_- - 2u_-d_+u_-\}.
\end{aligned}$$

The normalization factor should be changed to $\sqrt{1/18}$.

9.5 Exercises

1. Calculate the dimensions of the irreducible representations denoted by the following Young patterns for the SU(3) group and for the SU(6) group, respectively:

$$[3], \quad [2,1], \quad [3,3], \quad [4,2], \quad [5,1].$$

2. In terms of the block weight diagram of the irreducible representation $[3,1]$ of SU(3) (see Fig. 8.6), calculate the orthonormal basis tensors in this representation $[3,1]$ as expansions with respect to the standard tensor Young tableaux. Then, write the expansion of the basis tensor for the highest weight state with respect to the basis tensors $\boldsymbol{\theta}_{abcd}$ in the tensor subspace $\mathcal{Y}_2^{[3,1]}\mathcal{T}$, where $\mathcal{T}$ is the tensor space of rank 4 for the SU(3) group and the standard Young tableau $\mathcal{Y}_2^{[3,1]}$ is $\begin{array}{|c|c|c|}\hline 1 & 2 & 4 \\ \hline 3 \\ \cline{1-1} \end{array}$.

3. Expand the Gel'fand bases in the irreducible representation $[3,3]$ of the SU(3) group with respect to the standard tensor Young tableaux by making use of its block weight diagram given in Fig. 8.4.

4. Express each Gel'fand bases in the representations $[1]$, $[1,1]$ and $[2,1,1,1]$ of the SU(5) group by the standard tensor Young tableau, respectively, and calculate the nonvanishing matrix elements for the lowering operators F_μ.

5. Draw the block weight diagrams for the representations $[5,1]$ and $[4,2]$ of the Lie group SU(3) (see Ex. 2 and Ex. 3 in subsection 9.2.2), respectively.

6. Draw the planar weight diagram for the representations $[3,1]$ and $[3,2]$ of SU(3), respectively.

7. Calculate the Clebsch–Gordan series for the following direct product representations, and compare their dimensions by Eq. (9.30) for the SU(3) group and for the SU(6) group, respectively:

(a) $[2,1]\times[3,0]$, (b) $[3,0]\times[3,0]$, (c) $[3,0]\times[3,3]$, (d) $[4,2]\times[2,1]$.

8. A neutron is composed of one u quark and two d quarks. Construct the wave function of a neutron with spin $S_3 = -1/2$, satisfying the correct permutation symmetry among the identical particles.

9. Transform the following traceless mixed tensor representations of the SU(6) group into the covariant tensor representations, respectively, and calculate their dimensions:

$$(1)\ \ [3]\backslash[2,1]^*, \qquad (2)\ \ [3,2,1]\backslash[3,3]^*, \qquad (3)\ \ [4,3,1]\backslash[3,2]^*.$$

10. Prove the identity where T_A is the generator of SU(N):

$$\sum_{A=1}^{N^2-1} (T_A)_{ac}\,(T_A)_{bd} = \frac{1}{2}\delta_a^d\delta_b^c - \frac{1}{2N}\delta_a^c\delta_b^d\,.$$

Chapter 10

REAL ORTHOGONAL GROUPS

In this chapter we will study the tensor representations and the spinor representations of the SO(N) groups, and calculate the basis functions in those irreducible representations.

10.1 Tensor Representations of SO(N)

The tensor representations of SO(N) are its single-valued representations. In this section, the decomposition of a tensor space of SO(N) is studied and the orthonormal irreducible basis tensors are calculated.

10.1.1 *Tensors of* SO(N)

Similar to the tensors of SU(N), a tensor of rank n of SO(N) has N^n components and transforms in $R \in \text{SO}(N)$,

$$\boldsymbol{T}_{a_1\cdots a_n} \xrightarrow{R} (O_R\boldsymbol{T})_{a_1\cdots a_n} = \sum_{b_1\cdots b_n} R_{a_1b_1}\cdots R_{a_nb_n}\boldsymbol{T}_{b_1\cdots b_n}. \qquad (10.1)$$

A basis tensor $\boldsymbol{\theta}_{d_1\cdots d_n}$ contains only one nonvanishing component which is equal to 1, (see Eq. (9.16))

$$(\boldsymbol{\theta}_{d_1\cdots d_n})_{a_1\cdots a_n} = (\boldsymbol{\theta}_{d_1}\times\cdots\times\boldsymbol{\theta}_{d_n})_{a_1\cdots a_n} = \delta_{d_1a_1}\delta_{d_2a_2}\cdots\delta_{d_na_n}, \qquad (10.2)$$

$$O_R\boldsymbol{\theta}_{d_1\cdots d_n} = \sum_{b_1\cdots b_n}\boldsymbol{\theta}_{b_1\cdots b_n}R_{b_1d_1}\cdots R_{b_nd_n}. \qquad (10.3)$$

Any tensor can be expanded with respect to the basis tensors,

$$\boldsymbol{T}_{a_1\cdots a_n} = \sum_{d_1\cdots d_n}(\boldsymbol{\theta}_{d_1\cdots d_n})_{a_1\cdots a_n}T_{d_1\cdots d_n} = T_{a_1\cdots a_n}. \qquad (10.4)$$

Remind that $\boldsymbol{T}_{a_1\cdots a_n}$ and $T_{a_1\cdots a_n}$ are different in the SO(N) transformation although they are equal in value.

In a permutation of S_n the tensors and the basis tensors are transformed as Eqs. (9.4) and (9.18), respectively. **The tensor space is an invariant space both in SO(N) and in S_n.** The SO(N) transformation commutes with the permutation (Weyl reciprocity), so that the tensor space can be decomposed by the projection of the Young operators.

The main difference of the tensors of SO(N) from the tensors of SU(N) is that the transformation matrix $R \in \mathrm{SO}(N) \subset \mathrm{SU}(N)$ is real. As a result, the tensors of SO(N) have the following new characteristics. First, the real part and the imaginary part of a tensor of SO(N) transform separately in Eq. (10.1) so that **only the real tensors are needed to be studied.** Second, there is no difference between a covariant tensor and a contravariant tensor for the SO(N) transformations. **The contraction of a tensor are accomplished between any two tensor indices.** Before the projection of a Young operator the tensor space should be decomposed into the direct sum of a series of traceless tensor subspaces with different ranks, each of which are left invariant in SO(N). Third, let $\mathcal{T}$ denote the traceless tensor space of rank n. After the projection of a Young operator, $\mathcal{T}_{\mu}^{[\lambda]} = \mathcal{Y}_{\mu}^{[\lambda]}\mathcal{T}$ **is a traceless tensor subspace with the given permutation symmetry.** A similar proof to that in subsection 9.3.3 shows that $\mathcal{T}_{\mu}^{[\lambda]}$ **is a null space if the sum of the numbers of boxes in the first two columns of the Young pattern $[\lambda]$ is larger than N.** Fourth, when the row number m of the Young pattern $[\lambda]$ is larger than $N/2$, we can define the dual basis tensor. Assuming that the first column in the tensor Young tableau $\mathcal{Y}_{\mu}^{[\lambda]}\boldsymbol{\theta}_{a_1\cdots a_n} \equiv \psi_{d_1\cdots d_m c\cdots}$ is filled by $d_1\cdots d_m$, we define the dual tensor Young tableau $[{}^*\psi]_{b_1\cdots b_{N-m}c\cdots}$ of $\psi_{d_1\cdots d_m c\cdots}$:

$$[{}^*\psi]_{b_1\cdots b_{N-m}c\cdots} = \frac{1}{m!}\sum_{d_1\cdots d_m} \epsilon_{b_1\cdots b_{N-m}d_m\cdots d_1}\psi_{d_1\cdots d_m c\cdots}\,. \tag{10.5}$$

Its inverse transformation is

$$\begin{aligned}
&\frac{1}{(N-m)!}\sum_{b_1\cdots b_{N-m}} \epsilon_{a_1\cdots a_m b_{N-m}\cdots b_1}\,[{}^*\psi]_{b_1\cdots b_{N-m}c\cdots} \\
&= \frac{1}{m!(N-m)!}\sum_{b_1\cdots b_{N-m}d_1\cdots d_m} \epsilon_{a_1\cdots a_m b_{N-m}\cdots b_1}\epsilon_{b_1\cdots b_{N-m}d_m\cdots d_1}\psi_{d_1\cdots d_m c\cdots} \\
&= (-1)^{N(N-1)/2}\psi_{a_1\cdots a_m c\cdots}.
\end{aligned} \tag{10.6}$$

The correspondence between the pair of two mutually dual basis tensors is one-to-one. The difference of two sets of basis tensors is only in the arranged order and some possible signs. Following the proof for Eq. (9.74), we prove that a traceless tensor subspace $\mathcal{T}_{\mu}^{[\lambda]}$, where the row number m of the Young pattern $[\lambda]$ is larger than $N/2$, is equivalent to a traceless tensor subspace $\mathcal{T}_{\nu}^{[\lambda']}$, where **the row number of the Young pattern $[\lambda']$ is $N-m<N/2$,**

$$[\lambda'] \simeq [\lambda], \qquad \lambda'_j = \begin{cases} \lambda_j, & j \leqslant (N-m), \\ 0, & j > (N-m), \end{cases} \qquad N/2 < m \leqslant N. \quad (10.7)$$

Fifth, when $N=2\ell$ is even, the Young pattern $[\lambda]$, whose row number is ℓ, is the same as its dual Young pattern $[\lambda']$, which is called **the self-dual Young pattern.** In order to remove the factor $(-1)^{N(N-1)/2}=(-1)^{\ell}$ in Eq. (10.6), we introduce a factor i^{ℓ} in the dual relation (10.5),

$$\begin{aligned} [{}^{*}\psi]_{d_1\cdots d_\ell c\cdots} &= \frac{\mathrm{i}^{\ell}}{\ell!} \sum_{d_{\ell+1}\cdots d_{2\ell}} \epsilon_{d_1\cdots d_\ell d_{\ell+1}\cdots d_{2\ell}} \psi_{d_{2\ell}\cdots d_{\ell+1}c\cdots}, \\ \psi_{d_1\cdots d_\ell c\cdots} &= \frac{\mathrm{i}^{\ell}}{\ell!} \sum_{d_{\ell+1}\cdots d_{2\ell}} \epsilon_{d_1\cdots d_\ell d_{\ell+1}\cdots d_{2\ell}} [{}^{*}\psi]_{d_{2\ell}\cdots d_{\ell+1}c\cdots}, \end{aligned} \quad (10.8)$$

where the first column in the tensor Young tableau $\mathcal{Y}_{\mu}^{[\lambda]}\boldsymbol{\theta}_{a_1\cdots a_n}=\psi_{d_1\cdots d_\ell c\cdots}$ is filled by $d_1\cdots d_\ell$. We define

$$\phi^{\pm}_{d_1\cdots d_\ell c\cdots} = \frac{1}{2}\left\{\psi_{d_1\cdots d_\ell c\cdots} \pm [{}^{*}\psi]_{d_1\cdots d_\ell c\cdots}\right\}. \quad (10.9)$$

$\phi^{+}_{d_1\cdots d_\ell d\cdots}$ is left invariant in the dual transformation and is called the **self-dual basis tensor.** $\phi^{-}_{d_1\cdots d_\ell c\cdots}$ changes its sign in the dual transformation and is called the **anti-self-dual basis tensor**. For example,

$$\begin{aligned} \phi^{\pm}_{135\cdots(2\ell-1)c\cdots} &= \frac{1}{2}\left\{\psi_{135\cdots(2\ell-1)c\cdots} \pm \mathrm{i}^{\ell}\psi_{246\cdots(2\ell)c\cdots}\right\}, \\ \phi^{\pm}_{235\cdots(2\ell-1)c\cdots} &= \frac{1}{2}\left\{\psi_{235\cdots(2\ell-1)c\cdots} \mp \mathrm{i}^{\ell}\psi_{146\cdots(2\ell)c\cdots}\right\}, \end{aligned} \quad (10.10)$$

where $\epsilon_{135\cdots(2\ell-1)(2\ell)\cdots 642}=1$. Thus, when the row number of $[\lambda]$ is equal to $N/2$, **the representation space $\mathcal{T}_{\mu}^{[\lambda]}$ is decomposed into the self-dual and the anti-self-dual tensor subspaces with the same dimension.** Note that the combinations by the Young operators and the dual transformations (10.5) and (10.10) are all real except that the dual transformation (10.10) with $N=4m+2$ is complex.

In summary, the traceless tensor subspace $\mathcal{T}_{\mu}^{[\lambda]}$ is a null space if the sum of the numbers of boxes in the first two columns of the Young pattern

$[\lambda]$ is larger than N. $\mathcal{T}_\mu^{[\lambda]}$ corresponds to a representation $[\lambda]$ of SO(N), where the row number of $[\lambda]$ is less than $N/2$. When the row number of $[\lambda]$ is larger than $N/2$, two dual Young patterns (see Eq. (10.7)) $[\lambda]$ and $[\lambda']$ correspond to equivalent representations. When the row number ℓ of $[\lambda]$ is equal to $N/2$, $\mathcal{T}_\mu^{[\lambda]}$ is decomposed into the self-dual and the anti-self-dual tensor subspaces, $\mathcal{T}_\mu^{[(+)\lambda]}$ and $\mathcal{T}_\mu^{[(-)\lambda]}$, corresponding to the representation $[(\pm)\lambda]$, respectively. The highest weights of the representations $[\lambda]$ and $[(\pm)\lambda]$ will be calculated later and proved to be irreducible (similar to Theorem 9.3). In fact, there is no further constraint to construct a nontrivial invariant subspace in their representation spaces. **All the irreducible representations are real except for $[(\pm)\lambda]$ when $N = 4m + 2$.**

As far as the orthonormal irreducible basis tensors of SO(N) is concerned, there are two problems. One is **how to decompose the standard tensor Young tableaux into a sum of the traceless basis tensors.** The decomposition is straightforward, but tedious. The second is **how to combine the basis tensors such that they are the common eigenfunctions of H_j and orthonormal to each other.**

For SU(N), the standard tensor Young tableaux are the common eigenfunctions of H_j but not orthonormal. Because the highest weight is simple, the orthonormal basis tensors for SU(N) can be obtained from the highest weight state by the lowering operators F_μ in terms of the method of the block weight diagram and the Gel'fand's method. Generalizing this method to SO(N), we have **to find the common eigenstates of H_μ in the self-representation** of SO(N), and then, **to find the highest weight state in the representations** $[\lambda]$ and $[(\pm)\lambda]$. The orthonormal basis tensors for SO(N) can be calculated from the highest weight state by the lowering operators F_μ in terms of **the method of the block weight diagram and the generalized Gel'fand's method**. The calculated orthonormal basis tensors must be traceless, just like that for the highest weight state.

In the self-representation of SO(N), the generators T_{ab} satisfy

$$\begin{aligned} (T_{ab})_{cd} &= -\mathrm{i}\left\{\delta_{ac}\delta_{bd} - \delta_{ad}\delta_{bc}\right\}, \\ [T_{ab},\, T_{cd}] &= -\mathrm{i}\left\{\delta_{bc}T_{ad} + \delta_{ad}T_{bc} - \delta_{bd}T_{ac} - \delta_{ac}T_{bd}\right\}, \end{aligned} \tag{10.11}$$

where the generators in the Cartan subalgebra are $H_j = T_{(2j-1)(2j)}$, $1 \leqslant j \leqslant N/2$.

10.1.2 *Irreducible Basis Tensors of* SO$(2\ell + 1)$

The Lie algebra of SO$(2\ell + 1)$ is B_ℓ. The simple roots of SO$(2\ell + 1)$ are

$$\boldsymbol{r}_\mu = \boldsymbol{e}_\mu - \boldsymbol{e}_{\mu+1}, \qquad 1 \leqslant \mu \leqslant \ell - 1, \qquad \boldsymbol{r}_\ell = \boldsymbol{e}_\ell. \tag{10.12}$$

$\boldsymbol{r}_\mu$ are the longer roots with $d_\mu = 1$ and $\boldsymbol{r}_\ell$ is the shorter root with $d_\ell = 1/2$. From the definition (8.1), the Chevalley bases of SO$(2\ell+1)$ in the self-representation are $(1 \leqslant \mu \leqslant \ell - 1)$

$$\begin{aligned}
H_\mu &= T_{(2\mu-1)(2\mu)} - T_{(2\mu+1)(2\mu+2)}, \qquad H_\ell = 2T_{(2\ell-1)(2\ell)}, \\
E_\mu &= \tfrac{1}{2}\left\{T_{(2\mu)(2\mu+1)} - \mathrm{i}T_{(2\mu-1)(2\mu+1)} - \mathrm{i}T_{(2\mu)(2\mu+2)} - T_{(2\mu-1)(2\mu+2)}\right\}, \\
F_\mu &= \tfrac{1}{2}\left\{T_{(2\mu)(2\mu+1)} + \mathrm{i}T_{(2\mu-1)(2\mu+1)} + \mathrm{i}T_{(2\mu)(2\mu+2)} - T_{(2\mu-1)(2\mu+2)}\right\}, \\
E_\ell &= T_{(2\ell)(2\ell+1)} - \mathrm{i}T_{(2\ell-1)(2\ell+1)}, \qquad F_\ell = T_{(2\ell)(2\ell+1)} + \mathrm{i}T_{(2\ell-1)(2\ell+1)}.
\end{aligned} \tag{10.13}$$

$\boldsymbol{\theta}_a$ is not the common eigenvector of H_μ. Generalizing the spherical harmonic basis vectors (5.184) for SO(3), we define the **spherical harmonic basis vectors** in the self-representation of SO$(2\ell+1)$

$$\boldsymbol{\phi}_\alpha = \begin{cases} (-1)^{\ell-\alpha+1}\sqrt{1/2}\,(\boldsymbol{\theta}_{2\alpha-1} + \mathrm{i}\boldsymbol{\theta}_{2\alpha}), & 1 \leqslant \alpha \leqslant \ell, \\ \boldsymbol{\theta}_{2\ell+1}, & \alpha = \ell + 1, \\ \sqrt{1/2}\,(\boldsymbol{\theta}_{4\ell-2\alpha+3} - \mathrm{i}\boldsymbol{\theta}_{4\ell-2\alpha+4}), & \ell + 2 \leqslant \alpha \leqslant 2\ell + 1. \end{cases} \tag{10.14}$$

The spherical harmonic basis vectors $\boldsymbol{\phi}_\alpha$ are orthonormal and complete. The matrix elements of the Chevalley bases in $\boldsymbol{\phi}_\alpha$ are

$$\begin{aligned}
H_\mu \boldsymbol{\phi}_\nu &= \left[\delta_{\nu\mu} - \delta_{\nu(\mu+1)} + \delta_{\nu(2\ell-\mu+1)} - \delta_{\nu(2\ell-\mu+2)}\right] \boldsymbol{\phi}_\nu, \\
H_\ell \boldsymbol{\phi}_\ell &= 2\delta_{\nu\ell}\boldsymbol{\phi}_\ell - 2\delta_{\nu(\ell+2)}\boldsymbol{\phi}_{\ell+2}, \\
E_\mu \boldsymbol{\phi}_\nu &= \delta_{\nu(\mu+1)}\boldsymbol{\phi}_\mu + \delta_{\nu(2\ell-\mu+2)}\boldsymbol{\phi}_{2\ell-\mu+1}, \\
F_\mu \boldsymbol{\phi}_\nu &= \delta_{\nu\mu}\boldsymbol{\phi}_{\mu+1} + \delta_{\nu(2\ell-\mu+1)}\boldsymbol{\phi}_{2\ell-\mu+2}, \\
E_\ell \boldsymbol{\phi}_\nu &= \sqrt{2}\delta_{\nu(\ell+1)}\boldsymbol{\phi}_\ell + \sqrt{2}\delta_{\nu(\ell+2)}\boldsymbol{\phi}_{\ell+1}, \\
F_\ell \boldsymbol{\phi}_\nu &= \sqrt{2}\delta_{\nu\ell}\boldsymbol{\phi}_{\ell+1} + \sqrt{2}\delta_{\nu(\ell+1)}\boldsymbol{\phi}_{\ell+2},
\end{aligned} \tag{10.15}$$

where $1 \leqslant \mu \leqslant \ell - 1$. Namely, H_μ and H_ℓ are diagonal in the spherical harmonic basis vectors $\boldsymbol{\phi}_\alpha$:

$$\begin{aligned}
H_\mu &= \text{diag}\{\underbrace{0, \cdots, 0}_{\mu-1}, 1, -1, \underbrace{0, \cdots, 0}_{2\ell-2\mu-1}, 1, -1, \underbrace{0, \cdots, 0}_{\mu-1}\}, \\
H_\ell &= \text{diag}\{\underbrace{0, \cdots, 0}_{\ell-1}, 2, 0, -2, \underbrace{0, \cdots, 0}_{\ell-1}\}.
\end{aligned}$$

The spherical harmonic basis tensor $\boldsymbol{\phi}_{\alpha_1\cdots\alpha_n}$ of rank n for SO$(2\ell+1)$ is the direct product of n spherical harmonic basis vectors $\boldsymbol{\phi}_{\alpha_1} \times \cdots \times \boldsymbol{\phi}_{\alpha_n}$.

The standard tensor Young tableaux $\mathcal{Y}_\mu^{[\lambda]}\phi_{\alpha_1\cdots\alpha_n}$ are **the common eigenstates of H_μ, but generally not orthonormal and traceless.** The eigenvalue of H_μ in the standard tensor Young tableau $\mathcal{Y}_\mu^{[\lambda]}\phi_{\alpha_1\cdots\alpha_n}$ is equal to the number of the digits μ and $(2\ell-\mu+1)$ in the tableau, minus the number of $(\mu+1)$ and $(2\ell-\mu+2)$. The eigenvalue of H_ℓ in $\mathcal{Y}_\mu^{[\lambda]}\phi_{\alpha_1\cdots\alpha_n}$ is equal to the number of ℓ in the tableau, minus the number of $\ell+2$, and then multiplied with 2. The standard tensor Young tableau $\mathcal{Y}_\mu^{[\lambda]}\phi_{\alpha_1\cdots\alpha_n}$ in the action of F_μ becomes the sum of all possible tensor Young tableaux, each of which is obtained from the original one by replacing one filled digit μ with the digit $(\mu+1)$, or by replacing one filled digit $(2\ell-\mu+1)$ with the digit $(2\ell-\mu+2)$. $\mathcal{Y}_\mu^{[\lambda]}\phi_{\alpha_1\cdots\alpha_n}$ in the action of F_ℓ becomes the sum, multiplied with a factor $\sqrt{2}$, of all possible tensor Young tableaux, each of which is obtained from the original one by replacing one filled digit ℓ with the digit $(\ell+1)$ or by replacing one filled digit $(\ell+1)$ with $(\ell+2)$. The actions of $E_\mu(E_\ell)$ are opposite to that of $F_\mu(F_\ell)$. **The obtained tensor Young tableaux may be not standard, but they can be transformed to the sum of the standard tensor Young tableaux by** Eqs. (9.22–23).

Two standard tensor Young tableaux with different sets of the filled digits are orthogonal to each other. For an irreducible representation $[\lambda]$ of SO$(2\ell+1)$, where the row number of $[\lambda]$ is not larger than ℓ, **the highest weight state corresponds to the standard tensor Young tableau where each box in the αth row is filled with the digit α** because every raising operator E_μ annihilates it. The highest weight $\boldsymbol{M}=\sum_\mu \boldsymbol{w}_\mu M_\mu$ is calculated from Eq. (10.15),

$$M_\mu=\lambda_\mu-\lambda_{\mu+1}, \qquad 1\leqslant\mu<\ell, \qquad M_\ell=2\lambda_\ell. \tag{10.16}$$

The tensor representation $[\lambda]$ of SO$(2\ell+1)$, where M_ℓ is even, is a single-valued representation. It will be known later that the representation with odd M_ℓ is a double-valued representation, called the spinor one.

Although the standard tensor Young tableaux is generally not traceless, **the standard tensor Young tableau with the highest weight is traceless** because it only contains ϕ_α with $\alpha<\ell+1$ (see Eq. (10.14)). For example, the tensor basis $\boldsymbol{\theta}_1\boldsymbol{\theta}_1$ is not traceless, but $\boldsymbol{\phi}_1\boldsymbol{\phi}_1$ is traceless. Since the highest weight is simple, the highest weight state is orthogonal to any other state in the irreducible representation. Generalizing the method of finding the orthonormal basis tensors in SU(N), we are able to find the remaining orthonormal and traceless basis tensors in $[\lambda]$ of SO$(2\ell+1)$ from the highest weight state by the lowering operators F_μ in terms of the method of the block weight diagram and the generalized Gel'fand's method. The multiplicity of a weight in the representation can be obtained by counting the number of the traceless tensor Young tableaux with this weight.

We give three examples for calculating the orthonormal basis states in the highest weight representations in SO(7) whose Lie algebra is B_3 (see subsection 8.3.3).

Ex. 1. The irreducible representation $[\lambda] = [2,0,0] = [2]$ of SO(7).

From Eq. (10.16), the highest weight is $\boldsymbol{M} = 2\boldsymbol{w}_1 = (2,0,0)$. In the brief symbols, $|\boldsymbol{M},\boldsymbol{m}\rangle = |\boldsymbol{m}\rangle$, the highest weight state is

$$|\boldsymbol{M}\rangle = |(2,0,0)\rangle = \left|\begin{matrix}2&&0&&0\\&2&&0&\\&&2&&\end{matrix}\right\rangle^{1}_{0} = \begin{array}{|c|c|}\hline 1&1\\\hline\end{array} = 2\phi_{11}.$$

Two typical standard tensor Young tableaux in this representation are

$$\begin{aligned}\begin{array}{|c|c|}\hline \mathrm{a}&\mathrm{a}\\\hline\end{array} &= \mathcal{Y}^{[2]}\phi_{aa} = 2\phi_{aa},\\ \sqrt{2}\begin{array}{|c|c|}\hline \mathrm{a}&\mathrm{b}\\\hline\end{array} &= \sqrt{2}\ \mathcal{Y}^{[2]}\phi_{ab} = \sqrt{2}\ \{\phi_{ab}+\phi_{ba}\},\end{aligned}$$

where $a \neq b$. The highest weight state $|(2,0,0)\rangle$ is also the highest weight state of an A_2-sixlet with $\boldsymbol{M}^{(2)} = (2,0)$:

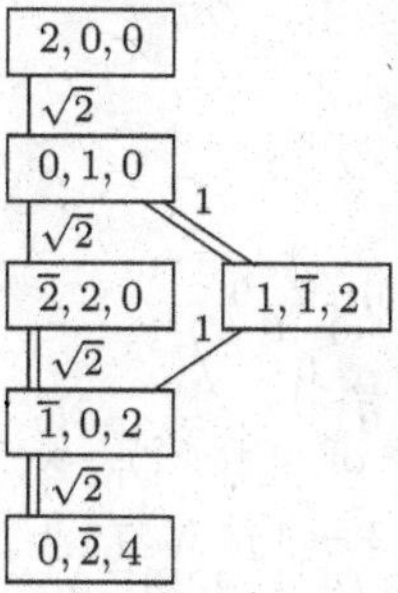

$$\left|\begin{matrix}2&&0&&0\\&2&&0&\\&&2&&\end{matrix}\right\rangle^{1}_{0} = \begin{array}{|c|c|}\hline 1&1\\\hline\end{array}$$

$$\left|\begin{matrix}2&&0&&0\\&2&&0&\\&&1&&\end{matrix}\right\rangle^{2}_{0} = \sqrt{2}\begin{array}{|c|c|}\hline 1&2\\\hline\end{array}$$

$$\left|\begin{matrix}2&&0&&0\\&2&&0&\\&&0&&\end{matrix}\right\rangle^{3}_{0} = \begin{array}{|c|c|}\hline 2&2\\\hline\end{array}, \qquad \left|\begin{matrix}2&&0&&0\\&1&&0&\\&&1&&\end{matrix}\right\rangle^{3}_{2} = \sqrt{2}\begin{array}{|c|c|}\hline 1&3\\\hline\end{array}$$

$$\left|\begin{matrix}2&&0&&0\\&1&&0&\\&&0&&\end{matrix}\right\rangle^{4}_{2} = \sqrt{2}\begin{array}{|c|c|}\hline 2&3\\\hline\end{array}$$

$$\left|\begin{matrix}2&&0&&0\\&0&&0&\\&&0&&\end{matrix}\right\rangle^{5}_{4} = \begin{array}{|c|c|}\hline 3&3\\\hline\end{array}$$

Two $\mathcal{A}_3$-triplets are constructed from the states $\left|\begin{matrix}2&&0&&0\\&1&&0&\\&&1&&\end{matrix}\right\rangle^{3}_{2}$ and $\left|\begin{matrix}2&&0&&0\\&1&&0&\\&&0&&\end{matrix}\right\rangle^{4}_{2}$ due to the parallel principle.

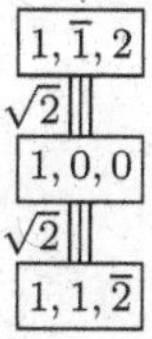

$$\left|\begin{matrix}2&&0&&0\\&1&&0&\\&&1&&\end{matrix}\right\rangle^{3}_{2} = \sqrt{2}\begin{array}{|c|c|}\hline 1&3\\\hline\end{array}$$

$$\left|\begin{matrix}1&&0&&0\\&1&&0&\\&&1&&\end{matrix}\right\rangle^{4}_{0} = \sqrt{2}\begin{array}{|c|c|}\hline 1&4\\\hline\end{array}$$

$$\left|\begin{matrix}1&&0&&\bar{1}\\&1&&0&\\&&1&&\end{matrix}\right\rangle^{5}_{\bar{2}} = \sqrt{2}\begin{array}{|c|c|}\hline 1&5\\\hline\end{array}$$

$$\left|\begin{matrix}2&&0&&0\\&1&&0&\\&&0&&\end{matrix}\right\rangle^{4}_{2} = \sqrt{2}\begin{array}{|c|c|}\hline 2&3\\\hline\end{array}$$

$$\left|\begin{matrix}1&&0&&0\\&1&&0&\\&&0&&\end{matrix}\right\rangle^{5}_{0} = \sqrt{2}\begin{array}{|c|c|}\hline 2&4\\\hline\end{array}$$

$$\left|\begin{matrix}1&&0&&\bar{1}\\&1&&0&\\&&0&&\end{matrix}\right\rangle^{6}_{\bar{2}} = \sqrt{2}\begin{array}{|c|c|}\hline 2&5\\\hline\end{array}$$

Two states $\left|\begin{smallmatrix}1&&0&&0\\&1&&0&\\&&1&&\end{smallmatrix}\right\rangle^{4}_{0}$ and $\left|\begin{smallmatrix}1&&0&&\bar{1}\\&1&&0&\\&&1&&\end{smallmatrix}\right\rangle^{5}_{\bar{2}}$ are the highest weight states of an A_2-triplet with $\boldsymbol{M}^{(2)}=(1,0)$ and an A_2-octet with $\boldsymbol{M}^{(2)}=(1,1)$, respectively:

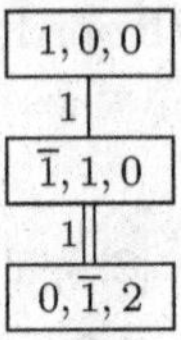

$$\left|\begin{smallmatrix}1&&0&&0\\&1&&0&\\&&1&&\end{smallmatrix}\right\rangle^{4}_{0}=\sqrt{2}\,\begin{array}{|c|c|}\hline 1&4\\\hline\end{array}$$

$$\left|\begin{smallmatrix}1&&0&&0\\&1&&0&\\&&0&&\end{smallmatrix}\right\rangle^{5}_{0}=\sqrt{2}\,\begin{array}{|c|c|}\hline 2&4\\\hline\end{array}$$

$$\left|\begin{smallmatrix}1&&0&&0\\&0&&0&\\&&0&&\end{smallmatrix}\right\rangle^{6}_{2}=\sqrt{2}\,\begin{array}{|c|c|}\hline 3&4\\\hline\end{array}$$

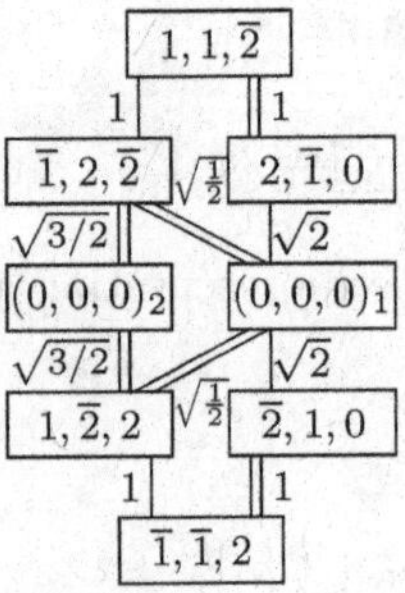

$$\left|\begin{smallmatrix}1&&0&&\bar{1}\\&1&&0&\\&&1&&\end{smallmatrix}\right\rangle^{5}_{\bar{2}}=\sqrt{2}\,\begin{array}{|c|c|}\hline 1&5\\\hline\end{array}$$

$$\left|\begin{smallmatrix}1&&0&&\bar{1}\\&1&&0&\\&&0&&\end{smallmatrix}\right\rangle^{6}_{\bar{2}}=\sqrt{2}\,\begin{array}{|c|c|}\hline 2&5\\\hline\end{array},\qquad \left|\begin{smallmatrix}1&&0&&\bar{1}\\&1&&\bar{1}&\\&&1&&\end{smallmatrix}\right\rangle^{6}_{0}=\sqrt{2}\,\begin{array}{|c|c|}\hline 1&6\\\hline\end{array}$$

$$\left|\begin{smallmatrix}1&&0&&\bar{1}\\&0&&0&\\&&0&&\end{smallmatrix}\right\rangle^{7}_{0},\qquad \left|\begin{smallmatrix}1&&0&&\bar{1}\\&1&&\bar{1}&\\&&0&&\end{smallmatrix}\right\rangle^{7}_{0}=\begin{array}{|c|c|}\hline 1&7\\\hline\end{array}+\begin{array}{|c|c|}\hline 2&6\\\hline\end{array}$$

$$\left|\begin{smallmatrix}1&&0&&\bar{1}\\&0&&\bar{1}&\\&&0&&\end{smallmatrix}\right\rangle^{8}_{2}=\sqrt{2}\,\begin{array}{|c|c|}\hline 3&6\\\hline\end{array},\qquad \left|\begin{smallmatrix}1&&0&&\bar{1}\\&1&&\bar{1}&\\&&\bar{1}&&\end{smallmatrix}\right\rangle^{8}_{0}=\sqrt{2}\,\begin{array}{|c|c|}\hline 2&7\\\hline\end{array}$$

$$\left|\begin{smallmatrix}1&&0&&\bar{1}\\&0&&\bar{1}&\\&&\bar{1}&&\end{smallmatrix}\right\rangle^{9}_{2}=\sqrt{2}\,\begin{array}{|c|c|}\hline 3&7\\\hline\end{array}$$

$$|(0,0,0)_2\rangle=\left|\begin{smallmatrix}1&&0&&\bar{1}\\&0&&0&\\&&0&&\end{smallmatrix}\right\rangle^{7}_{0}=\frac{1}{\sqrt{3}}\left\{-\begin{array}{|c|c|}\hline 1&7\\\hline\end{array}+\begin{array}{|c|c|}\hline 2&6\\\hline\end{array}+2\begin{array}{|c|c|}\hline 3&5\\\hline\end{array}\right\}.$$

The state $\left|\begin{smallmatrix}2&&0&&0\\&0&&0&\\&&0&&\end{smallmatrix}\right\rangle^{5}_{4}$ in the A_2-sixlet with $\boldsymbol{M}^{(2)}=(2,0)$ is the highest weight state of an $\mathcal{A}_3$-quintuplet (see Eq. (8.7)). We calculate the $\mathcal{A}_3$-quintuplet by Eqs. (8.25-28). Letting

$$F_3\left|\begin{matrix}2&&0&&0\\&0&&0&\\&&0&&\end{matrix}\right\rangle^{5}_{4}=2\left|\begin{matrix}1&&0&&0\\&0&&0&\\&&0&&\end{matrix}\right\rangle^{6}_{2},$$

$$F_3\left|\begin{matrix}1&&0&&0\\&0&&0&\\&&0&&\end{matrix}\right\rangle^{6}_{2}=a_1\left|\begin{matrix}1&&0&&\bar{1}\\&0&&0&\\&&0&&\end{matrix}\right\rangle^{7}_{0}+a_2\left|\begin{matrix}0&&0&&0\\&0&&0&\\&&0&&\end{matrix}\right\rangle^{7}_{0},$$

where a_2 is real and positive by choosing the phase of $\left|\begin{smallmatrix}0&&0&&0\\&0&&0&\\&&0&&\end{smallmatrix}\right\rangle^{7}_{0}$, we have

$$\begin{aligned}E_2F_3\left|\begin{matrix}1&&0&&0\\&0&&0&\\&&0&&\end{matrix}\right\rangle^{6}_{2}&=a_1E_2\left|\begin{matrix}1&&0&&\bar{1}\\&0&&0&\\&&0&&\end{matrix}\right\rangle^{7}_{0}=a_1\sqrt{\frac{3}{2}}\left|\begin{matrix}1&&0&&\bar{1}\\&1&&0&\\&&0&&\end{matrix}\right\rangle^{6}_{\bar{2}}\\&=F_3E_2\left|\begin{matrix}1&&0&&0\\&0&&0&\\&&0&&\end{matrix}\right\rangle^{6}_{2}=F_3\left|\begin{matrix}1&&0&&0\\&1&&0&\\&&0&&\end{matrix}\right\rangle^{5}_{0}=\sqrt{2}\left|\begin{matrix}1&&0&&\bar{1}\\&1&&0&\\&&0&&\end{matrix}\right\rangle^{6}_{\bar{2}},\\E_3F_3\left|\begin{matrix}1&&0&&0\\&0&&0&\\&&0&&\end{matrix}\right\rangle^{6}_{2}&=(a_1^2+a_2^2)\left|\begin{matrix}1&&0&&0\\&0&&0&\\&&0&&\end{matrix}\right\rangle^{6}_{2}\\&=(F_3E_3+H_3)\left|\begin{matrix}1&&0&&0\\&0&&0&\\&&0&&\end{matrix}\right\rangle^{6}_{2}=(4+2)\left|\begin{matrix}1&&0&&0\\&0&&0&\\&&0&&\end{matrix}\right\rangle^{6}_{2}.\end{aligned}$$

Thus, $a_1 = \sqrt{4/3}$ and $a_2 = \sqrt{4+2-4/3} = \sqrt{14/3}$.

The block weight diagram of the $\mathcal{A}_3$-quintuplet is as follows.

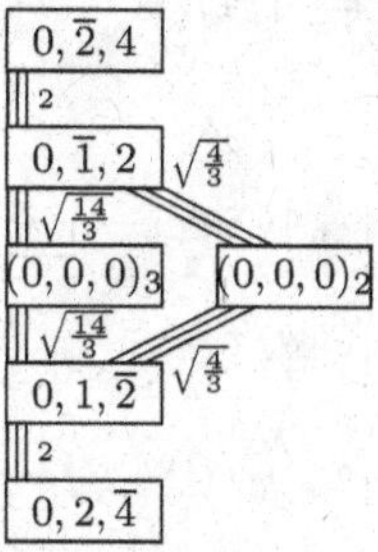

$$\left|{}^{2}{}_{0}{}^{0}_{0}{}_{0}{}^{0}\right\rangle^{5}_{4} = \boxed{3 \mid 3}$$

$$\left|{}^{1}{}_{0}{}^{0}_{0}{}_{0}{}^{0}\right\rangle^{6}_{2} = \sqrt{2}\,\boxed{3 \mid 4}$$

$$\left|{}^{0}{}_{0}{}^{0}_{0}{}_{0}{}^{0}\right\rangle^{7}_{0}, \quad \left|{}^{1}{}_{0}{}^{0}_{0}{}_{0}{}^{\overline{1}}\right\rangle^{7}_{0}$$

$$\left|{}^{0}{}_{0}{}^{0}_{0}{}_{0}{}^{\overline{1}}\right\rangle^{8}_{\overline{2}} = \sqrt{2}\,\boxed{4 \mid 5}$$

$$\left|{}^{0}{}_{0}{}^{0}_{0}{}_{0}{}^{\overline{2}}\right\rangle^{9}_{\overline{4}} = \boxed{5 \mid 5}$$

$$|(0,0,0)_3\rangle = \left|{}^{0}{}_{0}{}^{0}_{0}{}_{0}{}^{0}\right\rangle^{7}_{0} = \sqrt{\tfrac{2}{21}}\left\{\boxed{1 \mid 7} - \boxed{2 \mid 6} + \boxed{3 \mid 5} + 3\,\boxed{4 \mid 4}\right\},$$

$$|(0,0,0)_2\rangle = \left|{}^{1}{}_{0}{}^{0}_{0}{}_{0}{}^{\overline{1}}\right\rangle^{7}_{0} = \tfrac{1}{\sqrt{3}}\left\{-\boxed{1 \mid 7} + \boxed{2 \mid 6} + 2\,\boxed{3 \mid 5}\right\}.$$

The state $|(0,0,0)_3\rangle = \left|{}^{0}{}_{0}{}^{0}_{0}{}_{0}{}^{0}\right\rangle^{7}_{0}$ spans an A_2-singlet. Two states $\left|{}^{0}{}_{0}{}^{0}_{0}{}_{0}{}^{\overline{1}}\right\rangle^{8}_{\overline{2}}$ and $\left|{}^{0}{}_{0}{}^{0}_{0}{}_{0}{}^{\overline{2}}\right\rangle^{9}_{\overline{4}}$ are the highest weight states of an A_2-triplet with $\boldsymbol{M}^{(2)} = (0,1)$ and an A_2-sixptet with $\boldsymbol{M}^{(2)} = (0,2)$, respectively.

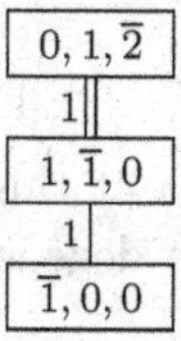

$$\left|{}^{0}{}_{0}{}^{0}_{0}{}_{0}{}^{\overline{1}}\right\rangle^{8}_{\overline{2}} = \sqrt{2}\,\boxed{4 \mid 5}$$

$$\left|{}^{0}{}_{0}{}^{0}_{0}{}_{\overline{1}}{}^{\overline{1}}\right\rangle^{9}_{0} = \sqrt{2}\,\boxed{4 \mid 6}$$

$$\left|{}^{0}{}_{0}{}^{0}_{\overline{1}}{}_{\overline{1}}{}^{\overline{1}}\right\rangle^{10}_{0} = \sqrt{2}\,\boxed{4 \mid 7}$$

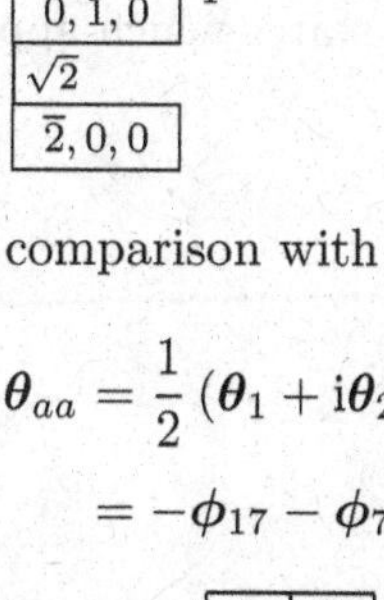

$$\left|{}^{0}{}_{0}{}^{0}_{0}{}_{0}{}^{\overline{2}}\right\rangle^{9}_{\overline{4}} = \boxed{5 \mid 5}$$

$$\left|{}^{0}{}_{0}{}^{0}_{0}{}_{\overline{1}}{}^{\overline{2}}\right\rangle^{10}_{\overline{2}} = \sqrt{2}\,\boxed{5 \mid 6}$$

$$\left|{}^{0}{}_{0}{}^{0}_{0}{}_{\overline{2}}{}^{\overline{2}}\right\rangle^{11}_{0} = \boxed{6 \mid 6}, \quad \left|{}^{0}{}_{0}{}^{0}_{\overline{1}}{}_{\overline{1}}{}^{\overline{2}}\right\rangle^{11}_{\overline{2}} = \sqrt{2}\,\boxed{5 \mid 7}$$

$$\left|{}^{0}{}_{0}{}^{0}_{\overline{1}}{}_{\overline{2}}{}^{\overline{2}}\right\rangle^{12}_{0} = \sqrt{2}\,\boxed{6 \mid 7}$$

$$\left|{}^{0}{}_{0}{}^{0}_{\overline{2}}{}_{\overline{2}}{}^{\overline{2}}\right\rangle^{13}_{0} = \boxed{7 \mid 7}$$

In comparison with trace tensor

$$\begin{aligned}
\sum_{a=1}^{7} \boldsymbol{\theta}_{aa} &= \frac{1}{2}\left(\boldsymbol{\theta}_1 + \mathrm{i}\boldsymbol{\theta}_2\right)\left(\boldsymbol{\theta}_1 - \mathrm{i}\boldsymbol{\theta}_2\right) + \frac{1}{2}\left(\boldsymbol{\theta}_1 - \mathrm{i}\boldsymbol{\theta}_2\right)\left(\boldsymbol{\theta}_1 + \mathrm{i}\boldsymbol{\theta}_2\right) + \cdots + \boldsymbol{\theta}_7\boldsymbol{\theta}_7 \\
&= -\boldsymbol{\phi}_{17} - \boldsymbol{\phi}_{71} + \boldsymbol{\phi}_{26} + \boldsymbol{\phi}_{62} - \boldsymbol{\phi}_{35} - \boldsymbol{\phi}_{53} + \boldsymbol{\phi}_{44} \\
&= -\boxed{1 \mid 7} + \boxed{2 \mid 6} - \boxed{3 \mid 5} + \frac{1}{2}\boxed{4 \mid 4},
\end{aligned} \tag{10.17}$$

three states $|(0,0,0)_1\rangle$, $|(0,0,0)_2\rangle$, and $|(0,0,0)_3\rangle$ all are traceless.

Due to parallel principle, two $\mathcal{A}_3$-triplets are constructed from the states $\left|{}^{1}{}_{0}{}^{0}_{0}{}_{\bar{1}}{}^{\bar{1}}\right\rangle_2^{8}$ and $\left|{}^{1}{}_{0}{}^{0}_{\bar{1}}{}_{\bar{1}}{}^{\bar{1}}\right\rangle_2^{9}$ in the A_2-octet with $\boldsymbol{M}^{(2)} = (1,1)$.

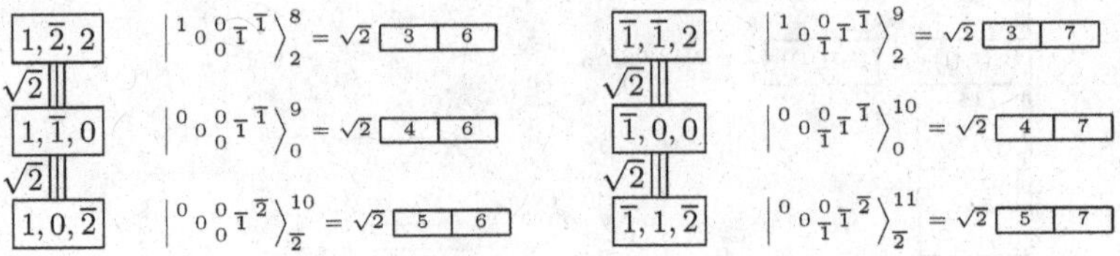

The highest weight representation $D^{\boldsymbol{M}}$ with $\boldsymbol{M} = 2\boldsymbol{w}_1 = (2,0,0)$ of B_3 is 27-dimensional, and contains 6 A_2-multiplets with $A^{(2)} = (2,0)$, $(1,0)$, $(1,1)$, $(0,0)$, $(0,1)$, and $(0,2)$, respectively. This representation contains three simple dominant weights $(2,0,0)$, $(0,1,0)$, $(1,0,0)$ and one triple dominant weight $(0,0,0)$.

Ex. 2. The irreducible representation $[\lambda] = [4,4,2]$ of SO(7).

From Eq. (10.16) the highest weight of this representation is $\boldsymbol{M} = 2\boldsymbol{w}_2 + 2\boldsymbol{w}_3 = (0,2,2)$, where the double degeneracy of the A_2-multiplet occurs in the action of F_3 (see Fig. 8.10).

Since $F_2\ \phi_\ell = \sqrt{2}\ \phi_{\ell+1}$ and $F_2\ \phi_{\ell+1} = \sqrt{2}\ \phi_{\ell+2}$, and **both $\phi_{\ell+1}$ and $\phi_{\ell+2}$ are annihilated by each raising operator of the subalgebra** $A_{\ell-1}$, this is the reason why the double degeneracy of the $A_{\ell-1}$-multiplet may occur in some highest weight representations of B_ℓ. When the double degeneracy occurs, **one has to define one state and then to calculate the another by** Eq. (8.9) which is orthonormal to the first one. Now, we calculate the tensor Young tableaux for the nine states which appear in Fig. 8.10:

$$\begin{aligned}
|(0,2,2)\rangle &= \left|\begin{smallmatrix}4&&4&&2\\&4&&4&\\&&4&&\end{smallmatrix}\right\rangle_2^1 = \begin{array}{|c|c|c|}\hline 1&1&1\\\hline 2&2&2\\\hline 3\\\cline{1-1}\end{array}\,,\\
|(1,0,4)\rangle &= \left|\begin{smallmatrix}4&&4&&2\\&4&&3&\\&&4&&\end{smallmatrix}\right\rangle_4^2 = \sqrt{\tfrac{1}{2}}F_2\,|(0,2,2)\rangle = \sqrt{2}\begin{array}{|c|c|c|}\hline 1&1&1\\\hline 2&2&3\\\hline 3\\\cline{1-1}\end{array}\,,\\
|(0,3,0)\rangle &= \left|\begin{smallmatrix}4&&4&&1\\&4&&4&\\&&4&&\end{smallmatrix}\right\rangle_0^2 = \sqrt{\tfrac{1}{2}}F_3\,|(0,2,2)\rangle = \begin{array}{|c|c|c|}\hline 1&1&1\\\hline 2&2&2\\\hline 4\\\cline{1-1}\end{array}\,,\\
|(0,4,\overline{2})\rangle &= \left|\begin{smallmatrix}4&&4&&0\\&4&&4&\\&&4&&\end{smallmatrix}\right\rangle_{\overline{2}}^3 = \sqrt{\tfrac{1}{2}}F_3\,|(0,3,0)\rangle = \begin{array}{|c|c|c|}\hline 1&1&1\\\hline 2&2&2\\\hline 5\\\cline{1-1}\end{array}\,,\\
|(1,1,2)_1\rangle &= \left|\begin{smallmatrix}4&&4&&1\\&4&&3&\\&&4&&\end{smallmatrix}\right\rangle_2^3 = \sqrt{\tfrac{1}{3}}F_2\,|(0,3,0)\rangle\\
&= \sqrt{3}\begin{array}{|c|c|c|}\hline 1&1&1\\\hline 2&2&3\\\hline 4\\\cline{1-1}\end{array} - \sqrt{\tfrac{1}{3}}\begin{array}{|c|c|c|}\hline 1&1&1\\\hline 2&2&4\\\hline 3\\\cline{1-1}\end{array}\,,
\end{aligned}$$

$$\begin{aligned}
|(1,1,2)_2\rangle &= \left|\begin{smallmatrix}4&&3&&2\\&4&&3&\\&&4&&\end{smallmatrix}\right\rangle_2^3 = \sqrt{\tfrac{3}{8}}\left\{F_3\,|(1,0,4)\rangle - \sqrt{\tfrac{4}{3}}\,|(1,1,2)_1\rangle\right\}\\
&= \sqrt{\tfrac{3}{4}}F_3\begin{array}{|c|c|c|}\hline 1&1&1\\\hline 2&2&3\\\hline 3\\\cline{1-1}\end{array} - \sqrt{\tfrac{3}{2}}\begin{array}{|c|c|c|}\hline 1&1&1\\\hline 2&2&3\\\hline 4\\\cline{1-1}\end{array} + \sqrt{\tfrac{1}{6}}\begin{array}{|c|c|c|}\hline 1&1&1\\\hline 2&2&4\\\hline 3\\\cline{1-1}\end{array}\\
&= \sqrt{\tfrac{8}{3}}\begin{array}{|c|c|c|}\hline 1&1&1\\\hline 2&2&4\\\hline 3\\\cline{1-1}\end{array}\,,\\
|(1,2,0)_1\rangle &= \left|\begin{smallmatrix}4&&4&&0\\&4&&3&\\&&4&&\end{smallmatrix}\right\rangle_0^4 = \frac{1}{2}F_2\,|(0,4,\overline{2})\rangle = \frac{1}{2}F_2\begin{array}{|c|c|c|}\hline 1&1&1\\\hline 2&2&2\\\hline 5\\\cline{1-1}\end{array}\\
&= \frac{1}{2}\begin{array}{|c|c|c|}\hline 1&1&1\\\hline 2&2&2\\\hline 6\\\cline{1-1}\end{array} + \frac{3}{2}\begin{array}{|c|c|c|}\hline 1&1&1\\\hline 2&2&3\\\hline 5\\\cline{1-1}\end{array} - \frac{1}{2}\begin{array}{|c|c|c|}\hline 1&1&1\\\hline 2&2&5\\\hline 3\\\cline{1-1}\end{array}\,.
\end{aligned}$$

We define

$$\begin{aligned}
|(1,2,0)_2\rangle &= \left|\begin{smallmatrix}4&&3&&1\\&4&&3&\\&&4&&\end{smallmatrix}\right\rangle_0^4 = \sqrt{\frac{3}{14}}F_3\,|(1,1,2)_2\rangle\\
&= \sqrt{\frac{4}{7}}F_3\begin{array}{|c|c|c|}\hline 1&1&1\\\hline 2&2&4\\\hline 3\\\cline{1-1}\end{array} = \sqrt{\frac{8}{7}}\left\{\begin{array}{|c|c|c|}\hline 1&1&1\\\hline 2&2&4\\\hline 4\\\cline{1-1}\end{array} + \begin{array}{|c|c|c|}\hline 1&1&1\\\hline 2&2&5\\\hline 3\\\cline{1-1}\end{array}\right\}.
\end{aligned}$$

Then,

$$
\begin{aligned}
|(1,2,0)_3\rangle &= \left|\begin{smallmatrix}4&&3&&1\\&4&&3&\\&&4&&\end{smallmatrix}\right\rangle'^{\,4}_{0} \\
&= \sqrt{\frac{14}{15}}\left\{F_3\,|(1,1,2)_1\rangle - \sqrt{\frac{3}{2}}\,|(1,2,0)_1\rangle - \sqrt{\frac{16}{21}}\,|(1,2,0)_2\rangle\right\} \\
&= \frac{14}{3\sqrt{35}}\left\{3\,\begin{array}{|c|c|c|}\hline 1&1&1\\\hline 2&2&3\\\hline 5\\\cline{1-1}\end{array} + 2\,\begin{array}{|c|c|c|}\hline 1&1&1\\\hline 2&2&4\\\hline 4\\\cline{1-1}\end{array} - \begin{array}{|c|c|c|}\hline 1&1&1\\\hline 2&2&5\\\hline 3\\\cline{1-1}\end{array}\right\} \\
&\quad - \frac{7}{2\sqrt{35}}\left\{\begin{array}{|c|c|c|}\hline 1&1&1\\\hline 2&2&2\\\hline 6\\\cline{1-1}\end{array} + 3\,\begin{array}{|c|c|c|}\hline 1&1&1\\\hline 2&2&3\\\hline 5\\\cline{1-1}\end{array} - \begin{array}{|c|c|c|}\hline 1&1&1\\\hline 2&2&5\\\hline 3\\\cline{1-1}\end{array}\right\} \\
&\quad - \frac{16}{3\sqrt{35}}\left\{\begin{array}{|c|c|c|}\hline 1&1&1\\\hline 2&2&4\\\hline 4\\\cline{1-1}\end{array} + \begin{array}{|c|c|c|}\hline 1&1&1\\\hline 2&2&5\\\hline 3\\\cline{1-1}\end{array}\right\} \\
&= \frac{1}{2\sqrt{35}}\left\{-7\,\begin{array}{|c|c|c|}\hline 1&1&1\\\hline 2&2&2\\\hline 6\\\cline{1-1}\end{array} + 8\,\begin{array}{|c|c|c|}\hline 1&1&1\\\hline 2&2&4\\\hline 4\\\cline{1-1}\end{array} + 7\,\begin{array}{|c|c|c|}\hline 1&1&1\\\hline 2&2&3\\\hline 5\\\cline{1-1}\end{array}\right. \\
&\qquad \left. - 13\,\begin{array}{|c|c|c|}\hline 1&1&1\\\hline 2&2&5\\\hline 3\\\cline{1-1}\end{array}\right\}.
\end{aligned}
$$

Ex. 3. The irreducible representation $[\lambda] = [2,1,0] = [2,1]$ of SO(7).

From Eq. (10.16) the highest weight of this representation is $\boldsymbol{M} = \boldsymbol{w}_1 + \boldsymbol{w}_2 = (1,1,0)$. This representation $(1,1,0)$ contains four A_2-octets, four A_2-sixlets, two A_2-15lets, six A_2-triplets, and one A_2-singlet, where some A_2-multiplets are **double degeneracy**. This representation contains two simple dominant weights $(1,1,0)$ and $(2,0,0)$, two double dominant weights $(0,1,0)$ and $(0,0,2)$, and two quintuple dominant weights $(1,0,0)$ and $(0,0,0)$, whose sizes of Weyl orbits are 24, 6, 12, 8, 6 and 1, respectively. The dimension of this representation is 105:

$$
\begin{aligned}
105 &= 4\times 8 + 4\times 6 + 2\times 15 + 6\times 3 + 1\times 1 \\
&= 24 + 6 + 2\times 12 + 2\times 8 + 5\times 6 + 5\times 1.
\end{aligned}
$$

The detailed calculation for the A_2-multiplets is left to the readers as exercise (see Prob. 3). We are going to calculate only two double degeneracies (levels 6 and 7) of $\mathcal{A}_3$-multiplets in some detail and to give the main results for the remaining calculations.

In the tensor subspace $\mathcal{Y}_1^{[2,1]}\mathcal{T}$, $\mathcal{Y}_1^{[2,1]} = \begin{array}{|c|c|}\hline 1&2\\\hline 3\\\cline{1-1}\end{array}$, we list the typical tensor Young tableaux which are normalized to $\sqrt{6}$, where $a \neq b \neq c \neq a$.

$$\begin{array}{|c|c|}\hline a & a \\ \hline b \\ \cline{1-1}\end{array} = \mathcal{Y}_1^{[2,1]} \phi_{aab} = 2\phi_{aab} - \phi_{baa} - \phi_{aba},$$

$$\sqrt{\frac{3}{2}}\,\begin{array}{|c|c|}\hline a & b \\ \hline c \\ \cline{1-1}\end{array} = \sqrt{\frac{3}{2}}\,\mathcal{Y}_1^{[2,1]} \phi_{abc} = \sqrt{\frac{3}{2}}\left\{\phi_{abc} + \phi_{bac} - \phi_{cba} - \phi_{bca}\right\},$$

$$\sqrt{\frac{3}{2}}\left\{\begin{array}{|c|c|}\hline a & b \\ \hline c \\ \cline{1-1}\end{array} - \begin{array}{|c|c|}\hline a & c \\ \hline b \\ \cline{1-1}\end{array}\right\} = \sqrt{\frac{3}{2}}\left\{\phi_{abc} + \phi_{bac} - \phi_{acb} - \phi_{cab}\right\},$$

$$\sqrt{\frac{1}{2}}\left\{2\begin{array}{|c|c|}\hline a & b \\ \hline c \\ \cline{1-1}\end{array} - \begin{array}{|c|c|}\hline a & c \\ \hline b \\ \cline{1-1}\end{array}\right\}, \qquad \sqrt{\frac{1}{2}}\left\{\begin{array}{|c|c|}\hline a & b \\ \hline c \\ \cline{1-1}\end{array} + \begin{array}{|c|c|}\hline a & c \\ \hline b \\ \cline{1-1}\end{array}\right\},$$

$$\sqrt{\frac{1}{14}}\left\{4\begin{array}{|c|c|}\hline a & b \\ \hline c \\ \cline{1-1}\end{array} - 5\begin{array}{|c|c|}\hline a & c \\ \hline b \\ \cline{1-1}\end{array}\right\}, \qquad \sqrt{\frac{1}{26}}\left\{7\begin{array}{|c|c|}\hline a & b \\ \hline c \\ \cline{1-1}\end{array} - 5\begin{array}{|c|c|}\hline a & c \\ \hline b \\ \cline{1-1}\end{array}\right\}.$$

The highest weight states of the seventeen A_2-multiplets in the representation $D^{\boldsymbol{M}}$ of SO(7) with $\boldsymbol{M} = (1,1,0)$ are listed as follows:

$$|(1,1,0)\rangle = \left|\begin{array}{ccccc} 2 & & 1 & & 0 \\ & 2 & & 1 & \\ & & 2 & & \end{array}\right\rangle_0^1 = \begin{array}{|c|c|}\hline 1 & 1 \\ \hline 2 \\ \cline{1-1}\end{array},$$

$$|(2,0,0)\rangle = \left|\begin{array}{ccccc} 2 & & 0 & & 0 \\ & 2 & & 0 & \\ & & 2 & & \end{array}\right\rangle_0^3 = \begin{array}{|c|c|}\hline 1 & 1 \\ \hline 4 \\ \cline{1-1}\end{array},$$

$$|(2,1,\overline{2})\rangle = \left|\begin{array}{ccccc} 2 & & 0 & & \overline{1} \\ & 2 & & 0 & \\ & & 2 & & \end{array}\right\rangle_{\overline{2}}^4 = \begin{array}{|c|c|}\hline 1 & 1 \\ \hline 5 \\ \cline{1-1}\end{array},$$

$$|(0,1,0)_2\rangle = \left|\begin{array}{ccccc} 1 & & 1 & & 0 \\ & 1 & & 1 & \\ & & 1 & & \end{array}\right\rangle_0^4 = \sqrt{\frac{3}{2}}\,\begin{array}{|c|c|}\hline 1 & 4 \\ \hline 2 \\ \cline{1-1}\end{array},$$

$$|(0,2,\overline{2})_2\rangle = \left|\begin{array}{ccccc} 1 & & 1 & & \overline{1} \\ & 1 & & 1 & \\ & & 1 & & \end{array}\right\rangle_{\overline{2}}^5 = \sqrt{\frac{3}{2}}\,\begin{array}{|c|c|}\hline 1 & 5 \\ \hline 2 \\ \cline{1-1}\end{array},$$

$$|(1,0,0)_4\rangle = \left|\begin{array}{ccccc} 1 & & 0 & & 0 \\ & 1 & & 0 & \\ & & 1 & & \end{array}\right\rangle_0^6 = \frac{\sqrt{3}}{4}\left\{2\begin{array}{|c|c|}\hline 1 & 4 \\ \hline 4 \\ \cline{1-1}\end{array} + \begin{array}{|c|c|}\hline 1 & 5 \\ \hline 3 \\ \cline{1-1}\end{array} - \begin{array}{|c|c|}\hline 1 & 6 \\ \hline 2 \\ \cline{1-1}\end{array}\right\},$$

$$|(1,0,0)_5\rangle = \left|\begin{array}{ccccc}1&&0&&0\\&1&&0&\\&&1&&\end{array}\right\rangle^{\prime\,6}_{0} = \frac{1}{4\sqrt{3}}\left\{4\,\begin{array}{|c|c|}\hline 1&1\\\hline \multicolumn{1}{|c|}{7}\\\cline{1-1}\end{array} - 4\,\begin{array}{|c|c|}\hline 1&2\\\hline \multicolumn{1}{|c|}{6}\\\cline{1-1}\end{array} + 5\,\begin{array}{|c|c|}\hline 1&6\\\hline \multicolumn{1}{|c|}{2}\\\cline{1-1}\end{array}\right.$$
$$\left. +4\,\begin{array}{|c|c|}\hline 1&3\\\hline \multicolumn{1}{|c|}{5}\\\cline{1-1}\end{array} - 5\,\begin{array}{|c|c|}\hline 1&5\\\hline \multicolumn{1}{|c|}{3}\\\cline{1-1}\end{array} + 2\,\begin{array}{|c|c|}\hline 1&4\\\hline \multicolumn{1}{|c|}{4}\\\cline{1-1}\end{array}\right\},$$

$$|(1,1,\overline{2})_1\rangle = \left|\begin{array}{ccccc}1&&0&&\overline{1}\\&1&&0&\\&&1&&\end{array}\right\rangle^{7}_{\overline{2}} = \sqrt{2}\,\begin{array}{|c|c|}\hline 1&4\\\hline \multicolumn{1}{|c|}{5}\\\cline{1-1}\end{array} - \sqrt{\frac{1}{2}}\,\begin{array}{|c|c|}\hline 1&5\\\hline \multicolumn{1}{|c|}{4}\\\cline{1-1}\end{array},$$

$$|(1,1,\overline{2})_2\rangle = \left|\begin{array}{ccccc}1&&0&&\overline{1}\\&1&&0&\\&&1&&\end{array}\right\rangle^{\prime\,7}_{\overline{2}} = \sqrt{\frac{3}{2}}\,\begin{array}{|c|c|}\hline 1&5\\\hline \multicolumn{1}{|c|}{4}\\\cline{1-1}\end{array},$$

$$|(1,2,\overline{4})_2\rangle = \left|\begin{array}{ccccc}1&&0&&\overline{2}\\&1&&0&\\&&1&&\end{array}\right\rangle^{8}_{\overline{4}} = \begin{array}{|c|c|}\hline 1&5\\\hline \multicolumn{1}{|c|}{5}\\\cline{1-1}\end{array},$$

$$|(0,0,0)_5\rangle = \left|\begin{array}{ccccc}0&&0&&0\\&0&&0&\\&&0&&\end{array}\right\rangle^{9}_{0} = \sqrt{\frac{1}{2}}\left\{\begin{array}{|c|c|}\hline 1&4\\\hline \multicolumn{1}{|c|}{7}\\\cline{1-1}\end{array} - \begin{array}{|c|c|}\hline 2&4\\\hline \multicolumn{1}{|c|}{6}\\\cline{1-1}\end{array} + \begin{array}{|c|c|}\hline 3&4\\\hline \multicolumn{1}{|c|}{5}\\\cline{1-1}\end{array}\right\},$$

$$|(2,0,\overline{2})_2\rangle = \left|\begin{array}{ccccc}1&&\overline{1}&&\overline{1}\\&1&&\overline{1}&\\&&1&&\end{array}\right\rangle^{9}_{\overline{2}} = \sqrt{\frac{3}{2}}\left\{\begin{array}{|c|c|}\hline 1&5\\\hline \multicolumn{1}{|c|}{6}\\\cline{1-1}\end{array} - \begin{array}{|c|c|}\hline 1&6\\\hline \multicolumn{1}{|c|}{5}\\\cline{1-1}\end{array}\right\},$$

$$|(0,1,\overline{2})_4\rangle = \left|\begin{array}{ccccc}0&&0&&\overline{1}\\&0&&0&\\&&0&&\end{array}\right\rangle^{10}_{\overline{2}}$$
$$= \sqrt{\frac{3}{10}}\left\{\begin{array}{|c|c|}\hline 1&5\\\hline \multicolumn{1}{|c|}{7}\\\cline{1-1}\end{array} - \begin{array}{|c|c|}\hline 2&5\\\hline \multicolumn{1}{|c|}{6}\\\cline{1-1}\end{array} + \begin{array}{|c|c|}\hline 3&5\\\hline \multicolumn{1}{|c|}{5}\\\cline{1-1}\end{array} + \begin{array}{|c|c|}\hline 4&4\\\hline \multicolumn{1}{|c|}{5}\\\cline{1-1}\end{array}\right\},$$

$$|(0,1,\overline{2})_5\rangle = \left|\begin{array}{ccccc}0&&0&&\overline{1}\\&0&&0&\\&&0&&\end{array}\right\rangle^{\prime\,10}_{\overline{2}} = \sqrt{\frac{1}{120}}\left\{-7\,\begin{array}{|c|c|}\hline 1&5\\\hline \multicolumn{1}{|c|}{7}\\\cline{1-1}\end{array} + 5\,\begin{array}{|c|c|}\hline 1&7\\\hline \multicolumn{1}{|c|}{5}\\\cline{1-1}\end{array}\right.$$
$$\left. +7\,\begin{array}{|c|c|}\hline 2&5\\\hline \multicolumn{1}{|c|}{6}\\\cline{1-1}\end{array} - 5\,\begin{array}{|c|c|}\hline 2&6\\\hline \multicolumn{1}{|c|}{5}\\\cline{1-1}\end{array} - 2\,\begin{array}{|c|c|}\hline 3&5\\\hline \multicolumn{1}{|c|}{5}\\\cline{1-1}\end{array} + 8\,\begin{array}{|c|c|}\hline 4&4\\\hline \multicolumn{1}{|c|}{5}\\\cline{1-1}\end{array}\right\},$$

$$|(0,2,\overline{4})\rangle = \left|\begin{array}{ccccc}0&&0&&\overline{2}\\&0&&0&\\&&0&&\end{array}\right\rangle^{11}_{\overline{4}} = \begin{array}{|c|c|}\hline 4&5\\\hline \multicolumn{1}{|c|}{5}\\\cline{1-1}\end{array},$$

$$|(1,0,\overline{2})_2\rangle = \left|\begin{array}{ccccc}0&&\overline{1}&&\overline{1}\\&0&&\overline{1}&\\&&0&&\end{array}\right\rangle^{12}_{\overline{2}} = \sqrt{\frac{3}{2}}\left\{\begin{array}{|c|c|}\hline 4&5\\\hline \multicolumn{1}{|c|}{6}\\\cline{1-1}\end{array} - \begin{array}{|c|c|}\hline 4&6\\\hline \multicolumn{1}{|c|}{5}\\\cline{1-1}\end{array}\right\},$$

$$|(1,1,\overline{4})\rangle = \left|\begin{array}{ccccc}0&&\overline{1}&&\overline{2}\\&0&&\overline{1}&\\&&0&&\end{array}\right\rangle^{13}_{\overline{4}} = \begin{array}{|c|c|}\hline 5&5\\\hline \multicolumn{1}{|c|}{6}\\\cline{1-1}\end{array}.$$

The highest weight state of $D^{\boldsymbol{M}}$ of SO(7) with $\boldsymbol{M} = (1,1,0)$ is that of an A_2-octet with $\boldsymbol{M}^{(2)} = (1,1)$, where there are six basis states satisfy the condition (8.7) with $\mu = 3$. Four $\mathcal{A}_3$-triplets and two $\mathcal{A}_3$-quintuplets are constructed from them.

$$\begin{array}{ll}
\boxed{2,\overline{1},2} & \left|\begin{array}{ccccc}2&&1&&0\\&2&&0&\\&&a&&\end{array}\right\rangle^{2}_{2} = \begin{array}{|c|c|}\hline 1&1\\\hline \multicolumn{1}{|c|}{3}\\\cline{1-1}\end{array} \\
\sqrt{2}\,\Vert & \\
\boxed{2,0,0} & \left|\begin{array}{ccccc}2&&0&&0\\&2&&0&\\&&a&&\end{array}\right\rangle^{3}_{0} = \begin{array}{|c|c|}\hline 1&1\\\hline \multicolumn{1}{|c|}{4}\\\cline{1-1}\end{array} \\
\sqrt{2}\,\Vert & \\
\boxed{2,1,\overline{2}} & \left|\begin{array}{ccccc}2&&0&&\overline{1}\\&2&&0&\\&&a&&\end{array}\right\rangle^{4}_{\overline{2}} = \begin{array}{|c|c|}\hline 1&1\\\hline \multicolumn{1}{|c|}{5}\\\cline{1-1}\end{array}
\end{array}
\qquad
\begin{array}{ll}
\boxed{(0,0,2)_2} & \left|\begin{array}{ccccc}2&&1&&0\\&1&&1&\\&&1&&\end{array}\right\rangle^{3}_{2} = \sqrt{\frac{3}{2}}\,\begin{array}{|c|c|}\hline 1&3\\\hline \multicolumn{1}{|c|}{2}\\\cline{1-1}\end{array} \\
\sqrt{2}\,\Vert & \\
\boxed{(0,1,0)_2} & \left|\begin{array}{ccccc}1&&1&&0\\&1&&1&\\&&1&&\end{array}\right\rangle^{4}_{0} = \sqrt{\frac{3}{2}}\,\begin{array}{|c|c|}\hline 1&4\\\hline \multicolumn{1}{|c|}{2}\\\cline{1-1}\end{array} \\
\sqrt{2}\,\Vert & \\
\boxed{(0,2,\overline{2})_2} & \left|\begin{array}{ccccc}1&&1&&\overline{1}\\&1&&1&\\&&1&&\end{array}\right\rangle^{5}_{\overline{2}} = \sqrt{\frac{3}{2}}\,\begin{array}{|c|c|}\hline 1&5\\\hline \multicolumn{1}{|c|}{2}\\\cline{1-1}\end{array}
\end{array}$$

where $a = 2$. There are two similar $\mathcal{A}_3$-triplets with $a = 1$ and 0 due to the parallel principle.

The $\mathcal{A}_3$-quintuplets need to be calculated in some detail in terms of Eqs. (8.25-28), because **the double degeneracies of the highest weight states of A_2-multiplets occur** there. The highest weight state of one $\mathcal{A}_3$-quintuplet in the A_2-octet is $|(1,\overline{2},4)\rangle = \left|\begin{smallmatrix}2&&1&&0\\&1&&0&\\&&1&&\end{smallmatrix}\right\rangle^4_4 = \begin{array}{|c|c|}\hline 1&3\\\hline 3&\multicolumn{1}{c}{}\\\cline{1-1}\end{array}$. Letting

$$F_3\left|\begin{smallmatrix}2&&1&&0\\&1&&0&\\&&1&&\end{smallmatrix}\right\rangle^4_4 = a_1\left|\begin{smallmatrix}2&&0&&0\\&1&&0&\\&&1&&\end{smallmatrix}\right\rangle^5_2 + a_2\left|\begin{smallmatrix}1&&1&&0\\&1&&0&\\&&1&&\end{smallmatrix}\right\rangle^5_2$$

we obtain (see Fig. 8.2, Fig. 8.3, and Fig. 8.5),

$$\begin{aligned}
E_2F_3\left|\begin{smallmatrix}2&&1&&0\\&1&&0&\\&&1&&\end{smallmatrix}\right\rangle^4_4 &= E_2\left\{a_1\left|\begin{smallmatrix}2&&0&&0\\&1&&0&\\&&1&&\end{smallmatrix}\right\rangle^5_2 + a_2\left|\begin{smallmatrix}1&&1&&0\\&1&&0&\\&&1&&\end{smallmatrix}\right\rangle^5_2\right\}\\
&= a_1\left|\begin{smallmatrix}2&&0&&0\\&2&&0&\\&&1&&\end{smallmatrix}\right\rangle^4_0 + a_2\left|\begin{smallmatrix}1&&1&&0\\&1&&1&\\&&1&&\end{smallmatrix}\right\rangle^4_0\\
&= F_3E_2\left|\begin{smallmatrix}2&&1&&0\\&1&&0&\\&&1&&\end{smallmatrix}\right\rangle^4_4 = \sqrt{\tfrac{1}{2}}F_3\left|\begin{smallmatrix}2&&1&&0\\&2&&0&\\&&1&&\end{smallmatrix}\right\rangle^3_2 + \sqrt{\tfrac{3}{2}}F_3\left|\begin{smallmatrix}2&&1&&0\\&1&&1&\\&&1&&\end{smallmatrix}\right\rangle^3_2\\
&= \left|\begin{smallmatrix}2&&0&&0\\&2&&0&\\&&1&&\end{smallmatrix}\right\rangle^4_0 + \sqrt{3}\left|\begin{smallmatrix}1&&1&&0\\&1&&1&\\&&1&&\end{smallmatrix}\right\rangle^4_0.
\end{aligned}$$

Thus, $a_1 = 1$, $a_2 = \sqrt{3}$, and

$$\begin{aligned}
F_3\left|\begin{smallmatrix}2&&1&&0\\&1&&0&\\&&1&&\end{smallmatrix}\right\rangle^4_4 &= F_3\begin{array}{|c|c|}\hline 1&3\\\hline 3&\multicolumn{1}{c}{}\\\cline{1-1}\end{array} = \sqrt{2}\left\{\begin{array}{|c|c|}\hline 1&3\\\hline 4&\multicolumn{1}{c}{}\\\cline{1-1}\end{array} + \begin{array}{|c|c|}\hline 1&4\\\hline 3&\multicolumn{1}{c}{}\\\cline{1-1}\end{array}\right\}\\
&= \left|\begin{smallmatrix}2&&0&&0\\&1&&0&\\&&1&&\end{smallmatrix}\right\rangle^5_2 + \sqrt{3}\left|\begin{smallmatrix}1&&1&&0\\&1&&0&\\&&1&&\end{smallmatrix}\right\rangle^5_2\\
&= 1\times\left\{\sqrt{2}\begin{array}{|c|c|}\hline 1&3\\\hline 4&\multicolumn{1}{c}{}\\\cline{1-1}\end{array} - \sqrt{\tfrac{1}{2}}\begin{array}{|c|c|}\hline 1&4\\\hline 3&\multicolumn{1}{c}{}\\\cline{1-1}\end{array}\right\} + \sqrt{3}\times\sqrt{\tfrac{3}{2}}\begin{array}{|c|c|}\hline 1&4\\\hline 3&\multicolumn{1}{c}{}\\\cline{1-1}\end{array}.
\end{aligned}$$

Letting

$$\begin{aligned}
F_3\left|\begin{smallmatrix}1&&1&&0\\&1&&0&\\&&1&&\end{smallmatrix}\right\rangle^5_2 &= b_1\left|\begin{smallmatrix}1&&1&&\overline{1}\\&1&&0&\\&&1&&\end{smallmatrix}\right\rangle^6_0 + b_2\left|\begin{smallmatrix}1&&0&&0\\&1&&0&\\&&1&&\end{smallmatrix}\right\rangle^6_0,\\
F_3\left|\begin{smallmatrix}2&&0&&0\\&1&&0&\\&&1&&\end{smallmatrix}\right\rangle^5_2 &= b_3\left|\begin{smallmatrix}2&&0&&\overline{1}\\&1&&0&\\&&1&&\end{smallmatrix}\right\rangle^6_0 + b_4\left|\begin{smallmatrix}1&&0&&0\\&1&&0&\\&&1&&\end{smallmatrix}\right\rangle^6_0 + \cdots,
\end{aligned}$$

we obtain (see Fig. 8.2, Fig. 8.3, and Fig. 8.6),

$$\begin{aligned}
E_2F_3\left|\begin{smallmatrix}1&&1&&0\\&1&&0&\\&&1&&\end{smallmatrix}\right\rangle^5_2 &= b_1E_2\left|\begin{smallmatrix}1&&1&&\overline{1}\\&1&&0&\\&&1&&\end{smallmatrix}\right\rangle^6_0 = \sqrt{2}\,b_1\left|\begin{smallmatrix}1&&1&&\overline{1}\\&1&&1&\\&&1&&\end{smallmatrix}\right\rangle^5_{\overline{2}}\\
&= F_3E_2\left|\begin{smallmatrix}1&&1&&0\\&1&&0&\\&&1&&\end{smallmatrix}\right\rangle^5_2 = F_3\left|\begin{smallmatrix}1&&1&&0\\&1&&1&\\&&1&&\end{smallmatrix}\right\rangle^4_0 = \sqrt{2}\left|\begin{smallmatrix}1&&1&&\overline{1}\\&1&&1&\\&&1&&\end{smallmatrix}\right\rangle^5_{\overline{2}},
\end{aligned}$$

$$\begin{aligned}
E_2F_3 \left|\begin{smallmatrix} 2 && 0 && 0 \\ & 1 && 0 & \\ && 1 && \end{smallmatrix}\right\rangle^{5}_{2} &= b_3 E_2 \left|\begin{smallmatrix} 2 && 0 && \bar{1} \\ & 1 && 0 & \\ && 1 && \end{smallmatrix}\right\rangle^{6}_{0} = \sqrt{\frac{4}{3}}\, b_3 \left|\begin{smallmatrix} 2 && 0 && \bar{1} \\ & 2 && 0 & \\ && 1 && \end{smallmatrix}\right\rangle^{5}_{\bar{2}} \\
&= F_3E_2 \left|\begin{smallmatrix} 2 && 0 && 0 \\ & 1 && 0 & \\ && 1 && \end{smallmatrix}\right\rangle^{5}_{2} = F_3 \left|\begin{smallmatrix} 2 && 0 && 0 \\ & 2 && 0 & \\ && 1 && \end{smallmatrix}\right\rangle^{4}_{0} = \sqrt{2} \left|\begin{smallmatrix} 2 && 0 && \bar{1} \\ & 2 && 0 & \\ && 1 && \end{smallmatrix}\right\rangle^{5}_{\bar{2}}, \\
E_3F_3 \left|\begin{smallmatrix} 1 && 1 && 0 \\ & 1 && 0 & \\ && 1 && \end{smallmatrix}\right\rangle^{5}_{2} &= \left(b_1^2+b_2^2\right) \left|\begin{smallmatrix} 1 && 1 && 0 \\ & 1 && 0 & \\ && 1 && \end{smallmatrix}\right\rangle^{5}_{2} + b_2b_4 \left|\begin{smallmatrix} 2 && 0 && 0 \\ & 1 && 0 & \\ && 1 && \end{smallmatrix}\right\rangle^{5}_{2} \\
&= \left(F_3E_3+H_3\right) \left|\begin{smallmatrix} 1 && 1 && 0 \\ & 1 && 0 & \\ && 1 && \end{smallmatrix}\right\rangle^{5}_{2} = \left(a_2^2+2\right) \left|\begin{smallmatrix} 1 && 1 && 0 \\ & 1 && 0 & \\ && 1 && \end{smallmatrix}\right\rangle^{5}_{2} + a_2a_1 \left|\begin{smallmatrix} 2 && 0 && 0 \\ & 1 && 0 & \\ && 1 && \end{smallmatrix}\right\rangle^{5}_{2}.
\end{aligned}$$

Choosing the phase of $\left|\begin{smallmatrix} 1 && 0 && 0 \\ & 1 && 0 & \\ && 1 && \end{smallmatrix}\right\rangle^{6}_{0}$ such that b_2 is real and positive, we obtain $b_1 = 1$, $b_3 = \sqrt{3/2}$, $b_2 = \sqrt{3+2-1} = 2$ and $b_4 = \sqrt{3}/2$. Then, the confliction occurs:

$$\begin{aligned}
E_3F_3 \left|\begin{smallmatrix} 2 && 0 && 0 \\ & 1 && 0 & \\ && 1 && \end{smallmatrix}\right\rangle^{5}_{2} &= \left(b_3^2+b_4^2\right) \left|\begin{smallmatrix} 2 && 0 && 0 \\ & 1 && 0 & \\ && 1 && \end{smallmatrix}\right\rangle^{5}_{2} + b_4b_2 \left|\begin{smallmatrix} 1 && 1 && 0 \\ & 1 && 0 & \\ && 1 && \end{smallmatrix}\right\rangle^{5}_{2} \\
&\neq \left(F_3E_3+H_3\right) \left|\begin{smallmatrix} 2 && 0 && 0 \\ & 1 && 0 & \\ && 1 && \end{smallmatrix}\right\rangle^{5}_{2} = \left(a_1^2+2\right) \left|\begin{smallmatrix} 2 && 0 && 0 \\ & 1 && 0 & \\ && 1 && \end{smallmatrix}\right\rangle^{5}_{2} + a_1a_2 \left|\begin{smallmatrix} 1 && 1 && 0 \\ & 1 && 0 & \\ && 1 && \end{smallmatrix}\right\rangle^{5}_{2}, \\
b_3^2+b_4^2 &= \frac{3}{2}+\frac{3}{4} = \frac{9}{4}, \qquad \left(a_1^2+2\right) = 3 = \frac{9}{4}+\frac{3}{4}.
\end{aligned}$$

It means that another highest weight state $\left|\begin{smallmatrix} 1 && 0 && 0 \\ & 1 && 0 & \\ && 1 && \end{smallmatrix}\right\rangle'^{6}_{0}$ with the coefficient $\sqrt{3}/2$ has to occur in $F_3 \left|\begin{smallmatrix} 2 && 0 && 0 \\ & 1 && 0 & \\ && 1 && \end{smallmatrix}\right\rangle^{5}_{2}$.

$$\begin{aligned}
F_3 \left|\begin{smallmatrix} 1 && 1 && 0 \\ & 1 && 0 & \\ && 1 && \end{smallmatrix}\right\rangle^{5}_{2} &= F_3 \sqrt{\tfrac{3}{2}} \begin{array}{|c|c|}\hline 1 & 4 \\ \hline 3 \\ \cline{1-1}\end{array} = \sqrt{3}\left\{ \begin{array}{|c|c|}\hline 1 & 4 \\ \hline 4 \\ \cline{1-1}\end{array} + \begin{array}{|c|c|}\hline 1 & 5 \\ \hline 3 \\ \cline{1-1}\end{array} \right\} \\
&= \left|\begin{smallmatrix} 1 && 1 && \bar{1} \\ & 1 && 0 & \\ && 1 && \end{smallmatrix}\right\rangle^{6}_{0} + 2 \left|\begin{smallmatrix} 1 && 0 && 0 \\ & 1 && 0 & \\ && 1 && \end{smallmatrix}\right\rangle^{6}_{0} = 1 \times \tfrac{\sqrt{3}}{2} \left\{ \begin{array}{|c|c|}\hline 1 & 5 \\ \hline 3 \\ \cline{1-1}\end{array} + \begin{array}{|c|c|}\hline 1 & 6 \\ \hline 2 \\ \cline{1-1}\end{array} \right\} \\
&\quad + 2 \times \frac{\sqrt{3}}{4} \left\{ 2 \begin{array}{|c|c|}\hline 1 & 4 \\ \hline 4 \\ \cline{1-1}\end{array} + \begin{array}{|c|c|}\hline 1 & 5 \\ \hline 3 \\ \cline{1-1}\end{array} - \begin{array}{|c|c|}\hline 1 & 6 \\ \hline 2 \\ \cline{1-1}\end{array} \right\}, \\
\left|\begin{smallmatrix} 1 && 0 && 0 \\ & 1 && 0 & \\ && 1 && \end{smallmatrix}\right\rangle'^{6}_{0} &= \frac{2}{\sqrt{3}} \left\{ F_3 \left|\begin{smallmatrix} 2 && 0 && 0 \\ & 1 && 0 & \\ && 1 && \end{smallmatrix}\right\rangle^{5}_{2} - \sqrt{\frac{3}{2}} \left|\begin{smallmatrix} 2 && 0 && \bar{1} \\ & 1 && 0 & \\ && 1 && \end{smallmatrix}\right\rangle^{6}_{0} - \frac{\sqrt{3}}{2} \left|\begin{smallmatrix} 1 && 0 && 0 \\ & 1 && 0 & \\ && 1 && \end{smallmatrix}\right\rangle^{6}_{0} \right\} \\
&= \frac{2}{\sqrt{3}} F_3 \left\{ \sqrt{2} \begin{array}{|c|c|}\hline 1 & 3 \\ \hline 4 \\ \cline{1-1}\end{array} - \frac{1}{\sqrt{2}} \begin{array}{|c|c|}\hline 1 & 4 \\ \hline 3 \\ \cline{1-1}\end{array} \right\} - \sqrt{2} \left|\begin{smallmatrix} 2 && 0 && \bar{1} \\ & 1 && 0 & \\ && 1 && \end{smallmatrix}\right\rangle^{6}_{0} - \left|\begin{smallmatrix} 1 && 0 && 0 \\ & 1 && 0 & \\ && 1 && \end{smallmatrix}\right\rangle^{6}_{0}
\end{aligned}$$

$$
\begin{aligned}
&= \frac{2}{\sqrt{3}}\left\{2\,\begin{array}{|c|c|}\hline 1 & 3 \\ \hline 5 & \multicolumn{1}{c}{} \\ \cline{1-1}\end{array} - \begin{array}{|c|c|}\hline 1 & 5 \\ \hline 3 & \multicolumn{1}{c}{} \\ \cline{1-1}\end{array} + \begin{array}{|c|c|}\hline 1 & 4 \\ \hline 4 & \multicolumn{1}{c}{} \\ \cline{1-1}\end{array}\right\} - \sqrt{2}\times\frac{1}{2\sqrt{6}}\left\{-2\,\begin{array}{|c|c|}\hline 1 & 1 \\ \hline 7 & \multicolumn{1}{c}{} \\ \cline{1-1}\end{array}\right. \\
&\quad \left. +2\,\begin{array}{|c|c|}\hline 1 & 2 \\ \hline 6 & \multicolumn{1}{c}{} \\ \cline{1-1}\end{array} - \begin{array}{|c|c|}\hline 1 & 6 \\ \hline 2 & \multicolumn{1}{c}{} \\ \cline{1-1}\end{array} + 6\,\begin{array}{|c|c|}\hline 1 & 3 \\ \hline 5 & \multicolumn{1}{c}{} \\ \cline{1-1}\end{array} - 3\,\begin{array}{|c|c|}\hline 1 & 5 \\ \hline 3 & \multicolumn{1}{c}{} \\ \cline{1-1}\end{array}\right\} \\
&\quad - \frac{\sqrt{3}}{4}\left\{2\,\begin{array}{|c|c|}\hline 1 & 4 \\ \hline 4 & \multicolumn{1}{c}{} \\ \cline{1-1}\end{array} + \begin{array}{|c|c|}\hline 1 & 5 \\ \hline 3 & \multicolumn{1}{c}{} \\ \cline{1-1}\end{array} - \begin{array}{|c|c|}\hline 1 & 6 \\ \hline 2 & \multicolumn{1}{c}{} \\ \cline{1-1}\end{array}\right\} \\
&= \frac{1}{4\sqrt{3}}\left\{4\,\begin{array}{|c|c|}\hline 1 & 1 \\ \hline 7 & \multicolumn{1}{c}{} \\ \cline{1-1}\end{array} - 4\,\begin{array}{|c|c|}\hline 1 & 2 \\ \hline 6 & \multicolumn{1}{c}{} \\ \cline{1-1}\end{array} + 5\,\begin{array}{|c|c|}\hline 1 & 6 \\ \hline 2 & \multicolumn{1}{c}{} \\ \cline{1-1}\end{array} + 4\,\begin{array}{|c|c|}\hline 1 & 3 \\ \hline 5 & \multicolumn{1}{c}{} \\ \cline{1-1}\end{array}\right. \\
&\quad \left. - 5\,\begin{array}{|c|c|}\hline 1 & 5 \\ \hline 3 & \multicolumn{1}{c}{} \\ \cline{1-1}\end{array} + 2\,\begin{array}{|c|c|}\hline 1 & 4 \\ \hline 4 & \multicolumn{1}{c}{} \\ \cline{1-1}\end{array}\right\}.
\end{aligned}
$$

The only possible basis state obtained by the action of F_3 on each of all four states $\left|\begin{smallmatrix} 2 & & 0 & & \bar{1} \\ & 1 & & 0 & \\ & & 1 & & \end{smallmatrix}\right\rangle^6_0$, $\left|\begin{smallmatrix} 1 & & 1 & & \bar{1} \\ & 1 & & 0 & \\ & & 1 & & \end{smallmatrix}\right\rangle^6_0$, $\left|\begin{smallmatrix} 1 & & 0 & & 0 \\ & 1 & & 0 & \\ & & 1 & & \end{smallmatrix}\right\rangle^6_0$, and $\left|\begin{smallmatrix} 1 & & 0 & & 0 \\ & 1 & & 0 & \\ & & 1 & & \end{smallmatrix}\right\rangle'^6_0$ in the $\mathcal{A}_3$-quintuplet is $\left|\begin{smallmatrix} 1 & & 0 & & \bar{1} \\ & 1 & & 0 & \\ & & 1 & & \end{smallmatrix}\right\rangle^7_{\bar{2}}$, which is the highest weight state of an A$_2$-octet with $M^{(2)} = (1,1)$. We are going to show that the double degeneracy of $\left|\begin{smallmatrix} 1 & & 0 & & \bar{1} \\ & 1 & & 0 & \\ & & 1 & & \end{smallmatrix}\right\rangle^7_{\bar{2}}$ has to occur. In fact, letting

$$
\begin{aligned}
F_3\left|\begin{smallmatrix} 2 & & 0 & & \bar{1} \\ & 1 & & 0 & \\ & & 1 & & \end{smallmatrix}\right\rangle^6_0 &= c_1\left|\begin{smallmatrix} 1 & & 0 & & \bar{1} \\ & 1 & & 0 & \\ & & 1 & & \end{smallmatrix}\right\rangle^7_{\bar{2}}, \\
F_3\left|\begin{smallmatrix} 1 & & 1 & & \bar{1} \\ & 1 & & 0 & \\ & & 1 & & \end{smallmatrix}\right\rangle^6_0 &= c_2\left|\begin{smallmatrix} 1 & & 0 & & \bar{1} \\ & 1 & & 0 & \\ & & 1 & & \end{smallmatrix}\right\rangle^7_{\bar{2}} + c_3\left|\begin{smallmatrix} 1 & & 0 & & \bar{1} \\ & 1 & & 0 & \\ & & 1 & & \end{smallmatrix}\right\rangle'^7_{\bar{2}}, \\
F_3\left|\begin{smallmatrix} 1 & & 0 & & 0 \\ & 1 & & 0 & \\ & & 1 & & \end{smallmatrix}\right\rangle^6_0 &= c_4\left|\begin{smallmatrix} 1 & & 0 & & \bar{1} \\ & 1 & & 0 & \\ & & 1 & & \end{smallmatrix}\right\rangle^7_{\bar{2}} + c_5\left|\begin{smallmatrix} 1 & & 0 & & \bar{1} \\ & 1 & & 0 & \\ & & 1 & & \end{smallmatrix}\right\rangle'^7_{\bar{2}}, \\
F_3\left|\begin{smallmatrix} 1 & & 0 & & 0 \\ & 1 & & 0 & \\ & & 1 & & \end{smallmatrix}\right\rangle'^6_0 &= c_6\left|\begin{smallmatrix} 1 & & 0 & & \bar{1} \\ & 1 & & 0 & \\ & & 1 & & \end{smallmatrix}\right\rangle^7_{\bar{2}} + c_7\left|\begin{smallmatrix} 1 & & 0 & & \bar{1} \\ & 1 & & 0 & \\ & & 1 & & \end{smallmatrix}\right\rangle'^7_{\bar{2}},
\end{aligned}
$$

where c_1 is real and positive by choosing the phases of $\left|\begin{smallmatrix} 1 & & 0 & & \bar{1} \\ & 1 & & 0 & \\ & & 1 & & \end{smallmatrix}\right\rangle^7_{\bar{2}}$, we obtain

$$
\begin{aligned}
E_3F_3\left|\begin{smallmatrix} 2 & & 0 & & \bar{1} \\ & 1 & & 0 & \\ & & 1 & & \end{smallmatrix}\right\rangle^6_0 &= c_1E_3\left|\begin{smallmatrix} 1 & & 0 & & \bar{1} \\ & 1 & & 0 & \\ & & 1 & & \end{smallmatrix}\right\rangle^7_{\bar{2}} \\
&= c_1\left\{c_1\left|\begin{smallmatrix} 2 & & 0 & & \bar{1} \\ & 1 & & 0 & \\ & & 1 & & \end{smallmatrix}\right\rangle^6_0 + c_2\left|\begin{smallmatrix} 1 & & 1 & & \bar{1} \\ & 1 & & 0 & \\ & & 1 & & \end{smallmatrix}\right\rangle^6_0 + c_4\left|\begin{smallmatrix} 1 & & 0 & & 0 \\ & 1 & & 0 & \\ & & 1 & & \end{smallmatrix}\right\rangle^6_0 + c_6\left|\begin{smallmatrix} 1 & & 0 & & 0 \\ & 1 & & 0 & \\ & & 1 & & \end{smallmatrix}\right\rangle'^6_0\right\} \\
&= (F_3E_3 + H_3)\left|\begin{smallmatrix} 2 & & 0 & & \bar{1} \\ & 1 & & 0 & \\ & & 1 & & \end{smallmatrix}\right\rangle^6_0 = \sqrt{\tfrac{3}{2}}\,F_3\left|\begin{smallmatrix} 2 & & 0 & & 0 \\ & 1 & & 0 & \\ & & 1 & & \end{smallmatrix}\right\rangle^5_2 \\
&= \tfrac{3}{2}\left|\begin{smallmatrix} 2 & & 0 & & \bar{1} \\ & 1 & & 0 & \\ & & 1 & & \end{smallmatrix}\right\rangle^6_0 + \tfrac{3}{2\sqrt{2}}\left|\begin{smallmatrix} 1 & & 0 & & 0 \\ & 1 & & 0 & \\ & & 1 & & \end{smallmatrix}\right\rangle^6_0 + \tfrac{3}{2\sqrt{2}}\left|\begin{smallmatrix} 1 & & 0 & & 0 \\ & 1 & & 0 & \\ & & 1 & & \end{smallmatrix}\right\rangle'^6_0.
\end{aligned}
$$

Thus, $c_1 = \sqrt{3/2}$, $c_2 = 0$, and $c_4 = c_6 = \sqrt{3}/2$. Furthermore, choosing the phase of $\left|\begin{smallmatrix} 1 & & 0 & & \bar{1} \\ & 1 & & 0 & \\ & & 1 & & \end{smallmatrix}\right\rangle'^7_{\bar{2}}$ such that c_3 is real and positive, we have

$$\begin{aligned}
E_3F_3\left|\begin{smallmatrix}1&&1&&\bar1\\&1&&0&\\&&1&&\end{smallmatrix}\right\rangle^6_0 &= c_3E_3\left|\begin{smallmatrix}1&&0&&\bar1\\&1&&0&\\&&1&&\end{smallmatrix}\right\rangle'^7_{\bar2}\\
&= c_3\left\{c_3\left|\begin{smallmatrix}1&&1&&\bar1\\&1&&0&\\&&1&&\end{smallmatrix}\right\rangle^6_0 + c_5\left|\begin{smallmatrix}1&&0&&0\\&1&&0&\\&&1&&\end{smallmatrix}\right\rangle^6_0 + c_7\left|\begin{smallmatrix}1&&0&&0\\&1&&0&\\&&1&&\end{smallmatrix}\right\rangle'^6_0\right\}\\
&= (F_3E_3+H_3)\left|\begin{smallmatrix}1&&1&&\bar1\\&1&&0&\\&&1&&\end{smallmatrix}\right\rangle^6_0 = \left|\begin{smallmatrix}1&&1&&\bar1\\&1&&0&\\&&1&&\end{smallmatrix}\right\rangle^6_0 + 2\left|\begin{smallmatrix}1&&0&&0\\&1&&0&\\&&1&&\end{smallmatrix}\right\rangle^6_0.
\end{aligned}$$

Then, $c_3=1$, $c_5=2$, and $c_7=0$. The block weight diagram of this $\mathcal{A}_3$-quintuplet with the highest weight state $\left|\begin{smallmatrix}2&&1&&0\\&1&&0&\\&&a&&\end{smallmatrix}\right\rangle^4_4$ where $a=1$ is given as follows.

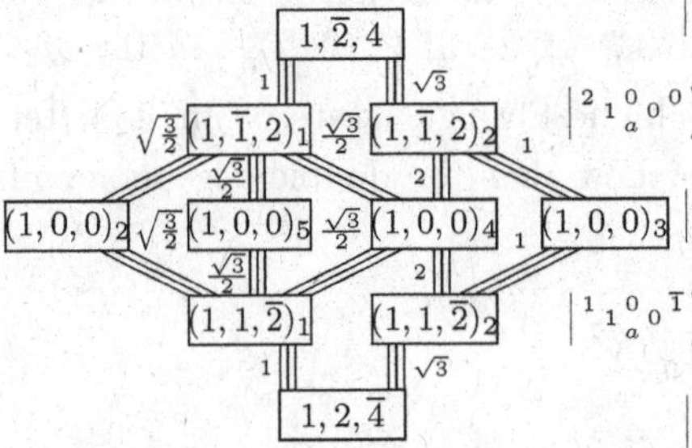

$$\left|\begin{smallmatrix}2&&1&&0\\&1&&0&\\&&a&&\end{smallmatrix}\right\rangle^4_4 = \begin{array}{|c|c|}\hline 1&3\\\hline 3\\\cline{1-1}\end{array}$$

$$\left|\begin{smallmatrix}2&&0&&0\\&1&&0&\\&&a&&\end{smallmatrix}\right\rangle^5_2,\quad \left|\begin{smallmatrix}1&&1&&0\\&1&&0&\\&&a&&\end{smallmatrix}\right\rangle^5_2 = \sqrt{\tfrac{3}{2}}\begin{array}{|c|c|}\hline 1&4\\\hline 3\\\cline{1-1}\end{array}$$

$$\left|\begin{smallmatrix}2&&0&&\bar1\\&1&&0&\\&&a&&\end{smallmatrix}\right\rangle^6_0,\ \left|\begin{smallmatrix}1&&0&&0\\&1&&0&\\&&a&&\end{smallmatrix}\right\rangle'^6_0,\ \left|\begin{smallmatrix}1&&0&&0\\&1&&0&\\&&a&&\end{smallmatrix}\right\rangle^6_0,\ \left|\begin{smallmatrix}1&&1&&\bar1\\&1&&0&\\&&a&&\end{smallmatrix}\right\rangle^6_0$$

$$\left|\begin{smallmatrix}1&&0&&\bar1\\&1&&0&\\&&a&&\end{smallmatrix}\right\rangle^7_{\bar2},\quad \left|\begin{smallmatrix}1&&0&&\bar1\\&1&&0&\\&&a&&\end{smallmatrix}\right\rangle'^7_{\bar2} = \sqrt{\tfrac{3}{2}}\begin{array}{|c|c|}\hline 1&5\\\hline 4\\\cline{1-1}\end{array}$$

$$\left|\begin{smallmatrix}1&&0&&\bar2\\&1&&0&\\&&a&&\end{smallmatrix}\right\rangle^8_{\bar4} = \begin{array}{|c|c|}\hline 1&5\\\hline 5\\\cline{1-1}\end{array}$$

$$\left|\begin{smallmatrix}2&&0&&0\\&1&&0&\\&&a&&\end{smallmatrix}\right\rangle^5_2 = \sqrt{2}\begin{array}{|c|c|}\hline 1&3\\\hline 4\\\cline{1-1}\end{array} - \sqrt{\tfrac{1}{2}}\begin{array}{|c|c|}\hline 1&4\\\hline 3\\\cline{1-1}\end{array}$$

$$\left|\begin{smallmatrix}2&&0&&\bar1\\&1&&0&\\&&a&&\end{smallmatrix}\right\rangle^6_0 = \frac{1}{2\sqrt6}\left\{-2\begin{array}{|c|c|}\hline 1&1\\\hline 7\\\cline{1-1}\end{array} + 2\begin{array}{|c|c|}\hline 1&2\\\hline 6\\\cline{1-1}\end{array} - \begin{array}{|c|c|}\hline 1&6\\\hline 2\\\cline{1-1}\end{array} + 6\begin{array}{|c|c|}\hline 1&3\\\hline 5\\\cline{1-1}\end{array} - 3\begin{array}{|c|c|}\hline 1&5\\\hline 3\\\cline{1-1}\end{array}\right\}$$

$$\begin{aligned}\left|\begin{smallmatrix}1&&0&&0\\&1&&0&\\&&a&&\end{smallmatrix}\right\rangle'^6_0 = \frac{1}{4\sqrt3}\Bigg\{&4\begin{array}{|c|c|}\hline 1&1\\\hline 7\\\cline{1-1}\end{array} - 4\begin{array}{|c|c|}\hline 1&2\\\hline 6\\\cline{1-1}\end{array} + 5\begin{array}{|c|c|}\hline 1&6\\\hline 2\\\cline{1-1}\end{array} + 4\begin{array}{|c|c|}\hline 1&3\\\hline 5\\\cline{1-1}\end{array}\\
&-5\begin{array}{|c|c|}\hline 1&5\\\hline 3\\\cline{1-1}\end{array} + 2\begin{array}{|c|c|}\hline 1&4\\\hline 4\\\cline{1-1}\end{array}\Bigg\}\end{aligned}$$

$$\left|\begin{smallmatrix}1&&0&&0\\&1&&0&\\&&a&&\end{smallmatrix}\right\rangle^6_0 = \frac{\sqrt3}{4}\left\{2\begin{array}{|c|c|}\hline 1&4\\\hline 4\\\cline{1-1}\end{array} + \begin{array}{|c|c|}\hline 1&5\\\hline 3\\\cline{1-1}\end{array} - \begin{array}{|c|c|}\hline 1&6\\\hline 2\\\cline{1-1}\end{array}\right\}$$

$$\left|\begin{smallmatrix}1&&1&&\bar1\\&1&&0&\\&&a&&\end{smallmatrix}\right\rangle^6_0 = \frac{\sqrt3}{2}\left\{\begin{array}{|c|c|}\hline 1&5\\\hline 3\\\cline{1-1}\end{array} + \begin{array}{|c|c|}\hline 1&6\\\hline 2\\\cline{1-1}\end{array}\right\}$$

$$\left|\begin{smallmatrix}1&&0&&\bar1\\&1&&0&\\&&a&&\end{smallmatrix}\right\rangle^7_{\bar2} = \sqrt2\begin{array}{|c|c|}\hline 1&4\\\hline 5\\\cline{1-1}\end{array} - \sqrt{\tfrac12}\begin{array}{|c|c|}\hline 1&5\\\hline 4\\\cline{1-1}\end{array}$$

There is a similar $\mathcal{A}_3$-quintuplet with $a=0$ due to the parallel principle. $E_3F_3 = F_3E_3 + H_3$ holds evidently when it applies to the states $\left|\begin{smallmatrix}1&&0&&0\\&1&&0&\\&&1&&\end{smallmatrix}\right\rangle^6_0$ and $\left|\begin{smallmatrix}1&&0&&0\\&1&&0&\\&&1&&\end{smallmatrix}\right\rangle'^6_0$. The last state $\left|\begin{smallmatrix}1&&0&&\bar2\\&1&&0&\\&&1&&\end{smallmatrix}\right\rangle^8_{\bar4}$ is calculated as follows:

$$\begin{aligned}
F_3\left|\begin{smallmatrix}1&&0&&\bar1\\&1&&0&\\&&1&&\end{smallmatrix}\right\rangle^7_{\bar2} &= F_3\left\{\sqrt2\begin{array}{|c|c|}\hline 1&4\\\hline 5\\\cline{1-1}\end{array} - \sqrt{\tfrac12}\begin{array}{|c|c|}\hline 1&5\\\hline 4\\\cline{1-1}\end{array}\right\} = \begin{array}{|c|c|}\hline 1&5\\\hline 5\\\cline{1-1}\end{array} = \left|\begin{smallmatrix}1&&0&&\bar2\\&1&&0&\\&&1&&\end{smallmatrix}\right\rangle^8_{\bar4},\\
F_3\left|\begin{smallmatrix}1&&0&&\bar1\\&1&&0&\\&&1&&\end{smallmatrix}\right\rangle'^7_{\bar2} &= F_3\sqrt{\frac32}\begin{array}{|c|c|}\hline 1&5\\\hline 4\\\cline{1-1}\end{array} = \sqrt3\begin{array}{|c|c|}\hline 1&5\\\hline 5\\\cline{1-1}\end{array} = \sqrt3\left|\begin{smallmatrix}1&&0&&\bar2\\&1&&0&\\&&1&&\end{smallmatrix}\right\rangle^8_{\bar4}.
\end{aligned}$$

The remaining $\mathcal{A}_3$-multiplets are as follows.

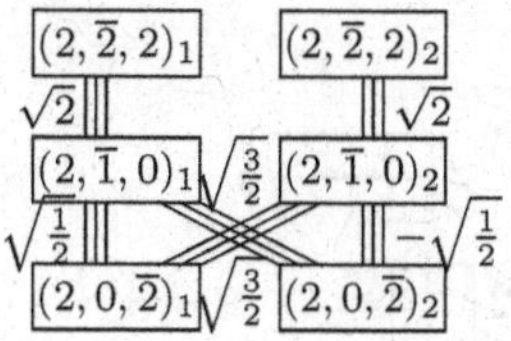

$$\left|\begin{smallmatrix}2&&0&&\bar{1}\\&1&&\bar{1}&\\&&a&&\end{smallmatrix}\right\rangle^{7}_{2},\quad \left|\begin{smallmatrix}1&&1&&\bar{1}\\&1&&\bar{1}&\\&&a&&\end{smallmatrix}\right\rangle^{7}_{2}=\sqrt{\tfrac{3}{2}}\begin{array}{|c|c|}\hline 1&6\\\hline 3\\\cline{1-1}\end{array}$$

$$\left|\begin{smallmatrix}1&&0&&\bar{1}\\&1&&\bar{1}&\\&&a&&\end{smallmatrix}\right\rangle^{8}_{0},\quad \left|\begin{smallmatrix}1&&0&&\bar{1}\\&1&&\bar{1}&\\&&a&&\end{smallmatrix}\right\rangle^{\prime\,8}_{0}=\sqrt{\tfrac{3}{2}}\begin{array}{|c|c|}\hline 1&6\\\hline 4\\\cline{1-1}\end{array}$$

$$\left|\begin{smallmatrix}1&&0&&\bar{2}\\&1&&\bar{1}&\\&&a&&\end{smallmatrix}\right\rangle^{9}_{\bar{2}},\quad \left|\begin{smallmatrix}1&&\bar{1}&&\bar{1}\\&1&&\bar{1}&\\&&a&&\end{smallmatrix}\right\rangle^{9}_{\bar{2}}=\sqrt{\tfrac{3}{2}}\left\{\begin{array}{|c|c|}\hline 1&5\\\hline 6\\\cline{1-1}\end{array}-\begin{array}{|c|c|}\hline 1&6\\\hline 5\\\cline{1-1}\end{array}\right\}$$

$$\left|\begin{smallmatrix}2&&0&&\bar{1}\\&1&&\bar{1}&\\&&a&&\end{smallmatrix}\right\rangle^{7}_{2}=\sqrt{\tfrac{1}{2}}\left\{2\begin{array}{|c|c|}\hline 1&3\\\hline 6\\\cline{1-1}\end{array}-\begin{array}{|c|c|}\hline 1&6\\\hline 3\\\cline{1-1}\end{array}\right\},$$

$$\left|\begin{smallmatrix}1&&0&&\bar{1}\\&1&&\bar{1}&\\&&a&&\end{smallmatrix}\right\rangle^{8}_{0}=\sqrt{\tfrac{1}{2}}\left\{2\begin{array}{|c|c|}\hline 1&4\\\hline 6\\\cline{1-1}\end{array}-\begin{array}{|c|c|}\hline 1&6\\\hline 4\\\cline{1-1}\end{array}\right\},$$

$$\left|\begin{smallmatrix}1&&0&&\bar{2}\\&1&&\bar{1}&\\&&a&&\end{smallmatrix}\right\rangle^{9}_{\bar{2}}=\sqrt{\tfrac{1}{2}}\left\{\begin{array}{|c|c|}\hline 1&5\\\hline 6\\\cline{1-1}\end{array}+\begin{array}{|c|c|}\hline 1&6\\\hline 5\\\cline{1-1}\end{array}\right\}$$

where $a = 1$. There are two similar $\mathcal{A}_3$-multiplets with $a = 0$ and $\bar{1}$ due to the parallel principle.

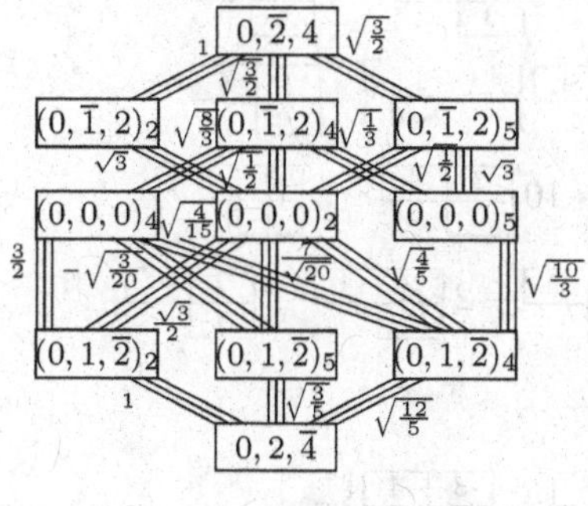

$$\left|\begin{smallmatrix}2&&0&&0\\&0&&0&\\&&0&&\end{smallmatrix}\right\rangle^{7}_{4}=\begin{array}{|c|c|}\hline 3&3\\\hline 4\\\cline{1-1}\end{array}$$

$$\left|\begin{smallmatrix}2&&0&&\bar{1}\\&0&&0&\\&&0&&\end{smallmatrix}\right\rangle^{8}_{2},\quad \left|\begin{smallmatrix}1&&0&&0\\&0&&0&\\&&0&&\end{smallmatrix}\right\rangle^{8}_{2},\quad \left|\begin{smallmatrix}1&&0&&0\\&0&&0&\\&&0&&\end{smallmatrix}\right\rangle^{\prime\,8}_{2}$$

$$\left|\begin{smallmatrix}1&&0&&\bar{1}\\&0&&0&\\&&0&&\end{smallmatrix}\right\rangle^{\prime\,9}_{0},\quad \left|\begin{smallmatrix}1&&0&&\bar{1}\\&0&&0&\\&&0&&\end{smallmatrix}\right\rangle^{9}_{0},\quad \left|\begin{smallmatrix}0&&0&&0\\&0&&0&\\&&0&&\end{smallmatrix}\right\rangle^{9}_{0}$$

$$\left|\begin{smallmatrix}1&&0&&\bar{2}\\&0&&0&\\&&0&&\end{smallmatrix}\right\rangle^{10}_{\bar{2}},\quad \left|\begin{smallmatrix}0&&0&&\bar{1}\\&0&&0&\\&&0&&\end{smallmatrix}\right\rangle^{\prime\,10}_{\bar{2}},\quad \left|\begin{smallmatrix}0&&0&&\bar{1}\\&0&&0&\\&&0&&\end{smallmatrix}\right\rangle^{10}_{\bar{2}}$$

$$\left|\begin{smallmatrix}0&&0&&\bar{2}\\&0&&0&\\&&0&&\end{smallmatrix}\right\rangle^{11}_{\bar{4}}=\begin{array}{|c|c|}\hline 4&5\\\hline 5\\\cline{1-1}\end{array}$$

$$\left|\begin{smallmatrix}2&&0&&\bar{1}\\&0&&0&\\&&0&&\end{smallmatrix}\right\rangle^{8}_{2}=\frac{1}{2\sqrt{2}}\left\{-2\begin{array}{|c|c|}\hline 1&3\\\hline 7\\\cline{1-1}\end{array}+\begin{array}{|c|c|}\hline 1&7\\\hline 3\\\cline{1-1}\end{array}+2\begin{array}{|c|c|}\hline 2&3\\\hline 6\\\cline{1-1}\end{array}-\begin{array}{|c|c|}\hline 2&6\\\hline 3\\\cline{1-1}\end{array}+2\begin{array}{|c|c|}\hline 3&3\\\hline 5\\\cline{1-1}\end{array}\right\},$$

$$\left|\begin{smallmatrix}1&&0&&0\\&0&&0&\\&&0&&\end{smallmatrix}\right\rangle^{8}_{2}=\frac{\sqrt{3}}{4}\left\{-\begin{array}{|c|c|}\hline 1&7\\\hline 3\\\cline{1-1}\end{array}+\begin{array}{|c|c|}\hline 2&6\\\hline 3\\\cline{1-1}\end{array}+2\begin{array}{|c|c|}\hline 3&4\\\hline 4\\\cline{1-1}\end{array}\right\},$$

$$\left|\begin{smallmatrix}1&&0&&0\\&0&&0&\\&&0&&\end{smallmatrix}\right\rangle^{\prime\,8}_{2}=\frac{1}{4\sqrt{3}}\left\{4\begin{array}{|c|c|}\hline 1&3\\\hline 7\\\cline{1-1}\end{array}+\begin{array}{|c|c|}\hline 1&7\\\hline 3\\\cline{1-1}\end{array}-4\begin{array}{|c|c|}\hline 2&3\\\hline 6\\\cline{1-1}\end{array}-\begin{array}{|c|c|}\hline 2&6\\\hline 3\\\cline{1-1}\end{array}+4\begin{array}{|c|c|}\hline 3&3\\\hline 5\\\cline{1-1}\end{array}+2\begin{array}{|c|c|}\hline 3&4\\\hline 4\\\cline{1-1}\end{array}\right\},$$

$$\left|\begin{smallmatrix}1&&0&&\bar{1}\\&0&&0&\\&&0&&\end{smallmatrix}\right\rangle^{\prime\,9}_{0}=\frac{1}{2}\left\{-\begin{array}{|c|c|}\hline 1&7\\\hline 4\\\cline{1-1}\end{array}+\begin{array}{|c|c|}\hline 2&6\\\hline 4\\\cline{1-1}\end{array}+2\begin{array}{|c|c|}\hline 3&5\\\hline 4\\\cline{1-1}\end{array}\right\},$$

$$\left|\begin{smallmatrix}1&&0&&\bar{1}\\&0&&0&\\&&0&&\end{smallmatrix}\right\rangle^{9}_{0}=\frac{1}{2\sqrt{3}}\left\{-2\begin{array}{|c|c|}\hline 1&4\\\hline 7\\\cline{1-1}\end{array}+\begin{array}{|c|c|}\hline 1&7\\\hline 4\\\cline{1-1}\end{array}+2\begin{array}{|c|c|}\hline 2&4\\\hline 6\\\cline{1-1}\end{array}-\begin{array}{|c|c|}\hline 2&6\\\hline 4\\\cline{1-1}\end{array}+4\begin{array}{|c|c|}\hline 3&4\\\hline 5\\\cline{1-1}\end{array}-2\begin{array}{|c|c|}\hline 3&5\\\hline 4\\\cline{1-1}\end{array}\right\},$$

$$\left|\begin{smallmatrix}0&&0&&0\\&0&&0&\\&&0&&\end{smallmatrix}\right\rangle^{9}_{0}=\frac{1}{\sqrt{2}}\left\{\begin{array}{|c|c|}\hline 1&4\\\hline 7\\\cline{1-1}\end{array}-\begin{array}{|c|c|}\hline 2&4\\\hline 6\\\cline{1-1}\end{array}+\begin{array}{|c|c|}\hline 3&4\\\hline 5\\\cline{1-1}\end{array}\right\},$$

$$\left|\begin{smallmatrix}1&&0&&\bar{2}\\&0&&0&\\&&0&&\end{smallmatrix}\right\rangle^{10}_{\bar{2}}=\frac{1}{2\sqrt{2}}\left\{-\begin{array}{|c|c|}\hline 1&5\\\hline 7\\\cline{1-1}\end{array}-\begin{array}{|c|c|}\hline 1&7\\\hline 5\\\cline{1-1}\end{array}+\begin{array}{|c|c|}\hline 2&5\\\hline 6\\\cline{1-1}\end{array}+\begin{array}{|c|c|}\hline 2&6\\\hline 5\\\cline{1-1}\end{array}+2\begin{array}{|c|c|}\hline 3&5\\\hline 5\\\cline{1-1}\end{array}\right\},$$

$$\begin{aligned}\left|\begin{smallmatrix}0&&0&&\bar{1}\\&0&&0&\\&&0&&\end{smallmatrix}\right\rangle^{\prime\,10}_{\bar{2}}=\sqrt{\tfrac{1}{120}}\Bigg\{&-7\begin{array}{|c|c|}\hline 1&5\\\hline 7\\\cline{1-1}\end{array}+5\begin{array}{|c|c|}\hline 1&7\\\hline 5\\\cline{1-1}\end{array}+7\begin{array}{|c|c|}\hline 2&5\\\hline 6\\\cline{1-1}\end{array}-5\begin{array}{|c|c|}\hline 2&6\\\hline 5\\\cline{1-1}\end{array}\\&-2\begin{array}{|c|c|}\hline 3&5\\\hline 5\\\cline{1-1}\end{array}+8\begin{array}{|c|c|}\hline 4&4\\\hline 5\\\cline{1-1}\end{array}\Bigg\},\end{aligned}$$

$$\left|\begin{smallmatrix}0&&0&&\bar{1}\\&0&&0&\\&&0&&\end{smallmatrix}\right\rangle^{10}_{\bar{2}}=\sqrt{\tfrac{3}{10}}\left\{\begin{array}{|c|c|}\hline 1&5\\\hline 7\\\cline{1-1}\end{array}-\begin{array}{|c|c|}\hline 2&5\\\hline 6\\\cline{1-1}\end{array}+\begin{array}{|c|c|}\hline 3&5\\\hline 5\\\cline{1-1}\end{array}+\begin{array}{|c|c|}\hline 4&4\\\hline 5\\\cline{1-1}\end{array}\right\},$$

The method of the tensor Young tableaux can be used to check the calculation, for example:

$$\begin{aligned}
&F_3 \left|\begin{smallmatrix}1 & & 0 & & 0\\ & 0 & & 0 & \\ & & 0 & & \end{smallmatrix}\right\rangle^{8}_{2} = F_3\, \frac{\sqrt{3}}{4}\left\{-\begin{array}{|c|c|}\hline 1&7\\\hline 3\\\cline{1-1}\end{array}+\begin{array}{|c|c|}\hline 2&6\\\hline 3\\\cline{1-1}\end{array}+2\begin{array}{|c|c|}\hline 3&4\\\hline 4\\\cline{1-1}\end{array}\right\}\\
&=\sqrt{\tfrac{3}{8}}\left\{-\begin{array}{|c|c|}\hline 1&7\\\hline 4\\\cline{1-1}\end{array}+\begin{array}{|c|c|}\hline 2&6\\\hline 4\\\cline{1-1}\end{array}+2\begin{array}{|c|c|}\hline 3&4\\\hline 5\\\cline{1-1}\end{array}+2\begin{array}{|c|c|}\hline 3&5\\\hline 4\\\cline{1-1}\end{array}\right\}\\
&=\sqrt{\tfrac{8}{3}}\times\tfrac{1}{2}\left\{-\begin{array}{|c|c|}\hline 1&7\\\hline 4\\\cline{1-1}\end{array}+\begin{array}{|c|c|}\hline 2&6\\\hline 4\\\cline{1-1}\end{array}+2\begin{array}{|c|c|}\hline 3&5\\\hline 4\\\cline{1-1}\end{array}\right\}\\
&\quad+\sqrt{\tfrac{1}{2}}\times\frac{1}{2\sqrt{3}}\left\{-2\begin{array}{|c|c|}\hline 1&4\\\hline 7\\\cline{1-1}\end{array}+\begin{array}{|c|c|}\hline 1&7\\\hline 4\\\cline{1-1}\end{array}+2\begin{array}{|c|c|}\hline 2&4\\\hline 6\\\cline{1-1}\end{array}-\begin{array}{|c|c|}\hline 2&6\\\hline 4\\\cline{1-1}\end{array}\right.\\
&\quad\left.+4\begin{array}{|c|c|}\hline 3&4\\\hline 5\\\cline{1-1}\end{array}-2\begin{array}{|c|c|}\hline 3&5\\\hline 4\\\cline{1-1}\end{array}\right\}\\
&\quad+\sqrt{\tfrac{1}{3}}\times\sqrt{\tfrac{1}{2}}\left\{\begin{array}{|c|c|}\hline 1&4\\\hline 7\\\cline{1-1}\end{array}-\begin{array}{|c|c|}\hline 2&4\\\hline 6\\\cline{1-1}\end{array}+\begin{array}{|c|c|}\hline 3&4\\\hline 5\\\cline{1-1}\end{array}\right\},
\end{aligned}$$

$$\begin{aligned}
&F_3 \left|\begin{smallmatrix}1 & & 0 & & 0\\ & 0 & & 0 & \\ & & 0 & & \end{smallmatrix}\right\rangle^{\prime 8}_{2} = F_3\, \frac{1}{4\sqrt{3}}\left\{4\begin{array}{|c|c|}\hline 1&3\\\hline 7\\\cline{1-1}\end{array}+\begin{array}{|c|c|}\hline 1&7\\\hline 3\\\cline{1-1}\end{array}-4\begin{array}{|c|c|}\hline 2&3\\\hline 6\\\cline{1-1}\end{array}\right.\\
&\quad\left.-\begin{array}{|c|c|}\hline 2&6\\\hline 3\\\cline{1-1}\end{array}+4\begin{array}{|c|c|}\hline 3&3\\\hline 5\\\cline{1-1}\end{array}+2\begin{array}{|c|c|}\hline 3&4\\\hline 4\\\cline{1-1}\end{array}\right\}=\frac{1}{2\sqrt{6}}\left\{4\begin{array}{|c|c|}\hline 1&4\\\hline 7\\\cline{1-1}\end{array}\right.\\
&\quad\left.+\begin{array}{|c|c|}\hline 1&7\\\hline 4\\\cline{1-1}\end{array}-4\begin{array}{|c|c|}\hline 2&4\\\hline 6\\\cline{1-1}\end{array}-\begin{array}{|c|c|}\hline 2&6\\\hline 4\\\cline{1-1}\end{array}+10\begin{array}{|c|c|}\hline 3&4\\\hline 5\\\cline{1-1}\end{array}-2\begin{array}{|c|c|}\hline 3&5\\\hline 4\\\cline{1-1}\end{array}\right\}\\
&=\sqrt{\tfrac{1}{2}}\times\frac{1}{2\sqrt{3}}\left\{-2\begin{array}{|c|c|}\hline 1&4\\\hline 7\\\cline{1-1}\end{array}+\begin{array}{|c|c|}\hline 1&7\\\hline 4\\\cline{1-1}\end{array}+2\begin{array}{|c|c|}\hline 2&4\\\hline 6\\\cline{1-1}\end{array}-\begin{array}{|c|c|}\hline 2&6\\\hline 4\\\cline{1-1}\end{array}\right.\\
&\quad\left.+4\begin{array}{|c|c|}\hline 3&4\\\hline 5\\\cline{1-1}\end{array}-2\begin{array}{|c|c|}\hline 3&5\\\hline 4\\\cline{1-1}\end{array}\right\}\\
&\quad+\sqrt{3}\times\sqrt{\tfrac{1}{2}}\left\{\begin{array}{|c|c|}\hline 1&4\\\hline 7\\\cline{1-1}\end{array}-\begin{array}{|c|c|}\hline 2&4\\\hline 6\\\cline{1-1}\end{array}+\begin{array}{|c|c|}\hline 3&4\\\hline 5\\\cline{1-1}\end{array}\right\},
\end{aligned}$$

$$\begin{aligned}
&F_3 \left|\begin{smallmatrix}1 & & 0 & & \bar{1}\\ & 0 & & 0 & \\ & & 0 & & \end{smallmatrix}\right\rangle^{\prime 9}_{0} = F_3\, \tfrac{1}{2}\left\{-\begin{array}{|c|c|}\hline 1&7\\\hline 4\\\cline{1-1}\end{array}+\begin{array}{|c|c|}\hline 2&6\\\hline 4\\\cline{1-1}\end{array}+2\begin{array}{|c|c|}\hline 3&5\\\hline 4\\\cline{1-1}\end{array}\right\}\\
&=\sqrt{\tfrac{1}{2}}\left\{-\begin{array}{|c|c|}\hline 1&7\\\hline 5\\\cline{1-1}\end{array}+\begin{array}{|c|c|}\hline 2&6\\\hline 5\\\cline{1-1}\end{array}+2\begin{array}{|c|c|}\hline 3&5\\\hline 5\\\cline{1-1}\end{array}\right\}\\
&=\tfrac{3}{2}\times\frac{1}{2\sqrt{2}}\left\{-\begin{array}{|c|c|}\hline 1&5\\\hline 7\\\cline{1-1}\end{array}-\begin{array}{|c|c|}\hline 1&7\\\hline 5\\\cline{1-1}\end{array}+\begin{array}{|c|c|}\hline 2&5\\\hline 6\\\cline{1-1}\end{array}+\begin{array}{|c|c|}\hline 2&6\\\hline 5\\\cline{1-1}\end{array}+2\begin{array}{|c|c|}\hline 3&5\\\hline 5\\\cline{1-1}\end{array}\right\}\\
&\quad-\sqrt{\tfrac{3}{20}}\times\sqrt{\tfrac{1}{120}}\left\{-7\begin{array}{|c|c|}\hline 1&5\\\hline 7\\\cline{1-1}\end{array}+5\begin{array}{|c|c|}\hline 1&7\\\hline 5\\\cline{1-1}\end{array}+7\begin{array}{|c|c|}\hline 2&5\\\hline 6\\\cline{1-1}\end{array}-5\begin{array}{|c|c|}\hline 2&6\\\hline 5\\\cline{1-1}\end{array}\right.\\
&\quad\left.-2\begin{array}{|c|c|}\hline 3&5\\\hline 5\\\cline{1-1}\end{array}+8\begin{array}{|c|c|}\hline 4&4\\\hline 5\\\cline{1-1}\end{array}\right\}\\
&\quad+\sqrt{\tfrac{4}{15}}\times\sqrt{\tfrac{3}{10}}\left\{\begin{array}{|c|c|}\hline 1&5\\\hline 7\\\cline{1-1}\end{array}-\begin{array}{|c|c|}\hline 2&5\\\hline 6\\\cline{1-1}\end{array}+\begin{array}{|c|c|}\hline 3&5\\\hline 5\\\cline{1-1}\end{array}+\begin{array}{|c|c|}\hline 4&4\\\hline 5\\\cline{1-1}\end{array}\right\},
\end{aligned}$$

$$\begin{aligned}
F_3\left|\begin{smallmatrix}1&&0&&\overline{1}\\&0&&0&\\&&0&&\end{smallmatrix}\right\rangle_0^9 &= F_3\,\frac{1}{2\sqrt{3}}\left\{-2\begin{array}{|c|c|}\hline 1&4\\\hline 7\\\cline{1-1}\end{array}+\begin{array}{|c|c|}\hline 1&7\\\hline 4\\\cline{1-1}\end{array}+2\begin{array}{|c|c|}\hline 2&4\\\hline 6\\\cline{1-1}\end{array}-\begin{array}{|c|c|}\hline 2&6\\\hline 4\\\cline{1-1}\end{array}\right.\\
&\quad\left.+4\begin{array}{|c|c|}\hline 3&4\\\hline 5\\\cline{1-1}\end{array}-2\begin{array}{|c|c|}\hline 3&5\\\hline 4\\\cline{1-1}\end{array}\right\}=\sqrt{\tfrac{1}{6}}\left\{-2\begin{array}{|c|c|}\hline 1&5\\\hline 7\\\cline{1-1}\end{array}+\begin{array}{|c|c|}\hline 1&7\\\hline 5\\\cline{1-1}\end{array}\right.\\
&\quad\left.+2\begin{array}{|c|c|}\hline 2&5\\\hline 6\\\cline{1-1}\end{array}-\begin{array}{|c|c|}\hline 2&6\\\hline 5\\\cline{1-1}\end{array}+2\begin{array}{|c|c|}\hline 3&5\\\hline 5\\\cline{1-1}\end{array}+4\begin{array}{|c|c|}\hline 4&4\\\hline 5\\\cline{1-1}\end{array}\right\}\\
&=\frac{\sqrt{3}}{2}\times\frac{1}{2\sqrt{2}}\left\{-\begin{array}{|c|c|}\hline 1&5\\\hline 7\\\cline{1-1}\end{array}-\begin{array}{|c|c|}\hline 1&7\\\hline 5\\\cline{1-1}\end{array}+\begin{array}{|c|c|}\hline 2&5\\\hline 6\\\cline{1-1}\end{array}+\begin{array}{|c|c|}\hline 2&6\\\hline 5\\\cline{1-1}\end{array}+2\begin{array}{|c|c|}\hline 3&5\\\hline 5\\\cline{1-1}\end{array}\right\}\\
&\quad+\frac{7}{\sqrt{20}}\times\sqrt{\frac{1}{120}}\left\{-7\begin{array}{|c|c|}\hline 1&5\\\hline 7\\\cline{1-1}\end{array}+5\begin{array}{|c|c|}\hline 1&7\\\hline 5\\\cline{1-1}\end{array}+7\begin{array}{|c|c|}\hline 2&5\\\hline 6\\\cline{1-1}\end{array}-5\begin{array}{|c|c|}\hline 2&6\\\hline 5\\\cline{1-1}\end{array}\right.\\
&\quad\left.-2\begin{array}{|c|c|}\hline 3&5\\\hline 5\\\cline{1-1}\end{array}+8\begin{array}{|c|c|}\hline 4&4\\\hline 5\\\cline{1-1}\end{array}\right\}\\
&\quad+\sqrt{\frac{4}{5}}\times\sqrt{\frac{3}{10}}\left\{\begin{array}{|c|c|}\hline 1&5\\\hline 7\\\cline{1-1}\end{array}-\begin{array}{|c|c|}\hline 2&5\\\hline 6\\\cline{1-1}\end{array}+\begin{array}{|c|c|}\hline 3&5\\\hline 5\\\cline{1-1}\end{array}+\begin{array}{|c|c|}\hline 4&4\\\hline 5\\\cline{1-1}\end{array}\right\}.
\end{aligned}$$

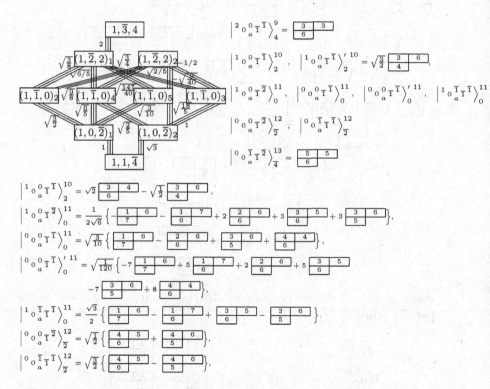

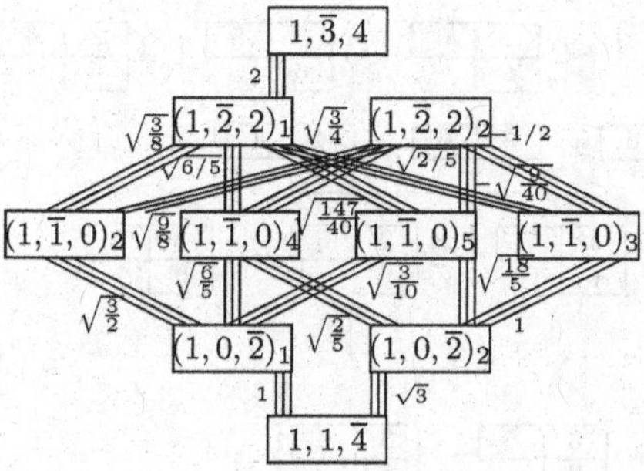

$$\left|\begin{smallmatrix}2&&0&&\overline{1}\\&0&&\overline{1}&\\&&a&&\end{smallmatrix}\right\rangle_4^9=\begin{array}{|c|c|}\hline 3&3\\\hline 6\\\cline{1-1}\end{array}$$

$$\left|\begin{smallmatrix}1&&0&&\overline{1}\\&0&&\overline{1}&\\&&a&&\end{smallmatrix}\right\rangle_2^{10},\quad \left|\begin{smallmatrix}1&&0&&\overline{1}\\&0&&\overline{1}&\\&&a&&\end{smallmatrix}\right\rangle_2^{\prime\,10}=\sqrt{\tfrac{3}{2}}\begin{array}{|c|c|}\hline 3&6\\\hline 4\\\cline{1-1}\end{array},$$

$$\left|\begin{smallmatrix}1&&0&&\overline{2}\\&0&&\overline{1}&\\&&a&&\end{smallmatrix}\right\rangle_0^{11},\quad \left|\begin{smallmatrix}0&&0&&\overline{1}\\&0&&\overline{1}&\\&&a&&\end{smallmatrix}\right\rangle_0^{11},\quad \left|\begin{smallmatrix}0&&0&&\overline{1}\\&0&&\overline{1}&\\&&a&&\end{smallmatrix}\right\rangle_0^{\prime\,11},\quad \left|\begin{smallmatrix}1&&\overline{1}&&\overline{1}\\&0&&\overline{1}&\\&&a&&\end{smallmatrix}\right\rangle_0^{11}$$

$$\left|\begin{smallmatrix}0&&0&&\overline{2}\\&0&&\overline{1}&\\&&a&&\end{smallmatrix}\right\rangle_{\overline{2}}^{12},\quad \left|\begin{smallmatrix}0&&\overline{1}&&\overline{1}\\&0&&\overline{1}&\\&&a&&\end{smallmatrix}\right\rangle_{\overline{2}}^{12}$$

$$\left|\begin{smallmatrix}0&&\overline{1}&&\overline{2}\\&0&&\overline{1}&\\&&a&&\end{smallmatrix}\right\rangle_{\overline{4}}^{13}=\begin{array}{|c|c|}\hline 5&5\\\hline 6\\\cline{1-1}\end{array}$$

$$\begin{aligned}
\left|\begin{smallmatrix}1&&0&&\overline{1}\\&0&&\overline{1}&\\&&a&&\end{smallmatrix}\right\rangle_2^{10}&=\sqrt{2}\begin{array}{|c|c|}\hline 3&4\\\hline 6\\\cline{1-1}\end{array}-\sqrt{\tfrac{1}{2}}\begin{array}{|c|c|}\hline 3&6\\\hline 4\\\cline{1-1}\end{array},\\
\left|\begin{smallmatrix}1&&0&&\overline{2}\\&0&&\overline{1}&\\&&a&&\end{smallmatrix}\right\rangle_0^{11}&=\frac{1}{2\sqrt{6}}\left\{-\begin{array}{|c|c|}\hline 1&6\\\hline 7\\\cline{1-1}\end{array}-\begin{array}{|c|c|}\hline 1&7\\\hline 6\\\cline{1-1}\end{array}+2\begin{array}{|c|c|}\hline 2&6\\\hline 6\\\cline{1-1}\end{array}+3\begin{array}{|c|c|}\hline 3&5\\\hline 6\\\cline{1-1}\end{array}+3\begin{array}{|c|c|}\hline 3&6\\\hline 5\\\cline{1-1}\end{array}\right\},\\
\left|\begin{smallmatrix}0&&0&&\overline{1}\\&0&&\overline{1}&\\&&a&&\end{smallmatrix}\right\rangle_0^{11}&=\sqrt{\tfrac{3}{10}}\left\{\begin{array}{|c|c|}\hline 1&6\\\hline 7\\\cline{1-1}\end{array}-\begin{array}{|c|c|}\hline 2&6\\\hline 6\\\cline{1-1}\end{array}+\begin{array}{|c|c|}\hline 3&6\\\hline 5\\\cline{1-1}\end{array}+\begin{array}{|c|c|}\hline 4&4\\\hline 6\\\cline{1-1}\end{array}\right\},\\
\left|\begin{smallmatrix}0&&0&&\overline{1}\\&0&&\overline{1}&\\&&a&&\end{smallmatrix}\right\rangle_0^{\prime\,11}&=\sqrt{\tfrac{1}{120}}\left\{-7\begin{array}{|c|c|}\hline 1&6\\\hline 7\\\cline{1-1}\end{array}+5\begin{array}{|c|c|}\hline 1&7\\\hline 6\\\cline{1-1}\end{array}+2\begin{array}{|c|c|}\hline 2&6\\\hline 6\\\cline{1-1}\end{array}+5\begin{array}{|c|c|}\hline 3&5\\\hline 6\\\cline{1-1}\end{array}\right.\\
&\qquad\left.-7\begin{array}{|c|c|}\hline 3&6\\\hline 5\\\cline{1-1}\end{array}+8\begin{array}{|c|c|}\hline 4&4\\\hline 6\\\cline{1-1}\end{array}\right\},\\
\left|\begin{smallmatrix}1&&\overline{1}&&\overline{1}\\&0&&\overline{1}&\\&&a&&\end{smallmatrix}\right\rangle_0^{11}&=\frac{\sqrt{3}}{2}\left\{\begin{array}{|c|c|}\hline 1&6\\\hline 7\\\cline{1-1}\end{array}-\begin{array}{|c|c|}\hline 1&7\\\hline 6\\\cline{1-1}\end{array}+\begin{array}{|c|c|}\hline 3&5\\\hline 6\\\cline{1-1}\end{array}-\begin{array}{|c|c|}\hline 3&6\\\hline 5\\\cline{1-1}\end{array}\right\}.\\
\left|\begin{smallmatrix}0&&0&&\overline{2}\\&0&&\overline{1}&\\&&a&&\end{smallmatrix}\right\rangle_{\overline{2}}^{12}&=\sqrt{\tfrac{1}{2}}\left\{\begin{array}{|c|c|}\hline 4&5\\\hline 6\\\cline{1-1}\end{array}+\begin{array}{|c|c|}\hline 4&6\\\hline 5\\\cline{1-1}\end{array}\right\},\\
\left|\begin{smallmatrix}0&&\overline{1}&&\overline{1}\\&0&&\overline{1}&\\&&a&&\end{smallmatrix}\right\rangle_{\overline{2}}^{12}&=\sqrt{\tfrac{3}{2}}\left\{\begin{array}{|c|c|}\hline 4&5\\\hline 6\\\cline{1-1}\end{array}-\begin{array}{|c|c|}\hline 4&6\\\hline 5\\\cline{1-1}\end{array}\right\},
\end{aligned}$$

where $a = 0$. There is a similar $\mathcal{A}_3$-quintuplet with $a = \overline{1}$ due to the parallel principle. As a check, we have

$$
\begin{aligned}
F_3 \left|\begin{smallmatrix}1&&0&&\overline{1}\\&0&&\overline{1}&\\&&0&&\end{smallmatrix}\right\rangle^{10}_{2} &= F_3 \left\{\sqrt{2}\,\young(34,6) - \sqrt{\tfrac{1}{2}}\,\young(36,4)\right\} \\
&= 2\,\young(35,6) - \young(36,5) + 2\,\young(44,6) \\
&= \sqrt{\tfrac{3}{8}} \times \tfrac{1}{2\sqrt{6}} \left\{-\young(16,7) - \young(17,6) + 2\,\young(26,6) + 3\,\young(35,6) + 3\,\young(36,5)\right\} \\
&\quad + \sqrt{\tfrac{6}{5}} \times \sqrt{\tfrac{3}{10}} \left\{\young(16,7) - \young(26,6) + \young(36,5) + \young(44,6)\right\} \\
&\quad + \sqrt{\tfrac{147}{40}} \times \sqrt{\tfrac{1}{120}} \left\{-7\,\young(16,7) + 5\,\young(17,6) + 2\,\young(26,6) + 5\,\young(35,6)\right. \\
&\qquad \left. -7\,\young(36,5) + 8\,\young(44,6)\right\} \\
&\quad + \sqrt{\tfrac{3}{4}} \times \sqrt{\tfrac{3}{4}} \left\{\young(16,7) - \young(17,6) + \young(35,6) - \young(36,5)\right\},
\end{aligned}
$$

$$
\begin{aligned}
F_3 \left|\begin{smallmatrix}1&&0&&\overline{1}\\&0&&\overline{1}&\\&&0&&\end{smallmatrix}\right\rangle'^{\,10}_{2} &= F_3 \left\{\sqrt{\tfrac{3}{2}}\,\young(36,4)\right\} = \sqrt{3}\,\young(36,5) \\
&= \sqrt{\tfrac{9}{8}} \times \tfrac{1}{2\sqrt{6}} \left\{-\young(16,7) - \young(17,6) + 2\,\young(26,6) + 3\,\young(35,6) + 3\,\young(36,5)\right\} \\
&\quad + \sqrt{\tfrac{2}{5}} \times \sqrt{\tfrac{3}{10}} \left\{\young(16,7) - \young(26,6) + \young(36,5) + \young(44,6)\right\} \\
&\quad - \sqrt{\tfrac{9}{40}} \times \sqrt{\tfrac{1}{120}} \left\{-7\,\young(16,7) + 5\,\young(17,6) + 2\,\young(26,6) + 5\,\young(35,6)\right. \\
&\qquad \left. -7\,\young(36,5) + 8\,\young(44,6)\right\} \\
&\quad - \tfrac{1}{2} \times \sqrt{\tfrac{3}{4}} \left\{\young(16,7) - \young(17,6) + \young(35,6) - \young(36,5)\right\}.
\end{aligned}
$$

$$
\begin{array}{cl}
\boxed{2,\overline{3},2} & \left|\begin{smallmatrix}1&&0&&\overline{2}\\&0&&\overline{2}&\\&&a&&\end{smallmatrix}\right\rangle^{12}_{2} = \young(36,6) \\
\sqrt{2}\,\Vert & \\
\boxed{2,\overline{2},0} & \left|\begin{smallmatrix}0&&0&&\overline{2}\\&0&&\overline{2}&\\&&a&&\end{smallmatrix}\right\rangle^{13}_{0} = \young(46,6) \\
\sqrt{2}\,\Vert & \\
\boxed{2,\overline{1},\overline{2}} & \left|\begin{smallmatrix}0&&\overline{1}&&\overline{2}\\&0&&\overline{2}&\\&&a&&\end{smallmatrix}\right\rangle^{14}_{\overline{2}} = \young(56,6)
\end{array}
$$

where $a = 0$. There are two similar $\mathcal{A}_3$-triplets with $\overline{1}$, and $\overline{2}$ due to the parallel principle.

$$
\begin{array}{cl}
\boxed{(0,\overline{2},2)_2} & \left|\begin{smallmatrix}1&&\overline{1}&&\overline{1}\\&\overline{1}&&\overline{1}&\\&&\overline{1}&&\end{smallmatrix}\right\rangle^{13}_{2} = \sqrt{\tfrac{3}{2}}\left\{\young(36,7) - \young(37,6)\right\} \\
\sqrt{2}\,\Vert & \\
\boxed{(0,\overline{1},0)_2} & \left|\begin{smallmatrix}0&&\overline{1}&&\overline{1}\\&\overline{1}&&\overline{1}&\\&&\overline{1}&&\end{smallmatrix}\right\rangle^{14}_{0} = \sqrt{\tfrac{3}{2}}\left\{\young(46,7) - \young(47,6)\right\} \\
\sqrt{2}\,\Vert & \\
\boxed{(0,0,\overline{2})_2} & \left|\begin{smallmatrix}0&&\overline{1}&&\overline{2}\\&\overline{1}&&\overline{1}&\\&&\overline{1}&&\end{smallmatrix}\right\rangle^{15}_{\overline{2}} = \sqrt{\tfrac{3}{2}}\left\{\young(56,7) - \young(57,6)\right\}
\end{array}
$$

10.1.3 *Irreducible Basis Tensors of* SO(2ℓ)

The Lie algebra of SO(2ℓ) is D_ℓ. The simple roots of SO(2ℓ) are

$$\boldsymbol{r}_\mu = \boldsymbol{e}_\mu - \boldsymbol{e}_{\mu+1}, \qquad 1 \leqslant \mu \leqslant \ell - 1, \qquad \boldsymbol{r}_\ell = \boldsymbol{e}_{\ell-1} + \boldsymbol{e}_\ell. \tag{10.18}$$

The lengths of all simple roots are the same, $d_\mu = 1$. From the definition (8.1), the Chevalley bases of SO(2ℓ) in the self-representation are the same as those of SO(2ℓ + 1) (see Eq. (10.13)) except for $\mu = \ell$,

$$\begin{aligned}
H_\ell &= T_{(2\ell-3)(2\ell-2)} + T_{(2\ell-1)(2\ell)},\\
E_\ell &= \tfrac{1}{2}\left\{T_{(2\ell-2)(2\ell-1)} - iT_{(2\ell-3)(2\ell-1)} + iT_{(2\ell-2)(2\ell)} + T_{(2\ell-3)(2\ell)}\right\},\\
F_\ell &= \tfrac{1}{2}\left\{T_{(2\ell-2)(2\ell-1)} + iT_{(2\ell-3)(2\ell-1)} - iT_{(2\ell-2)(2\ell)} + T_{(2\ell-3)(2\ell)}\right\}.
\end{aligned} \tag{10.19}$$

$\boldsymbol{\theta}_a$ is not the common eigenvector of H_μ. We define the **spherical harmonic basis vectors** in the self-representation of SO(2ℓ), which are the generalization of those for SO(4) (see Eq. (11.4)),

$$\boldsymbol{\phi}_\alpha = \begin{cases} (-1)^{\ell-\alpha}\sqrt{1/2}\,(\boldsymbol{\theta}_{2\alpha-1} + i\boldsymbol{\theta}_{2\alpha}), & 1 \leqslant \alpha \leqslant \ell,\\ \sqrt{1/2}\,(\boldsymbol{\theta}_{4\ell-2\alpha+1} - i\boldsymbol{\theta}_{4\ell-2\alpha+2}), & \ell+1 \leqslant \alpha \leqslant 2\ell. \end{cases} \tag{10.20}$$

The spherical harmonic basis vectors $\boldsymbol{\phi}_\alpha$ are orthonormal and complete. The matrix elements of the Chevalley bases in $\boldsymbol{\phi}_\alpha$ are

$$\begin{aligned}
H_\mu \boldsymbol{\phi}_\nu &= \delta_{\nu\mu}\boldsymbol{\phi}_\mu - \delta_{\nu(\mu+1)}\boldsymbol{\phi}_{\mu+1} + \delta_{\nu(2\ell-\mu)}\boldsymbol{\phi}_{2\ell-\mu} - \delta_{\nu(2\ell-\mu+1)}\boldsymbol{\phi}_{2\ell-\mu+1},\\
H_\ell \boldsymbol{\phi}_\nu &= \delta_{\nu(\ell-1)}\boldsymbol{\phi}_{\ell-1} + \delta_{\nu\ell}\boldsymbol{\phi}_\ell - \delta_{\nu(\ell+1)}\boldsymbol{\phi}_{\ell+1} - \delta_{\nu(\ell+2)}\boldsymbol{\phi}_{\ell+2},\\
E_\mu \boldsymbol{\phi}_\nu &= \delta_{\nu(\mu+1)}\boldsymbol{\phi}_\mu + \delta_{\nu(2\ell-\mu+1)}\boldsymbol{\phi}_{2\ell-\mu},\\
E_\ell \boldsymbol{\phi}_\nu &= \delta_{\nu(\ell+1)}\boldsymbol{\phi}_{\ell-1} + \delta_{\nu(\ell+2)}\boldsymbol{\phi}_\ell,\\
F_\mu \boldsymbol{\phi}_\nu &= \delta_{\nu\mu}\boldsymbol{\phi}_{\mu+1} + \delta_{\nu(2\ell-\mu)}\boldsymbol{\phi}_{2\ell-\mu+1},\\
F_\ell \boldsymbol{\phi}_\nu &= \delta_{\nu(\ell-1)}\boldsymbol{\phi}_{\ell+1} + \delta_{\nu\ell}\boldsymbol{\phi}_{\ell+2},
\end{aligned} \tag{10.21}$$

where $1 \leqslant \mu \leqslant \ell - 1$. Namely, the diagonal matrices of H_μ and H_ℓ in the spherical harmonic basis vectors $\boldsymbol{\phi}_\alpha$ are

$$\begin{aligned}
H_\mu &= \text{diag}\{\underbrace{0,\cdots,0}_{\mu-1},1,-1,\underbrace{0,\cdots,0}_{2\ell-2\mu-2},1,-1,\underbrace{0,\cdots,0}_{\mu-1}\},\\
H_\ell &= \text{diag}\{\underbrace{0,\cdots,0}_{\ell-2},1,1,-1,-1,\underbrace{0,\cdots,0}_{\ell-2}\}.
\end{aligned}$$

The spherical harmonic basis tensor $\boldsymbol{\phi}_{\alpha_1\cdots\alpha_n}$ of rank n for SO(2ℓ) is the direct product of n spherical harmonic basis vectors $\boldsymbol{\phi}_{\alpha_1} \times \cdots \times \boldsymbol{\phi}_{\alpha_n}$. The standard tensor Young tableaux $\mathcal{Y}_\mu^{[\lambda]}\phi_{\alpha_1\cdots\alpha_n}$ are the common eigenstates of H_μ, but generally not orthonormal and traceless. **The eigenvalue of H_μ** in the standard tensor Young tableau $\mathcal{Y}_\mu^{[\lambda]}\phi_{\alpha_1\cdots\alpha_n}$ is equal to **the number of the digits μ and $(2\ell-\mu)$ in the tableau, minus the number of $(\mu+1)$ and $(2\ell-\mu+1)$. The eigenvalue of H_ℓ** in $\mathcal{Y}_\mu^{[\lambda]}\phi_{\alpha_1\cdots\alpha_n}$ is equal **to the number of the digits $(\ell-1)$ and ℓ in the tableau, minus the number of $(\ell+1)$ and $(\ell+2)$**. The eigenvalues constitute the weight $\boldsymbol{m}$ of the standard tensor Young tableau. **The standard tensor Young tableau $\mathcal{Y}_\mu^{[\lambda]}\phi_{\alpha_1\cdots\alpha_n}$ in the action of F_μ** becomes the sum of all possible tensor Young tableaux, each of which is obtained from the original one by **replacing one filled digit μ with the digit $(\mu+1)$, or by replacing one filled digit $(2\ell-\mu)$ with the digit $(2\ell-\mu+1)$. $\mathcal{Y}_\mu^{[\lambda]}\phi_{\alpha_1\cdots\alpha_n}$ in the action of F_ℓ** becomes the sum of all possible tensor Young tableaux, each of which is obtained from the original one by **replacing one filled digit $(\ell-1)$ with the digit $(\ell+1)$ or by replacing one filled digit ℓ with the digit $(\ell+2)$**. The actions of $E_\mu(E_\ell)$ are opposite to that of $F_\mu(F_\ell)$. The obtained tensor Young tableaux may be not standard, but they can be transformed to the sum of the standard tensor Young tableaux by Eqs. (9.22–23).

The standard tensor Young tableaux with different weights are orthogonal to each other. For an irreducible representation $[\lambda]$, where the row number of $[\lambda]$ is not larger than ℓ, or $[(+)\lambda]$ of SO(2ℓ), the highest weight state corresponds to the standard tensor Young tableau where each box in the αth row is filled with the digit α because every raising operator E_μ annihilates it. In the standard tensor Young tableau with the highest weight of the representation $[(-)\lambda]$, the box in the αth row is filled with the digit α, but the box in the ℓth row is filled with the digit $(\ell+1)$. The highest weight $\boldsymbol{M}=\sum_\mu \boldsymbol{w}_\mu M_\mu$ is calculated from Eq. (10.21),

$$\begin{aligned}
&M_\mu=\lambda_\mu-\lambda_{\mu+1}, && 1\leqslant\mu<\ell-1,\\
&M_{\ell-1}=M_\ell=\lambda_{\ell-1}, && \text{if}\ \ \lambda_\ell=0,\\
&M_{\ell-1}=\lambda_{\ell-1}-\lambda_\ell,\ \ M_\ell=\lambda_{\ell-1}+\lambda_\ell, && \text{for}\ \ [(+)\lambda],\\
&M_{\ell-1}=\lambda_{\ell-1}+\lambda_\ell,\ \ M_\ell=\lambda_{\ell-1}-\lambda_\ell, && \text{for}\ \ [(-)\lambda].
\end{aligned} \tag{10.22}$$

The tensor representation $[\lambda]$ of SO(2ℓ) is single-valued where $(M_{\ell-1}+M_\ell)$ is even. As shown later, the representation with odd $(M_{\ell-1}+M_\ell)$ is a double-valued representation, called the spinor one.

Although the standard tensor Young tableau is generally not traceless, **the standard tensor Young tableau with the highest weight is**

traceless because ϕ_α and $\phi_{2\ell-\alpha+1}$ do not appear in that tensor Young tableau simultaneously (see Eq. (10.20)). Since the highest weight is simple, the highest weight state is orthogonal to any other standard tensor Young tableau in the irreducible representation. One is able to find the remaining orthonormal and traceless basis tensors in the irreducible representation of SO(2ℓ) **from the highest weight state by the lowering operators F_μ in terms of the method of the block weight diagrams and the generalized Gel'fand's method** (see subsection 8.3.4). The multiplicity of a weight in the representation can be calculated by counting the number of the traceless tensor Young tableaux with this weight.

Based on the definition (10.9), we need to judge from the highest weight state which one in $[(\pm)\lambda]$ corresponds to the self-dual tensor representation or the anti-self-dual one. For simplicity we discuss the tensor representation $[(+)\lambda] = [(+)1^\ell]$, whose highest weight state is

$$\begin{aligned}
\mathcal{Y}^{[1^\ell]}\phi_{12\cdots\ell} &= (-1)^{\ell(\ell-1)/2}2^{-\ell/2}\mathcal{Y}^{[1^\ell]}\left\{(\boldsymbol{\theta}_1 + \mathrm{i}\boldsymbol{\theta}_2) \times \cdots \times (\boldsymbol{\theta}_{2\ell-1} + \mathrm{i}\boldsymbol{\theta}_{2\ell})\right\} \\
&= (-1)^{\ell(\ell-1)/2}2^{-\ell/2}\left\{\left[\mathcal{Y}^{[1^\ell]}\boldsymbol{\theta}_{135\cdots(2\ell-1)} + \mathrm{i}^\ell \mathcal{Y}^{[1^\ell]}\boldsymbol{\theta}_{246\cdots(2\ell)}\right]\right. \\
&\quad \left. + \mathrm{i}\left[\mathcal{Y}^{[1^\ell]}\boldsymbol{\theta}_{235\cdots(2\ell-1)} - \mathrm{i}^\ell \mathcal{Y}^{[1^\ell]}\boldsymbol{\theta}_{146\cdots(2\ell)}\right] + \cdots\right\}.
\end{aligned}$$

Thus, $[(+)\lambda]$ **denotes the self-dual representation and** $[(-)\lambda]$ **denotes the anti-self-dual one**, where the row number of $[\lambda]$ is ℓ.

We give two examples for calculating the orthonormal basis states in the highest weight representations in SO(8) whose Lie algebra is D_4.

Ex. 1. The irreducible representation $[\lambda] = [2,0,0,0] = [2]$.

From Eq. (10.22), the highest weight of this representation is $\boldsymbol{M} = 2\boldsymbol{w}_1 = (2,0,0,0)$. In the brief symbols, $|\boldsymbol{M},\boldsymbol{m}\rangle = |\boldsymbol{m}\rangle$, the highest weight state is

$$|\boldsymbol{M}\rangle = |(2,0,0,0)\rangle = \left|\begin{smallmatrix} 2 & & 0 & & 0 & & 0 \\ & 2 & & 0 & & 0 & \\ & & 2 & & 0 & & \\ & & & 2 & & & \end{smallmatrix}\right\rangle_0^1 = \begin{array}{|c|c|}\hline 1 & 1 \\ \hline\end{array} = 2\phi_{11}.$$

Two typical standard tensor Young tableaux in this representation are

$$\begin{aligned}
&\begin{array}{|c|c|}\hline a & a \\ \hline\end{array} = \mathcal{Y}^{[2]}\phi_{aa} = 2\phi_{aa}, \\
&\sqrt{2}\begin{array}{|c|c|}\hline a & b \\ \hline\end{array} = \sqrt{2}\ \mathcal{Y}^{[2]}\phi_{ab} = \sqrt{2}\ \{\phi_{ab} + \phi_{ba}\},
\end{aligned}$$

where $a \neq b$. The highest weight state $|(2,0,0,0)\rangle$ is also the highest weight state of an A_3-decuplet with $\boldsymbol{M}^{(3)} = (2,0,0)$:

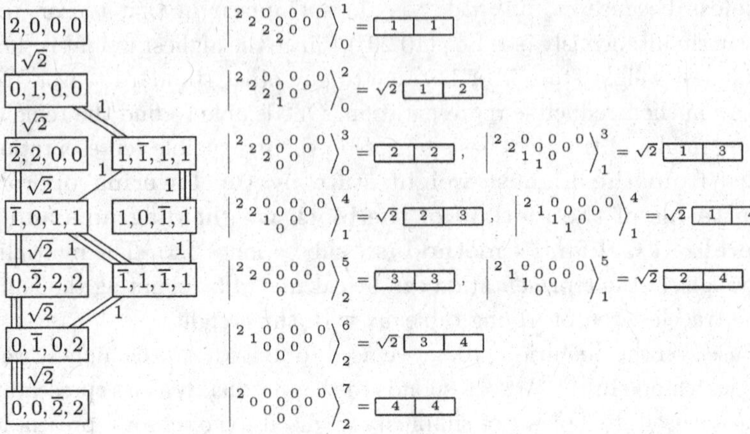

$$\boxed{2,0,0,0}$$

$\sqrt{2}$

$$\boxed{0,1,0,0}$$

$\sqrt{2}$

$$\boxed{\bar{2},2,0,0} \quad \boxed{1,\bar{1},1,1}$$

$\|\sqrt{2}$

$$\boxed{\bar{1},0,1,1} \quad \boxed{1,0,\bar{1},1}$$

$\sqrt{2}$

$$\boxed{0,\bar{2},2,2} \quad \boxed{\bar{1},1,\bar{1},1}$$

$\|\sqrt{2}$

$$\boxed{0,\bar{1},0,2}$$

$\|\sqrt{2}$

$$\boxed{0,0,\bar{2},2}$$

$$\left|{}^{2}\,{}_{2}{}^{2}_{2}{}^{0}_{0}{}^{0}\,{}^{0}_{0}\,{}^{0}\right\rangle^{1}_{0} = \boxed{1\;|\;1}$$

$$\left|{}^{2}\,{}_{2}{}^{2}_{1}{}^{0}_{0}{}^{0}\,{}^{0}_{0}\,{}^{0}\right\rangle^{2}_{0} = \sqrt{2}\,\boxed{1\;|\;2}$$

$$\left|{}^{2}\,{}_{2}{}^{2}_{0}{}^{0}_{0}{}^{0}\,{}^{0}_{0}\,{}^{0}\right\rangle^{3}_{0} = \boxed{2\;|\;2}\,, \qquad \left|{}^{2}\,{}_{2}{}^{0}_{1}{}^{0}_{0}{}^{0}\,{}^{0}_{0}\,{}^{0}\right\rangle^{3}_{1} = \sqrt{2}\,\boxed{1\;|\;3}$$

$$\left|{}^{2}\,{}_{2}{}^{0}_{1}{}^{0}_{0}{}^{0}\,{}^{0}_{0}\,{}^{0}\right\rangle^{4}_{1} = \sqrt{2}\,\boxed{2\;|\;3}\,, \qquad \left|{}^{2}\,{}_{1}{}^{0}_{1}{}^{0}_{0}{}^{0}\,{}^{0}_{0}\,{}^{0}\right\rangle^{4}_{1} = \sqrt{2}\,\boxed{1\;|\;4}$$

$$\left|{}^{2}\,{}_{2}{}^{0}_{0}{}^{0}_{0}{}^{0}\,{}^{0}_{0}\,{}^{0}\right\rangle^{5}_{2} = \boxed{3\;|\;3}\,, \qquad \left|{}^{2}\,{}_{1}{}^{0}_{1}{}^{0}_{0}{}^{0}\,{}^{0}_{0}\,{}^{0}\right\rangle^{5}_{1} = \sqrt{2}\,\boxed{2\;|\;4}$$

$$\left|{}^{2}\,{}_{1}{}^{0}_{0}{}^{0}_{0}{}^{0}\,{}^{0}_{0}\,{}^{0}\right\rangle^{6}_{2} = \sqrt{2}\,\boxed{3\;|\;4}$$

$$\left|{}^{2}\,{}_{0}{}^{0}_{0}{}^{0}_{0}{}^{0}\,{}^{0}_{0}\,{}^{0}\right\rangle^{7}_{2} = \boxed{4\;|\;4}$$

There are seven basis states in the A$_3$-decuplet with $\boldsymbol{M}^{(3)} = (2,0,0)$ satisfying the condition (8.7) with $\mu = 4$, so that the $\mathcal{A}_4$-multiplets are constructed. We construct four $\mathcal{A}_4$-doublets from four states $|(1,\overline{1},1,1)\rangle$, $|(\overline{1},0,1,1)\rangle$, $|(1,0,\overline{1},1)\rangle$ and $(\overline{1},1,\overline{1},1)$ by the parallel principle. The partners in those $\mathcal{A}_4$-doublets belong to the A$_3$-15plet with $\boldsymbol{M}^{(3)} = (1,0,1)$, including its highest weight state $\left|\begin{smallmatrix}1 & & 0 & & 0 & & \overline{1}\\ & 1 & & 0 & & 0 & \\ & & 1 & & 0 & & \\ & & & 1 & & & \end{smallmatrix}\right\rangle^{4}_{\overline{1}}$.

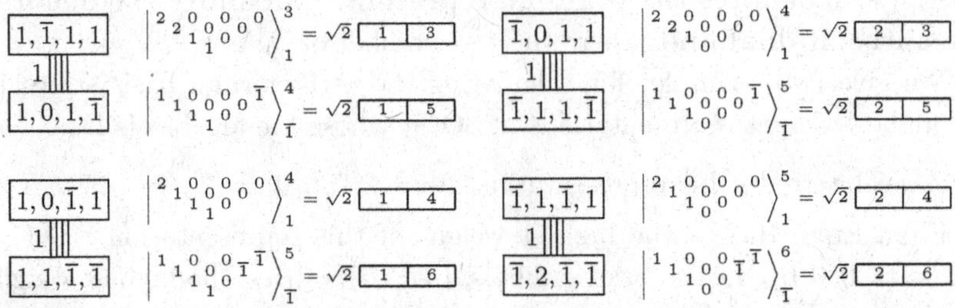

We construct three $\mathcal{A}_4$-triplets from three states $|(0,\overline{2},2,2)\rangle$, $|(0,\overline{1},0,2)\rangle$ and $|(0,0,\overline{2},2)\rangle$ in terms of Eqs. (8.25-28) and the parallel principle. The first two $\mathcal{A}_4$-triplets are easy to construct.

$$\boxed{0,\bar{2},2,2} \qquad \left|{}^{2}\,{}_{0}{}^{2}_{0}{}^{0}_{0}{}^{0}\,{}^{0}_{0}\,{}^{0}\right\rangle^{5}_{2} = \boxed{3\;|\;3} \qquad \boxed{0,0,\bar{2},2} \qquad \left|{}^{2}\,{}_{0}{}^{0}_{0}{}^{0}_{0}{}^{0}\,{}^{0}_{0}\,{}^{0}\right\rangle^{7}_{2} = \boxed{4\;|\;4}$$

$\sqrt{2}\|\|$ $\sqrt{2}\|\|$

$$\boxed{0,\bar{1},2,0} \qquad \left|{}^{1}\,{}_{0}{}^{0}_{0}{}^{0}_{0}{}^{0}\,{}^{\bar{1}}_{0}\,{}^{0}\right\rangle^{6}_{0} = \sqrt{2}\,\boxed{3\;|\;5} \qquad \boxed{0,1,\bar{2},0} \qquad \left|{}^{1}\,{}_{0}{}^{0}_{0}{}^{0}_{0}{}^{0}\,{}^{\bar{1}}_{0}\,{}^{\bar{1}}\right\rangle^{8}_{0} = \sqrt{2}\,\boxed{4\;|\;6}$$

$\sqrt{2}\|\|$ $\sqrt{2}\|\|$

$$\boxed{0,0,2,\bar{2}} \qquad \left|{}^{0}\,{}_{0}{}^{0}_{0}{}^{0}_{0}{}^{0}\,{}^{\bar{2}}_{0}\,{}^{0}\right\rangle^{7}_{\bar{2}} = \boxed{5\;|\;5} \qquad \boxed{0,2,\bar{2},\bar{2}} \qquad \left|{}^{0}\,{}_{0}{}^{0}_{0}{}^{0}_{0}{}^{0}\,{}^{\bar{2}}_{0}\,{}^{\bar{2}}\right\rangle^{9}_{\bar{2}} = \boxed{6\;|\;6}$$

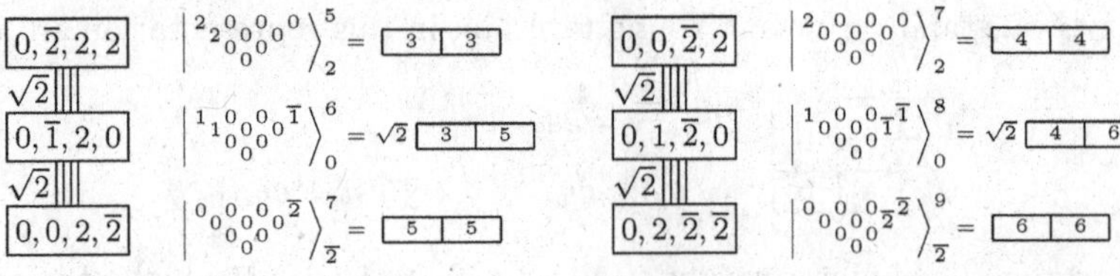

The state $\left|\begin{smallmatrix}0 & & 0 & & 0 & & \overline{2}\\ & 0 & & 0 & & 0 & \\ & & 0 & & 0 & & \\ & & & 0 & & & \end{smallmatrix}\right\rangle^{7}_{\overline{2}} = \boxed{5 \mid 5}$ is the highest weight state of the A$_3$-

decuplet with $M^{(3)} = (0, 0, 2)$. The calculation of the last $\mathcal{A}_4$-triplet has to know the knowledge on the A$_3$-15plet with $M^{(3)} = (1, 0, 1)$, so that we calculate the A$_3$-15plet and A$_3$-decuplet first.

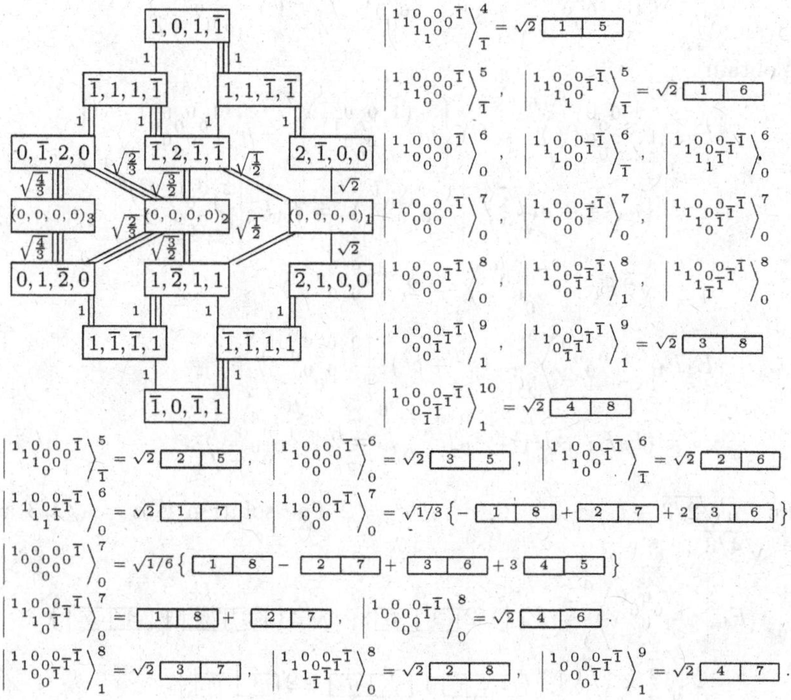

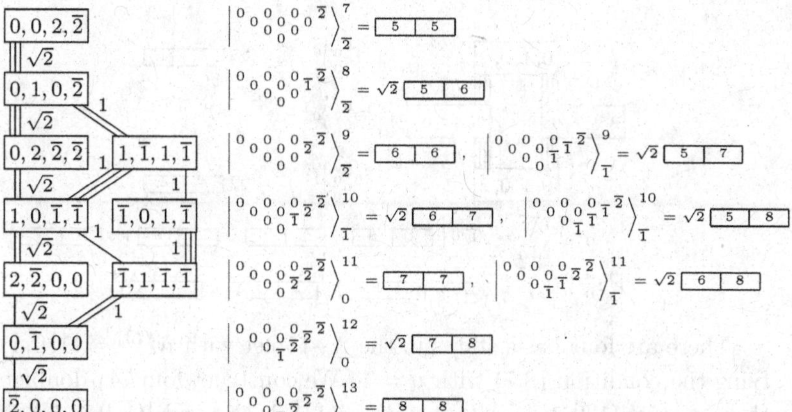

Now, we return to calculate the $\mathcal{A}_4$-triplet from $\left|\begin{smallmatrix}2&0&0&0\\1&0&0\\0&0\\0\end{smallmatrix}\right\rangle^{6}_{2}$ by Eqs. (8.25-28). Letting

$$F_4\left|\begin{smallmatrix}2&0&0&0\\1&0&0\\0&0\\0\end{smallmatrix}\right\rangle^{6}_{2}=a\left|\begin{smallmatrix}1&0&0&\overline{1}\\1&0&\overline{1}\\0&0\\0\end{smallmatrix}\right\rangle^{7}_{0}+b\left|\begin{smallmatrix}1&0&0&\overline{1}\\0&0&0\\0&0\\0\end{smallmatrix}\right\rangle^{7}_{0},$$

we obtain

$$\begin{aligned}
E_3F_4\left|\begin{smallmatrix}2&0&0&0\\1&0&0\\0&0\\0\end{smallmatrix}\right\rangle^{6}_{2}&=E_3\left\{a\left|\begin{smallmatrix}1&0&0&\overline{1}\\1&0&\overline{1}\\0&0\\0\end{smallmatrix}\right\rangle^{7}_{0}+b\left|\begin{smallmatrix}1&0&0&\overline{1}\\0&0&0\\0&0\\0\end{smallmatrix}\right\rangle^{7}_{0}\right\}\\
&=\left(a\sqrt{\tfrac{2}{3}}+b\sqrt{\tfrac{4}{3}}\right)\left|\begin{smallmatrix}1&0&0&\overline{1}\\1&0&0\\0&0\\0\end{smallmatrix}\right\rangle^{6}_{0}=F_4E_3\left|\begin{smallmatrix}2&0&0&0\\1&0&0\\0&0\\0\end{smallmatrix}\right\rangle^{6}_{2}\\
&=\sqrt{2}F_4\left|\begin{smallmatrix}2&0&0&0\\2&0&0\\0&0\\0\end{smallmatrix}\right\rangle^{5}_{2}=2\left|\begin{smallmatrix}1&0&0&\overline{1}\\1&0&0\\0&0\\0\end{smallmatrix}\right\rangle^{6}_{0},\\
E_4F_4\left|\begin{smallmatrix}2&0&0&0\\1&0&0\\0&0\\0\end{smallmatrix}\right\rangle^{6}_{2}&=\left(a^2+b^2\right)\left|\begin{smallmatrix}2&0&0&0\\1&0&0\\0&0\\0\end{smallmatrix}\right\rangle^{6}_{2}\\
&=\left(F_4E_4+H_4\right)\left|\begin{smallmatrix}2&0&0&0\\1&0&0\\0&0\\0\end{smallmatrix}\right\rangle^{6}_{2}=2\left|\begin{smallmatrix}2&0&0&0\\1&0&0\\0&0\\0\end{smallmatrix}\right\rangle^{6}_{2}.
\end{aligned}$$

Thus, $a\sqrt{2/3}+b\sqrt{4/3}=2$, $a^2+b^2=2$ The solution is $a=\sqrt{2/3}$ and $b=\sqrt{4/3}$.

$$\begin{aligned}
F_4\left|\begin{smallmatrix}2&0&0&0\\1&0&0\\0&0\\0\end{smallmatrix}\right\rangle^{6}_{2}&=F_4\left\{\sqrt{2}\,\begin{array}{|c|c|}\hline 3&4\\\hline\end{array}\right\}=\sqrt{2}\left\{\begin{array}{|c|c|}\hline 3&6\\\hline\end{array}+\begin{array}{|c|c|}\hline 4&5\\\hline\end{array}\right\}\\
&=\sqrt{\tfrac{2}{3}}\times\sqrt{\tfrac{1}{3}}\left\{-\begin{array}{|c|c|}\hline 1&8\\\hline\end{array}+\begin{array}{|c|c|}\hline 2&7\\\hline\end{array}+2\begin{array}{|c|c|}\hline 3&6\\\hline\end{array}\right\}\\
&\quad+\sqrt{\tfrac{4}{3}}\times\sqrt{\tfrac{1}{6}}\left\{\begin{array}{|c|c|}\hline 1&8\\\hline\end{array}-\begin{array}{|c|c|}\hline 2&7\\\hline\end{array}+\begin{array}{|c|c|}\hline 3&6\\\hline\end{array}+3\begin{array}{|c|c|}\hline 4&5\\\hline\end{array}\right\}.
\end{aligned}$$

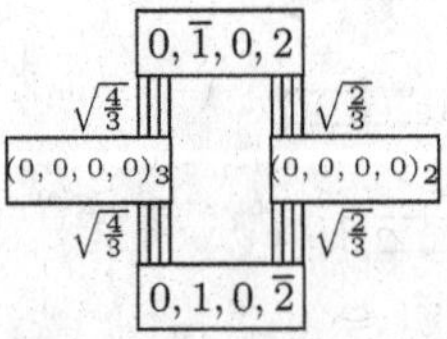

$$\left|\begin{smallmatrix}2&0&0&0\\1&0&0\\0&0\\0\end{smallmatrix}\right\rangle^{6}_{2}=\sqrt{2}\,\begin{array}{|c|c|}\hline 3&4\\\hline\end{array}$$

$$\left|\begin{smallmatrix}1&0&0&\overline{1}\\0&0&0\\0&0\\0\end{smallmatrix}\right\rangle^{7}_{0},\quad\left|\begin{smallmatrix}1&0&0&\overline{1}\\1&0&\overline{1}\\0&0\\0\end{smallmatrix}\right\rangle^{7}_{0}$$

$$\left|\begin{smallmatrix}0&0&0&\overline{2}\\0&0&\overline{1}\\0&0\\0\end{smallmatrix}\right\rangle^{8}_{\overline{2}}=\sqrt{2}\,\begin{array}{|c|c|}\hline 5&6\\\hline\end{array}$$

$$\left|\begin{smallmatrix}1&0&0&\overline{1}\\0&0&0\\0&0\\0\end{smallmatrix}\right\rangle^{7}_{0}=\sqrt{1/6}\left\{\begin{array}{|c|c|}\hline 1&8\\\hline\end{array}-\begin{array}{|c|c|}\hline 2&7\\\hline\end{array}+\begin{array}{|c|c|}\hline 3&6\\\hline\end{array}+3\begin{array}{|c|c|}\hline 4&5\\\hline\end{array}\right\}$$

$$\left|\begin{smallmatrix}1&0&0&\overline{1}\\1&0&\overline{1}\\0&0\\0\end{smallmatrix}\right\rangle^{7}_{0}=\sqrt{1/3}\left\{-\begin{array}{|c|c|}\hline 1&8\\\hline\end{array}+\begin{array}{|c|c|}\hline 2&7\\\hline\end{array}+2\begin{array}{|c|c|}\hline 3&6\\\hline\end{array}\right\}.$$

There are four basis states in the A_3-15plet with $\boldsymbol{M}^{(3)}=(1,0,1)$ satisfying the condition (8.7) with $\mu=4$. We construct four $\mathcal{A}_4$-doublets from those states $|(1,\overline{2},1,1)\rangle$, $|(\overline{1},\overline{1},1,1)\rangle$, $|(1,\overline{1},\overline{1},1)\rangle$, and $|(\overline{1},0,\overline{1},1)\rangle$ (see the parallel principle), where the partners all belong to the A_3-decuplet with $\boldsymbol{M}^{(3)} = (0, 0, 2)$.

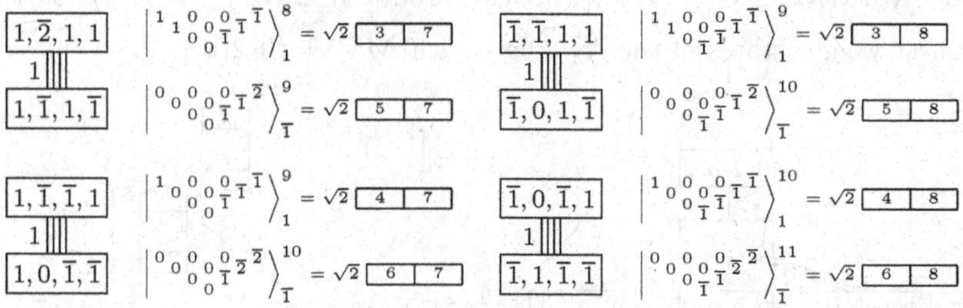

In summary, the representation $\boldsymbol{M} = (2, 0, 0, 0)$ of SO(8) is 35-dimensional. It contains an A_3-decuplet with $\boldsymbol{M}^{(3)} = (2, 0, 0)$, an A_3-15plet with $\boldsymbol{M}^{(3)} = (1, 0, 1)$, and an A_3-decuplet with $\boldsymbol{M}^{(3)} = (0, 0, 2)$. There are two simple dominant weights $(2, 0, 0, 0)$ and $(0, 1, 0, 0)$ and one triplet dominant weight $(0, 0, 0, 0)$ in this representation, where their sizes of Weyl orbits are 8, 24, and 1, respectively. Remind that three basis states with the weight $(0, 0, 0, 0)$ all are traceless and orthogonal to the trace tensor

$$\sum_{a=1}^{8} \boldsymbol{\theta}_{aa} = -\boxed{1 \mid 8} + \boxed{2 \mid 7} - \boxed{3 \mid 6} + \boxed{4 \mid 5}\,. \qquad (10.23)$$

Ex. 2. The self-dual representation $[(+)1, 1, 1, 1]$ of SO(8).

From Eq. (10.22), the highest weight is $\boldsymbol{M} = 2\boldsymbol{w}_4 = (0, 0, 0, 2)$. The self-dual representation is real (see subsection 10.1.1) and its highest weight state is

$$|\boldsymbol{M}\rangle = |(0, 0, 0, 2)\rangle = \left|\begin{smallmatrix} 2 & & 2 & & 2 & & 2 \\ & 2 & & 2 & & 2 & \\ & & 2 & & 2 & & \\ & & & 2 & & & \end{smallmatrix}\right\rangle_{2}^{1} = \begin{array}{|c|}\hline 1\\\hline 2\\\hline 3\\\hline 4\\\hline\end{array}.$$

The highest weight state $|(0, 0, 0, 2)\rangle$ spans an A_3-singlet with $\boldsymbol{M}^{(3)} = (0, 0, 0)$ and is also the highest weight state of an $\mathcal{A}_4$-triplet:

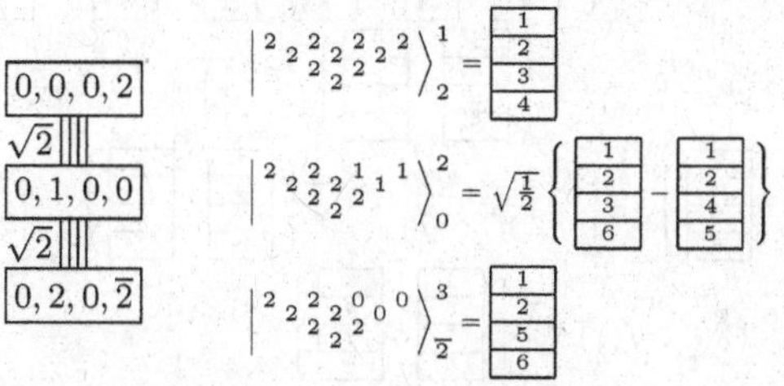

The produced state $\left|\begin{smallmatrix}2&&2&&1&&1\\&2&&2&&1&\\&&2&&2&&\\&&&2&&&\end{smallmatrix}\right\rangle^{2}_{0}$ is the highest weight states of the A_3-sixlet with $\boldsymbol{M}^{(3)}=(0,1,0)$, and the produced state $\left|\begin{smallmatrix}2&&2&&0&&0\\&2&&2&&0&\\&&2&&2&&\\&&&2&&&\end{smallmatrix}\right\rangle^{3}_{\overline{2}}$ is the highest weight states of the A_3-20lets with $\boldsymbol{M}^{(3)}=(0,2,0)$.

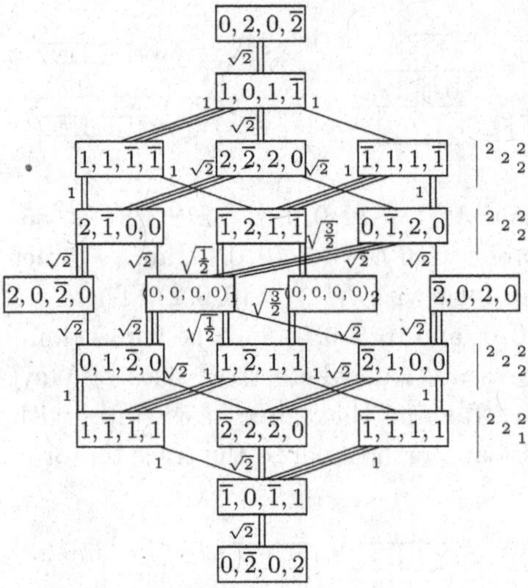

$$\left|\begin{smallmatrix}2&&2&&0&&0\\&2&&2&&0&\\&&2&&2&&\\&&&2&&&\end{smallmatrix}\right\rangle^{3}_{\overline{2}}=\begin{array}{|c|}\hline 1\\\hline 2\\\hline 5\\\hline 6\\\hline\end{array}$$

$$\left|\begin{smallmatrix}2&&2&&0&&0\\&2&&2&&0&\\&&2&&1&&\\&&&2&&&\end{smallmatrix}\right\rangle^{4}_{\overline{1}}=\sqrt{\tfrac{1}{2}}\left\{\begin{array}{|c|}\hline 1\\\hline 2\\\hline 5\\\hline 7\\\hline\end{array}+\begin{array}{|c|}\hline 1\\\hline 3\\\hline 5\\\hline 6\\\hline\end{array}\right\}$$

$$\left|\begin{smallmatrix}2&&2&&0&&0\\&2&&1&&0&\\&&2&&1&&\\&&&2&&&\end{smallmatrix}\right\rangle^{5}_{\overline{1}},\quad \left|\begin{smallmatrix}2&&2&&0&&0\\&2&&2&&0&\\&&2&&0&&\\&&&2&&&\end{smallmatrix}\right\rangle^{5}_{0},\quad \left|\begin{smallmatrix}2&&2&&0&&0\\&2&&2&&0&\\&&2&&1&&\\&&&1&&&\end{smallmatrix}\right\rangle^{5}_{\overline{1}}$$

$$\left|\begin{smallmatrix}2&&2&&0&&0\\&2&&1&&0&\\&&2&&0&&\\&&&2&&&\end{smallmatrix}\right\rangle^{6}_{0},\quad \left|\begin{smallmatrix}2&&2&&0&&0\\&2&&1&&0&\\&&2&&1&&\\&&&1&&&\end{smallmatrix}\right\rangle^{6}_{\overline{1}},\quad \left|\begin{smallmatrix}2&&2&&0&&0\\&2&&2&&0&\\&&2&&0&&\\&&&1&&&\end{smallmatrix}\right\rangle^{6}_{0}$$

$$\left|\begin{smallmatrix}2&&2&&0&&0\\&2&&0&&0&\\&&2&&0&&\\&&&2&&&\end{smallmatrix}\right\rangle^{7}_{0},\ A\ ,\ B\ ,\quad \left|\begin{smallmatrix}2&&2&&0&&0\\&2&&2&&0&\\&&2&&0&&\\&&&0&&&\end{smallmatrix}\right\rangle^{7}_{0}$$

$$\left|\begin{smallmatrix}2&&2&&0&&0\\&2&&0&&0&\\&&2&&0&&\\&&&1&&&\end{smallmatrix}\right\rangle^{8}_{0},\quad \left|\begin{smallmatrix}2&&2&&0&&0\\&2&&1&&0&\\&&1&&0&&\\&&&1&&&\end{smallmatrix}\right\rangle^{8}_{1},\quad \left|\begin{smallmatrix}2&&2&&0&&0\\&2&&1&&0&\\&&2&&0&&\\&&&0&&&\end{smallmatrix}\right\rangle^{8}_{0}$$

$$\left|\begin{smallmatrix}2&&2&&0&&0\\&2&&0&&0&\\&&1&&0&&\\&&&1&&&\end{smallmatrix}\right\rangle^{9}_{1},\quad \left|\begin{smallmatrix}2&&2&&0&&0\\&2&&0&&0&\\&&2&&0&&\\&&&0&&&\end{smallmatrix}\right\rangle^{9}_{0},\quad \left|\begin{smallmatrix}2&&2&&0&&0\\&2&&1&&0&\\&&1&&0&&\\&&&0&&&\end{smallmatrix}\right\rangle^{9}_{1}$$

$$\left|\begin{smallmatrix}2&&2&&0&&0\\&2&&0&&0&\\&&1&&0&&\\&&&0&&&\end{smallmatrix}\right\rangle^{10}_{1}=\sqrt{\tfrac{1}{2}}\left\{\begin{array}{|c|}\hline 2\\\hline 4\\\hline 7\\\hline 8\\\hline\end{array}+\begin{array}{|c|}\hline 3\\\hline 4\\\hline 6\\\hline 8\\\hline\end{array}\right\}$$

$$\left|\begin{smallmatrix}2&&2&&0&&0\\&2&&0&&0&\\&&0&&0&&\\&&&0&&&\end{smallmatrix}\right\rangle^{11}_{2}=\begin{array}{|c|}\hline 3\\\hline 4\\\hline 7\\\hline 8\\\hline\end{array}$$

$$\left|\begin{smallmatrix}2&&2&&0&&0\\&2&&1&&0&\\&&2&&1&&\\&&&2&&&\end{smallmatrix}\right\rangle^{5}_{\overline{1}}=\sqrt{\tfrac{1}{2}}\left\{\begin{array}{|c|}\hline 1\\\hline 2\\\hline 6\\\hline 7\\\hline\end{array}+\begin{array}{|c|}\hline 1\\\hline 4\\\hline 5\\\hline 6\\\hline\end{array}\right\},\quad \left|\begin{smallmatrix}2&&2&&0&&0\\&2&&2&&0&\\&&2&&0&&\\&&&2&&&\end{smallmatrix}\right\rangle^{5}_{0}=\begin{array}{|c|}\hline 1\\\hline 3\\\hline 5\\\hline 7\\\hline\end{array},$$

$$\left|\begin{smallmatrix}2&&2&&0&&0\\&2&&2&&0&\\&&2&&1&&\\&&&1&&&\end{smallmatrix}\right\rangle^{5}_{\overline{1}}=\sqrt{\tfrac{1}{2}}\left\{\begin{array}{|c|}\hline 1\\\hline 2\\\hline 5\\\hline 8\\\hline\end{array}+\begin{array}{|c|}\hline 2\\\hline 3\\\hline 5\\\hline 6\\\hline\end{array}\right\},\quad \left|\begin{smallmatrix}2&&2&&0&&0\\&2&&1&&0&\\&&2&&0&&\\&&&2&&&\end{smallmatrix}\right\rangle^{6}_{0}=\sqrt{\tfrac{1}{2}}\left\{\begin{array}{|c|}\hline 1\\\hline 3\\\hline 6\\\hline 7\\\hline\end{array}+\begin{array}{|c|}\hline 1\\\hline 4\\\hline 5\\\hline 7\\\hline\end{array}\right\},$$

$$\left|\begin{smallmatrix}2&&2&&0&&0\\&2&&1&&0&\\&&2&&1&&\\&&&1&&&\end{smallmatrix}\right\rangle^{6}_{\overline{1}}=\sqrt{\tfrac{1}{2}}\left\{\begin{array}{|c|}\hline 1\\\hline 2\\\hline 6\\\hline 8\\\hline\end{array}+\begin{array}{|c|}\hline 2\\\hline 4\\\hline 5\\\hline 6\\\hline\end{array}\right\},\quad \left|\begin{smallmatrix}2&&2&&0&&0\\&2&&2&&0&\\&&2&&0&&\\&&&1&&&\end{smallmatrix}\right\rangle^{6}_{0}=\sqrt{\tfrac{1}{2}}\left\{\begin{array}{|c|}\hline 1\\\hline 3\\\hline 5\\\hline 8\\\hline\end{array}+\begin{array}{|c|}\hline 2\\\hline 3\\\hline 5\\\hline 7\\\hline\end{array}\right\}$$

$$\left|\begin{smallmatrix}2&&2&&0&&0\\&2&&0&&0&\\&&2&&0&&\\&&&2&&&\end{smallmatrix}\right\rangle^{7}_{0}=\begin{array}{|c|}\hline 1\\\hline 4\\\hline 6\\\hline 7\\\hline\end{array},\quad A=|(0,0,0,0)_1\rangle=\left|\begin{smallmatrix}2&&2&&0&&0\\&2&&1&&0&\\&&2&&0&&\\&&&1&&&\end{smallmatrix}\right\rangle^{7}_{0}=\tfrac{1}{2}\left\{\begin{array}{|c|}\hline 1\\\hline 3\\\hline 6\\\hline 8\\\hline\end{array}+\begin{array}{|c|}\hline 1\\\hline 4\\\hline 5\\\hline 8\\\hline\end{array}+\begin{array}{|c|}\hline 2\\\hline 3\\\hline 6\\\hline 7\\\hline\end{array}+\begin{array}{|c|}\hline 2\\\hline 4\\\hline 5\\\hline 7\\\hline\end{array}\right\}$$

$$B=|(0,0,0,0)_2\rangle=\left|\begin{smallmatrix}2&&2&&0&&0\\&2&&1&&0&\\&&1&&1&&\\&&&1&&&\end{smallmatrix}\right\rangle^{7}_{0}=\sqrt{\tfrac{1}{12}}\left\{2\begin{array}{|c|}\hline 1\\\hline 2\\\hline 7\\\hline 8\\\hline\end{array}+\begin{array}{|c|}\hline 1\\\hline 3\\\hline 6\\\hline 8\\\hline\end{array}-\begin{array}{|c|}\hline 1\\\hline 4\\\hline 5\\\hline 8\\\hline\end{array}-\begin{array}{|c|}\hline 2\\\hline 3\\\hline 6\\\hline 7\\\hline\end{array}+\begin{array}{|c|}\hline 2\\\hline 4\\\hline 5\\\hline 7\\\hline\end{array}+2\begin{array}{|c|}\hline 3\\\hline 4\\\hline 5\\\hline 6\\\hline\end{array}\right\},$$

$$\left|\begin{smallmatrix}2&&2&&0&&0\\&2&&2&&0&\\&&2&&0&&\\&&&0&&&\end{smallmatrix}\right\rangle^{7}_{0}=\begin{array}{|c|}\hline 2\\\hline 3\\\hline 5\\\hline 8\\\hline\end{array},\quad \left|\begin{smallmatrix}2&&2&&0&&0\\&2&&0&&0&\\&&2&&0&&\\&&&1&&&\end{smallmatrix}\right\rangle^{8}_{0}=\sqrt{\tfrac{1}{2}}\left\{\begin{array}{|c|}\hline 1\\\hline 4\\\hline 6\\\hline 8\\\hline\end{array}+\begin{array}{|c|}\hline 2\\\hline 4\\\hline 6\\\hline 7\\\hline\end{array}\right\},\quad \left|\begin{smallmatrix}2&&2&&0&&0\\&2&&1&&0&\\&&1&&0&&\\&&&1&&&\end{smallmatrix}\right\rangle^{8}_{1}=\sqrt{\tfrac{1}{2}}\left\{\begin{array}{|c|}\hline 1\\\hline 3\\\hline 7\\\hline 8\\\hline\end{array}+\begin{array}{|c|}\hline 3\\\hline 4\\\hline 5\\\hline 7\\\hline\end{array}\right\},$$

$$\left|\begin{smallmatrix}2&&2&&0&&0\\&2&&1&&0&\\&&2&&0&&\\&&&0&&&\end{smallmatrix}\right\rangle^{8}_{0}=\sqrt{\tfrac{1}{2}}\left\{\begin{array}{|c|}\hline 2\\\hline 3\\\hline 6\\\hline 8\\\hline\end{array}+\begin{array}{|c|}\hline 2\\\hline 4\\\hline 5\\\hline 8\\\hline\end{array}\right\},\quad \left|\begin{smallmatrix}2&&2&&0&&0\\&2&&0&&0&\\&&1&&0&&\\&&&1&&&\end{smallmatrix}\right\rangle^{9}_{1}=\sqrt{\tfrac{1}{2}}\left\{\begin{array}{|c|}\hline 1\\\hline 4\\\hline 7\\\hline 8\\\hline\end{array}+\begin{array}{|c|}\hline 3\\\hline 4\\\hline 6\\\hline 7\\\hline\end{array}\right\}$$

$$\left|\begin{smallmatrix}2&&2&&0&&0\\&2&&0&&0&\\&&2&&0&&\\&&&0&&&\end{smallmatrix}\right\rangle^{9}_{0}=\begin{array}{|c|}\hline 2\\\hline 4\\\hline 6\\\hline 8\\\hline\end{array},\quad \left|\begin{smallmatrix}2&&2&&0&&0\\&2&&1&&0&\\&&1&&0&&\\&&&0&&&\end{smallmatrix}\right\rangle^{9}_{1}=\sqrt{\tfrac{1}{2}}\left\{\begin{array}{|c|}\hline 2\\\hline 3\\\hline 7\\\hline 8\\\hline\end{array}+\begin{array}{|c|}\hline 3\\\hline 4\\\hline 5\\\hline 8\\\hline\end{array}\right\}.$$

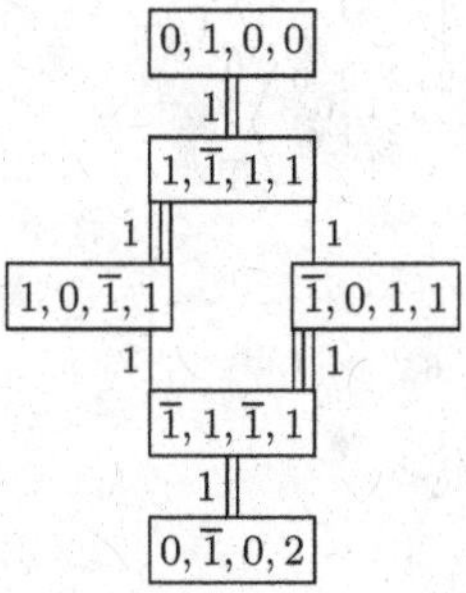

$$\left|\begin{smallmatrix} 2&&2&&1&&1\\ &2&&2&&1&\\ &&2&&2&&\\ &&&2&&& \end{smallmatrix}\right\rangle_0^2 = \sqrt{\tfrac{1}{2}}\left\{\begin{array}{|c|}\hline 1\\\hline 2\\\hline 3\\\hline 6\\\hline\end{array} - \begin{array}{|c|}\hline 1\\\hline 2\\\hline 4\\\hline 5\\\hline\end{array}\right\}$$

$$\left|\begin{smallmatrix} 2&&2&&1&&1\\ &2&&2&&1&\\ &&2&&1&&\\ &&&2&&& \end{smallmatrix}\right\rangle_1^3 = \sqrt{\tfrac{1}{2}}\left\{\begin{array}{|c|}\hline 1\\\hline 2\\\hline 3\\\hline 7\\\hline\end{array} - \begin{array}{|c|}\hline 1\\\hline 3\\\hline 4\\\hline 5\\\hline\end{array}\right\}$$

$$\left|\begin{smallmatrix} 2&&2&&1&&1\\ &2&&1&&1&\\ &&2&&1&&\\ &&&2&&& \end{smallmatrix}\right\rangle_1^4, \quad \left|\begin{smallmatrix} 2&&2&&1&&1\\ &2&&2&&1&\\ &&2&&1&&\\ &&&1&&& \end{smallmatrix}\right\rangle_1^4 = \sqrt{\tfrac{1}{2}}\left\{\begin{array}{|c|}\hline 1\\\hline 2\\\hline 3\\\hline 8\\\hline\end{array} - \begin{array}{|c|}\hline 2\\\hline 3\\\hline 4\\\hline 5\\\hline\end{array}\right\}$$

$$\left|\begin{smallmatrix} 2&&2&&1&&1\\ &2&&1&&1&\\ &&2&&1&&\\ &&&1&&& \end{smallmatrix}\right\rangle_1^5 = \sqrt{\tfrac{1}{2}}\left\{\begin{array}{|c|}\hline 1\\\hline 2\\\hline 4\\\hline 8\\\hline\end{array} - \begin{array}{|c|}\hline 2\\\hline 3\\\hline 4\\\hline 6\\\hline\end{array}\right\}$$

$$\left|\begin{smallmatrix} 2&&2&&1&&1\\ &2&&1&&1&\\ &&1&&1&&\\ &&&1&&& \end{smallmatrix}\right\rangle_2^6 = \sqrt{\tfrac{1}{2}}\left\{\begin{array}{|c|}\hline 1\\\hline 3\\\hline 4\\\hline 8\\\hline\end{array} - \begin{array}{|c|}\hline 2\\\hline 3\\\hline 4\\\hline 7\\\hline\end{array}\right\}$$

$$|1,0,\overline{1},1\rangle = \left|\begin{smallmatrix} 2&&2&&1&&1\\ &2&&1&&1&\\ &&2&&1&&\\ &&&2&&& \end{smallmatrix}\right\rangle_1^4 = \sqrt{\tfrac{1}{2}}\left\{\begin{array}{|c|}\hline 1\\\hline 2\\\hline 4\\\hline 7\\\hline\end{array} - \begin{array}{|c|}\hline 1\\\hline 3\\\hline 4\\\hline 6\\\hline\end{array}\right\}.$$

Four basis states $|(1,\overline{1},1,1)\rangle$, $|(\overline{1},0,1,1)\rangle$, $|(1,0,\overline{1},1)\rangle$, and $|(\overline{1},1,\overline{1},1)\rangle$ in the A$_3$-sixlet with $\boldsymbol{M}^{(3)} = (0,1,0)$ satisfy the condition (8.7) with $\mu = 4$. Four $\mathcal{A}_4$-doublets are constructed from those states due to the parallel principle, where the partners all belong to the A$_3$-20let with $\boldsymbol{M}^{(3)} = (0,2,0)$.

$\boxed{1,\overline{1},1,1}$ $\quad \left|\begin{smallmatrix} 2&&2&&1&&1\\ &2&&2&&1&\\ &&2&&1&&\\ &&&2&&& \end{smallmatrix}\right\rangle_1^3 = \sqrt{\tfrac{1}{2}}\left\{\begin{array}{|c|}\hline 1\\\hline 2\\\hline 3\\\hline 7\\\hline\end{array} - \begin{array}{|c|}\hline 1\\\hline 3\\\hline 4\\\hline 5\\\hline\end{array}\right\}$

1

$\boxed{1,0,1,\overline{1}}$ $\quad \left|\begin{smallmatrix} 2&&2&&0&&0\\ &2&&2&&0&\\ &&2&&1&&\\ &&&2&&& \end{smallmatrix}\right\rangle_{\overline{1}}^4 = \sqrt{\tfrac{1}{2}}\left\{\begin{array}{|c|}\hline 1\\\hline 2\\\hline 5\\\hline 7\\\hline\end{array} + \begin{array}{|c|}\hline 1\\\hline 3\\\hline 5\\\hline 6\\\hline\end{array}\right\}$

$\boxed{\overline{1},0,1,1}$ $\quad \left|\begin{smallmatrix} 2&&2&&1&&1\\ &2&&2&&1&\\ &&2&&1&&\\ &&&1&&& \end{smallmatrix}\right\rangle_1^4 = \sqrt{\tfrac{1}{2}}\left\{\begin{array}{|c|}\hline 1\\\hline 2\\\hline 3\\\hline 8\\\hline\end{array} - \begin{array}{|c|}\hline 2\\\hline 3\\\hline 4\\\hline 5\\\hline\end{array}\right\}$

1

$\boxed{\overline{1},1,1,\overline{1}}$ $\quad \left|\begin{smallmatrix} 2&&2&&0&&0\\ &2&&2&&0&\\ &&2&&1&&\\ &&&1&&& \end{smallmatrix}\right\rangle_{\overline{1}}^5 = \sqrt{\tfrac{1}{2}}\left\{\begin{array}{|c|}\hline 1\\\hline 2\\\hline 5\\\hline 8\\\hline\end{array} + \begin{array}{|c|}\hline 2\\\hline 3\\\hline 5\\\hline 6\\\hline\end{array}\right\}$

$\boxed{1,0,\overline{1},1}$ $\quad \left|\begin{smallmatrix} 2&&2&&1&&1\\ &2&&1&&1&\\ &&2&&1&&\\ &&&2&&& \end{smallmatrix}\right\rangle_1^4 = \sqrt{\tfrac{1}{2}}\left\{\begin{array}{|c|}\hline 1\\\hline 2\\\hline 4\\\hline 7\\\hline\end{array} - \begin{array}{|c|}\hline 1\\\hline 3\\\hline 4\\\hline 6\\\hline\end{array}\right\}$

1

$\boxed{1,1,\overline{1},\overline{1}}$ $\quad \left|\begin{smallmatrix} 2&&2&&0&&0\\ &2&&1&&0&\\ &&2&&1&&\\ &&&2&&& \end{smallmatrix}\right\rangle_{\overline{1}}^5 = \sqrt{\tfrac{1}{2}}\left\{\begin{array}{|c|}\hline 1\\\hline 2\\\hline 6\\\hline 7\\\hline\end{array} + \begin{array}{|c|}\hline 1\\\hline 4\\\hline 5\\\hline 6\\\hline\end{array}\right\}$

$\boxed{\overline{1},1,\overline{1},1}$ $\quad \left|\begin{smallmatrix} 2&&2&&1&&1\\ &2&&1&&1&\\ &&2&&1&&\\ &&&1&&& \end{smallmatrix}\right\rangle_1^5 = \sqrt{\tfrac{1}{2}}\left\{\begin{array}{|c|}\hline 1\\\hline 2\\\hline 4\\\hline 8\\\hline\end{array} - \begin{array}{|c|}\hline 2\\\hline 3\\\hline 4\\\hline 6\\\hline\end{array}\right\}$

1

$\boxed{\overline{1},2,\overline{1},\overline{1}}$ $\quad \left|\begin{smallmatrix} 2&&2&&0&&0\\ &2&&1&&0&\\ &&2&&1&&\\ &&&1&&& \end{smallmatrix}\right\rangle_{\overline{1}}^6 = \sqrt{\tfrac{1}{2}}\left\{\begin{array}{|c|}\hline 1\\\hline 2\\\hline 6\\\hline 8\\\hline\end{array} + \begin{array}{|c|}\hline 2\\\hline 4\\\hline 5\\\hline 6\\\hline\end{array}\right\}$

The basis state $|(0,\overline{1},0,2)\rangle$ in the A$_3$-sixlet with $\boldsymbol{M}^{(3)} = (0,1,0)$ satisfies the condition (8.7) with $\mu = 4$, and an $\mathcal{A}_4$-triplet is constructed. Letting

$$F_4\,|(0,\overline{1},0,2)\rangle = F_4\left|\begin{smallmatrix} 2&&2&&1&&1\\ &2&&1&&1&\\ &&1&&1&&\\ &&&1&&& \end{smallmatrix}\right\rangle_2^6 = a\left|\begin{smallmatrix} 2&&2&&0&&0\\ &2&&1&&0&\\ &&1&&1&&\\ &&&1&&& \end{smallmatrix}\right\rangle_0^7 + b\left|\begin{smallmatrix} 1&&1&&1&&1\\ &1&&1&&1&\\ &&1&&1&&\\ &&&1&&& \end{smallmatrix}\right\rangle_0^7,$$

where b is real and positive by choosing the phase of the state $\left|\begin{smallmatrix} 1&&1&&1&&1\\ &1&&1&&1&\\ &&1&&1&&\\ &&&1&&& \end{smallmatrix}\right\rangle_0^7$, we have

$$\begin{aligned}
E_2F_4\left|\begin{smallmatrix}2\ 2\ 1\ 1\\2\ 1\ 1\\1\ 1\\1\end{smallmatrix}\right\rangle_2^6 &= aE_2\left|\begin{smallmatrix}2\ 2\ 0\ 0\\2\ 1\ 0\\1\ 1\\1\end{smallmatrix}\right\rangle_0^7 = a\sqrt{\tfrac{3}{2}}\left|\begin{smallmatrix}2\ 2\ 0\ 0\\2\ 1\ 0\\2\ 1\\1\end{smallmatrix}\right\rangle_{\overline{1}}^6\\
&= F_4E_2\left|\begin{smallmatrix}2\ 2\ 1\ 1\\2\ 1\ 1\\1\ 1\\1\end{smallmatrix}\right\rangle_2^6 = F_4\left|\begin{smallmatrix}2\ 2\ 1\ 1\\2\ 1\ 1\\2\ 1\\1\end{smallmatrix}\right\rangle_1^5 = \left|\begin{smallmatrix}2\ 2\ 0\ 0\\2\ 1\ 0\\2\ 1\\1\end{smallmatrix}\right\rangle_{\overline{1}}^6,\\
E_4F_4\left|\begin{smallmatrix}2\ 2\ 1\ 1\\2\ 1\ 1\\1\ 1\\1\end{smallmatrix}\right\rangle_2^6 &= (a^2+b^2)\left|\begin{smallmatrix}2\ 2\ 1\ 1\\2\ 1\ 1\\1\ 1\\1\end{smallmatrix}\right\rangle_2^6\\
&= (F_4E_4+H_4)\left|\begin{smallmatrix}2\ 2\ 1\ 1\\2\ 1\ 1\\1\ 1\\1\end{smallmatrix}\right\rangle_2^6 = 2\left|\begin{smallmatrix}2\ 2\ 1\ 1\\2\ 1\ 1\\1\ 1\\1\end{smallmatrix}\right\rangle_2^6.
\end{aligned}$$

Thus, $a=\sqrt{2/3}$, $b=\sqrt{2-2/3}=\sqrt{4/3}$, and

$$\begin{aligned}
\left|\begin{smallmatrix}1\ 1\ 1\ 1\\1\ 1\ 1\\1\ 1\\1\end{smallmatrix}\right\rangle_0^7 &= \sqrt{\tfrac{3}{4}}\left\{F_4\left|\begin{smallmatrix}2\ 2\ 1\ 1\\2\ 1\ 1\\1\ 1\\1\end{smallmatrix}\right\rangle_2^6 - \sqrt{\tfrac{2}{3}}\left|\begin{smallmatrix}2\ 2\ 0\ 0\\2\ 1\ 0\\1\ 1\\1\end{smallmatrix}\right\rangle_0^7\right\}\\
&= \sqrt{\tfrac{3}{4}}\times\sqrt{\tfrac{1}{2}}\left\{\begin{array}{|c|}\hline 1\\\hline 3\\\hline 6\\\hline 8\\\hline\end{array} - \begin{array}{|c|}\hline 1\\\hline 4\\\hline 5\\\hline 8\\\hline\end{array} - \begin{array}{|c|}\hline 2\\\hline 3\\\hline 6\\\hline 7\\\hline\end{array} + \begin{array}{|c|}\hline 2\\\hline 4\\\hline 5\\\hline 7\\\hline\end{array}\right\}\\
&\quad - \sqrt{\tfrac{1}{2}}\times\sqrt{\tfrac{1}{12}}\left\{2\begin{array}{|c|}\hline 1\\\hline 2\\\hline 7\\\hline 8\\\hline\end{array} + \begin{array}{|c|}\hline 1\\\hline 3\\\hline 6\\\hline 8\\\hline\end{array} - \begin{array}{|c|}\hline 1\\\hline 4\\\hline 5\\\hline 8\\\hline\end{array} - \begin{array}{|c|}\hline 2\\\hline 3\\\hline 6\\\hline 7\\\hline\end{array} + \begin{array}{|c|}\hline 2\\\hline 4\\\hline 5\\\hline 7\\\hline\end{array} + 2\begin{array}{|c|}\hline 3\\\hline 4\\\hline 5\\\hline 6\\\hline\end{array}\right\}\\
&= \sqrt{\tfrac{1}{6}}\left\{-\begin{array}{|c|}\hline 1\\\hline 2\\\hline 7\\\hline 8\\\hline\end{array} + \begin{array}{|c|}\hline 1\\\hline 3\\\hline 6\\\hline 8\\\hline\end{array} - \begin{array}{|c|}\hline 1\\\hline 4\\\hline 5\\\hline 8\\\hline\end{array} - \begin{array}{|c|}\hline 2\\\hline 3\\\hline 6\\\hline 7\\\hline\end{array} + \begin{array}{|c|}\hline 2\\\hline 4\\\hline 5\\\hline 7\\\hline\end{array} - \begin{array}{|c|}\hline 3\\\hline 4\\\hline 5\\\hline 6\\\hline\end{array}\right\}.
\end{aligned}$$

The block weight diagram for the $\mathcal{A}_4$-triplet with the highest weight state $|(0,\overline{1},0,2)\rangle$ is listed as follows.

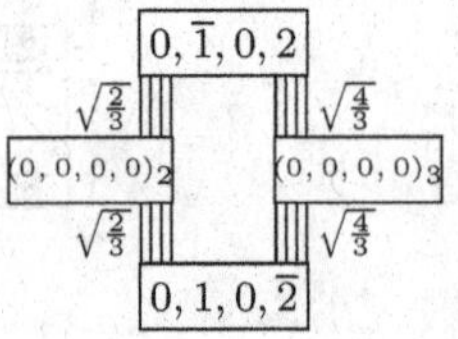

$$\left|\begin{smallmatrix}2\ 2\ 1\ 1\\2\ 1\ 1\\1\ 1\\1\end{smallmatrix}\right\rangle_2^6 = \sqrt{\tfrac{1}{2}}\left\{\begin{array}{|c|}\hline 1\\\hline 3\\\hline 4\\\hline 8\\\hline\end{array} - \begin{array}{|c|}\hline 2\\\hline 3\\\hline 4\\\hline 7\\\hline\end{array}\right\}.$$

$$\left|\begin{smallmatrix}2\ 2\ 0\ 0\\2\ 1\ 0\\1\ 1\\1\end{smallmatrix}\right\rangle_0^7, \quad \left|\begin{smallmatrix}1\ 1\ 1\ 1\\1\ 1\ 1\\1\ 1\\1\end{smallmatrix}\right\rangle_0^7$$

$$\left|\begin{smallmatrix}1\ 1\ 0\ 0\\1\ 1\ 0\\1\ 1\\1\end{smallmatrix}\right\rangle_{\overline{2}}^8 = \sqrt{\tfrac{1}{2}}\left\{\begin{array}{|c|}\hline 1\\\hline 5\\\hline 6\\\hline 8\\\hline\end{array} - \begin{array}{|c|}\hline 2\\\hline 5\\\hline 6\\\hline 7\\\hline\end{array}\right\}$$

$$\left|\begin{smallmatrix}2\ 2\ 0\ 0\\2\ 1\ 0\\1\ 1\\1\end{smallmatrix}\right\rangle_0^7 = \sqrt{\tfrac{1}{12}}\left\{2\begin{array}{|c|}\hline 1\\\hline 2\\\hline 7\\\hline 8\\\hline\end{array} + \begin{array}{|c|}\hline 1\\\hline 3\\\hline 6\\\hline 8\\\hline\end{array} - \begin{array}{|c|}\hline 1\\\hline 4\\\hline 5\\\hline 8\\\hline\end{array} - \begin{array}{|c|}\hline 2\\\hline 3\\\hline 6\\\hline 7\\\hline\end{array} + \begin{array}{|c|}\hline 2\\\hline 4\\\hline 5\\\hline 7\\\hline\end{array} + 2\begin{array}{|c|}\hline 3\\\hline 4\\\hline 5\\\hline 6\\\hline\end{array}\right\},$$

$$\left|\begin{smallmatrix}1\ 1\ 1\ 1\\1\ 1\ 1\\1\ 1\\1\end{smallmatrix}\right\rangle_0^7 = \sqrt{\tfrac{1}{6}}\left\{-\begin{array}{|c|}\hline 1\\\hline 2\\\hline 7\\\hline 8\\\hline\end{array} + \begin{array}{|c|}\hline 1\\\hline 3\\\hline 6\\\hline 8\\\hline\end{array} - \begin{array}{|c|}\hline 1\\\hline 4\\\hline 5\\\hline 8\\\hline\end{array} - \begin{array}{|c|}\hline 2\\\hline 3\\\hline 6\\\hline 7\\\hline\end{array} + \begin{array}{|c|}\hline 2\\\hline 4\\\hline 5\\\hline 7\\\hline\end{array} - \begin{array}{|c|}\hline 3\\\hline 4\\\hline 5\\\hline 6\\\hline\end{array}\right\}.$$

The state $|(0,0,0,0)_3\rangle$ spans an A_3-singlet with $\boldsymbol{M}^{(3)}=0$ and the state $|(0,1,0,\overline{2})\rangle$ is the highest weight state of the A_3-sixlet with $\boldsymbol{M}^{(3)}=(0,1,0)$, from which an A_3-sixlet with $\boldsymbol{M}^{(3)}=(0,1,0)$ is constructed.

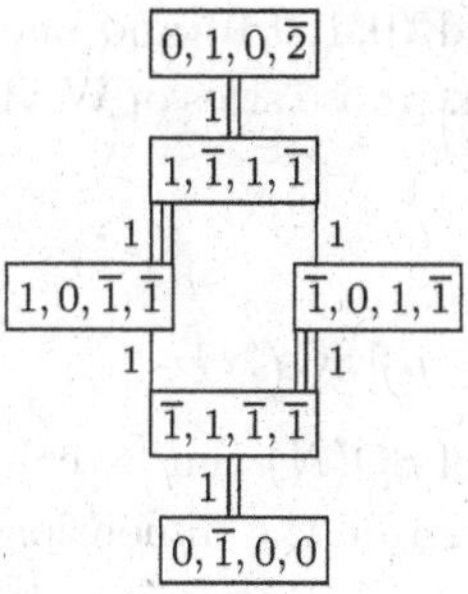

$$\left|\begin{smallmatrix}1&1&0&0\\1&1&0\\1&1\\1\end{smallmatrix}\right\rangle^{8}_{\overline{2}} = \sqrt{\tfrac{1}{2}}\left\{\begin{array}{|c|}\hline 1\\\hline 5\\\hline 6\\\hline 8\\\hline\end{array} - \begin{array}{|c|}\hline 2\\\hline 5\\\hline 6\\\hline 7\\\hline\end{array}\right\}$$

$$\left|\begin{smallmatrix}1&1&0&0\\1&1&0\\1&0\\1\end{smallmatrix}\right\rangle^{9}_{\overline{1}} = \sqrt{\tfrac{1}{2}}\left\{\begin{array}{|c|}\hline 1\\\hline 5\\\hline 7\\\hline 8\\\hline\end{array} - \begin{array}{|c|}\hline 3\\\hline 5\\\hline 6\\\hline 7\\\hline\end{array}\right\}$$

$$\left|\begin{smallmatrix}1&1&0&0\\1&0&0\\1&0\\1\end{smallmatrix}\right\rangle^{10}_{\overline{1}}, \quad \left|\begin{smallmatrix}1&1&0&0\\1&1&0\\1&0\\0\end{smallmatrix}\right\rangle^{10}_{\overline{1}} = \sqrt{\tfrac{1}{2}}\left\{\begin{array}{|c|}\hline 2\\\hline 5\\\hline 7\\\hline 8\\\hline\end{array} - \begin{array}{|c|}\hline 3\\\hline 5\\\hline 6\\\hline 8\\\hline\end{array}\right\}$$

$$\left|\begin{smallmatrix}1&1&0&0\\1&0&0\\1&0\\0\end{smallmatrix}\right\rangle^{11}_{\overline{1}} = \sqrt{\tfrac{1}{2}}\left\{\begin{array}{|c|}\hline 2\\\hline 6\\\hline 7\\\hline 8\\\hline\end{array} - \begin{array}{|c|}\hline 4\\\hline 5\\\hline 6\\\hline 8\\\hline\end{array}\right\}$$

$$\left|\begin{smallmatrix}1&1&0&0\\1&0&0\\0&0\\0\end{smallmatrix}\right\rangle^{12}_{0} = \sqrt{\tfrac{1}{2}}\left\{\begin{array}{|c|}\hline 3\\\hline 6\\\hline 7\\\hline 8\\\hline\end{array} - \begin{array}{|c|}\hline 4\\\hline 5\\\hline 7\\\hline 8\\\hline\end{array}\right\}$$

$$|1,0,\overline{1},1\rangle = \left|\begin{smallmatrix}1&1&0&0\\1&0&0\\1&0\\1\end{smallmatrix}\right\rangle^{10}_{\overline{1}} = \sqrt{\tfrac{1}{2}}\left\{\begin{array}{|c|}\hline 1\\\hline 6\\\hline 7\\\hline 8\\\hline\end{array} - \begin{array}{|c|}\hline 4\\\hline 5\\\hline 6\\\hline 7\\\hline\end{array}\right\}.$$

Five basis states $|(1,\overline{2},1,1)\rangle$, $|(\overline{1},\overline{1},1,1)\rangle$, $|(1,\overline{1},\overline{1},1)\rangle$, $|(\overline{1},0,\overline{1},1)\rangle$, and $|(0,\overline{2},0,2)\rangle$ in the A_3-20let with $\boldsymbol{M}^{(3)} = (0,2,0)$ satisfy the condition (8.7) with $\mu = 4$. Four $\mathcal{A}_4$-doublets are constructed from the first four states and one $\mathcal{A}_4$-triplet is constructed from the last state. All partners in those $\mathcal{A}_4$-multiplets belong to the A_3-sixlet with $\boldsymbol{M}^{(3)} = (0,1,0)$, except for the last state $|(0,0,0,\overline{2})\rangle$, which spans an A_3-singlet and is the lowest weight state in the self-dual representation $\boldsymbol{M} = (0,0,0,2)$ of SO(8).

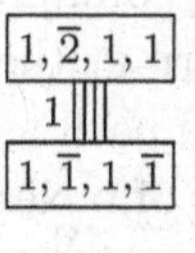

$$\left|\begin{smallmatrix}2&2&0&0\\2&1&0\\1&0\\1\end{smallmatrix}\right\rangle^{8}_{1} = \sqrt{\tfrac{1}{2}}\left\{\begin{array}{|c|}\hline 1\\\hline 3\\\hline 7\\\hline 8\\\hline\end{array} - \begin{array}{|c|}\hline 3\\\hline 4\\\hline 5\\\hline 7\\\hline\end{array}\right\}$$

$$\left|\begin{smallmatrix}1&1&0&0\\1&1&0\\1&0\\1\end{smallmatrix}\right\rangle^{9}_{\overline{1}} = \sqrt{\tfrac{1}{2}}\left\{\begin{array}{|c|}\hline 1\\\hline 5\\\hline 7\\\hline 8\\\hline\end{array} + \begin{array}{|c|}\hline 3\\\hline 5\\\hline 6\\\hline 7\\\hline\end{array}\right\}$$

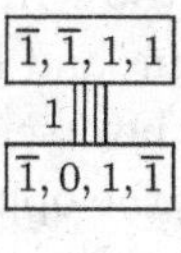

$$\left|\begin{smallmatrix}2&2&0&0\\2&1&0\\1&0\\0\end{smallmatrix}\right\rangle^{9}_{1} = \sqrt{\tfrac{1}{2}}\left\{\begin{array}{|c|}\hline 2\\\hline 3\\\hline 7\\\hline 8\\\hline\end{array} - \begin{array}{|c|}\hline 3\\\hline 4\\\hline 5\\\hline 8\\\hline\end{array}\right\}$$

$$\left|\begin{smallmatrix}1&1&0&0\\1&1&0\\1&0\\0\end{smallmatrix}\right\rangle^{10}_{\overline{1}} = \sqrt{\tfrac{1}{2}}\left\{\begin{array}{|c|}\hline 2\\\hline 5\\\hline 7\\\hline 8\\\hline\end{array} + \begin{array}{|c|}\hline 3\\\hline 5\\\hline 6\\\hline 8\\\hline\end{array}\right\}$$

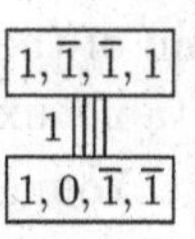

$$\left|\begin{smallmatrix}2&2&0&0\\2&0&0\\1&0\\1\end{smallmatrix}\right\rangle^{9}_{1} = \sqrt{\tfrac{1}{2}}\left\{\begin{array}{|c|}\hline 1\\\hline 4\\\hline 7\\\hline 8\\\hline\end{array} - \begin{array}{|c|}\hline 3\\\hline 4\\\hline 6\\\hline 7\\\hline\end{array}\right\}$$

$$\left|\begin{smallmatrix}1&1&0&0\\1&0&0\\1&0\\1\end{smallmatrix}\right\rangle^{10}_{\overline{1}} = \sqrt{\tfrac{1}{2}}\left\{\begin{array}{|c|}\hline 1\\\hline 6\\\hline 7\\\hline 8\\\hline\end{array} + \begin{array}{|c|}\hline 4\\\hline 5\\\hline 6\\\hline 7\\\hline\end{array}\right\}$$

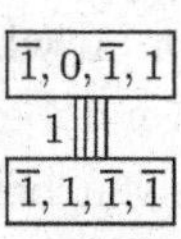

$$\left|\begin{smallmatrix}2&2&0&0\\2&0&0\\1&0\\0\end{smallmatrix}\right\rangle^{10}_{1} = \sqrt{\tfrac{1}{2}}\left\{\begin{array}{|c|}\hline 2\\\hline 4\\\hline 7\\\hline 8\\\hline\end{array} - \begin{array}{|c|}\hline 3\\\hline 4\\\hline 6\\\hline 8\\\hline\end{array}\right\}$$

$$\left|\begin{smallmatrix}1&1&0&0\\1&0&0\\1&0\\0\end{smallmatrix}\right\rangle^{11}_{\overline{1}} = \sqrt{\tfrac{1}{2}}\left\{\begin{array}{|c|}\hline 2\\\hline 6\\\hline 7\\\hline 8\\\hline\end{array} + \begin{array}{|c|}\hline 4\\\hline 5\\\hline 6\\\hline 8\\\hline\end{array}\right\}$$

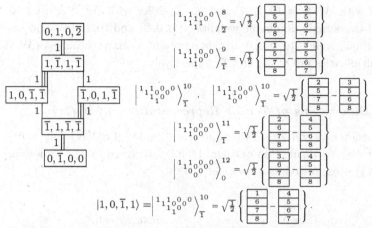

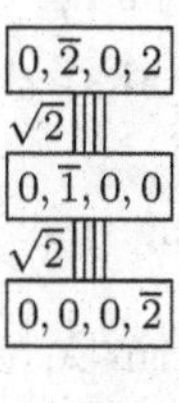

$$\left|\begin{smallmatrix}2&2&0&0\\2&0&0\\0&0\\0\end{smallmatrix}\right\rangle^{11}_{2} = \begin{array}{|c|}\hline 3\\\hline 4\\\hline 7\\\hline 8\\\hline\end{array}$$

$$\left|\begin{smallmatrix}1&1&0&0\\1&0&0\\0&0\\0\end{smallmatrix}\right\rangle^{12}_{0} = \sqrt{\tfrac{1}{2}}\left\{\begin{array}{|c|}\hline 3\\\hline 6\\\hline 7\\\hline 8\\\hline\end{array} - \begin{array}{|c|}\hline 4\\\hline 5\\\hline 7\\\hline 8\\\hline\end{array}\right\}$$

$$\left|\begin{smallmatrix}0&0&0&0\\0&0&0\\0&0\\0\end{smallmatrix}\right\rangle^{13}_{\overline{2}} = \begin{array}{|c|}\hline 5\\\hline 6\\\hline 7\\\hline 8\\\hline\end{array}$$

In summary, the self-dual representation $\boldsymbol{M} = (0,0,0,2)$ of SO(8) is 35-dimensional. It contains three A_3-singlets with $\boldsymbol{M}^{(3)} = (0,0,0)$, two

A_3-sixlets with $M^{(3)} = (0,1,0)$, and an A_3-20let with $M^{(3)} = (0,2,0)$. There are two simple dominant weights $(0,0,0,2)$ and $(0,1,0,0)$ and one triple dominant weight $(0,0,0,0)$ in this representation, whose sizes of Weyl orbits multiplicities are 8, 24, and 1, respectively.

10.1.4 *Dimensions of Tensor Representations of* SO*(N)*

The dimension $d_{[\lambda]}(\mathrm{SO}(N))$ of the representation $[\lambda]$ of $\mathrm{SO}(N)$ can be calculated by the hook rule [Ma and Dai (1982)]. In this rule the dimension $d_{[\lambda]}(\mathrm{SO}(N))$ is expressed as a quotient,

$$
\begin{aligned}
d_{[\lambda]}(\mathrm{SO}(N)) &= \frac{Y_T^{[\lambda]}}{Y_h^{[\lambda]}}, \qquad &\text{when } \lambda_\ell = 0, \\
d_{[(\pm)\lambda]}(\mathrm{SO}(2\ell)) &= \frac{Y_T^{[\lambda]}}{2Y_h^{[\lambda]}}, \qquad &\text{when } \lambda_\ell \neq 0,
\end{aligned}
\tag{10.24}
$$

where the symbols $Y_T^{[\lambda]}$ in the numerator and $Y_h^{[\lambda]}$ in the denominator are defined as follows. The **hook path** $(i,\ j)$ **in the Young pattern** $[\lambda]$ is defined to be a path which enters the Young pattern at the rightmost of the ith row, goes leftward in the i row, turns downward at the j column, goes downward in the j column, and leaves from the Young pattern at the bottom of the j column. The inverse hook path $\overline{(i,\ j)}$ is the same path as the hook path $(i,\ j)$ but along the opposite direction. The number of boxes contained in the hook path $(i,\ j)$ is called the **hook number** h_{ij}. $Y_h^{[\lambda]}$ is a tableau of the Young pattern $[\lambda]$ where the box in the jth column of the ith row is filled with the hook number h_{ij}. Define a series of the tableaux $Y_{T_a}^{[\lambda]}$ recursively by the rule given below. $Y_T^{[\lambda]}$ is a tableau of the Young pattern $[\lambda]$ where each box is filled with the sum of the digits which are respectively filled in the same box of each tableau $Y_{T_a}^{[\lambda]}$ in the series. **The symbol $Y_T^{[\lambda]}$ means the product of the filled digits in it, so does the symbol $Y_h^{[\lambda]}$.**

The tableaux $Y_{T_a}^{[\lambda]}$ are defined by the following rule:

(a) $Y_{T_0}^{[\lambda]}$ is a tableau of the Young pattern $[\lambda]$ where the box in the jth column of the ith row is filled with the digit $(N+j-i)$.

(b) Let $[\lambda^{(1)}] = [\lambda]$. Beginning with $[\lambda^{(1)}]$, we define recursively the Young pattern $[\lambda^{(a)}]$ by removing the first row and the first column of the Young pattern $[\lambda^{(a-1)}]$ until $[\lambda^{(a)}]$ contains less than two columns.

(c) If $[\lambda^{(a)}]$ contains more than one column, define $Y_{T_a}^{[\lambda]}$ to be a tableau of the Young pattern $[\lambda]$ where the boxes in the first $(a-1)$ row and in the first $(a-1)$ column are filled with 0, and the remaining part of the Young pattern is nothing but $[\lambda^{(a)}]$. Let $[\lambda^{(a)}]$ have r rows. Fill the first r boxes along the hook path (1, 1) of the Young pattern $[\lambda^{(a)}]$, beginning with the box on the rightmost, with the digits $(\lambda_1^{(a)}-1)$, $(\lambda_2^{(a)}-1)$, $\cdots$, $(\lambda_r^{(a)}-1)$, box by box, and fill the first $(\lambda_i^{(a)}-1)$ boxes in each inverse hook path $\overline{(i,\ 1)}$ of the Young pattern $[\lambda^{(a)}]$, $1 \leqslant i \leqslant r$, with -1. The remaining boxes are filled with 0. If a few -1 are filled in the same box, the digits are summed. The sum of all filled digits in the pattern $Y_{T_a}^{[\lambda]}$ with $a > 0$ is 0.

The calculation method (10.24) is explained through some examples.

Ex. 1 The dimension of the representation $[3,3,3]$ of SO(7).

$$Y_T^{[3,3,3]} = \begin{array}{|c|c|c|}\hline 7 & 8 & 9 \\ \hline 6 & 7 & 8 \\ \hline 5 & 6 & 7 \\ \hline \end{array} + \begin{array}{|c|c|c|}\hline 2 & 2 & 2 \\ \hline -2 & & \\ \hline -3 & -1 & \\ \hline \end{array} + \begin{array}{|c|c|c|}\hline & & \\ \hline & 1 & 1 \\ \hline & -2 & \\ \hline \end{array} = \begin{array}{|c|c|c|}\hline 9 & 10 & 11 \\ \hline 4 & 8 & 9 \\ \hline 2 & 3 & 7 \\ \hline \end{array},$$

$$d_{[3,3,3]}(\mathrm{SO}(7)) = \frac{\begin{array}{|c|c|c|}\hline 9 & 10 & 11 \\ \hline 4 & 8 & 9 \\ \hline 2 & 3 & 7 \\ \hline \end{array}}{\begin{array}{|c|c|c|}\hline 5 & 4 & 3 \\ \hline 4 & 3 & 2 \\ \hline 3 & 2 & 1 \\ \hline \end{array}} = 11 \times 9 \times 7 \times 2 = 1386.$$

Ex. 2 The representation of one-row Young pattern $[n]$ of SO(N).

$$\begin{aligned} d_{[n]}(\mathrm{SO}(N)) &= \frac{\begin{array}{|c|c|c|c|}\hline N & N+1 & \cdots & N+n-1 \\ \hline \end{array} + \begin{array}{|c|c|c|c|}\hline -1 & \cdots & -1 & n-1 \\ \hline \end{array}}{\begin{array}{|c|c|c|c|}\hline n & n-1 & \cdots & 1 \\ \hline \end{array}} \\ &= \begin{array}{|c|c|c|c|c|}\hline N-1 & N & \cdots & N+n-3 & N+2n-2 \\ \hline \end{array} \;/n! \\ &= \frac{(N+n-3)!(N+2n-2)}{(N-2)!n!}. \end{aligned} \tag{10.25}$$

Thus, $d_{[n]}(\mathrm{SO}(3)) = 2n+1$, $d_{[n]}(\mathrm{SO}(4)) = (n+1)^2$, and $d_{[n]}(\mathrm{SO}(5)) = (n+1)(n+2)(2n+3)/6$.

Ex. 3 The representation of two-row Young pattern $[n, m]$ of SO(N).

$$
\begin{aligned}
Y_T^{[n,m]} &= \begin{array}{|c|c|c|c|c|c|c|c|}\hline N & N+1 & \cdots & N+m-2 & N+m-1 & \cdots & N+n-2 & N+n-1 \\ \hline N-1 & N & \cdots & N+m-3 & N+m-2 \\ \cline{1-5}\end{array} \\
&+ \begin{array}{|c|c|c|c|c|c|c|c|c|}\hline -1 & -1 & \cdots & -1 & -1 & \cdots & -1 & m-1 & n-1 \\ \hline -2 & -1 & \cdots & -1 & \\ \cline{1-5}\end{array} \\
&+ \begin{array}{|c|c|c|c|c|c|c|}\hline \ & \ & \ & \ & \ & \ & \ \\ \hline \ & -1 & \cdots & -1 & m-2 \\ \cline{1-5}\end{array} \\
&= \begin{array}{|c|c|c|c|c|c|c|c|}\hline N-1 & \ldots & N+m-3 & N+m-2 & \ldots & N+n-4 & N+n+m-3 & N+2n-2 \\ \hline N-3 & \ldots & N+m-5 & N+2m-4 \\ \cline{1-4}\end{array}, \\
Y_h^{[n,m]} &= \begin{array}{|c|c|c|c|c|c|}\hline n+1 & \cdots & n-m+2 & n-m & \cdots & 1 \\ \hline m & \cdots & 1 \\ \cline{1-3}\end{array},
\end{aligned}
$$

$$
\begin{aligned}
d_{[n,m]}(\mathrm{SO}(N)) = &\frac{(n-m+1)(N+n-4)!(N+m-5)!}{(n+1)!m!(N-2)!(N-4)!} \\
&\times (N+n+m-3)(N+2n-2)(N+2m-4).
\end{aligned} \tag{10.26}
$$

Thus, $d_{[(\pm)n,m]}(SO(4)) = (n-m+1)(n+m+1)$ and $d_{[n,m]}(SO(5)) = (n-m+1)(n+m+2)(2n+3)(2m+1)/6$.

For the representation with one-column Young pattern $[1^n]$, there is no traceless condition so that when $n < N/2$,

$$
d_{[1^n]}(\mathrm{SO}(N)) = d_{[1^n]}(\mathrm{SU}(N)) = \binom{N}{n} = \frac{N!}{n!(N-n)!}. \tag{10.27}
$$

10.1.5 *Adjoint Representation of* SO*(N)*

As shown in subsection 7.5.6, the highest weight of the adjoint representation of SO(N) is $\boldsymbol{M} = \boldsymbol{w}_2$, corresponding to the Young pattern $[1, 1]$. In this subsection we are going to discuss the adjoint representation of SO(N) by replacement of tensors.

The $N(N-1)/2$ generators T_{ab} in the self-representation of SO(N) construct the complete bases of N-dimensional antisymmetric matrices. Let T_A denote T_{ab} for convenience, $1 \leqslant A \leqslant N(N-1)/2$. Tr $(T_A T_B) = 2\delta_{AB}$. For $R \in$SO(N), one has from Eq. (7.8)

$$
RT_A R^{-1} = \sum_{B+1}^{N(N-1)/2} T_B D^{\mathrm{ad}}_{BA}(R)\ . \tag{10.28}
$$

The antisymmetric tensor $\boldsymbol{T}_{ab}$ of rank 2 of SO(N) satisfies a similar relation in the SO(N) transformation R

$$(O_R \boldsymbol{T})_{ab} = \sum_{cd} R_{ac} \boldsymbol{T}_{cd} \left(R^{-1}\right)_{db} = \left(R \boldsymbol{T} R^{-1}\right)_{ab}.$$

$\boldsymbol{T}_{ab}$ is an antisymmetric matrix and can be expanded with respect to $(T_A)_{ab}$, where the coefficient $\boldsymbol{F}_A$ is a tensor,

$$\boldsymbol{T}_{ab} = \sum_{A=1}^{N(N-1)/2} (T_A)_{ab} \boldsymbol{F}_A, \qquad \boldsymbol{F}_A = \frac{1}{2} \sum_{ab} (T_A)_{ba} \boldsymbol{T}_{ab}. \tag{10.29}$$

Thus, $\boldsymbol{F}_A$ transforms in the SO(N) transformation R according to the adjoint representation of SO(N)

$$\begin{aligned}
(O_R \boldsymbol{T})_{ab} &= \sum_B (T_B)_{ab} O_R \boldsymbol{F}_B \\
&= \left(R \boldsymbol{T} R^{-1}\right)_{ab} = \sum_A \left(R T_A R^{-1}\right)_{ab} \boldsymbol{F}_A \\
&= \sum_B (T_B)_{ab} \left\{ \sum_A D^{\text{ad}}_{BA}(R) \boldsymbol{F}_A \right\}, \\
(O_R \boldsymbol{F})_B &= \sum_A D^{\text{ad}}_{BA}(R) \boldsymbol{F}_A.
\end{aligned} \tag{10.30}$$

The adjoint representation of SO(N) **is equivalent to the antisymmetric tensor representation** $[1, 1]$ **of rank 2**. It is irreducible when $N = 3$ or $N > 4$ so that SO(N) is a simple Lie group except for $N = 2$ and 4. For SO(3), $[1, 1] \simeq [1]$. The SO(2) group is abelian. The adjoint representation of SO(4) is reducible, and the SO(4) group is the homomorphic image of the direct product of two SU(2) (see Eq. (11.7)).

10.1.6 *Tensor Representations of* O(N)

The group O(N) is a mixed Lie group, whose group space falls into two disjoint regions corresponding to det $R = 1$ and det $R = -1$. Its invariant subgroup SO(N) has a connected group space corresponding to det $R = 1$. The set of elements related to the other connected piece where det $R = -1$ is the coset of SO(N). **The property of** O(N) **can be characterized completely by** SO(N) **and a representative element in the coset.**

When $N = 2\ell + 1$ is odd, $\sigma = -\mathbf{1}$ is usually chosen to be the representative element in the coset. σ commutes with every element in O($2\ell + 1$) and $\sigma^2 = \mathbf{1}$ so that the representation matrix $D(\sigma)$ in an irreducible representation of O($2\ell + 1$) is a constant matrix

$$D(\sigma) = c\mathbf{1}, \qquad D(\sigma)^2 = \mathbf{1}, \qquad c = \pm 1. \tag{10.31}$$

The induced representation of each irreducible representation $D^{[\lambda]}$ of SO$(2\ell+1)$ contains two irreducible representations $D^{[\lambda]\pm}$ of O$(2\ell+1)$):

$$D^{[\lambda]\pm}(R) = D^{[\lambda]}(R), \quad R \in \mathrm{SO}(2\ell+1), \qquad D^{[\lambda]\pm}(\sigma) = \pm\mathbf{1}. \qquad (10.32)$$

Two irreducible representations $D^{[\lambda]\pm}(\mathrm{O}(N))$ are inequivalent because the characters of σ in two representations are different.

When $N = 2\ell$ is even, $\sigma = -\mathbf{1} \in \mathrm{SO}(2\ell)$, and the diagonal matrix τ is usually chosen to be the representative element in the coset, where the diagonal elements of τ are 1 except for $\tau_{NN} = -1$. $\tau^2 = \mathbf{1}$, but τ does not commute with some elements in O(2ℓ). Any tensor Young tableau $\mathcal{Y}_\mu^{[\lambda]}\boldsymbol{\theta}_{a_1\cdots a_n}$ is an eigentensor of τ with the eigenvalue 1 or -1 depending on whether the number of filled digits N in the tensor Young tableau is even or odd. In the spherical harmonic basis tensors, τ interchanges the filled digits ℓ and $(\ell+1)$ in the tensor Young tableau $\mathcal{Y}_\mu^{[\lambda]}\phi_{\alpha_1\cdots\alpha_n}$. Thus, **the representation matrix $D^{[\lambda]}(\tau)$ is known.**

The induced representation of each irreducible representation $D^{[\lambda]}$ of SO(2ℓ), where the row number of $[\lambda]$ is less than ℓ, contains two irreducible representations $D^{[\lambda]\pm}$ of O(2ℓ):

$$D^{[\lambda]\pm}(R) = D^{[\lambda]}(R), \quad R \in \mathrm{O}(2\ell), \qquad D^{[\lambda]\pm}(\tau) = \pm D^{[\lambda]}(\tau). \qquad (10.33)$$

Two representations $D^{[\lambda]\pm}(\mathrm{O}(N))$ are inequivalent because the characters of τ in two representations are different.

When the row number of $[\lambda]$ is $\ell = N/2$, there are two inequivalent irreducible representations $D^{[(\pm)\lambda]}$ of SO(2ℓ) whose basis tensors are given in Eq. (10.9). Two terms in Eq. (10.9) contain different numbers of the subscripts N such that τ changes the tensor Young tableau in $[(\pm)\lambda]$ to that in $[(\mp)\lambda]$, namely, the representation spaces of both $D^{[(\pm)\lambda]}(\mathrm{SO}(2\ell))$ are not invariant in O(2ℓ). Only their direct sum is an invariant space corresponding to an irreducible representation $D^{[\lambda]}$ of O(2ℓ),

$$\begin{aligned} &D^{[\lambda]}(R) = D^{[(+)\lambda]}(R) \oplus D^{[(-)\lambda]}(R), \qquad && R \in \mathrm{SO}(2\ell), \\ &D^{[\lambda]}(\tau) = \begin{pmatrix} \mathbf{0} & X \\ Y & \mathbf{0} \end{pmatrix}, && XY = YX = \mathbf{1}, \end{aligned} \qquad (10.34)$$

where the representation matrix $D^{[\lambda]}(\tau)$ can be calculated by interchanging the filled digits ℓ and $(\ell+1)$ in the tensor Young tableau $\mathcal{Y}_\mu^{[\lambda]}\phi_{\alpha_1\cdots\alpha_n}$. Two representations by changing the sign of $D^{[\lambda]}(\tau)$ are equivalent because they can be related by a similarity transformation $Z = \left(\begin{smallmatrix} 1 & 0 \\ 0 & -1 \end{smallmatrix}\right)$.

10.2 Γ Matrix Groups

Dirac introduced four γ matrices, which are the generalization of the Pauli matrices. Similar to the Pauli matrices, four γ matrices also satisfy the anticommutative relations. In terms of the tool of the γ matrices, Dirac established the equation of motion for the relativistic particle with spin 1/2, called the Dirac equation. In the language of group theory, Dirac found the spinor representation of the Lorentz group. The tool of the γ matrices is generalized for finding the spinor representations of SO(N). The set of products of the γ matrices forms the matrix group Γ. Its group algebra is called the Clifford algebra in mathematics.

10.2.1 *Property of* Γ *Matrix Groups*

Define N matrices γ_a satisfying the anticommutative relations

$$\{\gamma_a\ ,\ \gamma_b\} = \gamma_a\gamma_b + \gamma_b\gamma_a = 2\delta_{ab}\mathbf{1}, \quad 1 \leqslant a \leqslant N, \quad 1 \leqslant b \leqslant N, \tag{10.35}$$

namely, $\gamma_a^2 = \mathbf{1}$ and $\gamma_a\gamma_b = -\gamma_b\gamma_a$ when $a \neq b$. **The inverse of a product of γ_a matrices is the same product but in the opposite order.** Two γ_b with the same subscript in products of γ_a matrices can be moved together and eliminated. The set of all different products of γ_a matrices, in the multiplication rule of matrices, satisfies four axioms of a group and forms a group, denoted by Γ_N. Γ_N is a finite matrix group.

Choose a faithful irreducible unitary representation of Γ_N to be its self-representation. As shown in Eq. (10.46), the representation does exist. Thus, from Eq. (10.35) γ_a **is unitary and hermitian,**

$$\gamma_a^\dagger = \gamma_a^{-1} = \gamma_a. \tag{10.36}$$

The eigenvalue of γ_a is 1 or −1. Let

$$\gamma_\chi^{(N)} = \gamma_1\gamma_2\cdots\gamma_N, \qquad \left(\gamma_\chi^{(N)}\right)^2 = (-1)^{N(N-1)/2}\,\mathbf{1}. \tag{10.37}$$

When N is odd, $\gamma_\chi^{(N)}$ commutes with each γ_a matrix so that it is a constant matrix owing to the Schur theorem,

$$\gamma_\chi^{(N)} = \begin{cases} \pm\mathbf{1}, & N = 4m+1, \\ \pm i\mathbf{1}, & N = 4m-1. \end{cases} \tag{10.38}$$

Since $\gamma_1\gamma_2\gamma_1\gamma_2 = -\mathbf{1}$, the diagonal matrix $\gamma_\chi^{(4m+1)}$ is not a new element in Γ_{4m+1}. Two groups with different $\gamma_\chi^{(4m+1)}$ are isomorphic through a one-to-one correspondence, say

$$\gamma_a \longleftrightarrow \gamma'_a, \qquad 1 \leqslant a \leqslant 4m, \qquad \gamma_{4m+1} \longleftrightarrow -\gamma'_{4m+1}. \tag{10.39}$$

Furthermore, for a given $\gamma_\chi^{(4m+1)}$, γ_{4m+1} in Γ_{4m+1} can be expressed as a product of other γ_a matrices. Thus, all elements both in Γ_{4m} and in Γ_{4m+1} can be expressed as the products of matrices γ_a, $1 \leqslant a \leqslant 4m$ so that they are isomorphic. On the other hand, since $\gamma_\chi^{(4m-1)}$ is equal to either $\mathrm{i}\mathbf{1}$ or $-\mathrm{i}\mathbf{1}$, Γ_{4m-1} is isomorphic onto a group composed of Γ_{4m-2} and $\mathrm{i}\Gamma_{4m-2}$,

$$\Gamma_{4m+1} \approx \Gamma_{4m}, \qquad \Gamma_{4m-1} \approx \{\Gamma_{4m-2}, \ \mathrm{i}\Gamma_{4m-2}\}. \tag{10.40}$$

10.2.2 *The Case $N = 2\ell$*

First, we calculate the order $g^{(2\ell)}$ of $\Gamma_{2\ell}$. Obviously, if $R \in \Gamma_{2\ell}$, $-R$ belongs to $\Gamma_{2\ell}$, too. Choosing one element in each pair of elements $\pm R$, we obtain a set $\Gamma'_{2\ell}$ containing $g^{(2\ell)}/2$ elements. Let S_n denote a product of n different γ_a. Since the number of different S_n contained in the set $\Gamma'_{2\ell}$ is equal to the combinatorial number of n among 2ℓ, then

$$g^{(2\ell)} = 2\sum_{n=0}^{2\ell} \binom{2\ell}{n} = 2(1+1)^{2\ell} = 2^{2\ell+1}. \tag{10.41}$$

Second, for any element $S_n \in \Gamma_{2\ell}$ except for $\pm\mathbf{1}$, one is able to find a matrix γ_a which is anticommutable with S_n. In fact, if n is even and γ_a appears in the product S_n, we have $\gamma_a S_n = -S_n \gamma_a$. If n is odd, there must exist at least one γ_a which does not appear in the product S_n so that $\gamma_a S_n = -S_n \gamma_a$. Thus, **the trace of S_n vanishes**,

$$\mathrm{Tr}\ S_n = \mathrm{Tr}\left(\gamma_a^2 S_n\right) = -\mathrm{Tr}\left(\gamma_a S_n \gamma_a\right) = -\mathrm{Tr}\ S_n = 0.$$

Namely, the character of the element S in the self-representation of $\Gamma_{2\ell}$ is

$$\chi(S) = \begin{cases} \pm d^{(2\ell)} & \text{when } S = \pm\mathbf{1}, \\ 0 & \text{when } S \neq \pm\mathbf{1}, \end{cases} \tag{10.42}$$

where $d^{(2\ell)}$ is the dimension of γ_a. Since the self-representation of $\Gamma_{2\ell}$ is irreducible,

$$2\left(d^{(2\ell)}\right)^2 = \sum_{S\in\Gamma_{2\ell}} |\chi(S)|^2 = g^{(2\ell)} = 2^{2\ell+1},$$
$$d^{(2\ell)} = 2^\ell. \tag{10.43}$$

Due to Eqs. (10.35) and (10.42), $\det \gamma_a = 1$ when $\ell > 1$.

Third, since $\gamma_\chi^{(2\ell)}$ is anticommutable with every γ_a, we define $\gamma_f^{(2\ell)} = (-\mathrm{i})^\ell \gamma_\chi^{(2\ell)}$ such that $\gamma_f^{(2\ell)}$ satisfies Eq. (10.35):

$$\gamma_f^{(2\ell)} = (-\mathrm{i})^\ell \gamma_\chi^{(2\ell)} = (-\mathrm{i})^\ell \gamma_1 \gamma_2 \cdots \gamma_{2\ell}, \qquad \left(\gamma_f^{(2\ell)}\right)^2 = \mathbf{1}. \tag{10.44}$$

In fact, $\gamma_f^{(2\ell)}$ **may be defined to be the matrix** $\gamma_{2\ell+1}$ **in** $\Gamma_{2\ell+1}$.

Fourth, **the matrices in the set** $\Gamma'_{2\ell}$ **are linearly independent.** Otherwise, there is a linear relation $\sum_S C(S)S = 0$, where $S \in \Gamma'_{2\ell}$. Multiplying it with $R^{-1}/d^{(2\ell)}$ and taking the trace, one obtains that any coefficient $C(R) = 0$. Thus, the set $\Gamma'_{2\ell}$ contains $2^{2\ell}$ linearly independent matrices of dimension $d^{(2\ell)} = 2^\ell$ so that they constitute a complete set of basis matrices. Any matrix M of dimension $d^{(2\ell)}$ can be expanded with respect to $S \in \Gamma'_{2\ell}$,

$$M = \sum_{S \in \Gamma'_{2\ell}} C(S)S, \qquad C(S) = \frac{1}{d^{(2\ell)}} \mathrm{Tr}\left(S^{-1}M\right). \tag{10.45}$$

Fifth, due to Eq. (10.37), $\pm S$ construct a class. $\mathbf{1}$ and $-\mathbf{1}$ construct two classes, respectively. **The** $\Gamma_{2\ell}$ **group contains** $2^{2\ell}+1$ **classes.** It is a one-dimensional representation that arbitrarily chosen n matrices γ_a correspond to 1 and the remaining matrices γ_b correspond to -1. The number of the one-dimensional inequivalent representations is

$$\sum_{n=0}^{2\ell} \binom{2\ell}{n} = 2^{2\ell}.$$

The remaining irreducible representation of $\Gamma_{2\ell}$ has to be $d^{(2\ell)}$-dimensional, which is faithful. The γ_a matrices in this representation is called **the irreducible** γ_a **matrices**. The irreducible γ_a matrices may be chosen as the direct product of ℓ two-dimensional matrices, which are the Pauli matrices σ_a and the unit matrix $\mathbf{1}$ [Georgi (1982)]:

$$\begin{aligned} \gamma_{2n-1} &= \underbrace{\mathbf{1} \times \cdots \times \mathbf{1}}_{n-1} \times \sigma_1 \times \underbrace{\sigma_3 \times \cdots \times \sigma_3}_{\ell-n}, \\ \gamma_{2n} &= \underbrace{\mathbf{1} \times \cdots \times \mathbf{1}}_{n-1} \times \sigma_2 \times \underbrace{\sigma_3 \times \cdots \times \sigma_3}_{\ell-n}, \\ \gamma_f^{(2\ell)} &= \underbrace{\sigma_3 \times \cdots \times \sigma_3}_{\ell}. \end{aligned} \tag{10.46}$$

$\gamma_f^{(2\ell)}$ is a diagonal matrix where the diagonal elements are ± 1.

At last, there is an equivalent theorem for the γ_a matrices.

Theorem 10.1 (The equivalent theorem) Two sets of $d^{(2\ell)}$-dimensional matrices γ_a and $\overline{\gamma}_a$, both of which satisfy the anticommutative relation (10.35) with $N = 2\ell$, are equivalent

$$\overline{\gamma}_a = X^{-1}\gamma_a X , \qquad 1 \leqslant a \leqslant 2\ell. \tag{10.47}$$

The similarity transformation matrix X is determined up to a constant factor. If the determinant of X is restricted to be 1, there are $d^{(2\ell)}$ choices by a factor: $\exp\left(-\mathrm{i}2n\pi/d^{(2\ell)}\right)$, $0 \leqslant n < d^{(2\ell)}$.

Proof The irreducible representations of $\Gamma_{2\ell}$ constructed by two sets of matrices γ_a are equivalent because the characters of any element $S \in \Gamma_{2\ell}$ are equal to each other (see Eq. (10.38)).

If there are two similarity transformation matrices X and Y,

$$\overline{\gamma}_a = X^{-1}\gamma_a X, \qquad \overline{\gamma}_a = Y^{-1}\gamma_a Y,$$

then, YX^{-1} can commute with every γ_a so that $Y = cX$. □

The charge conjugation matrix $C^{(2\ell)}$, which is commonly used in particle physics, is defined based on this theorem. Define $\overline{\gamma}_a = -\left(\gamma_a\right)^T$, where T denotes the transpose of the matrix. Both γ_a and $\overline{\gamma}_a$ satisfy the anticommutative relation (10.35), so that $\overline{\gamma}_a$ are equivalent to γ_a. Choose a unimodular and unitary similarity transforation matrix $C^{(2\ell)}$:

$$\begin{aligned}
&\left(C^{(2\ell)}\right)^{-1}\gamma_a C^{(2\ell)} = -\left(\gamma_a\right)^T ,\\
&\left(C^{(2\ell)}\right)^{\dagger} C^{(2\ell)} = \mathbf{1}, \qquad \det C^{(2\ell)} = 1,\\
&\left(C^{(2\ell)}\right)^{-1}\gamma_f^{(2\ell)} C^{(2\ell)} = (-\mathrm{i})^{\ell}\left(\gamma_1\right)^T\left(\gamma_2\right)^T\cdots\left(\gamma_{2\ell}\right)^T\\
&\quad = (-1)^{\ell}\left(\gamma_f^{(2\ell)}\right)^T .
\end{aligned} \tag{10.48}$$

Taking the transpose of Eq. (10.48), one has

$$\begin{aligned}
\gamma_a &= -\left(C^{(2\ell)}\right)^T\gamma_a^T\left[\left(C^{(2\ell)}\right)^{-1}\right]^T\\
&= \left[\left(C^{(2\ell)}\right)^T\left(C^{(2\ell)}\right)^{-1}\right]\gamma_a\left[\left(C^{(2\ell)}\right)^T\left(C^{(2\ell)}\right)^{-1}\right]^{-1}.
\end{aligned}$$

Thus, $\left(C^{(2\ell)}\right)^T\left(C^{(2\ell)}\right)^{-1} = \lambda^{(2\ell)}\mathbf{1}$, $\left(C^{(2\ell)}\right)^T = \lambda^{(2\ell)}C^{(2\ell)}$, and

$$C^{(2\ell)} = \lambda^{(2\ell)}\left(C^{(2\ell)}\right)^T = \left(\lambda^{(2\ell)}\right)^2 C^{(2\ell)}, \qquad \lambda^{(2\ell)} = \pm 1.$$

The constant $\lambda^{(2\ell)}$ can be determined as follows. Remind that $\Gamma'_{2\ell}$ is a complete set of $d^{(2\ell)}$-dimensional matrices and is composed of S_n, $0 \leqslant n \leqslant$

2ℓ, where S_n is a product of n different γ_a matrices. Since

$$\left(S_n C^{(2\ell)}\right)^T = \lambda^{(2\ell)} C^{(2\ell)} \left(S_n\right)^T = \lambda^{(2\ell)} (-1)^{n(n+1)/2} \left(S_n C^{(2\ell)}\right), \tag{10.49}$$

$S_n C^{(2\ell)}$ is either symmetric or antisymmetric. The number of S_n as well as $S_n C^{(2\ell)}$ is the combinatorial number of n among 2ℓ. Because the number of the symmetric matrices of dimension $d^{(2\ell)}$ is larger than that of the antisymmetric ones, $S_\ell C^{(2\ell)}$ has to be symmetric,

$$\left(C^{(2\ell)}\right)^T = \lambda^{(2\ell)} C^{(2\ell)}, \qquad \lambda^{(2\ell)} = (-1)^{\ell(\ell+1)/2} . \tag{10.50}$$

The charge conjugation matrix $C^{(2\ell)}$ satisfies Eqs. (10.48) and (10.50). Choosing the γ_a in (10.46), we have

$$\begin{aligned} C^{(4m)} &= \underbrace{(\sigma_1 \times \sigma_2) \times (\sigma_1 \times \sigma_2) \times \cdots \times (\sigma_1 \times \sigma_2)}_{m}, \\ C^{(4m+2)} &= \sigma_2 \times C^{(4m)}. \end{aligned} \tag{10.51}$$

In the particle physics, the strong space-time inversion matrix $B^{(2\ell)}$ is also used,

$$\begin{aligned} & B^{(2\ell)} = \gamma_f^{(2\ell)} C^{(2\ell)} = (-1)^\ell \lambda^{(2\ell)} \left(B^{(2\ell)}\right)^T, \qquad \det B^{(2\ell)} = 1 \\ & \left(B^{(2\ell)}\right)^{-1} \gamma_a B^{(2\ell)} = \left(\gamma_a\right)^T, \qquad \left(B^{(2\ell)}\right)^\dagger B^{(2\ell)} = \mathbf{1}. \end{aligned} \tag{10.52}$$

10.2.3 The Case $N = 2\ell + 1$

Since $\gamma_f^{(2\ell)}$ and (2ℓ) matrices γ_a in $\Gamma_{2\ell}$, $1 \leqslant a \leqslant 2\ell$, satisfy the antisymmetric relations (10.35), they can be defined to be the $(2\ell+1)$ matrices γ_a in $\Gamma_{2\ell+1}$. In this definition, $\gamma_\chi^{(2\ell+1)}$ in $\Gamma_{2\ell+1}$ has been chosen,

$$\gamma_{2\ell+1} = \gamma_f^{(2\ell)}, \qquad \gamma_\chi^{(2\ell+1)} = \gamma_1 \cdots \gamma_{2\ell+1} = (\mathrm{i})^\ell \mathbf{1}. \tag{10.53}$$

Obviously, the dimension $d^{(2\ell+1)}$ of the matrices in $\Gamma_{2\ell+1}$ is the same as $d^{(2\ell)}$ in $\Gamma_{2\ell}$,

$$d^{(2\ell+1)} = d^{(2\ell)} = 2^\ell. \tag{10.54}$$

When N is odd, the equivalent theorem has to be modified because the multiplication rule of elements in $\Gamma_{2\ell+1}$ includes Eq. (10.37). A similarity transformation cannot change the sign of $\gamma_\chi^{(2\ell+1)}$. Namely, the equivalent condition for two sets of γ_a and $\overline{\gamma}_a$ has to include a new condition $\gamma_\chi = \overline{\gamma}_\chi$, in addition to those given in Theorem 10.1.

Letting $\overline{\gamma}_a = -\left(\gamma_a\right)^T$, we have

$$\begin{aligned}\overline{\gamma}_\chi^{(2\ell+1)} &= \overline{\gamma}_1 \cdots \overline{\gamma}_{2\ell+1} = -\left\{\gamma_{2\ell+1} \cdots \gamma_1\right\}^T \\ &= (-1)^{\ell+1}\left\{\gamma_\chi^{(2\ell+1)}\right\}^T = (-1)^{\ell+1}\gamma_\chi^{(2\ell+1)},\end{aligned} \tag{10.55}$$

namely, $C^{(4m-1)}$ satisfying Eq. (10.48) exists, but $C^{(4m+1)}$ does not. In the same reason, $B^{(4m+1)}$ satisfying Eq. (10.52) exists, but $B^{(4m-1)}$ does not. In fact, due to Eq. (10.48), $\gamma_{2\ell+1} = \gamma_f^{(2\ell)}$ satisfies Eq. (10.47) when $N = 4m-1$, but does not when $N = 4m+1$. Thus,

$$\begin{aligned}&C^{(4m-1)} = C^{(4m-2)}, && \left(C^{(4m-1)}\right)^T = (-1)^m C^{(4m-1)}, \\ &B^{(4m+1)} = B^{(4m)} = \gamma_f^{(4m)} C^{(4m)}, && \left(B^{(4m+1)}\right)^T = (-1)^m B^{(4m+1)}.\end{aligned} \tag{10.56}$$

$C^{(4m-1)}$ and $B^{(4m+1)}$ have the same symmetry in transpose.

10.3 Spinor Representations of SO(N)

10.3.1 *Covering Groups of* SO*(N)*

From a set of N irreducible unitary matrices γ_a which satisfy the anticommutative relation (10.35), we define

$$\overline{\gamma}_a = \sum_{b=1}^N R_{ab}\gamma_b, \qquad R \in \text{SO}(N). \tag{10.57}$$

Since R is a real orthogonal matrix, $\overline{\gamma}_a$ satisfy

$$\overline{\gamma}_a\overline{\gamma}_b + \overline{\gamma}_b\overline{\gamma}_a = \sum_{cd} R_{ac}R_{bd}\left\{\gamma_c\gamma_d + \gamma_d\gamma_c\right\} = 2\sum_c R_{ac}R_{bc}\mathbf{1} = 2\delta_{ab}\mathbf{1}.$$

Due to Eq. (10.35) and $\sum_a R_{ja}R_{ka} = \delta_{jk}$,

$$\begin{aligned}\sum_{a_1a_2} R_{ja_1}R_{ka_2}\gamma_{a_1}\gamma_{a_2} &= \frac{1}{2}\sum_{a_1,a_2} R_{ja_1}R_{ka_2}\left(\gamma_{a_1}\gamma_{a_2} - \gamma_{a_2}\gamma_{a_1}\right), \quad j \neq k, \\ \overline{\gamma}_1\overline{\gamma}_2\cdots\overline{\gamma}_N &= \sum_{a_1\cdots a_N} R_{1a_1}\cdots R_{Na_N}\gamma_{a_1}\gamma_{a_2}\cdots\gamma_{a_N} \\ &= \sum_{a_1\cdots a_N} R_{1a_1}\cdots R_{Na_N}\epsilon_{a_1\cdots a_N}\gamma_1\gamma_2\cdots\gamma_N \\ &= (\det\ R)\,\gamma_1\gamma_2\cdots\gamma_N = \gamma_1\gamma_2\cdots\gamma_N.\end{aligned}$$

From the equivalent theorem, γ_a and $\overline{\gamma}_a$ can be related through a unitary similarity transformation $D(R)$ with determinant 1,

$$D(R)^{-1}\gamma_a D(R) = \sum_{d=1}^{N} R_{ad}\gamma_d, \qquad \det\, D(R) = 1, \tag{10.58}$$

where $D(R)$ is determined up to a constant

$$\exp\left(-\mathrm{i}2n\pi/d^{(N)}\right), \qquad 0 \leqslant n < d^{(N)}. \tag{10.59}$$

The set of $D(R)$ defined in Eq. (10.58), in the multiplication rule of matrices, satisfies four axioms of a group and forms a Lie group G'_N. There is a $d^{(N)}$-to-one correspondence between the elements in G'_N and the elements in SO(N), and the correspondence is left invariant in the multiplication of elements. Therefore, SO(N) is the homomorphic image of G'_N. Since the group space of SO(N) is doubly-connected, the correspondence between SO(N) and its covering group has to be one-to-two. Namely, the group space of G'_N must fall into several disjoint pieces, where the piece containing the identity E forms an invariant subgroup G_N of G'_N. **G_N is a connected Lie group and is the covering group of** SO(N). Based on the property of the infinitesimal elements, a discontinuous condition will be found to pick up G_N from G'_N.

Let R denote an infinitesimal element of SO(N). Expand R and $D(R)$ with respect to the infinitesimal parameters ω_{ab},

$$R_{cd} = \delta_{cd} - \mathrm{i}\sum_{a<b} \omega_{ab}\left(T_{ab}\right)_{cd} = \delta_{cd} - \omega_{cd},$$
$$D(R) = \mathbf{1} - \mathrm{i}\sum_{a<b} \omega_{ab} S_{ab},$$

where T_{ab} are the generators in the self-representation of SO(N) given in Eq. (10.11) and S_{ab} are the generators in G_N. From Eq. (10.58) we have

$$[\gamma_c,\ S_{ab}] = \sum_d \left(T_{ab}\right)_{cd}\gamma_d = -\mathrm{i}\left\{\delta_{ac}\gamma_b - \delta_{bc}\gamma_a\right\}. \tag{10.60}$$

The solution is

$$S_{ab} = \frac{1}{4\mathrm{i}}\left(\gamma_a\gamma_b - \gamma_b\gamma_a\right). \tag{10.61}$$

S_{ab} is hermitian because $D(R)$ is unitary.

For simplifying the notations, we define

$$C = \begin{cases} B^{(N)} & \text{when } N = (4m+1), \\ C^{(N)} & \text{when } N \neq (4m+1). \end{cases} \tag{10.62}$$

Thus, $C^{-1}S_{ab}C = -\left(S_{ab}\right)^T = -S_{ab}^*$,

$$C^{-1}D(R)C = \left\{D(R^{-1})\right\}^T = D(R)^*. \qquad (10.63)$$

The discontinuous condition (10.63) restricts the factor in $D(R)$ (see Eq. (10.59)) such that there is a two-to-one correspondence between $\pm D(R)$ in G_N and R in $\mathrm{SO}(N)$ through the relations (10.58) and (10.63). Namely, **G_N is the covering group of $\mathrm{SO}(N)$**,

$$\mathrm{G}_N \sim \mathrm{SO(N)}. \qquad (10.64)$$

When $N = 3$, $\mathrm{G}_3 \approx \mathrm{SU}(2)$. G_N is called **the fundamental spinor representation** of $\mathrm{SO}(N)$, or briefly, the spinor representation, denoted by $D^{[s]}(\mathrm{SO}(N))$. S_{ab} is called the **spinor angular momentum operators**. The irreducible tensor representation $[\lambda]$ is a single-valued representation of $\mathrm{SO}(N)$, but a non-faithful one of G_N. The faithful representation of G_N is a double-valued one of $\mathrm{SO}(N)$.

Since the products S_n span a complete set of the $d^{(N)}$-dimensional matrices, it can be decided by checking the commutative relations of S_n with the generators S_{ab} whether there is a nonconstant matrix commutable with all S_{ab}. The result is that only $\gamma_\chi^{(N)}$ is commutable with all S_{ab}. $\gamma_\chi^{(2\ell+1)}$ is a constant matrix so that the **fundamental spinor representation $D^{[s]}(\mathrm{SO}(2\ell+1))$ is irreducible**. Due to Eqs. (10.63) and (10.56), $D^{[s]}(\mathrm{SO}(N))$ is self-conjugate when $N = 2\ell+1$ and is real when $N = 8k\pm 1$.

$\gamma_\chi^{(2\ell)}$ is not a constant matrix so that the **fundamental spinor representation $D^{[s]}(\mathrm{SO}(2\ell))$ is reducible.** Through a similarity transformation X, $X^{-1}\gamma_f^{(2\ell)}X = \sigma_3 \times \mathbf{1}$ (see Eq. (10.46)) and $D^{[s]}(\mathrm{SO}(2\ell))$ is decomposed into the direct sum of two irreducible representations,

$$X^{-1}D^{[s]}(R)X = \begin{pmatrix} D^{[+s]}(R) & 0 \\ 0 & D^{[-s]}(R) \end{pmatrix}. \qquad (10.65)$$

Two representations $D^{[\pm s]}(\mathrm{SO}(2\ell))$ can be proved to be inequivalent by reduction to absurdity. In fact, if $Z^{-1}D^{[-s]}(R)Z = D^{[+s]}(R)$ and $Y = \mathbf{1} \oplus Z$, then all generators $(XY)^{-1}S_{ab}XY$ commute with $\sigma_1 \times \mathbf{1}$, but their product does not commute with it,

$$2^\ell (XY)^{-1}\left(S_{12}S_{34}\cdots S_{(2\ell-1)(2\ell)}\right)XY = Y^{-1}\left[X^{-1}\gamma_f^{(2\ell)}X\right]Y = \sigma_3 \times \mathbf{1}.$$

It leads to contradiction.

Introduce two projective operators $P_\pm$,

$$
\begin{aligned}
&P_{\pm} = \frac{1}{2}\left(\mathbf{1} \pm \gamma_f^{(2\ell)}\right), && P_{\pm} D^{[s]}(R) = D^{[s]}(R) P_{\pm}, \\
&X^{-1} P_{+} X = \begin{pmatrix} \mathbf{1} & 0 \\ 0 & 0 \end{pmatrix}, && X^{-1} P_{+} D^{[s]}(R) X = \begin{pmatrix} D^{[+s]}(R) & 0 \\ 0 & 0 \end{pmatrix}, \\
&X^{-1} P_{-} X = \begin{pmatrix} 0 & 0 \\ 0 & \mathbf{1} \end{pmatrix}, && X^{-1} P_{-} D^{[s]}(R) X = \begin{pmatrix} 0 & 0 \\ 0 & D^{[-s]}(R) \end{pmatrix}.
\end{aligned}
\tag{10.66}
$$

From Eq. (10.48) we have

$$
C^{-1} D^{[s]}(R) P_{\pm} C = \begin{cases} D^{[s]}(R)^* P_{\pm} & \text{when } N = 4m, \\ D^{[s]}(R)^* P_{\mp} & \text{when } N = 4m+2. \end{cases}
\tag{10.67}
$$

Two inequivalent representations $D^{[\pm s]}(R)$ are conjugate to each other when $N = 4m+2$, are self-conjugate when $N = 4m$, and are real when $N = 8k$ owing to Eq. (10.50). The dimension of the irreducible spinor representations of SO(N) is

$$
d_{[s]}[\mathrm{SO}(2\ell+1)] = 2^{\ell}, \qquad d_{[\pm s]}[\mathrm{SO}(2\ell)] = 2^{\ell-1}.
\tag{10.68}
$$

10.3.2 *Fundamental Spinors of* SO(*N*)

In an SO(N) transformation R, $\mathbf{\Psi}$ is called the fundamental spinor of SO(N) if it transforms by the fundamental spinor representation $D^{[s]}(R)$:

$$
\left(O_R \mathbf{\Psi}\right)_{\mu} = \sum_{\nu} D_{\mu\nu}^{[s]}(R) \mathbf{\Psi}_{\nu}, \qquad O_R \mathbf{\Psi} = D^{[s]}(R) \mathbf{\Psi},
\tag{10.69}
$$

where $\mathbf{\Psi}$ is a column matrix with $d_{[s]}$ components.

The Chevalley bases $H_{\mu}(S)$, $E_{\mu}(S)$ and $F_{\mu}(S)$ with respect to the spinor angular momentum can be obtained from Eqs. (10.13) and (10.19) by replacing T_{ab} with S_{ab}. In the chosen forms of γ_a given in Eq. (10.46), the Chevalley bases for the SO($2\ell+1$) group (B$_\ell$ Lie algebra) are

$$
\begin{aligned}
H_{\mu}(S) &= \underbrace{\mathbf{1} \times \cdots \times \mathbf{1}}_{\mu-1} \times \frac{1}{2}\left\{\sigma_3 \times \mathbf{1} - \mathbf{1} \times \sigma_3\right\} \times \underbrace{\mathbf{1} \times \cdots \times \mathbf{1}}_{\ell-\mu-1}, \\
H_{\ell}(S) &= \underbrace{\mathbf{1} \times \cdots \times \mathbf{1}}_{\ell-1} \times \sigma_3, \\
E_{\mu}(S) &= \underbrace{\mathbf{1} \times \cdots \times \mathbf{1}}_{\mu-1} \times \left\{\sigma_{+} \times \sigma_{-}\right\} \times \underbrace{\mathbf{1} \times \cdots \times \mathbf{1}}_{\ell-\mu-1} = F_{\mu}(S)^T, \\
E_{\ell}(S) &= \underbrace{\sigma_3 \times \cdots \times \sigma_3}_{\ell-1} \times \sigma_{+} = F_{\ell}(S)^T,
\end{aligned}
\tag{10.70}
$$

where $1 \leqslant \mu < \ell$. The Chevalley bases for the SO(2ℓ) group (D_ℓ Lie algebra) are the same as those for SO($2\ell+1$) except for $\mu = \ell$,

$$\begin{aligned} H_\ell(S) &= \underbrace{\mathbf{1} \times \cdots \times \mathbf{1}}_{\ell-2} \times \frac{1}{2}\left\{\sigma_3 \times \mathbf{1} + \mathbf{1} \times \sigma_3\right\}, \\ E_\ell(S) &= -\underbrace{\mathbf{1} \times \cdots \times \mathbf{1}}_{\ell-2} \times \left\{\sigma_+ \times \sigma_+\right\} = F_\ell(S)^T. \end{aligned} \tag{10.71}$$

The basis spinor $\chi(\boldsymbol{m})$ of SO(N) is also expressed as a direct product of ℓ two-dimensional basis spinors χ_α,

$$\begin{aligned} \chi(\boldsymbol{m}) &= \chi_{\alpha_1\alpha_2\cdots\alpha_\ell} = \chi_{\alpha_1} \times \chi_{\alpha_2} \times \cdots \times \chi_{\alpha_\ell}, \\ \chi_1 &= \begin{pmatrix} 1 \\ 0 \end{pmatrix}, \qquad \chi_2 = \begin{pmatrix} 0 \\ 1 \end{pmatrix}. \end{aligned} \tag{10.72}$$

When N is even, the fundamental spinor space is decomposed into two subspaces by the project operators $P_\pm$, $\boldsymbol{\Psi}_\pm = P_\pm \boldsymbol{\Psi}$, corresponding to irreducible spinor representations $D^{[\pm s]}$. **The basis spinor in the representation space of $D^{[+s]}$ contains even number of factors χ_2, and that of $D^{[-s]}$ contains odd number of χ_2.** The highest weight states $\chi(\boldsymbol{M})$ and their highest weights $\boldsymbol{M}$ are

$$\begin{aligned} &\chi_{\underbrace{1\cdots1}_{\ell-1}1}, && \boldsymbol{M} = (\underbrace{0,\cdots,0}_{\ell-1},1), && [s] \text{ of SO}(2\ell+1), \\ &\chi_{\underbrace{1\cdots1}_{\ell-1}1}, && \boldsymbol{M} = (\underbrace{0,\cdots,0}_{\ell-2},0,1), && [+s] \text{ of SO}(2\ell), \\ &\chi_{\underbrace{1\cdots1}_{\ell-1}2}, && \boldsymbol{M} = (\underbrace{0,\cdots,0}_{\ell-2},1,0), && [-s] \text{ of SO}(2\ell). \end{aligned} \tag{10.73}$$

The remaining basis states can be calculated by the applications of lowering operators $F_\mu(S)$.

10.3.3 *Direct Products of Spinor Representations*

The spinor representation is unitary so that

$$O_R \boldsymbol{\Psi}^\dagger = \boldsymbol{\Psi}^\dagger D^{[s]}(R)^{-1}. \tag{10.74}$$

$$\begin{aligned} \boldsymbol{\Psi}^\dagger \boldsymbol{\Psi} &= \sum_\mu \boldsymbol{\Psi}_\mu^* \boldsymbol{\Psi}_\mu = \sum_{\mu\nu} \boldsymbol{\Psi}_\mu^* \delta_{\mu\nu} \boldsymbol{\Psi}_\nu, \\ O_R\left(\boldsymbol{\Psi}^\dagger \boldsymbol{\Psi}\right) &= \boldsymbol{\Psi}^\dagger D^{[s]}(R)^{-1} D^{[s]}(R) \boldsymbol{\Psi} = \boldsymbol{\Psi}^\dagger \boldsymbol{\Psi}. \end{aligned} \tag{10.75}$$

$\mathbf{\Psi}^*\mathbf{\Psi}$ is left invariant in the SO(N) transformations. It is a scalar of SO(N). In the language of group theory, the products of $\mathbf{\Psi}^*_\mu$ and $\mathbf{\Psi}_\nu$ span an invariant space, corresponding to the direct product representation $D^{[s]*} \times D^{[s]}$ of SO(N). **The decomposition of $D^{[s]*} \times D^{[s]}$ contains an identical representation, denoted by $[\lambda] = [0]$, where the Clebsch–Gordan coefficients are $\delta_{\mu\nu}$.** Generally,

$$\begin{aligned} O_R\left(\mathbf{\Psi}^\dagger \gamma_{a_1} \cdots \gamma_{a_n} \mathbf{\Psi}\right) &= \mathbf{\Psi}^\dagger D^{[s]}(R)^{-1} \gamma_{a_1} \cdots \gamma_{a_n} D^{[s]}(R) \mathbf{\Psi} \\ &= \sum_{b_1 \cdots b_n} R_{a_1 b_1} \cdots R_{a_n b_n} \mathbf{\Psi}^\dagger \gamma_{b_1} \cdots \gamma_{b_n} \mathbf{\Psi}. \end{aligned} \tag{10.76}$$

$\mathbf{\Psi}^\dagger \gamma_{a_1} \cdots \gamma_{a_n} \mathbf{\Psi}$ is an antisymmetric tensor of rank n of SO(N) corresponding to the Young pattern $[1^n]$. n is obviously not larger than N, otherwise the repetitive γ_a can be moved together and eliminated.

When $N = 2\ell + 1$, $\gamma_f^{(2\ell+1)}$ is a constant matrix so that the product of $(N - n)$ matrices γ_a can be changed to a product of n matrices γ_a. Thus, the rank n of the tensor (10.76) is less than $N/2$, and the Clebsch–Gordan series is

$$[s]^* \times [s] \simeq [s] \times [s] \simeq [0] \oplus [1] \oplus [1^2] \oplus \cdots \oplus [1^\ell], \quad \text{for SO}(2\ell+1). \tag{10.77}$$

The matrix elements of the product of γ_a are the Clebsch–Gordan coefficients. The highest weight of $[1^n]$ is $\boldsymbol{M} = \boldsymbol{w}_n$ when $1 \leqslant n < \ell$, and the highest weight of $[1^\ell]$ is $\boldsymbol{M} = 2\boldsymbol{w}_\ell = (0, \cdots, 0, 2)$ (see Eq. (10.15)), which is also the highest weight in the product space (see Eq. (10.71)).

When $N = 2\ell$, due to the property of the projective operators $P_\pm$,

$$\begin{aligned} &P_+ P_- = P_- P_+ = 0, \qquad P_\pm P_\pm = P_\pm, \qquad \gamma_f^{(2\ell)} P_\pm = \pm P_\pm, \\ &P_\mp \gamma_{a_1} \cdots \gamma_{a_{2m}} P_\pm = 0, \qquad P_\pm \gamma_{a_1} \cdots \gamma_{a_{2m+1}} P_\pm = 0, \end{aligned} \tag{10.78}$$

the product of $(N - n)$ matrices γ_a can still be changed to a product of n matrices γ_a. If $n = \ell$, one has $\gamma_1 \gamma_2 \cdots \gamma_\ell \gamma_f^{(2\ell)} = (-\mathrm{i})^\ell \gamma_{2\ell} \gamma_{2\ell-1} \cdots \gamma_{\ell+1}$,

$$\gamma_1 \gamma_2 \cdots \gamma_\ell P_\pm = \frac{1}{2} \left\{\gamma_1 \gamma_2 \cdots \gamma_\ell \pm (-\mathrm{i})^\ell \gamma_{2\ell} \gamma_{2\ell-1} \cdots \gamma_{\ell+1}\right\} P_\pm. \tag{10.79}$$

If $N = 4m$,

$$\begin{aligned} &[\pm s]^* \times [\pm s] \simeq [\pm s] \times [\pm s] \simeq [0] \oplus [1^2] \oplus [1^4] \oplus \cdots \oplus [(\pm)1^{2m}], \\ &[\mp s]^* \times [\pm s] \simeq [\mp s] \times [\pm s] \simeq [1] \oplus [1^3] \oplus [1^5] \oplus \cdots \oplus [1^{2m-1}]. \end{aligned} \tag{10.80}$$

If $N = 4m + 2$,

$$\begin{aligned}
&[\pm s]^* \times [\pm s] \simeq [\mp s] \times [\pm s] \simeq [0] \oplus [1^2] \oplus [1^4] \oplus \cdots \oplus [1^{2m}] \\
&[\mp s]^* \times [\pm s] \simeq [\pm s] \times [\pm s] \simeq [1] \oplus [1^3] \oplus [1^5] \oplus \cdots \oplus [(\pm)1^{2m+1}].
\end{aligned} \tag{10.81}$$

By calculating the highest weight in the product space, **the self-dual representation $[(+)1^\ell]$ with the highest weight $M = (0, \cdots, 0, 0, 2)$ occurs in the decomposition of the direct product** $[+s] \times [+s]$, the anti-self-dual one $[(-)1^\ell]$ with the highest weight $M = (0, \cdots, 0, 2, 0)$ occurs in the decomposition of the direct product $[-s] \times [-s]$, and the antisymmetric tensor representations $[1^{\ell-1}]$ with the highest weights $M = (0, \cdots, 0, 1, 1)$ occurs in the decomposition of the direct product $[\mp s] \times [\pm s]$.

Remind that, **the combination** (10.79) **of the self-dual tensor components is different, when $N = 2\ell = 4m + 2$, to the combination** (10.10) **for the self-dual basis tensor** by a sign $(-1)^\ell = -1$. It occurs in Eq. (5.184), too. We write the decomposition of an antisymmetric tensor $\boldsymbol{T}$ of rank 5 with respect to SO(10) (D_5 Lie algebra) into self-dual tensor and anti-self-dual tensor in some detail, where the subscript "10" is denoted by "0".

$$\begin{aligned}
\boldsymbol{T} &= \sum_{(a)} \boldsymbol{\theta}_{a_1 \cdots a_5} T_{a_1 \cdots a_5} \\
&= \frac{1}{2}\left(\boldsymbol{\theta}_{12345} + \mathrm{i}^5 \boldsymbol{\theta}_{09876}\right) \frac{1}{2}\left(T_{12345} + (-\mathrm{i})^5 T_{09876}\right) \\
&\quad + \frac{1}{2}\left(\boldsymbol{\theta}_{09876} - \mathrm{i}^5 \boldsymbol{\theta}_{12345}\right) \frac{1}{2}\left(T_{09876} - (-\mathrm{i})^5 T_{12345}\right) + \cdots \\
&\quad + \frac{1}{2}\left(\boldsymbol{\theta}_{12345} - \mathrm{i}^5 \boldsymbol{\theta}_{09876}\right) \frac{1}{2}\left(T_{12345} - (-\mathrm{i})^5 T_{09876}\right) \\
&\quad + \frac{1}{2}\left(\boldsymbol{\theta}_{09876} + \mathrm{i}^5 \boldsymbol{\theta}_{12345}\right) \frac{1}{2}\left(T_{09876} + (-\mathrm{i})^5 T_{12345}\right) + \cdots ,
\end{aligned}$$

where $\epsilon_{0987654321} = -1$.

10.3.4 *Spinor Representations of Higher Ranks*

In the SO(3) group, $D^{1/2}$ is the fundamental spinor representation, and the spinor representations $D^{\ell+1/2}$ of higher ranks can be obtained from the decomposition of the direct product of the fundamental spinor representation and a tensor representation,

$$D^{1/2} \times D^\ell \simeq D^{\ell+1/2} \oplus D^{\ell-1/2}.$$

The spinor representations of higher ranks of SO(N) can be obtained in the same way.

A spinor $\boldsymbol{\Psi}_{a_1\cdots a_n}$ with the tensor indices is called a spin-tensor if it transforms in $R \in \mathrm{SO}(N)$ as follows

$$(O_R\boldsymbol{\Psi})_{a_1\cdots a_n} = \sum_{b_1\cdots b_n} R_{a_1b_1}\cdots R_{a_nb_n}D^{[s]}(R)\boldsymbol{\Psi}_{b_1\cdots b_n}. \tag{10.82}$$

The spin-tensor $\boldsymbol{\Psi}_{a_1\cdots a_n}$ can be simply looked as the direct product of a tensor and a spin: $\boldsymbol{\Psi}_{a_1\cdots a_n} = \boldsymbol{T}_{a_1\cdots a_n}\boldsymbol{\Psi}$. The tensor part $\boldsymbol{T}_{a_1\cdots a_n}$ can be decomposed into a direct sum of the traceless tensors with different ranks and each traceless tensor subspace can be decomposed by the projection of the Young operators. Thus, the reduced subspace of the traceless tensor part $\boldsymbol{T}_{a_1\cdots a_n}$ is denoted by a Young pattern $[\lambda]$ or $[(\pm)\lambda]$ where the row number of $[\lambda]$ is not larger than $N/2$. However, the direct product of the irreducible tensor representation $[\lambda]$ or $[(\pm)\lambda]$ and the fundamental spinor representation $[s]$ is still reducible. It is required to find a new constraint to pick up the irreducible subspace like the subspace of $D^{\ell+1/2}$ for SO(3). The constraint comes from the so-called **trace of the second kind of the spin-tensor** which is left invariant in the SO(N) transformations:

$$\begin{aligned}
\boldsymbol{\Phi}_{a_1\cdots a_{i-1}a_{i+1}\cdots a_n} &= \sum_{b=1}^{N} \boldsymbol{T}_{a_1\cdots a_{i-1}ba_{i+1}\cdots a_n}\gamma_b\boldsymbol{\Psi}, \\
(O_R\boldsymbol{\Phi})_{a_1\cdots a_{i-1}a_{i+1}\cdots a_n} &= \sum_{b_1\cdots \hat{b_i}\cdots b_n b'} R_{a_1b_1}\cdots R_{a_nb_n}\boldsymbol{T}_{b_1\cdots b_{i-1}b'b_{i+1}\cdots b_n} \\
&\quad \cdot \left[\sum_b \gamma_b R_{bb'}\right] D^{[s]}(R)\boldsymbol{\Psi} \\
&= \sum_{b_1\cdots \hat{b_i}\cdots b_n} R_{a_1b_1}\cdots R_{a_nb_n}D^{[s]}(R)\left[\sum_{b'}\boldsymbol{T}_{b_1\cdots b_{i-1}b'b_{i+1}\cdots b_n}\gamma_{b'}\boldsymbol{\Psi}\right] \\
&= \sum_{b_1\cdots \hat{b_i}\cdots b_n} R_{a_1b_1}\cdots R_{a_nb_n}D^{[s]}(R)\boldsymbol{\Phi}_{b_1\cdots b_{i-1}b_{i+1}\cdots b_n},
\end{aligned}$$

where $\hat{b_i}$ means no summation over b_i. The irreducible subspace of SO(N) contained in the spin-tensor space, in addition to the projection of a Young operator, satisfies the usual traceless conditions of tensors and the traceless conditions of the second kind:

$$\sum_b \boldsymbol{T}_{a\cdots b\cdots b\cdots c}\boldsymbol{\Psi} = 0, \qquad \sum_b \boldsymbol{T}_{a\cdots b\cdots c}\gamma_b\boldsymbol{\Psi} = 0. \tag{10.83}$$

The highest weight state of an irreducible representation satisfies both traceless conditions.

The irreducible representation is denoted by $[s, \lambda]$ for SO$(2\ell+1)$, whose highest weight $\boldsymbol{M}$ is equal to the sum of the highest weights for two representation $[\lambda]$ and $[s]$:

$$\begin{aligned}&[s] \times [\lambda] \simeq [s, \lambda] \oplus \cdots \\ &\boldsymbol{M} = [(\lambda_1 - \lambda_2), \cdots, (\lambda_{\ell-1} - \lambda_\ell), (2\lambda_\ell + 1)],\end{aligned} \tag{10.84}$$

and the irreducible representation is denoted by $[\pm s, \lambda]$ for SO(2ℓ), which has the highest weight $\boldsymbol{M}$:

$$\begin{aligned}&[+s] \times [\lambda] \quad \text{or} \quad [+s] \times [(+)\lambda] \simeq [+s, \lambda] \oplus \cdots \\ &\boldsymbol{M} = [(\lambda_1 - \lambda_2), \cdots, (\lambda_{\ell-1} - \lambda_\ell), (\lambda_{\ell-1} + \lambda_\ell + 1)], \\ &[-s] \times [\lambda] \quad \text{or} \quad [-s] \times [(-)\lambda] \simeq [-s, \lambda] \oplus \cdots \\ &\boldsymbol{M} = [(\lambda_1 - \lambda_2), \cdots, (\lambda_{\ell-1} + \lambda_\ell + 1), (\lambda_{\ell-1} - \lambda_\ell)], \\ &[+s] \times [(-)\lambda] \simeq [-s, \lambda_1, \lambda_2, \cdots, \lambda_{\ell-1}, (\lambda_\ell - 1)] \oplus \cdots \\ &\boldsymbol{M} = [(\lambda_1 - \lambda_2), \cdots, (\lambda_{\ell-1} + \lambda_\ell), (\lambda_{\ell-1} - \lambda_\ell + 1)], \\ &[-s] \times [(+)\lambda] \simeq [+s, \lambda_1, \lambda_2, \cdots, \lambda_{\ell-1}, (\lambda_\ell - 1)] \oplus \cdots \\ &\boldsymbol{M} = [(\lambda_1 - \lambda_2), \cdots, (\lambda_{\ell-1} - \lambda_\ell + 1), (\lambda_{\ell-1} + \lambda_\ell)].\end{aligned} \tag{10.85}$$

These irreducible representations $[s, \lambda]$ of SO$(2\ell+1)$ and $[\pm s, \lambda]$ of SO(2ℓ) are called the spinor representations of higher ranks. Remind that **the row number of the Young pattern $[\lambda]$ in the spinor representation of higher rank is not larger than ℓ**, otherwise the space is null. For example, in the space of the spinor representation $[s, 1^n]$ of SO$(2\ell+1)$, the number of spin-tensors before taking the traceless conditions is $d^{(2\ell+1)}$ times the combinatorial number of n among $(2\ell+1)$. There is no traceless condition of the first kind. The number of traceless conditions of the second kind is $d^{(2\ell+1)}$ times the combinatorial number of $(n-1)$ among $(2\ell+1)$. Thus, the space is null when $n > \ell$ because the number of the traceless conditions is not less than the number of tensors.

The remaining representations in the Clebsch–Gordan series (10.84) and (10.85) are calculated by the method of dominant weight diagram. For example, when $[\lambda] = [\lambda_1]$ is a one-row Young diagram, one has

$$\begin{aligned}&SO(2\ell+1): \quad [s] \times [\lambda_1] \simeq [s, \lambda_1] \oplus [s, \lambda_1 - 1], \\ &SO(2\ell): \quad [\pm s] \times [\lambda_1] \simeq [\pm s, \lambda_1] \oplus [\mp s, \lambda_1 - 1].\end{aligned} \tag{10.86}$$

$[\mp s, \lambda_1 - 1]$ appears in the second equation because the factor γ_b in Eq. (10.83) anti-commutes with γ_f in $P_\pm$.

We calculate the orthonormal basis states in the spin-tensor representation $[+s, 1]$ of SO(8) (D_4 Lie algebra) as example. From Eq. (10.85), the highest weight is $\boldsymbol{M} = \boldsymbol{w}_1 + \boldsymbol{w}_4 = (1, 0, 0, 1)$. In the brief symbols, $|\boldsymbol{M}, \boldsymbol{m}\rangle = |\boldsymbol{m}\rangle$, the highest weight state is

$$|\boldsymbol{M}\rangle = |(1,0,0,1)\rangle = \left|\begin{smallmatrix}2&&1&&1&&1\\&2&&1&&1&\\&&2&&1&&\\&&&2&&&\end{smallmatrix}\right\rangle_1^1 = \boxed{\ 1\ }_{1111} = -\sqrt{\frac{1}{2}}\ (\boldsymbol{\theta}_1 + \mathrm{i}\boldsymbol{\theta}_2)\,\chi_{1111}.$$

This state satisfies the traceless conditions of the second kind:

$$-\sqrt{\frac{1}{2}}\sum_a \left[(\boldsymbol{\theta}_1)_a + \mathrm{i}\,(\boldsymbol{\theta}_2)_a\right]\gamma_a\chi_{1111} = -\sqrt{\frac{1}{2}}(\gamma_1 + \mathrm{i}\gamma_2)\,\chi_{1111} = 0.$$

The highest weight state $|(1,0,0,1)\rangle$ is the highest weight state of an A_3-quadruplet with $\boldsymbol{M}^{(3)} = (1,0,0)$. All four basis states in this A_3-quadruplet satisfy the condition (8.7) with $\mu = 4$. Two $\mathcal{A}_4$-doublets are easy to construct from the states $|(1,0,0,1)\rangle$ and $(\overline{1},1,0,1)\rangle$, and the partners in these two $\mathcal{A}_4$-doublets belong to an A_3-20let with $\boldsymbol{M}^{(3)} = (1,1,0)$. Two $\mathcal{A}_4$-triplets are constructed from the states $(0,\overline{1},1,2)\rangle$ and $(0,0,\overline{1},2)\rangle$ in terms of Eqs. (8.25-28) where the knowledge of this A_3-20let with $\boldsymbol{M}^{(3)} = (1,1,0)$ is used.

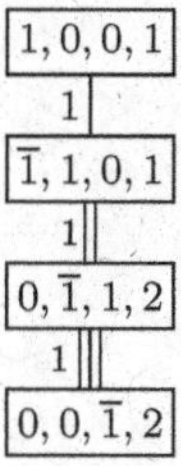

$$\left|\begin{smallmatrix}2&&1&&1&&1\\&2&&1&&1&\\&&2&&1&&\\&&&2&&&\end{smallmatrix}\right\rangle_1^1 = \boxed{\ 1\ }_{1111}$$

$$\left|\begin{smallmatrix}2&&1&&1&&1\\&2&&1&&1&\\&&2&&1&&\\&&&1&&&\end{smallmatrix}\right\rangle_1^2 = \boxed{\ 2\ }_{1111}$$

$$\left|\begin{smallmatrix}2&&1&&1&&1\\&2&&1&&1&\\&&1&&1&&\\&&&1&&&\end{smallmatrix}\right\rangle_2^3 = \boxed{\ 3\ }_{1111}$$

$$\left|\begin{smallmatrix}2&&1&&1&&1\\&1&&1&&1&\\&&1&&1&&\\&&&1&&&\end{smallmatrix}\right\rangle_2^4 = \boxed{\ 4\ }_{1111}$$

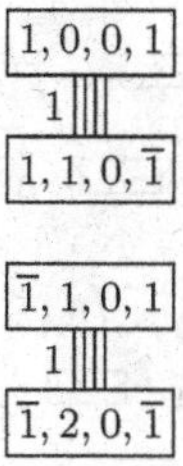

$$\left|\begin{smallmatrix}2&&1&&1&&1\\&2&&1&&1&\\&&2&&1&&\\&&&2&&&\end{smallmatrix}\right\rangle_1^1 = \boxed{\ 1\ }_{1111}$$

$$\left|\begin{smallmatrix}2&&1&&0&&0\\&2&&1&&0&\\&&2&&1&&\\&&&2&&&\end{smallmatrix}\right\rangle_{\overline{1}}^2 = \boxed{\ 1\ }_{1122}$$

$$\left|\begin{smallmatrix}2&&1&&1&&1\\&2&&1&&1&\\&&2&&1&&\\&&&1&&&\end{smallmatrix}\right\rangle_1^2 = \boxed{\ 2\ }_{1111}$$

$$\left|\begin{smallmatrix}2&&1&&0&&0\\&2&&1&&0&\\&&2&&1&&\\&&&1&&&\end{smallmatrix}\right\rangle_{\overline{1}}^3 = \boxed{\ 2\ }_{1122}$$

Letting

$$F_4\left|\begin{smallmatrix}2&&1&&1&&1\\&2&&1&&1&\\&&1&&1&&\\&&&1&&&\end{smallmatrix}\right\rangle_2^3 = a_1\left|\begin{smallmatrix}2&&1&&0&&0\\&2&&1&&0&\\&&1&&1&&\\&&&1&&&\end{smallmatrix}\right\rangle_0^4 + a_2\left|\begin{smallmatrix}1&&1&&1&&0\\&1&&1&&1&\\&&1&&1&&\\&&&1&&&\end{smallmatrix}\right\rangle_0^4,$$

we obtain (see the block weight diagram of A_3-20let with $\boldsymbol{M}^{(3)} = (1,1,0)$)

$$\begin{aligned}
E_2F_4\left|\begin{smallmatrix}2&&1&&1&&1\\&2&&1&&1&\\&&1&&1&&\\&&&1&&&\end{smallmatrix}\right\rangle_2^3 &= a_1E_2\left|\begin{smallmatrix}2&&1&&0&&0\\&2&&1&&0&\\&&1&&1&&\\&&&1&&&\end{smallmatrix}\right\rangle_0^4 = a_1\sqrt{\frac{3}{2}}\left|\begin{smallmatrix}2&&1&&0&&0\\&2&&1&&0&\\&&2&&1&&\\&&&1&&&\end{smallmatrix}\right\rangle_{\overline{1}}^3\\
= F_4E_2\left|\begin{smallmatrix}2&&1&&1&&1\\&2&&1&&1&\\&&1&&1&&\\&&&1&&&\end{smallmatrix}\right\rangle_2^3 &= F_4\left|\begin{smallmatrix}2&&1&&1&&1\\&2&&1&&1&\\&&2&&1&&\\&&&1&&&\end{smallmatrix}\right\rangle_1^2 = \left|\begin{smallmatrix}2&&1&&0&&0\\&2&&1&&0&\\&&2&&1&&\\&&&1&&&\end{smallmatrix}\right\rangle_{\overline{1}}^3.
\end{aligned}$$

Choosing the phase of the state $\left|\begin{smallmatrix}1&&1&&1&&0\\&1&&1&&1&\\&&1&&1&&\\&&&1&&&\end{smallmatrix}\right\rangle_0^4$ such that a_2 is real and positive, we have $a_1 = \sqrt{2/3}$ and $a_2 = \sqrt{2 - 2/3} = \sqrt{4/3}$. By Eq. (8.9) again we complete the calculation for the $\mathcal{A}_4$-triplet. Due to the parallel principle, another $\mathcal{A}_4$-triplet can be constructed from the state $(0,0,\overline{1},2))\rangle$. Their block weight diagrams and basis states are given as follows.

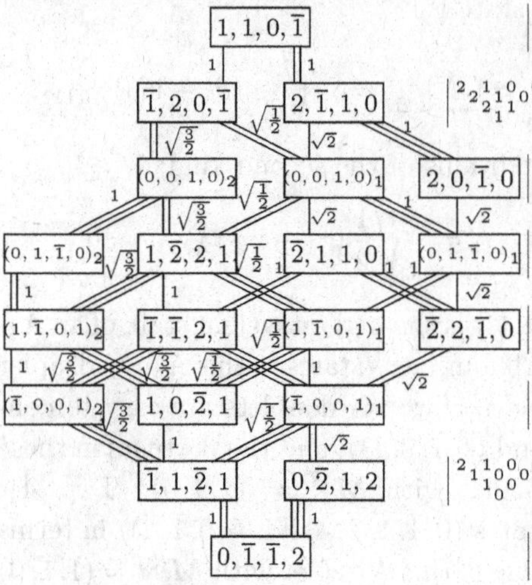

$$\left|\begin{smallmatrix}2&1&0&0\\2&1&0\\2&1\\2\end{smallmatrix}\right\rangle^{2}_{\overline{1}} = \boxed{1}_{1122}$$

$$\left|\begin{smallmatrix}2&1&0&0\\2&1&0\\2&1\\1\end{smallmatrix}\right\rangle^{3}_{\overline{1}} = \boxed{2}_{1122}, \quad \left|\begin{smallmatrix}2&1&0&0\\2&1&0\\2&0\\2\end{smallmatrix}\right\rangle^{3}_{0} = \boxed{1}_{1212}$$

$$\left|\begin{smallmatrix}2&1&0&0\\2&1&0\\1&1\\1\end{smallmatrix}\right\rangle^{4}_{0}, \quad \left|\begin{smallmatrix}2&1&0&0\\2&1&0\\2&0\\1\end{smallmatrix}\right\rangle^{4}_{0}, \quad \left|\begin{smallmatrix}2&1&0&0\\2&0&0\\2&0\\2\end{smallmatrix}\right\rangle^{4}_{0}$$

$$\left|\begin{smallmatrix}2&1&0&0\\1&1&0\\1&1\\1\end{smallmatrix}\right\rangle^{5}_{0}, \quad \left|\begin{smallmatrix}2&1&0&0\\2&1&0\\1&0\\1\end{smallmatrix}\right\rangle^{5}_{1}, \quad \left|\begin{smallmatrix}2&1&0&0\\2&1&0\\2&0\\0\end{smallmatrix}\right\rangle^{5}_{0}, \ \mathrm{A}$$

$$\left|\begin{smallmatrix}2&1&0&0\\1&1&0\\1&0\\1\end{smallmatrix}\right\rangle^{6}_{1}, \quad \left|\begin{smallmatrix}2&1&0&0\\2&1&0\\1&0\\0\end{smallmatrix}\right\rangle^{6}_{1}, \quad \left|\begin{smallmatrix}2&1&0&0\\2&0&0\\1&0\\1\end{smallmatrix}\right\rangle^{6}_{1}, \ B$$

$$\left|\begin{smallmatrix}2&1&0&0\\1&1&0\\1&0\\0\end{smallmatrix}\right\rangle^{7}_{1}, \quad \left|\begin{smallmatrix}2&1&0&0\\1&0&0\\1&0\\1\end{smallmatrix}\right\rangle^{7}_{1}, \quad \left|\begin{smallmatrix}2&1&0&0\\2&0&0\\1&0\\0\end{smallmatrix}\right\rangle^{7}_{1}$$

$$\left|\begin{smallmatrix}2&1&0&0\\1&0&0\\1&0\\0\end{smallmatrix}\right\rangle^{8}_{1} = \boxed{4}_{2121}, \quad \left|\begin{smallmatrix}2&1&0&0\\2&0&0\\0&0\\0\end{smallmatrix}\right\rangle^{8}_{2} = \boxed{3}_{2211}$$

$$\left|\begin{smallmatrix}2&1&0&0\\1&0&0\\0&0\\0\end{smallmatrix}\right\rangle^{9}_{2} = \boxed{4}_{2211}$$

$$\left|\begin{smallmatrix}2&1&0&0\\2&1&0\\1&1\\1\end{smallmatrix}\right\rangle^{4}_{0} = \sqrt{\tfrac{1}{6}}\left\{-\boxed{1}_{2112} + \boxed{2}_{1212} + 2\,\boxed{3}_{1122}\right\},$$

$$\left|\begin{smallmatrix}2&1&0&0\\2&1&0\\2&0\\1\end{smallmatrix}\right\rangle^{4}_{0} = \sqrt{\tfrac{1}{2}}\left\{\boxed{1}_{2112} + \boxed{2}_{1212}\right\}, \quad \left|\begin{smallmatrix}2&1&0&0\\2&0&0\\2&0\\2\end{smallmatrix}\right\rangle^{4}_{0} = \boxed{1}_{1221},$$

$$\left|\begin{smallmatrix}2&1&0&0\\1&1&0\\1&1\\1\end{smallmatrix}\right\rangle^{5}_{0} = \sqrt{\tfrac{1}{6}}\left\{-\boxed{1}_{2121} + \boxed{2}_{1221} + 2\,\boxed{4}_{1122}\right\},$$

$$\left|\begin{smallmatrix}2&1&0&0\\2&1&0\\1&0\\1\end{smallmatrix}\right\rangle^{5}_{1} = \boxed{3}_{1212}, \quad \left|\begin{smallmatrix}2&1&0&0\\2&1&0\\2&0\\0\end{smallmatrix}\right\rangle^{5}_{0} = \boxed{2}_{2112},$$

$$A = \left|\begin{smallmatrix}2&1&0&0\\2&0&0\\2&0\\1\end{smallmatrix}\right\rangle^{5}_{0} = \sqrt{\tfrac{1}{2}}\left\{\boxed{1}_{2121} + \boxed{2}_{1221}\right\},$$

$$\left|\begin{smallmatrix}2&1&0&0\\1&1&0\\1&0\\1\end{smallmatrix}\right\rangle^{6}_{1} = \sqrt{\tfrac{1}{6}}\left\{-\boxed{1}_{2211} + \boxed{3}_{1221} + 2\,\boxed{4}_{1212}\right\}, \quad \left|\begin{smallmatrix}2&1&0&0\\2&1&0\\1&0\\0\end{smallmatrix}\right\rangle^{6}_{1} = \boxed{3}_{2112},$$

$$\left|\begin{smallmatrix}2&1&0&0\\2&0&0\\1&0\\1\end{smallmatrix}\right\rangle^{6}_{1} = \sqrt{\tfrac{1}{2}}\left\{\boxed{1}_{2211} + \boxed{3}_{1221}\right\}, \quad B = \left|\begin{smallmatrix}2&1&0&0\\2&0&0\\2&0\\0\end{smallmatrix}\right\rangle^{6}_{0} = \boxed{2}_{2121},$$

$$\left|\begin{smallmatrix}2&1&0&0\\1&1&0\\1&0\\0\end{smallmatrix}\right\rangle^{7}_{1} = \sqrt{\tfrac{1}{6}}\left\{-\boxed{2}_{2211} + \boxed{3}_{2121} + 2\,\boxed{4}_{2112}\right\},$$

$$\left|\begin{smallmatrix}2&1&0&0\\1&0&0\\1&0\\1\end{smallmatrix}\right\rangle^{7}_{1} = \boxed{4}_{1221}, \quad \left|\begin{smallmatrix}2&1&0&0\\2&0&0\\1&0\\0\end{smallmatrix}\right\rangle^{7}_{1} = \sqrt{\tfrac{1}{2}}\left\{\boxed{2}_{2211} + \boxed{3}_{2121}\right\}.$$

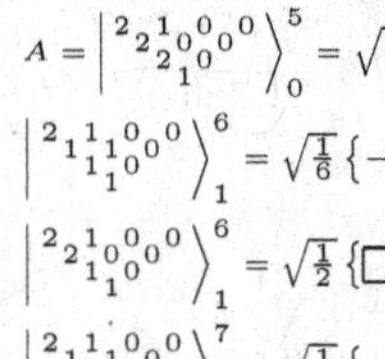

$$\left|\begin{smallmatrix}2&1&1&1\\2&1&1\\1&1\\1\end{smallmatrix}\right\rangle^{3}_{2} = \boxed{3}_{1111}$$

$$\left|\begin{smallmatrix}2&1&0&0\\2&1&0\\1&1\\1\end{smallmatrix}\right\rangle^{4}_{0} = \sqrt{\tfrac{1}{6}}\left\{-\boxed{1}_{2112} + \boxed{2}_{1212} + 2\,\boxed{3}_{1122}\right\}, \quad \left|\begin{smallmatrix}1&1&1&0\\1&1&1\\1&1\\1\end{smallmatrix}\right\rangle^{4}_{0}$$

$$\left|\begin{smallmatrix}1&1&0&\overline{1}\\1&1&0\\1&1\\1\end{smallmatrix}\right\rangle^{5}_{\overline{2}} = \boxed{5}_{1122}$$

$$\left|\begin{smallmatrix}1&1&1&0\\1&1&1\\1&1\\1\end{smallmatrix}\right\rangle^{4}_{0} = \sqrt{\tfrac{1}{12}}\left\{\boxed{1}_{2112} - \boxed{2}_{1212} + \boxed{3}_{1122} + 3\,\boxed{5}_{1111}\right\}$$

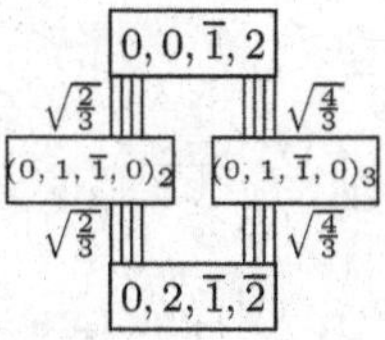

$$\left|\begin{smallmatrix} 2 & 1 & 1 & 1 \\ 1 & 1 & 1 \\ 1 & 1 \\ 1 \end{smallmatrix}\right\rangle_{2}^{4} = \boxed{4}_{1111}$$

$$\left|\begin{smallmatrix} 2 & 1 & 0 & 0 \\ 1 & 1 & 0 \\ 1 & 1 \\ 1 \end{smallmatrix}\right\rangle_{0}^{5} = \sqrt{\tfrac{1}{6}}\left\{-\boxed{1}_{2121} + \boxed{2}_{1221} + 2\,\boxed{4}_{1122}\right\}, \quad \left|\begin{smallmatrix} 1 & 1 & 1 & 0 \\ 1 & 1 & 0 \\ 1 & 1 \\ 1 \end{smallmatrix}\right\rangle_{0}^{5}$$

$$\left|\begin{smallmatrix} 1 & 1 & 0 & \bar{1} \\ 1 & 1 & \bar{1} \\ 1 & 1 \\ 1 \end{smallmatrix}\right\rangle_{\bar{2}}^{6} = \boxed{6}_{1122}$$

$$\left|\begin{smallmatrix} 1 & 1 & 1 & 0 \\ 1 & 1 & 0 \\ 1 & 1 \\ 1 \end{smallmatrix}\right\rangle_{0}^{4} = \sqrt{\tfrac{1}{12}}\left\{\boxed{1}_{2121} - \boxed{2}_{1221} + \boxed{4}_{1122} + 3\,\boxed{6}_{1111}\right\}$$

The states $\left|\begin{smallmatrix} 1 & 1 & 1 & 0 \\ 1 & 1 & 1 \\ 1 & 1 \\ 1 \end{smallmatrix}\right\rangle_{0}^{4}$ and $\left|\begin{smallmatrix} 1 & 1 & 0 & \bar{1} \\ 1 & 1 & 0 \\ 1 & 1 \\ 1 \end{smallmatrix}\right\rangle_{\bar{2}}^{5}$ in the $\mathcal{A}_4$-triplet are the highest weight states of an A_3-quadruplet with $\boldsymbol{M}^{(3)} = (0, 0, 1)$ and an A_3-20let with $\boldsymbol{M}^{(3)} = (0, 1, 1)$, respectively. The calculation results of the block weight diagram and the basis states of the A_3-20let and the A_3-quadruplet are given as follows.

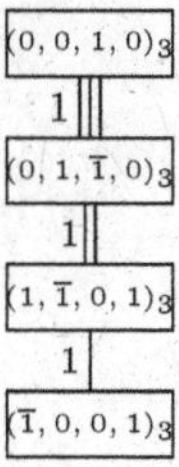

$$\left|\begin{smallmatrix} 1 & 1 & 1 & 0 \\ 1 & 1 & 1 \\ 1 & 1 \\ 1 \end{smallmatrix}\right\rangle_{0}^{4} = \sqrt{\tfrac{1}{12}}\left\{\boxed{1}_{2112} - \boxed{2}_{1212} + \boxed{3}_{1122} + 3\,\boxed{5}_{1111}\right\}$$

$$\left|\begin{smallmatrix} 1 & 1 & 1 & 0 \\ 1 & 1 & 0 \\ 1 & 1 \\ 1 \end{smallmatrix}\right\rangle_{0}^{5} = \sqrt{\tfrac{1}{12}}\left\{\boxed{1}_{2121} - \boxed{2}_{1221} + \boxed{4}_{1122} + 3\,\boxed{6}_{1111}\right\}$$

$$\left|\begin{smallmatrix} 1 & 1 & 1 & 0 \\ 1 & 1 & 0 \\ 1 & 0 \\ 1 \end{smallmatrix}\right\rangle_{1}^{6} = \sqrt{\tfrac{1}{12}}\left\{\boxed{1}_{2211} - \boxed{3}_{1221} + \boxed{4}_{1212} + 3\,\boxed{7}_{1111}\right\}$$

$$\left|\begin{smallmatrix} 1 & 1 & 1 & 0 \\ 1 & 1 & 0 \\ 1 & 0 \\ 0 \end{smallmatrix}\right\rangle_{1}^{7} = \sqrt{\tfrac{1}{12}}\left\{\boxed{2}_{2211} - \boxed{3}_{2121} + \boxed{4}_{2112} + 3\,\boxed{8}_{1111}\right\}$$

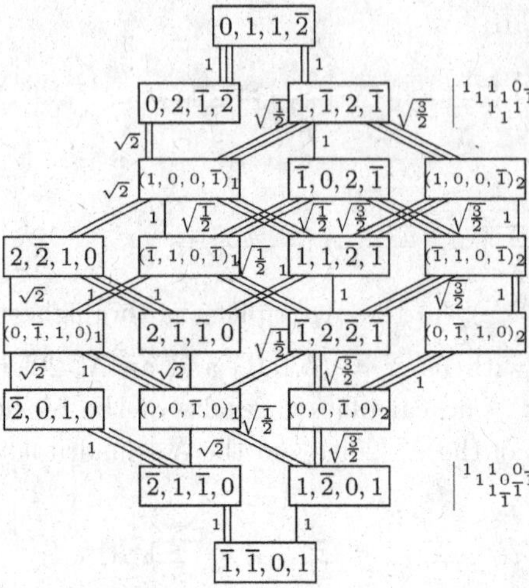

$$\left|\begin{smallmatrix}1&1&0&\bar1\\1&1&0\\1&1\\1\end{smallmatrix}\right\rangle^{5}_{\bar2}=\boxed{5}_{1122}$$

$$\left|\begin{smallmatrix}1&1&0&\bar1\\1&1&\bar1\\1&1\\1\end{smallmatrix}\right\rangle^{6}_{\bar2}=\boxed{6}_{1122},\quad\left|\begin{smallmatrix}1&1&0&\bar1\\1&1&0\\1&0\\1\end{smallmatrix}\right\rangle^{6}_{\bar1}=\boxed{5}_{1212}$$

$$\left|\begin{smallmatrix}1&1&0&\bar1\\1&1&\bar1\\1&0\\1\end{smallmatrix}\right\rangle^{7}_{\bar1},\quad\left|\begin{smallmatrix}1&1&0&\bar1\\1&1&0\\1&0\\0\end{smallmatrix}\right\rangle^{7}_{\bar1},\quad\left|\begin{smallmatrix}1&1&0&\bar1\\1&0&0\\1&0\\1\end{smallmatrix}\right\rangle^{7}_{\bar1}$$

$$\left|\begin{smallmatrix}1&1&0&\bar1\\1&1&\bar1\\1&\bar1\\1\end{smallmatrix}\right\rangle^{8}_{0},\quad\left|\begin{smallmatrix}1&1&0&\bar1\\1&1&\bar1\\1&0\\0\end{smallmatrix}\right\rangle^{8}_{\bar1},\quad\left|\begin{smallmatrix}1&1&0&\bar1\\1&0&\bar1\\1&0\\1\end{smallmatrix}\right\rangle^{8}_{\bar1},\ \mathrm{A}$$

$$\left|\begin{smallmatrix}1&1&0&\bar1\\1&1&\bar1\\1&\bar1\\0\end{smallmatrix}\right\rangle^{9}_{0},\quad\left|\begin{smallmatrix}1&1&0&\bar1\\1&0&\bar1\\1&\bar1\\1\end{smallmatrix}\right\rangle^{9}_{0},\quad\left|\begin{smallmatrix}1&1&0&\bar1\\1&0&\bar1\\1&0\\0\end{smallmatrix}\right\rangle^{9}_{\bar1},\ B$$

$$\left|\begin{smallmatrix}1&1&0&\bar1\\1&1&\bar1\\1&\bar1\\\bar1\end{smallmatrix}\right\rangle^{10}_{0},\quad\left|\begin{smallmatrix}1&1&0&\bar1\\1&0&\bar1\\1&\bar1\\0\end{smallmatrix}\right\rangle^{10}_{0},\quad\left|\begin{smallmatrix}1&1&0&\bar1\\1&0&\bar1\\0&0\\0\end{smallmatrix}\right\rangle^{10}_{0}$$

$$\left|\begin{smallmatrix}1&1&0&\bar1\\1&0&\bar1\\1&\bar1\\\bar1\end{smallmatrix}\right\rangle^{11}_{0}=\boxed{8}_{2121},\quad\left|\begin{smallmatrix}1&1&0&\bar1\\1&0&\bar1\\0&\bar1\\0\end{smallmatrix}\right\rangle^{11}_{1}=\boxed{7}_{2211}$$

$$\left|\begin{smallmatrix}1&1&0&\bar1\\1&0&\bar1\\0&\bar1\\\bar1\end{smallmatrix}\right\rangle^{12}_{1}=\boxed{8}_{2211}$$

$$\left|\begin{smallmatrix}1&1&0&\bar1\\1&1&\bar1\\1&0\\1\end{smallmatrix}\right\rangle^{7}_{\bar1}=\sqrt{\tfrac{1}{2}}\left\{\boxed{6}_{1212}+\boxed{7}_{1122}\right\},\quad\left|\begin{smallmatrix}1&1&0&\bar1\\1&1&0\\1&0\\0\end{smallmatrix}\right\rangle^{7}_{\bar1}=\boxed{5}_{2112},$$

$$\left|\begin{smallmatrix}1&1&0&\bar1\\1&0&0\\1&0\\1\end{smallmatrix}\right\rangle^{7}_{\bar1}=\sqrt{\tfrac{1}{6}}\left\{2\,\boxed{5}_{1221}+\boxed{6}_{1212}-\boxed{7}_{1122}\right\}$$

$$\left|\begin{smallmatrix}1&1&0&\bar1\\1&1&\bar1\\1&\bar1\\1\end{smallmatrix}\right\rangle^{8}_{0}=\boxed{7}_{1212},\quad\left|\begin{smallmatrix}1&1&0&\bar1\\1&1&\bar1\\1&0\\0\end{smallmatrix}\right\rangle^{8}_{\bar1}=\sqrt{\tfrac{1}{2}}\left\{\boxed{6}_{2112}+\boxed{8}_{1122}\right\}$$

$$\left|\begin{smallmatrix}1&1&0&\bar1\\1&0&\bar1\\1&0\\1\end{smallmatrix}\right\rangle^{8}_{\bar1}=\boxed{6}_{1221},\quad A=\left|\begin{smallmatrix}1&1&0&\bar1\\1&0&0\\1&0\\0\end{smallmatrix}\right\rangle^{8}_{\bar1}=\sqrt{\tfrac{1}{6}}\left\{2\,\boxed{5}_{2121}+\boxed{6}_{2112}-\boxed{8}_{1122}\right\},$$

$$\left|\begin{smallmatrix}1&1&0&\bar1\\1&1&\bar1\\1&\bar1\\0\end{smallmatrix}\right\rangle^{9}_{0}=\sqrt{\tfrac{1}{2}}\left\{\boxed{7}_{2112}+\boxed{8}_{1212}\right\},\quad\left|\begin{smallmatrix}1&1&0&\bar1\\1&0&\bar1\\1&\bar1\\1\end{smallmatrix}\right\rangle^{9}_{0}=\boxed{7}_{1221},$$

$$\left|\begin{smallmatrix}1&1&0&\bar1\\1&0&\bar1\\1&0\\0\end{smallmatrix}\right\rangle^{9}_{\bar1}=\boxed{6}_{2121},\quad B=\left|\begin{smallmatrix}1&1&0&\bar1\\1&0&0\\0&0\\0\end{smallmatrix}\right\rangle^{9}_{0}=\sqrt{\tfrac{1}{6}}\left\{2\,\boxed{5}_{2211}+\boxed{7}_{2112}-\boxed{8}_{1212}\right\},$$

$$\left|\begin{smallmatrix}1&1&0&\bar1\\1&1&\bar1\\1&\bar1\\\bar1\end{smallmatrix}\right\rangle^{10}_{0}=\boxed{8}_{2112},\quad\left|\begin{smallmatrix}1&1&0&\bar1\\1&0&\bar1\\1&\bar1\\0\end{smallmatrix}\right\rangle^{10}_{0}=\sqrt{\tfrac{1}{2}}\left\{\boxed{7}_{2121}+\boxed{8}_{1221}\right\},$$

$$\left|\begin{smallmatrix}1&1&0&\bar1\\1&0&\bar1\\0&0\\0\end{smallmatrix}\right\rangle^{10}_{0}=\sqrt{\tfrac{1}{6}}\left\{2\,\boxed{6}_{2211}+\boxed{7}_{2121}-\boxed{8}_{1221}\right\}.$$

Eight states in the A_3-20let with $M^{(3)}=(1,1,0)$ and two states in the A_3-quadruplet with $M^{(3)}=(0,0,1)$ satisfy the condition (8.7) with $\mu=4$ and they are the highest weight states of the $\mathcal{A}_4$-doublets. The weights of four states $|(1,\bar2,2,1)\rangle=\boxed{3}_{1212}$, $|(\bar1,\bar1,2,1)\rangle=\boxed{3}_{2112}$, $|(1,0,\bar2,1)\rangle=\boxed{4}_{1221}$, and $|(\bar1,1,\bar2,1)\rangle=\boxed{4}_{2121}$ are single so that four $\mathcal{A}_4$-doublets from them are easy to calculate by Eq. (8.9) and the parallel principle.

$$F_4\left|\begin{smallmatrix}2&1&0&0\\2&1&0\\1&0\\1\end{smallmatrix}\right\rangle^{5}_{1}=F_4\,\boxed{3}_{1212}=\boxed{5}_{1212}=\left|\begin{smallmatrix}1&1&0&\bar1\\1&1&0\\1&0\\1\end{smallmatrix}\right\rangle^{6}_{\bar1},$$

$$F_4\left|\begin{smallmatrix}2&1&0&0\\ 2&1&0\\ 1&0\\ 0\end{smallmatrix}\right\rangle^{6}_{1}=F_4\,\boxed{3}_{\,2112}=\boxed{5}_{\,2112}=\left|\begin{smallmatrix}1&1&0&\bar{1}\\ 1&1&0\\ 1&0\\ 0\end{smallmatrix}\right\rangle^{7}_{\bar{1}},$$

$$F_4\left|\begin{smallmatrix}2&1&0&0\\ 1&0&0\\ 1&0\\ 1\end{smallmatrix}\right\rangle^{7}_{1}=F_4\,\boxed{4}_{\,1221}=\boxed{6}_{\,1221}=\left|\begin{smallmatrix}1&1&0&\bar{1}\\ 1&0&\bar{1}\\ 1&0\\ 1\end{smallmatrix}\right\rangle^{8}_{\bar{1}},$$

$$F_4\left|\begin{smallmatrix}2&1&0&0\\ 1&0&0\\ 1&0\\ 0\end{smallmatrix}\right\rangle^{8}_{1}=F_4\,\boxed{4}_{\,2121}=\boxed{6}_{\,2121}=\left|\begin{smallmatrix}1&1&0&\bar{1}\\ 1&0&\bar{1}\\ 1&0\\ 0\end{smallmatrix}\right\rangle^{9}_{\bar{1}}.$$

The $\mathcal{A}_4$-doublets from the remaining six basis states are overlapping. We have to calculate those $\mathcal{A}_4$-doublets carefully by Eqs. (8.25-28), and then, by the parallel principle. Letting

$$F_4\left|\begin{smallmatrix}2&1&0&0\\ 2&0&0\\ 1&0\\ 1\end{smallmatrix}\right\rangle^{6}_{1}=b_1\left|\begin{smallmatrix}1&1&0&\bar{1}\\ 1&0&0\\ 1&0\\ 1\end{smallmatrix}\right\rangle^{7}_{\bar{1}}+b_2\left|\begin{smallmatrix}1&0&0&0\\ 1&0&0\\ 1&0\\ 1\end{smallmatrix}\right\rangle^{7}_{\bar{1}},$$

$$F_4\left|\begin{smallmatrix}2&1&0&0\\ 1&1&0\\ 1&0\\ 1\end{smallmatrix}\right\rangle^{6}_{1}=b_3\left|\begin{smallmatrix}1&1&0&\bar{1}\\ 1&0&0\\ 1&0\\ 1\end{smallmatrix}\right\rangle^{7}_{\bar{1}}+b_4\left|\begin{smallmatrix}1&1&0&\bar{1}\\ 1&1&\bar{1}\\ 1&0\\ 1\end{smallmatrix}\right\rangle^{7}_{\bar{1}}+b_5\left|\begin{smallmatrix}1&0&0&0\\ 1&0&0\\ 1&0\\ 1\end{smallmatrix}\right\rangle^{7}_{\bar{1}},$$

$$F_4\left|\begin{smallmatrix}1&1&1&0\\ 1&1&0\\ 1&0\\ 1\end{smallmatrix}\right\rangle^{6}_{1}=b_6\left|\begin{smallmatrix}1&1&0&\bar{1}\\ 1&0&0\\ 1&0\\ 1\end{smallmatrix}\right\rangle^{7}_{\bar{1}}+b_7\left|\begin{smallmatrix}1&1&0&\bar{1}\\ 1&1&\bar{1}\\ 1&0\\ 1\end{smallmatrix}\right\rangle^{7}_{\bar{1}}+b_8\left|\begin{smallmatrix}1&0&0&0\\ 1&0&0\\ 1&0\\ 1\end{smallmatrix}\right\rangle^{7}_{\bar{1}},$$

and choosing the phase of the state $\left|\begin{smallmatrix}1&0&0&0\\ 1&0&0\\ 1&0\\ 1\end{smallmatrix}\right\rangle^{7}_{\bar{1}}$ such that b_2 is real and positive, we have

$$E_3F_4\left|\begin{smallmatrix}2&1&0&0\\ 2&0&0\\ 1&0\\ 1\end{smallmatrix}\right\rangle^{6}_{1}=\sqrt{\tfrac{3}{2}}\,b_1\left|\begin{smallmatrix}1&1&0&\bar{1}\\ 1&1&0\\ 1&0\\ 1\end{smallmatrix}\right\rangle^{6}_{\bar{1}}$$
$$=F_4E_3\left|\begin{smallmatrix}2&1&0&0\\ 2&0&0\\ 1&0\\ 1\end{smallmatrix}\right\rangle^{6}_{1}=\sqrt{\tfrac{1}{2}}\,F_4\left|\begin{smallmatrix}2&1&0&0\\ 2&1&0\\ 1&0\\ 1\end{smallmatrix}\right\rangle^{5}_{1}=\sqrt{\tfrac{1}{2}}\left|\begin{smallmatrix}1&1&0&\bar{1}\\ 1&1&0\\ 1&0\\ 1\end{smallmatrix}\right\rangle^{6}_{\bar{1}},$$

$$E_3F_4\left|\begin{smallmatrix}2&1&0&0\\ 1&1&0\\ 1&0\\ 1\end{smallmatrix}\right\rangle^{6}_{1}=\sqrt{\tfrac{3}{2}}\,b_3\left|\begin{smallmatrix}1&1&0&\bar{1}\\ 1&1&0\\ 1&0\\ 1\end{smallmatrix}\right\rangle^{6}_{\bar{1}}+\sqrt{\tfrac{1}{2}}\,b_4\left|\begin{smallmatrix}1&1&0&\bar{1}\\ 1&1&0\\ 1&0\\ 1\end{smallmatrix}\right\rangle^{6}_{\bar{1}}$$
$$=F_4E_3\left|\begin{smallmatrix}2&1&0&0\\ 1&1&0\\ 1&0\\ 1\end{smallmatrix}\right\rangle^{6}_{1}=\sqrt{\tfrac{3}{2}}\,F_4\left|\begin{smallmatrix}2&1&0&0\\ 2&1&0\\ 1&0\\ 1\end{smallmatrix}\right\rangle^{5}_{1}=\sqrt{\tfrac{3}{2}}\left|\begin{smallmatrix}1&1&0&\bar{1}\\ 1&1&0\\ 1&0\\ 1\end{smallmatrix}\right\rangle^{6}_{\bar{1}},$$

$$E_2F_4\left|\begin{smallmatrix}2&1&0&0\\ 1&1&0\\ 1&0\\ 1\end{smallmatrix}\right\rangle^{6}_{1}=\sqrt{2}\,b_4\left|\begin{smallmatrix}1&1&0&\bar{1}\\ 1&1&\bar{1}\\ 1&1\\ 1\end{smallmatrix}\right\rangle^{6}_{\bar{2}}$$
$$=F_4E_2\left|\begin{smallmatrix}2&1&0&0\\ 1&1&0\\ 1&0\\ 1\end{smallmatrix}\right\rangle^{6}_{1}=F_4\left|\begin{smallmatrix}2&1&0&0\\ 1&1&0\\ 1&1\\ 1\end{smallmatrix}\right\rangle^{5}_{0}=\sqrt{\tfrac{2}{3}}\left|\begin{smallmatrix}1&1&0&\bar{1}\\ 1&1&\bar{1}\\ 1&1\\ 1\end{smallmatrix}\right\rangle^{6}_{\bar{2}},$$

$$E_3F_4\left|\begin{smallmatrix}1&1&1&0\\ 1&1&0\\ 1&0\\ 1\end{smallmatrix}\right\rangle^{6}_{1}=\sqrt{\tfrac{3}{2}}\,b_6\left|\begin{smallmatrix}1&1&0&\bar{1}\\ 1&1&0\\ 1&0\\ 1\end{smallmatrix}\right\rangle^{6}_{\bar{1}}+\sqrt{\tfrac{1}{2}}\,b_7\left|\begin{smallmatrix}1&1&0&\bar{1}\\ 1&1&0\\ 1&0\\ 1\end{smallmatrix}\right\rangle^{6}_{\bar{1}}$$

$$
\begin{aligned}
&= F_4E_3 \left|\begin{smallmatrix} 1&&1&&1&&0\\ &1&&1&&0&\\ &&1&&0&&\\ &&&1&&& \end{smallmatrix}\right\rangle^{6}_{1} = 0, \\
E_2F_4 &\left|\begin{smallmatrix} 1&&1&&1&&0\\ &1&&1&&0&\\ &&1&&0&&\\ &&&1&&& \end{smallmatrix}\right\rangle^{6}_{1} = \sqrt{2}\, b_7 \left|\begin{smallmatrix} 1&&1&&0&&\overline{1}\\ &1&&1&&\overline{1}&\\ &&1&&1&&\\ &&&1&&& \end{smallmatrix}\right\rangle^{6}_{\overline{2}} \\
&= F_4E_2 \left|\begin{smallmatrix} 1&&1&&1&&0\\ &1&&1&&0&\\ &&1&&0&&\\ &&&1&&& \end{smallmatrix}\right\rangle^{6}_{1} = F_4 \left|\begin{smallmatrix} 1&&1&&1&&0\\ &1&&1&&0&\\ &&1&&1&&\\ &&&1&&& \end{smallmatrix}\right\rangle^{5}_{0} = \sqrt{\tfrac{4}{3}} \left|\begin{smallmatrix} 1&&1&&0&&\overline{1}\\ &1&&1&&\overline{1}&\\ &&1&&1&&\\ &&&1&&& \end{smallmatrix}\right\rangle^{6}_{\overline{2}},
\end{aligned}
$$

Thus, $b_1 = b_4 = \sqrt{1/3}$, $b_3 = \sqrt{2/3}(\sqrt{3/2} - \sqrt{1/2}b_4) = 2/3$, $b_7 = \sqrt{2/3}$, and $b_6 = -\sqrt{2/3} \times \sqrt{1/2}b_7 = -\sqrt{2}/3$. Furthermore,

$$
\begin{aligned}
E_4F_4 &\left|\begin{smallmatrix} 2&&1&&0&&0\\ &2&&0&&0&\\ &&1&&0&&\\ &&&1&&& \end{smallmatrix}\right\rangle^{6}_{1} = \left(b_1^2 + b_2^2\right) \left|\begin{smallmatrix} 2&&1&&0&&0\\ &2&&0&&0&\\ &&1&&0&&\\ &&&1&&& \end{smallmatrix}\right\rangle^{6}_{1} + (b_1b_3 + b_2b_5) \left|\begin{smallmatrix} 2&&1&&0&&0\\ &1&&1&&0&\\ &&1&&0&&\\ &&&1&&& \end{smallmatrix}\right\rangle^{6}_{1} \\
&= (F_4E_4 + H_4) \left|\begin{smallmatrix} 2&&1&&0&&0\\ &2&&0&&0&\\ &&1&&0&&\\ &&&1&&& \end{smallmatrix}\right\rangle^{6}_{1} = \left|\begin{smallmatrix} 2&&1&&0&&0\\ &2&&0&&0&\\ &&1&&0&&\\ &&&1&&& \end{smallmatrix}\right\rangle^{6}_{1}, \\
E_4F_4 &\left|\begin{smallmatrix} 2&&1&&0&&0\\ &1&&1&&0&\\ &&1&&0&&\\ &&&1&&& \end{smallmatrix}\right\rangle^{6}_{1} = (b_3b_1 + b_5b_2) \left|\begin{smallmatrix} 2&&1&&0&&0\\ &2&&0&&0&\\ &&1&&0&&\\ &&&1&&& \end{smallmatrix}\right\rangle^{6}_{1} + \left(b_3^2 + b_4^2 + b_5^2\right) \left|\begin{smallmatrix} 2&&1&&0&&0\\ &1&&1&&0&\\ &&1&&0&&\\ &&&1&&& \end{smallmatrix}\right\rangle^{6}_{1} \\
&\quad + (b_3b_6 + b_4b_7 + b_5b_8) \left|\begin{smallmatrix} 1&&1&&1&&0\\ &1&&1&&0&\\ &&1&&0&&\\ &&&1&&& \end{smallmatrix}\right\rangle^{6}_{1} \\
&= (F_4E_4 + H_4) \left|\begin{smallmatrix} 2&&1&&0&&0\\ &1&&1&&0&\\ &&1&&0&&\\ &&&1&&& \end{smallmatrix}\right\rangle^{6}_{1} = \left|\begin{smallmatrix} 2&&1&&0&&0\\ &1&&1&&0&\\ &&1&&0&&\\ &&&1&&& \end{smallmatrix}\right\rangle^{6}_{1}, \\
E_4F_4 &\left|\begin{smallmatrix} 1&&1&&1&&0\\ &1&&1&&0&\\ &&1&&0&&\\ &&&1&&& \end{smallmatrix}\right\rangle^{6}_{1} = (b_6b_1 + b_8b_2) \left|\begin{smallmatrix} 2&&1&&0&&0\\ &2&&0&&0&\\ &&1&&0&&\\ &&&1&&& \end{smallmatrix}\right\rangle^{6}_{1} \\
&\quad + (b_6b_3 + b_7b_4 + b_8b_5) \left|\begin{smallmatrix} 2&&1&&0&&0\\ &1&&1&&0&\\ &&1&&0&&\\ &&&1&&& \end{smallmatrix}\right\rangle^{6}_{1} + \left(b_6^2 + b_7^2 + b_8^2\right) \left|\begin{smallmatrix} 1&&1&&1&&0\\ &1&&1&&0&\\ &&1&&0&&\\ &&&1&&& \end{smallmatrix}\right\rangle^{6}_{1} \\
&= (F_4E_4 + H_4) \left|\begin{smallmatrix} 1&&1&&1&&0\\ &1&&1&&0&\\ &&1&&0&&\\ &&&1&&& \end{smallmatrix}\right\rangle^{6}_{1} = \left|\begin{smallmatrix} 1&&1&&1&&0\\ &1&&1&&0&\\ &&1&&0&&\\ &&&1&&& \end{smallmatrix}\right\rangle^{6}_{1}.
\end{aligned}
$$

Thus, $b_2 = \sqrt{1 - b_1^2} = \sqrt{2/3}$, $b_5 = -b_1b_3/b_2 = -\sqrt{2}/3$, $b_8 = -b_6b_1/b_2 = 1/3$, and

$$
\begin{aligned}
|(1,0,0,\overline{1})_3\rangle &= \left|\begin{smallmatrix} 1&&0&&0&&0\\ &1&&0&&0&\\ &&1&&0&&\\ &&&1&&& \end{smallmatrix}\right\rangle^{7}_{\overline{1}} = \sqrt{\tfrac{3}{2}} \left\{ F_4 \,|(1,\overline{1},0,1)_1\rangle - \sqrt{\tfrac{1}{3}}\, |(1,0,0,\overline{1})_2\rangle \right\} \\
&= \sqrt{\tfrac{3}{2}} F_4 \left|\begin{smallmatrix} 2&&1&&0&&0\\ &2&&0&&0&\\ &&1&&0&&\\ &&&1&&& \end{smallmatrix}\right\rangle^{6}_{1} - \sqrt{\tfrac{1}{2}} \left|\begin{smallmatrix} 1&&1&&0&&\overline{1}\\ &1&&0&&0&\\ &&1&&0&&\\ &&&1&&& \end{smallmatrix}\right\rangle^{7}_{\overline{1}}
\end{aligned}
$$

$$
\begin{aligned}
&= \sqrt{\tfrac{3}{2}} \times \sqrt{\tfrac{1}{2}} \left\{ \boxed{1}_{2222} + \boxed{5}_{1221} \right\} \\
&\quad - \sqrt{\tfrac{1}{2}} \times \sqrt{\tfrac{1}{6}} \left\{ 2\,\boxed{5}_{1221} + \boxed{6}_{1212} - \boxed{7}_{1122} \right\} \\
&= \sqrt{\tfrac{1}{12}} \left\{ 3\,\boxed{1}_{2222} + \boxed{5}_{1221} - \boxed{6}_{1212} + \boxed{7}_{1122} \right\}.
\end{aligned}
$$

The state $\left|\begin{smallmatrix}1\;0\;0\;0\\1\;0\;0\\1\;0\\1\end{smallmatrix}\right\rangle^{7}_{\overline{1}}$ is the highest weight state of an A_3-quadruplet with $\boldsymbol{M}^{(3)} = (1,0,0)$, from which the A_3-quadruplet is constructed.

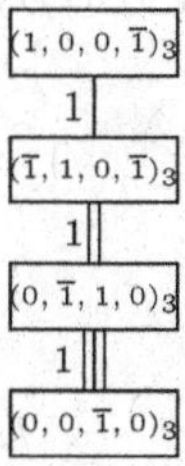

$$
\begin{aligned}
\left|\begin{smallmatrix}1\;0\;0\;0\\1\;0\;0\\1\;0\\1\end{smallmatrix}\right\rangle^{7}_{\overline{1}} &= \sqrt{\tfrac{1}{12}} \left\{ 3\,\boxed{1}_{2222} + \boxed{5}_{1221} - \boxed{6}_{1212} + \boxed{7}_{1122} \right\} \\
\left|\begin{smallmatrix}1\;0\;0\;0\\1\;0\;0\\1\;0\\0\end{smallmatrix}\right\rangle^{8}_{\overline{1}} &= \sqrt{\tfrac{1}{12}} \left\{ 3\,\boxed{2}_{2222} + \boxed{5}_{2121} - \boxed{6}_{2112} + \boxed{8}_{1122} \right\} \\
\left|\begin{smallmatrix}1\;0\;0\;0\\1\;0\;0\\0\;0\\0\end{smallmatrix}\right\rangle^{9}_{0} &= \sqrt{\tfrac{1}{12}} \left\{ 3\,\boxed{3}_{2222} + \boxed{5}_{2211} - \boxed{7}_{2112} + \boxed{8}_{1212} \right\} \\
\left|\begin{smallmatrix}1\;0\;0\;0\\0\;0\;0\\0\;0\\0\end{smallmatrix}\right\rangle^{10}_{0} &= \sqrt{\tfrac{1}{12}} \left\{ 3\,\boxed{4}_{2222} + \boxed{6}_{2211} - \boxed{7}_{2121} + \boxed{8}_{1221} \right\}
\end{aligned}
$$

The three $\mathcal{A}_4$-doublets are shown by the block weight diagram:

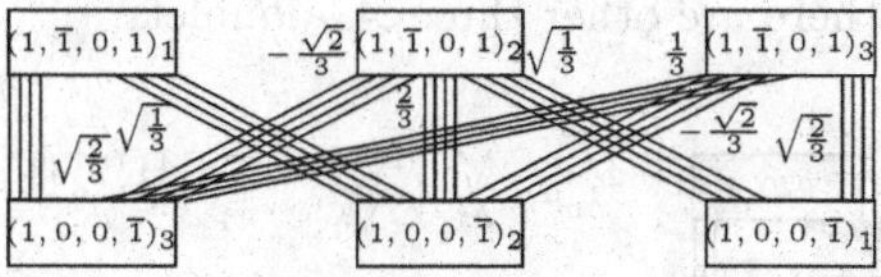

$$
\left|\begin{smallmatrix}2\;1\;0\;0\\2\;0\;0\\1\;0\\1\end{smallmatrix}\right\rangle^{6}_{1},\ \left|\begin{smallmatrix}2\;1\;0\;0\\1\;1\;0\\1\;0\\1\end{smallmatrix}\right\rangle^{6}_{1},\ \left|\begin{smallmatrix}1\;1\;1\;0\\1\;1\;0\\1\;0\\1\end{smallmatrix}\right\rangle^{6}_{1}
$$

$$
\left|\begin{smallmatrix}1\;0\;0\;0\\1\;0\;0\\1\;0\\1\end{smallmatrix}\right\rangle^{7}_{\overline{1}},\ \left|\begin{smallmatrix}1\;1\;0\;\overline{1}\\1\;0\;0\\1\;0\\1\end{smallmatrix}\right\rangle^{7}_{\overline{1}},\ \left|\begin{smallmatrix}1\;1\;0\;\overline{1}\\1\;1\;\overline{1}\\1\;0\\1\end{smallmatrix}\right\rangle^{7}_{\overline{1}}
$$

$$
\begin{aligned}
\left|\begin{smallmatrix}2\;1\;0\;0\\2\;0\;0\\1\;0\\1\end{smallmatrix}\right\rangle^{6}_{1} &= \sqrt{\tfrac{1}{2}} \left\{ \boxed{1}_{2211} + \boxed{3}_{1221} \right\} \\
\left|\begin{smallmatrix}2\;1\;0\;0\\1\;1\;0\\1\;0\\1\end{smallmatrix}\right\rangle^{6}_{1} &= \sqrt{\tfrac{1}{6}} \left\{ -\boxed{1}_{2211} + \boxed{3}_{1221} + 2\,\boxed{4}_{1212} \right\} \\
\left|\begin{smallmatrix}1\;1\;1\;0\\1\;1\;0\\1\;0\\1\end{smallmatrix}\right\rangle^{6}_{1} &= \sqrt{\tfrac{1}{12}} \left\{ \boxed{1}_{2211} - \boxed{3}_{1221} + \boxed{4}_{1212} + 3\,\boxed{7}_{1111} \right\} \\
\left|\begin{smallmatrix}1\;0\;0\;0\\1\;0\;0\\1\;0\\1\end{smallmatrix}\right\rangle^{7}_{\overline{1}} &= \sqrt{\tfrac{1}{12}} \left\{ 3\,\boxed{1}_{2222} + \boxed{5}_{1221} - \boxed{6}_{1212} + \boxed{7}_{1122} \right\} \\
\left|\begin{smallmatrix}1\;1\;0\;\overline{1}\\1\;0\;0\\1\;0\\1\end{smallmatrix}\right\rangle^{7}_{\overline{1}} &= \sqrt{\tfrac{1}{6}} \left\{ 2\,\boxed{5}_{1221} + \boxed{6}_{1212} - \boxed{7}_{1122} \right\} \\
\left|\begin{smallmatrix}1\;1\;0\;\overline{1}\\1\;1\;\overline{1}\\1\;0\\1\end{smallmatrix}\right\rangle^{7}_{\overline{1}} &= \sqrt{\tfrac{1}{2}} \left\{ \boxed{6}_{1212} + \boxed{7}_{1122} \right\}
\end{aligned}
$$

In addition to the formula for calculating $|(1,0,0,\overline{1})_3\rangle$, other two formulas with the parameters b_j can be checked by the basis spin-tensors.

$$
F_4 \left|\begin{smallmatrix}2\;1\;0\;0\\1\;1\;0\\1\;0\\1\end{smallmatrix}\right\rangle^{6}_{1} = \sqrt{\tfrac{1}{6}} \left\{ -\boxed{1}_{2222} + \boxed{5}_{1221} + 2\,\boxed{6}_{1212} \right\}
$$

$$
\begin{aligned}
&= \frac{2}{3}\left|\begin{array}{l} 1\ 1\ 0\ \bar{1} \\ \ 1\ 0\ 0 \\ \ \ 1\ 0 \\ \ \ \ 1 \end{array}\right\rangle_{\bar{1}}^{7} + \sqrt{\frac{1}{3}}\left|\begin{array}{l} 1\ 1\ 0\ \bar{1} \\ \ 1\ 1\ \bar{1} \\ \ \ 1\ 0 \\ \ \ \ 1 \end{array}\right\rangle_{\bar{1}}^{7} - \frac{\sqrt{2}}{3}\left|\begin{array}{l} 1\ 0\ 0\ 0 \\ \ 1\ 0\ 0 \\ \ \ 1\ 0 \\ \ \ \ 1 \end{array}\right\rangle_{\bar{1}}^{7}, \\
&= \frac{2}{3} \times \sqrt{\frac{1}{6}}\left\{2\,\boxed{5}_{1221} + \boxed{6}_{1212} - \boxed{7}_{1122}\right\} \\
&\quad + \sqrt{\frac{1}{3}} \times \sqrt{\frac{1}{2}}\left\{\boxed{6}_{1212} + \boxed{7}_{1122}\right\} \\
&\quad - \frac{\sqrt{2}}{3} \times \sqrt{\frac{1}{12}}\left\{3\,\boxed{1}_{2222} + \boxed{5}_{1221} - \boxed{6}_{1212} + \boxed{7}_{1122}\right\},
\end{aligned}
$$

$$
\begin{aligned}
F_4\left|\begin{array}{l} 1\ 1\ 1\ 0 \\ \ 1\ 1\ 0 \\ \ \ 1\ 0 \\ \ \ \ 1 \end{array}\right\rangle_{1}^{6} &= \sqrt{\frac{1}{12}}\left\{\boxed{1}_{2222} - \boxed{5}_{1221} + \boxed{6}_{1212} + 3\,\boxed{7}_{1122}\right\} \\
&= -\frac{\sqrt{2}}{3}\left|\begin{array}{l} 1\ 1\ 0\ \bar{1} \\ \ 1\ 0\ 0 \\ \ \ 1\ 0 \\ \ \ \ 1 \end{array}\right\rangle_{\bar{1}}^{7} + \sqrt{\frac{2}{3}}\left|\begin{array}{l} 1\ 1\ 0\ \bar{1} \\ \ 1\ 1\ \bar{1} \\ \ \ 1\ 0 \\ \ \ \ 1 \end{array}\right\rangle_{\bar{1}}^{7} + \frac{1}{3}\left|\begin{array}{l} 1\ 0\ 0\ 0 \\ \ 1\ 0\ 0 \\ \ \ 1\ 0 \\ \ \ \ 1 \end{array}\right\rangle_{\bar{1}}^{7}, \\
&= -\frac{\sqrt{2}}{3} \times \sqrt{\frac{1}{6}}\left\{2\,\boxed{5}_{1221} + \boxed{6}_{1212} - \boxed{7}_{1122}\right\} \\
&\quad + \sqrt{\frac{2}{3}} \times \sqrt{\frac{1}{2}}\left\{\boxed{6}_{1212} + \boxed{7}_{1122}\right\} \\
&\quad + \frac{1}{3} \times \sqrt{\frac{1}{12}}\left\{3\,\boxed{1}_{2222} + \boxed{5}_{1221} - \boxed{6}_{1212} + \boxed{7}_{1122}\right\}.
\end{aligned}
$$

Due to the parallel principle, there are other three $\mathcal{A}_4$-doublets.

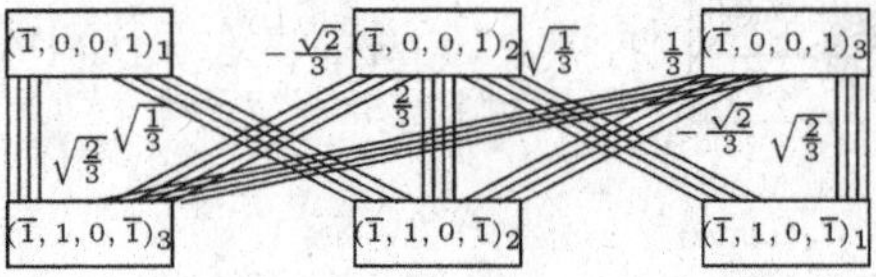

$$
\left|\begin{array}{l} 2\ 1\ 0\ 0 \\ \ 2\ 0\ 0 \\ \ \ 1\ 0 \\ \ \ \ 0 \end{array}\right\rangle_{1}^{7}, \left|\begin{array}{l} 2\ 1\ 0\ 0 \\ \ 1\ 1\ 0 \\ \ \ 1\ 0 \\ \ \ \ 0 \end{array}\right\rangle_{1}^{7}, \left|\begin{array}{l} 1\ 1\ 1\ 0 \\ \ 1\ 1\ 0 \\ \ \ 1\ 0 \\ \ \ \ 0 \end{array}\right\rangle_{1}^{7}
$$

$$
\left|\begin{array}{l} 1\ 0\ 0\ 0 \\ \ 1\ 0\ 0 \\ \ \ 1\ 0 \\ \ \ \ 0 \end{array}\right\rangle_{\bar{1}}^{8}, \left|\begin{array}{l} 1\ 1\ 0\ \bar{1} \\ \ 1\ 0\ 0 \\ \ \ 1\ 0 \\ \ \ \ 0 \end{array}\right\rangle_{\bar{1}}^{8}, \left|\begin{array}{l} 1\ 1\ 0\ \bar{1} \\ \ 1\ 1\ \bar{1} \\ \ \ 1\ 0 \\ \ \ \ 0 \end{array}\right\rangle_{\bar{1}}^{8}
$$

$$
\begin{aligned}
\left|\begin{array}{l} 2\ 1\ 0\ 0 \\ \ 2\ 0\ 0 \\ \ \ 1\ 0 \\ \ \ \ 0 \end{array}\right\rangle_{1}^{7} &= \sqrt{\frac{1}{2}}\left\{\boxed{2}_{2211} + \boxed{3}_{2121}\right\} \\
\left|\begin{array}{l} 2\ 1\ 0\ 0 \\ \ 1\ 1\ 0 \\ \ \ 1\ 0 \\ \ \ \ 0 \end{array}\right\rangle_{1}^{7} &= \sqrt{\frac{1}{6}}\left\{-\boxed{2}_{2211} + \boxed{3}_{2121} + 2\,\boxed{4}_{2112}\right\} \\
\left|\begin{array}{l} 1\ 1\ 1\ 0 \\ \ 1\ 1\ 0 \\ \ \ 1\ 0 \\ \ \ \ 0 \end{array}\right\rangle_{1}^{7} &= \sqrt{\frac{1}{12}}\left\{\boxed{2}_{2211} - \boxed{3}_{2121} + \boxed{4}_{2112} + 3\,\boxed{8}_{1111}\right\} \\
\left|\begin{array}{l} 1\ 0\ 0\ 0 \\ \ 1\ 0\ 0 \\ \ \ 1\ 0 \\ \ \ \ 0 \end{array}\right\rangle_{\bar{1}}^{8} &= \sqrt{\frac{1}{12}}\left\{3\,\boxed{2}_{2222} + \boxed{5}_{2121} - \boxed{6}_{2112} + \boxed{8}_{1122}\right\} \\
\left|\begin{array}{l} 1\ 1\ 0\ \bar{1} \\ \ 1\ 0\ 0 \\ \ \ 1\ 0 \\ \ \ \ 0 \end{array}\right\rangle_{\bar{1}}^{8} &= \sqrt{\frac{1}{6}}\left\{2\,\boxed{5}_{2121} + \boxed{6}_{2112} - \boxed{8}_{1122}\right\} \\
\left|\begin{array}{l} 1\ 1\ 0\ \bar{1} \\ \ 1\ 1\ \bar{1} \\ \ \ 1\ 0 \\ \ \ \ 0 \end{array}\right\rangle_{\bar{1}}^{8} &= \sqrt{\frac{1}{2}}\left\{\boxed{6}_{2112} + \boxed{8}_{1122}\right\}
\end{aligned}
$$

Two states $\left|\begin{array}{l} 2\ 1\ 0\ 0 \\ \ 2\ 0\ 0 \\ \ \ 0\ 0 \\ \ \ \ 0 \end{array}\right\rangle_{2}^{8}$ and $\left|\begin{array}{l} 2\ 1\ 0\ 0 \\ \ 1\ 0\ 0 \\ \ \ 0\ 0 \\ \ \ \ 0 \end{array}\right\rangle_{2}^{9}$ in the A_3-20let with $M^{(3)} = (1,1,0)$ satisfy the condition (8.7) with $\mu = 4$, and they are the highest weight states of the $\mathcal{A}_4$-triplets. In terms of Eqs. (8.25-28), letting

$$F_4\left|\begin{matrix}2\ 1\ 0\ 0\\ 2\ 0\ 0\\ 0\ 0\\ 0\end{matrix}\right\rangle_2^8 = c_1\left|\begin{matrix}1\ 1\ 0\ \overline{1}\\ 1\ 0\ 0\\ 0\ 0\\ 0\end{matrix}\right\rangle_0^9 + c_2\left|\begin{matrix}1\ 0\ 0\ 0\\ 1\ 0\ 0\\ 0\ 0\\ 0\end{matrix}\right\rangle_0^9,$$

we have

$$\begin{aligned}E_2F_4\left|\begin{matrix}2\ 1\ 0\ 0\\ 2\ 0\ 0\\ 0\ 0\\ 0\end{matrix}\right\rangle_2^8 &= c_1\left|\begin{matrix}1\ 1\ 0\ \overline{1}\\ 1\ 0\ 0\\ 1\ 0\\ 0\end{matrix}\right\rangle_{\overline{1}}^8 + c_2\left|\begin{matrix}1\ 0\ 0\ 0\\ 1\ 0\ 0\\ 1\ 0\\ 0\end{matrix}\right\rangle_{\overline{1}}^8\\ &= F_4E_2\left|\begin{matrix}2\ 1\ 0\ 0\\ 2\ 0\ 0\\ 0\ 0\\ 0\end{matrix}\right\rangle_2^8 = \sqrt{2}F_4\left|\begin{matrix}2\ 1\ 0\ 0\\ 2\ 0\ 0\\ 1\ 0\\ 0\end{matrix}\right\rangle_1^7\\ &= \sqrt{\tfrac{2}{3}}\left|\begin{matrix}1\ 1\ 0\ \overline{1}\\ 1\ 0\ 0\\ 1\ 0\\ 0\end{matrix}\right\rangle_{\overline{1}}^8 + \sqrt{\tfrac{4}{3}}\left|\begin{matrix}1\ 0\ 0\ 0\\ 1\ 0\ 0\\ 1\ 0\\ 0\end{matrix}\right\rangle_{\overline{1}}^8.\end{aligned}$$

Thus, $c_1 = \sqrt{2/3}$ and $c_2 = \sqrt{4/3}$.

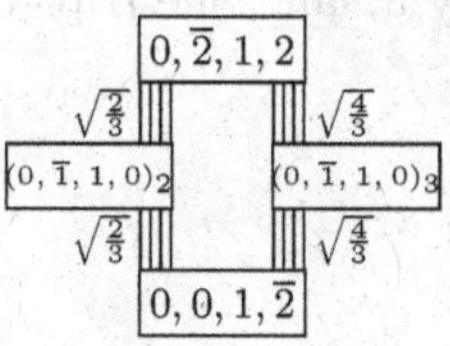

$$\left|\begin{matrix}2\ 1\ 0\ 0\\ 2\ 0\ 0\\ 0\ 0\\ 0\end{matrix}\right\rangle_2^8 = \boxed{3}_{2211}$$

$$\left|\begin{matrix}1\ 1\ 0\ \overline{1}\\ 1\ 0\ 0\\ 0\ 0\\ 0\end{matrix}\right\rangle_0^9 = \sqrt{\tfrac{1}{6}}\left\{2\,\boxed{5}_{2211} + \boxed{7}_{2112} - \boxed{8}_{1212}\right\},\quad \left|\begin{matrix}1\ 0\ 0\ 0\\ 1\ 0\ 0\\ 0\ 0\\ 0\end{matrix}\right\rangle_0^9$$

$$\left|\begin{matrix}0\ 0\ 0\ \overline{1}\\ 0\ 0\ 0\\ 0\ 0\\ 0\end{matrix}\right\rangle_{\overline{2}}^{10} = \boxed{5}_{2222}$$

$$\left|\begin{matrix}1\ 0\ 0\ 0\\ 1\ 0\ 0\\ 0\ 0\\ 0\end{matrix}\right\rangle_0^9 = \sqrt{\tfrac{1}{12}}\left\{3\,\boxed{3}_{2222} + \boxed{5}_{2211} - \boxed{7}_{2112} + \boxed{8}_{1212}\right\}$$

Due to the parallel principle, there is another $\mathcal{A}_4$-triplet.

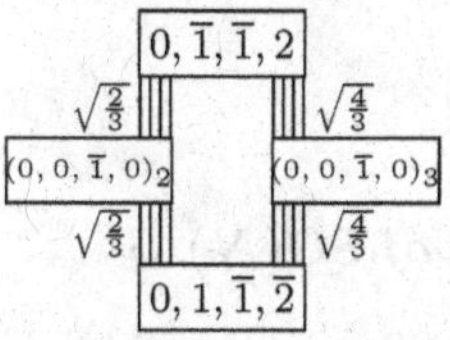

$$\left|\begin{matrix}2\ 1\ 0\ 0\\ 1\ 0\ 0\\ 0\ 0\\ 0\end{matrix}\right\rangle_2^9 = \boxed{4}_{2211}$$

$$\left|\begin{matrix}1\ 1\ 0\ \overline{1}\\ 1\ 0\ \overline{1}\\ 0\ 0\\ 0\end{matrix}\right\rangle_0^{10} = \sqrt{\tfrac{1}{6}}\left\{2\,\boxed{6}_{2211} + \boxed{7}_{2121} - \boxed{8}_{1221}\right\},\quad \left|\begin{matrix}1\ 0\ 0\ 0\\ 0\ 0\ 0\\ 0\ 0\\ 0\end{matrix}\right\rangle_0^{10}$$

$$\left|\begin{matrix}0\ 0\ 0\ \overline{1}\\ 0\ 0\ \overline{1}\\ 0\ 0\\ 0\end{matrix}\right\rangle_{\overline{2}}^{11} = \boxed{6}_{2222}$$

$$\left|\begin{matrix}1\ 0\ 0\ 0\\ 0\ 0\ 0\\ 0\ 0\\ 0\end{matrix}\right\rangle_0^{10} = \sqrt{\tfrac{1}{12}}\left\{3\,\boxed{4}_{2222} + \boxed{6}_{2211} - \boxed{7}_{2121} + \boxed{8}_{1221}\right\}$$

The state $\left|\begin{matrix}0\ 0\ 0\ \overline{1}\\ 0\ 0\ 0\\ 0\ 0\\ 0\end{matrix}\right\rangle_{\overline{2}}^{10}$ is the highest weight state of an A_3-quadruplet with $\boldsymbol{M}^{(3)} = (0,0,1)$, from which the A_3-quadruplet is constructed.

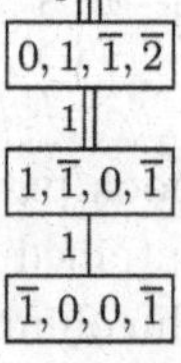

$$\left|\begin{matrix}0\ 0\ 0\ \overline{1}\\ 0\ 0\ 0\\ 0\ 0\\ 0\end{matrix}\right\rangle_{\overline{2}}^{10} = \boxed{5}_{2222}$$

$$\left|\begin{matrix}0\ 0\ 0\ \overline{1}\\ 0\ 0\ \overline{1}\\ 0\ 0\\ 0\end{matrix}\right\rangle_{\overline{2}}^{11} = \boxed{6}_{2222}$$

$$\left|\begin{matrix}0\ 0\ 0\ \overline{1}\\ 0\ 0\ \overline{1}\\ 0\ \overline{1}\\ 0\end{matrix}\right\rangle_{\overline{1}}^{12} = \boxed{7}_{2222}$$

$$\left|\begin{matrix}0\ 0\ 0\ \overline{1}\\ 0\ 0\ \overline{1}\\ 0\ \overline{1}\\ \overline{1}\end{matrix}\right\rangle_{\overline{1}}^{13} = \boxed{8}_{2222}$$

The state $\left|\begin{matrix} 0 & 0 & 0 & \bar{1} \\ 0 & 0 & \bar{1} \\ 0 & \bar{1} \\ \bar{1} \end{matrix}\right\rangle^{13}_{\bar{1}}$ is the lowest weight state in this representation. Two states $\left|\begin{matrix} 1 & 1 & 0 & \bar{1} \\ 1 & 0 & \bar{1} \\ 0 & \bar{1} \\ 0 \end{matrix}\right\rangle^{11}_{1}$ and $\left|\begin{matrix} 1 & 1 & 0 & \bar{1} \\ 1 & 0 & \bar{1} \\ 0 & \bar{1} \\ \bar{1} \end{matrix}\right\rangle^{12}_{1}$ in the A_3-20let with $\boldsymbol{M}^{(3)} = (0,1,1)$ satisfy the condition (8.7) with $\mu = 4$ and two $\mathcal{A}_4$-doublets are constructed from them (the parallel principle).

$$\begin{array}{c} \boxed{1,\bar{2},0,1} \\ 1 \Big\| \\ \boxed{1,\bar{1},0,\bar{1}} \end{array} \quad \begin{array}{l} \left|\begin{matrix} 1 & 1 & 0 & \bar{1} \\ 1 & 0 & \bar{1} \\ 0 & \bar{1} \\ 0 \end{matrix}\right\rangle^{11}_{1} = \boxed{\ 7\ }_{2211} \\[2ex] \left|\begin{matrix} 0 & 0 & 0 & \bar{1} \\ 0 & 0 & \bar{1} \\ 0 & \bar{1} \\ 0 \end{matrix}\right\rangle^{12}_{\bar{1}} = \boxed{\ 7\ }_{2222} \end{array} \qquad \begin{array}{c} \boxed{\bar{1},\bar{1},0,1} \\ 1 \Big\| \\ \boxed{\bar{1},0,0,\bar{1}} \end{array} \quad \begin{array}{l} \left|\begin{matrix} 1 & 1 & 0 & \bar{1} \\ 1 & 0 & \bar{1} \\ 0 & \bar{1} \\ \bar{1} \end{matrix}\right\rangle^{12}_{1} = \boxed{\ 8\ }_{2211} \\[2ex] \left|\begin{matrix} 0 & 0 & 0 & \bar{1} \\ 0 & 0 & \bar{1} \\ 0 & \bar{1} \\ \bar{1} \end{matrix}\right\rangle^{13}_{\bar{1}} = \boxed{\ 8\ }_{2222} \end{array}$$

In summary, the representation $\boldsymbol{M} = (1,0,0,1)$ of SO(8) contains four A_3-quadruplets with $\boldsymbol{M}^{(3)} = (1,0,0)$, and $(0,0,1)$ and two A_3-20lets with $\boldsymbol{M}^{(3)} = (1,1,0)$, and $(0,1,1)$, respectively. Its dimension is 56. There are one single dominant weight $(1,0,0,1)$ with multiplicity 32 and one triplet dominant weight $(0,0,1,0)$ with multiplicity 8.

$$\begin{aligned}
&\left|\begin{matrix} 2 & 1 & 1 & 1 \\ 2 & 1 & 1 \\ 2 & 1 \\ 2 \end{matrix}\right\rangle^{1}_{1} = \boxed{\ 1\ }_{1111}\,, \qquad \left|\begin{matrix} 0 & 0 & 0 & \bar{1} \\ 0 & 0 & 0 \\ 0 & 0 \\ 0 \end{matrix}\right\rangle^{10}_{\bar{2}} = \boxed{\ 5\ }_{2222}\,, \\
&\left|\begin{matrix} 1 & 1 & 1 & 0 \\ 1 & 1 & 1 \\ 1 & 1 \\ 1 \end{matrix}\right\rangle^{4}_{0} = \sqrt{\tfrac{1}{12}}\left\{\boxed{\ 1\ }_{2112} - \boxed{\ 2\ }_{1212} + \boxed{\ 3\ }_{1122} + 3\,\boxed{\ 5\ }_{1111}\right\}, \\
&\left|\begin{matrix} 1 & 0 & 0 & 0 \\ 1 & 0 & 0 \\ 1 & 0 \\ 1 \end{matrix}\right\rangle^{7}_{\bar{1}} = \sqrt{\tfrac{1}{12}}\left\{3\,\boxed{\ 1\ }_{2222} + \boxed{\ 5\ }_{1221} - \boxed{\ 6\ }_{1212} + \boxed{\ 7\ }_{1122}\right\}, \\
&\left|\begin{matrix} 2 & 1 & 0 & 0 \\ 2 & 1 & 0 \\ 2 & 1 \\ 2 \end{matrix}\right\rangle^{2}_{\bar{1}} = \boxed{\ 1\ }_{1122}\,, \qquad \left|\begin{matrix} 1 & 1 & 0 & \bar{1} \\ 1 & 1 & 0 \\ 1 & 1 \\ 1 \end{matrix}\right\rangle^{5}_{\bar{2}} = \boxed{\ 5\ }_{1122}\,.
\end{aligned}$$

10.3.5 *Dimensions of Spinor Representations of* SO(N)

The dimension of a spinor representation $[s,\lambda]$ of $SO(2\ell+1)$ or $[\pm s,\lambda]$ of $SO(2\ell)$ can be calculated by the hook rule [Dai (1983)]. In this rule the dimension is expressed as a quotient multiplied with the dimension of the fundamental spinor representation, where the numerator and the denominator of the quotient are denoted by the symbols $Y_S^{[\lambda]}$ and $Y_h^{[\lambda]}$, respectively:

$$d_{[s,\lambda]}(SO(2\ell+1)) = 2^{\ell}\frac{Y_S^{[\lambda]}}{Y_h^{[\lambda]}}, \qquad d_{[\pm s,\lambda]}(SO(2\ell)) = 2^{\ell-1}\frac{Y_S^{[\lambda]}}{Y_h^{[\lambda]}}. \tag{10.87}$$

The concepts of a hook path $(i,\ j)$ and an inverse hook path $\overline{(i,\ j)}$ are discussed in subsection 10.1.4. The number of boxes contained in the hook path $(i,\ j)$ is the hook number h_{ij} of the box in the jth column of the ith row. $Y_h^{[\lambda]}$ is a tableau of the Young pattern $[\lambda]$ where the box in the jth column of the ith row is filled with the hook number h_{ij}. Define a series of tableaux $Y_{S_a}^{[\lambda]}$ recursively by the rule given below. $Y_S^{[\lambda]}$ is a tableau of the Young pattern $[\lambda]$ where each box is filled with the sum of the digits which are respectively filled in the same box of each tableau $Y_{S_a}^{[\lambda]}$ in the series. The symbol $Y_S^{[\lambda]}$ means the product of the filled digits in it, so does the symbol $Y_h^{[\lambda]}$.

The tableaux $Y_{S_a}^{[\lambda]}$ are defined by the following rule:

(a) $Y_{S_0}^{[\lambda]}$ is a tableau of the Young pattern $[\lambda]$ where the box in the jth column of the ith row is filled with the digit $(N-1+j-i)$.

(b) Let $[\lambda^{(1)}]=[\lambda]$. Beginning with $[\lambda^{(1)}]$, we define recursively the Young pattern $[\lambda^{(a)}]$ by removing the first row and the first column of the Young pattern $[\lambda^{(a-1)}]$ until $[\lambda^{(a)}]$ contains less than two rows.

(c) If $[\lambda^{(a)}]$ contains more than one row, define $Y_{S_a}^{[\lambda]}$ to be a tableau of the Young pattern $[\lambda]$ where the boxes in the first $(a-1)$ row and in the first $(a-1)$ column are filled with 0, and the remaining part of the Young pattern is nothing but $[\lambda^{(a)}]$. Let $[\lambda^{(a)}]$ have r rows. Fill the first $(r-1)$ boxes along the hook path (1, 1) of the Young pattern $[\lambda^{(a)}]$, beginning with the box on the rightmost, with the digits $\lambda_2^{(a)}$, $\lambda_3^{(a)}$, $\cdots$, $\lambda_r^{(a)}$, box by box, and fill the first $\lambda_i^{(a)}$ boxes in each inverse hook path $\overline{(i,\ 1)}$ of the Young pattern $[\lambda^{(a)}]$, $2\leqslant i\leqslant r$, with -1. The remaining boxes are filled with 0. If a few -1 are filled in the same box, the digits are summed. The sum of all filled digits in the pattern $Y_{S_a}^{[\lambda]}$ with $a>0$ is 0.

Ex. 1 Dimension of the representation $[+s,3,3,3]$ of SO(8).

$$Y_S^{[3,3,3]}=\begin{array}{|c|c|c|}\hline 7&8&9\\\hline 6&7&8\\\hline 5&6&7\\\hline\end{array}+\begin{array}{|c|c|c|}\hline &3&3\\\hline -1&-1&\\\hline -2&-1&-1\\\hline\end{array}+\begin{array}{|c|c|c|}\hline &&\\\hline &&2\\\hline &-1&-1\\\hline\end{array}$$

$$=\begin{array}{|c|c|c|}\hline 7&11&12\\\hline 5&6&10\\\hline 3&4&5\\\hline\end{array},$$

$$Y_h^{[3,3,3]}=\begin{array}{|c|c|c|}\hline 5&4&3\\\hline 4&3&2\\\hline 3&2&1\\\hline\end{array},$$

$$d_{[+s(3,3,3)]}(\mathrm{SO}(8))=\ 2^3\times 11\times 7\times 5^2=15400.$$

Ex. 2 Dimension of the representations $[\pm s, n]$ of SO(2ℓ) and $[s,n]$ of SO$(2\ell+1)$.

$$Y_S^{[n]} = \begin{array}{|c|c|c|c|}\hline N-1 & N & \cdots & N+n-2 \\ \hline \end{array} = \frac{(N+n-2)!}{(N-2)!},$$

$$Y_h^{[n]} = \begin{array}{|c|c|c|c|}\hline n & n-1 & \cdots & 1 \\ \hline \end{array} = n!,$$

$$\begin{aligned} d_{[\pm s,n]}(\mathrm{SO}(2\ell)) &= 2^{\ell-1}\frac{(2\ell+n-2)!}{n!(2\ell-2)!} = 2^{\ell-1}\binom{2\ell+n-2}{n}, \\ d_{[s,n]}(\mathrm{SO}(2\ell+1)) &= 2^{\ell}\frac{(2\ell+n-1)!}{n!(2\ell-1)!} = 2^{\ell}\binom{2\ell+n-1}{n}. \end{aligned} \tag{10.88}$$

Ex. 3 Dimension of the representations $[\pm s, 1^n]$ of SO(2ℓ) and $[s,n]$ of SO$(2\ell+1)$.

$$Y_S^{[1^n]} = \begin{array}{|c|}\hline N-1 \\ \hline N-2 \\ \hline \vdots \\ \hline N-n+1 \\ \hline N-n \\ \hline \end{array} + \begin{array}{|c|}\hline 1 \\ \hline 1 \\ \hline \vdots \\ \hline 1 \\ \hline -n+1 \\ \hline \end{array} = \begin{array}{|c|}\hline N \\ \hline N-1 \\ \hline \vdots \\ \hline N-n+2 \\ \hline N-2n+1 \\ \hline \end{array},$$

$$\begin{aligned} d_{[\pm s,1^n]}(SO(2\ell)) &= 2^{\ell-1}\,\frac{(2\ell)!(2\ell-2n+1)}{n!(2\ell-n+1)!}, \\ d_{[\pm s,1^n]}(SO(2\ell+1)) &= 2^{\ell}\,\frac{(2\ell+1)!(2\ell-2n+2)}{n!(2\ell-n+2)!}. \end{aligned} \tag{10.89}$$

10.4 Rotational Symmetry in N-Dimensional Space

10.4.1 *Orbital Angular Momentum Operators*

The relations between the rectangular coordinates x_a and the spherical coordinates r and φ_b in an N-dimensional space are

$$\begin{aligned} x_1 &= r\cos\varphi_1 \sin\varphi_2 \cdots \sin\varphi_{N-1}, \\ x_2 &= r\sin\varphi_1 \sin\varphi_2 \cdots \sin\varphi_{N-1}, \\ x_b &= r\cos\varphi_{b-1} \sin\varphi_b \cdots \sin\varphi_{N-1}, \qquad 3 \leqslant b \leqslant N-1, \\ x_N &= r\cos\varphi_{N-1}, \\ \sum_{a=1}^{N} x_a^2 &= r^2. \end{aligned} \tag{10.90}$$

The unit vector along $\boldsymbol{x}$ is usually denoted by $\hat{\boldsymbol{x}} = \boldsymbol{x}/r$. The volume element of the configuration space is

$$\prod_{a=1}^{N} dx_a = r^{N-1} dr d\Omega, \qquad d\Omega = \prod_{a=1}^{N-1} (\sin\varphi_a)^{a-1}\, d\varphi_a,$$
$$0 \leqslant r < \infty, \quad -\pi \leqslant \varphi_1 \leqslant \pi, \quad 0 \leqslant \varphi_b \leqslant \pi, \quad 2 \leqslant b \leqslant N-1. \tag{10.91}$$

The orbital angular momentum operators L_{ab} are the generators of the transformation operators P_R for the scalar function, $R \in$SO(N),

$$L_{ab} = -L_{ba} = -\mathrm{i}x_a \frac{\partial}{\partial x_b} + \mathrm{i}x_b \frac{\partial}{\partial x_a}, \qquad L^2 = \sum_{a<b=2}^{N} L_{ab}^2, \tag{10.92}$$

where the natural units are used, $\hbar = c = 1$. From the second Lie theorem, L_{ab} satisfy the same commutation relations as Eq. (10.11) of the generators T_{ab} in the self-representation of SO(N). L^2 is the Casimir operator of order 2 calculated by Eq. (7.133).

10.4.2 *Spherical Harmonic Functions*

The Chevalley bases $H_\mu(L)$, $E_\mu(L)$, and $F_\mu(L)$ can be obtained from Eqs. (10.13) and (10.19) by replacing T_{ab} with L_{ab}. **The spherical harmonic function $Y_{\boldsymbol{m}}^{[\lambda]}(\hat{\boldsymbol{x}})$ in the N-dimensional real space are defined as the basis functions of the irreducible representation $D^{[\lambda]}$** of SO(N). Since there is only one coordinate vector $\boldsymbol{x}$, $[\lambda]$ has to be the one-row Young pattern $[\lambda] = [\lambda_1, 0, \cdots, 0] = [\lambda_1]$:

$$P_R Y_{\boldsymbol{m}}^{[\lambda_1]}(\hat{\boldsymbol{x}}) = \sum_{\boldsymbol{m}'} Y_{\boldsymbol{m}'}^{[\lambda_1]}(\hat{\boldsymbol{x}}) D_{\boldsymbol{m}'\boldsymbol{m}}^{[\lambda_1]}(R), \qquad R \in \mathrm{SO}(N). \tag{10.93}$$

As usual, by making $D^{[\lambda_1]}(H_\mu(L))$ diagonal, $Y_{\boldsymbol{m}}^{[\lambda_1]}(\hat{\boldsymbol{x}})$ are the common eigenfunctions of $H_\mu(L)$ and L^2:

$$H_\mu(L) Y_{\boldsymbol{m}}^{[\lambda_1]}(\hat{\boldsymbol{x}}) = m_\mu Y_{\boldsymbol{m}}^{[\lambda_1]}(\hat{\boldsymbol{x}}), \tag{10.94}$$

We cannot use the method in subsection 5.4.4 to calculate $Y_{\boldsymbol{m}}^{[\lambda_1]}(\hat{\boldsymbol{x}})$ because we do not have the analytic expressions for $D_{\boldsymbol{m}}^{[\lambda_1]}(R)$. It is a good choice to calculate $Y_{\boldsymbol{m}}^{[\lambda_1]}(\hat{\boldsymbol{x}})$ through the **harmonic polynomials** $\mathcal{Y}_{\boldsymbol{m}}^{[\lambda_1]}(\boldsymbol{x}) = r^{\lambda_1} Y_{\boldsymbol{m}}^{[\lambda_1]}(\hat{\boldsymbol{x}})$ (see subsection 5.4.5). Because

$$(O_R \boldsymbol{x})_d = \sum_{c=1}^{N} R_{dc} x_c = \sum_{c=1}^{N} x_c R_{cd}, \qquad O_R \boldsymbol{\theta}_d = \sum_{c=1}^{N} \boldsymbol{\theta}_c R_{cd},$$

the common eigenfunctions X_α of $H_\mu(L)$ can be obtained by following the combination from $\boldsymbol{\theta}_a$ to $\boldsymbol{\phi}_\alpha$ (see Eq. (10.15) for SO($2\ell+1$) and Eq. (10.20) for SO(2ℓ)). **Note the different sign and the factor $\sqrt{2}$ in the definition of X_α.** We define for the SO($2\ell+1$) group

$$\begin{aligned}
&X_\alpha = \begin{cases} x_{2\alpha-1} + \mathrm{i}x_{2\alpha}, & 1 \leqslant \alpha \leqslant \ell, \\ x_{2\ell+1}, & \alpha = \ell+1, \\ x_{4\ell-2\alpha+3} - \mathrm{i}x_{4\ell-2\alpha+4}, & \ell+2 \leqslant \alpha \leqslant 2\ell+1, \end{cases} \\
&H_\mu(L)X_\nu = \left[\delta_{\nu\mu} - \delta_{\nu(\mu+1)} + \delta_{\nu(2\ell-\mu+1)} - \delta_{\nu(2\ell-\mu+2)}\right] X_\nu, \\
&H_\ell(L)X_\nu = \left[2\delta_{\nu\ell} - 2\delta_{\nu(\ell+2)}\right] X_\nu, \\
&E_\mu(L)X_\nu = -\delta_{\nu(\mu+1)}X_\mu + \delta_{\nu(2\ell-\mu+2)}X_{2\ell-\mu+1}, \\
&E_\ell(L)X_\nu = -\delta_{\nu(\ell+1)}X_\ell + \delta_{\nu(\ell+2)}X_{\ell+1}, \\
&F_\mu(L)X_\nu = -\delta_{\nu\mu}X_{\mu+1} + \delta_{\nu(2\ell-\mu+1)}X_{2\ell-\mu+2}, \\
&F_\ell(L)X_\nu = -\delta_{\nu\ell}X_{\ell+1} + \delta_{\nu(\ell+1)}X_{\ell+2},
\end{aligned} \tag{10.95}$$

and for the SO(2ℓ) group

$$\begin{aligned}
&X_\alpha = \begin{cases} x_{2\alpha-1} + \mathrm{i}x_{2\alpha}, & 1 \leqslant \alpha \leqslant \ell, \\ x_{4\ell-2\alpha+1} - \mathrm{i}x_{4\ell-2\alpha+2}, & \ell+1 \leqslant \alpha \leqslant 2\ell, \end{cases} \\
&H_\mu(L)X_\nu = \left(\delta_{\nu\mu} - \delta_{\nu(\mu+1)} + \delta_{\nu(2\ell-\mu)} - \delta_{\nu(2\ell-\mu+1)}\right) X_\nu, \\
&H_\ell(L)X_\nu = \left(\delta_{\nu(\ell-1)} + \delta_{\nu\ell} - \delta_{\nu(\ell+1)} - \delta_{\nu(\ell+2)}\right) X_\nu, \\
&E_\mu(L)X_\nu = -\delta_{\nu(\mu+1)}X_\mu + \delta_{\nu(2\ell-\mu+1)}X_{2\ell-\mu}, \\
&E_\ell(L)X_\nu = -\delta_{\nu(\ell+1)}X_{\ell-1} + \delta_{\nu(\ell+2)}X_\ell, \\
&F_\mu(L)X_\nu = -\delta_{\nu\mu}X_{\mu+1} + \delta_{\nu(2\ell-\mu)}X_{2\ell-\mu+1}, \\
&F_\ell(L)X_\nu = -\delta_{\nu(\ell-1)}X_{\ell+1} + \delta_{\nu\ell}X_{\ell+2},
\end{aligned} \tag{10.96}$$

where $1 \leqslant \mu \leqslant \ell-1$.

When $\lambda_1 = 1$, the highest weight state $\mathcal{Y}^{[1]}_{(1,0\cdots,0)}(\hat{\boldsymbol{x}})$ is proportional to $X_1 = (x_1 + \mathrm{i}x_2)$:

$$\mathcal{Y}^{[1]}_{(1,0\cdots,0)}(\hat{\boldsymbol{x}}) = C_{N,1}\frac{(-1)^t}{\sqrt{2}}X_1, \quad t = \begin{cases} \ell, & N = 2\ell+1, \\ \ell-1, & N = 2\ell. \end{cases} \tag{10.97}$$

Generally, the highest weight state $Y^{[\lambda_1]}_{\boldsymbol{M}}(\hat{\boldsymbol{x}})$ with $\boldsymbol{M} = (\lambda_1, 0, \cdots, 0)$ is

$$\mathcal{Y}_{\boldsymbol{M}}^{[\lambda_1]}(\hat{\boldsymbol{x}}) = r^{\lambda_1} Y_{\boldsymbol{M}}^{[\lambda_1]}(\hat{\boldsymbol{x}}) = C_{N,\lambda_1}\left(\frac{(-1)^t X_1}{\sqrt{2}}\right)^{\lambda_1},$$

$$(C_{N,\lambda_1})^2 = \begin{cases} \dfrac{(2\lambda_1+2\ell-1)!}{2^{\lambda_1+2\ell}\pi^\ell \lambda_1!(\lambda_1+\ell-1)!}, & N = 2\ell+1, \\ \dfrac{2^{\lambda_1-1}(\lambda_1+\ell-1)!}{\pi^\ell \lambda_1!}, & N = 2\ell, \end{cases} \tag{10.98}$$

where the coefficient C_{N,λ_1} is calculated by the normalized condition

$$1 = \int d\Omega \, |Y_{\boldsymbol{M}}^{[\lambda_1]}(\hat{\boldsymbol{x}})|^2 = \frac{\pi \, (C_{N,\lambda_1})^2}{2^{\lambda_1-1}} \prod_{a=2}^{N-1} A_{2\lambda_1+a-1},$$

$$A_n = \int_0^\pi d\varphi \, (\sin\varphi)^n = \begin{cases} \pi \cdot \dfrac{1\cdot 3\cdot 5\cdots(n-1)}{2\cdot 4\cdot 6\cdots(n)}, & n \text{ is even}, \\ 2\cdot \dfrac{2\cdot 4\cdot 6\cdots(n-1)}{1\cdot 3\cdot 5\cdots(n)}, & n \text{ is odd}, \end{cases} \tag{10.99}$$

The remaining harmonic polynomials $\mathcal{Y}_{\boldsymbol{m}}^{[\lambda_1]}(\hat{\boldsymbol{x}})$ with the weight $\boldsymbol{m}$ can be calculated by the lowering operators $F_\mu(L)$. For example,

$$\begin{aligned} \mathcal{Y}_{(\lambda_1-2,1,0\cdots,0)}^{[\lambda_1]}(\hat{\boldsymbol{x}}) &= \lambda_1^{-1/2} F_1(L) \mathcal{Y}_{\boldsymbol{M}}^{[\lambda_1]}(\hat{\boldsymbol{x}}) \\ &= -\sqrt{\lambda_1} C_{N,\lambda_1}\left(\frac{(-1)^t}{\sqrt{2}}\right)^{\lambda_1} X_1^{\lambda_1-1} X_2, \quad N > 3. \end{aligned}$$

The harmonic polynomial $\mathcal{Y}_{\boldsymbol{M}}^{[\lambda_1]}(\boldsymbol{x})$ is a homogeneous polynomial of degree λ_1 with respect to x_a and satisfies the Laplace equation, so do $\mathcal{Y}_{\boldsymbol{m}}^{[\lambda_1]}(\boldsymbol{x})$:

$$\triangledown_{\boldsymbol{x}}^2 \mathcal{Y}_{\boldsymbol{m}}^{[\lambda_1]}(\boldsymbol{x}) = 0, \qquad \triangledown_{\boldsymbol{x}}^2 = \sum_{a=1}^N \frac{\partial^2}{\partial x_a^2}. \tag{10.100}$$

The eigenvalue $C_2([\lambda_1])$ of L^2 on $Y_{\boldsymbol{m}}^{[\lambda_1]}(\boldsymbol{x})$ is calculated by Eq. (7.133) where A^{-1} is given in subsection 7.5.6:

$$L^2 \, Y_{\boldsymbol{m}}^{[\lambda_1]}(\hat{\boldsymbol{x}}) = C_2([\lambda_1]) Y_{\boldsymbol{m}}^{[\lambda_1]}(\hat{\boldsymbol{x}}), \quad C_2([\lambda_1]) = \lambda_1(\lambda_1+2\ell-1). \tag{10.101}$$

10.4.3 *Schrödinger Equation for a Two-body System*

An isolated n-body quantum system is invariant in the translation of space-time and the spatial rotation. After separating the motion of the center-of-mass (see subsection 5.9), there is only one Jacobi coordinate vectors for a two-body system in N-dimensions, denoted by $\boldsymbol{R}_1 \equiv \boldsymbol{x}$ for simplicity. Remind that a factor of the square root of mass has been included in $\boldsymbol{R}_1$

(see Eq. (5.225)). The eigenfunction of angular momentum of the system has to be proportional to the harmonic functions $\mathcal{Y}_{\boldsymbol{m}}^{[\lambda_1]}(\hat{\boldsymbol{x}})$,

$$\psi_{\boldsymbol{m}}^{[\lambda_1]}(\boldsymbol{x}) = \phi^{[\lambda_1]}(r)\mathcal{Y}_{\boldsymbol{m}}^{[\lambda_1]}(\boldsymbol{x}), \tag{10.102}$$

where $\phi^{[\lambda_1]}(r)$ is the radial function.

$$\begin{aligned}
\bigtriangledown_{\boldsymbol{x}}^2 \left[\phi^{[\lambda_1]}(r)\mathcal{Y}_{\boldsymbol{M}}^{[\lambda_1]}(\hat{\boldsymbol{x}})\right] &= \mathcal{Y}_{\boldsymbol{M}}^{[\lambda_1]}(\boldsymbol{x}) \bigtriangledown_{\boldsymbol{x}}^2 \phi^{[\lambda_1]}(r) + 2 \bigtriangledown_{\boldsymbol{x}} \phi^{[\lambda_1]}(r) \cdot \bigtriangledown_{\boldsymbol{x}}\mathcal{Y}_{\boldsymbol{M}}^{[\lambda_1]}(\boldsymbol{x}) \\
&= \mathcal{Y}_{\boldsymbol{M}}^{[\lambda_1]}(\boldsymbol{x}) \left\{ r^{1-N}\frac{d}{dr}r^{N-1}\frac{d}{dr}\phi^{[\lambda_1]}(r) \right\} + 2\frac{d\phi^{[\lambda_1]}(r)}{dr}\frac{\boldsymbol{x}}{r} \cdot \bigtriangledown_{\boldsymbol{x}}\mathcal{Y}_{\boldsymbol{M}}^{[\lambda_1]}(\boldsymbol{x}) \\
&= \mathcal{Y}_{\boldsymbol{M}}^{[\lambda_1]}(\boldsymbol{x}) \left\{ \frac{d^2\phi^{[\lambda]}(r)}{dr^2} + \frac{N-1}{r}\frac{d\phi^{[\lambda_1]}(r)}{dr} \right\} + \frac{d\phi^{[\lambda_1]}(r)}{dr}\frac{2\lambda_1}{r^2}\mathcal{Y}_{\boldsymbol{M}}^{[\lambda_1]}(\boldsymbol{x}) \\
&= \mathcal{Y}_{\boldsymbol{M}}^{[\lambda_1]}(\hat{\boldsymbol{x}}) \left\{ \frac{d^2\phi^{[\lambda_1]}(r)}{dr^2} + \frac{N+2\lambda_1-1}{r}\frac{d\phi^{[\lambda_1]}(r)}{dr} \right\}.
\end{aligned}$$

Substituting $\psi_{\boldsymbol{M}}^{[\lambda_1]}(\boldsymbol{x})$ into the Schrödinger equation in the coordinate system of the center-of-mass

$$-\frac{1}{2} \bigtriangledown_{\boldsymbol{x}}^2 \psi_{\boldsymbol{M}}^{[\lambda_1]}(\boldsymbol{x}) + V(r)\psi_{\boldsymbol{M}}^{[\lambda_1]}(\boldsymbol{x}) = E\psi_{\boldsymbol{M}}^{[\lambda_1]}(\boldsymbol{x}), \tag{10.103}$$

we obtain the radial equation

$$\begin{aligned}
-\frac{1}{2}\left\{ \frac{d^2\phi^{[\lambda_1]}(r)}{dr^2} + \frac{N+2\lambda_1-1}{r}\frac{d\phi^{[\lambda_1]}(r)}{dr} \right\} \\
= \{E - V(r)\} \phi^{[\lambda_1]}(r).
\end{aligned} \tag{10.104}$$

10.4.4 *Schrödinger Equation for a Three-body System*

After separating the motion of the center-of-mass (see subsection 5.9), there are two Jacobi coordinate vectors in a three-body system (see Eq. (5.225)). $\boldsymbol{R}_1 \equiv \boldsymbol{x}$ describes the mass-weighted separation from the first particle to the center-of-mass of the last two particles, and $\boldsymbol{R}_2 \equiv \boldsymbol{y}$ describes the mass-weighted separation of the last two particles. There are $3N$ degrees of freedom for a three-body system. N degrees of freedom describe the motion of center-of-mass. A SO(N) rotation R transforms $\boldsymbol{x}$ along the Nth axis and $\boldsymbol{y}$ on the $x_N x_{N-1}$-plane with the positive x_N-component where the rotation in the remaining $(N-2)$ subspace is free. Thus R contains $N(N-1)/2 - (N-2)(N-3)/2 = 2N-3$ degrees of freedom which describe the global rotation of the system. The remaining three degrees of freedom describe the internal motion. The internal variables are denoted by $\xi_1 = \boldsymbol{x} \cdot \boldsymbol{x}$, $\xi_2 = \boldsymbol{y} \cdot \boldsymbol{y}$, and $\xi_3 = \boldsymbol{x} \cdot \boldsymbol{y}$, which are left invariant in the global rotation.

Since there are two coordinate vectors, the angular momentum states of the system are described by a two-row Young pattern $[\lambda] = [\lambda_1, \lambda_2]$, $\lambda_1 \geqslant \lambda_2 \geqslant 0$. The eigenfunction of the angular momentum is expressed as the combination of the products of $\mathcal{Y}^{[q]}_{\boldsymbol{m}}(\boldsymbol{x})$ and $\mathcal{Y}^{[p]}_{\boldsymbol{m}'}(\boldsymbol{y})$, where $[q]$ and $[p]$ both are one-row Young patterns and $[\lambda_1, \lambda_2]$ is contained in the decomposition of $[q] \times [p]$.

The Clebsch–Gordan (CG) series of $[q] \times [p]$ of SO(N) consists of two parts. One is calculated by the Littlewood–Richardson rule, just like the decomposition of $[q] \times [p]$ of SU(N). The other comes from the trace operation between two indices belonging to two representations, respectively. Without loss of generality, we assume $q \geqslant p$,

$$\begin{aligned}
([q] \times [p])_{LR} &\simeq [q+p, 0] \oplus [q+p-1, 1] \oplus \cdots \oplus [q, p] \\
&\simeq \bigoplus_{s=0}^{p} [q+p-s, s], \\
[q] \times [p] &\simeq ([q] \times [p])_{LR} \oplus ([q-1] \times [p-1])_{LR} \oplus \cdots \\
&\quad \oplus ([q-p] \times [0])_{LR} \simeq \bigoplus_{s=0}^{p} \bigoplus_{t=0}^{p-s} [q+p-s-2t, s].
\end{aligned} \tag{10.105}$$

For SO(3), $s = 0$ or 1 and $[\lambda, 1] \simeq [\lambda, 0]$. For SO(4), the representation with $s \neq 0$ is reduced into the direct sum of a self-dual representation and an anti-self-dual representation.

As far as the independent eigenfunctions of the angular momentum is concerned, one has to exclude the functions in the form $(\boldsymbol{x} \cdot \boldsymbol{y})F(\boldsymbol{x}, \boldsymbol{y})$ because the factor $(\boldsymbol{x} \cdot \boldsymbol{y})$ is an internal variable and can be incorporated into the radial functions. The function, which belongs to a representation with $t \neq 0$ in the decomposition (10.105), is nothing but that in the form $(\boldsymbol{x} \cdot \boldsymbol{y})F(\boldsymbol{x}, \boldsymbol{y})$. Thus, **the independent eigenfunction of the angular momentum described by $[\lambda_1, \lambda_2]$ is expressed as the combination of $\mathcal{Y}^{[q]}_{\boldsymbol{m}}(\boldsymbol{x})\mathcal{Y}^{[\lambda_1+\lambda_2-q]}_{\boldsymbol{m}'}(\boldsymbol{y})$**, which is usually calculated by the Clebsch–Gordan coefficients. Fortunately, in the derivation of the radial equation, we only need to calculate the eigenfunctions $Q^{[\lambda_1,\lambda_2]}_q(\boldsymbol{x}, \boldsymbol{y})$ with the highest weight $\boldsymbol{M} = (\lambda_1 - \lambda_2, \lambda_2, 0, \cdots, 0)$ due to the spherical symmetry of the system.

The eigenfunction $Q^{[\lambda_1,\lambda_2]}_q(\boldsymbol{x}, \boldsymbol{y})$ of the angular momentum with the highest weight corresponds to the tensor Young tableau with the Young pattern $[\lambda_1, \lambda_2]$ where the box in the first row is filled by 1 and the boxes in the second row is filled by 2. Thus, $Q^{[\lambda_1,\lambda_2]}_q(\boldsymbol{x}, \boldsymbol{y})$ satisfies

$$\begin{aligned}
&H_1(L)Q_q^{[\lambda_1,\lambda_2]}(\boldsymbol{x},\boldsymbol{y}) = (\lambda_1-\lambda_2)Q_q^{[\lambda_1,\lambda_2]}(\boldsymbol{x},\boldsymbol{y}),\\
&H_2(L)Q_q^{[\lambda_1,\lambda_2]}(\boldsymbol{x},\boldsymbol{y}) = \lambda_2 Q_q^{[\lambda_1,\lambda_2]}(\boldsymbol{x},\boldsymbol{y}),\\
&H_\nu(L)Q_q^{[\lambda_1,\lambda_2]}(\boldsymbol{x},\boldsymbol{y}) = E_\mu(L)Q_q^{[\lambda_1,\lambda_2]}(\boldsymbol{x},\boldsymbol{y}) = 0,\\
&3\leqslant \nu \leqslant \ell, \qquad 1\leqslant \mu\leqslant \ell, \qquad L_{ab} = L_{ab}(\boldsymbol{x}) + L_{ab}(\boldsymbol{y}).
\end{aligned} \tag{10.106}$$

From Eqs. (10.14), (10.15), (10.20), (10.21), (10.95) and (10.96), we know

$$\begin{aligned}
&X_1 = x_1 + \mathrm{i}x_2, \quad X_2 = x_3 + \mathrm{i}x_4,\\
&H_\nu(L)X_1 = \delta_{\nu 1}X_1, \quad H_\nu(L)X_2 = -\delta_{\nu 1}X_2 + \delta_{\nu 2}X_2,\\
&E_\nu(L)X_1 = 0, \quad E_\nu(L)X_2 = -\delta_{\nu 1}X_1,
\end{aligned} \tag{10.107}$$

and those formulas obtained by replacing x_a by y_a and X_a by Y_a. Thus, for the angular momentum described by $[\lambda_1, \lambda_2]$, there are $(\lambda_1 - \lambda_2 + 1)$ linearly independent unnormalized eigenfunctions $Q_q^{[\lambda_1,\lambda_2]}(\boldsymbol{x},\boldsymbol{y})$ with the highest weight $\boldsymbol{M} = (\lambda_1-\lambda_2, \lambda_2, 0, \cdots, 0)$:

$$\begin{aligned}
&Q_q^{[\lambda_1,\lambda_2]}(\boldsymbol{x},\boldsymbol{y}) = \frac{X_1^{q-\lambda_2}Y_1^{\lambda_1-q}}{(q-\lambda_2)!(\lambda_1-q)!}\,(X_1Y_2 - X_2Y_1)^{\lambda_2},\\
&\lambda_2 \leqslant q \leqslant \lambda_1, \qquad N > 4,
\end{aligned} \tag{10.108}$$

where the coefficient is introduced for convenience. $Q_q^{[\lambda_1,\lambda_2]}(\boldsymbol{x},\boldsymbol{y})$ **is a homogeneous polynomial of degree** q **and degree** $(\lambda_1 + \lambda_2 - q)$ **with respect to the components of** $\boldsymbol{x}$ and $\boldsymbol{y}$, respectively, **and satisfies** Eq. (10.106) **and the Laplace equations**

$$\begin{aligned}
\nabla_{\boldsymbol{x}}^2 Q_q^{[\lambda_1,\lambda_2]}(\boldsymbol{x},\boldsymbol{y}) &= \nabla_{\boldsymbol{y}}^2 Q_q^{[\lambda_1,\lambda_2]}(\boldsymbol{x},\boldsymbol{y})\\
&= \nabla_{\boldsymbol{x}}\cdot\nabla_{\boldsymbol{y}} Q_q^{[\lambda_1,\lambda_2]}(\boldsymbol{x},\boldsymbol{y}) = 0.
\end{aligned} \tag{10.109}$$

For SO(3), λ_2 in Eq. (10.108) has to be 0 or 1, and $X_2 = x_3$, $Y_2 = y_3$. For SO(4), the representation $[\lambda_1, \lambda_2]$ is reduced into the direct sum of a self-dual representation $[(+)\lambda_1, \lambda_2]$ and an anti-self-dual representation $[(-)\lambda_1, \lambda_2]$. $Q_q^{[(+)\lambda_1,\lambda_2]}(\boldsymbol{x},\boldsymbol{y})$ has the same form as Eq. (10.108), but $Q_q^{[(-)\lambda_1,\lambda_2]}(\boldsymbol{x},\boldsymbol{y})$ is obtained from Eq. (10.108) by replacing X_2 and Y_2 with $X_3 = x_3 - \mathrm{i}x_4$ and $Y_3 = y_3 - \mathrm{i}y_4$, respectively.

The Schrödinger equation for a three-body system in the coordinate system of the center-of-mass is

$$-\frac{1}{2}\left\{\nabla_{\boldsymbol{x}}^2 + \nabla_{\boldsymbol{y}}^2\right\}\psi_{\boldsymbol{M}}^{[\lambda_1,\lambda_2]}(\boldsymbol{x},\boldsymbol{y}) = \left[E - V(\xi_1,\xi_2,\xi_3)\right]\psi_{\boldsymbol{M}}^{[\lambda_1,\lambda_2]}(\boldsymbol{x},\boldsymbol{y}). \tag{10.110}$$

Let

$$\psi_{\boldsymbol{M}}^{[\lambda_1,\lambda_2]}(\boldsymbol{x},\boldsymbol{y}) = \sum_{q=\lambda_2}^{\lambda_1} \phi_q^{[\lambda_1,\lambda_2]}(\xi_1,\xi_2,\xi_3) Q_q^{[\lambda_1,\lambda_2]}(\boldsymbol{x},\boldsymbol{y}). \tag{10.111}$$

The parity $\phi_q^{[\lambda_1,\lambda_2]}(\xi_1,\xi_2,\xi_3)$ is even, and the parities of $\psi_{\boldsymbol{M}}^{[\lambda_1,\lambda_2]}(\boldsymbol{x},\boldsymbol{y})$ and $Q_q^{[\lambda_1,\lambda_2]}(\boldsymbol{x},\boldsymbol{y})$ both are $(-1)^{\lambda_1+\lambda_2}$. The action of the Laplace operator on $\psi_{\boldsymbol{M}}^{[\lambda_1,\lambda_2]}(\boldsymbol{x},\boldsymbol{y})$ is divided into three parts. The first part is its action on the radial functions $\phi_q^{[\lambda_1,\lambda_2]}$, which can be calculated by replacement of variables (see Eq. (10.113)). The second part is its action on the basis eigenfunctions $Q_q^{[\lambda_1,\lambda_2]}(\boldsymbol{x},\boldsymbol{y})$ which vanishes. The third part is its mixed action

$$\begin{aligned}
&2\left\{\left(\frac{\partial}{\partial\xi_1}\phi_q^{[\lambda_1,\lambda_2]}\right)2\boldsymbol{x} + \left(\frac{\partial}{\partial\xi_3}\phi_q^{[\lambda_1,\lambda_2]}\right)\boldsymbol{y}\right\}\cdot\nabla_{\boldsymbol{x}}Q_q^{[\lambda_1,\lambda_2]}(\boldsymbol{x},\boldsymbol{y}) \\
&\quad + 2\left\{\left(\frac{\partial}{\partial\xi_2}\phi_q^{[\lambda_1,\lambda_2]}\right)2\boldsymbol{y} + \left(\frac{\partial}{\partial\xi_3}\phi_q^{[\lambda_1,\lambda_2]}\right)\boldsymbol{x}\right\}\cdot\nabla_{\boldsymbol{y}}Q_q^{[\lambda_1,\lambda_2]}(\boldsymbol{x},\boldsymbol{y}).
\end{aligned}$$

From Eq. (10.108) one has

$$\begin{aligned}
\boldsymbol{x}\cdot\nabla_{\boldsymbol{x}}Q_q^{[\lambda_1,\lambda_2]} &= qQ_q^{[\lambda_1,\lambda_2]}, \\
\boldsymbol{y}\cdot\nabla_{\boldsymbol{y}}Q_q^{[\lambda_1,\lambda_2]} &= (\lambda_1+\lambda_2-q)Q_q^{[\lambda_1,\lambda_2]}, \\
\boldsymbol{y}\cdot\nabla_{\boldsymbol{x}}Q_q^{[\lambda_1,\lambda_2]} &= (\lambda_1-q+1)Q_{q-1}^{[\lambda_1,\lambda_2]}, \\
\boldsymbol{x}\cdot\nabla_{\boldsymbol{y}}Q_q^{[\lambda_1,\lambda_2]} &= (q-\lambda_2+1)Q_{q+1}^{[\lambda_1,\lambda_2]}.
\end{aligned} \tag{10.112}$$

Thus, the general radial equation for the radial function $\phi_q^{[\lambda_1,\lambda_2]}$ is (see [Gu et al. (2001c)])

$$\begin{aligned}
&\left\{\nabla_{\boldsymbol{x}}^2+\nabla_{\boldsymbol{y}}^2\right\}\phi_q^{[\lambda_1,\lambda_2]} + 4q\frac{\partial}{\partial\xi_1}\phi_q^{[\lambda_1,\lambda_2]} + 4(\lambda_1+\lambda_2-q)\frac{\partial}{\partial\xi_2}\phi_q^{[\lambda_1,\lambda_2]} \\
&\quad + 2(\lambda_1-q)\frac{\partial}{\partial\xi_3}\phi_{q+1}^{[\lambda_1,\lambda_2]} + 2(q-\lambda_2)\frac{\partial}{\partial\xi_3}\phi_{q-1}^{[\lambda_1,\lambda_2]} \\
&\quad = -2\left(E-V\right)\phi_q^{[\lambda_1,\lambda_2]}, \\
&\left\{\nabla_{\boldsymbol{x}}^2+\nabla_{\boldsymbol{y}}^2\right\}\phi_q^{[\lambda_1,\lambda_2]} = \left\{4\xi_1\frac{\partial^2}{\partial\xi_1^2} + 4\xi_2\frac{\partial^2}{\partial\xi_2^2} + 2N\left(\frac{\partial}{\partial\xi_1}+\frac{\partial}{\partial\xi_2}\right)\right. \\
&\quad \left. + (\xi_1+\xi_2)\frac{\partial^2}{\partial\xi_3^2} + 4\xi_3\left(\frac{\partial^2}{\partial\xi_1\partial\xi_3}+\frac{\partial^2}{\partial\xi_2\partial\xi_3}\right)\right\}\phi_q^{[\lambda_1,\lambda_2]}.
\end{aligned} \tag{10.113}$$

This method can be generalized to a quantum multiple-body system (see [Gu et al. (2003b)]).

10.5 Exercises

1. Calculate the orthonormal basis tensors, expressed by the standard tensor Young tableaux and the generalized Gel'fand bases, in the irreducible representation $[2,1]$ of SO(5) with the highest weight $\boldsymbol{M} = (1,2)$.

2. Calculate the orthonormal basis tensors, expressed by the standard tensor Young tableaux and the generalized Gel'fand bases, in the irreducible representation $[2,2]$ of SO(5) with the highest weight $\boldsymbol{M} = (0,4)$.

3. Complete the calculating for the orthonormal basis tensors in the irreducible representation $[2,1]$ of SO(7) with the highest weight $\boldsymbol{M} = (1,1,0)$ (see subsection 10.1.2)

4. Calculate the orthonormal basis tensors, expressed by the standard tensor Young tableaux and the generalized Gel'fand bases, in the anti-self-dual representation space $[(-)1,1,1,1]$ of SO(8) with the highest weight $\boldsymbol{M} = (0,0,2,0)$.

5. Calculate the dimensions of the irreducible tensor representations of the SO(8) group denoted by the following Young patterns: (a) $[4,2]$, (b) $[3,2]$, (c) $[4,4]$, (d) $[(+)3,2,1,1]$, (e) $[(+)3,3,1,1]$.

6. Calculate all the basis spinors in the fundamental spinor representation $[s]$ of SO(7)

7. Calculate all the basis spinors in two fundamental spinor representations $[\pm s]$ of SO(8)

8. There are four $\mathcal{A}_4$-doublets in the space of the spin-tensor representation $[+s,1]$ of SO(8) (see pp. 565–568):

$$F_4 \left| \begin{array}{ccccccc} 2&&1&&0&&0\\ &2&&1&&0&\\ &&1&&0&&\\ &&&1&&& \end{array} \right\rangle_1^5 = \left| \begin{array}{ccccccc} 1&&1&&0&&\bar{1}\\ &1&&1&&0&\\ &&1&&0&&\\ &&&1&&& \end{array} \right\rangle_{\bar{1}}^6,$$

$$F_4 \left| \begin{array}{ccccccc} 2&&1&&0&&0\\ &2&&0&&0&\\ &&1&&0&&\\ &&&1&&& \end{array} \right\rangle_1^6 = \sqrt{\frac{1}{3}} \left| \begin{array}{ccccccc} 1&&1&&0&&\bar{1}\\ &1&&0&&0&\\ &&1&&0&&\\ &&&1&&& \end{array} \right\rangle_{\bar{1}}^7 + \sqrt{\frac{2}{3}} \left| \begin{array}{ccccccc} 1&&0&&0&&0\\ &1&&0&&0&\\ &&1&&0&&\\ &&&1&&& \end{array} \right\rangle_{\bar{1}}^7,$$

$$F_4 \left| \begin{array}{ccccccc} 2&&1&&0&&0\\ &1&&1&&0&\\ &&1&&0&&\\ &&&1&&& \end{array} \right\rangle_1^6 = \frac{2}{3} \left| \begin{array}{ccccccc} 1&&1&&0&&\bar{1}\\ &1&&0&&0&\\ &&1&&0&&\\ &&&1&&& \end{array} \right\rangle_{\bar{1}}^7 + \sqrt{\frac{1}{3}} \left| \begin{array}{ccccccc} 1&&1&&0&&\bar{1}\\ &1&&1&&\bar{1}&\\ &&1&&0&&\\ &&&1&&& \end{array} \right\rangle_{\bar{1}}^7 - \frac{\sqrt{2}}{3} \left| \begin{array}{ccccccc} 1&&0&&0&&0\\ &1&&0&&0&\\ &&1&&0&&\\ &&&1&&& \end{array} \right\rangle_{\bar{1}}^7,$$

$$F_4 \left| \begin{array}{ccccccc} 2&&1&&0&&0\\ &1&&0&&0&\\ &&1&&0&&\\ &&&1&&& \end{array} \right\rangle_1^7 = \left| \begin{array}{ccccccc} 1&&1&&0&&\bar{1}\\ &1&&0&&\bar{1}&\\ &&1&&0&&\\ &&&1&&& \end{array} \right\rangle_{\bar{1}}^8.$$

Please check that they are consistent in applying F_3 to them.

9. Calculate the orthonormal basis states, expressed by the standard tensor Young tableaux and the generalized Gel'fand bases, in the spin-tensor representation space $[-s, 1]$ of SO(8) with the highest weight $\boldsymbol{M} = (1, 0, 1, 0)$.

10. Calculate the dimension of the irreducible spinor representation of the SO(7) group denoted by the following Young patterns:

$$(1)\ [s, 4, 2], \qquad (2)\ [s, 3, 2], \qquad (3)\ [s, 4, 4],$$
$$(4)\ [s, 3, 1, 1], \qquad (5)\ [s, 3, 2, 2].$$

11. Calculate the harmonic polynomials $\mathcal{Y}^{[4]}_{(4,0)}$ in an 5-dimensional Euclidian space.

12. Calculate the harmonic polynomials $\mathcal{Y}^{[3]}_{(3,0,0,0)}$ in an 8-dimensional Euclidian space.

13. Calculate the expression of the harmonic polynomials $\mathcal{Y}^{[\lambda_1]}_{\boldsymbol{m}}(\boldsymbol{x})$ where $\lambda_1 = 1$, 2 and 3 for the SO(4) and SO(5) groups, respectively.

14. Change the definition (10.102) for the radial function as follows:

$$\psi^{[\lambda_1]}_{\boldsymbol{m}}(\boldsymbol{x}) = \phi^{[\lambda_1]}(r)\mathcal{Y}^{[\lambda_1]}_{\boldsymbol{m}}(\boldsymbol{x}) = \overline{\phi}^{[\lambda_1]}(r)Y^{[\lambda_1]}_{\boldsymbol{m}}(\hat{\boldsymbol{x}}).$$

Please calculate the radial equation for the new radial function $\overline{\phi}^{[\lambda_1]}(r)$ (see Eq. (10.113)).

Chapter 11

LORENTZ GROUPS

The proper Lorentz group L_p is a noncompact Lie group, and the SO(4) group is a compact Lie group. Two groups have the same Lie algebra but their real Lie algebras are different. In this chapter the irreducible representations of the proper Lorentz group is studied through those of the SO(4) group. The study of the Lorentz group provides a typical example in mathematics for studying the noncompact Lie groups.

11.1 The SO(4) Group

11.1.1 *Irreducible Representations of* SO*(4)*

The Lie algebra of SO(4) is $D_2 \approx A_1 \oplus A_1$. In this subsection the SO(4) group is evidently decomposed into a direct product of two SU(2) groups. Based on this decomposition, the parameters of SO(4) are chosen and the irreducible representations of SO(4) are calculated analytically.

Six generators of SO(4) in its self-representation are given in Eq. (7.80). Through a suitable combination one has

$$T_1^{(\pm)} = \frac{1}{2}\left(T_{23} \pm T_{14}\right) = \frac{1}{2}\begin{pmatrix} 0 & 0 & 0 & \mp \mathrm{i} \\ 0 & 0 & -\mathrm{i} & 0 \\ 0 & \mathrm{i} & 0 & 0 \\ \pm \mathrm{i} & 0 & 0 & 0 \end{pmatrix} \tag{11.1}$$

and its cyclic combination. $T_a^{(\pm)}$ can be expressed as

$$\begin{aligned} T_1^{(+)} &= \frac{1}{2}\sigma_2 \times \sigma_1, & T_2^{(+)} &= \frac{-1}{2}\sigma_2 \times \sigma_3, & T_3^{(+)} &= \frac{1}{2}\mathbf{1}_2 \times \sigma_2, \\ T_1^{(-)} &= \frac{-1}{2}\sigma_1 \times \sigma_2, & T_2^{(-)} &= \frac{-1}{2}\sigma_2 \times \mathbf{1}_2, & T_3^{(-)} &= \frac{1}{2}\sigma_3 \times \sigma_2. \end{aligned} \tag{11.2}$$

The generators are divided into two sets, each of which satisfies the commutative relations of generators in SU(2),

$$\left[T_a^{(\pm)},\, T_b^{(\pm)}\right] = \mathrm{i}\sum_{c=1}^{3} \epsilon_{abc} T_c^{(\pm)}, \qquad \left[T_a^{(+)},\, T_b^{(-)}\right] = 0. \tag{11.3}$$

Though a similarity transformation N, this property becomes clearer:

$$N^{-1}T_a^{(+)}N = (\sigma_a/2)\times \mathbf{1}_2, \qquad N^{-1}T_a^{(-)}N = \mathbf{1}_2 \times (\sigma_a/2)\,,$$
$$N = \frac{1}{\sqrt{2}}\begin{pmatrix} -1 & 0 & 0 & 1 \\ -\mathrm{i} & 0 & 0 & -\mathrm{i} \\ 0 & 1 & 1 & 0 \\ 0 & \mathrm{i} & -\mathrm{i} & 0 \end{pmatrix}. \tag{11.4}$$

Note that Eq. (10.20) is a generalization of Eq. (11.4). An arbitrary element R in SO(4) can be expressed as

$$\begin{aligned} R = &\ \exp\left(-\mathrm{i}\sum_{a<b}^{4} \omega_{ab}T_{ab}\right) = \exp\left\{-\mathrm{i}\sum_{a=1}^{3}\left(\omega_a^{(+)}T_a^{(+)} + \omega_a^{(-)}T_a^{(-)}\right)\right\} \\ = &\ \exp\left\{-\mathrm{i}\omega^{(+)}\hat{\boldsymbol{n}}^{(+)}\cdot \boldsymbol{T}^{(+)}\right\}\ \exp\left\{-\mathrm{i}\omega^{(-)}\hat{\boldsymbol{n}}^{(-)}\cdot \boldsymbol{T}^{(-)}\right\} \\ = &\ N\left\{u(\hat{\boldsymbol{n}}^{(+)},\omega^{(+)})\times u(\hat{\boldsymbol{n}}^{(-)},\omega^{(-)})\right\}N^{-1}, \end{aligned} \tag{11.5}$$

where

$$\begin{aligned} \omega_1^{(\pm)} &= \omega_{23} \pm \omega_{14} = \omega^{(\pm)}n_1^{(\pm)}, \\ \omega_2^{(\pm)} &= \omega_{31} \pm \omega_{24} = \omega^{(\pm)}n_2^{(\pm)}, \qquad \omega^{(\pm)} = \left\{\sum_{a=1}^{3}\left(\omega_a^{(\pm)}\right)^2\right\}^{1/2}. \\ \omega_3^{(\pm)} &= \omega_{12} \pm \omega_{34} = \omega^{(\pm)}n_3^{(\pm)}, \end{aligned} \tag{11.6}$$

Thus, R **is evidently written as the direct product of two unimodular unitary matrices of dimension 2**. R is left invariant if two unitary matrices change their signs simultaneously. Namely, Eq. (11.5) gives a one-to-two correspondence between the element R in SO(4) and the element $u\times u'$ in SU(2) $\times$ SU(2)$'$, and the correspondence leaves invariant in the multiplication of elements, so that

$$\mathrm{SU}(2)\otimes \mathrm{SU}(2)' \sim \mathrm{SO}(4). \tag{11.7}$$

The parameters of the group SO(4) are chosen to be those of SU(2) $\times$ SU(2)$'$ where the group space of SU(2)$'$ is shortened by half such that there is a one-to-one correspondence between the set of parameters and the group element of SO(4) at least in the region where the measure is not vanishing. Namely, the group space of SO(4) is

$$0 \leqslant \omega^{(+)} \leqslant 2\pi, \quad 0 \leqslant \omega^{(-)} \leqslant \pi, \quad 0 \leqslant \theta^{(\pm)} \leqslant \pi, \quad -\pi \leqslant \varphi^{(\pm)} \leqslant \pi, \tag{11.8}$$

where $\theta^{(\pm)}$ and $\varphi^{(\pm)}$ are the polar angle and the azimuthal angle of $\hat{\boldsymbol{n}}^{(\pm)}$, respectively. Since

$$R(\hat{\boldsymbol{n}}^{(+)}, \omega^{(+)}; \hat{\boldsymbol{n}}^{(-)}, \omega^{(-)}) = R(-\hat{\boldsymbol{n}}^{(+)}, (2\pi - \omega^{(+)}); -\hat{\boldsymbol{n}}^{(-)}, (2\pi - \omega^{(-)})),$$

the group space of SO(4) is doubly-connected:

$$R(\hat{\boldsymbol{n}}^{(+)}, \omega^{(+)}; \hat{\boldsymbol{n}}^{(-)}, \pi) = R(-\hat{\boldsymbol{n}}^{(+)}, (2\pi - \omega^{(+)}); -\hat{\boldsymbol{n}}^{(-)}, \pi). \tag{11.9}$$

The covering group of SO(4) **is** SU(2) × SU(2)′. The weight function for the group integral of SO(4) is

$$\begin{aligned} dR = & \left(8\pi^4\right)^{-1} \sin^2\left(\omega^{(+)}/2\right) \sin^2\left(\omega^{(-)}/2\right) \sin\theta^{(+)} \sin\theta^{(-)} \\ & \times\ d\omega^{(+)} d\omega^{(-)} d\theta^{(+)} d\theta^{(-)} d\varphi^{(+)} d\varphi^{(-)}. \end{aligned} \tag{11.10}$$

An irreducible representation D^{jk} of SO(4) is a direct product of representations of two SU(2), which is $(2j+1)(2k+1)$-dimensional,

$$D^{jk}\left(\hat{\boldsymbol{n}}^{(+)}, \omega^{(+)}; \hat{\boldsymbol{n}}^{(-)}, \omega^{(-)}\right) = D^{j}\left(\hat{\boldsymbol{n}}^{(+)}, \omega^{(+)}\right) \times D^{k}\left(\hat{\boldsymbol{n}}^{(-)}, \omega^{(-)}\right). \tag{11.11}$$

Its row (column) index is denoted by two letters $(\mu\nu)$ and its generator I^{jk}_{ab} is expressed in terms of I^{j}_{a} of SU(2),

$$\begin{aligned} & I_a^{jk(+)} = I_a^j \times \mathbf{1}_{2k+1}, && I_a^{jk(-)} = \mathbf{1}_{2j+1} \times I_a^k, \\ & I_{ab}^{jk} = \sum_{c=1}^{3} \epsilon_{abc}\left(I_c^{jk(+)} + I_c^{jk(-)}\right), && I_{a4}^{jk} = I_a^{jk(+)} - I_a^{jk(-)}. \end{aligned} \tag{11.12}$$

The decomposition of the direct product of two D^{jk} can be calculated by the Clebsch–Gordan series for SU(2) such as

$$\begin{aligned} \left(C^{j_1 j_2}\right)^{-1} \left(D^{j_1 0}(R) \times D^{j_2 0}(R)\right) \left(C^{j_1 j_2}\right) &= \bigoplus_{J=|j_1-j_2|}^{j_1+j_2} D^{J0}(R), \\ \left(C^{k_1 k_2}\right)^{-1} \left(D^{0 k_1}(R) \times D^{0 k_2}(R)\right) \left(C^{k_1 k_2}\right) &= \bigoplus_{K=|k_1-k_2|}^{k_1+k_2} D^{0K}(R), \\ D^{j_1 k_1}(R) \times D^{j_2 k_2}(R) &\simeq \bigoplus_{J=|j_1-j_2|}^{j_1+j_2} \bigoplus_{K=|k_1-k_2|}^{k_1+k_2} D^{JK}(R). \end{aligned} \tag{11.13}$$

In fact, $I_a^{jk(\pm)}$ are related directly with the Chevalley bases (see Eqs. (10.13) and (10.19)),

$$\begin{aligned} &H_1 = 2I_3^{jk(-)}, && H_2 = 2I_3^{jk(+)}, \\ &E_1 = I_1^{jk(-)} + \mathrm{i}I_2^{jk(-)}, && E_2 = I_1^{jk(+)} + \mathrm{i}I_2^{jk(+)}. \end{aligned} \tag{11.14}$$

The highest weight in D^{jk} is $M = (2k, 2j)$, and its Young pattern is

$$\begin{aligned} &D^{jj} \simeq [2j, 0], && j = k, \\ &D^{jk} \simeq [(+)(j+k), (j-k)], && j-k > 0 \text{ is integer}, \\ &D^{jk} \simeq [(-)(j+k), (k-j)], && k-j > 0 \text{ is integer}, \\ &D^{jk} \simeq [+s, (j+k-1/2), (j-k-1/2)], && j-k-1/2 > 0 \text{ is integer}, \\ &D^{jk} \simeq [-s, (j+k-1/2), (k-j-1/2)], && k-j-1/2 > 0 \text{ is integer}. \end{aligned} \tag{11.15}$$

The identical representation of SO(4) is D^{00}, and the self-representation of SO(4) is equivalent to $D^{\frac{1}{2}\frac{1}{2}}$. The self-dual and anti-self-dual antisymmetric tensor representation of rank two of SO(4) are D^{10} and D^{01}, respectively. In fact, from the definition (10.9) we have

$$\begin{aligned} \boldsymbol{\theta}_{12}^{\pm} &= \pm\boldsymbol{\theta}_{34}^{\pm} = \tfrac{1}{2}\ \{(\boldsymbol{\theta}_{12} - \boldsymbol{\theta}_{21}) \pm (\boldsymbol{\theta}_{34} - \boldsymbol{\theta}_{43})\}, \\ \boldsymbol{\theta}_{13}^{\pm} &= \pm\boldsymbol{\theta}_{42}^{\pm} = \tfrac{1}{2}\ \{(\boldsymbol{\theta}_{13} - \boldsymbol{\theta}_{31}) \pm (\boldsymbol{\theta}_{42} - \boldsymbol{\theta}_{24})\}, \\ \boldsymbol{\theta}_{14}^{\pm} &= \pm\boldsymbol{\theta}_{23}^{\pm} = \tfrac{1}{2}\ \{(\boldsymbol{\theta}_{14} - \boldsymbol{\theta}_{41}) \pm (\boldsymbol{\theta}_{23} - \boldsymbol{\theta}_{32})\}, \end{aligned}$$

$$\begin{aligned} \begin{array}{|c|}\hline 1 \\ \hline 2 \\ \hline \end{array} &= \mathcal{Y}^{[1,1]}\boldsymbol{\phi}_{12} = \boldsymbol{\phi}_{12} - \boldsymbol{\phi}_{21} \\ &= \tfrac{1}{2}\{-(\boldsymbol{\theta}_1 + \mathrm{i}\boldsymbol{\theta}_2)(\boldsymbol{\theta}_3 + \mathrm{i}\boldsymbol{\theta}_4) + (\boldsymbol{\theta}_3 + \mathrm{i}\boldsymbol{\theta}_4)(\boldsymbol{\theta}_1 + \mathrm{i}\boldsymbol{\theta}_2)\} = -\boldsymbol{\theta}_{13}^{+} - \mathrm{i}\boldsymbol{\theta}_{14}^{+}, \\ \sqrt{\tfrac{1}{2}}\, F_2 \begin{array}{|c|}\hline 1 \\ \hline 2 \\ \hline \end{array} &= \sqrt{\tfrac{1}{2}} \left(\begin{array}{|c|}\hline 1 \\ \hline 4 \\ \hline \end{array} - \begin{array}{|c|}\hline 2 \\ \hline 3 \\ \hline \end{array} \right) = \sqrt{\tfrac{1}{2}}\, \{\boldsymbol{\phi}_{14} - \boldsymbol{\phi}_{41} - \boldsymbol{\phi}_{23} + \boldsymbol{\phi}_{32}\} \\ &= \sqrt{\tfrac{1}{8}}\, \{-(\boldsymbol{\theta}_1 + \mathrm{i}\boldsymbol{\theta}_2)(\boldsymbol{\theta}_1 - \mathrm{i}\boldsymbol{\theta}_2) + (\boldsymbol{\theta}_1 - \mathrm{i}\boldsymbol{\theta}_2)(\boldsymbol{\theta}_1 + \mathrm{i}\boldsymbol{\theta}_2) \\ &\quad - (\boldsymbol{\theta}_3 + \mathrm{i}\boldsymbol{\theta}_4)(\boldsymbol{\theta}_3 - \mathrm{i}\boldsymbol{\theta}_4) + (\boldsymbol{\theta}_3 - \mathrm{i}\boldsymbol{\theta}_4)(\boldsymbol{\theta}_3 + \mathrm{i}\boldsymbol{\theta}_4)\} = \mathrm{i}\sqrt{2}\boldsymbol{\theta}_{12}^{+}, \\ \sqrt{\tfrac{1}{2}}\, F_2 &\left\{ \sqrt{\tfrac{1}{2}} \left(\begin{array}{|c|}\hline 1 \\ \hline 4 \\ \hline \end{array} - \begin{array}{|c|}\hline 2 \\ \hline 3 \\ \hline \end{array} \right) \right\} = \begin{array}{|c|}\hline 3 \\ \hline 4 \\ \hline \end{array} = \boldsymbol{\phi}_{34} - \boldsymbol{\phi}_{43} \\ &= \tfrac{1}{2}\{(\boldsymbol{\theta}_3 - \mathrm{i}\boldsymbol{\theta}_4)(\boldsymbol{\theta}_1 - \mathrm{i}\boldsymbol{\theta}_2) - (\boldsymbol{\theta}_1 - \mathrm{i}\boldsymbol{\theta}_2)(\boldsymbol{\theta}_3 - \mathrm{i}\boldsymbol{\theta}_4)\} = -\boldsymbol{\theta}_{13}^{+} + \mathrm{i}\boldsymbol{\theta}_{14}^{+}, \\ \begin{array}{|c|}\hline 1 \\ \hline 3 \\ \hline \end{array} &= \mathcal{Y}^{[1,1]}\boldsymbol{\phi}_{13} = \boldsymbol{\phi}_{13} - \boldsymbol{\phi}_{31} \\ &= \tfrac{1}{2}\{-(\boldsymbol{\theta}_1 + \mathrm{i}\boldsymbol{\theta}_2)(\boldsymbol{\theta}_3 - \mathrm{i}\boldsymbol{\theta}_4) + (\boldsymbol{\theta}_3 - \mathrm{i}\boldsymbol{\theta}_4)(\boldsymbol{\theta}_1 + \mathrm{i}\boldsymbol{\theta}_2)\} = -\boldsymbol{\theta}_{13}^{-} + \mathrm{i}\boldsymbol{\theta}_{14}^{-}, \end{aligned}$$

$$
\begin{aligned}
\sqrt{\tfrac{1}{2}}\, F_1 \begin{array}{|c|}\hline 1\\\hline 3\\\hline\end{array} &= \sqrt{\tfrac{1}{2}} \left(\begin{array}{|c|}\hline 1\\\hline 4\\\hline\end{array} + \begin{array}{|c|}\hline 2\\\hline 3\\\hline\end{array} \right) = \sqrt{\tfrac{1}{2}} \left\{ \phi_{14} - \phi_{41} + \phi_{23} - \phi_{32} \right\} \\
&= \sqrt{\tfrac{1}{8}} \left\{ - \left(\boldsymbol{\theta}_1 + \mathrm{i}\boldsymbol{\theta}_2\right)\left(\boldsymbol{\theta}_1 - \mathrm{i}\boldsymbol{\theta}_2\right) + \left(\boldsymbol{\theta}_1 - \mathrm{i}\boldsymbol{\theta}_2\right)\left(\boldsymbol{\theta}_1 + \mathrm{i}\boldsymbol{\theta}_2\right) \right. \\
&\quad \left. + \left(\boldsymbol{\theta}_3 + \mathrm{i}\boldsymbol{\theta}_4\right)\left(\boldsymbol{\theta}_3 - \mathrm{i}\boldsymbol{\theta}_4\right) - \left(\boldsymbol{\theta}_3 - \mathrm{i}\boldsymbol{\theta}_4\right)\left(\boldsymbol{\theta}_3 + \mathrm{i}\boldsymbol{\theta}_4\right) \right\} = \mathrm{i}\sqrt{2}\boldsymbol{\theta}^{-}_{12}, \\
\sqrt{\tfrac{1}{2}}\, F_1 \left\{ \sqrt{\tfrac{1}{2}} \left(\begin{array}{|c|}\hline 1\\\hline 4\\\hline\end{array} + \begin{array}{|c|}\hline 2\\\hline 3\\\hline\end{array} \right) \right\} &= \begin{array}{|c|}\hline 2\\\hline 4\\\hline\end{array} = \phi_{24} - \phi_{42} \\
&= \tfrac{1}{2} \left\{ \left(\boldsymbol{\theta}_3 + \mathrm{i}\boldsymbol{\theta}_4\right)\left(\boldsymbol{\theta}_1 - \mathrm{i}\boldsymbol{\theta}_2\right) - \left(\boldsymbol{\theta}_1 - \mathrm{i}\boldsymbol{\theta}_2\right)\left(\boldsymbol{\theta}_3 + \mathrm{i}\boldsymbol{\theta}_4\right) \right\} = -\boldsymbol{\theta}^{-}_{13} - \mathrm{i}\boldsymbol{\theta}^{-}_{14}.
\end{aligned}
$$

The fundamental spinor representation of SO(4) is $D^{\frac{1}{2}0} \oplus D^{0\frac{1}{2}}$ and its generators (see Eq. (10.45)) are

$$
\begin{aligned}
S_{23} &= \left(\sigma_2 \times \sigma_2\right)/2, & S_{14} &= -\left(\sigma_1 \times \sigma_1\right)/2, \\
S_{31} &= -\left(\sigma_1 \times \sigma_2\right)/2, & S_{24} &= -\left(\sigma_2 \times \sigma_1\right)/2, \\
S_{12} &= \left(\sigma_3 \times \mathbf{1}\right)/2, & S_{34} &= \left(\mathbf{1} \times \sigma_3\right)/2.
\end{aligned}
\tag{11.16}
$$

Through the combination (11.11), $S_a^{\pm}$ are the generators of $D^{\frac{1}{2}0}$ and $D^{0\frac{1}{2}}$, respectively. In fact, through a similarity transformation X, one has

$$
\begin{aligned}
X &= \begin{pmatrix} 1 & 0 & 0 & 0 \\ 0 & 0 & 1 & 0 \\ 0 & 0 & 0 & 1 \\ 0 & -1 & 0 & 0 \end{pmatrix}, \\
X^{-1} S_a^{+} X &= \begin{pmatrix} \sigma_a/2 & 0 \\ 0 & 0 \end{pmatrix}, \qquad X^{-1} S_a^{-} X = \begin{pmatrix} 0 & 0 \\ 0 & \sigma_a/2 \end{pmatrix}.
\end{aligned}
\tag{11.17}
$$

11.1.2 *Single-valued Representations of* O(4)

The group space of O(4) falls into two pieces depending on whether the determinant of the element R is 1 or -1. The piece with det $R = 1$ constructs its invariant subgroup SO(4). As discussed in subsection 10.1.6, the diagonal matrix τ with $\tau_{11} = \tau_{22} = \tau_{33} = 1$ and $\tau_{44} = -1$ is usually chosen to be the representative element in the coset. The transformation rule of generators in the action of τ can be calculated in the self-representation,

$$
\begin{aligned}
\tau T_{ab} \tau^{-1} &= T_{ab}\,, & \tau T_{a4} \tau^{-1} &= -T_{a4}, \\
\tau T_a^{(\pm)} \tau^{-1} &= T_a^{(\mp)}, & \tau \left(T^{(\pm)}\right)^2 \tau^{-1} &= \left(T^{(\mp)}\right)^2,
\end{aligned}
\tag{11.18}
$$

where

$$\left(T^{(\pm)}\right)^2 = \sum_{a=1}^{3} \left(T_a^{(\pm)}\right)^2 . \tag{11.19}$$

We only discuss the single-valued representation D^{jk} of SO(4) in this subsection, where $j+k$ is an integer. Let $\Psi_{\mu\nu}^{jk}$ denote the basis function belonging to the representation D^{jk} of SO(4). P_R is the transformation operator of SO(4) for scalar functions and L_{ab} are its generators. $L_a^{(\pm)}$ and $\left(L^{(\pm)}\right)^2$ can be obtained by Eqs. (11.1) and (11.19).

$$\begin{aligned} P_R \Psi_{\mu\nu}^{jk} &= \sum_{\mu'\nu'} \Psi_{\mu'\nu'}^{jk} D_{\mu'\nu',\mu\nu}^{jk}(R) \\ &= \sum_{\mu'\nu'} \Psi_{\mu'\nu'}^{jk} D_{\mu'\mu}^{j}(\hat{\boldsymbol{n}}^{(+)},\omega^{(+)}) D_{\nu'\nu}^{k}(\hat{\boldsymbol{n}}^{(-)},\omega^{(-)}), \end{aligned} \tag{11.20}$$

$\Psi_{\mu\nu}^{jk}$ is the common eigenfunction of $L_3^{(\pm)}$ and $\left(L^{(\pm)}\right)^2$,

$$\begin{aligned} &L_3^{(+)} \Psi_{\mu\nu}^{jk} = \mu \Psi_{\mu\nu}^{jk}, & &L_3^{(-)} \Psi_{\mu\nu}^{jk} = \nu \Psi_{\mu\nu}^{jk}, \\ &\left(L^{(+)}\right)^2 \Psi_{\mu\nu}^{jk} = j(j+1) \Psi_{\mu\nu}^{jk}, & &\left(L^{(-)}\right)^2 \Psi_{\mu\nu}^{jk} = k(k+1) \Psi_{\mu\nu}^{jk}. \end{aligned} \tag{11.21}$$

In the transformation of τ, $P_\tau \Psi_{\mu\nu}^{jk} = \Phi_{\nu\mu}^{kj}$ belongs to the representation D^{kj} of SO(4) and still is the common eigenfunction of $L_3^{(\pm)}$ and $\left(L^{(\pm)}\right)^2$ with interchanged eigenvalues,

$$\begin{aligned} &L_3^{(+)} \Phi_{\nu\mu}^{kj} = \nu \Phi_{\nu\mu}^{kj}, & &L_3^{(-)} \Phi_{\nu\mu}^{kj} = \mu \Phi_{\nu\mu}^{kj}, \\ &\left(L^{(+)}\right)^2 \Phi_{\nu\mu}^{kj} = k(k+1) \Phi_{\nu\mu}^{kj}, & &\left(L^{(-)}\right)^2 \Phi_{\nu\mu}^{kj} = j(j+1) \Phi_{\nu\mu}^{kj}. \end{aligned}$$

Due to $P_\tau^2 = P_E = \mathbf{1}$, $P_\tau \Phi_{\nu\mu}^{kj} = \Psi_{\mu\nu}^{jk}$.

When $j \neq k$, neither of the two representation spaces of D^{jk} and D^{kj} of SO(4) is invariant in O(4), only their direct sum is invariant. Namely, the subduced representation of an irreducible representation Δ^{jk} of O(4) with respect to the subgroup SO(4) is the direct sum of two irreducible representations D^{jk} and D^{kj} of SO(4). In addition to the indices μ and ν, a new index $\alpha = \pm$ has to be added for the row (column) indices of Δ^{jk} to distinguish the two representation subspaces. For $R \in$ SO(4),

$$\begin{aligned} &\Delta_{\mu\nu+,\mu'\nu'+}^{jk}(R) = D_{\mu\nu,\mu'\nu'}^{jk}(R), \qquad \Delta_{\nu\mu-,\nu'\mu'-}^{jk}(R) = D_{\nu\mu,\nu'\mu'}^{kj}(R), \\ &\Delta_{\mu\nu+,\nu'\mu'-}^{jk}(R) = \Delta_{\nu\mu-,\mu'\nu'+}^{jk}(R) = 0, \\ &\Delta_{\mu\nu\alpha,\nu'\mu'\beta}^{jk}(\tau) = \delta_{(-\alpha)\beta}\delta_{\mu\mu'}\delta_{\nu\nu'}. \end{aligned} \tag{11.22}$$

Changing the sign of $\Delta^{jk}(\tau)$ leads to an equivalent representation.

When $j = k$, we define

$$\psi_{\mu\nu}^{jj\pm} \sim \Psi_{\mu\nu}^{jj} \pm P_\tau \Psi_{\mu\nu}^{jj}, \qquad P_\tau \psi_{\mu\nu}^{jj\pm} = \pm \psi_{\nu\mu}^{jj\pm}.$$

Thus, the subduced representation of each irreducible representation $\Delta^{jj\pm}$ of O(4) with respect to SO(4) is D^{jj} of SO(4):

$$\Delta^{jj\pm}(R) = D^{jj}(R), \quad R \in \mathrm{SO}(4), \qquad \Delta_{\mu\nu,\nu'\mu'}^{jj\pm}(\tau) = \pm\delta_{\mu\mu'}\delta_{\nu\nu'}. \tag{11.23}$$

11.2 The Lorentz Groups

The transformation between two inertial systems in four-dimensional space-time is the Lorentz transformation. There are two commonly used sets of coordinates and metric tensors for the four-dimensional space-time. One is (x_0, x_1, x_2, x_3) with the Minkowski metric tensor $\eta = \mathrm{diag}(1, -1, -1, -1)$ where $x_0 = ct$ (see [Bjorken and Drell (1964)]). The other is (x_1, x_2, x_3, x_4) with the Euclidian metric tensor $\delta_{\mu\nu}$ where $x_4 = \mathrm{i}ct$ (see [Schiff (1968); Marshak et al. (1969)]). The Lorentz transformation matrices for two sets are related by a similarity transformation,

$$\mathcal{A}^T \eta \mathcal{A} = \eta, \qquad A^T A = \mathbf{1}, \qquad X^{-1}\mathcal{A}X = A,$$
$$X = \begin{pmatrix} 0 & 0 & 0 & -\mathrm{i} \\ 1 & 0 & 0 & 0 \\ 0 & 1 & 0 & 0 \\ 0 & 0 & 1 & 0 \end{pmatrix}, \qquad X^{-1} = \begin{pmatrix} 0 & 1 & 0 & 0 \\ 0 & 0 & 1 & 0 \\ 0 & 0 & 0 & 1 \\ \mathrm{i} & 0 & 0 & 0 \end{pmatrix}. \tag{11.24}$$

In this textbook we adopt the second set. The formula for the first set can be found in [Gu et al. (2002)].

11.2.1 *The Proper Lorentz Group and its Cosets*

Due to $A^T A = \mathbf{1}$, the elements of the matrix A satisfy the condition

$$\begin{array}{l} A_{ab} \text{ and } A_{44} \text{ are real}, \\ A_{a4} \text{ and } A_{4a} \text{ are imaginary}, \end{array} \qquad a \text{ and } b = 1,\ 2,\ 3. \tag{11.25}$$

This condition is left invariant in the product of two Lorentz transformations. The set of all such orthogonal matrices A, in the multiplication rule of matrices, constitutes the **homogeneous Lorentz group**, denoted by O(3,1) or L_h. The orthogonal condition $A^T A = \mathbf{1}$ gives

$$\det A = \pm 1, \qquad A_{44}^2 = 1 + \sum_{a=1}^{3} |A_{a4}|^2 \geqslant 1. \tag{11.26}$$

These two discontinuous constraints divide the group space of L_h into four disjointed pieces. The elements in the piece to which the identity E belongs constitute an invariant subgroup L_p of L_h, called the **proper Lorentz group**. The element in L_p satisfies

$$\det A = 1, \qquad A_{44} \geqslant 1. \tag{11.27}$$

Since A_{44} does not have its upper limit, the group space of L_p is an open region in the Euclidean space, and L_p **is a noncompact Lie group**. The representative elements in L_p and its three cosets are usually chosen to be the identity E, the space inversion σ, the time inversion τ, and the space-time inversion ρ:

$$\begin{aligned}
&E = \mathrm{diag}\,(1,\ 1,\ 1,\ 1) \in \mathrm{L}_+^\uparrow = \mathrm{L}_p, && \det A = 1, && A_{44} \geqslant 1,\\
&\sigma = \mathrm{diag}\,(-1,\ -1,\ -1,\ 1) \in \mathrm{L}_-^\uparrow, && \det A = -1, && A_{44} \geqslant 1,\\
&\tau = \mathrm{diag}\,(1,\ 1,\ 1,\ -1) \in \mathrm{L}_-^\downarrow, && \det A = -1, && A_{44} \leqslant -1,\\
&\rho = \mathrm{diag}\,(-1,\ -1,\ -1,\ -1) \in \mathrm{L}_+^\downarrow, && \det A = 1, && A_{44} \leqslant -1.
\end{aligned} \tag{11.28}$$

Four elements constitute the inversion group V_4 of order 4.

11.2.2 *Irreducible Representations of* L_p

Discuss the generators in the self-representation of L_p. Let A denote an infinitesimal element of L_p:

$$\begin{aligned}
&A = \mathbf{1} - \mathrm{i}\alpha X, && A^T = \mathbf{1} - \mathrm{i}\alpha X^T,\\
&\mathbf{1} = A^T A = \mathbf{1} - \mathrm{i}\alpha\left(X + X^T\right), && X^T = -X,\\
&1 = \det A = 1 - \mathrm{i}\alpha \mathrm{Tr} X, && \mathrm{Tr} X = 0.
\end{aligned}$$

Thus, X is a traceless antisymmetric matrix. Expand X with respect to the generators T_{ab} in the self-representation of SO(4):

$$\begin{aligned}
A &= \mathbf{1} - \mathrm{i} \sum_{a<b=2}^{3} \omega_{ab} T_{ab} - \mathrm{i} \sum_{a=1}^{3} \omega_{a4} T_{a4}\\
&= \mathbf{1} - \mathrm{i} \sum_{a=1}^{3} \left(\Omega_a T_a^{(+)} + \Omega_a^* T_a^{(-)}\right),
\end{aligned} \tag{11.29}$$

where ω_{ab} is real, ω_{a4} is imaginary, and

$$\Omega_a = \frac{1}{2}\sum_{b,c=1}^{3} \epsilon_{abc}\omega_{bc} + \omega_{a4}. \tag{11.30}$$

Except for the imaginary parameters, the generators in the self-representations of SO(4) and L_p are completely the same. **The finite-dimensional inequivalent irreducible representations of L_p are also denoted by $D^{jk}(L_p)$**, whose generators are the same as those for SO(4), given in Eq. (11.12). Two representations $D^{jk}(\mathrm{SO}(4))$ and $D^{jk}(\mathrm{L}_p)$ have the same dimensions and are single-valued or double-valued simultaneously. The decompositions of their direct product representations are also the same. However, **the global properties of the two groups are very different because the parameters ω_{a4} in L_p are imaginary**. The finite-dimensional irreducible representation D^{jk} of L_p, except for the identical representation, is not unitary. There are infinite-dimensional unitary representations of L_p [Adams et al. (1987)].

The transformation generated by T_{12} with the real parameter $\omega_{12} = \varphi$ (see the first Lie theorem 5.1) is obviously a pure rotation about the Z-axis, but now it is a four-dimensional matrix:

$$R(\boldsymbol{e}_3, \varphi) = \exp\{-\mathrm{i}\varphi T_{12}\} = \begin{pmatrix} \cos\varphi & -\sin\varphi & 0 & 0 \\ \sin\varphi & \cos\varphi & 0 & 0 \\ 0 & 0 & 1 & 0 \\ 0 & 0 & 0 & 1 \end{pmatrix} \in \mathrm{SO}(3). \tag{11.31}$$

The transformation generated by T_{34} with the imaginal parameter $\omega_{34} = \mathrm{i}\omega$ (see the first Lie theorem 5.1) is a Lorentz boost along the direction of the Z-axis with the relative velocity v:

$$A(\boldsymbol{e}_3, i\omega) = \exp\{-\mathrm{i}(\mathrm{i}\omega)T_{34}\} = \begin{pmatrix} 1 & 0 & 0 & 0 \\ 0 & 1 & 0 & 0 \\ 0 & 0 & \cosh\omega & -\mathrm{i}\sinh\omega \\ 0 & 0 & \mathrm{i}\sinh\omega & \cosh\omega \end{pmatrix}, \tag{11.32}$$
$$\cosh\omega = \left(1 - v^2\right)^{-1/2}, \quad \sinh\omega = v\left(1 - v^2\right)^{-1/2}, \quad v = \tanh\omega,$$

where the natural units $\hbar = c = 1$ are employed. In comparison with the Lorentz boost in the usual physical textbook where the system is fixed, the transformation matrix becomes its inverse because the viewpoint of transformation of the system is used here.

Similar to the Euler angles in SO(3), a new set of parameters of L_p is expected such that each element A of L_p is expanded as **a product of some pure rotations and a Lorentz boost along the Z-direction.**

From a given A matrix in L_p, calculate ω from $A_{44} = \cosh\omega$, and then, calculate the polar angle and the azimuthal angle of the unit vector $\hat{\boldsymbol{n}}(\theta,\varphi)$ in the three-dimensional space, $\hat{\boldsymbol{n}}(\theta,\varphi) = (\mathrm{i}/\sinh\omega)(A_{14}, A_{24}, A_{34})$,

$$\begin{aligned} &A_{44} = \cosh\omega, && \mathrm{i}A_{14}/\sinh\omega = \sin\theta\cos\varphi, \\ &\mathrm{i}A_{24}/\sinh\omega = \sin\theta\sin\varphi, && \mathrm{i}A_{34}/\sinh\omega = \cos\theta. \end{aligned} \tag{11.33}$$

Since

$$R(\boldsymbol{e}_3,\varphi)R(\boldsymbol{e}_2,\theta)A(\boldsymbol{e}_3,\mathrm{i}\omega)\begin{pmatrix}0\\0\\0\\1\end{pmatrix} = \begin{pmatrix}A_{14}\\A_{24}\\A_{34}\\A_{44}\end{pmatrix}, \tag{11.34}$$

$\{R(\boldsymbol{e}_3,\varphi)R(\boldsymbol{e}_2,\theta)A(\boldsymbol{e}_3,i\omega)\}^{-1} A$ is a pure rotation $R(\alpha,\beta,\gamma)$, where the Euler angles α, β, and γ can be calculated. Thus, **an arbitrary Lorentz transformation A in L_p is expressed as a product**

$$\begin{aligned} A = A(\varphi,\theta,\omega,\alpha,\beta,\gamma) &= R(\varphi,\theta,0)A(\boldsymbol{e}_3,\mathrm{i}\omega)R(\alpha,\beta,\gamma), \\ R(\alpha,\beta,\gamma) &= R(\boldsymbol{e}_3,\alpha)R(\boldsymbol{e}_2,\beta)R(\boldsymbol{e}_3,\gamma), \end{aligned} \tag{11.35}$$

where

$$\begin{aligned} &0 \leqslant \omega < \infty, && 0 \leqslant \theta \leqslant \pi, && 0 \leqslant \beta \leqslant \pi, \\ &-\pi \leqslant \varphi \leqslant \pi, && -\pi \leqslant \alpha \leqslant \pi, && -\pi \leqslant \gamma \leqslant \pi. \end{aligned} \tag{11.36}$$

The geometrical meaning of the expression (11.35) is evident. Two rotations in the two sides of $A(\boldsymbol{e}_3,\mathrm{i}\omega)$ transform two frames before and after the Lorentz transformation A such that two Z-axes are changed to the direction of the relative motion, and the remaining axes are changed to be parallel to each other, respectively. After two rotations, the Lorentz transformation A is simplified into $A(\boldsymbol{e}_3,\mathrm{i}\omega)$.

Due to Eq. (11.30), $D^{jk}(A)$ is the representation matrix of a pure rotation when Ω_a of $A \in L_p$ is real, and it is that of a Lorentz boost when Ω_a is imaginary:

$$\begin{aligned} D^{jk}(\alpha,\beta,\gamma) &= D^j(\alpha,\beta,\gamma) \times D^k(\alpha,\beta,\gamma), \\ D^{jk}(\boldsymbol{e}_3,\mathrm{i}\omega) &= \exp(\omega I_3^j) \times \exp(-\omega I_3^k), \end{aligned} \tag{11.37}$$

where I_3^j and I_3^k are diagonal. Thus,

$$\begin{aligned} D^{jk}_{\mu\nu,\mu'\nu'}(\varphi,\theta,\omega,\alpha,\beta,\gamma) = \sum_{\rho\tau} &e^{-\mathrm{i}(\mu+\nu)\varphi} d^j_{\mu\rho}(\theta) d^k_{\nu\tau}(\theta) e^{(\rho-\tau)\omega} \\ &\times e^{-\mathrm{i}(\rho+\tau)\alpha} d^j_{\rho\mu'}(\beta) d^k_{\tau\nu'}(\beta) e^{-\mathrm{i}(\mu'+\nu')\gamma}. \end{aligned} \tag{11.38}$$

11.2.3 *The Covering Group of* L_p

An exponential mapping exists for a matrix Lie group G if its every element can be written in an exponential function of matrix where the exponent belongs to its real Lie algebra. Equation (11.5) shows that there exists an exponential mapping of SO(4). Based on Eq. (11.5) the covering group of SO(4) is found. **The exponential mapping exists for each compact Lie group**, but is not necessary to exist for a noncompact Lie group. For example, the set of two-dimensional unimodular complex matrices, in the multiplication rule of matrices, constitutes a group SL$(2, C)$. SL$(2, C)$ is noncompact and its exponential mapping does not exist, because the following elements cannot be written in an exponential form in the condition that the exponent belongs to the real Lie algebra of SL$(2, C)$,

$$\begin{pmatrix} -1 & -2 \\ 0 & -1 \end{pmatrix} = -e^{\sigma_1 + i\sigma_2}, \qquad \begin{pmatrix} -1-i & -1 \\ -1 & -1+i \end{pmatrix} = -e^{\sigma_1 + i\sigma_3}.$$

In this subsection we are going to find the covering group of L_p and to show that every element in L_p can be written in an exponential function of matrix where the exponent belongs to its real Lie algebra, namely, **there exists an exponential mapping of** $\mathbf{L}_p$ although L_p is noncompact and some elements in L_p cannot be diagonalized.

The self-representation of SO(4) can be changed to $D^{\frac{1}{2}\frac{1}{2}}$ by the similarity transformation N (see Eq. (11.5)), so can that of L_p because the generators in the two representations are the same. Due to

$$\begin{aligned} &u(\boldsymbol{e}_a, \omega) = e^{-i\omega\sigma_a/2}, \qquad u(\boldsymbol{e}_3, i\omega) = e^{\omega\sigma_3/2}, \\ &D^{1/2}(\alpha, \beta, \gamma) = u(\alpha, \beta, \gamma) = e^{-i\alpha\sigma_3/2} e^{-i\beta\sigma_2/2} e^{-i\gamma\sigma_3/2}, \end{aligned}$$

we have

$$\begin{aligned} A = &\ A(\varphi, \theta, \omega, \alpha, \beta, \gamma) \\ = &\ N\left\{u(\varphi, \theta, 0) \times u(\varphi, \theta, 0)\right\} \left\{\exp\left(\omega\sigma_3/2\right) \times \exp\left(-\omega\sigma_3/2\right)\right\} \\ &\cdot \left\{u(\alpha, \beta, \gamma) \times u(\alpha, \beta, \gamma)\right\} N^{-1} \\ = &\ N\left\{M \times (\sigma_2 M^* \sigma_2)\right\} N^{-1}, \\ M = &\ u(\varphi, \theta, 0) \exp\left(\omega\sigma_3/2\right) u(\alpha, \beta, \gamma) \in \mathrm{SL}(2, C). \end{aligned} \tag{11.39}$$

The determinant of M is 1 so that M belongs to SL$(2, C)$. Equation (11.39) gives a map from each element A in L_p to at least one element M in SL$(2, C)$. If one element A in L_p maps two different elements M and M' in SL$(2, C)$ through Eq. (11.39), then,

$$\mathbf{1} = N^{-1}\left(AA^{-1}\right)N = MM'^{-1} \times \sigma_2 \left(MM'^{-1}\right)^* \sigma_2.$$

Namely, MM'^{-1} is a constant matrix $c\mathbf{1}$. Since det M = det $M' = 1$, one has $c^2 = 1$. **Each element A in L_p maps onto two elements $\pm M$ in** $\mathrm{SL}(2, C)$. In the following we are going to show that each pair of elements $\pm M$ in $\mathrm{SL}(2, C)$ maps onto one element A in L_p. Since the orders of $\mathrm{SL}(2, C)$ and L_p both are 6, there exists a two-to-one correspondence between two groups. The correspondence is left invariant in the multiplication of elements. L_p is the homomorphic image of $\mathrm{SL}(2, C)$. $\mathrm{SL}(2, C)$ **is the covering group of** L_p:

$$\mathrm{SL}(2, C) \sim \mathrm{L}_p. \tag{11.40}$$

Since det $M = 1$, the product of two eigenvalues of M is equal to 1. M can be diagonalized if its two eigenvalues are different,

$$Y^{-1}MY = \begin{pmatrix} \mathrm{e}^{-\mathrm{i}\tau} & 0 \\ 0 & \mathrm{e}^{\mathrm{i}\tau} \end{pmatrix} = \exp\left\{-\mathrm{i}\begin{pmatrix} \tau & 0 \\ 0 & -\tau \end{pmatrix}\right\}, \quad Y \in \mathrm{SL}(2, C). \tag{11.41}$$

Thus, $M = \exp\left(-\mathrm{i}\tau Y\sigma_3 Y^{-1}\right)$ and τ is a complex number with $-\pi <$ Re $\tau \leqslant \pi$ and Im $\tau > 0$. If two eigenvalues of M both are 1 but $M \neq \mathbf{1}$, M can be transformed to

$$Z^{-1}MZ = \begin{pmatrix} 1 & -2 \\ 0 & 1 \end{pmatrix} = \exp\left\{\begin{pmatrix} 0 & -2 \\ 0 & 0 \end{pmatrix}\right\}, \qquad Z \in \mathrm{SL}(2, C). \tag{11.42}$$

Thus, $M = \exp\left\{-Z\left(\sigma_1 + \mathrm{i}\sigma_2\right)Z^{-1}\right\}$. If two eigenvalues of M both are -1, we can diagonalize $-M$ instead of M. Therefore, **by removing the possible minus sign**, $M = \exp(-\mathrm{i}B)$ where B is a traceless matrix and can be expanded with respect to the Pauli matrices,

$$M = \exp\left(-\mathrm{i}\boldsymbol{\Omega} \cdot \boldsymbol{\sigma}/2\right). \tag{11.43}$$

Substituting Eq. (11.43) into (11.39), we obtain a matrix:

$$\begin{aligned}
&N\left\{\exp\left(-\mathrm{i}\boldsymbol{\Omega} \cdot \boldsymbol{\sigma}/2\right) \times \exp\left(-\mathrm{i}\boldsymbol{\Omega}^* \cdot \boldsymbol{\sigma}/2\right)\right\}N^{-1} \\
&\quad = \exp\left(-\mathrm{i}\boldsymbol{\Omega} \cdot \boldsymbol{T}^{(+)}\right)\exp\left(-\mathrm{i}\boldsymbol{\Omega}^* \cdot \boldsymbol{T}^{(-)}\right) \\
&\quad = \exp\left\{-\mathrm{i}\sum_{a=1}^{3}\left(\Omega_a T_a^{(+)} + \Omega_a^* T_a^{(-)}\right)\right\} \\
&\quad = \exp\left\{-\mathrm{i}\sum_{a<b}\omega_{ab}T_{ab} - \mathrm{i}\sum_{a=1}^{3}\omega_{a4}T_{a4}\right\},
\end{aligned} \tag{11.44}$$

where ω_{ab} is real and ω_{a4} is pure imaginary,

$$\omega_{ab} = \frac{1}{2}\sum_{c=1}^{3} \epsilon_{abc}\left(\Omega_c + \Omega_c^*\right), \qquad \omega_{a4} = \frac{1}{2}\left(\Omega_a - \Omega_a^*\right). \tag{11.45}$$

The exponent in Eq. (11.44) belongs to the real Lie algebra of L_p. So that due to the first Lie Therorem 5.1 the matrix given in Eq. (11.44) belongs to L_p and we have proved that two elements $\pm M$ in $\mathrm{SL}(2, C)$ maps onto one element A in L_p. Equation (11.44) also shows that there exists an exponential mapping of L_p. When the parameters are infinitesimal, Eq. (11.44) returns to Eq. (11.29). The representation matrix of A in D^{jk} is

$$\begin{aligned} D^{jk}(A) &= \exp\left\{-\mathrm{i}\sum_{a<b} \omega_{ab} I_{ab}^{jk} - \mathrm{i}\sum_{a=1}^{3} \omega_{a4} I_{a4}^{jk}\right\} \\ &= \exp\left\{-\mathrm{i}\sum_{a=1}^{3} \Omega_a I_a^j\right\} \times \exp\left\{-\mathrm{i}\sum_{a=1}^{3} \Omega_a^* I_a^k\right\}. \end{aligned} \tag{11.46}$$

In principle, **two sets of parameters can be transformed to each other**, although the transformation is quite complicated. In fact, from the parameters $(\varphi, \theta, \omega, \alpha, \beta, \gamma)$ one is able to calculate M by Eq. (11.39) and determines the parameters ω_{ab} and ω_{a4} by Eqs. (11.43) and (11.45). Conversely, from the parameters ω_{ab} and ω_{a4} one calculates M by Eq. (11.43) and A by Eq. (11.39), and then, determines the parameters $(\varphi, \theta, \omega, \alpha, \beta, \gamma)$ by Eqs. (11.33) and (11.35).

11.2.4 *Classes of* L_p

The classes of L_p can be studied from the classes of $\mathrm{SL}(2, C)$ owing to $\mathrm{SL}(2, C) \sim \mathrm{L}_p$. When two eigenvalues of $M \in \mathrm{SL}(2, C)$ are different, M is conjugate to an element given in Eq. (11.41), which corresponds to an element $A(\varphi, 0, \omega, 0, 0, 0)$ in L_p:

$$A = \begin{pmatrix} \cos\varphi & -\sin\varphi & 0 & 0 \\ \sin\varphi & \cos\varphi & 0 & 0 \\ 0 & 0 & \cosh\omega & -\mathrm{i}\sinh\omega \\ 0 & 0 & \mathrm{i}\sinh\omega & \cosh\omega \end{pmatrix} \qquad \begin{array}{l} -\pi \leqslant \varphi \leqslant \pi, \\ 0 \leqslant \omega < \infty. \end{array} \tag{11.47}$$

Since $\Omega_3 = 2\tau$, $\omega_{12} = \varphi = \tau + \tau^* = 2\mathrm{Re}\,\tau$ and $\omega_{34} = \mathrm{i}\omega = \tau - \tau^* = 2\mathrm{i}\ \mathrm{Im}\,\tau$. Two φ with difference 2π describe two elements $\pm M$ in $\mathrm{SL}(2, C)$ but one element $A(\varphi, 0, \omega, 0, 0, 0)$ in L_p. $\pm M$ belong to two classes in $\mathrm{SL}(2, C)$ but correspond to one class in L_p.

If two eigenvalues of M both are 1, there are four classes in SL$(2, C)$ but only two classes in L_p. Two classes $M = \pm 1$ in SL$(2, C)$ correspond to one class (identity) in L_p. All elements M, which conjugate to an element given in Eq. (11.42), constitute one class in SL$(2, C)$ and those elements $-M$ constitute another class in it. These two classes of SL$(2, C)$ both correspond to one class of L_p. Since

$$\begin{aligned} & u(-\pi, \pi/4, 0) \begin{pmatrix} \sqrt{2}+1 & 0 \\ 0 & \sqrt{2}-1 \end{pmatrix} u(\pi, 3\pi/4, 0) = \begin{pmatrix} 1 & -2 \\ 0 & 1 \end{pmatrix} \\ & = \exp\left\{ \begin{pmatrix} 0 & -2 \\ 0 & 0 \end{pmatrix} \right\} = \exp\{-\sigma_1 - \mathrm{i}\sigma_2\} = \exp\left\{ -\mathrm{i} \sum_{a=1}^{3} \Omega_a \sigma_a / 2 \right\}, \end{aligned}$$

where $\Omega_1 = -2\mathrm{i}$, $\Omega_2 = 2$, and $\Omega_3 = 0$, the representative element in this class of L_p is $A(-\pi, \pi/4, \omega, \pi, 3\pi/4, 0)$ with $\cosh\omega = 3$,

$$A = \mathrm{e}^{-\mathrm{i}2T_{31} - 2T_{14}} = \begin{pmatrix} 1 & 0 & 2 & 2\mathrm{i} \\ 0 & 1 & 0 & 0 \\ -2 & 0 & -1 & -2\mathrm{i} \\ -2\mathrm{i} & 0 & -2\mathrm{i} & 3 \end{pmatrix}. \tag{11.48}$$

This matrix A is a proper Lorentz transformation matrix where four eigenvalues all are 1.

11.2.5 *Irreducible Representations of* L_h

The irreducible representations of L_h can be obtained from those of L_p and the properties of the representative elements τ and ρ of two cosets $L_-^{\downarrow}$ and $L_+^{\downarrow}$. Another representative element $\sigma = \tau\rho$ in the coset $L_-^{\uparrow}$ is calculable. Since ρ commutes with every element in L_h and $\rho^2 = E$, it takes a constant matrix in an irreducible representation of L_h. The property of τ has been given in Eq. (11.18).

We introduce four irreducible representations of the inversion group V_4 of order 4,

$$\begin{aligned} & V^{(1)}(\tau) = V^{(2)}(\tau) = V^{(1)}(\rho) = V^{(4)}(\rho) = 1, \\ & V^{(3)}(\tau) = V^{(4)}(\tau) = V^{(2)}(\rho) = V^{(3)}(\rho) = -1. \end{aligned} \tag{11.49}$$

Corresponding to the single-valued representation $D^{jk}(L_p)$, where $j+k$ is integer, we calculate the irreducible representation $D^{jk}(L_h)$, just like those of O(4). When $j = k$, there are four irreducible representations $\Delta^{jj\lambda}(L_h)$ whose subduced representation with respect to L_p is $D^{jj}(L_p)$:

$$\begin{aligned}
&\Delta^{jj\lambda}(A) = D^{jj}(A), \qquad A \in \mathrm{L}_p, \\
&\Delta^{jj\lambda}_{\mu\nu,\nu'\mu'}(\tau) = V^{(\lambda)}(\tau)\delta_{\mu\mu'}\delta_{\nu\nu'}, \\
&\Delta^{jj\lambda}(\rho) = V^{(\lambda)}(\rho)\ \mathbf{1}, \qquad 1 \leqslant \lambda \leqslant 4.
\end{aligned} \tag{11.50}$$

When $j \neq k$ but $j+k$ is an integer, only the direct sum of two representation spaces of $D^{jk}(\mathrm{L}_p)$ and $D^{kj}(\mathrm{L}_p)$ is left invariant for the group L_h. There are two irreducible representations $\Delta^{jk\pm}(\mathrm{L}_h)$ whose subduced representation with respect to L_p is $D^{jk}(\mathrm{L}_p) \oplus D^{kj}(\mathrm{L}_p)$:

$$\begin{aligned}
&\Delta^{jk\pm}(A) = D^{jk}(A) \oplus D^{kj}(A), \qquad A \in \mathrm{L}_p, \qquad \Delta^{jk\pm}(\rho) = \pm\mathbf{1}, \\
&\Delta^{jk\pm}_{\mu\nu\alpha,\nu'\mu'\beta}(\tau) = \delta_{(-\alpha)\beta}\delta_{\mu\mu'}\delta_{\nu\nu'}, \qquad \alpha,\ \beta = \pm 1,
\end{aligned} \tag{11.51}$$

where two representation spaces of $D^{jk}(\mathrm{L}_p)$ and $D^{kj}(\mathrm{L}_p)$ are distinguished by the subscript α ($\alpha = \pm 1$). Changing the sign of $\Delta^{jk\pm}(\tau)$ leads to an equivalent representation.

For the double-valued representation $D^{jk}(\mathrm{L}_p)$, we only discuss the Dirac spinor representation $D(\mathrm{L}_h)$, whose subduced representation with respect to L_p is $D^{\frac{1}{2}0}(\mathrm{L}_p) \oplus D^{0\frac{1}{2}}(\mathrm{L}_p)$. Let

$$\overline{\gamma}_\mu = \sum_{\nu=1}^{4} A_{\mu\nu}\gamma_\nu, \qquad 1 \leqslant \mu \leqslant 4.$$

Since $A \in \mathrm{L}_h$ is an orthogonal matrix, $\overline{\gamma}_\mu$ also satisfy the anticommutative relations (10.35). From Theorem 10.1 two sets of γ_μ and $\overline{\gamma}_\mu$ are equivalent and they are related by a unimodular similarity transformation $D(A)$,

$$D(A)^{-1}\gamma_\mu D(A) = \sum_{\nu=1}^{4} A_{\mu\nu}\gamma_\nu, \qquad \det D(A) = 1. \tag{11.52}$$

The set of $D(A)$ forms a multiple-valued representation of L_h where the generators are

$$I_{\mu\nu} = \frac{-\mathrm{i}}{4}\left(\gamma_\mu\gamma_\nu - \gamma_\nu\gamma_\mu\right). \tag{11.53}$$

Introduce the charge conjugate matrix C

$$C^{-1}\gamma_\mu C = -\gamma_\mu^T, \quad C^\dagger C = \mathbf{1}, \quad C^T = -C, \quad \det C = 1. \tag{11.54}$$

Due to $C^{-1}I_{\mu\nu}C = -I_{\mu\nu}^T$, a new constraint is added to restrict the representation to be double-valued,

$$C^{-1}D(A)C = \left\{D(A^{-1})\right\}^T, \qquad A \in \mathrm{L}_p. \tag{11.55}$$

Remind that the right-hand side of Eq. (11.55) is not equal to $D(A)^*$. In physics, the representation matrices of four representative elements in L_h are restricted by

$$\begin{aligned} & C^{-1}D(A)C = \frac{A_{44}}{|A_{44}|}\left\{D(A^{-1})\right\}^T, \qquad A \in L_h, \\ & D(\sigma) = \pm i\gamma_4, \qquad D(\tau) = \pm\gamma_4\gamma_5, \qquad D(\rho) = \pm i\gamma_5 \, . \end{aligned} \tag{11.56}$$

There is a one-to-two correspondence between A in L_h and $\pm D(A)$, which satisfy Eqs. (11.52) and (11.56). The set of $D(A)$ is called the **Dirac spinor representation** which is the covering group of L_h. For L_p, the Dirac spinor representation is decomposed to two inequivalent irreducible representations of dimension 2, both of which are faithful representations of the covering group SL$(2, C)$ of L_p. **The Dirac spinor representation is not unitary.** We take the conjugate of Eq. (11.52):

$$\begin{aligned} D(A)^\dagger \gamma_a D(A^{-1})^\dagger &= \sum_{b=1}^{3} A_{ab}\gamma_b \; - \; A_{a4}\gamma_4 \, . \\ D(A)^\dagger \gamma_4 D(A^{-1})^\dagger &= -\sum_{b=1}^{3} A_{4b}\gamma_b \; + \; A_{44}\gamma_4 . \end{aligned}$$

By taking a similarity transformation γ_4, we obtain

$$\left\{\gamma_4 D(A)^\dagger \gamma_4\right\} \gamma_\mu \left\{\gamma_4 D(A)^\dagger \gamma_4\right\}^{-1} = \sum_{\nu=1}^{4} A_{\mu\nu}\gamma_\nu .$$

From Theorem 10.1 we have

$$\gamma_4 D(A)^\dagger \gamma_4 = cD(A)^{-1}.$$

The constant c is determined by Eq. (11.56):

$$\gamma_4 D(A)^\dagger \gamma_4 = \frac{A_{44}}{|A_{44}|}\, D(A)^{-1}. \tag{11.57}$$

This is the reason for that the Lorentz invariant constructed by the spinor field ψ is $\overline{\Psi}\,\Psi$ instead of $\Psi^\dagger\,\Psi$:

$$\overline{\Psi}\,\Psi = \Psi^\dagger \gamma_4 \Psi . \tag{11.58}$$

11.3 Dirac Equation in $(N + 1)$-dimensional Space-time

The Dirac equation in $(D + 1)$-dimensional space-time can be expressed as (see [Schiff (1968)])

$$\sum_{\mu=1}^{N+1} \gamma_\mu \left(\frac{\partial}{\partial x_\mu} - ieA_\mu \right) \boldsymbol{\Psi}(\boldsymbol{x}, t) + M\boldsymbol{\Psi}(\boldsymbol{x}, t) = 0, \tag{11.59}$$

where M is the mass of the particle, and $(N + 1)$ matrices γ_μ satisfy the anticommutative relation:

$$\gamma_\mu \gamma_\nu + \gamma_\nu \gamma_\mu = 2\delta_{\mu\nu} \mathbf{1}. \tag{11.60}$$

For simplicity, the natural units $\hbar = c = 1$ are employed. Discuss the special case where only the time component A_{N+1} is nonvanishing and it is spherically symmetric:

$$eA_{N+1} = iV(r), \qquad A_a = 0 \qquad \text{when } 1 \leqslant a \leqslant N. \tag{11.61}$$

The Hamiltonian $H(\boldsymbol{x})$ of the system is expressed as

$$\begin{aligned} &i\frac{\partial}{\partial t}\boldsymbol{\Psi}(\boldsymbol{x}, t) = H(\boldsymbol{x})\boldsymbol{\Psi}(\boldsymbol{x}, t), \\ &H(\boldsymbol{x}) = \sum_{a=1}^{N} \gamma_{N+1}\gamma_a \frac{\partial}{\partial x_a} + V(r) + \gamma_{N+1}M. \end{aligned} \tag{11.62}$$

The orbital angular momentum operator L_{ab} is given in Eq. (10.92). The spinor operator S_{ab} is given in Eq. (10.61). The total angular momentum operator is $J_{ab} = L_{ab} + S_{ab}$. There are three Casimir operators of order 2 for the total, orbital, and spinor wave functions, respectively,

$$J^2 = \sum_{a<b=2}^{N} J_{ab}^2, \qquad L^2 = \sum_{a<b=2}^{N} L_{ab}^2, \qquad S^2 = \sum_{a<b=2}^{N} S_{ab}^2. \tag{11.63}$$

Due to the spherical symmetry, J_{ab} commutes with the Hamiltonian $H(\boldsymbol{x})$. There is another conservative operator κ which commutes with both J_{ab} and $H(\boldsymbol{x})$ (see [Schiff (1968)])

$$\begin{aligned} \kappa &= -i\gamma_{N+1} \sum_{a<b=2}^{N} \gamma_a \gamma_b L_{ab} + \frac{1}{2}\gamma_{N+1}(N - 1) \\ &= \gamma_{N+1} \left\{ J^2 - L^2 - S^2 + (N - 1)/2 \right\}. \end{aligned} \tag{11.64}$$

The set of mutually commutable operators consists of $H(x)$, J^2, κ, S^2, **and the Chevalley bases** $H_\mu(J)$. Their common eigenfunctions span an irreducible representation space of SO(N). Due to the spherical symmetry, we only need **to calculate the eigenfunction with the highest weight**, which is the linear combination of the products of the radial function, the spherical harmonic functions $Y_{\boldsymbol{m}}^{[\lambda_1]}(\hat{\boldsymbol{x}})$ and the fundamental basis spinor $\chi(\boldsymbol{m}_s)$, combined by the Clebsch–Gordan coefficients,

We introduce a set of N unitary matrices β_a, satisfying

$$\beta_a\beta_b + \beta_b\beta_a = 2\delta_{ab}\mathbf{1}. \tag{11.65}$$

The explicit forms of β_a are the same as those γ_a given in Eq. (10.46). When $N = 2\ell+1$, $\beta_{2\ell+1}$ is given in Eq. (10.53).

(a) *The case of* $N = 2\ell + 1$

Define $(N+1)$ γ_a matrices satisfying the anticommutative relations (11.60)

$$\gamma_{N+1} = \sigma_3 \times \mathbf{1}, \qquad \gamma_a = \sigma_2 \times \beta_a, \qquad 1 \leqslant a \leqslant 2\ell+1. \tag{11.66}$$

The spinor operator S_{ab} and the κ operator become the block matrices and they commute with γ_{N+1}:

$$\begin{aligned} S_{ab} &= \mathbf{1} \times \overline{S}_{ab}, \qquad & \overline{S}_{ab} &= -\mathrm{i}\left(\beta_a\beta_b - \beta_b\beta_a\right)/4, \\ \kappa &= \sigma_3 \times \overline{\kappa}, & \overline{\kappa} &= -\mathrm{i}\sum_{a<b=2}^{N} \beta_a\beta_b L_{ab} + \frac{N-1}{2}. \end{aligned} \tag{11.67}$$

The relation between S_{ab} and $\overline{S}_{ab}$ is similar to that between the spinor operators for the Dirac spinors and those for the Pauli spinors in the usual 3 + 1 space-time.

At the level of the Pauli spinors, the fundamental spinor $\chi[\boldsymbol{m}]$ belongs to the fundamental spinor representation $[s]$ with the highest weight $(0, \cdots, 0, 1)$. Due to Eq. (10.86), **there are two sets of the wave functions belonging to the representation** $[j] \equiv [s, \lambda_1]$ of SO($2\ell+1$), called the spherical spinor functions (see Eq. (5.195)). Two sets of the spherical spinor functions come from the decompositions of $[s] \times [\lambda_1]$ and $[s] \times [\lambda_1 + 1]$, respectively. We are going to calculate the spherical spinor functions $Y_{\kappa,[j]}(\hat{\boldsymbol{x}})$ with the highest weight. The Clebsch–Gordan coefficients are calculated **from the condition that the highest weight state is annihilated by each raising operators** $E_\mu(J)$.

Before calculating the highest weight states, we first calculate the eigenvalues K of κ (see Eq. (11.64)) which relates to the casimir invariants of order 2. Substituting the Cartan matrix A^{-1} of the Lie algebra B_ℓ (see subsection 7.5.6) into Eq. (7.133), we obtain

$$\begin{aligned} &\boldsymbol{M}([j]) = (\lambda_1, 0, \cdots, 0, 1), \quad C_2([j]) = \lambda_1(\lambda_1 + 2\ell) + \ell(2\ell+1)/4, \\ &\boldsymbol{M}([\lambda_1]) = (\lambda_1, 0, \cdots, 0), \quad C_2([\lambda_1]) = \lambda_1(\lambda_1 + 2\ell - 1), \\ &\boldsymbol{M}([s]) = (0, \cdots, 0, 1), \qquad C_2([s]) = \ell(2\ell+1)/4. \end{aligned} \tag{11.68}$$

Then, due to Eq. (11.64),

$$\begin{aligned} &C_2([j]) - C_2([\lambda_1]) - C_2([s]) + \ell = \lambda_1 + \ell \equiv |K|, \\ &C_2([j]) - C_2([\lambda_1+1]) - C_2([s]) + \ell = -\lambda_1 - \ell = -|K|. \end{aligned} \tag{11.69}$$

The highest weight state $Y_{|K|,[j]}(\hat{\boldsymbol{x}})$, which comes from the decompositions of $[s] \times [\lambda_1]$, is easy to calculate:

$$\begin{aligned} Y_{|K|,[j]}(\hat{\boldsymbol{x}}) &= Y^{[\lambda_1]}_{(\lambda_1,0,\cdots,0)}(\hat{\boldsymbol{x}})\chi[(0,\cdots,0,1)] \\ &= C_{(2\ell+1),\lambda_1}\left\{\frac{(-1)^\ell X_1}{r\sqrt{2}}\right\}^{\lambda_1} \chi[(0,\cdots,0,1)] \\ &= C_{(2\ell+1),\lambda_1}\left\{\frac{(-1)^\ell (x_1 + \mathrm{i}x_2)}{r\sqrt{2}}\right\}^{\lambda_1} \chi_{1\cdots 1}, \end{aligned} \tag{11.70}$$

Another highest weight state $Y_{-|K|,[j]}(\hat{\boldsymbol{x}})$, which comes from the decompositions of $[s] \times [\lambda_1 + 1]$, is calculated by Eq. (10.95):

$$\begin{aligned} Y_{-|K|,[j]}(\hat{\boldsymbol{x}}) &= \frac{C_{(2\ell+1),(\lambda_1+1)}\sqrt{\lambda_1+1}}{r\sqrt{2\ell+2\lambda_1+1}}\left\{\frac{(-1)^\ell X_1}{r\sqrt{2}}\right\}^{\lambda_1} \\ &\quad \times \{X_{\ell+1}\chi[(0,\cdots,0,1)] + X_\ell\chi[(0,\cdots,0,1,\overline{1})] \\ &\quad + X_{\ell-1}\chi[(0,\cdots,0,1,\overline{1},1)] + \cdots \\ &\quad + X_2\chi[(1,\overline{1},0,\cdots,0,1)] + X_1\chi[(\overline{1},0,\cdots,0,1)]\} \\ &= r^{-1}C_{(2\ell+1),\lambda_1}\left\{\frac{(-1)^\ell(x_1+\mathrm{i}x_2)}{r\sqrt{2}}\right\}^{\lambda_1} \{x_{2\ell+1}\chi_{1\cdots 1} \\ &\quad +(x_{2\ell-1}+\mathrm{i}x_{2\ell})\chi_{1\cdots 12} + (x_{2\ell-3}+\mathrm{i}x_{2\ell-2})\chi_{1\cdots 121} + \cdots \\ &\quad + (x_3+\mathrm{i}x_4)\chi_{121\cdots 1} + (x_1+\mathrm{i}x_2)\chi_{21\cdots 1}\} \\ &= (\boldsymbol{\beta}\cdot\hat{\boldsymbol{x}})\, Y_{|K|,[j]}(\hat{\boldsymbol{x}}), \end{aligned} \tag{11.71}$$

where from Eq. (10.98) $C_{2\ell+1,(\lambda_1+1)}\sqrt{\dfrac{\lambda_1+1}{2\ell+1+2\lambda_1}} = C_{2\ell+1,\lambda_1}$, and

$$\boldsymbol{\beta}\cdot\hat{\boldsymbol{x}} = r^{-1}\sum_{a=1}^{N}\beta_a x_a, \qquad (\boldsymbol{\beta}\cdot\hat{\boldsymbol{x}})^2 = \mathbf{1}. \tag{11.72}$$

Thus,

$$\begin{aligned} \overline{\kappa} Y_{K,[j]}(\hat{\boldsymbol{x}}) &= K Y_{K,[j]}(\hat{\boldsymbol{x}}), \\ (\boldsymbol{\beta}\cdot\hat{\boldsymbol{x}})\, \boldsymbol{\phi}_{K,[j]}(\hat{\boldsymbol{x}}) &= \boldsymbol{\phi}_{-K,[j]}(\hat{\boldsymbol{x}}). \end{aligned} \tag{11.73}$$

This is the generalization of Eq. (5.196). Introducing

$$\boldsymbol{\beta}\cdot\bigtriangledown = \sum_{a=1}^{N}\beta_a\frac{\partial}{\partial x_a}, \quad (\boldsymbol{\beta}\cdot\hat{\boldsymbol{x}})(\boldsymbol{\beta}\cdot\bigtriangledown) = \frac{\partial}{\partial r} + \frac{\mathrm{i}}{r}\sum_{a<b}\beta_a\beta_b L_{ab}, \tag{11.74}$$

we have

$$\begin{aligned} &(\boldsymbol{\beta}\cdot\bigtriangledown) r^{-\ell} f(r) Y_{K,[j]}(\hat{\boldsymbol{x}}) \\ &= (\boldsymbol{\beta}\cdot\hat{\boldsymbol{x}})\left[\frac{\partial}{\partial r} + \frac{\mathrm{i}}{r}\sum_{a<b}\beta_a\beta_b L_{ab}\right] r^{-\ell} f(r) Y_{K,[j]}(\hat{\boldsymbol{x}}) \\ &= r^{-\ell}\left[\frac{df(r)}{dr} - \frac{Kf(r)}{r}\right] Y_{-K,[j]}(\hat{\boldsymbol{x}}). \end{aligned} \tag{11.75}$$

At the level of the Dirac spinors, the wave function $\boldsymbol{\Psi}_{K,[j]}(\boldsymbol{x})$ with the highest weight in the irreducible representation $[j]$ can be expressed as

$$\begin{aligned} &\boldsymbol{\Psi}_{K,[j]}(\boldsymbol{x},t) = r^{-\ell} e^{-\mathrm{i}Et}\begin{pmatrix} F_K(r) Y_{K,[j]}(\hat{\boldsymbol{x}}) \\ \mathrm{i} G_K(r) Y_{-K,[j]}(\hat{\boldsymbol{x}}) \end{pmatrix}, \\ &\kappa\Psi_{K,(j)}(\boldsymbol{x}) = K\Psi_{K,(j)}(\boldsymbol{x}), \qquad K = \pm(\lambda_1+\ell). \end{aligned} \tag{11.76}$$

The factor i in the radial function $\mathrm{i}G_K(r)$ comes from the definition (11.66) for γ_a (Note the factor σ_2). Substituting $\boldsymbol{\Psi}_{K,[j]}(\boldsymbol{x},t)$ into the Dirac equation (11.62) one obtains the radial equation

$$\begin{aligned} &\frac{dG_K(r)}{dr} + \frac{K}{r}G_K(r) = [E - V(r) - M]F_K(r), \\ &-\frac{dF_K(r)}{dr} + \frac{K}{r}F_K(r) = [E - V(r) + M]G_K(r). \end{aligned} \tag{11.77}$$

Equation (11.77) holds for the usual Dirac equation in 3 + 1-dimensional space-time.

(b) *The case of* $N = 2\ell$ *with* $\ell > 2$

The spinor representation of SO(2ℓ) is decomposed into two fundamental spinor representations $[\pm s]$ with the highest weights $(0,\cdots,0,1)$ and $(0,\cdots,0,1,0)$, respectively. Let

$$\gamma_{N+1}=\beta_{2\ell+1},\qquad \gamma_a=\beta_a,\qquad 1\leqslant a\leqslant 2\ell. \tag{11.78}$$

γ_{N+1} is a diagonal matrix with eigenvalues ± 1:

$$\begin{aligned}&\gamma_{N+1}\chi[(0,\cdots,0,1)]=\chi[(0,\cdots,0,1)],\\&\gamma_{N+1}\chi[(0,\cdots,0,1,0)]=-\chi[(0,\cdots,0,1,0)].\end{aligned} \tag{11.79}$$

Because the spinor operator S_{ab} and the operator κ commute with γ_{N+1}, they both become the block matrices.

At the level of the Pauli spinors, due to Eq. (10.86), **there are two sets of the wave functions belonging to each representation** $[j_\pm]=[\pm s,\lambda_1]$ of SO(2ℓ), called the spherical spinor functions. Two sets of spherical spinor functions come from the decompositions of $[\pm s]\times[\lambda_1]$ and $[\mp s]\times[\lambda_1+1]$, respectively. We are going to calculate the spherical spinor functions $Y_{K,[j_\pm]}(\hat{\boldsymbol{x}})$ with the highest weight. The Clebsch–Gordan coefficients are calculated from the condition that **the highest weight state is annihilated by each raising operators E_μ**.

Before calculating the highest weight states, we first calculate the eigenvalues K of κ (see Eq. (11.64)) which relates to the casimir invariants of order 2. Substituting the Cartan matrix A^{-1} of the Lie algebra D_ℓ (see subsection 7.5.6) into Eq. (7.133), we obtain

$$\begin{aligned}&\boldsymbol{M}([j_+])=(\lambda_1,0,\cdots,0,1),\qquad \boldsymbol{M}([j_-])=(\lambda_1,0,\cdots,0,1,0),\\&C_2([j_\pm])=\lambda_1(\lambda_1+2\ell-1)+\ell(2\ell-1)/4,\\&\boldsymbol{M}([\lambda_1])=(\lambda_1,0,\cdots,0),\qquad C_2([\lambda_1])=\lambda_1(\lambda_1+2\ell-2),\\&\boldsymbol{M}([+s])=(0,\cdots,0,1),\qquad \boldsymbol{M}([-s])=(0,\cdots,0,1,0),\\&C_2([\pm s])=\ell(2\ell-1)/4.\end{aligned} \tag{11.80}$$

Then,

$$\begin{aligned}&C_2([j_\pm])-C_2([\lambda_1])-C_2([\pm s])+\ell-1/2=\lambda_1+\ell-1/2=|K|,\\&C_2([j_\pm])-C_2([\lambda_1+1])-C_2([\pm s])+\ell-1/2=-\lambda_1-\ell+1/2=-|K|.\end{aligned} \tag{11.81}$$

Two highest weight states $Y_{|K|,[j_+]}(\hat{\boldsymbol{x}})$ and $Y_{|K|,[j_-]}(\hat{\boldsymbol{x}})$, which come from the decompositions of $[+s]\times[\lambda_1]$ and $[-s]\times[\lambda_1]$, respectively, are easy to calculate:

$$\begin{aligned}
Y_{|K|,[j_+]}(\hat{\boldsymbol{x}}) &= Y^{[\lambda_1]}_{(\lambda_1,0,\cdots,0)}(\hat{\boldsymbol{x}})\chi[(0,\cdots,0,1)] \\
&= C_{2\ell,\lambda_1}\left\{\frac{(-1)^{\ell-1}X_1}{r\sqrt{2}}\right\}^{\lambda_1}\chi[(0,\cdots,0,1)] \\
&= C_{2\ell,\lambda_1}\left\{\frac{(-1)^{\ell-1}(x_1+\mathrm{i}x_2)}{r\sqrt{2}}\right\}^{\lambda_1}\chi_{1\cdots 1}, \\
Y_{|K|,[j_-]}(\hat{\boldsymbol{x}}) &= Y^{[\lambda_1]}_{(\lambda_1,0,\cdots,0)}(\hat{\boldsymbol{x}})\chi[(0,\cdots,0,1,0)] \\
&= C_{2\ell,\lambda_1}\left\{\frac{(-1)^{\ell-1}X_1}{r\sqrt{2}}\right\}^{\lambda_1}\chi[(0,\cdots,0,1,0)] \\
&= C_{2\ell,\lambda_1}\left\{\frac{(-1)^{\ell-1}(x_1+\mathrm{i}x_2)}{r\sqrt{2}}\right\}^{\lambda_1}\chi_{1\cdots 1,2},
\end{aligned} \tag{11.82}$$

Two other highest weight state $Y_{-|K|,[j_+]}(\hat{\boldsymbol{x}})$ and $Y_{-|K|,[j_-]}(\hat{\boldsymbol{x}})$, which come from the decompositions of $[-s]\times[\lambda_1+1]$ and $[s]\times[\lambda_1+1]$, respectively, are calculated by Eq. (10.96):

$$\begin{aligned}
Y_{-|K|,[j_+]}(\hat{\boldsymbol{x}}) &= \frac{C_{2\ell,(\lambda_1+1)}\sqrt{\lambda_1+1}}{r\sqrt{2\ell+2\lambda_1}}\left\{\frac{(-1)^{\ell-1}X_1}{r\sqrt{2}}\right\}^{\lambda_1} \\
&\quad\times\left\{X_\ell\chi[(0,\cdots,0,1,0)]+X_{\ell-1}\chi[(0,\cdots,0,1,\overline{1},0)]\right. \\
&\quad+X_{\ell-2}\chi[(0,\cdots,0,1,\overline{1},0,1)]+\cdots \\
&\quad\left.+X_2\chi[(1,\overline{1},0,\cdots,0,1)]+X_1\chi[(\overline{1},0,\cdots,0,1)]\right\} \\
&= r^{-1}C_{2\ell,\lambda_1}\left\{\frac{(-1)^{\ell-1}(x_1+\mathrm{i}x_2)}{r\sqrt{2}}\right\}^{\lambda_1}\left\{(x_{2\ell-1}+\mathrm{i}x_{2\ell})\chi_{1\cdots 12}\right. \\
&\quad+(x_{2\ell-3}+\mathrm{i}x_{2\ell-2})\chi_{1\cdots 121}+(x_{2\ell-5}+\mathrm{i}x_{2\ell-4})\chi_{1\cdots 1211}+\cdots \\
&\quad\left.+(x_3+\mathrm{i}x_4)\chi_{121\cdots 1}+(x_1+\mathrm{i}x_2)\chi_{21\cdots 1}\right\} \\
&= (\boldsymbol{\beta}\cdot\hat{\boldsymbol{x}})\,Y_{|K|,[j_+]}(\hat{\boldsymbol{x}}),
\end{aligned}$$

$$\begin{aligned}
Y_{-|K|,[j_-]}(\hat{\boldsymbol{x}}) &= \frac{C_{2\ell,(\lambda_1+1)}\sqrt{\lambda_1+1}}{r\sqrt{2\ell+2\lambda_1}}\left\{\frac{(-1)^{\ell-1}X_1}{r\sqrt{2}}\right\}^{\lambda_1} \\
&\quad\times\left\{X_{\ell+1}\chi[(0,\cdots,0,1)]+X_{\ell-1}\chi[(0,\cdots,0,1,0,\overline{1})]\right. \\
&\quad+X_{\ell-2}\chi[(0,\cdots,0,1,\overline{1},1,0)]+\cdots \\
&\quad\left.+X_2\chi[(1,\overline{1},0,\cdots,1,0)]+X_1\chi[(\overline{1},0,\cdots,1,0)]\right\} \\
&= r^{-1}C_{2\ell,\lambda_1}\left\{\frac{(-1)^{\ell-1}(x_1+\mathrm{i}x_2)}{r\sqrt{2}}\right\}^{\lambda_1}\left\{(x_{2\ell-1}-\mathrm{i}x_{2\ell})\chi_{1\cdots 1}\right.
\end{aligned}$$

$$+ (x_{2\ell-3} + \mathrm{i}x_{2\ell-2})\chi_{1\cdots 122} + (x_{2\ell-5} + \mathrm{i}x_{2\ell-4})\chi_{1\cdots 1212} + \cdots$$
$$+ (x_3 + \mathrm{i}x_4)\chi_{121\cdots 12} + (x_1 + \mathrm{i}x_2)\chi_{21\cdots 12}\} \tag{11.83}$$
$$= (\boldsymbol{\beta} \cdot \hat{\boldsymbol{x}})\, Y_{|K|,[j_-]}(\hat{\boldsymbol{x}}),$$

From Eq. (11.74) we have

$$\begin{aligned}
&(\boldsymbol{\beta} \cdot \nabla)\, r^{-\ell+1/2} f(r) Y_{K,[j_\pm]}(\hat{\boldsymbol{x}}) \\
&\quad = (\boldsymbol{\beta} \cdot \hat{\boldsymbol{x}}) \left[\frac{\partial}{\partial r} + \frac{\mathrm{i}}{r} \sum_{a<b} \beta_a \beta_b L_{ab}\right] r^{-\ell+1/2} f(r) Y_{K,[j_\pm]}(\hat{\boldsymbol{x}}) \\
&\quad = r^{-\ell+1/2} \left[\frac{df(r)}{dr} - \frac{K f(r)}{r}\right] Y_{-K,[j_\pm]}(\hat{\boldsymbol{x}}).
\end{aligned} \tag{11.84}$$

At the level of the Dirac spinors, the wave functions $\boldsymbol{\Psi}_{K,[j_\pm]}(\boldsymbol{x})$ with the highest weights in the irreducible representation $[j_\pm]$ are

$$\begin{aligned}
&\boldsymbol{\Psi}_{|K|,[j_+]}(\hat{\boldsymbol{x}}) = r^{-\ell+1/2} e^{-\mathrm{i}Et} \left\{F_{|K|}(r) Y_{|K|,[j_+]}(\hat{\boldsymbol{x}}) + G_{|K|}(r) Y_{-|K|,[j_+]}(\hat{\boldsymbol{x}})\right\}, \\
&\boldsymbol{\Psi}_{-|K|,[j_-]}(\hat{\boldsymbol{x}}) \\
&\quad = r^{-\ell+1/2} e^{-\mathrm{i}Et} \left\{F_{-|K|}(r) Y_{-|K|,([j_-]}(\hat{\boldsymbol{x}}) + G_{-|K|}(r) Y_{|K|,[j_-]}(\hat{\boldsymbol{x}})\right\}, \\
&\kappa \boldsymbol{\Psi}_{\pm|K|,[j_\pm]}(\boldsymbol{x}) = \pm |K| \boldsymbol{\Psi}_{\pm|K|,[j_\pm]}(\boldsymbol{x}), \qquad |K| = \lambda_1 + \ell - 1/2.
\end{aligned} \tag{11.85}$$

Substituting $\boldsymbol{\Psi}_{K[j_\pm]}(\boldsymbol{x})$ into the Dirac equation (11.62) we obtain the radial equations, which are the same as Eq. (11.77). Their solutions for the Coulomb potential are given in [Gu et al. (2002)]. The Levinson theorem for the Dirac equation in $(N+1)$-dimensional space-time is given in [Gu et al. (2003e)]. When $N = 4$, SO(4) $\sim$ SU(2)$\times$SU(2), and the representations $[j_\pm]$ belong to two different SU(2) groups, respectively. When $N = 2$, the SO(2) group is an abelian group, and the radial equation (11.77) holds for SO(2) but $K = \pm 1/2, \pm 3/2, \cdots$ [Dong et al. (2003); Dong et al. (1998c)].

11.4 Exercises

1. Discuss the classes in the SO(4) group and calculate their characters in the irreducible representation D^{jk}.
2. Calculate six parameters φ, θ, ω, α, β, and γ of the following proper Lorentz transformation A, and write its representation matrix in the irreducible representation $D^{jk}(A)$ of the proper Lorentz group L_p:

$$A(\varphi,\theta,\omega,\alpha,\beta,\gamma)=\begin{pmatrix}1 & 0 & 0 & 0\\ 0 & \sqrt{3}/2 & (\cosh\omega)/2 & -\mathrm{i}(\sinh\omega)/2\\ 0 & -1/2 & \sqrt{3}(\cosh\omega)/2 & -\mathrm{i}\sqrt{3}(\sinh\omega)/2\\ 0 & 0 & i\sinh\omega & \cosh\omega\end{pmatrix}.$$

3. Please calculate the parameters ω_{ab} and ω_{a4} of the proper Lorentz transformation A given in the Prob. 2.

4. Please calculate the matrix $u(\boldsymbol{e}_3, \mathrm{i}\omega)=\mathrm{e}^{\omega\sigma_3/2}$ where cosh $\omega=3$.

5. Please prove that the operator κ given in Eq. (11.64) commutes with the Hamiltonian operator $H(\boldsymbol{x})$ given in Eq. (11.62).

6. Please check the radial equation (11.77) by substituting $\boldsymbol{\Psi}_{K,[j]}(\boldsymbol{x},t)$ given in Eq. (11.76) into the Dirac equation (11.62).

7. Please check the radial equation (11.77) by substituting $\boldsymbol{\Psi}_{K,[j\pm]}(\boldsymbol{x},t)$ given in Eq. (11.85) into the Dirac equation (11.62).

8. Calculate the Casimir invariants $C_2([j])$, $C_2([\lambda_1])$ and $C_2([s])$ given in Eq. (11.68) for the B_ℓ Lie algebra.

9. Calculate the Casimir invariants $C_2([j_\pm])$, $C_2([\lambda_1])$ and $C_2([\pm s])$ given in Eq. (11.80) for the D_ℓ Lie algebra.

Chapter 12

SYMPLECTIC GROUPS

In subsection 7.4.3 we have introduced the fundamental property of the USp(2ℓ) group and the Sp($2\ell, R$) group. In this chapter we will study the irreducible representations of USp(2ℓ) and their applications to physics.

12.1 Irreducible Representations of USp(2ℓ)

12.1.1 *Decomposition of the Tensor Space of* USp(2ℓ)

Let the element u in USp(2ℓ) denote a transformation matrix in a (2ℓ)-dimensional complex space,

$$x_a \xrightarrow{u} x'_a = \sum_b u_{ab} x_b, \qquad u \in \mathrm{USp}(2\ell), \tag{12.1}$$

where the index a is taken to be j or $\overline{j}$, $1 \leqslant j \leqslant \ell$, in the following order,

$$a = 1,\ \overline{1},\ 2,\ \overline{2},\ \cdots,\ \ell,\ \overline{\ell}. \tag{12.2}$$

The u matrix satisfies

$$u^T J u = J, \qquad u^\dagger = u^{-1}, \tag{12.3}$$

where

$$J_{ab} = \begin{cases} 1 & \text{when } a = j, \quad b = \overline{j}, \\ -1 & \text{when } a = \overline{j}, \quad b = j, \\ 0 & \text{the remaining cases,} \end{cases} \tag{12.4}$$

$$J = \mathbf{1}_\ell \times (\mathrm{i}\sigma_2) = -J^{-1} = -J^T, \qquad \det\ J = 1.$$

Thus, the self-representation of USp(2ℓ) is self-conjugate

$$u^* = J u J^{-1} = J^{-1} u J. \tag{12.5}$$

The tensor $\boldsymbol{T}_{a_1\cdots a_n}$ of rank n and the basis tensors $\boldsymbol{\theta}_{a_1\cdots a_n}$ with respect to USp(2ℓ) are defined as

$$\begin{aligned} \boldsymbol{T}_{a_1\cdots a_n} &\xrightarrow{u} (O_u\boldsymbol{T})_{a_1\cdots a_n} = \sum_{b_1\cdots b_n} u_{a_1b_1}\cdots u_{a_nb_n}\boldsymbol{T}_{b_1\cdots b_n}, \\ \boldsymbol{\theta}_{a_1\cdots a_n} &\xrightarrow{u} O_u\boldsymbol{\theta}_{a_1\cdots a_n} = \sum_{b_1\cdots b_n} \boldsymbol{\theta}_{b_1\cdots b_n}u_{b_1a_1}\cdots u_{b_na_n}. \end{aligned} \tag{12.6}$$

Similar the tensors of SO(N), there also exist two invariant tensors of USp(2ℓ): J_{ab} and $\epsilon_{a_1\cdots a_{2\ell}}$. Due to Eq. (12.5) **the contravariant tensor representation is equivalent to the covariant one.** The invariant tensor $\epsilon_{a_1\cdots a_{2\ell}}$ is irrelevant to the decomposition of the tensor space for USp(2ℓ). We are going to pay attention to the invariant tensor J_{ab}. Let $\boldsymbol{J}$ denote an antisymmetric tensor of rank 2 with components J_{ab},

$$(O_u\boldsymbol{J})_{ab} = \sum_{cd} u_{ac}u_{bd}\boldsymbol{J}_{cd} = \left(u\boldsymbol{J}u^T\right)_{ab} = \boldsymbol{J}_{ab}. \tag{12.7}$$

$\boldsymbol{J}$ is an invariant tensor of USp(2ℓ). Through a contraction of two antisymmetric indices j and $\overline{j}$, the trace tensor of a tensor of rank n for USp(2ℓ) is a tensor of rank $(n-2)$,

$$\begin{aligned} &\sum_{a_ra_s} (O_uJ_{a_ra_s}\boldsymbol{T})_{a_1\cdots a_{r-1}a_ra_{r+1}\cdots a_{s-1}a_sa_{s+1}\cdots a_n} \\ &\quad= \sum_{a_ra_sb_1\cdots b_n} (u_{a_rb_r}u_{a_sb_s}J_{a_ra_s})\, u_{a_1b_1}\cdots u_{a_{r-1}b_{r-1}}u_{a_{r+1}b_{r+1}}\cdots u_{a_{s-1}b_{s-1}} \\ &\quad\quad\times u_{a_{s+1}b_{s+1}}\cdots u_{a_nb_n}\boldsymbol{T}_{b_1\cdots b_{r-1}b_rb_{r+1}\cdots b_{s-1}b_sb_{s+1}\cdots b_n} \\ &\quad= \sum_{b_1\cdots b_{r-1}b_{r+1}\cdots b_{s-1}b_{s+1}\cdots b_n} u_{a_1b_1}\cdots u_{a_{r-1}b_{r-1}}u_{a_{r+1}b_{r+1}}\cdots u_{a_{s-1}b_{s-1}} \\ &\quad\quad\times u_{a_{s+1}b_{s+1}}\cdots u_{a_nb_n}\left(\sum_{b_rb_s} J_{b_rb_s}\boldsymbol{T}_{b_1\cdots b_{r-1}b_rb_{r+1}\cdots b_{s-1}b_sb_{s+1}\cdots}\right). \end{aligned}$$

The subspace of the trace tensors is invariant in USp(2ℓ). Similar to Eq. (9.66), a tensor space of USp(2ℓ) can be decomposed into the direct sum of a series of subspaces of traceless tensors with different ranks:

$$\sum_{ab} J_{ab}\boldsymbol{T}_{\cdots a\cdots b\cdots} = 0. \tag{12.8}$$

For example, a tensor of rank 2 is decomposed into a sum of a traceless tensor of rank 2 and a scalar

$$\boldsymbol{T}_{ab} = \left\{\boldsymbol{T}_{ab} - \boldsymbol{J}_{ab}\left(\frac{1}{2\ell}\sum_{cd} J_{cd}\boldsymbol{T}_{cd}\right)\right\} + \boldsymbol{J}_{ab}\left(\frac{1}{2\ell}\sum_{cd} J_{cd}\boldsymbol{T}_{cd}\right). \tag{12.9}$$

The Weyl reciprocity holds for the tensors of USp(2ℓ) so that the tensor space can be decomposed by the projection of the Young operators. Let $\mathcal{T}$ denote the space of traceless tensors of rank n of USp(2ℓ). Projecting by a Young operator $\mathcal{Y}_{\mu}^{[\lambda]}$, one obtains a traceless tensor subspace $\mathcal{Y}_{\mu}^{[\lambda]}\mathcal{T} = \mathcal{T}_{\mu}^{[\lambda]}$, where the basis tensors are taken to be the tensor Young tableaux.

We first study the condition whether the traceless tensor space $\mathcal{T}_{\mu}^{[\lambda]}$ is a null space or not. Without loss of generality, we assume that the row number of $[\lambda]$ is r, and there are m pairs of digits j and $\overline{j}$ in the first column of a given tensor Young tableau in $\mathcal{T}_{\mu}^{[\lambda]}$. Fixing the unpaired digits in the first column and the digits in the remaining columns, we change the pairs of digits in the first column from 1 to ℓ, where some tensor Young tableaux may be vanishing owing to antisymmetry of indices. The number of linearly independent tensor Young tableaux is the combinatorial number of m among $[\ell-(r-2m)]$. The traceless condition is written as Eq. (12.8), where the $(m-1)$ pairs of digits are fixed and only one pair of digits runs over from 1 to ℓ. The number of traceless conditions is the number of possible values of the $(m-1)$ pairs of digits, that is the combinatorial number of $(m-1)$ among $[\ell-(r-2m)]$. The traceless tensor space $\mathcal{T}_{\mu}^{[\lambda]}$ is a null space if the number of traceless conditions is not less than the number of independent tensors, namely, $[\ell-(r-2m)]/2 < m$, and then, $r > \ell$. Thus, the traceless tensor space $\mathcal{T}_{\mu}^{[\lambda]}$ is a null space if the row number r of the Young pattern $[\lambda]$ is larger than ℓ. **An irreducible representation of** USp(2ℓ) **is self-conjugate** and **denoted by the Young pattern** $[\lambda]$ **with the row number not larger than** ℓ (see Eq. (12.14)).

12.1.2 *Orthonormal Irreducible Basis Tensors of* USp(2ℓ)

The generators in the self-representation of USp(2ℓ), given in Eq. (7.102), satisfy the orthonormal condition $\text{Tr}(T_A T_B) = \delta_{AB}$. The Chevalley bases of USp(2ℓ) can be calculated by Eq. (8.1),

$$\begin{aligned}
H_{\mu} &= \left\{T_{\mu\mu}^{(1)} - T_{(\mu+1)(\mu+1)}^{(1)}\right\} \times \sigma_3 \ , \\
H_{\ell} &= T_{\ell\ell}^{(1)} \times \sigma_3, \\
E_{\mu} &= T_{\mu(\mu+1)}^{(1)} \times \sigma_3 + \text{i}T_{\mu(\mu+1)}^{(2)} \times \mathbf{1}_2 \ , \\
E_{\ell} &= T_{\ell\ell}^{(1)} \times (\sigma_1 + \text{i}\sigma_2)/2, \\
F_{\mu} &= T_{\mu(\mu+1)}^{(1)} \times \sigma_3 - \text{i}T_{\mu(\mu+1)}^{(2)} \times \mathbf{1}_2 \ , \\
F_{\ell} &= T_{\ell\ell}^{(1)} \times (\sigma_1 - \text{i}\sigma_2)/2,
\end{aligned} \tag{12.10}$$

where $1 \leqslant \mu < \ell$, $T^{(r)}_{jk}$ $(r = 1, 2)$ are the generators in SU(ℓ), and $\left(T^{(1)}_{jj}\right)_{ab} = \delta_{aj}\delta_{bj}$. The action of the generators on the basis vectors $\boldsymbol{\theta}_a$ are

$$\begin{aligned}
&H_\mu \boldsymbol{\theta}_\nu = \left(\delta_{\nu\mu} - \delta_{\nu(\mu+1)} + \delta_{\nu\overline{\mu+1}} - \delta_{\nu\overline{\mu}}\right)\boldsymbol{\theta}_\nu, \\
&H_\ell \boldsymbol{\theta}_\nu = \delta_{\nu\ell}\boldsymbol{\theta}_\ell - \delta_{\nu\overline{\ell}}\boldsymbol{\theta}_{\overline{\ell}}, \\
&E_\mu \boldsymbol{\theta}_\nu = \delta_{\nu(\mu+1)}\boldsymbol{\theta}_\mu - \delta_{\nu\overline{\mu}}\boldsymbol{\theta}_{\overline{\mu+1}}, \quad E_\ell \boldsymbol{\theta}_\nu = \delta_{\nu\overline{\ell}}\boldsymbol{\theta}_\ell, \\
&F_\mu \boldsymbol{\theta}_\nu = \delta_{\nu\mu}\boldsymbol{\theta}_{\mu+1} - \delta_{\nu\overline{\mu+1}}\boldsymbol{\theta}_{\overline{\mu}}, \quad F_\ell \boldsymbol{\theta}_\nu = \delta_{\nu\ell}\boldsymbol{\theta}_{\overline{\ell}}.
\end{aligned} \tag{12.11}$$

Namely, $\boldsymbol{\theta}_a$ is the common eigenvector of H_μ and H_ℓ, but its arranged order and the minus signs in the actions of E_μ and F_μ are not convenient for calculation. Define

$$\boldsymbol{\phi}_\mu = \boldsymbol{\theta}_\mu, \qquad \boldsymbol{\phi}_{\ell+\mu} = (-1)^{\mu+1}\boldsymbol{\theta}_{\overline{\ell-\mu+1}}, \qquad 1 \leqslant \mu \leqslant \ell, \tag{12.12}$$

Thus, Eq. (12.11) becomes

$$\begin{aligned}
&H_\mu \boldsymbol{\phi}_\nu = \left(\delta_{\nu\mu} - \delta_{\nu(\mu+1)} + \delta_{\nu(2\ell-\mu)} - \delta_{\nu(2\ell-\mu+1)}\right)\boldsymbol{\phi}_\nu, \\
&H_\ell \boldsymbol{\phi}_\nu = \delta_{\nu\ell}\boldsymbol{\theta}_\ell - \delta_{\nu(\ell+1)}\boldsymbol{\phi}_{\ell+1}, \\
&E_\mu \boldsymbol{\phi}_\nu = \delta_{\nu(\mu+1)}\boldsymbol{\theta}_\mu + \delta_{\nu(2\ell-\mu+1)}\boldsymbol{\phi}_{2\ell-\mu}, \quad E_\ell \boldsymbol{\theta}_\nu = \delta_{\nu(\ell+1)}\boldsymbol{\theta}_\ell, \\
&F_\mu \boldsymbol{\phi}_\nu = \delta_{\nu\mu}\boldsymbol{\phi}_{\mu+1} + \delta_{\nu(2\ell-\mu)}\boldsymbol{\phi}_{2\ell-\mu+1}, \quad F_\ell \boldsymbol{\phi}_\nu = \delta_{\nu\ell}\boldsymbol{\phi}_{\ell+1},
\end{aligned} \tag{12.13}$$

where $1 \leqslant \mu < \ell$.

The basis tensor $\boldsymbol{\phi}_{\alpha_1\cdots\alpha_n}$ is the direct product of the basis vectors $\boldsymbol{\phi}_{\alpha_i}$. The action of a generator on the basis tensor $\boldsymbol{\phi}_{\alpha_1\cdots\alpha_n}$ is equal to the sum of its action on each basis vector $\boldsymbol{\phi}_{\alpha_i}$ in the product. **The standard tensor Young tableaux $\mathcal{Y}^{[\lambda]}_\mu \boldsymbol{\phi}_{\alpha_1\cdots\alpha_n}$ are the common eigenstates of H_μ, but generally not orthonormal and traceless**. The eigenvalue of H_μ in the standard tensor Young tableau $\mathcal{Y}^{[\lambda]}_\mu \boldsymbol{\phi}_{\alpha_1\cdots\alpha_n}$ is the number of the digits μ and $(2\ell - \mu)$ filled in the tableau, minus the number of the digits $(\mu+1)$ and $(2\ell-\mu+1)$. The eigenvalue of H_ℓ in the standard tensor Young tableau is equal to the number of the digit ℓ in the tableau, minus the number of the digit $\ell + 1$. The eigenvalues constitutes the weight $\boldsymbol{m}$ of the standard tensor Young tableau. **Two standard tensor Young tableaux are orthogonal if their weights are different**. The action of F_μ on the standard tensor Young tableau is equal to the sum of all possible tensor Young tableaux, each of which is obtained from the original one by replacing one filled digit μ with the digit $(\mu + 1)$, or by replacing one filled digit $(2\ell - \mu)$ with the digit $(2\ell - \mu + 1)$. The action of F_ℓ on the standard tensor Young tableau is equal to the sum of all possible tensor Young tableaux, each of which is obtained from the original one by replacing one filled digit ℓ with the digit $(\ell+1)$. The actions of $E_\mu (E_\ell)$ are opposite to that of $F_\mu (F_\ell)$. **The obtained tensor Young tableaux may be not standard, but they can be transformed to the sum of the standard tensor Young tableaux by the relations** (9.22) **and** (9.23).

The row number of $[\lambda]$ in the traceless tensor subspace $\mathcal{T}_\mu^{[\lambda]}$ is not larger than ℓ, $[\lambda] = [\lambda_1, \lambda_2, \cdots, \lambda_\ell]$. Through a similar proof as that for Theorem 9.3, there is one and only one traceless standard tensor Young tableau in $\mathcal{T}_\mu^{[\lambda]}$, where each box in its αth row is filled with the digit α, which is annihilated by each raising operators E_μ (see Eq. (7.111)). This traceless standard tensor Young tableau describes the highest weight state in $\mathcal{T}_\mu^{[\lambda]}$ with the highest weight $\boldsymbol{M} = \sum_\mu \boldsymbol{w}_\mu M_\mu$:

$$M_\mu = \lambda_\mu - \lambda_{\mu+1}, \qquad M_\ell = \lambda_\ell, \qquad 1 \leqslant \mu < \ell. \tag{12.14}$$

It means that **the irreducible representation of** USp(2ℓ) **is denoted by the Young pattern** $[\lambda]$ **with the row number not larger than** ℓ. The standard tensor Young tableau which describes the highest weight state is obviously **traceless** because there is no digit larger than ℓ filled in this tableau. The remaining basis tensors in $\mathcal{T}_\mu^{[\lambda]}$ can be calculated from the standard tensor Young tableau with the highest weight by the lowering operators F_μ in the method of block weight diagram and the generalized Gel'fand's method. **The calculated orthonormal basis states are the linear combinations of the standard tensor Young tableaux, which are traceless.**

The irreducible representations of Sp($2\ell, R$) can be obtained from those of USp(2ℓ) by replacing some parameters to be pure imaginary (see the discussion below Eq. (7.102)) but leaving the generators invariant.

In subsection 8.3.2 we calculated the basis states of some highest weight representations of C_3 (the USp(6) group) by the method of the block weight diagram and the generalized Gel'fand's method. Now we calculate the expressions of those orthonormal basis states in terms of the linear combinations of the standard tensor Young tableaux through five examples. The simple roots $\boldsymbol{r}_\mu$ of USp(6) are expressed with respect to the fundamental dominant weights $\boldsymbol{w}_\nu$,

$$\boldsymbol{r}_1 = 2\boldsymbol{w}_1 - \boldsymbol{w}_2, \quad \boldsymbol{r}_2 = -\boldsymbol{w}_1 + 2\boldsymbol{w}_2 - \boldsymbol{w}_3, \quad \boldsymbol{r}_3 = -2\boldsymbol{w}_2 + 2\boldsymbol{w}_3. \tag{12.15}$$

Ex. 1. The self-representation with $\boldsymbol{M} = \boldsymbol{w}_1 = (1,0,0)$ of C_3.

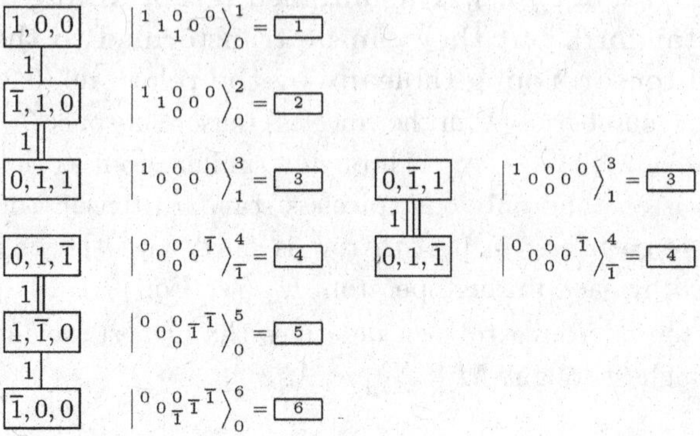

Ex. 2. The basic representation with $\boldsymbol{M} = \boldsymbol{w}_2 = (0,1,0)$ of C_3.

This representation contains two A_2-triplets and one A_2-octet with $\boldsymbol{M}^{(2)} = (0,1)$, $(1,0)$ and $(1,1)$, respectively. The highest weight state of this representation is also the highest weight state of an A_2-triplet with $\boldsymbol{M}^{(2)} = (0,1)$. Two $\mathcal{A}_3$-doublets are constructed from two states in this A_2-triplet due to the parallel principle.

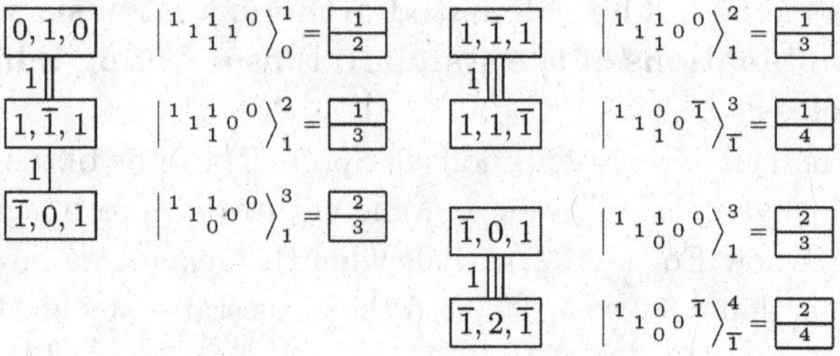

The state $\left|\,{}^{1}{}_{1}{}^{0}_{1}{}_{0}{}^{\bar{1}}\right\rangle^{3}_{\bar{1}}$ is the highest weight state of an A_2-octet with $\boldsymbol{M}^{(2)} = (1,1)$, from which the A_2-octet is constructed. Remind that two states $\left|\,{}^{1}{}_{1}{}^{0}_{0}{}_{\bar{1}}{}^{\bar{1}}\right\rangle^{5}_{0}$ and $\left|\,{}^{1}{}_{0}{}^{0}_{0}{}_{0}{}^{\bar{1}}\right\rangle^{5}_{0}$ in the A_2-octet are traceless and orthogonal to the trace tensor

$$\boxed{\begin{smallmatrix}1\\6\end{smallmatrix}} - \boxed{\begin{smallmatrix}2\\5\end{smallmatrix}} + \boxed{\begin{smallmatrix}3\\4\end{smallmatrix}}\,.$$

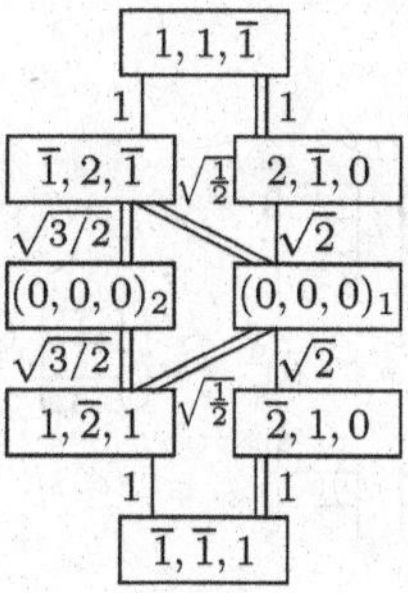
1, 1, 1̄
1
1
1̄, 2, 1̄
√½
2, 1̄, 0
√3/2
√2
(0, 0, 0)₂
(0, 0, 0)₁
√3/2
√2
1, 2̄, 1
√½
2̄, 1, 0
1
1
1̄, 1̄, 1

$$\left|\begin{smallmatrix}1&&0&&\bar1\\&1&&0&\\&&1&&\end{smallmatrix}\right\rangle^{3}_{\bar1}=\begin{array}{|c|}\hline 1\\\hline 4\\\hline\end{array}$$

$$\left|\begin{smallmatrix}1&&0&&\bar1\\&1&&0&\\&&0&&\end{smallmatrix}\right\rangle^{4}_{\bar1}=\begin{array}{|c|}\hline 2\\\hline 4\\\hline\end{array},\qquad \left|\begin{smallmatrix}1&&0&&\bar1\\&1&&\bar1&\\&&1&&\end{smallmatrix}\right\rangle^{4}_{0}=\begin{array}{|c|}\hline 1\\\hline 5\\\hline\end{array}$$

$$\left|\begin{smallmatrix}1&&0&&\bar1\\&0&&0&\\&&0&&\end{smallmatrix}\right\rangle^{5}_{0},\quad \left|\begin{smallmatrix}1&&0&&\bar1\\&1&&\bar1&\\&&0&&\end{smallmatrix}\right\rangle^{5}_{0}=\sqrt{\tfrac{1}{2}}\left\{\begin{array}{|c|}\hline 1\\\hline 6\\\hline\end{array}+\begin{array}{|c|}\hline 2\\\hline 5\\\hline\end{array}\right\}$$

$$\left|\begin{smallmatrix}1&&0&&\bar1\\&0&&\bar1&\\&&0&&\end{smallmatrix}\right\rangle^{6}_{1}=\begin{array}{|c|}\hline 3\\\hline 5\\\hline\end{array},\qquad \left|\begin{smallmatrix}1&&0&&\bar1\\&1&&\bar1&\\&&\bar1&&\end{smallmatrix}\right\rangle^{6}_{0}=\begin{array}{|c|}\hline 2\\\hline 6\\\hline\end{array}$$

$$\left|\begin{smallmatrix}1&&0&&\bar1\\&0&&\bar1&\\&&\bar1&&\end{smallmatrix}\right\rangle^{7}_{1}=\begin{array}{|c|}\hline 3\\\hline 6\\\hline\end{array}$$

$$\left|\begin{smallmatrix}1&&0&&\bar1\\&0&&0&\\&&0&&\end{smallmatrix}\right\rangle^{5}_{0}=\sqrt{\tfrac{1}{6}}\left\{-\begin{array}{|c|}\hline 1\\\hline 6\\\hline\end{array}+\begin{array}{|c|}\hline 2\\\hline 5\\\hline\end{array}+2\begin{array}{|c|}\hline 3\\\hline 4\\\hline\end{array}\right\}.$$

Two $\mathcal{A}_3$-doublets are constructed from two states $\left|\begin{smallmatrix}1&&0&&\bar1\\&0&&\bar1&\\&&0&&\end{smallmatrix}\right\rangle^{6}_{1}$ and $\left|\begin{smallmatrix}1&&0&&\bar1\\&0&&\bar1&\\&&\bar1&&\end{smallmatrix}\right\rangle^{7}_{1}$ in this A_2-octet due to the parallel principle. Then, an A_2-triplet with $\boldsymbol{M}^{(2)}=(1,0)$ follows.

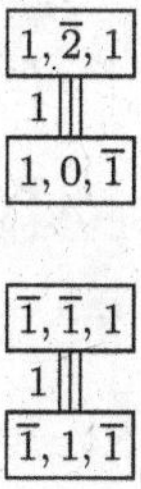
1, 2̄, 1
1
1, 0, 1̄
1̄, 1̄, 1
1
1̄, 1, 1̄

$$\left|\begin{smallmatrix}1&&0&&\bar1\\&0&&\bar1&\\&&0&&\end{smallmatrix}\right\rangle^{6}_{1}=\begin{array}{|c|}\hline 3\\\hline 5\\\hline\end{array}$$

$$\left|\begin{smallmatrix}0&&\bar1&&\bar1\\&0&&\bar1&\\&&0&&\end{smallmatrix}\right\rangle^{7}_{\bar1}=\begin{array}{|c|}\hline 4\\\hline 5\\\hline\end{array}$$

$$\left|\begin{smallmatrix}1&&0&&\bar1\\&0&&\bar1&\\&&\bar1&&\end{smallmatrix}\right\rangle^{7}_{1}=\begin{array}{|c|}\hline 3\\\hline 6\\\hline\end{array}$$

$$\left|\begin{smallmatrix}0&&\bar1&&\bar1\\&0&&\bar1&\\&&\bar1&&\end{smallmatrix}\right\rangle^{8}_{\bar1}=\begin{array}{|c|}\hline 4\\\hline 6\\\hline\end{array}$$

1, 0, 1̄
1
1̄, 1, 1̄
1
0, 1̄, 0

$$\left|\begin{smallmatrix}0&&\bar1&&\bar1\\&0&&\bar1&\\&&0&&\end{smallmatrix}\right\rangle^{7}_{\bar1}=\begin{array}{|c|}\hline 4\\\hline 5\\\hline\end{array}$$

$$\left|\begin{smallmatrix}0&&\bar1&&\bar1\\&0&&\bar1&\\&&\bar1&&\end{smallmatrix}\right\rangle^{8}_{\bar1}=\begin{array}{|c|}\hline 4\\\hline 6\\\hline\end{array}$$

$$\left|\begin{smallmatrix}0&&\bar1&&\bar1\\&\bar1&&\bar1&\\&&\bar1&&\end{smallmatrix}\right\rangle^{9}_{0}=\begin{array}{|c|}\hline 5\\\hline 6\\\hline\end{array}$$

Ex. 3. The basic representation with $\boldsymbol{M}=\boldsymbol{w}_3=(0,0,1)$ of C_3.

This representation contains two A_2-singlets with $\boldsymbol{M}^{(2)}=(0,0)$ and two A_2-sixlets with $\boldsymbol{M}^{(2)}=(0,2)$, and $(2,0)$, respectively. Its highest weight state spans the A_2-singlet and is also the highest weight state of an $\mathcal{A}_3$-doublet.

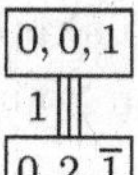
0, 0, 1
1
0, 2, 1̄

$$\left|\begin{smallmatrix}1&&1&&1\\&1&&1&\\&&1&&\end{smallmatrix}\right\rangle^{1}_{1}=\begin{array}{|c|}\hline 1\\\hline 2\\\hline 3\\\hline\end{array}$$

$$\left|\begin{smallmatrix}1&&1&&\bar1\\&1&&1&\\&&1&&\end{smallmatrix}\right\rangle^{2}_{\bar1}=\begin{array}{|c|}\hline 1\\\hline 2\\\hline 4\\\hline\end{array}$$

The basis state $\left|\begin{smallmatrix}1&&1&&\bar1\\&1&&1&\\&&1&&\end{smallmatrix}\right\rangle^{2}_{\bar1}$ is the highest weight state of an A_2-sixlet with $\boldsymbol{M}^{(2)}=(0,2)$, and the A_2-sixlet is constructed from it (see Fig. 8.3):

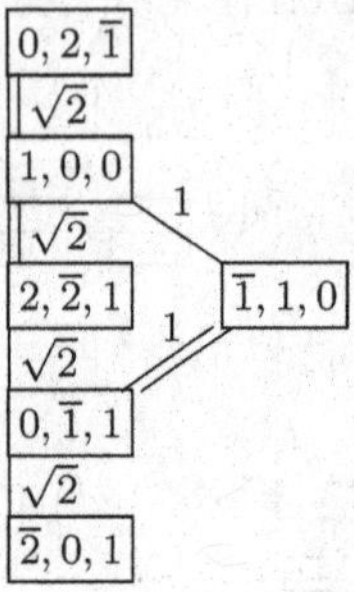

$$\left|\begin{smallmatrix}1&&1&&\bar{1}\\&1&&1&\\&&1&&\end{smallmatrix}\right\rangle^{2}_{\bar{1}}=\begin{array}{|c|}\hline 1\\\hline 2\\\hline 4\\\hline\end{array}$$

$$\left|\begin{smallmatrix}1&&1&&\bar{1}\\&1&&0&\\&&1&&\end{smallmatrix}\right\rangle^{3}_{0}=\sqrt{\tfrac{1}{2}}\left\{\begin{array}{|c|}\hline 1\\\hline 2\\\hline 5\\\hline\end{array}+\begin{array}{|c|}\hline 1\\\hline 3\\\hline 4\\\hline\end{array}\right\}$$

$$\left|\begin{smallmatrix}1&&1&&\bar{1}\\&1&&\bar{1}&\\&&1&&\end{smallmatrix}\right\rangle^{4}_{1}=\begin{array}{|c|}\hline 1\\\hline 3\\\hline 5\\\hline\end{array},\quad \left|\begin{smallmatrix}1&&1&&\bar{1}\\&1&&0&\\&&0&&\end{smallmatrix}\right\rangle^{4}_{0}=\sqrt{\tfrac{1}{2}}\left\{\begin{array}{|c|}\hline 1\\\hline 2\\\hline 6\\\hline\end{array}+\begin{array}{|c|}\hline 2\\\hline 3\\\hline 4\\\hline\end{array}\right\}$$

$$\left|\begin{smallmatrix}1&&1&&\bar{1}\\&1&&\bar{1}&\\&&0&&\end{smallmatrix}\right\rangle^{5}_{1}=\sqrt{\tfrac{1}{2}}\left\{\begin{array}{|c|}\hline 1\\\hline 3\\\hline 6\\\hline\end{array}+\begin{array}{|c|}\hline 2\\\hline 3\\\hline 5\\\hline\end{array}\right\}$$

$$\left|\begin{smallmatrix}1&&1&&\bar{1}\\&1&&\bar{1}&\\&&\bar{1}&&\end{smallmatrix}\right\rangle^{6}_{1}=\begin{array}{|c|}\hline 2\\\hline 3\\\hline 6\\\hline\end{array}$$

Three $\mathcal{A}_3$-doublets are constructed from three states in this A_2-sixlet due to the parallel principle. Then, an A_2-sixlet with $\boldsymbol{M}^{(2)}=(2,0)$ follows.

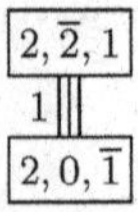

$$\left|\begin{smallmatrix}1&&1&&\bar{1}\\&1&&\bar{1}&\\&&1&&\end{smallmatrix}\right\rangle^{4}_{1}=\begin{array}{|c|}\hline 1\\\hline 3\\\hline 5\\\hline\end{array}$$

$$\left|\begin{smallmatrix}1&&\bar{1}&&\bar{1}\\&1&&\bar{1}&\\&&1&&\end{smallmatrix}\right\rangle^{5}_{\bar{1}}=\begin{array}{|c|}\hline 1\\\hline 4\\\hline 5\\\hline\end{array}$$

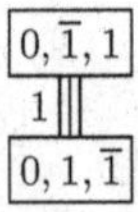

$$\left|\begin{smallmatrix}1&&1&&\bar{1}\\&1&&\bar{1}&\\&&0&&\end{smallmatrix}\right\rangle^{5}_{1}=\sqrt{\tfrac{1}{2}}\left\{\begin{array}{|c|}\hline 1\\\hline 3\\\hline 6\\\hline\end{array}+\begin{array}{|c|}\hline 2\\\hline 3\\\hline 5\\\hline\end{array}\right\}$$

$$\left|\begin{smallmatrix}1&&\bar{1}&&\bar{1}\\&1&&\bar{1}&\\&&0&&\end{smallmatrix}\right\rangle^{6}_{\bar{1}}=\sqrt{\tfrac{1}{2}}\left\{\begin{array}{|c|}\hline 1\\\hline 4\\\hline 6\\\hline\end{array}+\begin{array}{|c|}\hline 2\\\hline 4\\\hline 5\\\hline\end{array}\right\}$$

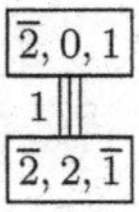

$$\left|\begin{smallmatrix}1&&1&&\bar{1}\\&1&&\bar{1}&\\&&\bar{1}&&\end{smallmatrix}\right\rangle^{6}_{1}=\begin{array}{|c|}\hline 2\\\hline 3\\\hline 6\\\hline\end{array}$$

$$\left|\begin{smallmatrix}1&&\bar{1}&&\bar{1}\\&1&&\bar{1}&\\&&\bar{1}&&\end{smallmatrix}\right\rangle^{7}_{\bar{1}}=\begin{array}{|c|}\hline 2\\\hline 4\\\hline 6\\\hline\end{array}$$

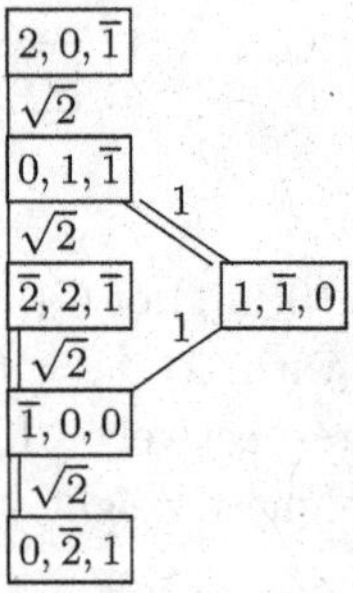

$$\left|\begin{smallmatrix}1&&\bar{1}&&\bar{1}\\&1&&\bar{1}&\\&&1&&\end{smallmatrix}\right\rangle^{5}_{\bar{1}}=\begin{array}{|c|}\hline 1\\\hline 4\\\hline 5\\\hline\end{array}$$

$$\left|\begin{smallmatrix}1&&\bar{1}&&\bar{1}\\&1&&\bar{1}&\\&&0&&\end{smallmatrix}\right\rangle^{6}_{\bar{1}}=\sqrt{\tfrac{1}{2}}\left\{\begin{array}{|c|}\hline 1\\\hline 4\\\hline 6\\\hline\end{array}+\begin{array}{|c|}\hline 2\\\hline 4\\\hline 5\\\hline\end{array}\right\}$$

$$\left|\begin{smallmatrix}1&&\bar{1}&&\bar{1}\\&1&&\bar{1}&\\&&\bar{1}&&\end{smallmatrix}\right\rangle^{7}_{\bar{1}}=\begin{array}{|c|}\hline 2\\\hline 4\\\hline 6\\\hline\end{array},\quad \left|\begin{smallmatrix}1&&\bar{1}&&\bar{1}\\&0&&\bar{1}&\\&&0&&\end{smallmatrix}\right\rangle^{7}_{0}=\sqrt{\tfrac{1}{2}}\left\{\begin{array}{|c|}\hline 1\\\hline 5\\\hline 6\\\hline\end{array}+\begin{array}{|c|}\hline 3\\\hline 4\\\hline 5\\\hline\end{array}\right\}$$

$$\left|\begin{smallmatrix}1&&\bar{1}&&\bar{1}\\&0&&\bar{1}&\\&&\bar{1}&&\end{smallmatrix}\right\rangle^{8}_{0}=\sqrt{\tfrac{1}{2}}\left\{\begin{array}{|c|}\hline 2\\\hline 5\\\hline 6\\\hline\end{array}+\begin{array}{|c|}\hline 3\\\hline 4\\\hline 6\\\hline\end{array}\right\}$$

$$\left|\begin{smallmatrix}1&&\bar{1}&&\bar{1}\\&\bar{1}&&\bar{1}&\\&&\bar{1}&&\end{smallmatrix}\right\rangle^{9}_{1}=\begin{array}{|c|}\hline 3\\\hline 5\\\hline 6\\\hline\end{array}$$

The last state $\left|\begin{smallmatrix}1&&\bar{1}&&\bar{1}\\&\bar{1}&&\bar{1}&\\&&\bar{1}&&\end{smallmatrix}\right\rangle^{9}_{1}$ is the highest weight state of an $\mathcal{A}_3$-doublet. In the $\mathcal{A}_3$-doublet the lowest weight state of this representation follows and it spans an A_2-singlet. Remind that all tensor Young tableaux for the states in this representation are traceless.

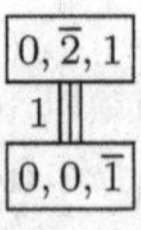

$$\left|\begin{smallmatrix}1&&\bar{1}&&\bar{1}\\&\bar{1}&&\bar{1}&\\&&\bar{1}&&\end{smallmatrix}\right\rangle^{9}_{1}=\begin{array}{|c|}\hline 3\\\hline 5\\\hline 6\\\hline\end{array}$$

$$\left|\begin{smallmatrix}\bar{1}&&\bar{1}&&\bar{1}\\&\bar{1}&&\bar{1}&\\&&\bar{1}&&\end{smallmatrix}\right\rangle^{10}_{\bar{1}}=\begin{array}{|c|}\hline 4\\\hline 5\\\hline 6\\\hline\end{array}$$

Ex. 4. The adjoint representation with $\boldsymbol{M} = 2\boldsymbol{w}_1 = (2,0,0)$ of C_3.

The highest weight state of this representation of C_3 is also the highest weight state of the A_2-sixlet with $\boldsymbol{M}^{(2)} = (2,0)$.

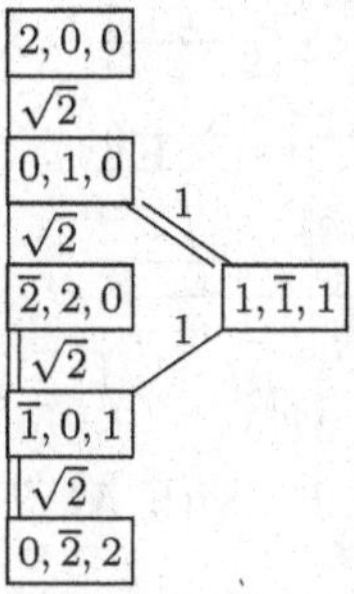

$$\left|{}^{2}\,{}_{2}\,{}^{0}_{2}\,{}_{0}\,{}^{0}\right\rangle^{1}_{0} = \boxed{1\,|\,1}$$

$$\left|{}^{2}\,{}_{2}\,{}^{0}_{1}\,{}_{0}\,{}^{0}\right\rangle^{2}_{0} = \sqrt{2}\,\boxed{1\,|\,2}$$

$$\left|{}^{2}\,{}_{2}\,{}^{0}_{0}\,{}_{0}\,{}^{0}\right\rangle^{3}_{0} = \boxed{2\,|\,2}, \qquad \left|{}^{2}\,{}_{1}\,{}^{0}_{1}\,{}_{0}\,{}^{0}\right\rangle^{3}_{1} = \sqrt{2}\,\boxed{1\,|\,3}$$

$$\left|{}^{2}\,{}_{1}\,{}^{0}_{0}\,{}_{0}\,{}^{0}\right\rangle^{4}_{1} = \sqrt{2}\,\boxed{2\,|\,3}$$

$$\left|{}^{2}\,{}_{0}\,{}^{0}_{0}\,{}_{0}\,{}^{0}\right\rangle^{5}_{2} = \boxed{3\,|\,3}$$

The states $\left|{}^{2}\,{}_{1}\,{}^{0}_{1}\,{}_{0}\,{}^{0}\right\rangle^{3}_{1}$ and $\left|{}^{2}\,{}_{1}\,{}^{0}_{0}\,{}_{0}\,{}^{0}\right\rangle^{4}_{1}$ satisfy the condition (8.7) with $\mu = 3$. Due to the parallel principle, two $\mathcal{A}_3$-doublets are constructed from these two states, where the highest weight state of an A_2-octet with $\boldsymbol{M}^{(2)} = (1,1)$ appears.

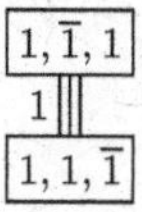

$$\left|{}^{2}\,{}_{1}\,{}^{0}_{1}\,{}_{0}\,{}^{0}\right\rangle^{3}_{1} = \sqrt{2}\,\boxed{1\,|\,3}, \qquad \left|{}^{1}\,{}_{1}\,{}^{0}_{1}\,{}_{0}\,{}^{\bar{1}}\right\rangle^{4}_{\bar{1}} = \sqrt{2}\,\boxed{1\,|\,4}$$

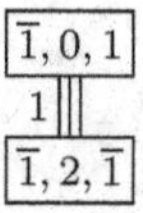

$$\left|{}^{2}\,{}_{1}\,{}^{0}_{0}\,{}_{0}\,{}^{0}\right\rangle^{4}_{1} = \sqrt{2}\,\boxed{2\,|\,3}, \qquad \left|{}^{1}\,{}_{1}\,{}^{0}_{0}\,{}_{0}\,{}^{\bar{1}}\right\rangle^{5}_{\bar{1}} = \sqrt{2}\,\boxed{2\,|\,4}$$

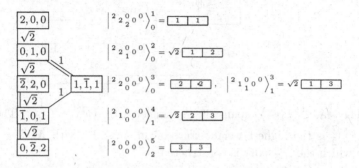

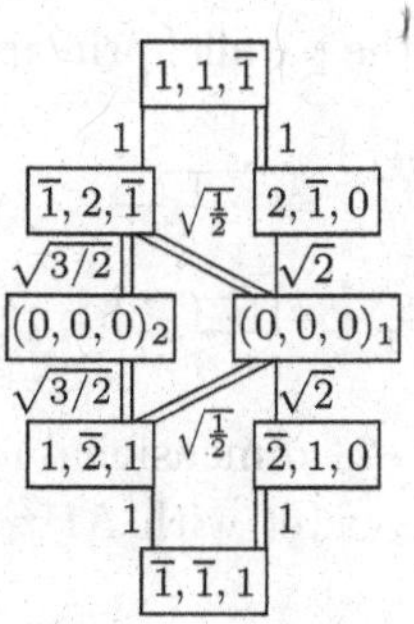

$$\left|{}^{1}\,{}_{1}\,{}^{0}_{1}\,{}_{0}\,{}^{\bar{1}}\right\rangle^{4}_{\bar{1}} = \sqrt{2}\,\boxed{1\,|\,4}$$

$$\left|{}^{1}\,{}_{1}\,{}^{0}_{0}\,{}_{0}\,{}^{\bar{1}}\right\rangle^{5}_{\bar{1}} = \sqrt{2}\,\boxed{2\,|\,4}, \qquad \left|{}^{1}\,{}_{1}\,{}^{0}_{1}\,{}_{\bar{1}}\,{}^{\bar{1}}\right\rangle^{5}_{0} = \sqrt{2}\,\boxed{1\,|\,5}$$

$$\left|{}^{1}\,{}_{0}\,{}^{0}_{0}\,{}_{0}\,{}^{\bar{1}}\right\rangle^{6}_{0}, \qquad \left|{}^{1}\,{}_{1}\,{}^{0}_{0}\,{}_{\bar{1}}\,{}^{\bar{1}}\right\rangle^{6}_{0} = \boxed{1\,|\,6} + \boxed{2\,|\,5}$$

$$\left|{}^{1}\,{}_{0}\,{}^{0}_{0}\,{}_{\bar{1}}\,{}^{\bar{1}}\right\rangle^{7}_{1} = \sqrt{2}\,\boxed{3\,|\,5}, \qquad \left|{}^{1}\,{}_{1}\,{}^{0}_{\bar{1}}\,{}_{\bar{1}}\,{}^{\bar{1}}\right\rangle^{7}_{0} = \sqrt{2}\,\boxed{2\,|\,6}$$

$$\left|{}^{1}\,{}_{0}\,{}^{0}_{\bar{1}}\,{}_{\bar{1}}\,{}^{\bar{1}}\right\rangle^{8}_{1} = \sqrt{2}\,\boxed{3\,|\,6}$$

$$\left|{}^{1}\,{}_{0}\,{}^{0}_{0}\,{}_{0}\,{}^{\bar{1}}\right\rangle^{6}_{0} = \sqrt{\tfrac{1}{3}}\left\{-\boxed{1\,|\,6} + \boxed{2\,|\,5} + 2\,\boxed{3\,|\,4}\right\}$$

The state $\left|{}^{2}\,{}_{0}\,{}^{0}_{0}\,{}_{0}\,{}^{0}\right\rangle^{5}_{2}$ in the A_2-sixlet with $\boldsymbol{M}^{(2)} = (2,0)$ satisfies the condition (8.7) with $\mu = 3$, and an $\mathcal{A}_3$-triplet is constructed from it in terms of Eqs. (8.25-28).

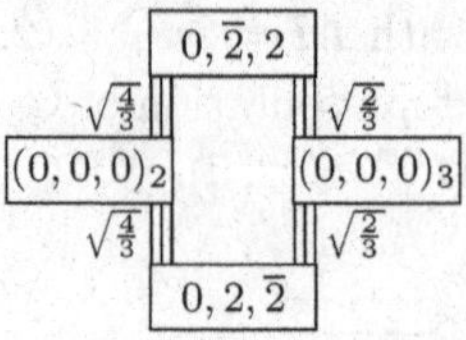

$$\left|{}^{2}\,{}_{0}\,{}^{0}_{0}\,{}_{0}\,{}^{0}\right\rangle^{5}_{2} = \begin{array}{|c|c|}\hline 3 & 3 \\ \hline\end{array}$$

$$\left|{}^{1}\,{}_{0}\,{}^{0}_{0}\,{}_{0}\,{}^{\bar{1}}\right\rangle^{6}_{0}, \quad \left|{}^{0}\,{}_{0}\,{}^{0}_{0}\,{}_{0}\,{}^{0}\right\rangle^{6}_{0}$$

$$\left|{}^{0}\,{}_{0}\,{}^{0}_{0}\,{}_{0}\,{}^{\bar{2}}\right\rangle^{7}_{\bar{2}} = \begin{array}{|c|c|}\hline 4 & 4 \\ \hline\end{array}$$

$$\left|{}^{1}\,{}_{0}\,{}^{0}_{0}\,{}_{0}\,{}^{\bar{1}}\right\rangle^{6}_{0} = \sqrt{\tfrac{1}{3}}\left\{-\begin{array}{|c|c|}\hline 1 & 6 \\ \hline\end{array} + \begin{array}{|c|c|}\hline 2 & 5 \\ \hline\end{array} + 2\begin{array}{|c|c|}\hline 3 & 4 \\ \hline\end{array}\right\}$$

$$\left|{}^{0}\,{}_{0}\,{}^{0}_{0}\,{}_{0}\,{}^{0}\right\rangle^{6}_{0} = \sqrt{\tfrac{2}{3}}\left\{\begin{array}{|c|c|}\hline 1 & 6 \\ \hline\end{array} - \begin{array}{|c|c|}\hline 2 & 5 \\ \hline\end{array} + \begin{array}{|c|c|}\hline 3 & 4 \\ \hline\end{array}\right\}$$

The state $\left|{}^{0}\,{}_{0}\,{}^{0}_{0}\,{}_{0}\,{}^{0}\right\rangle^{6}_{0}$ spans the A_2-singlet with $\boldsymbol{M}^{(2)} = (0,0)$. The state $\left|{}^{0}\,{}_{0}\,{}^{0}_{0}\,{}_{0}\,{}^{\bar{2}}\right\rangle^{7}_{\bar{2}}$ is the highest weight state of an A_2-sixlet with $\boldsymbol{M}^{(2)} = (0,2)$, from which the A_2-sixlet is constructed.

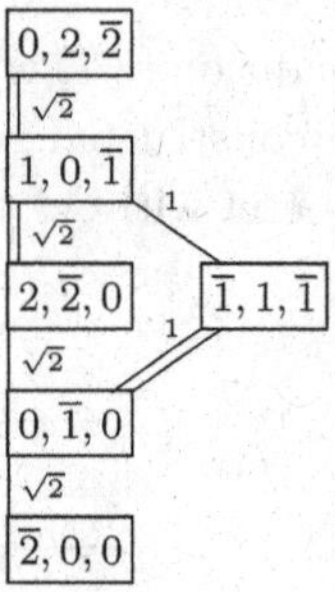

$$\left|{}^{0}\,{}_{0}\,{}^{0}_{0}\,{}_{0}\,{}^{\bar{2}}\right\rangle^{7}_{\bar{2}} = \begin{array}{|c|c|}\hline 4 & 4 \\ \hline\end{array}$$

$$\left|{}^{0}\,{}_{0}\,{}^{0}_{0}\,{}_{\bar{1}}\,{}^{\bar{2}}\right\rangle^{8}_{\bar{1}} = \sqrt{2}\begin{array}{|c|c|}\hline 4 & 5 \\ \hline\end{array}$$

$$\left|{}^{0}\,{}_{0}\,{}^{0}_{0}\,{}_{\bar{2}}\,{}^{\bar{2}}\right\rangle^{9}_{0} = \begin{array}{|c|c|}\hline 5 & 5 \\ \hline\end{array}, \quad \left|{}^{0}\,{}_{0}\,{}^{0}_{\bar{1}}\,{}_{\bar{1}}\,{}^{\bar{2}}\right\rangle^{9}_{\bar{1}} = \sqrt{2}\begin{array}{|c|c|}\hline 4 & 6 \\ \hline\end{array}$$

$$\left|{}^{0}\,{}_{0}\,{}^{0}_{\bar{1}}\,{}_{\bar{2}}\,{}^{\bar{2}}\right\rangle^{10}_{0} = \sqrt{2}\begin{array}{|c|c|}\hline 5 & 6 \\ \hline\end{array}$$

$$\left|{}^{0}\,{}_{0}\,{}^{0}_{\bar{2}}\,{}_{\bar{2}}\,{}^{\bar{2}}\right\rangle^{11}_{0} = \begin{array}{|c|c|}\hline 6 & 6 \\ \hline\end{array}$$

Two states $\left|{}^{1}\,{}_{0}\,{}^{0}_{0}\,{}_{\bar{1}}\,{}^{\bar{1}}\right\rangle^{7}_{1}$ and $\left|{}^{1}\,{}_{0}\,{}^{0}_{\bar{1}}\,{}_{\bar{1}}\,{}^{\bar{1}}\right\rangle^{8}_{1}$ in the A_2-octet with $\boldsymbol{M}^{(2)} = (1,1)$ are the highest weight states of two $\mathcal{A}_3$-doublet due to the parallel principle.

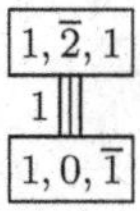

$$\left|{}^{1}\,{}_{0}\,{}^{0}_{0}\,{}_{\bar{1}}\,{}^{\bar{1}}\right\rangle^{7}_{1} = \sqrt{2}\begin{array}{|c|c|}\hline 3 & 5 \\ \hline\end{array}, \quad \left|{}^{0}\,{}_{0}\,{}^{0}_{0}\,{}_{\bar{1}}\,{}^{\bar{2}}\right\rangle^{8}_{\bar{1}} = \sqrt{2}\begin{array}{|c|c|}\hline 4 & 5 \\ \hline\end{array}$$

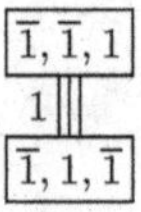

$$\left|{}^{1}\,{}_{0}\,{}^{0}_{\bar{1}}\,{}_{\bar{1}}\,{}^{\bar{1}}\right\rangle^{8}_{1} = \sqrt{2}\begin{array}{|c|c|}\hline 3 & 6 \\ \hline\end{array}, \quad \left|{}^{0}\,{}_{0}\,{}^{0}_{\bar{1}}\,{}_{\bar{1}}\,{}^{\bar{2}}\right\rangle^{9}_{\bar{1}} = \sqrt{2}\begin{array}{|c|c|}\hline 4 & 6 \\ \hline\end{array}$$

The adjoint representation with $\boldsymbol{M} = 2\boldsymbol{w}_1$ of C_3 is 21-dimensional and it contains two A_2-sixlets, one A_2-singlet, and one A_2-octet with $\boldsymbol{M}^{(2)} = (2,0)$, $(0,2)$, $(0,0)$, and $(1,1)$, respectively.

Ex. 5. The representation with $\boldsymbol{M} = \boldsymbol{w}_1 + \boldsymbol{w}_2 = (1,1,0)$ of C_3.

Since we have calculated the orthonormal basis states of this representation in subsection 8.3.2 in some detail. Here we only list the tensor Young tableaux for the highest weight states of eight A_2-multiplets and for some $\mathcal{A}_3$-multiplets. The detailed calculation for the remaining orthonormal basis states are left as exercise for readers (see prob. 3).

$$|(1,1,0)\rangle = \left|\begin{smallmatrix}2&&1&&0\\&2&&1&\\&&2&&\end{smallmatrix}\right\rangle^{1}_{0} = \begin{array}{|c|c|}\hline 1&1\\\hline 2\\\cline{1-1}\end{array}\,,$$

$$|(2,1,\overline{1})\rangle = \left|\begin{smallmatrix}2&&0&&\overline{1}\\&2&&0&\\&&2&&\end{smallmatrix}\right\rangle^{3}_{\overline{1}} = \begin{array}{|c|c|}\hline 1&1\\\hline 4\\\cline{1-1}\end{array}\,,$$

$$|(0,2,\overline{1})_2\rangle = \left|\begin{smallmatrix}1&&1&&\overline{1}\\&1&&1&\\&&1&&\end{smallmatrix}\right\rangle^{4}_{\overline{1}} = \sqrt{\tfrac{3}{2}}\begin{array}{|c|c|}\hline 1&4\\\hline 2\\\cline{1-1}\end{array}\,,$$

$$|(1,0,0)_4\rangle = \left|\begin{smallmatrix}1&&0&&0\\&1&&0&\\&&1&&\end{smallmatrix}\right\rangle^{5}_{0} = \sqrt{\tfrac{1}{56}}\left\{2\begin{array}{|c|c|}\hline 1&1\\\hline 6\\\cline{1-1}\end{array} - 2\begin{array}{|c|c|}\hline 1&2\\\hline 5\\\cline{1-1}\end{array} - 5\begin{array}{|c|c|}\hline 1&5\\\hline 2\\\cline{1-1}\end{array} + 2\begin{array}{|c|c|}\hline 1&3\\\hline 4\\\cline{1-1}\end{array} + 5\begin{array}{|c|c|}\hline 1&4\\\hline 3\\\cline{1-1}\end{array}\right\}\,,$$

$$|(1,2,\overline{2})\rangle = \left|\begin{smallmatrix}1&&0&&\overline{2}\\&1&&0&\\&&1&&\end{smallmatrix}\right\rangle^{6}_{\overline{2}} = \begin{array}{|c|c|}\hline 1&4\\\hline 4\\\cline{1-1}\end{array}\,,$$

$$|(2,0,\overline{1})_2\rangle = \left|\begin{smallmatrix}1&&\overline{1}&&\overline{1}\\&1&&\overline{1}&\\&&1&&\end{smallmatrix}\right\rangle^{7}_{\overline{1}} = \sqrt{\tfrac{3}{2}}\left\{\begin{array}{|c|c|}\hline 1&4\\\hline 5\\\cline{1-1}\end{array} - \begin{array}{|c|c|}\hline 1&5\\\hline 4\\\cline{1-1}\end{array}\right\}\,,$$

$$|(0,1,\overline{1})_4\rangle = \left|\begin{smallmatrix}0&&0&&\overline{1}\\&0&&0&\\&&0&&\end{smallmatrix}\right\rangle^{8}_{\overline{1}} = \sqrt{\tfrac{1}{56}}\left\{-5\begin{array}{|c|c|}\hline 1&4\\\hline 6\\\cline{1-1}\end{array} + 7\begin{array}{|c|c|}\hline 1&6\\\hline 4\\\cline{1-1}\end{array} + 5\begin{array}{|c|c|}\hline 2&4\\\hline 5\\\cline{1-1}\end{array} - 7\begin{array}{|c|c|}\hline 2&5\\\hline 4\\\cline{1-1}\end{array} + 2\begin{array}{|c|c|}\hline 3&4\\\hline 4\\\cline{1-1}\end{array}\right\}\,,$$

$$|(1,1,\overline{2})\rangle = \left|\begin{smallmatrix}0&&\overline{1}&&\overline{2}\\&0&&\overline{1}&\\&&0&&\end{smallmatrix}\right\rangle^{10}_{\overline{2}} = \begin{array}{|c|c|}\hline 4&4\\\hline 5\\\cline{1-1}\end{array}\,.$$

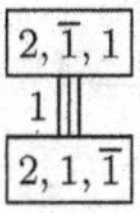

$$\left|\begin{smallmatrix}2&&1&&0\\&2&&0&\\&&2&&\end{smallmatrix}\right\rangle^{2}_{1} = \begin{array}{|c|c|}\hline 1&1\\\hline 3\\\cline{1-1}\end{array} \qquad \left|\begin{smallmatrix}2&&0&&\overline{1}\\&2&&0&\\&&2&&\end{smallmatrix}\right\rangle^{3}_{\overline{1}} = \begin{array}{|c|c|}\hline 1&1\\\hline 4\\\cline{1-1}\end{array}$$

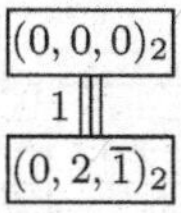

$$\left|\begin{smallmatrix}2&&1&&0\\&1&&1&\\&&1&&\end{smallmatrix}\right\rangle^{3}_{1} = \sqrt{\tfrac{3}{2}}\begin{array}{|c|c|}\hline 1&3\\\hline 2\\\cline{1-1}\end{array} \qquad \left|\begin{smallmatrix}1&&1&&\overline{1}\\&1&&1&\\&&1&&\end{smallmatrix}\right\rangle^{4}_{\overline{1}} = \sqrt{\tfrac{3}{2}}\begin{array}{|c|c|}\hline 1&4\\\hline 2\\\cline{1-1}\end{array}$$

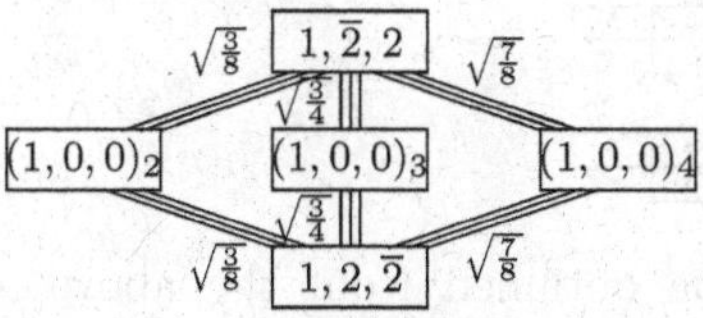

$$\left|\begin{smallmatrix}2&&1&&0\\&1&&0&\\&&1&&\end{smallmatrix}\right\rangle^{4}_{2} = \begin{array}{|c|c|}\hline 1&3\\\hline 3\\\cline{1-1}\end{array}$$

$$\left|\begin{smallmatrix}2&&0&&\overline{1}\\&1&&0&\\&&1&&\end{smallmatrix}\right\rangle^{5}_{0}, \quad \left|\begin{smallmatrix}1&&1&&\overline{1}\\&1&&0&\\&&1&&\end{smallmatrix}\right\rangle^{5}_{0}, \quad \left|\begin{smallmatrix}1&&0&&0\\&1&&0&\\&&1&&\end{smallmatrix}\right\rangle^{5}_{0}$$

$$\left|\begin{smallmatrix}1&&0&&\overline{2}\\&1&&0&\\&&1&&\end{smallmatrix}\right\rangle^{6}_{\overline{2}} = \begin{array}{|c|c|}\hline 1&4\\\hline 4\\\cline{1-1}\end{array}$$

$$\left|\begin{smallmatrix}2&&0&&\overline{1}\\&1&&0&\\&&1&&\end{smallmatrix}\right\rangle^{5}_{0} = \sqrt{\tfrac{1}{24}}\left\{-2\begin{array}{|c|c|}\hline 1&1\\\hline 6\\\cline{1-1}\end{array} + 2\begin{array}{|c|c|}\hline 1&2\\\hline 5\\\cline{1-1}\end{array} - \begin{array}{|c|c|}\hline 1&5\\\hline 2\\\cline{1-1}\end{array} + 6\begin{array}{|c|c|}\hline 1&3\\\hline 4\\\cline{1-1}\end{array} - 3\begin{array}{|c|c|}\hline 1&4\\\hline 3\\\cline{1-1}\end{array}\right\},$$

$$\left|\begin{smallmatrix}1&&1&&\overline{1}\\&1&&0&\\&&1&&\end{smallmatrix}\right\rangle^{5}_{0} = \sqrt{\tfrac{3}{4}}\left\{\begin{array}{|c|c|}\hline 1&4\\\hline 3\\\cline{1-1}\end{array} + \begin{array}{|c|c|}\hline 1&5\\\hline 2\\\cline{1-1}\end{array}\right\},$$

$$\left|\begin{smallmatrix}1&&0&&0\\&1&&0&\\&&1&&\end{smallmatrix}\right\rangle^{5}_{0} = \sqrt{\tfrac{1}{56}}\left\{2\begin{array}{|c|c|}\hline 1&1\\\hline 6\\\cline{1-1}\end{array} - 2\begin{array}{|c|c|}\hline 1&2\\\hline 5\\\cline{1-1}\end{array} - 5\begin{array}{|c|c|}\hline 1&5\\\hline 2\\\cline{1-1}\end{array} + 2\begin{array}{|c|c|}\hline 1&3\\\hline 4\\\cline{1-1}\end{array} + 5\begin{array}{|c|c|}\hline 1&4\\\hline 3\\\cline{1-1}\end{array}\right\}.$$

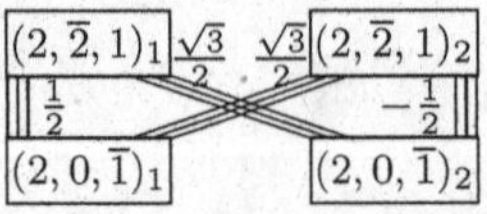

$$\left|\begin{smallmatrix}2&&0&&\bar1\\&1&&\bar1&\\&&1&&\end{smallmatrix}\right\rangle^6_1=\sqrt{\tfrac{1}{2}}\left\{2\begin{array}{|c|c|}\hline 1&3\\\hline 5&\multicolumn{1}{c}{}\\\cline{1-1}\end{array}-\begin{array}{|c|c|}\hline 1&5\\\hline 3&\multicolumn{1}{c}{}\\\cline{1-1}\end{array}\right\},\quad\left|\begin{smallmatrix}1&&1&&\bar1\\&1&&\bar1&\\&&1&&\end{smallmatrix}\right\rangle^6_1$$

$$\left|\begin{smallmatrix}1&&0&&\bar2\\&1&&\bar1&\\&&1&&\end{smallmatrix}\right\rangle^7_{\bar1}=\sqrt{\tfrac{1}{2}}\left\{\begin{array}{|c|c|}\hline 1&4\\\hline 5&\multicolumn{1}{c}{}\\\cline{1-1}\end{array}+\begin{array}{|c|c|}\hline 1&5\\\hline 4&\multicolumn{1}{c}{}\\\cline{1-1}\end{array}\right\},\quad\left|\begin{smallmatrix}1&&\bar1&&\bar1\\&1&&\bar1&\\&&1&&\end{smallmatrix}\right\rangle^7_{\bar1}$$

$$\left|\begin{smallmatrix}1&&1&&\bar1\\&1&&\bar1&\\&&1&&\end{smallmatrix}\right\rangle^6_1=\sqrt{\tfrac{3}{2}}\begin{array}{|c|c|}\hline 1&5\\\hline 3&\multicolumn{1}{c}{}\\\cline{1-1}\end{array},\qquad\left|\begin{smallmatrix}1&&\bar1&&\bar1\\&1&&\bar1&\\&&1&&\end{smallmatrix}\right\rangle^7_{\bar1}=\sqrt{\tfrac{3}{2}}\left\{\begin{array}{|c|c|}\hline 1&4\\\hline 5&\multicolumn{1}{c}{}\\\cline{1-1}\end{array}-\begin{array}{|c|c|}\hline 1&5\\\hline 4&\multicolumn{1}{c}{}\\\cline{1-1}\end{array}\right\},$$

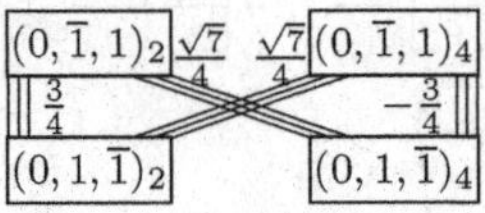

$$\left|\begin{smallmatrix}2&&0&&\bar1\\&0&&0&\\&&0&&\end{smallmatrix}\right\rangle^7_1,\quad\left|\begin{smallmatrix}1&&0&&0\\&0&&0&\\&&0&&\end{smallmatrix}\right\rangle^7_1$$

$$\left|\begin{smallmatrix}1&&0&&\bar2\\&0&&0&\\&&0&&\end{smallmatrix}\right\rangle^8_{\bar1},\quad\left|\begin{smallmatrix}0&&0&&\bar1\\&0&&0&\\&&0&&\end{smallmatrix}\right\rangle^8_{\bar1}$$

$$\left|\begin{smallmatrix}2&&0&&\bar1\\&0&&0&\\&&0&&\end{smallmatrix}\right\rangle^7_1=\sqrt{\tfrac{1}{8}}\left\{-2\begin{array}{|c|c|}\hline 1&3\\\hline 6&\multicolumn{1}{c}{}\\\cline{1-1}\end{array}+\begin{array}{|c|c|}\hline 1&6\\\hline 3&\multicolumn{1}{c}{}\\\cline{1-1}\end{array}+2\begin{array}{|c|c|}\hline 2&3\\\hline 5&\multicolumn{1}{c}{}\\\cline{1-1}\end{array}-\begin{array}{|c|c|}\hline 2&5\\\hline 3&\multicolumn{1}{c}{}\\\cline{1-1}\end{array}+2\begin{array}{|c|c|}\hline 3&3\\\hline 4&\multicolumn{1}{c}{}\\\cline{1-1}\end{array}\right\},$$

$$\left|\begin{smallmatrix}1&&0&&0\\&0&&0&\\&&0&&\end{smallmatrix}\right\rangle^7_1=\sqrt{\tfrac{1}{56}}\left\{2\begin{array}{|c|c|}\hline 1&3\\\hline 6&\multicolumn{1}{c}{}\\\cline{1-1}\end{array}-7\begin{array}{|c|c|}\hline 1&6\\\hline 3&\multicolumn{1}{c}{}\\\cline{1-1}\end{array}-2\begin{array}{|c|c|}\hline 2&3\\\hline 5&\multicolumn{1}{c}{}\\\cline{1-1}\end{array}+7\begin{array}{|c|c|}\hline 2&5\\\hline 3&\multicolumn{1}{c}{}\\\cline{1-1}\end{array}+2\begin{array}{|c|c|}\hline 3&3\\\hline 4&\multicolumn{1}{c}{}\\\cline{1-1}\end{array}\right\},$$

$$\left|\begin{smallmatrix}1&&0&&\bar2\\&0&&0&\\&&0&&\end{smallmatrix}\right\rangle^8_{\bar1}=\sqrt{\tfrac{1}{8}}\left\{-\begin{array}{|c|c|}\hline 1&4\\\hline 6&\multicolumn{1}{c}{}\\\cline{1-1}\end{array}-\begin{array}{|c|c|}\hline 1&6\\\hline 4&\multicolumn{1}{c}{}\\\cline{1-1}\end{array}+\begin{array}{|c|c|}\hline 2&4\\\hline 5&\multicolumn{1}{c}{}\\\cline{1-1}\end{array}+\begin{array}{|c|c|}\hline 2&5\\\hline 4&\multicolumn{1}{c}{}\\\cline{1-1}\end{array}+2\begin{array}{|c|c|}\hline 3&4\\\hline 4&\multicolumn{1}{c}{}\\\cline{1-1}\end{array}\right\},$$

$$\left|\begin{smallmatrix}0&&0&&\bar1\\&0&&0&\\&&0&&\end{smallmatrix}\right\rangle^8_{\bar1}=\sqrt{\tfrac{1}{56}}\left\{-5\begin{array}{|c|c|}\hline 1&4\\\hline 6&\multicolumn{1}{c}{}\\\cline{1-1}\end{array}+7\begin{array}{|c|c|}\hline 1&6\\\hline 4&\multicolumn{1}{c}{}\\\cline{1-1}\end{array}+5\begin{array}{|c|c|}\hline 2&4\\\hline 5&\multicolumn{1}{c}{}\\\cline{1-1}\end{array}-7\begin{array}{|c|c|}\hline 2&5\\\hline 4&\multicolumn{1}{c}{}\\\cline{1-1}\end{array}+2\begin{array}{|c|c|}\hline 3&4\\\hline 4&\multicolumn{1}{c}{}\\\cline{1-1}\end{array}\right\},$$

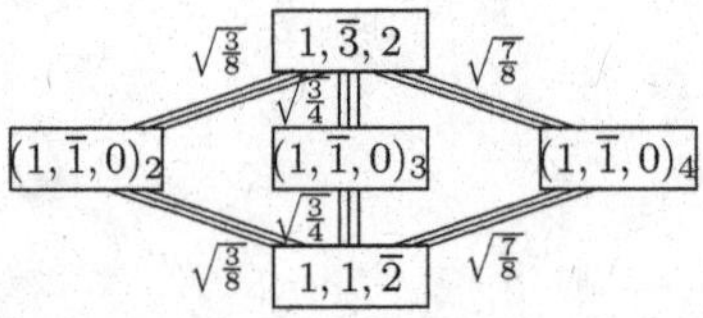

$$\left|\begin{smallmatrix}2&&0&&\bar1\\&0&&\bar1&\\&&0&&\end{smallmatrix}\right\rangle^8_2=\begin{array}{|c|c|}\hline 3&3\\\hline 5&\multicolumn{1}{c}{}\\\cline{1-1}\end{array}$$

$$\left|\begin{smallmatrix}1&&0&&\bar2\\&0&&\bar1&\\&&0&&\end{smallmatrix}\right\rangle^9_0,\quad\left|\begin{smallmatrix}1&&\bar1&&\bar1\\&0&&\bar1&\\&&0&&\end{smallmatrix}\right\rangle^9_0,\quad\left|\begin{smallmatrix}0&&0&&\bar1\\&0&&\bar1&\\&&0&&\end{smallmatrix}\right\rangle^9_0$$

$$\left|\begin{smallmatrix}0&&\bar1&&\bar2\\&0&&\bar1&\\&&0&&\end{smallmatrix}\right\rangle^{10}_{\bar2}=\begin{array}{|c|c|}\hline 4&4\\\hline 5&\multicolumn{1}{c}{}\\\cline{1-1}\end{array}$$

$$\left|\begin{smallmatrix}1&&0&&\bar2\\&0&&\bar1&\\&&0&&\end{smallmatrix}\right\rangle^9_0=\sqrt{\tfrac{1}{24}}\left\{-\begin{array}{|c|c|}\hline 1&5\\\hline 6&\multicolumn{1}{c}{}\\\cline{1-1}\end{array}-\begin{array}{|c|c|}\hline 1&6\\\hline 5&\multicolumn{1}{c}{}\\\cline{1-1}\end{array}+2\begin{array}{|c|c|}\hline 2&5\\\hline 5&\multicolumn{1}{c}{}\\\cline{1-1}\end{array}+3\begin{array}{|c|c|}\hline 3&4\\\hline 5&\multicolumn{1}{c}{}\\\cline{1-1}\end{array}+3\begin{array}{|c|c|}\hline 3&5\\\hline 4&\multicolumn{1}{c}{}\\\cline{1-1}\end{array}\right\},$$

$$\left|\begin{smallmatrix}1&&\bar1&&\bar1\\&0&&\bar1&\\&&0&&\end{smallmatrix}\right\rangle^9_0=\sqrt{\tfrac{3}{4}}\left\{\begin{array}{|c|c|}\hline 1&5\\\hline 6&\multicolumn{1}{c}{}\\\cline{1-1}\end{array}-\begin{array}{|c|c|}\hline 1&6\\\hline 5&\multicolumn{1}{c}{}\\\cline{1-1}\end{array}+\begin{array}{|c|c|}\hline 3&4\\\hline 5&\multicolumn{1}{c}{}\\\cline{1-1}\end{array}-\begin{array}{|c|c|}\hline 3&5\\\hline 4&\multicolumn{1}{c}{}\\\cline{1-1}\end{array}\right\},$$

$$\left|\begin{smallmatrix}0&&0&&\bar1\\&0&&\bar1&\\&&0&&\end{smallmatrix}\right\rangle^9_0=\sqrt{\tfrac{1}{56}}\left\{-5\begin{array}{|c|c|}\hline 1&5\\\hline 6&\multicolumn{1}{c}{}\\\cline{1-1}\end{array}+7\begin{array}{|c|c|}\hline 1&6\\\hline 5&\multicolumn{1}{c}{}\\\cline{1-1}\end{array}-2\begin{array}{|c|c|}\hline 2&5\\\hline 5&\multicolumn{1}{c}{}\\\cline{1-1}\end{array}+7\begin{array}{|c|c|}\hline 3&4\\\hline 5&\multicolumn{1}{c}{}\\\cline{1-1}\end{array}-5\begin{array}{|c|c|}\hline 3&5\\\hline 4&\multicolumn{1}{c}{}\\\cline{1-1}\end{array}\right\},$$

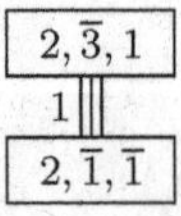

$$\left|\begin{smallmatrix}1&&0&&\bar2\\&0&&\bar2&\\&&0&&\end{smallmatrix}\right\rangle^{10}_1=\begin{array}{|c|c|}\hline 3&5\\\hline 5&\multicolumn{1}{c}{}\\\cline{1-1}\end{array}$$

$$\left|\begin{smallmatrix}0&&\bar1&&\bar2\\&0&&\bar2&\\&&0&&\end{smallmatrix}\right\rangle^{11}_{\bar1}=\begin{array}{|c|c|}\hline 4&5\\\hline 5&\multicolumn{1}{c}{}\\\cline{1-1}\end{array}$$

The remaining $\mathcal{A}_3$-multiplets can be obtained from the above $\mathcal{A}_3$-multiplets by the parallel principle.

12.1.3 *Dimensions of Irreducible Representations*

The dimension of an irreducible representation $[\lambda]$ of $\mathrm{USp}(2\ell)$ can be calculated by the hook rule. In this rule, the dimension is expressed as a quotient, where the numerator and the denominator are denoted by the symbols $Y_P^{[\lambda]}$ and $Y_h^{[\lambda]}$, respectively:

$$d_{[\lambda]}(\text{Sp}(2\ell)) = \frac{Y_P^{[\lambda]}}{Y_h^{[\lambda]}}. \tag{12.16}$$

We still use the concept of the hook path $(i,\ j)$ and the inverse hook path $\overline{(i,\ j)}$ in the Young pattern $[\lambda]$ as given in subsection 10.1.4. The number of boxes contained in the hook path $(i,\ j)$ is the hook number h_{ij} of the box in the jth column of the ith row. $Y_h^{[\lambda]}$ is a tableau of the Young pattern $[\lambda]$ where the box in the jth column of the ith row is filled with the hook number h_{ij}. Define a series of the tableaux $Y_{P_a}^{[\lambda]}$ recursively by the rule given below. $Y_P^{[\lambda]}$ is a tableau of the Young pattern $[\lambda]$ where each box is filled with the sum of the digits which are respectively filled in the same box of each tableau $Y_{P_a}^{[\lambda]}$ in the series. The symbol $Y_P^{[\lambda]}$ means the product of the filled digits in it, so does the symbol $Y_h^{[\lambda]}$.

The tableaux $Y_{P_a}^{[\lambda]}$ are defined by the following rule:

(a) $Y_{P_0}^{[\lambda]}$ is a tableau of the Young pattern $[\lambda]$ where the box in the jth column of the ith row is filled with the digit $(2\ell + j - i)$.
(b) Let $[\lambda^{(1)}] = [\lambda]$. Beginning with $[\lambda^{(1)}]$, we define recursively the Young pattern $[\lambda^{(a)}]$ by removing the first row and the first column of the Young pattern $[\lambda^{(a-1)}]$ until $[\lambda^{(a)}]$ contains less than two rows.
(c) If $[\lambda^{(a)}]$ contains more than one row, define $Y_{P_a}^{[\lambda]}$ to be a tableau of the Young pattern $[\lambda]$ where the boxes in the first $(a-1)$ rows and in the first $(a-1)$ columns are filled with 0, and the remaining part of the Young pattern is nothing but $[\lambda^{(a)}]$. Let $[\lambda^{(a)}]$ have r rows. Fill the first $(r-1)$ boxes along the hook path $(1,\ 1)$ of the Young pattern $[\lambda^{(a)}]$, beginning with the box on the rightmost, with the digits $\lambda_2^{(a)}, \lambda_3^{(a)}, \cdots, \lambda_r^{(a)}$, box by box, and fill the first $\lambda_i^{(a)}$ boxes in each inverse hook path $\overline{(i,\ 1)}$ of the Young pattern $[\lambda^{(a)}]$, $2 \leqslant i \leqslant r$, with -1. The remaining boxes are filled with 0. If a few -1 are filled in the same box, the digits are summed. The sum of all filled digits in the pattern $Y_{P_a}^{[\lambda]}$ with $a > 0$ is 0.

The calculation method (12.16) is explained through some examples.

Ex. 1 The dimension of the representation [3, 3, 3] of USp(6).

$$Y_P^{[3,3,3]} = \begin{array}{|c|c|c|}\hline 6 & 7 & 8 \\ \hline 5 & 6 & 7 \\ \hline 4 & 5 & 6 \\ \hline \end{array} + \begin{array}{|c|c|c|}\hline & 3 & 3 \\ \hline -1 & -1 & \\ \hline -2 & -1 & -1 \\ \hline \end{array} + \begin{array}{|c|c|c|}\hline & & \\ \hline & & 2 \\ \hline & -1 & -1 \\ \hline \end{array} = \begin{array}{|c|c|c|}\hline 6 & 10 & 11 \\ \hline 4 & 5 & 9 \\ \hline 2 & 3 & 4 \\ \hline \end{array},$$

$$d_{[3,3,3]}[\mathrm{USp}(6)] = \frac{\begin{array}{|c|c|c|}\hline 6 & 10 & 11 \\ \hline 4 & 5 & 9 \\ \hline 2 & 3 & 4 \\ \hline \end{array}}{\begin{array}{|c|c|c|}\hline 5 & 4 & 3 \\ \hline 4 & 3 & 2 \\ \hline 3 & 2 & 1 \\ \hline \end{array}} = 11 \times 5 \times 3 \times 2 = 330.$$

Ex. 2 The representation of one-row Young pattern $[n]$ of USp(2ℓ). The tensors in the representation are totally symmetric, so that all standard tensor Young tableaux are traceless.

$$d_{[n]}[\mathrm{USp}(2\ell)] = d_{[n]}[\mathrm{SU}(2\ell)] = \binom{n+2\ell-1}{n}. \tag{12.17}$$

Ex. 3 The representation of one-column Young pattern $[1^n]$ of USp(2ℓ).

$$Y_P^{[1^n]} = \begin{array}{|c|}\hline 2\ell \\ \hline 2\ell-1 \\ \hline \vdots \\ \hline 2\ell-n+2 \\ \hline 2\ell-n+1 \\ \hline \end{array} + \begin{array}{|c|}\hline 1 \\ \hline 1 \\ \hline \vdots \\ \hline 1 \\ \hline -n+1 \\ \hline \end{array} = \begin{array}{|c|}\hline 2\ell+1 \\ \hline 2\ell \\ \hline \vdots \\ \hline 2\ell-n+3 \\ \hline 2\ell-2n+2 \\ \hline \end{array},$$

$$d_{[1^n]}[\mathrm{USp}(2\ell)] = \frac{(2\ell+1)!(2\ell-2n+2)}{n!(2\ell-n+2)!}. \tag{12.18}$$

Ex. 4 The representation of two-row Young pattern $[n, m]$ of USp(2ℓ).

$$Y_P^{[n,m]} = \begin{array}{|c|c|c|c|c|c|}\hline 2\ell & \cdots & 2\ell+m-1 & \cdots & 2\ell+n-2 & 2\ell+n-1 \\ \hline 2\ell-1 & \cdots & 2\ell+m-2 & & & \\ \hline \end{array}$$

$$+ \begin{array}{|c|c|c|c|c|c|}\hline 0 & \cdots & 0 & \cdots & 0 & m \\ \hline -1 & \cdots & -1 & & & \\ \hline \end{array}$$

$$= \begin{array}{|c|c|c|c|c|c|}\hline 2\ell & \cdots & 2\ell+m-1 & \cdots & 2\ell+n-2 & 2\ell+n+m-1 \\ \hline 2\ell-2 & \cdots & 2\ell+m-3 & & & \\ \hline \end{array},$$

$$Y_h^{[n,m]} = \begin{array}{|c|c|c|c|c|c|}\hline n+1 & \cdots & n-m+2 & n-m & \cdots & 1 \\ \hline m & \cdots & 1 & & & \\ \hline \end{array},$$

$$d_{[n,m]}[\mathrm{USp}(2\ell)] = \frac{(n-m+1)(2\ell+n+m-1)(2\ell+n-2)!(2\ell+m-3)!}{(n+1)!m!(2\ell-1)!(2\ell-3)!}. \tag{12.19}$$

For the groups USp(4) and USp(6), one has

$$\begin{aligned} d_{[n,m]}[\mathrm{USp}(4)] &= (n-m+1)(n+m+3)(n+2)(m+1)/6, \\ d_{[n,m]}[\mathrm{USp}(6)] &= (n-m+1)(n+m+5)(n+4)(n+3)(n+2) \\ &\quad \times (m+3)(m+2)(m+1)/720. \end{aligned}$$

12.2 Physical Application

The Hamiltonian equation of a classical system with ℓ degrees of freedom is

$$\frac{dq_j}{dt} = \frac{\partial H}{\partial p_j}, \qquad \frac{dp_j}{dt} = -\frac{\partial H}{\partial q_j}. \tag{12.20}$$

Arranging the coordinates q_j and the momentums p_j in the order,

$$x_a = (q_1,\ p_1,\ q_2,\ p_2,\ \cdots,\ q_\ell,\ p_\ell), \tag{12.21}$$

we obtain the coordinates x_a in the (2ℓ)-dimensional phase space. The Hamiltonian equation can be expressed in a unified form

$$\frac{dx_a}{dt} = \sum_b J_{ab} \frac{\partial H}{\partial x_b}, \qquad \frac{dx}{dt} = J\frac{\partial H}{\partial x}, \tag{12.22}$$

called the **symplectic form of the Hamiltonian equation**. If dx_a satisfy the Hamiltonian equation (12.22), after the symplectic transformation

$$dz_a = \sum_b R_{ab} dx_b\,, \qquad R_{ab} = \frac{\partial z_a}{\partial x_b}, \qquad R^T J R = J, \tag{12.23}$$

dz_a still satisfy the Hamiltonian equation (12.22),

$$\begin{aligned} \frac{dz_a}{dt} &= \sum_b R_{ab}\frac{dx_b}{dt} = \sum_{bc} R_{ab} J_{bc} \sum_r \frac{\partial z_r}{\partial x_c}\frac{\partial H}{\partial z_r} \\ &= \sum_r \left(RJR^T\right)_{ar} \frac{\partial H}{\partial z_r} = \sum_r J_{ar}\frac{\partial H}{\partial z_r}\,. \end{aligned}$$

The method of Runge–Kutta is commonly used in the numerical calculations by computer. This method does not reflect the characteristic of the equation of motion so that the calculation error will be accumulated. If the calculation is repeated in a tremendous number, the accumulated error will make a big deviation of the calculation data from the real orbit, for example, the calculation in cyclotron reaction and in satellites. If each step in the numerical calculation reflects the characteristic of the Hamiltonian equation, say each step satisfies the symplectic transformation,

$$z_a = z_a^{(0)} + \tau \sum_b J_{ab} \frac{\partial H(x)}{\partial z_b}, \qquad x_a = \left(z_a + z_a^{(0)}\right)/2, \tag{12.24}$$

where τ is the length of the step, the accumulated error will decrease greatly. The group led by Professor Kang Feng studied deeply this problem. Please see his paper [Feng (1991)] in detail.

12.3 Exercises

1. Calculate the positive roots of USp(6) from the weights of the states in its adjoint representation (see Ex. 4 in subsection 12.1.2).
2. Calculate the tensor Young tableaux for all orthonormal basis states in the representation of C_3 with the highest weight $\boldsymbol{M} = 2\boldsymbol{w}_2 = (0, 2, 0)$.
3. Calculate the tensor Young tableaux for all orthonormal basis states in the representation of C_3 with the highest weight $\boldsymbol{M} = 2\boldsymbol{w}_2 = (0, 0, 2)$.
4. Calculate the tensor Young tableaux for all orthonormal basis states in the representation of C_3 with the highest weight $\boldsymbol{M} = \boldsymbol{w}_1 + \boldsymbol{w}_3 = (1, 0, 1)$.
5. Calculate the tensor Young tableaux for all orthonormal basis states in the representation of C_3 with the highest weight $\boldsymbol{M} = \boldsymbol{w}_2 + \boldsymbol{w}_3 = (0, 1, 1)$.
6. Calculate the dimensions of the irreducible representations of the USp(6) group denoted by the following Young patterns:

   (1) $[4, 2]$, (2) $[3, 2]$, (3) $[4, 4]$, (4) $[3, 3, 2]$, (5) $[4, 4, 3]$.

Bibliography

Adams, B. G., Cizek, J. and Paldus, J. (1987). Lie algebraic methods and their applications to simple quantum systems, *Advances in Quantum Chemistry*, Vol. **19**, Academic Press, New York.

Andrews, G. E. (1976). The Theory of Partitions, Encyclopedia of Mathematics and Its Applications, Vol. **2**, Ed. Gian-Carlo Rota, Addison-Wesley.

Baird, G. E. and Biedenharm, L. C. (1976). On the representations of the semisimple Lie groups, II, *J. Math. Phys.* **4**, 1449.

Bayman, B. F. (1960). Some Lectures on Groups and their Applications to Spectroscopy, Nordita.

Berenson, R. and Birman, J. L. (1975). Clebsch–Gordan coefficients for crystal space group, *J. Math. Phys.* **16**, 227.

Bethe, H.A. (1929). Splitting of terms in crystals, *Annln Phys.* **3**, 133 (in German); translated by Cracknell, A.P. *Applied Group theory*, Pergamon Press, Oxford.

Biedenharn, L. C., Giovannini, A. and Louck, J. D. (1970). Canonical definition of Wigner coefficients in $U(n)$, *J. Math. Phys.* **11**, 2368.

Biedenharn, L. C. and Louck, J. D. (1981). Angular Momentum in Quantum Physics, Theory and Application, *Encyclopedia of Mathematics and its Application*, Vol. **8**, Ed. G. C. Rota, Addison-Wesley, Massachusetts.

Bjorken, J. D. and Drell, S. D. (1964). Relativistic Quantum Mechanics, McGraw-Hill Book Co., New York.

Bloch, F. (1928). Quantum mechanics of electrons in crystal lattices, *Z. Phys.* **52**, 555–600 (in German).

Boerner, H. (1963). Representations of Groups, North-Holland, Amsterdam.

Bourbaki, N. (1989). Elements of Mathematics, Lie Groups and Lie Algebras, Springer-Verlag, New York.

Bradley, C. J. and Cracknell, A. P. (1972). The Mathematical Theory of Symmetry in Solids, Clarendon Press, Oxford.

Bremner, M. R., Moody, R. V. and Patera, J. (1985). Tables of Dominant Weight Multiplicities for Representations of Simple Lie Algebras, Pure and Applied Mathematics, A Series of Monographs and Textbooks 90, Marcel Dekker, New York.

Burns, G. and Glazer, A. M. (1978). Space Groups for Solid State Scientists, Academic Press, New York.

Chen, J. Q., Wang, P. N., Lu Z. M. and Wu, X. B. (1987). Tables of the Clebsch–Gordan, Racah and Subduction Coefficients of $SU(n)$ Groups, World Scientific, Singapore.

Chen, J. Q., Ping, J. L. and Wang, F. (2002). Group Representation Theory for Physicists, 2nd edition, World Scientific, Singapore.

Chen, J. Q. and Ping, J. L. (1997). Algebraic expressions for irreducible bases of icosahedral group, *J. Math. Phys.* **38**, 387.

Cotton, F. A. (1971). Chemical Applications of Group Theory, Wiley, New York.

de Swart, J. J. (1963). The octet model and its Clebsch–Gordan coefficients, *Rev. Mod. Phys.* **35**, 916.

Dai A. Y. (1983). A graphic rule for dimensions of irreducible spinor representations of SO(N), J. Lanzhou Univ. (Natural Sciences) **19**, No. 2, 33 (in Chinese).

Deng Y. F. and Yang, C. N. (1992). Eigenvalues and eigenfunctions of the Hückel Hamiltonian for Carbon-60, *Phys. Lett. A* **170**, 116.

Dirac, P. A. M. (1958). The Principle of Quantum Mechanics, Clarendon Press, Oxford.

Dong, S. H., Hou, X. W. and Ma, Z. Q. (1998a). Irreducible bases and correlations of spin states for double point groups, *Inter. J. Theor. Phys.* **37**, 841.

Dong, S. H., Xie, M. and Ma, Z. Q. (1998b). Irreducible bases in icosahedral group space, *Inter. J. Theor. Phys.* **37**, 2135.

Dong, S. H., Hou, X. W. and Ma, Z. Q. (1998c). Relativistic Levinson theorem in two dimensions, *Phys. Rev. A* **58**, 2160.

Dong, S. H., Hou, X. W. and Ma, Z. Q. (2001). Correlations of spin states for icosahedral double group, *Inter. J. Theor. Phys.* **40**, 569.

Dong, S. H. and Ma, Z. Q. (2003). Exact solutions to the Dirac equation with a Coulomb potential in $2 + 1$ dimensions, *Phys. Lett. A* **312**, 78.

Duan, B., Gu, X. Y. and Ma, Z. Q. (2001). Precise calculation for energy levels of a helium atom in P states, *Phys. Lett. A* **283**, 229.

Duan, B., Gu, X. Y. and Ma, Z. Q. (2001). Energy levels in D waves for a helium atom, *Phys. Rev.* **A 64**, 012102.

Duan, B., Gu, X. Y. and Ma, Z. Q. (2002). Numerical calculation of energies of some excited states in a helium atom, *Eur. Phys. J. D* **19**, 9.

Dynkin, E. B. (1947). The structure of semisimple algebras, *Usp. Mat. Nauk. (N. S.)*, **2**, 59. Transl. in *Am. Math. Soc. Transl. (I)*, **9**, 308, 1962.

Eckart, C. (1934). The kinetic energy of polyatomic molecules, *Phys. Rev.* **46**, 383.

Edmonds, A. R. (1957). Augular Momentum in Quantum Mechanics, Princeton University Press, Princeton.

Elliott J. P. and Dawber, P. G. (1979). Symmetry in Physics, McMillan Press, London.

Feng Kang (1991). The Hamiltonian way for computing Hamiltonian dynamics, *Applied and Industrial Mathematics*, Ed. R. Spigler, Kluwer Academic Publishers, p. 17.

Fronsdal, C. (1963). Group theory and applications to particle physics, 1962, *Brandies Lectures*, Vol. **1**, 427. Ed. K. W. Ford, Benjamin, New York.

Gao, S. S. (1992). Group Theory and its Applications in Particle Physics, Higher Education Press, Beijing (in Chinese).

Gel'fand, I. M., Minlos, R. A. and Shapiro, Z. Ya. (1963). Representations of the Rotation and Lorentz Groups and Their Applications, Transl. from Russian by G. Cummins and T. Boddington, Pergamon Press, New York.

Gel'fand, I. M. and Tsetlin, M. L. (1950). Finite-dimensional representations of the group of unimodular matrices, Dokl. Akad. Nauk. SSSR **71** 825 (in Russian). English transl. in: I. M. Gelfand, "Collected papers". Vol. II, Springer-Verlag, Berlin, 1988, 653–656.

Gell-Mann, M. and Ne'eman, Y. (1964). The Eightfold Way, Benjamin, New York.

Georgi, H. (1982). Lie Algebras in Particle Physics, Benjamin, New York.

Gilmore, R. (1974). Lie Groups, Lie Algebras and Some of Their Applications, Wiley, New York.

Girardeau, (1960). Relationship between systems of impenetrable bosons and fermions in one dimension, *J. Math. Phys.* **1**, 516.

Gradshteyn, I. S. and Ryzhik, I. M. (2007) Table of Integrals, Series, and Products, 7th Ed., Editors: A. Jeffrey and D. Zwillinger (Academic Press).

Gu, X. Y., Duan, B. and Ma, Z. Q. (2001a). Conservation of angular momentum and separation of global rotation in a quantum N-body system, *Phys. Lett. A* **281**, 168.

Gu, X. Y., Duan, B. and Ma, Z. Q. (2001b). Independent eigenstates of angular momentum in a quantum N-body system, *Phys. Rev. A* **64**, 042108(1–14).

Gu, X. Y., Duan, B. and Ma, Z. Q. (2001c). Quantum three-body system in D dimensions, *J. Math. Phys.* **43**, 2895.

Gu, X. Y., Ma, Z. Q. and Dong, S. H. (2002). Exact solutions to the Dirac equation for a Coulomb potential in $D+1$ dimensions, *Inter. J. Mod. Phys. E* **11**, 335.

Gu, X. Y., Ma, Z. Q. and Duan, B. (2003a). Interdimensional degeneracies for a quantum three-body system in D dimensions, *Phys. Lett. A* **307**, 55.

Gu, X. Y., Ma, Z. Q. and Sun, J. Q. (2003b). Quantum four-body system in D dimensions, *J. Math. Phys.* **44**,3763.

Gu, X. Y., Ma, Z. Q. and Sun, J. Q. (2003c). Interdimensional degeneracies in a quantum isolated four-body system, *Phys. Lett. A* **314**, 156.

Gu, X. Y., Ma, Z. Q. and Sun, J. Q. (2003d). Interdimensional degeneracies in a quantum N-body system, *Europhys. Lett.* **64**, 586.

Gu, X. Y., Ma, Z. Q. and Dong, S. H. (2003e). The Levinson theorem for the Dirac equation in $D+1$ dimensions, *Phys. Rev. A* **67**, 062715(1–12).

Guan, L. M., Chen. S., Wang Y. P., and Ma, Z. Q., (2009). Exact solution of infinitely strongly interacting Fermi gases in tight waveguides, *Phys. Rev. Lett.* **102**, 160402.

Hamermesh, M. (1962). Group Theory and its Application to Physical Problems, Addison-Wesley, Massachusetts.

Han, Q. Z. and Sun, H. Z. (1987). Group Theory, Peking University Press, Beijing (in chinese).

Heine, V. (1960). Group Theory in Quantum Mechanics, Pergamon Press, London.

Hirschfelder, J. O. and Wigner, E. P. (1935). Separation of rotational coordinates from the Schrödinger equation for N particles, *Proc. Natl. Acad. Sci. U.S.A.* **21**, 113.

Hou, Bo-Yu, Hou, Bo-Yuan and Ma Z. Q. (1990a). Clebsch–Gordan coefficients, Racah coefficients and braiding fusion of quantum $s\ell(2)$ enveloping algebra I, *Commun. Theor. Phys.* **13**, 181.

Hou, Bo-Yu, Hou, Bo-Yuan and Ma Z. Q. (1990b). Clebsch–Gordan coefficients, Racah coefficients and braiding fusion of quantum $s\ell(2)$ enveloping algebra II, *Commun. Theor. Phys.* **13**, 341.

Hou, Bo-Yuan and Hou, Bo-Yu (1997). Differential Geometry for Physicists, World Scientific, Singapore.

Hou, X. W., Xie, M., Dong, S. H. and Ma, Z. Q. (1998). Overtone spectra and intensities of tetrahedral molecules in boson-realization models, *Ann. Phys. (N.Y.)* **263**, 340.

Hsiang, W. T. and Hsiang, W. Y. (1998). On the reduction of the Schrödinger's equation of three-body problem to a system of linear algebraic equations, preprint.

Hsiang, W. Y. (1998). On the kinematic geometry of many body system, preprint.

Itzykson, C. and Nauenberg, M. (1966). Unitary groups: Representations and decompositions, *Rev. Mod. Phys.* **38**, 95.

Joshi, A. W. (1977). Elements of Group Theory for Physicists, Wiley.

Kastant, B. (1959). A formula for the multiplicity of a weight, Transactions of the American Mathematical Society **93** 53.

Kazdan, D. and Lusztig, G. (1979). Representations of Coxeter groups and Hecke algebras, *Invent. Math.* **53**, 165.

Kogan, E. and Nazarov, V.U. (2012). Symmetry classification of energy bands in graphene, *Phys. Rev.* **B85**, 115418.

Koster, G. F. (1957). Space Groups and Their Representations in Solid State Physics, Eds. F. Seitz and D. Turnbull, Academic Press, New York, **5**, 174.

Kovalev, O. V. (1961). Irreducible Representations of Space Groups, translated from Russian by A. M. Gross, Gordon & Breach.

Lipkin, H. J. (1965). Lie Groups for Pedestrians, North-Holland, Amsterdam.

Littlewood, D. E. (1958). The Theory of Group Characters, Oxford University Press, Oxford.

Liu, F., Ping, J. L. and Chen, J. Q. (1990). Application of the eigenfunction method to the icosahedral group, *J. Math. Phys.* **31**, 1065.

Ma, Z. Q. (1993). Yang–Baxter Equation and Quantum Enveloping Algebras, World Scientific, Singapore.

Ma, Z. Q. and Dai, A. Y. (1982). A graphic rule for dimensions of irreducible tensor representations of SO(N), *J. Lanzhou Univ.* (Natural Sciences) **18**, No. 2, 97 (in Chinese).

Ma, Z. Q. and Gu, X. Y. (2004). Problems & Solutions in Group Theory for Physicists, World Scientific, Singapore.

Ma, Z. Q., Hou, X. W., and Xie, M., (1996). Boson-realization model for the

vibrational spectra of tetrahedral molecules, *Phys. Rev. A* **53**, 2173.
Ma, Z. Q. and Yan, Z. C., (2015). The rotational invariants constructed by the products of three harmornic polynomials, *Chin. Phys.* **C39**, 063104.
Ma, Z. Q. and Yang, C. N., (2010). Spin 1/2 Fermions in 1D Harmonic Trap with Repulsive Delta Function Interparticle Interaction, *Chin. Phys. Lett.* **27**, 080501.
Marshak, R. E., Riazuddin, and Ryan, C. P. (1969). Theory of Weak Interactions in Particle Physcs, John Wiley & Sons, Inc., New York.
Miller, Jr. W., (1972). Symmetry Groups and Their Applications, Academic Press, New York.
Molev, A. I., (2006). Gelfand–Tsetlin bases for classical Lie algebras, Handbook of Algebra, Ed. by M. Hazewinkel, Elservier.
Opechowski, W. (1940) On the 'double' crystallographic groups, *Physica's Grav.* **7**, 552; translated by Cracknell, A.P. *Applied Group theory*, Pergamon Press, Oxford.
Racah, G. (1951). Group Theory and Spectroscopy, Lecture Notes in Princeton.
Ren S. Y. (2017). Electronic States in Crystals of Finite Size, Quantum confinement of Bloch waves, Second Ed. Springer, New York.
Roman, P. (1964). Theory of Elementary Particles, North-Holland, Amsterdam.
Rose, M. E. (1957). Elementary Theory of Angular Momentum, Wiley, New York.
Salam, A. (1963). The Formalism of Lie Groups, in *Theoretical Physics*, Director: A. Salam, International Atomic Energy Agency, Vienna, 173.
Schiff, L. I. (1968). Quantum Mechanics, Third Edition, McGraw-Hill, New York.
Schwartz, C. (1961). Lamb shift in the helium atom, *Phys. Rev.* **123**, 1700.
Serre, J. P. (1965). Lie Algebras and Lie Groups, Benjamin, New York.
Tinkham, M. (1964). Group Theory and Quantum Mechanics, McGraw-Hill, New York.
Tong, D. M., Zhu, C. J. and Ma, Z. Q. (1992). Irreducible representations of braid groups, *J. Math. Phys.* **33**, 2660.
Tung, W. K. (1985). Group Theory in Physics, World Scientific, Singapore.
Wang Y. H. and Ma, Z. Q., (2017). Spin 1/2 fermion gas in 1D harmonic trap with attractive delta function interaction, *Chin. Phys. Lett.* **34**, 020501.
Weyl, H. (1931). The Theory of Groups and Quantum Mechanics, translated from German by H. P. Robertson, Dover Publications.
Weyl, H. (1946). The Classical Groups, Princeton University Press, Princeton.
Wigner, E. P. (1959). Group Theory and its Applications to the Quantum Mechanics of Atomic Spectra, Academic Press, New York.
Wybourne, B. G. (1974). Classical Groups for Physicists, Wiley, New York.
Yamanouchi, T. (1937). On the construction of unitary irreducible representation of the symmetric group, *Proc. Phys. Math. Soc. Jpn* **19**, 436.
Zachariasen, W. H. (1951). Theory of X ray diffraction in crystals, (John-Wiley & Sons, Inc. Charman & Hall, Ltd. London).
Zou, P. C. and Huang, Y. C. (1995). Proof of "irreducible postulation" and its applications, *High Ener. Phys. Nucl. Phys.* **19**, 375.

Index

Printed in the United States
By Bookmasters